Lecture Notes in Electrical Engineering

Volume 718

The book series *Lecture Notes in Electrical Engineering* (LNEE) publishes the latest developments in Electrical Engineering - quickly, informally and in high quality. While original research reported in proceedings and monographs has traditionally formed the core of LNEE, we also encourage authors to submit books devoted to supporting student education and professional training in the various fields and applications areas of electrical engineering. The series cover classical and emerging topics concerning:

- Communication Engineering, Information Theory and Networks
- Electronics Engineering and Microelectronics
- Signal, Image and Speech Processing
- Wireless and Mobile Communication
- Circuits and Systems
- Energy Systems, Power Electronics and Electrical Machines
- Electro-optical Engineering
- Instrumentation Engineering
- Avionics Engineering
- Control Systems
- Internet-of-Things and Cybersecurity
- Biomedical Devices, MEMS and NEMS

For general information about this book series, comments or suggestions, please contact leontina.dicecco@springer.com.

To submit a proposal or request further information, please contact the Publishing Editor in your country:

China

Jasmine Dou, Editor (jasmine.dou@springer.com)

India, Japan, Rest of Asia

Swati Meherishi, Editorial Director (Swati.Meherishi@springer.com)

Southeast Asia, Australia, New Zealand

Ramesh Nath Premnath, Editor (ramesh.premnath@springernature.com)

USA, Canada:

Michael Luby, Senior Editor (michael.luby@springer.com)

All other Countries:

Leontina Di Cecco, Senior Editor (leontina.dicecco@springer.com)

**** This series is indexed by EI Compendex and Scopus databases. ****

More information about this series at http://www.springer.com/series/7818

Yusheng Xue · Yuping Zheng · Anjan Bose
Editors

Proceedings of 2020 International Top-Level Forum on Engineering Science and Technology Development Strategy and The 5th PURPLE MOUNTAIN FORUM (PMF2020)

Springer

Editors
Yusheng Xue
NARI Group Co., Ltd.
Nanjing, Jiangsu, China

Yuping Zheng
NARI Group Co., Ltd.
Nanjing, Jiangsu, China

Anjan Bose
School of EECS
Washington State University
Pullman, WA, USA

ISSN 1876-1100 ISSN 1876-1119 (electronic)
Lecture Notes in Electrical Engineering
ISBN 978-981-15-9745-9 ISBN 978-981-15-9746-6 (eBook)
https://doi.org/10.1007/978-981-15-9746-6

This Springer imprint is published by the registered company Springer Nature Singapore Pte Ltd.
The registered company address is: 152 Beach Road, #21-01/04 Gateway East, Singapore 189721, Singapore

Contents

Protection Comparative Analysis of Typical Phase-Shifting Transformers

Xiaoyu Zhang[1(✉)], Qiaogen Gu[1], HaiDong Huang[2], and Xiao Cheng[1]

[1] NR Electric Co., Ltd., Nanjing, China
zhangxiaoyu@nrec.com
[2] State Grid Jiangsu Electric Power Co., Ltd., Nanjing, China

Abstract. There are many kinds of phase-shifting transformer (PST) because of the wide variety of applications. Therefore, the differential protection of the PST are different. This paper analyzes the differential protection of PST and sum up the typical type of it. Based on two fundamental differential protection principles, this paper proposes a series of differential protection configuration for typical PST. The protection scope, fault type and matching relation of different protection are analyzed base on theoretical analysis and RTDS simulation.

Keywords: PST · Magnetic balance differential protection · Electrical balance differential protection · Parallel transformer · Series transformer

1 Introduction

Phase-shifting transformers (PST) control the steady state power flow and voltage of a transmission line by stringing horizontal or vertical voltages into the transmission line and changing the amplitude and phase of the voltage at the installation point [1]. With the development of the power system, the structure of the power grid is becoming more and more complex. It has become a huge AC/DC hybrid system consisting of multiple types of power supplies, multiple types of loads, and power devices in different voltage levels. This puts forward higher and higher requirements for reliable and economic operation of power grid [2]. As an effective method to control the power flow, PST can avoid overload and underload conditions, increase the transmission capacity of existing transmission lines, and reduce transmission costs. The new type of PST can also improve the dynamic stability of the power system [3]. PST have been widely used in foreign high voltage grids [1].

At present, there are related researches on the use of PST to solve system transient and dynamic problems. For example, how to improve the transient stability of the system; how to alleviate the cross flow which leads to the out-of-step of the tie line; and how to inhibit the switch overload and damped oscillation caused by the sudden increase in the power of line after a fault [4].

However, China has not adopted PST technology wildly, and there is less research on its protection. Based on the structure of PST, this paper presents the differential protection configuration schemes for different PST, and analyzes the protection scope and reflected fault types of each differential protection through RTDS simulation.

Y. Xue et al. (Eds.): PMF 2020, LNEE 718, pp. 1–14, 2021.
https://doi.org/10.1007/978-981-15-9746-6_1

2 Basic Types of PST

There are many types of PST, the primary structure is complex and diverse, so the secondary protection configuration also needs to be different depending on the different types.

2.1 Basic Principle of PST

Although there are great differences in the structure and control method of PST, the principle of the adjusting method of the power flow is the same. The basic principle is to inject a voltage vector with a certain angle to the input voltage to shift the phase angle of the output voltage [5]. A phase shift is generated between the primary side (power supply) and the secondary side (load) of the transformer, so the power flow can be controlled.

For a transmission line which connect two AC systems, the power can be expressed as follows [4] (Fig. 1):

$$P_{SL} = \frac{U_S U_L}{X_{SL}} \sin(\delta_S - \delta_L) \quad (1)$$

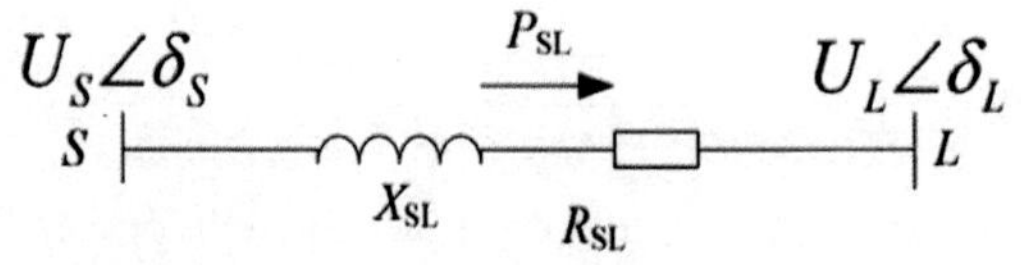

Fig. 1. Power before PST is installed

Where P_{SL} is the active power from node S to node L; U_s, U_L, U_s, U_L are bus voltage amplitude and phase of node S and L respectively; X_{SL} is the reactance value of the line.

It can be seen from formula (1) that the active power of the line is mainly determined by the node voltage, phase angle and line impedance. The power can be changed by controlling the voltage, impedance and phase angle of the line.

Figure 2 shows the results with PST installed on the S side of the line.

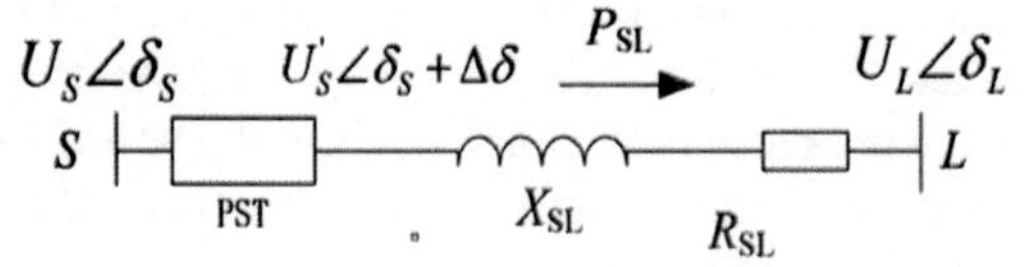

Fig. 2. Power after PST is installed

Ignoring the line resistance, the active power of the line after installing the phase-shift transformer is:

$$P_{SL} = \frac{U_S' U_L}{X_{SL}} \sin(\delta_S - \delta_L + \Delta\delta) \tag{2}$$

Where, U_S' is the voltage after compensating; $\Delta\delta$ is the compensation phase angle of PST; The compensation of PST changes the amplitude and phase angle of the line voltage, so as to regulate the line power.

2.2 Asymmetric Type and Symmetric Type

According to the electrical characteristics after compensation, PST can be divided into two types: asymmetrical type and symmetrical type.

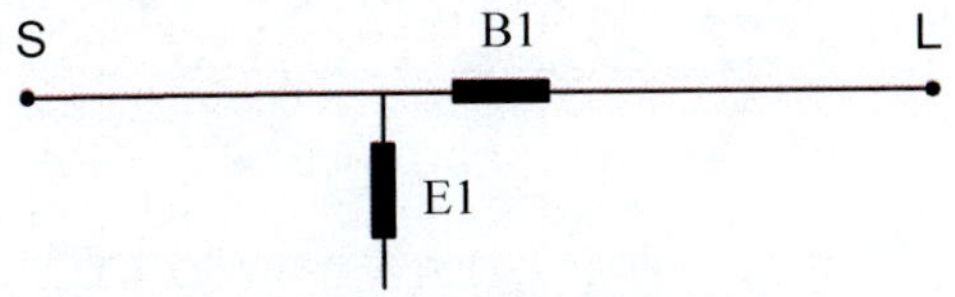

Fig. 3. Schematic diagram of asymmetry PST

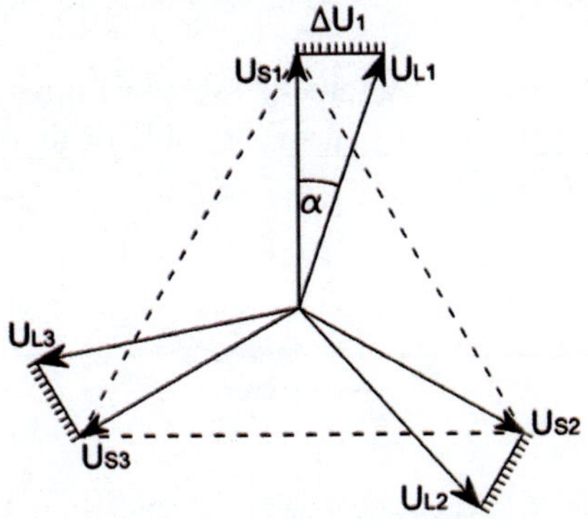

Fig. 4. Voltage vector of asymmetry PST

For the asymmetric type, the phase angle and amplitude of the output voltage are both changed compared with the input voltage. The voltage of excitation winding E1 is taken from one side of series winding B (Figs. 3 and 4).

For the symmetrical type, the phase angle of the output voltage is changed compared with the input voltage, but the amplitude is the same. The voltage of excitation winding E1 is taken from the middle of two series windings B1 and B2 (Figs. 5 and 6).

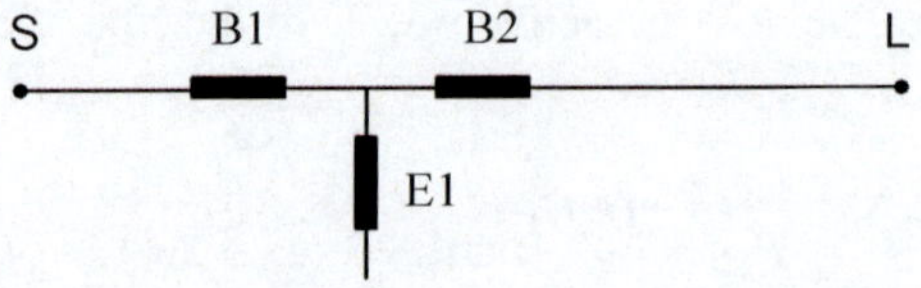

Fig. 5. Schematic diagram of symmetry PST

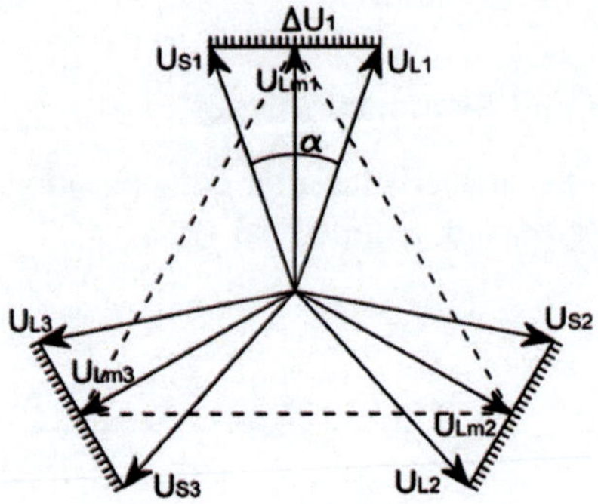

Fig. 6. Voltage vector of symmetry PST

2.3 Single Core Type and Double Core Type

According to primary structure, PST can be divided into two types: single core type and double core type.

The single core PTS is based on a three-phase transformer, and the phase-shifting is realized by proper wiring mode. The secondary side of the series winding is directly connected with the line (Fig. 7).

Fig. 7. Schematic diagram of single-core PST

The double core PST is based on two independent transformers, the parallel transformer (ET) is used to adjust the voltage extracted from the system, and the series transformer (BT) is used to inject the voltage into the line (Fig. 8).

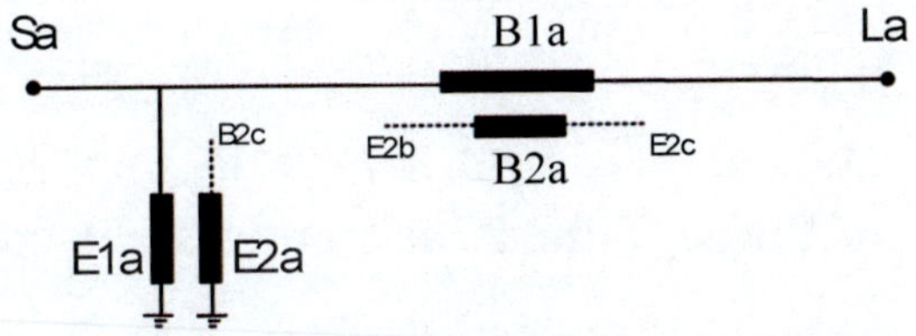

Fig. 8. Schematic diagram of double core PST

2.4 Continuous Controlled Type and Step Controlled Type

According to the result of voltage regulation, PST can be divided into two types: continuous controlled type and step controlled type.

The continuous controlled PST realizes the continuous adjustment of the compensation voltage by adjusting the tap of the same winding (Fig. 9).

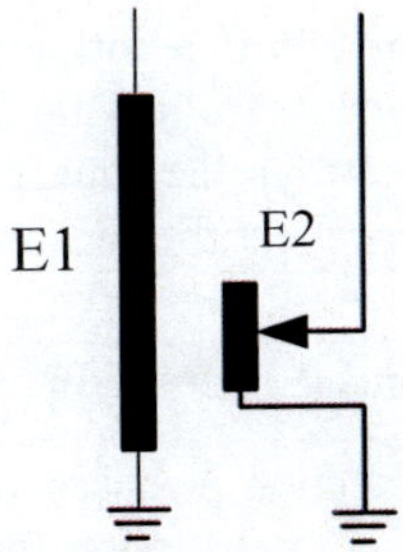

Fig. 9. Schematic diagram of continuous controlled PST

The step controlled PST realizes the adjustment of the compensation voltage by dividing the secondary winding of transformer into differential windings (the ratio of each winding can be the same or different according to the demand). Through the different windings put into use which are controlled by the control system, the input

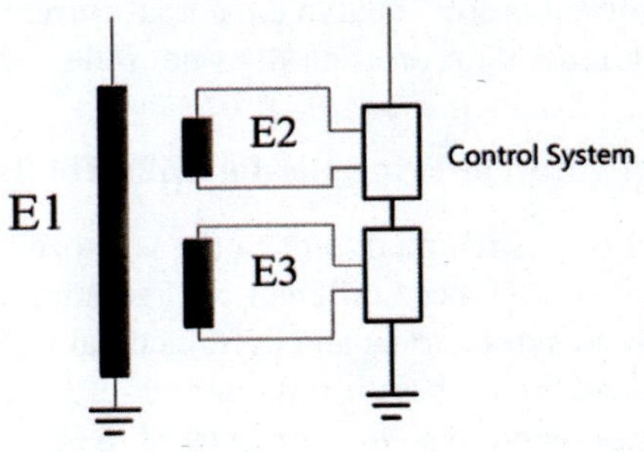

Fig. 10. Schematic diagram of step controlled PST

voltage is changed step by step, thereby the compensation voltage is also changed step by step (Fig. 10).

3 Basic Types of Phase-Shifted Differential Protection

There are two basic principles of differential protection applied to PST.

3.1 Magnetic Balance Differential Protection

Magnetic balance, also known as flux balance or ampere turn balance, is the basic theoretical of general transformer differential protection. For two winding transformer of Y/Y connection, the calculation formula of differential current is as follows:

$$I_d = KI_1 + I_2 \tag{3}$$

Where I_d is the different current. I_1 and I_2 are primary currents of the two windings, K is the ratio of the transformer.

PST usually consists of one or two transformers, each of which can form a magnetic balance differential protection to reflect the internal fault of the transformer. Because the turn to turn fault destroys the balance of magnetic flux, the magnetic balance differential protection can reflect the turn to turn fault.

3.2 Electrical Balance Differential Protection

Electrical balance differential protection is based on Kirchhoff's first law. For a T-connected line, the difference current calculation formula is as follows:

$$I_d = I_1 + I_2 + I_3 \tag{4}$$

Where I_d is the different current. I_1, I_2 and I_3 are primary currents of the T-connected line.

To achieve the purpose of phase-shifting, the winding of PST is usually connected in a certain mode. Besides the magnetic connection, there is also a direct circuit connection between each winding. Therefore, the differential protection based on Kirchhoff's first law can be configured.

Because the input current is still equal to the output current during turn to turn fault, the electrical balance differential protection does not reflect the turn to turn fault.

3.3 Differential Configuration Principles for Different Types of PST

The configuration of PST differential protection is closely related to its primary structure. Different types of PST need different configuration of differential protection.

The difference between symmetrical and asymmetrical type is the different location of excitation voltage. In addition, there is only one series winding for asymmetric type and there are two series windings for symmetrical type. There is little difference between them. The symmetrical type has two windings, so the protection is slightly complicated.

The structure of single core type and double core type are essentially different. Because there is only one transformer in single core PST, only one differential protection based on magnetic balance principle is needed. The double core PST consists of two transformers, which can be configured with two magnetic balance differential protections according to the conditions. Moreover, due to the different connection mode, both of them can be configured with differential protection of electrical balance principle, but the protection range of electrical balance differential protection is different. According to the installation position of CT, only the primary side of the transformer or both the primary and secondary sides of the transformer can be protected for the single core PST. Generally, the primary winding of the series transformer and the primary winding of the parallel transformer can be protected for the double core PST.

The main difference between continuous controlled type and step controlled type is magnetic balance differential protection. The parallel transformer in the continuous controlled PST is difficult to be equipped with normal transformer differential protection because of its large variation range of ratio. But for the step controlled type, the ratios of each windings are fixed, so if CTs are installed in each winding, the parallel transformer can be equipped with magnetic balance differential protection.

4 Differential Protection Configuration Scheme

The configuration of differential protection is closely related to the primary structure of PST and the configuration of CT. The following three typical phase-shifting transformers are taken as examples to analyze the protection configuration of different type of PST. The protection of other types of PST can be configured refer to the example, except some special ones that are not analyzed in this paper.

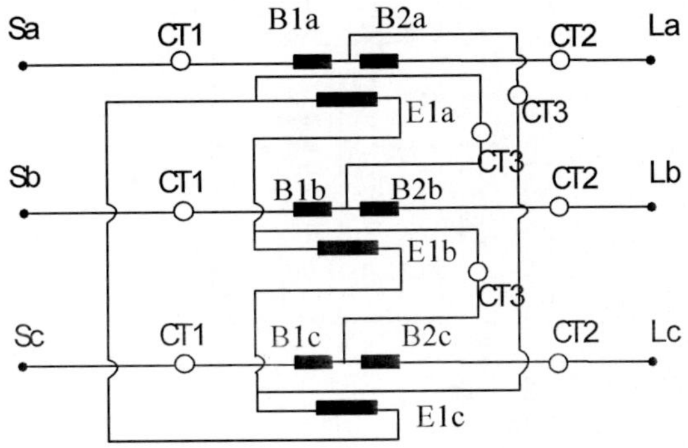

Fig. 11. Wiring diagram of single-core symmetry PST

The following analysis is based on the typical CT configuration. According to different applications, manufacturers, the location and numbers of CT will be different. The corresponding protection configuration and protection scope also need to be adjusted accordingly.

4.1 Single-Core Symmetrical PST

Single core symmetrical PST is a common phase-shifting transformer with simple structure and low cost. Its internal wiring structure is as follows (Fig. 11):

The configuration of the protection for single core PST differential is simple. Generally, only electrical balance differential protection is provided. Some faults need to be reflected by backup protection and gas protection. The specific configuration and protection scope are as follows (Table 1).

Table 1. Protection configuration of single-core symmetry PST

Type of protection	CT used	Protection scope
Electrical balance differential protection	CT1, CT2, CT3	Series windings B1, B2 and the lines between them

4.2 Double Core Asymmetric Continuous Controlled PST

Double core asymmetric continuous controlled PST has the advantages of double core phase-shifting transformer. Moreover, the voltage regulation mode is relatively simple, and no complex control circuit is required. Its wiring mode is as follows (Fig. 12):

Because of the continuous control of the voltage, only series transformer is configured with magnetic balance differential protection, and its protection principle is

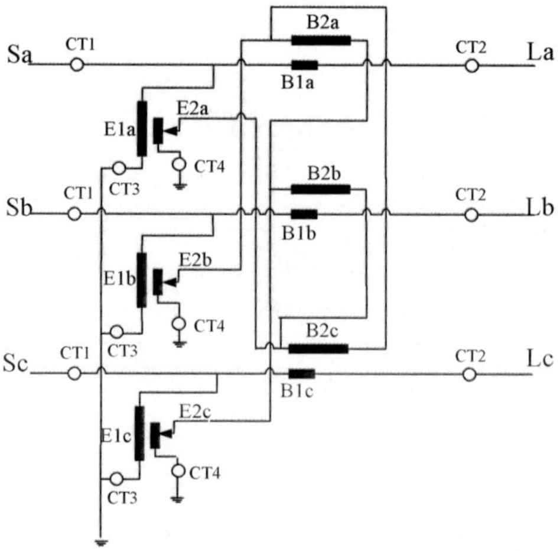

Fig. 12. Wiring diagram of double core asymmetry continuous control PST

Table 2. Protection configuration of double core asymmetry continuous controlled PST

Type of protection	CT used	Protection scope
Electrical balance differential protection	CT1, CT2, CT3	Series windings B1, Parallel windings E1 and the lines between them
Magnetic balance differential protection of Series transformer	CT1, CT2, CT4	Series windings B1, B2, Parallel windings E2 and the lines between them

similar to that of two winding Y/D transformer. Besides, an electrical balance differential protection is configured to protect the primary winding of the two transformers. The differential protection is configured as follows [6] (Table 2).

4.3 Double Core Symmetrical Step Controlled PST

The structure of the double core symmetrical step controlled PST is complex, and it needs independent control system to adjust the voltage. Its structure is as follows (Fig. 13):

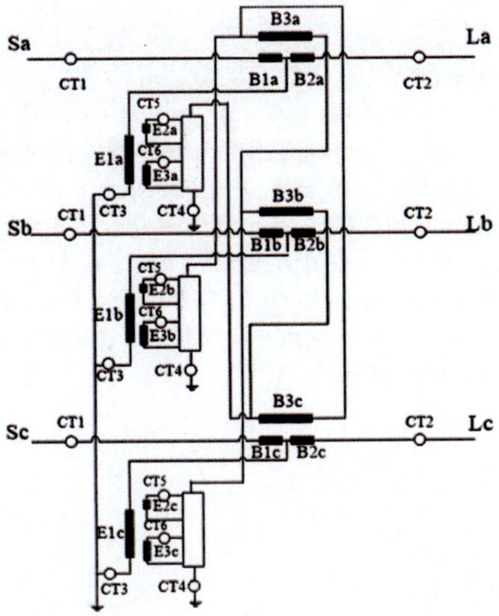

Fig. 13. Wiring diagram of double core symmetry step controlled PST

Independent magnetic balance differential protection can be configured for series transformer and parallel transformer in the double core symmetrical step controlled PST. The protection principle of parallel transformer is similar to that of Y/Y/Y connected transformer, and for the series transformer it is similar to that of Y/Y/D connected transformer. In addition, the electrical balance differential protection can also be configured to protect the primary winding of two transformers. The differential protection configuration is as follows (Table 3):

Table 3. Protection configuration of double core symmetry step controlled PST

Type of protection	CT used	Protection Scope
Electrical balance differential protection	CT1, CT2, CT3	Series windings B1, B2, Parallel windings E1 and the lines between them
Magnetic balance differential protection of Series transformer	CT1, CT2, CT4	Series windings B1, B2, Parallel windings E2, E3 and the lines between them
Magnetic balance differential protection of Parallel transformer	CT3, CT5, CT6	Parallel windings E1, E2, E3 and the lines between them

5 Simulation Analysis

There are many kinds of PST, which are difficult to discuss one by one. In this chapter, the most complex one, double core symmetrical step controlled PST is taken as the research object to establish the RTDS simulation model and the protection range and operation characteristics of different differential are analyzed.

Phase A ground fault occurs between CT1 and B1a, which can be reflected by electrical balance differential and magnetic balance differential of series transformer. The waveforms are as follows (Fig. 14):

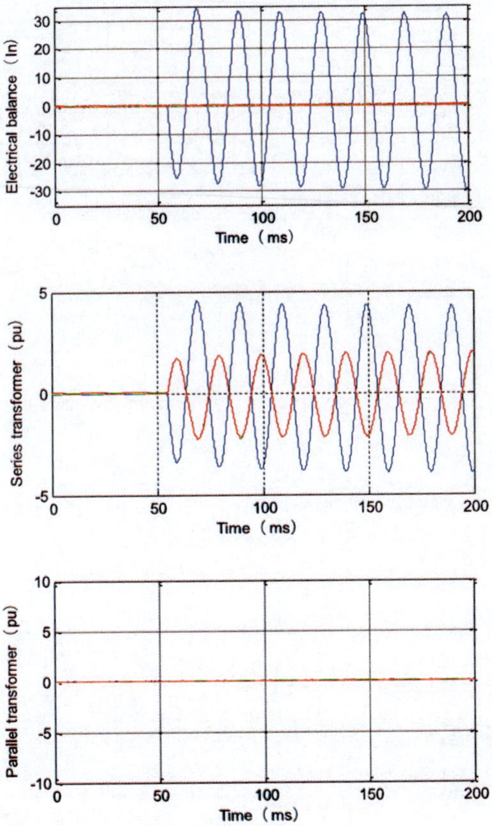

Fig. 14. Differential current with phase A fault between CT1 and B1a

100% turn to turn fault occurs in the parallel transformer winding E2, which can be reflected by the magnetic balance differential protection of the parallel transformer. The waveforms are as follows (Fig. 15):

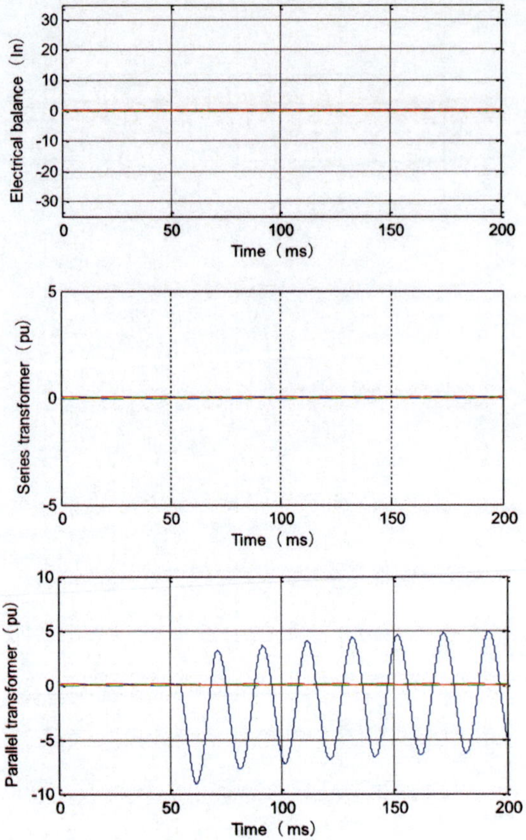

Fig. 15. Differential current with inter-turn fault of E2

15% turn to turn fault occurs in series transformer winding B3, which can be reflected by the magnetic balance differential of series transformer. The waveforms are as follows (Fig. 16).

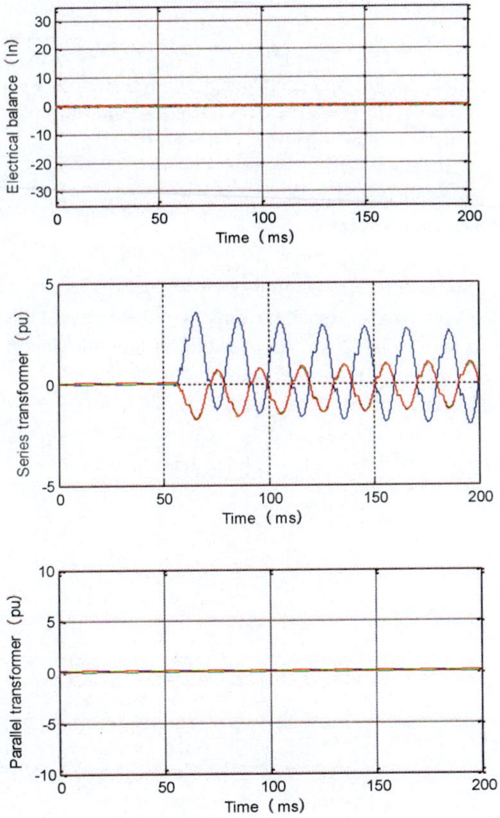

Fig. 16. Differential current with inter-turn fault of B3

It can also be seen from the above waveforms that the three differential protection have their own protection ranges and reflect different fault types. Each differential protection cooperates with each other to form a complete PST protection.

6 Conclusion

From the point of view of protection, the classification of PST is studied in this paper, and the differential protection configuration scheme of typical PST is put forward. Through theoretical analysis and simulation, the scope and coordination of the differential protection are verified. It should be noted that the CT configuration will be different according to different engineering applications, so the actual protection range of each protection may be changed. Besides, some special PST outside the discussion scope of this paper need further study because of their complex structure.

References

1. Zheng, B., Xiang, Z., Ban, L., Ni, S.: Electromagnetic transient analysis on static phase shifter applied in UHV power grid. In: Power System Technology, pp. 445–449 (2012)
2. Zhu, F.: Modeling of the voltage injecting phase-shifting transformers. Power Syst. Clean Energy **28**(2), 7–10 (2012)
3. Zhao, X., Guo, M.: Study of installation of phase-shifting transformers for East China power grid. East China Electr. Power **34**(11) (2006)
4. Yu, H., Zhou, F., Yang, Z.: Parameter design and steady state characteristic analysis of TCPSTs in a EHV-grid. Electr. Power **46**(11), 36–41 (2013)
5. Zhao, X.: Research on improving transmission capacity of east china 500 kV power grid by phase-shifting transformer technology. In: Academic Annual Meeting of ShangHai Society for electrical engineering, pp. 103–106 (2006)
6. Gajie, Z., Podboj, M., Traven, B., Krašovec, A.: When existing recommendations for pst protection can let you down. In: 11th IET International Conference on Developments in Power System Protection 2012, pp. 218–224 (2012)

Process-Level Network Analysis and Optimization Scheme Research in Smart Substation

Gui Yang[1,2](✉), Xin Huang[2], Xiao Cheng[1], and Qin Li[2]

[1] NR Electric Co., Ltd., Nanjing 211100, China
yangg@nrec.com
[2] State Key Laboratory of Smart Grid Protection and Control (NARI Group Corporation), Nanjing 211106, China

Abstract. Heavy TRAFFIC load and transmission congestion related to GOOSE/SV messages transmitted sometimes happened in process-level network of smart substation, and this may cause unpredictable problems. Theoretical analysis and research of GOOSE/SV messages transmission over process-level network is conducted in the article, and the characteristics of the network. The correctness of theoretical analysis is verified by comparing the results between OPNET simulation and theoretical analysis, and the maximum delay of GOOSE/SV message transmission at 18 bays is analyzed and quantified. Also suggestions for process-layer networking are given to be the reference of theoretical analysis foundation and basis for smart substation process-level construction.

Keywords: Traffic analysis · Network optimization · Flow control · Switch · Process-level · Smart substation

1 Introduction

With the popularization and application of smart substations, a large number of substations use process-level networks to transmit SV and GOOSE message.

The services transmitted at the process-level network have the characteristics of large flow, low delay, and high importance. How to improve the real-time performance of the process-level network message transmission and ensure that the SV/GOOSE messages are delivered to the corresponding device at the fastest speed has become a key topic of current research.

At present, quite a few references have analyzed and researched the transmission characteristics of SV and GOOSE message network messages. References [2–4] have performed simulation analysis on process-level network traffic. References [5, 6] have analyzed process-level network storms and other abnormal traffic. Reference [7] carried out network performance testing and analysis on process-level network traffic. Reference [8] analyzed the characteristics of the process-level network message and provided the communication configuration. References [9–11] analyzed the fault location at the process-level. References [12, 13] proposed a scheme based on flow control to

Y. Xue et al. (Eds.): PMF 2020, LNEE 718, pp. 15–26, 2021.
https://doi.org/10.1007/978-981-15-9746-6_2

improve the reliability of network transmission in smart substations. References [14, 15] described current fault location method of process-level network, on the perspective of actual operation and maintenance of substations. The above references have made some progress in the research direction of relevant process-level, but there is no relevant research on the level of the maximum delay that SV and GOOSE messages can be transmitted on the process layer network.

2 Network Analysis of Process-Level

2.1 Networking Solution Analysis

There are two possible solutions for the smart substation process-level networking. The first solution uses the SV and GOOSE respective networking schemes, and the second solution uses the SV and GOOSE common networking schemes. Solution one needs to establish a process-level network for SV and GOOSE respectively. The first solution has the disadvantages of multiple network devices and high network costs compared with second solution. Therefore, only a few substations have adopted respective networking scheme in smart substaion, and they are no longer popularized and applied. Compared with the existing process-level network, the second solution no longer adopts protection to transmit SV message through optical fiber directly. It has the characteristics of basically the same networking cost and equivalent network equipment usage with current network. Therefore, the following section focuses on the analysis of SV and GOOSE common network transmission.

In order to reduce the network transmission delay and meet the transmission bandwidth requirements, the number of cascades in the process-level network is generally controlled within three cascades, and the cascade ports between the central switch and the bay switches use gigabit ports. A feasible networking scheme is shown in Fig. 1, SV and GOOSE common network transmission adopts a three-level cascade networking scheme. Central switch uses full Gigabit optical ports, and the bay switch uses 100M optical ports and Gigabit cascade optical ports. Figure 1 shows a typical process-level switch configuration and networking diagram. The process-level network central switch is configured with 16 Gigabit optical ports, which are used to implement the access of bay switches, sniffer and other equipment. The bay switch (1) are configured with 16 100M optical ports and 4 Giga optical ports, which can meet the access of IEDs with 3–4 bays. The bay switch (2) is configured with 6 100M optical ports and 2 Gigabit optical ports, which can meet IEDs access of one bay.

The following chapter analyzes the port traffic and characteristics of three differently configured switches. At the same time, in order to reflect the actual typical application scenarios of smart substations, the analysis will be based on the GOOSE and SV packet length in typical scenarios.

2.2 Process-Level Message Characteristics Analysis

Smart substation constructed in China, the process-level network is only used to transmit SV and GOOSE messages. SV generally uses a frequency of 4000 frames per

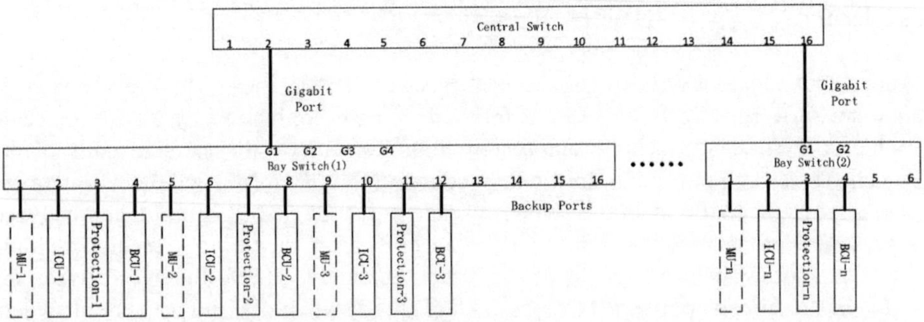

Fig. 1. Process-level network diagram.

second for message transmission, and the length of each SV message is generally around 250 Byte. The length of the GOOSE message varies greatly depending on the content carried. The maximum length can reach 1518 bytes, and there are also dozens of bytes in shorter cases.

3 IED Output Flow Analysis

At present, the mainstream IED in smart substations uses 100M optical ports. The busbar protection, transformer protection, line protection, BCU, ICU, MU and other devices are equipped with 100M optical ports. The sniffer supports both 100M and 1000M optical ports. The length and number of GOOSE and SV packets sent by each device are shown in the following table.

Table 1. Packet length of various types of devices

Name	Length of GOOSE (Byte)	Length of SV (Byte)
ICU (GOOSE1)	311	–
ICU (GOOSE2)	361	–
ICU (GOOSE3)	395	–
ICU (GOOSE4)	1062	–
MU	–	250
Line protection	200	–
BCU	270	–
Bus protection	280	–
transformer protection	206	–

The values listed in the table are the packet's length for GOOSE and SV configuration of each IED in typical applications.

4 Process-Level Network Traffic Analysis

After GOOSE and SV messages are sent by each IED in the process-level network, they are forwarded according to the forwarding path configured by VLAN or static multicast. Nodes with a large amount of traffic appear on the cascade ports of the switch. The switch port traffic for line protection, BCU, ICU, MU and other equipment access is much smaller than the switch cascade port traffic, so it is no longer analyzed separately. For bus protection and transformer protection, they need to receive multiple GOOSE and SV, which need to be analyzed separately. The following analyzes the traffic of the cascade ports of the bay switch (1), bay switch (2), and the central switch.

4.1 Bay-Switch (2) Traffic Analysis

Bay switch (2) has 6 100 M optical ports (ports 1–6) and 2 Gigabit/100M optical ports (ports G1–G2). The bay switch (2) cascade port is a Gigabit optical port. It is used to access single-bay protection, BCU, ICU and MU, etc.

Figure 2 is a schematic diagram of uplink traffic and downlink traffic of the bay Switch (2). In the following, the upstream traffic and downstream traffic of the cascade port of the switch are analyzed separately.

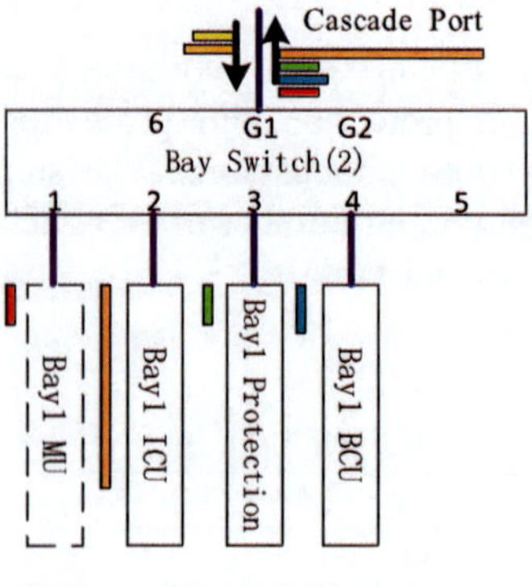

Fig. 2. Diagram of the traffic of bay-switch (2)

Upstream traffic analysis: Since sniffer is deployed on the central switch, the messages sent by the protection, BCU, ICU, and MU in the bay are transmitted to the central switch through the cascade port for sniffer surveillance. ICU adopts single-port access to the bay-switch, and GOOSE messages of the four control blocks are sequentially sent to the bay-switch. There is no problem with queuing the packets of the same device on the switch cascade port. However, in extreme cases, there is congestion in the packets sent by the protection, BCU, ICU, and MU through the cascade port. The extreme queue depth is calculated according to the following formula:

$$\mathrm{L} = \sum_{i=1}^{n} \mathrm{L}_i \tag{1}$$

Refer to (1), Where L is the extreme queuing depth of the port (in bytes), and Li is the length of the packets participating in the queuing.

Packet queuing on the cascade port of the bay-switch (2) can be instantiated into the following formula by formula (1):

$$\mathrm{L} = Lp + Lb + Li\,\mathrm{max} + Lm \tag{2}$$

Refer to (2), Where Lp is the length of the GOOSE packet sent by line protection, Lb is the length of the GOOSE packet sent by BCU, Limax is the length of the longest GOOSE packet sent by ICU, and Lm is the length of the SV packet sent by MU. Bring the data in Table 1 into formula (2) and calculate the value of L as 1782 Byte.

Therefore, in extreme cases, the buffer space of the cascade port of the bay-switch is larger than 1782 Byte, so no packet loss will occur. Because the interval for sending SV messages is 250 μs, only the number of bytes that can be transmitted within 250 μs of the Gigabit cascade port is greater than the number of bytes required for the message to ensure that no cumulative queuing occurs. Calculate the number of bytes sent at 250 μs on the Gigabit port. The calculation formula is:

$$S = v * t/B \tag{3}$$

Refer to (3), Where S is the number of bytes transmitted within a specified time, v is the port rate, which is 1.25 Gbps for a Gigabit port, t is the transmission time in seconds, and B is the length of a byte (8bit). Therefore, Use formula (3) to calculate the number of bytes S that can be transmitted at 250 μs as 39.06 kByte. It is much larger than the number of bytes that need to be transmitted, so there is no accumulation queuing. Therefore, the transmission delay of the cascade port of the bay-switch (2) can be calculated according to the following formula:

$$T = \frac{l}{v} * B + Ts \tag{4}$$

Refer to (4), Where T represents the maximum transmission delay of the cascade port (in seconds), L is the number of bytes transmitted by the cascade port, v is the rate of the cascade port, B is the number of bits in a byte, and Ts is Store and Forward delay (S&F) of the bay-switch (2). S&F is calculated based on a typical value of 10 μs. Bring L, v, B into formula (4) and calculate the T value as 21.41 μs.

According to formula (1), the total length of the packets sent by all bay IEDs in the 250 μs is calculated as 2.849 kByte. According to the calculation result of the 250 μs interval, it is known that there is no queuing phenomenon across the SV message transmission interval.

When an avalanche occurs, GOOSE messages are sent at intervals of 2 ms, 2 ms, 4 ms, and 8 ms. The total length of the messages sent by all devices within the 2 m interval is calculated as 4.599 kByte according to formula (1). The number of bytes that can be transmitted by the cascade port in a 2 ms time interval is calculated as 312.5 kByte according to formula (3).

Therefore, the number of bytes that need to be transmitted within 2 m is much smaller than the number of bytes that can be transmitted by the cascade port, and no cumulative queuing across 2 ms will occur. The maximum queuing depth is consistent with the 1782 Byte analyzed above, and the maximum transmission delay is consistent with the 21.41 μs delay analyzed above.

Downstream traffic analysis: Downstream messages mainly come from bus protection and transformer protection. In extreme cases, the queue depth is calculated according to formula (1) to obtain the number of queued bytes as 486 Byte.

Far less than the number of bytes that can be transmitted by the cascade port in 2 ms, the transmission delay calculated according to formula (4) is 13.12 μs. From the above analysis, the maximum uplink delay of the cascade port is 21.41 μs, and the maximum downlink delay is 13.12 μs.

4.2 Bay-Switch (1) Traffic Analysis

The bay switch (1) has 16 100M optical ports (ports 1–16) and 4 Gigabit/100M optical ports (ports G1–G4). Access to three intervals of protection, BCU, ICU, and MU. Figure 3 is a schematic diagram of uplink traffic and downlink traffic of bay switch (1). In the following, the upstream traffic and downstream traffic of the cascade port of the switch are analyzed separately.

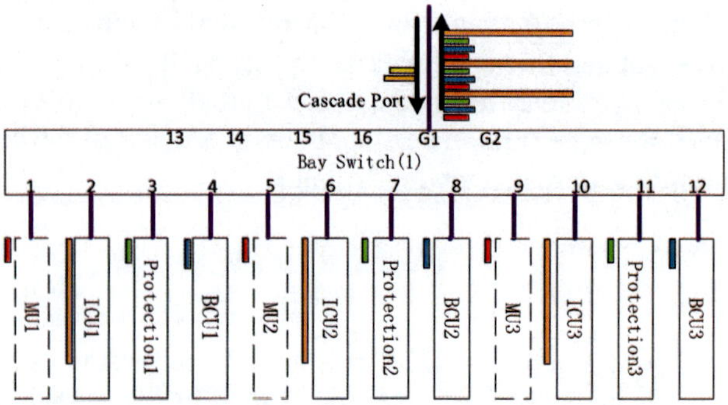

Fig. 3. Diagram of the traffic of the bay-switch (1)

Upstream traffic analysis: Since the sniffer is deployed on the central switch, the information sent by the protection, MCU, ICU and MU in the bay must be uploaded to the central switch through the cascade port for Message monitoring. The ICU uses a single port to access the bay switch, so the GOOSE messages of the four control blocks are sequentially sent to the bay switch, and there is no problem that the messages of the device are aligned on the cascade port of the switch. However, in extreme cases, the protection, MCU, ICU and MU connected to the same switch have congestion on the packets sent by the cascade port. The extreme queuing depth is calculated according to the instantiation of formula (1).

According to the method in Sect. 3.1, the value of the length L of the cascade port packet is calculated to be 5346 Byte. The calculation result of the transmission delay of the cascade port of the bay switch (1) is 44.22 μs. Calculate the total length of packets sent by all devices in the 250 μs interval to 8.547 kByte. Therefore, the cascading port packet transmission will not occur over 250 μs cumulative queuing phenomenon, that is, there is no queuing phenomenon across the SV packet sending interval. When an avalanche occurs, the total length of packets sent by all devices within a 2 m interval is 13.797 kByte.

Therefore, the cascading port will not have cumulative queuing across 2 ms, that is, there will be no queuing across the GOOSE message sending interval. The maximum queuing depth is consistent with the 5346 Byte analyzed above, and the maximum transmission delay is consistent with the 44.22 μs delay analyzed above.

4.3 Central Switch Traffic Analysis

As shown in Fig. 4, the central switch has 16 Gigabit optical ports (ports 1–16), which are used to access devices such as bay switches and Sniffer. The uplink traffic of the central switch's cascading port is the packets sent by the bay switches, while the downlink traffic is mainly packets sent by the transformer protection and bus protection. From the analysis of the bay switch, the upstream traffic is greater than the downstream packet traffic, and the upstream traffic of the central switch is all connected to the Sniffer port for monitoring.

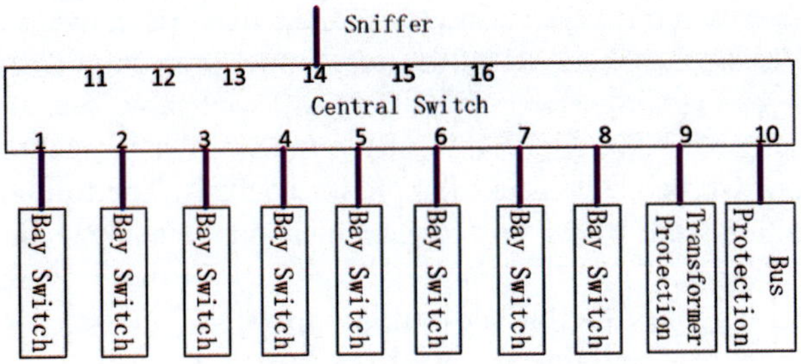

Fig. 4. Networking diagram of central switch

Therefore, analyzing the congestion of Sniffer port can represent the extreme queuing scenario of the central switch. The length of the queued packets on Sniffer port can be calculated according to the following formula:

$$L = \sum_{1}^{M} \sum_{1}^{N} Li \tag{5}$$

Where Li is the length of a packet sent by a device accessed by the bay switch, N is the number of devices accessed by the bay switch, and M is the number of bay switches.

According to the calculation of a typical substation with a maximum of 18 bays, the central switch needs to be connected with 6 bay switches (1) or 18 bay switches (2). Calculation based on the maximum concurrent transmission packet length of each bay of 1782 Byte. In extreme cases, all bay switches will send 1782 Byte (bay switch (2)) or 5346 Byte (bay switch (1)) to the central switch at the same time. Formula (5) calculates that the queueing depth of the port connected to the smart recorder is 32.076 kByte.

Therefore, the queue depth is 32.076 kByte, which can transmit 39.06 kByte within 250 μs. According to formula (4), the transmission time of the port connected to the sniffer is 205.29 μs. The 250 μs non-cumulative queuing transmission can achieve a maximum access bays of 21, which can meet the process-level networking requirements of a typical substation.

4.4 IED Traffic Analysis

The output message length of the IED has been analyzed in Sect. 2. The downstream messages of busbar protection and transformer protection have been explained in the analysis of the bay switch traffic. The number of GOOSE control blocks received by line protection, BCU, ICU and MU is significantly smaller than that of busbar protection and transformer protection. Therefore, this section mainly analyzes the received traffic of busbar protection and transformer protection.

According to the analysis of 18 bays in a typical substation, the busbar protection needs to receive data sending from 18 bays and a bus voltage MU, which measn 19 SV messages and 36 GOOSE messages. Considering transformer protection of the three-volume transformer and voltage level networking respectively, each voltage level needs to receive no more than three SV messages and ten GOOSE messages. The number of GOOSE and SV messages received by the transformer protection is significantly lower than that of the busbar protection. Therefore, the applicable scenario of the busbar protection can also meet the application of the transformer protection. Therefore, the busbar protection will be used as an example for analysis. The following uses Gigabit port communication and 100M port communication to analyze the bus protection respectively.

First the total length of the busbar protection received message is 11.95 kByte by calculation, and the transmission delay of the busbar protection is 76.48 μs. When cascading 100M ports, it is necessary to pay attention to the total traffic of busbar protection receiving packets of 180.8 Mbps. According to the typical plan that the flow of each port is less than 60%, the busbar protection needs to provide 4 ports. SV/GOOSE messages are evenly distributed to each port. Each port needs to input 5 SV messages and 9 GOOSE messages. Therefore, the maximum queuing depth of the port is 3.05 kByte. The maximum delay is 244 μs. Therefore, in extreme cases, the bus-protected ports will not be queued across the SV transmission interval.

4.5 Summary

From the analysis of the above bay-switch (1), bay-switch (2), central switch, and busbar protection, the key technical indicators that affect the network scale are the maximum transmission delay of the network and the buffer capacity of the switch port. The bay-switch (1) ports require a buffer size of 5346 Byte in the extreme case, the interval switch (2) ports require a buffer size of 1782 Byte in the extreme case, and the central switch ports require the buffer size of 32.076 kByte in the extreme case. The 100M port needs a cache size of 11.95 kByte, and the 100M port needs a cache size of 3.05 kByte.

From the perspective of the performance of the current mainstream power system process layer switches, the port cache of 100M switches is greater than 13 kByte, and the port cache of Gigabit switches is greater than 125 kByte. In terms of cache capacity, it can meet the process network networking requirements of a maximum of 18 bays.

5 Analysis of Process-Layer Network Transmission Delay

If the bay-switch (2) is used for networking, the equipment of 18 bays must be cascaded in three cascades. An example of networking is shown in Fig. 5.

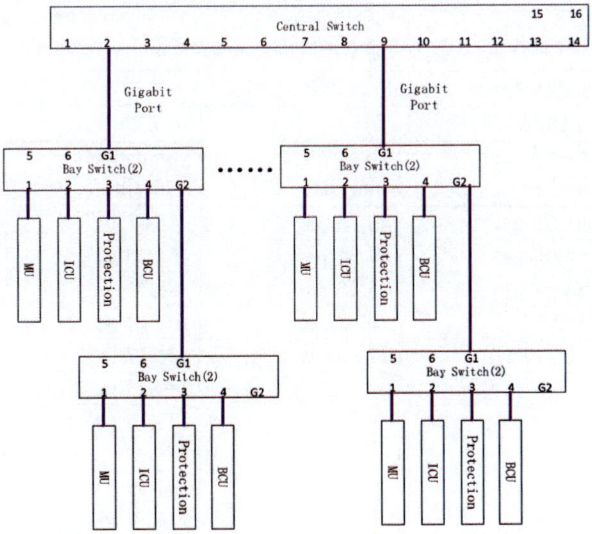

Fig. 5. Networking diagram of Bay-switch cascade

SV messages from the MU are still used for analysis, and the total transmission time to the sniffer is calculated to be 289.51 μs. Using the data in Table 1 to calculate the total transmission time to the busbar protection 100M port is 328.22 μs, which is far below the threshold of 2 ms.

From the above calculations, it can be known that with different bay switch networking, the final transmission delay difference is only in the S&F delay inside the switch introduced by the number of cascades.

Another feature that needs attention is that when the message is calculated according to the extreme delay in the transmission, the key that affects the transmission delay is the length of the packet sent from the sending device. The longer length, the greater delay. When the delay of reaching the Sniffer is calculated according to the longest GOOSE packet of ICU, the calculated total transmission time is 354.47 μs.

From the perspective of network transmission delay, the maximum packet transmission delay of 18 bays does not exceed 400 μs. When the network scale is further expanded to 30 bays, the maximum network delay can still be controlled within 1 ms. Due to space reasons, the calculation is no longer carried out here, but the cache capacity requirements of network equipment will further increase.

6 Simulation

In order to verify the correctness of the above theoretical analysis, the simulation results with 5 IEDs and 10 IEDs sending messages simultaneously are compared with the results of theoretical calculations. The delay in the table is the CT delay of the switch, and the S&F delay of the switch is calculated according to 10 μs (Tables 2 and 3).

Table 2. 100 Mbps port delay analysis and comparison

100M bandwidth, 218 Byte, sampling period 250 μs				
Message delay	5 IEDs		10 IEDs	
	Simulation analysis	Theoretical analysis	Simulation analysis	Theoretical analysis
IED1	45.01 μs	45.20 μs	45.01 μs	45.20 μs
IED2	64.05 μs	64.24 μs	64.05 μs	64.24 μs
IED3	83.09 μs	83.28 μs	83.09 μs	83.28 μs
IED4	102.13 μs	102.32 μs	102.13 μs	102.32 μs
IED5	121.17 μs	121.36 μs	121.17 μs	121.36 μs
IED6	–	–	140.21 μs	140.40 μs
IED7	–	–	159.25 μs	159.44 μs
IED8	–	–	178.29 μs	178.48 μs
IED9	–	–	197.33 μs	197.52 μs
IED10	–	–	216.37 μs	216.56 μs

From the simulation results of 100M and 1000M ports, there is only a fixed deviation between theoretical calculation and the simulation results,. The deviation is very small, which can indicate that the theoretical analysis results are credible.

7 Network Optimization Suggestions

From the analysis in Sects. 3, 4 and 5, for network equipment, the two important indicators that affect network transmission performance are the buffer capacity of the switch and the transmission delay. The buffering capability directly determines whether

Table 3. 1000 Mbps port delay analysis and comparison

1000M bandwidth, 218 Byte, sampling period 250 μs				
Message delay	5 IEDs		10 IEDs	
	Simulation analysis	Theoretical analysis	Simulation analysis	Theoretical analysis
IED1	12.47 μs	12.00 μs	12.50 μs	12.00 μs
IED2	14.47 μs	14.00 μs	14.50 μs	14.00 μs
IED3	16.47 μs	16.00 μs	16.50 μs	16.00 μs
IED4	18.47 μs	18.00 μs	18.50 μs	18.00 μs
IED5	20.47 μs	20.00 μs	20.50 μs	20.00 μs
IED6	–	–	22.50 μs	22.00 μs
IED7	–	–	24.50 μs	24.00 μs
IED8	–	–	26.50 μs	26.00 μs
IED9	–	–	28.50 μs	28.00 μs
IED10	–	–	30.50 μs	30.00 μs

packet loss occurs, and the transmission delay determines whether it meets the application of protecting services. The main factors affecting the delay are the packet length, the number of cascaded switches, and the port speed of the device. Since the packet length is directly related to the service, this article does not discuss it. Therefore, it is recommended to reduce the number of cascades and set the port rate as high as possible. To obtain the best network performance, access as many bays as possible. The networking recommendations are as follows:

(1) The total network traffic is controlled within 500 Mbps
(2) Transmission of network packets should be controlled within 4 switches as much as possible.
(3) It is recommended that devices that receive more packets across the bays, such as busbar protection and transformer protection, use Gigabit ports for communication.
(4) When cross-space equipment such as busbar protection and transformer protection must use 100M ports for communication, multi-port reception processing is used. The traffic of each port is controlled within 60 Mbps, and the traffic of the GOOSE message is calculated according to the 2 ms sending interval.
(5) Gigabit ports are recommended for Sniffer to access the central switch.

8 Conclusion

The paper analyzes the maximum network transmission delay and the switch port buffer capacity requirements of the process-layer network SV and GOOSE common network transmission. Through quantitative analysis, the port buffer capacity and the maximum transmission delay time that the network transmission may introduced are clear given. Based on this, the feasibility of SV and GOOSE common network transmission and the network scale and scheme are analyzed. From the analysis, it is theoretically feasible to use SV and GOOSE common network transmission. At the same time, the process-layer network optimization suggestions are given, which provides a theoretical analysis basis for the future smart substation networking, and helps to plan the process-layer networking scheme with quantitative data.

References

1. IEC/IEEE 61850 Communication networks and systems for power utility automation
2. Zhang, J., Zhang, J., Ni, M., et al.: Generic message source and flow calculation model for communication network in smart substation. Autom. Electr. Power Syst. (43), 147–155 (2019)
3. Zhang, Y., Cai, Z., Long, P., et al.: Real-time fault diagnosing models and method for communication network in smart substation. Power Syst. Technol. (40), 1851–1857 (2016)
4. Chen, Q., Chen, Z., Huang, Y., et al.: Calculation of distribution of message and traffic load for process bus network in smart substation. Power Syst. Protect. Control (44), 117–121 (2016)
5. Zhou, C., Wu, H., He, Z., et al.: Prevention of merging unit cascade network stomr. Southern Power Syst. Technol. (10), 44–48, 69 (2016)
6. Zhang, J., Zhang, J., Yang, C., et al.: Abnormal traffic detection on process layer network of smart substation based on cyber physical fusion. Autom. Electr. Power Syst. (43), 173–181 (2019)
7. Liu, H., Zuo, Q., Zhang, B.: Process level network performance testing and analysis in smart substation. Power Syst. Protect. Control (40), 112–116 (2012)
8. Liu, W., Wang, H., Zhang, Y.: Study on message characteristics and communication configuration of process layer network of intelligent substation. Power Syst. Protect. Control (42), 110–115 (2014)
9. Yi, Y., Liu, Y., Chen, Y., et al.: Research of network communication fault location technique in smart substation based on comprehensive information judgment. Power Syst. Protect. Control (44), 135–140 (2016)
10. Li, C., Luo, L., Wang, D., et al.: Design and implementation of network monitoring and fault location system for process layer network in smart substation. Electr. Power Eng. Technol. (38), 117–122, 141 (2019)
11. Zhou, H., Huang, J., Zhang, C., et al.: Petri net based fault diagnosis for GOOSE circuits of smart substation. Southern Power Syst. Technol. (11), 49–56 (2017)
12. Zhao, M., Lv, H., Yang, G., et al.: Data transmission reliability promotion scheme in smart substation based on flow control. Power Syst. Protect. Control (47), 141–146 (2019)
13. Yang, G., Lv, H., Yuan, Z., et al.: Research and realization of intelligent substation process leve network flow control and sysnchronization method. Power Syst. Protect. Control (43), 70–74 (2015)
14. Xu, G., Xia, X., An, Y., et al.: Research on fault diagnosis of secondary circuit of relay protection in intelligent substation. Electr. Technol. (4), 41–42 (2019)
15. Zhang, Y., Li, D., Wen, F., et al.: Process level fault diagnosis in IEC 61850 based smart substations. Electr. Power Constr. (39), 42–48 (2018)

Attitude Monitoring Method of Transmission Tower in Goaf Based on Multi-source Sensing and Information Fusion

Hongbin Zai(✉), Bing Song, and Jianjun Dong

State Grid Jincheng Power Supply Company, Jincheng 048000, Shanxi, China
qazw3182@163.com

Abstract. In view of the problems of incline, translation and lodging of high-voltage transmission tower in goaf, a method of transmission tower attitude monitoring based on multi-source sensor and information fusion is proposed to ensure the transmission tower in normal state. Firstly, the angle sensor and gyroscope sensor are used to collect the data of transmission tower, and the Beidou deformation monitoring system is designed to build the transmission tower attitude monitoring model. Then, the calibration based on Beidou difference code and multi parameter smooth difference can dynamically reflect the attitude change of transmission tower and further improve the accuracy of monitoring. Finally, based on the design of the transmission tower attitude monitoring system, the field data are collected and analyzed to demonstrate the system performance. The results show that the proposed method can accurately monitor the movement and inclination of transmission towers within the allowable error range, so as to ensure the safe and stable operation of the power grid.

Keywords: Goaf · Transmission tower · Attitude monitoring · Multi-source sensing · Beidou differential code · Multi parameter smooth

1 Introduction

In recent years, electric power as a modern basic industry has developed rapidly. As the UHV transmission line across regions and basins passes through complex terrain, many mined out areas and frequent mountain movement, it will lead to tower tilt, settlement, displacement, etc., and then tower collapse, as well as various conditions affecting the normal operation of the line such as line tension and sag change [1, 2]. Moreover, the towers of UHV transmission lines are huge and interrelated, and the tension between them is very large. The inclination or deformation of the towers may aggravate the collapse of the goaf. Therefore, the key to the management and protection of transmission lines is to monitor the translation, inclination and lodging of transmission towers and take corresponding measures [3, 4].

As a method to describe the spatial position information of an object, spatial coordinate is an important basis to determine the spatial position of an object, and spatial coordinate measurement facilities play an important role in the digital information age. There are four existing positioning systems: Global Positioning System

Y. Xue et al. (Eds.): PMF 2020, LNEE 718, pp. 27–39, 2021.
https://doi.org/10.1007/978-981-15-9746-6_3

(GPS), GLONASS system, Galileo system and Beidou satellite navigation and positioning system (BDS) [5]. With the development of satellite positioning and space measurement technology, the possibility of accurate positioning of an object is becoming higher and higher, especially when using Beidou differential technology, the positioning accuracy can reach centimeter level or even millimeter level, which provides the possibility for resource exploration and accurate positioning [6].

How to deal with the spatiotemporal correlation that affects the accuracy of differential positioning is the biggest difficulty faced by Beidou navigation system data processing at present; the idea of dealing with local differential data is to interpolate the error correction number of flow station, and the main methods at present include weighted average method, linear interpolation method and the least square error method [7]. But the positioning accuracy of these methods is not very high. Ref. [8] only the accuracy and stability of the difference code deviation solution of BDS provided by the German Aerospace Center are analyzed, and compared with the stability of the difference code of GPS satellite, but the performance difference of the deviation data is not demonstrated.

In addition, there are a lot of scientific research achievements in the research of transmission tower attitude monitoring methods and systems. Ref. [9], the mathematical model of transmission tower attitude is derived. Combined with the parametric simulation of virtual prototype, the load between transmission tower and surrounding rock is studied, and the method of tower monitoring is proposed, but the specific monitoring of angle sensor is not studied in depth. Ref. [10, 11], the angle sensor is used to measure the lodging of the transmission tower, and the influence of noise and other interference factors is eliminated by changing the position of the sensor and Kalman filter algorithm. However, the state that the transmission tower appears the inclination at the same time in the transverse and longitudinal directions of the base is not demonstrated and analyzed. Ref. [12] proposed a method of attitude monitoring based on spatial multi-source heterogeneous data is proposed, which integrates a variety of sensor information to achieve the monitoring of tower attitude in typhoon, but the accuracy of monitoring needs to be further improved.

In order to solve the problem that the accuracy of transmission tower monitoring in the existing methods is not high, a method of transmission tower attitude monitoring in goaf based on multi-source sensing and information fusion is proposed. Combined with the existing space measurement technology and other electronic technology, it can monitor the mountain movement and transmission tower displacement, and predict and control all kinds of potential power safety accidents. The main innovations of the proposed method are summarized as follows:

(1) Due to the low accuracy of the data collected by a single sensor, the tilt sensor and gyroscope sensor are used in the proposed Beidou deformation monitoring system to collect the data of transmission tower, and the adaptive weighted fusion algorithm is used to process the sensor data, so as to comprehensively monitor the attitude state of transmission tower.
(2) In order to further improve the accuracy of monitoring, the proposed method combines the actual situation of the transmission tower in the mined out area, and uses Beidou differential code and multi parameter smooth differential to calibrate

the monitoring data, so as to dynamically reflect the attitude or translation trend of the transmission tower.

(3) Using Beidou differential technology, sensor technology and wireless communication technology, a monitoring and analysis system composed of continuous operating reference system (CORS) server system, monitoring flow station and upper computer server is designed, which can monitor the parallel movement, vertical movement, tilt and collapse of transmission tower.

2 Proposed Model of Transmission Tower Attitude Monitoring

The proposed method takes the BDS as the reference point to establish the transmission tower attitude monitoring model which integrates the data of inclination sensor and gyroscope sensor.

2.1 Beidou Deformation Monitoring System

Using the advanced Beidou satellite positioning technology and high-precision baseline solution method, the Beidou deformation monitoring system (as shown in Fig. 1) is constructed to realize the real-time three-dimensional coordinate solution of the power tower and its surrounding geology, obtain the horizontal displacement and settlement of the monitoring point, analyze the deformation development trend based on the monitoring data, and prepare for the follow-up early warning work [13, 14].

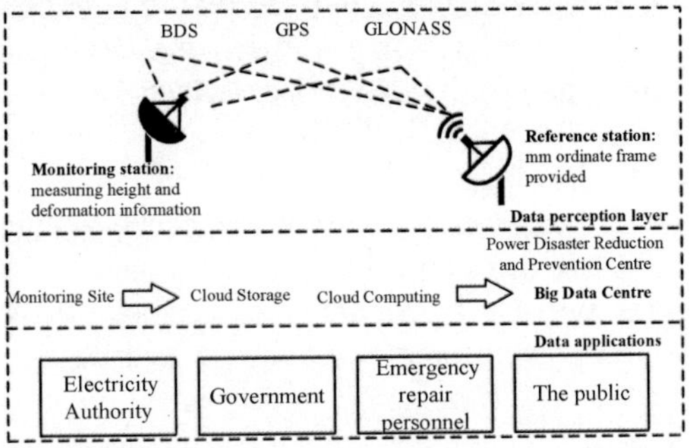

Fig. 1. Structure of Beidou deformation monitoring system

In the Beidou deformation monitoring system, the transmission tower deformation sensor sends the collected data of the tower horizontal tilt, vertical tilt, composite tilt and so on to the monitoring center through the wireless communication equipment. The monitoring center stores, displays and makes statistical reports on the status parameters, and analyzes the design parameters of the tower itself to complete the multi parameter warning function of the tower tilt. Among them, the monitoring object of the system is the environmental geological disaster monitoring around the transmission line (landslide, surface subsidence, collapse, etc.). the monitoring means are Beidou monitoring system (base station equipment, monitoring point equipment, etc.), tilt sensor, gyroscope and other sensors; the data communication is wireless network, Beidou short message, optical fiber, etc. the data is pushed to the transportation inspection department and relevant business management department.

1) Datum station: The reference station realizes the functions of satellite positioning data tracking, acquisition, transmission and system reliability monitoring, which mainly consists of outdoor equipment and indoor equipment. The indoor equipment includes Beidou data receiver, network equipment, network configuration and power distribution equipment of the reference station, and the outdoor equipment includes choke antenna and lightning protector. The main function of outdoor equipment is to receive satellite data and protect the reference station equipment.
2) *Data communication*: At present, long-distance data transmission mainly includes mobile network, optical fiber communication, communication cable, satellite, wireless network, etc. Because the monitoring system has the characteristics of remote control, remote management, real-time automatic monitoring, data two-way communication, and according to the data transmission mode of the system, Beidou is used for data communication (Beidou data pre solution mode is used).

2.2 Deformation Monitoring Point Combined with Multi-source Sensor

Through the deep integration technology research and equipment development based on Beidou system and multiple sensor technologies, the application scope of the Internet of things can be effectively improved to meet the comprehensive monitoring and supervision of environmental geology around the transmission line and the information of social service needs. Among them, the deformation monitoring point sensors include tilt sensor and gyroscope sensor, which have the advantages of small volume, good performance and high stability, and are widely used in the research of low-cost attitude monitoring.

Due to the weak anti-interference ability of gyroscope to temperature and so on, it is easy to produce random error and zero point error. In the case of long-time measurement, a single gyroscope can't guarantee the accuracy of angle measurement. The angle sensor uses the component force of gravity on each axis to get the angle. Under the state of static and uniform motion, the measurement angle is stable and accurate, but it is easy to have installation deviation in actual use, and the coupling effect between the two measurement axes of the angle sensor. The single use of acceleration sensor can't meet the requirements of accurate measurement. Therefore, the proposed method combines the angle sensor and gyroscope to collect the data of transmission tower, and uses the adaptive weighted fusion algorithm to solve the attitude angle of tower.

1) Solution of attitude angle of inclination sensor: The dual axis inclination sensor of MCO svt626-90 is selected, with resolution of 0.001°, accuracy of 0.05°, temperature drift of 0.008°/°C, which can be ignored. voltage output range is 0–10 V,

corresponding measurement angle range is −90°–90°, voltage output and angle conversion formula is as follows:

$$\theta = 18U - 90 \tag{1}$$

Where, θ is the output angle value of the inclination sensor; U is the voltage output value of the inclination sensor.

2) The solution of attitude angle of gyroscope: The gyroscope is installed on the top and base of the transmission tower. According to the working principle of the gyroscope, the angular velocity of the component can be calculated. This angular velocity is obtained by the carrier relative to the reference coordinate system, which can not be used directly. Instead, it needs to use the attitude solution to get the angle through the angular velocity integration [15].

When describing the attitude of transmission tower, it is necessary to define the coordinate system related to the attitude description. The reference coordinate system is the northeast sky coordinate system, and the carrier coordinate system is the transmission tower coordinate system. The gyroscope is positioned on the top of the transmission tower. The reference coordinate system is constant, and the two coordinate systems are coincident in the initial state. The carrier coordinate system can be seen as the composite result after the attitude in the reference coordinate system rotates around the yaw angle α, roll angle β and pitch angle γ in turn.

The initial coordinate system o-xyz obtains the final coordinate system o-$x_3y_3z_3$ through three consecutive basic rotations. After three basic rotations, the matrix is as follows:

$$\begin{aligned} C_0^3 &= C_2^3 C_1^2 C_0^1 \\ &= \begin{bmatrix} \cos\gamma\cos\alpha - \sin\gamma\sin\beta\sin\alpha & \cos\gamma\sin\alpha + \sin\gamma\sin\beta & -\sin\gamma\cos\beta \\ -\cos\beta\sin\alpha & \cos\beta\cos\alpha & \sin\beta \\ \sin\gamma\cos\alpha + \cos\gamma\sin\beta\sin\alpha & \sin\gamma\sin\alpha - \cos\gamma\sin\beta & \cos\gamma\cos\beta \end{bmatrix} \end{aligned} \tag{2}$$

The attitude transformation matrix from the final coordinate system o-$x_3y_3z_3$ to the original coordinate system o-XYZ, then the attitude angle of the transmission tower can be calculated from the attitude transformation matrix.

$$C_3^0 = \begin{bmatrix} T_{11} & T_{12} & T_{13} \\ T_{21} & T_{22} & T_{23} \\ T_{31} & T_{32} & T_{33} \end{bmatrix} \tag{3}$$

From $T_{32} = \sin\beta$, $T_{12} = -\cos\beta\sin\alpha$, $T_{22} = \cos\beta\cos\alpha$, $T_{31} = -\sin\gamma\cos\beta$ and $T_{31} = \cos\gamma\cos\beta$, the yaw angle α, roll angle β and pitch angle γ can be obtained:

$$\alpha = \arctan\left(-\frac{T_{12}}{T_{22}}\right), \beta = \arcsin T_{32}, \gamma = \arctan\left(-\frac{T_{31}}{T_{33}}\right) \tag{4}$$

3) Adaptive weighted fusion algorithm: By using the adaptive weighted fusion algorithm, the data from a single sensor are adaptively searched for their corresponding weight coefficients to achieve the best fusion effect for different types of sensors. Without any prior knowledge of the original measurement data, each sensor data complements each other to further eliminate the situation of inaccurate measurement caused by external factors.

Each sensor works independently, and the measured angle value is interfered by noise, vibration and other factors, so the height value calculated by the optimal angle value has randomness, $H_i \sim (\mu_i, \sigma_i)^2$, where μ_i is the expected value and σ_i^2 is the variance. In the case of N sensors, the variance of each sensor is $\sigma_1^2, \sigma_2^2, \cdots, \sigma_z^2$, the angle value calculated by each sensor is $\bar{\vartheta}_1, \bar{\vartheta}_2, \cdots, \bar{\vartheta}_z$, and its weighting factors are $W_1, W_2, \cdots, W_z$ respectively, then the fused $\hat{\vartheta}$ value meets the following formula:

$$\hat{\vartheta} = \sum_{i=1}^{N} W_i \bar{\vartheta}_i, \sum_{i=1}^{N} W_i = 1 \tag{5}$$

The total variance of N sensor angle values is:

$$\sigma^2 = \sum_{i=1}^{N} W_i^2 \sigma_i^2 \tag{6}$$

In the case of minimum total variance, the optimal weighting factor can be calculated as $W_i = 1 \Big/ \sigma_i^2 \sum_{i=1}^{N} 1/\sigma_i^2$. Therefore, the support height obtained by weighted adaptive fusion is:

$$\hat{\vartheta} = \frac{\sum_{i=1}^{N} \bar{\vartheta}_i}{\sigma_i^2} \Bigg/ \sum_{i=1}^{N} \frac{1}{\sigma_i^2} \tag{7}$$

3 Data Calibration Based on Beidou Difference Code and Multi Parameter Smooth Difference

3.1 Beidou Differential Code

The precise processing of BDS differential code deviation is one of the preconditions for global positioning service. At present, there are two types of BDS differential code deviation data, namely, the difference code deviation of Deutsche Zentrum fiir Luft und Raumfahrt (DLR) and the difference code deviation of Chinese Academy of Sciences (CAS).

DLR and CAS have the same method to deal with the intra frequency deviation of BDS satellite. They use two kinds of code observation measurements on the same frequency to estimate the integrated intra frequency deviation of satellite and receiver:

$$D_{L_{k,x}-L_{k,z}} = \frac{1}{N}\sum_{i=1}^{N}\left(P_{k,x}^{i} - P_{k,z}^{i}\right) \tag{8}$$

Where, $P_{k,x}^{i}$ and $P_{k,z}^{i}$ is the observation value of two different codes of the i th epoch on the frequency, and N is the total number of epoch in the calculation period.

The proposed method uses CAS to deal with the inter frequency deviation of BDS, and simultaneously estimates the parameters of the Ionospheric Total Electron Content Model and BDS satellite receiver differential code deviation using the measured data of the monitoring station. The main calculation process is as follows:

$$\bar{P}_{4,m} = 40.3\left(\frac{1}{f_1^2} - \frac{1}{f_2^2}\right)\delta_{STEC} + c(B^s + B^r) + \varepsilon \tag{9}$$

Where, $\bar{P}_{4,m}$ is the observed value of carrier phase smoothing pseudo range, δ_{STEC} is the total electron content of oblique ionosphere extracted based on carrier phase smoothing pseudo range method, B^s, B^r is the hardware delay of satellite and receiver respectively, f_1, f_2 is the frequency of two kinds of satellite signals respectively, c is the speed of light, ε is the measurement noise.

3.2 Multi Parameter Smooth Difference Calibration

With the development of modern space technology and electronic technology, continuous operation CORS is widely used in the field of positioning and navigation. The working process of CORS is shown in Fig. 2.

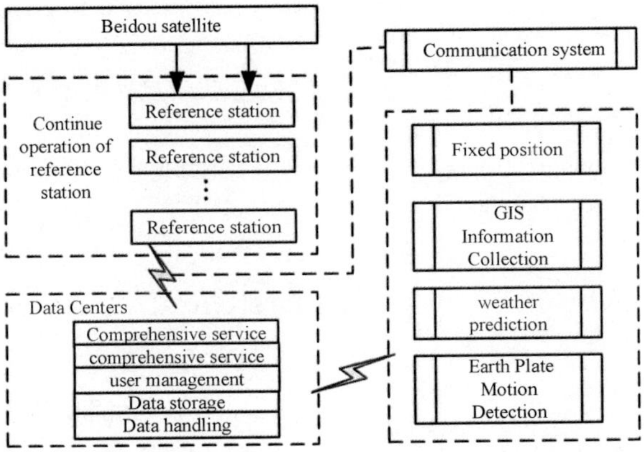

Fig. 2. Working process diagram of CORS

CORS is a modern information infrastructure to acquire spatial data and geographical features dynamically, continuously, rapidly and accurately. In the process of system positioning measurement, assuming that the ambiguity of the carrier signal does not change in the whole cycle, the BDS receiver measurement process is to continuously record the carrier signal phase. If the BDS receiver observes the same satellite in a certain period of time, and considers the multi parameter smoothing of pseudo range and phase, there are:

$$\begin{aligned} \rho_1 = (N+\varphi_1)\lambda, \rho_2 = (N+\varphi_2)\lambda \\ \cdots\cdots, \rho_j = (N+\varphi_j)\lambda \end{aligned} \tag{10}$$

Where, ρ_j is the corrected pseudo range for the j th time, φ_j is the fractional phase of the carrier signal observed for the j th time, N is the ambiguity of the carrier signal, and λ is the wavelength of the carrier signal.

j pseudo range after smoothing:

$$\rho_j = \frac{1}{j}\sum_{k=1}^{j}(\rho_k - \lambda\varphi_k) \tag{11}$$

Considering the pseudo distance difference measurement error and their weight coefficients, and considering the multi parameter smoothing of pseudo distance and phase, taking ρ_j as the user measurement distance ρ_b^i, a new multi parameter smoothing algorithm formula for differential transmission line tower positioning is obtained:

$$\begin{aligned} & K_1R_b^i + \left(\frac{1}{j}\sum_{k=1}^{j}(\rho_k - \lambda\varphi_k) - K_1\frac{1}{j_b}\sum_{k_b=1}^{j_b}(\rho_{k_b} - \lambda\varphi_{k_b})\right) \\ & + K_1\lambda(\varphi_j - \varphi_{jb}) + K_2\Delta\rho_b^i \\ & = \sqrt{(X^i - X_u)^2 + (Y^i - Y_u)^2 + (Z^i - Z_u)^2} + \Delta d\rho \\ & \Delta\rho_b^i = R_b^i - \frac{1}{j_b}\sum(\rho_{k_b} - \lambda\varphi_{k_b}) - \lambda\varphi_{j_b} \\ & = \sqrt{(X^i - X_u)^2 + (Y^i - Y_u)^2 + (Z^i - Z_u)^2} + \Delta d\rho \end{aligned} \tag{12}$$

Where, R_b^i is the actual distance between the reference station and the satellite, and K_1, K_2 is the weight coefficient.

4 Proposed Attitude Monitoring System of Transmission Tower

Aiming at the problem that the high-voltage transmission tower in goaf is easy to cause tilt, translation and lodging accidents, a transmission tower attitude monitoring and analysis system based on multi-source sensor and information fusion is developed. The system is composed of CORS service system, monitoring flow station and upper computer server software. By using remote CORS service technology, Beidou carrier phase difference technology, combined with existing space measurement technology and other electronic technology, it can effectively reduce transmission system accidents caused by tilt, settlement, displacement and other factors of transmission tower.

CORS service system is a local area network system composed of several continuous operation reference stations to provide Beidou navigation and positioning related services for users. It is mainly responsible for the transmission of the BDS system correction data to the monitoring flow station. The monitoring flow station is the monitoring terminal of the transmission tower posture monitoring and analysis system in the goaf. It is the core of the whole system. It is used to measure the settlement of the transmission tower and installed on the monitoring point of the transmission tower. It is mainly responsible for remote monitoring CORS service system communication, Beidou difference calculation is realized to get the three-dimensional coordinate information of transmission tower, and the data package is sent to the upper computer server through the communication module after calibration with the collected tilt angle. The upper computer server is the user's interface, which is responsible for receiving the data packets sent by the monitoring flow station, and storing, analyzing, displaying and warning them. It can not only display the current attitude of the transmission tower in three dimensions, but also analyze the data.

The monitoring flow station consists of solar power supply system, 32-bit microcontroller control system, Beidou Positioning System, communication system and other parts. The solar power supply system mainly consists of three parts: solar panel, charge and discharge control circuit and polymer lithium battery. The control system of 32-bit microcontroller is mainly composed of central processor, A/D conversion circuit, control circuit, peripheral signal processing circuit and calibration operation circuit based on ARM9 architecture.

5 Experimental Results and Analysis

On the developed system platform, after a period of debugging and testing, the monitoring flow station is installed on a transmission tower with serious goaf. During the installation process, the shell of the monitoring flow station is kept horizontal as far as possible. After the monitoring flow station is installed, start the upper computer server of the monitoring center, configure the translation warning value, tilt warning value, IP address and port number, and then start to receive the monitoring flow station data.

When the upper computer server software receives the first eight groups of data packets, the tower reference coordinate information is calculated by eliminating the abnormal value algorithm. After the completion of the reference coordinate configuration, the upper computer server starts to record the translation distance and tilt angle of the current transmission tower, and displays the horizontal translation distance, vertical translation distance and tilt angle of the current tower on the main interface of the server software. After receiving a certain amount of data, the upper computer server can click the data center to view all the data in the database or the attitude related data of a certain tower, or generate a report. In the monitoring data of mobile station with equipment No. 001, the data of a certain period of time is randomly selected for analysis.

5.1 Monitoring Results

The data detected by the monitoring flow station with equipment No. 001 from 10:00 to 16:00 on October 25, 2019, in which the collection period is 20 min. During this period, the relevant attitude information such as horizontal movement distance, vertical movement distance, X-direction inclination angle change and Y-direction inclination angle change of the transmission tower is plotted, as shown in Fig. 3.

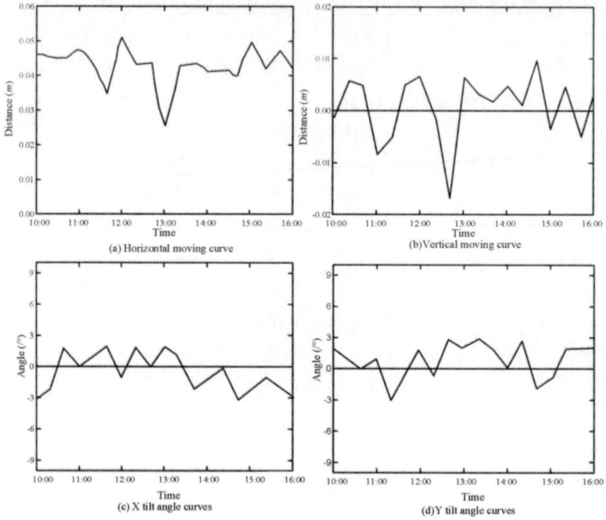

Fig. 3. Monitoring results of transmission tower attitude in goaf

It can be seen from the figure above that the transmission tower has moved a certain distance in both horizontal and vertical directions, but the amplitude is not very large, and the inclination angle also fluctuates within a certain range, and the cause of small fluctuation may be wind power, etc. It can be seen that the proposed method and system can accurately monitor the attitude of transmission tower in goaf and grasp the attitude of tower in time.

At the same time, the maintenance personnel can set the transmission tower translation or vertical movement distance warning value and tilt angle warning value through the upper computer server. When the detected transmission tower posture parameter exceeds the warning value, the upper computer server will send an alarm in time. The proposed method and system can effectively reduce the power accidents caused by the tilt, settlement, displacement and other factors of transmission towers, and greatly reduce the number of inspections, and realize the digital state management of transmission lines.

5.2 Comparative Analysis with Other Methods

In addition, the proposed monitoring method is compared with the methods in Ref. [9], Ref. [10] and Ref. [12] in terms of monitoring accuracy, and the results are shown in Fig. 4. The monitoring flow station with equipment No. 001 is in 2019, and the monitoring flow station with equipment No. 001 is in the data detected from 0:00 to 24:00 on August 25, 2019, in which the collection period is 24 h.

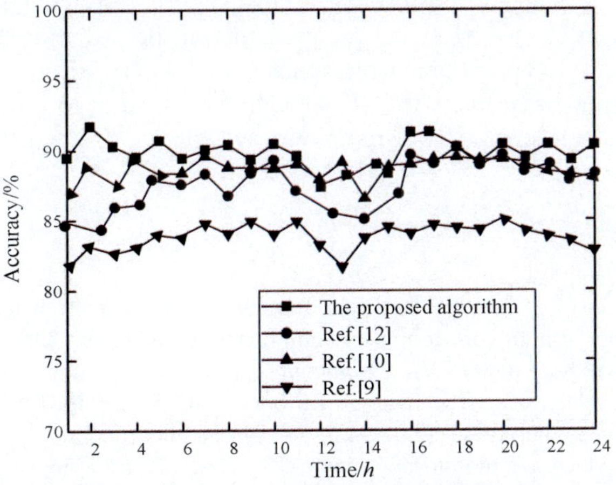

Fig. 4. Comparison results of different methods

As can be seen from Fig. 4, compared with other methods, the proposed method has the highest monitoring accuracy, about 92%. And at night, because it uses multi-source sensors such as gyroscope, and carries out information fusion, the accuracy will not decline significantly. Other methods are sensitive to light, so the accuracy of monitoring at night will be reduced. The experimental time is in summer, thunderstorm is easy to occur in summer afternoon, heavy rain has a great impact on the performance of the sensor, so the monitoring accuracy is significantly reduced, but the proposed method has the smallest drop. It can be seen that the proposed method has good performance in different weather conditions and is suitable for the complex environment of goaf.

6 Conclusion

Aiming at the problem that the high-voltage transmission tower in the goaf is prone to tilt, translation and lodging accidents, a method of transmission tower attitude monitoring based on multi-source sensor and information fusion is developed. Using the remote CORS service technology, Beidou differential code and multi parameter smooth differential technology, combined with the existing space measurement technology and other electronic technology, the transmission tower attitude monitoring system is constructed, and the system is analyzed experimentally. The results show that the proposed method can monitor the attitude of transmission tower within a certain accuracy range, and transmit it to the upper computer server, which is convenient for the staff to directly monitor. The system realizes the digital management of transmission lines, effectively reduces the transmission system accidents caused by the tilt, settlement, displacement and other factors of transmission towers, effectively improves the working efficiency and ensures the normal operation of the power grid.

The proposed method has achieved the expected goal, but the stability still needs further observation, and the positioning accuracy needs further improvement. In the future work, according to the actual requirements of the power system and its application field, we will apply the achievements of modern electronic technology to develop more products in line with the actual situation, so as to provide the necessary guarantee for the safe operation of the power system.

References

1. Cheng, L., Wu, G.: Obstacles detection and depth estimation from monocular vision for inspection robot of high voltage transmission line. Clust. Comput. **22**(2), 2611–2627 (2019)
2. Zhu, K., Lee, W.K., Pong, P.W.T.: Non-contact voltage monitoring of HVDC transmission lines based on electromagnetic fields. IEEE Sens. J. **19**(8), 3121–3129 (2019)
3. Stringer, K.L., Azuero, A., Ott, C., et al.: Feasibility and acceptability of real-time antiretroviral adherence monitoring among depressed women living with HIV in the deep south of the US. AIDS Behav. **23**(5), 1306–1314 (2019)
4. Liu, S., Wang, B., Gao, K.L., et al.: Target detection method of aerial patrol image based on R-FCN. Autom. Electr. Power Syst. **43**(13), 162–172 (2019)
5. Cheng, F., Lu, W., Liu, C., et al.: A genome-wide positioning systems network algorithm for in silico drug repurposing. Nat. Commun. **10**(1), 1–14 (2019)
6. Specht, C., Pawelski, J., Smolarek, L., et al.: Assessment of the positioning accuracy of DGPS and EGNOS systems in the bay of Gdansk using maritime dynamic measurements. J. Navigat. **72**(3), 575–587 (2019)
7. McDuie, F., Casazza, M.L., Keiter, D., et al.: Moving at the speed of flight: dabbling duck-movement rates and the relationship with electronic tracking interval. Wildlife Res. **46**(6), 533–543 (2019)
8. Grekousis, G., Gialis, S.: More flexible yet less developed? Spatio-temporal analysis of labor flexibilization and gross domestic product in crisis-hit European Union regions. Soc. Indicat. Res. **143**(2), 505–524 (2019)
9. Baik, H., Valenzuela, J.: Unmanned aircraft system path planning for visually inspecting electric transmission towers. J. Intell. Robot. Syst. **95**(3–4), 1097–1111 (2019)
10. Shi, M.J., Qin, K.Y., Li, K., et al.: Design and test of multi UAV autonomous cooperative line patrol for high voltage transmission lines. Autom. Electr. Power Syst. **41**(10), 117–122 (2017)
11. Wang, R., Tian, J., Wu, F., et al.: Optimisation electric field inverse solving algorithm for non-destructive detection of voltage on transmission lines. IET Gener. Transm. Distrib. **13** (23), 5423–5430 (2019)
12. Hou, H., Yu, S.W., Xiao, X., et al.: Risk assessment of transmission tower under typhoon based on spatial multi-source heterogeneous data. Autom. Electr. Power Syst. **44**(10), 127–137 (2020)

13. Wang, L., Li, L., Qiu, R.: Edge computing-based differential positioning method for BeiDou navigation satellite system. TIIS **13**(1), 69–85 (2019)
14. Popov, S.G., Zaborovsky, V.S., Kurochkin, L.M., et al.: Method of dynamic selection of satellite navigation system in the autonomous mode of positioning. Trudy SPIIRAN **18**(2), 302–325 (2019)
15. Zhang, C., Mao, Y., Zhou, X., et al.: Application of the Wiener filter for intensity noise reduction in fiber optic gyroscopes. Appl. Opt. **58**(20), 5568–5574 (2019)

Check for updates

Empirical Study of the Influence of AC/DC Touch Fault on Transformer Differential Protection

Qiaogen Gu(✉), Hang Lv, Xiaoyu Zhang, Lulu Cheng, and Xiao Cheng

State Key Laboratory of Smart Grid Protection and Control, Nanjing 211106, China
guqg@nrec.com

Abstract. In practical engineering, the cross and crossing of AC and DC lines are more common at present. The communication lines are complex, especially in the South. With many DC projects, the cross over phenomenon of AC and DC in the China Southern Power Grid is more obvious. This paper is based on the practical PSCAD simulation model of Jin-Zhong converter station, built a simulation model of AC and DC concomitant systems, and selecte 220 kV and 500 kV communication systems as typical scenarios for analysis by simulation of metal touch faults outside the AC transformer protection range. According to the simulation waveform, The characteristic analysis results of transformer differential protection are given in the end.

Keywords: Acrossing AC/DC lines · AC/DC touch fault · Transformer differential protection

1 Introduction

1.1 A Subsection Sample

Due to the geographical distance between the power production center and the load center in China, it is easy to occur that the capacity of the power supply side cannot be delivered, but the demand side is short of power. In order to solve the problem of efficient power transmission, China will build a power backbone network based on UHVAC and UHVDC technology. UHV power transmission is a good technical means to solve the problem of large capacity and long-distance power transmission, but at the same time, the power corridor of UHV power transmission needs to occupy increasingly tense land resources. Therefore, the sharing corridor of UHV AC and DC lines is an inevitable trend [1, 2]. Although there is no same tower layout of AC/DC lines in the current engineering application, it is not ruled out that this kind of transmission layout may appear in the future. Different from the AC/DC independent transmission mode, a series of new problems will arise if the AC/DC lines are erected on the same tower, or the AC/DC transmission lines are close to and cross each other [3, 4]. For example, the operating AC lines will produce electromagnetic coupling effect on the adjacent DC lines, resulting in the mixed phenomenon of AC electric field and DC electric field,

Y. Xue et al. (Eds.): PMF 2020, LNEE 718, pp. 40–52, 2021.
https://doi.org/10.1007/978-981-15-9746-6_4

which may lead to touching fault which occurs between the AC/DC transmission lines after long-term operation.

When touch fault happens between the AC grid and DC grid, the impact of the overlapping of DC fault current and AC fault current on the traditional relay protection is rare at present, especially for the component protection which is easily affected by DC current, such as transformer protection, it is necessary to conduct an in-depth study on whether it will be saturated and lead to protection malfunction [7]. In this paper, from the perspective of transformer differential protection, according to the actual project simulation model, the waveform characteristics of AC transformer differential protection under AC-DC line impact fault are simulated, and the adaptability of conventional transformer differential protection under such fault conditions is discussed.

2 Fault Process of AC/DC Line Impact Fault

For the neutral point grounding system of AC power grid, analyze the fault current after AC/DC touch fault, as shown in Fig. 1.

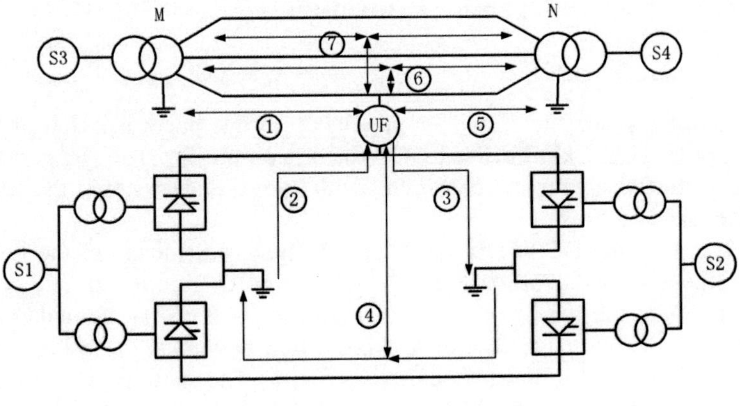

Fig. 1. Current schematic diagram of AC/DC touch fault

As shown in Fig. 1, UF is the equivalent fault power supply. The main flow of the fault current is shown as 1, 2, 3, 5 in the figure; In addition, the fault current will also flow through the phase inductance between the DC bipolar lines, as shown in the flow direction at 4; The fault current will also flow through the phase to phase mutual inductance of the AC line, as shown as 6,7 in the figure.

For the AC power grid with neutral point grounded, as shown in Fig. 1, the AC power grid composed of M-terminal and N-terminal, if an AC-DC line-ground fault occurs, the fault current will flow through the ground point of the fault point, at this time, the DC component flowing into the AC system is small. In case of AC-DC touch line ungrounded fault, the fault current will flow through the neutral point of the AC system, resulting in a large DC fault current flowing through the AC grid, which will affect the normal operation of grid equipment and relay protection.

3 Impact Assessment of Transformer Protection

When the DC current flows into the transformer, it will produce DC bias in the transformer, which may cause saturation of the core flux of the transformer [6–8]. This will not only damage the safety of the transformer, but also further affect the adaptability of the differential protection of the transformer, resulting in misoperation of the protection and further expand the accident.

At present, it is common for AC transmission lines and DC transmission lines to cross over in actual projects. The possible fault points are mainly on transmission towers and transmission lines. The transformer is usually located in the substation, the equipment occupies a relatively small area, the transformer differential protection scope is only limited to the transformer body and the lead part, and the probability of line collision fault is small. Therefore, this paper focuses on the analysis of the anti maloperation characteristics of the transformer differential protection after the line contact fault, without considering that the line contact fault occurs in the transformer differential protection area.

4 Simulation of Fault Characteristics of AC 220 kV Transformer Differential Protection

4.1 Simulation Model

The DC transmission system used for simulation in this paper is based on the actual operation of the ±500 kV Jinzhong DC transmission project. The DC part simulation model constructed is highly consistent with the actual system; the engineering parameters are as follows:

The length of the DC line is 1100 km. The main parameters of the tower are as follows: 4-bundle conductor, the height of the two poles from the ground is 30 m, the distance is 20 m, the height of the two ground wires from the ground is 41 m, the distance is 20 m; the valve control and protection system is the actual control and protection system used in Jinzhong converter station. The starting value of transformer differential protection is set to 0.3 ie, the inrush current locking mode is set to second harmonic locking, and set the inrush current locking value as 15%.

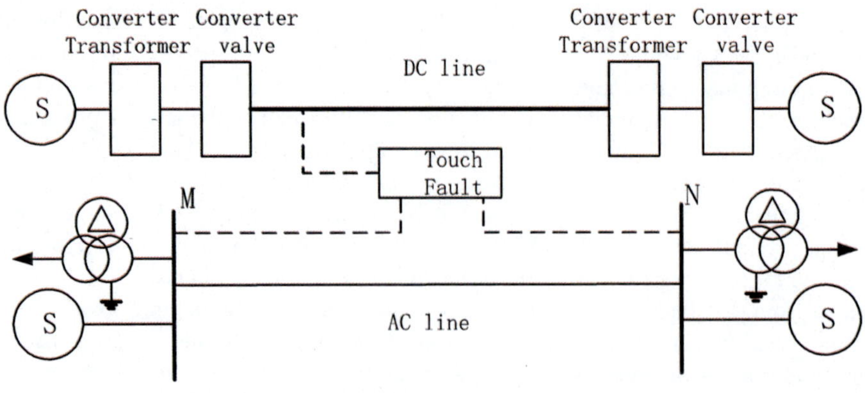

Fig. 2. Schematic diagram of AC/DC touch fault

Schematic diagram of AC and DC line collision fault is shown in Fig. 2.
The analysis methods used in this article include:

1. According to PSCAD DC model applied in Jinzhong converter station, the simulation system of AC-DC line collision fault is built;
2. Based on the basic principles of relay protection, the mathematical model of simulating the actual relay protection elements is built by using MATLAB software to simulate the vertical differential protection commonly used in UHV and UHV power grids. Based on various working conditions and the fault waveforms obtained by simulation, the adaptability of conventional differential protection in AC-DC line collision is analyzed to check whether the action of differential protection is reliable.

The transformer differential protection analyzed in this paper uses the three-fold brake characteristic, the expression is as follows:

$$\begin{cases} I_d > I_{cdqd} & I_r \leq 0.5I_e \\ I_d > K_{b1}[I_r - 0.5I_e] + I_{cdqd} & 0.5I_e \leq I_r \leq 6I_e \\ I_d > K_{b2}[I_r - 6I_e] + K_{b1}[5.5I_e] + I_{cdqd} & I_r > 6I_e \\ I_r = \frac{1}{2}\sum\limits_{i=1}^{m} |I_i| & \\ I_d = \left|\sum\limits_{i=1}^{m} I_i\right| & \end{cases}$$

Where: I_e is the rated current of transformer, I_d is differential current, I_r is constrained current. K_{b1} can take 0.5 as the typical value, K_{b2} can take 0.75 as the typical value.

The simulation model in this paper sets the DC system startup time to 12 s to ensure that the system is stable before the fault. Set the fault start time to the 12th second, the duration of the fault is 1 s; after the fault disappears, the system recovery time lasts for 7 s. Set the simulation time to 20 s in total. The analysis graph shown below omits the first 10 s of the DC system startup process during the simulation process.

Although the fault point of the AC/DC touch line itself is not grounded, due to the collision with the DC line, the DC system is grounded through the ground electrode line, so the AC short-circuit current can form a loop [8]. However, because this kind of fault is different from the common AC ground fault, it is obvious that the short-circuit impedance of AC fault circuit is greater than that of the common AC ground fault. Figure 3 shows the comparison of short-circuit current in the case of short-circuit fault of M-side line.

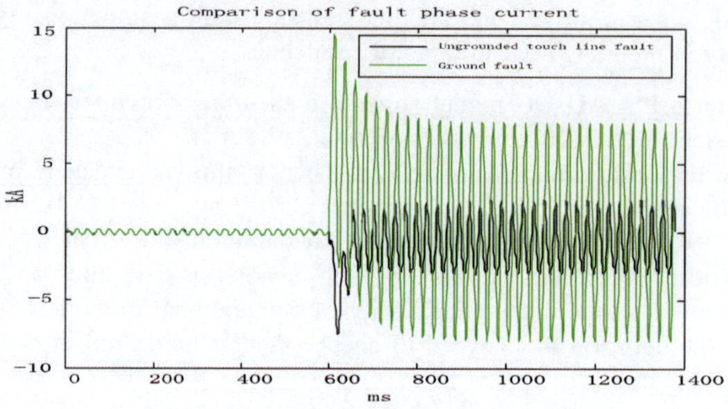

Fig. 3. Comparison of line fault current

It can be seen from Fig. 3 that due to the effect of short-circuit impedance, under the same conditions, the short-circuit current of the line in the case of an ungrounded line-to-line fault is significantly smaller than that of a general ground fault. The simulation results given in [9] are valid under certain line protection settings. However, for more general setting of line protection settings, and considering the adverse effects of DC short-circuit current on transformers [10, 11], the reliability and sensitivity of line protection will inevitably decrease [12–15]. This paper does not further discuss the issue of line protection. Based on the AC line protection, there is still a large uncertainty [16, 17]. It is only assumed that during the 1 s simulation process, the line protection does not operate and the line switch does not open.

According to the valve control and protection strategy, the DC system will enter the self-restarting process when the line protection protection action occurs after a fault occurs. If the contact point is a permanent fault, this process will gradually increase the impact of the DC short-circuit current on the AC system.

Analyzing this process, it can be foreseen that the valve control system is shut down quickly after a fault, and the transformer flows through non-periodic components for a short time, and the transformer is less saturated. Because the fault is permanent, during the subsequent transient process of the valve control system, several short-term non-periodic components repeatedly flow through the transformer winding, causing the transformer saturation to gradually deepen.

In the case of simulating DC protection input, a permanent ungrounded line fault (ABC three-phase fault) outside the transformer area occurs.

Calculate the transformer differential current under this fault. The waveform characteristics are shown in Fig. 4.

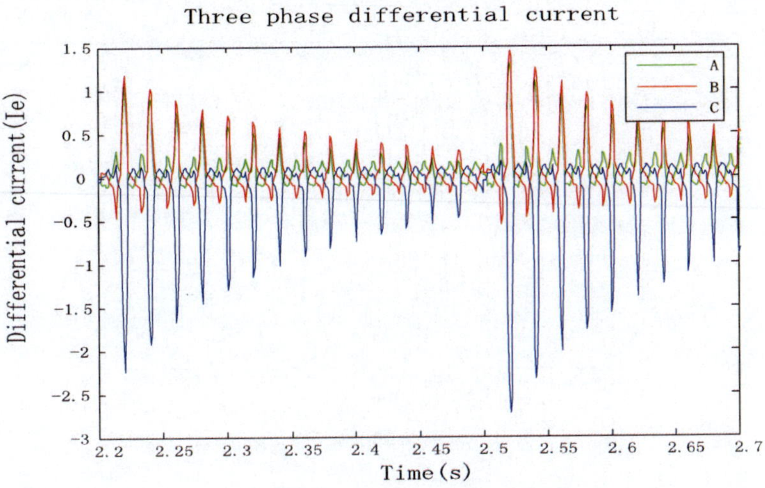

Fig. 4. Waveform of differential flow under this fault

Due to the rapid protection action of the DC system, aperiodic components only flow through the transformer in a short time, the transformer is less saturated, and the differential current is less than the differential threshold. As can be seen in Fig. 4, as the valve control system is re-energized multiple times, the differential flow gradually rises, and it is likely to reach the operating threshold. However, in this process, due to the high content of the second harmonic in the differential current, the inrush current criterion can play the role of braking differential protection.

4.2 Transformer Differential Protection in the Event of a Remote Touch Line Fault

Analyze the failure process. Assuming that there are grounded neutral points on both the M and N sides, when the line-contact fault occurs near the N side, the fault point-M side and the fault point-N side form a parallel circuit. When the fault point is closer to the N side, the line impedance serially connected to the fault point-M side is larger, and the DC component flowing into the M side is smaller.

When the simulated line fault occurred at the far end (200 km) of the AC transmission line, an ABC three-phase line fault outside the transformer area occurred.

After the fault, the three-phase current on the high voltage side was reduced and no significant DC offset was detected. After the system was put back into operation, the transformer showed no signs of saturation. The differential differential protection differential current during the fault is shown in Fig. 5.

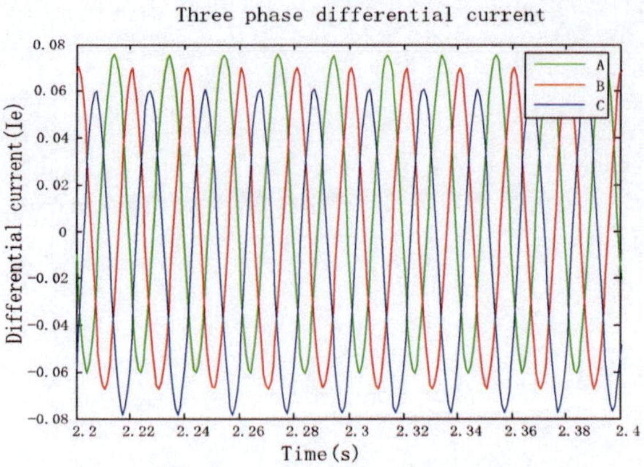

Fig. 5. Waveform of differential protection flow

It can be seen from Fig. 5 that when the AC/DC line fault point occurs at the far end of the transmission line, the flow of DC short-circuit current follows Kirchhoff's law. It chooses a circuit with a smaller impedance to flow into the ground, that is, the short-circuit current in this fault, it flows into the earth through the neutral point at the far end of the line, so there is basically no DC bias on the transformer on the local side, so the transformer differential protection does not produce a significant differential current.

4.3 Ungrounded Wire-to-Line Fault Outside the Transformer Near-End Area in the Event of Valve Protection Failure

The purpose of analyzing valve protection failure in AC/DC line impact fault is to simulate the situation of DC short circuit current passing through transformer winding for a long time.

Calculate the transformer differential current under this fault and perform harmonic analysis. The differential current waveform within 200 ms after the fault starts is shown in Fig. 6-A; the calculation of the effective value of the differential current during the

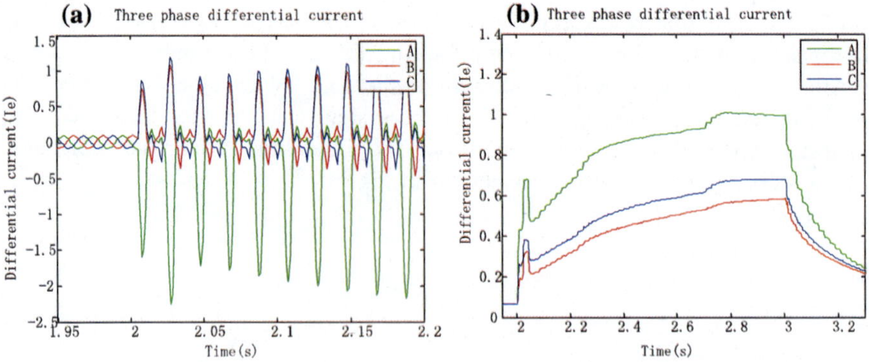

Fig. 6. A. Waveforms of differential protection flow B. Amplitude of differential protection flow

entire fault is shown in Fig. 6-B. It can be seen from the figure that after the fault, the transformer enters saturation, and the second harmonic content of the transformer winding current and differential current is rich, as shown in Fig. 7.

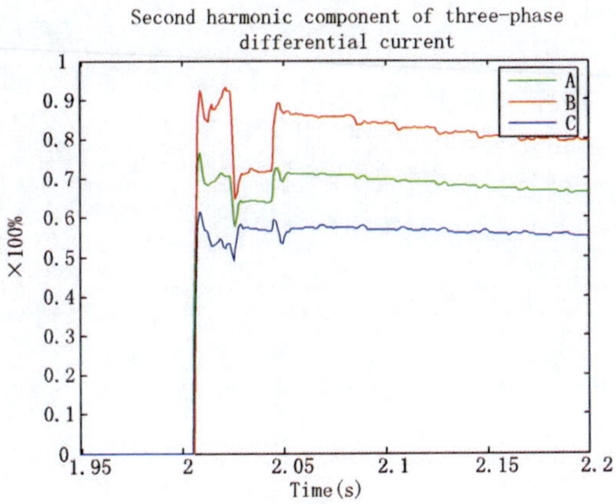

Fig. 7. The harmonic components of the fault current

After calculation, the differential differential current has satisfied the differential protection braking characteristic curve during the fault and can be operated in theory. However, due to the high second harmonic content, the inrush current characteristics are obvious. effect. In addition, the differential current did not reach the quick-set value.

5 Fault Characteristics of 500 kV Transformer Differential Protection

5.1 Line-to-Line Fault Near the Main Transformer with Parallel High Reactance

Consider the case where the 500 kV transmission line contains a parallel high-impedance connection. In the M-N transmission line shown in Fig. 2, a parallel high-impedance is added. Since there is a ground point with a small reactance through a neutral point for the high reactance, from the perspective of the transformer, it is equivalent to connecting a ground point before the neutral point of the transformer.

It is simulated that when the valve control protection is not put into operation, an ungrounded wire-faulting fault occurs in the AB phase outside the booster transformer zone.

Waveform characteristics: After a fault, the AC short-circuit current flows through the 500 kV short-circuit point. The direction of AC Short-Circuit Current on each side presents typical characteristics of out of zone faults. At the same time, the DC short-circuit current flows directly into the ground through the neutral point of the parallel high reactance, and the DC current flowing through the transformer is extremely small.

Calculate the difference current at the moment of failure. The difference current waveform within 200 ms after the failure starts is shown in Fig. 8:

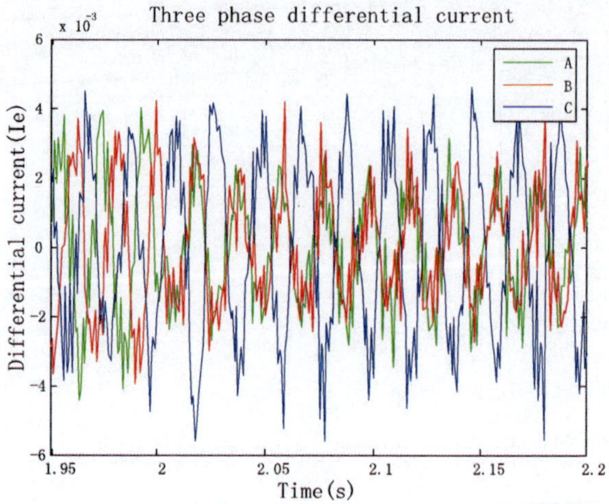

Fig. 8. Waveform of differential protection flow

As can be seen from Fig. 8, the differential current is very small, and the differential protection will not malfunction.

5.2 Line-to-Line Fault Near the Main Transformer Without Parallel High Reactance

The autotransformer has a large rated capacity, a high voltage level, and a larger core cross-sectional area. Therefore, the autotransformer has better resistance to saturation for the same amount of DC faults as the short-circuit capacity. When a line fault occurs outside the transformer area, the medium-voltage side power supply will provide an AC short-circuit current, which flows to the short-circuit point, and has a penetrating characteristic as a whole; the DC short-circuit current will directly flow into the ground through the common winding coil.

Simulate that the control protection of simulation valve is not put into operation, phase a ungrounded touch line fault occurs outside the contact transformer area. After the fault, the AC short-circuit current flows through the 500 kV short-circuit point, which is a characteristic of a fault outside the zone. After the fault, the transformer A-phase current has a certain DC bias condition. About 500 ms after the fault, the transformer gradually shows steady-state saturation.

Calculate the transformer differential current under this fault. The waveform characteristics are shown in Fig. 9:

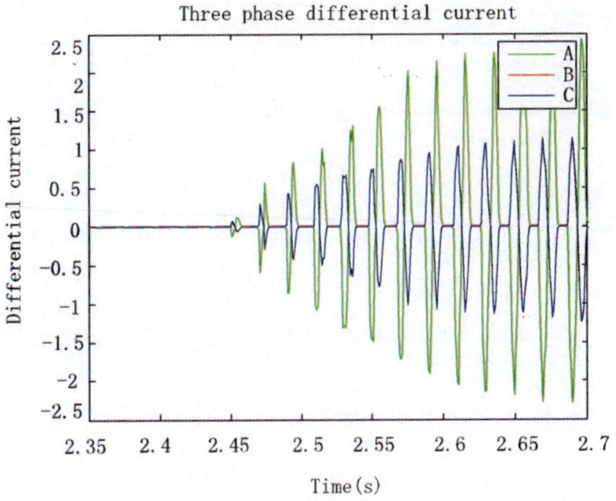

Fig. 9. Waveform of differential protection flow

As can be seen in Fig. 9, the transformer does not enter the saturated state for a period of time after the line fault occurs. During this period, the transformer differential current is very small; about 500 ms after the fault until the fault is removed, the transformer gradually enters the saturated state. At the same time, after the fault is removed, when the system recovers, the transformer generates a restoring inrush current, and the generated differential current crosses the differential threshold to meet the operating conditions of the differential protection. However, at this time, the differential current waveform is on the side of the time axis, and the harmonic content is high. The excitation inrush current criterion and the CT saturation criterion can make the differential protection reliably brake.

6 Analysis of Simulation Results

The differential protection characteristics of AC voltage and DC line fault under different voltage levels and system operation conditions are summarized, as shown in Table 1 and Table 2.

Table 1. Ungrounded AC/DC Touch Fault outside 220 kV transformer protection range

Fault type	Saturation condition	Ratio characteristic	Inrush characteristics
Near-end ABC three-phase touch line fault	With the valve reclosing, there is a lower saturation characteristic	Satisfactory	Can be blocked by inrush current
Remote ABC three-phase line touching fault	Basically no DC bias, unsaturated	Unsatisfactory	–
Near end failure when valve protection fails	Will be saturated after failure, with high saturation degree	Satisfactory	Can be blocked by inrush current

Table 2. Ungrounded AC/DC Touch Fault outside 500 kV transformer protection range

Fault type	Saturation condition	Ratio characteristic	Inrush characteristics
Connected in parallel high reactance	Unsaturated	Unsatisfactory	–
Near-end single-phase line fault	Does not saturate immediately, produces a restoring surge when the system recovers	Satisfactory	Can be blocked by inrush current
Near-end ABC three-phase touch line fault	Saturation characteristics are slight	Unsatisfactory	Can be blocked by inrush current

According to the simulation results above, it can be seen that after the AC/DC line fault, when the DC short-circuit current flows through the transformer winding and flows out through the transformer neutral point, the transformer core is liable to saturate the core due to DC bias. The factors that cause the DC bias of the transformer include: system impedance, valve control and protection characteristics, transformer capacity and structure. Specifically, the following conclusions can be drawn:

1. Low-voltage level, small-capacity transformers are easy to saturate, generating differential current can cross the brake threshold, and differentially enter the action zone. The high-voltage transformer will enter saturation slowly, and large differential current will be produced during recovery.
2. The degree to which the transformer is affected by DC bias depends on the impedance distribution of the system. The more it flows into the transformer, the easier it is to saturate. When the fault point is far away from the target transformer, the transformer is less affected. After the transformer is saturated, the inrush current criterion can play a certain braking role.
3. When a parallel high reactance is installed near the fault point, the DC short-circuit current flows into the ground through the high reactance neutral point, and the transformer is less affected. Similarly, in the event of a ground fault, the DC short-circuit current directly enters the ground, and the transformer is not affected.
4. After the AC/DC line-to-line fault occurs, the transformer near the fault point should arrange the demagnetization maintenance work of the transformer core as soon as possible.

After the transformer enters the saturation state, the core excitation resistance decreases, and the primary side excitation current increases. The imbalance of the switching current on the primary and secondary sides causes the transformer differential protection to be pushed to the action area. To sum up, the braking current of the weak system is small, and it is easy to enter the differential protection action zone after saturation. The braking current of the strong system is large, and certain braking characteristics can be guaranteed after saturation.

For low-voltage and small-capacity transformers near AC and DC lines on the same tower, the transformer differential protection setting should be appropriately increased to increase the differential protection braking interval and enhance the anti-error characteristics.

7 Summary

Based on the PSCAD simulation model applied in the actual DC engineering, a fault simulation model of AC and DC coexistence is established, and the characteristics of AC and DC transmission line collision faults in common scenarios of 220 kV and 500 kV are simulated and analyzed. Simulations show that when AC and DC line-to-line faults occur in AC systems with lower voltage levels, the AC system's transformer easily enters the saturation state, which brings the risk of transformer differential protection malfunction. For high-voltage AC transformers, although the transformer will not quickly enter the saturation state, the DC short-circuit current generated by the DC magnetic flux cannot be ignored, and it should also be paid attention to. AC and DC line faults do not essentially leave the analysis scope of short circuit calculations in power systems. Changes in system structure and composition will also affect the distribution and flow direction of short circuit currents.

It should be noted that there is still much to be done in the study of AC and DC line-to-line faults. For example, when the actual power grid fails, the fault characteristics are more complicated than the experimental simulation. After the current transformer and voltage transformer are affected by the DC bias, Which may affect its transmission characteristics, and then further adversely affect differential protection, impedance protection, etc. In addition, the interaction between the fault non-periodic component of the AC system and the DC short-circuit current after the fault needs further research.

References

1. Wang, J., Wang, Z., Li, X.: Simulation to study the dynamic performance of multi-infeed AC/DC power systems including UHVDC. Autom. Electr. Power Syst. **31**(11), 97–102 (2007)
2. Ren, S., He, W., Yang, F., et al.: Analysis and calculation for the capacitive coupling of the ac and dc transmission lines on one tower. High Volt. Apparatus **48**(4), 39–43 (2012)
3. Wu, P., Tian, M., Lu, Y., et al.: Influence of JIN-SU HTDC to transformer DC magnetic bias IN Jiangsu power grid. Jiangsu Electr. Eng. **33**(1), 5–9 (2014)
4. Wang, C., Song, G., Tang, H., et al.: Adaptability analysis of distance protection in power system integration with wind farms. Autom. Electr. Power Syst. (22), 10–15 (2015)
5. Zhu, S.: Principle and technology of relay protection in high voltage grid system. China Electric Power Press (2005)
6. Zhao, G., Ji, C., Zhang, H., et al.: China electric power research institute. Electr. Power Eng. Technol. **35**(1), 66–69 (2016). Nantong Power Supply Company
7. Huang, S., Shen, H., Liu, W., et al.: Effectof AC/DC interconnected network on transformer protection and its counter measures. Autom. Electr. Power Syst. **39**(23) (2015)

8. Liu, C., Ying, F., Zhou, N., et al.: Impact on DC bias magnetic of the power transformers by grounding electrode current of ±1100 kV Guquan convertor station. Electr. Power Eng. Technol. **37**(3), 145–150 (2018)
9. Xue, H., Wang, J., Zhang, S., et al.: Touching fault analysis and clearing action strategy for AC/DC crossed transmission lines. Autom. Electr. Power Syst. **41**(24), 150–157 (2017)
10. Yang, Z., Chen, W.: Research on the method of measuring residual flux in the core of current transformer. Electr. Power Eng. Technol. **36**(3), 38–42 (2018)
11. Liao, W., Zhou, Z., Zhan, R., et al.: Construction of the dynamic model test platform for hybrid operation of multi type current transformer and its influence on differential protection. Autom. Electr. Power Syst. **45**(22), 83–89 (2017)
12. Li, X., Jiang, W., Li, T.: Influence of Same-tower AC/DC hybrid transmission lines on DC bias of converter transformer. Autom. Electr. Power Syst. **35**(11), 87–92 (2011)
13. Hong, P.: Distance protection and impedance protection. Jiangsu Electr. Eng. **20**(1), 48–49 (2001)
14. Yang, S., He, S.: Improving method of action characteristics of distance protection and its application. Jiangsu Electr. Eng. **31**(4), 39–45 (2012)
15. Liu, H., Wei, F., Yu, B., et al.: Improvement on pilot zero sequence directional protection to avoid miss operation due to zero sequence weak-infeed and its simulation. Autom. Electr. Power Syst. **33**(22), 75–78 (2009)
16. Wang, G., Shen, J., Jiang, T.: Analysis of a 220 kV line protection failure action. Electron. Technol. Softw. Eng. (21), 224–225 (2016)
17. Suonan, J., Xu, Q., Song, G., et al.: Adaptive earth fault distance relay. Autom. Electr. Power Syst. **29**(17), 54–58 (2005)

Research on Reliability Improvement Scheme of Network in Smart Substation

Bing Shen[1], Zhihao Lu[1], Hang Lv[2,3](✉), Gui Yang[2,3], and Li Li[2,3]

[1] State Grid Shanghai Municipal Electric Power Company, Shanghai 200122, China
[2] State Key Laboratory of Smart Grid Protection and Control, Nanjing 211106, Jiangsu, China
[3] NR Electric Co., Ltd., Nanjing 211106, Jiangsu, China
lvh@nrec.com

Abstract. The data sharing based on process level network is the development trend of the digitalization of the secondary circuit in smart substation. However, abnormal problems of data transmission such as network storms bring certain hidden dangers to the safe and stable operation of the secondary equipment. In this paper, the main problems affecting the reliability of data transmission are analyzed, and the message management technology based on data flow control is proposed, which can effectively improve the reliability of data transmission. Aiming at the problem of cross-network data exchange, a solution based on cross-network port flow control is also proposed.

Keywords: Smart substation · Data transmission · Reliability · Flow control

1 Introduction

The main characteristics of smart substations are the digitization of the station-wide information, the communication platform networking, and information platforms sharing [1, 2]. The "direct sampling and direct tripping" method is widely used in China, which avoids the data transmission problems caused by abnormal switches [3, 4]. However, the method exacerbates the reliability problems caused by abnormal data links, especially for cross-bay equipment such as busbar protection. The reliability of multiple point-to-point transmission is worse than that of the dual network transmission, and the high power consumption and high heat generation of the device due to the large number of network ports affect the long-term reliability of the equipment. Therefore, replacing the secondary circuit with the process level network is an inevitable development trend, which is conducive to more convenient data sharing, simplified optical fiber circuits, and increased data transmission redundancy [5–9].

While networking brings convenience to data sharing in smart substation, but also brings certain hidden dangers to the safe operation of substations. Abnormal problems of data transmission such as network storms are among the most serious problems, which not only cause network congestion, but also seriously affect the real-time performance and reliability of data transmission, bring challenges to the reliable operation

Y. Xue et al. (Eds.): PMF 2020, LNEE 718, pp. 53–63, 2021.
https://doi.org/10.1007/978-981-15-9746-6_5

of secondary equipment such as protection and control, and seriously affect the safe and stable operation of the smart substation.

In this paper, the main causes of abnormal data in smart substation networks will be analyzed, and the specific solutions to improve the reliability of data transmission will be proposed. In the smart substation, two sets of independent networks are configured for the dual devices, but the data exchange requirement between the single set of devices and the dual devices often exists. Based on the analysis of the current status of cross-network data exchange, the scheme base on flow control is proposed to realize convenient data sharing and ensure the reliability of data transmission.

2 Analysis of Abnormal Data Transmission

The process level network is the basis for essential information transmission and sharing in smart substations, and its performance determines the safe and reliable operation of smart substations and even the power grid. In smart substation, the information flow of the process level network not only carries the operation status data of the power grid such as current sampling and the status of the circuit breaker, but also the control commands such as tripping and blocking of the secondary equipment. The performance of the process level network can be characterized as the distribution characteristics of the information flow, real-time performance, reliability, and synchronization [10–12]. In the case of abnormal network data flow caused by equipment abnormality, abnormal conditions such as network congestion and data loss may occur in the process level network, which will seriously affect the real-time performance and reliability of the information flow, and bring challenge to the reliable operation of the secondary systems such as relay protection.

The manifestations of network data transmission problems are different, and the causes of network anomalies are not only related to the network topology and data transmission methods, but also to the status of the switch equipment and IED(Intelligent Electronic Device) equipment related to the process level network [13–22]. There are several reasons that cause network data abnormality as follows.

1) Network loop

Although the star network has been adopted as the standard network topology in the smart station standard in China, and it's required that each network port of the secondary device use an independent Ethernet adapter. However, during the implementation of the project, if the network adapters of each port in certain secondary devices of are not completely independent, such as: using HUB to expand multiple network ports, and it is possible to cause a network loop when these devices are bridged in a star dual networks. Although it looks like a star network architecture, in fact, the interconnection of these ports constitutes a network loop. A large number of broadcast packets will be repeatedly forwarded in the network, which may cause network storms, and then causes crash of the network.

As shown in Fig. 1, equipment 1 and equipment 2 have two ports connected to switch 1 and switch 2, respectively. If independent Ethernet adapters are used in the different network ports of the secondary equipment, there is no problem in the star network structure. However, if the two ports of the device are actually ports expanded by a HUB chip, the two devices and two switches constitute a network loop, which risks causing network storms.

As an example, network storms once occurred in the station level network of an operating 220 kV smart station. It was found that the different ports of the recorder were not independent, when the recorders are connected to the dual networks using different ports, the interconnection of these ports constitutes the network loop. A large number of broadcast packets are repeatedly forwarded in the station level network, which affects the communication of the station level network of entire station.

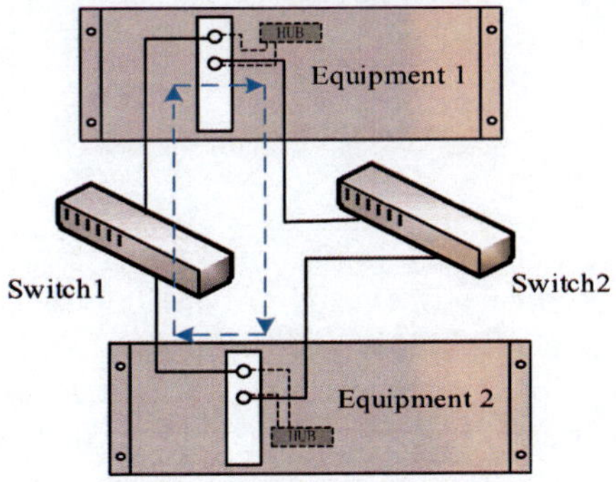

Fig. 1. Network loop caused by the non-independence of the equipment port

2) The failure of network hardware in the device
When a certain port or network card in the device is damaged, a large number of abnormal packets may be sent to the network, which will cause a large number of useless packets to be generated occupying a large amount of bandwidth. In severe cases, a network storm or avalanche will occur.

3) The defect of application
For example, writing a large number of messages into the message buffer uncontrollably may be trigged by the defect of the application in secondary device, which may also cause a large number of messages to be sent from the defective device, causing a network storm or avalanche.

VLAN (Virtual Local Area Network), static multicast, and GMRP (GARP Multicast Registration Protocol) all have a limited traffic management functions, but they cannot effectively solve the problem of interference among multiple packets in the same physical link. For example, when a GOOSE (Generic Object Oriented Substation Event) message is abnormal and a storm occurs, it may reach tens of megabytes or more in an instant. At this time, the bandwidth of the network device may be occupied by the abnormal GOOSE message, resulting in the impact of other normal messages such as packet loss, which may lead to abnormal communication on the entire network and expand the scope of influence.

As shown in Fig. 2, IED1 and IED2 send SMV (Sampled Value) and GOOSE messages to IED3, respectively. Under normal circumstances, both GOOSE and SMV messages can be received normally.

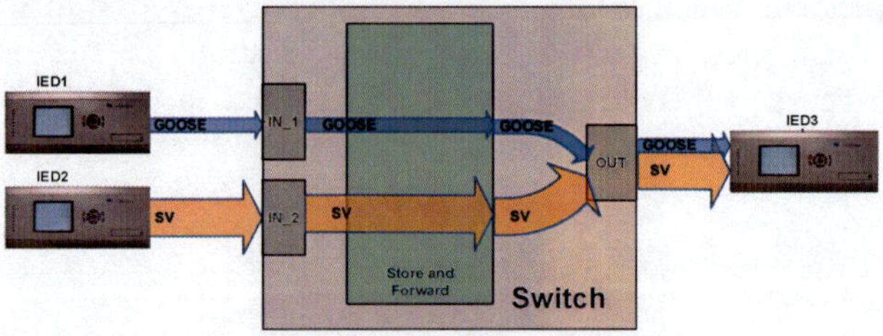

Fig. 2. Normal flow of the switch

As shown in Fig. 3, when there is abnormal traffic in the GOOSE message sent from IED1, if the priority of the SMV and GOOSE messages is set to the same, the switch will lose packets at the IED3 port because the traffic exceeds the designed bandwidth, normal SMV packets from IED2 are affected by IED1 abnormal traffic packets and frame loss occurs, resulting in incomplete SMV data received by IED3.

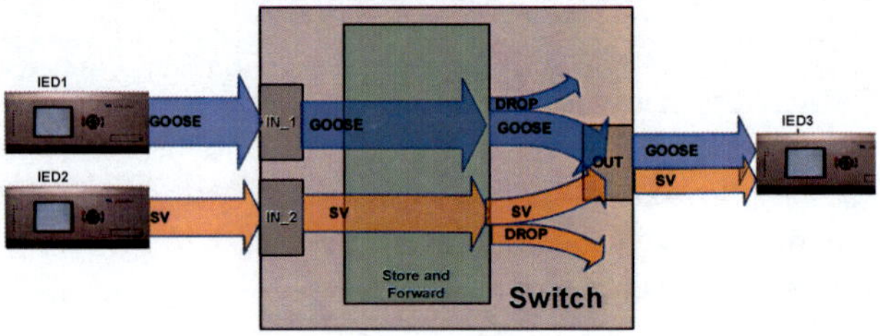

Fig. 3. Traffic in the event of network storm

3 Cross-Network Data Exchange

In some applications, a single set of equipment needs to obtain data from dual-set equipment (or data exchange between two sets of equipment), which inevitably encounters the problem of obtaining data across the dual networks. For example, a single set of BATS (backup automatic transfer switch) receives the blocking BATS signals from dual sets of transformer protection.

In order to improve the reliability of the process level network, it's required by the Chinese national standard "General specification for protection relay in smart substations": "The dual process level networks equipped for the dual protections should follow the principle of independence, when one network is abnormal or exits, it should not affect the operation of the other network".

At present, there are mainly two schemes to realize data exchange between single set device and dual-set devices, one is connecting the single set device to dual networks, and the other is connecting the single set device to the second set of devices by point-to-point fiber. Next, the operational risks of these two schemes will be analysed with the case of the single set BATS receiving the block signals from dual sets of transformer protection. In the following figures, the solid red and blue lines represent optical fiber links, and the dashed red and blue lines represent data flows.

1) Connecting the single set device to dual networks
A single set of device is connected to the dual networks using different network ports, as shown in Fig. 4. This solution has the risk of causing abnormalities in the dual networks at the same time. The more cross-network devices, the greater the risk. Especially some inconspicuous devices are easily ignored, which may lead to network abnormalities.

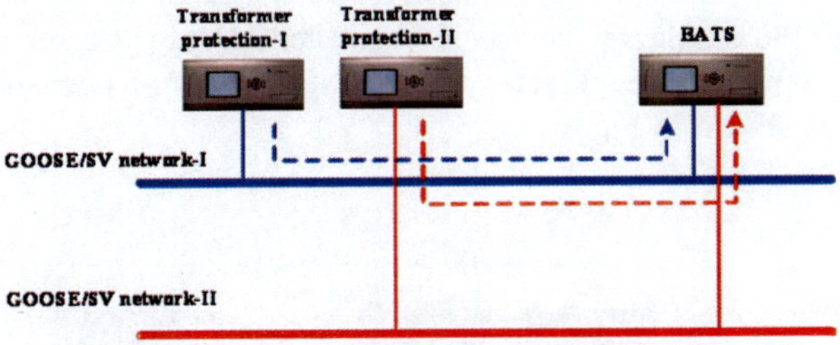

Fig. 4. Single set equipment across dual networks

As mentioned earlier, if the network adapters of different ports of the device are not independent, an invisible network loop may be formed. That is a typical case of network abnormality caused by equipment across the network.

In the Chinese national standard "General Technical Conditions for Relay Protection of Smart Substations", it is required that "devices should use independent data interface controllers when they are connected to different networks". But even if independent data interface controllers are used as required, there are still software operations problems such as abnormalities may cause the device to send a large amount of data to the dual networks simultaneously, which may affect the dual sets of process level networks at the same time.

Based on this solution, if flow control is added to the access point of the cross-network device, as shown in Fig. 5, it can prevent the abnormality of the dual-set networks caused by the cross-network device. However, any cross-network device is a risk point, and the flow control of all cross-network devices must be done to ensure the reliability of the entire process level network. Therefore, the risk points of this solution are scattered and difficult to control.

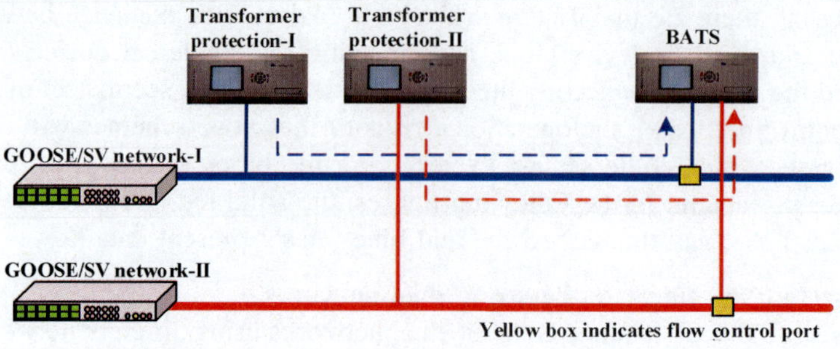

Fig. 5. Single set equipment across dual networks with flow control

2) Single set device communicate with the second set of devices using point-to-point method

In this scheme, the single set device is connected to the first set of process level networks, and communicates directly with the second set of devices using point-to-point method, as shown in Fig. 6.

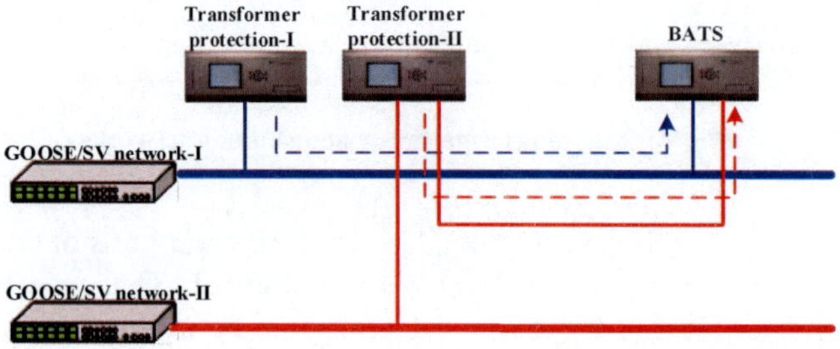

Fig. 6. Single set device communicates with the second set of devices using point-to-point

This solution requires the device to have more network ports. In some cases, the number of network ports may increase a lot, which will increase difficulties in the design of the device. The increase in communication links may also cause the overall reliability of the device to decrease.

4 Measures to Improve the Reliability of Data Transmission

1) Message management technology based on data flow control

VLAN, static multicast, and GMRP all have limited amount of traffic management functions [23–25], but they cannot effectively solve the problem of mutual interference between messages in Fig. 3.

The network flow control can realize the precise management of the network by the switch, which can solve the problem of mutual influence between the network messages and eliminate the hidden dangers to the safe operation of the secondary equipment caused by the network storm.

As shown in Fig. 7, as the data flow control function is enabled in the switch, when abnormal flow occurs in the GOOSE messages of IED1, the internal flow control policy of the switch is activated, and the GOOSE messages flow is limited to the threshold value. The SMV messages received by IED3 will not be affected by abnormal traffic of IED1.

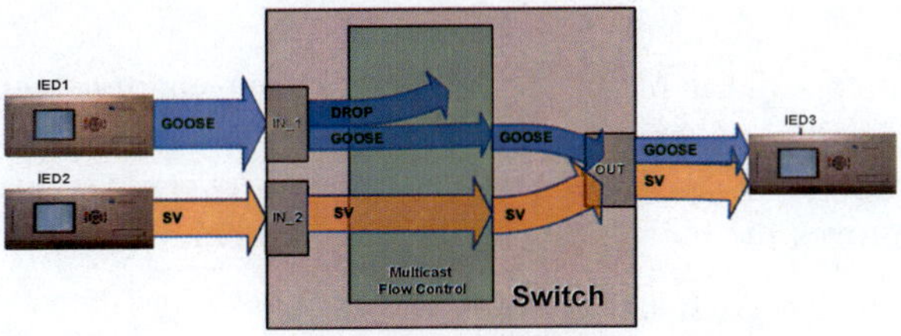

Fig. 7. Traffic in network storm case with flow control

Network storms are mainly caused by network loops and the damage of network card. There are three types of messages involved: broadcast messages, multicast messages, and unicast messages with unknown addresses. No matter what type of data packet is, when a network storm occurs, there will be a large number of repeated packets on the network, that is, the storm packets are repetitive. Network storm monitoring and filtering algorithms are mainly designed based on the characteristics of message repetition. As long as large number of duplicate messages are detected in the network in a short time, the occurrence of the network storm can be determined.

The specific implementation scheme of the message management technology based on data flow control is as follows:

a) Add FPGA (Field Programmable Gate Array) between switch processing chip and PHY (port physical layer), the message management strategy is sent from processing chip to FPGA, and mainly implemented by FPGA;
b) The FPGA registers the MAC (Media Access Control) address and APPID information of legitimate network messages in the white list of received messages, and registers the MAC address and APPID information of illegal network messages in the black list of blocked messages. The FPGA implements procession of network packets based on message management policies, such as receiving legitimate network packets, discarding illegal network packets and alarming, and discarding or forwarding network packets that are neither in the whitelist nor in the blacklist.
c) The FPGA program collects traffic statistics for different network packet classifications, discards network packets which exceed the traffic monitoring thresholds, and alerts. The SMV message is 4000 frames per second and the message length is

fixed, the traffic of SMV message is basically stable. Therefore, the flow control threshold of SMV messages need only add a small margin to the theoretical flow of SMV messages. The GOOSE message maintains one heartbeat message per 5 s in the case of no status change, and burst traffic occurs in the case of any status change. The flow control threshold needs to be set according to the system's most severe burst situation, which may reach the maximum burst flow. A typical SMV flow threshold is around 10M and a GOOSE flow threshold is around 2M.

d) The FPGA puts the CRC (Cyclic Redundancy Check) check code of the received network message into the data cache. If multiple same CRC check codes appear in the given data cache window, it indicates that there is a storm message in the network. The FPGA may discard storm messages based on message management policy.

Network flow control can realize the precise management of network packets by the switches. It effectively solves the problem that the abnormal messages occupy the network bandwidth, and avoids mutual influence between the network messages, significantly improves the reliability of data transmission.

2) Cross-network data exchange solution
In order to realize a single set of equipment obtaining data from a dual-set equipment without affecting the reliability of the dual networks, a cross-network port flow control scheme is proposed.

Effective isolation between dual networks is realized by network flow control device. Both the ports of the network flow control device and the access ports of the dual networks use flow control method (as shown in Fig. 8). It can be seen that there are four flow control points. In the event of an abnormality in the switches with access point or in the network flow control device, it is still possible to ensure that "when one network is abnormal or exits, it should not affect the operation of the other network".

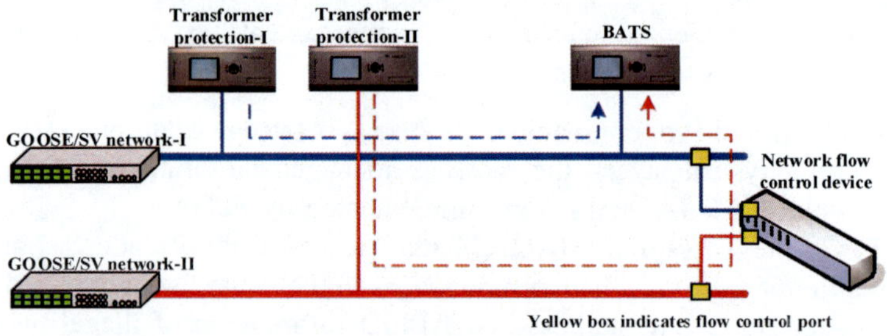

Fig. 8. Scheme of network device based on flow control

By using this solution, on the one hand, flexible data sharing between devices is realized; on the other hand, the relative independence of the dual networks is ensured by flow control of the cross-network ports, the abnormality of one network will not affect the stable operation of another network.

In order to verify the effectiveness of this solution, a test system is set up, as shown in Fig. 9. The BATS device is connected to network-I, and the blocking BATS signal from the second set of transformer protections is obtained via network flow control device. The threshold of the four flow control ports are all set to 20M.

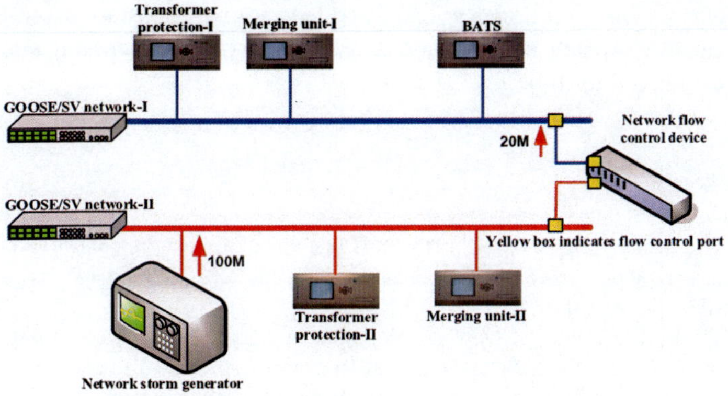

Fig. 9. Network storm test system

Test 1: Disable the protection functions of the transformer protection-I, and inject fault current to simulate a power system fault. The transformer protection-II operates to eliminate the fault. The BATS device receives the blocking BATS signal of the main transformer protection-II and executes related function normally. The test results show that this scheme realizes data exchange between devices on dual networks.

Test 2: In order to test if there is mutual influence between the dual networks due to abnormal traffic, a network storm generator is connected to the second network to simulate the subscribed messages of BATS including GOOSE storm/avalanche messages from transformer protection-II and SV storm messages.

Inject 100M of GOOSE storm/avalanche message and SV storm message to network-II, it is observed that the traffic of the access point at network-I is limited to less than 20M. Inject fault current to the dual stand-alone merging units to simulate power system fault, transformer protection-I operates normally to eliminate the fault, while transformer protection-II refuses to trip due to network congestion. The test results show that the flow control scheme ensures the relative independence of the dual networks, the abnormality of one network will not affect the stable operation of another network.

5 Conclusion

This paper analyzes in detail the main causes of network abnormal data in smart substation, and proposes a message management technology based on data flow control, which solves the problem of mutual influence between the network messages. Combined with white list filtering, network storms suppression, abnormal message identification and other functions, this scheme is conducive to improving the reliability of data transmission in smart substation.

Aiming at the needs of cross-network data exchange in smart substation, a cross-network port flow control scheme is proposed. Both flexible data sharing between cross-network devices and relative independence of the dual networks can be achieved by this scheme. Abnormality in one network will not affect the security and stability of another network. This solution has been applied in smart substation projects such as Gezhouba 500 kV switch station, and the secondary equipment on site have been working well for a long time.

References

1. Zheng, Y., Wang, D., Zhou, Z., et al.: Hierarchical protection control system of smart substations. J. Mod. Power Syst. Clean Energy **2**(3), 282–288 (2014). https://doi.org/10.1007/s40565-014-0070-2
2. Zhang, B., Hao, Z., Bo, Z.: New development in relay protection for smart grid. Protect. Control Mod. Power Syst. **1**(2), 27–33 (2016). https://doi.org/10.1186/s41601-016-0025-x
3. Han, B., Wang, Q., Sun, Z., et al.: Research on relay protection tripping modes in smart substation. Electr. Power **45**(8), 24–27 (2012)
4. Wang, T., Xie, M., Sun, Y., et al.: Analysis of reliability for relay protection systems in smart substation. Power Syst. Protect. Control **43**(6), 58–66 (2015)
5. Wang, W., Liu, M., Zhao, X., et al.: Hierarchical protection control system of smart substations. J. Mod. Power Syst. Clean Energy **2**(4), 438–443 (2014). https://doi.org/10.1007/s40565-014-0073-z
6. Fan, C., Ni, Y., Dou, R., et al.: Analysis of network scheme for process level in smart substation. Autom. Electr. Power Syst. **35**(18), 67–71 (2011)
7. Wang, S., Qiu, Y., Hou, W., et al.: Discussion of GOOSE network trip mode for smart substation. Autom. Electr. Power Syst. **39**(18), 140–144 (2015). https://doi.org/10.7500/AEPS20150204013
8. Zhao, J., Xu, C., Gao, Z., et al.: A networking technology for sampled values in smart substations based on distributed synchronization method. Autom. Electr. Power Syst. (24), 60–65, 75 (2013). https://doi.org/10.7500/aeps201302033
9. Li, H., Liu, H., Zhao, Y.: Improved process-level networking scheme of smart substation. Electr. Power Autom. Equip. **37**(3), 218–223 (2017). https://doi.org/10.16081/j.issn.1006-6047.2017.03.034
10. Chen, Q., Chen, Z., Huang, Y., et al.: Calculation of distribution of message and traffic load for process bus network in smart substation. Power Syst. Protect. Control **44**(9), 117–121 (2016)
11. Xu, C., Zhang, Y., Ge, Y.: Analysis and study on smart substation process level network data stream. Electr. Power **46**(8), 153–159 (2013)
12. Lyu, H., Chen, J., Yang, G., et al.: Synchronous sampling method based on measurement of switch data transmission delay. Autom. Electr. Power Syst. **40**(9), 124–128 (2016). https://doi.org/10.7500/AEPS20150520013
13. Zhang, J., Zhang, J., Yang, C., et al.: Abnormal traffic detection on process layer network of smart substation based on cyber physical fusion. Autom. Electr. Power Syst. **43**(14), 173–181 (2019). https://doi.org/10.7500/aeps20180611008

14. Zhu, L., Chen, X., Gao, Z.: Prevention of digital substation broadcast storm. Electr. Power Autom. Equip. **30**(11), 98–101 (2010)
15. Ouyang, F., Liu, H., Zhao, Y., et al.: Analysis on communication network congestion occurred in smart substation and preventive measures. Power Syst. Technol. **35**(11), 7–11 (2011)
16. Hao, W., Yang, Q., Li, W.: FARIMA model based analysis of communication traffic anomaly in smart substation. Autom. Electr. Power Syst. **43**(1), 158–167 (2019). https://doi.org/10.7500/AEPS20180705012
17. Zhou, H., Zheng, Y., Yao, J., et al.: Network packet control technology for secondary equipments in smart substation. Autom. Electr. Power Syst. (19), 96–100, 131 (2015). https://doi.org/10.7500/aeps20150115004
18. Huang, T., Liu, P., Xin, G., et al.: Analysis and treatment of process-level network storm in smart substation. Electr. Power **47**(11), 112–115, 120 (2014)
19. Yao, L., Zou, L., Wang, S.: Strategies on how to use smart substation relaying to deal with the network storm. Electr. Autom. **38**(5), 60–63 (2016)
20. Huang, S., Ma, W., Chen, J., et al.: Research on the network storm monitoring and filtering algorithm in smart substation. Power Syst. Protect. Control **41**(18), 68–72 (2013)
21. Wang, Z., Du, Z., Dong, Z., et al.: Study of smart substation network storm suppression technology. J. Electr. Eng. **11**(2), 40–46 (2016)
22. Yu, Y., Zhou, C., Zhan, R., et al.: Research on network stress causes and test method of relay protection in smart substation. Power Syst. Protect. Control **45**(18), 156–162 (2017)
23. Zhang, H., Eng, X.: Design and practice of process level network in substation based on GMRP. Comput. Eng. Design (7), 1681–1685 (2015). https://doi.org/10.16208/j.issn1000-7024.2015.07.001
24. Liu, J., Tian, L.: Research on VLAN division based communication network scheme of intelligent substation. Mod. Electron. Tech. **40**(20), 173–177 (2017). https://doi.org/10.16652/j.issn.1004-373x.2017.20.049
25. Lu, Y., Song, W., Yu, T., et al.: Research on reliability of the process level data in one network on smart substation. Electr. Power Sci. Eng. **28**(6), 51–55 (2012)

Short-Term Wind Power Prediction Based on Integration of Feature Set Mining and Two-Stage XGBoost

Ke Li(✉), Dongchen Huang, Zibin Tao, Haowen Li, Huan Xiong, and Yedong Du

NARI Technology Development Co. Ltd., Shenzhen Branch Company, Shenzhen 518054, China
like@sgepri.sgcc.com.cn

Abstract. Differences in geographic environment, climatic conditions, construction time, etc. lead to different available data bases for different wind farms, which increases the difficulty of wind power prediction methods with universality, making it not conducive to engineering applications. Aiming at the problems above and starting from two aspects of feature engineering and model construction, we propose a short-term wind power prediction method that integrates feature set mining and two-stage XGBoost. To begin with, a feature set mining method based on feature extraction and selection is designed. After extracting a large number of features, a feature selection method based on partial maximum information coefficient is further used for feature screening to obtain feature set composed of significant features without redundancy. Based on the above, a two-stage XGBoost model is constructed to achieve short-term wind power prediction. Using two wind farms with large differences as cases to verify, the results show that the proposed method can effectively improve the power prediction accuracy with universality in a certain degree.

Keywords: Short-term wind power prediction · Feature set mining · Two-stage XGBoost

1 Introduction

As a kind of green renewable energy that could be commercialized on a large scale, wind power has discontinuity and volatility with unobvious cycle law. When a large amount of wind power is integrated into the power grid, it may have adverse effects on the safe and stable operation of the power grid [1–3]. Therefore, in large-scale wind power integration, accurate wind power prediction plays an indispensable role.

Up till now, scholars all over the world have carried out a lot of exploration around the improvement of wind power prediction accuracy. Artificial Neural Network (ANN) [4–6] and Support Vector Machine (SVM) [7–9] are two classic wind power prediction models. ANN could adapt to self-learning and approximate any nonlinear mapping with arbitrary precision, which is suitable for describing the complex nonlinear characteristics of wind power prediction models. But it also has a trend of over-fitting. In comparison, SVM is highly generalized and not easily over-fitting. However,

Y. Xue et al. (Eds.): PMF 2020, LNEE 718, pp. 64–81, 2021.
https://doi.org/10.1007/978-981-15-9746-6_6

there is no obvious improvement when the training set has a sharp increase of examples.

In recent years, the eXtreme Gradient Boosting (XGBoost) [10] algorithm has attracted a lot of attention due to its excellent performance in a series of big data algorithm competitions such as Kaggle and KDD. The model not only performs better than other traditional algorithm in accuracy, but also supports parallel operation so that the training time of the model could be reduced. At present, the model has been widely applied to various prediction fields, including bank failure prediction [11], energy consumption prediction [12], new energy prediction [13–15], etc. Although deep neural networks have become more popular in the field of new energy prediction [16–18], XGBoost still has its unique advantages in scenarios with limited training samples, short training time and lack of parameter tuning knowledge. Compared with deep neural networks, XGBoost is able to process tabular data better and has stronger interpretability. Moreover, it has the advantages of no need to normalize the input data [19]. Therefore, we use XGBoost to construct wind power prediction model in this paper.

In addition to model construction, many scholars have chosen the selection of similar training samples [20, 21] as breakthrough points for research in recent years. Attaching importance to the refined selection of similar training samples and mining the similarity between training samples and samples to be predicted can enhance the representativeness and adaptability of prediction model to specific meteorological conditions. However, it requires a large number of samples. In fact, there are not so many samples for those wind farms with short construction time. Furthermore, the similarity measurement is time-consuming whether it is in training or prediction stage.

There are also some scholars putting the research focus on the feature engineering, such as feature extraction and feature selection. In terms of feature extraction, reference [22] analyzes the influence of wind speed on wind power generation and extracts UP-DOWN-features of wind speed. Reference [23] introduced turbulence intensity variables in meteorological data so as to better characterize wind speed fluctuation. According to the theory of multifractal spectrum, reference [24] uses variable time windows to dynamically extract features from wind speed series. The works mentioned above have certain effects, while using rich artificial experience to extract a small number of features is the common point of them. However, in practical engineering, the extracted features still pose a certain risk to the general effectiveness of the wind farms since climatic conditions, geographical features, wind farm layout, etc. of wind farms are different to some extent. In terms of feature selection, reference [25] uses the Pearson coefficient to simply filter the five features including wind speed, wind direction, temperature, atmospheric density and wind power before the model was constructed. However, the improvement is not obvious since the number of features is too limited. Reference [26] added uncanonical features on the basis of traditional meteorological features before applying a feature selection method. However, in practical engineering, the acquisition of numerical weather prediction (NWP) data and measured data related to uncanonical features requires huge costs. As to the present works, the mining of significant and meaningful feature set for wind power prediction is not sufficient, which are also lack of the consideration of universality.

Based on the above analysis, a feature set mining method combining feature extraction and selection is designed. After expanding the common meteorological features in the field of wind power prediction in single time dimension, a time series mining tool–tsfresh [27] is used to further automatically extract a comprehensive number of time series features. Besides, a feature selection algorithm based on partial maximum information coefficient (PMIC) is proposed. By introducing partial information, it is helpful to further excavate the optimal feature set that has more correlation with wind power without redundancy. Finally, the feature set mining method is combined with two-stage XGBoost for prediction model construction. We use two wind farms with large differences as cases for verification analysis. The results show that the feature set mining method proposed is able to excavate a non-redundant feature set composed of significant features. In additions, the integration of feature set mining with the XGBoost model including the two-stage regression of wind speed correction and power prediction can further improve the power prediction accuracy.

2 Feature Set Mining

For mining non-redundant feature set composed of significant features, we design an algorithm that combines the extraction of a comprehensive number of features with feature selection based on PMIC. On the basis of expanding the common meteorological features in single time dimension, tsfresh is employed to further automatically extract time series features of different time scales, thereby generating a comprehensive number of features. To get a feature set composed of significant features without redundancy, a feature selection based on PMIC is performed.

2.1 Features Extraction

2.1.1 Single Time Features Extraction

In the research of short-term wind power prediction, wind speed, wind direction, temperature, humidity and air pressure are the most commonly used meteorological features. In this paper, we extend these features by feature extraction while the result is shown in Table 1. For convenience, we call them ST features.

2.1.2 Time Series Feature Extraction

Based on the ST features mentioned in Sect. 2.1.1, we employ tsfresh to further extract time series features, which allows for over 700 time series features to be generated from time series data. For convenience, we call them TF features. The complete list of these features is available in the documentation of tsfresh (https://tsfresh.readthedocs.io/en/latest/). Additionally, tsfresh provides a feature selection function to filter out irrelevant features. However, in our cases, there are still too many TF features left after performing the feature selection operation provided by tsfresh.

Table 1. The set of ST features

Feature symbol	Feature description
V	Wind speed
H	Air humidity
T	Air temperature
θ	Direction angle
A	Atmospheric pressure
$\cos\theta$	Cosine of wind direction angle
$\sin\theta$	Sinusoidal of wind direction angle
$V\cos\theta$	The product of V and $\cos\theta$
$V\sin\theta$	The product of V and $\sin\theta$
C_V	The variation of wind speed compared to the previous time
C_D	The variation of wind direction compared to the previous time

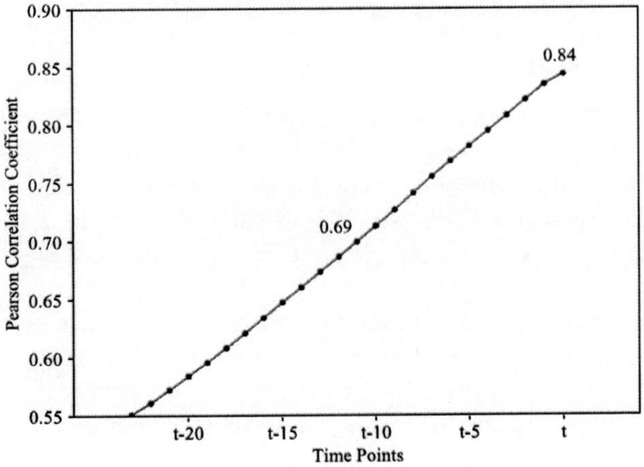

Fig. 1. The Pearson correlation coefficient of measured wind speed of the sample and wind power of wind farm A located in East China

In the process of extracting TF features, we restrict the length of time series by using correlation analysis of wind speed and wind power in order to avoid unnecessary computation cost. Thanks to the fact that wind speed is the most important factor that affecting wind power. Taking wind farm A located in East China for example for the principle illustration, as shown in Fig. 1, the horizontal axis represents the time difference between the sample and the current time t, and the vertical axis represents the Pearson correlation coefficient of measured wind speed of the sample and wind power. Generally speaking, the larger the time difference, the smaller the correlation coefficient. We note the first time point greater than 80% of the correlation coefficient at time t as $t - l$. In the scope of time $t - l$ to time t, we extract different scale of time series. For the example shown above, the time scope is from $t - 10$ to t, namely, the length of time series for feature extraction is restrict to 10. As shown in Fig. 2, the terminal of time series is always be time *t*. In practice, we use the sliding window technology to extract TF features, while a scale of time series is called a window size.

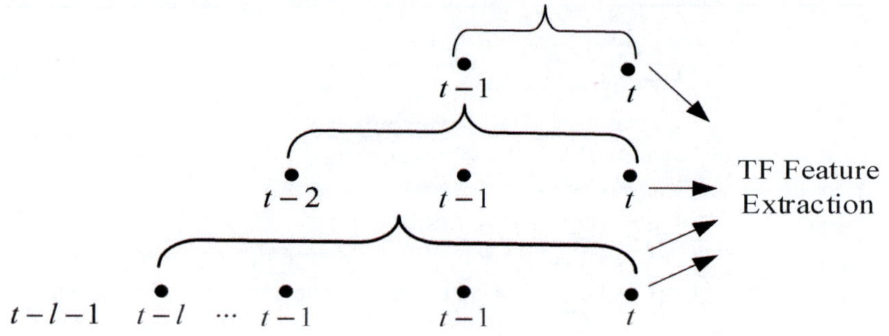

Fig. 2. The illustration of multi-scale time series feature extraction

2.2 Feature Selection

In the feature extraction stage, we only select features for TF features by feature selection function provided by tsfresh, while process does not take the ST features into account. In addition, the number of features remaining at this time is still considerable, which affects the generalization performance and computational efficiency of the model. Therefore, we propose a feature selection algorithm based on PMIC.

The maximum information coefficient (MIC) [28] is a statistic to measure associations between two variables, which could capture not only the linear and nonlinear relationship, but also the potential non-functioned relationship. It is based on the theory of mutual information (MI). Given variables $X = \{x_i, i = 1, 2, \ldots, n\}$ and $Y = \{y_i, i = 1, 2, \ldots, n\}$, where n is the sample size, the MI between X and Y is defined as

$$MI(X, Y) = \sum_{x \in X} \sum_{y \in Y} f(x, y) \log \frac{f(x, y)}{f(x) f(y)} \tag{1}$$

where $f(x, y)$ denotes the joint probability density function of X and Y, while $f(x)$ and $f(y)$ denote the marginal probability density functions of X and Y respectively, using Gaussian function to estimate the related probability densities. At this point, the MIC can be further obtained.

Given a dataset $D = \{(x_i, y_i),\ i = 1, 2, \ldots, n\}$. Suppose that the x-values of X and y-values of Y are partitioned into x bins and y bins respectively, generating a pair of partitions called an x-by-y grid G. For a specific dataset D, there are different kinds of grid partition schemes. The MI is calculated for each scheme, and the maximum value of MI under different schemes is taken as the MI of partition G. Therefore, MIC for set D under the partition G could be defined:

$$I^*(D, x, y) = \max_G I(D|G) \tag{2}$$

where $D|_G$ denotes the distribution of D on G, $I(D|_G)$ denotes MI of the probability distribution, $\max_G$ denotes traversing all possible x-by-y grid G. Different values of $I^*(D, x, y)$ can be obtained for different x-by-y partition schemes, which could be regularized and composed as a characteristic matrix.

$$M(D)_{xy} = \frac{I^*(D, x, y)}{\log_{\min}(x, y)} \tag{3}$$

The MIC is defined as

$$MIC(D) = \max_{xy < B(n)} \frac{I * (D, x, y)}{\log_2 \min(x, y)} \tag{4}$$

where $B(n)$ denotes the upper value of x-by-y grid. In the article of Reshelf et al. the authors suggest setting ε as 0.6, that is $B(n) = n^{0.6}$.

However, there obviously exists different degrees of interdependency among the large number of meteorological features related to this paper. Hence, we further introduce PMI on the basis of the MIC to transform the MIC into PMIC to eliminate the adverse effects of interdependency on feature selection.

Let X and Z be input variables in a multi-input system, and Y be an output variable. If X and Z has interdependency, there will be some deviation of the calculation of the MIC between X and Y. Therefore, we use the conditional expectation $m_X(z)$ and $m_Y(z)$ to discard the information of Z included in X and Y, which are denoted as U and V.

$$m_X(z) = E(x|Z = z) = \frac{\sum_{i=1}^{n} x_i f(z)}{\sum_{i=1}^{n} f(z)} \tag{5}$$

$$U = X - m_X(Z) \tag{6}$$

$$V = Y - m_Y(Z) \tag{7}$$

Where $f(z)$ denotes the marginal probability density function of Z. Then the PMIC of X and Y is denoted as

$$PMIC(X, Y) = MIC(U, V) \tag{8}$$

In this paper, Akaike Information Criterion (AIC) is used as the end condition of variable selection, namely

$$T_{\mathrm{AIC}} = n \log\left(\frac{1}{n} \sum_{i=1}^{n} r_i^2\right) + 2(p + 1)^2 \tag{9}$$

Where r_i is the Y regression residual calculated from the selected variables; p is the number of selected variables. With the screening of variables, the AIC value T_{AIC} decreases continuously. When T_{AIC} reaches the minimum value, the optimal independent variable set is screened.

Suppose the input variable set is F_S, the input variable is Y, the optimal input variable set is S, and F_S is the candidate variable corresponding to the largest PMIC value. The process of PMIC variable selection algorithm is as follows:

1) Initialize S to an empty set.
2) Calculate the maximum information coefficient $MIC(F_i, Y)$ of each variable in F_S and Y.
3) Select F_S that corresponding to the maximal value of $MIC(F_i, Y)$.
4) Calculate T_{AIC} value and move F_S into S.
5) If $F \neq \phi$, calculate $V = Y - m_Y(S)$, and for each $F_j \in F$, calculate $U = F_j - m_{F_j}(S)$.
6) Select F_S that corresponding to the maximal value of $MIC(U, V)$.
7) Update the T_{AIC}. If the T_{AIC} decreases, move F_S into S and return to step 5, otherwise terminate the screening.

3 Two-Stage XGBoost Model Constructing with Feature Set Mining

3.1 XGBoost Algorithm

As an optimization of boosting algorithm, XGBoost has higher model accuracy and outstanding generalization ability. In additions, it supports parallel operations, which is helpful to train the model efficiently.

Given a dataset $D = \{(x_i, y_i), i = 1, \ldots, n\}, (x_i \in R^m, y_i \in R)$ with n samples and m features, a tree ensemble model $\varphi(x_i)$ can be represented by K addictive CART regression functions $f_k(x_i)$.

$$\hat{y}_i = \varphi(x_i) = \sum_{i=1}^{K} f_k(x_i), f_k \in F \tag{10}$$

where f denotes a decision tree function of set F. In order to learn the function of (5), it is necessary to minimize the following objective function.

$$obj(\varphi) = \sum_{i=1}^{n} l(\hat{y}_i, y_i) + \sum_{k=1}^{K} \Omega(f_k) \tag{11}$$

where l is used to measure the error between predictive value $\hat{y}_i$ and actual value y_i. Since the wind power prediction is a regression problem, the function $l(\hat{y}_i, y_i)$ is commonly be replaced by square loss function $(\hat{y}_i - y_i)^2$. Additionally, the Ω is a regularization term to penalize the complexity pf the model to avoid over-fitting, which is defined as

$$\Omega(f_k) = \gamma T + \frac{1}{2}\lambda\sum_{j=1}^{T} w_j^2 \tag{12}$$

where T denotes the number of leaves in the tree, ω denotes the weight of each leaf, γ is the coefficient of T and λ is the coefficient of $\sum_{j=1}^{T} w_j^2$.

It is difficult to employ traditional optimization method such as stochastic gradient descent to optimize XGBoost model. However, it can be trained in an addictive manner. Adding a decision tree for each step based on the decision trees set of the previous step, the objective function can be written as

$$obj(\varphi)^t = \sum_{i=1}^{n} l\left(\hat{y}_i, y_i^{(t-1)} + f_t(x_i)\right) + \Omega(f_t) + C \tag{13}$$

Using the second order Taylor expansion to approximate it with remove of the constant term, the objective function can be written as

$$obj(\varphi)^t \simeq \sum_{i=1}^{n} [l(\hat{y}_i, y_i^{(t-1)}) + g_i f_t(x_i) + \frac{1}{2}h_i f_t^2(x_i)] + \Omega(f_t) \tag{14}$$

where $g_i = \partial l(y_i, \hat{y}_i^{(t-1)})/\partial \hat{y}_i^{(t-1)}$ and $h_i = \partial^2 l(y_i, \hat{y}_i^{(t-1)})/\partial(\hat{y}_i^{t-1})^2$ are first and second order gradient statistics on the objective function. The measure of using first and second order gradient statistics helps the model converge faster on the training set.

Combine the formula (12) and (14), and let $G_j = \sum_{i \in I_j} g_i$ and $H_j = \sum_{i \in I_j} h_i$, the above formula can be rewritten as

$$obj(\varphi)^t = \sum_{j=1}^{T} [G_j\omega_j + \frac{1}{2}(H_j + \lambda)\omega_j^2] + \gamma \tag{15}$$

For a confirmed tree structure q, the corresponding optimized objective function value is

$$obj(\varphi)^* = -\frac{1}{2}\sum_{j=1}^{T}\frac{G_j^2}{H_j + \lambda} + \gamma T \tag{16}$$

Let $\partial obj(\varphi)^t/\partial w_j = 0$, the optimal weight ω_j^* of leaf is

$$\omega_j^* = \frac{-G_j}{H_j + \lambda} \tag{17}$$

Formula (16) can be used to measure the quality of a tree structure q. Normally it is impossible to enumerate all the possible tree structures. Thus, a greedy algorithm is applied to searching for the optimal tree structure. Suppose I_L and I_R are instances subsets of left and right leaf nodes after splitting, Let $I = I_L \cup I_R$, the information gain for node split is given by

$$\text{Gain} = \frac{1}{2}\left[\frac{G_L^2}{H_L + \lambda} + \frac{G_R^2}{H_R + \lambda} - \frac{(G_L + G_R)^2}{H_L + H_R + \lambda}\right] - \gamma \tag{18}$$

Formula (18) can be a measurement to evaluate the split candidates.

3.2 Two-Stage XGBoost Model Constructing with Feature Set Mining

Due to the influence of surface roughness, there is a difference between the actual wind speed of the wind farm and NWP wind speed [29]. Therefore, in this paper, XGBoost is not only used for short-term wind power regression prediction, but also used to correct the NWP wind speed. Integrated with the feature set mining method mentioned above, a two-stage XGBoost model is constructed.

In the first stage, historical measured wind speed and XGBoost model are used to correct the predicted wind speed. The procedure is shown in Fig. 3. In the training stage, the feature set mining method mentioned above is used to explore the optimal feature subset from historical NWP data. Then, the XGBoost model is trained with the optimal feature subset as input, while the output is historical measured wind speed. In the prediction stage, the optimal feature subset is first extracted from the NWP data of the prediction day. Using the feature subset as input, predicted wind speed is corrected by the XGBoost model obtained in the training stage.

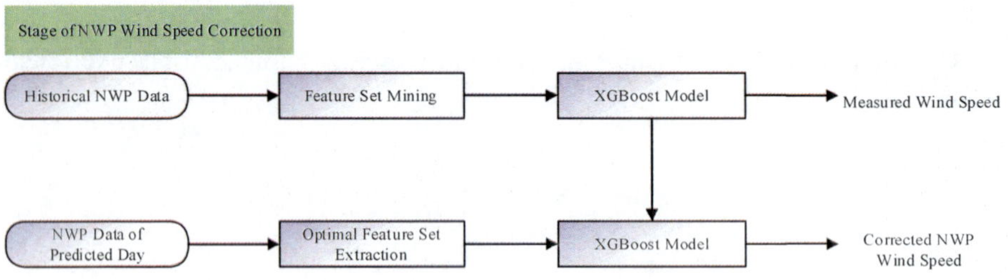

Fig. 3. XGBoost model construction for NWP wind speed correction

In the second stage, the XGBoost model is used again for power prediction. The procedure is shown in Fig. 4. In the training stage, the feature set mining method mentioned above is used to explore the optimal feature subset from historical measured data. Taking the optimal feature subset as input and the measured wind power as output, a XGBoost model is trained. In the prediction stage, the optimal feature subset is extracted from the "corrected predicted wind speed + other NWP data", the XGBoost obtained in the training stage is used to further make prediction of wind power in the predicted days.

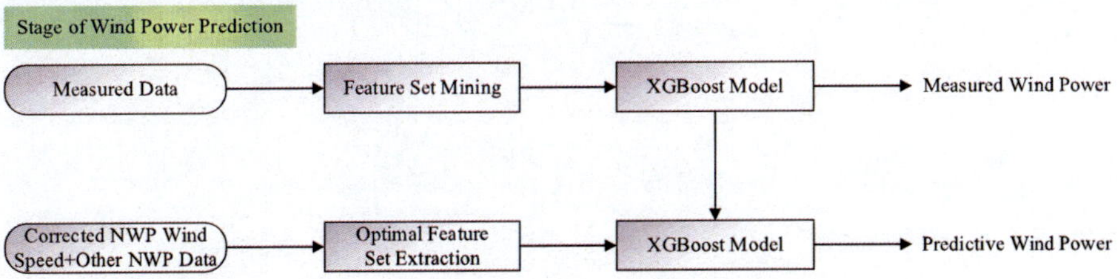

Fig. 4. XGBoost model construction for wind power prediction

4 Case Study

4.1 Dataset and Data Preprocessing

In order to demonstrate the validity of the method proposed in this paper, the measured data and historical NWP data of a wind farm A in East China and a wind farm B in Northwest China are used for relevant experiments. The data of the two wind farms are shown in Table 2. While the average wind speed of wind farm A is relatively small, and the fluctuation is relatively stable. The wind farm B has a larger average wind speed and strong volatility. Compared with wind farm B, wind farm A has more data available. For the wind power prediction model, the original input in training stage is the measured data of wind speed, wind direction, temperature, humidity and pressure, while the output is measured power. The original input in testing stage is the NWP data, and the output is predictive power. For short-term power prediction in this paper, the numerical weather forecast data for the following day is released by the forecasting provider at 0 o'clock the previous day.

Table 2. The data overview of wind farms A and B

Wind farm	Training set	Testing set	Sampling interval	Installed capacity
A	2018/05/01–2019/09/29	2019/10/01–2020/01/31	15 min	49.5 MW
B	2018/10/01–2019/01/30	2019/02/01–2019/03/31	15 min	200 MW

The data preprocessing method in [30] was used to preprocess the relevant data.

4.2 Evaluation Indexes for Wind Power Prediction

Refer to the document of the China State Administration of Energy [31], we use three key indexes to evaluate the performance of wind power prediction, which are root mean square error e_{RMSE}, mean absolute error e_{MAE} and pass rate Q_{R}. The corresponding formulas are defined as follows:

$$e_{\mathrm{RMSE}} = \frac{1}{S_{\mathrm{op}}} \sqrt{\frac{1}{N} \sum_{i=1}^{N} (P_i - P_i^f)^2} \tag{19}$$

$$e_{\mathrm{MAE}} = \frac{1}{N} \sum_{i=1}^{N} \left(\frac{\left| P_i - P_i^f \right|}{S_{\mathrm{op}}} \right) \tag{20}$$

$$Q_{\mathrm{R}} = \frac{1}{N} \sum_{i=1}^{N} B_k \tag{21}$$

$$B_k = \begin{cases} 1 & \frac{\left|P_i - P_i^f\right|}{S_{\mathrm{op}}} < 0.25 \\ 0 & \frac{\left|P_i - P_i^f\right|}{S_{\mathrm{op}}} \geq 0.25 \end{cases} \tag{22}$$

where P_i denotes the predictive value, P_i^f denotes the measured value, S_{op} denotes the installed capacity, N denotes the number of testing samples.

4.3 The Process of Feature Set Mining

To begin with, we use Pearson correlation coefficient the restrict the window sizes of time series for feature extraction. According to Fig. 1 and Fig. 5, the window sizes of time series for feature extraction is limited to 11, while the window sizes of time series for feature extraction is limited to 9.

Table 3. The number of features of wind farms A and B after feature extraction.

Wind farm	NWP wind correction stage	Wind power prediction stage
A	8977	11613
B	6197	8275

Table 3 shows the number of features of wind farms A and B after feature extraction in the NWP wind speed correction stage and power prediction stage respectively. It can be seen that the feature extraction method described in this paper can greatly enrich the feature set. On the other hand, it also explains the necessity of restricting the window sizes to avoid more expensive computational cost.

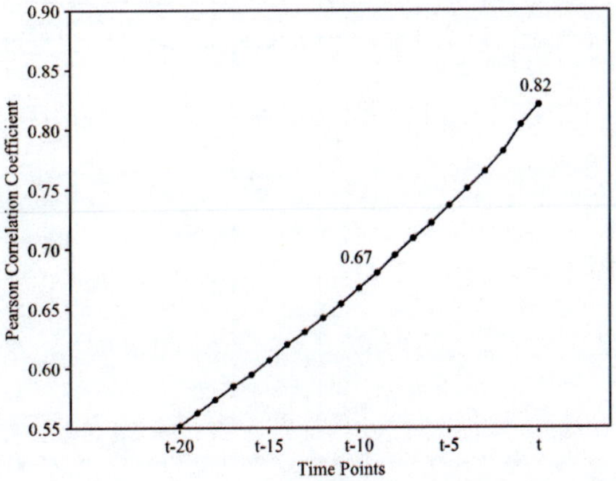

Fig. 5. The Pearson correlation coefficient of measured wind speed of the sample and wind power of wind farm B located in Northwest China

In order to get the most significant features and eliminate redundancy, it is necessary to use the feature selection method to optimize the above feature set. The corresponding selection process is shown in Fig. 6 and Fig. 7. In the stage of NWP wind speed correction: For wind farm A, when the 11th feature is selected, the T_{AIC} value is the smallest, which is −94651. For wind farm B, when the 8th feature is selected, the T_{AIC} value is the smallest, which is −25402. In the stage of wind power prediction: For wind farm A, when the 16th feature is selected, the T_{AIC} value is the smallest, which is −97252. For wind farm B, when the 13th feature is selected, the T_{AIC} value is the smallest, which is −27159. The obtained optimal feature sets are shown in Table 4 and Table 5.

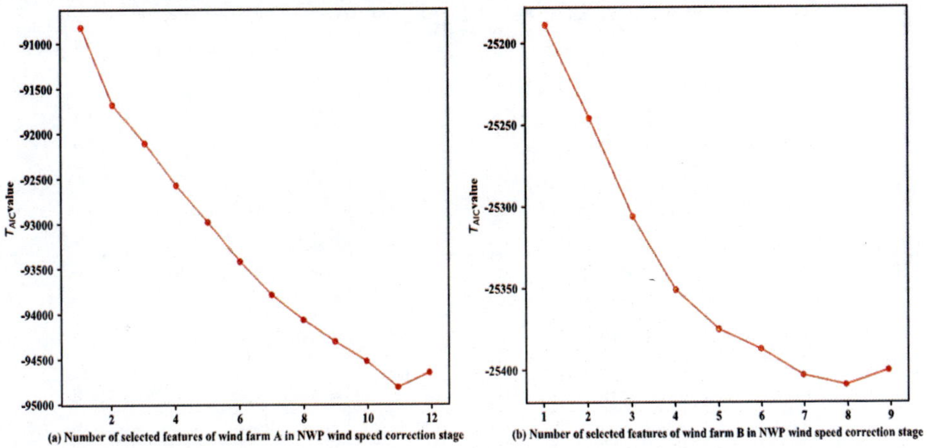

Fig. 6. Number of selected features of wind farm A and wind farm B in NWP Wind speed correction stage

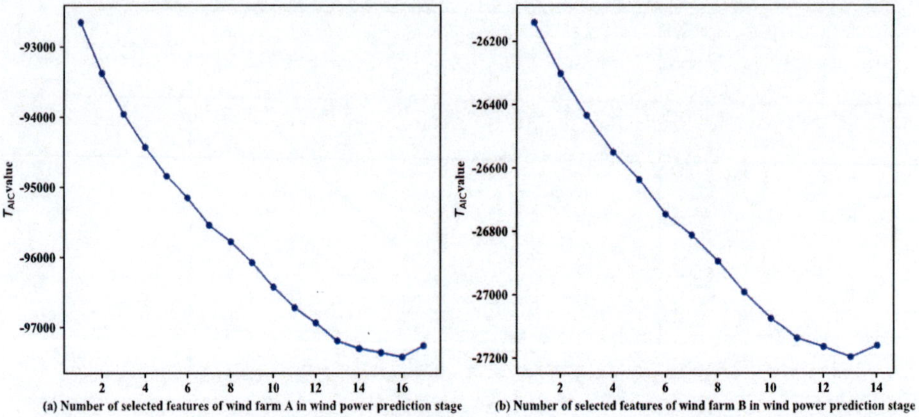

Fig. 7. Number of selected features of wind farm A and wind farm B in wind power prediction stage

Table 4. The optimal feature set of wind farm A and B for NWP data correction

Wind farm	ST feature	TF feature	Parameter of TF feature	Window size
A	V	sum_values	–	11
	θ	cwt_coefficients	coeff:2, w:20, width (2, 5, 10, 20)	10
	θ	cwt_coefficients	coeff:0, w:2, width(2, 5, 10, 20)	9
	$V\cos\theta$	maximum	–	11
	A	cwt_coefficients	coeff:9, w:5, width(2, 5, 10, 20)	11
	H	sum_of_reoccurring_data_points	–	9
	H	cwt_coefficients	coeff:0, w:5, width(2, 5, 10, 20)	4
	T	quantile	q:0.1	11
	$V\sin\theta$	maximum	–	8
	C_V	quantile	q:0.8	9
	$V\sin\theta$	change_quantiles	ql:1, qh:0.4, isabs: False, f_agg: "var"	11
B	V	cwt_coefficients	coeff:0, w:2, width(2, 5, 10, 20)	6
	H	cwt_coefficients	coeff:0, w:2, width(2, 5, 10, 20)	9
	θ	minimum	–	8
	V	linear_trend	attr: "slope"	9
	$V\sin\theta$	median	–	9
	A	minimum	–	6
	$V\sin\theta$	abs_energy	–	8
	C_V	quantile	q:0.4	4

Table 5. The optimal feature set of wind farm A and B for wind power prediction

Windfarm	ST feature	TF feature	Parameter of TF feature	Window size
A	V	quantile	q:0.9	4
	V	c3	lag:3	8
	A	cwt_coefficients	coeff:0, w:2, width (2, 5, 10, 20)	9
	H	cwt_coefficients	coeff:1, w:2, width (2, 5, 10, 20)	7
	T	fft_coefficient	coeff:0, attr: "abs"	3
	θ	–	–	–
	C_V	quantile	q:0.1	7
	T	change_quantiles	ql:0, qh:0.8, isabs: True, f_agg: "var"	11
	C_V	quantile	q:0.1	4
	T	linear_trend	attr: "stderr"	9
	A	max_langevin_fixed_point	m:3 r:30	5
	A	change_quantiles	ql:1, qh:0.4, isabs: False, f_agg: "var"	11
	H	linear_trend	attr: "stderr"	1
	C_V	quantile	q:0.4	3
	T	standard_deviation	–	11
	C_V	quantile	q:0.3	4
B	V	quantile	q:0.8	9
	θ	quantile	q:0.9	4
	V	c3	lag:3	7
	C_V	fft_coefficient	coeff:2, attr: "abs"	6
	θ	c3	lag:2	5
	H	cwt_coefficients	coeff:7, w:2, width (2, 5, 10, 20)	8
	θ	minimum	–	9
	$V \cos \theta$	quantile	q:0.9	5
	θ	variation_coefficient	–	9
	C_V	quantile	q:0.3	7
	V	–	–	–
	C_V	change_quantiles	ql:0, qh:0.4, isabs: True, f_agg: "mean"	8
	T	sum_of_reoccurring_data_point	–	9

The following points can be summarized from these tables:

(1) A comprehensive number of extracted features are greatly simplified after feature selection filtering, and the optimal feature set is basically composed of new features, which illustrates the rationality of the combination of feature extraction and feature extraction.
(2) Features such as $V\cos\theta$, C_V, etc. appear in the optimal feature set, indicating the necessity of ST Feature extension. And also, almost all the feature appear in the optimal feature set are TF feature, indicating the necessity of TF feature extraction.
(3) The window sizes of the TF features in the optimal feature set are different, indicating that it is not proper to simply extract features with a specific window size. Additionally, the features with different window sizes have to be filtered by feature selection at the same time.

4.4 The Efficiency of Two-Stage XGBoost with Feature Set Mining

In order to verify the effectiveness of the method in this paper, SVM and Long Short-Term Memory Network (LSTM) are introduced for experimental comparison, which are both popular prediction models. In order to make the algorithm more automatic and avoid future data leakage to affect the model parameters, time series split cross-validation shown in Fig. 8 and grid search are used for tuning the model parameters of XGBoost, SVM and LSTM as shown in Table 6, while the size of the cross fold K is set to 4.

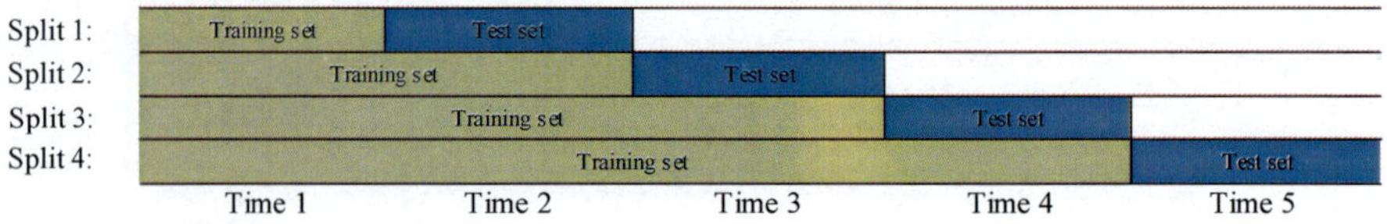

Fig. 8. The schematic diagram of time series split cross-validation

Table 6. The main parameters of SVM, LSTM and XGBoost

Model	Parameters
SVM	kernel, c, gamma
LSTM	learning_rate, num_layers, num_steps, hidden_size, keep_prob
XGBoost	n_estimators, max_depth, min_child_wight, gamma, reg_lambda

Using the evaluation indicators listed in Sect. 4.2 to evaluate different prediction models, the results are shown in Table 7. We first compare the models without feature set mining. As to wind farm A, the results of LSTM is close to XGBoost, while SVM obtains worst results since it has less advantages when the amount of training data is enough. As to wind farm B, the accuracy of LSTM is worst since it is easy to overfit when the amount of training data is insufficient. Compared with SVM and LSTM, XGBoost has better accuracy. The reasons are as follows: In terms of learning ability, XGBoost is an ensemble learning model composed of several decision trees. The number of decision trees can theoretically be arbitrarily large, which means that XGBoost can fit training samples more flexibly and learn more rich information from the sample. In addition, the measurement of the second-order Taylor expansion helps to make the model more easily converge on the training set. In terms of generalization ability, the XGBoost algorithm controls the number of leaf nodes and the weight of the leaf nodes in a single tree on the regular term to avoid overfitting, which improve the generalization ability of the model outside the training set.

Compared XGBoost with FM-XGBoost, the latter has better accuracy performance. As for wind farm A, the e_{RMSE} and e_{MAE} of FM-XGBoost decreased by 3.12% and 2.48% respectively compared to XGBoost, and the Q_R increased by 3.47%. As for wind farm B, the e_{RMSE} and e_{MAE} of FM-XGBoost also decreased by 3.09% and 2.39% respectively compared to XGBoost, and the Q_R increased by 2.87%. The comparison verifies the effectiveness of the feature set mining method in this paper.

Compared FM-XGBoost with TS-FM-XGBoost, whether it is wind farm A or wind farm B, the e_{RMSE} and e_{MAE} of TS-FM-XGBoost are less than FM-XGBoost, and the Q_R is also higher than FM-XGBoost. The result shows that the NWP wind speed correction should be performed before the power prediction, which can further improve the prediction effect.

Table 7. Evaluation indexes comparison for different models of wind farms A and B

Wind farm	Model	$e_{\mathrm{RMSE}}/\%$	$e_{\mathrm{MAE}}/\%$	$Q_R/\%$
A	SVM	13.02	9.21	84.58
	LSTM	12.93	9.16	84.91
	XGBoost	12.37	9.01	85.15
	FM-XGBoost	9.25	6.53	88.62
	TS-FM-XGBoost	8.51	5.76	89.43
B	SVM	15.61	10.98	82.83
	LSTM	16.33	11.65	82.42
	XGBoost	14.06	10.18	84.16
	FM-XGBoost	10.97	7.79	87.03
	TS-FM-XGBoost	9.95	6.86	88.34

5 Conclusion

In the practical application about short-term wind power prediction, there are two significant problems to be taken into consideration:

(1) For different wind farms, the geographic environment, climatic conditions, construction time, etc. are different, which makes the data available different. It is necessary to consider the universality problem in the research field of wind power prediction.

(2) The model accuracy problem.

To solve the problem above, a short-term wind power prediction method based on integration of feature set mining and two-stage XGBoost is proposed in this paper. Using data of two wind farms with large differences for verification. The experimental results show that the overall method can improve the accuracy of short-term wind power prediction. Additionally, there are several conclusions can be drawn.

(1) The combination of extracting a comprehensive number of features and feature selection considering partial relationship helps to obtain new and significant feature set without redundancy.
(2) As to the feature extraction part, it is necessary to perform the ST feature extension and TF feature extraction with the restriction the window sizes.
(3) TF features with different window sizes needs to be filter by feature selection at the same time.
(4) Correcting the NWP wind speed before wind power prediction helps to further improve the predictive accuracy.

In the next work, we will consider optimizing the efficiency of the feature set mining part in this paper.

References

1. Qian, Z., Pei, Y., Cao, L., et al.: Review of wind power forecasting method. High Volt. Eng. **42**(4), 1047–1060 (2016)
2. Wu, W.Z., Qiao, Y., Lu, Z.X., et al.: Method and prospects for probabilistic forecasting of wind power. Autom. Electr. Power Syst. **41**(18), 167–175 (2017)
3. Vargas, S.A., Esteves, G.R.T., Maria, P.M., et al.: Wind power generation: a review and a research agenda. J. Cleaner Prod. **218**(1), 850–870 (2019)
4. Peng, H.W., Liu, F.R., Yang, X.F., et al.: Short term wind power forecasting based on artificial neural network. Acta Energiae Solaris Sinica **32**(8), 1245–1250 (2011)
5. Lee, D., Baldick, R.: Short-term wind power ensemble prediction based on Gaussian processes and neural networks. IEEE Trans. Smart Grid **5**(1), 501–510 (2014)
6. Li, S., Wang, P., Goel, L., et al.: Wind power forecasting using neural network ensembles with feature selection. IEEE Trans. Sustain. Energy **6**(4), 1447–1456 (2015)
7. Ye, L., Liu, P.: Combined model based on EMD-SVM for short-term wind power prediction. Proc. Chin. Soc. Electr. Eng. **31**(31), 102–108 (2011)
8. Ding, Z.Y., Yang, P., Yang, X., et al.: Wind power prediction method based on sequential time clustering support vector machine. Autom. Electr. Power Syst. **36**(14), 131–135 (2012)
9. Liu, Y., Shi, J., Yang, Y., et al.: Short-term wind-power prediction based on wavelet transform–support vector machine and statistic-characteristics analysis. IEEE Trans. Ind. Appl. **48**(4), 1136–1141 (2019)
10. Chen, T., Guestrin, C.: XGBoost: a scalable tree boosting system. In: Proceedings of the 22nd ACM SIGKDD International Conference on Knowledge Discovery and Data Mining, pp. 785–794. ACM, New York (2016)
11. Carmona, P., Climent, F., Momparler, A., et al.: Predicting failure in the U.S. banking sector: an extreme gradient boosting approach. Int. Rev. Econ. Finan. **61**, 304–323 (2019)

12. Touzani, S., Granderson, J., Fernandes, S., et al.: Gradient boosting machine for modeling the energy consumption of commercial buildings. Energy Build. **158**, 1533–1543 (2018)
13. Zheng, H., Wu, Y., et al.: A XGBoost model with weather similarity analysis and feature engineering for short-term wind power forecasting. Appl. Sci. **9**(15), 3019 (2019)
14. Demolli, H., Dokuz, A.S., Ecemis, A., et al.: Wind power forecasting based on daily wind speed data using machine learning algorithms. Energy Convers. Manag. **198**, 11823 (2019)
15. Torres-Barrán, A., Alonso, Á., Dorronsoro, J.R.: Regression tree ensembles for wind energy and solar radiation prediction. Neurocomputing **326**, 151–160 (2019)
16. Niu, Z.W., Yu, Z.Y., Li, B., et al.: Short-term wind power forecasting model based on deep gated recurrent unit neural network. Electr. Power Autom. Equip. **38**(5), 43–49 (2018)
17. Khodayar, M., Wang, J.: Spatio-Temporal graph deep neural network for short-term wind speed forecasting. IEEE Trans. Sustain. Energy **10**(2), 670–681 (2018)
18. Li, P.T., Zhou, K.L., Lu, X.H., et al.: A hybrid deep learning model for short-term PV power forecasting. Appl. Energy **259**, 114216 (2020)
19. He, L.: Depth Understanding XGBoost: Efficient and Advanced Machine Learning Algorithms, 1st edn. China Machine Press, Beijing (2020)
20. Yan, J., Xu, C.Z., Liu, Y.Q., et al.: Short-term wind power prediction method based on wind speed cloud model in similar day. Autom. Electr. Power Syst. **42**(6), 53–59 (2018)
21. Xiong, Y.D., Liu, K.P., Qin, L., et al.: Short-term wind power prediction method based on dynamic wind power weather division of time sequence data. Power Syst. Technol. **43**(9), 3353–3359 (2019)
22. Ye, X.L., Chen, H., Guo, X.J., et al.: Short-term wind power prediction based on the UP-DOWN-features of wind speed. Power Syst. Prot. Control **44**(19), 56–62 (2018)
23. Huang, R., Du, W.J., Wang, H.F.: Short-term prediction of wind power considering turbulence intensity. Power Syst. Technol. **42**(6), 1907–1914 (2018)
24. Ye, L., Teng, J.Z., Lan, H.B., et al.: Combined prediction for short-term wind power based on variable time window and feature extraction. Autom. Electr. Power Syst. **41**(17), 29–36 (2017)
25. Zhu, Q.M., Li, H.Y., Wang, Z.Q., et al.: Short-term wind power forecasting based on LSTM. Power Syst. Technol. **41**(12), 3797–3802 (2017)
26. Song, J.K., Peng, Y.G., Cai, H.D., et al.: Research of short-term wind power forecasting considering multi-location NWP, Uncanonical Feature. Power Syst. Technol. **42**(10), 3234–3242 (2018)
27. Christ, M., Braun, N., Neuffer, J., et al.: Time series feature extraction on basis of scalable hypothesis tests (tsfresh – a python package). Neurocomputing **307**, 72–77 (2018)
28. Reshef, D.N., Reshef, Y.A., Finucane, H.K., et al.: Detecting novel associations in large data sets. Science **334**(6062), 1518–1520 (2011)
29. Peng, X.L., Xiong, L., Wen, J.Y., et al.: A summary of the state of the art for short-term and ultra-short-term wind power prediction of regions. Proc. Chin. Soc. Electr. Eng. **36**(23), 6315–6326 (2016)
30. Zhou, Y.H., Zhang, G.J., Li, K., et al.: Prediction on Ultra-short-term wind power based on optimal predictive time length. Guangdong Electr. Power **28**(8), 19–22 (2015)
31. QGDW 10588—2015. Function specification of wind power forecasting. State Grid Corporation of China, Beijing (2015)

Research on Comprehensive Evaluation Method of Reconstruction Plan of Distribution Area Based on Ubiquitous Power Internet of Things

Geng Lu[1(✉)], Yan Yang[1], and Zhou Jie[2]

[1] China Energy Engineering Group Jiangsu Power Design Institute Co., Ltd., Nanjing 211102, China
1046339056@qq.com
[2] State Grid Jibei Electric Economic Research Institute, Beijing 100045, China

Abstract. In order to solve the problem of diverse construction indicators of the ubiquitous electric power Internet of Things in distribution transformer service area and the lack of unified evaluation standards, a research on comprehensive evaluation method of reconstruction plan of distribution area based on ubiquitous electricity internet of things is carried out. First of all, combined with the current situations such as non-shared operation and distribution data in distribution transformer service area, duplication of terminal collection equipment functions, and heavy operation and maintenance workload, based on the ubiquitous Internet of Things construction needs in distribution system, the technical structure of the Internet of Things in distribution transformer service area is proposed and the two typical transformation schemes – edge-end separation type and edge-end fusion type are summarized from them. After that, a comprehensive evaluation index system for the construction of smart stations was established from three aspects: the level of operation and distribution, the quality of customer service, and the level of smart things. Second, the DEMATEL-ANP-entropy weight method and the improved gray correlation analysis method are used to determine the weight of each assessment factor and the index scoring function is used to obtain a comprehensive scoring result. Finally, combined with a demonstration project of the power distribution Internet of Things in a city in Jiangsu Province, a comprehensive evaluation of its distribution transformer service area reconstruction plan is performed to verify the feasibility of the proposed comprehensive evaluation method.

Keywords: Ubiquitous power Internet of Things · Distribution transformer service area · DEMATEL-ANP-entropy weight method · Improved gray correlation analysis method · Index system

1 Introduction

State Grid Corporation put forward the strategic goal of "three types of two networks and world-class" in 2019, and the ubiquitous electric power Internet of Things is its main component [1]. The low-voltage distribution network is usually managed by the

Y. Xue et al. (Eds.): PMF 2020, LNEE 718, pp. 82–98, 2021.
https://doi.org/10.1007/978-981-15-9746-6_7

area (distribution transformer service area) of the power supply range of the distribution transformer [2]. "Outline of Ubiquitous Electricity Internet of Things Construction" of State Grid Corporation emphasizes the "optimization and promotion of operation and distribution" business [3]. Since the distribution transformer service area is also the junction of marketing and equipment expertise. The two parties install concentrators and distribution transformer terminals on the side of the distribution transformer service area based on business needs, which will lead to problems such as unshared business data, duplicate functions of terminal acquisition equipment, and heavy operation and maintenance workload [4, 5]. At the same time, the current ubiquitous electric power Internet of Things construction has a variety of indicators, and the evaluation criteria for the transformation plan in the distribution transformer service area are not uniform, which is not conducive to the comparison and selection of plans and decision-making.

This article conducts in-depth research and analysis on the above issues. Based on the ubiquitous IoT construction needs in distribution system, this paper studies the transformation plan of the distribution transformer service area, and analyzes two on-site implementation options: edge-end separation type and edge-end fusion type. In addition, a comprehensive evaluation index system for the effectiveness of smart station construction is constructed from three aspects: the level of operation and distribution, the quality of customer service, and the level of smart Internet of Things. The weight of each evaluation factor is determined through DEMATEL-ANP-entropy weight method and improved gray correlation method. Finally, the index score function is used to calculate the comprehensive score result, which can objectively reflect the construction effect of the smart distribution transformer service area and make the choice of the scheme more reasonable. Also it is of great significance to promote the construction of the power distribution Internet of Things.

2 Reconstruction Plan of Distribution Transformer Service Area Based on Ubiquitous Power Internet of Things

2.1 Current Situation

At present, due to the different frequency of data requirements of the equipment in the distribution transformer service area and the diversified concerns of business departments, the frequency of data collection for distribution transformers, feeders, branch boxes, and household meters in the station area is 5 min, 15 min, and hours. Due to the frequency problem during data application, the cross-professional application of data lacks a credible foundation for multi-data fusion. The original power distribution automation system, power consumption information collection system, moving ring system, and inspection robot system were independent of each other, and the data integration was not achieved. The current situation of low-voltage fault location in the distribution transformer service area is slow, and repairs are not timely, resulting in long power outages and low user satisfaction [6–8].

In order to realize the automatic collection of electricity meter data, the marketing majors deploy gateway electricity meters, collectors, and smart meters in low-voltage stations. The concentrator aggregats the meter data and connects the collected data to the power information collection system through communication methods such as wireless public network/private network (see Fig. 1).

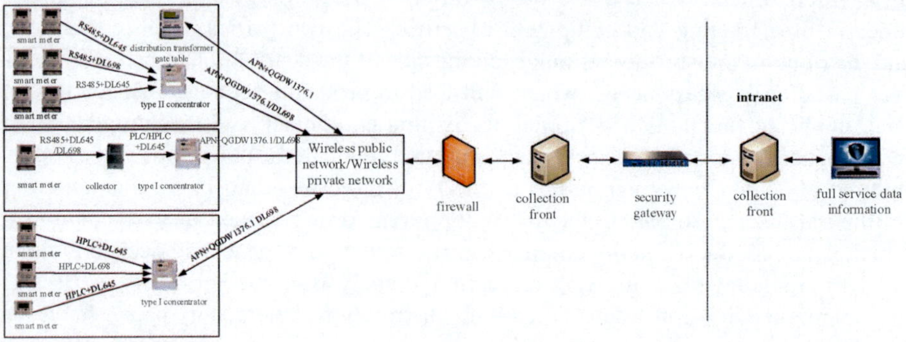

Fig. 1. Electricity information collection system for marketing professionals.

In order to realize the intelligent monitoring of the transformer, distribution room environment, and low-voltage lines in the station area, sensing terminals such as low-voltage branch monitoring units, low-voltage fault sensors, and environmental sensors are deployed in the low-voltage station area. The intelligent distribution terminal TTU aggregates the sensing data through the wireless public network/private network and other communication methods, the collected data is connected to the new generation distribution automation master station system in the management information area (see Fig. 2).

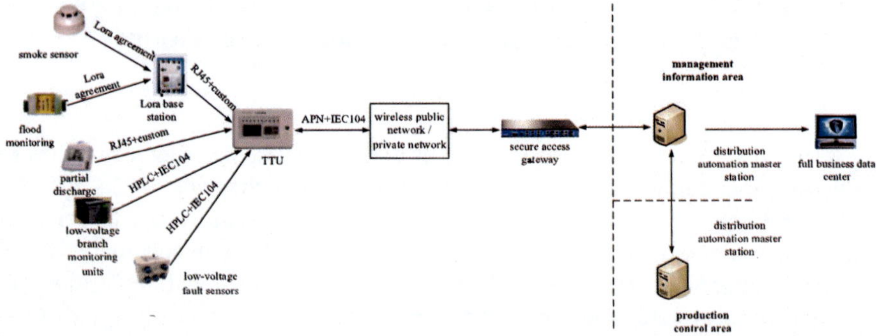

Fig. 2. Equipment professional intelligent distribution transformer monitoring system.

2.2 Transformation Plan

According to the "cloud-network-edge-terminal" technology architecture, aiming at the integration of equipment on the distribution transformer service area and the penetration of business distribution data, through the deployment of new concentrators and edge IoT agents. User power consumption information, environmental awareness and equipment status data are extensively accessed to upgrade the electric power wireless communication network, build a unified IoT-based management platform and implement advanced repairs such as active repairs, lean management of line losses in stations, active inspection of stations, and accurate positioning of low-voltage faults (see Fig. 3).

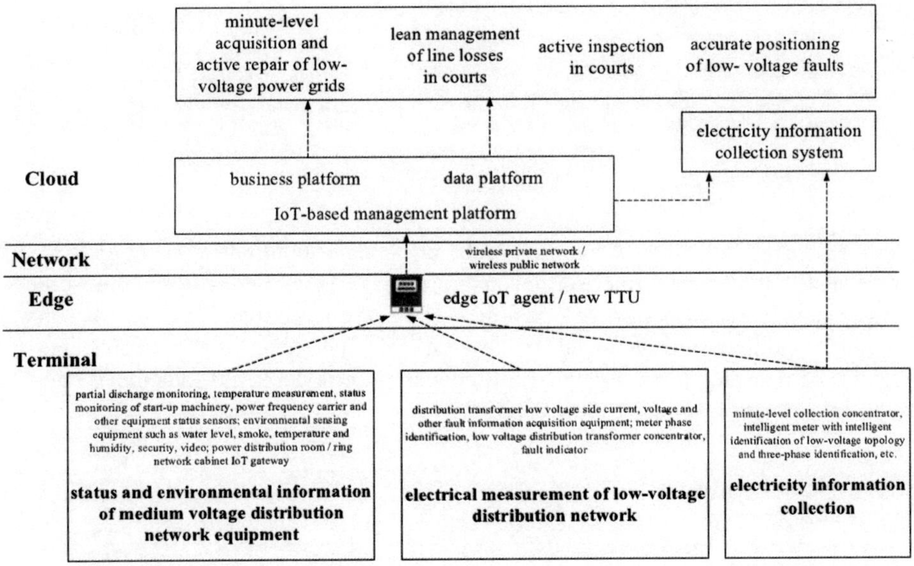

Fig. 3. IoT technology architecture in distribution transformer service area.

The demonstration project of power distribution Internet of Things in Jiangsu Province is investigated and the implementation plan for the transformation of the distribution station area is summarized well, which is mainly divided into two types: edge-end separation type and edge-end integration type.

Edge-Separated Solution. According to the principle of configuring one edge IoT agent in one power distribution room, an edge IoT agent is deployed in one of the stations in the power distribution room to access the mobile ring host and the new type II concentrator (see Fig. 4). The new type II concentrator has a one-shot, double-shot, minute-level collection function. In order to ensure that the new concentrator access scheme does not affect the operation of the existing mining system, the new type II concentrator has been submitted to the provincial metrology center for testing and obtained the network asset number.

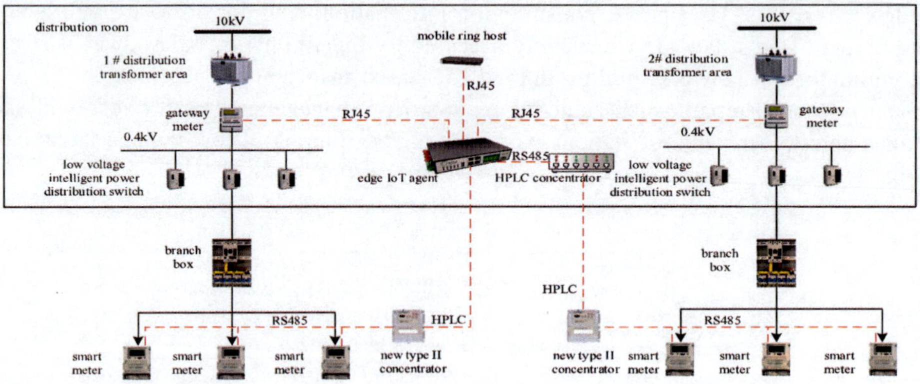

Fig. 4. Edge-separated solution

Edge-Blending Solution. According to the principle of configuring an edge IoT agent in the distribution transformer service area, a new TTU with edge computing function in each station area or perform a function upgrade of the original TTU that has software upgrade conditions are deployed as the edge side IoT agent at the station area (see Fig. 5). At the same time, a new type II concentrator is deployed to achieve one-shot, two-shot, minute-level collection functions.

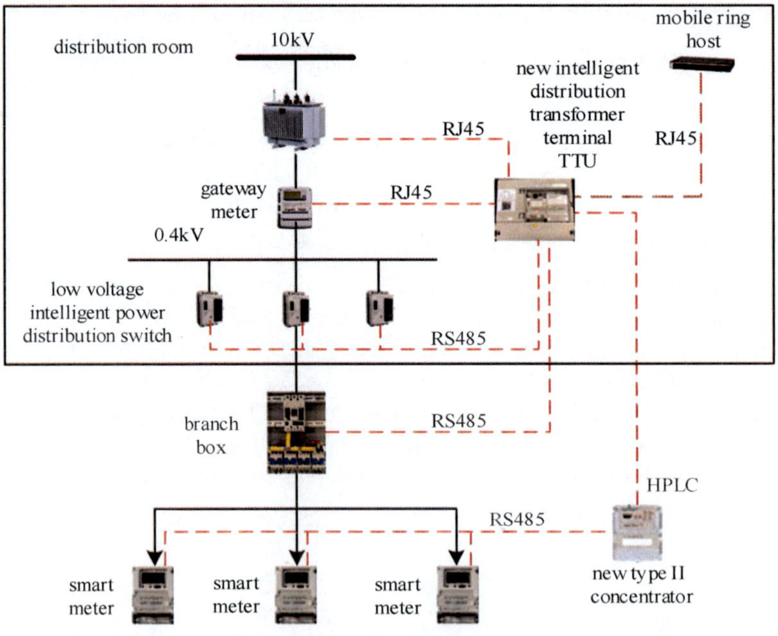

Fig. 5. Edge-blending solution.

3 Index System

The comprehensive evaluation index system for transformation scheme of the distribution transformer service area based on ubiquitous power internet of things proposed in this paper establishes evaluation criteria from three aspects: operational level, customer service quality and smart IoT level. Also it establishes specific indicators under each criterion layer.

3.1 Operational Allocation Level Indicators

0.4 kV Negative Loss Rate. Corresponding distribution transformer refers to the monthly line loss rate for the same period in distribution transformer service is not more than −1% and power supply gate table in distribution transformer service area is complete, monthly power supply is greater than 300 kWh or monthly sales is more than 450 kWh, the number of corresponding distribution transformers in the distribution transformer service area whose power loss is less than −90 kWh.

0.4 kV Negative Loss rate = the Number of Corresponding Distribution Transformers in the Distribution Transformer Service area/10 kV Negative Loss Rate × 100%.

Timeliness of Reporting the Power Outage (on) of Smart Meters (Station-Wide Meter). The number of smart meter (station-wide meter) power-off (power-on) events reported in a timely manner refers to the number of events that distribution automation, power consumption information collection and other systems upload the distribution transformer trip (off) events to power supply service command system and 95598 business support system within 5 min after 10 kV distribution transformer trip (off).

Timeliness of reporting the power outage (on) of smart meters (station-wide meter) = number of smart meter (station-wide meter) power-off (power-on) events reported in a timely manner/number of 10 kV distribution transformers for planned, temporary, and fault outages × 100%.

Consistency Rate of Online Data and Field. Whether the information on marketing business applications, equipment asset lean management, power grid geographic information platform substations, lines, distribution transformers, user accounts, and graphic topology are consistent with the site.

Agreement rate = Amount of online data consistent with the field/Number of online data × 100%.

3.2 Customer Service Quality Indicators

User Average Outage Time. Data statistics range is 'City center + Urban area + Town + Countryside' (Eliminate the influence of uncontrollable factors such as major policy adjustments and natural disasters).

User's average outage time = $\sum$ (Duration of each power outage × Number of users per power outage)/Total number of users during the statistical period.

Analysis of Outage Information for Households. Realizing analysis of outage information for households refers to the amount of outage information submitted to outage equipment (distribution transformers) and complete user list in a timely manner during the statistical period. Among them: the equipment name, equipment identification, and equipment type of the equipment affected by the power outage are not empty, and the marketing application system can automatically associate the corresponding files; there are valid corresponding files in the user name and user code in the user list affected by the power outage. When the power outage (sending) information is first reported, the user status should be normal power users.

Analysis of outage information for households = the amount of outage information to realize analysis for households/total number of outage information × 100%.

Probability of Outage Notification to Households Precisely. The number of messages of informing users of power outages actively refers to the number of times to inform users of power outages through online channels such as 'Online State Grid', SMS, etc. according to the time requirements of the power outage announcement in accordance with *State Grid Corporation 95598 Customer Service Business Management Measures*. Repeated notifications from various channels shall be combined and counted as one.

Probability of outage notification to households precisely = the number of messages of informing users of power outages actively/total number of users affected by the power outage × 100%.

3.3 Smart IoT Level Indicator

Practical Level of Power Equipment IoT Equipment. The online rate of sensing terminals and node equipment = the number of sensing terminals and node equipment that have not been offline for 3 consecutive days/the total number of sensing terminals and node equipment in the area × 100%.

Access Rate of Perceived Terminal. Perceived terminals include smart energy meters, transformers, acquisition terminals, meter boxes, etc. access rate of sensing terminals = number of sensing terminals accessing the IoT management platform/total number of sensing terminals × 100%.

Accuracy of Fault Investigation and Judgment. Accuracy of fault investigation and judgment = repair orders accurately judged/total number of repair orders × 100%.

4 Comprehensive Evaluation Method

In the comprehensive evaluation index system for the transformation plan of the distribution transformer service area based on the ubiquitous electric power Internet of Things, the relationship between the three criteria of the level of operation and distribution, customer service quality, and smart Internet of Things is complex and ambiguous. In view of this, this paper uses the DEMATEL-ANP-entropy weight method to comprehensively consider subjective and objective information to calculate the weights of each criterion layer, and then use the improved gray correlation analysis method to find out the weights of each index in the criterion. Finally, the comprehensive scores of different wiring modes can be calculated according to the index weight, so as to provide decision makers with a basis for selection.

4.1 DEMATEL-ANP-Entropy Qeight Method

DEMATEL method, can be used to simplify the system structure analysis process, which means using "decision and evaluation experiments" to screen the main elements of complex system. At the same time, the network analysis method (Analytic Network Process, ANP) can fully consider the interaction and dominance of the elements, so as to solve the actual decision-making problem [9]. Therefore, this paper uses the network analysis method based on DEMATEL to determine the subjective weights of the three criteria.

Assume that there are n criteria in the network layer of ANP, which are represented by $C_1, C_2, \cdots, C_n$, and there are dependencies between the criteria, and y_{ji} is the importance of the criterion $C_j (j \neq i)$ to C_i. Taking $C_i \quad (i = 1, 2, \cdots, n)$ as a criterion in turn, comparing C_i with other index values except itself to obtain the corresponding weight matrix.According to the characteristic root method, a weight vector $W_i = \left[w_1^{(i)} \ \cdots \ w_{i-1}^{(i)} \ w_{i+1}^{(i)} \ \cdots \ w_n^{(i)}\right]^T$ with C_i as a criterion is obtained, and the weight matrix is shown in formula (1):

$$\begin{bmatrix} 1 & \cdots & \frac{y_{1i}}{y_{(i-1)i}} & \frac{y_{1i}}{y_{(i+1)i}} & \cdots & \frac{y_{1i}}{y_{ni}} \\ \vdots & \ddots & \vdots & \vdots & \ddots & \vdots \\ \frac{y_{(i-1)i}}{y_{1i}} & \cdots & 1 & \frac{y_{(i-1)i}}{y_{(i+1)i}} & \cdots & \frac{y_{(i-1)i}}{y_{ni}} \\ \frac{y_{(i+1)i}}{y_{1i}} & \cdots & \frac{y_{(i+1)i}}{y_{(i-1)i}} & 1 & \cdots & \frac{y_{(i+1)i}}{y_{ni}} \\ \vdots & \cdots & \vdots & \vdots & \ddots & \vdots \\ \frac{y_{ni}}{y_{1i}} & \cdots & \frac{y_{ni}}{y_{(i-1)i}} & \frac{y_{ni}}{y_{(i+1)i}} & \cdots & 1 \end{bmatrix} \tag{1}$$

Under each criterion layer, the criterion does not have an influence relationship on itself, that is, the diagonal element of the matrix is 0, and the direct influence matrix W_d is obtained, see Eq. (2):

$$W_d = \begin{bmatrix} 0 & w_1^{(2)} & \cdots & w_1^{(i)} & \cdots & w_1^{(n)} \\ w_2^{(1)} & 0 & \cdots & w_2^{(i)} & \cdots & w_2^{(n)} \\ \vdots & \vdots & \ddots & \vdots & \ddots & \vdots \\ w_i^{(1)} & w_i^{(2)} & \cdots & 0 & \cdots & w_i^{(n)} \\ \vdots & \vdots & \ddots & \vdots & \ddots & \vdots \\ w_n^{(i)} & w_n^{(2)} & \cdots & w_n^{(i)} & \cdots & 0 \end{bmatrix} \tag{2}$$

To find the limitable comprehensive influence matrix for the direct influence matrix W_{d}, see Eq. (3):

$$W = \lim_{n \to \infty} (W_{\mathrm{d}})^n \tag{3}$$

The elements of each row of the matrix tend to a certain stable value, and the non-zero stable value of each row is the corresponding subjective weight value w_{si} of each element, then the subjective weight set is

$$w_s = \{w_{si} | 1 \leq i \leq n\} \tag{4}$$

The entropy weight method is a comprehensive evaluation method that can be used for multiple objects and multiple indicators. The evaluation results are mainly based on objective data and are almost free from subjective influence. In order to avoid the extreme sensitivity of the index difference in the entropy weight method, which leads to the extreme situation that the index is too small in the weight distribution, this article uses the anti-entropy weight method to determine the objective weight. Assume that there are m evaluation objects, n evaluation indexes, and an index evaluation matrix $X = (x_{ij})_{n \times m}$, in which $x_{ij} (i = 1, 2, \cdots, n; j = 1, 2, \cdots, m)$ is an index value. The steps of the anti-entropy weight method to obtain the index weight are as follows:

Determine the anti-entropy of each indicator, as shown in Eq. (5).

$$h_i = -\sum_{j=1}^{m} r_{ij} \cdot \ln(1 - r_{ij}) \tag{5}$$

In the formula, $r_{ij} = x_{ij} \Big/ \sum_{j=1}^{m} x_{ij}$.

According to h_i obtained by the above formula, weight normalization processing is performed to obtain w_{oi}, as shown in formula (6).

$$w_{oi} = h_i \Big/ \sum_{i=1}^{n} h_i \tag{6}$$

Then the objective weight set is

$$w_o = \{w_{oi} | 1 \leq i \leq n\} \tag{7}$$

The relative importance of subjective and objective weights varies according to the criteria. Let the relative importance levels of subjective and objective weights be expressed as α_i and β_i, respectively. Combined with the basic ideas of matrix estimation theory, the subjective and objective weight important coefficients α_i and β_i of each criterion are finally calculated. The formula is shown in formula (8).

$$\begin{cases} \alpha_i = w_{si}/(w_{si} + w_{oi}) \\ \beta_i = w_{oi}/(w_{si} + w_{oi}) \end{cases}, 1 \leq i \leq n \tag{8}$$

Using the obtained subjective weight set, objective weight set, and the relative importance coefficients of subjective and objective weights, the combined weight can be finally calculated. The formula is shown in formula (9).

$$w_i = \frac{\alpha_i w_{si} + \beta_i w_{oi}}{\sum_{i=1}^{m} (\alpha_i w_{si} + \beta_i w_{oi})} \tag{9}$$

4.2 Improved Grey Correlation Method

The improved gray correlation method fully combines subjective information of expert experience judgment values with a scientific simple mathematical model, and overcomes the shortcomings of traditional gray correlation analysis methods that are easily affected by the value of the resolution coefficient in the calculation process. The calculation method and steps are as follows:

Step 1: Determine the initial weight value.

Set m experts to make empirical judgments on n evaluation index weights to determine the target, and form a judgment data column, see formula (10):

$$X = [X_1, X_2, \ldots, X_n]^T = \begin{bmatrix} x_{11} & x_{12} & \cdots & x_{1m} \\ x_{21} & x_{22} & \cdots & x_{2m} \\ \vdots & \vdots & \ddots & \vdots \\ x_{n1} & x_{n2} & \cdots & x_{nm} \end{bmatrix} \tag{10}$$

The character x_{nm} stands for the weight value in formulas (10)–(12). The range of x_{nm} is [0, 1].

Step 2: Determine the reference sequence.

The maximum weight value is selected from each column of X as the reference weight value to form a reference data column X_0.

$$X_0 = [x_{01}, x_{02}, \ldots, x_{0m}] \tag{11}$$

Find the distance between each indicator sequence $X_1, X_2, \ldots, X_n$ and the reference data X_0.

$$D_{0i} = \sum_{k=1}^{m} (x_{0k} - x_{ik})^2 \tag{12}$$

Step 3: Find the weight of each indicator.

$$\tilde{w}_i = 1/(1 + D_{0i}) \tag{13}$$

Step 4: Find the normalized weight of each indicator.

$$\overline{w_i} = \tilde{w}_i \Bigg/ (\sum_{i=1}^{n} \tilde{w}_i) \tag{14}$$

Finally, the comprehensive score of different wiring modes can be calculated according to the index weight.

4.3 Implementation of Comprehensive Assessment Method

The implementation process of the comprehensive evaluation method for the construction of smart stations based on the ubiquitous electric power Internet of Things is shown in Fig. 6. First, determine the weight of the criterion layer of the evaluation system, then determine the weights of the indicators of the evaluation system, and finally establish an index system scoring function. The index weight calculates the comprehensive score result (see Fig. 6).

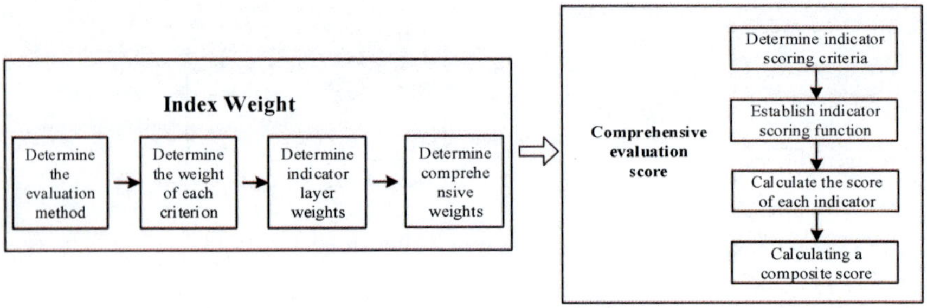

Fig. 6. Integrated assessment method implementation process.

Weight of the Criterion Layer of the Evaluation System. By investigating the evaluation results of the construction results of the ubiquitous electric power Internet of Things demonstration area in Jiangsu Province, the empirical judgments on the mutual direct influence between the macro demand indicators are made, and the three criteria C_1, C_2, C_3 are used to represent the level of business allocation, customer service quality, and smart IoT level. The layer index, using the 1–5 scale method, can be divided into five basic scales: general, slightly important, obviously important, strongly important, and extremely important. The corresponding numerical scales are given to 1, 3, 5, 7, and 9 respectively. Measured to get the impact relationship matrix, as shown in Table 1:

Table 1. Criterion layer indicator influence relationship matrix.

Scale	C_1	C_2	C_3
C_1	/	5	7
C_2	5	/	9
C_3	7	9	/

According to formulas (1)–(4), the subjective weight vector formula can be obtained as shown in (15):

$$w_s = [0.378 \quad 0.252 \quad 0.37]^T \tag{15}$$

When determining the objective weights, the anti-entropy weight method is used for calculation. First, through the survey results, refer to the initial weighted experience scores of the macro demand indicators during the planning of the three ubiquitous power IoT demonstration areas. Finally, the initial judgment matrix X shown in formula (16) is obtained. Observation shows that the elements of each row of matrix X respectively represent the set of weighted opinions on a certain indicator during the planning of each demonstration area, and the elements of each column represent the set of weighted opinions on all indicators during the planning of each demonstration area. The sum of the column elements is 1.

$$X = \begin{bmatrix} 0.42 & 0.38 & 0.35 \\ 0.22 & 0.30 & 0.27 \\ 0.36 & 0.32 & 0.38 \end{bmatrix} \tag{16}$$

By performing calculations according to formulas (5), (6), and (7), the objective weight vector of each index of the normalized standard layer can be obtained w_o, as shown in formula (17).

$$w_0 = [0.3318 \quad 0.3368 \quad 0.3314]^T \tag{17}$$

Based on the subjective and objective weight vectors of each criterion layer indicator, and according to the combined weight calculation methods of formulas (8) and (9), the final combined weight vector of each indicator can be obtained, as shown in formula (18).

$$w_i = [0.3533 \quad 0.2979 \quad 0.3487]^T \tag{18}$$

It can be seen from the weight calculation results that the more basic the demand, the larger the index weight value, and the greater its impact on the evaluation of the wiring mode, which verifies the rationality of the demand hierarchy in the previous article. The final combination weight effectively integrates the subjective and objective information of the indicator, and plays a role in optimizing the indicator weight.

Weights of Indicators in the Evaluation System. By investigating relevant professional departments' understanding of the importance of indicators for the construction of smart stations, and assigning relevant weights to each indicator (Table 2).

Table 2. Evaluation set of indicators under the security criteria.

Criterion layer	Indicator layer	1	2	3
Operational allocation level	0.4 kV Negative Loss Rate	0.42	0.41	0.40
	Timeliness of reporting the power outage (on) of smart meters (station-wide meter)	0.35	0.34	0.36
	Consistency rate of online data and field	0.23	0.25	0.24
Customer service quality	User average outage time	0.52	0.50	0.48
	Analysis of outage information for households	0.23	0.25	0.27
	Probability of outage notification to households precisely	0.25	0.25	0.25
Smart IoT level	Practical level of power equipment IoT equipment	0.31	0.30	0.32
	Access rate of perceived terminal	0.48	0.45	0.46
	Accuracy of fault investigation and judgment	0.21	0.25	0.22

According to the improved grey correlation method, the weights of the indicators in the indicator layer are solved and multiplied by the weights of the indicators in the criterion layer to obtain the weight values of each indicator in the transformation plan of the distribution station area, as shown in Table 3.

Table 3. Each indicator weight value.

Serial number	Criterion layer	Indicator layer	Weight
1	Operational allocation level	0.4 kV Negative Loss Rate	0.1449
2		Timeliness of reporting the power outage (on) of smart meters (station-wide meter)	0.1201
3		Consistency rate of online data and field	0.0883
4	Customer service quality	User average outage time	0.1490
5		Analysis of outage information for households	0.0745
6		Probability of outage notification to households precisely	0.0745
7	Smart IoT level	Practical level of power equipment IoT equipment	0.1081
8		Access rate of perceived terminal	0.1604
9		Accuracy of fault investigation and judgment	0.0802

Index System Scoring Function. According to the different types of indicators and the ideal values of indicators, refer to the survey results and expert opinions to formulate reasonable scoring standards. Scoring is expressed on a percentage basis. Benefit-type indicators increase their scores as the value of the indicator increases, and cost-type indicators increase their scores as the value of the indicator decreases. The corresponding relationship between the index value and the index score is shown in Table 4.

Table 4. Rating criteria for the indicator system.

Indicator		Grade					
		0	20	40	60	80	100
0.4 kV Negative Loss Rate	Cost type	0.6%	0.5%	0.4%	0.3%	0.2%	0.1%
Timeliness of reporting the power outage (on) of smart meters (station-wide meter)	Benefit type	50%	60%	70%	80%	90%	100%
Consistency rate of online data and field	Benefit type	75%	80%	85%	90%	95%	100%
User average outage time	Cost type	12h	10h	8h	6h	4h	2h
Analysis of outage information for households	Benefit type	50%	60%	70%	80%	90%	100%
Probability of outage notification to households precisely	Benefit type	50%	60%	70%	80%	90%	100%
Practical level of power equipment IoT equipment	Benefit type	75%	80%	85%	90%	95%	100%
Access rate of perceived terminal	Benefit type	90%	92%	94%	96%	98%	100%
Accuracy of fault investigation and judgment	Benefit type	50%	60%	70%	80%	90%	100%

MATLAB software and Curve Fitting Tool are used to perform curve fitting on discrete smart station construction effectiveness scoring standards, and obtain the middle smart station construction effectiveness index system scoring function, as shown in Table 5.

Table 5. Evaluation indicator system scoring function.

Serial number	Criterion layer	Indicator layer	Unit	Scoring function
1	Operational allocation level	0.4 kV Negative Loss Rate	%	y = −200x + 120
2		Timeliness of reporting the power outage (on) of smart meters (station-wide meter)	%	y = 2x − 100
3		Consistency rate of online data and field	%	y = 4x − 300
4	Customer service quality	User average outage time	h	y = −10x + 120
5		Analysis of outage information for households	%	y = 2x − 100
6		Probability of outage notification to households precisely	%	y = 2x − 100
7	Smart IoT level	Practical level of power equipment IoT equipment	%	y = 4x − 300
8		Access rate of perceived terminal	%	y = 10x − 900
9		Accuracy of fault investigation and judgment	%	y = 2x − 100

5 Engineering Case Analysis

Take a demonstration project of the power distribution IoT in a city in Jiangsu Province as an example. This project selects a power distribution cell for the construction of smart stations. In stations where new TTUs have been deployed (with edge computing functions through software upgrades), the edge-end fusion solution is preferred, and the TTU function is upgraded to serve as the edge IoT agent at the station side., Minute-level collection. In the station area where the new TTU is not deployed, the edge-separated solution is preferred. Each edge distribution unit is equipped with an edge IoT agent device, which reduces the total investment in the smart station area. A new type II concentrator is also deployed.

The calculation results of the project's expected effectiveness indicators combined with the various professional departments are as follows (Table 6):

Table 6. Reliability calculation result data.

Serial number	Criterion layer	Indicator layer	Results before transformation	Results after transformation
1	Operational allocation level	0.4 kV Negative Loss Rate	0.3%	0.1%
2		Timeliness of reporting the power outage (on) of smart meters (station-wide meter)	85%	95%
3		Consistency rate of online data and field	85%	98%
4	Customer service quality	User average outage time	5h	2h
5		Analysis of outage information for households	80%	98%
6		Probability of outage notification to households precisely	80%	98%
7	Smart IoT level	Practical level of power equipment IoT equipment	90%	98%
8		Access rate of perceived terminal	96%	100%
9		Accuracy of fault investigation and judgment	80%	98%

Bring the index data into the corresponding index scoring function in Table 5 to get the score of each index, and then combine the weights to calculate the comprehensive score result. The comprehensive evaluation results of the power distribution IoT demonstration project are shown in Table 7. It can be known that the comprehensive rating of the transformation plan proposed in this paper is relatively high, and the scores have been greatly improved in three aspects: the level of business and distribution, customer service quality, and the level of smart things, which is consistent with the actual project and verifies the feasibility and effectiveness of the evaluation system.

Table 7. Comprehensive evaluation results

Criterion layer	Operational allocation level	Customer service quality	Smart IoT level	Comprehensive score
Score before transformation	20.6	19.4	20.9	60.9
Score after transformation	33.4	29.2	33.7	96.3

6 Conclusion

This paper proposes a comprehensive evaluation method for the transformation scheme of distribution station areas based on the ubiquitous electric power Internet of Things, which combines the DEMATEL-ANP-inverse entropy weight method and the improved gray correlation method.

Based on the analysis of the existing problems and needs for reconstruction in the distribution transformer service area, the transformation scheme of the distribution station area based on the ubiquitous power Internet of Things is studied, and the technical framework and field implementation plan of the distribution network of the distribution area are proposed, which provides a theoretical basis for establishing a reasonable comprehensive evaluation system.

A comprehensive evaluation method of the distribution station area reconstruction plan based on the ubiquitous power Internet of things is proposed. The analysis of the effectiveness of the construction of the power network of the distribution transformer service area is from the perspectives of the level of operation and distribution, the quality of customer service, and the level of the smart Internet of Thing which is helpful for decision makers to choose the appropriate construction or transformation plan for the distribution transformer service area.

The index evaluation system for the transformation plan of the distribution transformer service area is established based on the ubiquitous electricity internet of things. It not only combines the expert's empirical knowledge with theoretical knowledge, but also combines qualitative analysis with quantitative calculation, making the evaluation method more scientific and practical.

References

1. Wang, Z., Zhao, H., Zhu, M., et al.: Application of power wireless private network in ubiquitous power internet of things. Electr. Power **25**(12), 27–38 (2019)
2. Lü, J., Luan, W., Liu, R., et al.: Architecture of distribution internet of things based on widespread sensing & software defined technology. Power Syst. Technol. **42**(10), 3110–3117 (2018)
3. Gao, B., Li, F., Tang, R., et al.: Distribution side smart sensing device design of electric internet of things. Hebei Electr. Power **38**(06), 11–13+30 (2019)
4. Ma, H., Liu, L., Tao, P., et al.: Comprehensive evaluation research of distribution network substation area based on electric internet of things. Hebei Electr. Power **38**(06), 7–10 (2019)
5. Lü, J., Sheng, W., Liu, R., et al.: Design and application of power distribution internet of things. High Voltage Eng. **45**(06), 1681–1688 (2019)
6. He, F., Chen, J., Li, Q., et al.: Application and development of internet of things in smart grid. Power Syst. Protect. Control **48**(03), 58–69 (2020)
7. Zhang, Y., Yang, T., Meng, G.: Review and prospect of ubiquitous power internet of things in smart distribution system. Electr. Power Constr. **40**(06), 1–12 (2019)
8. Shen, B., Cai, Z., Dai, G., et al.: Communication analysis of information collection services in smart distribution system towards ubiquitous power internet of things. Electr. Power Constr. **40**(09), 27–34 (2019)
9. Han, J., Xie, Z., Geng, L., et al.: Research on comprehensive evaluation method of high reliability wiring mode for medium voltage distribution network. Electr. Power **53**(03), 91–100 (2020)

Flexible Control Strategy of DC Bus for Hybrid AC/DC Microgrid with Electric Vehicle

Jingsong Wang[1(✉)], Peng Yun Li[2], Chengming Gong[1], Donghui Li[2], and Chunyang Fan[2]

[1] DERI Energy Research Institute, Nanjing, China
wangjingsong@deri.energy
[2] State Grid Jiangsu Electric Power Company Nantong Power Supply Company, Nantong, China

Abstract. As a new type of microgrid structure, hybrid AC/DC microgrid can efficiently consume photovoltaic-based distributed renewable energy, fitting for the circumstances where electric vehicles work as the main load. Unlike the voltage of AC bus, the DC bus voltage of the hybrid AC/DC microgrid has typical low inertia guaranteed by the power electronic equipment connected to it. Thus, how to enhance the DC bus inertia of hybrid AC/DC microgrid system and improve the stability of DC bus voltage become particularly important. To solve these two problems, this paper presents a flexible method of coordinated control of microgrid bus voltages. The P/U droop characteristics of DC ports of power electronic equipment, such as energy storage and charging and discharging equipment for electric vehicles are taken into account, meanwhile the power reserve rate and energy reserve rate of various distributed renewable energy are comprehensively considered. Under the use of curve translation and other adjustment methods flexibly, DC bus voltage stability is maximized in different working conditions and stable operation of the hybrid AC/DC microgrid system is guaranteed. The validity of the proposed method is also verified by establishing a matlab/simulink simulation system.

Keywords: Electric vehicles · Hybrid AC/DC microgrid · DC bus · Stability control

1 Introduction

Global warming has become one of the biggest challenges human beings face. Promoting the use of electric vehicles sheds new light on the reduction of CO2 emissions [1–3]. Aiming to promote the development of electric vehicles in China, *Guidelines on Electric Vehicle Charging Infrastructure Development* (2015–2020) was issued by the National Development and Reform Commission as early as 2015. At present, however, the primary energy form of China is still mainly coal. Thus, it does no help in energy conservation and emission reduction by charging electric vehicles directly connected to the power grid [4, 5].

Y. Xue et al. (Eds.): PMF 2020, LNEE 718, pp. 99–112, 2021.
https://doi.org/10.1007/978-981-15-9746-6_8

To realize charging energy vehicles with renewable energy, photovoltaic system and energy storage are built in charging stations, forming a microgrid system with electric vehicles [6]. In order to raise the efficiency of renewable energy consumption, not only DC bus is applied to realize the interaction between electric vehicles and energy storage, but also the microgrid mentioned above construction is employed to build the hybrid AC/DC microgrid system [7].

The stability of the DC bus in hybrid AC/DC microgrid system is the basis of the stable operation of the microgrid system [8–10]. At present, in actual hybrid AC/DC microgrid projects with electric vehicles, usually only one device connected to DC bus is selected as the voltage source, and other devices operate in the current source mode. As a result, the stability of the DC bus is completely limited to the device's power range and stability while the reliability of the system is low. Therefore, how to improve the stability margin of the DC bus becomes particularly important.

The AC droop control method has been widely used in the field of AC system frequency and voltage control, and applied in some engineering pilots [11–13]. Prior researches on DC droop voltage control methods applied to DC bus have not been able to demonstrate how to realize flexible and stable control of DC bus in a hybrid AC/DC system, nor how to apply it in engineering when considering comprehensively the flexibility of the charging and discharging of electric vehicles, energy storage, and distributed renewable energy [14–17]. Therefore, researches and validations are still needed.

Based on the research of the existing DC bus droop control method, this paper proposes a DC bus voltage flexible control method of the AC/DC hybrid microgrid with direct participation of electric vehicles. A matlab/simulink simulation system is built to verify the effectiveness of this method which is applied in actual projects.

2 Hybrid AC/DC Microgrid System with Electric Vehicles

The traditional microgrid system has mainly AC bus, through which renewable energy power generators, energy storage, load and power grid exchange energy. The main load of the microgrid system studied in this paper is DC charging, and the main renewable energy power generation unit of the microgrid is photovoltaic system, which is also the most efficient conversion to DC. Therefore, the AC/DC hybrid microgrid with DC bus for energy interaction is more suitable for the microgrid system with electric vehicles. The flexible control strategy of DC bus for hybrid AC/DC microgrid is accomplished by power electronic equipment and microgrid controllers connected to DC bus. The detailed description is as follows.

The structure of hybrid AC/DC microgrid system with electric vehicles is shown in Fig. 1. The primary equipment of the microgrid includes PV power generation units, energy storage units, charging and discharging equipment for electric vehicles, four-quadrant AC/DC converters (short for AC/DC), power distribution switches and loads. The main secondary equipment includes a microgrid controller and microgrid energy cloud software.

2.1 Primary System Structure

The microgrid system studied in this paper is composed of two parts: AC part and DC part. Within the microgrid, the photovoltaic system, energy storage and charging and discharging equipment of the DC part exchange energy through the DC bus. For two-way energy interaction, the AC part is connected to the large power grid through an AC switch, and there is also some AC load on AC side.

2.2 Secondary System Structure

The secondary system consists of two parts: a microgrid controller and microgrid energy cloud software.

The microgrid controller collects DC bus voltage, PV, energy storage, AC/DC, and charging and discharging equipment output current through DC Hall sensors.

The microgrid controller communicates with the microgrid energy cloud through Internet. Since the stability control issue mainly discussed in this paper is mainly completed in the microgrid controller and various power electronic devices, the microgrid energy cloud is not described much.

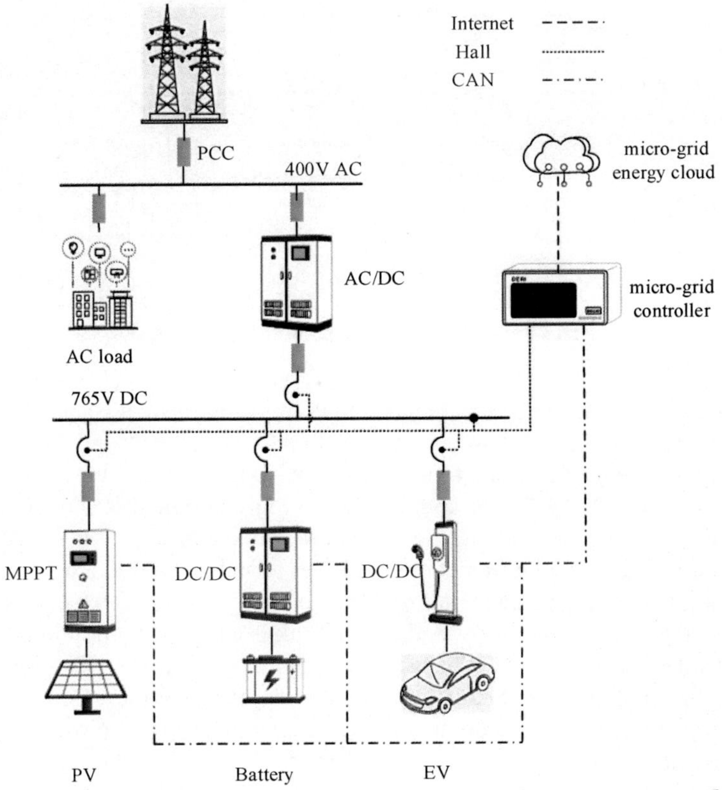

Fig. 1. The system structure of hybrid AC/DC microgrid

3 Flexible Control Strategy of DC Bus

3.1 Power Electronic Equipment Control

There are two types of power electronic equipment connected to DC bus: AC/DC and DC/DC. AC/DC is used to realize the bidirectional power interaction between AC and DC system, and DC/DC to interact among the power electric devices, energy storage, electric vehicles and photovoltaic system.

DCDC. The topology and control loop of DC/DC are displayed in Fig. 2.

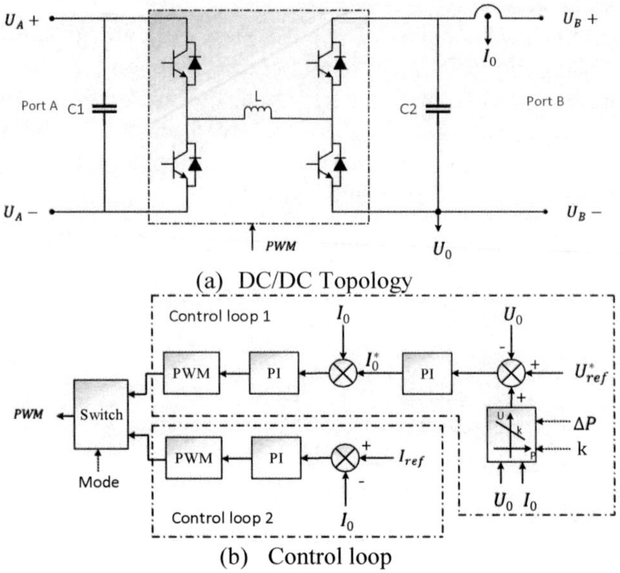

Fig. 2. The topology and control loop of DC/DC

The DC/DC applied in this paper uses a two-port four-switch Buck-Boost topology method to achieve bidirectional conversion. Port A is connected to electric vehicles, energy storage devices and other equipment, and port B to DC bus. This paper only considers the control of DC bus, and the control logic of port B is shown in Fig. 2 (b).

It can be seen that the control logic of port B is composed of two logics, and the control output selection of different control loops is realized through a switch. Between them, control loop 1 guarantees the stability of output voltage by means of voltage outer loop and current inner loop. A drooping link is added to control loop 1 outside to ensure that the port voltage will change linearly with output power. The control logic of the droop loop will be described in detail in the following sections. Control loop 2, as a current loop, ensures that the output current is output in accordance with the set current. The current loop can not support the bus voltage.

AC/DC. The topology and control loop of AC/DC are displayed in Fig. 3.

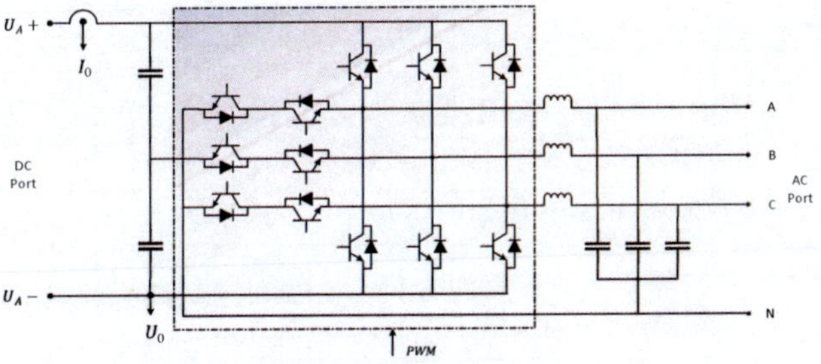

(c) AC/DC Topology

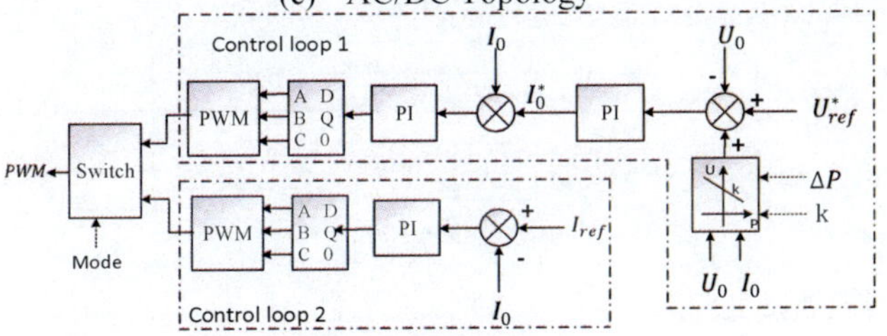

(d) Control loop.

Fig. 3. The topology and control loop of AC/DC

AC/DC equipment adopts T-type three-level topology, which can realize bidirectional power interaction. Figure 3 (b) gives the control logic of DC port, which is similar to DC/DC and won't be detailed here.

Droop Control. The droop control loop is written:

$$U_0^* = U_{ref}^* - k(P_{out} - \Delta P) \tag{1}$$

where U_0^* is the reference voltage of the voltage outer loop, U_{ref}^* is the rated voltage set by the power electronic device, k is the droop coefficient, P_{out} is the current output power, and ΔP is the power change rate adjusted by the microgrid controller.

The value of ΔP and k is set. When ΔP is set to 0, the actual reference voltage U_0^* of the voltage outer loop will decrease as the output current increases, which shows the droop control characteristic. When ΔP is not 0, not only U_0^* of the voltage outer loop decreases as the output current increases, but the rated voltage U_{ref}^* set by the power electronic device also changes, i.e., the droop curve has shifted.

3.2 Flexible Control Strategy of DC Bus Voltage

The flexible control strategy of bus voltage is realized by microgrid controllers, and its control logic diagram is illustrated in Fig. 4.

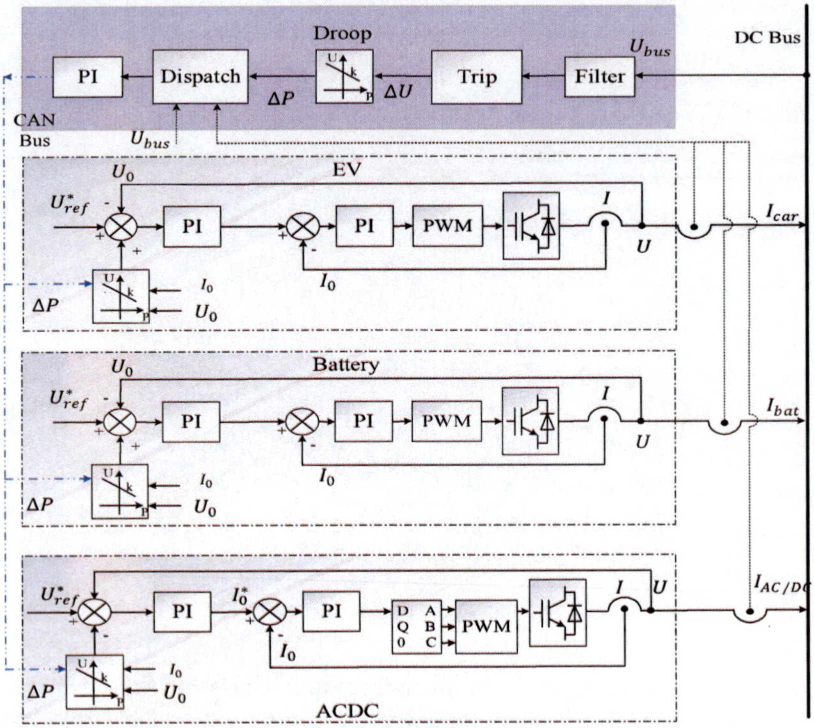

Fig. 4. The diagram of flexible control logic block

The core idea of flexible control is to calculate the shortfalls of DC bus system power through the change of DC bus voltage, and select the appropriate power electronic equipment to compensate for it, and finally achieve the purpose of DC bus voltage stabilization.

Figure 4 presents a logic diagram of three-node system in bus flexible control where the control method can be decomposed into three parts: trip logic, system droop curve computational logic, and power distribution logic. Elaborations respectively are as follows.

Trip Logic. The bus voltage trip logic triggers the process and output difference of current voltage. The voltage trip criterion is written:

$$|\mathrm{U} - \mathrm{U}_n| > \Delta\mathrm{U}_{set} \tag{2}$$

where U is bus voltage, U_n is the bus rated voltage, ΔU_{set} is the voltage deviation set-value. The setting principle is that the equipment connected to DC bus can operate normally within this range.

At this time, the output voltage fluctuation value is $\Delta U = U - U_n$.

Droop Curve Computational Logic. The main function of the droop curve calculation logic is to calculate the droop curve of microgrid system in real time, and output the unbalanced power value of the system according to the voltage value output by the start-up logic.

For example, in Fig. 4, the control system is composed of two DC/DC and one AC/DC in parallel. Each power electronic device has a U/P droop due to its virtual impedance, so the virtual impedance of the DC bus system is parallel connection of the impedances of three power electronic modules.

It should be noted that only when the device is running in voltage source mode with droop loop can the virtual impedance of the device be calculated into the system impedance. In other words, virtual impedance of the device should not be calculated in parallel with DC bus system when the device is running in current source mode or in a fault state.

Power Distribution Logic. The droop curve calculation logic has calculated the system power fluctuation value, and the power allocation logic allows the user to achieve the final power allocation. The calculation method is as follows:

(1) Each drooping node can be artificially set a weight coefficient δ ($\delta \in [0, 1]$). Increase the setting value if you want the node to participate in the adjustment first;
(2) The device is in a locked state, that is, if the node state is abnormal and the regulated power cannot be output, then the device is locked and will not participate in adjustment;
(3) Calculate the power reserve rate. Take P_n as the rated output power of the power electronic equipment participating in the control, and P as the current output power, then the power reserve rate of the droop device at this moment is:

$$\beta = |P_n - P|/P_n \quad (3)$$

4. Calculate the energy reserve rate. For the energy storage device, the energy reserve rate is calculated as follows:

$$\gamma = (\mathrm{W}_n - \mathrm{W})/\mathrm{W}_n \quad (4)$$

Where W_n is the rated capacity of energy storage. W is the current capacity value of energy storage. For actual control system, the energy reserve rate can be directly replaced by the stored energy SOC (State-of-charge) value ([0, 100]). For non-energy storage power electronic equipment set $\gamma = 100$;

5. For a non-locked device, the competition coefficient is:

$$\partial = \delta \times \beta \times \gamma \quad (5)$$

For the locked device, $\partial = 0$;

6. Sort the competition coefficients of each drooping node in descending order;
7. The total power fluctuation value calculated by the droop curve computational logic is distributed according to the competition coefficient calculated in step 5). The principle of allocation is to prioritize drooping nodes, the number of which can be more than one, that are ranked first. After allocating to each node through the CAN bus, this control is completed.

3.3 Impact Analysis Droop Factor of Stability

The value of the droop coefficient will affect the stability of the system. Literature [18] and literature [19] discussed the stability of multi-machine parallel connection in detail using the small signal stability method. This paper uses a graphical method to qualitatively analyze the impact of different droop coefficients on system operation.

Take two modules in parallel as an example, as shown in the following figure:

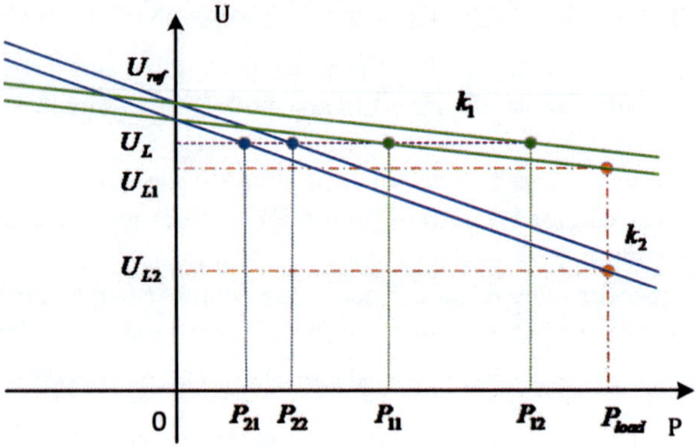

Fig. 5. Diagram of the effect of droop parameters on stability

Test case 1 is the combination of two modules with droop gain k_1. Test case 2 is the parallel connection of two modules with droop gain k_2. The k_1 is less than k_2.

It can be seen that when the bus voltage changes from U_{ref} to U_L, the power difference of test case 1 is $P_{22} - P_{21}$, and the power difference of test case 2 is $P_{12} - P_{11}$. It can be seen that a larger droop gain has better current sharing performance.

In addition, when the system power output P_{load}, the voltage of test case 1 drops to U_{L1}, and the voltage of test case 2 drops to U_{L2}. This means that a small droop gain has better stability.

4 Test Verification

4.1 Test System

In order to verify the effectiveness of the method described, a test system was built based in matlab/simulink. Figure 5 shows the schematic diagram (Fig. 6).

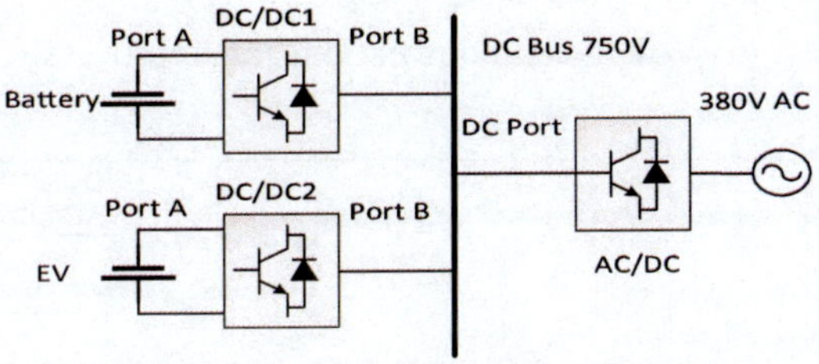

Fig. 6. Test system structure.

The initial equipment parameters and control parameters of each distributed power supply are displayed in the table below (Table 1).

Table 1. System parameters

Device	$U_n(V)$	$P_n(kW)$	Soc	k(V/kW)	δ
AC/DC	750	60	/	1	1
Battery	750	15	50	4	1
EV	750	15	50	4	1

In the table: U_n is rated voltage, P_n is the rated capacity of distributed power supply, and SOC is the power of each distributed power supply. For systems without energy storage, the default SOC is 1, k is the droop coefficient of the distributed power supply, δ is manually set weight coefficient. The range of stable start-up voltage was [748, 752] V.

4.2 Test Cases

Case 1: Electric Vehicles Charging in Grid-Connected Mode. The initial test conditions were: DC/DC1 stabilized the voltage on port B; DC/DC2 worked in current source mode, and AC/DC in stable DC port voltage. After the test had started, DC/DC2 charged the electric vehicle at a constant power of 15 kW. The test results are as follows (Fig. 7).

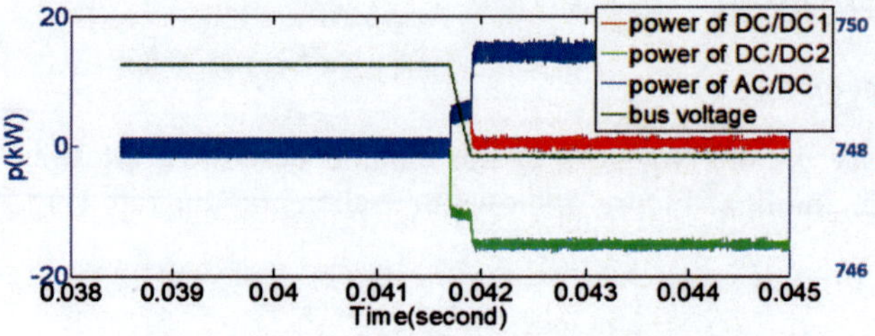

(a) Voltage and power adjustment process.

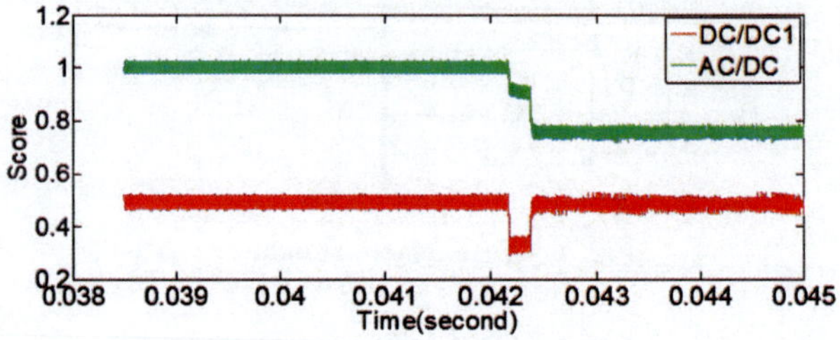

(b) Competition coefficient.

Fig. 7. Test results of case 1.

It is apparent that when DC/DC2 was charged at a constant power of 15 kW, bus voltage started to decline due to the drooping effects of DC/DC1 and AC/DC ports. When the voltage was below 748 V, the flexible control strategy of bus voltage came into effect. As can be seen from the competition scores of DC/DC1 and AC/DC in Figure (b), AC/DC has always had a high competition coefficient. The system added AC/DC power output to supplement the 15 kW charging power shortage. DC/DC1 did not output power under steady state.

In conventional engineering projects, energy storage will operate as a current source, and AC/DC stabilizes the bus voltage. The test results for this condition are as follows (Fig. 8):

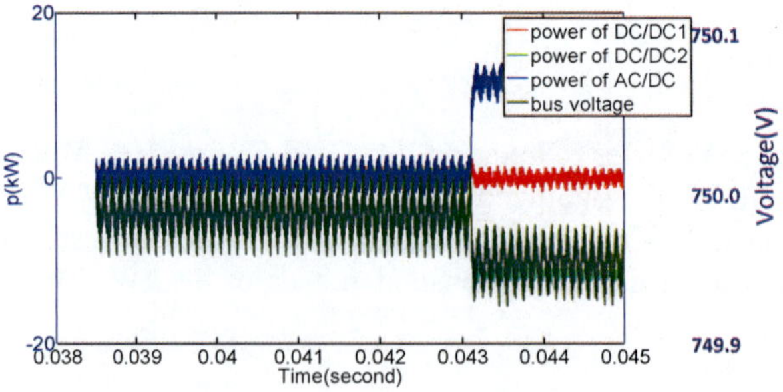

Fig. 8. Test results of engineering conditions

It can be seen that in the conventional engineering projects, DC/DC1 will not participate in the system stability adjustment, and the power imbalance of the system is adjusted by AC/DC. Since there is no droop, the voltage fluctuation is small.

Case 2: Charging Regulation Process in Off-grid State. The initial test conditions were: DC/DC1 stabilized the voltage on port B; DC/DC2 worked in the current source mode, and AC/DC in current source mode; DC/DC1 maintained stable DC bus voltage. After the test had started, DC/DC2 charged the electric vehicle at a constant power of 15 kW. The test results are as follows (Fig. 9).

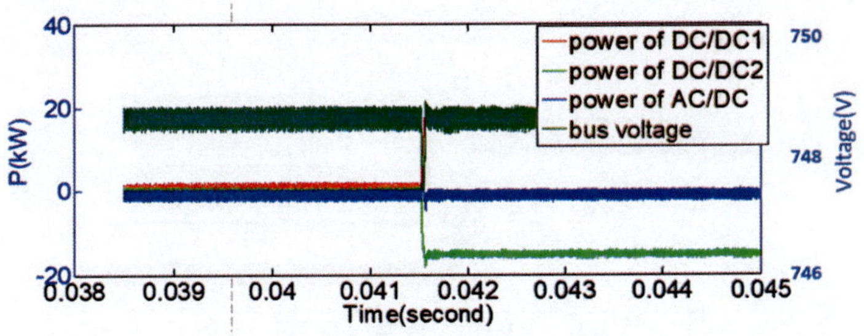

Fig. 9. Test results of case 2.

Because the system only had one voltage source, when DC/DC2 was charged at a constant power of 15 kW, the DC bus voltage descended due to the droop control of DC/DC1. After the voltage was under 748 V, DC/DC1 increased the power output to 15 kW, and the range of bus voltage was brought back to [748, 752] V, while the system entered steady state operation.

Under normal engineering conditions, the bus voltage is supported by DC/DC1, and the test data is as follows (Fig. 10):

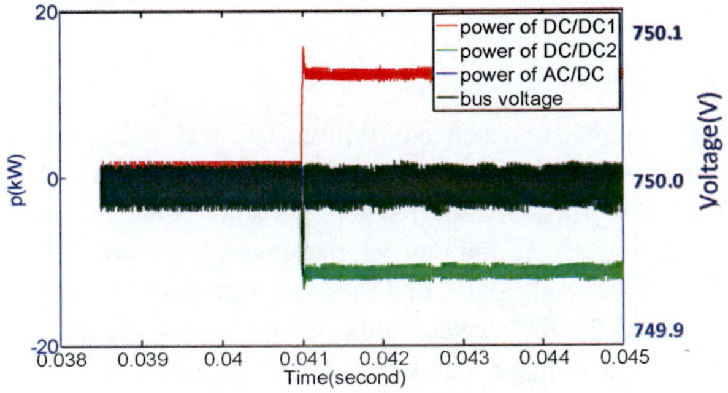

Fig. 10. Test results of engineering conditions

Since this working condition has only one source of stable DC bus, the test result under this working condition is basically consistent with the above result. The advanced nature of this solution can only be reflected when multiple voltage sources stabilize the bus at the same time.

Case 3: Supporting AC Load in Off-grid State. The initial test conditions were: DC/DC1 stabilized the voltage on port B with DC/DC2; AC/DC worked in current source mode. After the test had started, AC/DC generated power to the AC grid at a constant power of 15 kW. The test results are as follows (Fig. 11).

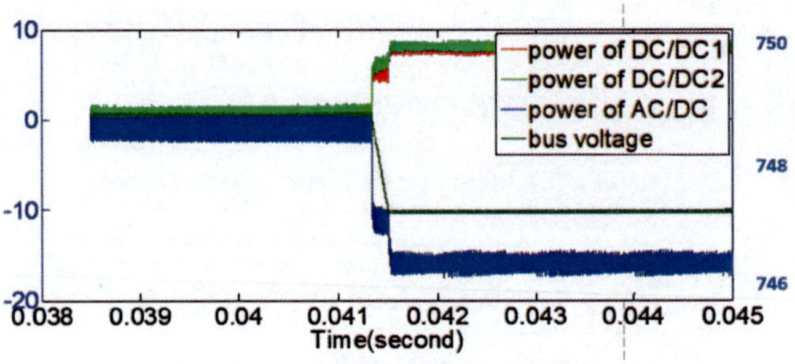

(a) Voltage and power adjustment process.

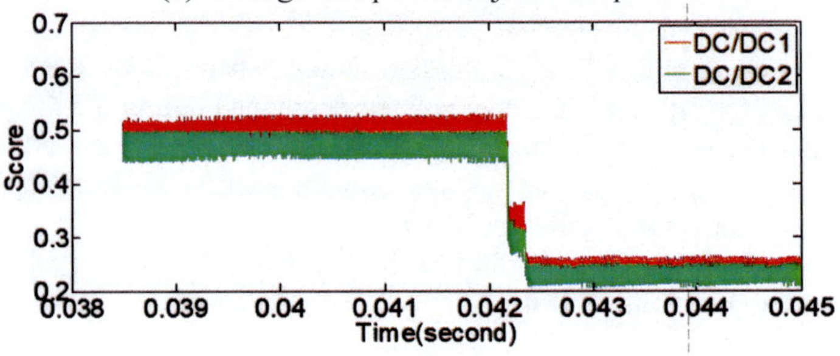

(b) Competition coefficient.

Fig. 11. Test results of case 3.

Similar to the previous two test conditions, DC bus voltage began to drop at the beginning stage due to the droop effect of DC/DC1 and DC/DC2 ports. The flexible control strategy of bus voltage started when the start value was lower. DC/DC1 and DC/DC2 shared the same original setting parameters. It can also be seen from Figure (b) that the two devices basically had the same competition coefficient. The average output of DC/DC1 and DC/DC2 was about 7.5 kW. As a result, the system entered a steady state and the bus voltage was stable at the set interval.

The bus voltage is supported by DC/DC1 in normal engineering conditions. The test data is as follows (Fig. 12):

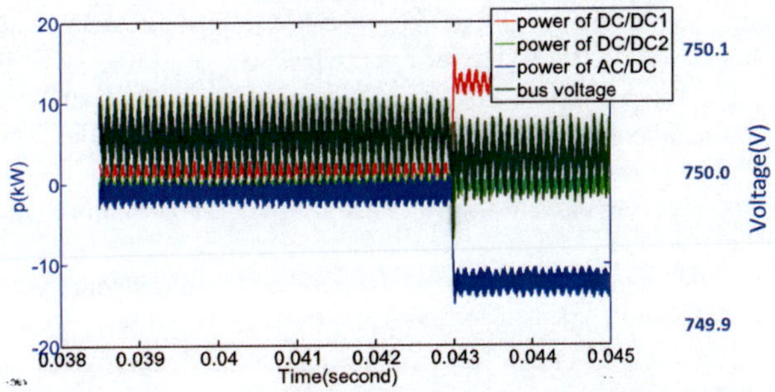

Fig. 12. Test results of engineering conditions

It can be seen that the power imbalance of the system is adjusted by DC/DC1, and the system will lose its stability after the AC power exceeds the maximum output power of DC/DC1.

From these three results, it can be seen that compared with power electronic devices without droop loops, when sudden power changes occur, power electronic devices could automatically distribute power according to their respective virtual impedance, and jointly support the stability of the bus; and the flexible control strategy of bus voltage could coordinate power electronic devices in different operating modes such as grid-connected mode and off-grid mode, it could also select devices with strong adjustment capabilities and keep the system voltage in the set range. Thus, the inertia and the power quality of DC bus are greatly improved.

5 Conclusion

This article focuses on the stability of the DC bus voltage of a hybrid AC/DC microgrid system with electric vehicles. The P/U droop characteristics of DC ports of power electronic equipment such as devices of energy storage, charging and discharging equipment for electric vehicles are fully utilized, meanwhile, the power reserve rate and energy reserve rate of different types of distributed energy are considered. Employing flexible adjustment methods like curve translation, the inertia of DC bus is improved, the stability of the microgrid system voltage is effectively guaranteed, and the power quality of DC system is promoted. The method described in this article has been widely used in charging system in the TELD company, integrating photovoltaics energy storage and charging for electric vehicles, which guarantees the safe and stable operation of the system and brings great economic benefits.

References

1. Ahmed, M.A., El-Sharkawy, M.R., et al.: Remote monitoring of electric vehicle charging stations in smart campus parking lot. J. Mod. Power Syst. Clean Energy **8**(1), 124–132 (2020)

2. Chen, T., Zhang, X.-P., Wang, J., et al.: A review on electric vehicle charging infrastructure development in the UK. J. Mod. Power Syst. Clean Energy **8**(2), 193–205 (2020)
3. Lu, Z., Jiang, C., Li, X., Zhao, H., et al.: Low-carbon benefit analysis on coordinating investment for clean energy and electric vehicle charging station. Trans. China Electr. Technol. **31**(9), 163–171 (2016)
4. Zhou, T., et al.: Electric vehicle-to-grid technology based on microgrid. Autom. Electr. Power Syst. **42**(3), 98–117 (2018)
5. Atia, R., Yamada, N.: More accurate sizing of renewable energy sources under high levels of electric vehicle integration. Renew. Energy **81**, 918–925 (2015)
6. Li, X., Guo, L., Wang, C., et al.: Key technologies of DC microgrids: an overview. Proc. CSEE **36**(1), 2–17 (2016)
7. Chen, A.: Coordination control and mode switching strategy for hybrid AC/DC microgrid with multi-bus structure. Autom. Electr. Power Syst. **42**(17), 175–183 (2018)
8. Zhou, T., et al.: Analysis on dynamic performance of droop control for multi-terminal VSC based DC distribution system. Autom. Electr. Power Syst. **43**(2), 89–96 (2019)
9. Bi, D., Fan, Z., Xie, D., et al.: Autonomous control strategy of island DC microgrid. Power Syst. Technol. **39**(4), 886–891 (2015). (in Chinese)
10. Sun, F., et al.: Optimized dispatching and control method of droop parameters with assurance of small-signal stability in DC distribution network. Autom. Electr. Power Syst. **42** (3), 48–55 (2018)
11. Mohammed, A., Refaat, S.S., Bayhan, S., Abu-Rub, H.: AC microgrid control and management strategies: evaluation and review. IEEE Power Electron. Mag. **6**(2), 18–31 (2019)
12. Liang, H., Zheng, C., Gao, Y., Li, P.: Research on improved droop control strategy for microgrid. Proc. CSEE **37**(17), 4901–4910 (2017)
13. Gao, F., Bozhko, S.B., Costabeber, A., et al.: Control design and voltage stability analysis of a droop-controlled electrical power system for more electric aircraft. IEEE Trans. Ind. Electron. **64**(12), 9271–9281 (2017)
14. Li, C., Chaudhary, S.K., Savaghebi, M., et al.: Power flow analysis for low-voltage AC and DC microgrids considering droop control and virtual impedance. IEEE Trans. Smart Grid **8** (6), 2754–2764 (2017)
15. Zhang, Y., Li, Y.W.: Energy management strategy for supercapacitor in droop-controlled DC microgrid using virtual impedance. IEEE Trans. Power Electron. **32**(4), 2704–2716 (2017)
16. Meng, L., Dragicevic, T., Vasquez, J.C., et al.: Tertiary and secondary control levels for efficiency optimization and system damping in droop controlled DC-DC converters. IEEE Trans. Smart Grid **6**(6), 2615–2626 (2015)
17. Li, C., Bosio, F., Chen, F., et al.: Economic dispatch for operating cost minimization under real time pricing in droop controlled DC microgrid. IEEE J. Emerg. Sel. Top. Power Electron. **5**(1), 587–595 (2017)
18. Jin, Y., et al.: Small-signal stability analysis of DC distribution network adopting droop control. Autom. Electr. Power Syst. **40**(14), 78–85 (2016)
19. Li, X., Li, Z., et al.: Flexible control and stability analysis of AC/DC microgrids clusters. Proc. CSEE **39**(20), 5948–5961 (2019)

Influence of Different New Energy Penetration Rates on Distance Protection

Tao Huang(✉), Qingchun Zhao, Hua Xie, Jinfeng Lu, and Hongxi Zhang

NR Electric Co. Ltd., Nanjing 211102, China
huangtao4@nrec.com

Abstract. The integration of new energy into power grid brings a series of problems to relay protection. The influence of system impedance ratio (SIR) on distance protection was introduced firstly. Then grid simulation models integrated with photovoltaic and wind power respectively were built. Based on simulation results, the characteristics of system impedance ratio under different penetration rates of new energy were analyzed. It is pointed out that the system impedance ratio increases gradually with the increase of new energy penetration rate. Then the action performances of distance protection under different new energy penetration rates were studied. It is found that under a fixed capacitor voltage transformer (CVT) error, when the new energy penetration rate is greater than a certain value, the line distance protection will have an overreach. Some countermeasures were also put forward in the paper.

Keywords: New energy · Distance protection · System impedance ratio · Penetration rate · Capacitor voltage transformer

1 Introduction

In recent years, the impact of large-scale and rapid development of new energy on relay protection has been widely concerned. Due to the special transient characteristics of new energy [1, 2], the traditional relay protection principles have lost their best working conditions, resulting in the deterioration of operation performance. In recent years, a lot of research has been done on the impact of new energy on differential protection [3], distance protection [4, 5], phase selection components [6] and direction components [7].

System impedance ratio (SIR) is an important factor to determine the performance of distance protection [8, 9]. When capacitor voltage transformer (CVT) is used, because of the existence of CVT transient error [10–12], large SIR is easy to cause an overreach of distance protection. In fact, the impedance characteristics of new energy resources are quite different from those of traditional power resources [7]. With the change of the penetration rate of new energy resources, the SIR characteristics will change and its impact on distance protection will also be different. However, there is little research on this area so far.

In this paper, simulation models of grid integrated with photovoltaic and wind power respectively are built. Based on simulation analysis, the change rules of SIR with

Y. Xue et al. (Eds.): PMF 2020, LNEE 718, pp. 113–123, 2021.
https://doi.org/10.1007/978-981-15-9746-6_9

different new energy penetration rates are studied. Then the operation performance of distance protection is analyzed considering the transient error of CVT.

2 Mechanism of SIR Effect on Distance Protection

Figure 1 shows a simplified diagram of power network, where Z_s is the impedance of system, Z_{lk} the line impedance from bus M to fault point F, $\dot{U}_\varphi$ and $\dot{I}_\varphi$ the voltage and current at the protection installation respectively, $\varphi = A, B, C$.

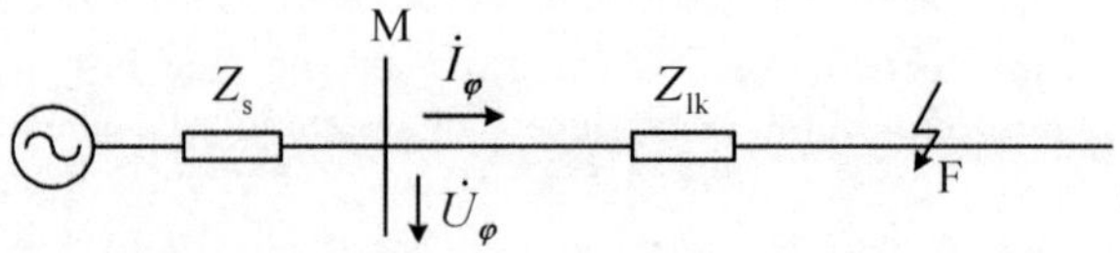

Fig. 1. Simplified diagram of system network.

It is assumed that the voltage and current measured by the protection device are $\dot{U}'_\varphi$ and $\dot{I}'_\varphi$ respectively, and there is no error in the measured current, while the measured voltage has amplitude error ρ_φ and phase error θ_φ due to CVT transient. The expression of ρ_φ and θ_φ is as follows:

$$\begin{cases} \rho_\varphi = \frac{|\dot{U}'_\varphi| - |\dot{U}_\varphi|}{|\dot{U}_\varphi|} \\ \theta_\varphi = \arg(\frac{\dot{U}'_\varphi}{\dot{U}_\varphi}) \end{cases} \tag{1}$$

Let $K_\varphi = (1 + \rho_\varphi)e^{j\theta_\varphi}$, then the voltage and current measured by protection will be:

$$\begin{cases} \dot{U}'_\varphi = K_\varphi \dot{U}_\varphi \\ \dot{I}'_\varphi = \dot{I}_\varphi \end{cases} \tag{2}$$

Take the BC two-phase short-circuit at point F in Fig. 1 as an example for analysis. The fault phasor diagram is shown in Fig. 2, where $\dot{E}_\varphi$ is the system voltage and $\dot{U}_F$ the fault point voltage.

According to Fig. 2, there is the following equation [8]:

$$\begin{cases} \dot{U}_B = \dot{U}_F + \frac{1}{2}\dot{U}_{BC} \\ \dot{U}_C = \dot{U}_F - \frac{1}{2}\dot{U}_{BC} \end{cases} \tag{3}$$

And the line SIR value can be calculated using the formula:

$$\mathrm{SIR} = \frac{Z_s}{Z_{lk}} \tag{4}$$

Substituting $\dot{E}_A = -2\dot{U}_F$ and $\dot{E}_{BC} = \dot{U}_{BC}(1+\mathrm{SIR})$ into formula (3) and according to formula (2), it can be obtained that:

$$K_{BC} = \frac{\dot{U}'_{BC}}{\dot{U}_{BC}} = \frac{1}{2}(K_B + K_C) - j(K_B - K_C)\frac{1+\mathrm{SIR}}{2\sqrt{3}} \tag{5}$$

Where the relationship $\dot{E}_{BC} = \sqrt{3}\dot{E}_A e^{-j\frac{\pi}{2}}$ is used.

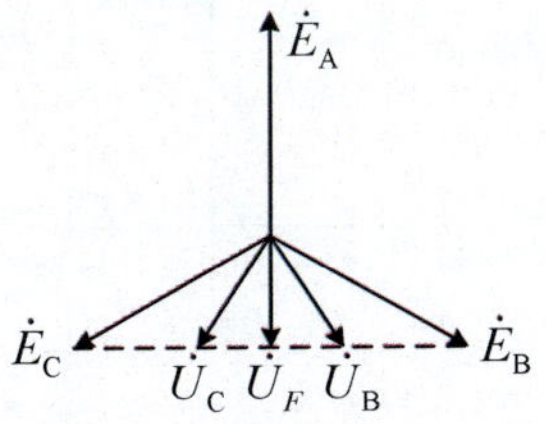

Fig. 2. Fault phasor diagram under phase B-C fault.

Therefore, after considering the CVT transient error, the impedance measured by protection Z_{BC} and the actual fault impedance Z_{lk} meet the relationship:

$$Z_{BC} = \frac{\dot{U}'_{BC}}{\dot{I}'_{BC}} = \frac{K_{BC}\dot{U}_{BC}}{\dot{I}_{BC}} = K_{BC}Z_{lk} \tag{6}$$

According to formula (5) and formula (6), under certain CVT transient error K_B and K_C, the error of measuring impedance is only related to SIR. When SIR increases, the error of measuring impedance also increases.

3 Change Rules of SIR Under Different New Energy Penetration Rates

Generally, the impedance of new energy resource is different from that of traditional power resource, which leads to a difference of line SIR property. In this section, change rules of SIR under different new energy penetration rates will be studied through simulation analysis.

3.1 Simulation Model

Using EMTDC/PSCAD simulation platform, two power grid simulation models integrated photovoltaic and wind power shown in Fig. 3 and Fig. 4 are established respectively. The grid power in Fig. 3 is composed of photovoltaic (PV) station and synchronous generator (SG), and the grid power in Fig. 4 is composed of doubly fed induction generator (DFIG)-based wind power and SG, with the total capacity of 300 MW. After each power resource is boosted to 220 kV through step-up transformer, it supplies power to the load through transmission line. The positive-sequence parameter of transmission line is Z1 = (1.14 + j14.78) Ω and the zero-sequence parameter is Z0 = (12.47 + j46.55) Ω.

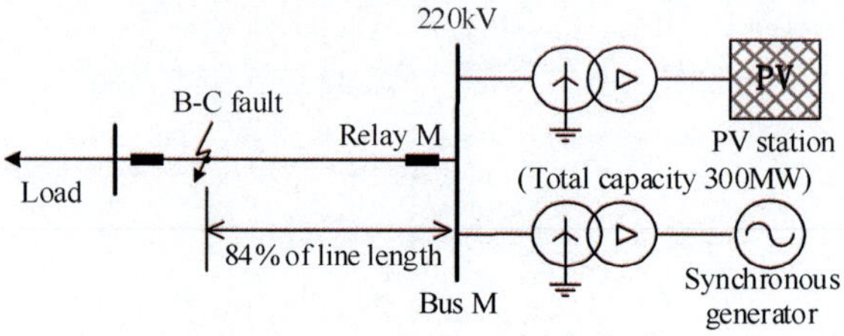

Fig. 3. Model of power grid integrated with PV station.

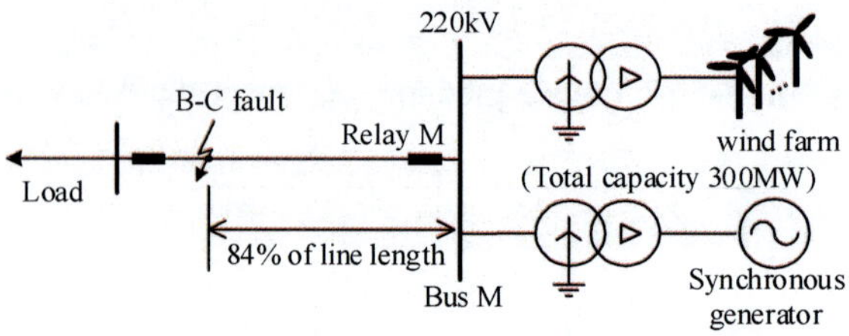

Fig. 4. Model of power grid integrated with wind farm.

The calculation of SIR needs to know the impedance of the system power resource. However, PV inverter and DFIG have no equivalent circuit, their impedances cannot be given directly. In this paper, a measured method is used to estimate SIR. The specific method is as follows: In Fig. 3 and Fig. 4, suppose there is no load in normal operation, and three-phase metallic short-circuit fault is set at the outlet of relay M. Assuming that the effective value of steady-state short-circuit phase-to-phase current measured by relay M is I_m, and the unit is kA, then the system power impedance behind relay M can be calculated by the following formula:

$$Z_s \approx \frac{220}{I_m} \quad (7)$$

In fact, the calculation method of formula (7) refers to the method of calculating the system impedance at the back side of the relay protection, that is, the system impedance is calculated by dividing the voltage fault component by the current fault component, $Z_s = |\Delta\dot{U}/\Delta\dot{I}|$. When a three-phase short-circuit fault occurs at the outlet of relay M, the voltage fault component is the rated voltage, and the current fault component is the short-circuit current since there is no load before fault.

3.2 Change Rules of SIR

Keep the total power supply capacity of 300 MW in Fig. 3 and Fig. 4 unchanged, and gradually increase the capacity of photovoltaic and wind power. It means that the penetration of new energy in the system power supply is increase. Selecting Z_{lk} equals to 0.84*Z1, thus the system impedance Z_s and SIR value observed at relay M can be calculated according to formula (7) and formula (4). In Fig. 5 and Table 1, the simulation results of Z_s and SIR under different PV penetration rates are presented. The results under different DFIG wind power penetration rates are given in Fig. 6 and Table 2 where DFIG is equipped with crowbar protection and Fig. 7 and Table 3 where DFIG without crowbar protection.

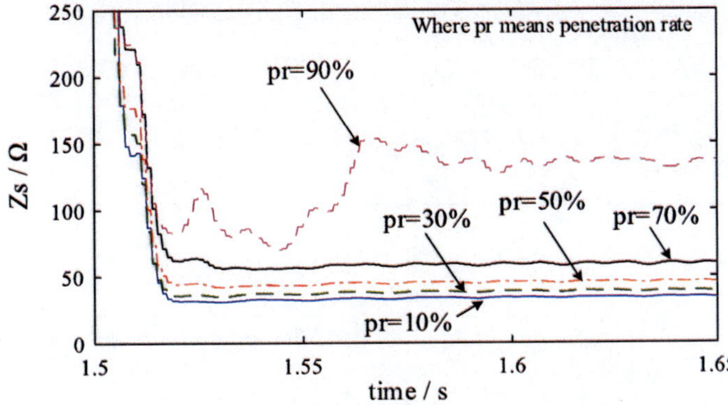

Fig. 5. Waveforms of Z_s under different PV penetration rates.

From the simulation results above, it can be seen that with the gradual increase of new energy penetration rate, the system impedance Z_s and line SIR increase gradually, and when penetration rate is below 60%, the SIR growth rate is relatively slow, but when penetration rate exceeds 60%, the SIR growth rate increases rapidly. SIR value when penetration rate is 100% is 3.7–6 times higher than the penetration rate is 0.

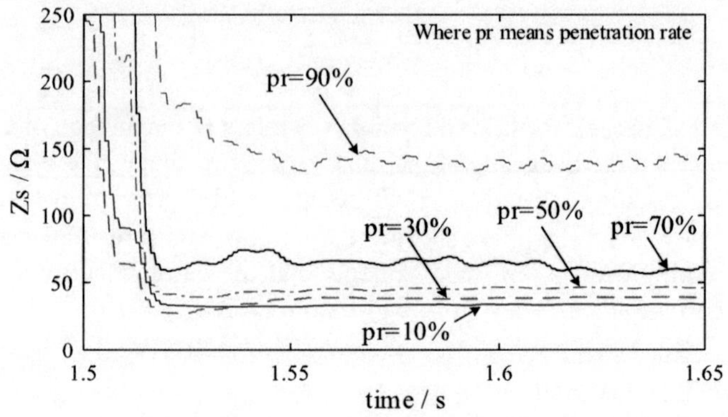

Fig. 6. Waveforms of Z_s under different wind power penetration rates (with crowbar protection).

Table 1. SIR under different photovoltaic penetration rates.

Capacity of PV station (MW)	Capacity of SG (MW)	Penetration rate (%)	SIR
0	300	0	2.5
30	270	10	2.7
60	240	20	2.9
90	210	30	3.1
120	180	40	3.3
150	150	50	3.6
180	120	60	3.9
210	90	70	4.8
240	60	80	7.8
270	30	90	10.5
300	0	100	14.4

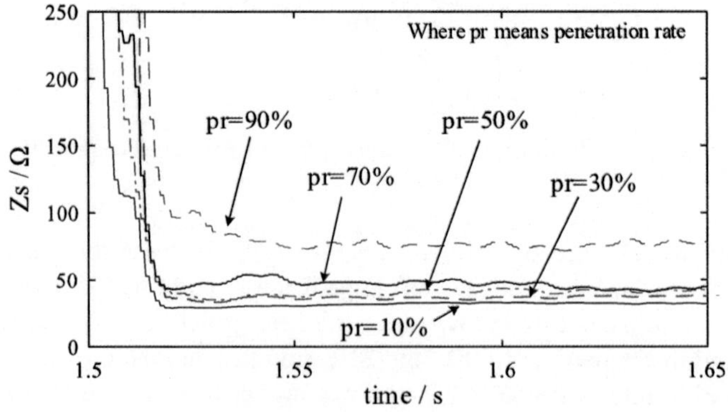

Fig. 7. Waveforms of Z_s under different wind power penetration rates (no crowbar protection).

Table 2. SIR under different wind power penetration rates (with crowbar protection).

Capacity of PV station (MW)	Capacity of SG (MW)	Penetration rate (%)	SIR
0	300	0	2.5
30	270	10	2.7
60	240	20	2.9
90	210	30	3.1
120	180	40	3.3
150	150	50	3.7
180	120	60	4.0
210	90	70	5.3
240	60	80	8.6
270	30	90	11.3
300	0	100	15.1

Table 3. SIR under different wind power penetration rates (no crowbar protection).

Capacity of PV station (MW)	Capacity of SG (MW)	Penetration rate (%)	SIR
0	300	0	2.5
30	270	10	2.6
60	240	20	2.7
90	210	30	2.9
120	180	40	3.1
150	150	50	3.2
180	120	60	3.5
210	90	70	3.8
240	60	80	4.3
270	30	90	6.1
300	0	100	9.2

The simulation results show that under the same capacity, the equivalent impedance of new energy is much larger than that of synchronous machine, which leads to the increase of SIR with the increase of new energy penetration rate. When CVT error is considered, the distance protection performance will be affected differently.

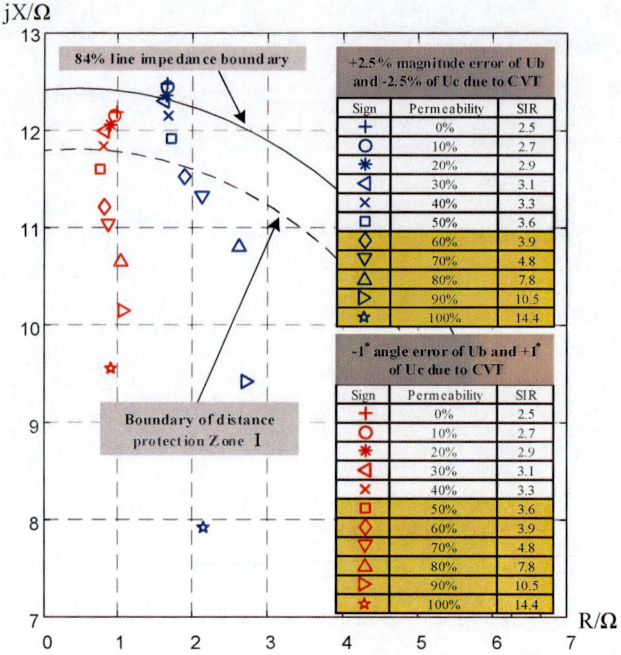

Fig. 8. Change track of Z_{bc} under different PV penetration rates.

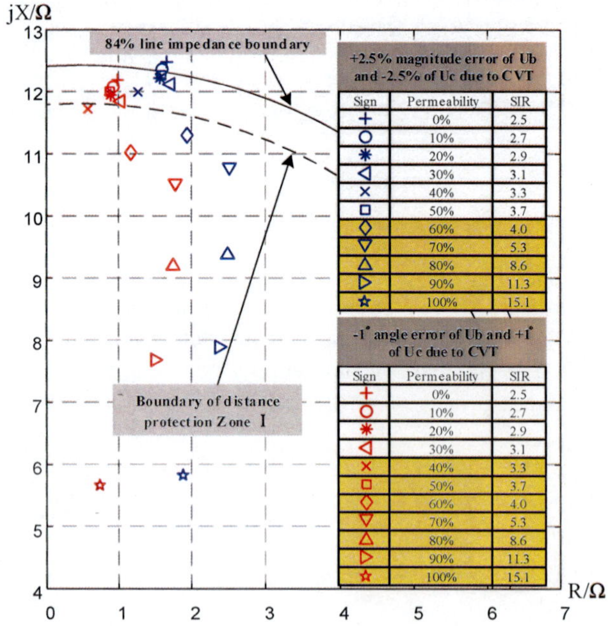

Fig. 9. Change track of Z_{bc} under different wind power penetration rates.

4 Distance Protection Effected by Different New Energy Penetration Rates

This section selects phase B-to-phase C fault at 84% of line length in Fig. 3 and Fig. 4 to analyze. The setting value of distance protection zone I is 80% of line length, and generally a ± 5% error is allowed in impedance measurement, that is to say, when fault occurs at 84% of line length, considering a −5% maximum error, the impedance measured by protection is exactly equal to the setting value. If the measured impedance exceeds the −5% accuracy limit, distance protection will have an overreach.

The transient error of CVT in simulation considers two situations: one is amplitude error, assuming that amplitude error of phase B voltage is +2.5%, and amplitude error of phase C voltage is −2.5%. The other is phase error, assuming that angle error of phase B voltage is −1°, and angle error of phase C voltage is +1°.

Figure 8 and Fig. 9 respectively show the change track of the measured impedance Z_{bc} at relay M in Fig. 3 and Fig. 4 under different PV and wind power penetration rates considering CVT transient error.

It can be seen from Fig. 8 and Fig. 9 that with the increase of new energy penetration rate, the error of distance protection measured impedance increases gradually. Considering 5% allowable error, under the set CVT amplitude error, when PV and wind power penetration rate exceeds 60%, the measured fault impedance will enter the range of distance protection zone I, resulting in a maloperation. Under the set CVT phase error, when PV penetration rate is more than 50% and wind power penetration rate is more than 40%, distance protection will have an overreach.

From the above simulation results, it can be seen that with the continuous increase of new energy capacity, the system impedance ratio also increases gradually. When CVT is adopted, the distance protection may have a maloperation. In order to eliminate the impact of high penetration rate of new energy on distance protection, the following measures are recommended:

A) In the system with high penetration rate of new energy, it is suggested to use electromagnetic voltage transformer with better transient characteristics, which can improve the accuracy of voltage measurement and reduce the impact of larger SIR on distance protection.
B) Automatically adjust distance protection setting according to SIR. Distance protection calculates the equivalent impedance of system in real time using measured fault components of voltage and current, and divides the calculated system impedance and line impedance as the estimated value of SIR. When the estimated SIR is over a certain value, the setting of distance protection zone I will be adjusted adaptively. On the one hand, the action time of distance protection zone I can be properly extended to prevent it operating faster than the pilot protection of adjacent line due to a serious transient overreach. On the other hand, through off-line simulation, the relationship between different SIR and impedance measurement error under the most serious CVT transient error can be obtained, then the maximum impedance measurement error under current SIR can be calculated according

to this relationship, so as to automatically adjust the setting threshold of distance protection zone I, which can effectively avoid the adverse impact of larger SIR value. The method proposed above can be regarded as a research work in the future.

5 Conclusions

As the equivalent impedances of photovoltaic and wind power are quite different from that of synchronous machine, with the increase of new energy penetration rate, system impedance ratio also gradually increases. Under a certain CVT error, the error brought by SIR to distance protection also increases. From the simulation results, when the new energy penetration rate reaches 40%, it may cause the distance protection overreach. Therefore, with the continuous development of new energy, it is suggested to take corresponding measures simultaneously to improve the action performance of distance protection.

References

1. Lopez, J., Sanchis, P., Robam, X., et al.: Dynamic behavior of the doubly fed induction generator during three-phase voltage dips. IEEE Trans. Energy Convers. **22**(3), 709–717 (2007)
2. El-Naggar, A., Erlich, I.: Fault current contribution analysis of doubly fed induction generator-based wind turbines. IEEE Trans. Energy Convers. **30**(3), 874–882 (2015)
3. Wei, W., Zhang, L.M., Gao, B.T., et al: Frequency inconsistency in DFIG-based wind farm during outgoing transmission line faults and its effect on longitudinal differential protection. In: The 4th Annual IEEE International Conference on Cyber Technology in Automation, Control and Intelligent, Hong Kong, China, pp. 25–30. IEEE (2014)
4. Hooshyar, A., Azzouz, M.A., El-Saadany, E.F.: Distance protection of lines connected to induction generator-based wind farms during balanced faults. IEEE Trans. Sustain. Energy **5** (4), 1193–1203 (2014)
5. Hooshyar, A., Azzouz, M.A., EI-Saadany, E.F.: Distance protection of lines emanating from full-scale converter-interfaced renewable energy power plants – part II: solution description and evaluation. IEEE Trans. Power Deliv. **30**(4), 1781–1791 (2015)
6. Huang, T., Lu, Y.P.: Improved superimposed current phase selector of wind farm with crowbar system. In: IEEE PES General Meeting | Conference & Exposition, National Harbor, MD, USA, pp. 1–5. IEEE (2014)
7. Huang, T., Lu, Y.P., Cai, C.: Analysis of phase angle characteristics of DFIG equivalent sequence superimposed impedances and its impact on fault components based on direction relay. Proc. CSEE **36**(14), 3929–3939 (2016)
8. He, B.T., Zhou, C.X.: Error analysis of distance relay protection in phase-to-phase short circuit fault. Electr. Power Autom. Equip. **21**(11), 12–13 (2001)
9. Singh, M., Vishnuvardhan, T., Srivani, S.G.: Adaptive protection coordination scheme for power networks under penetration of distributed energy resources. IET Gener. Transm. Distrib. **10**(15), 3919–3929 (2016)

10. Ahmadi, S.A., Sanaye-Pasand, M., Davarpanah, M.: Preventing maloperation of distance protection due to CCVT transients. IET Gener. Transm. Distrib. **13**(13), 2828–2835 (2019)
11. He, B., Li, Y.Q., Bo, Z.Q.: An adaptive distance relay based on transient error estimation of CVT. IEEE Trans. Power Deliv. **21**(4), 1856–1861 (2006)
12. Davarpanah, M., Sanaye-Pasand, M., Ajaei, F.B.: Compensation of CVT increased error and its impacts on distance relays. IEEE Trans. Power Deliv. **27**(3), 1670–1677 (2012)

A Cluster Charging Strategy for Electric Vehicles Considering the Willingness of EVs' Owners

Xiangyong Zeng(✉) and Wendi Zheng

School of Electrical Engineering and Automation, Fuzhou University, Fuzhou 350108, China
zhengwendi@fzu.edu.cn

Abstract. This paper presents an optimal dispatching model for the charging problem of massive electric vehicles (EVs) connected to the power grid, takes charging discount to incent EVs' owners, and the benefits of power grid and electric vehicle aggregators (EVAs) are both token into consideration. The EVs' cluster charging model is established to avoid the dimension explosion of variables, then a new incentive mechanism based on charging discount is proposed, and the probabilistic model of owners' willingness based on the Weber-Fechner Law of psychology under that mechanism is constructed to estimate the willingness of EVs' owners. Then, the non-cooperative Stackelberg game model between power grid and EVAs is established to reduce the difference between the load peak and valley while balancing the benefits of EVAs and the power grid. Finally, an example based on the presented optimal dispatching model considering massive EVs is put forward, the result of example verify the validity and practicality of the proposed model in reducing the peak-valley difference while balancing the willingness of EVs' owners and the benefits of EVAs.

Keywords: Massive EVs · Weber-Fechner law · The willingness of owners · Non-cooperative stackelberg game

1 Introduction

As the result of the fossil energy crisis and the air pollution caused by fossil energy, EVs have attracted the attention of various research institutions and government departments for its energy-saving and environmental characteristics. Therefore, countries around the world have issued relevant policies to promote the development of EVs industry [1]. However, large-scale EVs connected to the distribution network for disorderly charging will lead to the "peak plus peak" of power load, which will harm the safety and economic operation of power system [2, 3].

In order to reduce the harm caused by massive EVs connected to the distribution network, and give full play to its load transfer characteristic to lower the load peak-valley difference and reduce network loss, the orderly charging strategy of EVs has been studied in many literatures. In reference [4–6], the optimal charging dispatching model is established with the objective functions of minimizing the distribution network loss [4], EVs' charging cost [5], and power purchase cost [6], and using heuristic

Y. Xue et al. (Eds.): PMF 2020, LNEE 718, pp. 124–135, 2021.
https://doi.org/10.1007/978-981-15-9746-6_10

algorithm to solve the optimal charging strategy for each EV. The single EV model mentioned above can lead to the dimension explosion of variables with massive EVs connected to power grid, which is not conducive to calculation. Massive EVs are divided into clusters with different priorities according to the expected finish time of charging, the charging and discharging strategy of EVs' cluster is solved hierarchically [7]. Reference [8] proposes a control strategy of demand response for the cluster of EVs that can change the adjustable capacity boundary of EVs.

On the other hand, the charging facilities of EVs are built by independent investors, which make EVAs can independently dispatch its own charging power as a new stakeholder, and pursue the maximization of its own benefit. Based on the EVs' multi-agent management model, reference [9] establishes the optimal decision model of the EVs' agent based on the game theory. Reference [10] considers the uncertainty of charging demand for EVs, a robust optimal model for cooperation and non-cooperation Stackelberg game between the power grid and EVAs is established. Reference [11] proposes a massive EVs dynamic non-cooperative scheduling model that takes into account multiple EVAs stakeholders under real-time dynamic electricity price. However, the above papers ignore the absolute control right of EVs' owners over EVs, and the charging dispatching of EVs is a process of demand response, which needs to consider the benefits of owners and adopts reasonable incentive mechanism. Reference [12] establishes a multi-objective bi-level optimization model considering the satisfaction of EVs' owners and the economic operation of the power grid. Reference [13] establishes EVs' owners' satisfaction model which comprehensively considers charging time and charging discount, and proposes a distributed real-time collaborative optimization model considering the benefits of the power grid, EVAs and EVs' owners. The above methods require the owners to provide preference index, departure time and so on, and the owners' satisfaction model is set up subjectively, which can not fully reflect the owner's independent choice; In addition, this kind of owners-side model is a comprehensive evaluation of the benefit of all EVs' owners, which cannot guarantee the charging fairness of each EVs. Further, because EVAs fully refers to the departure time provided by the owner to formulate the charging strategy, it can cause more loss of owners in the situation that the EVs aren't charged when the owners need to use EVs in an emergency.

To solve the above problems, the dispatching model of EVs is established in this paper and a novel charging incentive mechanism based on charging discount is proposed, under this mechanism, the probabilistic model of owners' willingness based on the Weber-Fechner Law of psychology is constructed to estimate the willingness of EVs' owners. Then a non-cooperative Stackelberg game model balancing the benefits of power grid and EVAs is established and solved by differential evolution algorithm.

2 The Architecture of EV Charging Optimization

In this paper, the architecture diagram of the EVAs' charging optimization is shown in Fig. 1, which includes distribution system operator (DSO) and EVAs. Among them, DSO obtains power from the upper power grid to supply conventional load and EVAs' load, and it can also guide EVAs to transfer load for cutting peak and filling the valley through reasonable time-of-use electricity price strategy. EVAs are the administrative agencies of charging equipment within a certain area, they can be a set of charging piles in scattered residents area or a large shopping mall parking lot. When the massive charging equipment under an area aggregated into a stakeholder, we need to make sure that all the charging equipment in that area belongs to an investor or company. Thus, EVAs can formulate a reasonable charging strategy to maximize its overall benefits as an independent stakeholder. EVs in residential or commercial office areas can be charged for a long time and has strong scheduling ability for slow charging model [14]. Therefore, the converged EVAs are a collection of charging devices built by the same investor in residential and commercial office areas in a region in this paper.

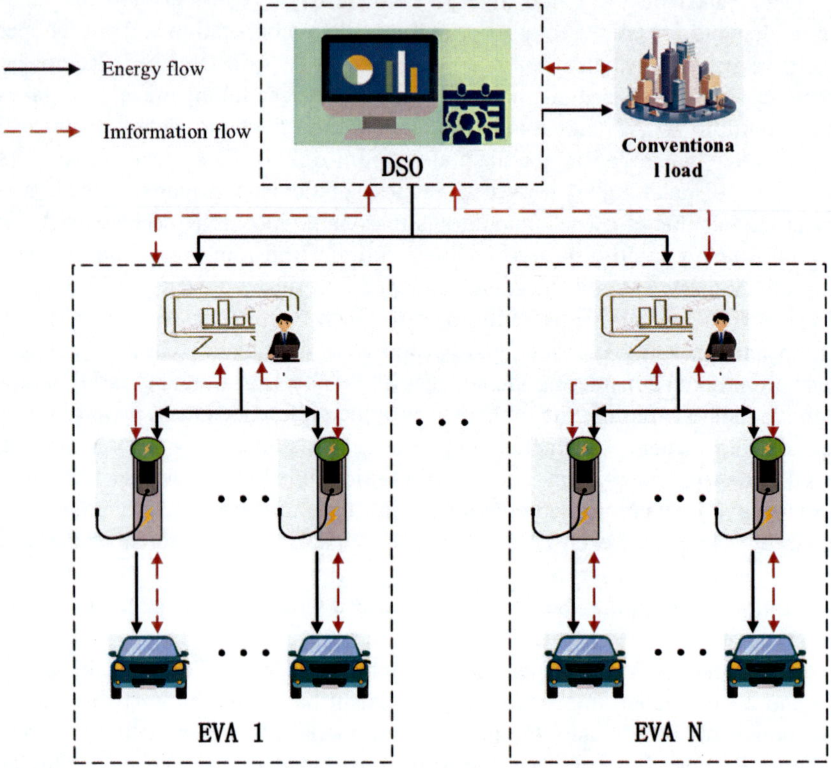

Fig. 1. Architecture diagram of the optimization model

In the Fig. 1, DSO and EVAs are nodes for information flow and energy flow, with the capacity of information processing and computing. Data transmission between DSO and EVAs can be realized by the high-speed communication such as 5G network and optical fiber communication. Among them, DSO mainly issues price information to EVAs, and accepts the EVAs' charging plan to adjust electricity price. EVAs receive the electricity price information issued by DSO, make their own charging strategy according to the price information. Without considering the EVAs' charging guidance for owners, there is no need for the long-distance communication between EVAs and EV, the EVs' owners can obtain the real-time charging discount from the screen of charging device. Of course, if there is communication between EVAs and EV, EVAs can take the result of charging guidance as the prediction information which can provide a reference for charging scheduling.

3 Electric Vehicle Charging Model

3.1 Probability Model of EV Access Time and Driven Mileage

It is assumed that EVs have the same driving characteristics as ordinary cars, and the charging access time of EVs is the return time of the last going out. Collecting actual data and using maximum likelihood estimation method to approximate EV's charging access time and daily driven mileage, we can obtain the following normal distribution and lognormal distribution [15].

The access time of EVs conforms to the following normal distribution:

$$f_S(x) = \begin{cases} \frac{1}{\sigma_S\sqrt{2\pi}}\exp\left[\frac{(x-\mu_S)^2}{2\sigma_S^2}\right], & (\mu_S - 12) < x \leq 24 \\ \frac{1}{\sigma_S\sqrt{2\pi}}\exp\left[\frac{(x+24-\mu_S)^2}{2\sigma_S^2}\right], & 0 < x \leq (\mu_S - 12) \end{cases} \tag{1}$$

From (1), $\mu_S = 17.6$; $\sigma_S = 3.4$

The daily driven mileage of EVs conforms to the following normal distribution is shown as follow.

$$f_D(x) = \frac{1}{x\sigma_D\sqrt{2\pi}}\exp\left[-\frac{(\ln x - \mu_D)^2}{2\sigma_D^2}\right] \tag{2}$$

From (2), $\mu_D = 3.20$; $\sigma_D = 0.88$.

3.2 EV Charging Model and EV Cluster Charging Model

It is considered that the charging power of EVs is constant during each scheduling period, and the charging pile ensures that the EV is fully charged before the EVs' departure period. The charging model of single EV is shown in the following formula (3).

$$\begin{cases} d_i = S_i W_{100}/100 \\ \eta \sum\limits_{t=t_{start}}^{t_{end}} P_{EV.t}^{i} \Delta t = d_i \\ P_{EV.t}^{i} = 0, \quad t \notin \left[t_i^{start}, t_i^{end}\right] \\ 0 \leq P_{EV.t}^{i} \leq P_{EV.\max}^{i} \end{cases} \tag{3}$$

From (3), t_i^{start}, t_i^{end} are the access and departure time of the ith EV; d_i is the charging demand of ith EV; S_i is the daily driven mileage of ith EV; W_{100} is power consumption per 100 km; η is charging efficiency of EV; $P_{EV.t}^i$ is charging power of ith EV at time t; $P_{EV.\max}^i$ is the maximum charging power of ith EV.

The number of variables in the above model is $O(\sum_{t=1}^{T}|\mathcal{I}(t)|)$, and $|\mathcal{I}(t)|$ represents the number of EVs charging on the charging pile at time t. With the continuous development of EV technology, a large number of EVs have begun to connect to charging piles, which increase the magnitude of $|\mathcal{I}(t)|$, and will cause the dimension explosion of variables. Therefore, this paper aggregates EVs that access or depart at the same period and establishes a cluster scheduling model for EVs as shown in Eq. (4).

$$\begin{cases} e_{\min.t} \le e_t \le e_{\max.t} \\ e_t = \eta \sum_{t=t_0}^{t} P_{\mathrm{FEV},t}\Delta t \\ e_{\min.t} = \sum_{t=t_0}^{t} \sum_{i\in\left\{i|t_{EV.i}^{end}=t\right\}} d_i \\ e_{\max.t} = \sum_{t=t_0}^{t} \sum_{i\in\left\{i|t_{EV.i}^{start}=t\right\}} d_i \\ 0 \le e_t - e_{t-1} \le P_{\mathrm{FEV,max}}\Delta t \end{cases} \tag{4}$$

From (4), e_t is the cumulative electric energy injected into the EV by the EVAs from the initial time t_0 to the time t, $e_{\min.t}$, $e_{\max.t}$ is the upper and lower limit of the cumulative injected energy from the initial time t_0 to the time t; $t_{EV.i}^{start}$, $t_{EV.i}^{end}$ is access time and departure time of ith EV, $P_{\mathrm{FEV},t}$ is the total charging power of EVAs at time t; $P_{\mathrm{FEV,max}}$ is the upper limit of the charging power of EVAs. The model described by Eq. (4) focuses on the characteristics of the EVs' cluster and ignores the characteristics of single EV, reference [16] pointed out that the cluster charging strategy that satisfies the model described by Eq. (4) must be able to find a power allocation strategy that satisfies the model described by Eq. (3).

3.3 The Probability Model of EV Owners Based on Weber-Fechner Law

In actually situations, EVs' owners have the absolute control right of EVs, which makes EVAs cannot dispatch EVs randomly. In this section, we consider the basic charging price of EV charging piles invariable, but there are two charging modes designed for EVs' owners to choose:

Mode 1: No dispatching mode, in this mode, EVAs directly charges EVs with maximum charging power, EVs' owners pay EVAs for basic charging price;

Mode 2: Dispatching mode, in this mode, EVAs can schedule the charging power and time of EVs randomly, but it ensures that EVs are fully charged in k periods, and give charging discount to EVs' owners.

The above mechanism allows EVs' owners to participate in orderly charging dispatching completely according to their own willingness without providing any information, which accords with the commercial practicality.

The above charging mechanism makes EVAs need to estimate the selection of EVs' owners for maximizing its benefits. For this purpose, the Weber-Fechner Law (W-F) is introduced to estimate owners' willingness. This law can accurately expresses the functional relationship between the amount of human response s and the amount of objective stimulus I, and is widely used in the field of psychology.

The W-F law states that the magnitude of sensation is directly proportional to the logarithm of the stimulus intensity, the stimulus intensity increases by geometric progression, and the sensory intensity increases by arithmetic progression [17]. Which can be expressed as follow.

$$s = k\ln(I) + s_0 \tag{5}$$

From (5), s is the amount of human response; I is the amount of objective stimulus; k is weber coefficient; s_0 is stimulus constant; This paper believes that the owners' willingness to choose is affected by the amount of the charging discount, and make the charging account φ as the objective stimulus, the human response is the probability of owners choosing no dispatch mode (mode 1). The probability owners' willingness model is established as follows:

$$P(\varphi) = \begin{cases} P_0, & \varphi \leq I_0 \\ \frac{k_0 \ln(\varphi) + s_0}{s_0}, & I_0 \leq \varphi \leq I_z \\ 1, & \varphi \geq I_z \end{cases} \tag{6}$$

From (6), $P(I)$ is the probability of EV owner choosing no dispatch mode (mode 1) under objective stimulus φ; I_0 is the amount of smallest sensible; P_0 is the maximum response, usually $P_0 > 0$, it means that the willingness of owners not entirely influenced by discount; I_z is the maximum limit of stimulation, usually $I_z = 1$,it means that EVs' owners don't want to accept the charging dispatching of EVAs at all without charging account.

4 Non-cooperative Stackelberg Game Optimization Model

This paper assumes that EVAs purchase electricity from DSO by time-of-use electricity price. Under the time-of-use electricity price, EVAs use charging discount mechanism to incent EV owners and dispatch the charging power of EVs to maximize its own benefits.

4.1 DSO Dispatching Model

The disorderly charging of EVs will make the load appear to the severe situation of "peak plus peak". Therefore, the distribution network hopes to make full use of the load transfer characteristics of EVs, and relies on time-of-use electricity prices to incent EVAs to achieve the purpose of peak clipping and valley filling. Therefore, the objective function of the DSO can be shown in (7):

$$\begin{aligned} C_M &= \min \sum_{t=1}^{T} (P_{load}^{t} - P_{load}^{avg})^2 \\ P_{load}^{avg} &= \frac{1}{T} \sum_{t=1}^{T} P_{load}^{t} \end{aligned} \tag{7}$$

From (7), P_{load}^{t} is the amount of conventional load plus EVAs' load at time t; T is the number of optimal dispatching period.

4.2 EVA Scheduling Model

The goal of EVAs is to incent EV owners to participate in scheduling through charging discounts, and then to shift the peak and valley of EV load to reduce the total cost. Its objective function is:

$$C_{EVA} = \min(C_{WX} + C_Z + C_{\varphi}) \tag{8}$$

From (8), C_{WX} is the EV charging cost of disorderly charging part that select mode 1; C_Z is EV charging cost of orderly charging part that select mode 2; C_{φ} the charging discount cost of orderly charging part that select mode 2.

$$\begin{aligned} C_{WX} &= \sum_{t=1}^{T} p_{cs}^{t} |\mathcal{I}_{WX}(t)| P_{EV}^{\max} \Delta t \\ C_Z &= \sum_{t=1}^{T} p_{cs}^{t} P_{EVA}^{t} \Delta t \\ C_{\varphi} &= \sum_{t=1}^{T} \Big(\sum_{i \in \mathcal{I}_{start}(t)} \frac{d_i}{\eta} \varphi_t p_{EV} \Big) \end{aligned} \tag{9}$$

In (9), p_{cs}^{t} is the electricity price at time t; $|\mathcal{I}_{WX}(t)|$ is the cluster of EVs that charg at time t and select mode 1; $P_{EV}^{\max}$ is maximum charging power of EV; P_{EVA}^{t} is cluster charging power of EVs that select mode 2 at time t; $\mathcal{I}_{start}(t)$ is the cluster of EV that assess to charging pile at time t; φ_t is charge discount for EVs that assess to charging pile at time t; p_{EV} is basic charging electricity price of charging pile.

4.3 Model Solving

In the above non-cooperative Stackelberg game model, the EVA solves its optimal strategy under the time-of-use electricity price p_{cb}^{t} issued by DSO. The EVA side solution model can be expressed as:

$$\left\{P_{EVA}^{t*}, \varphi_t^*\right\} = \underset{P_{cb}^{t}, \varphi_t^*}{\arg\min}\, C_{EVA}\left(p_{cb}^{t}, P_{EVA}^{t}, \varphi_t\right) \tag{10}$$

The DSO guides the EVA to actively schedule EV load by setting the time-sale electricity price p_{cb}^{t}. The DSO side solution model can be expressed as follow.

$$\left\{p_{cb}^{t*}\right\} = \underset{p_{cb}^{t}}{\arg\min}\, C_M\left(p_{cb}^{t}, P_{EVA}^{t*}, \varphi_t^*\right) \tag{11}$$

The above non-cooperative Stackelberg game model is a two-level programming problem. The upper-layer DSO finds out the optimal time-of-use electricity price p_{cs}^{t}, and the lower-level EVAs look for the optimal charging discount φ_t and charging power P_{EVA}^{t} under the time-of-use electricity price. In this paper, the differential evolution algorithm is used to search for the most optimal electricity price strategy in the upper layer, taking individuals in the population as an internal electricity price strategy for DSO. The EVAs calculates the optimal strategy under each individuals of electricity price in parallel, and feedback results to DSO. According to the feedback from EVAs, DSO calculates the fitness of each price strategy; through population variation, crossover, and selection, obtaining the optimal internal electricity price strategy for DSO. The specific solution steps are as follows:

(1) Input system parameters, including conventional load forecast data, EV forecast data, algorithm parameters and power distribution system parameters.
(2) The upper-layer DSO randomly generates initial populations which take electricity prices strategy as individuals.
(3) DSO delivers all the individuals to EVAs. According to (8)–(10), EVA calculates its own optimal dispatch strategy under each electricity price strategy by multi-population genetic algorithm in parallel, and feedback the result to DSO.
(4) Under the result feed from EVAs, DSO calculates the corresponding objective function of each individuals for (7) and (11), then extracting the best individual
(5) DSO makes sure whether the maximum iteration generation or the maximum retentive generation of the optimal value is met If it is satisfied, the optimal electricity price strategy is output, and EVAs calculate the corresponding charging strategy. If it is not satisfied, DSO generates a new population through population variation, crossover, and selection, and repeat (3)–(5) until it is satisfied.

5 Example Analysis

This paper takes the charging optimization of massive EVs under the typical working day in a certain distribution area as an example to verify the effectiveness and practicality of the proposed model. Assuming that the whole area is divided into 3 EVAs according to region, each EVAs has totally 4000 EVs assessed in a day; the maximal cluster charging power of each EVA area is 2.4 kW; the efficiency of the charging pile η is 0.95; the maximum charging power of each EV is 6 kW, and the power consumption per 100 km is 12.5 kW; Optimal interval $\Delta t = 15$ min, The optimal period is one day, a total of 96 optimization periods; The basic electricity price of the charging equipment is 1.5 yuan/kW. The level of time-of-use electricity price from DSO are respectively 0.4 yuan/kWh, 0.7 yuan/kWh and 1.0 yuan/kWh.

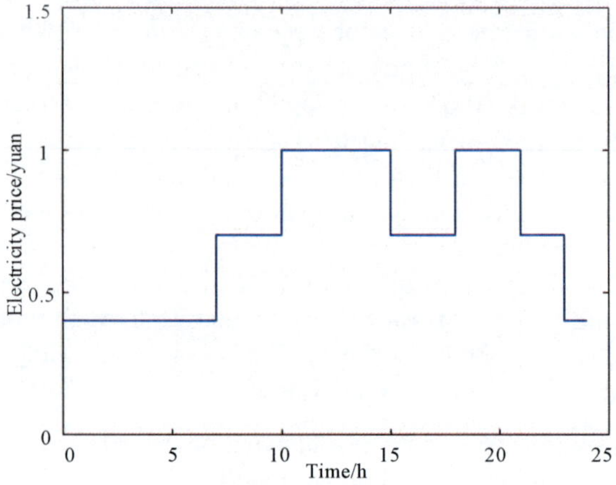

Fig. 2. The curve graph of DSO time-sharing electricity price.

Figure 2 Show the the curve graph of DSO time-sharing electricity price after Non-cooperative Stackelberg game. In order to verify the peak-cutting and valley-filling effects of the proposed model, Fig. 3 shows the result of EVs' load transfer after charging optimization in one day. In Fig. 3, we can see that the conventional load is at the peak of the day at about 18 o'clock, and the disorderly charging of EVs leads to the problem of "peak plus peak" for people return from work at this time. The conventional load is at a low point during the 0:00–7:00, at the same time, the EVs are almost at the state of fully charging after 0:00 under disorderly charging, which cause the total load still at a valley. Through the proposed model of charging optimization, we can see that the EVs' load shift from the peak to valley, reducing the total load of difference between valley and peak, and the total load variance is decreased from 121.01 to 61.33 after charging optimization, which means that the model can reduce the load fluctuations and improve the operating conditions of power system effectively.

Table 1 shows the operation cost of EVAs before and after charging dispatching. It can be seen from the table that EVAs needs additional incentive costs, but the cost of

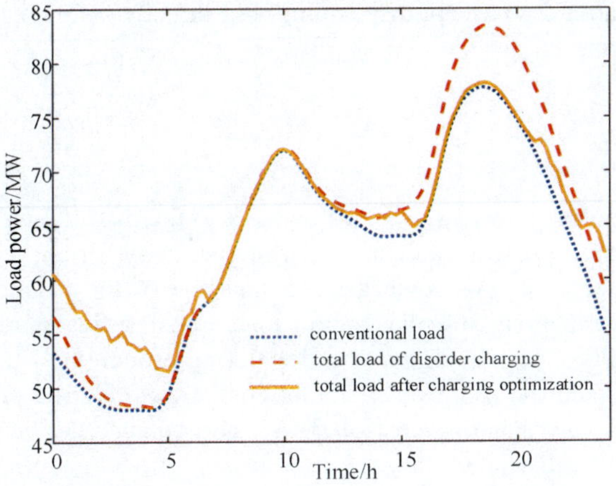

Fig. 3. Performance of load transfer for charging optimization model

Table 1. The operating cost of each EVAs.

		C_{WX}(yuan)	C_Z(yuan)	C_φ(yuan)	Total cos t (yuan)	Numbers of EVs with dispatching
EVA1	Disorderly charging	13902	0	0	13902	0
	Charging optimization	4469	2616	4769	11854	2463
EVA2	Disorderly charging	14079	0	0	14079	0
	Charging optimization	4864	2447	4674	11985	2380
EVA3	Disorderly charging	13988	0	0	13988	0
	Charging optimization	4591	2558	4731	11880	2421

power purchase from the grid is greatly reduced through EV load transfer, which leads to a reduction of total operating costs by about 15%. This shows that the proposed model can take into account the benefit of both DSO and EVAs, making EVAs voluntarily provide charging discounts to incentivize car owners, which eventually maximize EVAs own benefits through load transfer. In the example, the optimal discount strategy of EVAs incent about 60% of EVs' owners to choose mode 2 which means that they are willing to accept charging scheduling for getting discount, It shows that there is no need to give discount compensation for all EVs when EVAs are used as independent stakeholder to maximize the benefit of EVAs. And this paper takes the willingness of EVs' owners into consideration, EVAs only can dispatch the EVs whose owners choose mode 2 to accept dispatching, but directly fully charge the EVs whose owners select mode 1.

6 Conclusion

For the optimal problem of massive EVs connect to the distribution network to charge in cluster, this paper proposes a cluster charging strategy for electric vehicles considering the willingness of EVs' owners. The analysis of the example shows that the proposed model can give full play to the load transfer characteristics of EVs, and realize the function of peak cutting and valley filling through the guidance of electricity price from DSO, and the result of operating cost shows that the proposed model can effectively reduce the operating cost of EVAs, and balance the benefits of both DSO and EVAs. On the other hand, this paper proposes a novel charging mechanism with multi-modes, and uses the W-F law of psychology to estimate the willingness of owners, which can reflect owners independent choice and accords with its commercial practicality. Further, more optional modes with different fully charging time can be added for the EVs' owners to choose flexibly in practice, and we can update the information of EVs and calculate the optimal strategy in real-time to avoid the loss caused by the error of estimation.

References

1. Junhua, Z., Fushuan, W., Aiming, Y., et al.: Impacts of electric vehicles on power systems as well as the associated dispatching and control problem. Autom. Electr. Power Syst. **35**(14), 2–10 (2011)
2. Li, M., Zhao, B.: Charging and discharging decision-making model of electric vehicles and influence analysis of electricity price. Autom. Electr. Power Syst. **39**(15), 75–81 (2015)
3. Pieltain Ferna, N.L., Gomez, S.R.T., Cossent, R., et al.: Assessment of the impact of plug-in electric vehicles on distribution networks. IEEE Trans. Power Syst. **26**(1), 206–213 (2011)
4. Yang, X., Bai, X., Li, P., et al.: Charging optimization of massive electric vehicles in distribution network. Electr. Power Autom. Equipment **35**(6), 31–36 (2015)
5. Turker, H., Bacha, S.: Optimal minimization of plug-in electric vehicle charging cost with vehicle-to-home and vehicle-to-grid concepts. IEEE Trans. Veh. Technol. **67**(11), 10281–10292 (2018)
6. Sortomme, E., El-Sharkawi, M.A.: Optimal scheduling of vehicle-to-grid energy and ancillary services. IEEE Trans. Smart Grid **3**(1), 351–359 (2012)
7. Luo, Z., Hu, Z., Song, Y., et al.: Study on charging load modeling and coordinated charging of electric vehicles under battery swapping modes. Proc. CSEE **32**(31), 1–10 (2012)
8. Lixing, Li., Shihong, M., Dandan, S., et al.: Dynamic games of complete information in active distribution network with multi-stakeholder participation. Trans. China Electrotech. Soc. **33**(15), 3449–3509 (2018)
9. Li, M., Song, Y., Yan, Z., et al.: Research on game model and algorithm for electric vehicle aggregations. Power Syst. Technol. **38**(6), 1512–1517 (2014)
10. Yang, H., Xie, X., Vasilakos, A.V.: Non cooperative and cooperative optimization of electric vehicle charging under demand uncertainty: a robust Stackelberg game. IEEE Trans. Veh. Technol. **65**(3), 1043–1058 (2016)

11. Chen, L., Pan, Z., Yu, T., et al.: Real-time optimal dispatch for large-scale electric vehicles based on dynamic non-cooperative game theory. Autom. Electr. Power Syst. **43**(24), 32–40 (2019)
12. Wang, X., Zhao, J., Wang, K., et al.: Multi-objective bi-level electric vehicle charging optimization considering user satisfaction degree and distribution grid security. Power Syst. Technol. **41**(7), 2165–2172 (2017)
13. Pan, Z., Yu, T., Wang, K.: Decentralized coordinated dispatch for real-time optimization of massive electric vehicles considering various interests. Proc. CSEE **39**(12), 3528–3541 (2019)
14. Junhua, Z., Fushuan, W., Aimin, Y., et al.: Impacts of electric vehicles on power systems as well as the associated dispatching and control problem. Autom. Electr. Power Syst. **35**(14), 2–10 (2011)
15. Tian, L., Shi, S., Jia, Z.: A statistical model for charging power demand of electric vehicles. Power System Technol. **34**(11), 126–130 (2010)
16. Tang, W., Zhang, Y.: A model predictive control approach for low-complexity electric vehicle charging scheduling: optimality and scalability. IEEE Trans. Power Syst. **32**(2), 1050–1063 (2017)
17. Wei, H., Li., N., Li., W., et al.: Bi-level joint optimization dispatch of active distribution network considering the participation of multi-stakeholder. Proc. CSEE **37**(12), 3418–3428 (2017)
18. Xiaodong, Y., Shuaijie, R., Youbing, Z., et al.: Schedulable ability model and priority-based intraday scheduling strategy for electric vehicle. Autom. Electr. Power Syst. **41**(02), 84–93 (2017)
19. Hou, H., Fan, H., Su, S., et al.: Multilateral win-win strategy for charging service of electric vehicle. Autom. Electr. Power Syst. **41**(19), 66–73 (2017)

Research on State Estimation Method of Unscented Kalman Distribution Network Considering Measurement Correlation

Jianxiong Nie[1(✉)] and Wendi Zheng[1,2]

[1] Fuzhou University, Fuzhou 350116, China
[2] Fujian Smart Electrical Engineering Technology Research Center, Fuzhou 350108, China

Abstract. Distribution system state estimation is one of the key elements for distribution network monitoring activities. It estimates the voltage amplitude and phase angle of each node on the basis of redundant measurement information. This paper presents a dynamic state update method considering the influence of the correlation among different measurement data. Firstly, the correlation between the amplitude and phase measured by the same transformer is analyzed. Secondly, the covariance matrix among the measurement data can be obtained by using unscented transformation method, and is applied to the unscented Kalman filter (UKF) for calculating the current state. Finally, we assess the performance of the proposed method on an IEEE-33 distribution network. The results indicate that the method can achieve a satisfactory estimation accuracy and avoids obtaining suitable measurement error covariance only relying on experience.

Keywords: Distribution network · Measurement correlation · Dynamic state estimation · Unscented kalamn filter · Covariance matrix

1 Introduction

State estimation is the core component of Distribution Management System (DMS). It provides reliable data support for advanced online application functions [1, 2]. Therefore, improving the accuracy and reliability of state estimation is very important for the safe and efficient operation of the distribution network.

Dynamic state estimation is mainly based on Extend Kalman filter (EKF) [3] and UKF [4] algorithms. EKF inevitably has linearization errors due to linear approximation of nonlinear system equations. Base on Unscented Transformation (UT), UKF obtains the statistical characteristics of the nonlinear transformation and has higher accuracy and stability for filter. Reference [5] proposes a robust generalized maximum likelihood UKF algorithm for the problem that the statistical characteristics of system process noise is difficult to obtain. Reference [6] introduces the UKF square root form (square root UKF, SRUKF) that can be used to enhance the numerical stability of UKF. The adaptive UKF approach is described in [7] by constructing adaptive factors.

Y. Xue et al. (Eds.): PMF 2020, LNEE 718, pp. 136–145, 2021.
https://doi.org/10.1007/978-981-15-9746-6_11

Through a lot of simulations and tests, Eduardo Caro and other scholars find that the measurement errors of the supervisory control and data acquisition (SCADA) basically obey the Gaussian distribution, but they are not independent of each other [8–11]. However, the correlation among different measurements is usually neglected and the analysis of its impact on estimated results is lacking. Therefore, the UKF with measurement correlation(UKF- correlation) is proposed in this paper.

The paper is organized as follows. In Sect. 2 the principle of UKF is presented. In Sect. 3 this paper analyzes the correlation in measurement and calculates the covariance matrix of measurement. Finally, the conclusion is drawn in Sect. 4.

2 Principle of UKF

The distribution network is a high-dimensional nonlinear system. The nonlinear state transition and measurement function can be expressed in the following form:

$$\boldsymbol{x}_k = \boldsymbol{f}(x_{k-1}) + \boldsymbol{q}_k \tag{1}$$

$$\boldsymbol{y}_k = h(\boldsymbol{x}_k) + \boldsymbol{v}_k \tag{2}$$

2.1 UT Transformation

The UT transformation approximates a nonlinear function by approximating its probability density distribution. This paper uses the symmetric proportional modified sampling method [12] to construct the dimensional state quantity Sigma points and calculate corresponding weights.

$$\begin{cases} \mathbf{x}^s_{1,k-1} = \mathbf{x}_{k-1} \\ \mathbf{x}^s_{i,k-1} = \mathbf{x}_{k-1} + \left[\sqrt{(n+\lambda)\mathbf{P}_{k-1}}\right]_i \quad i = 2, 3, \cdots, n+1 \\ \mathbf{x}^s_{i,k-1} = \mathbf{x}_{k-1} - \left[\sqrt{(n+\lambda)\mathbf{P}_{k-1}}\right]_i \\ \qquad\qquad i = n+2, n+3, \cdots, 2n+1 \end{cases} \tag{3}$$

Where $\boldsymbol{x}_{k-1}$ represents the n dimension state at time $k-1$, $\boldsymbol{x}^s_{i,k-1}$ is Sigma sampling points of $\boldsymbol{x}_{k-1}$; $\boldsymbol{P}_{k-1}$ is covariance matrix of $\boldsymbol{x}_{k-1}$; $\lambda = \alpha^2(n+k_r) - n$, $0 < \alpha \leq 1$, is the scale factor used to reduce the total prediction error, α is the proportional parameter that controls the distribution of the Sigma point.; k_r to ensure $(n+\lambda)\boldsymbol{P}$ maintains positive semidefiniteness. The weights of each Sigma point are:

$$\begin{cases} W_1^{(m)} = \frac{\lambda}{n+\lambda} \\ W_i^{(m)} = \frac{\lambda}{2(n+\lambda)} \quad i = 2, 3, \cdots, 2n+1 \\ W_1^{(c)} = \frac{\lambda}{n+\lambda} + 1 - \alpha^2 + \beta \\ W_i^{(c)} = \frac{\lambda}{2(n+\lambda)} \quad i = 2, 3, \cdots, 2n+1 \end{cases} \tag{4}$$

Where W_i^m is mean weight; W_i^c is covariance weight.

2.2 Implementation of the UKF Algorithm

1) evaluating the sigma points $\bar{\boldsymbol{x}}_{k|k-1}$ with the dynamic model function and estimating the predicted state mean $\bar{\boldsymbol{x}}_{k|k-1}$

$$\boldsymbol{x}_{i,k|k-1} = f\left(\boldsymbol{x}_{i,k-1}^{s}\right) + \boldsymbol{q}_{k-1} \quad i = 1, 2, \cdots, 2n+1 \tag{5}$$

$$\bar{\boldsymbol{x}}_{k|k-1} = \sum_{i=1}^{2n+1} W_i^{(m)} \boldsymbol{x}_{i,k|k-1} \tag{6}$$

$$\boldsymbol{P}_{k|k-1} = \sum_{i=1}^{2n+1} W_i^{(c)} \left[\left(\boldsymbol{x}_{i,k|k-1} - \bar{\boldsymbol{x}}_{k|k-1}\right)\left(\boldsymbol{x}_{i,k|k-1} - \bar{\boldsymbol{x}}_{k|k-1}\right)^{\mathrm{T}}\right] + \boldsymbol{Q}_k \tag{7}$$

2) evaluating the propagated sigma points $\boldsymbol{y}_{i,k|k-1}$ with measurement function and estimating the predicted measurement $\hat{\boldsymbol{y}}_{k|k-1}$

$$\boldsymbol{y}_{i,k|k-1} = h\left(\boldsymbol{x}_{i,k|k-1}\right) \quad i = 1, 2, \cdots, 2n+1 \tag{8}$$

$$\hat{\boldsymbol{y}}_{k|k-1} = \sum_{i=1}^{2n+1} W_i^{(m)} \boldsymbol{y}_{i,k|k-1} \tag{9}$$

3) Calculating the Kalman gain $\boldsymbol{K}_k$

$$\boldsymbol{S}_k = \sum_{i=1}^{2n+1} W_i^{(c)} \left[\left(\boldsymbol{y}_{i,k|k-1} - \hat{\boldsymbol{y}}_{k|k-1}\right)\left(\boldsymbol{y}_{i,k|k-1} - \hat{\boldsymbol{y}}_{k|k-1}\right)^{\mathrm{T}}\right] + \boldsymbol{R}_k \tag{10}$$

$$\boldsymbol{C}_k = \sum_{i=1}^{2n+1} W_i^{(c)} \left[\left(\boldsymbol{x}_{i,k|k-1} - \hat{\boldsymbol{x}}_{k|k-1}\right)\left(\boldsymbol{y}_{i,k|k-1} - \hat{\boldsymbol{y}}_{k|k-1}\right)^{\mathrm{T}}\right] \tag{11}$$

$$\boldsymbol{K}_k = \boldsymbol{C}_k \boldsymbol{S}_k^{-1} \tag{12}$$

4) Updating state and covariance

$$\hat{\boldsymbol{x}}_k = \hat{\boldsymbol{x}}_{k|k-1} + \boldsymbol{K}_k(\boldsymbol{y}_k - \hat{\boldsymbol{y}}_{k|k-1}) \tag{13}$$

$$\boldsymbol{P}_k = \boldsymbol{P}_{k|k-1} - \boldsymbol{K}_k \boldsymbol{S}_k \boldsymbol{K}_k^{\mathrm{T}} \tag{14}$$

3 Research on UKF Considering Instrument Transformer Measurement Correlation

3.1 Correlation Analysis

The intermediate measurements first are measured by voltage transformer and current transformer, and convert into the power measurements by the following formulation:

$$P_i = U_i I_i \cos(\varphi_{U_i} - \varphi_{I_i}) \tag{15}$$

$$Q_i = U_i I_i \sin(\varphi_{U_i} - \varphi_{I_i}) \tag{16}$$

$$P_{ij} = U_i I_{ij} \cos(\varphi_{U_i} - \varphi_{I_{ij}}) \tag{17}$$

$$Q_{ij} = U_i I_{ij} \sin(\varphi_{U_i} - \varphi_{I_{ij}}) \tag{18}$$

Where U_i and φ_{U_i} are the voltage amplitude and phase angle in the node i; I_i and φ_{I_i} are current amplitude and phase angle in the node i; I_{ij} and $\varphi_{I_{ij}}$ are the magnitude and phase angle of the current measured on the branch ij.

For the same instrument transformer, the measured voltage amplitude and angle will be affected by the error of each other, and there is no direct correlation between different instrument transformers. So the covariance matrix of the intermediate measurements is:

$$\begin{aligned} &\operatorname{cov}([U_i, \varphi_{U_i}, I_i, \varphi_{I_i}, I_{ij}, \varphi_{I_{ij}}]) \\ &\quad = \begin{bmatrix} \boldsymbol{D}_1 & 0 & 0 \\ 0 & \boldsymbol{D}_2 & 0 \\ 0 & 0 & \boldsymbol{D}_3 \end{bmatrix} \end{aligned} \tag{19}$$

$$\boldsymbol{D}_1 = \begin{bmatrix} \sigma_{U_i}^2 & \sigma_{\varphi_{U_i}} \sigma_{U_i} \rho_U^{\mathrm{TS}} \\ \sigma_{\varphi_{U_i}} \sigma_{U_i} \rho_U^{\mathrm{TS}} & \sigma_{\varphi_{U_i}}^2 \end{bmatrix}$$

$$D_2 = \begin{bmatrix} \sigma_{I_i}^2 & \sigma_{\varphi_{I_i}}\sigma_{I_i}\rho_I^{TS} \\ \sigma_{\varphi_{I_i}}\sigma_{I_i}\rho_I^{TS} & \sigma_{\varphi_{I_i}}^2 \end{bmatrix}$$

$$D_3 = \begin{bmatrix} \sigma_{I_{ij}}^2 & \sigma_{\varphi_{I_{ij}}}\sigma_{I_{ij}}\rho_I^{TS} \\ \sigma_{\varphi_{I_{ij}}}\sigma_{I_{ij}}\rho_I^{TS} & \sigma_{\varphi_{I_{ij}}}^2 \end{bmatrix}.$$

Where σ_{U_i}, $\sigma_{\varphi_{U_i}}$, σ_{I_i}, $\sigma_{\varphi_{I_i}}$, $\sigma_{I_{ij}}$ and $\sigma_{\varphi_{I_{ij}}}$ are the standard deviation of U_i, φ_{U_i}, I_i, φ_{I_i}, I_{ij}, $\varphi_{I_{ij}}$; ρ_U^{TS} and ρ_I^{TS} are correlation coefficient between the amplitude and phase angle measured by the same instrument transformer. The covariance matrix of intermediate measurements is no longer a diagonal matrix, and its off-diagonal elements show a correlation among the measurements.

3.2 Calculating Covariance Matrix Among Power Measurements

From Eqs. (15) to (18), we can know that there is a nonlinear relationship between the intermediate measurements by the instrument transformer and power measurements. Therefore, the covariance matrix of power measurement is obtained through follows:

1) symmetric sampling method to obtain the sampling point matrix:

$$\mathbf{Z}_i = \bar{z} + (\sqrt{m_z \mathbf{P}_z}) \tag{20}$$

$$\mathbf{Z}_i = \bar{z} - (\sqrt{m_z \mathbf{P}_z}) \tag{21}$$

Where $\bar{z}$, m_z and $\mathbf{P}_Z$ are the mean, number, and covariance matrix of intermediate measurements.

2) Calculating sampling points of power measurements

$$\mathbf{Y}_i = f(\mathbf{Z}_i) \tag{22}$$

3) Calculating the covariance matrix $\overline{\mathbf{R}}$ of power measurements

$$\bar{y} = \frac{1}{2m_z}\sum_{i=1}^{2m_z}(\mathbf{Y}_i) \tag{23}$$

$$\overline{\mathbf{R}} = \frac{1}{2m_y}\sum_{i=1}^{2m_y}[(\mathbf{Y}_i - \bar{y})(\mathbf{Y}_i - \bar{y})^{\mathrm{T}}] \tag{24}$$

Where $\overline{\mathbf{R}}$ is the covariance matrix of the power measurement. The diagonal element can reflect the measurement accuracy, while the off-diagonal element represents the correlation between the measurements. It is applied to the UKF method to replace the measurement error variance matrix $\boldsymbol{R}_k$.

The steps of the entire algorithm are as follows:

step 1: Input the initial value of state and covariance matrix;

step 2: Use Eqs. (3)–(4) to obtain the sampling points and weights of UKF algorithm.

step 3: Obtain the predicted state and predicted measurement through Eqs. (5)–(9), and use Eqs. (20)–(24) to calculate the covariance matrix $\overline{\boldsymbol{R}}$ with Correlation;

step 4: Calculate the Kalman gain through Eqs. (10)–(12);

step 5: Finally, based on the measurement $\boldsymbol{y}_k$ and gain $\boldsymbol{K}_k$, the state is corrected and its variance is updated through Eqs. (13)–(14).

4 Calculation Example and Result Analysis

In order to verify the effectiveness and correctness of the proposed algorithm, this paper uses IEEE-33 distribution network [13, 14] to test the algorithm and makes the injected power of all nodes fluctuate randomly.

The method combining Latin hypercube and Cholesky decomposition [15, 16] is used to simulate the guassian noises with correlation measured by voltage transformer and current transformer. These noises are added to the true values of voltage and current obtained by the power flow calculation, and power measurements are calculated through formulas (15) to (18).

After estimating the voltage amplitude and phase angle by the algorithm proposed in this paper and traditional UKF algorithm, we can draw a curve with the estimated value and the real value for bus 3 to compare the accuracy under different algorithms. It can be seen from the Fig. 1 and Fig. 2 that the algorithm of this paper improves the performance of dynamic state estimation after considering the correlation of measurement values.

It can be seen that the estimation results of UKF-correlation are obviously better than that of UKF algorithm. Because the noise covariance matrix of UKF is a diagonal constant matrix, which neither considers the correlation of measurement nor adapts to the time-varying noise. Comparatively, UKF-correlation obtains noise covariance matrix through real-time computing. The method makes it possible for estimation to get accurate results when the load fluctuates greatly.

To quantitatively compare the estimation results, we use root-mean square-error (RMSE) as index

$$R_{MSE}(k) = \sqrt{\frac{1}{N}\sum_{i=1}^{N}\left(\hat{y}_{k,i} - y_{k,i}^{t}\right)^2} \tag{25}$$

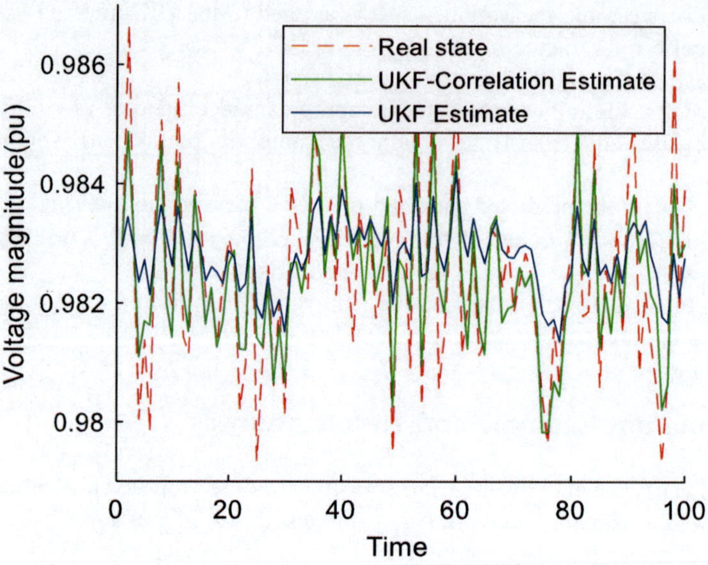

Fig. 1. Comparison results of estimation the voltage magnitude at bus 3 in the IEEE-33 bus system.

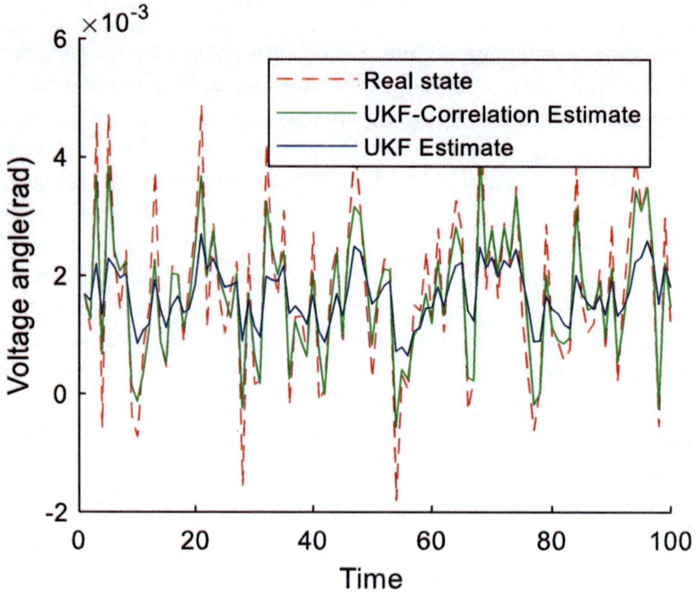

Fig. 2. Comparison results of estimation the voltage angle at bus 3 in the IEEE-33 bus system.

Where N is number of states, $\hat{y}_{k,i}$ and $y'_{k,i}$ are the estimated value and real value of the state at the k moment (Fig. 3, Fig. 4 and Table 1).

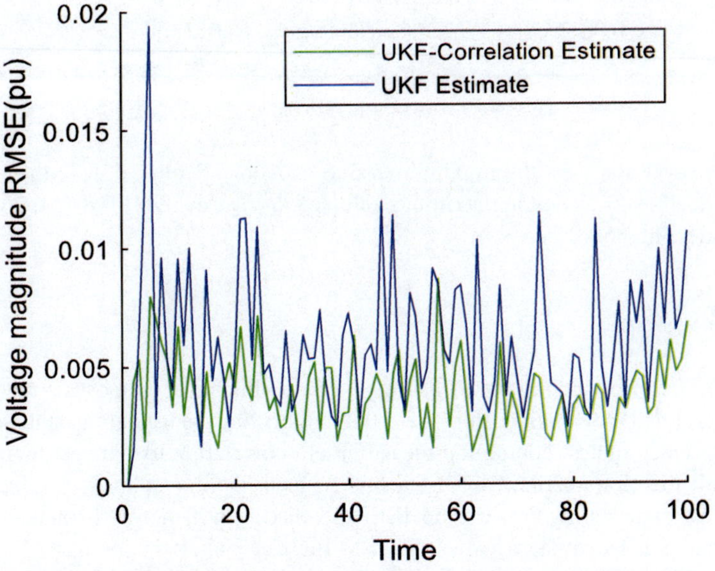

Fig. 3. RMSE Comparison of estimation the voltage magnitude

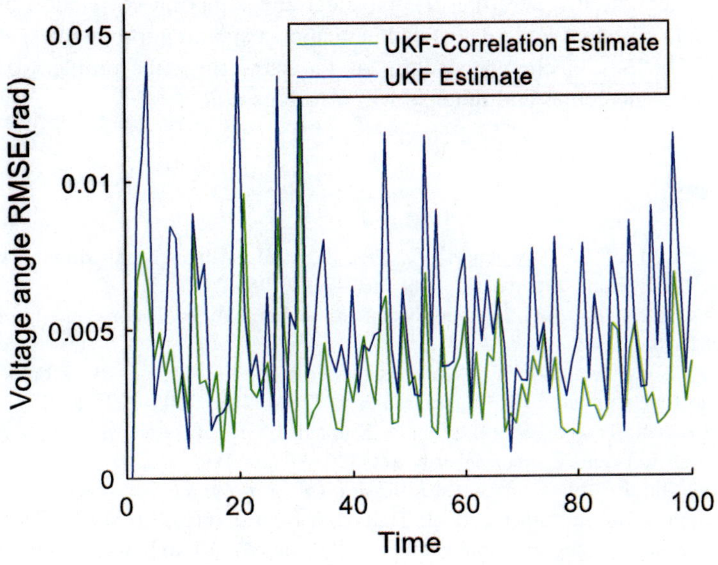

Fig. 4. RMSE Comparison of estimation the voltage angle

Table 1. Performance comparison of UKF and UKF-Correlation

Method	Maximum of R_{UMSE}	Mean of R_{UMSE}	Maximum of $R_{\theta MSE}$	Mean of $R_{\theta MSE}$
UKF	0.0194	0.0061	0.0151	0.0053
UKF-Correlation	0.0087	0.0038	0.0121	0.0025

Comparison analysis, the maximum value of R_{UMSE} declines 55%; the mean value of R_{UMSE} declines 37.7%; the maximum value of $R_{\theta MSE}$ declines 19.8%; the mean value of $R_{\theta MSE}$ declines 52.8%.

5 Conclusion

Based on correlation among the measurement data, this paper proposes a method fusing the measurement correlation into the UKF algorithm. In dynamic state estimation, using covariance matrix containing measurement correlation to state estimation is more in line with the characteristics of the actual measurement system. It can be seen from the result of state estimation on the IEEE-33 distribution network that the proposed method makes the estimated value closer to the true value.

In the actual measurement system, there is some bad data objectively. The traditional bad data detection methods include target maxima detection, weighted residual detection, and pseudo-measurement detection, which cannot guarantee that all the bad data can be identified. Therefore, the physical and temporal correlation between the measured data will be used to detect bad data in next step to improve the sensitivity and success rate of the detection algorithm. At the same time, the problem of false data injection attack detection and defense will also be studied.

References

1. Meliopoulos, A.P.S., Polymeneas, E., Tan, Z., et al.: Advanced distribution management system. IEEE Trans. Smart Grid **4**(4), 2109–2117 (2013)
2. Abur, A., Exposito, A.G.: Power System State Estimation: Theory and Implementation. Marcel Dekker, New York (2004)
3. Zhao, J., Netto, M., Mili, L.: A robust iterated extended Kalman fllter for power system dynamic state estimation. IEEE Trans. Power Syst. **32**(4), 3205–3216 (2017)
4. Valverde, G., Terzija, V.: Unscented Kalman filter for power system dynamic state estimation. IETGen. Transm. Distrib. **5**(1), 29–37 (2011)
5. Zhao, J., Mili, L.:Robust unscented Kalman fllter for power system dynamic state estimation with unknown noise statistics. IEEE Trans. Smart Grid **10**(2), 1215–1224 (2019)
6. Wei, Z., Sun, G., Pang, B.: Application of UKF and SRUKF to power system dynamic state estimation. Proc. CSEE **31**(16), 74–80 (2011)
7. Sun, J., Liu, M., Deng, L.,Ying, L., Li, Z.: State estimation of distribution network based on AUKF. Power Syst. Protection Control **46**(11), 1–7 (2018)

8. Caro, E., Conejo, A.J., Minguez, R.: Power system state estimation considering measurement dependencies. IEEE Trans. Power Syst. **24**(4), 1875–1885 (2009)
9. Caro, E., Morales, J.M., Conejo, A.J., et al.: Calculation of measurement correlations using point estimate. IEEE Trans. Power Delivery **25**(4), 2095–2103 (2010)
10. Caro, E., Conejo, A.J., Minguez, R., et al.: Multiple bad data identification considering measurement dependencies. IEEE Trans. Power Syst. **26**(4), 1953–1961 (2011)
11. Caro, E., Valverde, G.: Impact of transformer correlations in state estimation using the unscented transformation. IEEE Trans. Power Syst. **29**(1), 368–376 (2014)
12. Zhao, H., Tian, T.: Dynamic state estimation for power system based on an adaptive unscented Kalman filter. Power Syst. Technol. (2014)
13. Baran, M.E., Wu, F.F.: Optimal capacitor placement on radial distribution systems. IEEE Trans. Power Delivery **41**, 725–734 (1989)
14. Meng, Y., Yan, Z., Xu, X., et al.: Global sensitivity analysis of distribution network state estimation and its application. Autom. Electric Power Syst. **44**(02), 114–126 (2020)
15. Yu, H., Chung, C., Wong, K., et al.: Probabilistic load flow evaluation with hybrid latin hypercube sampling and Cholesky decomposition. IEEE Trans. Power Syst. **24**(2), 661–667 (2009)
16. Huna, Y., Xu, Q., Bian, H., et al.: Cumulant method based on Latin hypercube sampling for calculating probabilistic power flow. Electric Power Autom. Equipment **36**(11), 7–112–119 (2016)

Micro-grid Dispatch Decision-Making Method Based on Adjustable Robust Optimization Algorithm

Shaopeng Li(✉), Bin Zhou, Hongxiang Pan, and Feng Liang

NARI Group Corporation (State Grid Electric Power Research Institute), Nanjing 211106, China
lishaopeng@sgepri.sgcc.com.cn

Abstract. Vehicle to Grid technology (V2G) can provide stable support for the grid connection of renewable energy, and the nearby consumption of renewable energy and electric vehicle (EV) in the form of microgrid can also effectively alleviate the impact on the grid shock and calm load fluctuation. In order to solve the uncertainties of electric vehicle charging and discharging power and renewable energy output power, this paper establishes an inner and outer double-layer model firstly. The inner layer model uses Monte Carlo method and interval prediction theory to perform uncertain variable modeling and analysis. The outer layer model obtains uncertain variables with volatility in the inner layer model for scheduling optimization; secondly, in order to prevent the overly conservative nature of traditional robust optimization (RO), an adjustable robust optimization algorithm is proposed to quantify the system's economy and robustness. Finally, the effectiveness of the proposed method is verified by using a residential grid-connected microgrid as a simulation scenario.

Keywords: Vehicle-to-grid electric vehicle · Interval prediction theory · Dual theory · Two-layer model · Robust optimization algorithm · Adjustable robust optimization algorithm

1 Introduction

Environmental problems caused by the shortage of fossil energy reserves and pollution emissions have brought new challenges to human society. With the proposal of the energy interconnection architecture, clean energy with wind and light as the leading factor brings new ideas for promoting energy interconnection due to its environmental protection and low cost. However, due to large fluctuations in the output of wind turbines and photovoltaic generations, which poses great challenges to the safe and stable operation of the power grid [1, 2]. Therefore, how to suppress the power fluctuation of the grid and improve the capacity of renewable energy consumption is an urgent problem to be solved. The development of the microgrid introduces new ideas for solving this problem. The microgrid has the advantages of integrating a distributed energy resources, reducing long-distance transmission losses of renewable energy, and absorbing renewable energy nearby [3, 4]. V2G electric vehicles are used as mobile energy storage units in the microgrid. On the one hand, it can further support the safe

Y. Xue et al. (Eds.): PMF 2020, LNEE 718, pp. 146–160, 2021.
https://doi.org/10.1007/978-981-15-9746-6_12

and stable operation of the main grid by virtue of the duality of "source and charge" [5, 6]. On the one hand, the daily mileage of electric vehicle users is uncertain. According to the Monte Carlo simulation, most of the parking time of electric vehicles is concentrated in the peak period of electricity consumption [7, 8]. While the high proportion of low-cost wind resources at night has not been absorbed; on the other hand, V2G technology has strict requirements on the depth of charge and discharge of electric vehicles. Therefore, it is important to guide electric vehicles to charge and discharge in a reasonable and orderly manner, and to solve the interference problem of a large scale of uncertain factors in the microgrid. From the perspective of power grid economy, Shi et al. established an optimal dispatch model with distributed output units and electric vehicles, and introduced a robust optimization algorithm to solve the problem of interference of sources and uncertain variables [9]. However, the two-way flow characteristics of EV in V2G networks are not considered in the article. In this paper, an adjustable robust optimization algorithm is introduced, which can effectively integrate distributed energy sources, avoids the excessively conservative characteristics of traditional robust optimization, and makes decision makers find a compromise between robustness and economy.

2 Scheduling Model

2.1 Inner Uncertain Variable Model

1) Uncertainty modeling analysis of electric vehicles.
   The daily mileage of electric vehicles ds can be modeled as a probability density function that follows a logarithmic normal distribution:

$$h(ds;\mu,\sigma)=\frac{1}{ds\sqrt{2\pi\sigma^2}}e^{-\frac{(\ln ds-\mu)^2}{2\sigma^2}} \tag{1}$$

According to the *ds-SOC* conversion method introduced in [9], the fluctuation interval expressed by the mean and deviation of SOC_n is obtained, as shown in formula (2):

$$U_{soc}=\left\{SOC_n^G=\overline{SOC_n^G}+\hat{SOC_n^G}\right\} \tag{2}$$

In addition, the discharge behavior of electric vehicles is limited to avoid the impact of deep charging and discharging on the life of the battery:

$$SOC_n=\begin{cases} SOC_d, & SOC_n > SOC_r \\ SOC_c, & SOC_n \le SOC_r \end{cases} \tag{3}$$

Where d and c represent the two states of electric vehicle discharge and charge, respectively, and SOC_r represents the rated limit of discharge. In order to meet the needs of users traveling the next day, the total predicted charge and discharge capacity of n electric vehicles can be expressed as:

$$\begin{cases} \sum_{d} PEV_d = |SOC_d|_1 E, & when\ SOC > SOC_r \\ \sum_{c} PEV_c = \begin{pmatrix} |SOC_c|_1 + \\ Num_d(SOC_r - SOC_d) \end{pmatrix} E, & when\ SOC \le SOC_r \end{cases} \tag{4}$$

The random behavior of charging and discharging characteristic of electric vehicles can be modeled as an uncertain set, as shown in formula (5):

$$U_{d/c} = \left\{ \begin{array}{l} PEV^{G}_{d/c,t}\left(SOC^{G}_{d/c}\right) = \overline{PEV^{G}_{d/c,t}} + \hat{PEV}^{G}_{d/c,t} : \\ \underline{\hat{PEV}^{G}_{d/c,t}} \le \hat{PEV}^{G}_{d/c,t} \le \overline{\hat{PEV}^{G}_{d/c,t}} \end{array} \right\} \tag{5}$$

2) Uncertainty modeling analysis of wind turbines.
   The wind speed model can be modeled as a probability density function that follows the Rayleigh distribution, as shown in Eq. (6):

$$f_{wind}(V) = \frac{Z}{\sigma^2} \exp\left(-\frac{Z^2}{2\sigma^2}\right) \tag{6}$$

Among them, Z and σ are the distribution parameters of Rayleigh distribution. Assuming that the predicted wind speed value is equal to the mean value of the Rayleigh distribution, the parameters of the Rayleigh distribution can be calculated, and then the wind power-wind power calculation method described in [9] is used to represent the fluctuation of wind power as an uncertain set:

$$PWT = \left\{ PWT^{G}_{l,t} = \overline{PWT^{G}_{l,t}} + \hat{PWT}^{G}_{l,t} : \underline{\hat{PWT}^{G}_{l,t}} \le \hat{PWT}^{G}_{l,t} \le \overline{\hat{PWT}^{G}_{l,t}} \right\} \tag{7}$$

Among them, G represents the predicted value, $\overline{PWT^{G}_{l,t}}$ and $\hat{PWT}^{G}_{l,t}$ represent the mean and deviation of the wind power prediction interval, the magnitude of the deviation is limited by the upper and lower limit values.
This paper introduces the quantization method to calculate the robustness of the system through the probability that the rotation reserve constraint is violated, see the literature [9] for a detailed description.

2.2 Outer Scheduling Model

The outer model is a multi-objective planning model. From the perspective of economy and environmental benefits, the effect of quantifying economy is achieved through scheduling optimization. The distributed power sources in the microgrid include diesel engines (DE), wind turbines (WT) and V2G electric vehicles, and the objective function is shown in function (8):

1) Objective function 1: Minimize operating costs:

$$Minf_1(x) = C_1 = \sum_{t=1}^{T} \left\{ \sum_{i} \left[C_f(P_{i,t}) + C_{OM}(P_{i,t}) + C_{bat}(PEV_t) \right] + C_{grid,t} \right\} \quad (8)$$

The formula mainly includes the fuel and operation and maintenance costs of the distributed power supply, where i represents the type of power supply, including DE generators and WT, $P_{i,t}$ represents the output power of various types of power supply at time t; C_f represents the fuel cost function, C_{om} represents the operation Dimensional cost function, $C_{grid,t}$ represents the transmission cost with the main grid[10,11], C_{bat} represents the battery depreciation cost of electric vehicles[12], the transmission cost function between the main grid and the microgrid is shown in Eq. (9):

$$C_{grid,t} = P_{grid,t} M_t \quad (9)$$

Among them, $P_{grid,t}$ is the transmission power between the main grid and the microgrid at time t, and M_t respectively represent the electricity price of the microgrid at time t.

2) Objective function 2: Minimize environmental governance costs:

$$Minf_2(x) = C_2 = \sum_{t=1}^{T} \left(\sum_{i=1}^{N} \sum_{h}^{H} (C_h u_{i,h}) P_{i,t} + \sum_{h}^{H} (C_h u_{grid,t}) P_{grid,t} \right) \quad (10)$$

Among them, h represents the type of pollution, including CO_2, SO_2, NO_X, $u_{i,h}$ represents the amount of pollution caused by each distributed power supply, $u_{grid,t}$ represents the amount of pollution caused by transmission between power grids [13].

2.3 Constraint

The real-time output of the wind turbine should be constrained by the predicted value. The constraints are shown in formula (11):

$$0 \le PWT_{l,t} \le PWT_{l,t}^{G} \quad (11)$$

In the formula (11), l is the number of wind turbines, and $PWT_{l,t}$ represents the output power of the l-th wind turbine at time t. The real-time output constraint and climbing constraint of DE is:

$$P_{i,t}^{\min} \leq P_{i,t} \leq P_{i,t}^{\max} \tag{12}$$

$$P_{i,t-1} - P_{i,down} \leq P_{i,t} \leq P_{i,up} + P_{i,t-1} \tag{13}$$

In the formula (13), $P_{i,t}$ should satisfy the upper and lower limit constraints of $P_{i,t}^{\min}$ and $P_{i,t}^{\max}$, and $P_{i,up}$ and $P_{i,down}$ are the upper and lower limits of the climbing constraint. According to the power transmission protocol, the transmission power between the microgrid and the main grid cannot exceed the limit.

$$P_{buy}^{\min} \leq P_{grid,t}^{+} \leq P_{buy}^{\max} \tag{14}$$

$$P_{sell}^{\min} \leq P_{grid,t}^{-} \leq P_{sell}^{\max} \tag{15}$$

In V2G networks, power balance constraints are as follows:

$$\sum_i PDE_{i,t} + \sum_l PWT_{l,t} + \left|P_{grid,t}\right| + \sum_d PEV_{d,t} - \sum_c PEV_{c,t} = P_{load,t} \tag{16}$$

In the formula (16), $P_{load,t}$ represents the resident load at time t. and the electric vehicle charging power constraints are as follows:

$$0 \leq PEV_{k,t} \leq PEV_{k,t}^{\max} \tag{17}$$

$PEV_{k,t}^{\max}$ represent the maximum charging power of $PEV_{k,t}$ at time t, the rotation spare constraints are as follows:

$$\begin{aligned} &\sum_{\mathrm{i}} PDE_{i,t} + \sum_l PWT_{l,t}^{G} + \sum_d PEV_{d,t}^{G}\left(SOC_{d,t}^{G}\right) \geq \\ &(1 + L_t)\left(\sum_c PEV_{c,t}^{G}\left(SOC_{c,t}^{G}\right) + P_{load,t}\right) \end{aligned} \tag{18}$$

3 Scheduling Application of Adjustable Robust Optimization Algorithm

3.1 Robust Optimization Algorithm

The robust optimization algorithm needs to construct the worst case of the scheduling system, that is, the case where the wind resource power generation accounted for the lowest, and the electric vehicle charging load is the largest, we will use the model (19) to represent this worst case [14]:

$$F = worst\left\{ \overset{\wedge}{PWT}_{l,t}^{G} + \sum_{d=1}^{D} \overset{\wedge}{PEV}_{d,t}^{G} - (1+L_t)\sum_{c=1}^{C} \overset{\wedge}{PEV}_{c,t}^{G} \right\} \tag{19}$$

Due to the volatility of multiple uncertain variables, the planning model at this time is a semi-infinite optimization problem, and the huge solution space increases the difficulty of solving the model [15].

3.2 Robust Equal Conversion

According to the strong duality theory, the duality problem (20) of the model (19) is also feasible and bounded, and the target results are consistent:

$$Min\left\{ -\sum_{l=1}^{L} \underline{\overset{\wedge}{PWT}_{l,t}^{G}}\alpha_t + \sum_{l=1}^{L} \overline{\overset{\wedge}{PWT}_{l,t}^{G}}\beta_t - \sum_{k=1}^{K} \underline{\overset{\wedge}{PEV}_{k,t}^{G}}\gamma_t + \sum_{k=1}^{K} \overline{\overset{\wedge}{PEV}_{k,t}^{G}}\delta_t \right\}$$

$$s.t. \begin{array}{c} -\alpha_t + \beta_t \geq 1 \\ -\gamma_t + \delta_t \geq 1 \\ -v_t + \lambda_t \geq -1 - L_t \\ \alpha_t, \beta_t, \gamma_t, \delta_t, v_t, \lambda_t \geq 0 \end{array} \tag{20}$$

Among them, $\alpha, \beta, \gamma, \delta, v, \lambda$ are the deviation size control coefficient, the original fluctuation variable is converted by duality, expressed in the form of upper and lower bounds, which reduces the difficulty of solving the model.

In order to embody the core content of the adjustable robust optimization algorithm and simplify the scheduling model, this paper uses the weighted average method to deal with multi-objective programming problems, w_1 and w_2 are the weight coefficients. In summary, the robust optimal scheduling model is reconstructed as a model (21):

$$\underset{P_{i,t}, PEV_{n,t}}{Min}\left\{ \underset{PWT_{l,t}^{G}, SOC_{n}^{G}}{Worst}\ w_1 C_1 + w_2 C_2 \right\} \tag{21}$$

s.t. (11)–(17), (20).

Constructing the worst set is the core idea of robust optimization, but this method is too conservative, so we introduce an adjustable robust optimization algorithm to improve system economy.

3.3 Flow of Adjustable Robust Optimization Algorithm

We define Γ as the uncertainty coefficient, which is used to represent the number of uncertain variables. We take WT as an example, assume that each uncertain variable is defined as $\alpha \in [-1, 1]$, which is the scheduling interval coefficient indicating the size of the interval. Our goal is to ensure the stable operation of the system under the condition of fluctuation of uncertain variables with a very high probability. When

$\alpha = 1$, it means that the uncertain variable fluctuates throughout the interval, and $\alpha = 0$ means that there is no volatility in the current variable, the change process during the period is the application of adjustable robust optimization algorithm, as shown in Fig. 1.

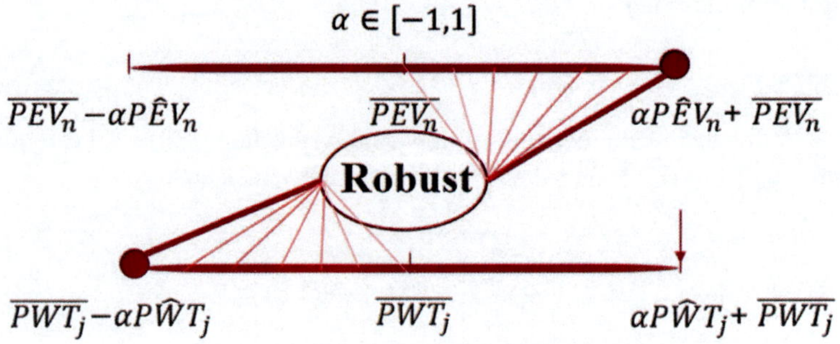

Fig. 1. Schematic diagram of fluctuation range adjustment.

The core idea of adjustable robust optimization is to control the magnitude of the fluctuation of uncertain variables through Γ. if $\Gamma = 15$, it means that there is no fluctuation of uncertain variables; if $\Gamma = 23.4$, it means that 23 uncertain variables fluctuate in the entire range, and one decimal fluctuates within the range of 0.4 times [16], the process is shown in Fig. 2:

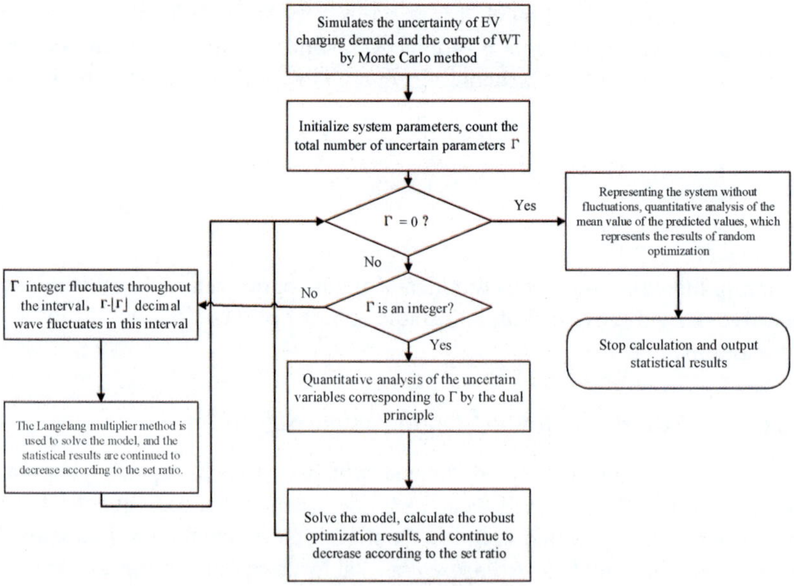

Fig. 2. Flow chart of adjustable robust optimization algorithm.

4 Simulation

In order to confirm the effectiveness of the proposed scheme, this chapter takes the grid-connected microgrid in the residential area as a simulation scenario. The distributed energy includes one DE, four WT, and one hundred V2G electric vehicles. Typical parameter information, model parameters and capacity are selected Limit [17], the simulation period is set to 16:00–04:00, and the simulation interval is 30 min. w1/w2 is set to 0.6/0.4, the scheduling model is selected by the quadratic programming method with faster convergence speed.

4.1 Comparative Analysis of Stochastic Optimization and Robust Optimization

Figure 3 is a graph of the optimization results of the stochastic optimization method. The uncertain variables in the inner model are selected for unbiased fluctuations. The optimization results show that during the peak hours of electricity consumption in residential areas, electric vehicles are discharged to the microgrid, and wind energy is preferentially absorbed at night In order to reduce the high fuel cost of the DE due to the rotation reserve, the total cost of the system under random optimization is 10324.2 $ This strategy effectively improves the stability and economy of power grid operation, but at this time the system has no rules to avoid the interference of uncertain variables.

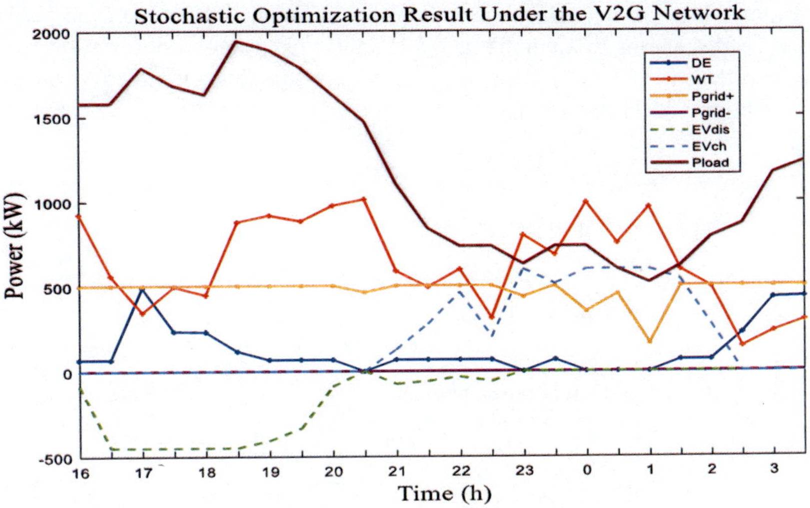

Fig. 3. Diagram of stochastic optimization.

Compared with stochastic optimization, robust optimization emphasizes the robustness of the system, as shown in Fig. 4, considering the worst scheduling system and the largest fluctuation of uncertain variables, but its total cost is as high as 19642.0 $ Cost, and therefore has an overly conservative character.

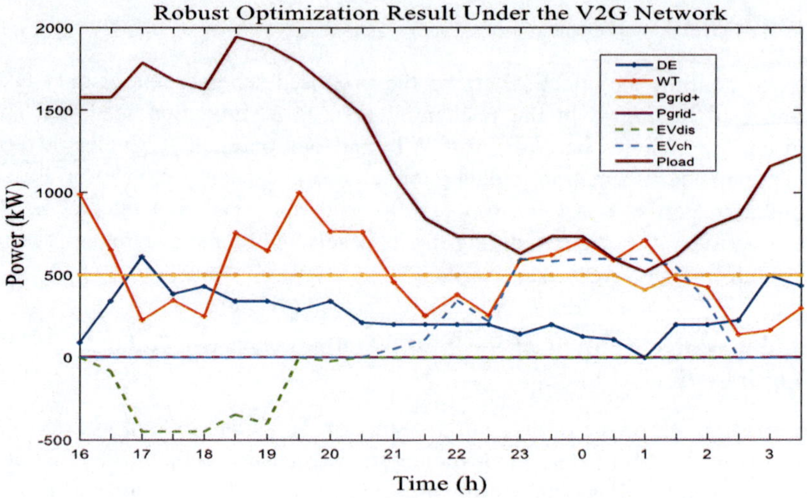

Fig. 4. Diagram of robust optimization results.

4.2 The Effect of Adjustable Robust Optimization on the Effect of Peak Cutting and Valley Filling

Figure 5 selects the optimization results of typical uncertain variable coefficients, draws the integrated load curve, and marks relevant trend change information and peak-valley difference information to demonstrate the adjustable robust optimization algorithm under the premise of ensuring load balance. It has a significant trend effect on the effect of peak cutting and valley filling:

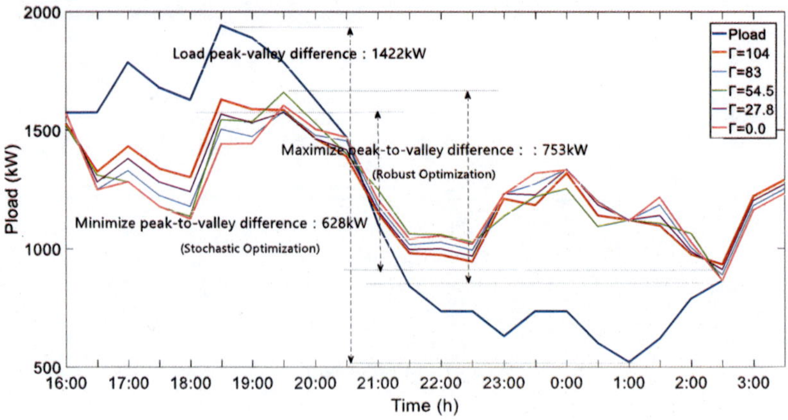

Fig. 5. Diagram of Impact of adjustable robust optimization on peak clipping and valley filling effect.

4.3 The Effect of Adjustable Robust Optimization on Distributed Unit Output

$\Gamma = 104$ represents that all variables fluctuate within the fluctuation range, and the robustness of the scheduling system is optimal at this time, $\Gamma = 0$ represents that there are no uncertain variables, and the process of gradient change during the period is an adjustable robust optimization algorithm In order to reflect the influence of the algorithm on the output of the distributed power supply, the typical power change trend graphs of two output units are selected for display, as shown in Fig. 6, 7, 8 and 9.

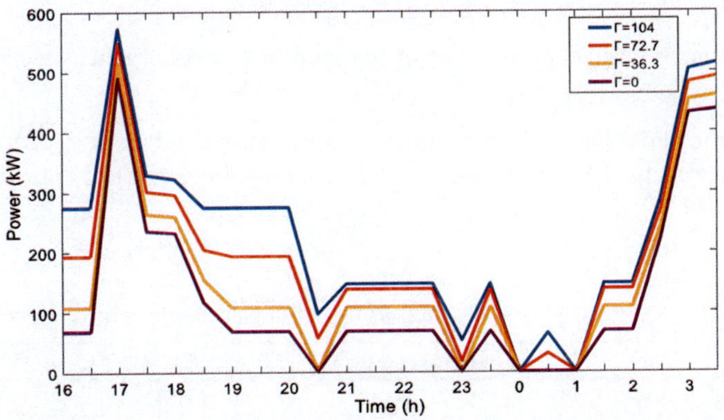

Fig. 6. Schematic diagram of the effect of adjustable robust optimization algorithm on DE.

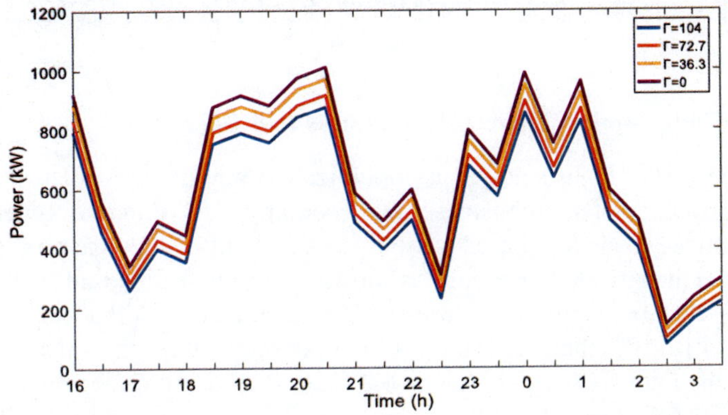

Fig. 7. Schematic diagram of the effect of adjustable robust optimization algorithm on WT.

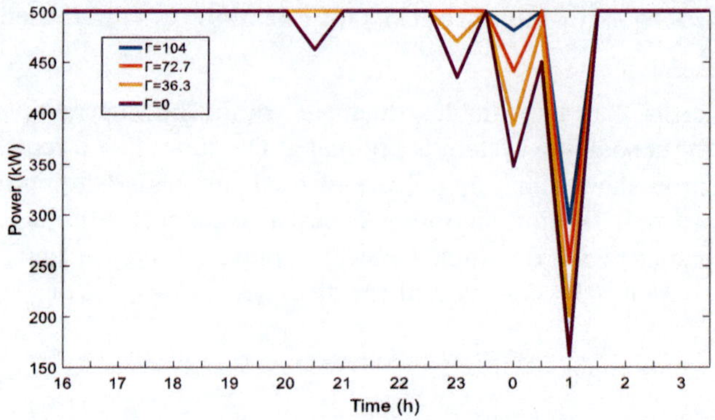

Fig. 8. Schematic diagram of the effect of adjustable robust optimization algorithm on $P_{grid,t}$.

In order to show the advantages of the algorithm, we will express the impact of the algorithm on each distributed energy and economy in numerical form, as shown in Table 1.

Table 1. Multi-objective dispatch RO result under the dynamic parameter Γ.

Γ	DE	WT	Pgrid+	PEV_d	PEV_c	C_1	C_2	C_{total}
	kW					USD		
Γ = 104	6320.1	12207.7	11544.6	2703.7	4524.7	18994.6	647.4	19642.0
Γ = 81.8	5804.9	12704.0	11490.2	2821.9	4570.1	16813.9	625.7	17439.6
Γ = 63.6	5298.4	13201.4	11436.5	3019.8	4696.2	14779.6	604.3	15383.9
Γ = 27.2	4309.2	14194.7	11327.3	3326.1	4789.0	11701.2	562.3	12263.5
Γ = 0.0	3454.5	15021.8	11174.0	3492.6	5019.2	9785.5	538.7	10324.2

4.4 The Compromise Between Robustness and Economy

The essence of the adjustable robust optimization algorithm is to quantify the relationship between system robustness and economy, and improve system economy. Users can make decisions based on the needs of different scenarios. Consider the situation with the largest fluctuations as shown in Fig. 9, then optimize. The robustness of the later scheduling system is close to 100%, that is, the CVP is close to zero. If the CVP is set to 10.13%, then according to the proposed method, the value of Γ will be set to 17, and the total cost at this time is $ 11220.4, reducing the operating cost from $ 19642.0 to 42.9% (Fig. 10).

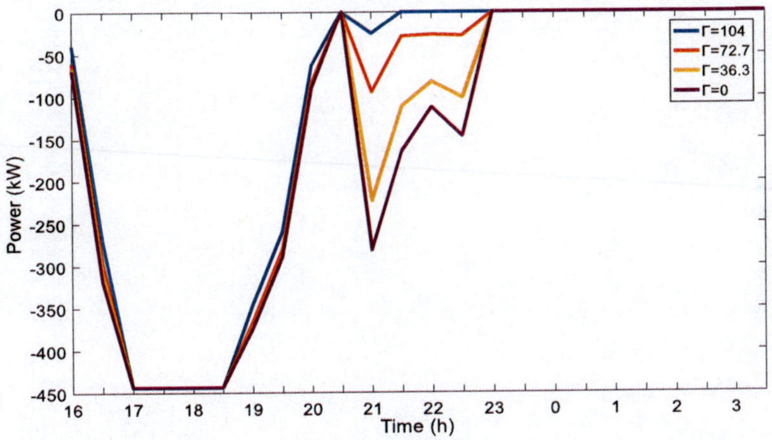

Fig. 9. Schematic diagram of the effect of adjustable robust optimization algorithm on EV discharging.

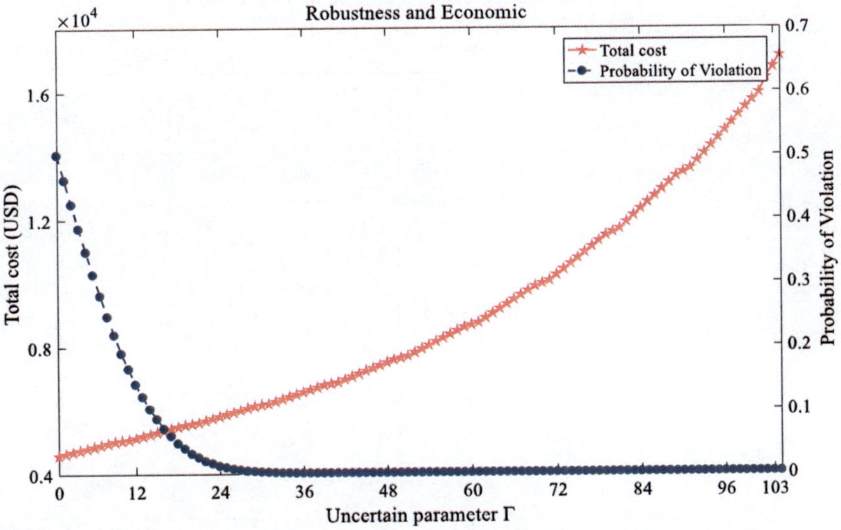

Fig. 10. Diagram of Impact of adjustable robust optimization on peak clipping and valley filling effect.

4.5 Sensitivity Analysis

In order to provide decision-making solutions for decision makers, this paper carries out a sensitivity analysis of the uncertain variable number Γ to the economics of the scheduling model, and quantifies C_1 and C_2 to reflect the impact of the adjustable robust optimization algorithm on the system economics, as shown in Fig. 11 and Fig. 12.

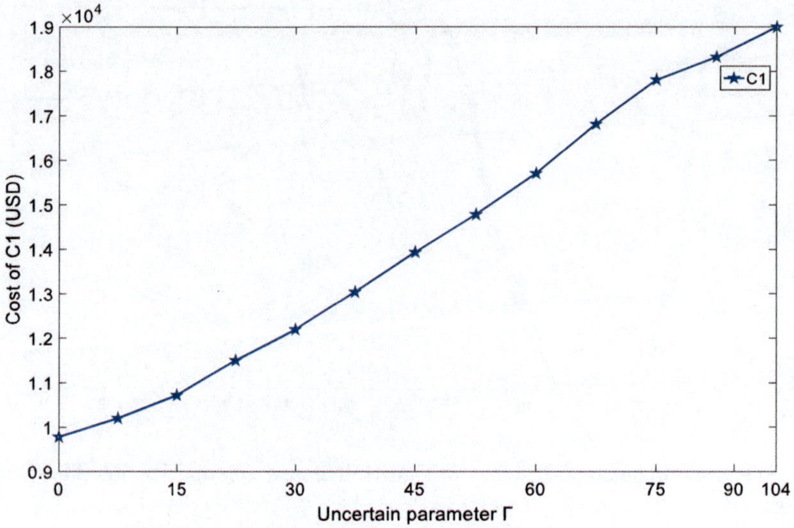

Fig. 11. Influence of uncertain parameter Γ on C_1.

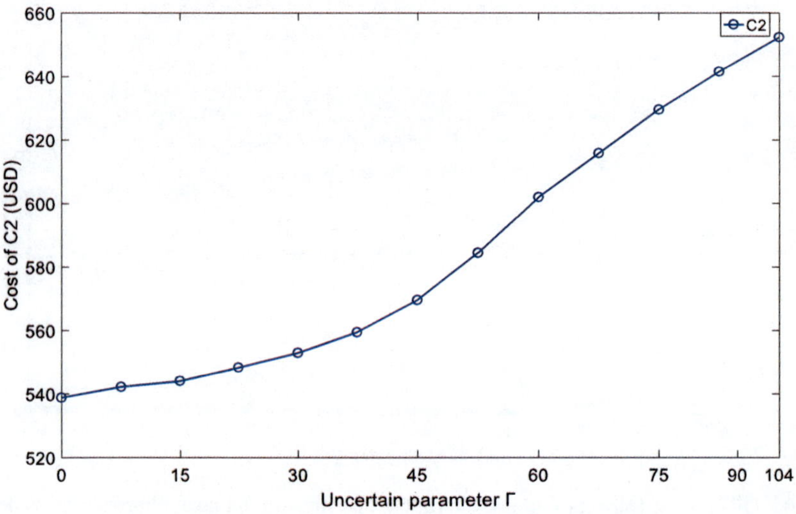

Fig. 12. Influence of uncertain parameter Γ on C_2.

Figure 13 is a weight analysis chart of w_1/w_2. The decision maker can formulate the ratio of w_1 and w_2 according to different scenarios and wishes.

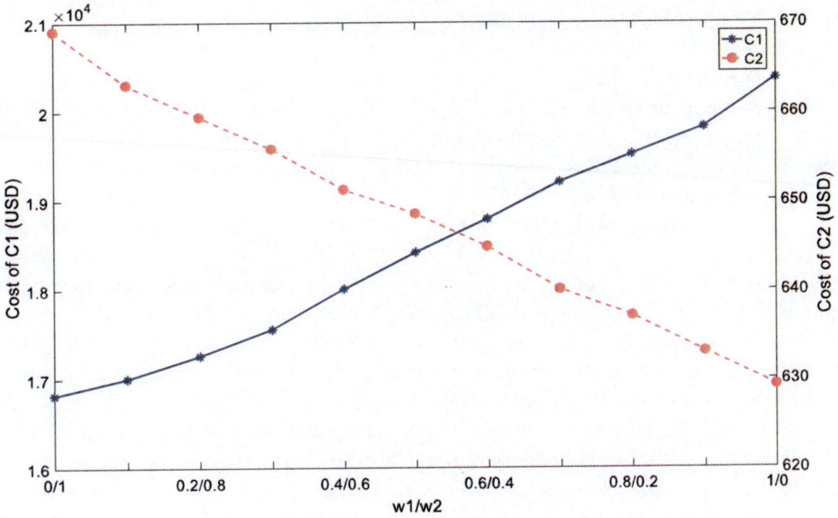

Fig. 13. Diagram of weight analysis.

5 Conclusion

In this paper, from the perspective of the uneconomical and unstable problems in the microgrid, based on the adjustable robust optimization algorithm, the construction of the internal and external double-layer models with uncertain variable coefficients as the control variables is optimized, which improves the economics and stability of micro-grid operation, the main conclusions are as follows:

(1) First of all, the phenomenon of peaking on load peaks in peak hours of residential areas is effectively resolved by stochastic optimization strategy, and at the same time, the high proportion of wind resources at night is absorbed nearby, ensuring the safe and economic operation of the microgrid.

(2) In order to improve the robustness of the system, this paper makes the scheduling system immune to the influence of uncertain variable interference based on the theory of robust optimization modeling, but it pays a certain economic loss as the price.

(3) In order to solve the excessively conservative characteristics of traditional robust optimization, an adjustable robust optimization algorithm is introduced. By adjusting the deviation of the uncertain variable, the effect of the uncertain variable coefficient on the economic benefits in the microgrid is elaborated and quantified. The economics of the system provides decision-makers with a basis for decision-making.

References

1. Guo, Z., Li, G., Zhou, M., et al.: Two-stage robust optimal scheduling of regional integrated energy system considering network constraints and uncertainties in source and load. Power Syst. Technol. **43**(09), 3090–3100 (2019)
2. Prudhviraj, D., Kiran, P.B.S., Pindoriya, N.M.: Stochastic energy management of microgrid with nodal pricing. J. Modern Power Syst. Clean Energy **8**(1), 102–110 (2020)
3. An, R., Ji, M., Zhang, S.: Global warming and obesity: a systematic review. Obes. Rev. **19** (2), 1–3 (2018)
4. Qiu, H., Gu, W., Xu, Y., et al.: Tri-level mixed-integer optimization for two-stage microgrid dispatch with multi-uncertainties. IEEE Trans. Power Syst. **PP**(99), 1 (2020)
5. Carfora, A., Pansini, R.V., Romano, A.A., et al.: Renewable energy development and green public policies complementarities: the case of developed and developing countries. Renewable Energy **115**, 741–749 (2018)
6. Lu, Y., Chang, R., Lim, S.: Crowdfunding for solar photovoltaics development: a review and forecast. Renew. Sustain. Energy Rev. **93**, 439–450 (2018)
7. Bi, R., Liu, R., Ding, M., et al.: Renewable energy generation cluster partition method aiming at improving accommodation capacity. Proc. Chinese Soc. Electr. Eng. **39**(22), 6583–6592 (2019)
8. Karthik, N., Parvathy, A.K., Arul, R., et al. Economic load dispatch in a microgrid using Interior Search Algorithm.In: Innovations in Power & Advanced Computing Technologies (2019)
9. Ruifeng, S.H.I., Shaopeng, L.I., Changhao, S.U.N., et al.: Adjustable robust optimization algorithm for residential microgrid multi-dispatch strategy with consideration of wind power and electric vehicles. Energies **11**(8), 2050 (2018)
10. Tabatabaee, S., Mortazavi, S.S., Niknam, T.: Stochastic scheduling of local distribution systems considering high penetration of plug-in electric vehicles and renewable energy sources. Energy **121**, 480–490 (2017)
11. Qiu, H., Gu, W., Xu, Y., et al.: Interval-partitioned uncertainty constrained robust dispatch for AC/DC hybrid microgrids with uncontrollable renewable generators. IEEE Trans. Smart Grid **10**(4), 4603–4614 (2019)
12. Ma, Z., Zou, S., Ran, L., et al.: Efficient decentralized coordination of large-scale plug-in electric vehicle charging ☆. Automatica **69**(C), 35–47 (2016)
13. Lu, X., Zhou, K., Yang, S.: Multi-objective optimal dispatch of microgrid containing electric vehicles. J. Cleaner Production **165**, 1572–1581 (2017)
14. Ruifeng, S., Shaopeng, L., Changhao, S., Kwang, L.: Multi-dispatch for grid-connected microgrid with robust optimization algorithm. IFAC PapersOnline **51**(27) (2018)
15. Bertsimas, D., Sim, M.: The price of robustness. Oper. Res. **52**(1), 35–53 (2004)
16. Zhou, Z., Sun, Z., Shi, Z., et al.: Robust energy scheduling in vehicle-to-grid networks. IEEE Netw. **31**(2), 30–37 (2017)
17. Shi, R., Li, S., Zhang, P., et al.: Integration of renewable energy sources and electric vehicles in V2G network with adjustable robust optimization. Renew. Energy **153**, 1067–1080 (2020)

Research of Backward Compatibility of Conductive High Power Charging (HPC) Technology for Electric Vehicle

Yuan Sun(✉), Jiande Ye, Fuzhong Dan, and Wei Zhang

NARI Group Corporation, No.19 Chengxin Avenue, Nanjing, China
sunyuan3@sgepri.sgcc.com.cn

Abstract. The high power charging technology using the new vehicle coupler leads to the compatibility problem of the new and old charging interfaces. In order to solve the backward compatibility of conductive HPC, a new adaptor used to transfer GB 2015 connector to HPC connector was introduced. The structure of HPC coupler and the new control pilot circuit using adaptor were both described in this paper. Meanwhile, tests of thermal and mechanical performance of cable assembly using adaptor were conducted. The results showed that the mechanical strength and the temperature rise of terminals of cable assembly using adaptor can meet the requirements defined in relevant standards. And the temperature sensors installed in GB connector will reflect the over temperature signal of vehicle inlet with a little delay. The paper verified the feasibility of the adaptor and solved the compatibility of old and new interfaces and will promote the commercial application of HPC technology.

Keywords: Conductive high power charging · Adaptor · Cable assembly · Test

1 Introduction

With the continuous development of China's new energy automobile industry, the range of electric vehicles has become one of the key points to constrain the development of the industry. At present, China is continuously improving the range requirements of electric vehicles. Considering the improvement of battery energy density is limited as the development status of battery technology, the improvement of battery capacity is one of the feasible solutions. However, the charging time is bound to increase if the charging current does not change with the increase of battery capacity. In order to shorten charging time, improve travel efficiency and user experience, research of high power charging technology has been carried out at home and abroad [1].

At abroad, tesla has built V2 and V3 super power charging stations, which greatly reduces the charging time. In China, China Electricity Council led the establishment of the work group on high power charging technology and standards development in 2017, which preliminarily determined the development route and key points of high power charging. Meanwhile, it was decided that a new charging interface adapted for HPC would be designed. In 2018, the adoption of a seven pins design for the new

Y. Xue et al. (Eds.): PMF 2020, LNEE 718, pp. 161–173, 2021.
https://doi.org/10.1007/978-981-15-9746-6_13

interface was accepted inside the working group, as well as the new design of communication protocol, control pilot circuit and charging backward compatibility. In 2019, demonstration projects of high power charging stations have been completed in Beijing, Changzhou, Jinan and other places led by state grid corporation of China.

The new charging interface adapts to the development of HPC technology and solves the relevant problems of GB 2015 vehicle coupler [2], and releases the charging current limitation of GB 2015. However, it usually takes a long time to upgrade electric vehicle hardware, so the old and new interfaces will coexist for a certain period of time. The backward compatibility problem of charging was considered in parallel when the working group was established as well as the design of the adaptor, which will be helpful for demonstration project construction, relevant HPC standards compilation and subsequently commercial operation.

2 HPC Vehicle Coupler

2.1 Definition of HPC and ChaoJi

Currently, there is no unified definition of high power charging in the world. It is generally believed that a certain high charging rate [3] can be called high power charging. The charging rate is proportional to the charging current while inversely proportional to the battery capacity. It is necessary to increase the charging current or reduce the battery capacity in order to improve the charging rate. Currently, China keeps raising the range requirements of electric vehicles and has set different subsidy policy according to the range. Considering the current development level of battery technology, improving battery capacity is one of the effective ways to increase the range of electric vehicle. Therefore, the charging current must be increased if the charging rate is to be increased on the premise of the increase of battery capacity.

At present, the conductive dc vehicle coupler used in the market satisfies the standard requirements of GB/T 20234.1-2015 and GB/T 20234.3-2015, hereinafter referred to as GB2015 coupler. GB/T 20234.3-2015 clearly stipulates that the rated output voltage is lower than 1000 V and rated output current is lower than 250 A [4]. In order to enhance the charging current, increase the charging rate and shorten the charging time, it is urgent to develop a new coupler that meets the requirements of high power charging. Compared to normal charging, a new high power charging control pilot circuit should be designed and supplementary messages should be added to the communication protocols in GB/T 27930-2015.

After the consultation of China Electricity Council and Japan CHAdeMO association, the high power charging coupler is named as ChaoJi. Through ChaoJi coupler, the high power charger can be connected to electric vehicles supporting high power charging, which enables connection and disconnection of electrical circuit [5, 6]. Figure 1 shows the definition of the boundary and contact pins of ChaoJi coupler. Compared to GB2015 coupler, ChaoJi coupler removed the A+ and A- auxiliary power contact pins, reducing the original 9 pins to 7 pins. ChaoJi coupler is designed for output voltage not exceeding 1500 VDC and output current not exceeding 800 A.

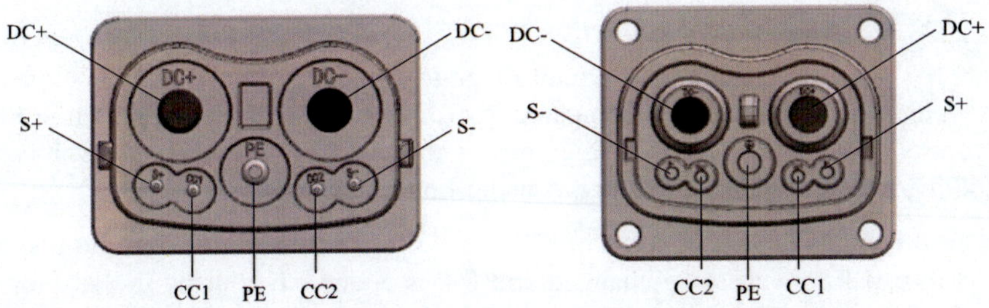

Fig. 1. Boundary of ChaoJi

2.2 Connection

Now, most dc chargers operated in the market are normal chargers equipped with GB2015 coupler. With the emergence of electric vehicles supporting high power charging and high power chargers, there will be the coexistence of normal power charger (NPC), NPC vehicles, high power charger and HPC vehicles. In view of the present quantity of normal charger and NPC vehicle, the four are bound to coexist for a period of time. As shown in Fig. 2, it is worth studying how to be compatible with each other and improve the utilization rate and charging efficiency of the existing chargers.

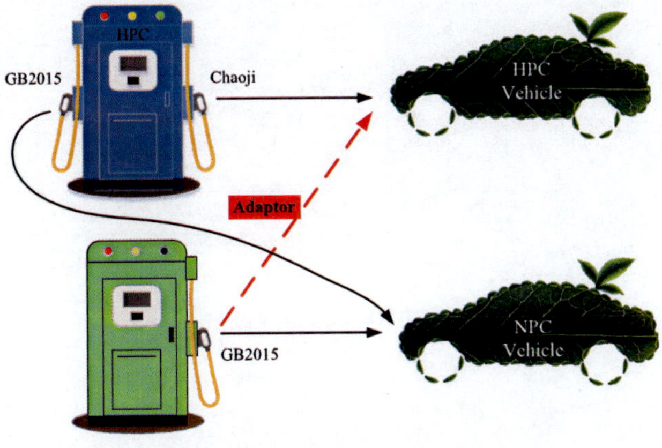

Fig. 2. Compatibility of HPC and NPC

Considering the combined mode shown in Fig. 2, the newly built HPC charger is equipped with ChaoJi and GB2015 coupler to be compatible with HPC vehicle and NPC vehicle. For the NPC charger already built, connecting the HPC vehicle using the GB2015 coupler has become an urgent problem to be solved. This paper introduces an adaptor that converts GB2015 connector to ChaoJi connector. Tests and simulation of temperature rise and structural mechanical strength of the cable assembly had been done to solve the compatibility problem between HPC vehicle and NPC.

2.3 Control Pilot Circuit

Figure 4 shows the control pilot circuit diagram connecting the GB2015 coupler and HPC vehicle using an adaptor. Compared to GB2015 [7] control pilot shown in Fig. 3, the control pilot circuit supporting high power charging is improved on the vehicle side as follow, which is indicated in Fig. 4 using botted rectangle.

(a) In order to keep the voltage of CC1 as 4V after full connection, the resistance value of R4 needs to be changed and R4′ is added when using the adaptor;
(b) To be compatible with CCS charging mode, diode D1 is added;
(c) Add optional switch S2.

The adaptor physically converts the GB2015 connector to ChaoJi connector, and the resistors R' and Rc' solve the compatibility between the old control pilot circuit and HPC vehicle. For different types of charger (CHAdeMO or CCS), charging compatibility can be achieved simply by changing the resistance values of R' and Rc' in the adaptor.

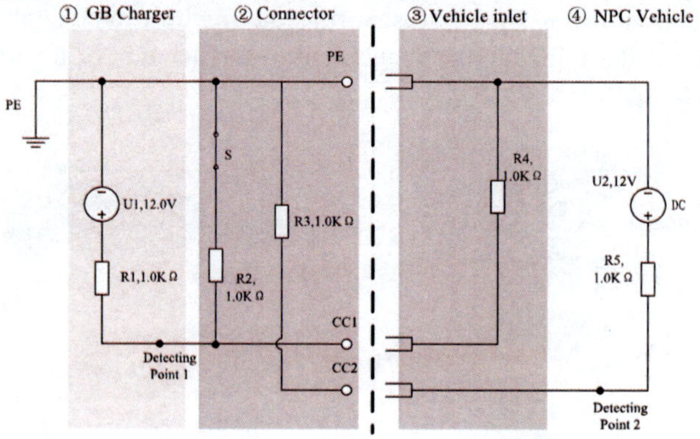

Fig. 3. GB2015 control pilot circuit

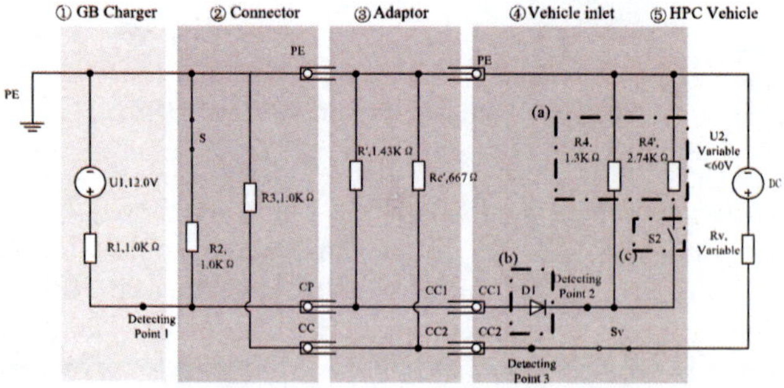

Fig. 4. Control pilot circuit adapted for GB2015 using adaptor

The charging sequence using an adaptor is shown in Fig. 5. The charging sequence of the charger complies with the standard of GB2015. Due to the charging reservation function, sleep/standby and wake processes are added in the process of electric vehicles that support high power charging. At the same time, since the circuit was designed to be compatible with CHAdeMO and CCS charging modes, the type of communication protocol will be judged to identify the type of charger after identifying the voltage of detection point 3.

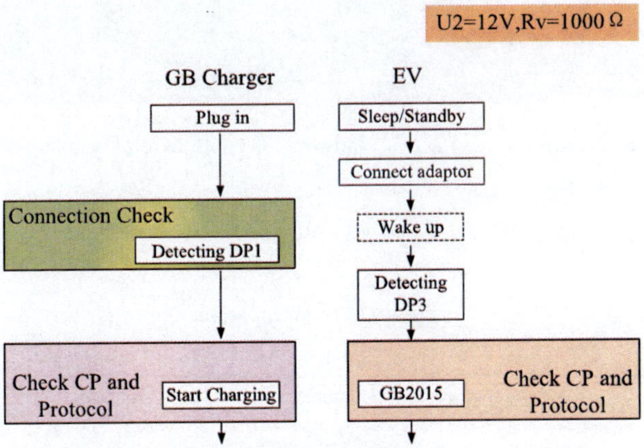

Fig. 5. Charging sequence

In the charging procedure, the parameters and states of each detection point and switches are shown in Table 1.

Table 1. Parameters of detecting points and switches

Type	Unit	Nominal	Minimum	Maximum
DP1	V	5.99	5.47	6.55
DP2	V	5.29	4.79	5.85
DP3	V	12.00	11.40	12.60
Switch S2	Close			
Switch Sv	Open			

3 Analysis of Thermal Performance

3.1 Test Objects and Apparatus

In order to evaluate the thermal performance of the cable assembly using the adaptor, a test system shown in Fig. 6 is built to conduct temperature rise test on the cable assembly.

Fig. 6. Temperature rise test platform for cable assembly using adaptor

The test apparatus used are listed in Table 2.

Table 2. Test apparatus

No	Name	Type	Parameter
1	DC temperature rise test system	Ametek	Output 0–10V 0–800 A
2	Omega thermocouple	K type	500 °C f.s
3	Heater	STG-2000W	Output 0–300 V 0–2000 W
4	Temperature acquisition unit	HIOKI LR8510	-
5	Wireless data logger	HIOKI LR8410-30	-

At present, ChaoJi coupler and the adaptor are still in the design and verification stage. Samples used in the demonstration project are all 3D printed. The test objects are listed in Table 3.

Table 3. Test objects

No	Name	Electric parameter	Note
1	GB 2015 cable assembly	DC + /DC- cross-sectional area 70 mm^2, length 5 m	-
2	Adaptor	ChaoJi DC + /DC- cross-sectional diameter 10 mm^2	3D printed
3	ChaoJi coupler	-	3D printed
4	ChaoJi test gauge	-	-
5	Cable in the vehicle	Length 2.5 m cross-sectional area 95 mm^2	-

3.2 Test Environment

According to the test requirements of cable assembly used for high power charging in IEC 62196-3-1 [8], the test was conducted in two groups, respectively the connection mode (1) The normal temperature rise test of GB2015 connector, adaptor and ChaoJi vehicle inlet connection combination. And connection mode (2) The over temperature test of GB2015 connector, adaptor, ChaoJi test gauge connection combination. The purpose of the normal temperature rise test is to verify whether the temperature rise of the terminals in the connector, the adaptor and the vehicle inlet does not exceed 50 K [8] under the specified test current using the adaptor, and the maximum temperature at each measuring point does not exceed 95 °C. Over temperature rise test intends to verify if the temperature sensor installed in the connector would be able to detect the temperature change happened in the vehicle inlet when overheating occurs in the inlet and at the same time the temperature sensor installed in in inlet fail to work. If it is, the charger should reduce the charging current or stop charging in time to prevent accident happen. In that case, the charger has the ability to provide backup protection for vehicles. The distribution of the temperature sensor and the two connection modes are shown in Fig. 7 and Fig. 8 respectively.

The test was carried out in the lab with the ambient temperature of 21.5 °C, and the sampling interval of the temperature logging unit was set to be 10 s. Now SGCC mainly uses 60 kW integral charger and 120 kW split type charger with multiple interfaces to build charging stations, and the output voltage range is 200–500 V and the rated output current is 120 A and 240 A respectively. The maximum output current of the 120 kW split type charger is still 120 A when two cars are charging at the same time. Considering the above, selecting 125 A as the test current and setting the output current of the test system as 125 A to start the test.

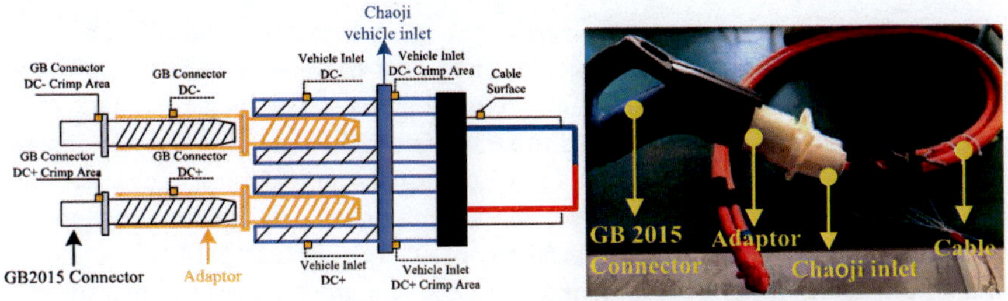

Fig. 7. Cable assembly using GB connector, adaptor and ChaoJi vehicle inlet

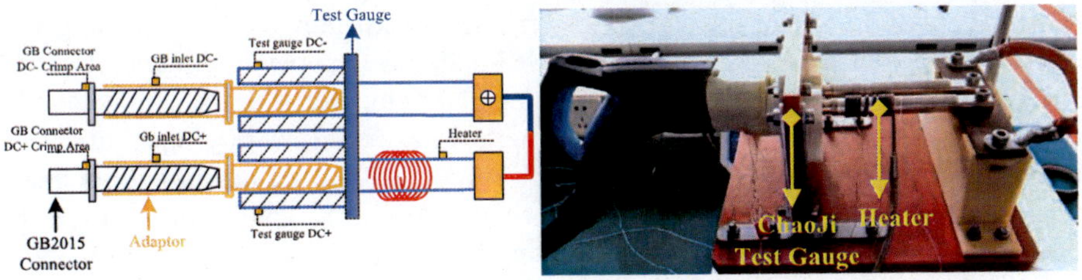

Fig. 8. Test gauge for cable assembly using GB connector and adaptor

3.3 Analysis of Test Results

Figure 9 indicates that after 2 h of the test current 125 A, all the measuring points have reached heat balance (temperature change less than 0.1 °C measured at an interval of 10 min for 3 consecutive times). The temperature measured on the DC+/DC- pins in the adaptor GB side is higher than other parts. The highest temperature is 39.65 °C, which appears in the DC- pins of the adaptor GB side, lower than the standard requirement of 95 °C, and the temperature rises is 18.15 K, less than the standard requirement of 50 K. It can be concluded that after using the adaptor, the temperature rise of the terminals can meet the standard requirements under the test current.

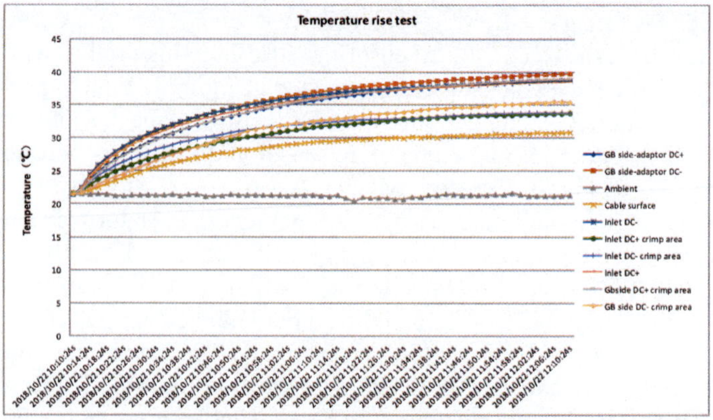

Fig. 9. Temperature rise test

Figure 10 shows the over temperature test of connection mode 2. The test was conducted after each temperature point reached heat balance. Set the output power of the heater to 45 V/300 W manually, and the heating resistance located on DC + pins of test gauge starts to work at 11:25:34 s to simulate overheating signal. The temperature trend of each measurement point in Fig. 10 obviously indicate that each measurement point reaches heat balance again after about 50 min, and the maximum temperature does not exceed the allowable temperature of 95 °C. At 12:15:14 s, set the output power of the heater to 60 V/400 W and at 12:19:14 s, the temperature measured at the inlet has reached to 94.55 °C, which is the maximum allowable temperature of the standard and the test is stopped.

From the trend of temperature change in Fig. 10, the following conclusions can be drawn: (1) with the use of the adaptor, when local overheating occurs, the temperature sensors in the vehicle inlet and connector will both have temperature mutation, and the change trend is the same. The change of temperature measured in the connector is less than the temperature measured in the vehicle inlet because the temperature sensor located in the connector is further from the hot spot, and the temperature difference ΔT is not linear. (2) Since the temperature sensor located in the connector is further from the hot spot, the temperature change measured in the connector is slightly behind the temperature measured in the vehicle inlet. But the delay Δt is not obvious because the good thermal conductivity of copper. (3) If overheat is not serious when using the adaptor, the cable assembly can still reach heat balance at a certain ambient temperature and meanwhile the temperature and temperature rise of the terminals will not exceed the standard requirements.

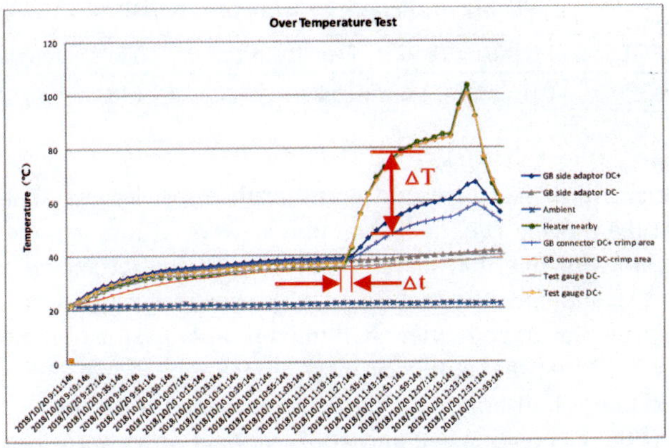

Fig. 10. Over temperature signaling test

4 Analysis of Mechanical Strength

4.1 Coupler Modeling

Considering the material of the test object, the structural strength is verified by simulation. When using the adaptor, the direct connection between the connector and vehicle is changed to an indirect connection, as shown in Fig. 11, and the physical connection length is increased. The torque exerts on the vehicle inlet is bigger if a same force is added to the connector when using the adaptor. The connector, adaptor and vehicle inlet models were established by Solidworks [10] to evaluate the impact of the use of adaptor upon the interface. In order to verify if the cable assembly using adaptor will meet the requirements of strength design, ANSYS [11] is used to analyze the stress of interface.

The coupler model was simplified and small features in the model were removed in order to reduce the workload of finite element analysis.

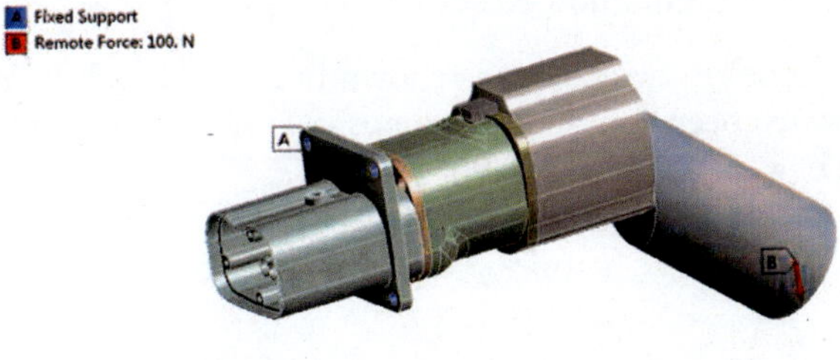

Fig. 11. Simplified model of charging coupler

4.2 Static Analysis

The simplified 3D model is imported into ANSYS, and the material is defined as nylon, and the following assumptions are made: 1) the material direction is. isotropy; 2) model gravity is considered; 3) the deformation is small. The main material parameters are shown in Table 4 below.

Table 4. Parameter table of material

Material	MOE(pa)	Poisson's ratio μ	Density ρ (g/cm^3)	Yield strength Re (Mpa)
Nylon	8.5E9	0.4	1540	148

An automatic grid division method with hexahedron element as the main element was adopted for the model. The maximum unit size was 3 mm and the total number of units was 120,980. For the analysis of a single part, the tetrahedral mesh automatic division method is adopted, the maximum unit size is 1.5 mm, the local details are 0.5mm, the total number of converter shell units is 114776, the total number of vehicle inlet shell units is 63758, the assembly mesh division is shown in Fig. 12.

The four mounting bolt holes of ChaoJi vehicle inlet were fully constrained in the XYZ direction (Fig. 11 position A), and 100N vertical downward force was applied at the rear of GB connector (Fig. 11 position B) to analyze the stress of the adaptor and ChaoJi vehicle inlet.

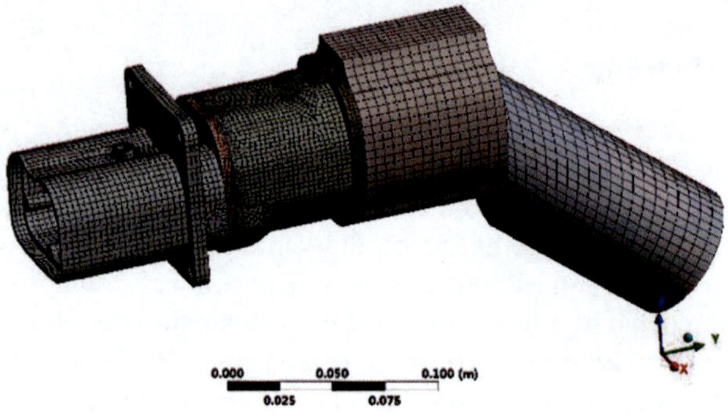

Fig. 12. Meshing diagram of the model

4.3 Analysis of the Simulation Results

The stress distribution of the adaptor is shown in Fig. 13, and the maximum stress is about 51 MPa, which occurs in the groove position of GB inlet. The material is nylon, and the safety factor S = 148/51 = 2.90.

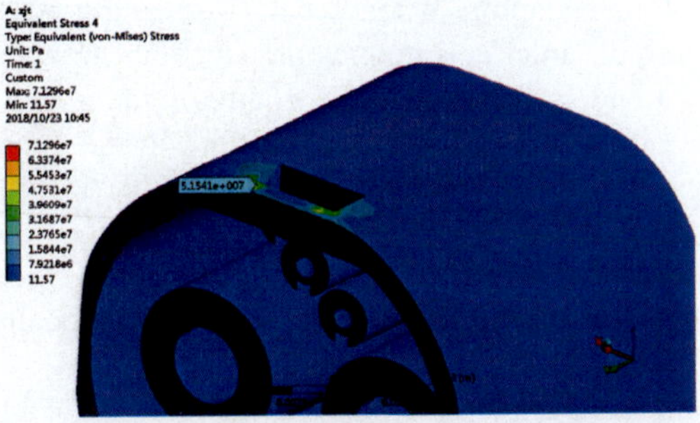

Fig. 13. Stress distribution of adaptor

The stress distribution of ChaoJi vehicle inlet is shown in Fig. 14. The maximum stress is about 25 MPa, which occurs at the bolt hole, and the material is nylon. The safety factor S = 148/25 = 5.92.

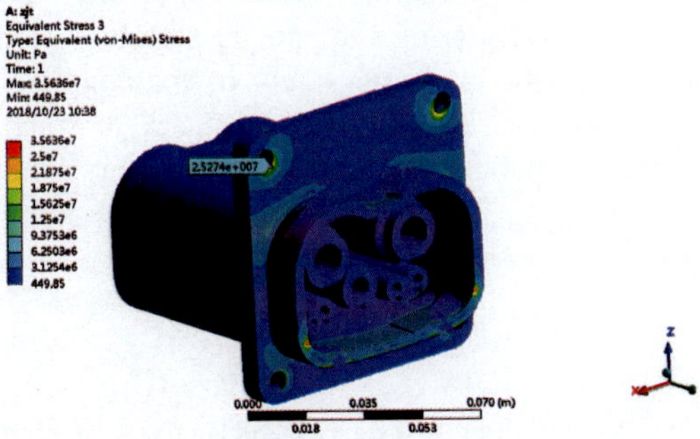

Fig. 14. Stress distribution of ChaoJi inlet

From the analysis of the above simulation results, it can be concluded that when using the adaptor, the maximum stress of the converter and ChaoJi vehicle inlet is about 51 MPa and 25 MPa, which occurs at the groove location of the GB inlet and the bolt hole of HPC vehicle inlet respectively. The safety factor is both greater than 1, which can meet the strength design requirements.

5 Conclusion

This paper introduces the boundary of the HPC coupler, and studies the connection combination using the adaptor about control pilot circuit design, thermal performance and structural strength. The results show that:

(1) The temperature rise of the terminals of cable assembly using adaptor can meet the standard requirements under the dc 125 A test current.

(2) With the use of the adaptor, when local overheating occurs, the temperature sensors in the vehicle inlet and connector will both have temperature mutation, and the change trend is the same. The change of temperature measured in the connector is less than the temperature measured in the vehicle inlet because the temperature sensor located in the connector is further from the hot spot, and the temperature difference ΔT is not linear. Meanwhile, since the temperature sensor located in the connector is further from the hot spot, the temperature change measured in the connector is slightly behind the temperature measured in the vehicle inlet. But the delay Δt is not obvious.
(3) If overheat is not serious when using the adaptor, the cable assembly can still reach heat balance at a certain ambient temperature and meanwhile the temperature and temperature rise of the terminals will not exceed the standard requirements.
(4) By using the adaptor, the maximum stress of the converter and ChaoJi vehicle inlet is about 51 MPa and 25 MPa, which occurs at the groove location of the GB inlet and the bolt hole of HPC vehicle inlet respectively. The safety factor is both greater than 1, which can meet the strength design requirements.

This paper mainly analyzes the thermal performance and structural strength of the adaptor in the initial state. Considering the special structure of the adaptor, it is particularly important to ensure the charging safety of the adaptor in any state. Such as after a period of use, the thermal performance in the wear state, and considering the performance in the high temperature, low temperature environment is worth further study.

References

1. Yang, L., Chuan, L., Xin, H.: Future research and analysis of high power electric vehicle power charging technology in China. Electronic Electric **5**, 001–006 (2018)
2. GB/T 20234.1–2015 Connection set for conductive charging of electric vehicle-part 1:general requirements[S]. Standardization administration of the P.R.C (2015)
3. Huaihe, S., Shubin, Y., Xiaohong, C.: Factors affecting the high rate charge discharge performance of Li-ion batteries. Chinese J. Power Sources **6**(133), 443–447 (2009)
4. GB/T 20234.3–2015 Connection set for conductive charging of electric vehicle—part 3:DC charging coupler[S]. Standardization administration of the P.R.C (2015)
5. Sang, L., Xu, H., Guan, X.: Experiment design on control pilot circuit of AC charging coupler for electric vehicle. Electr. Measur. Instrum. **2**(50), 112–115 (2013)
6. Jia, J., Ni, F.: A discussion on standardization of electric vehicle's charging coupler. Autom. Electric Power Syst. **35**(8), 76–79 (2011)
7. GB/T 18487.1–2015 Electric Vehicle conductive charging system-part1: General requirements[S]. Standardization administration of the P.R.C (2016)
8. IEC 62196–3–1 Plugs, Socket-outlets, vehicle connectors and vehicle inlets- conductive charging of electric vehicles Part 3–1: Vehicle connector, vehicle inlet and cable assembly intended to be used with a thermal management system for DC charging [S]. Switzerland: International Electro technical Commission
9. GB/T 11918.1–2014 Plug socket outlet and couplers for industrial purpose-part1:General requirements[S]. Standardization administration of the P.R.C (2016)

10. Zhao, F., Yang, X., Zhao N.: SolidWorks 2018 Chinese mechanical design from entry to mastery.1st ch. Post and Telecom Press Co., Ltd, Beijing (2018)
11. Saeed, M.: Finite element analysis theory and application with ANSYS. 3nd ch. Publish House of Electronics Industry, Beijing (2015)

Technology for Improving Backup Protection of Extreme Working Conditions Based on Station Information

Yulong Wang[1,2(✉)], Mingliang Jin[3], Qingchun Zhao[1,2], Jun Xie[3], Xiangwen Chen[3], Xiaochun Xu[1,2], Yanping Long[3], and Guangwu Dai[1,2]

[1] NR Electric Co. Ltd., Nanjing 211102, China
wangyulong@nrec.com
[2] NR Engineering Technology Co. Ltd., Nanjing 211102, China
[3] Central China Branch of State Grid Corporation of China, Wuhan 430000, China

Abstract. In substations, there is an extreme working condition in which the line interval loses the double main protection and a short-circuit fault occurs again. In this extreme working condition, the backup protection may have problems of slow fault removal or misstepping or refusal of overstepping. This paper analyzes the limitations of the configuration settings of the existing backup distance protection and zero-sequence protection for the above extreme conditions caused by the blocking of protection equipment, communication equipment failure and channel damage. According to the characteristics of different extreme working conditions, an extreme working condition recognition logic based on the shared information of the station domain is proposed. Combining the identification results of the station domain, the backup protection acceleration trigger mechanism of the substation under extreme operating conditions and the passive protection trigger acceleration mechanism of the corresponding interval of the neighboring substation are proposed. After the acceleration is triggered, the operation time of the backup protection is shortened, and the range of the backup protection corresponding to the interval between the neighboring substations is adjusted to realize the remote backup. A simulation platform was built on the RTDS platform in combination with the switching device of the station domain and the optimized line protection. The simulation results verified that the optimized backup protection can quickly and orderly remove the fault when the short circuit fault occurs after the main protection is lost due to different extreme conditions. The level of safety and stability of the UHV transmission system has been improved.

Keywords: Extreme working conditions · Station domain · Backup protection · Acceleration · Loss of main protection

Y. Xue et al. (Eds.): PMF 2020, LNEE 718, pp. 174–186, 2021.
https://doi.org/10.1007/978-981-15-9746-6_14

1 Introduction

With the construction of high-voltage transmission lines and the rapid development of communication technology, the transmission line generally adopts a double differential configuration mode. When the main protection configuration is strengthened, the timing of backup protection can be simplified and not fully coordinated [1, 2]. However, under extreme operating conditions such as protection blocking caused by protection against DC power loss in the power grid, loss of multiplex channel communication equipment, and communication channel failure, the extreme main protection of the substation single, multiple, or even the entire station line may be lost. These extremes short-circuit faults on the line under working conditions often need to be removed through backup protection. If the backup protection is simplified or mismatched, the backup protection will refuse to act or skip actions [3–6], which will cause the stability of the power grid and the safety of the equipment.

Domestic and international researches have been carried out on the optimization of related backup protection under extreme operating conditions. The literature [7] has studied the strategy of simply accelerating the backup protection section II when the line transmission channel is abnormal while ensuring that the delay of the adjacent line backup protection meets the coordination relationship, but there is also the risk of accelerating overstepping misoperation of the faulty adjacent line backup protection in the line area when the multi-interval or the total station line channel is abnormal. Literature [8] studied the identification of DC power loss in total station protection and put forward the concept of regional protection master station-substation protection system. This regional protection requires the use of additional communication channels to achieve data transmission to achieve rapid removal of regional distance protection and achieve cost high communication requirements.

In view of the extreme power grid conditions where the line differential main protection is lost, the switch information in the substation is used to identify the extreme conditions that lead to the loss of differential protection, and then the accelerated coordination strategy for backup protection is completed, and distance protection is achieved according to the main protection lost conditions adaptive range adjustment strategy. The optimized solution eliminates the problems of far-delay action of backup protection, the risk of disordered misoperation and insufficient protection sensitivity, and improves the reliability of backup protection to quickly remove faults.

2 Adaptability Analysis of Backup Protection

2.1 Backup Protection Setting Principle

In the ultra-high voltage line setting specification, for the line of 220 kV–750 kV voltage level, the setting is generally based on the near reserve and near-backup principle. When setting the distance protection section II, first consider the sensitivity, and cooperate with the adjacent line section I or longitudinal protection. If the time is considered to be in conjunction with the general time setting value of 0.5 s–1.7 s, the simplified section will set the operation time consistency of section II [9, 10]. The distance protection section III is set according to sufficient sensitivity, and cooperates with the adjacent line distance section II or section III. Only when the conditions permit, the combination of remote and near is adopted, and the sensitivity coefficient of remote backup is not required, and the action time avoids the maximum oscillation period, generally 1.7 s–4.0 s [9–11].

The sensitive section of the zero sequence protection is set according to the sensitivity of the metallic fault at the end of the line. At the same time, it cooperates with the sensitive section of the zero sequence protection of the adjacent line, and the time is not less than 1.0 s. The end of zero sequence protection is used to supplement the backup distance protection. The setting value is usually 300 A. The action time is set according to the end of the zero sequence overcurrent of the adjacent line. The starting time avoids the longest non-full phase running time [9–11].

2.2 Analysis of the Problem of Backup Protection When the Main Protection is Lost

According to the analysis of the existing setting principles, the backup protection setting has a far action delay when it is matched, and the misselection of the mismatch setting timing cannot meet the protection scope and selectivity requirements, which in turn leads to mismatch or overstepping action to the lower line bus in complex grid extreme conditions malfunction. In Fig. 1, lines L1–L3 adopt dual configuration line differential protection, and at the same time the backup protection of each line is configured according to the principle of simplification. Assuming extreme operating conditions such as device lockout, communication equipment abnormality, and channel failure caused by protection power failure at substation B, a backup protection adaptability analysis is performed.

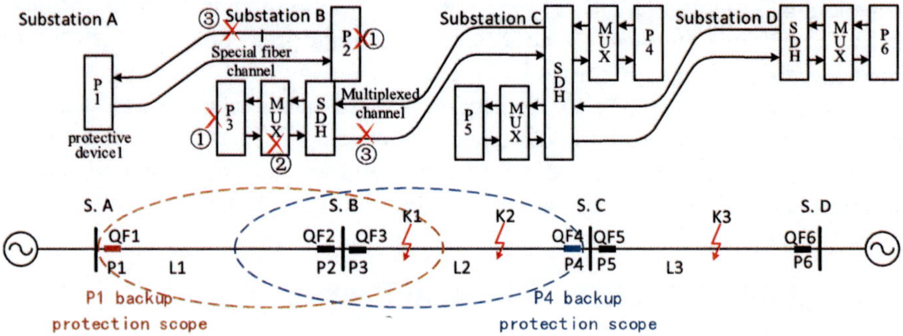

Fig. 1. Schematic diagram of power transmission system protection configuration.

(1) Substation B protection device blocking.
When protection P3 occurs DC power loss in substation B, the differential main protection is lost, as shown by ① in Fig. 1. When a fault occurs in line L2, it is hoped that P1 and P4 will quickly remove the fault. As shown in Fig. 1, if the fault point K1 is within the backup protection range of the protection P1, if there is no

special treatment, normal coordination delay action will be performed, and the fault removal will be slower, which may cause serious burnout of the device. If the fault point is at K2, P4 can remove the fault, but P1 cannot be removed according to the near backup setting, which may cause the protection to refuse to operate. When the whole station protection is blocked, there is also the problem that the backup protection of the adjacent station refuses to move or the action is slow.

(2) Substation B communication equipment is abnormal.
When the substation B communication DC power loss line L2 adopts the multiplexing channel, the line loses the main protection, as shown in ② in Fig. 1. If the K1 fault point is within the range of the near-backup section II, the action time is not less than 0.5 s, and when the multi-level difference is involved, the action delay is longer, which is not conducive to the rapid removal of the fault. However, due to the loss of the main protection, the line L2 fault may enter the near-backup protection range of the adjacent lines P1 and P6 at the same time. When the time setting is consistent or mismatched, the adjacent lines will misstep and expand the fault removal range. If all station line protections are multiplexed channels, the communication equipment is abnormal and the main station line main protection is lost. The near-backup protection time is set according to the step difference and the action delay is long. If the action time acceleration is shortened to the same, it may cause the loss of the main protection phase. Adjacent line near-backup protection overstepping action expands the scope of fault removal and affects the safety of power equipment. For dedicated fiber differentials, the main protection is not affected.

(3) Channel failure

The substation B line L2 channel fault and the fault in the area are within the range of backup section II, as shown in ③ in Fig. 1. When the multi-level difference is involved, the protection action delay is long, which is not conducive to the rapid removal of the fault. When all substation B channels are faulty, the main protection of lines L1 and L2 is lost. The fault of line L1 or L2 may enter the near-backup protection range of the adjacent line. When the time setting is consistent or mismatched, the adjacent line will be misstepped and expand the fault. The resection range, and the resection time may be longer.

Blocking of protection device, abnormal communication equipment, and channel failure will cause the main protection to be lost and output an abnormally signal for the channel. When the line protection uses a dedicated optical fiber or multiplexed channel, the channel quality judged by the protection is determined. In the differential protection, the channel error occurs when the data check fails in the received frame. The line differential protection requires a channel error rate of 10^{-3}–10^{-5}. In order to ensure high sensitivity, the current line protection channel error rate is set at a fixed value 10^{-5}. In addition to data verification, When the received data frame header is completely wrong, it is considered that there is no valid frame. When the channel has no valid frames or the channel error exceeds 10^{-5}, the main protection and output channel abnormal signal are blocked after a certain time delay. This delay is 200 ms according to the guaranteed discrimination reliability. There is no delay output when the protection device is locked. After the channel quality is restored, the main protection and channel anomaly signals are restored after 1 s widening. This widening time is greater than the blocking time of the protection device when the setting value is modified. The main protection blocking and channel anomaly discrimination logic are shown in Fig. 2.

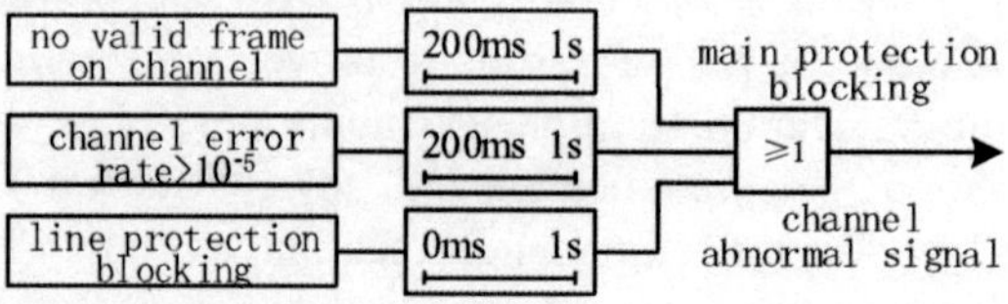

Fig. 2. Differential main protection blocking and reset logic.

3 Optimized Scheme for Backup Protection

Regardless of the line differential protection caused by protection blocking, communication equipment abnormality or channel failure, according to the existing backup protection setting scheme, there is a risk of long action delay, rejection or misleading.

The design specifications of the substation require the protection power supply and the communication power supply to be independent, because there may be different DC power loss under extreme working conditions, so the full substation switching quantity acquisition equipment adopts the protection power supply and channel power supply dual power supply mode for the discrimination of extreme working conditions.

3.1 Judging Extreme Conditions

(1) Judgment of protection blocking in the substation

When the protection device is blocked, the dual line protection at each interval outputs the "device blocking" and "channel abnormal" signals. If the full substation protection fails, the DC screen in the substation outputs the "protection DC loss" signal.based on these signals, it is determined that the line protection device in the substation has failed. Its discriminating logic is shown in Fig. 3.

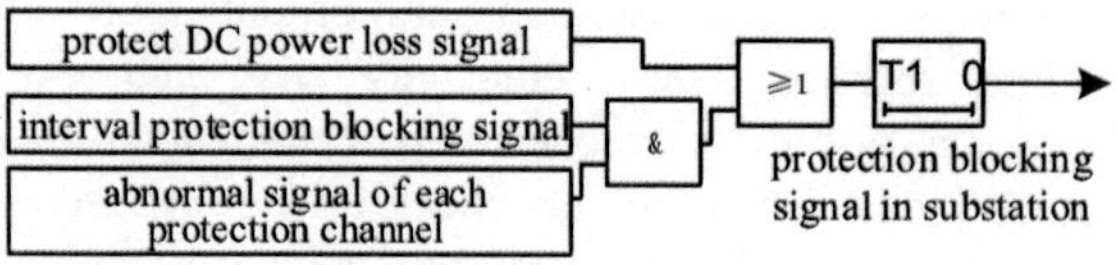

Fig. 3. Discrimination of blocking of line protection interval in substation.

In Fig. 3, the switching information of the substation domain meets the delay time T1 and outputs the "protection blocking signal in the substation", which is output to each interval line protection device. T1 avoids the restart time of the protection device to modify the fixed value, which is generally less than 500 ms.

(2) Abnormal identification of communication equipment in substation

When the multiplex channel MUX communication is blocked, the dual MUX equipment in each interval outputs the "communication device blocking" signal, and the corresponding channel protection device outputs the "channel abnormal" signal. If the communication power of the entire substation loses power, the communication DC screen in the substation outputs "communication power loss" signal to identify the blocking of communication equipment in the substation. The discrimination logic is shown in Fig. 4.

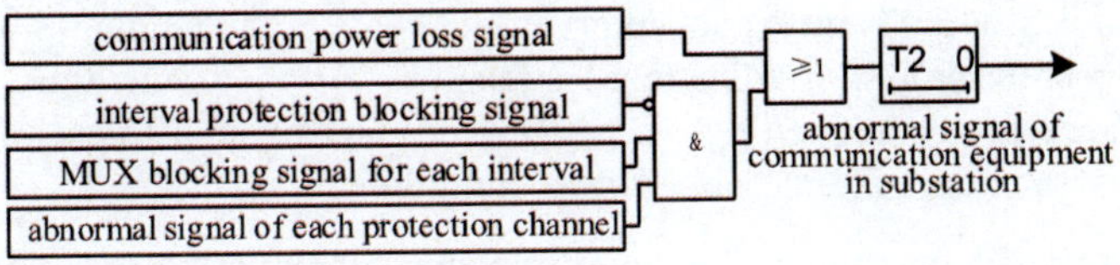

Fig. 4. Abnormal discrimination of line communication equipment in the substation.

In Fig. 4, T2 is the confirmation delay of the communication abnormal signal in the station, avoiding the restart time of the MUX device. The abnormality of the communication equipment makes the main protection of the line lost but the logic of the backup protection function of the protection device is normal.

(3) Channel fault identification

When the external channel is abnormal and then short-circuited, it is assumed that the double-channel or multi-circuit operation between the two stations is considered to be a single-interval channel failure. If the substation is connected to multiple substations, the multi-interval channel abnormality is considered to be the multi-interval channel failure. The discriminant logic is shown in Fig. 5.

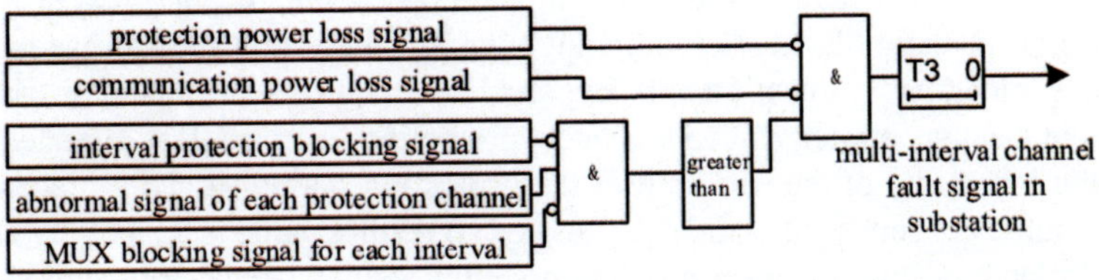

Fig. 5. Fault identification of multi-interval line channels in the substation.

In Fig. 5, T3 is the confirmation delay of the multi-interval channel fault signal in the station, generally 1 s. The channel fault is eliminated due to the device blocking and the abnormal communication device. The channel's main protection is lost when the channel fails, but the protection device's own near-backup protection function logic is normal.

Through the collection and logic judgment of the switching value in the substation, it can distinguish between protection blocking, abnormal communication equipment and channel failure, laying a foundation for the configuration and optimization of backup protection of line protection.

3.2 Accelerated Self-adjustment Strategy for Backup Protection

In order to ensure versatility, the "backup self-adjusting control word" is added to the line protection to switch the function on the standard line protection device. Adding the input of "protection blocking signal in substation", "abnormal signal of communication equipment in substation" and "multi-interval line channel fault signal in substation" to realize different backup self-adjustment strategies. At the same time, the "remote backup distance protection setting" is added for adaptive adjustment of the distance protection range under extreme power grid conditions.

(1) Backup protection strategy for line protection blocking in substation

Taking the protection blocking of substation B in Fig. 1 as an example, with the substation B full station protection blocking or partial protection blocking, adaptive acceleration of substation A protection P1 and station C protection P4 backup protection is realized.

The line protection that the main protection in the substation does not lose when the partition protection of the substation B is blocked will transmit the received "blocking protection signal in the substation" to the opposite side through the optical fiber to trigger the optimization of the backup protection of the opposite side. The abnormal and extreme conditions in the substation are judged by passive adaptive acceleration, so the passive adaptive acceleration trigger logic of the backup protection of the neighboring station is shown in Fig. 6.

Combined with the protection passive acceleration self-adjustment trigger logic in Fig. 5, the conditions are that the interval line protection device is not blocked, there is no protection blocking in the substation, the communication equipment is abnormal, and the multi-interval channel fault, and the interval channel is abnormal or received in the opposite substation Protection blocking signal. When substation B has interval line protection blocking, protection of P1 and P4 requires an adaptive acceleration adjustment backup strategy. The distance protection section II accelerates to t_{z2} to ensure the sensitivity of the end fault of the protection line itself. The adjacent line bus fault has sufficient sensitivity, and the distance III setting value is adaptively changed to the set "remote backup distance protection setting value". Accelerate zero sequence II section to t_{L2}, accelerate zero sequence III section to t_{L4}, in order to ensure the fast removal of short circuit faults in the area and the operation time of the main protection and failure protection of multiple normal adjacent lines, $t_{z2} = t_{L2} = 0.5$ s , t_{z4} hide the longest oscillation time over the line is 1.8 s, and t_{L4} and t_{z4} retain a time difference, its value takes 2 s.

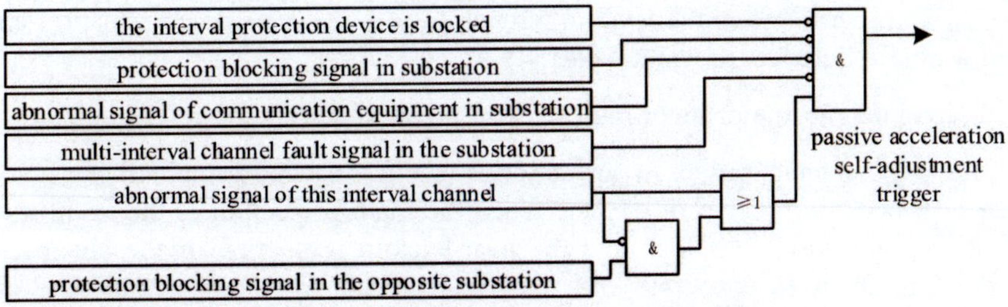

Fig. 6. Protection passive acceleration self-adjustment trigger logic.

(2) Backup protection strategy for abnormal blocking of communication equipment in substations

Taking the MUX failure of substation B in Fig. 1 as an example, regardless of the loss of the main protection of the substation B's total station line or part of the interval line, the substation B communication equipment abnormal interval line near-backup protection and the loss of main protection adjacent line near-backup protection action time need to cooperate and accelerate. The line protection device in substation B will receive the "abnormal signal of the line communication equipment in the substation", combined with the channel abnormal signal of the line protection device, to actively trigger acceleration automatically. The protection device performs passive adaptive acceleration triggering, and the logic is in Fig. 6.

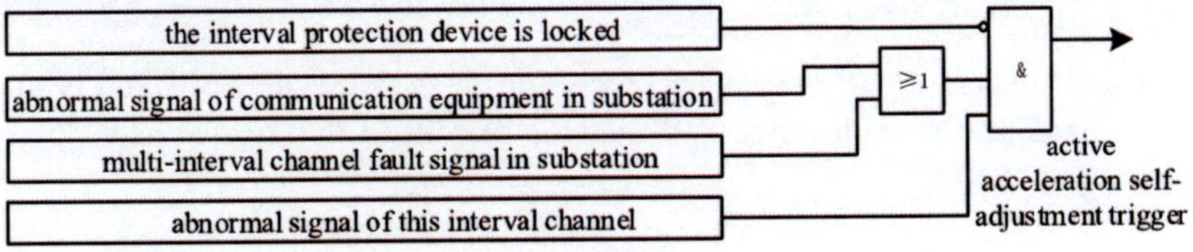

Fig. 7. Protection active acceleration self-adjustment trigger logic.

Taking the active acceleration self-adjustment trigger condition of Fig. 7 as the interval device is not blocked but the device channel is abnormal, and the communication equipment in the substation is abnormal or there are other interval channels in the substation. Combined with the protection passive and active acceleration self-adjustment trigger logic in Figs. 6 and 7, in Fig. 1 it is assumed that when the lines in substation B are multiplexed channels and the communication power supply is powered off, protection P1 cooperates with protection P3, protection P4 cooperates with protection P2, protection P1, P2, P3, P4 all need adaptive acceleration. The zero sequence protection stage II accelerates to t_{L1}, and the zero sequence protection stage III accelerates to t_{L3}. The adaptive acceleration strategy for protecting P1 and P4 is the same as in Fig. 6. In order to ensure the rapid elimination of the fault in the area and the near-backup protection coordination characteristics of the adjacent lines, the action time coordination relationship is $t_{z1} < t_{z2} < t_{z3} < t_{z4}$, $t_{L1} < t_{L2} < t_{L3} < t_{L4}$, $t_{z3} < t_{L3}$,

$t_{z4} < t_{L4}$. In order to retain the level difference $t_{z1} = t_{L1} = 0.3$ s, $t_{z3} = 1.6$ s, t_{L3} and t_{z3} retain a time difference, its value takes 1.8 s.

(3) Multi-interval line channel fault backup protection strategy in substation

Taking the channel failure of substation B in Fig. 1 as an example, there is no loss of the near-backup function of the line when the main protection of the multi-interval line in the substation B is lost, and the near-backup protection of the channel fault interval line needs to cooperate with the near-backup protection action time of the adjacent line that lost the main protection speed up protection action time. The line protection device in substation B will receive the "multi-interval line channel fault signal in station", combined with the channel abnormal signal of the line protection device, to perform active adaptive acceleration trigger. The adaptive acceleration trigger logic is the same as that in Fig. 7. Passive adaptive acceleration triggering on the opposite line protection device with abnormal channel is the same as that in Fig. 6. Combined with the protection passive and active acceleration self-adjustment trigger logic in Figs. 6 and 7, the multi-interval line channel fault backup protection strategy in the substation is the same as the communication device in the substation abnormally blocking backup protection strategy, but passive adaptation is performed only when the single-interval channel fails. Aiming at the extreme working conditions such as line protection device blocking, communication equipment abnormality and channel failure in the power grid, the active and passive acceleration self-adjustment trigger logic of near-backup protection is proposed. From the backup retention strategy of Fig. 6 and Fig. 7, when the line that loses the main protection fails, the near-backup protection can expedite the removal of the fault. At the same time, it ensures that the action time and protection range between the protection device in the substation and the protection in the opposite substation after the main protection of the multi-interval line is lost. The risk of expanding the fault range and the risk of refusal due to mismatch of backup protection are avoided.

4 Simulation Test

The above-mentioned backup protection acceleration self-adjusting trigger logic is implemented on standard ultra-high voltage line protection. At the same time, the distance, zero-sequence backup action time and action range of the line protection are self-adjusted according to the trigger logic. Identify the device under extreme conditions, and carry out systematic simulation tests on the protection logic and backup protection strategy.

According to Fig. 1, a simulation test environment is established. Each substation is equipped with a dual-power station domain switch quantity acquisition device. The unit positive sequence impedance of lines L1–L3 is $Z_1 = 0.02 + j0.28\,\Omega/\text{km}$, and the unit zero sequence impedance is $Z_0 = 0.19 + j0.84\,\Omega/\text{km}$.The lengths of lines L1–L3 are 100 km, 40 km and 60 km respectively. The protection P1–P6 are all put into the differential protection function, the backup protection distance I is set according to 70% reliability, the distance II is set according to the simplified consideration of sensitivity, and the other settings are set according to cooperation. The protection once is set in Table 1.

Table 1. Primary value of simulation test protection setting.

Setting value	Unit	P1	P2	P3	P4	P5	P6
Distance I	Ω	19.6	19.6	7.9	7.9	11.8	11.8
Distance = II	Ω	37.9	37.9	16.8	16.8	23.6	23.6
	s	0.8	0.8	0.8	0.8	0.8	0.8
Distance = III	Ω	38.9	38.9	27.3	35.5	27.6	27.6
	s	2.2	1.8	2.0	2.0	1.8	2.2
Zero sequence II	A	860	600	720	720	600	860
	s	1.6	1.2	1.4	1.4	1.2	1.6
Zero sequence III	A	300	300	300	300	300	300
	s	2.8	2.4	2.6	2.6	2.4	2.8
Romote black protection	Ω	51.1	51.1	36.5	51.1	36.5	36.5

Short circuit faults at different locations on line L2 under different extreme conditions of analog substation B. The location of the selected fault point is 20%, 80% and 100% of the total length of the line at the protection P2 installation, and the backup protection of different fault points P1–P6 before and after the optimization of the action, through the "backup self-adjusting control word" to switch back and forth to achieve the comparison before and after optimization, the action results are shown in Table 2, Table 3 and Table 4. In addition, it simulates the action behavior of substation B under different extreme conditions when the L3 line is faulty.

Table 2. Protection action when L2 line has a short circuit fault at 20%.

Fault point	Extreme conditions	Protection action	P1 ms	P2 ms	P3 ms	P4 ms	P5 ms	P6 ms
Short circuit fault at 20%	P3 protection lockout	Before optimization	824	\	\	822	\	\
		Optimized	528	\	\	523	\	\
	P2,P3 protection lockout	Before optimization	825	\	\	822	\	\
		Optimized	526	\	\	522	\	\
	L2 line MUX failure	Before optimization	\	\	22	824	\	\
		Optimized	\	\	21	524	\	\
	Full failure of MUX during L1 multiplex	Before optimization	\	\	23	825	\	\
		Optimized	\	\	24	523	\	\
	L2 channel failure	Before optimization	\	\	22	822	\	\
		Optimized	\	\	22	523	\	\
	L1,L2 channel failure	Before optimization	\	\	23	824	\	\
		Optimized	\	\	24	523	\	\

Table 3. Protection action when L2 line has a short circuit fault at 80%.

Fault point	Extreme conditions	Protection action	P1 ms	P2 ms	P3 ms	P4 ms	P5 ms	P6 ms
Short circuit fault at 80%	P3 protection lockout	Before optimization	829	\	\	19	\	\
		Optimized	529	\	\	20	\	\
	P2,P3 protection lockout	Before optimization	826	\	\	19	\	\
		Optimized	526	\	\	21	\	\
	L2 line MUX failure	Before optimization	829 false action	\	823	21	\	\
		Optimized	\	\	324	20	\	\
	Full failure of MUX during L1 multiplex	Before optimization	829 false action	\	824	20	\	\
		Optimized	\	\	323	22	\	\
	L2 channel failure	Before optimization	828 false action	\	824	21	\	\
		Optimized	\	\	523	21	\	\
	L1,L2 channel failure	Before optimization	829 false action	\	822	22	\	\
		Optimized	\	\	322	21	\	\

Table 4. Protection action when L2 line has a short circuit fault at 100%.

Fault point	Extreme conditions	Protection action	P1 ms	P2 ms	P3 ms	P4 ms	P5 ms	P6 ms
Short circuit fault at 100%	P3 protection lockout	Before optimization	1613 slow action	\	\	23	\	\
		Optimized	515	\	\	20	\	\
	P2,P3 protection lockout	Before optimization	1616 slow action	\	\	22	\	\
		Optimized	516	\	\	20	\	\
	L2 line MUX failure	Before optimization	\	\	828	19	\	\
		Optimized	\	\	326	20	\	\
	Full failure of MUX during L1 multiplex	Before optimization	\	\	827	22	\	\
		Optimized	\	\	326	21	\	\
	L2 channel failure	Before optimization	\	\	827	20	\	\
		Optimized	\	\	525	21	\	\
	L1,L2 channel failure	Before optimization	\	\	828	21	\	\
		Optimized	\	\	326	20	\	\

Under different extreme working conditions of substation B, protection P5 and P6 are reliably cut off by differential protection when line L3 has a short-circuit fault, and protection P1–P4 are not activated. The protection action results from Table 2, Table 3 and Table 4 show that before optimization, according to the existing backup protection principle and setting principle, the line fault backup protection under different extreme conditions has the problems of long fault removal, overstepping or refusal to operate. Identify the grid protection blocking, communication equipment abnormalities and communication channel faults under extreme operating conditions based on the switching value of the station domain. The optimized backup protection action speeds up and reduces the risk of backup protection overstepping action when line faults occur after extreme operating conditions.

5 Conclusions

Aiming at the loss of the main protection of the line in the substation due to the extreme operating conditions of the power grid, the paper proposes an extreme working condition identification method based on station domain information, near-backup and remote-backup protection accelerated self-adjustment strategy. Introducing dual power supply domain switching devices in substations to realize the discrimination of the main protection loss of single-interval and multi-interval lines caused by protection device blocking, communication equipment abnormality and channel failure, and then provide external decision-making information for emergency protection strategies. According to the discrimination of the extreme working conditions in the station, an active and passive adaptive backup protection acceleration trigger mechanism for line protection is proposed to realize the line protection of the main protection and the line protection of the neighboring station due to the loss of extreme protection in the station. The time is reasonably coordinated, and the simulation test shows that the emergency strategy of near-backup protection and remote-backup protection shortens the fault removal time of the fault line and the remote backup adjacent line, and avoids the long action time of the adjacent line or the occurrence of overstepping misoperation.

References

1. Zhao, J., Zhang, G., Tong, X.: Self-adaptive tripping strategy based on Dijkstra algorithm for wide-area protection. Power Syst. Prot. Control **46**(16), 127–134 (2018)
2. Li, H., Xie, J., Wang, Y., et al.: Adaptive-setting and optimization of performance for EHV transmission line protection device. Autom. Electric Power Syst. **42**(17), 148–155 (2018)
3. Gai, W.: Research on Distributed Wide-Area Back-Up Protection System for High-Voltage Transmission System. Shangdong University, Jinan (2017)
4. Yu, F., Booth, C., Dysko, A., Hong, Q.: Wide-area backup protection and protection performance analysis scheme using PMU data. Int. J. Electr. Power Energy Syst. 110 (2019)
5. Ahmadinia, M., Sadeh, J.: A modified wide-area backup protection scheme for shunt-compensated transmission lines. Electric Power Syst. Res. 183 (2020)

6. Liu, K., Cao, R., Dong, X.: Solution to avoid unreasonable action of remote backup protection under post-fault overload condition. Guangdong Electric Power **28**(07), 78–83,108 (2015)
7. Yang, X., Liu, G., Zhu, X., et al.: Technology of accelerating back-up protection of transmission line zone ii with abnormal fiber channel. Autom. Electric Power Syst. **38**(04), 123–126 (2014)
8. Dai, G., Xie, H., Xu, X., et al.: Power loss area distance protection based on substation's secondary DC power loss. Jiangsu Electr. Eng. **44**(06), 117–121 (2016)
9. Zhang, Y., Gong, R., Wang, L.: Discussion on the coordination of backup protection relay settings in the 220–750 kV network. Power Syst. Protection Control **43**(21), 136–142 (2015)
10. DL/T 559-2019 Setting guide for 200 kV–750 kV power system protection equipment[S] (2019)
11. Chen, M., Wang, H., He, B.: A simplified method of distance relay zone-II setting calculation and a region-based backup relaying scheme. Autom. Electric Power Syst.**39**(07), 151–156,163 (2015)

Real-Time Transaction Prediction Method Based on Generalized Tie-Line

Yantao Zhang[1,2(✉)], Jingxiao Jiang[1], Mengfu Tu[1,2], Fan Xu[1,2], and Wen Wang[1,2]

[1] NARI Group Corporation (State Grid Electric Power Research Institute), Nanjing 211106, Jiangsu, China
zhangyantao@sgepri.sgcc.com.cn

[2] State Key Laboratory of Smart Grid Protection and Control, Nanjing 211106, Jiangsu, China

Abstract. In order to cope with the rapid development of renewable energy in the Northwest Power Grid, it is an urgent need to predict the renewable energy consumption capacity and transaction needs of the provinces, and promote the timely consumption of renewable energy within and between provinces. Therefore, this paper proposes a real-time transaction prediction method based on the generalized tie-line. Based on the definition of generalized tie lines, a multi-objective optimization model is established. The optimization objectives include positive and negative reserve of the provincial power grid and generalized positive and negative reserve of the main and sub-control areas. The method analyzes the power transmission and reception capacity and power transmission and reception needs of the main and sub-control areas in the next few hours, calculates the power purchase and sales capacity among the provinces, and automatically generates intra-provincial replacement and inter-provincial transaction plan based on the transaction rule library to enhance the intelligence level of renewable energy consumption in the Northwest Power Grid.

Keywords: Real-time Transaction prediction · Generalized tie-line · Inter-provincial transaction · Intro-provincial replacement · New energy consumption

1 Introduction

The development of new energy is a general trend of world energy development and an inevitable choice for all countries in the world [1]. In recent years, China's renewable energy has developed rapidly, and its installed capacity has reached the world's first. At the same time, the abandoned problem of wind and photovoltaic becomes more and more serious. In order to solve the unreasonable domestic power supply structure and accept more renewable energy such as wind energy, many experts have proposed effective methods for renewable energy consumption from different perspectives. From the perspective of grid construction and managers, it proposes an implementation model of the "two alternatives" [2] strategy that focuses on "clean alternatives" and supplemented by "electrical energy substitutions" to promote the local consumption of

Y. Xue et al. (Eds.): PMF 2020, LNEE 718, pp. 187–201, 2021.
https://doi.org/10.1007/978-981-15-9746-6_15

new energy. At the same time, improve the direct electricity purchase mechanism for large users [3, 4] and propose corresponding incentives to promote the on-site consumption of new energy. From the perspective of renewable energy optimization technology for smart grid operation and control system, multi-cycle [5, 6] and multi-energy [7, 8] coordinated optimization technologies are proposed. Multi-cycle coordinated optimization is a renewable energy coordinated optimized scheduling method based on multiple time scales, including annual/monthly new energy power plan formulation, day-to-day and intra-day rolling schedule scheduling, and gradually reduced the impact of new energy forecast errors by multi-cycle coordinated optimization, to achieve the maximum safe consumption of renewable energy. Multi-energy coordinated optimization is a scheduling method based on the coordinated optimization of multiple types of energy units. The types of units include hydropower, pumped storage, solar energy, wind power, thermal power, nuclear power, etc. This method make full use of the operating characteristics of different types of generating units to optimize and adjust the output of the units to achieve full consumption of renewable energy. From the perspective of reverse distribution of power generation and consumption, the renewable energy-rich provincial power grids cannot be fully consumed in the province. For this reason, State Grid Corporation of China issued the "Notice on Doing a Good Job in the Preparation of the Cross-Provincial Spot Market for Rich Renewable Energy". Many experts and scholars have proposed cross-provincial UHV AC/DC optimization technology [9, 10], as well as cross-provincial transaction demand and acceptance capacity assessment analysis technology [11] to promote the cross-regional consumption of affluent renewable energy. However, the cross-regional renewable energy consumption lacks a perfect market mechanism [12] and incentives, which cannot mobilize the enthusiasm of new energy consumption. The cross-regional channel margin and the acceptance capacity of the end grid limit the cross-regional renewable energy transactions. Therefore, the current consumption of renewable energy is mainly completed within the regional grid.

This paper takes the Northwest China Grid as an example to explain how to realize the maximum consumption of renewable energy among provinces within the regional power grid. In order to promote the preferential consumption of renewable energy in the northwest power grid, the Northwest Real-Time Trading Platform was officially launched at the end of 2014. Intra-day transaction signing, the dispatcher manually judges the transaction volume based on the intra-day standby situation, which results in large execution deviations and cross-section flow limitations due to untimely transaction signing or improper transaction volume signing. After more than three years operation of the platform, users have put forward higher requirements for the level of transaction management. It is necessary to introduce real-time transaction prediction methods to replace artificial experience judgment, to analyze the safe and effective generalized tie-line reserve and inter-provincial purchase and sale capacity and demand of each main and sub-control area in real time, and to provide timely transaction prompts for provinces with insufficient consumption capacity. Promote the renewable energy consumption through the intelligent generation of the intra-provincial generalized tie-line plan replacement and inter-provincial transaction proposals, and further improve the intelligent level of renewable energy consumption within and between provinces.

2 Analysis of Real-Time Transaction Prediction Characteristics of Generalized Tie-Line Model

The Northwest Regional Power Grid divides five provinces into ten control areas based on the different dispatch nature of generator. There are five main and sub-control areas, named after the provincial power grid. Units in the main control area are directly dispatched by the Northwest Regional Power Grid. Units in the sub control area are directly dispatched by each provincial power grid. From the above definition, it can be seen that the electricity demand of each province is shared by the main control area and the sub-control area units. In order to reflect the "open fair and just" dispatching principles and ensure the balanced generation of units in the main and sub-control areas, the concept of the generalized tie-lines scheduling for the main and sub-control areas was introduced [13].

The real-time power generation plan of Northwest Regional Power Grid adopts the operation mode of tracking the generalized tie-line plan, which determines the prediction method of real-time transaction prediction. Real-time transaction prediction needs to take the main and sub-control areas as the main body of the forecast, analyze the power transmission and receiving capabilities and power transmission and reception needs of the main and sub-control areas based on the generalized tie-line plan of the main and sub-control areas, and then calculate the purchasing and selling capacity and demand of each province based on the above four types of prediction results, and provide basic information for intra-provincial replacement and inter-provincial transactions.

Intra-provincial replacement is mainly based on the power transmission and receiving capabilities and power transmission and reception needs the main and sub-control areas, and amending the intraday generalized line plan for main and sub-control area. The real-time power generation plan adjusts the output of the thermal power units in the main and sub-control areas by tracking the revised intra-day generalized tie line plan for the main and sub-control areas to ensure the maximum consumption of new energy in the province.

The intelligent generation of inter-provincial transactions depends on the purchase and sale capacity of each province, the demand for power purchase and sale in each province, and the inter-provincial transaction rules. At the same time, it is necessary to consider the safety of inter-provincial sections. After the transaction is generated, the northwest regional power grid revises the provincial geographic tie line plan and the generalized tie line plan of the main and sub-control areas to provide the latest boundary conditions for the regional power grid and provincial power grid real-time power generation plan.

3 Real-Time Transaction Prediction

The real-time transaction prediction is based on basic data such as ultra-short-term system load forecasting in each province, ultra-short-term new energy forecasting, generalized tie-line scheduling for each main or sub-control area, and unit operating parameter information, etc., comprehensively considering various types of grid operation constraints, with a period of 15 min rolling forecast of the power purchase and sales capacity and demand for power purchase and sale in each province in the next 4 h. At the same time, it analyzes the power transmission and reception capabilities and power transmission and reception requirements of the main and sub-control areas to provide a basis for the intelligent generation of intra-provincial replacement and inter-provincial transactions.

3.1 Mathematical Model

The real-time transaction prediction model takes the maximum positive and negative reserves of each province and the generalized positive and negative reserve costs of the main sub-control areas as the optimization objectives, calculates the positive and negative reserves of each province and the generalized positive and negative reserves of the main and sub control areas. Each province needs two optimization models to support the positive and negative reserve types, which are called positive and negative reserve optimization models. The objective functions are:

$$\max F_{A}^{u} = \min_{t} R_{A,t}^{u} + G_{A}^{u} \tag{1}$$

$$\max F_{A}^{d} = \min_{t} R_{A,t}^{d} + G_{A}^{d} \tag{2}$$

Where F_{A}^{u} is the optimization objective of the positive reserve optimization model of province A; F_{A}^{d} is the optimization objective of the negative reserve optimization model of province A; $R_{A,t}^{u}$ represents the positive reserve of province A at period t; $R_{A,t}^{d}$ represents the negative reserve of province A at period t; G_{A}^{u} represents the generalized positive reserve cost of province A; G_{A}^{d} represents the generalized negative reserve cost of province A.

$R_{A,t}^{u}$ and $R_{A,t}^{d}$ can be further expressed as:

$$R_{A,t}^{u} = \sum_{k \in S_A} p_{k,t} + \sum_{j \in J_A} \gamma_{j,t} - L_{A,t} \tag{3}$$

$$R_{A,t}^{d} = L_{A,t} - \sum_{k \in S_A} p_{k,t} - \sum_{j \in J_A} \gamma_{j,t} \tag{4}$$

Where S_A is the collection of units of province A; $p_{k,t}$ represents power output of unit k at period t; J_A is the collection of tie-lines of province A; $\gamma_{j,t}$ represents active power of tie-line j at period t; $L_{A,t}$ represents system demand at period t of province A.

Assuming that the positive and negative reserve reserved at period t of province A are respectively expressed as $V_{A,t}^{u}$ and $V_{A,t}^{d}$, then the power purchase and sale capacity and the power purchase and sale demand can be expressed as:

$$\begin{array}{ll} F_{A,t}^{sc} = R_{A,t}^{u} - V_{A,t}^{u} & R_{A,t}^{u} - V_{A,t}^{u} > 0 \\ F_{A}^{pn} = \left| R_{A,t}^{u} - V_{A,t}^{u} \right| & R_{A,t}^{u} - V_{A,t}^{u} < 0 \\ F_{A}^{pc} = R_{A,t}^{d} - V_{A,t}^{d} & R_{A,t}^{d} - V_{A,t}^{d} > 0 \\ F_{A}^{sn} = \left| R_{A,t}^{d} - V_{A,t}^{d} \right| & R_{A,t}^{d} - V_{A,t}^{d} < 0 \end{array} \tag{5}$$

Where $F_{A,t}^{sc}$ represents power sale capacity of province A; F_{A}^{pn} represents the power purchase demand for province A; F_{A}^{pc} represents power purchase capacity of province A; F_{A}^{sn} represents the power sale demand for province A.

The constraints of the positive and negative reserve optimization models are exactly the same. Constraint conditions include maximum and minimum power constraints, ramp up and down rate constraints, unit reserve constraints, unit group reserve constraints, unit group power constraints, unit output smooth constraints [14], network security constraints, etc. These conventional constraints will not be elaborated one by one in this paper, only the characteristic constraints of the positive and negative reserve optimization models will be described.

$$\left(L_{A,t} - N_{m,t}\right) \times \left(P_{m,\max} / \left(P_{m,\max} + P_{n,\max}\right)\right) - \left(T_{m,t} + W_{m,t}\right) = G_{m,t} + G_{m,t}^{r} \tag{6}$$

Where province A is divided into sub-control area m and main control area n according to the characteristics of unit dispatch nature; $L_{A,t}$ represents the ultra-short-term system load forecast at period t of province A; $N_{m,t}$ represents the ultra-short-term new energy forecast at period t of sub-control area m; $P_{m,\max}$ is the maximum output capacity of the sub-control area m; $P_{n,\max}$ is the maximum output capacity of the main control area n; $T_{m,t}$ is the total output of the thermal power units at time t of the sub-control area m; $W_{m,t}$ is the total output of the hydro power units at time t of the sub-control area m; $G_{m,t}$ is the intraday generalized tie-line scheduling at time t of the sub-control area m; $G_{m,t}^{r}$ is the intraday generalized tie-line scheduling constraint relaxation at time t of the sub-control area m.

The components in Eq. (6) can be further expressed as:

$$\begin{aligned} N_{m,t} &= \sum_{d \in S_m^d} p_{d,t} \\ T_{m,t} &= \sum_{i \in S_m^i} p_{i,t} \\ W_{m,t} &= \sum_{w \in S_m^w} p_{w,t} \\ G_{m,t}^{r} &= P_{m,t}^{+} - P_{m,t}^{-} \end{aligned} \tag{7}$$

Where $p_{d,t}$ represents the output of the wind power unit d at time t; S_m^d is the collection of new energy units of the sub-control area m; $p_{i,t}$ represents the output of the thermal power unit i at time t; S_m^i is the collection of thermal units of the sub-control area m; $p_{w,t}$ represents the output of the hydro power unit w at time t; S_m^w is the collection of hydro units of the sub-control area m; $P_{m,t}^{+}$ and $P_{m,t}^{-}$ are the upward and downward nonnegative relaxation variables for the generalized tie-line plan constraint at time t of the sub-control area m.

Intraday generalized tie-line plan constraints of the main control area:

$$\left(L_{A,t} - N_{m,t}\right) \times \left(P_{n,\max} / \left(P_{m,\max} + P_{n,\max}\right)\right) - \left(T_{n,t} + W_{n,t} + N_{n,t}\right) = G_{n,t} + G^{r}_{n,t} \quad (8)$$

Where $T_{n,t}$ is the total output of the thermal power units at time t of the main control area n; $W_{n,t}$ is the total output of the hydro power units at time t of the main control area n; $N_{n,t}$ represents the ultra-short-term new energy forecast at time t of main control area n; $G_{n,t}$ is the intraday generalized tie-line scheduling at time *t* of the main control area *n*; $G^{r}_{n,t}$ is the intraday generalized tie-line plan constraint relaxation variable at time *t* of the main control area *n*.

The components in Eq. (8) can be further expressed as:

$$\begin{aligned} N_{n,t} &= \sum_{d \in S_n^d} p_{d,t} \\ T_{n,t} &= \sum_{i \in S_n^i} p_{i,t} \\ W_{n,t} &= \sum_{w \in S_{\mathrm{n}}^w} p_{w,t} \\ G^{r}_{n,t} &= P^{+}_{n,t} - P^{-}_{n,t} \end{aligned} \quad (9)$$

Where S_n^d is the collection of new energy units of the main control area n; S_n^i is the collection of thermal units of the main control area n; $P^{+}_{n,t}$ and $P^{-}_{n,t}$ are the upward and downward adjustment nonnegative relaxation variables for the generalized tie-line scheduling constraint at time t of the main control area n.

Since the sum of the generalized tie-line plan of the main and sub-control areas is equal to the geographic tie-line plan [13], the relationship between the positive and negative reserve of province and the generalized positive and negative reserve of the main and sub-control areas can be expressed as:

$$R^{u}_{A,t} = G^{r}_{m,t} + G^{r}_{n,t} \quad (10)$$

$$R^{d}_{A,t} = G^{r}_{m,t} + G^{r}_{n,t} \quad (11)$$

Equation (10) is the relationship between the positive reserve of province A and the positive reserve of the main and sub-control areas in the positive reserve optimization model. If $G^{r}_{m,t}$ is greater than zero, it indicates the increased power generation capacity (power transmission capacity) of the sub-control area m at time t, otherwise, it indicates the power demand (insufficient power generation) of the sub-control area m at time t; If $G^{r}_{n,t}$ is greater than zero, it indicates the increased power generation capacity (power transmission capacity) of the main control area n at time t, otherwise, it indicates the power demand (insufficient power generation) of the main control area n at time t. Equation (11) is the relationship between the negative reserve of province A and the negative reserve of the main and sub-control areas in the negative reserve optimization model.

The generalized tie-line reserve cost in Eqs. (1) and (2) has a certain functional relationship with the generalized tie-line reserve of the main and sub-control areas in Eqs. (10) and (11). Due to the same processing method in the positive and negative reserve models, for convenience of expression, only the positive reserve model will be described below.

From the Eqs. (7) and (9) in the positive reserve model, we can see that the generalized tie-line reserve of the main and sub-control areas can be expressed as the difference between the positive and negative relaxation variables of the generalized tie-line plan constraint. Taking the sub-control area m as an example, the upwards and downwards relaxation variables correspond to different types of cost piecewise functions, as shown in Fig. 1 and Fig. 2.

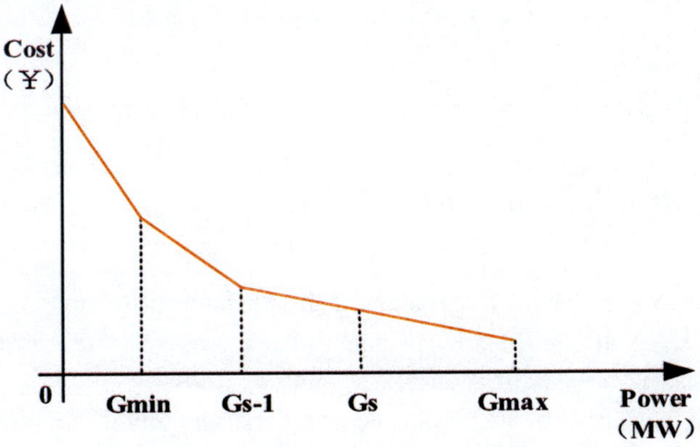

Fig. 1. Cost piecewise functions 1

If $G^r_{m,t}$ is greater than zero, it indicates the power transmission capacity of the sub-control area m at time t. From Eq. (7), $P^+_{m,t}$ is greater than zero, and $P^-_{m,t}$ is equal to zero. In order to achieve the maximum optimization goal of the generalized positive reserve of sub-control area m, $P^+_{m,t}$ uses Fig. 1 as the cost piecewise function 1.

If $G^r_{m,t}$ is less than zero, it indicates the power demand of the sub-control area m at time t. From Eq. (7), $P^+_{m,t}$ is equal to zero, and $P^-_{m,t}$ is greater than zero. In order to achieve the maximum optimization goal of the generalized positive reserve of sub-control area m, $P^-_{m,t}$ uses Fig. 2 as the cost piecewise function 2.

The generalized positive reserve costs are composed of two parts: main and sub-control areas. In order to achieve a broad distribution of positive reserve between the main and sub-control areas according to the proportion of the installed capacity, the main control area n uses a cost piecewise method similar to the sub-control area. The only difference is that the ratio of the output interval length of the main control zone n cost segmentation function 1, 2 to the sub-control zone m segmentation function 1, 2 is equal to the installed capacity ratio of the main and sub-control zone.

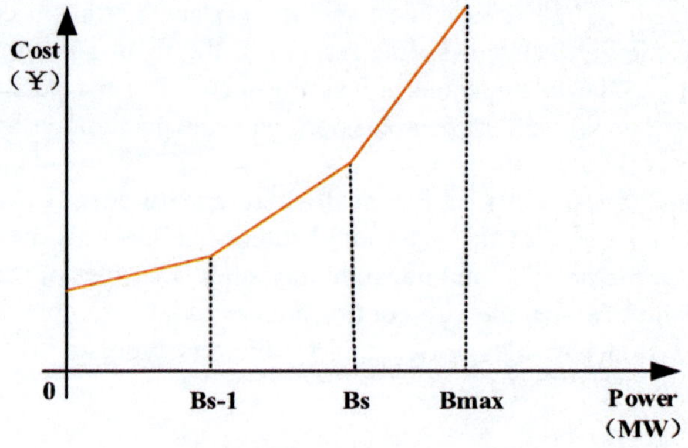

Fig. 2. Cost piecewise functions 2

The generalized positive reserve costs can be further expressed as:

$$G_A^u = \sum_{s=1}^{S}\left[\left(b_{m,s,t}^{+} + b_{n,s,t}^{+}\right)\lambda_s\right] + \sum_{s=1}^{S}\left[\left(b_{m,s,t}^{-} + b_{n,s,t}^{-}\right)\mu_s\right] \tag{12}$$

Where S is the total number of segments of the piecewise function; $b_{m,s,t}^{+}$ and $b_{n,s,t}^{+}$ are the output of the sub-control area m and the main control area n at time t on the s segment of the piecewise function 1; $b_{m,s,t}^{-}$ and $b_{n,s,t}^{-}$ are the output of the sub-control area m and the main control area n at time t on the s segment of the piecewise function 2; λ_s is the incremental cost of the s segment of the cost piecewise function 1; μ_s is the incremental cost of the s segment of the cost piecewise function 2.

The positive and negative adjustment relaxation variables of the generalized tie-line plan constraint of the sub-control areas can be further expressed as:

$$\begin{aligned} P_{m,t}^{+} &= \sum_{s=1}^{S} b_{m,s,t}^{+} \\ P_{m,t}^{-} &= \sum_{s=1}^{S} b_{m,s,t}^{-} \\ 0 \le b_{m,s,t}^{+} &\le B_{m,s}^{+} - B_{m,s-1}^{+} \\ 0 \le b_{m,s,t}^{-} &\le B_{m,s}^{-} - B_{m,s-1}^{-} \end{aligned} \tag{13}$$

Where $B_{m,s}^{+}$ is the right end point power of the sth segment of the cost segment function 1 of the sub-control area m. $B_{m,s}^{-}$ is the right end point power of the sth segment of the cost segment function 2 of the sub-control area m. The formula for the main control area is similar to Eq. (13).

3.2 Model Analysis

Since the inter-provincial transactions are generally smooth curves, the maximum and minimum targets are introduced into the optimization objectives, see Eqs. (1) and (2) to maximize the lowest point of the power purchase capacity of the inter-provincial transactions. Therefore, the optimization model is a nonlinear model with a maximum and minimum optimization goal. It is necessary to perform linear transformation on the maximum and minimum goals to achieve linearization of the model. The specific conversion is as follows:

$$\max F_A^u = \min_t R_{A,t}^u + G_A^u \Rightarrow \begin{array}{c} \max F_A^u = Z_A^u + G_A^u \\ Z_A^u \leq R_{A,t}^u \end{array} \tag{14}$$

$$\max F_A^d = \min_t R_{A,t}^d + G_A^d \Rightarrow \begin{array}{c} \max F_A^d = Z_A^d + G_A^d \\ Z_A^d \leq R_{A,t}^d \end{array} \tag{15}$$

Combine formulas (3)–(5), (10) and (11) to calculate the inter-provincial purchasing power and selling capacity and the power demand for purchasing and selling. At the same time, it obtains transmission and reception capabilities and transmission and reception requirements of each main and sub-control area, which provides basic data for inter-provincial transactions and intra-provincial replacement.

According to the definition of the generalized tie-line plan of the main and sub-control areas, the generalized tie-line plan constraints (6) to (9) are introduced. According to the positive and negative directions of the generalized tie-line relaxation of the main and sub-control areas, two different types of cost segmentation curves are constructed. The constraints of (12) to (13) are used to achieve the optimization goal of generalized reserve balance of the main and sub-control areas and maximum system reserve.

The real-time transaction prediction model based on the generalized tie-line is a linear programming model. The algorithm theory is mature, and the CPLEX optimization algorithm package is used for solution. The calculation efficiency is high and the actual operation requirements are met.

If the replacement of the main and sub-control areas in the province cannot achieve full consumption of renewable energy, inter-provincial transactions need to be implemented to achieve the cross-province renewable energy consumption. Inter-provincial transaction intelligence generation is based on the basic information of each province's inter-provincial purchasing and selling capacity and demand in the next four hours. It complies with the transaction rule base, fully considers the inter-provincial cross-section stability limits, intelligently generate inter-provincial transaction plans, and providing auxiliary decision-making for the inter-provincial consumption of renewable energy.

4 Case Study

This paper takes some provincial power grids in the northwest power grid as examples to verify and analyze the established model.

4.1 Analysis of Positive Reserve of the Main and Sun-control Areas

The installed capacity percentage of Ningxia main and sub-control areas is 0.35: 0.65. Therefore, the Ningxia positive and negative reserve models try to make the generalized positive(negative) of the main and sub-control areas proportional to each other as long as the positive and negative reserve of province are guaranteed to be maximum. From zero to four on June 26, 2017, the relationship between negative reserve in Ningxia and the generalized negative reserve in the main and sub-control areas is shown in Fig. 3.

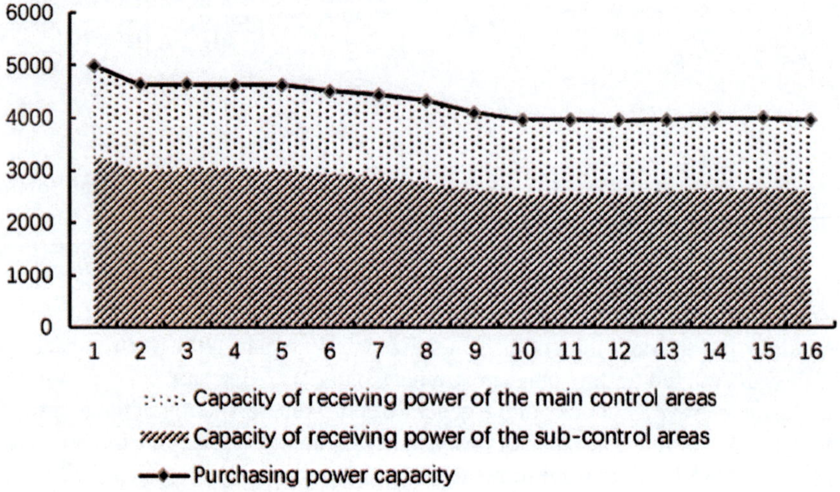

Fig. 3. The relationship between negative reserve of province and the generalized reserve of main and sub-control areas in Ningxia

It can be seen from Fig. 3 that the Ningxia negative reserve is equal to the sum of generalized negative reserve of the main and sub-control area. The negative reserve of the Ningxia can also be called Ningxia purchasing power capacity, and the reserve of the main and sub-control areas is also called the capacity of receiving power of the main and sub-control areas.

The relationship between the positive reserve of the Ningxia system and the positive reserve of the main and sub-control areas is similar and will not be repeated here.

4.2 Replacement of the Main and Sub-control Areas

The intraday generalized tie-line plan for the main and sub-control areas is generated based on the percentage of installed capacity in the main and sub-control areas, but the ratio of the start-up capacity of the main and sub-control areas may not be the same as the installed proportion. Considering various reasons such as line maintenance, unit limitation, and network overload, the generalized reserve ratio of the main and sub-control areas is often different from the installed capacity ratio. Therefore, when the power of new energy reaches a certain level, it is necessary to modify the generalized tie line plan through the replacement of the main and sub-control areas to promote the consumption of new energy.

Taking Ningxia as an analysis scenario from 8:30 to 12:30 on June 23, 2017, the renewable energy output rise rapidly, the system load changed steadily, and the geographical contact line plan was stable. Taking 8:30 new energy and system load as the base value, the new energy change trend, the system load change trend and the new energy change trend deduct the load change trend as shown in Fig. 4.

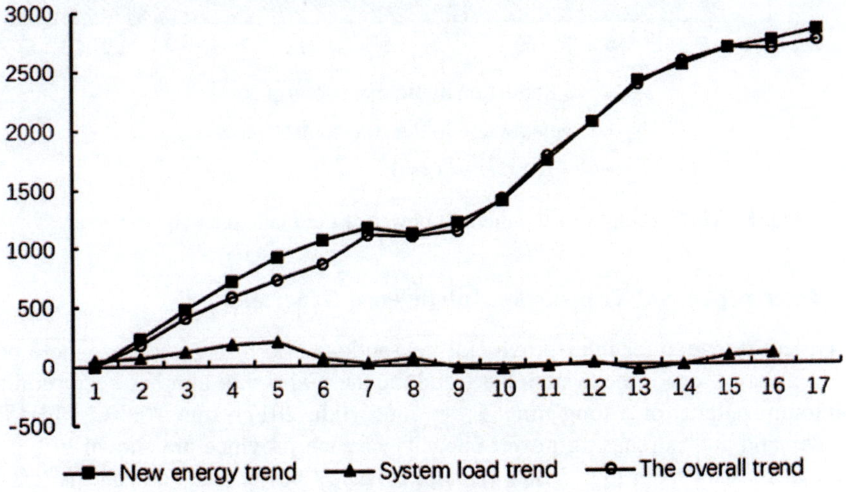

Fig. 4. System load, new energy and its overall trend

Figure 4 shows the overall change trend in the future, which means that the main and sub-control areas need to reduce the output of thermal power units to accept new energy. If the output of the thermal power unit is reduced according to the percentage of installed capacity in the main and sub-control areas, the reduction of the thermal power in the main and sub-control areas is shown in Fig. 5.

At 8:30, the generalized negative reserve values of the main and sub-control areas are 1534 and 1356 respectively. The sub-control area's negative reserve adjustment amount reaches 1543 at 11:30, which exceeds the sub-control area's generalized reserve. At the moment, the amount of adjustment in the main control area is 830, and there is sufficient space for the adjustment in the main control area. Therefore, new energy consumption can be achieved by modifying the generalized tie-line plan of the main and sub-area through intra-provincial replacement.

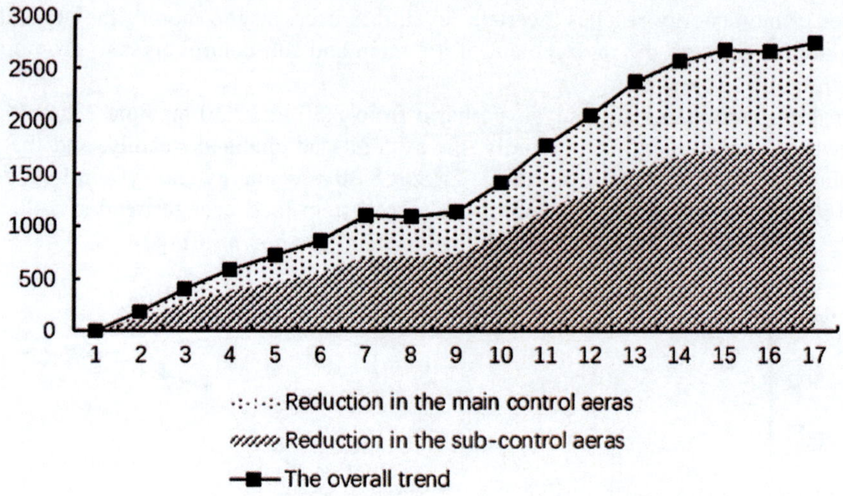

Fig. 5. The reduction of the thermal power of the main and sub-control areas

4.3 Inter-provincial Transaction Intelligence Generation

The typical province for inter-provincial transactions is Gansu Province, where power of new energy exceeds the system load, and the thermal power unit has been running at a minimum output for a long time. From June 10th, 2017, from 10:00 to 13:45, the sales demand and purchasing power capacity of each province are shown in Fig. 6.

It can be seen from Fig. 6 that Gansu Province's power sales demand is less than the purchasing power capacity of Ningxia and Shanxi Province. Therefore, the nearest principle in the transaction rule database (a transaction sequence, Ningxia, Shanxi, Xinjiang, Qinghai). First, a transaction is generated with Ningxia Province, but inter-provincial cross-section safety margin needs to be considered. The sections that Gansu needs to pass through to Ningxia include the four-fish section (Dunyu, Gaoyu, Heyu, Quanyu) and the south-to-north transmission section. The margins of each section are shown in Fig. 7.

It can be seen from Fig. 7 that the Heyu section margin cannot meet Gansu's electricity sales demand. Therefore, Gansu and Ningxia generated the first transaction to ensure the safety of the Heyu section. According to the principle of proximity, the second transaction is Gansu to Shanxi. However, Gansu to Shanxi also passed through the four-yu section, so no transaction could be generated. According to the principle of proximity, continue to generate the second transaction Gansu sent to Xinjiang, Gansu's remaining transaction demand is less than Xinjiang's power purchase capacity, and meet the requirements of inter-provincial cross-section, the second transaction is successfully generated and the transaction is completed.

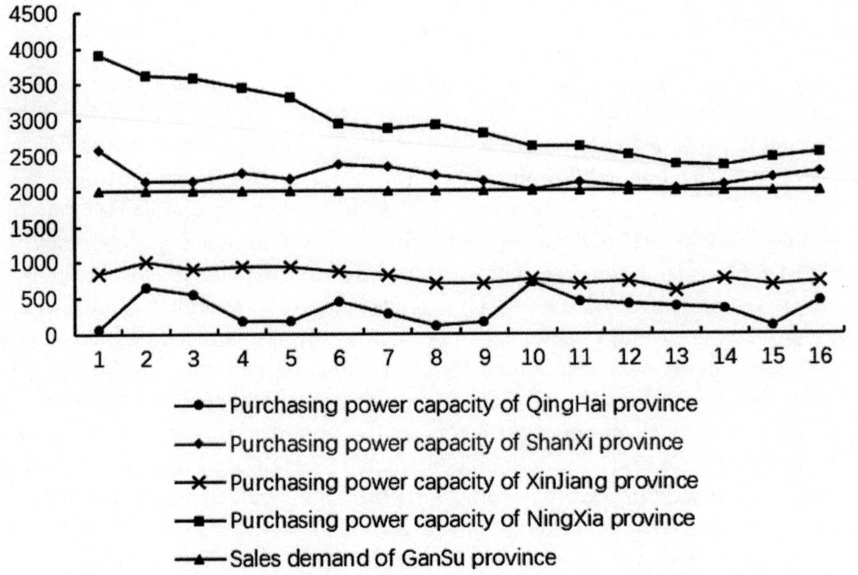

Fig. 6. The purchasing power capacity and sales demand of each province

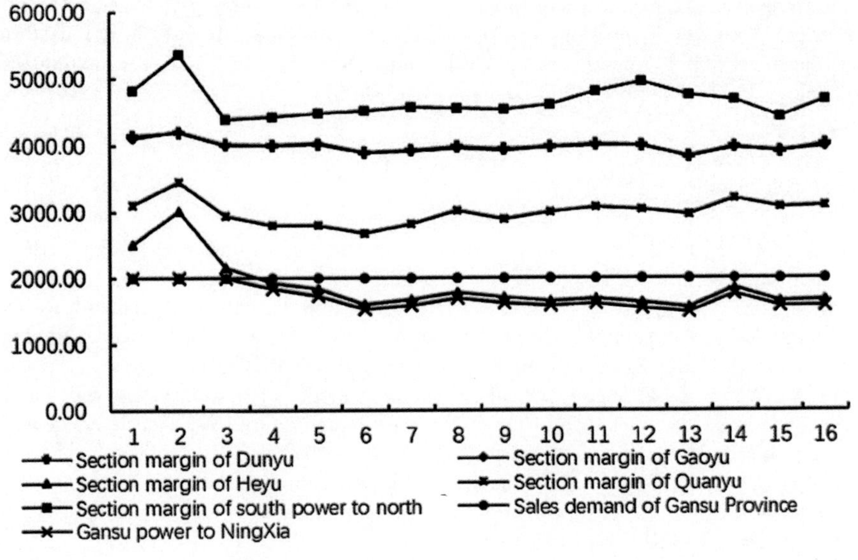

Fig. 7. Inter-provincial section margin of Gansu power to Ningxia

5 Conclusion

This paper proposes a real-time transaction prediction method based on the generalized tie-line in view of the problem of wind and photovoltaic abandoned caused by the rapid development of new energy sources in the Northwest Power Grid. This forecast model analyzes the power transmission and reception capacity and demand of the main and sub-control areas in the next few hours. At the same time, it calculates the purchasing and selling capacity and demand of each province, and automatically generates intra-provincial replacement and inter-provincial transaction plan based on the transaction rule database, promotes the largest consumption of new energy within and among provinces, increases the intelligence level of northwest new energy consumption. Through the construction of this method and analysis of examples, we can know that the effectiveness of positive and negative reserves needs to be considered in intra-provincial replacement and inter-provincial transactions, and the effectiveness of positive and negative reserves needs to be constructed to optimize the goal of maximizing reserves and considering various types of grid Safe operation constraints, through adjustment of unit output, generalized tie line plan and other variables, finally get effective results. The real-time transaction prediction system based on this paper has been applied in the Northwest Power Grid. The actual application effect is good, and it can be promoted in similar regional power grids.

Acknowledgement. This work is supported by Nanjing NARI Group Corporation Open Project: Research on Key Technology of Clearing and Simulation Analysis of bilaterally Competitive Electricity Spot Market, State Grid Corporation Science and Technology Project (Technology Project Number: 1200-201940420A-0-0-00).

References

1. F, Li., W, Q., H, S., et al.: Smart transmission grid: vision and framework . IEEE Trans Smart Grid **1**(2), 168–177 (2010)
2. Ming, L.I., Diangang, H.U., Youxue, Z.H.O.U., et al.: Research and Practice of Renewable Energy Local Consumption Mode in Gansu Province Based on "Double Alternative" Strategy [J]. Power System Technology **40**(10), 2991–2997 (2016)
3. Li, Y., Wang, X., Zhang, W., et al.: Scheduling model of power system with renewable energy and transaction mode of direct electricity purchase by large consumers considering network security constraints. J. Eng. **2019**(16), 899–904 (2019)
4. Zhang, X., Yang, T., Ren, J., et al.: Large consumer direct-purchasing mechanism and its application in northwest power grid considering renewable energy consumption. Electric Power Constr. **38**(01), 138–143 (2017)
5. Xie, L., Zhang, Y., Wang, G., et al.: Coordination method and closed-loop control of medium-short term multi-cycle generation scheduling. Autom. Electric Power Syst. **38**(10), 126–130 (2014)
6. Liu, C., Huang, Y., Zhang, N., et al.: Renewable energy dispatching based on smart grid dispatching and control system platform. Autom. Electric Power Syst. **39**(1), 159–163 (2015)

7. Bschorer, S., Kuschke, M., Strunz, K., et al.: Object-oriented modeling for planning and control of multi-energy systems . CSEE J. Power Energy Syst. **5**, 355–364 (2019)
8. L, Li., M, Tu., Q, D., et al.: Two-stage optimization method for generation output scheduling with large-scale wind power integration . Autom. Electric Power Syst. **38**(9), 45–52 (2014)
9. Zhong, H., Xia, Q., Ding, M., et al.: A new mode of HVDC tie-line operation optimization for maximizing renewable energy accommodation. Autom. Electric Power Syst. **39**(3), 36–42 (2015)
10. M, Xu., L, Wu., H, L., et al.: Multi-objective optimal scheduling strategy for wind power, PV and pumped storage plant in VSC-HVDC grid. J. Eng. **2019**(16), 3017–3021 (2019)
11. Yuan, B., Xu, Z., Liu, J., et al.:Research and application of multi-area new energy integration capability evaluation model. Electric Power **52**(12), 132–139 (2019)
12. Shu, Y., Zhang, Z., Guo, J., et al.: Study on key factors and solution of renewable energy accommodation. Proc. CESS **37**(1), 1–8 (2017)
13. Zhang, Y., Ren, J., Sun, X., et al.: Two-stage real-time generation scheduling model based on coordination of regional and provincial power grid generalized tie-line. Autom. Electric Power Syst. **39**(6), 128–133 (2015)
14. F, Xu., L, Li., Z, C., et al.: Generation scheduling model and application with fluctuation reduction of unit output. Autom. Electric Power Syst. **36**(5), 45–50 (2012)

An Integrated System of Line Protection and Fault Location for Submarine Hybrid Line

Xiaochun Xu[1(✉)], Gong Zhang[2], Haijun Shen[2], Bingyang Wu[2], and Hao Tan[1]

[1] NR Electirc Co., Ltd., China, Nanjing, People's Republic of China
xuxc@nrec.com
[2] State Grid Zhejiang Electric Power Co., Ltd. Construction Company, Hangzhou, People's Republic of China

Abstract. Based on the structure analysis and fault analysis of the hybrid line, this paper proposes an integrated system of protection and fault location which is suitable for the hybrid line. The system is composed of line protection relays on the ends of the hybrid line and section positioning devices on the ends of the cable. It collects three phase currents on both sides of the cable by using flexible optical current transformer which suits to be installed on the joint points. It calculates the differential current at the ends of the cable line through optical fiber and discriminates the fault is whether in or out of the cable line section. Also, it sends the result to the protection devices on both sides of the hybrid line. When the fault is in the cable, the reclosing function on both sides of the line is blocked immediately. In addition, the line protection relays on both sides integrate the traveling wave fault locating function. With receiving section signal, traveling wave fault location can locate the fault point accurately. With lots of simulation tests, it can be proved that the integrated system proposed in this paper can realize the hybrid line protection and precise fault location of the hybrid line. It can greatly improve the power supply reliability of the hybrid lines. At present, the system has been operated in 500 kV cross-linked polyethylene Zhoushan submarine cable hybrid line.

Keywords: Hybrid transmission line · Flexible optical current transformer · Traveling wave fault location

1 Introduction

With rapid development of power grid, lots of high-voltage cable-overhead hybrid transmission lines have been applied. When there is a fault on the hybrid line, it is difficult to search the exact fault location, especially with submarine cables. Therefore, fast and accuracy fault location is the key point to speed up the repair of transmission lines and reduce the outage time [1]. Besides, most of the overhead line faults are transient faults, while most of the cable faults are permanent faults caused by defects during production process or human factors. In order to improve the reliability of power supply, automatic reclosing is usually used in common overhead line protection. But for the permanent fault on the cable, auto reclosing will cause another impact on

Y. Xue et al. (Eds.): PMF 2020, LNEE 718, pp. 202–209, 2021.
https://doi.org/10.1007/978-981-15-9746-6_16

the system and affect the stability of the power system. This requires accurate fault section location of cable-overhead hybrid line. When the fault occurs in the overhead line section, reclosing function is on, but when the fault occurs in the cable section, reclosing function is blocked. At present, the method of fault location based on impedance method cannot guarantee the accuracy of fault location. When the result of impedance distance measurement is used as the judgment basis of reclosing, malop-eration could happen.

According to the different principles of fault location, fault location algorithms can be divided into two types: impedance method and traveling wave method [2]. Due to the influence of line parameters and measurement accuracy, the impedance method usually has a large deviation for hybrid lines; the traveling wave method has the advantages of high accuracy, not affected by the transition resistance [3–5], etc., but when traveling wave location is applied to hybrid lines There is also the problem of uneven propagation of traveling wave velocity, which will affect the final results [6, 7].

In order to solve the problem of accuracy fault location in hybrid transmission line, a fault location system is proposed in this paper. It can distinguish the sections of the hybrid line accurately, and the protection relays can distinguish auto reclosing strategy adaptively according to the results of the section positioning. In addition, on the basis of the section positioning, the line protection relay integrated traveling wave location module can greatly improve the accuracy of fault location.

2 Protection and Fault Location Scheme for Hybrid Lines

Take a typical three section hybrid line as an example, as shown in Fig. 1:

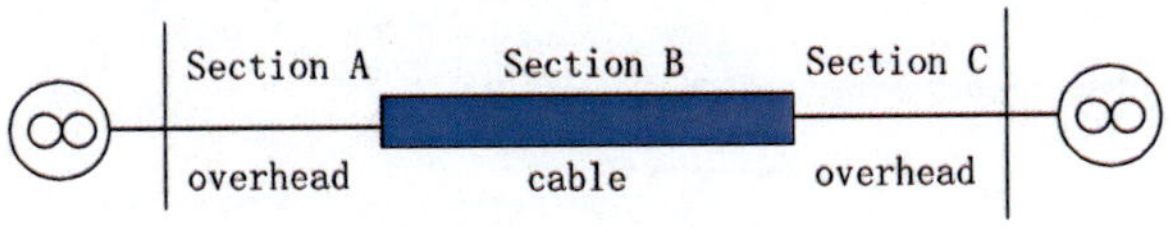

Fig. 1. Typical three section hybrid line

In order to realize the accurate fault section positioning and improve the fault location performance of the protection relays on the ends of the hybrid line, this paper proposes a scheme of accurate section positioning and fault location system based on flexible optical current transformer (OCT).

The fault section precise positioning module integrates the functions of current collection, data synchronization and section positioning. The whole system adopts a compact and centralized development mode, which can meet the requirements of outdoor installation of cable terminal station.

Section positioning devices are installed at both ends of the cable. They complete section discrimination by collecting the currents at both ends of the cable line. The section positioning device collects the data of the optical current transformer at the joint point of the hybrid line. Also, it uses optical fiber to realize the interaction of the

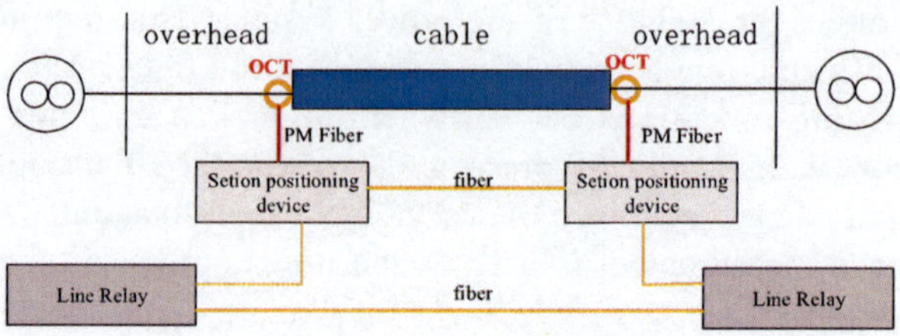

Fig. 2. Configuration of fault location system for hybrid line

currents on both sides of the cable and the calculation of the differential current. If the fault point is in the cable section, the section positioning device will send the blocking reclosing signal to the line protection relay on both sides through the communication optical fiber. To ensure the universality, the blocking reclosing signal will be transmitted to the goose network in the power station through the optical fiber (Fig. 2).

3 Research and Design of the Fault Location System

3.1 Integrated Design of the Section Positioning Device

For the normal application of optical current transformer, each phase of current needs to be equipped with a separate sampling unit, which is connected with polarization maintaining optical fiber to receive current data. Then, the signal is converted into original 4–8 sampling signal to the merging unit. At last, the synchronized data after merging unit is sent to the corresponding protection relays.

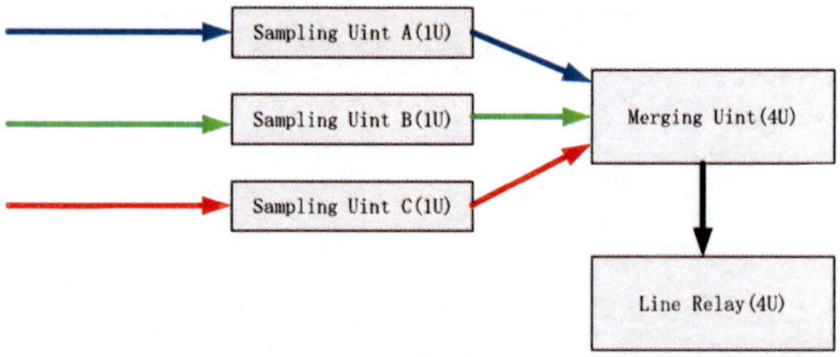

Fig. 3. Traditional application scenario of optical current transformer

At the cable joint of the hybrid line, devices are usually operating in severely disturbance environment. Also, the installation site of the equipment is limited. It is impossible to adopt the installation mode of conventional substation. As shown in Fig. 3, the normal application of optical current transformer requires 11U panel and cabinet space for installation, involving a large number of devices and high power consumption. The devices are connected through optical fiber with low reliability. Therefore, it is necessary to integrate the functions of relevant devices and reduce unnecessary links of devices. It can improve the stability of the whole system.

The device can receive the data of optical current transformer directly from polarization maintaining fiber. It converts the sampling data and complete the function of three-phase current synchronization.

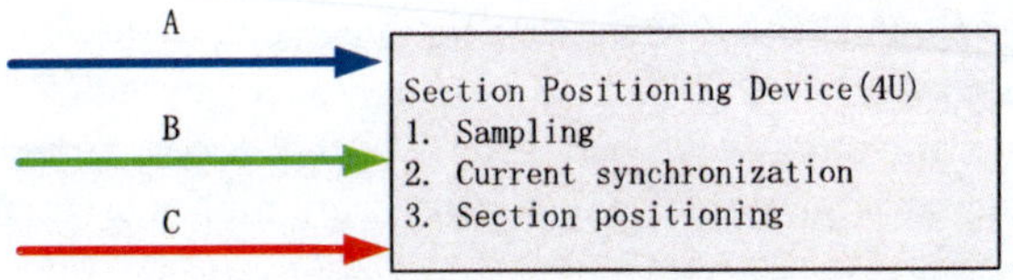

Fig. 4. Functional structure of the fault location device

As shown in Fig. 4, the high integration fault section positioning device proposed in this paper integrates the functions of data acquisition, data synchronization and section positioning in 4U space (compressing 11u device space into 4U device space) to meet the application requirements of miniaturization and local installation. Data exchange is completed by high reliability backplane bus. It greatly improves the integration between function modules and the overall reliability of the system.

3.2 Principles of Fault Section Positioning

The synchronization of three-phase current is completed by interpolation method. Then, the synchronized three-phase current data is directly sent to the section positioning module from the device backplane. The section positioning module transmits the synchronized current data to the device on the opposite side through the optical fiber with HDLC protocol. The section positioning device on both sides completes the data synchronization of the currents on both sides of the cable section by ping-pong principle.

Fault section positioning mainly depends on the differential principle based on Kirchhoff's law. In order to reflect various fault types and operation modes of transmission lines, the section positioning device is equipped with the following differential relays:

1) High sensitivity variable current differential relay. It can quickly distinguish whether the fault is in the section. It is not affected by the load current;
2) Two stage of steady state current differential relays. It can not only ensure the sensitivity, but also the reliability;
3) Zero sequence current differential relay with short time delay. It reflects fault with high resistance.

Once any of the above differential relays operates, the fault is considered to be in the protected cable section. At the same time, the section positioning device sends the "fault in section" goose signal to the subscribers at the ends of the hybrid line.

3.3 Fault Location System

For hybrid transmission lines, the parameters and traveling wave propagation velocity are not uniform. According to the conventional location measurement method, there will be a large deviation, especially for the impedance method, which has great relationship with line parameter and measurement accuracy.

Although the traveling wave method has high accuracy generally, it also has the problem of uneven traveling wave parameters. In this paper, a compensation method based on fault section is proposed, which can ensure high measurement accuracy.

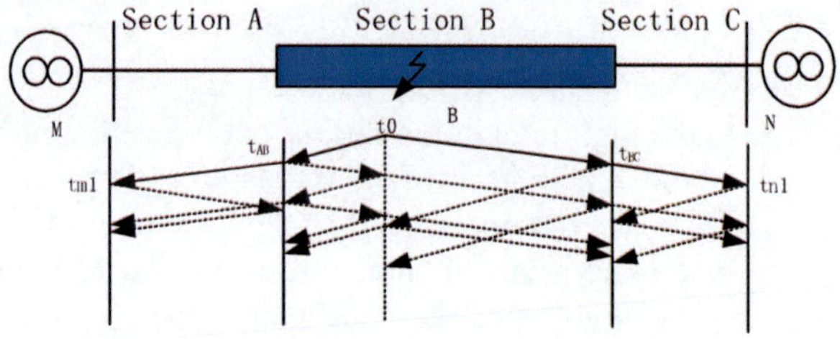

Fig. 5. Traveling wave scheme on hybridline

The basic steps of fault location for hybrid lines are as follows:

1) According to the result of section positioning device, the fault section is determined.
2) According to the traveling wave transmission time of the adjacent section, the time difference between the traveling wave arriving at both sides of the fault section is deduced.
3) Finally, fault location is based on the principle of conventional two terminal traveling wave.

As the hybrid line shown in Fig. 5, the length of the separate section is L_A, L_B, L_C, corresponding propagation velocity is v_A, v_B, v_C.

Assuming that fault point is located in section B, according to the transmission time of traveling wave in section A and section C, the time of the initial traveling wave reaching the connection point AB(t_{AB}) and BC(t_{BC}) can be calculated, as shown in formula (1):

$$\begin{cases} t_{AB} = t_{m1} - L_A/v_A \\ t_{BC} = t_{n1} - L_C/v_C \end{cases} \tag{1}$$

Where, t_{m1} and t_{n1} is the time when the fault traveling wave first reaches the protection relays on both sides of the hybrid line.

In this case, the distance between the fault point and the end side M can be obtained by using the two terminal traveling wave method for line section B, as shown in formula (2):

$$x = L_A + 0.5[(t_{AB} - t_{BC})v_B + L_B] \tag{2}$$

When the fault point is located in section A or section C, the fault position can be calculated by analogy.

4 Simulation and Application

Limited by the power amplifier of the real-time simulation system, the simulation test is divided into two parts: one is the real-time simulation system based on RTDS, focusing on the section positioning section and auto reclose logic; the other is the off-line simulation system based on PSCAD, with the simulation sampling rate of 1MHz as the waveform source of the fault location system. Both offline and online systems use unified models (Fig. 6).

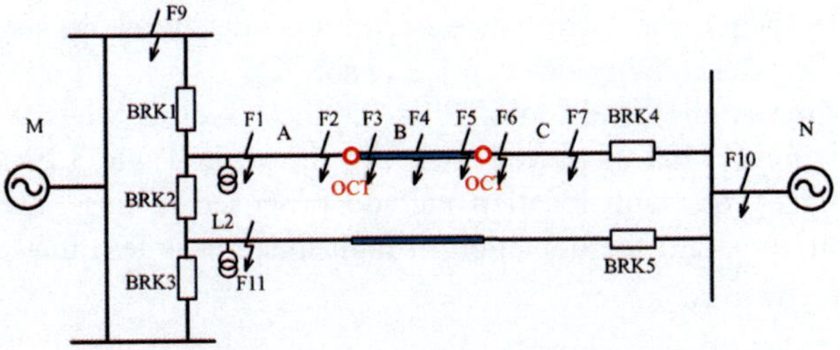

Fig. 6. Testing environment

The test system adopts 500 kV hybrid line based on the Zhoushan hybrid transmission line, with 3 km of overhead line in Section A, 17 km of submarine cable line in Section B and 33 km of overhead line in section C. The line parameters are shown in Table 1:

Table 1. Simulation parameters of the overhead line

	Value	Unit
Resistance(+)	0.02486	Ohm/km
Inductance(+)	0.266	Ohm/km
Capacitance(+)	0.0135	μF/km
Resistance(0)	0.01948	Ohm
Inductance(0)	0.07462	Deg
Capacitance(0)	7.57e-3	Ohm

Table 2. Simulation parameters of the submarine cable line

	Value	Unit
Resistance(+)	0.0195	Ohm/km
Inductance(+)	0.07464	Ohm/km
Capacitance(+)	0.171e-3	μF/km
Resistance(0)	0.01948	Ohm
Inductance(0)	0.07462	Deg
Capacitance(0)	0.171e-3	Ohm

Fault points F1–F7 are respectively set in Section A, B and C and F9–F11 are set out of the hybrid line. The performance of the devices under different types of faults in various operation modes is verified. The section positioning device can accurately detect the faults in and out of the section B. When the fault occurs in section B, the section positioning device can reliably send the section positioning signal to the reclosing device. When the disturbance occurs outside the cable section, the section positioning device can reliably not operate (Table 2).

Using the simulation waveform of PSCAD, the traveling wave waveform of each fault location of hybrid line is played back. As shown in Table 3 (where R represents transient resistance), the fault location method proposed in this paper can locate the fault point accurately, and the deviation of fault location is less than 300 m.

Table 3. Fault location results of hybrid line

Fault	Fault Section	Location(km)	R(Ω)	Deviation (km)
F1	A	0.3	0.1	0.12
F1	A	0.3	300	0.23
F2	A	2.3	0.1	0.05
F4	B	10.8	0.1	0.233
F4	B	10.8	300	0.198
F6	C	23.3	0.1	0.156
F6	C	23.3	300	−0.257
F7	C	50.85	0.1	−0.269
F7	C	50.85	300	−0.271

This integrated system has been operated in 500 kV cross-linked polyethylene Zhoushan submarine cable hybrid line.

Line relays are settled in ZH station and ZS station. The section positioning device are settled in ZH terminal and ZS terminal, where the OCTs are located at the joint point of the submarine cable. The system has been keeping in good situation (Fig. 7).

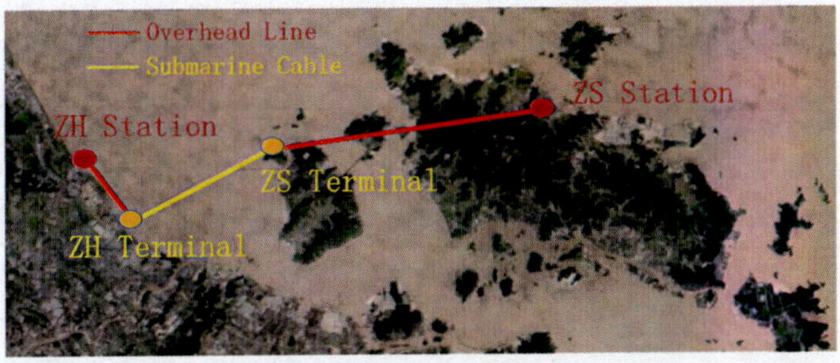

Fig. 7. Zhoushan submarine cable hybrid line

5 Conclusion

In this paper, a fault location system based on flexible optical current transformer is proposed for cable-overhead hybrid line. The system can accurately locate the fault section. It provides the feasibility for the auto reclosing of hybrid lines. When the fault is on the cable, it can broadcast the blocking reclosing signal to the reclosing device. It improves the reliability of reclosing and improves the reliability of power supply of the hybrid line. At the same time, the traveling wave location system based on the section positioning device can get high-precision fault location on the basis of the fault section.

References

1. Yang, L., Wang, B., Dong, X.: Overview of fault location methods in high voltage direct current transmission lines. Autom. Electr. Power Syst. **42**(08), 185–191 (2018)
2. Qi, Z., Zhuang, S., He, X., Huang, Z.: Traveling wave fault location technology for hybrid lines in distribution network based on disturbance signal injection at Busbar. Autom. Electr. Power Syst. **43**(13), 124–133 (2019)
3. Shu, H., Song, J., Tian, X.: Fault location of single-terminal traveling wave for unequal-length double-circuit lines based on traveling wave propagation path. Autom. Electr. Power Syst. **42** (17), 140–149 (2018)
4. Gao, H., Jiang, S., He, J.: Sampling synchronization methods in digital current differential protection. Autom. Electr. Power Syst. **20**(9), 46–49 (1996)
5. Quan, Y., Wang, X., Yang, M., et al.: Two new algorithms and the IR robustness for two-terminal fault location on HV transmission line. Autom. Electr. Power Syst. **24**(10), 28–32 (2000)
6. Zhang, Y., Wang, H., Li, T., Lin, Z., Chang, Y.: Combined Single-end Fault Location Method for LCC-VSC Hybrid HVDC Transmission Line. Autom. Electr. Power Syst. **43**(21), 187–199 (2019)
7. Dai, Z., Liu, N., Zhang, C., Wang, J.: Pilot Protection for High-resistance Grounded Fault of UHVDC Transmission Line Based on Correlation Coefficient of Current Variable. Autom. Electr. Power Syst. **43**(19), 156–163 (2019)

A Multi-scale Attention Networks for Substation Equipment Image Defect Detection

Ke Zhang[1,2](✉), Wenli Huang[2], Houqiang Li[1], and Rui Li[3]

[1] School of Information Science and Technology, University of Science and Technology of China, Hefei 230026, China
zhangke2@sgepri.sgcc.com.cn

[2] Anhui NARI Jiyuan Power Grid Technology Co., Ltd., State Grid Power Research Institute, Hefei, China

[3] Institute of Intelligent Machines, and Hefei Institute of Physical Science, Chinese Academy of Sciences, Hefei 230031, China

Abstract. With the continuous development of pattern recognition technology especially the development of convolutional neural network (CNN), the applications of defect monitoring in substations are widely utilized in industrial circumstances. However, current image analytic methods based on CNN achieved low accuracy and poor robustness due to different scales, non-rigid and complex background of defects. In order to solve these problems, a multi-scale attention networks (MAN) for substation equipment image defect detection is proposed in this paper. In our approach, we firstly design an attention module to extract the global and local features over depth and spatial position activation levels and then three columns attention networks are developed for extracting multi-scale features of defects. Finally, the multi-scale feature maps are fused by using fully connected neural network. The experimental results show that our approach achieves satisfactory results and outweighs the state-of-the-art methods in our defect dataset.

Keywords: Substation equipment · Defect detection · Multi-scale · Attention network · Deep learning

1 Introduction

With the increasing demand for electricity, the defect detection in substation is more important for maintaining the safe operation of power grid. However, there are many types of defects in substation, the manual detection is very time-consuming, inefficient and error-prone, which can affect the detection efficiency of defects. Due to the development artificial of intelligence and computer vision, many researchers have focused on using these technologies to automatically detect and recognize the defects in substation. Yu et al. [1] used image processing and artificial neural network for recognizing the defects in substation. To improve the efficiency of data processing, Zheng et al. [2] proposed an intelligent substation operation, inspection management and controlling system based on artificial intelligence. Their method can enhance the

Y. Xue et al. (Eds.): PMF 2020, LNEE 718, pp. 210–220, 2021.
https://doi.org/10.1007/978-981-15-9746-6_17

analysis and process ability of defect detection. To automatically read and analyze the indication of measuring instruments in substation equipment, Zheng et al. [3] developed a novel automatic reading approach for the indication of analog measuring instruments, which has robustness to camera angle and illumination. To measure the power line tension and sag, Baker et al. [4] used smartphone to assess feasibility of ground-based imaging with Hough transform techniques and line tracing methods. In order to automatically recognize the power equipment, Wang et al. [5] presented a method of recognizing the power equipment based on computer vision. Experiments showed that their method achieved good recognition accuracy. Despite that all above methods achieved good results, they might mainly pay attention to hand-crafted feature extraction or classifiers, there are many constraints by using these methods. In the practical application, the complexity of image background, non-rigid, brightness, different scales and arbitrary direction are the major challenges in feature extraction and classification.

Due to the excellent performance of deep learning in the field of computer vision, many research scholars are committed to using deep learning frameworks to solve defect detection problems. In order to detect the defects of power equipment, Xie et al. [6] extracted the features by using ZF-net, VGG-16 and ResNet-101, the experiment showed that their method had achieved a high detection rate. Liang et al. [7] used Faster R-CNN to detect the defects on power transmission, their proposed method could effectively improve the detection rate of defects. To detect the defects of powerline equipment, Zahid et al. [8] proposed a robust inspection system based on CNN, the experimental results showed that their method achieved 98% accuracy. Li et al. [9] proposed a new method, which applied random forest to classify the electric equipment images by using features extracted with deep convolutional neural network. Experiments showed that their method could provide a new solution for intelligent analysis and recognition of unstructured mass electric equipment images. To detect the foreign object detection (FOD) for securing power transmission lines, Zhang et al. [10] proposed a new deep learning network named RCNN4SPTL, experimental results showed that compared with Faster-RCNN, their method improved the detection speed and recognition accuracy. To improve the accuracy of defects in insulators, Filho et al. [11] used convolutional neural network and multi-task learning (MTL) to detect and recognize the defects, and their proposed methodology could achieve an accuracy of 85% for defect detection. Besides, there are many experts who have conducted research on the defect detection of insulators [12–16]. Although these methods could achieve satisfied performance, they cannot accurately detect the defects in substation due to different scales, complex background and uneven illumination.

In complex substation environment, the defect detection has many constraints: 1) Different scales of defects in images may result lower detection rate. As Fig. 1(a) shows, the scales of "smoking", "nest" and "respirator discoloration" are very tiny, and it is difficult to detect tiny defects via CNN. 2) The defects in substation contain many non-rigid defects, such as "equipment oil", "ground oil" and "metal corrosion" as shown in Fig. 1(b). In different scenes, the defects that belong to the same class have great differences in appearance and shape. Thus, it is difficult to detect these defects because of the characteristic of non-rigid. 3) The environment of substation is very complex, and it is difficult for detectors to effectively leverage appearance information to distinguish the objects from surrounding background or similar objects. As Fig. 1(c) shows, the nest hidden in the background makes it difficult to detect. To address these problems, in this paper, we developed a multi-scale attention networks (MAN) for substation equipment image defect detection, which can improve the detection accuracy of defects in substations. To address scale problem, we designed three columns convolutional neural networks, each column of the network used different-sized convolution kernels to detect defects at different scales. And to tackle non-rigid characteristic and complex environment problem, the attention module is added in each column network. Our attention module could extract the global and local features at the same time, so that it can reduce the effect of uneven non-rigid and complex environment.

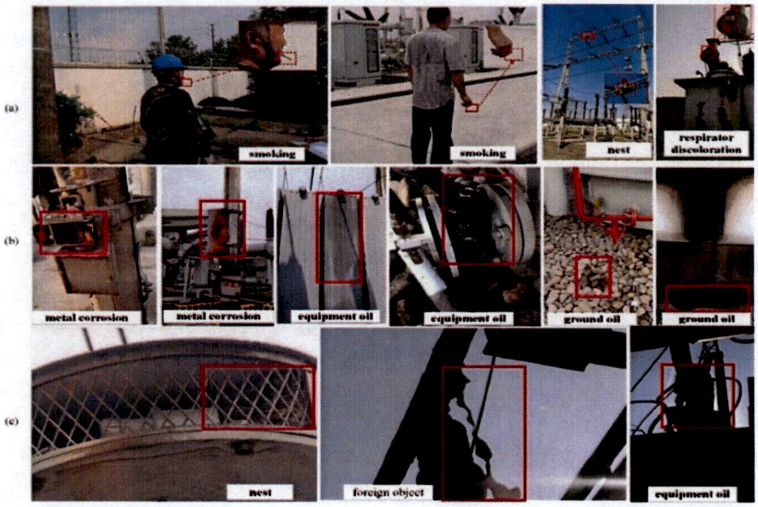

Fig. 1. The constraints of defect detection in substation equipment image. (a) the tiny defects (b) non-rigid defects (c) complex background

2 Materials and Methods

2.1 Image Acquisition and Preprocessing

All the images analyzed in this paper were collected in China. These images were captured by HD camera, inspection robot and mobile phone. In our dataset, there are 3000 images and 12 kinds of defects, which have various attributes: tiny defects, non-rigid, complex background. The dataset is randomly divided into training subset and validation subset according to ratio of 9:1. After image acquisition, the images are firstly annotated by power specialist, and then data cleaning would be used to guarantee the consistency of these data. In this paper, the data augmentation strategy is used including: mirroring, rotation, random cropping and random noise.

2.2 Method

Due to different scales, non-rigid and complex background of defects, the defect detection and recognition based on deep learning methods are not satisfactory. Thus, to improve the accuracy of defects, we proposed a multi-scale attention networks (MAN) for substation equipment image defect detection as shown in Fig. 2. Our multi-scale attention networks consist three different attention subnets. The features of defects contain global and local features. The global features are the representation of the whole object, which could be extracted via channel attention module. The local features contain much semantic information, which could be extracted via spatial attention module. Thus, in this paper, each sub-network is composed of some attention blocks, and an attention block contains a channel attention module and a spatial attention module. Finally, the output features are refined by the attention module.

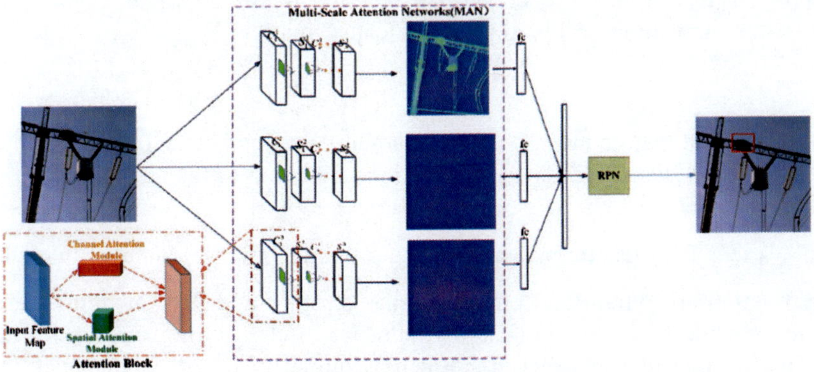

Fig. 2. Framework of defect detection based on multi-scale attention networks.

2.2.1 Attention Module

Channel Attention Module: Generally speaking, in general convolutional neural network, the feature maps output by the convolution layer are obtained by using different convolution kernels as filters, and the number of convolution kernels corresponds to the number of channels C_g. In order to obtain the activation vector at the channel level of the global feature map, a channel attention module is developed to obtain the activation vectors. The channel attention module should avoid the influence of spatial information as much as possible, so as to extract the channel activation vectors more effectively. The specific process is as follows:

Step 1: The feature map X_g is obtained through the convolution kernels.
Step 2: Global pooling is used to eliminate the influence of X_g on height and width. The global pooling F_{gp} can be defined:

$$z_c = F_{gp}(x_c) = \frac{1}{H_g \times W_g} \sum_{i=1}^{H_g} \sum_{j=1}^{W_g} x_c(i,j) \tag{1}$$

H_g and W_g are the height and width of X_g. x_c is one of the channel feature maps in X_g, z_c is the result of average pooling processing for x_c. The results of z_c of each channel feature map are connected together to form a 1- dimensional vector Z_C. Thus, the influence of spatial information is eliminated via the average pooling processing.

Step 3: Two consecutive full-connections are adopted to Z_C for further feature extraction and weight learning:

$$U_c = \delta_2\left(W_2^T \delta_1\left(W_1^T Z_C + b_1\right) + b_2\right) \tag{2}$$

W_1 and W_2 are the weight of the two fully connected layers, and b_1 and b_2 are the offsets. δ_1 is the activation function of ReLU. δ_2 is the activation function of Sigmoid.

Step 4: Complete the calibration of the feature map:

$$\tilde{x}_c = x_c \cdot u_c \tag{3}$$

the input global feature map X_g and the output activation vector U_c are weighted and averaged at the channel level to complete the calibration of the feature map. u_c is one of elements in U_c.

Thus, $\tilde{x}_c$ is the final feature map after passing through channel attention module.

Spatial Attention Module: Contrary to the channel attention module, the spatial attention module needs to eliminate the influence of channel information for learning the global feature of the defect in substation equipment image. The spatial attention module is trained through supervised learning by using the annotations of defect that are in the spatial position. In the standard convolutional neural network, the smaller receptive field of the convolution operation has always been the limitation that it is difficult to feel more context under small objects. Thus, a new type of mask-supervised learning method is proposed to learn the spatial activation matrix. The specific training process is as follows:

Step 1: The feature map X_g is obtained through the convolution kernels.

Step 2: Global convolution $Z_s(i,j)$ is adopted to eliminate the influence on channels of X_g. The global convolution can be defined:

$$Z_s(i,j) = F_{gc}\left(X_g\right) = \sum_{c=1}^{C_g} \sum_{m} \sum_{n} X_g(i+m, j+n, c) K(m,n,c) + b \tag{4}$$

H_g and W_g are the height and width of X_g, $i \in H_g, j \in W_g$. $F_{gc}\left(X_g\right)$ is the global convolution function, K is the convolution kernel of the global convolution operation, and the size of K is $m \times n \times C_g$. b is the offset.

Step 3: Dilated convolution feature map U_s is used to refine the spatial single-channel feature map:

$$U_s = \delta_2(K_2 * \delta_1(K_1 * Z_S + b_1) + b_2) \tag{5}$$

In this paper, two sets of dilated convolutions with different convolution kernel sizes are used to learn the activation matrix for gradually expanding the spatial receptive field K_1 and K_2 are the convolution kernels of two dilated convolutional layers, b_1 and b_2 are the offsets. δ_1 is the activation function of ReLU. δ_2 is the activation function of Sigmoid. $*$ is dilated convolution operation.

Step 4: Calculate the loss function $PCE(U_s, U_{gt})$:

$$PCE(U_s, U_{gt}) = \frac{1}{H_g \times W_g} \sum_{i=1}^{H_g} \sum_{j=1}^{W_g} -U_{gt}(i,j) log U_s(i,j) \tag{6}$$

U_{gt} is the activation map generated by ground truth.

Step 5: Enhance the spatial sensitivity of feature maps to defects:

$$\tilde{x}_s(i,j,c) = x_g(i,j,c) \cdot exp(u_s(i,j)) \tag{7}$$

In order to maintain the information of the original global feature map, we do not use pure multiplication by position, and exponential calculation on each channel is used to highlight the object information.

Attention Module: As shown in Fig. 3, feature map from each level is input into our proposed attention module to refine the features, which involves two branches for depth and spatial level activation. Thus, the final feature maps are the sum of the two branches:

$$\tilde{X}_g = \tilde{X}_c + \tilde{X}_s$$

Finally, the global and local feature maps are obtained via attention module.

2.2.2 Multi-scale Networks

In order to detect the defects at different scales, in our multi-scale networks, convolution filters of different sizes are used to extract feature maps. The larger filters are more effective for detecting the defects of larger size and the defects of smaller size could be easily detected by smaller filters. As Fig. 3 shown, in this paper, our multi-scale networks contain three parallel CNN, whose structures of columns are similar except for the sizes and numbers of filters. In order to reduce the dimensionality of the features output by the convolutional layer and prevent the CNN from learning over-fitting, max pooling and Rectified linear unit (ReLU) are adopted in this paper. To reduce the computational complexity of CNN, we use less filter. In this paper, softmax is used as the loss function:

$$L_{Softmax} = \sum_{i}^{N} \alpha_c L_{Softmax-cls} + \alpha_l L_{Softmax-loc} \tag{8}$$

N is the number of training image, α_c and α_l are the weights of classification layer and location layer. $L_{Softmax-cls}$ is the loss function of classification layer and $L_{Softmax-loc}$ is the loss function of location layer. $L_{Softmax-cls}$ can be calculated:

$$L_{Softmax-cls} = -\sum_{k=1}^{p} 1[y = 1] log P_k \tag{9}$$

$L_{Softmax-loc}$ can be calculated:

$$L_{Softmax-loc} = \sum_{i=1}^{4} smooth_{L1}(t^u, v) \tag{10}$$

2.3 Evaluation Metrics

In order to effectively evaluate the experimental results, we use Average Precision (AP) as the evaluation metric in this paper. AP is one of the often used evaluation metrics in the field of object detection. It can be calculated from the Precision-Recall (PR) curve:

$$AP(c) = \int PR(c) \tag{11}$$

in which c is the class name.

PR is calculated by:

$$Precision(c) = \frac{\#TP(c)}{\#TP(c) + \#FP(c)},\ Recall(c) = \frac{\#TP(c)}{\#TP(c) + \#FN(c)} \tag{12}$$

in which *TP* is the number of true positive, *FP* is the number of false positive and *FN* is the number of false negative.

And the mean Average Precision (mAP) could be calculated by:

$$mAP = \frac{1}{|C|} \sum_{c \in C} AP(c) \tag{13}$$

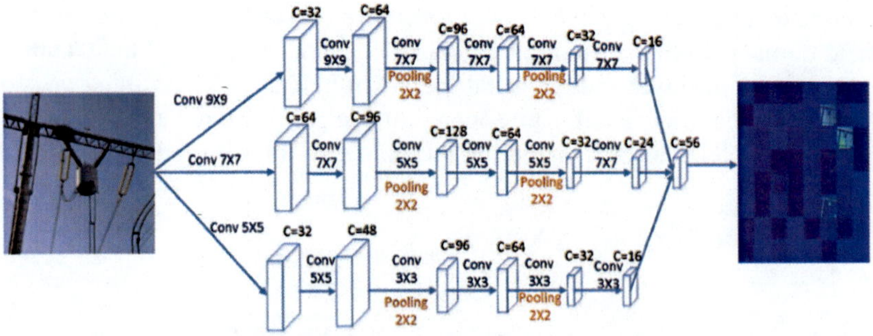

Fig. 3. Framework of multi-scale networks

3 Experiment

3.1 Experimental Settings

In order to verify that our multi-scale attention networks could be adopted for detecting defects in substations, some experimental settings are used to finish experiment. In our experiment, the deep learning framework of "mmdetection" with Python API 3.7 is applied in this paper and run on 12 GB GTX 1080Ti GPU (https://github.com/open-mmlab/mmdetection). In the training step, Batch formation is adopted for preventing the model from overfitting. The fine-tune strategy is used to make the model converge as soon as possible. And the learning rate is set to 0.0025, it would be divided by 10 when the 'epoch' is 8 and 11. In our experiment, the size of mini-batch is set to 2.

3.2 Experimental Results

In this paper, five state-of-the-art methods are experimented, the experimental results are shown in Table 1, and the best results have been bolded. Through Table 1, we firstly find that multi-scale attention networks (MAN) have the highest detection accuracy. YOLOv3 [17], RefineDet [18], M2Det [19] and Cascade-RCNN [20] all achieved the multi-scale object detection by getting features from pyramid networks. The experimental results show that among all of the state-of-the-art approaches, the best detection result is MAN, the mAP is 61.6%. Compared with other four methods, our method separately improved the mAP by 5.4%, 4.8%, 2.6% and 1.8%. Thus, the results show that our attention module and multi-scale CNN could increase the accuracy of defect detection in substations.

Furthermore, there are many tiny and non-rigid defects in our dataset. For instance, the size of "smoking" category is less than 5% in an image. Compared with other state-of-the-art algorithms, our proposed method could improve the AP by 4.9%. Except for tiny defects, there are also many non-rigid defects such as "metal corrosion", "ground oil" and "equipment oil". These defects vary greatly, and it is difficult to learn effective features through the CNN, thus, the AP of these classes is lower than others. However, thanks to the attention modules, the results show that our method can improve the Ap of these defects more than 2%. The effects of multi-scale attention networks (MAN) are experimented as Fig. 4 shown. The experiment show that the attention module could improve the detection rate of tiny and non-rigid defects.

Table 1. Result of state-of-the-art methods.

Class name/method	YOLOv3 (AP)	RefineDet (AP)	M2Det (AP)	Cascade RCNN (AP)	Ours
Without helmet	88.6%	87.9%	89.4%	89.7%	**90.9%**
Without work clothes	86.3%	87.8%	87.6%	87.1%	**88.7%**
Respirator discoloration	89.2%	89.6%	90.3%	90.8%	**91.9%**
Damaged oil seal	42.7%	43.9%	45.6%	46.3%	**47.5%**
Metal corrosion	28.8%	29.9%	33.4%	34.3%	**36.7%**
Meter abnormal	58.3%	59.1%	63.2%	64.7%	**67.6%**
Ground oil	35.9%	36.2%	38.6%	39.3%	**41.5%**
Foreign objects	70.4%	72.3%	74.6%	75.9%	**76.8%**
Smoking	24.0%	25.3%	29.6%	30.8%	**35.7%**
Nest	80.5%	81.6%	82.4%	83.1%	**84.9%**
Equipment oil	27.9%	26.8%	29.7%	30.2%	**33.5%**
Climbing	41.3%	40.6%	42.7%	43.9%	**44.8%**
mAP	56.2%	56.8%	59.0%	59.8%	**61.6%**

Fig. 4. The result of MAN

4 Conclusion

For improving the recognition accuracy of defects with tiny size and non-rigid, in this paper, we propose a multi-scale attention networks method. The experiments show that our method could achieve satisfactory results for detecting the defects in substation equipment images. And our method could be extended to other defects detection in power fields.

References

1. Yu, H., Wang, L., Wang, K.: Application of image processing technology and neural network in power equipment identification. China Comput. Commun. **2012**(10), 122–123 (2012)
2. Zheng, T., Tian, Q., Wu, J., Zhao, D., Gao, Y.: Application of artificial intelligence algorithm in unattended substation. Digit. Technol. Appl. **2019**(37), 8–12 (2019)
3. Zheng, C., Wang, S., Zhang, Y., et al.: A robust and automatic recognition system of analog instruments in power system by using computer vision. Measurement **2016**(92), 413–420 (2016)
4. Baker, L., Mills, S., Langlotz, T., Rathbone, C.: Power line detection using hough transform and line tracing techniques. In: 2016 International Conference on Image and Vision Computing, New Zealand (IVCNZ), pp. 1–6 (2016)
5. Wang, H., Meng, F.: Research on power equipment recognition method based on image processing. EURASIP J. Image Video Process. **2019**(57), 1–11 (2019)
6. Xie, X., Pang, C.: Power equipment image recognition based on deep learning. Int. J. Sci. **6** (9), 22–27 (2019)
7. Tao, X., Zhang, D., Wang, Z., Liu, X., Zhang, H., Xu, D.: Detection of power line insulator defects using aerial images analyzed with convolutional neural networks. IEEE Trans. Syst. Man Cybern. Syst. 1–713 (2018)
8. Zahid, A.S., Unsang, P., Sang, W.L., Nam, J.J., Minhee, C., Chanuk, L., Jang, H.S.: Robust powerline equipment inspection system based on a convolutional neural network. Sensors **18** (3837), 1–25 (2018)
9. Li, J., Wang, Q., Li, M.: Electric equipment image recognition based on deep learning and random forest. High Volt. Eng. **43**, 3705–3711 (2017)
10. Zhang, W., Liu, X., Yuan, J.: RCNN-based foreign object detection for securing power transmission lines. Procedia Comput. Sci. **147**, 331–337 (2019)
11. Filho, E., Prates, R.M., Marotta, A.P.: Insulator visual non-conformity detection in overhead power distribution lines using deep learning. Comput. Electr. Eng. **78**, 343–355 (2019)
12. Liang, H., Zuo, C., Wei, W.: Detection and evaluation method of transmission line defects based on deep learning **2020**(8), 38448–38458 (2020)
13. Bushra, J., Giuseppe, R.L., Massimo, M., Davide, M., Maria, A.P., Andrea, B.: Fault detection in power equipment via an unmanned aerial system using multi modal data. Sensors **19**(1034), 1–15 (2019)
14. Zhao, Z., Xu, G., Qi, Y., Liu, N., Zhang, T.: Multi-patch deep features for power line insulator status classification from aerial images. In: Proceedings of the 2016 International Joint Conference on Neural Networks, IJCNN 2016, Vancouver, BC, Canada, 24–29 July 2016, pp. 3187–3194 (2016)
15. Zhai, Y., Chen, R., Yang, Q., Li, X., Zhao, Z.: Insulator fault detection based on spatial morphological features of aerial images. IEEE Access **2018**(6), 35316–35326 (2018)
16. Zhao, Z., Zhen, Z., Zhang, L., Qi, Y., Kong, Y., Zhang, K.: Insulator detection method in inspection image based on improved faster R-CNN. Energies **12**(7), 1–15 (2019)
17. Redmon, J., Farhadi, A.: Yolov3: an incremental improvement (2018). https://arxiv.org/abs/1804.02767

18. Zhang, S., Wen, L., Bian, X.: Single-shot refinement neural network for object detection. In: 2017 IEEE Conference on Computer Vision and Pattern Recognition (CVPR) (2017)
19. Zhao, Q., Sheng, T., Wang, Y., Tang, Z., Chen, Y., Cai, L., Ling, H.: M2Det: a single-shot object detector based on multi-level feature pyramid network. In: 2018 IEEE Conference on Computer Vision and Pattern Recognition (CVPR) (2018)
20. Cai, Z., Vasconcelos, N.: Cascade R-CNN: delving into high quality object detection. In: 2018 IEEE Conference on Computer Vision and Pattern Recognition (CVPR) (2018)
21. Joseph, R., Ali, F.: YOLO9000: better, faster, stronger. In: 2016. IEEE Conference on Computer Vision and Pattern Recognition (CVPR) (2016)
22. Li, X., Zhao, B., Lu, X.: MAM-RNN: multi-level attention model based RNN for video captioning. In: 2017 International Joint Conferences on Artificial Intelligence (IJCAI), pp. 2208–2214 (2017)
23. Redmon, J., Divvala, S., Girshick, R., Farhadi, A.: You only look once: real-time object detection. In: 2016 IEEE Conference on Computer Vision and Pattern Recognition (CVPR), pp. 779–788 (2016)

Study on Propagation Process and Signal Acquisition of Traveling Wave for HVDC Transmission Line

Yulin Chen[1,2(✉)], Yucan Zhao[2], Guangquan Xian[2], Jianfeng Zhang[2], and Jie Zhang[2]

[1] State Key Laboratory of Smart Grid Protection and Control, Nanjing 211106, Jiangsu, China
chenyl@nrec.com

[2] Nanjing NR Electric Co. Ltd., Nanjing 211106, Jiangsu, China

Abstract. Traveling wave fault location is very important for high voltage direct current (HVDC) transmission lines. The attenuation and distortion of traveling wave in HVDC transmission lines and secondary circuits will have non-ignorable negative impact on traveling wave fault location. In this paper, the attenuation and dispersion characteristics of traveling wave at different fault distance are simulated and analyzed. The further attenuation and distortion of traveling wave in the secondary circuit are studied by physical model. The problems of traveling wave reflection caused by the secondary cable and signal interference caused by strong electromagnetic interference are analyzed. Through the analysis of the whole process of traveling wave propagation and acquisition, the necessity of on-site acquisition of traveling wave signal for HVDC transmission line is proved. In this paper, the system scheme and key technology of on-site acquisition of traveling wave signal for HVDC line are given, and the technical feasibility of this scheme is stated.

Keywords: HVDC transmission line · Secondary circuit · Traveling wave acquisition · Traveling wave fault location

1 Introduction

High voltage direct current (HVDC) transmission lines play an important role in energy transmission for the power system. HVDC transmission lines span a wide area, the length of which can reach 2000–3000 km. The geographical conditions along the lines are complex, and line faults are easy to occur. The traveling wave fault location (TWFL) of HVDC line is of great significance to locate the fault point quickly and shorten the outage time [1]. Compared with the AC transmission line, HVDC transmission lines have two important differences on TWFL issue. Firstly, the length of HVDC line is much larger than AC line, so the attenuation and dispersion of traveling wave is more serious. Secondly, electromagnetic CT can't be used on HVDC line to obtain the current traveling wave directly. In most actual projects, a small current transformer (CT) is equipped on the grounding circuit of a capacitive equipment to obtain the current traveling wave, which in fact is the differential of line voltage

Y. Xue et al. (Eds.): PMF 2020, LNEE 718, pp. 221–231, 2021.
https://doi.org/10.1007/978-981-15-9746-6_18

traveling wave. Furthermore, according to the traditional method, the sensed current also needs to propagate along a complex second circuit before it is sampled by high speed A/D module. The second circuit includes a transmission cable, and another small CT at in the fault location device. Due to the long-distance cable transmission of small signal, the traveling wave will undergo more attenuation and dispersion, and can be easily interfered by strong electromagnetic noise, such as switch operation and valve switching in the HVDC station [2, 3], which have significant negative impact on TWFL.

For the above problem, main factors that cause traveling wave dispersion on HVDC lines and their influence on traveling wave protection are studied in [4], but the influence from secondary circuit is not considered. Considering the whole process of traveling wave transmission, the influence of secondary cable on traveling wave oscillation frequency is studied with lumped parameter model in [4], the influence of the second cable is simulated with ATP cable model for the AC transmission line in [5], and the simulation method and test technology including small CT and secondary cable for HVDC line are proposed in [6]. In [7], the waveform distortions at different fault distance of 500 m, 1 km and 2 km are compared. About the problem of wave velocity variation and waveform dispersion and their influence on TWFL, [8–10] use different methods to define the traveling wave arriving time and match the corresponding traveling wave velocity, so as to improve the accuracy of TWFL.

No matter what kind of TWFL is adopted, it is necessary to obtain high-quality traveling wave data first. In this paper, the whole process of traveling wave transmission of HVDC transmission line is considered. The transmission characteristics of traveling wave on HVDC line and coupling capacitor are analyzed by PSCAD simulation. A physical verification of the distortion of traveling wave in the secondary circuit is carried out by building a secondary circuit test system with coaxial cable. The research shows that the problem of attenuation and dispersion of traveling wave in long-distance HVDC transmission lines is prominent, and the distortion and interference introduced by the secondary circuit have a significant negative impact on TWFL, so it is reasonably necessary to sample the traveling wave signal of DC transmission lines in situ to get high-quality traveling wave data. This paper presents a system of on-site acquisition and its key technologies, and states the technical feasibility of this system.

2 Distortion Simulation of Traveling Wave on HVDC Transmission Line

2.1 Traditional Acquisition Method of Traveling Wave for HVDC Line

For HVDC transmission line, electromagnetic PT or CT can't be used to obtain signal. For most HVDC projects, the traveling wave is obtained indirectly by sensing the surge current of the shunt capacitive equipment, as shown in the figure below.

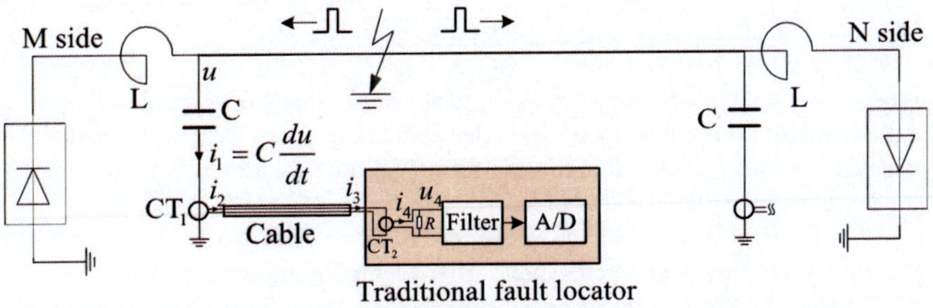

Fig. 1. Schematic diagram of traveling wave acquisition for HVDC line.

In the figure above, L is the smoothing inductor, with inductance of 20–200 mH. C is a capacitive device, which can be a PLC noise filter or a capacitor specially equipped for TWFL, with capacity of 10–100 nF in general. u is the voltage traveling wave at the terminal of the HVEC line caused by line fault, and i_1 is the current traveling wave flowing through the capacitor C.

The relation of voltage traveling wave u and current traveling wave i_1 is shown in the following equation.

$$i_1 = C\frac{du}{dt} \tag{1}$$

The equation means that voltage traveling wave can be obtained indirectly through measuring the surge current of the capacitor. With traditional wave acquisition method, a small current transformer, labeled as CT_1 in the figure, is used to sense the current wave flowing through capacitor C, and its output current signal i_2 is transmitted to a secondary cable. At the end of the cable, signal i_2 is changed to i_3, due to the non-ideal transmission characteristics of the cable and the wave refraction caused by wave impedance discontinuity at the cable terminal. Inside the fault location device, a smaller CT, labeled as CT_2 in the figure above, transfers i_3 to i_4. Then, the traveling wave signal i_4 is transferred to voltage u_4 by a resistor, then filtered by a band-pass filter, and sampled by an A/D module.

It can be seen that the traditional way of obtaining the traveling wave signal of HVDC line has many transmission process. During the whole process, there will be signal attenuation and distortion at each transmission section. Their influences on traveling wave signal acquisition and TWFL is analyzed in detail as follows.

2.2 Attenuation and Dispersion Mechanism of Traveling Wave

If the wave impedance discontinuity factors, such as line tower, line sag and so on, is ignored, HVDC transmission line can be modelled as a lossy uniform transmission line. The propagation characteristics of lossy uniform transmission lines can be expressed by transmission constant parameter γ, shown as follows.

$$\gamma = \alpha + j\beta = \sqrt{(R_0 + j\omega L_0)(G_0 + j\omega C_0)} \quad (2)$$

Where, α is attenuation constant, β is phase-shift constant, R_0 is distributed resistance, L_0 is distributed inductance, G_0 is distributed conductivity, and C_0 is distributed capacitance. R_0 and L_0 will change with frequency due to skin effect, so α and β are also both frequency-related parameters.

Due to the skin effect, the traveling wave has higher wave speed and greater attenuation with higher frequency. Therefore, when traveling wave propagates a certain distance, the waveform will decay and distort. If the fault is far from the signal detector, the attenuation and dispersion of traveling wave on the transmission line will be very prominent. In such situation, the amplitude of traveling wave will be greatly reduced, and the wave-front will become quite gentle, which will significantly increase the difficulty of TWFL.

The voltage change at the fault point can be regarded as an ideal step signal, but at the line terminal the signal will change to a gradual step signal due to the attenuation and dispersion effect of traveling wave. Furthermore, the traveling wave reflection will occur at the terminals of the HVDC line with a frequency-related reflection coefficient due to inductor L and capacitor C. The reflection will cause further distortion to the traveling wave. The distorted voltage traveling wave signal is differentiated by capacitor C, and the current wave will show as a narrow pulse waveform.

2.3 Simulation of Traveling Wave on HVDC Transmission Line

From the above analysis, it can be seen that the transmission process of traveling wave is very complex, especially considering the influence of inductor L and capacitor C at the HVDC transmission terminals. In order to quantitatively analyze the effect of traveling wave distortion induced by transmission line, inductor L and capacitor C, PSCAD modeling and simulation are carried out for a ±800 kV HVDC transmission system shown in Fig. 1. The simulation step is 0.1 μs, and the maximum length of the HVDC line is 3000 km. The reactance of L is 20 mH, the capacitance of C is 10 nF. The physical parameters of the transmission line are cited from [11], and is shown in the table below (Table 1).

Table 1. Physical parameters of HVDC transmission line.

Object	Item	Value	Unit
Pole wire	Sub-conductor number	6	/
	Sub-conductor spacing	0.45	m
	Sub-conductor radius	33.6	mm
	DC resistance	0.04633	Ω/km
	Conductor spacing	22	m
	Conductor height	33.5	m
	Conductor sag	16	m

(continued)

Table 1. *(continued)*

Object	Item	Value	Unit
Ground wire	Sub-conductor number	1	/
	Sub-conductor radius	17.5	mm
	DC resistance	0.7098	Ω/km
	Conductor spacing	18.2	m
	Conductor height	50	m
	Conductor sag	11	m
Ground	Ground resistivity	1000	Ω·m

Faults are applied at different positions of the HVDC line, ranging from 2 km to 3000 km, and the traveling wave current flowing through capacitor C is measured. In order to facilitate comparative analysis, the measured current traveling wave under different fault distance are aligned according to their actual arriving time, shown as follows.

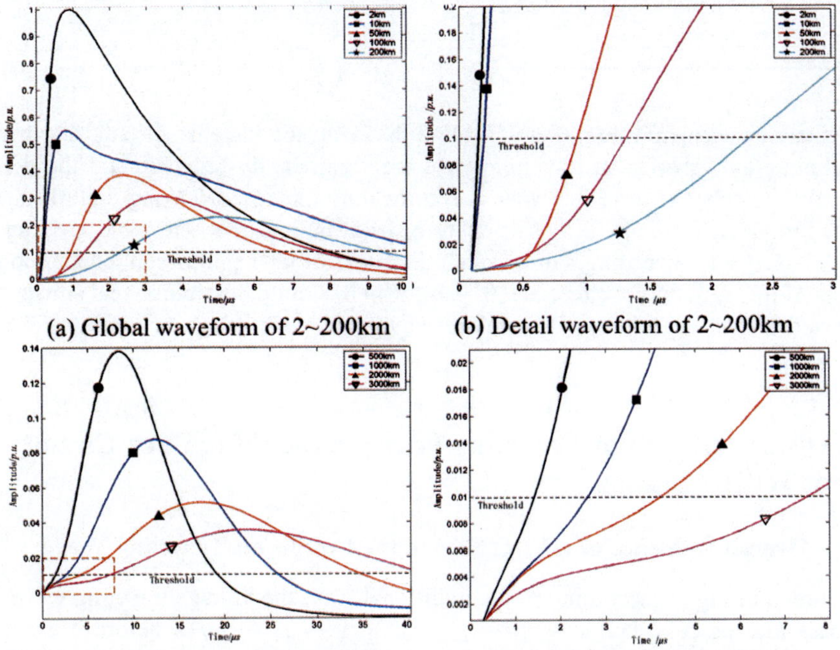

(a) Global waveform of 2~200km (b) Detail waveform of 2~200km

(c) Global waveform of 500~3000km (d) Detail waveform of 500~3000km

Fig. 2. Traveling waves with different fault distance.

It can be seen from Fig. 2(a) and (c) that with the increase of fault distance, the amplitude of traveling wave current in capacitor C decreases accordingly, and the surge width increases gradually. It can be seen from Fig. 2(b) and (d) that the detectable arriving time of traveling wave also increases significantly with the fault distance, supposing a certain threshold is used to detect the arrival time of traveling wave. The delay of wave peak and detectable arriving time under different fault distance are shown as follows.

Table 2. Traveling wave amplitude and delay of different fault distance.

Fault distance (km)	Peak value (p.u.)	Peak delay (μs)	Detect threshold (p.u.)	Arriving delay (μs)
2	1.000	0.9	0.1	0.20
10	0.550	0.8	0.1	0.25
50	0.381	2.3	0.1	0.95
100	0.302	3.3	0.1	1.33
200	0.226	5.0	0.1	2.30
500	0.138	8.4	0.01	1.56
1000	0.088	12.4	0.01	2.62
2000	0.0512	17.5	0.01	4.29
3000	0.0353	22.4	0.01	7.58

It can be seen from Fig. 2 and Table 2 that with the increase of fault distance, the amplitude and gradient of traveling wave are significantly reduced, and the detected arrival time delay of traveling wave is significantly increased. Taking fault distance of 2000 km as an example, the peak delay is 17.5 μs, and the detectable arriving time delay is 4.29 μs, supposing a quite small threshold of 0.01 p.u. is applied and no noise exists. If the signal noise exceeds 0.01 p.u., which is quite possible in real situation, the detected arrival time delay of traveling wave can reach 10 μs or more, referring to Fig. 2(c).

3 Distortion Test of Traveling Wave in the Secondary Circuit of HVDC Station

3.1 Overall Influence of HVDC Secondary Circuit on Traveling Wave

As shown in Fig. 1, according to the traditional way, the voltage traveling wave u has to pass four process, i.e., $u \rightarrow i_1 \rightarrow i_2 \rightarrow i_3 \rightarrow i_4 \rightarrow u_4 \rightarrow \mathrm{A/D}$, before it is sampled by A/D module. In every link, because of the non-ideal characteristics of the transmission equipment or transmission cable, the traveling wave waveform will be attenuated and distorted.

The length of transmission cable, related to signal i_2 and i_3, can be as long as 1000 m or even more [6]. Due to the impedance mismatch at the two ends of the cable, the reflection interference of the traveling wave will be caused in the process. In addition, because the cable laying environment is usually very complex, the weak traveling wave is easy to be disturbed by the strong electromagnetic noise around the transmission cable. The longer the transmission cable path is, the more serious the problem of attenuation, dispersion and interference will be.

3.2 Influence Test of HVDC Secondary Circuit on Traveling Wave

The transmission cable of the secondary circuit is usually KVVP2-22 control cable (copper core, PVC insulated, and sheathed steel strip armored) or SYWV-75-9 coaxial cable (copper core, polyethylene physical foaming insulation). SYWV-75-9 coaxial cable has much better high frequency characteristics than that of KVVP2-22 control cable, so it is more suitable for transmission of traveling wave. The characteristic parameters of SYWV-75-9 coaxial cable include: its wave impedance is 75 Ω, its wave speed is 250 m/μs, and the attenuation of 5 MHz signal per 100 m is less than 1.2 dB.

In order to study the traveling wave transmission characteristics of the whole secondary circuit of HVDC station, the test system shown in the following figure is built with a SYWV -75-9 coaxial cable of 350 m (Fig. 3).

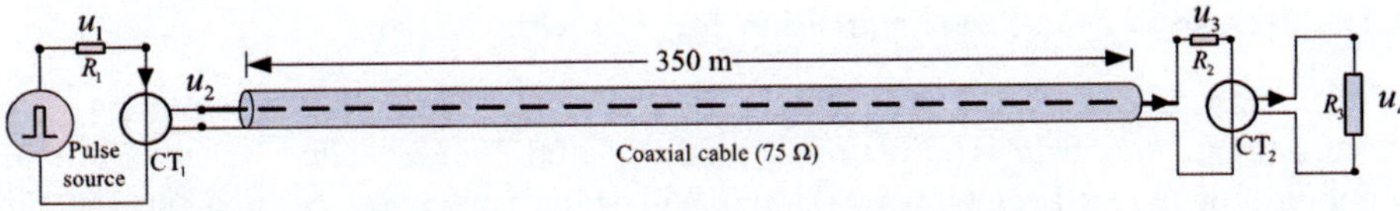

Fig. 3. Test system diagram of secondary circuit in HVDC station.

In the figure, the characteristics of current sensors CT_1 and CT_2 are similar to those used in field projects. The traveling wave pulse width is 1 μs. The signal output resistance of the pulse source is 75 Ω. The transformation ratio of CT_1 is 6:200, and the transformation ratio of CT_2 is 1:200. The sampling resistor R_2 is 2 Ω, and the load resistance at the output side of CT_2 is 150 Ω.

A four-channel oscilloscope is used to measure the voltage signals at different position, and the measured waveforms are shown as follows (Fig. 4).

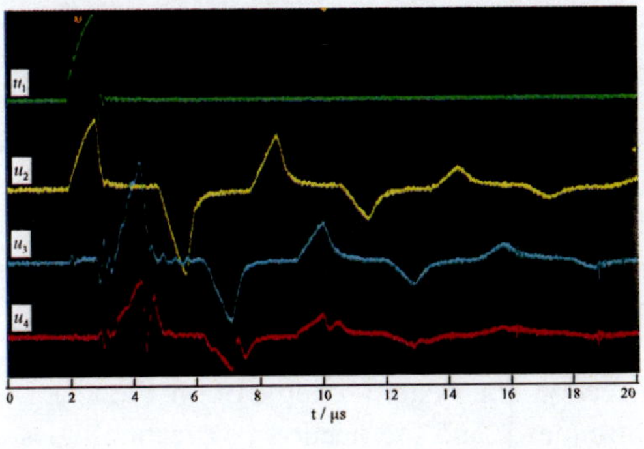

Fig. 4. Wave forms of traveling wave in secondary circuit test system.

In the figure above, the amplitude of each channel is properly scaled so as to achieve the best display effect. From the above figure, the following test results can be obtained:

(1) Due to the discontinuous wave impedance at both ends of the cable, the traveling wave produces multiple reflections on the cable.
(2) The reflected wave on the cable will not be fed back to the input side of transformer CT_1 due to the transfer characteristic of CT_1.
(3) For the traveling wave on the cable, CT_1 shows the characteristics of high inductance (positive reflection for the voltage traveling wave), and CT_2 shows the characteristics of short circuit (negative reflection for the voltage traveling wave).
(4) The more conversion and transmission process the secondary circuit has, the lower amplitude and gradient the traveling wave will have, and the more difficult it will be to identify the wave front accurately and reliably.

4 On-Site Acquisition Method of Traveling Wave Signal

4.1 Necessity of On-Site Acquisition for Traveling Wave

HVDC transmission lines are usually much longer than AC transmission lines. According to the results of previous studies, after long-distance transmission, the amplitude of fault traveling wave signal will significantly reduce, and the traveling wave front will also significantly slow down. Furthermore, the conversion and transmission process of the secondary circuit of HVDC station will lead to further attenuation and distortion of traveling wave signal, and cause complex line wave refraction and reflection behavior due to the discontinuity of wave impedance. The normal operation of converter valve and switching operation in HVDC converter station will produce strong electromagnetic interference, and the traveling wave can be easily interfered by these strong electromagnetic noises during the long-distance cable transmission process.

With weak traveling wave amplitude and strong electrical noise, it will not only make the triggering threshold difficult to set (the threshold is either too low or too high), but also seriously affect the accurate calibration of traveling wave arriving time. Therefore, high-precision acquisition of traveling wave is the premise of reliable and accurate TWFL for HVDC line.

According to the above analysis, if the transmission process of secondary circuit can be reduced, then the acquisition accuracy and signal-noise-ratio (SNR) of traveling wave signal can be improved. In particular, for the system shown in Fig. 1, on-site sampling of traveling wave i_2 and digital transmission of the data can be a good solution. By canceling the following transmission cables and other process, the distortion and attenuation of traveling wave caused by secondary circuit can be avoided, the electromagnetic interference and wave refraction problem caused by secondary circuit can be avoided.

In the following section, the system scheme of on-site acquisition of traveling wave for HVDC line is introduced, and the needed key technologies are discussed.

4.2 System Scheme of On-Site Acquisition of Traveling Wave

In this paper, the following system scheme is recommended for on-site acquisition of traveling wave for HVDC transmission lines.

(1) A small current transformer CT_1 is equipped on the grounding wire of the capacitive equipment C.
(2) A traveling wave acquisition and transmission module is configured locally, which is synchronized with PPS or IRIG-B synchronization signal.
(3) The module synchronously samples the output signal of CT_1, and the sampling rate is not less than 1 MHz.
(4) The module uses optical fiber to continuously output traveling wave data based on 100 Mbps or Gigabit Ethernet.
(5) A locator host is installed in the HVDC control room to receive the traveling wave data sent by the acquisition module.
(6) Based on the received traveling wave data, the host performs start-up calculation, data storage and TWFL.

In this scheme, the function of the on-site traveling wave acquisition module is relatively fixed and simple, so it can be designed as simple and reliable as possible, which is convenient for its operation and daily maintenance.

It can be seen that the acquisition module depends only on power supply and optical-fiber time source, which can be offered in field projects. The Ethernet output link and the time source can be realized by different optical fiber cores in one fiber cable. So the system scheme of on-site acquisition has enough practical feasibility for HVDC projects.

4.3 Key Technologies of On-Site Acquisition of Traveling Wave Signal

High Speed Synchronous Acquisition Technology
Based on accurate GPS time source, highly stable crystal oscillator and time-keeping technology, the synchronization accuracy of traveling wave acquisition module can reach 100 ns, and the corresponding location error is only 30 m.

Furthermore, based on the high-precision internal clock, the acquisition module can control the A/D sampling sequence, so that the number of sampling points per second is strictly same with the sampling rate setting, and the sampling time of the first point is accurately aligned with the PPS time. With these technologies, it is not only convenient to align the traveling wave data from different acquisition modules in a station, but also convenient to align the data between stations.

High Speed Data Transmission Technology
The sampling rate required for TWFL is generally not less than 1 MHz. In order to achieve a higher location resolution and to smooth the white noise, a higher sampling rate can be used, such as 2 MHz, 5 MHz or 10 MHz. On the other hand, in order to achieve redundant sampling and realize different range sampling, the acquisition module can use two channels to sample the traveling wave of one line.

Supposing each sampling data is 16 bits, if one channel of traveling wave is sampled at the rate of 1 MHz, then the data flow bandwidth is about 16 Mbps. If double channels are sampled at 20 MHz, then the data flow is about 320 Mbps. To realize continuous transmission of traveling wave data, 100 Mbps or Gigabit Ethernet can be used according to the actual data flow bandwidth.

To transmit traveling wave data through Ethernet, it is necessary to gather the traveling wave data in a period of time, such as 100 μs, and arrange them into one Ethernet report, so the Ethernet frame rate is in a reasonable range, like within 100 k frames per second. Each frame of Ethernet message has a time tag to indicate the first sampled data in the frame.

High Speed Data Processing Technology

The locator host receives the traveling wave data based on Ethernet, and realizes data buffer, triggering calculation, data storage and TWFL. Because of the huge amount of traveling wave data, the host needs to use Field Programmable Gate Array (FPGA) to decode the message frame, cache the data in cycle, and carry out triggering calculation. Generally, the depth of data cache should be more than 4 ms, which should be longer then the time delay of triggering algorithm.

If a trigger acts, FPGA will carry the cached data to another cyclic memory, which can save tens or hundreds of triggered data blocks, so the host has enough time to analyze them one by one in case of dense wave triggering scenarios. The traveling wave ranging host simultaneously carries out a slower but more reliable triggering algorithm based on larger time window, so as to check the primary triggering results and reduce false triggering. Based on this two-level buffer technology and triggering check technology, the locator device can both reduce the dead time of wave recording, and enhance the sensitivity and reliability of traveling wave recording.

Test Prototype Development

The technical feasibility of this system has been verified by a primary test prototype. This acquisition unit is designed as a compact device, suitable to be equipped in the HDVC yard. It can sample two channels of traveling wave at 5 MHz, 14 bit. The traveling data can be reliably transmitted by the unit and received by the locator host through Ethernet, with a frame rate of 50 kHz. The acquisition unit can work in harsh outdoor environment from −40 °C to +70 °C.

5 Conclusion

The traveling wave attenuation and distortion on HVDC line and its secondary circuit will have significant negative influence on the traditional traveling wave sampling and transmission method. This paper aims to study the traveling wave propagation process of HVDC line and offer a more accurate and reliable signal acquisition and transmission method for HVDC line. PSCAD simulation of a HVDC transmission line is carried out with the fault distance ranging from 2 km to 3000 km, and it shows that after long-distance transmission, the fault traveling wave of the HVDC line will have significant attenuation and distortion, the traveling wave front will become gentle, the peak time and the detectable arriving time of the traveling wave will lag significantly.

It is shown that the secondary circuit of HVDC station will cause further attenuation and distortion on the traveling wave. Furthermore, the wave reflection will be caused by wave impedance discontinuity at the ends of transmission cable. Long distance secondary cable transmission can also cause the pollution of the weak traveling wave by strong electrical noise, which will cause enormous trouble on traveling wave triggering and fault location. This paper takes into account the whole propagation process of traveling wave in HVDC system, and proves it is necessary to sample the traveling wave of HVDC transmission line locally, and transmit the digital data through optical fiber. The system scheme and key technologies is introduced in this paper, and the technical feasibility of this scheme is proved by a test prototype.

Acknowledgement. This work was supported by State Grid Electric Power Research Institute Co., Ltd with Contract No. JS2000448.

References

1. Yang, L., Wang, B., Dong, X.: Overview of fault location methods in high voltage direct current transmission lines. Autom. Electr. Power Syst. **42**(08), 185–191 (2018)
2. Wu, H., Li, H., Liu, J.: Research on the conducted voltage interference in the secondary circuit of an UHVDC converter station. Electr. Measur. Instrumen. **55**(20), 142–147 (2018)
3. Luo, H., Le, J., Mao, T., et al.: Conducted electromagnetic interference of secondary equipment circuit in ±800 kV UHVDC converter station. Electr. Power Autom. Equip. **38** (01), 173–178 (2018)
4. Fei, X.U., Xinzhou, D.O.N.G.: Travelling wave transfer characteristics of a substation secondary circuit. Tsinghua Univ. (Sci. Technol.) **55**(02), 251–256 (2015)
5. Wang, S., Dong, X., Shi, S., Liu, C.: Study on current travelling wave transmission characteristics in secondary control cable connected with CT. Phys. Procedia, **33** (2012)
6. Zhen, W., Chen, W., Chen, P., et al.: Simulation method and test technology of transmission line traveling wave. Electr. Power Autom. Equip. **31**(06), 74–79 (2011)
7. Han, K., Cai, Z., He, Z., et al.: Propagation characteristic of fault traveling wave on HVDC line and its influence on HVDC line traveling wave protection. Power Syst. Protect. Control **41**(21), 20–25 (2013)
8. Zhang, Y., Xu, M., Liu, Y., et al.: A novel travelling wave fault location algorithm for HVDC transmission lines considering variation characteristics of wave speed. South. Power Syst. Technol. **35**(07), 227–232 (2011)
9. Chen, S., Zhang, J., Bi, G., et al.: A fault location method based on high frequency attenuation characteristic of UHVDC transmission line. Power Syst. Protect. Control **42**(10), 77–83 (2014)
10. Xu, M., Cai, Z., Liu, Y., et al.: A novel fault location method for HVDC transmission line based on the broadband travelling wave information. Trans. China Electrotech. Soc. **28**(01), 259–265 (2013)
11. Quan, Y., Yang, M., Wang, X., et al.: Accuracy modeling and simulation of Yunan-Guangdong ± 800 kV UHVDC transmission system. J. Kunming Univ. Sci. Technol. (Nat. Sci. Ed.), **37**(02), 43–48, 60 (2012)

Economic Analysis of Thermal Storage Boiler Cluster Participating in Renewable Energy Consumption

Ling Dong[1], Zheng Zhang[2](✉), Jian Li[1], Dongliang Xie[2], and Yanhe Li[1]

[1] State Grid Qinghai Electric Power Company, Xining 810008, China
[2] NARI Technology Development Co., Ltd., Nanjing 211106, China
zzhengnjnu@163.com

Abstract. In view of the limited peak shaving reserve capacity of thermal power units during the winter heating period in the Three North Areas of China, the serious imbalance between power generation and consumption leads to the problem of wind curtailment. This paper proposes an optimized control method for wind power consumption based on thermal storage boilers. In order to solve the problem of on-site wind power consumption, the government in the northern region of China is promoting the application of thermal storage boilers heating to increase the power-to-heat load of the power grid. In this paper, we study the optimal strategy of thermal storage boilers in the indirect power supply mode of wind farms. With the goal of minimizing the total coal cost and wind curtailment risk, the complex problem with continuous and discrete variables is transformed into a mixed integer linear optimization problem. The analysis of the calculation example shows that the configuration of the thermal storage boilers can decouple the constraint of ordering power by heat, making the interaction between electricity and heat more effective.

Keywords: Thermal storage boilers · Wind power consumption · Electricity-heat interaction

1 Introduction

In recent years, the installed capacity of wind power in China has increased year by year, and the problem of wind curtailment has also become more serious [1–3]. According to data released by the National Energy Administration, in 2019, the country's wind curtailment was 16.9 billion kWh, and Xinjiang, Gansu and Inner Mongolia accounted for 81% of the country's wind curtailment [4]. Wind power consumption has become an urgent problem to be solved for the sustainable development of the renewable energy power industry. Among them, the problem of huge wind curtailment during the heating period in the Three North Areas (referring to the northeast, north, northwest) is particularly prominent.

Wind power generation can replace conventional thermal power to some extent. This approach brings economic and environmental benefits to the system, but it also brings challenges to the system's peak shaving reserve capability [5]. The power and

Y. Xue et al. (Eds.): PMF 2020, LNEE 718, pp. 232–243, 2021.
https://doi.org/10.1007/978-981-15-9746-6_19

heat supply of the Three North Areas during winter heating period is mutually restricted, and ordering power by heat further limits the ability of thermal power units to participate in deep peak shaving and reserve [6]. Moreover, the electric load at night is low while the heat load and wind power output are high, which aggravates seasonal wind curtailment. If the electric boiler is turned on during the wind curtailment period, it can decouple the constraint of ordering power by heat to a certain degree.

In order to solve the problem of wind power consumption in China, the government has introduced a series of measures to promote the heating of wind power [7, 8]. Ref. [9] systematically summarized the methods for alleviating the contradiction between wind power and heating units on the grid, and proposed the use of heat storage to improve the space-time optimization capability of integrated energy systems. In ref. [10], the thermal power plant is equipped with electric boilers to meet the heating demand while absorbing excess wind power. Ref. [11, 12] study the operation strategy of thermal power plants equipped with heat storage tanks to improve the capacity of wind power consumption during heating period. In ref. [13], heat storage is incorporated into the active power dispatching system of the power system that includes wind power. This method improves the capacity of wind power consumption by controlling the heat storage link. Ref. [14] compares the deployment position of boilers in the heat network, and the calculation example shows that the benefit is better on the demand side. Ref. [15] constructs an electric heating system scheduling model and compares the impact of different electric boiler start-stop strategies on economic benefits. However, few researchers consider the economic impact of configuring thermal storage boiler on an electric heating system from the perspective of power and thermal balance. Therefore, in this paper, we take the minimum of the sum of coal consumption and wind curtailment as the objective function to maximize the economic benefit of power-heat decoupling.

The thermal storage boiler is easy to install in the majority of cities, urban-rural junctions and rural areas, especially some areas which aren't covered by the central heating pipe network. This paper focuses on the economic impact of the thermal storage boiler heating of power system. Firstly, it discusses the mechanism of thermal storage boilers participating in the electro-thermal interaction, and studies its heat storage control strategy and builds a mathematical model. Secondly, it takes the minimum of coal consumption and wind curtailment as the objective function within a dispatching period. Finally, through a simulation example, we analyze and verify the impact of the configuration of thermal storage boiler on the economy and flexibility of the system.

2 Modeling of Thermal Storage Boilers in Power Grid Dispatch

2.1 Formalization of Thermal Storage Boilers Participating in Electricity-Heat Interaction

At present, most wind farms are planned to construct in remote suburbs. If thermal storage boilers are near the wind farm, they won't be easy to heat for users. Moreover, the investment of direct power supply from the wind farm to the boilers in the new line is relatively huge. Therefore, the indirect power supply mode is generally adopted at present, that is, all wind power is transmitted to the power grid, and the boiler cluster can receive power from the grid for heating. The indirect power supply mode is shown in Fig. 1.

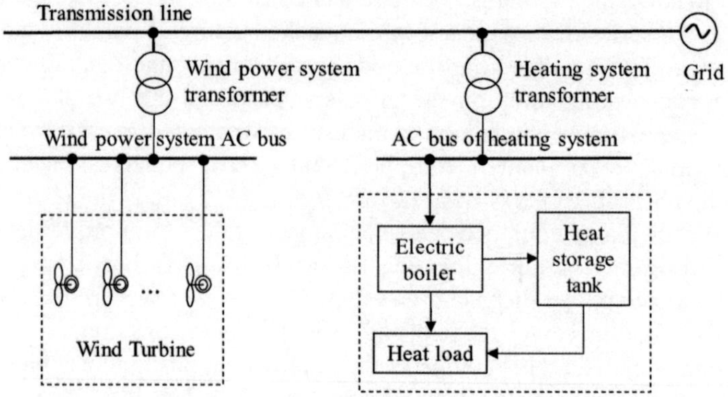

Fig. 1. Indirect power supply mode of thermal storage boilers

The thermal storage boiler is an energy storage system in a broad sense [16, 17]. The device can not only directly participate in heating, but also convert the electrical energy of one period to the thermal energy of another period, which can shift the load. In the clean heating demonstration project, the heat supplier stores heat with low price and releases heat with high price. This measure not only saves the cost of electricity, but also promotes the peak-shaving and valley-filling of the grid.

2.2 Mathematical Model of the Thermal Storage Boiler

The heat generated by the electric boiler is divided into two parts, one is for direct supply to users, and the other is stored in a heat storage tank. The boiler system is not only regarded as a heating unit, added to the heat network, but also regarded as an electric load, which is accumulated in the power demand. The thermal storage boiler model in the grid system is shown in Eq. (1).

$$P_{B.H} = P_B \eta_B \tag{1}$$

Where, P_B is the electric power of boiler, $P_{B.H}$ is the thermal power of boiler, η_B is the efficiency of power-heat.

The heat balance of thermal storage boilers is shown in Eq. (2).

$$P_{B.H} + P_S = D_H \tag{2}$$

Where, P_S is the thermal power of tank, D_H is the heat load.

The heat storage tanks in different states of the heat network (heat storage and heat release) have different thermal power, and the heat storage tanks participate in the heating system as in Eq. (3).

$$P_S = \begin{cases} P_{L.\max} & D_H < P_{B.H} \\ 0 & D_H = P_{B.H} \\ -P_{G.\max} & D_H > P_{B.H} \end{cases} \tag{3}$$

Where, P_S is the output value of heat storage tank, positive value indicates heat storage, negative value indicates heat release; $P_{L.\max}$ is the maximum heat storage power of the tank, $P_{G.\max}$ is the maximum heat release power of the tank.

3 Dispatching Optimization Model of Electricity-Heat Interactive System

3.1 Objective Function

In this paper, the minimum value of coal cost and wind curtailment risk is set as the optimization goal to evaluate the economic benefit of the system. The grid preferentially consumes wind power, and then dispatches conventional units and thermal storage boilers, so that the system can operate flexibly and economically on the premise of satisfying operational reliability. The objective function is shown in Eq. (4).

$$\min F = \sum_{t=1}^{T} \left(\lambda_f (\sum_{i=1}^{N_{tp}} f_{i,t}^{tp}(P_{i,t}^{tp}) + \sum_{i=1}^{N_{chp}} f_{i,t}^{chp}(P_{i,t}^{chp}, Q_{i,t}^{chp})) + \sum_{i=1}^{N_{tp}} S_{i,t}^{tp} + \lambda_w \sum_{i=1}^{N_w} (P_{i,t}^{wind} - P_{i,t}^{w}) \right) \tag{4}$$

Where, T is the number of optimization period, N_{tp} is the number of pure condensing units, N_{chp} is the number of thermoelectric units, N_w is the number of wind farms, $f_{i,t}^{tp}$ is the coal consumption function of pure condensing unit, $f_{i,t}^{chp}$ is the coal consumption function of thermoelectric unit, $P_{i,t}^{tp}$ is the electric power of pure condensing unit, $P_{i,t}^{chp}$ is the electric power of thermoelectric units, $Q_{i,t}^{chp}$ is the heat power of thermoelectric units, $S_{i,t}^{tp}$ is the startup and shutdown cost of pure condensing unit, $P_{i,t}^{wind}$ is the predicted wind power, $P_{i,t}^{w}$ is the actual wind power, λ_f is the coal price, λ_w is the curtailed wind price.

3.2 Restrictions

The electric power balance constraint is shown in Eq. (5).

$$\sum_{i=1}^{N_{tp}} P_i^{tp}(t) + \sum_{i=1}^{N_{chp}} P_i^{chp}(t) + \sum_{i=1}^{N_w} P_i^{w}(t) = P_{load}(t) + \sum_{i=1}^{N_b} P_i^{b}(t) \tag{5}$$

Where, $P_{load}(t)$ is the electric load, $P_i^b(t)$ is the electric boiler power.

The heat power balance constraint is shown in Eq. (6).

$$\sum_{i=1}^{N_{chp}} Q_i^{chp}(t) + \sum_{i=1}^{N_b} Q_i^b(t) + \sum_{i=1}^{N_{bs}} Q_i^{bs}(t) = Q_{load}(t) \tag{6}$$

Where, N_{bs} is the total number of tanks, $Q_i^b(t)$ is the heat power of electric boiler at time t, $Q_i^{bs}(t)$ is the heat power of tanks at time t, $Q_{load}(t)$ is the heat load at time t.

The pure condensing unit output constraint is shown in Eq. (7).

$$P_{i,\min}^{tp} \le P_i^{tp}(t) \le P_{i.\max}^{tp} \tag{7}$$

Where, $P_{i.\max}^{tp}$ is the maximum electric power of unit, $P_{i.\min}^{tp}$ is the minimum electric power of unit.

The ramping power constraint of pure condensing unit is as follows (8).

$$-R_{i.d}^{tp} \le P_i^{tp}(t) - P_i^{tp}(t-1) \le R_{i.u}^{tp} \tag{8}$$

Where, $R_{i.u}^{tp}$ is the power for ramping up, $R_{i.d}^{tp}$ is the power for ramping down.

The constraints of the on and off time of pure condensing unit are as follows Eq. (9) and (10).

$$(U_i(t-1) - U_i(t))(T_i^{on}(t-1) - M_i^{on}) \ge 0 \tag{9}$$

$$(U_i(t) - U_i(t-1))(T_i^{off}(t-1) - M_i^{off}) \ge 0 \tag{10}$$

Where, $U_i(t)$ is the on and off state of unit i in period t, 0 means off, 1 means on; $T_i^{on}(t-1)$ and $T_i^{off}(t-1)$ represent the running time and shutdown time of unit i in period t; M_i^{on} and M_i^{off} is the minimum running time and shutdown time of unit i.

The constraints of ordering power by heat are shown in Eq. (11)

$$\begin{cases} a_L + Q_i^{chp}(t) b_L = P_{i.\min}^{chp}(t) \\ a_H + Q_i^{chp}(t) b_H = P_{i.\max}^{chp}(t) \end{cases} \tag{11}$$

Where, a_L, a_H, b_L and b_H represent fitting coefficient.

The constraint of the heat power of unit is shown in Eq. (12).

$$Q_{i.\min}^{chp} \le Q_i^{chp}(t) \le Q_{i.\max}^{chp} \tag{12}$$

Where, $Q_{i.\max}^{chp}$ and $Q_{i.\min}^{chp}$ respectively represent upper and lower limits of heat power.

The constraint of the electric power of unit is shown in Eq. (13).

$$P_{i.\min}^{chp}(t) \leq P_i^{chp}(t) \leq P_{i.\max}^{chp}(t) \tag{13}$$

Where, $P_{i.\max}^{chp}(t)$ and $P_{i.\max}^{chp}(t)$ respectively represent upper and lower limits of electric power at time t.

The wind power constraint is shown in Eq. (14).

$$0 \leq P_i^{w}(t) \leq P_{i.\max}^{wind}(t) \tag{14}$$

Where, $P_{i.\max}^{wind}(t)$ represents the maximum of wind power at time t.

The constraints of thermal storage boilers are shown in Eq. (15) to (17).

$$0 \leq P_B(t) \leq P_{B.\max} \tag{15}$$

$$-Q_{S.\max} \leq Q_S(t) \leq Q_{S.\max} \tag{16}$$

$$E_{S.\min} \leq E(t) \leq E_{S.\max} \tag{17}$$

Where, $P_{B.\max}$ and $Q_{S.\max}$ represent the maximum electric power of boiler and the maximum heat power of tank; $E_{S.\min}$ and $E_{S.\max}$ are the minimum and maximum heat storage capacity of tank.

3.3 Solving Algorithm

The mathematical model of system optimization scheduling established in this paper is more complicated. The problem not only includes the continuous and discrete variables, but also is related to the inter-period constraints. It is transformed into a mixed integer linear optimization problem. The MILP method can directly optimize the unit commitment problem globally, and it's a relatively mature commercial software. In this paper we first establish a reasonable and efficient mixed integer programming model, and then directly calls CPLEX to solve the model. All simulations were run on a computer with an Intel® Core™ i7-7700HQ CPU @2.80 GHz and 8 GB of RAM.

4 Example Analysis

4.1 Simulation Parameter Setting

The parameters in this paper are simulated based on the actual ratio of power grid in a province in Northeast China. The main parameters of the equipment are shown in Table 1. The characteristic parameters of thermoelectric units are shown in Table 2, in which the unit commitment only considers pure condensing units. The predicted output of wind power in winter is shown in Fig. 2, and the typical electricity and heat load curves are shown in Fig. 3. The penalty price for wind curtailment is 200 yuan/MWh.

Table 1. Main parameters of the equipment

Device	Parameter	
Thermoelectric unit	Lower limit of electric power/MW	65+0.18D
	Upper limit of electric power/MW	100–0.05D
	Lower limit of extraction heating/(*t*/*h*)	30
	Upper limit of extraction heating/(*t*/*h*)	120
Pure condensing unit	Lower limit of electric power/MW	50
	Upper limit of electric power/MW	100
Thermal storage boiler	Rated thermal power of electric boiler/MW	100
	Rated thermal power of thermal storage tank/MW	100
	Minimum heat capacity of heat storage tank/MWh	0
	Maximum heat capacity of heat storage tank/MWh	400
	Boot time	9 pm–7 am

Table 2. The coal consumption's fitting coefficient of thermal power unit

Unit	x_1	y_1	z_1	x_2	y_2	z_2
#1	3.000	0.4195	0.123	0.0008	0.000173	0.00029
#2	2.987	0.4125	0.120	0.0009	0.000176	0.00022
#3	3.102	0.4185	0.122	0.0008	0.000175	0.00028
#4	3.105	0.4254	0.128	0.0007	0.000166	0.00032
#5	2.989	0.4187	0.124	0.0008	0.000172	0.00030
#6	3.015	0.4176	0.125	0.0009	0.000175	0.00031
#7	2.999	0.4234	0.127	0.0008	0.000178	0.00021
#8	3.089	a	0.123	0.0007	0.000176	0.00026
#9	5.9653	0.2608	0.0008	/	/	/
#10	4.9369	0.2762	0.0009	/	/	/
#11	3.000	0.3200	0.0013	/	/	/
#12	2.3298	0.3101	0.0015	/	/	/

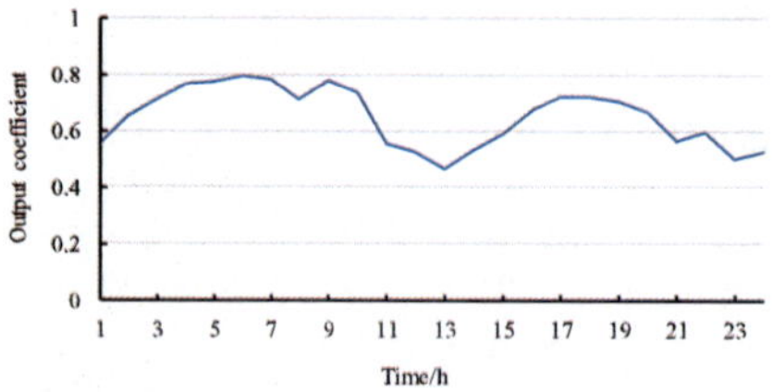

Fig. 2. The forecast output coefficient of typical daily wind power

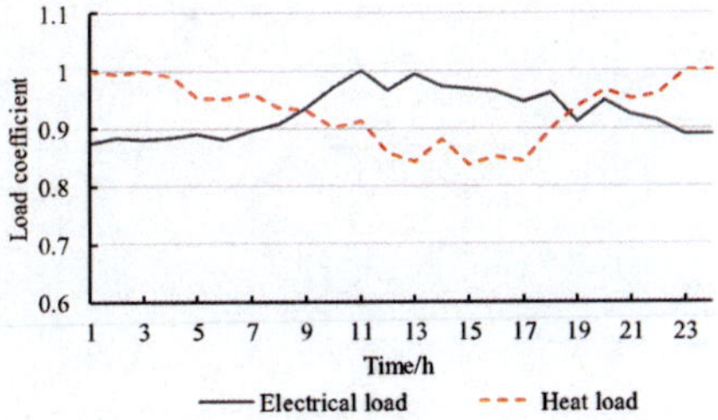

Fig. 3. The coefficient of typical daily electric load and heat load

4.2 Comparison of Thermoelectric Units' Operating Cost

According to the simulation results, the curve of the electric and heat power under different operating modes is drawn. Among them, Fig. 4 and Fig. 5 are system heat load distribution diagrams, and Fig. 6 is a comparison curve of electric power in different operation modes.

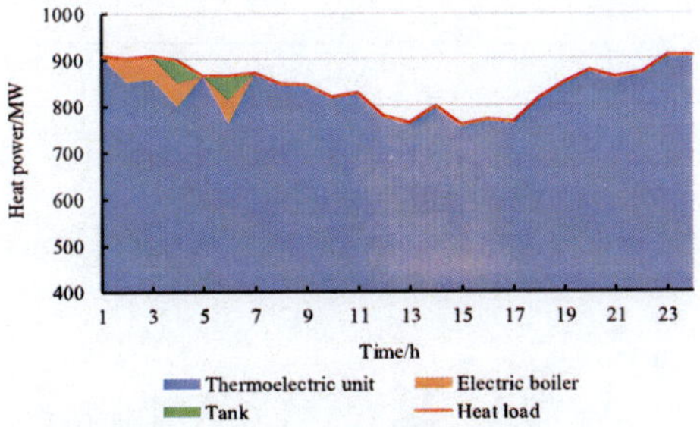

Fig. 4. The heat load distribution of half-configuration system

As can be seen from Fig. 4, the boiler effectively relieves the heating pressure of the thermoelectric unit. However, due to its insufficient configuration, the degree of electro-thermal coupling is shallow. From 3 to 7 o'clock, the electric boiler and the heat storage tank simultaneously heat to increase the wind power consumption. In Fig. 5, the boiler is fully configured. The electric boiler supplies heat from 3 to 7 o'clock, and the tank participates in heating from 6 to 10 o'clock. Taking into account the constraint of ordering power by heat, the peak reduction capability of the thermoelectric unit is released. It can be seen from Fig. 6 that the full-capacity boiler makes the unit's electric power decrease more, which frees up 37.6 MWh of consuming space than the half-rate configuration, and also reduces coal consumption by 18.3t.

4.3 Comparison of Renewable Energy Consumption

Under different operation modes of thermal storage boiler configuration, 24-hour power generation statistics are in the calculation example. Their outputs are shown in Fig. 7, 8, and 9.

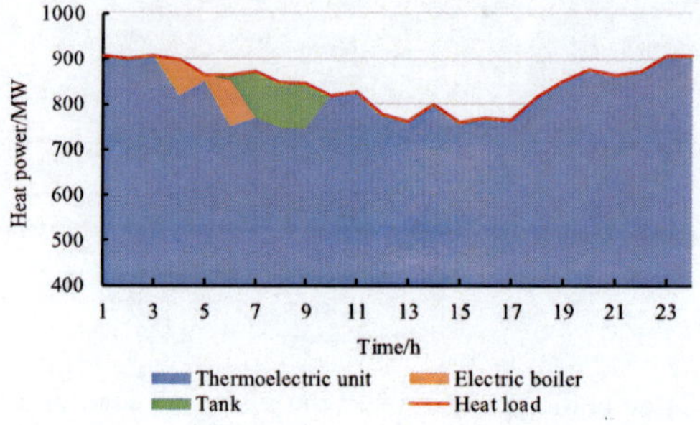

Fig. 5. The heat load distribution of full-configuration system

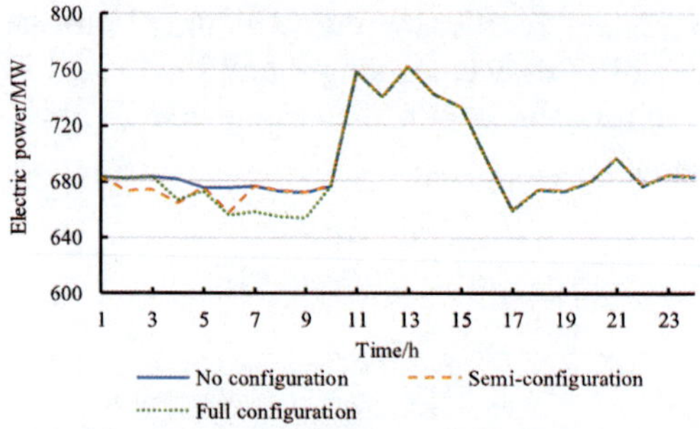

Fig. 6. The electric power comparison curve of thermoelectric unit

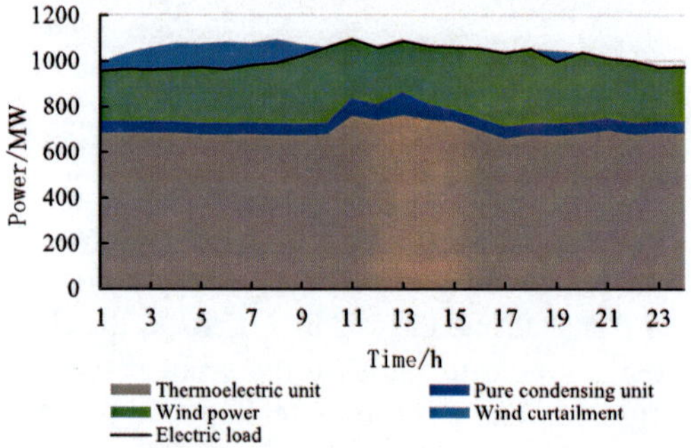

Fig. 7. The scenario operating condition without thermal storage boiler

Figure 7 shows the 24-h output statistics of various power supplies in the system. It can be found that the electric energy are mainly supplied from thermoelectric units and wind farms. The system's peaking power supply rely on thermoelectric units, but it is constrained by ordering power by heat and minimum technical effort. When the electric load in the daytime is high, the system preferentially consumes wind power and thermoelectric units participate in peak shaving. At night, the electric load is low, the thermoelectric units have been adjusted to the minimum technical output, which still

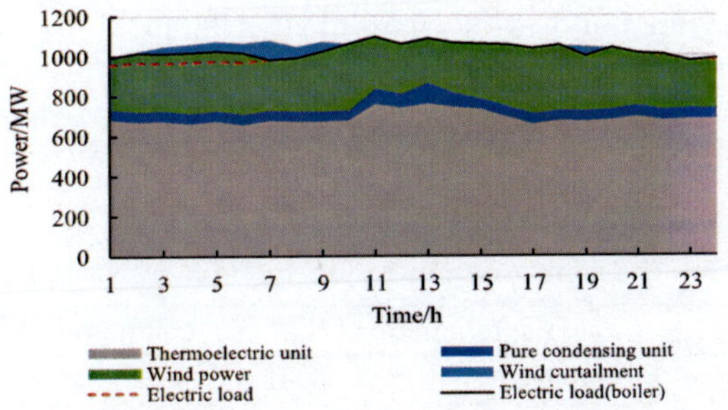

Fig. 8. The scenario operating condition with semi-configured thermal storage boiler

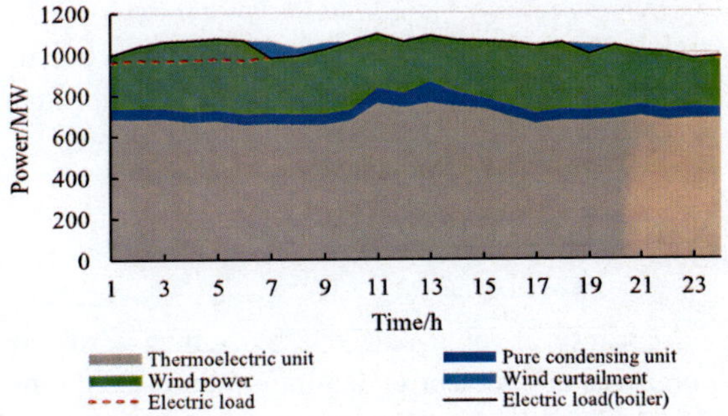

Fig. 9. The scenario operating condition with fully configured thermal storage boiler

has to cause a large amount of wind curtailment. From the above situation, power system and thermal system are interactive rigid energy systems. To a certain extent, thermal systems put pressure on the flexibility of power system regulation. There is a serious wind curtailment situation in the entire network.

As shown in Fig. 8 and 9, in the same scenario, the thermal storage boiler is configured so that the electric heating system can be decoupled and optimized. In the operating mode of fully configured thermal storage boiler, the change of wind power is most obvious. The electric boiler is turned on at night to increase the electrical load. Compared with the non-configuration and semi-configured boilers, the system respectively consumes 611.8 MWh and 301.8 MWh of wind curtailment. The electric boiler operates at full power from 1 to 7 o'clock, and the excess heat can be stored in the heat storage tank. It should be noted that due to the limitation of the start-up period of the boiler, even in the full configuration, there is still a small amount of wind curtailment from 7:00 to 10:00 and 18:00 to 20:00.

4.4 Comparison of Optimization Results

Table 3 shows the comparison result of coal consumption and wind curtailment in different operation modes. When the configuration of the boiler is insufficient (such as half-rate configuration), the economic cost of the system will be reduced to a certain extent. However, the amount of wind curtailment is 443.5 MWh, and there is still some space for decline. With fully configured boilers, the wind curtailment dropped to 141.7 MWh, and the wind curtailment rate was only 3.4%. As shown in Fig. 7, there is no curtailment during the start-up period of the boiler.

Table 3. The comparison of results in different operating modes

	No configuration	Half-rate configuration	Full configuration
Coal consumption(*t*)	6951.7	6938.9	6920.6
Wind curtailment (MWh)	753.5	443.5	141.7
Wind curtailment rate (%)	12.1	6.2%	3.4

In this paper, the peak shaving ability of the thermoelectric unit is improved by configuring thermal storage boilers. Its operation mode is better than the mode without boilers, which is conducive to power-heat effective interaction.

5 Conclusions

In this paper, the uncertainty of renewable energy output is solved as a deterministic problem. We have constructed an electric heating system scheduling model configured with thermal storage boilers to accommodate wind power heating. The model is transformed into a mixed integer linear model. We can use the commercial solver CPLEX to optimize the problem. Comparing the economics of different boilers configuration schemes, we get the following conclusions:

(1) The configuration scale of thermal storage boilers determines the power-heat decoupling degree. Compared with the half-rate configuration, the full configuration greatly reduces coal consumption and wind curtailment.
(2) The solution makes the thermal system contribute to wind power consumption. By transferring part of the heat load to wind power load, it has a certain coal-saving benefit. The amount of wind curtailment consumed determines the amount of coal saved.

Acknowledgement. This work was supported by Science and Technology Project of the State Grid Qinghai Electric Power Corporation of China “Study on the boundary condition of full clean power supply in Qinghai Power Grid”. The encoding of the project is “52280019004X”.

References

1. State Electricity Regulatory Commission. Regulatory report on wind power consumption in key regions (2012 No. 10). State Electricity Regulatory Commission, Beijing (2012)
2. Sun, H., Xu, J., Sun, Y., et al.: Active power optimization scheduling of wind farm based on mixed-integer linear programming. Autom. Electr. Power Syst. **40**(22), 27–33 (2016)

3. Pang, X., Zhang, N., Zhang, C., et al.: Study on high penetration wind power curtailment characteristic considering the system regulation constraint. Renew. Ener. Resour. **36**(12), 120–125 (2018)
4. National Energy Administration: Clean energy is more beautiful [EB/OL]. http://www.nea.gov.cn/2020-03/03/c_138838993.htm. Accessed 03 Mar 2020/04 Apr 2020
5. Wang, D., Chen, Z., Tu, M., et al.: Reserve capacity calculation considering large scale wind power integration. Autom. Electr. Power Syst. **36**(21), 24–28 (2012)
6. Zhang, Y., Lyu, Q., Li, Y., et al.: Analysis on operation flexibility of combined heat and power plant with four improved power-heat decoupling schemes. Autom. Electr. Power Syst. **44**(02), 164–177 (2020)
7. Wang, N., Wang, J., He, S.: Cross-border accommodation method and transmission scheme of Jiuquan wind power. Autom. Electr. Power Syst. **35**(22), 82–89 (2011)
8. Deng, T., Tian, L., Liu, J.: A control method of heat supply units for improving frequency control and peak load regulation ability with thermal storage in heat supply net. Proc. Chin. Soc. Electr. Eng. **35**(14), 162–169 (2015)
9. David, T., Aikaterini, F.: Do liberalized electricity markets help or hinder CHP and district heating. Ener. Policy **36**(4), 1448–1456 (2008)
10. Lyu, Q., Jiang, H., Chen, Y., et al.: Wind power accommodation by combined heat and power plant with electric boiler and its national economic evaluation. Autom. Electr. Power Syst. **38**(01), 6–12 (2014)
11. Lyu, Q., Wang, H., Chen, T., et al.: Operation strategies of heat accumulator in combined heat and power plant with uncertain wind power. Autom. Electr. Power Syst. **39**(14), 23–29 (2015)
12. Yu, Y., Chen, H., Jiang, X., et al.: Operation strategy for heat storage tank to improve wind power accommodation. Autom. Electr. Power Syst. **41**(7), 37–43 (2017)
13. Chen, L., Xu, F., Wang, X., et al.: Implementation and effect of thermal storage in improving wind power accommodation. Proc. Chin. Soc. Electr. Eng. **35**(17), 4283–4290 (2015)
14. Deng, J., Hu, L., Li, J.: Analysis on mechanism of curtailed wind power accommodation and its economic operation based on electric boiler for peak-load regulation at secondary heat supply network. Autom. Electr. Power Syst. **40**(18), 41–47 (2016)
15. Guo, F., Hu, L., Zhou, S.: Dispatching model of wind power accommodation based on heat storage electric boiler for peak-load regulation in secondary heat supply network. Autom. Electr. Power Syst. **42**(19), 50–56 (2018)
16. Zhang, Y.: Research on optimal operation of energy storage cooperative thermal storage electric boiler. North China Electric Power University (Beijing) (2018)
17. Yang, X., Ye, T., Zhang, Y.: A novel optimization model for combined wind power accommodation and electric boiler with thermal storage. Int. J. Ener. Res. **43**(12), 6494–6509 (2019)

Integrated Scheduling Model of Generation and Transmission for Promoting Clean Energy Accommodation

Fan Xu(✉), Rongzhang Cao, Mengfu Tu, Yantao Zhang, and Dandan Zhang

State Key Laboratory of Smart Grid Protection and Control, Nari Group Corporation, Nanjing 211106, Jiangsu, China
xufan@sgepri.sgcc.com.cn

Abstract. With the rapid development of ultra-high voltage AC and DC power supply and the increase of large-capacity power bases, China's power system shows obvious integration characteristics of power balance in the whole power grid, which poses new challenges to the existing model of scheduling plan with provincial balance and weakly coupled dispatching at all levels. For this reason, an integrated optimization model of power generation and transmission planning for the entire power grid is proposed, which is suitable for engineering practice and takes into account the characteristics of multi-energy generation complementation on the generation side, transmission operation characteristics of ultra-high voltage lines and flexible load response characteristics on the demand side. The model of multi-energy complementation of generation side is established, which can reduce the fluctuation of clean energy and increase the accommodation of clean energy. The flexible load response model of the demand side is established, which can coordinate with the power adjustment of the power generation to increase the accommodation space of clean energy. The power model of transmission channel is established to connect the energy base with the load center, which can consider the distribution characteristics of the power grid globally and reduce the overall operation cost of the grid. Based on the actual data of a cross-regional interconnected power grid, the validity of the model is verified.

Keywords: Ultra-high voltage transmission · Complementary characteristics · Flexible loads · Integrated scheduling

1 Introduction

In order to alleviate the ecological and environmental problems caused by fossil energy consumption, China has made great efforts to develop clean energy in recent years. At present, the installed capacity of clean energy has ranked first in the world, and the total consumption has also achieved significant growth. China has built a number of large-scale clean energy power generation bases in the southwest with abundant water energy resources and in the north with concentrated wind energy resources, and transmitted clean power to the economically developed east central load center through UHV

Y. Xue et al. (Eds.): PMF 2020, LNEE 718, pp. 244–261, 2021.
https://doi.org/10.1007/978-981-15-9746-6_20

AC/DC tie line. The rapid construction of UHV AC/DC lines and large clean energy power generation base makes the sending grid and receiving grid, the superior and subordinate dispatching organizations closely coupled, which shows the obvious characteristics of the integration of power balance of the whole network. The traditional dispatching mode of provincial balance and weak coupling of dispatching at all levels has been difficult to meet the needs of large-scale clean energy consumption across provinces and regions under the new situation.

In view of this, scholars at home and abroad have carried out relevant research. Tie line is the link of interaction and coordination among multi-regional power grids. With the increase of power exchange of interconnected power grid, the optimization model of tie line operation is increasingly refined. In [1], a refined model for generation and transmission scheduling of DC cross-regional interconnected power grid is established, which improves the wind power consumption capacity of the interconnected power grid and reduces the operation cost of the grid. [2] introduces the virtual cost of DC tie line power adjustment, establishes the DC power adjustment model, and realizes the coordination and optimization of cross-regional DC tie lines, conventional units and new energy units. The DC tie line power can be adjusted continuously and flexibly, but in the actual dispatching operation, in order to avoid the frequent fluctuation of DC tie line power, it is generally required that the DC power regulation should be greater than a certain threshold value. Based on the principle of power equivalent superposition, the DC tie line power is equivalent to multiple generating units [3], multiple transmission units [1] or multiple power feasible states [4]. The DC power is modeled as a discrete variable to realize the stepped operation, and to meet the various operation constraints of DC tie line, to achieve the coordination and optimization with other types of units.

Limited by the natural attributes of wind and solar energy, wind power and photovoltaic power generation have volatility and randomness. The research shows that by guiding the flexible load to adopt the best load schedule, giving full play to the flexible response ability of the flexible load [5, 6], supplemented by the coordination of multiple power sources such as thermal power, hydropower, etc., the clean energy consumption can be effectively improved [7–10], the peak load regulation pressure of the system can be eased [11], and the power supply cost of the large power grid can be reduced. At present, the research on coordination and complementary characteristics of renewable energy and conventional energy has been mature, and has been applied in practical projects [12–14].

On the basis of the existing research results, based on the actual project demand of large-scale clean energy consumption across provinces and regions, considering the current situation of the power grid with enhanced regional coupling under the background of UHV AC/DC line construction, this paper proposes a joint optimization model of power generation scheduling considering the generation, transmission, and flexible load resources of the whole grid. The model comprehensively considers the complementary characteristics of multi-energy generation at the generation side, the transmission operation characteristics of UHV tie lines and the flexible load response characteristics at the load side: the complementary characteristics of multi-energy sources at the generation side can reduce the power fluctuation of clean energy; the flexible load response at the load side can cooperate with the generation side to increase the consumption for clean consumption; UHV tie lines can be considered to reduce the overall operation cost of the power grid. The model can coordinate all provincial power grids, transfer all kinds of resources of the whole grid, enhance the utilization capacity of UHV transmission channel, promote the overall resource sharing, and improve the clean energy consumption of large power grid.

The model in this paper has certain engineering application value. Compared with the traditional provincial scheduling model, this model can reduce the cost of power generation and meet the actual engineering application requirements.

2 Unified Dispatching Mode of Power Resources

2.1 Current Provincial Dispatching Mode

Under the current provincial load balance mechanism in China, each provincial dispatching organization is responsible for the load balance and unit power generation schedules within the provincial control area. In general, the power transmission schedule of cross province AC/DC tie lines is formulated by the superior dispatching organization and sent to the subordinate dispatching organizations. The subordinate dispatching organization uses this as the boundary condition to prepare the power generation schedules in the control area, as shown in Fig. 1.

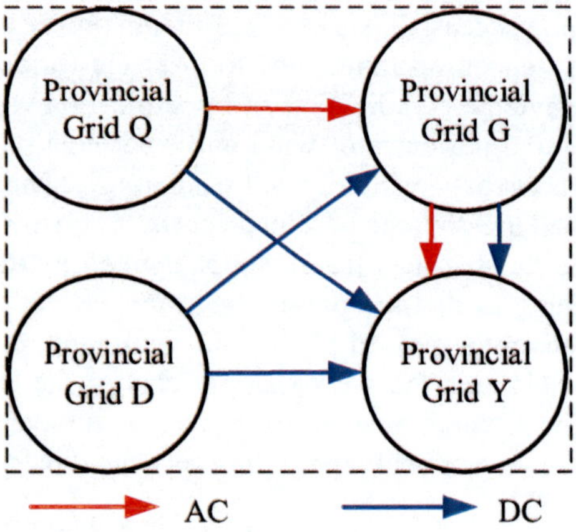

Fig. 1. Schematic diagram of provincial scheduling

When the installed capacity of clean energy is small, this mode is easy to manage the cross region and cross province power exchange, and easy to achieve the power balance of province. However, when the capacity of clean energy increases, the disadvantages of this mode gradually appear. First of all, when the superior dispatching organization arranges the transmission schedules of AC/DC tie lines, it generally meets the operation requirements of clean energy power generation and transmission. The subordinate dispatching organization, taking this as the boundary condition, may not be able to make the feasible generation schedule, which needs to be fed back to the superior dispatching organization to adjust the power schedules of tie lines. Secondly, the schedule of tie line takes less account of the actual power demand of the receiving grid. Even if the receiving grid dispatching organization makes the power schedule after iteration with the superior dispatching organization, it may also face greater operating pressure in the real-time dispatching stage. Thirdly, the clean energy power generation base is connected to the sending grid, but generally does not participate in the power balance of the sending grid. If the tie line power is taken as the equivalent load, and the power generation of the clean energy base is included in the power balance of the sending grid, the power supply cost of the local load of the sending grid can be reduced. In addition, due to the lack of comprehensive consideration of the operation of the receiving grid when making the schedules of the AC/DC tie lines, it is impossible to schedule the optimal power supply of clean energy. More power supply will increase the peak load regulation pressure of the receiving grid, and less power supply will result in the waste of clean energy.

2.2 Unified Dispatching Mode of the Whole Grid

The unified dispatching mode of the whole network is based on the current two-level dispatching mode of national and provincial level, taking the inter provincial interconnection line power as a decision variable, and jointly optimizing with the units and flexible loads in each province, so that the interconnection line can coordinate and interact with the resources in the province. According to different operation control characteristics of AC and DC tie lines, DC tie lines can be used as controllable power units in the network, and AC tie lines can be used as transmission lines or conventional sections in the network, so that the whole network can be regarded as a unified control area for power generation resource scheduling, as shown in Fig. 2.

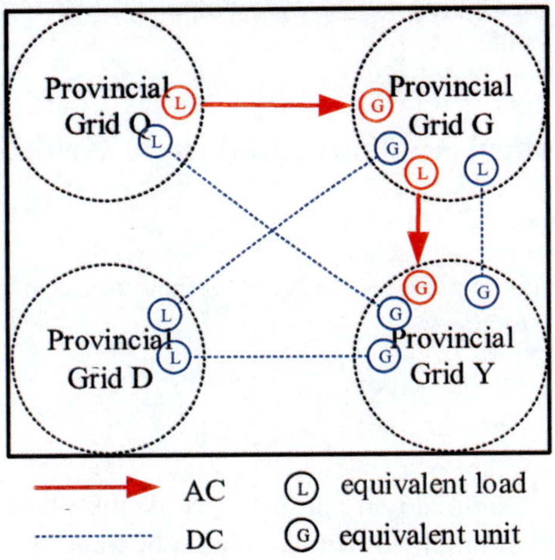

Fig. 2. Schematic diagram of scheduling in whole grid

For inter provincial DC tie line, its power can be controlled, and there is no direct coupling relationship between the power of units and the power of DC tie line. In essence, the power of DC tie line is an independent variable, and the DC power at sending grid and receiving grid are two variables in the whole network model. Because the DC power directions of the two grids are opposite, they cannot be reflected in the load balance constraints of the whole network. This problem can be solved by the node power balance model. However, in the current practical background, the completeness of bus load nodes and the accuracy of bus load forecasting can not be guaranteed, which makes the power balance equation of nodes difficult to be applied in engineering. Therefore, in this paper, the provincial power balance constraints are introduced into the whole network model, each province is regarded as an equivalent node, and the converter stations on both sides of the DC tie line are modeled as equivalent load and equivalent unit respectively, which are modeled in the provincial load balance constraints to obtain DC transmission power. For provincial AC tie line, it is a common AC transmission line in the whole network model, and its power flow value is essentially dependent variable, which is determined by the power grid topology and the power injection value of the whole network nodes. However, the characteristics of DC power require the introduction of provincial load balance constraints, and provincial balance constraints also require the equivalent treatment of AC tie line, so as to model in provincial load balance.

In the traditional provincial balance model, AC and DC power exist as the boundary condition of scheduling in the province, which is a fixed value in advance; in the whole network model, because of the operation characteristics of DC power, it is also necessary to introduce provincial load balance constraints, but the DC power is the same as the unit power in the province, which is also a decision variable, while the AC tie line power can not be directly controlled, but it can be indirectly controlled by adjusting the power of the whole network injection node, the AC and DC tie line can be used as the power channel to link the provincial power grid as a whole to optimize the power generation schedule.

3 The Mathematical Model of the Whole Grid Scheduling

3.1 Objective Function

The objective function is to achieve the lowest generation cost of the whole grid based on the priority of clean energy.

$$\min F = \sum_{i,t} c_{g,t} + \sum_{h,t} c_{h,t} + \sum_{m,t} c_{m,t} + \sum_{n,t} c_{h,t} + \sum_{w,t} \Delta_{w,t} \tag{1}$$

where $c_{g,t}$ is the cost of thermal unit i at time t; $c_{h,t}$ is the cost of hydropower unit h at time t; $c_{m,t}$ is the compensation cost of curtailable flexible load m at time t; $c_{n,t}$ is the transfer cost of transferable flexible load n at time t; $\Delta_{w,t}$ is the power curtailment penalty cost of clean energy unit w at time t.

3.2 The System Load Balance Model

The load balance constraint of the whole network is:

$$\sum_{i} p_{i,t} = \sum_{z} L_{z,t} + \sum_{m} l_{m,t} + \sum_{n} l_{n,t} + \sum_{k} \Delta p_{k,t} \tag{2}$$

where $p_{i,t}$ is the power of unit i at time t, including conventional thermal unit g, hydropower unit h, wind and solar energy unit w; $L_{z,t}$ is the system load forecast value deducting flexible load for control zone Z at time t; $l_{m,t}$ is the load of curtailable flexible load m at time t; $l_{n,t}$ is the load of transferable flexible load n at time t; $\Delta p_{k,t}$ is the power loss of AC/DC tie line at time t.

Due to the operation characteristics of tie line, it is necessary to introduce the constraint of load balance by province into its modeling:

$$\sum_{i \in z} p_{i,t} + \sum_{k \in z} p_{k,t}^{g} = L_{z,t} + \sum_{m \in z} l_{m,t} + \sum_{n \in z} l_{n,t} + \sum_{k \in z} p_{k,t}^{l} \tag{3}$$

where $p_{k,t}^{g}$ and $p_{k,t}^{l}$ are the power of equivalent unit and equivalent load of AC/DC tie line k at time t.

3.3 The Power Model of AC/DC Tie Line

The DC tie line power is controllable, which is a decision variable. However, for AC tie lines, the power at the sending grid is related to the grid topology and the injected power of nodes:

$$p_{a,t}^{l} = \sum_{i} p_{i,t} S_{i,a,t} - \sum_{l} l_{l,t} S_{l,a,t} \tag{4}$$

where is the power of AC tie line a at sending grid at time t; $S_{i,a,t}$ is the sensitivity of unit i to AC tie line a at time t, that is calculated by the network analysis module according to the whole network topology model; $l_{l,t}$ is the load of load node l at time t.

For the equivalent equipments on sending grid and receiving grid of the AC/DC tie lines, there are:

$$p_{k,t}^{l}(1 - \alpha_{k,t}) = p_{k,t}^{g} \tag{5}$$

$$\Delta p_{k,t} = p_{k,t}^{l} \alpha_{k,t} \tag{6}$$

where $\alpha_{k,t}$ is the loss coefficient of AC/DC tie line k [15].

The capacity limits constraints of AC/DC tie line [16] are:

$$P_{k,\min} \leq p_{k,t}^{l} \leq P_{k,\max} \tag{7}$$

where $P_{k,\min}$ and $P_{k,\max}$ are the minimum and maximum power limits of AC/DC tie line k respectively.

The constraint of ramp-up or ramp-down limits [2] of tie line:

$$-P_{k,down} \leq p_{k,t}^{l} - p_{k,t-1}^{l} \leq P_{k,up} \tag{8}$$

where $P_{k,up}$ and $P_{k,down}$ are the ramp-up or ramp-down limits of tie line k.

Transmission energy constraints for AC/DC tie line [1]:

$$Q_{k,\min} \leq \frac{\lambda}{60}\sum_{t} p_{k,t}^{l} \leq Q_{k,\max} \tag{9}$$

where λ is the minutes of a single period; $Q_{k,min}$ and $Q_{k,max}$ are the minimum and maximum energy limits of AC/DC tie line k during dispatching periods respectively.

The inter province exchange energy constraint [1] is:

$$Q_{(s,r),\min} \leq \frac{\lambda}{60}\sum_{t}\sum_{k\in(s,r)} p_{k,t}^{l} \leq Q_{(s,r),\max} \tag{10}$$

where $Q_{(s,r),min}$ and $Q_{(s,r),max}$ are the minimum and maximum of the exchange energy between the sending province s and the receiving province r respectively.

The power of the DC tie line adopts the stepped operation model, which is detailed in [4], and will not be described here. In this model, the total inter provincial exchange energy is considered, and the energy of the specific tie line is not limited. This processing method increases the optimization space of inter provincial resource allocation, and meets the requirements of inter provincial power transmission.

3.4 The Model of Flexible Load

(1) The model of curtailable flexible load

For air conditioning, lighting and other flexible loads, the load can be curtailed according to the compensation cost and system operation demand. According to the importance of the load, the load curtailment can be divided into segments, and each segment can be quoted a price differently according to its importance [17]. When load curtailment is needed, it generally starts from the unimportant load with low cost, so the cost of load curtailment is the piecewise linear incremental cost, and the load curtailment is the accumulated load of each segment [11]:

$$c_{m,t} = \sum_{l} b_{m,t,l} k_{m,t,l} \tag{11}$$

$$\Delta l_{m,t} = \sum_{l} b_{m,t,l} \tag{12}$$

where $\Delta l_{m,t}$ is the load curtailment of curtailable flexible load m at time t; $b_{m,t,l}$ is the curtailment in each segment of curtailable flexible load m at time t; $k_{m,t,l}$ is the price in each segment.

The load curtailment is less than the declared load curtailment:

$$\Delta l_{m,t} \leq \Delta L_{m,t} \tag{13}$$

where $\Delta L_{m,t}$ is the declared load curtailment of load m at time t.

After considering load curtailment, the actual load of flexible load is:

$$l_{m,t} = l_{m,t}^{0} - \Delta l_{m,t} \tag{14}$$

where $l_{m,t}^{0}$ is the original load of flexible load m at time t.

(2) The model of transferable flexible load

For electric vehicles and other transferable flexible loads, the power consumption can be flexibly adjusted during dispatching periods according to the transfer cost and system demand [5]. After the transferable load dispatching, the energy consumption remains unchanged:

$$\sum_{t \in T} l_{n,t} = \sum_{t \in T} l_{n,t}^{0} \tag{15}$$

where T is the power consumption time of transferable load; $l_{n,t}^{0}$ is the original load of load n at time t.

The transfer cost of the transferable flexible load is the adjustment cost of its deviation from the original power consumption [5]:

$$c_{n,t} = \lambda_n \left| l_{n,t} - l_{n,t}^{0} \right| \tag{16}$$

where λ_n is the transfer price of transferable flexible load n.

3.5 The Power Model of Generator Units

(1) The model of wind and solar energy units

Renewable energy such as wind and solar energy have priority consumption. When it cannot be fully consumed, it is allowed to curtail power, and introduce renewable energy curtailment punishment factor to ensure the maximum energy consumption of renewable energy [7].

The curtailment of renewable energy is:

$$p_{w,t}^{d} = \sum_{d} \Delta p_{w,d,t} \tag{17}$$

where $p_{w,t}^{d}$ is the power curtailment of renewable energy unit w at time t; $\Delta p_{w,d,t}$ is the power curtailment at segment d of renewable energy unit w at time t.

The penalty cost of renewable energy power curtailment is:

$$\Delta_{w,t} = \sum_{d} \rho_{w,d} \Delta p_{w,d,t} \tag{18}$$

where $\rho_{w,d}$ is the penalty price at segment d of renewable energy unit w.

After curtailment, the power of renewable energy unit is:

$$p_{w,t} = p^{0}_{w,t} - p^{d}_{w,t} \tag{19}$$

where $p^{0}_{w,t}$ is the power forecasting of renewable energy unit w at time t.

The complete generation scheduling model of renewable energy units such as wind and solar power have been detailed described in [7], which will not be described here.

(2) The model of hydropower units

The turbine of the hydropower unit will vibrate in the actual operation. Therefore, it is necessary to avoid the turbine working in the vibration area when scheduling the generating power of the hydropower unit.

The existence of the vibration region makes the operation region of the hydropower unit be several discontinuous feasible power operation intervals. So the power of hydropower unit can be expressed linearly as [18]:

$$p_{h,t} = \sum_{s} P^{d}_{h,s} e_{h,s,t} + \sum_{s} \delta_{h,s,t} \tag{20}$$

$$\sum_{s} e_{h,s,t} = u_{h,t} \tag{21}$$

where $p_{h,t}$ is the power of hydropower unit h at time t; $\delta_{h,s,t}$ is the power in operation interval s of hydropower unit h at time t; P^{d}_{h} is the lower power limit of the operation interval s; $e_{h,s,t}$ denotes whether the power of hydropower unit h is in operation interval s at time t, with 1 means in interval s and 0 means not in; $u_{h,t}$ is the on/off status of hydropower unit h at time t.

The generation cost of hydropower unit is:

$$c_{h,t} = \sum_{s} C^{d}_{h,s} e_{h,s,t} + \sum_{s} \delta_{h,s,t} \beta_{h,s,t} \tag{22}$$

where $c_{h,t}$ is the cost of hydropower unit h at time t; $C^{d}_{h,s}$ is the cost of hydropower unit h at the power $P^{d}_{h,s}$; $\beta_{h,s,t}$ is the incremental cost in operation interval s of hydropower unit h at time t.

The feasible power operation interval of the hydropower unit is:

$$s_{h,t} = \sum_{s} s\, e_{h,s,t} \tag{23}$$

where $s_{h,t}$ is the operation interval of hydropower unit h at time t. If the hydropower unit is shut down, it is considered to be in the operation interval 0.

Frequent power crossing through the vibration zone will increase the wear of the unit. Therefore, the maximum number of times of crossing the vibration zone should be considered.

A binary variable $\theta_{h,t}$ is introduced to indicate whether the hydropower unit h crosses the feasible power interval during the time t, then:

$$\theta_{h,t} \leq |s_{h,t} - s_{h,t-1}| \tag{24}$$

$$S_h \theta_{h,t} \geq |s_{h,t} - s_{h,t-1}| \tag{25}$$

where S_h is the number of power operation intervals of the hydropower unit h.

Thus, the maximum number of times of crossing the vibration zone can be obtained:

$$\sum_t \theta_{h,t} \leq \Delta_{h,\max} \tag{26}$$

where $\Delta_{h,\max}$ is the maximum number of times of crossing the vibration zone of hydropower unit h.

The generation power of the hydropower station is:

$$p_{k,t} = \sum_{h \in k} p_{h,t} \tag{27}$$

where $p_{k,t}$ is the generation power of the hydropower station k at time t.

For cascade hydropower stations in the same basin, it is necessary to consider the coupling relationship between the upstream and downstream hydropower stations. The flow coupling relationship between cascade hydropower stations is:

$$f_{k,t}^{out} = p_{k,t} b_{k,t} \tag{28}$$

$$f_{k,t}^{in} = f_{k-1,t}^{out} + f_{k,t}^{in0} \tag{29}$$

where $f_{k,t}^{out}$ and $f_{k,t}^{in}$ are outflow and inflow of cascade hydropower station k at time t; $b_{k,t}$ is the water consumption of hydropower station k at time t; $f_{k,t}^{In0}$ is the natural inflow of hydropower station k at time t.

The water level of the hydropower station is:

$$h_{k,t} = h_{k,t-1} + (f_{k,t}^{in} - f_{k,t}^{out})/S_k \tag{30}$$

where $h_{k,t}$ is the water level of hydropower station k at time t; S_k is the cross-sectional area of reservoir water of hydropower station k.

The water level constraint of hydropower station is:

$$H_{k,\min} \le h_{k,t} \le H_{k,\max} \tag{31}$$

where $H_{k,min}$ and $H_{k,max}$ are the lowest water level and the highest water level of the hydropower station k.

(3) The model of conventional thermal unit

Generation limits of conventional thermal unit is:

$$P_{g,\min} u_{g,t} \le p_{g,t} \le P_{g,\max} u_{g,t} \tag{32}$$

where $P_{g,min}$ and $P_{g,max}$ are the minimum and maximum power limits of thermal unit g respectively; is the on/off status of thermal unit g at time t.

The other constraints of generation scheduling model such as ramp up and ramp down rate limits of conventional units, system spinning reserve requirement and network security constraints have been detailed described in [19, 20], which are omitted here.

4 Case Study

Based on the actual data of a regional power system in China, the proposed model is verified and analyzed.

4.1 The Data of Power System

The regional power grid consists of 4 provincial power grids interconnected through 16 HV AC/DC tie line channels. The whole regional power grid includes 771 units with a total installed capacity of 212215 MW, including 344 thermal power units with an installed capacity of 126124 mw, 363 hydropower units with an installed capacity of 73981 mw, 17 wind power units with an installed capacity of 8300 mw, 47 solar units with an installed capacity of 3810 mw. There are 10 flexible loads in total, of which 6 can be curtailable and the power can be curtailed by 10500 mw; 4 can be transferred and the power can be transferred by 9500 mw. The maximum load on the day is at 12:00, 164765 MW, and the minimum load is at 6:00, 104189 MW. We made the optimization every 60 min over 24 h a day. In order to verify the effectiveness of the proposed model, based on the actual schedules of units and AC/DC tie lines, the following three conditions are constructed in this section, and the calculation results of each condition are compared and analyzed. Wind power and solar power prediction curves have a great impact on the calculation results [21–24]. The same wind and solar power prediction curve is used in all three conditions in this paper.

C1: The power schedules of units and AC/DC tie lines are fixed as the actual power.
C2: The power schedules of AC/DC tie lines are fixed as the actual power, and as the boundary condition of each provincial grid. The power generation schedule of each province is optimized.

C3: The model proposed in this paper is used to optimize the power schedules of units and AC/DC tie lines in the whole grid with the constant transmission energy during the day.

4.2 The Optimization Results

Using the unit cost parameter of C3, the actual generation cost of C1 is calculated, and the total generation cost of C1 is 568832487.93 yuan. C2 optimizes the generation schedules of four provincial grids respectively, among which the longest time-consuming provincial grid calculation time is 1052.13 s, and the other provincial grids are all within 300 s (because the parallel calculation mode can be adopted, C2 calculation time can be considered as 1052.13 s), and the total generation cost of the whole grid is 547162706.10 yuan. The calculation time of C3 is 3749.71 s, and the total generation cost of the whole grid is 536296932.41 yuan.

In the actual dispatching operation of each province, the dispatching mode with the lowest generation cost is not adopted, which leads to the high cost of the whole grid generation in C1. In C2, the actual transmission power schedules of AC/DC tie lines in C1 are used to optimize the generation schedules of units in the province, and the total generation cost is 3.81% lower than that in C1. Based on C2, C3 further optimizes the power of the AC/DC tie lines, so that the generation cost of the whole grid is 1.99% lower than C2, and 5.72% lower than C1..

(1) The schedules of flexible load

The flexible load is not scheduled in C1. C2 optimize generation schedules for each province. C3 uses the model proposed in this paper to optimize the schedules of the generation units, flexible loads and inter province AC/DC tie lines. The schedules of the flexible loads in the whole grid is shown in Fig. 3.

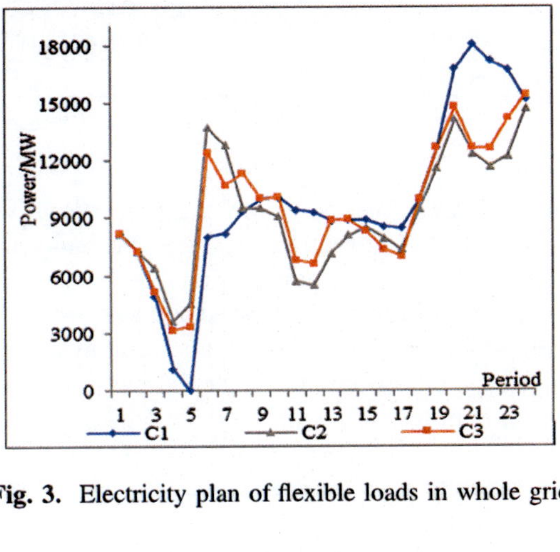

Fig. 3. Electricity plan of flexible loads in whole grid

It can be seen from Fig. 3 that the flexible load can be flexibly adjusted according to the overall situation of power generation and load of the whole grid, the curtailable flexible loads can curtailed the power consumption during the peak load period, and the transferable load can transfer the load between different time periods to achieve the effect of peak shaving and valley filling, reshape the load curve shape of the whole power system, make the system load curve more smooth and reduce the power supply cost of the whole grid. C1 is the actual operation of the system, and the flexible load is not scheduled. C2 and C3 have the same exchange energy of tie lines, but C3 can optimize the exchange power of tie lines, so there are more schedulable resources in the whole grid in C3 than in C2, which makes the adjustment range of flexible load in C3 smaller. The flexible load is adjusted greatly, so that the equivalent load curve of the whole network is stable. In C2, the flexible load is greatly adjusted, which makes the equivalent load of the whole grid relatively smooth. However, the tie line power of C2 cannot be further adjusted, resulting in the total generation cost of C2 higher than that of C3.

(2) Power generation analysis of multiple-type in the whole grid

The power generation of all types of units in the whole grid in three conditions is shown in Table 1.

Table 1. Generation capacity of multi-type units in whole grid

Energy type	Condition	Forecasted	C1	C2	C3
Wind	Energy/(MWh)	91449	59070	77352	80670
	Accommodation ratio/(%)	100.00	64.59	84.58	88.21
Solar	Energy/(MWh)	18150	15379	16667	17200
	Accommodation ratio/(%)	100.00	84.73	91.83	94.77
Hydro	Energy/(MWh)	–	1253891	1285535	1336001
	Accommodation ratio/(%)	–	70.62	72.40	75.24
Thermal	Energy/(MWh)	–	2035671	1969917	1921900
	Accommodation ratio/(%)	–	67.25	65.98	63.49

It can be seen from Table 1 that the accommodation ratio of wind and solar power in C1 is 64.59% and 84.73% respectively. In C2, the units in each province are optimized under the condition that the power of tie line remains unchanged, which can greatly improve the accommodation ratio of wind and solar power. In C3, the power of tie line is optimized, which further improves the accommodation ratio At the same time, through the optimization of unit power or tie line power, the hydropower consumption in the system is further improved, and the total generation cost of the whole grid is reduced.

The total power generation of wind and solar power in the whole grid in each condition is shown in Fig. 4.

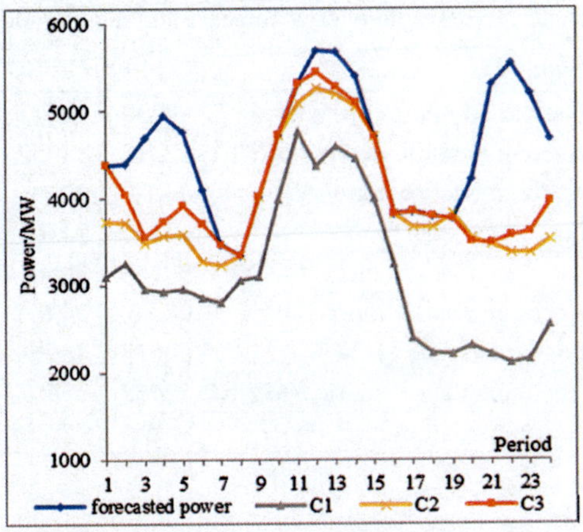

Fig. 4. Output of renewable energy in C3

C3 improves the accommodation ratio of renewable energy such as wind, solar and hydropower by reasonably scheduling the generation power of all types of units in the whole grid. Through the optimization of the power generation structure of the whole grid, the power supply cost of the large grid is significantly reduced. The generation power of wind, solar, hydropower and thermal power units in C3 is shown in Fig. 5. The wind power accommodation ratio is 88.21%, solar accommodation ratio is 94.77%, the daily average load rate of hydropower unit is 75.24%, and the daily average load rate of thermal power unit is 63.49%.

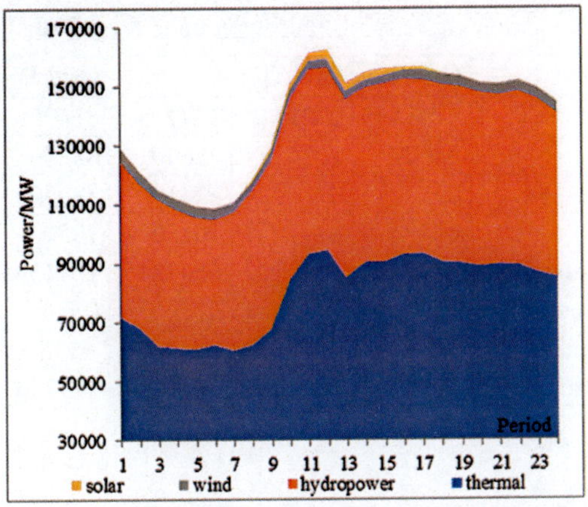

Fig. 5. Diagram of accumulative generation of multi-type units in C3

The forecasted energy and accommodation energy of wind and solar in C3 are shown in Table 2.

Table 2. Renewable energy forecast and accommodation

Energy type		Grid Q	Grid D	Sum
Wind	Forecasted energy/(MWh)	49938	41511	91449
	Accommodation energy/(MWh)	42318	38352	80670
	Accommodation ratio/(%)	84.74	92.39	88.21
Solar	Forecasted energy/(MWh)	10700	7450	18150
	Accommodation energy/(MWh)	10150	7050	17200
	Accommodation ratio/(%)	94.86	94.63	94.77
Sum	Forecasted energy/(MWh)	60638	48961	109599
	Accommodation energy/(MWh)	52468	45402	97870
	Accommodation ratio/(%)	86.53	92.73	89.30

It can be seen from Table 2 that the accommodation of wind and solar energy in the whole grid is 17200 MWh, and the accommodation ratio is 94.77%. In the valley load period, due to the low local load, a large number of external power transmission occupied the trans-provincial channel, resulting in a high curtailment of wind power. At 22:00, the curtailment ratio of wind power reached 35.64%.

In C3, the hydropower resources of provincial grid Q and D are rich, and a large amount of hydropower is directly or indirectly sent to the provincial grid Y of load center through provincial grid G. Provincial grid G is interconnected with provincial grid Q and Y through the AC tie lines, and the overall external performance is receiving power. See Table 3 for the power generation of multi-types of energy in the four provincial grids.

Table 3. Generation capacity of multi-type units for provincial grids in C3

Energy type	Grid Y	Grid G	Grid Q	Grid D	Sum
Thermal	1261335	228954	347915	83696	1921900
Hydro	24677	202628	226401	882295	1336001
Wind	0	0	42318	38352	80670
Solar	0	0	10150	7050	17200
Total generation	1286012	431582	626784	1011393	3355771
Energy received	707938	238264	0	0	946202
Energy sanded	0	−201638	−212564	−532000	−946202
Total consumption	1993950	468208	414220	479393	3355771

It can be seen from Table 3 that, provincial grid Y is the load center, and its external energy consumption accounts for 35.50% of the total load of the province; Grid G is a transfer grid, which mainly connects the sending grid and receiving grid, most of the power flow only flows through grid G; Provincial grid Q and D are the provinces sending out clean energy, and their outgoing energy accounts for 33.91%

and 52.60% of the total power generation of the province, respectively, of which the outgoing energy of provincial grid D has exceeded the load of the province, ensuring that clean energy is consumed in the whole grid.

(3) The energy flow in the whole grid

The overall flow of electric energy is from the western clean energy base to the eastern load center provincial grid, in which provincial grid D is a large province of hydropower resources, with 52.60% of the total generating energy transmitted across provinces; provincial grid Y is a large receiving end of the grid, with the external receiving energy accounting for 35.5% of the local load. The integrated optimization of the whole grid can effectively promote the accommodation of clean energy in a large space.

C1 and C2 have the same tie line power schedules. C3 optimizes the tie line power schedules with that the inter provincial exchange energy is unchanged. C3 considers the system load of the power grid at the sending and receiving end and the power generation of mutil-types of units, and reasonably uses the AC and DC channels between provinces to transmit the clean energy that could be curtailed at the sending end to the receiving end, which supports the power load during the peak period at the receiving end and improves the overall accommodation level of the clean energy of the whole grid.

C3 ensures that the exchange energy between provinces remains unchanged, but there is no requirement for the transmission power of the specific physical tie line, which further increases the optimization space of resource allocation, and promotes the more economical and reasonable schedules of the provincial AC/DC tie line channels.

5 Conclusions

In view of the practical background of the increasingly close coupling degree of dispatching at all levels under the rapid construction of HV AC/DC and large-scale energy bases, as well as the increasing practical demand of the accommodation of clean energy, this paper proposes an integrated optimization model for the dispatching schedule of the whole power grid. On the basis of considering the existing two-level dispatching mode and based on the practice, taking the tie line power as the coordination variable, the model optimizes the complementary characteristics of multi-type energy generation at the power generation side, the transmission operation characteristics of HV tie lines, and the flexible load response characteristics at the load side, which is in line with the dispatching mode of China's power grid with different levels and zones, and is helpful to improve the clean energy accommodation in large power grid. The integrated optimization model of the whole power grid can further improve the unified coordination and optimization ability of both sides of the generation source and load, change the traditional single scheduling mode of the power generation side, give full play to the flexible response ability of the schedulable load with the coordinated scheduling of generation, transmission and load, reduce the peak valley difference of the power grid, effectively improve the flexibility of the power grid regulation, and finally improve the accommodation of clean energy in the whole grid.

Acknowledgement. This work is supported by the Science and Technology Project of State Grid Corporation of China: Deepening Research on the Operation Mechanism of Inter Provincial Electricity Power Market.

References

1. Wang, B., Xia, Y., Xia, Q., et al.: Model and methods of generation and transmission scheduling of inter-regional power grid via HVDC tie-line. Autom. Electr. Power Syst. **40** (3), 8–13 (2016)
2. Han, H., Tu, M., Zhang, H., et al.: Day-ahead generation scheduling method considering adjustable HVDC plan and its analysis. Autom. Electric Power Syst. **39**(16), 38–143 (2015)
3. Zhong, H., Xia, Q., Ding, M., et al.: A new mode of HVDC tie-line operation optimization for maximizing renewable energy accommodation. Autom. Electr. Power Syst. **39**(3), 36–42 (2015)
4. Xu, F., Ding, Q., Han, H., et al.: Power optimization model and analysis of HVDC tie-line for promoting integration of inter-regional renewable energy accommodation. Autom. Electr. Power Syst. **41**(18), 152–159 (2017). https://doi.org/10.7500/AEPS20170401001
5. Xia, Y., Kang, C., Ning, B., et al.: A generation and load integrated scheduling on interaction mode on customer side. Autom. Electr. Power Syst. **36**(1), 17–23 (2012)
6. Yang, N., Wang, B., Liu, D., et al.: An integrated supply-demand stochastic optimization method considering large-scale wind power and flexible load. Proc. the CSEE **33**(16), 63–69 (2013)
7. Xu, F., Wang, Y., Yang, J., et al.: Generation scheduling model and application for wind-thermal power system considering security constraints. Autom. Electr. Power Syst. **38**(21), 114–120 (2014)
8. Chen, Q., Littler, T., Wang, H.: Optimal scheduling for coordinated wind and hydropower generation. Proc. CSEE **34**(34), 6074–6082 (2014)
9. Su, C., Shen, J., Wang, P., et al.: Coordinated dispatching method for wind-turbine-integrated power system with multi-type power sources based on power flexibility margin. Autom. Electr. Power Syst. **42**(17), 111–119 (2018). https://doi.org/10.7500/AEPS20171207001
10. Zeng, X., Liu, T., Li, Q., et al.: Short-term complementary optimal dispatch model of multi-source hybrid power system based on virtual power configuration strategy. Power Syst. Technol. **40**(5), 1379–1386 (2016)
11. Zhang, X., Jiang, J., Li, M., et al.: Multi-objective economic dispatch of power system with wind farms considering flexible load response. Autom. Electr. Power Syst. **41**(11), 61–67 (2017). https://doi.org/10.7500/aeps20160916001
12. Ye, L., Qu, X., Yao, Y., et al.: Analysis on intraday operation characteristics of hybrid wind-solar-hydro power generation system. Autom. Electr. Power Syst. **42**(4), 158–164 (2018). https://doi.org/10.7500/AEPS20170915010
13. Ren, L., Bai, Z., Yu, C., et al.: Research on active power control strategy for wind/photovoltaic/energy storage hybrid power system and its engineering application. Autom. Electr. Power Syst. **38**(7), 105–111 (2014)
14. Gao, M., Hui, D., Gao, Z., et al.: Presentation of national wind/photovoltaic/energy storage and transmission demonstration project and analysis of typical operation modes. Autom. Electr. Power Syst. **37**(1), 59–64 (2013)

15. Xu, D., Jiang, M., Zheng, X., et al.: Optimal modeling of dispatch scheduling for power grid with VSC-MTDC system. Autom. Electr. Power Syst. **41**(23), 22–28 (2017). https://doi.org/10.7500/aeps20170217003
16. Chen, C.L., Chen, Z.Y., Lee, T.Y.: Multi-area economic generation and reserve dispatch considering large-scale integration of wind power. Int. J. Electr. Power Ener. Syst. **55**(2), 171–178 (2014)
17. Wang, K., Yao, J., Yao, L., et al.: Survey of research on flexible loads scheduling technologies. Autom. Electr. Power Syst. **38**(20), 127–135 (2014)
18. Xu, F., Yao, J., Geng, J., Yang, Z.: Modeling and analysis of unit I/O characteristics based on mixed-integer programming. Autom. Electr. Power Syst. **34**(10), 45–50 (2010)
19. Lotfjou, A., Shahidehpour, M., Fu, Y., et al.: Security-constrained unit commitment with AC/DC transmission systems. IEEE Trans. Power Syst. **25**(1), 531–542 (2010)
20. Daneshi, H., Choobbari, L., Shahidehpour, M., et al.: Mixed integer programming method to solve security constrained unit commitment with restricted operating zone limits. In: Proceeding of IEEE International Conference on Electro/Information Technology, 18–20 May 2008, Ames, USA, pp. 187–192 (2008)
21. Ye, L., Zhang, C., Xue, H., Li, J.: Study of assessment on capability of wind power accommodation in regional power grids. Renew. Ener. **133**(Apr), 647–662 (2019)
22. Ye, L., Zhao, Y., Zeng, C., Zhang, C.: Short-term wind power prediction based on spatial model. Renew. Ener. **101**(Feb), 1067–1074 (2017)
23. Niu, D., Ji, H.: Quantitative analysis method for errors introduced by physical prediction model of wind power. Autom. Electr. Power Syst. **44**(8), 57–65 (2020)
24. Li, M., Lin, X., Zhang, Z., et al.: Interval prediction algorithm for ultra-short-term photovoltaic output and its application. Autom. Electr. Power Syst. **43**(3), 10–16 (2019)

An Active Power Comprehensive Control Scheme for Grid Connection of Renewable Energy

Yang Lei[1], Chang Chang[2(✉)], Zhang Jie[3], Xi Xinze[1], He Peng[1], Xiang Chuan[1], and He Xin[1]

[1] Electric Power Research Institute, Yunnan Power Grid Co., Ltd., Kunming 650217, Yunnan Province, China
[2] School of Electrical Engineering, Southwest Jiaotong University, Chengdu 611756, Sichuan Province, China
chris_e_mail@163.com
[3] Yunnan Power Dispatch Center, Kunming 650011, Yunnan Province, China

Abstract. As the proportion of renewable energy in the power system continues to increase, renewable energy plays an increasingly important role in the power grid. Renewable energy grid connection not only needs to improve its adaptability to the grid, but also needs to increase its active frequency and voltage regulation capabilities to improve the stability of the grid frequency and voltage at high penetration rates. This paper proposes an active power comprehensive control scheme and system for grid connection of renewable energy based on the current engineering status. It can effectively avoid the communication delay and out-of-sequence problems caused by traditional independent active control, and has the characteristics of convenient transformation and low investment cost. The scheme was verified to be feasible, through engineering transformation and field testing at a large-capacity photovoltaic power station in Yunnan Power Grid. And it proposes a highly scalable, economical, efficient and reliable reference scheme for power control of renewable energy grid connection.

Keywords: Fast power control · Yunnan Power Grid · Photovoltaic power generation

1 Foreword

Renewable energy has accounted for more than 50% of the operation mode of Yunnan Power Grid. After testing and verification in multiple regions, the active and reactive power of renewable energy is easy to decouple, so it has a superior control method compared to traditional generators. It has become a current development trend for renewable energy to participate in the power control of the system.

At present, there are many studies on the use of photovoltaic power stations to participate in the active power control of the power grid. Reference [1] theoretically analyzed the description function of primary frequency control in detail, established an analysis model with nonlinear links, and analyzed the limit range of gain parameters

Y. Xue et al. (Eds.): PMF 2020, LNEE 718, pp. 262–275, 2021.
https://doi.org/10.1007/978-981-15-9746-6_21

through nonlinear Nyquist theorem. It provides theoretical support for the determination of primary frequency control parameters. Reference [2] proposes a renewable energy inverter control strategy by analyzing the deficiencies of traditional renewable energy Automatic Generation Control (AGC), which can greatly reduce the response delay caused by signal transmission. Reference [3] estimates the cost and benefit of each energy storage device according to the model and indicators of dynamic reliability evaluation from the perspective of optimal economics, and then calculates the optimal plan based on the actual indicators of the photovoltaic power station. Reference [4] proposes a control strategy to achieve primary frequency control on the basis of active standby, and designs secondary frequency modulation based on different AGC control modes. In this way, differential adjustment is achieved.

The existing literature and standards are mainly for the research and improvement of the grid-connected adaptability and power control initiative of renewable energy. However, the control of renewable energy at high permeability will continue to be in the stage of continuous research and exploration. For example, the guideline for safety and stability of power systems GB 38755-2019 points out that the power supply should have the ability of primary frequency control, fast voltage regulation, and peak shaving, and should meet the requirements of relevant standards. The stations with higher proportion of renewable energy grid-connected power generation should provide the necessary inertia and short-circuit capacity support. In summary, the grid connection control requirements of renewable energy stations exist difference in different application scenarios, and the control scheme is also constantly being explored and improved. In order to adapt to the changing grid connection requirements and continuous engineering transformation, and to improve the adaptability, flexible expandability and economic reliability of renewable energy stations, this paper proposes a comprehensive control scheme and system for renewable energy grid connection. It effectively meets the continuous improvement of renewable energy grid control requirements, saves economic costs, and improves the reliability of each control logic. Through the engineering transformation and field test of a large-capacity photovoltaic power station in Yunnan Power Grid, the rapid response of the scheme to the system frequency change and the accurate and stable control of power are verified, so as to provide theoretical and engineering support to the scheme designed in this paper.

2 Current Power Control Strategy and Frequency Control Technical Specifications

2.1 Current Power Control Strategy

Yunnan Power Grid currently mainly controls the active power of the power grid by load shedding, DC frequency control, primary frequency control of traditional generators and AGC commands [5]. But its response speed is difficult to achieve immediate active power compensation, and the system frequency also cannot be stabilized quickly. At present, the response speed of each power control mode can be intuitively reflected through Table 1.

Table 1. Response time of current power control mode

Control method	Response time
Load shedding	0–3 s
DC frequency control	0–5 s
Traditional generators frequency adjustment	3–30 s
AGC instruction	$\geq$ 30 s
Start standby generators	$\geq$ 1 min

It can be seen from Table 1 that the fastest response is load shedding, however due to the large difference in load in each period, the part of the load that can be operated is not stable and is not conducive to cost control. Other control methods respond slowly, the faulty part is likely to cause more serious frequency fluctuations during a response. In order to maintain the stable operation of renewable energy, most of the renewable energy in Yunnan Power Grid is not equipped with frequency control function. Since the installed capacity of renewable energy in Yunnan Power Grid is already very large, if it can be used to participate in the frequency and active power control of the power grid, the safety and stability control pressure of Yunnan Power Grid can be greatly alleviated.

At present, most of the renewable energy that participate in AGC power control are also very slow in response, mainly due to the upper layer transmission of the communication path [6]. Currently, AGC-controlled photovoltaic power plants generally adopt a layered structure (see Fig. 1).

2.2 Primary Frequency Control Technical Specifications

According to the PV grid-connected technical specifications issued by the State Grid, the main parameters of the primary frequency control are regulated by the specifications [7–9]. The main technical indicators can now be summarized as shown in Table 2.

The adjustment rate δ is the reciprocal of the slope of the droop curve, from which the power of the droop curve can be expressed by Eq. (1).

$$P = P_0 - \frac{1}{\delta\%} \times P_N \times \frac{(f - f_d)}{f_N} \tag{1}$$

In the formula, f is the actual frequency of the system, f_N is the rated frequency, f_d is the dead zone of primary frequency control, P_N is the rated power, and P_o is the initial value of active power.

The parameters in the technical specifications are recommended values, and the specific situation still needs to be adapted to local conditions. When the dead zone is ±0.5 Hz and the load limit is 10% PN, the droop curve can be obtained (see Fig. 2).

When the photovoltaic power plant responds to frequency changes, it also needs to meet the dynamic response time. The main parameters can be briefly described as shown in Table 3.

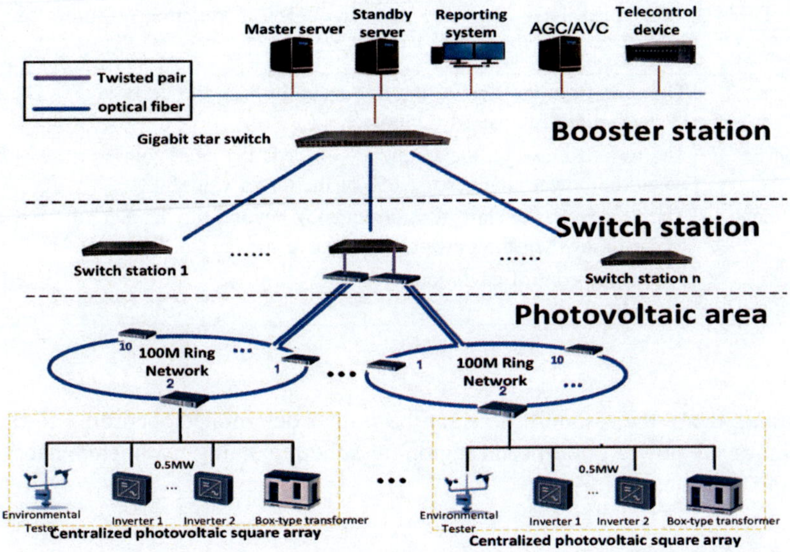

Fig. 1. Control structure diagram of photovoltaic power station. AGC needs to pass the control signal to the switch at first, and pass it down to the switch station, and then distribute it to each inverter in the photovoltaic power plant through the ring network, so as to issue the power control command. This process requires a response time of about 20 s. Obviously, current control strategies are difficult to meet the dynamic demands of the power grid.

Table 2. Technical index of photovoltaic primary frequency control

Parameter	Value
Dead zone (Hz)	−0.05–0.05
Adjustment rate δ	2%–3%
Maximum load limit (MW)	≥ 10% P_N
Minimum power to start frequency control (MW)	≥ 10% P_N

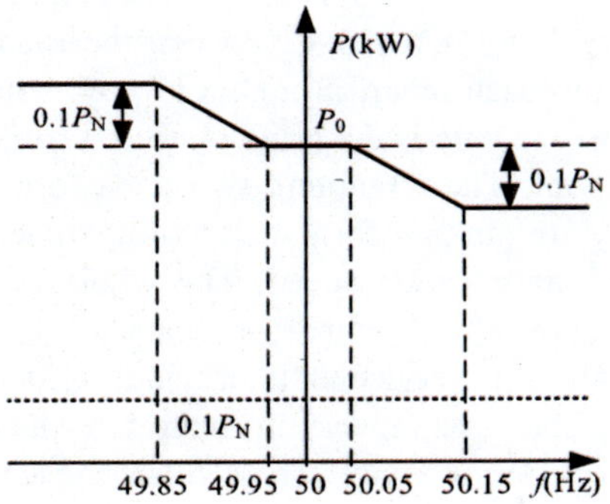

Fig. 2. Photovoltaic primary frequency control droop curve.

Table 3. Technical index of photovoltaic frequency modulation response time

Parameter	Meaning	Index
Lag time	The time from the start of the frequency beyond the dead zone to the initial response of the power	≤ 3 s
Response time	The time from when the frequency exceeds the dead zone to when the power changes to 90% of the target value	≤ 5 s
Adjustment time	The time from the start of the frequency beyond the dead zone to the time when the power converges to the control deviation	≤ 15 s
Control deviation	Frequency control deviation	$\pm 0.01\ P_N$

Among them, the response time can be further determined according to the actual situation of the power grid in each region, and the remaining parameters need to meet the response time.

3 Active Power Comprehensive Control Scheme for Grid Connection

In order to meet the function of renewable energy actively participating in system frequency control, the grid-connected comprehensive control system proposed in this paper consists of four main parts. They are comprehensive control device, monitoring service unit, fast communication optical fiber network and fast ring network communication control terminal. Among them, the grid-connected comprehensive control device mainly realizes fast active power control and emergency control, and provides frequency and power measurement functions for primary frequency control. The monitoring service unit can provide service functions such as AGC, AVC, primary frequency control, etc. It also has powerful communication, data record management, event recording or query, and human-computer interaction functions. The fast communication optical fiber network, to ensure the timeliness of fast-controlled communication, mainly provides special information and control command transmission network for fast power control. The fast ring network communication control terminal mainly realizes fast information exchange between the system and the inverter. The four parts are connected to each other through a layered structure (see Fig. 3).

Among the four parts, the core is the grid-connected comprehensive control device and monitoring service unit. The remaining two parts form a dedicated line for fast signal transmission, they are the transformation and upgrade of the original devices in order to improve the communication speed. The whole system mainly realizes two functions in terms of active power: primary frequency control and active power stability control. The monitoring service unit provides grid-connected comprehensive control device with frequency and power information of the current system. The grid-connected comprehensive control device can calculate the active power required by the system according to the current frequency deviation through an internally designed algorithm. Then the calculation result is transferred to the inverter of each photovoltaic

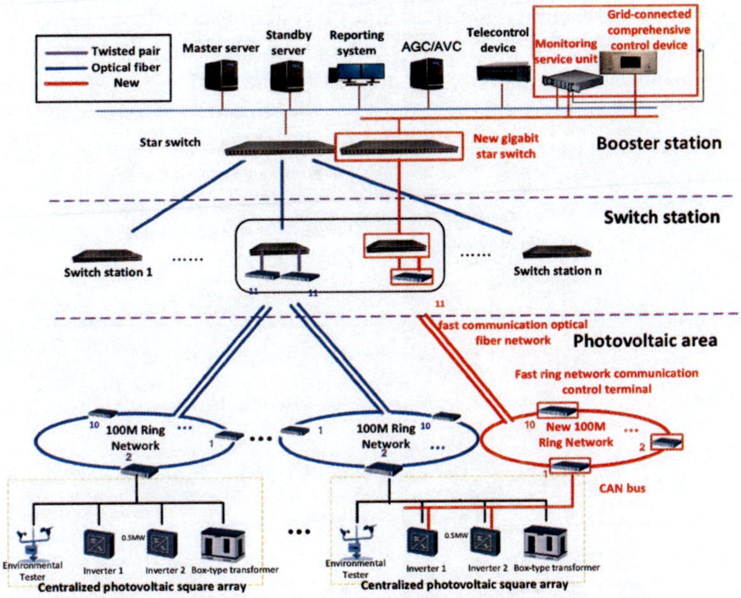

Fig. 3. Grid-connected comprehensive control system structure. The control structure can be divided into four parts: grid-connected comprehensive control device, monitoring service unit, fast communication optical fiber network, and fast ring network communication control terminal.

power station through the conversion signal down the fiber path, so as to realize the primary frequency control. In addition, when the AGC transmits the reference amount of active power output downward, it can also calculate the active output amount suitable for each photovoltaic unit according to the illumination and active status, so that the photovoltaic power station output close to the reference active power, so as to realize active power stability control function.

The grid-connected comprehensive control system is connected to the active power stability control system through a dedicated 2 M port. The stability control system uploads the entire station's switchable or adjustable power information, and then issues power control commands to the grid-connected integrated control system (see Fig. 4).

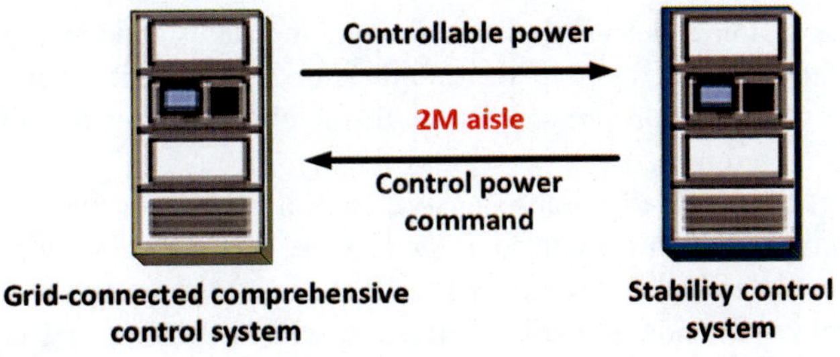

Fig. 4. Schematic diagram of data exchange between the integrated grid control system and the superior stability control system.

The grid-connected comprehensive control system established in Fig. 3 can communicate internally through the newly built optical fiber ring network to realize the downward transmission of superior commands (see Fig. 5).

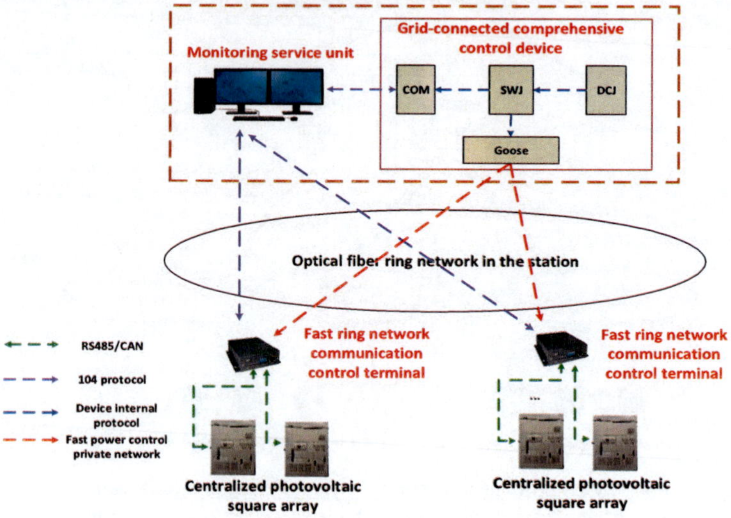

Fig. 5. Information exchange within the grid-connected comprehensive control system.

The strategy of this paper can detect the frequency state of the region through grid-connected comprehensive control device and monitoring service unit. When a frequency offset occurs in the system, grid-connected comprehensive control device can obtain the active compensation value through the frequency modulation algorithm. At the same time, the active power distribution amount is calculated according to the working status of each inverter of the photovoltaic power station. The frequency control function can be realized simultaneously by changing the active power adjustment value of the inverter. At the same time, the active power stability control command issued by the stability control system can accurately control the active power output of each inverter, thereby realizing accurate control of the active power.

The process of achieving fast active power control is as follows:

1) The inverter terminal detects the current operating status and the active power of the inverter, and uploads them to the monitoring service unit through the fast ring network communication control terminal, thereby realizing real-time state observation and recording.
2) After the grid-connected comprehensive control device receives the general command for fast active power control issued by the superior stability control system, it quickly calculates and decomposes it into the control commands of each inverter. The control commands of each inverter are sent to the fast ring network communication control terminal through the fast communication optical fiber network using the GOOSE protocol.

3) Using the fast ring network communication control terminal in the inverter area to make up the three fiber ring networks shown in Fig. 3. The fast ring network communication control terminal converts the GOOSE command into a CAN command acceptable to the inverter to control the inverter power output.
4) Each inverter sends active power according to the received instruction, and uploads the active power output to the monitoring service unit through the fast communication optical fiber network again. If there is any fault information, it is immediately uploaded to the control center for feedback.

4 Test Verification

The feasibility of the strategy in this paper is tested by conducting field tests at a large-capacity photovoltaic power station in Yunnan Power Grid. Mainly test the response of the photovoltaic power station under the frequency step of the system, the actual frequency disturbance, and the stability control command.

The test firstly confirms that the photovoltaic power generation unit is in a normal working state, and in a high-power output operating mode ($0.5P_N \leq P \leq 0.9P_N$). The operating effect of fast power control is verified by switching the detection signal of its grid connection point to a frequency disturbance signal (see Fig. 6).

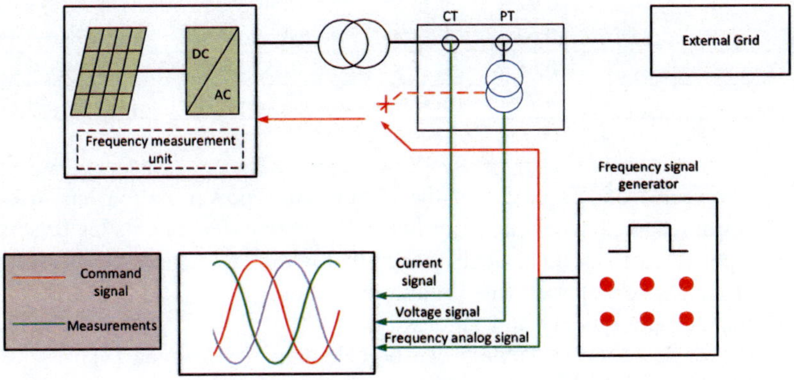

Fig. 6. Field test wiring diagram. Switch the measurement signal line at the PT point to a frequency simulation signal to simulate the frequency change when a fault occurs.

4.1 Frequency Step Disturbance Test

According to the wiring diagram shown in Fig. 6, the primary frequency control is fully tested by inputting two signals of one-sided frequency step and two-sided frequency step at the frequency signal generator. To better test the primary frequency control, the two signals are tested with two different primary frequency control parameters, as shown in Table 4.

Table 4. Primary frequency control parameter table for field test

Parameter name	Accuracy	Disturbance frequency	Cross-change frequency
Dead zone (Hz)	0.001	49.94–50.06	
Adjustment rate	0.1%	2.0%	
Rated power (MW)	0.01	30.00	
Load limiting	0.1	10.0%P_N	20.0%P_N
Control cycle (s)	0.01	5.00	

Since the photovoltaic power station does not achieve power-limited operation, it still maintains the MPPT working state under frequency disturbance [10, 11]. This field test mainly tests the step response during frequency disturbance. The following Table 5 is the response time of some tests.

Table 5. One-sided frequency step response schedule

Scrambling frequency (Hz)	Lag time	Response time	Adjustment time	Control deviation
50.1	0.7 s	1.39 s	2.29 s	0.57%
50.2	0.78	1.1	1.42	0.27%
50.4	1.0 s	1.62 s	1.9 s	0.2%
51	0.81 s	1.17 s	1.78 s	0.2%
51.3	0.72 s	1.32 s	1.46 s	0.3%
51.5	0.67 s	1.33 s	1.75 s	0.77%
51.8	0.88 s	1.27 s	1.58 s	0.37%
52	0.7 s	1.37 s	1.64 s	0.3%

It can be seen from Table 5 that the photovoltaic power station has made a frequency control response when the disturbance frequency step, and its lag time is within 1 s, the response time can be basically controlled within 1.5 s, and the adjustment time is also within 2 s. Compared with traditional control strategies, the strategy in this paper responds more quickly and efficiently.

The control deviation in Table 5 can be calculated by formula (2).

$$\Delta = \frac{|\Delta P_{\text{a}} - \Delta P_{\text{t}}|}{P_N} * 100\% \tag{2}$$

In the formula, $\varDelta P_t$ is the theoretical power offset, $\varDelta P_a$ is the actual power offset, and P_N is the rated power.

It can also be seen from Table 5 that the control deviations are all within 1%, and it can be seen that this strategy can not only complete primary frequency control quickly, but also meet the requirements of its control accuracy. The on-scatter frequency step response field test waveform is also recorded (see Fig. 7).

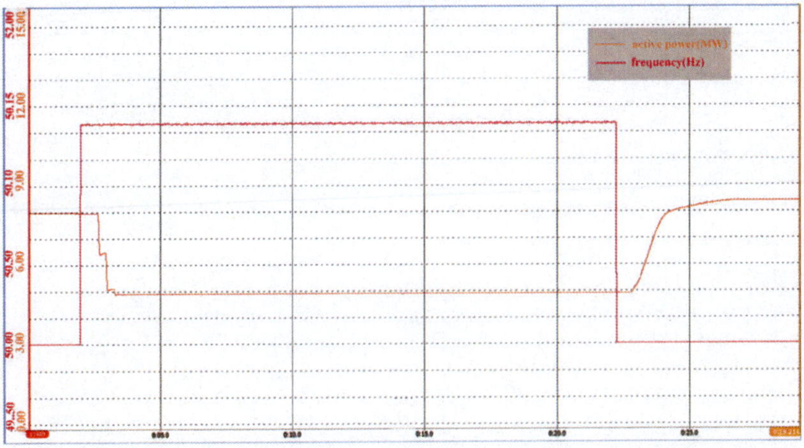

Fig. 7. One-sided frequency step response waveform (51.3 Hz).

In Fig. 7, The frequency-power waveform measured at the upset frequency of 51.3 Hz, where the purple line is the frequency step signal and the red line is the power response curve of the feeder. It can be seen that the active power starts to adjust at a certain time after the frequency crosses the dead zone. This is mainly a delay caused by logical judgment and communication interaction. When the frequency falls within the dead zone range, the photovoltaic power station immediately restores the MPPT state.

In order to test whether the response is normal when the frequency is stepped up and down, the input frequency is re-tested across the changing signal, and the response time is shown in Table 6.

Table 6. Two-sided frequency step response schedule

Cross-change frequency (Hz)	Lag time	Response time	Adjustment time	Control deviation
49 → 50.8	1.39 s	2.0 s	3.87 s	0.2%
49 → 51.5	1.35 s	1.92 s	3.71 s	0.43%
51.5 → 49	0.96 s	1.72 s	3.85 s	0.57%

In Fig. 8, the power response graph of the frequency changing from 50 Hz to 51.5 Hz and then to 49 Hz, and finally back to 50 Hz. It can be seen that when the frequency span changes, the low-band power adjustment amount is greater than the theoretical value. This is because when the frequency spans from high frequency to low frequency, the program records two events, and the power control is released from the high frequency into the dead zone., To restore the MPPT state, and then enter the low frequency from the dead zone. Due to the partial power release, the low-band power adjustment amount is the superposition of the released power and the theoretically calculated adjusted power.

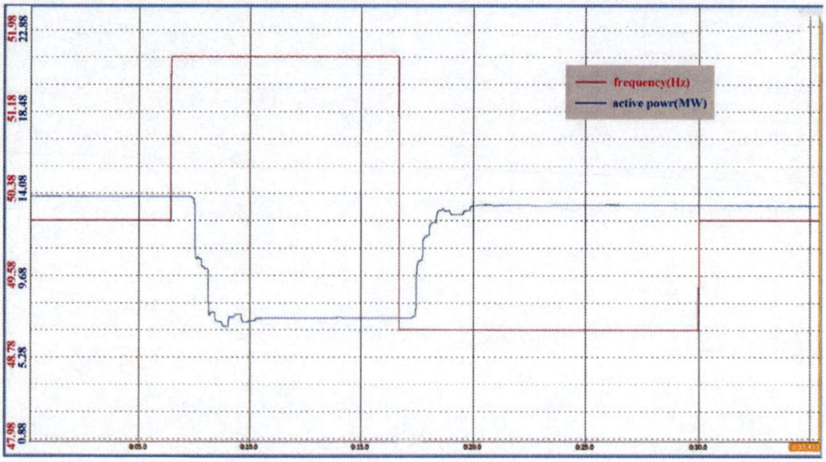

Fig. 8. Two-sided frequency step response waveform (51.5 Hz to 49.0 Hz).

4.2 Actual Frequency Disturbance Test

After the above test analysis, it can be seen that the fast power control system can make the photovoltaic power plant complete a frequency adjustment to suppress the frequency fluctuations. However, in practice, the frequency fluctuations are not simple step changes [12, 13]. This experiment records the frequency under the blocking of the Yunnan DC line Oscillate the waveform to the PMU to test the response of the fast power control in the actual waveform. Figure 7 shows the test waveform when a certain DC line is locked, and its frequency fluctuates between 50.24–50.05 Hz (see Fig. 9).

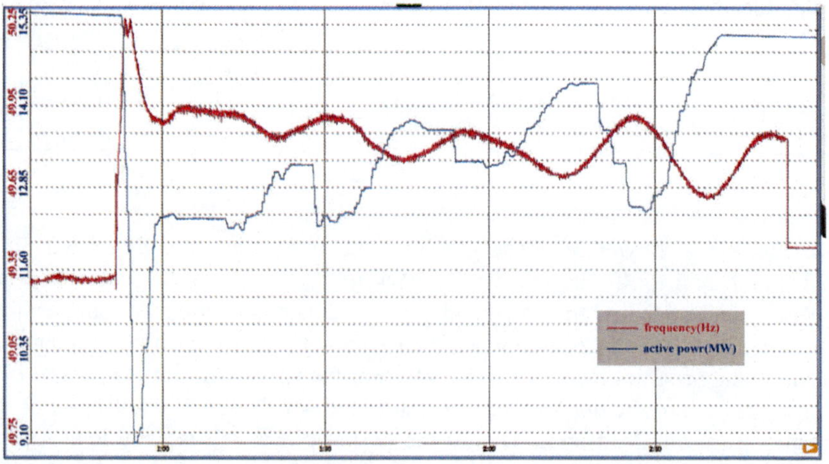

Fig. 9. The power response of the photovoltaic power station under actual frequency oscillation.

In Fig. 9, the purple line is the system frequency curve, and the blue line is the power response curve. After measurement, the highest frequency is 50.24 Hz, and the theoretical maximum adjustment value can be obtained by calculating formula (1).

$$\Delta P = \frac{(50.24 - 50.06)/50}{0.02} * 30 = 5.4(MW) \tag{3}$$

According to the recorded data, when the frequency crosses the dead zone, the active power is 15.23 MW, and the maximum adjustment value is 10.13 MW. It can be known that the actual adjustment value is 5.1 MW, from which the qualified rate of primary frequency control output response can be calculated as shown in calculation formula (4).

$$\frac{5.1}{5.4} * 100\% = 94.4\% \tag{4}$$

It can be seen that, the photovoltaic power can respond to the actual frequency oscillation in real time under the strategy, and its active output is very close to the theoretical value, thus verifying the practical feasibility of the control strategy.

4.3 Stability Control Command Response Test

At present, most Yunnan power grid photovoltaic power plants do not have power stability control capabilities, and this basic capability is very important for the power compensation of renewable energy sources under high penetration rate [14, 15]. This test passed the power command issued by the superior, and the test is fast. For the stability control capability of power control, Table 7 shows the implementation status of the active output of the photovoltaic power plant under each command (Fig. 10).

Table 7. Stability control command response schedule

Stability control command (MW)	Response time (s)	Adjustment time (s)	Actual adjustment (MW)	Control deviation
1	0.056	0.140	1.08	0.26%
5	0.039	0.162	4.98	0.07%
10	0.065	0.200	9.69	1.0%
15	0.067	0.219	14.42	1.9%
18	0.073	0.231	17.43	2.2%

It is not difficult to see from the results of the stability control test that when the stability control command changes from 1 MW–18 MW, the inverter power adjustment time is 0.14–0.231 s, and the 90% reduction time is 0.056–0.073 s. The actual power adjustment deviation is 0.26%−2.2%; The communication method and algorithm need to comprehensively consider the time of 90% drop, power adjustment time, and adjustment deviation.

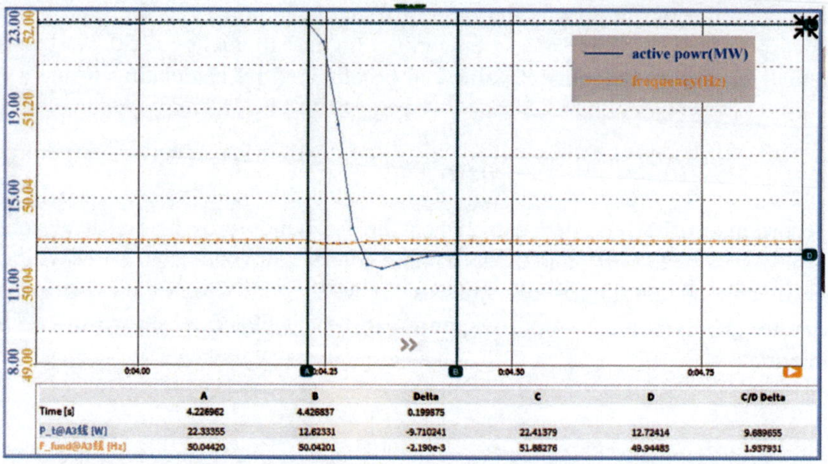

Fig. 10. Response waveform under 10 MW stability control command.

5 Conclusion

By analyzing the current power regulation status of Yunnan Power Grid and its renewable energy control system structure, this paper proposes a photovoltaic power stations grid-connected active power comprehensive control scheme. The specific work can be summarized as follows:

1) This strategy can realize the function of primary frequency modulation and active power stability control. At the same time, it is also flexible enough to facilitate the modification of control strategies, which can ensure the order of each function. The feasibility of the scheme was verified through on-site renovation and testing.
2) This strategy effectively shortens the communication time and improves the response speed of primary frequency control by setting up a fast ring network communication control terminal. The GOOSE message achieves active and stable control. The actual test shows that adjusting 90% of the response time is less than 73 ms, which can meet the control requirements of the stability control system.
3) The reliability of primary frequency modulation control scheme under different grid and system operating conditions is verified by inputting the actual grid frequency disturbance waveform. Field test shows that when the frequency crosses from high frequency to low frequency, the program records two events, and the MPPT state is restored when entering the dead zone. Due to the release of part of the power, the power adjustment amount in the low frequency band will be the superposition of the released power and the theoretically adjusted adjustment power, resulting in deviation of the control amount.
4) Field test shows that the adjustment time of active power stability control will increase with the increase of power adjustment, the deviation is less than 20 ms. At the same time, the power adjustment time is related to the communication method and algorithm, and the response time, adjustment time, and adjustment deviation

need to be comprehensively considered. Reducing the response time may cause the power adjustment time to increase and the adjustment deviation to increase.

The renewable energy grid-connected active power integrated control scheme proposed in this paper provides a flexible and efficient engineering solution for renewable energy to continuously adapt to the gradually changing control requirements of the power grid. It also meets the needs of different power grids and operating conditions. In the actual test, we also face the coordination problems of various functions of the program, such as the coordination of primary frequency control and stability control, and the meeting of high-cycle operating conditions within primary frequency control startup cycle. The scheme and system control need further study and improvement.

References

1. Zhao, D., Ma, J., Qian, H., et al.: Research on control gain of photovoltaic power station participating in primary frequency control of large power grid. Power Syst. Technol. **43**(2), 425–432 (2019)
2. Wang, S., Sun, G., Yu, C., et al.: Photovoltaic system-level rapid power regulation technology and its application. Chin. Soc. Electr. Eng. **38**(21), 6254–6263 (2018)
3. Xin, H., Liu, Y., Wang, Z., et al.: A new frequency regulation strategy for photovoltaic systems without energy storage. IEEE Trans. Sustain. Energy **4**(4), 985–993 (2013)
4. Duan, Z., Li, B., Huang, X., et al.: Photovoltaic power generation adapted to different AGC control modes participates in grid frequency regulation. Renew. Energy **38**(3), 373–379 (2020)
5. DL/T 1870–2018, Technical Specification for Power System Network Source Coordination. China Electric Power Press, Beijing (2018)
6. Sun, X., Liu, X., Cheng, L., et al.: Parameter setting scheme for fast frequency response of renewable energy sources in large power grids in northwestern China based on coordinated control of multi-frequency resources. Power Syst. Technol. **43**(5), 1760–1765 (2019)
7. Wang, S., Duan, S., Hou, W., et al.: Photovoltaic power plant layered distributed automatic power control technology. Autom. Electr. Power Syst. **40**(8), 126–132 (2016)
8. GB/T 19964-2012, Technical regulations for the connection of photovoltaic power stations to the power system. China Standard Press, Beijing (2013)
9. Q/CSG 1211002-2014, Technical specification for photovoltaic power station access to the power grid. China Southern Power Grid Co. Ltd. Guangzhou (2014)
10. Ding, M., Wang, W., Wang, X., et al.: Summary of the impact of large-scale photovoltaic power generation on power systems. Chin. Soc. Electr. Eng. **34**(1), 1–13 (2014)
11. Jiang, X., Ding, Y., Quan, S., et al.: Influence of photovoltaic penetration rate on power system frequency and control strategy. J. Wuhan Univ. **48**(5), 686–696 (2015)
12. Xu, P., Liu, F., Liu, B., et al.: Analysis, comparison and improvement of several MPPT methods for photovoltaic systems. Power Electron. Technol. **41**(5), 3–5 (2007)
13. Liu, J.: Basic problems of large-scale renewable energy power safe and efficient utilization. Chin. Soc. Electr. Eng. **33**(16), 1–8 (2013)
14. Chen, G., Li, M., Xu, T., et al.: Research on the technical bottleneck of renewable energy development. Chin. Soc. Electr. Eng. **37**(1), 20–26 (2017)
15. Jia, Q., Yan, G., Zhang, S., et al.: Dynamic coordination mechanism of grid frequency regulation with multiple photovoltaic generation units. Autom. Electr. Power Syst. **43**(24), 59–66 (2019)

Research and Engineering Practice on Voltage and Reactive Optimal Control of Magnetically Controlled Reactor

Dawei Xu[1,2(✉)], Tianhua Chen[1,2], He Li[3], Huijun Li[1,2], and Lei Du[1,2]

[1] NARI Group Corporation/State Grid Electric Power Research Institute, Nanjing 211106, China
xudawei@sgepri.sgcc.com.cn
[2] NARI Technology Co., Ltd., Nanjing 211106, China
[3] State Grid Tongliao Power Supply Co., Ltd., Tongliao 028000, China

Abstract. As a reactive power compensation device with stepless regulation, Magnetically controlled reactor (MCR) can realize continuous and smooth regulation of reactive power in the system, avoiding the problems of reactive over-compensation and operating over-voltage caused by traditional fixed-load switching reactor, which is an important mean to optimize voltage and reactive in power grid. This paper proposes a technical scheme for the MCR to participate in the AVC coordinated control. It elaborates the control principle and control strategy, which can effectively exert the rapid response and precise adjustment characteristics of the MCR, realize the rapid elimination of the voltage over-limit and the reactive power optimization of the whole network. The results of engineering practice show that the MCR can quickly and accurately respond to the control commands issued by the AVC system, and the optimization effect of voltage and reactive power is obvious, which proves that the control scheme is reliable and effective.

Keywords: MCR · AVC · Coordinated control · Engineering practice

1 Introduction

1.1 Research Background and Significance

Power system reactive and voltage optimization and control is an important means to increase the voltage pass rate, reduce network loss, and improve the operation safety of grid. AVC, full name as Automatic Voltage Control, is a combination of security and economy, which can achieve closed-loop control under security constraints. The power dispatch system with automatic voltage control function is recognized as the highest stage of the development of power dispatch control [1–3]. At present, the reactive power compensation devices widely used in substations of regional power grid are mainly parallel capacitor banks and reactor banks controlled by switches [4]. These traditional fixed-capacity devices have obvious disadvantages: the capacity is not adjustable and the regulation range is large, which are unable to adapt to complex system operating conditions and load changes in today's increasingly complex grid

Y. Xue et al. (Eds.): PMF 2020, LNEE 718, pp. 276–289, 2021.
https://doi.org/10.1007/978-981-15-9746-6_22

structure. With the development of ultra-high voltage technology, some 220 kV substations are prone to reactive power reverse transmission, and the large-scale use of cables in the transformation of urban power grids is likely to cause overvoltage problems due to capacitance effects at light or no load at the end of the line [5, 6]. Therefore, it is necessary to add adjustable voltage and reactive power control means in the system to meet the development needs of smart grid [7–9].

In recent years, MCR, full name as magnetically controlled reactor, has been more and more widely used in some large and medium-sized power grid systems. MCR overcomes the shortcomings of traditional saturable reactors, such as slow response speed, high loss, high noise, and rich harmonic content. It exhibits fast, flexible and continuous response characteristics, while the harmonic characteristics during operation have also been significantly improved. Compared with the switching-type reactor compensation device, the MCR avoids the limitation of the compensation accuracy caused by the stepped switching on the one hand, and does not cause frequent over or under compensation. On the other hand, it also avoids over-voltage, inrush current caused by frequent switching. These problems above-mentioned are likely to cause equipment breakdown and voltage operation reliability decreases. Therefore, on the basis of the existing reactive power compensation device, adding or replacing MCR will significantly improve the effect of reactive power compensation, realize the stepless adjustment of reactive power capacity, reduce the frequency of adjustment, save energy and reduce consumption. Compared with the existing reactive power compensation device, MCR is easy to maintain and operate, and has strong flexibility and adaptability [10]. This paper proposes a technical method and implementation plan for MCR participation in regional dispatch AVC coordinated control system, and explains the corresponding control objectives and strategies. Through the results of actual application on site, the method is feasible and effective, which lays a solid foundation for further deepening and application of the technical scheme in the future.

1.2 Introduction of AVC System

The AVC system in regional power grid generally adopts an integrated design with the intelligent dispatch control system. It obtains real-time data from SCADA and control models from network analysis application, performs centralized monitoring of substations, and controls the main transformer on-load voltage regulation device and reactive power compensation device to achieve closed-loop optimization of the voltage and reactive power of the entire network. According to the real-time position of switch and knife switch in SCADA, AVC system can automatically partition the network topology to realize regional automatic control of substation. The control flow chart of the system is shown in Fig. 1.

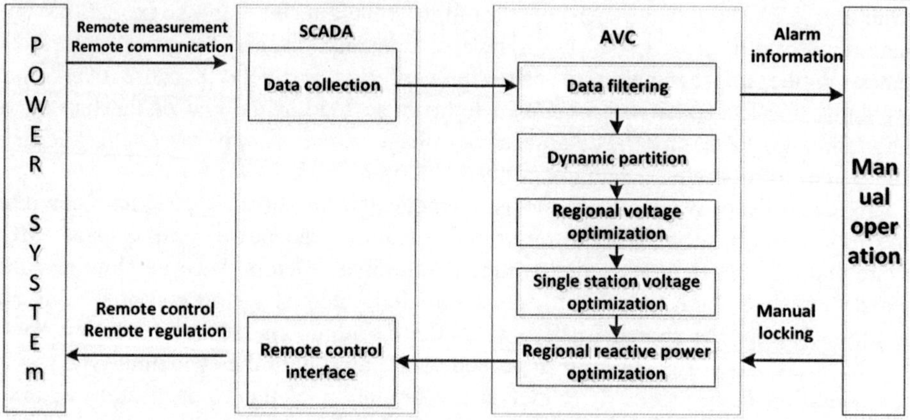

Fig. 1. The control flow chart of AVC system

AVC continuously improves the advancement on the basis of ensuring the practicality of the algorithm. It uses a mixed algorithm of analytical algorithm and expert rule to solve. The analytical algorithm includes real-time network loss sensitivity optimization analysis calculation. The kernel and hybrid algorithm ensure the robustness of the control system and the accuracy, speed and reliability of the control decision.

AVC adopts a substation/region two-level control system, which can set the substation to control locally or by region. In local control, it acts according to the basic principles of the nine-zone map, and in regional control, it participates in the optimization and coordination of the entire network. AVC has three control modes: regional voltage control, local voltage control and regional reactive power control, and automatically selects according to the reactive voltage distribution status. You can also manually set the cast/back control mode to ensure the flexibility of the control method.

1.3 Introduction of MCR

MCR is a direct-mounted inductive reactive power adjustment device, which is composed of the reactor body, thyristor control box and indoor control panel. The control panel collects the system voltage, current, active power, reactive power and other information in real time, and then changes the control angle of the thyristor according to different control targets, adjusts the saturation of the MCR core, and thus controls the MCR output capacity. The parallel MCR has a wide range of capacity changes and a large adjustment depth. It is continuously adjustable within the range of 0.5% to 110% of the rated capacity and allows long-term stable operation at any capacity [11, 12].

MCR has a variety of control modes such as set capacity mode, voltage control mode, reactive power flow control mode, power factor mode, and voltage and reactive power comprehensive control mode. Its control mode and control target value can be set locally or remotely. When the MCR is closed-loop running, it is steplessly adjusted according to the control target. It has fast response speed, strong coordination and control capabilities, and has the advantages of direct access to high-voltage systems, low harmonics and low losses, which is of great significance for improving the voltage quality and reducing the network loss [13, 14].

2 The Strategy of MCR Participating in AVC Coordinated Control

2.1 Topology Partition of MCR

The control of AVC follows the principle of hierarchical partitioning and on-site balance. Therefore, it is necessary to partition MCR according to its topological connection relationship in the power grid if MCR participates in the coordinated control of AVC system. MCR equipment is generally deployed on the low-voltage side of the substation. The typical topology structure is shown in Fig. 2.

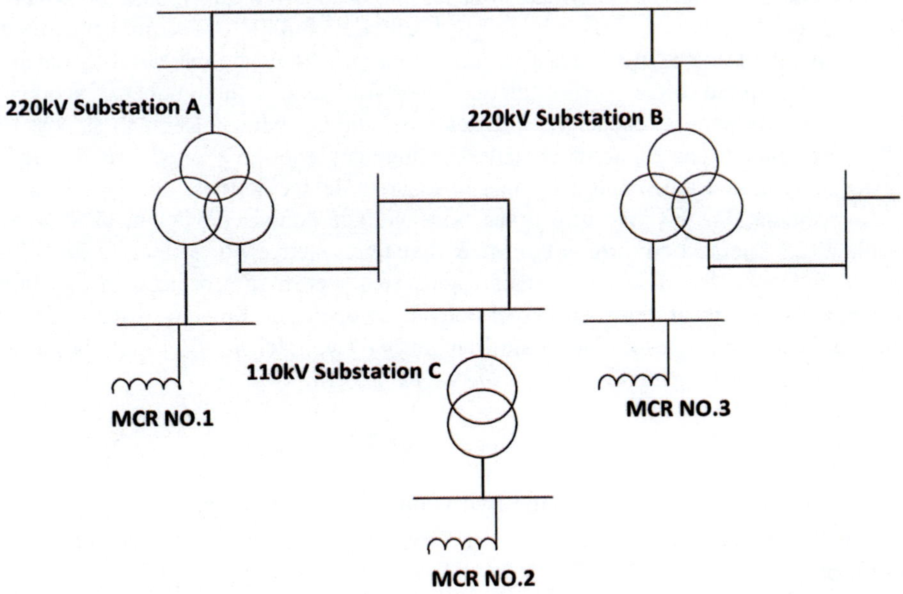

Fig. 2. Typical topological relationship diagram of MCR invloved in power grid

The structure of the regional power grid is generally radial, and the root node is a 220 kV substation. The lower voltage level substation is powered by the middle or low voltage side of the 220 kV substation, forming a tree structure. Based on such characteristics, the 220 kV substation can be used as a hub for topology partitioning, and the power grid can be divided into several independent regions that are decoupled from each other. The voltage and reactive power optimization in each region is relatively independent, thereby greatly reducing the complexity of the algorithm. The MCR equipment is divided into the corresponding substation according to its connection relationship. As shown in Fig. 2, MCR No. 1 and MCR No. 2 are divided into the same zone. The zone uses 220 kV substation A as the hub substation, and No. 3 MCR is divided into another zone. This zone takes 220 kV substation B as the hub substation.

2.2 Regional Control Algorithm

The traditional AVC system uses discrete devices such as capacitors or reactors as the control object, and satisfies the safe and economic operation of the voltage at the cost of the minimum number of device actions [15, 16]. In order to give full play to the advantages of MCR, this paper proposes a regional control algorithm which can consider the MCR participates in the coordinated adjustment of AVC. The optimization objective function is constructed as formula (1).

$$\mathrm{minF} = \beta_1 \sum\nolimits_{i=1}^{n} \|\Delta V_i\|^2 + \beta_2 \sum\nolimits_{j=1}^{m} \|\Theta_j\|^2 + \beta_3 \sum\nolimits_{r=1}^{d} \|K_{act}^{r}\|^2 \tag{1}$$

In formula (1), the minimum network loss is the final optimization goal, which is the minF on the left side of the equal sign. Under the condition of ensuring the power grid is qualified, the dynamic reactive power margin should be increased as much as possible to respond to the random fluctuation of voltage, and the number of actions of discrete devices such as capacitors or reactors should be reduced as much as possible. The constraints of control are the smallest voltage deviation, the largest MCR reactive power margin, and the smallest number of discrete device actions.

In formula (1), the first term is the node voltage control deviation, as shown in formula (2). The second term is the MCR dynamic reactive power margin index, the smaller the value, the greater the dynamic reactive power margin, which is shown in the formula (3). The third item is the total number of operation times of discrete devices such as capacitors, reactors and main transformer taps, etc. β_1, β_2, β_3 are weighting coefficients, and the general ones are: $\beta_1 > \beta_2 > \beta_3$.

$$\Delta \mathrm{V}_i = V_i^{\prime} - V_i^{opt} \tag{2}$$

In the formula (2), $V_i^{\prime}$ is the optimized voltage of node i, which is composed of the power flow balance equation as shown in formula (4), and V_i^{opt} is the voltage control target of node i.

$$\Theta_j = \frac{Q_j^{mcr} + \sum_{i=1}^{n} \left(\frac{\Delta V_i}{S_{i,j}^{mcr}} \right) - Q_{min}^{mcr}}{Q_{max}^{mcr} - Q_{min}^{mcr}} \tag{3}$$

In formula (3), Q_j^{mcr} is the current reactive power output of MCR, Q_{max}^{mcr} and Q_{min}^{mcr} are the maximum and minimum reactive power output of MCR respectively; ΔV_i is the voltage change at the node i. $S_{i,j}^{mcr}$ is the sensitivity coefficient of voltage change at the node to the reactive power change of MCR j.

$$V_i' = V_i + \sum_{c=1}^{n_{cp}} S_{i,c}^{cp}\left(N_{t,c}^{cp} - N_{t-1,c}^{cp}\right) + \sum_{r=1}^{n_{tr}} S_{i,j}^{tr}\left(N_{t,j}^{tr} - N_{t-1,j}^{tr}\right) + \sum_{d=1}^{n_{mcr}} S_{i,d}^{mcr}\left(Q_{t,d}^{mcr} - Q_{t-1,d}^{mcr}\right) \quad (4)$$

Formula (4) is the node voltage balance equation. Among them, V_i is the current voltage of the node i, $S_{i,c}^{cp}$, $S_{i,j}^{tr}$ and $S_{i,d}^{mcr}$ are the sensitivity coefficient of the voltage at the dominant node i to the switching state of the No.c capacitor group, the tap of No.r transformer, and the reactive power of No.d MCR, respectively.

In terms of specific algorithm implementation, firstly, the regional power grid is dynamically partitioned based on the topological relationship, and the strategy calculation is performed in units of partitions; secondly, the strategy preprocessing is performed according to the MCR equipment status (whether there is fault blocking, adjustable margin, etc.) in the partition, the controllable equipment is selected to participate in the optimization model calculation, based on the current SCADA real-time data and state estimation cross-section calculation results. By solving the objective function, the optimal control strategy suitable for the current partition. Finally, the control strategy is analyzed by voltage-reactive sensitivity analysis and transformed to a quantitative indicator that can be executed by MCR equipment.

2.3 Control Strategy and Control Mode

MCR automatically adjusts the inductive power emitted based on the results of the integrated calculation of the whole network by AVC to realize the control of voltage and reactive power at the connection point. The adjustment speed and control accuracy should meet the requirements of power system voltage regulation. Its main control strategy is shown as follows.

1. The control model adopts the method of AVC hierarchical partition and on-site balance, which is relatively mature and has rich on-site experience.
2. The control region of AVC is formed by using 220 kV substation as the root node. The MCR equipment is divided into corresponding regions to participate in the reactive power and voltage optimization of the area.
3. The AVC system deployed in the master station analyzes the reactive power regulation capability and balance of each zone on the premise of ensuring the MCR grid connection point bus voltage is qualified according to the limit issued by the higher-level power grid or the locally set limit. By calculating and determining the reactive power value that MCR equipment should output, AVC send the control instructions to the MCR controller at the substation.
4. After receiving the reactive power control value issued by the AVC system, the MCR controller at the substation first predicts the reactive power demand and adjustable range of the main network, and secondly calculates the adjustable reactive power margin of the MCR equipment, finally reasonably controls the reactive power output capacity of MCR, so as to give full play to the reactive power regulation capability of MCR.

AVC has two main control modes for MCR: automatic calculation and manual control. In the daily closed-loop operation of MCR, the automatic calculation mode is used. The AVC system automatically calculates the optimal MCR reactive output value according to the MCR coordinated control strategy and sends it to the MCR controller at the substation. In the MCR access commissioning phase or special holidays such as the Spring Festival, manual control mode can be used. The supervisor of the dispatch center can set the MCR reactive output value according to the reactive power demand of the whole network. When the reactive power output of the whole network is increased, the reactive power output of the MCR is reduced, on the contrary, when the reactive power of the whole network is under compensated, the reactive power output of the MCR is increased.

3 The Scheme of MCR Participating in AVC Coordinated Control

3.1 System Structure

The AVC system is generally deployed in the dispatching control system of the power grid. The real-time data collected by SCADA and the real-time operation mode provided by state estimation are analyzed and calculated to realize automatic monitoring of the bus voltage and reactive power flow of the power grid. The reactive power regulation equipment in the substation are controlled by AVC in 24-h online closed-loop way. Figure 3 shows the structure of a system where several MCRs participate in AVC coordinated control in a grid.

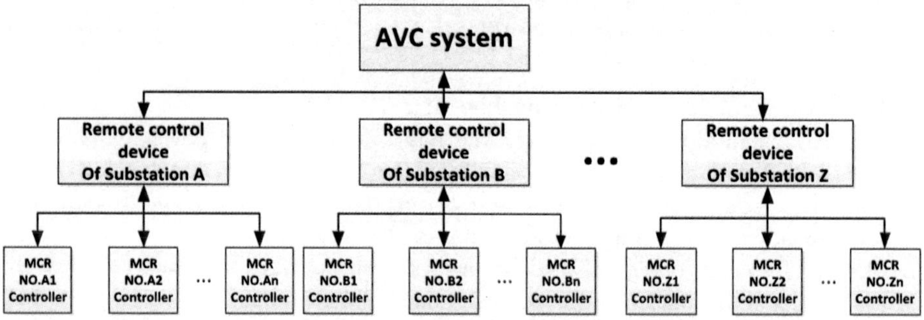

Fig. 3. System structure of MCR participating in AVC coordinated control

For the case of parallel operation of multiple MCR equipment, the AVC system will decouple the control time. When there is a need for reactive power adjustment, multiple MCR equipment will be rotated. The non-controlled MCR equipment will remain flat during this period status. When the grid-connected point has both MCR and ordinary capacitors, the resonant characteristics of inductive and capacitive reactive power will be considered, and the MCR will maintain the minimum capacity operation when the capacitor has been put in.

3.2 Technical Points

Based on the performance of the MCR itself, several technical points must be considered in the process of implementing the AVC system to directly control the MCR equipment.

3.2.1 Function Configuration

The AVC system adds MCR coordination related functional modules, and the substation is equipped with an MCR controller. The two implement data communication. The MCR controller uploads MCR control information in a periodic manner, such as the controllable state of the MCR, remote status of the MCR, inductive capacity can be increased, inductive capacity can be reduced, fault information, etc. AVC also sends the target value of reactive power control to the MCR controller in a periodic manner, either. And the target value is in signed floating point mode.

3.2.2 Communication Method

MCR adopts the RS485 serial communication method of half-duplex standard, and receives the AVC command of the master station through the remote control device at the substation. The AVC command is issued through the remote adjustment function of the IEC104 protocol. The substation remote control device realizes the MCR command conversion function by adding protocol conversion equipment.

3.2.3 Control Precision

The cycle of AVC system strategy calculation is between 10 and 60 s, and the control command is issued every 60 s. The MCR controller is assigned to the corresponding MCR equipment according to the reactive control command issued by AVC. The MCR automatically adjusts its reactive power output to ensure that the voltage of the MCR grid connection point and the reactive power of the main gateway port are near the set value. The response period is within 10 s.

3.2.4 Coordination of MCR and Fixed Capacitor or Reactor

Compared with fixed capacitor or reactor, MCR has the advantage of flexible response speed. The characteristic of the fixed capacitor or reactor is that the design capacity is relatively large, so when the fixed capacitor or reactor is input or exit, the voltage and reactive power of the connection point are significantly affected. When the substation has both MCR and fixed capacitor or reactor, the coordination between the two should be considered. This phenomenon cannot occur: MCR emits inductive reactive power, while the capacitor is input, or MCR emits capacitive reactive power, but reactor is input. It may cause resonance and damage the reactive equipment. For the above situation, the control strategy can be considered as follows: MCR as a priority adjustment means, and fixed capacitor or reactor as a backup supplementary adjustment means. When the MCR has been adjusted to full capacity and still cannot meet the voltage and reactive power optimization needs, the fixed capacitor reactor is then put into use. In terms of reactive power direction control, when the MCR needs to send inductive reactive power, but there is any capacitor that has been put in on the same connection point, the capacitor must be preferentially cut off, and vice versa, to ensure that the reactive power of the two is in the same direction at any time.

3.3 Interconnection Coordination Scheme

The AVC system considers the topology of MCR, performs partition and merge on the MCR, and then performs a coordinated control strategy. The main control principle of MCR is to maintain the same direction as the optimized control on the premise of ensuring the voltage of the grid connection point is qualified. The control flow is shown in Fig. 4.

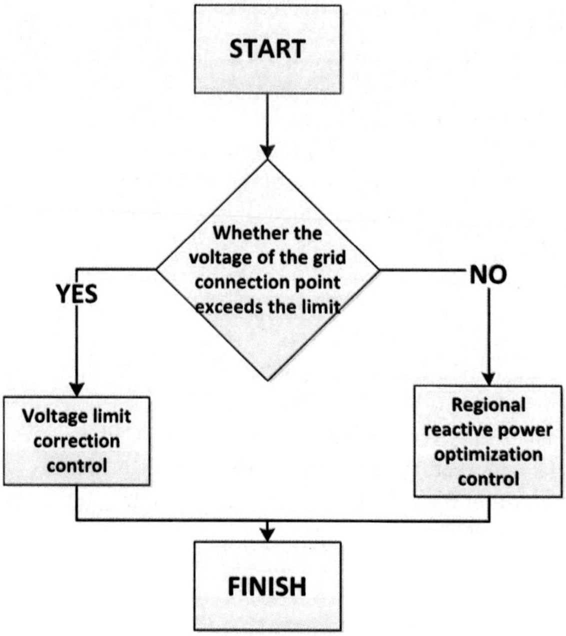

Fig. 4. Coordination control flow of MCR participating in AVC system

First MCR is partitioned according to the network topology. Secondly, AVC system starts to check whether the control target exceeds the limit. When the voltage of connection point exceeds the limit, the over-limit correction control is used. When there is sufficient reactive power regulation capability, the bus voltage is pulled into the normal control limit range. Otherwise, if the regulation ability is insufficient, AVC will also push MCR to reduce the bus voltage limit as far as possible. Finally, when there is no voltage exceeding the limit in the region, the MCR participates in the reactive power optimization in the whole network: According to the upper and lower reactive power thresholds issued by the upper-level power grid and considering the reactive power demand in the current region, the expected reactive power value and the required adjustment amount are obtained.

4 Example Application

In September 2017, the No. 2 MCR in 220 kV Wenjing substation in Zhejiang province were officially put into AVC closed-loop coordinated control after joint debugging test. AVC system generates a MCR reactive output target value every minute after calculation and analysis. Through the remote adjustment function based on the 104 protocol, the control command is delivered to the MCR controller established in the substation. After receiving the target value, the MCR controller can quickly adjust the reactive power output of the MCR equipment and the adjusted reactive power value is fed back to the AVC system.

According to the actual capacity of the MCR in the substation, the upper and lower limits of the MCR control in the AVC system are set to 0.1–9 MVar. At present, the No. 2 MCR in Wenjin Substation is under closed-loop operation, which is connected to the 35 kV bus for dynamic reactive power compensation, and participates in the AVC system interconnection coordination control. Figure 5 is the voltage curve of the grid connection bus of No. 2 MCR in Wenjin Substation on November 30, 2017, compared with the voltage curve of the same period last year. The yellow curve is the voltage curve after the MCR is put into operation, and the blue curve is the voltage curve before the MCR is put into operation.

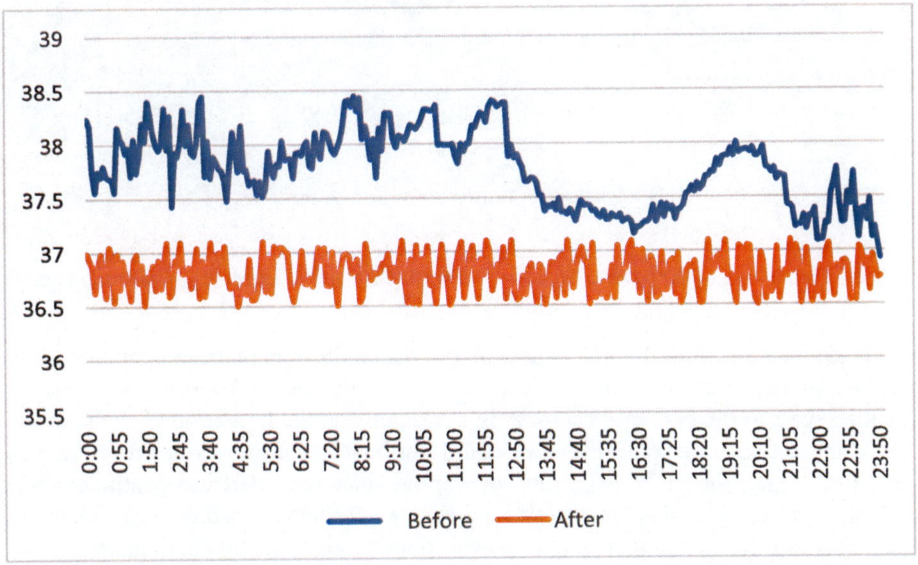

Fig. 5. Fluctuation of the busbar voltage before and after MCR closed-loop operation

As can be seen from Fig. 5, before the MCR is put into operation, the voltage of the 35 kV bus at the grid connection point fluctuates frequently, the voltage amplitude fluctuates greatly, and even approaches the upper limit. It is difficult to meet the requirements by switching existing reactors or capacitor. After MCR is put into operation, the 35 kV bus voltage fluctuations were significantly suppressed due to the continuous compensation output of the MCR, which ensured that the voltage was within the qualified range and was close to the rated voltage for normal operation.

Figure 6 is the reactive power curve of the main gateway port of the grid connection point of No. 2 MCR in Wenjin Substation on November 30, 2017, compared with the reactive power curve of the same period last year. The yellow curve is the reactive curve after the MCR is put into operation, and the blue curve is the reactive curve before the MCR is put into operation.

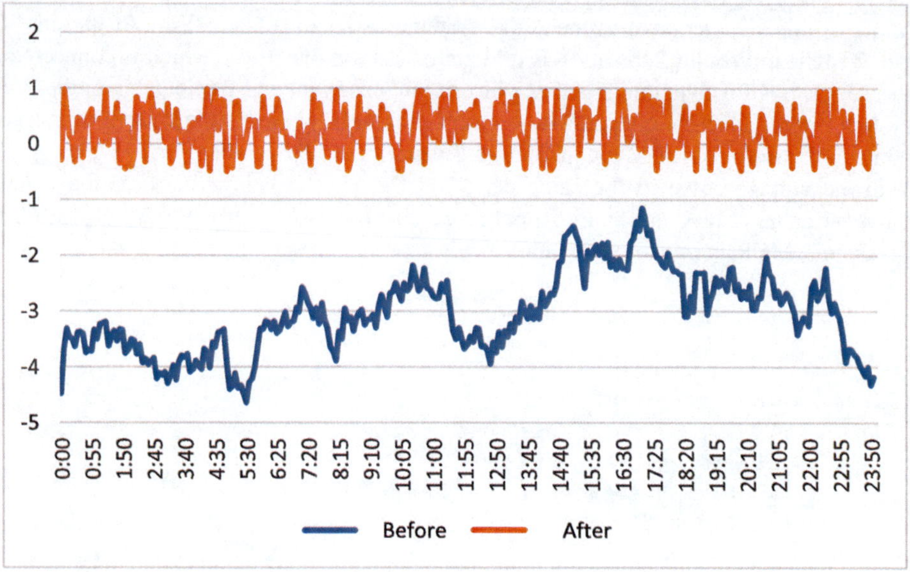

Fig. 6. Fluctuation of the gateway reactive power before and after MCR closed-loop operation

It can be seen from the Fig. 6 that before the MCR was put into operation, due to the lack of solid reactance configuration, the reactive power at the gate was in a state of sucking back to the system for a long time. Because of the limitation of the number of switching actions, the solid reactance could not handle the problem of reactive power feed back. After the MCR is put into operation, since the MCR can continuously and smoothly send out inductive reactive power, the 220 kV substation can maintain a relatively stable inductive reactive power output, and the fluctuation amplitude is reduced. On the other hand, it also avoids that the frequent switching of the capacitors or reactor group brings adverse effects to the safe and stable operation of the network and the equipment itself.

Figure 7 is a comparison curve of the control model calculation output and MCR action characteristic reactive output after the MCR participates in the AVC coordinated control. The blue curve is the model calculation output, and the red curve is the MCR action reactive power output.

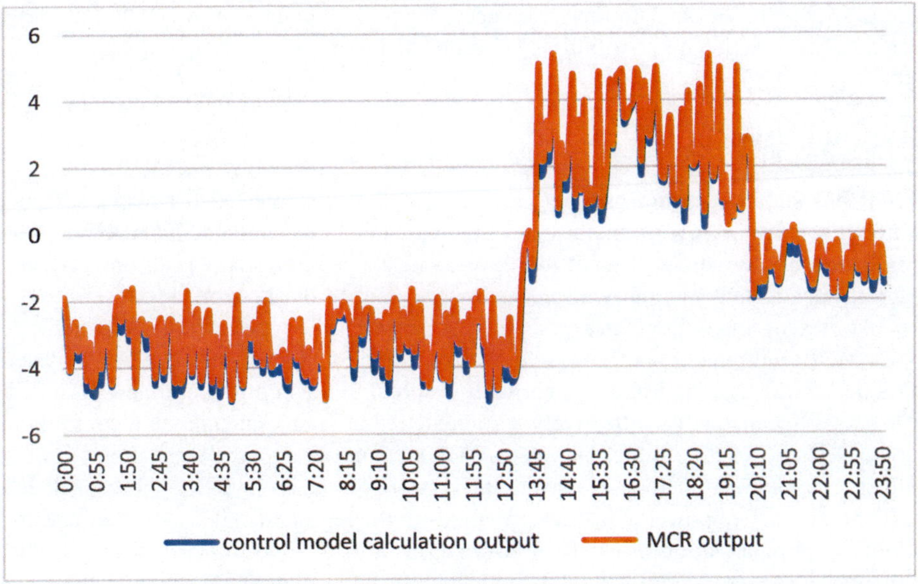

Fig. 7. The comparison of the control model calculation output and MCR output

It can be seen from Fig. 7 that before 13:50, when the voltage tends to exceed the upper limit, the model calculation output shows that the MCR emits inductive reactive power to smooth the upper trends of the voltage. After the voltage becomes stable, the MCR begins to emit some capacitive reactive power, and after 20:00, as the voltage rises again, the MCR again emits inductive reactive power. The actual reactive power output of the MCR is consistent with the model calculation output, with a deviation of less than 0.5 MVar, indicating that the controller at the MCR station receives the command of the AVC system is normal and can be adjusted sensitively, thus ensuring that the substation bus voltage is qualified.

Table 1 selects the number of operations of fixed-capacity reactive power compensation equipment (solid reactance, fixed-capacitance capacitor) for 5 days after the MCR is put into operation, compared with the number of equipment operations in the same period last year.

Table 1. Comparison of the action numbers before and after MCR is put into operation.

Date	1st day	2nd day	3rd day	4th day	5th day
Befor MCR is put into operation	6	8	8	7	6
After MCR is put into operation	2	2	3	2	1

As can be seen from Table 1, after the MCR is put into operation, due to the smooth adjustment characteristics and fast response speed of the MCR, the number of operations of the fixed-capacity reactive power compensation equipment is greatly reduced, which not only improves the accuracy and efficiency of adjustment, but also reduces the switch The possibility of equipment failure due to frequent actions.

5 Conclusion

MCR has stable operation, simple and convenient maintenance, fast response speed, smooth regulation, low harmonics and low loss. Its strong voltage and reactive power regulation ability can well meet the development needs of smart grid, and has great significance to improve the voltage quality and reduce the network loss.

At present, many large and medium-sized power grid systems have been equipped with MCR equipment, and have achieved coordinated control with the AVC system. Practice shows that the MCR can correctly respond to the control commands issued by the AVC system, and can effectively participate in the interconnection and coordination control. However, the limited number of MCR inputs in a single substation is not yet effective in reducing network losses in the region or even the system, and coordination with other reactive power adjustment methods in the region also needs further research. With the popularization of MCR in power grid, how to give full play to its excellent potential for reactive power regulation, deepen its coordination and optimization at the substation and inter-station level, and how to use MCR to improve the voltage quality and stability in the operating mode will be the focus of further attention and research.

Acknowledgment. This paper was supported by Science and Technology Project of State Grid Electric Power Research Institute <Reactive Power and Voltage Control Technology of Multi-Source and Multi-Object and Its Application>.

References

1. Sikander, A., Thakur, P.: A new control design strategy for automatic voltage regulator in power system. ISA Trans. **100**, 235–243 (2020)
2. Jumani, T.A., Mustafa, M.W., Hussain, Z., et al.: Jaya optimization algorithm for transient response and stability enhancement of a fractional-order PID based automatic voltage regulator system. Alexandria Eng. J. 145–153 (2020)
3. Anna, K., Sami, R., Pertti, J.: Coordinated voltage control in distribution networks including several distributed energy resources. IEEE Trans. Smart Grid **5**(4), 2010–2020 (2014)
4. Zhu, T., Nie, Y., Li, Z., et al.: Parameter estimation method of stray capacitance of capacitive voltage transformer based on improved particle swarm optimization algorithm. Autom. Electr. Power Syst. (04), 178–192 (2020)
5. Lyn, W., Xie, H., Xu, Q., et al.: Analysis on impact of AC filter sets on power grid harmonic in UHVDC converter station. Autom. Electr. Power Syst. (23), 217–225(2019)
6. Zhou, B., Yi, Y., Zhang, Y., et al.: Real-time slack optimal method for voltage and reactive control between provincial and prefectural power networks in heavy load flow mode. Autom. Electr. Power Syst. (11), 126–133 (2017)
7. Zhang, Z., Wang, C.: Coordinated voltage control strategy of wind farms based on hierarchical model predictive control. Autom. Electr. Power Syst. **43**(11), 34–42 (2019)

8. Zhao, J., Zhang, Z.: Heterogeneous decomposition based distributed reactive power optimization method for global transmission and distribution network. Autom. Electr. Power Syst. **43**(3), 108–114 (2019)
9. Guo, L., Zhang, D., Kou, X., et al.: Application analysis on magnetic valve control reactor based on fuzzy neural network control system. Power Capacitor React. Power Compensation (6), 28–32 (2017)
10. Liu, M., Li, Y., Gao, H., et al.: A fast adjustment method of magnetically controlled reactor with multi-tap. J. Magn. (4), 669–674 (2018)
11. Lei, X., Deng, Z.F., Xu, G., et al.: Research and practice of magnetically controlled shunt reactor. Proc. CSEE (34), 225–231 (2014)
12. Zheng, T., Zhao, Y., Jin, Y., et al.: Analysis on and protective schemes against control winding turn-to-turn fault of magnetic controllable shunt reactors. Autom. Electr. Power Syst. (10), 95–99 (2014)
13. Tao, L., Zheng, T., Zhao, Y.: Physical design of control windings for a three-phase magnetically controlled shunt reactor based on series-connected working windings. Autom. Electr. Power Syst. (19), 101–106 (2015)
14. Jin, F., Li, H., Yu, Y.: Discussion on control method of MCR type magnetic control reactor. Power Capacitor React. Power Compensation (5), 62–66 (2010)
15. Ye, H., Zhao, C., Tian, C.: The exploratory research of reactive power planning based on Magnetic controllable reactor. Yunnan Electr. Power (2), 7–11 (2014)
16. Chen, B., Song, J., Zhou, P., et al.: MCR and MSVC based voltage and reactive power combination control of different voltage level in 220 kV substation. Trans. China Electrotech. Soc. (16), 230–237 (2015)

Research and Application of Lightning Monitoring and Early Warning Network for Transmission Lines

Yue Chen[1,2,3](✉), Zhibo Jiang[1,2,3], Hantao Tao[1,2,3], Lei Zhang[1,2,3], and Dawei Wu[1,2,3]

[1] NARI Group Corporation Ltd., Nanjing, China
597265154@qq.com
[2] Wuhan NARI Limited Liability Company, State Grid Electric Power Research Institute, Wuhan, China
[3] Hubei Key Laboratory of Power Grid Lightning Risk Prevention, Wuhan, China

Abstract. The important transmission lines have the characteristics of large transmission capacity, long line, high tower, complex corridor terrain, which make the risk of lightning trip high. According to the demand of lightning early warning for transmission lines, the construction scheme of lightning early warning network for transmission lines is proposed. Firstly, the lightning risk level of transmission lines is evaluated, and the topological relationship between lightning movement trend and transmission line trend is considered. Combined with the operation and site selection characteristics of the early warning device, the distribution method of lightning early warning devices is proposed. A construction scheme of lightning monitoring and early warning network based on lightning strike risk assessment level of transmission lines is proposed, which provides technical support for the application of lightning early warning in power grid.

Keywords: Transmission line · Lightning warning · Distribution of early warning device · Construction of monitoring and early warning network

1 Introduction

Lightning disaster is listed as one of the ten natural disasters by the United Nations [1–3]. According to statistics, about 4000 people are killed or injured by lightning every year in China, which has caused direct economic losses to electric power, petrochemical, transportation and other industries by about 10 billion yuan. China's power grid is subjected to more than 350,000 lightning strikes annually, and the number of lightning strikes on transmission lines accounts for about 50% of the total number of trips [4–6]. Lightning can cause serious harm to the power grid, which directly affects the stability of power grid operation and the reliability of power supply and utilization

This work is supported by National Key Research and Development Project (2018YFF01010803).

Y. Xue et al. (Eds.): PMF 2020, LNEE 718, pp. 290–300, 2021.
https://doi.org/10.1007/978-981-15-9746-6_23

[7–9]. With expansion and upgrading of the power grid, the demand of raising the level of lightning protection is becoming more prominently in power grid.

The lightning protection performance has be improved by adding line arresters, reducing tower grounding resistance and strengthening insulation. Some results have been achieved in reducing the loss caused by lightning trip and fault [10–12]. However, this lightning protection mode is mainly aimed at the disposal and treatment after lightning stroke. A Pre-warning Technology for the risk of lightning stroke is urgently needed in power grid. Daniel analyzed the thunderstorm process of Terrassa in summer in northeastern Spain based on the data of atmospheric electric field and lightning location, and obtained that the method which combining the change of electric field polarity and electric field threshold can improve the lightning early warning effect [13]. Fang Biao obtained two lightning warning indexes by analyzing the correlation between lightning activity and radar echo cell in the northeast of Guizhou, the two indices that echo top(ET) of 35 dBz and 25 dBz reflectivity exceed the height of −10 °C and −20 °C layer had better lightning warning effect [14]. Wu Jian proposed the lightning early warning method based on the data of electric field strength and lightning location, the early warning of electric field strength is based on the threshold level of electric field strength and whether there is rapid change jitter in the initial stage of electric field curve, the lightning distance warning is based on the distance between the lightning location and the electric field instrument [15]. Liming calculated a number of early warning indicators by using meteorological data such as radar, atmospheric electric field and lightning location data, and realized thunderstorm recognition, tracking and extrapolation in three-dimensional space based on decision tree algorithm, and developed a lightning early warning platform [16].

At present, the lightning early warning method mainly depends on the atmospheric electric field, lightning location information, radar echo and other technical means, but it is not widely used in power grid. In order to solve this problem, the construction scheme of lightning monitoring and early warning network for transmission lines is proposed. The risk level of lightning damage of transmission line is evaluated based on the historical lightning trip fault and complex terrain conditions. The installation position of the equipment is selected in combination with the functions and performance characteristics of early warning devices such as atmospheric electric field instrument and radar. The construction of lightning monitoring and early warning network is realized.

2 The Assessment Method of Lightning Risk Level for Transmission Lines

Lightning activities and terrain environment are the main factors which affecting the lightning risk of transmission line. Because of the randomness of lightning strike, the distribution characteristics of lightning stroke faults near the line corridor should also be considered in the risk assessment of lightning damage.

In the grid area of 10 km around the transmission corridor, the lightning data in the past six years are selected as the sample for the statistical analysis of lightning density. The distribution characteristics of annual and average annual ground flash density of transmission lines are obtained by statistics. According to the lightning density, the lightning frequency is divided into four lightning hazard risk levels and seven lightning density levels from weak to strong [17, 18].

a) Level A —— lightning density < 0.78 times/(km2·a)
b) Level B1 —— 0.78 times/(km^2·a) ≤ lightning density < 2.0 times/(km^2·a)
c) Level B2 —— 2.0 times/(km^2·a) ≤ lightning density < 2.78 times/(km^2·a)
d) Level C1 —— 2.78 times/(km^2·a) ≤ lightning density < 5.0 times/(km^2·a)
e) Level C2 —— 5.0 times/(km^2·a) ≤ lightning density < 7.98 times/(km^2·a)
f) Level D1 —— 7.98 times/(km^2·a) ≤ lightning density < 11.0 times/(km^2·a)
g) Level D2 ——lightning density ≥ 11.0 times/(km^2·a)

Through the comprehensive analysis of the lightning density distribution characteristics, lightning fault distribution, altitude, terrain and other factors in the transmission line corridor, the lightning risk level for transmission lines is evaluated. The specific steps are as follows:

① The line corridor is taken as the center, and the grid is divided along the line within 10 km around, so the size of a grid is 20 km × 20 km.
② The lightning density of each grid is calculated, and the lightning density level is defined by according to the standard of the lightning distribution map. The corresponding relationship between the lightning density level and the lightning hazard risk level is: level A-level 1, level b1-level 1.5, level b2- level 2, level c1- level 2.5, level C2- level 3, level d1- level 3.5, level d2- level 4.
③ According to the operation experience, the number of lightning fault points for each grid in recent 6 years is counted. If the number of one grid is greater than zero, the risk level of that grid is increased by 1 level on the basis of ②. If the number of one grid is zero, but the numbers of lightning fault points for the 500 kV lines in the grid in recent three years is greater than zero, the risk level of that grid is increased by 0.5 level on the basis of ②.
④ The terrain of the transmission line section in the grid is investigated. If it is located at the top of the mountain, along the slope, across the valley or other harsh environment, the risk level of the grid will be increased by 1 level on the basis of ③.
⑤ The altitude of the transmission line section within the grid is investigated. If the altitude is more than 1000 m, the risk level of the grid will be increased by 0.5 on the basis of ④.
⑥ After the grid lightning hazard risk level of the line corridor considering the above factors are obtained, the value of the level is rounded up. If the level is more than 4, it is still calculated as level 4, and it is rewritten into Roman numeral form.

3 The Distribution Method of Lightning Early Warning Device in Power Grid

3.1 The Distribution Method of Atmospheric Electric Field Instrument

The atmospheric electric field instrument is a device for measuring the atmospheric electric field and its changes. In sunny days, the atmospheric electric field intensity is relatively small, with an average of 120 V/m on the land and 130 v/m on the sea [19]. Before the occurrence of lightning, the value of atmospheric electric field will appear obvious irregular vibration, and its absolute value will gradually increase. Therefore, the initial, growth, maturity and dissipation stages of thunderstorm can be judged by monitoring the change of atmospheric electric field. The development trend of thunderstorm activity and the possibility of disastrous weather such as lightning stroke can be warned, which is a technical means applied to lightning warning over a long period of time.

Providing real-time data of the atmospheric electric field in the range of 10–20 km for the early warning system is the main function of the atmospheric electric field instrument. The atmospheric electric field instrument can record, store and call the data of the atmospheric electric field remotely. The data can be transmitted to the server for comprehensive analysis of lightning warning in real time. The atmospheric electric field instrument is equipped with GPS antenna and high stability crystal oscillator to ensure accurate time for system measurement and calculation.

By studying the topological relationship between the lightning movement trend and the key sections of transmission lines, it is found that the proportion of lightning movement process along the line and in situ can reach 70% and the tangent movement only accounts for 30% for the east-west transmission lines. Therefore, most of the thunderstorm activities move from west to east [20]. In addition, it can be judged whether the thunderstorm activity trend is close to the line based on the monitoring data of radar and lightning location system. The effective monitoring radius of the atmospheric electric field instrument is 10 km–20 km. The monitoring coverage of thunderstorm activity along the transmission line and in the direction of vertical transmission line should be considered in the station deployment of the atmospheric electric field instrument. Therefore, it is recommended that the distance from vertical projection of station spacing to line for the east-west transmission lines is 15 km (Fig. 1).

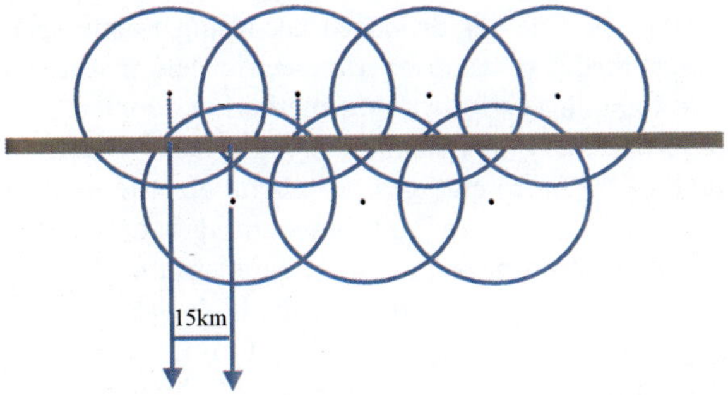

Fig. 1. The layout method of atmospheric electric field instrument

3.2 The Distribution Method of Radar Station

Radar can be used to predict the initial formation period and development period of thunderstorm. In the initial stage of thunderstorm formation, radar echo is used to characterize the state of thunderstorm formation. The radar echo can be used to judge whether there is thunderstorm cloud in 100–120 km around the radar. If there is thunderstorm cloud in 100–120 km around the radar, lightning early warning will be triggered. The radar data are the forecast factors of lightning early warning, such as radar echo reflectivity, velocity field, top height. The radar echo characteristics for lightning early warning are as follows: reflectivity is more than 40 dbz, zero radial velocity line is "s" distribution, echo top height needs to reach 6 km–9 km, etc. During the development and movement period of thunderstorm, optical flow extrapolation based on radar echo data is adopted. The change of thunderstorm clouds is excavated and the trend of lightning activity is predicted. When the prediction area of thunderstorm covers some tower positions of the transmission line, the lightning warning for transmission line will be issued.

The radar sends electromagnetic wave pulse to the air intermittently, and then receives the electromagnetic wave echo scattered by the meteorological target. The radar can be used to detect the spatial location and characteristics of meteorological targets in the range of 100–120 km around it, which provides important basic data for weather monitoring and early warning of small and medium scale lightning disaster. The monitoring radius of radar is 100 km–120 km. Because the detection efficiency of radar to the boundary is low, when multiple radar stations are combined for detection, overlapping coverage is required at the monitoring boundary. The recommended distance between radar stations is 150 km.

4 The Construction Method of Lightning Monitoring and Early Warning Network for Transmission Lines

The construction method of lightning monitoring and early warning network for transmission lines is shown in Fig. 2. Firstly, the risk level of lightning damage for transmission lines is evaluated according to the method described in Sect. 2. Then the function and performance of atmospheric electric field instrument and radar are considered. Finally, the construction scheme of lightning monitoring and early warning network for different risk levels is proposed according to site selection requirements and operation characteristics of atmospheric electric field instrument and radar.

The large-scale lightning early warning method is adopted for the low-risk transmission lines. It is necessary to install the atmospheric electric field instrument to monitor the change of the surrounding atmospheric electric field for the medium-risk transmission lines, which can realize lightning warning before thunderstorm matures. It is necessary to install the atmospheric electric field instrument and radar data to achieve high-precision lightning activity prediction for the high-risk transmission lines. For this reason, the construction scheme of lightning monitoring and early warning network for transmission lines is formulated. The scheme is as follows:

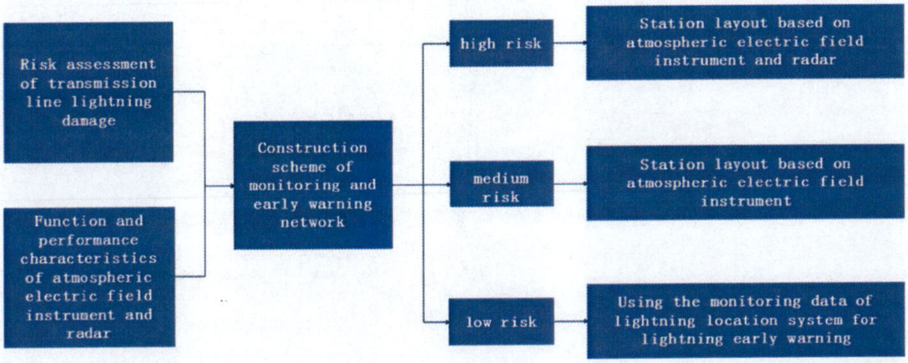

Fig. 2. Construction method of lightning monitoring and early warning network for transmission lines

The method based on atmospheric electric field instrument and radar is used for the lightning high risk section of transmission line. The atmospheric electric field instrument is used to monitor the change of atmospheric electric field, which can be used as the trigger and end conditions of lightning warning. The monitoring information of radar is used to identify, track and extrapolate thunderstorm. The false alarm rate can be reduced by combining with the early warning information of atmospheric electric field and radar.

The method based on atmospheric electric field instrument is used for the lightning medium risk section of transmission line. The atmospheric electric field instrument is used to monitor the electric field polarity, intensity, development and evolution between the thunderstorm cloud and the ground. The multi-level alarm value is set according to the installation position and terrain conditions of the atmospheric electric field instrument, which can realize the early warning during the thunderstorm occurrence and development period.

The method based on lightning location system is used for the lightning low risk section of transmission line. The lightning activities of transmission lines are calculated in real time and the trend of lightning movement is predicted, which can realize early warning of the thunderstorm for transmission lines in the development period.

5 Case Analysis

Within 10 km grid area around the transmission corridor, the lightning monitoring data in the past six years are selected as the sample of ground flash density statistical analysis. The annual and average annual lightning density distribution of transmission lines are obtained by statistics. The law of lightning activity in a transmission line corridor is analyzed. The average lightning density distribution, lightning fault points of transmission line corridor in recent six years and the lightning fault points of 500 kV line passing through provinces in recent three years are shown in Fig. 3.

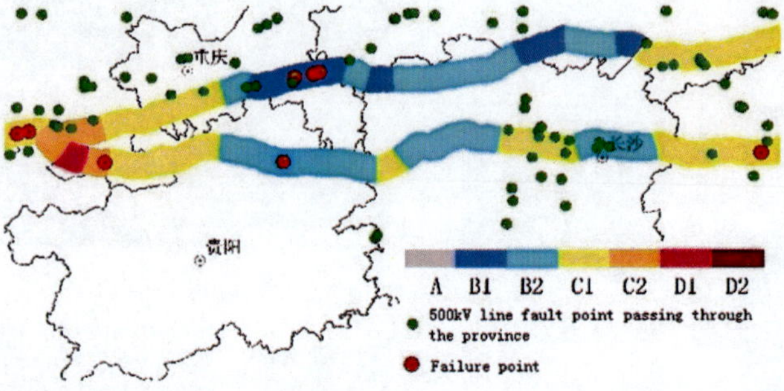

Fig. 3. Distribution of average ground flash density of transmission line in recent six years

The trend of lightning activity from west to east is "strong-weak-strong". The lightning density is large in Sichuan province, most of which are in C1 level. In Chongqing and Hunan province, the density of lightning is relatively small, at B1 and B2 levels respectively. The lightning density has gradually increased from B2 level to C1 level in Hubei Province. In Anhui province and Zhejiang province, the lightning density reaches the maximum value, most of which are above C1 level. The Annual ground flash density of a transmission line is shown in Fig. 4.

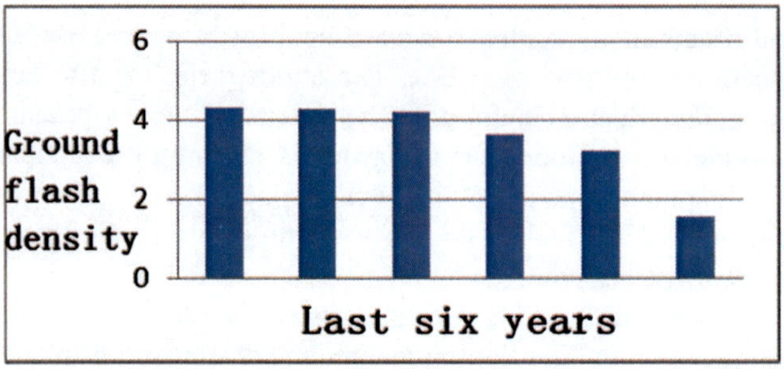

Fig. 4. Annual ground flash density of a transmission line

The lightning fault points of the transmission line are compared with the distribution diagram of the ground flash density. It is found that among the 8 lightning strike fault points, one is at the place with the ground flash density of C2 level, three are at the place with the ground flash density of C1 level, one is at the place with the ground flash density of B2 level, and three are at the place with the ground flash density of B1 level. The fault towers which located at the place with the ground flash density are located in the mountainous areas in the north of Guizhou and the south of Chongqing respectively. The altitude of those places is relatively high and the terrain conditions are complex, which are shown in Fig. 5.

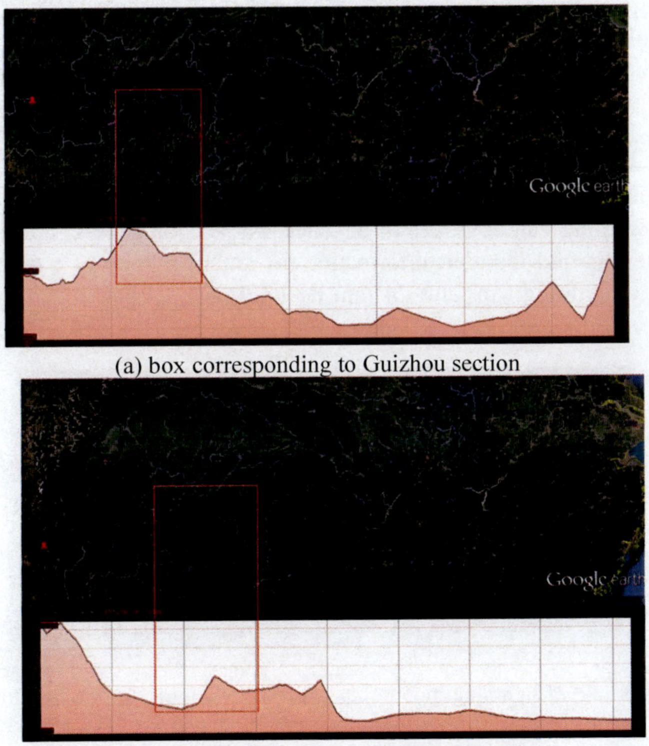

(a) box corresponding to Guizhou section

(b) box corresponding to Chongqing section

Fig. 5. Altitude of a transmission line corridor

After the distribution characteristics of lightning ground flash density in transmission line corridor are analyzed, the lightning fault distribution, altitude, terrain and other factors are considered. The risk level of each grid for transmission line corridor is evaluated, and the transmission line sections of the same grid are combined. The distribution characteristics of lightning hazard risk level along the transmission line are obtained, which are shown in Fig. 6.

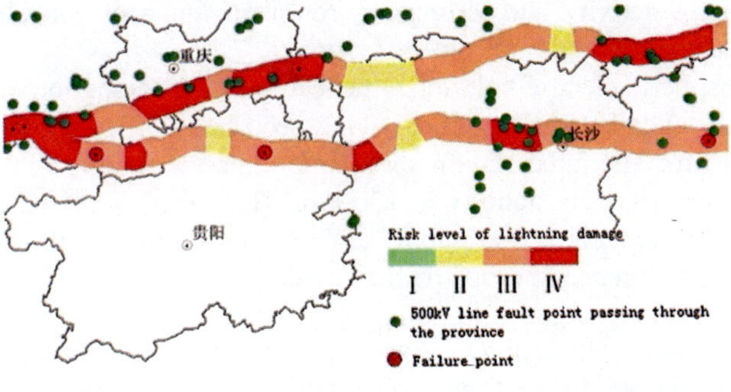

Fig. 6. Risk distribution of lightning damage on transmission lines

A lightning monitoring and early warning system is deployed according to the risk distribution of lightning damage in transmission lines. If the lightning ground flash density of the transmission line is high and there are many lightning faults, the atmospheric electric field instrument and radar are installed and deployed. If the transmission line is located at the place with medium and low lightning risk, the atmospheric electric field instrument is installed and deployed. After the lightning monitoring and early warning network for transmission lines is deployed, the lightning disaster risk warning can be issued in advance. At 5:02 on June 1, 2016, the lightning warning system issued red alert information for Anhui section of a transmission line. The line has a lightning trip fault 24 min later (Fig. 7).

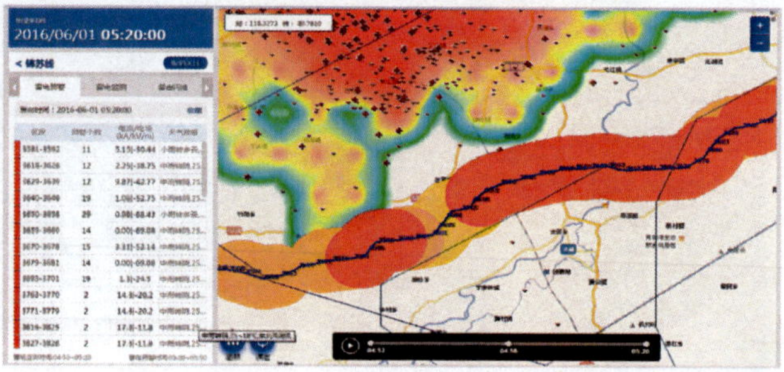

Fig. 7. Example of lightning monitoring and early warming for the transmission line.

6 Conclusion

The construction scheme of lightning monitoring and early warning network for transmission lines is studied in this paper. The risk assessment method of thunderstorm damage is studied. The operation characteristics and site selection principle of lightning early warning device in power grid are described. The construction scheme of monitoring and early warning network is proposed. The main points are as follows:

(1) The risk level of lightning damage for transmission lines should be evaluated before the construction of monitoring and early warning network. The risk assessment of lightning damage for transmission lines should not only consider the lightning activity and terrain environment, but also consider the lightning faults near the transmission line corridor.

(2) The atmospheric electric field instrument is deployed along the transmission line, and the distance between stations which projected vertically to the transmission line is 15 km. The radar station shall be arranged according to the principle that the distance between stations is 150 km. The location of the station shall be selected with reliable power supply and communication, and no serious electromagnetic interference and obstruction around.

(3) The construction process of lightning monitoring network for transmission lines is proposed. Firstly, the risk assessment method of lightning damage for transmission lines is studied. Then according to the function and performance

characteristics of the atmospheric electric field instrument and radar, the equipment installation locations are selected, the monitoring and early warning network are constructed.

References

1. Shu, C.: Design and implementation of integrated lightning warning management system based on WebGIS. Hubei University of Technology (2018)
2. Zhong, D., Gao, Q., Zhang, J., et al.: Development of on-line monitoring device for UHV AC metal oxide arrester. Insulators Surge Arresters (04), 120–123+127 (2018)
3. Liu, S., Zeng, W., Liao, W., et al.: Analysis of lightning transient characteristics and lightning protection measures of 35 kV overhead line. Insulators Surge Arresters **06**, 34–38 (2018)
4. Xie, S., Cao, S., Luo, S., et al.: Estimation on lightning shiolding porformanco of ±800 kV DC transmission linos based on loader propagation method. Insulators Surge Arresters (04), 87–93 (2019)
5. Qin, R., He, H., Sun, Z., et al.: Analysis of lightning parameters and study on the statistics method of ground flash density of distribution lines. Insulators Surge Arresters **03**, 134–139 (2019)
6. Li, Y.: Research and application of differential lightning protection technology for 500 kV transmission line in Shandong Province. Shandong University (2018)
7. Cao, W.: Research on lightning protection technology and measures of transmission lines in Inner Mongolia. North China Electric Power University (2019)
8. Zhuang, L.: Study on reliability improvement of distribution network in Haojiang District of Shantou. South China University of Technology (2018)
9. Li, J., He, G.: Research on comprehensive lightning protection technology of 220 kV transmission line in power system. Power Equip. Manag. **02**, 45–47 (2019)
10. Li, X.: Research and application of comprehensive measures of lightning protection for high voltage transmission lines. North China Electric Power University (2014)
11. Zhou, Y., Liu C., Zhou, S.: Comprehensive evaluation of lightning protection retrofitting measures for 500 kV transmission line based on extension analytic hierarchy process. Insulators Surge Arresters (05), 1–6+12 (2018)
12. Liu, Y.: Research on Lightning Protection Technology of High Voltage Transmission Line of Ordos District. Tianjin University (2018)
13. Daniel, A., Joan, M., Gloria, S., et al.: On the ligtning hazard warning using electrostatic field: analysis of summer thunderstorms in Spain. J. Elcclrostat. **67**, 507–512 (2009)
14. Fang, B., Zhang, Y., Yan, X.: Application research of radar data in lightning warning over northeast of Guizhou. Meteorol. Environ. Sci. **40**(03), 133–137 (2017)
15. Wu, J., Chen, Y., Zeng, Z.: Method of making a current pre-thundering report is to use the files of the ground electric field instrument and locating the lightning. Meteorol. Environ. Sci. **32**(01), 47–50 (2009)
16. Li, M.: Research and Design of the Lightning Warning Method and Platform Based on Muti-source Data. Chengdu University of Information Technology (2014)
17. Tan, Y.: Industry Application of Differentiation Technology in Lightning Protection and Analysis of Lightning Stroke Distribution in Inner Mongolia. Tianjin University (2016)

18. Yue, S.: Design and implementation of application system of "six prevention" thematic map of transmission line. University of Electronic Science and technology (2019)
19. Lu, B.: The Research On The Application Of The Data Of Atmospheric Electric Field In Lightning Warning. University of Electronic Science and technology (2012)
20. Wu, M., Wang, T., Zhao, C., et al.: A new description method of lightning activity around transmission line based on process. Power Syst. Technol. **42**(03), 989–995 (2018)

Research on Harmonic Resonance Mechanism and Characteristics for Power System with High Proportion of Renewable Energy

Jinhui Shi, Wei Chen(✉), Xiping Pei, Guoqiang Zheng, and Kun Wang

College of Electrical and Information Engineering, Lanzhou University of Technology, Lanzhou 730050, China
80658253@qq.com

Abstract. With the increasing proportion of renewable energy connected to the power grid, the harmonic resonance problem of the power grid becomes more complex. In this paper, a harmonic resonance analysis method combining frequency scanning and modal sensitivity is proposed. It solves the harmonic resonance problem caused by the high proportion of renewable energy into the power grid. Firstly, the frequency scanning method is used to determine the resonant frequency of the grid-connected system. Then the mode analysis and sensitivity analysis are combined to determine the resonance participation factors of different nodes and the sensitivity of each element. Finally, taking a grid-connected wind farm in Jiuquan as an analysis object, the time domain simulation in PSCAD/EMTDC is carried out to verify the effectiveness of this method. Furthermore, this paper analyzed the influence of reactive power compensation devices on resonance. These results show that through analyzing the harmonic resonance of grid-connected wind farms at the end of the power grid. It can provide a reference for the high proportion of new energy planning and harmonic resonance suppression.

Keywords: Harmonic resonance · Frequency scanning · Modal and sensitivity analysis · PSCAD/EMTDC modeling · Reactive compensation

1 Introduction

With the increasing scale of high proportion of renewable energy connected to the power grid, the proportion of power electronic devices is increasing, which causes the problem of harmonic resonance. Jiuquan area is a typical renewable energy high proportioned base. By the end of 2018, the proportion of wind power and photovoltaic was 71.4% and 24.9% respectively in Jiuquan area. The high proportion of large-scale renewable energy is mostly at the end of the power grid, and the installed capacity is large. Due to the weak structure of the grid terminal, there is a hybrid system of AC and DC [1–3]. Harmonic resonance will cause the reduction of power quality in power grid, even affect the safe operation of equipment and power grid [4–6].

Frequency scanning is a widely used method for harmonic resonance problem, which can determine the resonance frequency, but cannot detect the resonance position

Y. Xue et al. (Eds.): PMF 2020, LNEE 718, pp. 301–313, 2021.
https://doi.org/10.1007/978-981-15-9746-6_24

and the influence degree of each element on the resonance. [7] studied the harmonic level and characteristics of different electrical positions, including wind farm parallel network, collecting line bus and wind turbine. In reference [8], the frequency scanning method, small-signal analysis method and electromagnetic transient method are used to analyze the mechanism of sub-synchronous oscillations in wind farms. The harmonic resonance problem was analyzed in frequency domain based on the building of the wind farm harmonic model [9]. The node admittance matrix of the grid-connected system is established under the nonlinear factors such as multiple inverter control and time delay. The resonance frequency of the system is studied by using modal analysis in [10]. The modal analysis method can obtain the highest observation point and participation factor size [11, 12], but cannot determine the degree of each element participating in the resonance [13].

In view of the complex resonance problem caused by the high proportion of renewable energy connected to the power grid, considering the advantages and disadvantages of the frequency sweep method and the modal analysis method, this paper proposes a harmonic resonance analysis method combining frequency sweep and modal sensitivity. Taking a wind farm as the research object in Jiuquan area, Firstly, the resonance frequency point of the grid-connected wind farm was determined by frequency scanning method, using the modal sensitivity analysis method, the eigenvalues and eigenvectors obtained through decoupling were used to analyze the harmonic resonance participation factors and sensitivity of different nodes and components, which Improved the efficiency of the whole analysis process.

The validity of this method was verified using the time domain simulation under the PSCAD/EMTDC environment. Meanwhile, the system resonance was analyzed before and after the reactive power compensation device operated. Through analyzing the harmonic resonance frequency, participation factor and element sensitivity in wind farms, it provides a reference for the high proportion of new energy planning and harmonic resonance suppression.

2 Harmonic Model of the Grid-Connected Wind Farm

2.1 Harmonic Model of DFIG

DFIG is the main type of wind power generation [14–16]. The stator side of DFIG is directly connected to the power grid, while the rotor side is connected to the power grid through two power converters. Space vector control technology is used to realize bidirectional energy transmission between the rotor and the power grid. DC capacitor is mainly used to maintain the DC side voltage of the AC-DC-AC converter. When wind turbine operated, the harmonic current will be generated, which cause voltage distortion and power quality reduction. A large number of power electronic devices become the main harmonic source in wind farms. The harmonics become more and more serious with the increase of the wind turbines numbers. Under harmonic condition, for the hth harmonic, the equivalent model of DFIG is shown in Fig. 1:

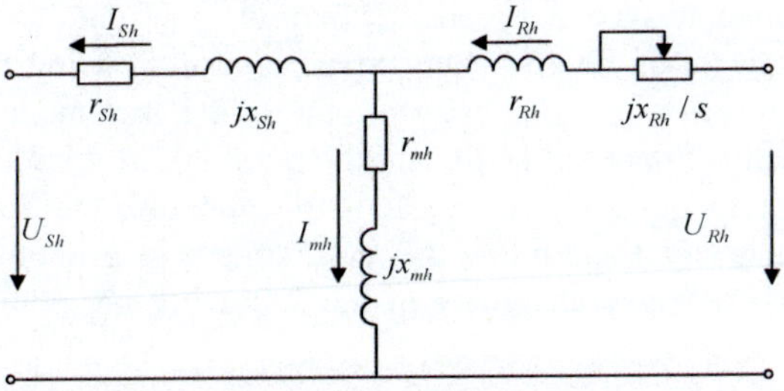

Fig. 1. The Harmonic model of DFIG

In Fig. 2, r_{Sh} and jx_{Sh} are stator leakage reactance for the hth harmonic, r_{Rh} and jx_{Rh}/s are rotor leakage reactance for the harmonic of h, r_{mh} and jx_{mh} are excitation leakage reactance for the harmonic of order h, U_{Sh} and U_{Rh} are stator electromotive force and rotor additional electromotive force for the hth harmonic respectively.

In large-scale wind farms, the harmonic emission of single wind turbines comes from power electronic converter, induction motor and power transformer. The harmonic impedance has a great influence on the resonance point of the power grid system. The system harmonic was enlarged by the resonance phenomenon, it will cause the harmonic level of the wind farm to be injected into the grid several times that of a single wind turbine. The simulation analysis is shown in Fig. 2 and Fig. 3:

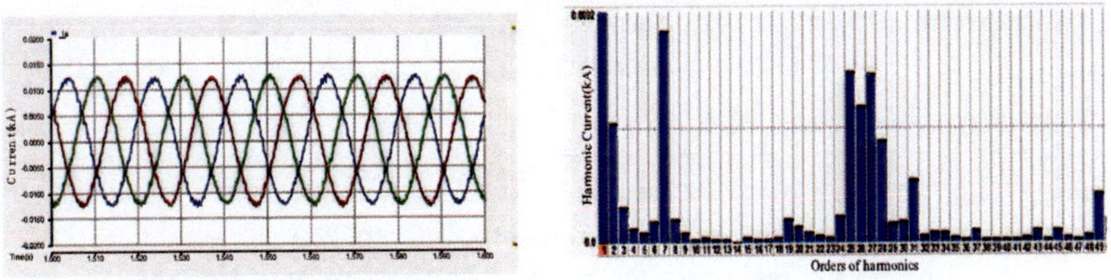

(a)Time-domain current (b)Harmonic current

Fig. 2. The Time-domain current and harmonic current of single wind turbine

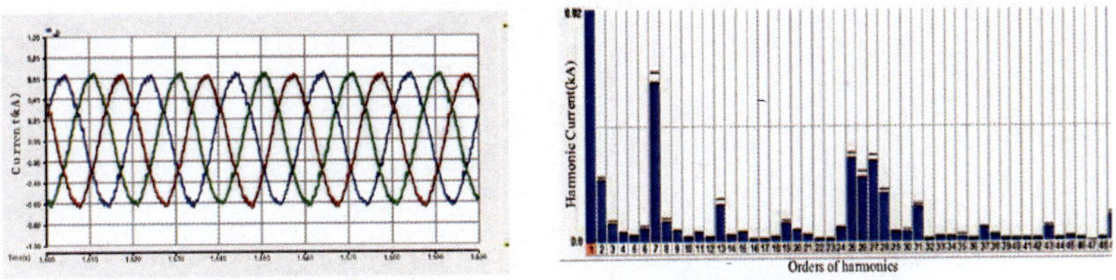

(a)Time-domain current (b)Harmonic current

Fig. 3. The Time-domain current and harmonic current of wind farm

In the simulation analysis experiment, the harmonic current of a single wind turbine and the harmonic current of the grid connection point of the wind farm were measured respectively. According to Fourier analysis, the odd harmonic content of the grid-connected wind farm is relatively large, and the harmonic content of each single fan is very small, which will not affect the system. Because the harmonic generated by a single wind turbine is small and meets the requirements of grid connection, this paper takes wind farm as the research object in the following harmonic resonance modal analysis.

2.2 The Harmonic Model of Grid Side

The grid-connected wind farm will cause harmonic impact on the grid, and the background harmonic of the grid will cause harmonic impact on the wind farm. The background harmonic of the grid is caused by various rectifier loads [17]. Because the wind turbine can not bear the background harmonic of the grid, the tripping accidents often occur, especially in the case of high voltage background harmonic, the harmonic resonance problem of the grid-connected wind farm becomes more serious [18, 19]. The grid harmonic model seen from the booster station is as follows:

$$Z_{sh} = jh \cdot X_s = jh\frac{U_0}{S}(U_{N1}/U_{N2})^2 \quad (1)$$

Where U_{N1} and U_{N2} are the rated voltage at the high voltage side and low voltage side of the step-up transformer respectively, U_0 is the open circuit voltage at the grid side, S is the short-circuit capacity of the parallel network.

2.3 Harmonic Modeling of SVG

Generally, The wind farm is equipped with a certain capacity of reactive power compensation device to improve the voltage support capacity and reactive power factor of the wind farm. SVG has become the first choice for wind farms due to its fast regulation speed and short response time. In this paper, the influence of SVG on grid-connected wind farm resonance is considered in the resonance analysis, so it is necessary to establish its harmonic model. SVG is mainly in two forms: voltage type and current type. Voltage type is a common type. The equivalent harmonic model of SVG is shown in Fig. 4.

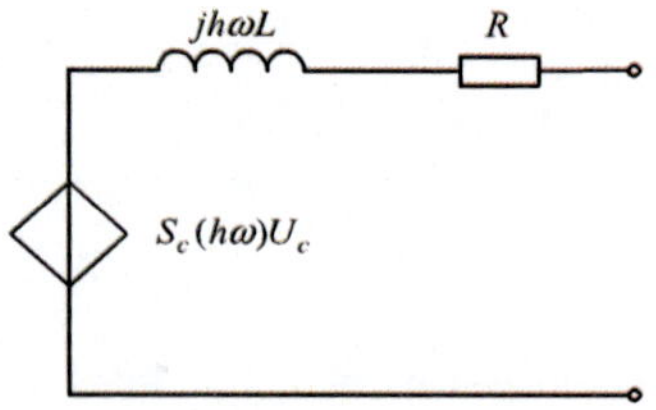

Fig. 4. The Harmonic model of SVG

In Fig. 4, $S_c(\omega)$ is the positive sequence switch function of SVG.

3 Modal Sensitivity Analysis Methods

3.1 The Principle of Modal Analysis

As a traditional resonance analysis method, Frequency domain method can calculate the system impedance spectrum of all nodes intuitively and quickly. It can determine the system harmonic resonance frequency, but cannot determine the location of resonance, the participation of each node and the influence of each device on the resonance.

The principle of modal analysis is that if the frequency of f resonance occurs in the system, the voltage and current equations of nodes in the system are:

$$V_f = Y_f^{-1} I_f \tag{2}$$

Where V_f is the node voltage matrix, I_f is node current matrix, and Y_f^{-1} is network admittance matrix at resonance frequency of f. According to [20–22], the admittance matrix of Y_f is decomposed into:

$$Y_f = L\Lambda T \tag{3}$$

L and T are eigenvector matrix, and, $L = T^{-1}$. Λ are diagonal eigenvalue matrix. Combining Eq. (2) and (3), Eq. (4) can be obtained:

$$V_f = L\Lambda^{-1} T I_f \tag{4}$$

Where Λ^{-1} is defined as the modal impedance. If $U_f = TV_f$ is defined as the modal voltage vector, and $J_f = TI_f$ as the modal current vector, then $U_f = \Lambda^{-1} J_f$.

$$\begin{bmatrix} U_{f1} \\ U_{f2} \\ \vdots \\ U_{fn} \end{bmatrix} = \begin{bmatrix} \lambda_{f1}^{-1} & 0 & 0 & 0 \\ 0 & \lambda_{f2}^{-1} & 0 & 0 \\ \vdots & \vdots & \vdots & \vdots \\ 0 & 0 & 0 & \lambda_{fn}^{-1} \end{bmatrix} \begin{bmatrix} J_{f1} \\ J_{f2} \\ \vdots \\ J_{fn} \end{bmatrix} \tag{5}$$

From Eq. (5), it is easy to get that when $\lambda_n = 0$ or λ_n very small, a small mode n injection current J_{fn} will lead to a large mode n voltage U_{fn}. At the same time, other mode voltage will not be affected, so it is easy to identify the resonance location in the model domain. The minimum eigenvalue is the key mode of harmonic resonance, and the left and right eigenvectors are the key eigenvectors.

If the resonance of mode 1 occurs in the system, the value of $1/\lambda_{f1}$ is much greater than that of other modes, there are:

$$\begin{bmatrix} V_{f1} \\ V_{f2} \\ \vdots \\ V_{fn} \end{bmatrix} = \lambda_{f1}^{-1} \begin{bmatrix} L_{11}T_{11} & L_{11}T_{12} & \cdots & L_{11}T_{1n} \\ L_{21}T_{11} & L_{21}T_{12} & \cdots & L_{21}T_{1n} \\ \vdots & \vdots & & \vdots \\ L_{n1}T_{11} & L_{n1}T_{11} & \cdots & L_{n1}T_{n1} \end{bmatrix} \begin{bmatrix} I_{f1} \\ I_{f2} \\ \vdots \\ I_{f3} \end{bmatrix} \tag{6}$$

The eigenvalue decomposition of nodal admittance matrix Y_f is carried out. The product of eigenvectors L and T is defined as the participation factor

$$PF = L_{mn}T_{nm} \tag{7}$$

Where m is the bus number, n is the model number, and PF is the contribution of each node in the system to the resonance under this model. The larger the participation factor is, the easier the resonance is to be activated and observed at this node.

The mode current J_{fn} is the linear mapping of the actual current in the n-th eigenvector direction.

$$J_{fn} = T_{n1}I_1 + T_{n2}I_2 + T_{n3}I_3 + \cdots + T_{nn}I_n \tag{8}$$

It can be known from the above formula that T_{n1} has the maximum value, node current I_1 has the maximum contribution to mode n current, and bus 1 is most likely to be activated and produce mode n resonance. If T_{n1} is zero, I_1 will not activate this mode resonance.

3.2 Sensitivity Analysis

Modal analysis can determine the size of influence factors, which can be used to estimate the contribution of each node, but it cannot determine the contribution of each component in the system harmonic resonance, while the sensitivity analysis method can make up for its shortcomings. Sensitivity analysis can obtain the influence degree of each element on harmonic resonance. When the system resonates at the frequency f, the nodal admittance matrix is decomposed. The corresponding elements of the eigenvector matrices L and T are multiplied, which indicates the degree of resonance participation of each node. The participation factor can only identify the nodes where resonance occurs, but it can't determine which devices are involved in the resonance center when resonance occurs, which is not conducive to the precise suppression of resonance. According to [23], if the eigenvalue of nodal admittance matrix Y_f is λ, then the partial derivative of λ for element Y_{ij} in row i and column j of Y_f is defined as element modal sensitivity, which is expressed as:

$$S_Y^{\lambda} = \frac{\partial \lambda}{\partial Y_{ij}} \tag{9}$$

In order to make the element sensitivity more comparable in a certain mode, it is necessary to normalize the sensitivity, which is expressed as:

$$\frac{\partial \lambda}{\partial Y_{ij}} \cdot \frac{Y_{ij}}{\lambda} = \frac{\partial \lambda}{\lambda} \Big/ \frac{\partial Y_{ij}}{Y_{ij}} \tag{10}$$

The normalized sensitivity indicates that when a certain element Y_{ij} in the admittance matrix has small change, the ratio of the eigenvalue λ to the variable Y_{ij} is the limit that ΔY_{ij} tends to zero.

When the admittance of parallel element in the system is $Y = G + jB$, the sensitivity of resonance eigenvalue λ to this element is expressed as follows:

$$\frac{\partial |\lambda|}{\partial G} = \frac{G(S_r \lambda_r + S_i \lambda_i)}{\lambda_r^2 + \lambda_i^2} \tag{11}$$

$$\frac{\partial |\lambda|}{\partial B} = \frac{B(S_r \lambda_i - S_i \lambda_r)}{\lambda_r^2 + \lambda_i^2} \tag{12}$$

S_r and S_i are the real part and virtual part of $S_{\lambda,ii}$ respectively; λ_r and λ_i are the real part and virtual part of resonance eigenvalue λ respectively. $\mu = \partial|\lambda|/\partial G$ and $v = \partial|\lambda|/\partial B$ represent the relative changes of resonance impedance and element parameters, respectively.

When the admittance of series element in the system is $Z = R + jX$, the sensitivity of resonance eigenvalue λ to this element is expressed as:

$$\frac{\partial |\lambda|}{\partial R} = \frac{\mu(X^2 - R^2) + 2vRX}{(R^2 + X^2)} \tag{13}$$

$$\frac{\partial |\lambda|}{\partial X} = \frac{v(X^2 - R^2) - 2\mu RX}{(R^2 + X^2)} \tag{14}$$

4 Example Analysis

The first new energy base is planned and constructed in Jiuquan. By the end of 2018, there are more than 78 wind farms with a total installed capacity of 9.15 million kilowatts, and more than 126 photovoltaic power stations with an installed capacity of 2 million kilowatts. However, which locate at the end of the power grid, with diverse energy sources and complicated grid structure. Covering a variety of new energy power generation forms such as wind energy, photovoltaics, solar thermal. The power output voltage grades mainly include AC 750 kV and DC ± 800 kV. The topological structure of new energy distribution in Jiuquan area is shown in Fig. 5:

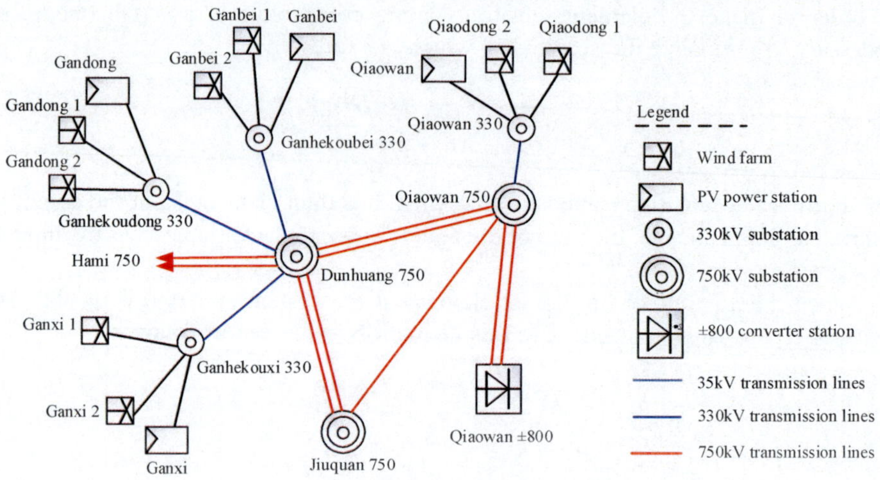

Fig. 5. The distribution of new energy in Jiuquan area

In this paper, a wind farm is selected as the research object in Jiuquan area, Using PSCAD/EMTDC, the simulation model of wind farm is established. the number of DFIG is 50, and the capacity of each DFIG is 2 MW. The output voltage of the DFIG is 0.69 kV, which is boosted to 35 kV through the box-type transformer, and it flows into the 35 kV bus from the primary transmission line. The voltage is raised to 330 kV through the booster transformer of the collection station and then connected to the large power grid. In the process of simulation, two sections of overhead lines are used. The length of the internal line of the wind farm is 30 km, and the output line is 100 km. The wiring diagram of the wind farm is shown in Fig. 6:

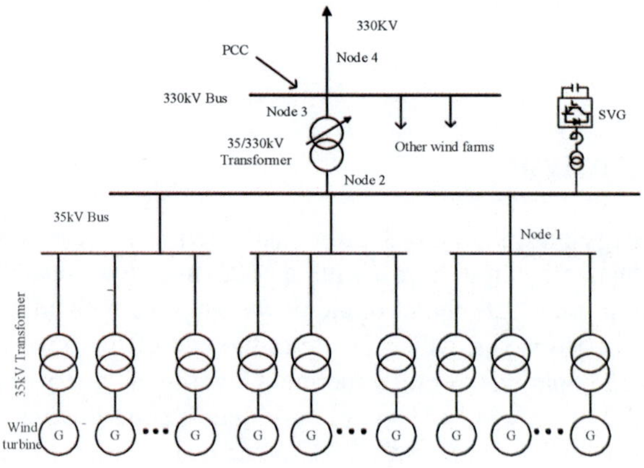

Fig. 6. The four node of test system

Main electrical equipment parameters of grid connected wind farm are shown in Table 1:

Table 1. The Main equipment parameters of grid-connected wind farms

Parameter	Parameter description
Wind turbine	DFIG,unit capacity 2 MW, Outlet voltage 0.69 kV, Rated current 1540 A
Box-type transformer	690 V/35 kV Rated capacity 2.35 MVA, Rated voltage 37 ± 2 * 2.5%/0.69 kV
Booster transformer	35 kV/330 kV,Rated capacity 400 MVA, Connection mode YNd11,
Transmission-line parameters	The resistance per kilometer 0.013 Ω,Reactance 0.029 Ω Susceptance 3.65×10^{-5} S
Reactive device	SVG:Rated voltage 35 kV,Rated capacity 30 Mvar,Connection mode: Y

4.1 Harmonic Resonance Analysis of Grid-Connected Wind Farm

The resonance frequency of the wind farm model is scanned, and the wind farm model is established by PSCAD/EMTDC. The resonance frequency point of wind farm is determined. Four nodes were scanned respectively, and the results of frequency impedance scanning are shown in Fig. 7:

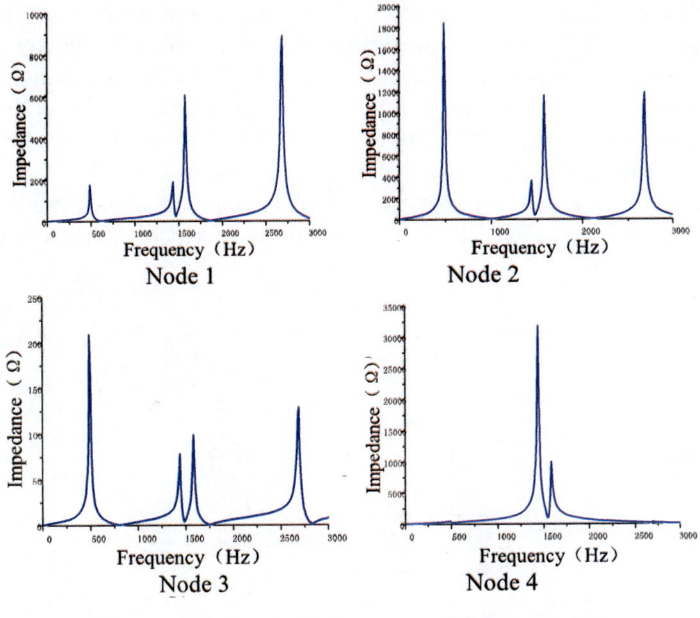

Fig. 7. The resonance frequency of different nodes

Through frequency scanning about four different nodes of grid-connected wind farms, it can be seen from Fig. 7 that there are four harmonic resonance points in grid-connected wind farms. Each node has a different resonance impedance value, but the magnitude of resonance frequency is the same. There are four resonance modes corresponding to the grid-connected wind farm, their resonance frequencies are 490 Hz, 1440 Hz, 1580 Hz and 2690 Hz.

4.2 Participation Factor and Sensitivity Analysis of Each Element

Based on the frequency scanning, the modal analysis method is used to analyze the participation factors of each node under four resonance modes. The analysis results are shown in Table 2:

Table 2. The Node participation factor in resonant mode

The Node participation factor				
Resonant frequency	490	1440	1580	2690
Node 1	0.2586	0.2877	0.2883	0.2901
Node 2	0.1725	0.1825	0.1855	0.1943
Node 3	0.1736	0.1839	0.1870	0.1985
Node 4	0.3824	0.5833	0.6861	0.7369

It can be known from the table that nodes 2 and 3 have similar participation factors under four resonance frequencies, that is, nodes 2 and 3 are affected in the same degree when resonance occurs. Node 4 has the largest participation factor under four resonance modes, which can be the Resonance Center under resonance rate. The resonance phenomenon is easily observed in the resonance center. Node 4 is directly connected to the large power grid and is easily influenced by background harmonics. Due to the installed reactive power compensation device in the low voltage side of the wind farm booster station, the participation factor of node 1 is larger than that of node 2 and node 3. So it is easier to activate harmonic resonance at nodes 1 and 4.

In this paper, the sensitivity analysis is carried out in combination with formulas (9) and (10) under 490 Hz, 1580 Hz and 2690 Hz resonant modes. The analysis results are shown in Fig. 8.

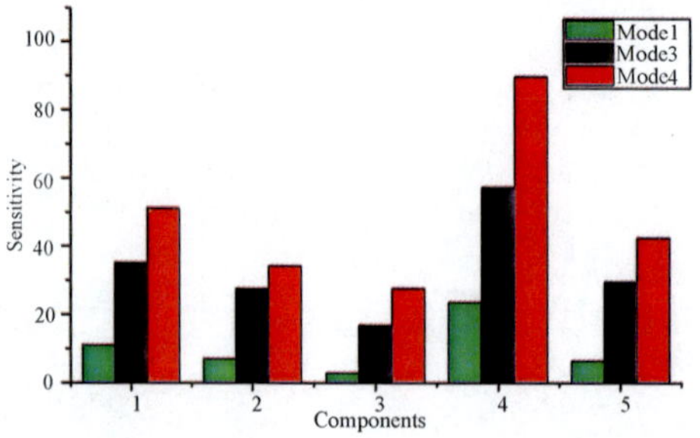

Fig. 8. The sensitivity of eigenvalue in three modes

It can be seen from Fig. 8 that element 1 is the sensitivity of equivalent impedance eigenvalue of the wind farm, element 2 and element 4 are the sensitivity of line impedance eigenvalue between nodes 1–2 and 3–4 respectively, element 3 is the sensitivity of main transformer eigenvalue of booster station, and element 5 is the sensitivity of reactive device eigenvalue. The greater the sensitivity value is, the greater the influence on resonance is.

4.3 Influence of Reactive Power Compensation Device on System Resonance

In order to improve the capacity of reactive power regulation and voltage control, wind farms are equipped with a corresponding proportion of reactive power compensation capacity. The impedance relationship of resonance frequency is compared and analyzed about the grid-connected wind farm, When SVG is put into operation before and after. Figure 9 shows the impedance relationship of resonance frequency when SVG is not put into operation, Fig. 7 shows the impedance relationship of resonance frequency when SVG is put into operation.

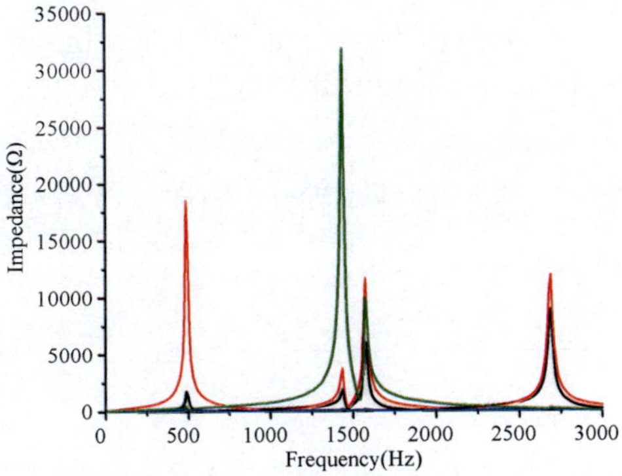

Fig. 9. The resonance impedance before SVG input

Table 3 shows the resonance impedance values of four nodes at 490 Hz, When SVG is put into operation before and after.

Table 3. The resonance point impedance comparison

490 Hz	SVG input	SVG output
Node 1	1738.3	1757.6
Node 2	17295.3	18424.4
Node 3	206.5	209.7
Node 4	435.8	438.3

Through comparative analysis, the reactive power compensation device will not change the point of resonance frequency, it will only change the resonance impedance value of each resonance point. After the dynamic reactive power compensation device is put into operation, the harmonic impedance values of resonance points 1, 2, 3 and 4 are reduced.

5 Conclusions

1) When studying the harmonic problem for the high proportion of new energy, the single wind turbine generates fewer harmonics, so the entire wind farm is taken as an analysis object.
2) Through frequency scanning, the frequency impedance relationship of each node can be obtained in the grid-connected system, and the key resonance of frequency can be obtained. The modal analysis only needs to decouple the key resonant modes, which reduces the amount of calculation.
3) The modal sensitivity analysis method can not only obtain the size of the participation factor of each node, but also give the sensitivity value of each element. The sensitivity analysis shows that the influence of wind farms and transmission line on resonance is relatively large.
4) The reactive power device will only cause the change of resonance impedance, but it will not change the resonance mode.

Acknowledgments. The is work was supported by the National Natural Science Foundation of China (No. 51767017 and No. 51867015) and Basic Research and Innovation Group Project of Gansu (No. 18JR3RA133).

References

1. Pérez, J., Cóbreces, S., Wang, X., et al.: A robust grid current controller with grid harmonic and filter resonance damping capabilities using a closed-loop admittance shaping. In: Applied Power Electronics Conference and Exposition (2017)
2. Tang, F., Guerrero, J.M., Vasquez, J.C., et al.: Distributed active synchronization strategy for microgrid seamless reconnection to the grid under unbalance and harmonic distortion. IEEE Trans. Smart Grid **6**(6), 2757–2769 (2017)
3. Cheng, H., Li, J., Wu, Y., et al.: Challenges and prospects for AC/DC transmission expansion planning considering high proportion of renewable energy. Autom. Electr. Power Syst. **41**(9), 19–27 (2017)
4. Hu, W., Sun, J., Ma, Q.: Analysis on interactive influences among multi grid-connected inverters. Power Syst. Technol. **38**(9), 2511–2518 (2014)
5. Kang, C., Yao, L.: Key scientific issues and theoretical research framework for power systems with high proportion of renewable energy. Autom. Electr. Power Syst. **41**(9), 2–11 (2017)
6. Lu, Z., Li, H., Qiao, Y.: Power system flexibility planning and challenges considering high proportion of renewable energy. Autom. Electr. Power Syst. **40**(13), 147–158 (2016)

7. Fan, Y., Zhang, J., Xie, J., et al.: Testing and analysis for harmonic resonance of wind farm. Autom. Electr. Power Syst. **40**(20), 147–151 (2016)
8. Qian, W., Hui, L.: Summary about mechanism and mitigation of SSR when wind power connecting grid. North China Electr. Power **6**, 10–15 (2014)
9. Cui, Y., Wang, X.: Modal frequency sensitivity for power system harmonic resonance analysis. IEEE Trans. Power Deliv. **27**(2), 1010–1017 (2012)
10. Liu, Y., Shuai, Z., Li, Y., et al.: Harmonic resonance modal analysis of multi-inverter grid-connected systems. Proc. CSEE **37**(14), 4156–4164 (2017)
11. Tan, M., Peng, X., Wang, T., et al.: Study on the harmonic resonance of grid-connected of wind farms and half-wave-length AC transmission system based on modal analysis method. High Volt. Eng. **44**(1), 90–98 (2018)
12. Wang, F., Duarte, J., Hendrix, M., et al.: Modeling and analysis of grid harmonic distortion impact of aggregated DG inverters. IEEE Trans. Power Electron. **26**(3), 786–797 (2011)
13. Zhang, M., He, Z., Gao, S.: Harmonic resonance modal sensitivity analysis for high-speed railway traction supply system considering different harmonic load model. Power Syst. Prot. Control **43**(20), 21–28 (2015)
14. Xie, X., Zhang, X., Liu, H., et al.: Characteristic analysis of subsynchronous resonance in practical wind farms connected to series-compensated transmissions. IEEE Trans. Energy Convers. **32**(3), 1117–1126 (2017)
15. Ma, S., Geng, H., Ma, J., et al.: An approachto establish detailed model of DFIG based wind farm. Trans. China Electro Tech. Soc. **32**(S1), 1–10 (2017)
16. Li, Q., Cheng, P., Wang, W.: Modeling and analysis of harmonic current in middle-frequency band in DFIG based wind turbines. Autom. Electr. Power Syst. **42**(18), 41–47 (2018)
17. Zhao, X., Wang, Q., Shao, B., et al.: Low voltage ride through control strategy and its analysis of doubly fed induction generator. Power Syst. Prot. Control **43**(16), 57–64 (2015)
18. Ebrahimzadeh, E., Blaabjerg Gae, F., Wang, X., et al.: Bus participation factor analysis for harmonic instability in power electronics based systems. IEEE Trans. Power Electron. **1**(1), 10341–10350 (2018)
19. Nie, C., Wang, Y., Wanjun, Lei, et al.: Harmonic resonance analysis and suppression method for transmission cables between renewable energy sources and power grid. Autom. Electr. Power Syst. **41**(17), 128–133 (2017)
20. Xu, W., Zhang, D.: A modal analysis method for harmonic resonance assessment. Proc. CSEE **25**(22), 89–93 (2005)
21. Yang, K., Bollen, M.H.J., Amaris, H., et al.: Decompositions of harmonic propagation in wind power plant. Electr. Power Syst. Res. **141**(12), 84–90 (2016)
22. Xing, F., Xu, Z., Wang, S.: Analysis and suppression of resonance problem in power system with unconventional generators. Autom. Electr. Power Syst. **43**(15), 71–79 (2019)
23. Tang, Z., Yang, H.: Series resonance analysis on grid-connected system of wind farm in remote areas. Autom. Electr. Power Syst. **40**(15), 120–125 (2016)

Research on Operation and Management of Decentralized Dispatching Agency for Multi-energy System Based on Block Chain Technology

Tianmeng Yang(✉), Fangyuan Yang, Xing Ji, Yi Liang, and Yong Wang

Economic Technical Research Institute of Liaoning Electric Co. Ltd., Shenyang, China
18240369315@163.com

Abstract. The integration of multi-energy systems is an important feature of energy internet system. The traditional energy systems, such as electric power, thermal power, gas and other energy systems, are operating separately. All kinds of energy systems have been separated and self-made in a long time. They are managed by different departments, and it is difficult to set up a centralized dispatching agency for operation management, which restricts the development of energy internet system greatly. Block chain technology has the main characteristics of decentralization and transparency. Therefore, a decentralization dispatching and operation control mechanism based on block chain technology for multi-energy fusion system is proposed in this paper. Through the establishment of decentralized dispatching agencies for operation management, it is proposed to share the information of energy supply and demand relationship between different energy systems and different regions in multiple energy systems. An optimal operation model of self-dispatching decision-making is established to obtain the results of energy self-dispatching transaction of each energy demand node. Finally, a simulation is conducted of a multi-energy system, and the calculation results verify the effectiveness and feasibility of the self-scheduling decision-making optimization model proposed in this paper.

Keywords: Energy internet · Multiple energy systems · Block chain · Self-scheduling decision · Marginal price

1 Introduction

The energy systems of traditional energy systems, such as electricity, heat and gas, are operating separately. With the rapid development of energy internet, the energy systems are closely integrated and optimized in production, conversion, storage, transportation, dispatch, control, management and use, so as to form a multi-energy system [1, 2]. All kinds of energy can flow flexibly in different physical systems through energy conversion equipment, which can realize flexible storage and cascade utilization

Y. Xue et al. (Eds.): PMF 2020, LNEE 718, pp. 314–326, 2021.
https://doi.org/10.1007/978-981-15-9746-6_25

of energy, and it can help improve the efficiency of energy conversion and utilization significantly.

At present, the energy demand is surging and the energy trading market is facing major changes. The development of multi-energy system has become the key to solve the problem of energy supply and demand. What's more, considering about the inconvenience of energy conversion, the opaque of energy trading information, the insecurity of user trading identity, and the cumbersome trading process, the multi-energy system cannot keep up with the expected pace of energy industry change. Block chain technology has the characteristics of de-centralization and no tampering of node data, which makes it have a broad application prospects in multi-energy system scheduling and operation and energy trading [3, 4].

To build a decentralized dispatching agency in multi-energy systems, and to adjust the demand of both suppliers and demanders in multi-energy systems through intelligent contracts effectively can simplify the energy trading system. Until now, there are few studies on block chains in multi-energy system scheduling, operation and trading. Due to the complexity of energy system, there are still many problems in the application of block chains in multi-energy system. In [5, 6], a block-chain-based framework for supply-demand interaction system is presented, which achieves the security of data at each node of the system. However, this framework has not been applied to deep research on energy trading. In [7, 8], a grid data federation storage system in the form of Federation chain is proposed, but less consideration was given to the security of data storage and transmission. In [9], a multi-energy complementary security transaction model for heterogeneous energy block chains based on block chain technology in distributed energy systems is proposed, but there is no in-depth study on multi-energy scheduling.

Base the analysis above, this paper proposes a decentralized dispatching agency for multi-energy system based on block chain technology, and its operation and management mode and energy trading mode is studied.

2 Research on the Related Principles of Block Chain Technology

2.1 The Characteristics of Block Chain Technology

Block chain technology is a chain structure that combines data blocks in accordance with certain rules. It has the characteristics of decentralization, de-trust, information cannot be tampered with, and open autonomy. It is essentially a reliable data ac-count maintained by all nodes. Block chains can be seen as distributed data accounts shared and maintained by each node in the system.

At present, block chain technology has been widely used in medical, financial, inter-net of things, energy and many other fields. According to the licensing authority, block chain can be divided into three types, which are public chain, private chain and alliance chain.

2.2 The Data Storage Structure of Block Chain in Multi-energy System

The data structure of block chain conclude of two parts, the head and body. The head mainly stores the information in the previous block linked to it, which is stored in hash value. The body mainly stores the information of energy conversion, usage and the amount and content of energy transactions. The data storage structure of block chain in multi-energy system is shown in Fig. 1.

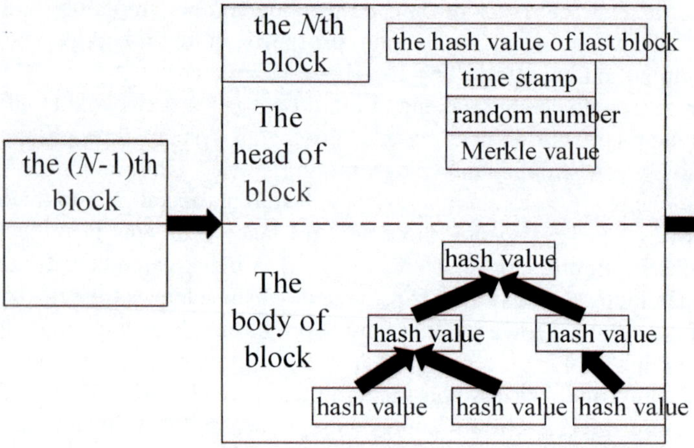

Fig. 1. Data storage structure diagram of block chain in multi-energy system

2.3 Intelligent Contract and Operation Mechanism on Block Chain

Intelligent contract is a kind of transaction protocol. It has the characteristics of no intermediary, self-verification and automatic execution of contract terms. With the development of block chain technology, intelligent contract based on block chain inheritable the advantages of decentralization, de-trust and not being tampered with [10].

The schematic diagram of the operation mechanism of the intelligent contract is shown in Fig. 2. Intelligent contract elements are mainly composed of the value and status. Generally, trigger scenarios and response rules are preset in the contract terms. After the completion of the common agreements and the signing of their respective transaction agreements within the system, the transaction initiated by the user is submitted, which is disseminated in the P2P network and stored in specific blocks of the block chain, and the contract address and contract interface of the transaction information are returned. When the user gets this information, he can initiate the transaction and call the corresponding transaction contract.

The miners in the block chain system will verify the transaction after receiving the instructions of creating contract and calling contract, execute the contract in the local execution environment, and automatically judge whether the current scenario meets the contract execution conditions according to the checking information of external data sources and world status. If the condition is satisfied, the transaction will be executed normally, the data will be packed into the main chain, and all nodes in the system will be updated.

What's more, the payment and transaction are undertaken by all nodes in the system, the centralized transaction institution is not required, which reduce the transaction costs. And the concrete transaction information are all restored in all nodes, which make sure the information is safe and can't be modified.

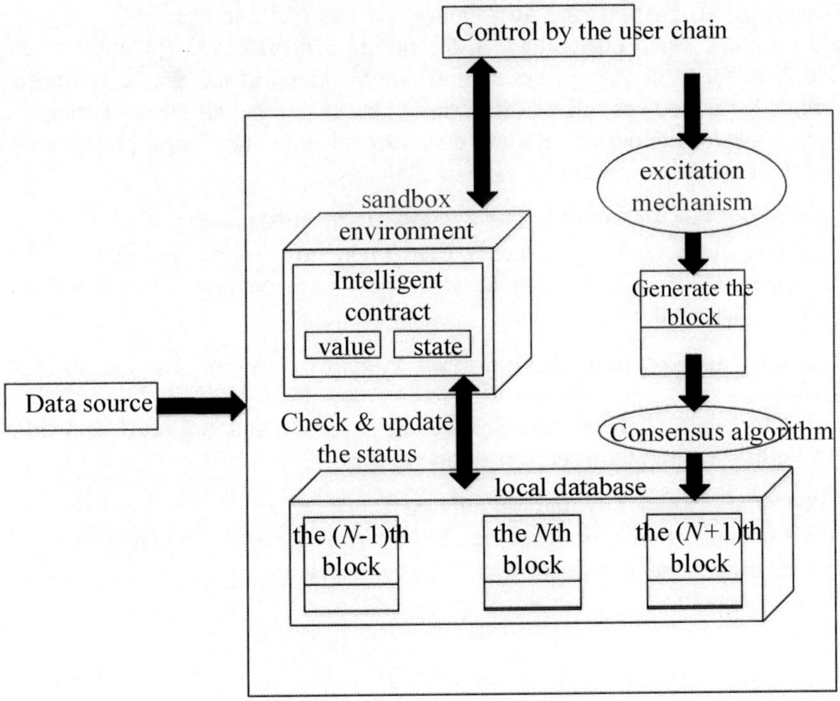

Fig. 2. Schematic diagram of operation mechanism of intelligent contract on block chain

3 Main Objectives of Multi-energy System Collaboration

3.1 Main Objectives of Multi-energy System Collaboration

The main objectives of multi-energy system collaboration can be summarized as follows:

1) All kinds of energy systems should be separated from the state of self-control and be able to accept the unified dispatch, operation and control management.
2) When all kinds of energy systems accept unified dispatching and trading, a unified security data platform should be based to ensure the security and reliability of data.

3) Collaborative work and trade among various energy systems should be supported by corresponding revenue incentive mechanism, so that the multi-energy system can operate normally and continuously.

3.2 Characteristics of Coordination of Multi-energy System Based on Block Chain Technology

1) Decentralized Dispatching and Operation Control Mechanism
Based on block chain technology, a decentralized dispatching operation mechanism can be built for multi-energy system, so that all kinds of energy do not need to be controlled by a fixed centralized dispatching agency, so that all kinds of energy can be free from the restrictions of regional distance and different energy prices, which can realize the optimal utilization of their own energy.

2) Guarantee The Reliability And Confidentiality of Data
Energy flow information and energy transaction information between multi-energy systems are confidential information, which need to be encrypted before uploading to the unified data platform.

3) Unified And Coordinated Operation Without Barriers Among Multi-energy Sources
It breaks the barriers between traditional energy sources, and makes all kinds of energy have a unified and coordinated operation platform.

4) Self-Collaboration Through Intelligent Contract
Intelligent contracts are introduced into multi-energy systems, so that the price of each node can trigger the intelligent contracts, and then they can be signed directly to achieve synergy.

3.3 Schematic Diagram of Cooperative Working Principle of Multi-energy System

The schematic diagram of cooperative work among multi-energy systems is shown in Fig. 3. In the traditional energy system, the power, thermodynamic, gas systems are all in discrete operation status. While in the future energy internet system, the energy production, transaction, storage, transportation, dispatch and control are all in close integration. In Fig. 3, the information transmission between the thermodynamic system and block chain system are the heating capacity, electricity price, heating price, and gas price. The information transmission between the power system and block chain system are the power generation and the electricity price. The information transmission between the gas system and block chain system are the quantity of gas supply and electricity price. All different energy systems sign the smart contract and execute automatic scheduling instructions according to the marginal price information.

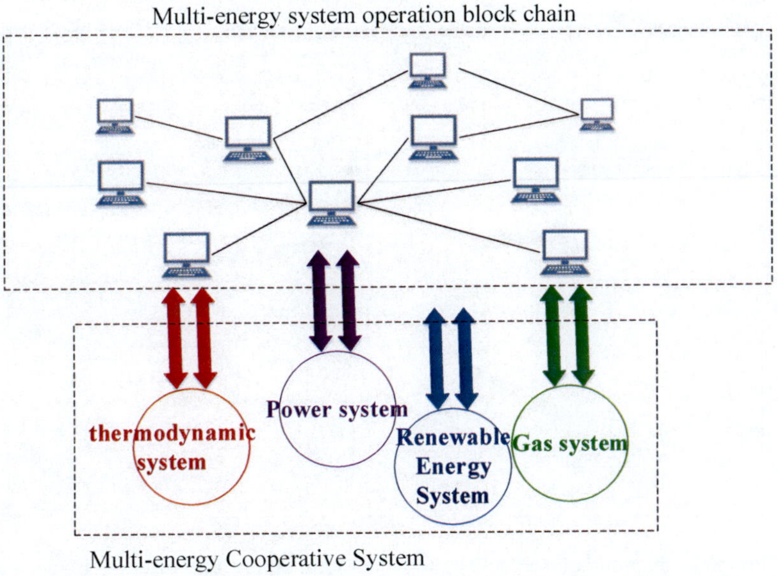

Fig. 3. The schematic diagram of cooperative working principle of multi-energy system

4 Main Objectives of Multi-energy System Collaboration Optimal Operation Model of Self-scheduling Decision-Making for Multi-energy System

4.1 Authors and Affiliations Optimal Operation Model of Self-scheduling Decision-Making for Multi-energy System

In this paper, an optimal operation and transaction model for multi-energy system with cross-energy self-scheduling decision-making is proposed, as is shown in Fig. 4.

1) Self-Scheduling Decision-Making Optimization Model for Multi-energy Systems

The model consists of eight elements, which are expressed as follows:

$$Y = \left(D_{\text{Energy}}, S_i, U_i, D_{\text{Dispatch}}, D_{\text{Transaction}}, D_{\text{Index}}, C_{\text{Energy}}\right) \tag{1}$$

$$D_{\text{Energy}} = \{D_{\text{Coal}}, D_{\text{PV}}, D_{\text{Wind}}, D_{\text{Gas}}, D_{\text{other}}\} \tag{2}$$

$$C_{\text{Energy}} = \{C_{\text{Energy-PoS}}, C_{\text{Dispatch-PoS}}, C_{\text{Transaction-PoS}}, C_{\text{Index-PoS}}\} \tag{3}$$

Where, D_{Energy} represents the energy supply block chain, mainly including coal thermal supply block chain D_{Coal}, photovoltaic power supply block chain D_{PV}, wind power supply block chain D_{Wind}, gas supply block chain D_{Gas} and other energy supply block chain D_{Other}; S_i represents the aggregation of various energy supply nodes; U_i represents all kinds of energy needs. Find the set of nodes; D_{Dispatch} represents the energy

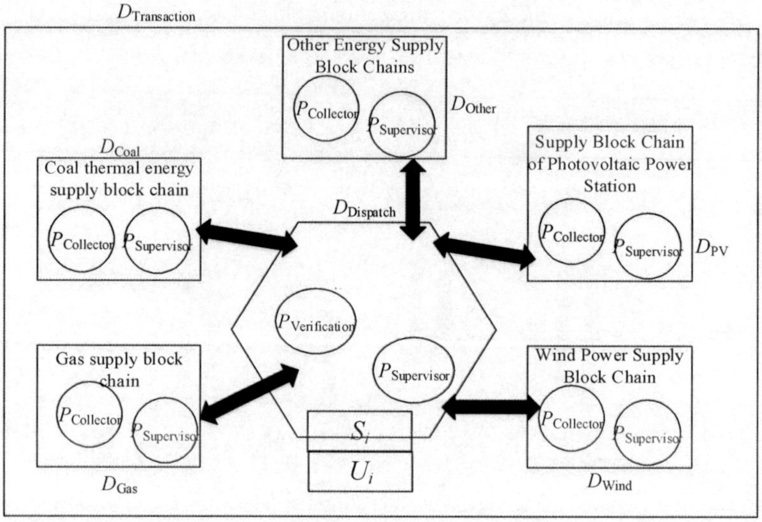

Fig. 4. Principle diagram of self-scheduling decision-making optimization model for multi-energy systems

dispatching block chain; $D_{Transaction}$ represents the energy trading block chain; D_{Index} represents the energy index trading block chain; C_{Energy} represents the energy dispatching operation and trading intelligent contract, mainly by the energy supply consensus algorithm $C_{Energy\text{-}PoS}$ based on PoS equity certification, and PoS equity certification based on PoS equity certification. $C_{Dispatch\text{-}PoS}$, $C_{Transaction\text{-}PoS}$ and $C_{Index\text{-}PoS}$ are the energy index consensus algorithms based on PoS equity certification.

2) Energy Supply Block Chain D_{Energy}

Energy supply block chain mainly includes coal thermal supply block chain D_{Coal}, photovoltaic power supply block chain D_{PV}, wind power supply block chain D_{Wind}, gas supply block chain D_{Gas} and other energy supply block chain D_{Other}. All types of energy supply block chains have the same data structure. The schematic diagram of the data structure is shown in Fig. 5.

The following is a detailed description of the coal thermal energy supply block chain, whose main components are as follows.

$$D_{Coal} = \left(P_{Collector}, P_{Superbisor}, S_i, C_{Energy-Pos}\right) \tag{4}$$

Where, $P_{Collector}$ represents the collector of block information, $P_{Supervisor}$ represents the supervisor of block information, S_i represents the supply node in the coal thermal energy supply chain, mainly composed of coal energy companies and thermal power companies, $C_{Energy\text{-}PoS}$ represents the consensus algorithm of energy supply based on PoS equity certification.

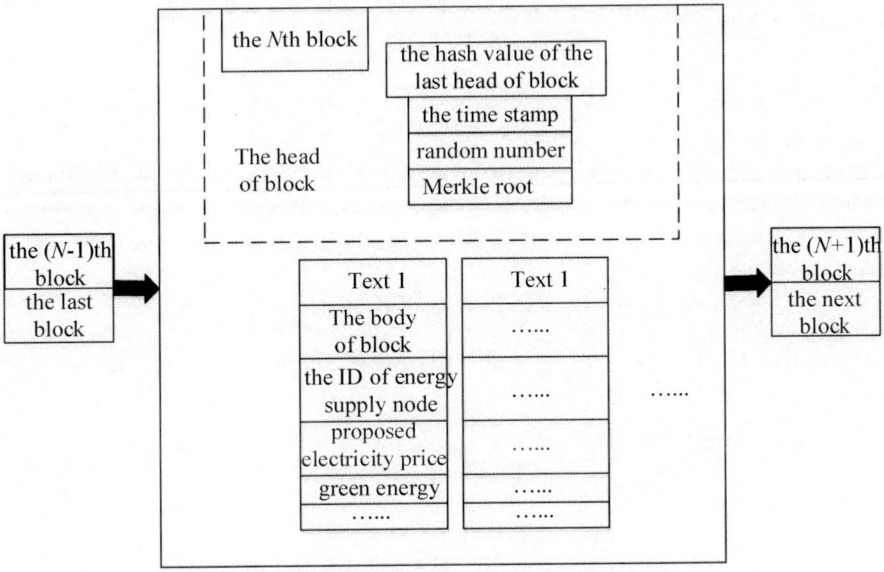

Fig. 5. Data structure diagram of energy supply block chain

Among them, block information collector and block information supervisor are very important roles in block chain system. As the collector of block information, it has all the information in the block chain. Its main task is to package and submit the newly generated blocks, and to execute allocation, scheduling, operation and transaction. As the supervisor of block information, he is mainly responsible for supervising the working mode in the block, supervising and reporting the illegal actions such as forged node information and forged transaction information found, and uploading the information of the illegal actions found to the block chain after encrypting. It is precisely because of the existence of the role of block information supervisor that the illegal actions are avoided.

$C_{\mathbf{Energy\text{-}PoS}}$ represents an energy supply consensus algorithm based on PoS equity certification. The main content of the algorithm is that the revenue of each node of both supply and demand comes from the reward brought by the successful exchange. Rewards are usually in the form of tokens, and interest can be accumulated over time through the holding of money by nodes. PoS-based proof of rights and interests has the advantages of low computational power and short consensus time, so it can avoid the hidden danger of over-centralization caused by the greater probability of miners' acquiring accounting rights, make the distribution of accounting rights of all nodes in the whole network balanced, and ensure the synchronous consistency of accounting books.

3) Energy Dispatching Block Chain $D_{\mathbf{Dispatch}}$

$$D_{\text{Dispatch}} = \left(U_i,\ C_{\text{Dispatch}-\text{PoS}}\right) \tag{5}$$

Where, U_i represents the energy demand nodes in the block chain, mainly including residential areas, office buildings, factories, electric vehicle charging stations, etc. $C_{\text{Dispatch-PoS}}$ represents the consensus algorithm of energy dispatching based on PoS rights certification.

$C_{\text{Dispatch-PoS}}$ is also based on PoS proof of rights. The energy scheduling consensus algorithm mainly packages and uploads the energy demand of all energy demand nodes in U_i to a new block for recording, so that the energy supply and energy demand can be matched efficiently.

4) Energy Trading Block Chain $D_{\mathbf{Tranaction\text{-}PoS}}$

$$D_{\text{Transaction}} = \left(U_i,\ C_{\text{Transaction}-\text{PoS}}\right) \tag{6}$$

Where, $C_{\text{Transaction-PoS}}$ represents an energy trading consensus algorithm based on PoS equity certification.

$C_{\text{Transaction-PoS}}$ is similar to $C_{\text{Dispatch-PoS}}$ in that it aims to efficiently match the energy demand and energy demand of all energy demand nodes in U_i to ensure an effective transaction between energy demand and supply.

5) Energy Index Trading Block Chain $D_{\mathbf{Index}}$

$$D_{\text{Index}} = \left(P_{\text{collector}}, P_{\text{Verification}}, C_{Index-PoS}\right) \tag{7}$$

Where, $P_{\text{Verification}}$ represents the verifier of block information and $C_{\text{Indec-PoS}}$ represents the consensus algorithm of energy index based on PoS proof of rights.

The block information verifier has higher authority than the block information collector and the block information supervisor.

4.2 Simulated Operation of Self-scheduling Decision-Making Optimization Model

In the simulation system, the energy supply nodes mainly include thermal power plants, thermal power plants, gas power plants and photovoltaic power plants, and energy demand nodes mainly include electric vehicle charging stations, residential buildings and shopping malls. Thermal power plants and thermal power plants belong to traditional energy sources, gas power plants and photovoltaic power plants belong to new energy sources. The purpose of self-scheduling decision-making optimization model is to achieve reasonable scheduling and effective complementarity of multiple energy sources.

Based on Energy-PoS, each energy supply node S_i submits its energy supply level, energy generation price and other information to their respective energy supply block chains, and then packages the information by their respective block information collectors, generates new blocks, and aggregates them into $D_{\text{Transaction}}$. On the other hand, each energy demand node U_i submits its energy demand information to D_{Dispatch} and $D_{\text{Transaction}}$, and submits it to $D_{\text{Transaction}}$. According to all the information submitted by the energy supplier and demander, the information verifier of the block gets the energy supply and use plan, and allocates the proportion of all kinds of energy rea-sonably.

5 Case Study

5.1 The Design of Multi-energy System

Based on the existing energy supply and demand information of a city, this paper makes some modifications to simulate a multi-energy collaborative system, and simulates the decentralized dispatching model of the multi-energy system proposed in this paper.

The energy suppliers and energy demanders in the system are shown in Table 1 and Table 2. And the generation cost of various energy sources can be found in [10–12].

Table 1. Energy supply node information

No.	Energy supply node	Energy output (MW·h)	Coal consumption g/kW·h
1	Thermal Power Plant G_1	500	352
2	Thermal Power Plant G_2	480	312
3	Gas power plant G_3	380	0
4	Photovoltaic power station G_4	40	0
5	Photovoltaic power station G_5	50	0
6	Wind farm G_6	60	0
7	Wind farm G_7	80	0

Table 2. Energy demand node information

No.	Energy supply node	Energy demand (MW·h)
1	Electric vehicle charging station L_1	80
2	Electric vehicle charging station L_2	100
3	Office Building L_3	280
4	Residential building L_4	340
5	Residential building L_5	350
6	Market L_6	365

5.2 Optimal Operation and Transaction Results of Multi-energy System Self-scheduling

Based on the multi-energy system self-scheduling decision-making optimization model proposed in this paper, the above multi-energy system is optimized and analyzed. The results of multi-energy simulation scheduling and trading are shown in Table 3.

Table 3. Energy supply node information

No.	Energy supply node	Energy supply node	Expected energy demand (MW·h)	Energy trading capacity (MW·h)	Trading price (RMB)
1	Thermal power Plant G_1	Residential building L_4	340	340	74120
2	Thermal Power Plant G_1	Market L_6	365	160	34880
3	Photovoltaic power station G_4	Electric vehicle charging station L_1	80	40	24000
4	Photovoltaic power station G_5	Electric vehicle charging station L_1	80	40	24000
5	Wind farm G_6	Electric vehicle charging station L_1	100	60	30000
6	Wind farm G_7	Electric vehicle charging station L_1	100	40	20000
7	Gas power plant G_3	Office Building L_3	340	340	95200
8	Gas power plant G_3	Residential building L_5	350	40	11200
9	Wind farm G_7	Market L_6	365	40	20000
10	Thermal Power Plant G_2	Residential building L_5	350	310	67580
11	Thermal Power Plant G_2	Market L_6	365	165	14170

From Table 3, it can be seen that there is a good match between the energy supply and demand nodes, and the new energy and clean energy in the system are also used efficiently. Wind power and photovoltaic power are sold to the charging station of electric vehicles in priority, which effectively reduces the thermal power generation, which is a traditional energy. It reduces the pollution to the environment. Also, the routine method is also conducted for compassion, the results show that the thermal power plants consume more coal and generate more power, while the wind farms and photovoltaic plants generate less power. After comparison, if the block chain technology didn't be used, the thermal power plants consumed at least $25 \times (256 \times 10^3$ $10^6) = 6.4$ t. And according to the quoted price of coal, it will expand more money about $6.4 \times 610 = 3904$ RMB. What's more, thanks to the block chain technology, more renewable energy is distributed and allocated more reasonably.

Through the self-scheduling decision-making optimization of the model, the energy output of the energy supply nodes in the system is more convenient, and the energy demand side is more transparent. The intelligent matching is achieved through the energy dispatching block chain and energy index trading chain, which verifies the effectiveness and feasibility of the self-scheduling decision-making optimization model.

6 Conclusion

In this paper, an optimal decision-making model of decentralized dispatching for multi-energy system based on block chain technology is proposed, and the optimal dispatching and trading process of energy is simulated. Through the establishment of decentralized dispatching agencies for operation management and the introduction of intelligent contracts into multi-energy transactions, not only to ensure the safety, reliability and confidentiality of energy dispatching and trading process in multi-energy systems, but also to ensure the rational matching of energy between energy supply nodes and energy demand nodes. Based on the research, the main advantages of block chain system are decentralized, de-trust, information cannot be tampered with and open autonomy, which are all verified in the paper. In the future, more in-depth research should be done in this area.

Acknowledgement. This work was supported in part by the Youth Promotion Fund "Research on load storage operation on load storage operation model of source network based on Grid Division".

References

1. Zeng, M., Cheng, J., Wang, Y., et al.: Primarily research for multi module cooperative autonomous mode of energy internet under blockchain framework. Proc. CSEE **37**(13), 3672–3681 (2017)
2. Zhang, N., Wang, Y., Kang, C., et al.: Blockchain technique in the energy internet: preliminary research framework and typical applications. Proc. CSEE **36**(15), 4011–4022 (2016)
3. Yuan, Y., Wang, F.: Blockchain: the state of the art and future trends. Zidonghua Xuebao/acta Autom. Sin. **42**(4), 481–494 (2016)

4. Zhao, X., Li, X., Zhan, L., et al.: Data integrity protection method for microorganism sampling robots based on blockchain technology. J. Huazhong univ. Sci. Tech. (Nat. Sci. Ed.) **43**(S1), 216–219 (2015)
5. Zeng, M., Cheng, J., Wang, Y., et al.: Primarily research for multi module cooperative autonomous mode of energy internet under blockchain framework. Proc. CSEE **37**(13), 3672–3681 (2017)
6. Lu, B., Lu, C., Cao, W., et al.: A preliminary study of block chain based automated demand response system. Proc. CSEE **37**(13), 3691–3702 (2017)
7. Wu, G., Zeng, B., Li, R., et al.: Research on the application of blockchain in the integrated demand response resource transaction. Proc. CSEE **37**(13), 3717–3728 (2017)
8. Wang, J., Zhou, N., Wang, Q., et al.: Electricity direct transaction mode and strategy in microgrid based on blockchain and continuous double auction mechanism. Proc. CSEE **38** (17), 5072–5084 (2018)
9. W, S., Gu, Z., Yang, X., et al.: Multienergy complementation and safety transaction model on heterogeneous energy blockchain. Power Syst. Technol. **21**(3), 321–330 (2019)
10. Ouyang, L., Wang, S., Yuan, Y., et al.: Blockchain-enabled smart conracts: architecture, applications and future trends. Acta Autom. Sin. **3**(4), 1–12 (2019)
11. Han, D., Zhang, C., Sun, W., et al.: Architecture design of intelligent power distribution trading platform based on blockchain technology. Power Syst. Autom. **43**(7), 132–142 (2019)
12. Chen, Q., Wang, K., Chen, S., et al.: Tradable energy system for distributed agents: architecture, mechanism design and key technologies. Power Syst. Autom. **32**(4), 140–148 (2018)

An Improved TET Method for Traveling-Waves Protection of VSC-HVDC

Guangbin Zhu, Yiqing Liu(✉), Gang Yu, and Peng Lv

School of Electrical Engineering, University of Jinan, Jinan, China
cse_liuyq@ujn.edu.cn

Abstract. As fault process of DC transmission line develops quickly, single-terminal protection is adopted for High Voltage Direct Current based on Voltage Source Converter (VSC-HVDC) for fast isolation of faults. Single-terminal protections generally need to identify initial arrival wave (IAW) of traveling-waves (TWs). IAW is identified by wavelet transform in conventional single-terminal traveling-wave protection. However, the identification accuracy of IAW by wavelet transform is low for VSC-HVDC. To increase identification accuracy of IAW, Transient Extraction transform (TET) with high time resolution is proposed to identify IAW for traveling-wave protection in this paper. Based on the characteristics of the IAW, the TET is improved to further increase the time resolution and identify the IAW more accurately. A simulation model of double-terminal VSC-HVDC is established in PSCAD/EMTDC to verify the proposed method. IAW identification accuracy of improved TET is compared with that of wavelet transform. The simulation results show that there is a higher accuracy for identification of IAW by the improved TET in different fault condition.

Keywords: Transient extraction transform · Voltage source converter high voltage direct current · Identification of initial arrival wave · High time resolution

1 Introduction

With the continuous development of power electronics technology, High Voltage Direct Current based on Voltage Source Converter (VSC-HVDC) has overcome many shortcomings of High Voltage Direct Current based on Line Commutated Converter (LCC-HVDC). Therefore, VSC-HVDC has been widely studied by scholars [1, 2].

When a fault occurs at DC transmission line of VSC-HVDC, the fault process develops quickly and the fault current increases rapidly [3]. Rapidly rise of fault current will cause the over-current of converters within a few milliseconds. In order to prevent damage to the power electronics of the converter, it is necessary for protections to quickly response within a few milliseconds.

This work was supported by Key Research and Development Program of Shandong Province (2019GGX103014), and Science and Technology Program of University of Jinan (XKY2020).

Y. Xue et al. (Eds.): PMF 2020, LNEE 718, pp. 327–342, 2021.
https://doi.org/10.1007/978-981-15-9746-6_26

VSC-HVDC protections can be divided into two categories, which are directional pilot protection and single-terminal protection. Although directional pilot protection can accurately detect fault location, the directional pilot protection needs to interact with double-terminal information [4]. Therefore, response speed of directional pilot protection is lower than that of single-terminal protections [5]. Because response speed of VSC-HVDC protection must be within a few milliseconds, single-terminal protection method is obviously more suitable for VSC-HVDC protection.

There are mainly the following types of single-terminal protection for VSC-HVDC, which are single-terminal protection based on boundary elements [6], traveling-waves protection (TWP) based on frequency spectrum characteristics [7], and TWP based on accurate identification of initial arrival wave (IAW) [8, 9].

Single-terminal protection based on boundary elements constructs protection criterion by electric parameters of energy-storage capacitance and current-limiting reactor that are called as boundary conditions of DC transmission lines. Single-terminal protection based on boundary elements is adopted only when the boundary conditions of DC transmission lines are obvious. TWP based on accurate identification of IAW does not require obvious boundary conditions. For TWP based on frequency spectrum characteristics, analysis of frequency domain is more complex than analysis of time domain, and analysis of frequency spectrum characteristics is not accurate by existing analysis method of frequency domain.

Therefore, among the above methods, TWP based on accurate identification of IAW is adopted for VSC-HVDC more widely. The accurate identification of IAW is the most important process in this kind of TWP methods [8].

Due to the frequency variation of DC transmission line parameters, the IAW will be distorted. The distortion of IAW will make the identification of IAW more difficult [9]. Therefore, IAW identification needs the time-frequency analysis method with higher accuracy.

At present, wavelet transform is already used for identifying IAW in TWPs for AC transmission lines and LCC-HVDC [10]. However, the time resolution of wavelet transform is low for TWP of VSC-HVDC protection. Thus it is urgent to study the method of higher time resolution for TWP of VSC-HVDC.

Transient Extraction Transform (TET) method that is a new time-frequency analysis method of higher time resolution is suitable for analyzing the abrupt-change signal in time domain. TET has been successfully applied in the vibration signal processing of the mechanical fault diagnosis, because there is the abrupt-change characteristic for vibration signal in the time domain. There is also obvious characteristic of abrupt-change for traveling-waves (TWs) signal in time domain. Therefore, TET can be used to analyze the TWs and accurately identify IAW.

The organization of this paper is as follows: in Sect. 2, problem of IAW identification is analyzed. In Sect. 3, TET is introduced. Considering the characteristics of TWs signal, the TET is improved to increase the identification accuracy of IAW. In Sect. 4, the simulation model of double-terminal VSC-HVDC is established and simulated in PSCAD. Data of simulation is analyzed by improved TET and wavelet transform in MATLAB respectively. Accuracy of IAW that is identified by improved TET and wavelet transform is compared in different fault conditions.

2 Analysis of IAW for VSC-HVDC

2.1 Structure of VSC-HVDC

The VSC-HVDC system mainly consists of AC grids, converter stations and DC transmission lines. A schematic diagram of VSC-HVDC is shown in Fig. 1. In Fig. 1, AC1 and AC2 are AC grids. VSC1 is the rectifying terminal and VSC2 is the inverting terminal. C_q is the energy-storage capacitance. L_{dc} is the current-limiting reactor of DC transmission line. In Fig. 1, MN represents the protected line. The protected line locates between L_{dc} of M-terminal and L_{dc} of N-terminal. R_M is the protection of M-terminal. u_{M+} and u_{M-} are the positive pole voltage and negative pole voltage of M-terminal respectively. i_{M+} and i_{M-} are the positive pole current and negative pole current of M-terminal respectively.

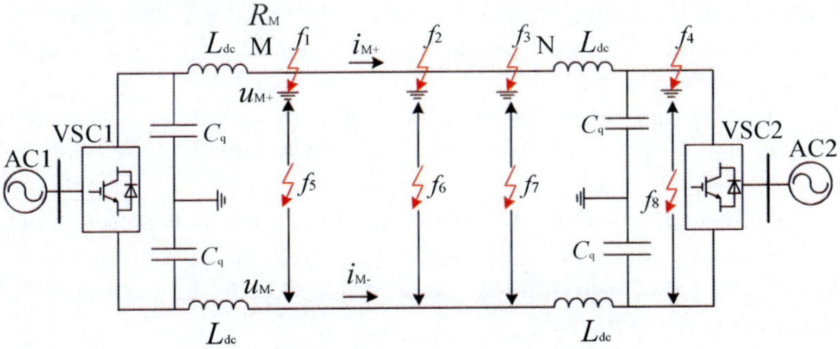

Fig. 1. Schematic diagram of VSC-HVDC

Typical faults of the DC transmission line are used to analysis and simulate the proposed method, which are shown in Fig. 1. f_1–f_8 are the fault locations. f_1 and f_5 locate at 0% of the protected line. f_2 and f_6 locate at 50% of the protected line. f_3 and f_7 locate at 100% of the protected line. f_1–f_3 are single-pole faults. f_5–f_7 are double-pole faults. In addition, f_4 and f_8 are also considered, which locate between L_{dc} of N-terminal and VSC2.

2.2 Phase-Mode Transformation Technology

There is electromagnetic coupling between positive pole and negative pole for the double-pole transmission lines of VSC-HVDC, which will result in difficulty of analysis and computation of TWs signal. Thus, phase-mode transformation will be used for decoupling the double-pole transmission lines in the proposed method.

Karenbell transformation that is a common phase-mode transformation in TWP is carried out, which is shown as:

$$\begin{bmatrix} i_{M0} \\ i_{M1} \end{bmatrix} = \frac{\sqrt{2}}{2}\begin{bmatrix} 1 & 1 \\ 1 & -1 \end{bmatrix}\begin{bmatrix} i_{M+} \\ i_{M-} \end{bmatrix} \quad (1)$$

where i_{M1} is the aerial-mode current and the subscript of 1 represents the variable that is related to the aerial-mode; i_{M0} is the zero-mode current and the subscript of 0 represents the variable that is related to the zero-mode.

Because IAW distortion of aerial-mode is much less than that of zero-mode, it is easier to accurately identify the IAW of aerial-mode than that of zero-mode. Therefore, the aerial-mode parameters are more suitable for the TWP based on accurate identification of IAW. Thus, TWs signal that is calculated and analyzed in the following of this paper is also aerial-mode current.

2.3 Propagation Analysis of TWs Signal

IAW identification is effected by TWs distortion. TWs distortion is caused by propagation of TWs signal. Therefore, propagation analysis of TWs signal is necessary for analysis of IAW identification.

When a fault occurs at the protected line in Fig. 1, namely f_1–f_3, the fault-induced TWs is a step signal that is shown as Fig. 2(a), which spreads to two-terminals of the line.

Because propagation speed of different frequency component is different at DC transmission line, R_M will receive different frequency component that is simultaneously generated at fault location in different time, which leads to the distortion of the TWs signal that is shown as Fig. 2(b).

According to Fig. 2(b), it is known that the longer distance that between fault location and R_M is, the more serious distortion of TWs signal is.

When a fault occurs at f_4, in addition to the propagation effect of the DC transmission line, IAW is also effected by the smoothing of L_{dc}. Due to effect of L_{dc}, distortion of IAW that is shown as Fig. 2(b) is more serious, which makes it more difficult than identification of IAW.

There is transition resistance when a fault occurs, which will effect IAW distortion that is shown as Fig. 2(b). In Fig. 2(b), IAW distortion is compared in different conditions that there is or not is transition resistance that is 100 Ω, when a fault occurs at f_2. According to Fig. 2(b), when the fault process exists transition resistance, the energy of IAW will diverge, and the rising of IAW become slow, namely the distortion of IAW will be more serious.

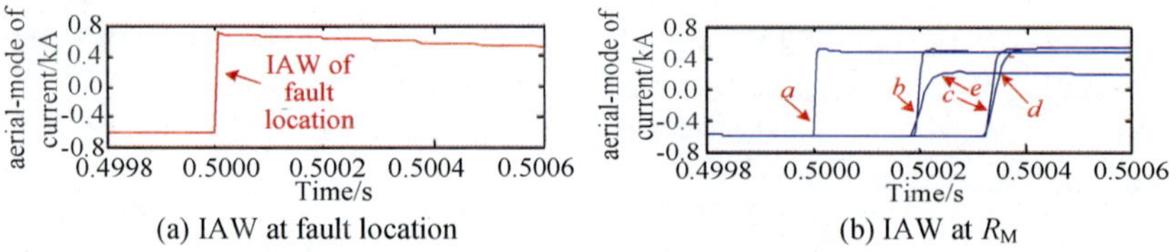

Fig. 2. IAW at the fault location and R_M

In Fig. 2(b), a–d represent IAW of TWs signal when faults occur at f_1–f_4 respectively, and transition resistance is 0 Ω. e represents IAW of TWs signal when a fault occurs at f_2, and transition resistance is 100 Ω.

By analysis of TWs signal propagation, the IAW that is detected by the R_M is distorted signal. According to Fig. 2(b), distorted degree of TWs signal is effected by fault location, and transition resistance will also make IAW more distorted. The more serious distortion of TWs signal is, the more difficult identification of IAW is. Thus, the time-frequency analysis method with high time resolution must be adopted to accurately identify IAW in different fault conditions. There is a high time resolution for TET that is mentioned in [11], and TET is suitable for analyzing abrupt-change signal. Thus, TET is considered to identify IAW for VSC-HVDC.

3 TET and Its Improvement

3.1 TET

An abrupt-change signal $s(t)$ is set to analyze TET:

$$s(t) = A\delta(t - t_0) \tag{2}$$

where $\delta(\bullet)$ is a pules abrupt-change function; t_0 is abrupt-change time of $s(t)$; A is coefficient of $\delta(\bullet)$.

The Short Time Fourier Transform (STFT) of $s(t)$ is $G(t, \omega)$. t is time, and ω is frequency. $G(t, \omega)$ is a function whose independent variable is t and ω.

According to the knowledge of STFT, it is known that the energy of signal analysis can be concentrated near the abrupt-change time by STFT. But because of the low time resolution of STFT, the abrupt-change time of signal can not be accurately detected. To increase time resolution, STFT is further processed.

According to the theory of STFT, the following equation can be obtained:

$$\partial_\omega G(t, \omega) = -jt_0 G(t, \omega) \tag{3}$$

where $\partial_\omega G(t,\omega) = \partial G(t,\omega)/\partial\omega$; j is an imaginary unit.

A function is constructed, and its independent variable is t and ω, as is shown in the following equation:

$$t_L(t, \omega) = -j\frac{\partial_\omega G(t, \omega)}{G(t, \omega)} \tag{4}$$

According to (3) and (4), when $G(t, \omega) \neq 0$, there is $t_L(t,\omega) = t_0$.

TET of $s(t)$ is as follows:

$$T_e(t, \omega) = G(t, \omega)\delta(t - t_L(t, \omega)) \tag{5}$$

where $T_e(t,\omega)$ is the result of TET. $\delta(t- t_L(t,\omega))$ is TET transformation factor. $T_e(t,\omega)$ is a function whose independent variable is t and ω.

The energy of STFT analysis is concentrated near abrupt-change time, namely $G(t, \omega) \neq 0$ near abrupt-change time, and there is $t_L(t,\omega) = t_0$. According to (5), when $t = t_L(t,\omega)$, there is $T_e(t,\omega) = G(t,\omega)$. So, when $t = t_0$, there is $T_e(t,\omega) = G(t,\omega)$.

TET is essentially post-processing of STFT. The result of STFT is multiplied with the TET transformation factor, which is essentially that energy of STFT analysis is only kept at the abrupt-change time t_0, and the divergence energy of time domain is removed, which improves the time resolution.

3.2 Performance Analysis of TET

To test performance of TET, an abrupt-change signal *h(t)* is simulated,

$$h(t) = \begin{cases} 1 & t \leq 0.001\,\text{s} \\ 1 - e^{100000(0.001-t)} & t > 0.001\,\text{s} \end{cases} \tag{6}$$

where t is the time; the decay time constant τ is 100000.

The abrupt-change time of $h(t)$ is t = 0.001 s, as shown in Fig. 3(a). $h(t)$ is similar to TWs signal, which exists a step abrupt-change. And its step abrupt-change is distorted; the degree of distortion is decided by decay time constant τ.

The simulation result that abrupt-change time is identified by TET is shown from Fig. 3(b) and Fig. 3(c).

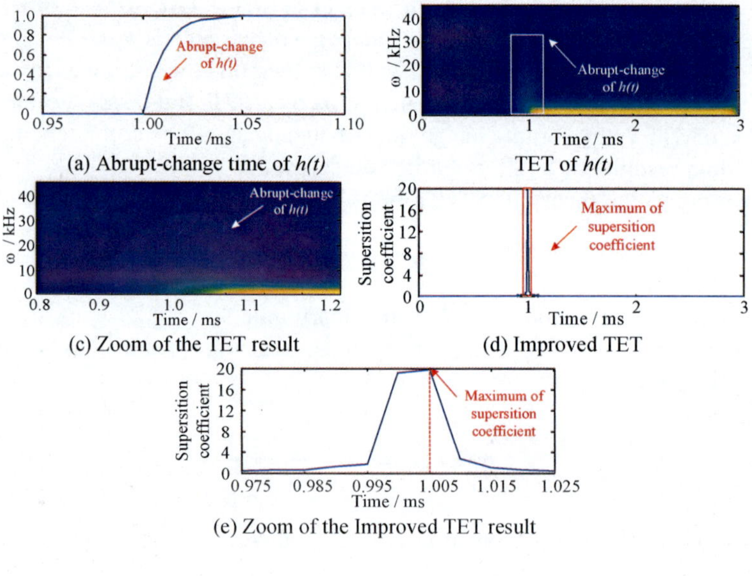

Fig. 3. Analysis of *h(t)*

According to Fig. 3(c), time domain energy near abrupt-change time is dispersive. It is difficult for identification of abrupt-change time to construct an accurate criterion. The time resolution of TET is still low.

It is further analyzed that the reason why the accuracy of abrupt-change time identified by TET is low. It is found that the target signal is pulse abrupt-change signal at the theory derivation process of TET. However, it can be known from Fig. 3(a) that there is the characteristics that is similar to the step abrupt-change signal for abrupt-change of $h(t)$. Therefore, it is not ideal simply that using the original TET transform to obtain the abrupt-change time of $h(t)$, and analysis process of TET needs to be improved.

3.3 Improved TET

Improved process of TET is following.

Step1:
In view of the feature that the TWs signal is a step signal, it is adopted that direct analysis of TWs signal is translated into analysis of average absolute distance of TWs signal.

The average absolute distance H_i of a given signal at sample point i is defined as follows:

$$H_i = \frac{|x_{i1} - \overline{x_i}| + |x_{i2} - \overline{x_i}| + \ldots\ldots + |x_{in} - \overline{x_i}|}{n} \tag{7}$$

where H_i is the average absolute distance of this signal at a sample point i, and H_i of all sample points of this signal constitute average absolute distance H of this signal. H_i is computed by n sample points. Sample point i is the midpoint of these n sample points. $\overline{x_i}$ is the average values of n sample points.

It can be known from the Eq. (7), H can represent the change degree of a signal. When the signal change is more drastic, H is larger. When IAW arrives at protection, namely TWs signal occurs abrupt-change, the TWs signal changes the most dramatically. In this time, H of TWs signal is the largest, and the pulse abrupt-change occurs, and the abrupt-change time of TWs signal corresponds to the pulse abrupt-change time of H one to one.

According to the above analysis, H is a pulse abrupt-change signal, which is suitable for being analyzed by TET, and the abrupt-change time corresponds to the abrupt-change time of TWs one to one. Therefore, the identification of IAW arrival time can be transformed into identification of the pulse abrupt-change time of the H.

To identify abrupt-change time of the H, H is analyzed by TET. The abrupt-change time of H is identified by analysis results of TET. However, through simulation, the time resolution of results of TET is still low, and it is still necessary to further increase the time resolution.

Step2:
To solve the problem that time resolution of TET results of the H is low, the TET results are post-processed. According to the time-frequency analysis theory, the signal

abrupt-change time contains the energy of many frequency bands, and superposition of these energy is higher than that of other time. The results of TET are superimposed in the direction of frequency domain at each time. The superposition results represent the sum of all frequency energies at each time, which is called the superposition coefficient. At the arrival time of IAW, the TWs signal occurs abrupt-change, and H also occurs abrupt-change. In this time, the frequency spectrum is abundant, so the energy superposition result is the largest.

To sum up, the time that the superposition coefficient is the maximum represents the arrival time of IAW.

After the improvement of the above two steps, the improved TET can obtain a higher time resolution and identify the IAW more accurately.

3.4 Accuracy Analysis of Improved TET

$h(t)$ is simulated by different time-frequency analysis method that includes TET, wavelet transform and improved TET. Simulation results are analyzed and compared. Through comparison, improvement effect of TET can be illustrated.

1) improved TET of *h(t)*

The simulation results that abrupt-change time are identified by improved TET are shown as Fig. 3(d) and Fig. 3(e).

According to Fig. 3(e), when t = 1.005 ms, superposition coefficient of improved TET is maximum value. It can be concluded that abrupt-change time that is identified by improved TET is t_1' = 1.005 ms.

In order to represent the identification accuracy of the time-frequency analysis method for abrupt-change time, an error that is represented by a deviation between theoretical value and the value that is identified by time-frequency analysis need to be is used. Compared with other errors, absolute error is most suitable to represent the measurement accuracy of a data, that is, to represent the deviation of the measured value from the theoretical value. Therefore, this paper uses absolute error to represent the accuracy of identifying abrupt-change time by the time-frequency analysis. The absolute error is defined as t_{OP}. t_{OP} is shown as:

$$t_{OP} = \left|t_1 - t_1'\right| \tag{8}$$

where t_{OP} represents the absolute error between the time that the abrupt-change is identified by time-frequency analysis method and theoretical time. t_1' is the time that the abrupt-change is identified by time-frequency analysis. t_1 is the theoretical abrupt-change time. When analyzed signal is $h(t)$, there is t_1 = 1.000 ms. When analyzed signal is TWs signal, t_1 is obtained by computation of theoretical wave velocity. The smaller t_{OP} is, the higher the accuracy is.

From above analysis, abrupt-change time that is identified by improved TET is t_1' = 1.005 ms. Theoretical abrupt-change time of $h(t)$ is t_1 = 1.000 ms. According to computation of Eq. (8), absolute error that is identified by improved TET is 5 μs.

2) Accuracy comparison of TET and improved TET

According to Fig. 3(c), after $h(t)$ is analyzed by TET, time domain energy is dispersive, and time resolution is low, which can not accurately identify abrupt-change time.

According to Fig. 3(e), analysis result of improved TET can identify abrupt-change time by a given analysis process, and the absolute error of the improved TET only is 5 μs.

Through comparing results of TET and results of improved TET in time domain, the following conclusions can be summarized that through improvement of TET analysis process, time resolution is increased, and abrupt-change time can be accurately identified. Accuracy of improved TET is better than accuracy of TET.

3) Accuracy comparison of wavelet transform and improved TET

To compare accuracy of improved TET and wavelet transform, $h(t)$ is analyzed by wavelet transform.

The simulation results of abrupt-change time that are identified by wavelet transform are shown as Fig. 4. The mother wavelet is db2. d_1–d_5 represent decomposition levels in different frequency band respectively. Figure 4(a), Fig. 4(c), Fig. 4(e), Fig. 4 (g), Fig. 4(i) represent wavelet transform coefficients of d_1–d_5 respectively. The frequency decreases from d_1 to d_5.

According to the theory of wavelet transform, time of modulus maxima of wavelet coefficients represents abrupt-change time of a signal. This time corresponds to t_1' of Eq. (8). Through analysis results of wavelet transform, t_1' can be obtained. t_1' of d_1–d_5 are 1.025 ms, 1.025 ms, 1.020 ms, 1.020 ms, and 1.020 ms respectively. Theoretical abrupt-change time of $h(t)$ is t_1 = 1.000 ms. According to Eq. (8), the absolute error of d_1–d_5 can be computed, which is 25 μs, 25 μs, 20 μs, 20 μs, 20 μs respectively. The smallest absolute error is 20 μs in different decomposition levels, while the largest absolute error is 25 μs in different decomposition levels

If absolute error of improved TET is smaller than the value that is the smallest absolute error of wavelet transform in different decomposition, accuracy of improved TET is higher than wavelet transform. Therefore, absolute error of improved TET is compared with the smallest absolute error of wavelet transform in different decomposition levels.

The t_{OP} of improved TET is 5 μs, while the smallest absolute error t_{OP} of wavelet transform is 20 μs. t_{OP} of improved TET is smaller than the smallest absolute error t_{OP} of wavelet transform. Thus, accuracy of improved TET is better than accuracy of wavelet transform.

From above analysis, through improvement, TET increases time resolution, and can identify abrupt-change time of a given signal. Accuracy of improved TET even is higher than accuracy of wavelet transform that is widely used to identify abrupt-change time.

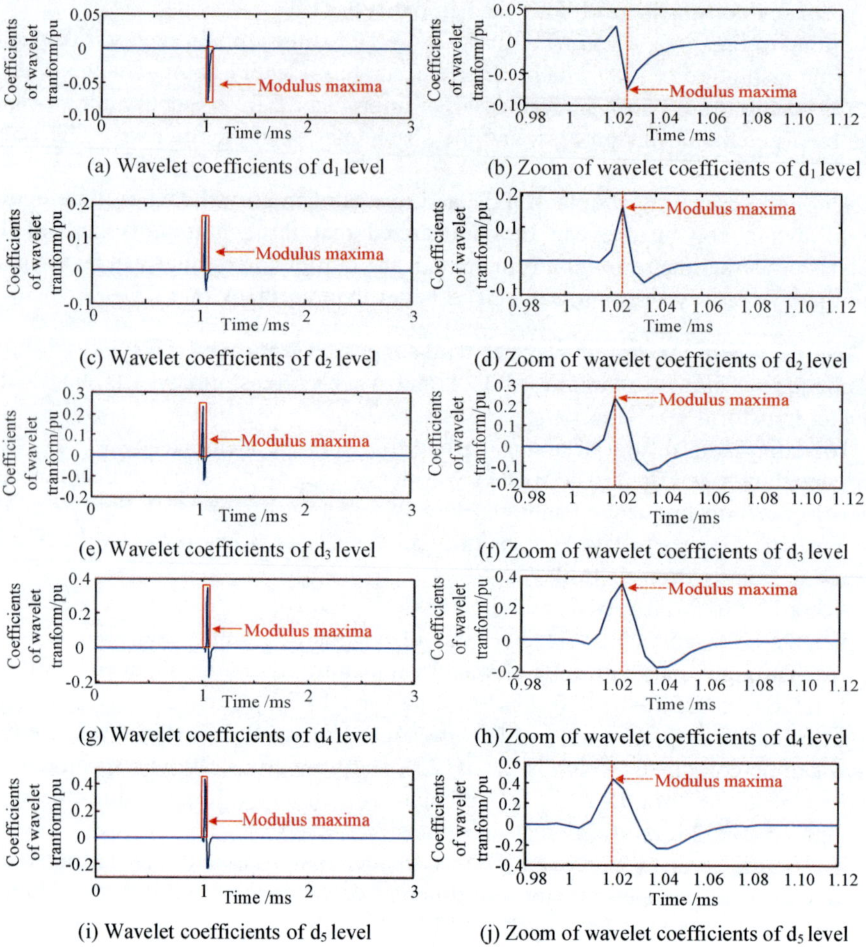

(a) Wavelet coefficients of d_1 level
(b) Zoom of wavelet coefficients of d_1 level
(c) Wavelet coefficients of d_2 level
(d) Zoom of wavelet coefficients of d_2 level
(e) Wavelet coefficients of d_3 level
(f) Zoom of wavelet coefficients of d_3 level
(g) Wavelet coefficients of d_4 level
(h) Zoom of wavelet coefficients of d_4 level
(i) Wavelet coefficients of d_5 level
(j) Zoom of wavelet coefficients of d_5 level

Fig. 4. Wavelet transform of *h(t)*

4 Simulation Test and Analysis

4.1 Simulation Model and Parameters

To compare IAW identification accuracy of improved TET and wavelet transform, simulations are carried out, and the results of simulation are analyzed and compared. Through comparison and analysis, performance of improved TET that is used for IAW identification can be illustrated.

PSCAD/EMTDC is used to establish the simulation model of VSC-HVDC system. According to Fig. 1, the simulation model of double-terminal VSC-HVDC is built in PSCAD. TWs signal is measured at R_M, and is analyzed by improved TET and wavelet transform in MATLAB.

The selection of converter is key for PSCAD simulation model. The converter of VSC-HVDC can be divided into two categories, which are two-level Voltage Source Converter (VSC) and Multilevel modular converter (MMC). The biggest difference of two-level VSC and MMCs is that there is lower switching frequency and simple construction for two-level VSC. Because it is not assumed in advance which type of converter it is in the derivation process of the proposed method, which can be known from analysis of Sect. 2.3 and Sect. 3.3.

However, considering the application of converter in the actual projects, the converter of PSCAD simulation model selects two-level VSC. Two-level VSC is more suitable for DC transmission lines than MMC and they are widely used in the actual projects of large-capacity transmission systems, because two-level VSCs possess simpler control strategies, lower investment costs and less converter losses. In addition, MMC are usually used in the actual projects of low-capacity transmission systems or distribution system. Because DC transmission lines of low-capacity transmission systems or distribution system are shorter than large-capacity transmission systems, traveling-wave protection is not suitable for these systems. Improved TET is used in traveling-wave protection. Therefore, simulation model based on two-level VSC is more valuable than simulation model based on MMC in simulation and analysis of identifying IAW by improved TET. From above all, the converter adopts two-level VSC in this paper. The control strategy of converter adopts direct current control.

The model parameters are as follows. The rated voltage of DC is 500 kV. The DC transmission line model adopts Frequency Dependent Model. The DC transmission line length is $l = 400$ km. The sampling rate is $f_0 = 200$ kHz. The fault occurs at $t_{p1} = 0.500$ s.

4.2 Theoretical Calculation of Wave Velocity

The theoretical time t_1 when the IAW arrives at R_M can be calculated by the theoretical wave velocity v_1. According to (8), absolute error t_{OP} is calculated by theoretical time t_1 and time t_1' that is obtained by simulation. Accuracy of time-frequency analysis method is represented by absolute error t_{OP}. Therefore, to represent accuracy of time-frequency analysis method, theoretical wave velocity v_1 must be calculated.

The DC transmission line adopts typical parameters of VSC-HVDC project. Per kilometer resistance of aerial-mode is $r_1 = 0.8945 \times 10^{-2}$ Ω/km. Per kilometer inductance of aerial-mode is $l_1 = 0.4292 \times 10^{-1}$ H/km. Per kilometer conductivity of aerial-mode is $g_1 = 0$ S/km. Per kilometer capacitance of aerial-mode is $c_1 = 0.4217 \times 10^{-5}$ F/km.

The wave velocity of aerial-mode can be calculated by typical parameters:

$$v_1 = \frac{\omega}{\sqrt{\left[\omega^2 l_1 c_1 - r_1 g_1 + \sqrt{(r_1^2 + \omega^2 l_1^2)(g_1^2 + \omega^2 c_1^2)}\right] \Big/ 2}} = 292826\text{km/s} \quad (9)$$

where $\omega = 2\pi f$; $f = 50$ Hz.

4.3 Simulation Results and Analysis

Considering different fault conditions that include fault locations, fault types and transition resistances, IAW identification accuracy that is identified by improved TET and wavelet transform is compared.

To compare accuracy of wavelet transform and improved TET, absolute error of wavelet transform and improved TET must be computed respectively. According to Eq. (8), t_{OP} is computed by theoretical time t_1 and the time that is obtained by simulation t_1'. Thus, t_1 must be computed by wave velocity. t_1' must be obtained by analysis of wavelet transform and improved TET respectively.

According to the theoretical wave velocity v_1 and distance between fault location and R_M, the theoretical time that the IAW arriving at R_M can be calculated as

$$t_1 = t_{p1} + \frac{l_d}{v_1} \tag{10}$$

where t_1 is the time that the IAW arriving at R_M by the theoretical calculation. t_{p1} is the time when a fault occurs. l_d is the distance between fault location and R_M.

According to (10), results of t_1 are 500.000 ms, 500.683 ms, and 501.364 ms when a fault occurs at 0%, 50%, and 100% of DC transmission line respectively.

The simulation results and their comparison are shown as Table 1. In Table 1, t_{w1}' is the time that the IAW is identified by wavelet transform. t_{e1}' is the time that the IAW is identified by improved TET. t_{OPw} is the absolute error between the time that the IAW is identified by wavelet transform and theoretical value. t_{OPe} is the absolute error between the time that the IAW is identified by improved TET and theoretical value.

According to analysis of Table 1, in the case of different fault locations, fault types and transition resistances, the absolute errors t_{OPe} are all lower than absolute errors t_{OPw}. Therefore, accuracy of improved TET is higher than accuracy of wavelet transform.

According to analysis of Table 1, the longer distance that between fault location and R_M is, the bigger transition resistances are, the bigger of absolute errors of wavelet transform is. Compared with the accuracy of identifying IAW by wavelet transform, accuracy of identifying IAW by improved TET is less 10 μs, and is hardly affected by fault conditions. Therefore, compared with the traditional method of identifying IAW by wavelet transform, accuracy of identifying IAW by improved TET is hardly affected by fault conditions, and there is better performance for improved TET.

Table 1. The comparison of absolute error in different fault conditions

Fault types	Fault location	Transition resistance/Ω	t_1/ms	t'_{e1}/ms	t_{OPe}/μs	t'_{w1}/ms	t_{OPw}/μs
Single pole fault	f_1	0	500.000	500.000	0	500.000	0
		100	500.000	500.000	0	500.000	0
		300	500.000	500.000	0	500.000	0
	f_2	0	500.683	500.685	2	500.665	18
		100	500.683	500.685	2	500.665	18
		300	500.683	500.690	7	500.660	23
	f_3	0	501.364	501.360	4	500.440	24
		100	501.364	501.360	4	500.440	24
		300	501.364	501.355	9	500.435	29
	f_4	0	501.364	501.355	9	500.330	34
		100	501.364	501.355	9	500.325	39
		300	501.364	501.355	9	500.325	39
Double pole fault	f_5	0	500.000	500.000	0	500.000	0
		100	500.000	500.000	0	500.000	0
		300	500.000	500.000	0	500.000	0
	f_6	0	500.683	500.685	2	500.665	18
		100	500.683	500.685	2	500.665	18
		300	500.683	500.690	7	500.660	23
	f_7	0	501.364	501.360	4	500.440	24
		100	501.364	501.360	4	500.440	24
		300	501.364	501.355	9	500.435	29
	f_8	0	501.364	501.355	9	500.330	34
		100	501.364	501.355	9	500.325	39
		300	501.364	501.355	9	500.325	39

4.4 Typical Simulation Cases

To clearly demonstrate that the accuracy of the improved TET is higher than that of the wavelet transform, a typical simulation as an example detailedly compares and analyzes accuracy of improved TET and accuracy of wavelet transform.

Fault condition of this simulation is that a fault occurs at f_2, and transition resistance is 0 Ω. According to Table 1, this typical fault can clearly compare accuracy of wavelet transform and accuracy of improved TET.

The simulation result that IAW is identified by improved TET at R_M is shown as Fig. 5.

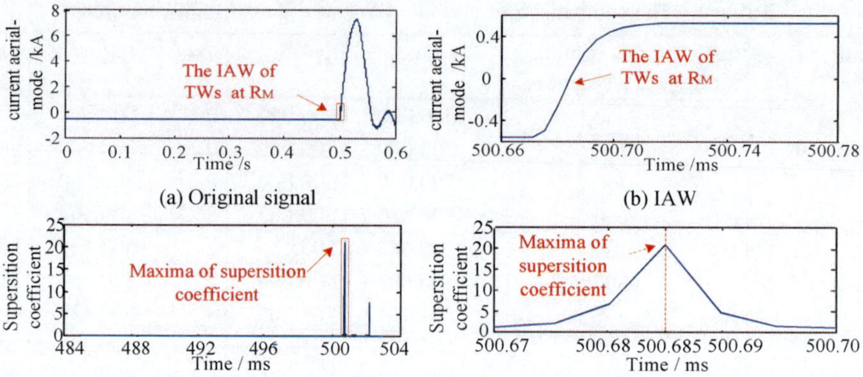

(a) Original signal

(b) IAW

(c) Improved TET

(d) Zoom of the Improved TET result

Fig. 5. Improved TET of TWs at R_M

According to Fig. 5(d), it can be concluded that time of identifying IAW by improved TET at R_M is t'_{e1} = 500.685 ms.

According to Eq. (10), there is t_1 = 500.683 ms, when a fault occurs at f_2.

According to compution of Eq. (8), absolute error of improved TET is 2 μs.

The simulation results that arrival time of IAW is identified by db2 wavelet transform are shown as Fig. 6.

The simulation results of wavelet transform can be obtained by Fig. 6. t'_{w1} is the time when IAW is identified by wavelet transform. t'_{w1} of d_1–d_5 are 500.665 ms, 500.660 ms, 500.655 ms, 500.660 ms, and 500.655 ms respectively. According to (10), there is t_1 = 500.683 ms, when a fault occurs at f_2. According to computation of Eq. (8), the smallest absolute error of wavelet transform is 18 μs in different decomposition levels.

From the above analysis, comparison of absolute error that IAW is identified by improved TET and wavelet transform is shown as Fig. 7.

It can be clearly known from Fig. 7 that t_{OPe} is smaller than t_{OPw}. It can be known that the accuracy of IAW identified by the improved TET is better than the wavelet transform.

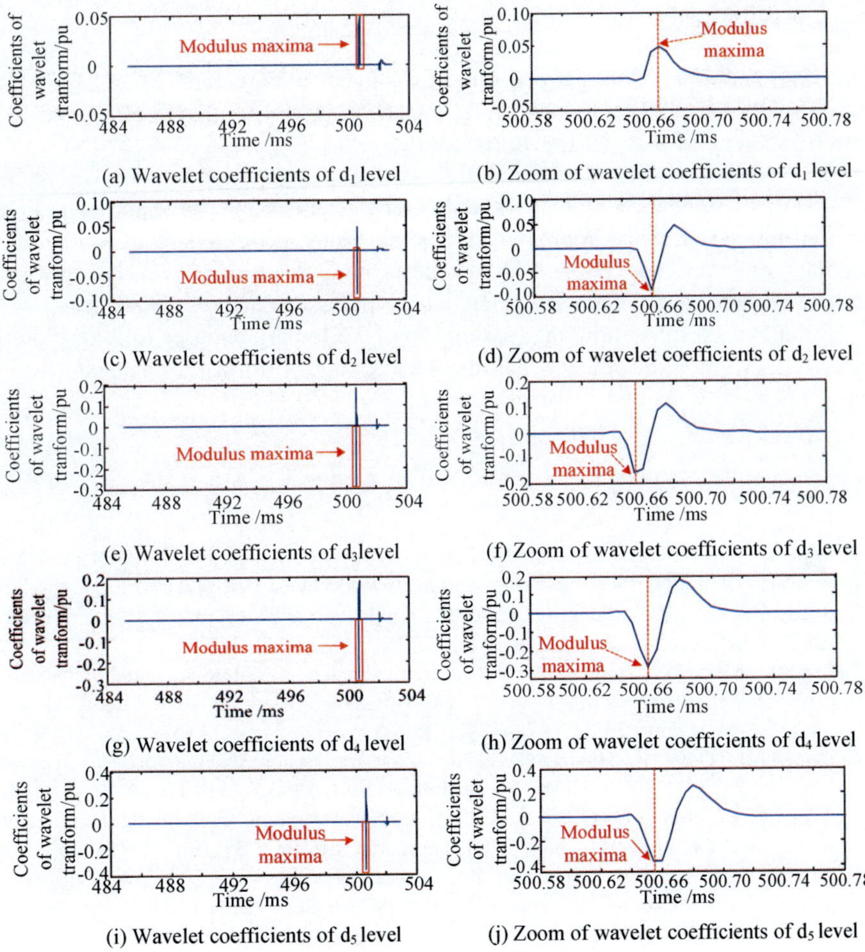

Fig. 6. Wavelet transform of TWs at the R_M

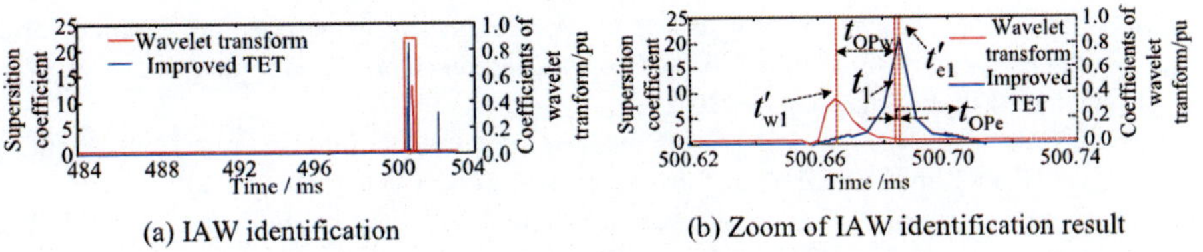

Fig. 7. Absolute error comparison of improved TET and wavelet transform

5 Conclusion

The single-terminal TWP generally needs to identify IAW for VSC-HVDC. Identification of IAW usually adopts wavelet transform in current TWP. However, identification accuracy of wavelet transform is low for TWP of VSC-HVDC.

To increase identification accuracy of IAW, this paper proposes a novel method for VSC-HVDC protection, which is that IAW is identified by improved TET. To demonstrate accuracy of improved TET, simulations that compare IAW identification accuracy of improved TET and IAW identification accuracy of wavelet transform are carried out in different fault locations, fault types and transition resistances.

Simulation results show that the improved TET that increases time resolution can identify IAW more accurately than the wavelet transform in different fault locations, fault types and transition resistances.

References

1. Guo, Q., Guo, H.P., Huang, L.B.: Stability influence of network voltage feedforward for VSC-HVDC under Weak network. Autom. Electr. Power Syst. **42**(14), 139–144 (2018)
2. Tang, G.F., He, Z.Y., Pang, H.: Research, application and development for engineering for VSC-HVDC. Autom. Electr. Power Syst. **37**(15), 3–14 (2013)
3. Han, X., Sima, W., Yang, M., et al.: Transient characteristics under ground and short-circuit faults in a ±500 kV MMC-based HVDC system With hybrid DC circuit breakers. IEEE Trans. Power Deliv. **33**(3), 1378–1387 (2018)
4. Zou, G.B., Qian, F., Huang, Q., et al.: A fast protection scheme for VSC based multi-terminal DC grid. Int. J. Electr. Power Energy Syst. **98**, 307–314 (2018)
5. Yang, J., Fletcher, J.E., O'Reilly, J.: Short-circuit and ground fault analyses and location in VSC-based DC network cables. IEEE Trans. Ind. Electron. **59**(10), 3827–3836 (2012)
6. Wang, X., Yang, S.Z., Xu, L., et al.: A transient voltage-based DC fault line protection scheme for MMC-based DC grid embedding DC breakers. IEEE Trans. Power Deliv. **34**(1), 334–345 (2019)
7. De Kerf, K., Srivastava, K., Reza, M., et al.: Wavelet-based protection strategy for DC faults in multi-terminal VSC HVDC systems. IET Gener. Transm. Distrib. **5**(4), 496–503 (2011)
8. Zhang, C.H., Song, G.B., Dong, X.Z.: The principle of single-terminal traveling wave protection for VSC-HVDC based on the fitting of initial traveling wave of fault current. Proc. CSEE **35**(12), 1–11 (2019)
9. Johannesson, N., Norrga, S., Wikströ, M.C.: Selective wave-front based protection algorithm for MTDC systems. In: 13th IET International Conference on Development in Power System Protection, Edinburgh, pp. 1–6 (2016)
10. Zhang, Y., Tai, N.L., Xu, B.: Fault analysis and traveling-wave protection scheme for bipolar HVDC lines. IEEE Trans. Power Deliv. **27**(3), 1583–1591 (2012)
11. Yu, G.: A concentrated time-frequency analysis tool for bearing fault diagnosis. IEEE Trans. Instrum. Meas. **41**(02), 7042–7053 (2019)

Non-intrusive Load Decomposition Based on Hidden Markov Model Considering Multiple Electrical Features

Zhukui Tan[1(✉)], Bin Liu[1], Saiqiu Tang[1], Xiao Su[2], and Weifeng Xu[2]

[1] Electric Power Research Institute of Guizhou Power Grid Co., Ltd., Guiyang 550002, China
1066048121@qq.com

[2] Suzhou Huatian Power Technology Co., Ltd., Suzhou 215000, China

Abstract. As the infrastructure of smart grid, the construction of Advanced Metering Infrastructure (AMI) has become increasingly important. Non-intrusive load decomposition (NILD) technology is one of the components of AMI, whose classic algorithms include Hidden Markov Model (HMM) and so on. However, the classic HMM only uses the single electrical features of the load, and it is not possible to extract more information about the load. In this paper, a binary parameters Hidden Markov Model based on active power and steady-state current is established to realize the utilization of multivariate electrical features information. In this paper, the sparse Viterbi algorithm is used to realize the state recognition of electrical appliances. Based on the principle of maximum likelihood estimation, the power decomposition of the load is achieved through the statistical features of the load state clustering. This method shows better load decomposition accuracy than the classic HMM algorithm, heuristic algorithm, and deep learning-based recognition method, which is verified by test examples.

Keywords: Non-intrusive load decomposition · Hidden markov model · Viterbi algorithm · Advanced metering infrastructure

1 Introduction

At present, the concept of smart grid is in the ascendant. More and more scholars and grid companies are participating in the theoretical research and practical exploration of smart grid. Advanced Metering Infrastructure (AMI) [1] is a key part of the smart grid construction. As one of the most important components of AMI, load monitoring and identification is the first step to realize smart grid [2]. Based on load monitoring and identification technology, power grid companies can grasp the load status and energy consumption, and then use big data and other technical methods to describe the load energy consumption pattern and user's power consumption behavior, so as to achieve demand response, optimize power resource allocation, and support the construction of smart grid.

With the establishment of AMI, non-intrusive load decomposition (NILD) was first proposed by Hart in the 1980s [3]. This technology only installs monitoring equipment

Y. Xue et al. (Eds.): PMF 2020, LNEE 718, pp. 343–357, 2021.
https://doi.org/10.1007/978-981-15-9746-6_28

at the user's electricity entrance to record electrical data, and then uses a decomposition algorithm to decompose the load to obtain the load status and energy consumption information of internal electrical appliances. At present, non-intrusive load decomposition technology is not mature, and scholars have proposed a variety of load decomposition algorithms. Reference [4] designed a normal distribution metric function that comprehensively considered the features of steady-state harmonic current and power, and proposed a load monitoring method based on an improved flock algorithm. Reference [5] constructed a current feature expression based on harmonics and combined power as the target function of the switching state identification of the equipment. The particle swarm optimization algorithm was used to search and implement NILD, but the accuracy rate of state recognition was low. Machine learning algorithms such as deep learning and Hidden Markov Models are also used in NILD. Reference [6] proposed a deep neural network to perform load disaggregation whose input of the network includes appliance states, power and time information; Reference [7] proposed a NILD model based on feature fusion and deep learning and overcame the disadvantage that V-I trajectory features cannot reflect the power of the equipment. The above two methods based on deep learning have a high recognition rate, but the model training time is long. Reference [8] uses the steady-state average power and steady-state differential power obtained by event detection as HMM observation values to realize load state recognition; Reference [9–12] adopts improved or expanded HMM model to make the speed of calculation and the accuracy of state recognition improve. The above improved HMM model has a higher recognition rate, but the model is complicated, and the use of unknown information and electrical features is insufficiently considered.

In view of the fact that the existing HMM application has a single observation value chain, this paper proposes a decomposition method of HMM based on binary parameters. This method first clusters and encodes the appliance states, then establishes the binary parameters HMM and the sparse Viterbi algorithm so that it can consider multiple electrical features and decompose the appliances more quickly. After calculating the state sequence, the maximum likelihood estimation is used to decompose the load power according to the statistical features of the clustering state of the appliance. The test examples demonstrate the effectiveness of the method.

2 Load Status Determination and Presentation

The selection of load features determines the physical description of the load state, and clustering is the method of determining the load state. The first step of the method described in this article is the selection of load features and clustering methods.

2.1 Load Feature Selection

At present, the load features selected by NILD's research can be roughly divided into transient features and steady features. The transient features are generally synthetic data, they are not actually collected by the measurement device, so the practicability is not strongly convinced. Therefore, this paper selects the steady-state electrical data as the load feature.

The steady-state electrical features mainly include active power and steady-state current. Since active power is an indicator of power decomposition calculation, NILD needs to give the decomposition value of load active power after state recognition is completed, active power is a feature adopted by most NILD research cases. However, the steady-state power fluctuates slightly, the steady-state current is not affected by voltage fluctuations, and the calculated load recognition accuracy is higher [13]. Therefore, this paper selects active power and steady-state current as load features.

2.2 State Clustering

Various types of loads have different operating states during operation due to their own electrical features or working conditions. From the perspective of load decomposition, electrical appliances can be divided into switch-type appliances, limited multi-state appliances, and continuous variable-state models according to the operating state [3]. The number of operating states of the first two types of appliances is countable. In theory, all their operating states can be obtained, but the working state of continuously variable state appliances is continuously changing. It is necessary to use a clustering algorithm to make their states with limited amounts.

This paper uses DBSCAN (Density-Based Spatial Clustering of Applications with Noise) clustering algorithm to extract and cluster electrical load features. The DBSCAN algorithm was proposed by Martin Ester et al. in 1996 and is a classic and representative density-based clustering method [14]. The core idea of DBSCAN is: starting from a selected core point, and continuously expanding to the area where the density is reachable, so as to obtain a maximum area containing the core point and the boundary point, and any two points in the area are connected by the density. Therefore, the number of clusters does not need to be determined in advance, and any number and shape of clusters can be identified in the data set containing noisy data [15]. The specific execution steps of the DBSCAN algorithm are as follows:

1) The user sets the radius parameter ε and the neighborhood density threshold *MinPts*.
2) Randomly draw a sample point P. If the number of points falling within the radius ε of the point is greater than or equal to *MinPts*, it will be included in the core point list, and the entire data set will be traversed, and the points with direct density will form corresponding temporary clusters.
3) For each temporary cluster, check whether the point is the core point. If so, merge the temporary cluster corresponding to the point with the current temporary cluster to obtain a new temporary cluster.
4) Repeat step 3 until every point in the current temporary cluster is not in the core point list, or the points with direct density are already in the temporary cluster, at this time the temporary cluster is upgraded to a cluster.
5) Repeat step 4 until all temporary clusters are processed.

2.3 State Decoding

In most studies, vectors are generally used to represent the current state of several electrical appliances at a time (hereinafter referred to as the total state for convenience). For example, assuming that the number of states of the three electrical appliances is 2, 3, 8, respectively, and the state at that time is respectively 0, 2, 6, then the total state can be expressed by the vector S = [0, 2, 6]. However, in HMM applications, the hidden state cannot be represented by a vector. To this end, this paper proposes a binary-based state coding method, which encodes the hidden state vectors of multiple appliances into a binary state value. With the above example, the specific steps are as follows:

1) Allocate digits. Determine the binary digits required for coding according to the number of electrical appliances. The number of states of the above three electrical appliances is 2, 3, 8 respectively, and the binary digits allocated to each electrical appliance are 1, 2, 3, respectively.
2) Determine the value. The binary status value is calculated according to the decimal status value of the electrical appliance at the current moment. The current three electrical appliances have decimal status values of 0, 2, and 6, respectively, and the binary status values are 0, 10, and 110, respectively.
3) Splicing representation. The obtained binary state values are spliced from high to low according to the sorting of electrical appliances, and the final result is obtained. The state value of the state vector after splicing at the current time is 010110.

The following table shows other examples of the encoding method (Table 1).

Table 1. State code of some electrical appliances

Current moment	Display method	Appliance 1	Appliance 2	Appliance 3	Total state
1	Decimal	1	1	3	000011
	Binary	0	00	011	
2	Decimal	1	2	4	001100
	Binary	0	01	100	
3	Decimal	2	3	5	110101
	Binary	1	10	101	
4	Decimal	2	1	8	100111
	Binary	1	00	111	

3 The Establishment of Binary Parameters HMM

After the determination and presentation of the load state, the HMM model can be established and the model parameters can be calculated.

3.1 Introduction to HMM

Hidden Markov Model is a probability model about time series [16]. In the HMM, there is a time series with unobservable values and an observable time series with values determined by the aforementioned time series. Time series with unobservable values are called implicit state series, and time series with observable values are called observation series. The model structure is shown in Fig. 1.

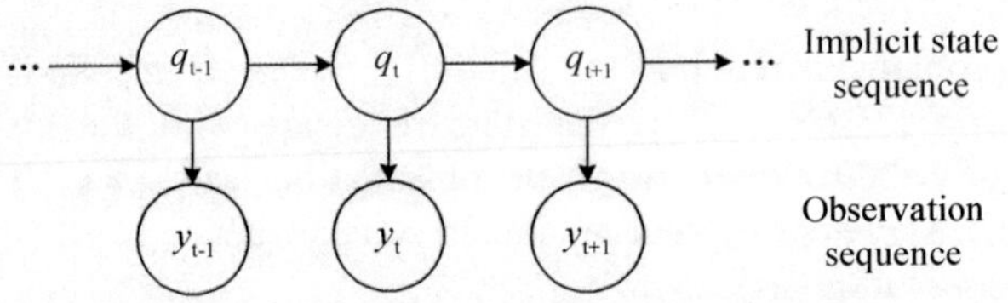

Fig. 1. Structure chart of Hidden Markov Model

An HMM model can be described by the following parameters:

1) Implicit state set S:

$$S = \{s_1, s_2, s_3, \ldots s_N\} \tag{1}$$

2) Observation state set V:

$$V = \{v_1, v_2, v_3, \ldots, v_M\} \tag{2}$$

3) State transition matrix:

$$\mathbf{A} = \left[a_{ij}\right]_{N \times N} \tag{3}$$

where $a_{ij} = P(q_{t+1} = s_j \mid q_t = s_i)$.

4) Output matrix:

$$\mathbf{B} = [b_{ik}]_{N \times M} \tag{4}$$

where $b_{ik} = P(y_t = v_k \mid q_t = s_i)$.

5) Initial probability matrix:

$$\boldsymbol{\pi} = [\pi_i]_{N \times 1} \tag{5}$$

where $\pi_i = P(q_1 = s_i)$.

It can be seen from the above formulas: the state transition probability matrix **A** and the initial state probability matrix $\boldsymbol{\pi}$ determine the hidden state sequence, and the observation probability matrix **B** determines how to generate the observation value from the hidden state. The number N and the number M of observation states are actually defined by the above three matrices, so the Hidden Markov Model can be expressed by $\lambda = \{\mathbf{A}, \mathbf{B}, \pi\}$.

HMM is used to solve the following three types of problems:

1) Probability calculation problem. Given the model $\lambda = \{\mathbf{A}, \mathbf{B}, \pi\}$ and the observation sequence $O = \{O_1 O_2 \ldots O_T\}$, calculate $P = (O|\lambda)$.

2) Learning problems. Given the observation sequence $O = \{O_1 O_2 \ldots O_T\}$, the model parameters $\lambda = \{\mathbf{A}, \mathbf{B}, \pi\}$ are estimated, that is to find the maximum value $P = (O|\lambda)$, the maximum likelihood estimation method is used to estimate the parameters.
3) Decoding problem. Knowing the model $\lambda = \{\mathbf{A}, \mathbf{B}, \pi\}$ and the observation sequence $O = \{O_1 O_2 \ldots O_T\}$, the state sequence with the highest conditional probability $P = (I|O)$ for a given observation sequence $I = \{I_1 I_2 \ldots I_T\}$ is obtained, that is, the corresponding hidden state sequence is predicted according to the given observation sequence.

3.2 Problem Modeling and Parameter Calculation

In the NILD study, the physical meaning of the two HMM time series is very clear: the implicit state sequence corresponds to the operating state of each appliance, and the observation sequence corresponds to the measurable electrical data. Therefore, the NILD problem is transformed into a problem which is to obtain the most likely hidden state sequence when the HMM model parameter and the observation sequence is given, that is, the decoding problem.

Further, this paper builds the NILD problem into the following HMM model and calculates its parameters:

1) Implicit state set S: In the NILD problem, S can be expressed as a set of combinations of operating states of each appliance, that is, a set of total states. The set is a complete arrangement of the operating states of each appliance, and the number of elements in the set is determined by the number of clusters of the status of each appliance, and its value is calculated using the state coding method introduced in Sect. 2.3.
2) Observation state set V: In the NILD problem, V represents the collection of specific electrical data recorded by the user's electricity entrance. In particular, the element of the general HMM model set V is the total active power, but the element in the set V of this paper is a vector composed of the total active power and the total steady-state current, $v_i = \left[P_i^L, I_i^L\right]$, which is also the origin of the binary parameterss in this paper.
3) State transition matrix **A**: a_{ij} refers to the probability of transition from the total state of each appliance at time t, $q_t = s_j$, to the total state at time $t + 1$, $q_{t+1} = s_j$. Calculation method:

$$a_{ij} = \frac{h_{ij}}{\sum_{j=1}^{N} h_{ij}} \tag{6}$$

where h_{ij} is the frequency of the total state transition from $q_t = s_i$ to $q_t = s_i$, and N is the total number of hidden states.
4) Output matrix $\mathbf{B}$:b_{ik} represents the probability that each appliance is in a general state $q_t = s_i$ at time t and the observed value is $y_t = v_k$. Calculation method:

$$b_{ik} = \frac{o_{ik}}{\sum_{k=1}^{M} o_{ik}} \tag{7}$$

where o_{ik} is the frequency of the total state $q_t = s_i$ at time t and the observed value is $y_t = v_k$, and M is the total number of observed values.

5) Initial probability matrix $\boldsymbol{\pi}$: π_i represents the probability that the overall state of each appliance is s_i at the initial moment. Calculation method:

$$\pi_i = \frac{d_i}{d} \tag{8}$$

among them, d is the total amount of data in the training set, d_i represents the frequency of occurrence of hidden states s_i in the training set.

It should be pointed out that, for the HMM modeling of multi-states appliances, a Factor Hidden Markov Model (FHMM) can be used. FHMM contains multiple hidden state chains, which can correspond to each appliance to be studied. However, related studies have shown that the state recognition accuracy of FHMM prediction is low [17]. Therefore, this article improves on the basis of the classic HMM model, cooperates with the state coding method described in Sect. 2.3 to convert the combined vector of each appliance's state into a binary value, and solves the problem that the hidden state of the HMM is difficult to be represented by a vector. Compared with the FHMM, the state transition matrix decoupling theoretically retains the correlation information between the state transitions of different appliances.

4 Load State and Power Decomposition Based on Binary Parameters HMM

After establishing the binary parameters HMM model for the NILD problem, the optimal hidden sequence can be predicted by the Viterbi algorithm, and the power decomposition model is established based on the maximum likelihood estimation to decompose the load power.

4.1 State Decomposition Based on Viterbi Algorithm

The Viterbi algorithm was proposed by the American Italian scientist Andrew Viterbi in 1967. The Viterbi algorithm is a dynamic programming algorithm used to solve the shortest path problem and is widely used in decoding and natural language processing [18].

The basic idea of the Viterbi algorithm is to start at $t = 1$ and recursively calculate the maximum probability of transitioning to each state i at time t:

$$\delta[t, i] = \max_j(\mathbf{B}[i, y_t] \bullet \delta[t-1, j] \bullet \mathbf{A}[j, i]) \tag{9}$$

and record the state from time *t-1* to the state with the highest probability of state i at time *t*:

$$\psi[t, i] = \arg\max_j(\delta[t-1, j] \bullet \mathbf{A}[j, i]) \tag{10}$$

after calculating the time T, find out the state $\max_i(\delta[T, i])$ which the probability belongs is the end point of the prediction sequence q_T^*. From this point, based on ψ the prediction point is gradually obtained from back to front to obtain the optimal prediction sequence $Q^* = (q_1^*, q_2^*, \ldots q_T^*)$.

Considering the sparseness of the state transition matrix and the observation matrix, the reference [19] proposed the sparse Viterbi algorithm only considers calculations where the state transition probability and the observation probability are not 0.

For a given observation sequence $Y = \{y_0 y_1, \ldots, y_T\}$ and implicit state sequence $Q = \{q_0 q_1, \ldots, q_T\}$, the specific steps of the improved Viterbi algorithm proposed in this paper are as follows:

1) Initialization:

$$\delta[0, i] = \pi[i] \bullet \mathbf{B}[i, y_0] \tag{11}$$

2) Recursive calculation:

$$\delta[t, i] = \max_j(\mathbf{B}[i, y_t] \bullet \delta[t-1, j] \bullet \mathbf{A}[j, i]) \tag{12}$$

$$\psi[t, i] = \arg\max_j(\delta[t-1, j] \bullet \mathbf{A}[j, i]) \tag{13}$$

3) Calculation of termination status:

$$p_T^* = \max_i(\delta[T, i]) \tag{14}$$

$$q_T^* = \arg\max_i(\delta[T, i]) \tag{15}$$

4) Optimal sequence backtracking:

$$q_t^* = \psi_{t+1}(q_{t+1}^*), \; t = T-1, T-2, \ldots, 0 \tag{16}$$

The sequence obtained at this time is the predicted optimal hidden state sequence $Q^* = (q_1^* q_2^*, \ldots, q_T^*)$.

4.2 Power Decomposition Based on Maximum Likelihood Estimation

The power of electrical appliances in a certain stable operating state fluctuates, and this fluctuation can be regarded as a random observation under a certain probability distribution [20, 21]. In this paper, the normal distribution is used to describe the randomness of power fluctuations during stable operation of the appliance, and it is used to calculate the power decomposition of the appliance.

The power decomposition calculation steps in this paper are: ① According to the average value and variance of the clusters of each electrical appliance sample, establish a normal distribution probability density function for each state of each electrical appliance. ② Establish the objective function based on maximum likelihood estimation, that is, to obtain the maximum value of the joint probability, and note that the sum of the power decomposition values of the electrical appliances at the same time should be equal to the constraint condition of the total power. The power decomposition calculation model is constructed as follows:

$$\begin{cases} f_{[i,j]}(x) = \frac{1}{\sqrt{2\pi}\sigma_{[i,j]}} \exp(-\frac{(x-\mu_{[i,j]})^2}{2\sigma_{[i,j]}^2}) \\ \max\limits_{P^{(1)},\ldots,P^{N}} \prod\limits_{i=1}^{N} f_{[i,j]}(P^{(i)}) \\ \text{s.t.} \ \sum\limits_{i=1}^{N} P^{i} = P^{L} \end{cases} \tag{17}$$

In the formula, $\sigma_{[i,j]}$ and $\mu_{[i,j]}$ refer to the standard deviation and mean value of the j-th cluster of the i-th appliance, respectively, N is the number of appliances, $P^{(i)}$ refers to the decomposed active power of each appliance, and P^{L} refers to the active power of the total load. $f_{[i,j]}\left(P^{(i)}\right)$ refers to the probability of $P^{(i)}$ when appliance i is in the j operating state. The above problem is a common convex quadratic programming problem after taking *ln()* on both sides of the objective function.

The complete flow of the non-intrusive load decomposition method proposed in this paper is shown in Fig. 2.

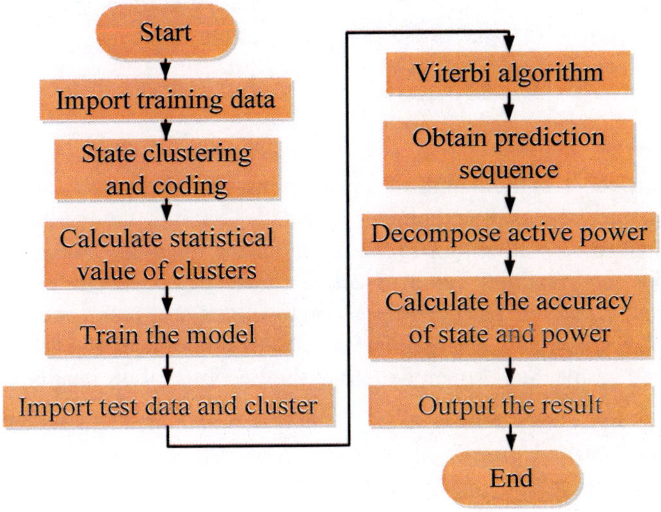

Fig. 2. Flow chart of non-intrusive load decomposition based on binary parameters HMM

5 Example Analysis

5.1 Experimental Method

This paper selects the public data set AMPds2 established by Canadian scholar Stephen Makonin to verify the method described in this article [22]. AMPds2 collects real electrical data of electrical equipment in a household, and records 11 electrical features including steady-state current and active power. The sampling frequency is 1/60 Hz, and the recording time is 2 years. It is suitable as a data set for analysis.

From the data set, select six kinds of appliances such as fireplace (WOE), clothes dryer (CDE), dishwasher (DWE), television (TVE), washing machine (CWE), and heat pump (HPE) for a total of 14400 sampling points in 10 days The active power and steady-state current data are divided into 10 groups according to time, which are recorded as test1-test10, and 10-fold cross-validation is performed. In each experiment, 9 groups of data are selected as training data, and one group is used as test data. Finally, the average of all the data groups is the result.

The PC configuration used in this article is: 16 GB RAM/Inter (R) Core (TM) i5-8300H@2.30 GHz, written in Python.

5.2 Decomposition Result

In this paper, the basic accuracy rate ACC_{state} is used to evaluate the accuracy of load state recognition, and the root mean square error ACC_{power} is used to evaluate the accuracy of load power decomposition:

$$ACC_{\text{state}} = \frac{\sum_{t=1}^{T} I\left(s_t = s_t^*\right)}{T} \tag{18}$$

$$ACC_{\text{power}} = \sqrt{\frac{1}{T}\sum_{t=1}^{T} \left(p_t - p_t^*\right)^2} \tag{19}$$

In the formula, T is the length of the sampling period, s_t and p_t are the actual state and actual power value of the appliance at time t, s_t^* and p_t^* are the predicted state and decomposition power value, and $I\ ()$ is the indicator function.

Select the superposition value of active power of six electrical appliances on a certain day as the test set. The decomposition results of heat pump and TV are shown in Figs. 3 and 4, respectively. The yellow line is the actual power and the green line is the decomposition result. See Appendix A for the decomposition results of all electrical appliances.

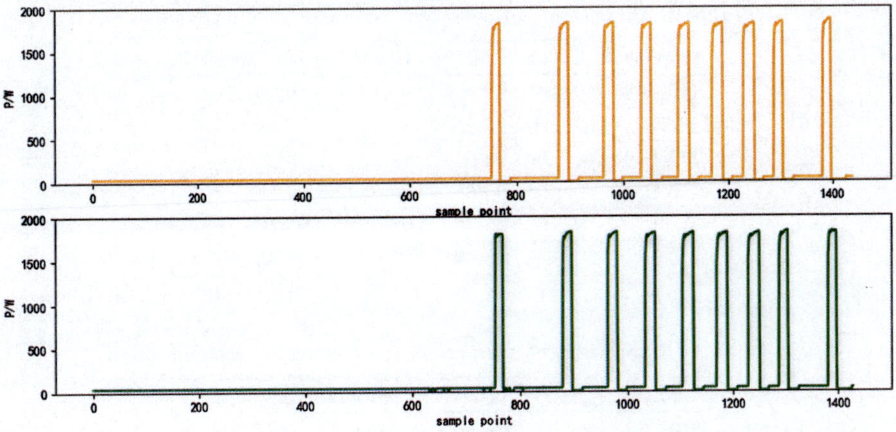

Fig. 3. Results of power decomposition (HPE)

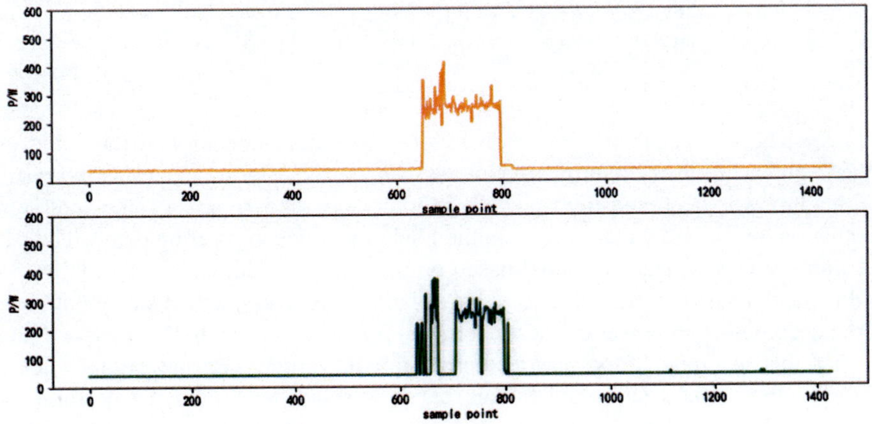

Fig. 4. Results of power decomposition (TVE)

This paper selects the decomposition method based on genetic algorithm (GA) proposed in reference [23], the decomposition method based on deep sequence translation model in reference [24] and the classic HMM [13] as comparison, chooses the same data to perform the 10-fold cross-validation calculation and uses the average value as result. The average accuracy rate of load state identification by the four methods, and the average accuracy rate of power decomposition are shown in Table 2 and Table 3, respectively. It can be seen that the load decomposition method based on the binary parameters HMM and the sparse Viterbi algorithm has a better effect on the state identification and power decomposition of the total load.

Table 2. Comparison of average accuracy of state recognition

Appliance	GA/%	RNN/%	HMM/%	This paper/%
HPE	96.82	97.23	97.12	99.79
TVE	93.41	95.56	94.44	97.69
DWE	91.86	94.55	93.22	94.34
CWE	94.28	94.65	94.38	94.27
WOE	92.77	95.85	93.25	95.10
CDE	95.17	96.40	97.67	98.46

Table 3. Comparison of average accuracy of power decomposition

Appliance	GA/W	RNN/W	HMM/W	This paper/W
HPE	23.54	15.24	8.69	4.97
TVE	73.76	68.79	69.71	30.09
DWE	61.02	55.68	52.90	45.38
CWE	92.56	84.62	82.29	44.52
WOE	66.93	49.28	50.33	25.24
CDE	64.88	54.46	67.39	58.37

It can be seen from the above results that the decomposition method based on genetic algorithm uses the typical value of load state clustering center as the result of the load power decomposition, which ignores the randomness of the load power fluctuation, resulting in a large error in the load power decomposition result. The power decomposition optimization model based on maximum likelihood estimation proposed in this paper considers and alleviates this volatility relatively, and ensures that the sum of the decomposition power of each electrical appliance is equal to the total load power, so that the accuracy of power decomposition is higher. For newer deep learning methods, the method described in this paper is superior to it in the decomposition results of clothes dryers, televisions and heat pumps, which also shows that improved HMM is still suitable for solving NILD problems. Compared with the classic HMM algorithm, the method described in this paper can use the multiple electrical features to extract the load state that can better reflect the load features, thereby simultaneously improving the accuracy of state recognition and power decomposition.

6 Conclusion

This paper proposes a non-intrusive load decomposition method based on the binary parameters Hidden Markov Model using load active power and steady-state current as features. This method first uses the DBSCAN clustering algorithm to extract the operating state of electrical appliances, proposes a binary-based encoding scheme to encode the total state, and then constructs a binary parameters Hidden Markov Model and calculates its parameters, using a sparse Viterbi algorithm to realize the state decomposition of electrical load, and finally combine maximum likelihood estimation to achieve load power decomposition according to the statistical values of electrical clusters. The results of the calculation examples show that the proposed method achieves better load decomposition accuracy than the classic HMM algorithm, heuristic algorithm, and deep learning-based recognition method. Considering the randomness of load power fluctuation, the method proposed in this paper is also valuable for the practical application of NILD.

In this work, the research on the transition between unknown state and new state is relatively preliminary, and the non-electrical features of load operation are not considered. The next research will focus on how to use non-electrical features, such as the working hours of electrical appliances, user behavior habits, study the treatment of newly added unknown electrical appliances, and propose a more practical non-intrusive load decomposition method.

Acknowledgement. The paper is supported by Science and technology project of China Southern Power Grid Co., Ltd. (Grant No. 066600KK52180051) and Special funds for basic business expenses of Central Universities (Grant No. D2192890).

Appendix

See Fig. 5

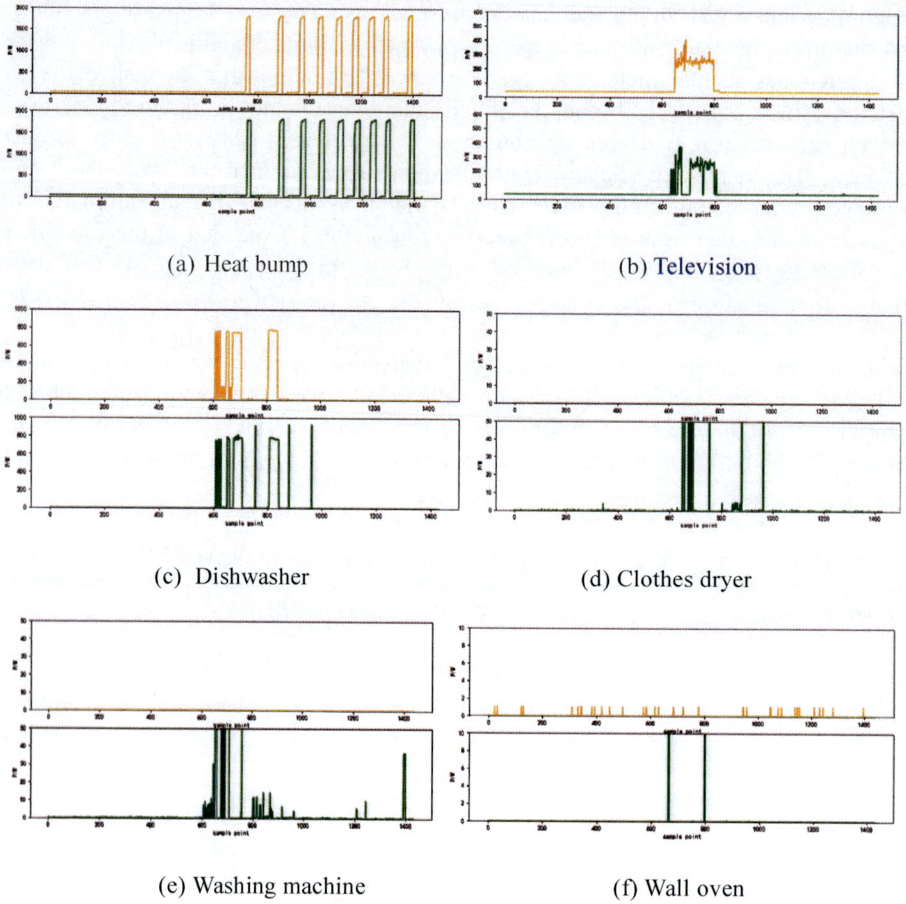

(a) Heat bump

(b) Television

(c) Dishwasher

(d) Clothes dryer

(e) Washing machine

(f) Wall oven

Fig. 5. Results of load decomposition (the yellow line is the actual power, and the green line is the decomposition result) (Color figure online)

References

1. Li, T.: Technical implications and development trends of flexible and interactive utilization of intelligent power. Autom. Electr. Power Syst. **36**(2), 11–17 (2012)
2. Yi-Xin, Y.U.: Technical composition of smart grid and its implementation sequence. South. Power Syst. Technol. **3**(2), 1–5 (2009)
3. Hart, G.W.: Nonintrusive appliance load monitoring. Proc. IEEE **80**(12), 1870–1891 (1992)
4. Xingjie, L., Meihan, C., Yuejuan, X.U.: Non-intrusive load monitoring based on improved chicken swarm optimization algorithm. Electr. Power Autom. Equip. **38**(5), 235–240 (2018)

5. Ruyi, L.I., Mingshan, H., Dongguo, Z., et al.: Optimized nonintrusive load decomposition method using particle swarm optimization algorithm **44**(8), 30–36 2016
6. Yan, X., Zhai, S., Wang, Z., et al.: Application of deep neural network in non-intrusive load disaggregation. Autom. Electr. Power Syst. **43**(1), 126–132 (2019)
7. Chen Haiwen, L., et al.: Non-intrusive load identification algorithm based on feature fusion and deep learning. Autom. Electr. Power Syst. **44**(9), 103–110 (2020)
8. Lu, T.: Research on Event-based HMM for Non-intrusive Electrical Load Monitoring Algorithm. Huazhong University of Science and Technology, Wuhan (2017)
9. Sun, X.: Research on Non-Intrusive Load Decomposition Based on FHMM. Zhengzhou University, Zhengzhou (2018)
10. Chen, H.: Study on Nonintrusive Load Monitoring Technology Based on HMM and Its Variants. Tianjin University, Tianjin (2016)
11. Ai-Jun, H., Yong-Sheng, W., Bo-Chuan, G.U., et al.: Non-intrusive Load Monitoring Method Based on Denoising Filtering and FHMM. Autom. Instr. **34**(11), 97–102 (2019)
12. Wu, Y.: Non-intrusive Load Decomposition Method based on Factorial Hidden Markov Model. North China Electric Power University (Beijing), Beijing (2019)
13. Dan, G.: Study on non intrusive load decomposition of residential electricity based on HMM. North China Electric Power University (Beijing), Beijing (2018)
14. Ester, M., Kriegel, H., Sander, J., et al.: A Density-Based Algorithm for Discovering Clusters in Large Spatial Databases with Noise. 20090721836-841
15. Feng, Z.: Research and Application of Clustering Algorithm based on DBSCAN. Jiangnan University, Wuxi (2016)
16. Berges, M., Goldman, E., Matthews, H.S., et al.: Learning systems for electric consumption of buildings. pp. 1–10 (2009)
17. Yixin, Y.U., Bo, L., Wenpeng, L.: Nonintrusive residential load monitoring and decomposition technology. South. Power Syst. Technol. **7**(4), 1–5 (2013)
18. Forney, G.D.: The viterbi algorithm. Proc. IEEE **61**(3), 268–278 (1973)
19. Makonin, S., Popowich, F., Bajić, I.V., et al.: Exploiting HMM sparsity to perform online real-time nonintrusive load monitoring. IEEE Trans. Smart Grid **7**(6), 2575–2585 (2016)
20. Liang, J., Ng, S., Kendall, G., et al.: Load signature study—part I: Basic concept, structure and methodology (2010)
21. Liang, J., Ng, S.K.K., Kendall, G., et al.: Load signature study—Part II: decomposition framework, simulation, and applications. IEEE Trans. Power Deliv. **25**(2), 561–569 (2010)
22. Makonin, S., Ellert, B., Bajić, I.V., et al.: Electricity, water, and natural gas consumption of a residential house in Canada from 2012 to 2014. Sci. Data **3**(1), 160037 (2016)
23. Xu, Q., Lou, O., Zheng, A., et al.: A non-intrusive load decomposition method based on affinity propagation and genetic algorithm optimization. Trans. China Electrotech. Soc. **33** (16), 3868–3878 (2018)
24. Ren, W., Xu, G.: Non-intrusive load decomposition method based on deep sequence translation model. Power Syst. Technol. **44**(01), 27–37 (2020)

Renewable Energy Accommodation Optimization Model Considering DC Tie Line Operation Conditions

Xiaoyang He[1], Zhi Zhou[1], Keming Zhan[1], and Jinming Yao[2](✉)

[1] State Grid Liaoning Electric Power Co. Ltd., Shenyang 110006, China
[2] North China Electric Power University, Baoding 071003, China
yaojinming97@163.com

Abstract. With the increasing scale of renewable energy in China, how to optimize the power regulation capacity of DC tie lines has become an important challenge for renewable energy accommodation in interconnected power grids. Therefore, this paper establishes a stair-like power optimization model for DC tie lines. This model can meet the constraints of conventional generating units and renewable energy generating units in the sending and receiving systems. The model can also meet the constraints of DC tie line, including DC power stair-like regulation, the times of regulation, constant daily exchange power and no reverse power regulation in short time, which can not only improve the flexibility of the DC tie line regulation, but also meet the safety requirements of the practical engineering operation. In addition, this model can take into account multiple DC tie lines of multiple interconnected power grids and the load complementarities in the sending and receiving systems, further promoting the accommodation of renewable energy in a larger space. Finally, based on the practical data of a provincial interconnected power grid in China, the effectiveness of this model was verified by the before-and-after optimized comparison by means of DC tie line power, conventional generating units output and proportion of renewable energy abandonment of wind and photovoltaic.

Keywords: Renewable energy accommodation · DC tie lines · Interconnected power grid · DC power Stair-like regulation

1 Introduction

Under the call of the strategy of sustainable development, China's installed capacity of renewable energy has increased year by year. As of the end of 2018, China's wind power installed capacity exceeded 180 GW, and photovoltaic installed capacity exceeded 170 GW, ranking first in the world [1–5]. With the rapid growth of renewable energy installed capacity, the problem of renewable energy accommodation is becoming increasingly prominent due to the random and volatile characteristics of wind and photovoltaic power [6, 7]. Among them, the "Three North Regions" (Northwest, Northeast, North China) where the renewable energy resources are abundant have relatively light loads and limited renewable energy accommodation capacity with serious phenomena of wind and photovoltaic abandonment, while in the

Y. Xue et al. (Eds.): PMF 2020, LNEE 718, pp. 358–373, 2021.
https://doi.org/10.1007/978-981-15-9746-6_29

east of China, where the renewable energy resources are scarce and the load is heavy, the energy supply is insufficient [8–11]. Therefore, it is necessary to study the improvement of renewable energy accommodation capacity through DC tie lines.

Related studies have analyzed the power optimization of interconnected DC tie lines. Literature [12] achieves the stair-like operation of DC tie line power by simulating the DC tie lines as the output and start-stop of equivalent conventional generating units, but the discrete power points reduce the flexibility of DC power regulation. Literature [13] improves the fixed output combination of the equivalent generating units of the DC tie lines. By simulating the power of the DC tie lines as a combination of transmission units with upper and lower limits, the flexibility of DC power regulation has been improved. Literature [14] based on the initial transmission plan of interconnected energy trading, considering the constraints of DC tie lines, renewable energy and conventional generating units to maximize the accommodation of renewable energy, but the model does not consider the DC power stair-like conditions and does not meet the practical operation need. Literature [15] considers the power demand in the receiving system, proposing a more flexible DC power step strategy. The literature establishes a conditional control section model and improves the level of renewable energy accommodation in a larger space. However, the model does not consider the load balance constraint of the sending system. Literature [16] considers the complementary characteristics of the load in sending and receiving systems and the renewable energy accommodation space. The literature establishes an interconnected DC power optimization model to achieve renewable energy accommodation. Without economic consideration, the model only optimizes the accommodation of renewable energy.

Existing research is mainly aimed at optimizing the configuration of various types of generating units in the sending system. The effect of improving the interconnected renewable energy accommodation has been achieved by optimizing the power of the DC tie lines. It is seldom considered that the needs in the receiving system for stable operation and the load complementarities in the interconnected systems [17–19]. Even if the receiving system is considered, it is still a simple optimization model with constraints, which is slightly different from the practical operation. This paper establishes a joint optimization model considering DC tie line operation conditions, conventional generating units and renewable energy of the interconnected power grids. Taking into account the power system security constraints and economy, the model can optimize multiple interconnected DC tie lines, promoting the accommodation of interconnected renewable energy and meet the practical engineering needs.

2 DC Tie Line Power Optimization Model

This paper promotes the interconnected renewable energy accommodation by establishing a power optimization model for DC tie lines.

2.1 Objective Function

With the minimum overall system cost as the objective function, the conventional generating units and renewable energy generating units (wind power and photovoltaic units) of the system have the lowest operating costs within the specified scheduling period. Among them, renewable energy power generation only incurs penalty costs when wind and photovoltaic abandonment occurs. Considering the objective function term generated by the power optimization of the DC tie lines, the objective function is:

$$\min F = \sum_{t=1}^{T}\sum_{j=1}^{N} C_j^t + \sum_{t=1}^{T}\sum_{e=1}^{E} C_e^t + \delta \tag{1}$$

where T represents the total running time of the dispatch; N represents the number of conventional generating units that the sending system participates in dispatching; E represents the number of renewable energy generating units in the sending system; C_j^t represents the power generation cost of the conventional unit j in time period t; C_e^t represents the penalty cost of wind abandoning and photovoltaic abandoning phenomenon of renewable energy unit e in time period t; δ represents the relevant objective function term of DC tie line power optimization.

2.2 Conventional Unit Related Constraints

Constraints on Upper and Lower Limits of Conventional Unit Output

$$X_j^t P_j^{\min} \le P_j^t \le X_j^t P_j^{\max} \tag{2}$$

where P_j^t represents the output of conventional unit j in time period t; $P_j^{\min}$ and $P_j^{\max}$ represent the lower and upper limits of the output of conventional unit j respectively; X_j^t represents a binary variable indicating whether the conventional unit j is in the running state, X_j^t equals 1 indicating that the unit is operating normally, and equals 0 indicating that the unit has been stopped.

Constraints on Climbing Rate of Conventional Generating Units

$$-R_j^{\text{down}}\Delta t \le P_j^t - P_j^{t-1} \le R_j^{up}\Delta t \tag{3}$$

where R_j^{down} and R_j^{up} represent the descending and climbing rates of the conventional unit j respectively; Δt represents the scheduling running time steps of the conventional unit j.

Minimum Start-Stop Time Constraints for Conventional Generating Units

$$\begin{aligned} (X_j^t - X_j^{t-1})(T_j^{\min up} - T_j^{t-1}) \geq 0 \\ (X_j^{t-1} - X_j^t)(T_j^{\min down} + T_j^{t-1}) \geq 0 \end{aligned} \tag{4}$$

where $T_j^{\min up}$ and $T_j^{\min down}$ represent the minimum continuous running time and minimum continuous downtime of the conventional unit j respectively; T_j^{t-1} represents the length of time that the conventional unit j has maintained the original state at time period $(t-1)$, in which the running state is positive and the shutdown state is negative.

2.3 Renewable Energy Unit Related Constraints

Renewable Energy Unit Output Constraints

$$\begin{aligned} P_{\mathrm{ep}}^t = P_{\mathrm{ea}}^t + P_{\mathrm{e}}^t \\ 0 \leq P_{\mathrm{ea}}^t \leq P_{\mathrm{ep}}^t \end{aligned} \tag{5}$$

where P_{e}^t represents the actual output of the renewable energy unit e in the time period t; P_{ep}^t represents the predicted output of the renewable energy unit e in the time period t; P_{ea}^t represents the amount of wind and photovoltaic abandoned by the renewable energy unit e during time period t.

Renewable Energy Unit Penalty Cost Constraints. With the increase of abandonment of wind and photovoltaic, the segmented penalty factors rapidly increases, and the cost of renewable energy power generation increases sharply. By controlling the size of the penalty factors in each segment, renewable energy generating units can better match the cost optimization of conventional generating units and promote the accommodation of renewable energy [20].

$$\begin{aligned} C_e^t = \sum_{l=1}^{L} w_e^l \Delta P_e^{l,t} \\ 0 \leq \Delta P_e^{l,t} \leq P_e^{l,t} - P_e^{l-1,t} \end{aligned} \tag{6}$$

where L represents the total number of segments of the penalty factors; w_e^l represents the penalty factor of the new energy unit e at the l segment of the penalty factor segmentation function; $\Delta P_e^{l,t}$ represents the corresponding power of the l segment of the renewable energy unit e in the time period t under the principle of piecewise accumulation; $P_e^{l,t}$ and $P_e^{l-1,t}$ represent the terminal powers of the renewable energy unit e at the l and ($l-1$) segments of the penalty factor piecewise function respectively.

2.4 Sending System Related Constraints

Load Balance Constraints in Sending System

$$\sum_{j=1}^{N} P_j^t + \sum_{e=1}^{E} P_e^t = P_L^t + \sum_{s=1}^{S} P_s^t \tag{7}$$

where S represents the number of DC tie lines; P_s^t represents the power transmitted by the s DC tie line in time period t; P_L^t represents the load predicted value in the sending system in time period t.

Reserve Capacity Constraints in Sending System. System reserve capacity constraints include positive spinning reserve capacity constraints and negative spinning reserve capacity constraints:

$$\begin{aligned} Sp &\leq \sum_{j=1}^{N} P_j^{max} - \sum_{j=1}^{N} P_j^t \\ Sn &\leq \sum_{j=1}^{N} P_j^t - \sum_{j=1}^{N} P_j^{min} \end{aligned} \tag{8}$$

where Sp and Sn represent the positive spinning reserve capacity and negative spinning reserve capacity respectively.

2.5 Receiving System Related Constraints

Power Grid Capacity Constraints in Receiving System

$$\sum_{i=1}^{I} P_{d,i}^{min} \leq P_d^t - \sum_{f=1}^{F} P_{d,f}^t - \sum_{s \in (d)} P_s^t \leq \sum_{i=1}^{I} P_{d,i}^{max} \tag{9}$$

where F represents the number of renewable energy generating units in the receiving system d; I represents the number of conventional generating units in the receiving system d participating in dispatching; $P_{d,i}^{min}$ and $P_{d,i}^{min}$ represent the minimum and maximum output of the conventional unit i in the receiving system d respectively; P_d^t represents the load predicted value in the receiving system d in time period t; $P_{d,f}^t$ represents the power predicted value of the renewable energy unit f in the receiving system d in time period t; $\sum_{s \in (d)} P_s^t$ represents the sum of the power values of the DC tie lines directly connected to the receiving system d in time period t.

Climbing Rate Constraints of Conventional Generating Units in Receiving System

$$
\begin{aligned}
P_{\mathrm{d}}^{cl,t} &= P_{d}^{t} - \sum_{f=1}^{F} P_{d,f}^{t} - \sum_{s\in(d)} P_{s}^{t} \\
-\sum_{i=1}^{I} R_{d,i}^{\mathrm{down}}\Delta t \le P_{\mathrm{d}}^{cl,t} &- P_{\mathrm{d}}^{cl,t-1} \le \sum_{i=1}^{I} R_{d,i}^{\mathrm{up}}\Delta t
\end{aligned}
\tag{10}
$$

where $P_{\mathrm{d}}^{cl,t}$ represents the net load in the receiving system d in the time period t; $R_{d,i}^{\mathrm{down}}$ and $R_{d,i}^{\mathrm{up}}$ represent the descending and climbing rates of the conventional unit i in the receiving system d respectively.

Reserve Capacity Constraints in Receiving System. The sending system has similar reserve capacity constraints as the receiving system:

$$
\begin{aligned}
Ssp &\le \sum_{i=1}^{I} P_{\mathrm{d},i}^{\mathrm{max}} - \sum_{i=1}^{I} P_{d,i}^{\mathrm{t}} \\
Ssn &\le \sum_{i=1}^{I} P_{d,i}^{\mathrm{t}} - \sum_{i=1}^{I} P_{d,i}^{\mathrm{min}}
\end{aligned}
\tag{11}
$$

where $P_{d,i}^{\mathrm{t}}$ represents the output of the conventional unit i in the receiving system d in the time period t; Ssp and Ssn represent the positive spinning reserve capacity and negative spinning reserve capacity in the receiving system d respectively.

2.6 DC Tie Line Related Constraints

Taking into account the characteristics of DC converter equipment regulation and receiving power grid coordination, the power transmission of the DC tie line adopts a stair-like operation mode. In order to more accurately establish a DC power stair-like model, this paper selects the stepped power of the DC link as the set of power that can be transmitted when the DC tie line runs.

$$
P_{\mathrm{s}}^{t} \in \{P_{s,1}, P_{s,2}, \cdots, P_{s,z}, \cdots, P_{s,Z}\} \tag{12}
$$

where Z represents the total amount of power states in the DC tie line power set; $P_{s,z}$ represents the power of the DC tie line s in the feasible state z in the power set.

Therefore, the DC tie line transmission power is expressed as

$$
P_{s}^{t} = \sum_{z=1}^{Z} P_{s,z} K_{s,z}^{t} \tag{13}
$$

where $K_{s,z}^{t}$ represents a binary variable indicating whether the DC tie line s is in the feasible state z in the power set during the time period t.

The status of the DC tie line power is expressed as

$$z_s^t = \sum_{z=1}^{Z} zK_{s,z}^t \tag{14}$$

where z_s^t is the power state of the DC tie line s in the time period t.

Considering that the DC tie line can only be in one state at the same time, it can be concluded that

$$\sum_{z=1}^{Z} K_{s,z}^t = X_s^t \tag{15}$$

where X_s^t represents a binary variable indicating whether the DC tie line s is in the running state in the time period t, X_s^t equals 1 indicating that the DC tie line is operating normally, and equals 0 indicating that the DC tie line has been stopped.

Constraints on No Reverse Power Regulation in Adjacent Time Periods

$$\begin{aligned} \Delta_{s,p}^t + \Delta_{s,n}^t &\leq 1 \\ \Delta_{s,p}^t + \Delta_{s,n}^{t+1} &\leq 1 \\ \Delta_{s,p}^{t+1} + \Delta_{s,n}^t &\leq 1 \end{aligned} \tag{16}$$

where $\Delta_{s,p}^t$ and $\Delta_{s,n}^t$ represent binary variables indicating whether the power of DC tie line s increases and decreases in time period t respectively.

In order to achieve the numerical optimization of $\Delta_{s,p}^t$ and $\Delta_{s,n}^t$, the binary auxiliary variables $v_{s,p}^t$ and $v_{s,n}^t$ are introduced as [21, 22]

$$\begin{aligned} P_s^t - P_s^{t-1} &\leq v_{s,p}^t P_s^{\max} \\ P_s^{t-1} - P_s^t &\leq v_{s,n}^t P_s^{\max} \\ \Delta_{s,p}^t &\geq v_{s,p}^t \\ \Delta_{s,n}^t &\geq v_{s,n}^t \\ \Delta_{s,p}^t &\leq \left| z_s^t - z_s^{t-1} \right| \\ \Delta_{s,p}^t &\leq \left| z_s^t - z_s^{t-1} \right| \end{aligned} \tag{17}$$

The absolute value constraint conditions appearing in the constraint conditions are non-linear forms, while the optimization model constructed in this paper is a linear model, so the absolute value conditions need to be linearized as [23]

$$\begin{aligned} \Delta_{s,p}^t &\leq z_{s,p}^t + z_{s,n}^t \\ \Delta_{s,n}^t &\leq z_{s,p}^t + z_{s,n}^t \\ z_s^t - z_s^{t-1} &= z_{s,p}^t - z_{s,n}^t \end{aligned} \tag{18}$$

where $z^t_{s,p}$ and $z^t_{s,n}$ are the two non-negative parameters introduced, which have no practical physical meaning.

In addition, due to the introduction of these two parameters, the relevant function terms need to be considered in the objective function as

$$\min\delta = \sum_{t=1}^{T}\sum_{s=1}^{S} z^t_{s,p} + z^t_{s,n} \tag{19}$$

Duration Constraints of DC Tie Line Power Stable Operation

$$\begin{gathered} \Delta^t_{\mathrm{s}} = \Delta^t_{\mathrm{s},p} + \Delta^t_{\mathrm{s,n}} \\ r^t_{\mathrm{s},p} - r^t_{\mathrm{s,n}} = \Delta^t_{\mathrm{s}} - \Delta^{t-1}_{\mathrm{s}} \\ r^t_{\mathrm{s},p} + r^t_{\mathrm{s,n}} \leq 1 \\ r^t_{\mathrm{s,n}} + \sum_{\alpha=t+1}^{t+q-1} r^{\alpha}_{\mathrm{s,p}} \leq 1 \end{gathered} \tag{20}$$

where Δ^t_{s} represents a binary variable indicating whether the power of the DC tie line s is regulated in the time period t; $r^t_{\mathrm{s},p}$ and $r^t_{\mathrm{s,n}}$ represent binary variables indicating whether the power of the DC tie line s starts to regulate and ends to regulate in the time period t respectively; q is the minimum duration of stable operation of the power of the DC tie line s.

Constraints on DC Tie Line Regulation Times

$$\sum_{t=1}^{T} r^t_{s,p} \leq N_s \tag{21}$$

where N_s represents the limit of the number of times that the DC tie line s can be regulated within the scheduling period.

Constraints on DC Tie Line Transmission Power

$$Q_S^{\min} \leq \sum_{t=1}^{T} \tau P^t_s \leq Q_S^{\max} \tag{22}$$

where $Q_S^{\min}$ and $Q_S^{\max}$ represent the minimum and maximum power delivered during the scheduled time period of the DC tie line s; τ represents the time interval of each time period.

At this point, the power optimization model of the DC tie lines has been established with the goal of promoting interconnected renewable energy accommodation. Due to the model contains multiple sets of binary variables, it is a mixed integer linear programming problem. Now the more mature mathematical optimization software CPLEX can be used to solve the model [24].

3 Case Analysis

The model is analyzed with the practical data of the typical day of a provincial interconnected power grid in China. In order to verify the superiority of the power optimization model of the DC tie lines proposed in this paper, the results of the following two modes are calculated and analyzed in this section:

Mode 1: DC tie line power is the practical operating value.

Mode 2: The power of the DC tie line is optimized on the premise that the daily exchange power is unchanged.

3.1 System Basic Data

The interconnected power grid consists of two regional power grids, the sending system and the receiving system, and the two regional power grids are connected by a DC tie line. The decision period of this case is 96 periods of one day, and the interval of each period is 15 min.

The sending system consists of 142 generating units with a total installed capacity of 15645.9 MW, including 50 conventional generating units with an installed capacity of 9133.2 MW; 38 wind turbines with an installed capacity of 4087.9 MW; 54 photovoltaic generating units with an installed capacity of 2424.8 MW. The receiving system consists of 130 generating units with an installed capacity of 28778 MW.

The load in the sending system reached the peak at the 30th and the 73rd time periods, which were 4757.46 MW and 4910.55 MW respectively, and reached the valley value at the 33rd time period, which was 4523.51 MW. The receiving system peaked at 23007 MW in the 44th time period. The load prediction curve in the receiving system is shown in Fig. 1:

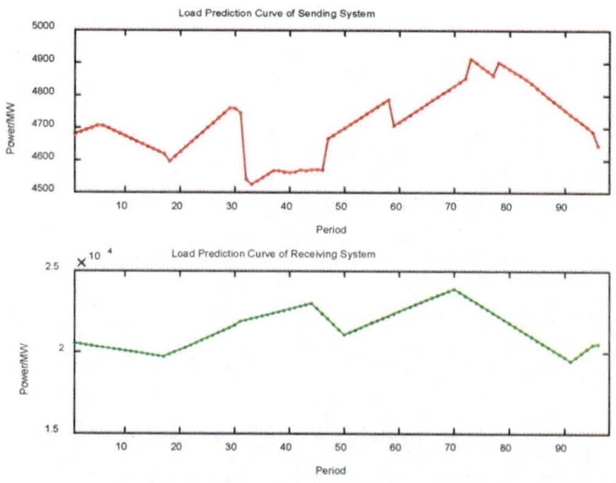

Fig. 1. Load prediction curve of sending and receiving systems

Among them, from the 1st to the 20th time period, the system load in the sending and receiving systems had a similar trend, but during the 30th to the 40th time period, the system load in the sending system had a rapid decline, while the system load in the receiving system steadily increased, and the system load in the sending and receiving systems had certain complementarities.

The total predicted output of renewable energy power in the sending system reached the minimum value of 1194.36 MW in the 28th time period and the maximum value of 2983.02 MW in the 58th time period. The predicted output of photovoltaic power generation reached the maximum value of 990.77 MW in the 56th time period, the predicted output of wind power generation reached the minimum value of 1190.97 MW in the 28th time period and the maximum value of 2005.88 MW in the 74th time period. The renewable energy power predicted curve in the sending system is shown in Fig. 2:

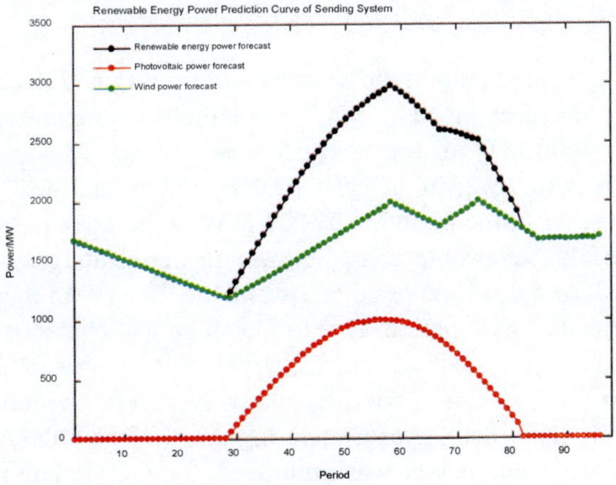

Fig. 2. Renewable energy power prediction curve of sending system

3.2 Analysis of Results

Comparative Analysis Before and After Power Optimization of DC Tie Line. In this paper, a power joint optimization model of conventional and renewable energy generating units and DC tie line is established. Under the premise that the daily exchange power of the DC tie line is unchanged the power regulation range of the tie line becomes larger and more flexible. The power comparison before and after the power optimization of the DC tie line is shown in Fig. 3:

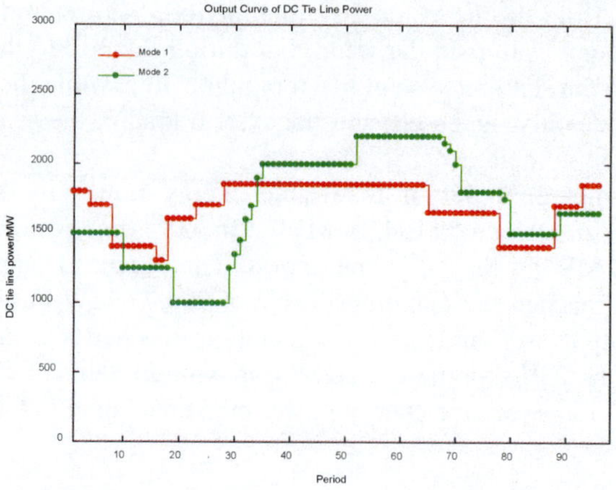

Fig. 3. Output curve of DC tie line power

As shown in Fig. 1, the load in the sending system dropped dramatically in the 31st time period with a decrease of 203.1 MW in one time period, and maintained at a low load level below 4600 MW in the next 15 time periods. Meanwhile, the receiving system in the same period was in the early peak period, with a steady rise in load, from 21895.5 MW in the 31st time period to 23007 MW at the peak in the 44th time period. At the same time, the renewable energy power in the sending system in Fig. 2 also continued to rise during this time periods, from 1411.72 MW in the 31st time period to 2375.08 MW in the 44th time period. Due to the numerical characteristics of the load in the sending and receiving systems and the renewable energy in the sending system during the 31st to the 44th time period, a relatively severe phenomenon of wind and photovoltaic abandonment will be caused during this time periods. As can be seen from Fig. 3, after the DC tie line power was optimized, the DC tie line transmission power increased from 1850 MW in the 31st time period to 2000 MW in the 35th time period. It transmitted renewable energy that was difficult to be accommodated by the sending system to the receiving system with an increased load, which enhanced the renewable energy accommodation capacity in the sending system.

Comparative Analysis of Wind and Photovoltaic Abandonment. Under the two modes, the output curves of wind power generation and photovoltaic power generation are shown in Fig. 4:

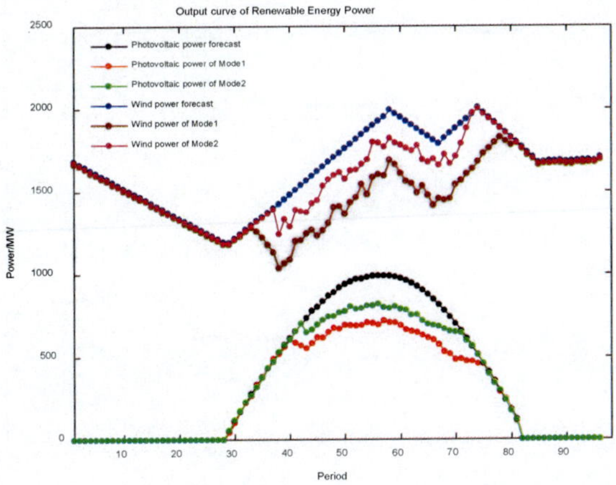

Fig. 4. Output curve of Renewable Energy Power

In the two modes, the power accommodation of wind power generation and photovoltaic power generation is shown in Table 1:

Table 1. Accommodated power of renewable energy

Types	Forecast power/ (MW·h)	Mode1		Mode2	
		Accommodation/ (MW·h)	Proportion/%	Accommodation/ (MW·h)	Proportion/%
Pv	6967.62	5642.22	80.98	5845.7	83.9
Wind	39435.67	36048.49	91.41	37809.23	95.88
Total	46403.29	42090.71	90.71	43654,93	94.08

It can be seen from Table 1 that by optimizing the power of the DC tie line through the model in this paper, the accommodation of renewable energy in the system increased by 1564.22 MW·h, and the accommodation proportion increased by 3.37%. Through the optimization of the power of the DC tie line, the renewable energy accommodation capacity in the sending system was effectively improved, while the load in the receiving system was large, and the comprehensive benefit was also enhanced by the renewable energy transferred from the sending system.

Comparative Analysis of the Conventional Unit in the Sending System. Under the two modes, the output curve of the conventional generating units in the sending system is shown in Fig. 5:

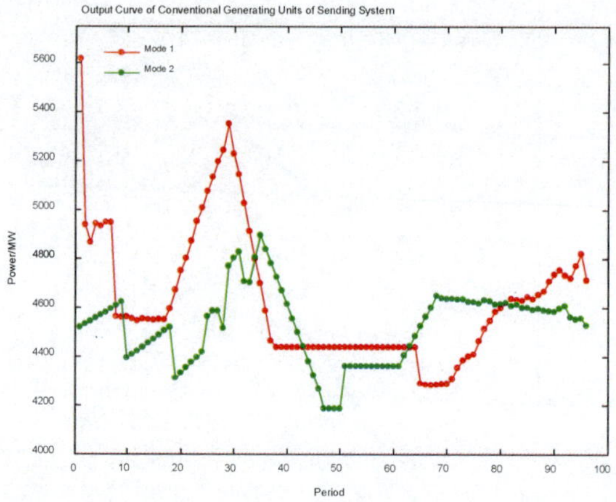

Fig. 5. Output curve of conventional generating units of sending system

Comparing the output of the conventional generating units in the two modes of Fig. 5, it can be seen that after the power optimization of the DC tie line, the output curve of the conventional generating units in the sending system was smoother, which had a certain effect of "peak shifting and valley filling". Combined with Fig. 1, it can be seen that the power of the DC tie line in mode 1 rose steadily from 1300 MW in the 16th time period to 1850 MW in the 28th time period, and remained unchanged for 1850 MW in the following period. However, the renewable energy output in the same time period reached the maximum of 1411.96 MW in the 16th time period, and remained below 1500 MW for a long time. Due to the high level of DC power output and low level of renewable energy power, the sending system required a high level of conventional unit output to maintain the balance, which rose rapidly from 4551.62 MW in the 16th time period and reached a peak of 5351.5 MW in the 29th time period. Mode 2 comprehensively considered the renewable energy output, system load and power of the DC tie line in the receiving and sending systems. When the renewable energy output was relatively low, the transmission power of the DC tie line was reduced, and a certain output margin of the conventional generating units was retained, thus promoting the accommodation of renewable energy in the interconnected power grid.

3.3 Case of Multiple DC Tie Lines

On the basis of the above example of DC tie line power optimization at the sending and receiving systems, further consideration is given to the interconnected grid with multiple DC tie lines across. Among them, area 3 is interconnected with area 2, area 1 and area 4 respectively, while other areas are not connected and there are three DC link lines. The comparison of the DC tie line power optimization before and after each area is shown in Fig. 6:

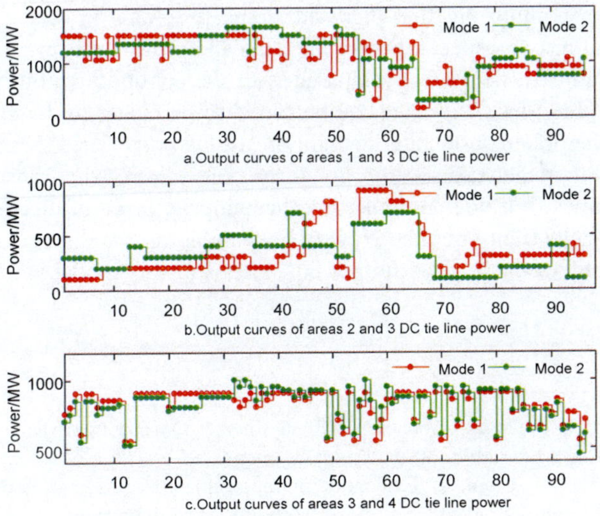

Fig. 6. Output curve of DC tie lines power

It can be seen from Fig. 6 that area three delivered DC power to other areas and area 3 had complementary load characteristics with area 1 and area 2. Starting from the 30th period, the transmission power of the DC tie line was increased to promote the accommodation of renewable energy, and the power regulation of the DC tie lines in the four areas was more flexible. The accommodation proportions of the four areas were increased by 1.21%, 2.37%, 2.94% and 1.45%, which further verified the feasibility of the DC tie line power optimization model proposed in this paper.

4 Conclusion

This paper proposes a stair-like power optimization model for DC tie lines with the goal of improving the capacity of renewable energy accommodation. The model comprehensively considers the constraints of the DC tie lines which including the stair-like power of the DC tie lines, constant daily exchange power and no reverse power regulation in short time. In addition, the related constraints of the conventional and renewable energy generating units in the interconnected systems also are considered in this model. The model improve the flexibility of the DC tie lines, which makes the sending system accommodate more renewable energy power, and supplement the power during the peak load period in the receiving system, improving the benefit of interconnected systems. Based on the practical data of a provincial interconnected power grid in China, the effectiveness of this model in promoting the renewable energy accommodation and improving the power grid economy is verified. The model in this paper can consider multiple DC tie lines of multiple interconnected power grids, which provides an effective reference for improving the accommodation of renewable energy in interconnected systems.

It should be pointed out that there are some problems in this model, which are mainly reflected in two aspects. Firstly, the stair-like power model of DC tie lines has some shortcomings in flexibility compared with the continuous power change model, so the power value of DC tie lines can be found to be closer to the continuous power change under the practical operating conditions of the power grid. Secondly, the model established in this paper takes into account the constraint of the transmission quantity of the DC tie lines, but does not take into account the price of electricity in different segments of the electricity market. How to better adapt to the demand of the electricity market is a problem that can be further considered in this model.

References

1. China Energy Research Association: China Energy Development Report 2018, 1st edn. China Building Materials Industry Press, Beijing (2019)
2. Wenqi, H., Gao, Q., Chen, Y.: Evaluation on renewable energy accommodation ability based on peak load regulation and capacity constraint. In: International Conference on Power System Technology (POWERCON), pp. 905–910. IEEE, Guangzhou (2018)
3. Zhengling, Z.: Research on situation and countermeasures of new energy integration in the 13th five-year plan period and its multi-scenario simulation. Electr. Power **51**(01), 2–9 (2018)
4. Mi, Z., Zhang, J., Liu, J., Zhao, Q., Zhang, F.: Decomposition algorithm of source-grid-load contribution for promoting renewable energy accommodation. Power Gener. Technol. **40** (06), 509–515 (2019)
5. Wang, H., Qin, H., Zhou, C., et al.: Cross-regional day-ahead to intra-day scheduling model considering forecasting uncertainty of renewable energy. Autom. Electr. Power Syst. **43**(19), 60–67 (2019)
6. Wu, D., Wang, J., Li, H., Wang, Z., Zhang, N., Kang, C.: Research on capacity and delivery sequence of interconnected northwestern-southwestern power grid towards renewable energy accommodation. Power Syst. Technol. **42**(07), 2103–2110 (2018)
7. Huang, W., Zhang, N., Yang, J., Wang, Y., Kang, C.: Optimal configuration planning of multi-energy systems considering distributed renewable energy. IEEE Trans. Smart Grid **10** (2), 1452–1464 (2019)
8. Li, M., Chen, G., Dong, C., Liang, Z., Wang, W., Fan, G.: Research on power balance of high proportion renewable energy system. Power Syst. Technol. **43**(11), 3979–3986 (2019)
9. Pei, Z., Wang, X., Dong, C., Xing, Y., Liang, Z.: Analysis of Impact of CHP plant on renewable energy accommodation in Northeast China and thermoelectric decoupling measures. Power Syst. Technol. **41**(06), 1786–1792 (2017)
10. Niu, D., Li, J., Wei, L., Chi, Y.: Study on technical factors analysis and overall evaluation method regarding wind power integration in trans-provincial power grid. Power Syst. Technol. **40**(04), 1087–1093 (2016)
11. Yuan, B., Xu, Z., Liu, J., Zong, J.: Research and application of multi-area new energy integration capability evaluation model. Power Syst. Technol. **52**(12), 132–139 (2019)
12. Zhong, H., Xia, Q., Ding, M., et al.: A new mode of HVDC tie-line operation optimization for maximizing renewable energy accommodation. Autom. Electr. Power Syst. **39**(3), 36–42 (2015)

13. Wang, B., Xia, Y., Xia, Q., et al.: Model and methods of generation and transmission scheduling of inter-regional power grid via HVDC tie-line. Autom. Electr. Power Syst. **40** (3), 8–13 (2016)
14. Han, H., Tu, M., Zhang, H., et al.: Day-ahead generation scheduling method considering adjustable HVDC plan and its analysis. Autom. Electr. Power Syst. **39**(16), 138–143 (2015)
15. Xu, F., Tu, M., Zhang, Y., et al.: Generation scheduling model of inter-regional HVDC generation and transmission system considering conditional sections. Autom. Electr. Power Syst. **43**(17), 176–183 (2019)
16. Dong, C., Liang, Z., Li, X., Wang, Y., Bai, J., Wang, S.: Study on power optimization of the trans-regional UHVDC delivery channels in promoting renewable energy accommodation capacity. Electr. Power **52**(04), 41–50 (2019)
17. Zhang, Z., Wang, W., Wang, Z., et al.: Impact analysis and application of cross-region HVDC delivery mode in renewable energy accommodation. Autom. Electr. Power Syst. **43** (11), 174–180 (2019)
18. Zeng, F., Li, G., Bie, Z., et al.: Decentralized dispatch method for multi-area interconnected power systems considering cross-area tie-line transaction strategy. Autom. Electr. Power Syst. **42**(16), 32–40 (2018)
19. Yang, J., Zhang, N., Wang, Y., et al.: Multi-energy system towards renewable energy accommodation: review and prospect. Autom. Electr. Power Syst. **42**(4), 11–24 (2018)
20. Xu, F., Tu, M., Li, L., et al.: Scheduling model and solution of integrated power generation in power grid for clean energy accommodation. Autom. Electr. Power Syst. **43**(19), 185–193 (2019)
21. Alawode, K.O., Jubril, A.M., Kehinde, L.O., Ogunbona, P.O.: Semidefinite programming solution of economic dispatch problem with non-smooth, non-convex cost functions. Electr. Power Syst. Res. **164**, 178–187 (2018)
22. Sun, X., Li, R.: Integer Programming, 1st edn. Science Press, Beijing (2010)
23. Xue, Y., Liu, D., Zhu, J., Taha, H.A., et al.: Introduction to Operations Research: Elementary, 8th edn. People's Posts and Telecommunications Press, Beijing (2008)
24. Cui, C., Cao, Y., Huang, Y., Li, P., Liu, C., Dai, Y.: Optimization model for starting-stopping curve of thermal power units. e-Sci. Technol. Appl. **6**(03), 79–85 (2015)

An Online Pre-decision Method for Security and Stability of Frequency and Voltage Emergency Regulation in AC/DC Power System

Qinfeng Ma[1], Xiaotong Xu[2](✉), Mingshun Liu[1], Feng Wu[2], Guosong Wang[1], Xiancheng Ren[2], and Yong Zhang[1]

[1] Power Dispatching Control Center of Guizhou Power Grid Co. Ltd., Beijing 550002, Guizhou, China
[2] NARI Group Corporation (State Grid Electric Power Research Institute), Nanjing 211106, China
66105768@qq.com

Abstract. There are risks of frequency and voltage security and stability due to the ongoing replacement of conventional synchronous power plants by intermittent generation and DC transmission. Prompt and effective emergency regulation after faults can prevent blackouts. An online pre-decision method for emergency regulation of frequency and voltage security and stability is proposed to avoid the risks caused by the insufficiency of offline-made accident plan. The decision-making model of frequency and voltage control aiming at the minimum cost of comprehensive control is constructed. The verification schemes are generated by enumerating measures and screened by the control quantity estimation. The pre-decision strategy considering frequency and voltage problems is carried out by iterative search. After the actual fault occurs, effective regulation action is determined according to the fault and the problem. The effectiveness of this method is verified by an example of a real power grid.

Keywords: Security and stability of frequency and voltage · System protection · Offline-made accident plan · Emergency regulation · Pre-decision

1 Introduction

Nowadays, there is a tendency of substituting the conventional synchronous power plants with renewable energy sources and HVDC power transmission. In 2019, the total DC capacity across east China reached 73.4 million kW, accounting for about 46 percent of the light load in east China during the flood season. Due to the high proportion of receiving power outside the zone and a large number of conventional power generators have been replaced, the inertia of the system has reduced and voltage support capacity has decreased. The risk of frequency and voltage limit violation is increasing [1–4]. Due to the large-scale access of renewable energy sources and the development of the power market, the deviation between the actual power grid operation mode and the off-line arrangement operation mode increases [5, 6]. In addition,

Y. Xue et al. (Eds.): PMF 2020, LNEE 718, pp. 374–387, 2021.
https://doi.org/10.1007/978-981-15-9746-6_30

the uncertainty of the actual operation state of controllable equipment such as generator and DC transmission, the accident plan based on off-line operation mode is difficult to ensure the security in actual power grid [7–10]. It is essential to make emergency regulation and optimization decisions based on online operation mode.

At present, the problem of voltage security has been solved through the calculation of online pre-decision and the matching of post-fault security and stability problems [11, 12]. However, frequency security and stability problem is not considered. Since DC power modulation, load cutting and other measures affect the system frequency and voltage at the same time, it is difficult to give pre-decision strategy considering both frequency and voltage problems. Reference [13] proposed a preventive control decision-making method, using the control measures performance index to optimize measures. Through the security problem decomposition coordination, parallel solution and combined-measure search are adopted to meet the multi-security constraints. This method also ignores the influence of active power measures on AC voltage and the influence of reactive power measures on system power flow, leading to deviation in the optimization results. When the fault occurs, the method of real-time decision calculation can provide a correction control strategy for dispatching operators [14]. However, due to the dynamic process of the power grid after failure and the time-consuming calculation of decision making, the correction control strategy cannot guarantee the accuracy and reliability of strategy [15, 16].

In order to make up for the risk of accident plan, and meet the requirements of frequency and voltage online calculation and emergency regulation, an online pre-decision method for emergency regulation of frequency and voltage security and stability is proposed to avoid the risks caused by the insufficiency of offline-made accident plan. The decision-making model of frequency and voltage control aiming at the minimum cost of comprehensive control is constructed. The verification schemes are generated by enumerating measures and screened by the control quantity estimation. The pre-decision strategy considering frequency and voltage problems is carried out by iterative search. After the actual fault occurs, effective regulation action is determined according to the fault and the problem. The effectiveness of this method is verified by an example of a real power grid.

2 Overall Scheme

According to the problem of frequency and voltage security and stability in multi-infeed DC receiving-end power grid, transient security problems are solved by system protection strategy, synchronous condenser, generator excitation control at present. The offline-made accident plan or real-time corrective control is adopted to solve the problem of frequency or voltage after the transient process calming to quasi-steady state. This paper proposes an online pre-decision method for emergency regulation of frequency and voltage security and stability. The overall scheme is shown in Fig. 1. Firstly, multi-source data such as state estimation and meteorological environment is obtained to generate the power grid operation mode and contingency set. The transient simulation is carried out with the on-duty control strategy of system protection and control model of DC reactive voltage. The accurate steady-state operation mode is obtained. On this basis, the effectiveness of the accident plan is evaluated. If the accident plan can solve the frequency or voltage problem effectively, it will be terminated. Otherwise, the online pre-decision of the frequency and voltage is supposed to be calculated to generate an emergency regulation strategy.

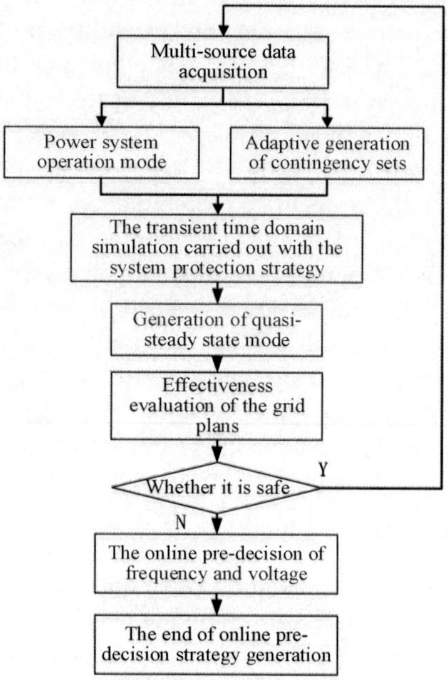

Fig. 1. Scheme of online pre-decision method for emergency regulation

One of the key problems of online pre-decision of frequency and voltage is the accuracy of the quasi-steady state mode after the actual fault. Therefore, it is necessary to identify the on-duty control strategy after the fault, simulate the second and the third line of defense, and generate the accurate power grid operation mode based on online monitoring and modeling of existing system protection. The effectiveness evaluation module of the accident plan obtains the disposal strategy by matching the contingency set, post-fault mode, disposal measures and stability requirements. The effectiveness evaluation module checks whether the power grid is safe after the measures are implemented. If the disposal strategy fails to solve the power grid problems effectively, the online pre-decision of frequency and voltage should be performed. The accident disposal measures suitable for the online operation mode should be generated as an online pre-decision strategy for emergency regulation. If the disposal measures are effective, the next round of multi-source data acquisition will be processed.

3 Identification of System Protection on-Duty Strategy

The emergency regulation system of frequency has put into operation in a real power grid [14, 17–19]. By means of multiple DC coordination control, rapid control of cutting pumped-storage, and load cutting control, the emergency regulation system solves the problem of frequency security and stability in multi-infeed DC receiving-end power grid under high-power disturbance. In order to obtain the accurate post-fault power grid operation mode, it is necessary to identify the on-duty control strategy of system protection based on online monitoring and modeling of system protection.

3.1 Online Monitoring of System Protection

The online monitoring of system protection has access to the control central station, multiple DC coordinated control station, pumping and storage control station, load cutting control station and each substation. Real-time monitoring of commissioning status, abnormal signs, action signals of device, to obtain the controllable quantity of each control device, and lay the foundation for the identification of on-duty control measures of post-fault system protection.

3.2 Online Modeling of System Protection Strategy

Online modeling of system protection strategy is a general description of the system protection strategy, designing a reasonable data structure and expressing the action logic of system protection device under different faults, which is an important part of on-duty strategy identification. Taking frequency control system in the real power grid as an example, the online strategy model includes the following contents accurately. The priority of control measures is DC modulation first, then pumping and storage, and finally fast load cutting. The above control measures are supposed to consider two kind of control modes: remote (feed-forward) control and local (feedback) control. In the remote control mode, the control quantity of each measure is calculated by the control central station according to the unbalanced power caused by the DC transmission fault and measure priority. Each control measure is distributed to the control substations. In the local control mode, each control substation detects the local bus voltage and implements DC modulation, pumping and storage, and load cutting when the bus voltage reaching the local fixed value.

3.3 Identification of On-duty Strategy

Identification of system protection on-duty strategy carries out online identification, analysis and transformation of off-line system protection strategy according to the power grid operation mode and contingency set, which is based on online monitoring of system protection and online modeling of strategy. The strategy is distributed to the control substations and execution stations or executed locally, in order to simulate the actions of system protection correctly. Furthermore, the post-fault power grid operation mode will be obtained, which lays the foundation for the effectiveness evaluation of emergency regulation and the calculation of pre-decision strategy.

4 Pre-decision Methods of Frequency and Voltage

For the situation that security and stability problems still exist after the implementation of system protection strategy and offline-made accident plan, a method of global frequency priority and local voltage second is adopted aiming at the minimum cost of comprehensive control. The online pre-decision for emergency regulation of frequency and voltage is carried out by the strategy of regulating the generator output, acting capacitor and reactor, cutting load and other measures.

4.1 Objective Function

The objective of pre-decision is to minimize the cost of comprehensive control. Formula (1) is the objective function. The main security constraints considered include steady-state frequency overruns, device overload/ cross-section overruns, and steady-state voltage overruns.

$$\begin{aligned} \min\{&\sum_{i=1}^{n_i}(C_{G.i}|\Delta P_{G.i}|) \\ &+\sum_{j=1}^{n_j}(C_{LP.j}|\Delta P_{L.j}|+C_{LQ.j}|\Delta Q_{L.j}|) \\ &+\sum_{k=1}^{n_k}(C_{D.k}|\Delta P_{D.k}|)+\sum_{l=1}^{n_l}(C_{X.l}|\Delta Q_{X.l}|)\} \end{aligned} \tag{1}$$

In the above formula, n_i represents the total number of active controllable generator measures, $C_{G.i}$ represents the cost of per unit active power of the generator i, $\Delta P_{G.i}$ represents the active power variation of the generator i, n_j represents the total number of controllable load measures, $C_{LP.j}$ represents the cost of per unit active power of the load j, $\Delta P_{L.j}$ represents the active power variation of the load j, $C_{LQ.j}$ represents the cost of per unit reactive power of the load j, $\Delta Q_{L.j}$ represents the reactive power variation of the load j, n_k represents the total number of controllable DC measures, $C_{D.k}$ represents the cost of per unit active power of the DC k, $\Delta P_{D.k}$ represents the active power variation of the DC k, n_l represents the total number of controllable capacitor and reactor, $C_{X.l}$ represents the cost of per unit active power of the capacitor and reactor l, $\Delta Q_{X.l}$ represents the reactive power variation of the capacitor and reactor l.

Formula (2) is the constraints of steady-state active power quota of the transmission facilities and the stability quota of the power transmission channel.

$$\begin{aligned} s.t.\ &P_{G.min} \le P_G \le P_{G.max} \\ &P_{L.min} \le P_L \\ &-P_{k.LMT} \le P_k \le P_{k.LMT} \\ &-P'_{m.LMT} \le P_m \le P_{m.LMT} \end{aligned} \tag{2}$$

In the above formula, $P_{G.max}$ and $P_{G.min}$ represent the maximum vector and minimum vector of the controllable range of steady-state active power of the controllable generator during the setting duration Δt_c (usually set to 3 min) after the contingency set, respectively. $P_{L.min}$ represents the minimum vector of the controllable range of steady-state active power of the controllable load during the setting duration Δt_c after the contingency set. $P_{k.LMT}$ represents the steady-state active power quota vector of the transmission facilities. $P_{m.LMT}$ and $P'_{m.LMT}$ represent the positive security and stability quota vector and the reverse security and stability quota vector of the power transmission channel used for security and stability monitoring after the contingency set, respectively.

Formula (3) is the security constraints of reactive power regulation and steady-state voltage.

$$\begin{aligned} s.t. \quad & Q_{X.min} \leq Q_X \leq Q_{X.max} \\ & V_{b.low} \leq V_b \leq V_{b.up} \end{aligned} \tag{3}$$

In the above formula, Q_X, $Q_{X.min}$, $Q_{X.max}$ represent the steady-state reactive power vector, steady-state reactive power minimum vector and steady-state reactive power maximum vector of the reactive power controllable node after the contingency set, respectively. V_b, $V_{b.low}$, $V_{b.up}$ represent the steady-state voltage vector, steady-state voltage minimum vector and steady-state voltage maximum vector after the contingency set, respectively.

Formula (4) is the constraint of steady-state frequency quota.

$$f_{cr.low} \leq f_r \leq f_{cr.up} \tag{4}$$

In the above formula, f_r, $f_{cr.low}$, $f_{cr.up}$ represent the steady-state frequency, lower limit and the upper limit of the steady-state frequency, respectively.

4.2 Measures Control Performance Index Calculation

The calculation methods of measures control performance index of generator active power adjustment of frequency security and stability are as follows:

$$Y_{gfd} = \frac{1}{J_1}\sum_{j_1=1}^{J_1}[(1-\eta_{tfd.j_1})^k y_{gfd.j_1}]/C_g \tag{5}$$

$$Y_{gfr} = \frac{1}{J_2}\sum_{j_2=1}^{J_2}[(1-\eta_{tfr.j_2})^k y_{gfr.j_2}]/C_g \tag{6}$$

In the above formulas, J_1 represents the number of key nodes/generators in W_{tfd}, $\eta_{tfd.j_1}$ represents the frequency drop safety margin of the key node/generator $j1$, J_2 represents the number of key nodes/generators in W_{tfr}, $\eta_{tfr.j_2}$ represents the frequency increase safety margin of the key node/generator $j2$, k is a parameter setting more than

1, $y_{gfd.j_1}$ represents the reciprocal of the electrical distance between the node where the generator is connected and the key node/generator node $j1$, $y_{gfr.j_2}$ represents the reciprocal of the electrical distance between the node where the generator is connected and the key node/generator node $j2$. W_{tfd} represents the frequency drop weakness pattern set, W_{tfr} represents the frequency increase weakness pattern set, C_g represents the cost of per unit active power of the generator.

The calculation methods of measures control performance index of capacitor and reactor and load of voltage security and stability are as follows:

$$X_{vl} = \frac{1}{J_3} \sum_{j_3=1}^{J_3} [(1 - \eta_{vl.j_3})^k s_{Qvl.j_3}] / C_x \tag{7}$$

$$X_{vu} = \frac{1}{J_4} \sum_{j_4=1}^{J_4} [(1 - \eta_{vu.j_4})^k s_{Qvu.j_4}] / C_x \tag{8}$$

In the above formulas, J_3 represents the number of key nodes in W_{vl}, $\eta_{vl.j_3}$ represents the voltage drop margin of the key node $j3$, $s_{Qvl.j_3}$ represents the voltage sensitivity of the capacitor/reactor commission/outage measures of the key node $j3$, J_4 represents the number of key nodes in W_{vu}, $\eta_{vu.j_4}$ represents the voltage rise margin of the key node $j4$, $s_{Qvu.j_4}$ represents the voltage sensitivity of the capacitor/reactor commission/outage measures of the key node $j4$. k is a parameter setting more than 1. W_{vl} represents the lower-limit voltage weakness pattern set, W_{vu} represents the upper-limit voltage weakness pattern set, C_x represents the cost of per unit reactive power of capacitor and reactor.

The calculation methods of frequency and voltage margin are as follows:

When the steady-state frequency is between the upper limit and the lower limit, the calculation method of frequency margin is as follows:

$$\eta_{fre} = \left(1 - \left| \frac{f - \frac{f_{\max} + f_{\min}}{2}}{\frac{f_{\max} - f_{\min}}{2}} \right| \right) \times 100\% \tag{9}$$

In the above formula, $f_{\max}$ represents the upper limit; $f_{\min}$ represents the lower limit; f represents steady-state frequency after fault.

When the steady-state frequency voltage is higher than the upper limit, the calculation method of frequency margin is as follows:

$$\eta_{fre} = \left(1 - \left| \frac{f}{f_{\max}} \right| \right) \times 100\% \tag{10}$$

When the steady-state frequency voltage is below the lower limit, the calculation method of frequency margin is as follows:

$$\eta_{vol} = \left(\left| \frac{f}{f_{\min}} \right| - 1 \right) \times 100\% \tag{11}$$

When the steady-state voltage is between the upper limit and the lower limit, the calculation method of voltage margin is as follows:

$$\eta_{vol} = \left(1 - \left| \frac{V - \frac{V_{\max} + V_{\min}}{2}}{\frac{V_{\max} - V_{\min}}{2}} \right| \right) \times 100\% \tag{12}$$

In the above formula, $V_{\max}$ represents the upper limit; $V_{\min}$ represents the lower limit; V represents actual voltage after fault.

When the steady-state voltage is higher than the upper limit, the calculation method of voltage margin is as follows:

$$\eta_{vol} = \left(1 - \left| \frac{V}{V_{\max}} \right| \right) \times 100\% \tag{13}$$

When the steady-state voltage is below the lower limit, the calculation method of voltage margin is as follows:

$$\eta_{vol} = \left(\left| \frac{V}{V_{\min}} \right| - 1 \right) \times 100\% \tag{14}$$

4.3 Online Strategy Searching Method

Based on the characteristics and influence range of frequency and voltage security and stability problems, the optimization decision is made according to the classification priority principle of frequency and voltage security and stability problems. The searching method of frequency and voltage emergency regulation strategy is shown in Fig. 2. Firstly, the optimization decision is made for the frequency safety problem which has a great influence on the global security and stability, and then the decision is made for the voltage safety problem. In the decision-making process, the optimal measures are searched according to the measures control performance index of frequency and voltage. If all measures in this round fail to meet the security and stability requirements of frequency and voltage, update the post-fault mode and controllable measures space. The search will be iterated until all security and stability constraints are met. After the actual fault occurs, the time effectiveness and reliability of control can be improved by matching the fault and problems and delivering the pre-decision strategy.

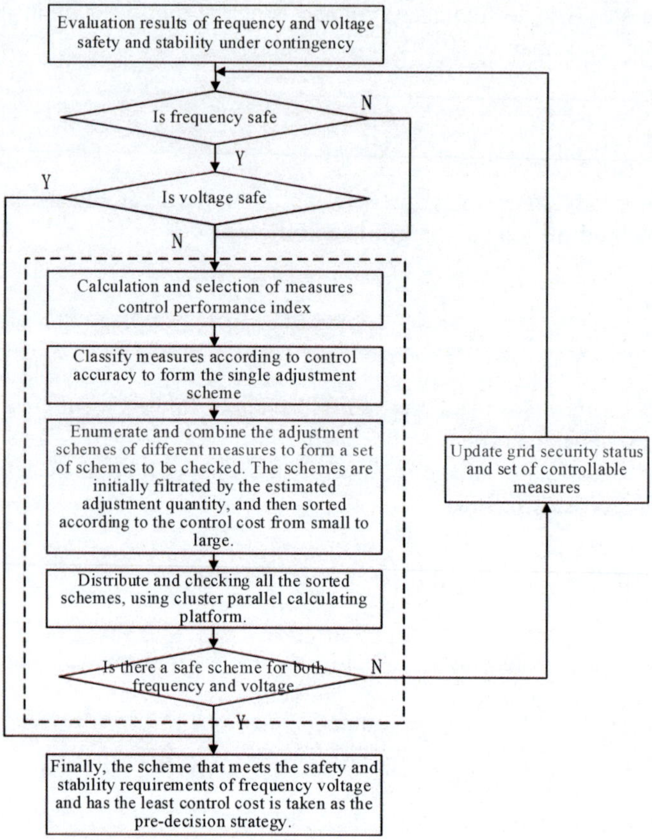

Fig. 2. Online pre-decision method for emergency regulation of frequency and voltage

The calculation speed of online pre-decision must be considered in the large scale AC/DC power system. The heuristic algorithm based on measures control performance index is usually used for searching optimal strategy, and the cluster parallel calculating method is used to improve the calculation efficiency. The strategy searching process shown in the dashed line box of Fig. 2 is explained as follows:

1) Calculate the measures control performance index of generator, load, DC transmission, capacitor and reactor and other measures of frequency and voltage security and stability, using formulas (5)–(8);
2) If there is a frequency security problem, it is necessary to estimate the adjustment amount P_{re} of the frequency safety problem according to the factor. For the case that the estimated adjustment amount is greater than the set threshold value (usually 2000 MW), adjust the N measures ranking first to the upper limit, and only enumerate and combine the M measures that are close to and exceed the estimated amount, so as to reduce the number of verification schemes. N and M are determined by the following methods: the adjustment amount of the first N measures is just greater than $P_{re} - \Delta P$ MW, and the adjustment amount of the first N + M

measures is just greater than $P_{re} + \Delta P$ MW; Otherwise, if there is only voltage security problem, it is necessary to calculate the estimated adjustment amount according to reactive power sensitivity, and screening measures are included in the subsequent grading enumeration and combination;

3) Classify measures according to control precision. For example, if the active power control precision of generator G1 is ΔP_G, the adjustment scheme of G1 is $P_{G,n} = P_0 + n \cdot \Delta P_G$, where, the post-adjustment active power of G1 is $P_{G,n}$, the initial active power is P_0, $n = 0, \pm 1, \pm 2\ldots$, $P_{G,n}$ is between the upper and lower limits. If the active power control precision of DC equipment D1 is ΔP_D, the adjustment scheme of D1 is $P_{D,m} = P_0 + m \cdot \Delta P_D$, where, the post-adjustment active power of D1 is P_D, the initial active power is P_0, $m = 0, \pm 1, \pm 2\ldots$, $P_{D,m}$ is between the upper and lower limits;
4) Enumerate and combine the adjustment schemes of different measures to form the schemes to be checked. Taking the combination of generator and DC equipment as an example, the schemes include ($P_{G,1}$, $P_{D,1}$), ($P_{G,1}$, $P_{D,2}$), ($P_{G,2}$, $P_{D,2}$), etc., which are sorted according to the control cost from small to large;
5) Distribute and check all the sorted schemes in turn using cluster parallel calculating platform. When there is a scheme both frequency and voltage are safe, the calculation will be terminated and the distribution of schemes will be suspended. Otherwise, the search will be iterated according to the frequency or voltage security and stability problems.

5 Simulation and Verification of Examples

Taking the time data of a regional grid on August 18, 2019 as an example, transmission power of BK DC bipolar was 5379 MW and transmission power of FF DC bipolar was 5353 MW under the actual power grid operation mode. There is a contingency set that BK and FF two-loop DC bipolar block faults. According to the real-time operation state detection information of the system protection, all the pumped-storage devices are in the generating state. Simulate the serious situation of the rejection of load cutting device. The DC power increased to 1562 MW urgently. The system steady-state frequency was 49.80 Hz after the fault. According to the method in this paper, the online frequency pre-decision calculation was started firstly, and the voltage over-limit assistant decision was made on this basis. The voltage and frequency limit information is shown in Table 1.

Table 1. Voltage and frequency out of limit

Monitoring components	Element type	Limit value kV/target value Hz	Quasi-steady current flow value
RH 500 kV II bus	bus	(540,510)	508.44 kV
JH 500 kV I bus	bus	(49.95,50.05)	49.80 Hz

According to formula (5), the measures control performance index is calculated. Only the results of the eight units with the higher index are shown in Table 2.

Table 2. Performance index of measure

Number	The unit	Adjustable amount/MW	Unit cost/10,000 yuan/MW	Index
1	TB G1	120.2	0.11	9.09
2	TB G4	119.4	0.11	9.09
3	THP G2	90.1	0.12	8.33
4	THP G3	90.0	0.12	8.33
5	THP G4	90.0	0.12	8.33
6	THP G6	88.9	0.12	8.33
7	FCJ G1	179.6	0.08	7.69
8	FCJ G2	181.2	0.08	7.69

The factor K for frequency is 3800 MW/0.1 Hz, and the estimated adjustment power P_{re} is 5700 MW. Since DC transmission has no adjustable space, enumeration combinations of generator measures are required. According to the method in Sect. 4.3, $N = 37$, $M = 4$, ΔP_G is 50 MW. There are 6 adjustment schemes around the estimated quantity shown in Table 3. The checking results of different control schemes is shown in Fig. 3.

The adjustment quantity of scheme 4 is 6002.8 MW, which is greater than 5990.1 MW of scheme 3. However, since the control cost of the generator is small, the total cost of scheme 4 is less than that of scheme 3. Scheme 4 satisfies the objective function and security constraints of formula (1)–(4). Therefore, scheme 4 was selected as the pre-decision measure of frequency. The system frequency was restored to 49.95 Hz by adding 6002.8 MW output of units of TB, THP and FCJ, etc.

Table 3. Details of different control schemes

Scheme number	Adjust amount/MW	Total cost/10,000 yuan	Frequency safety margin/%
1	5678.6	602.5	−2.71
2	5812.5	631.7	−1.21
3	5990.1	672.1	0.69
4	6002.8	642.7	1.69
5	6090.5	659.9	2.35
6	6240.4	681.4	3.49

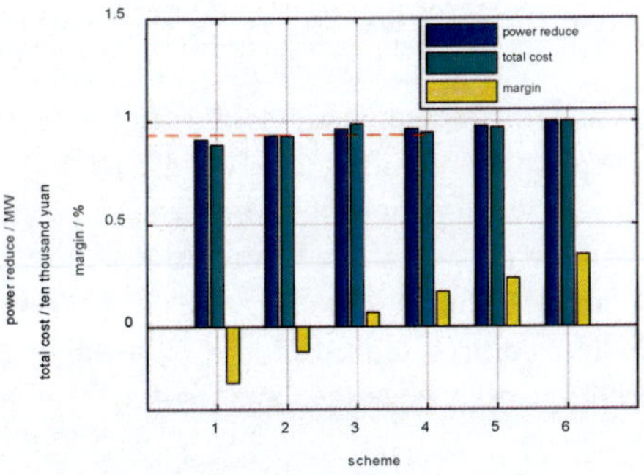

Fig. 3. Checking results of different control schemes

The frequency curve of JH station before and after the implementation of emergency regulation is shown in Fig. 4. The emergency regulation strategy was executed in about 40s after the fault. In actual operation, the strategy could be executed in batches by AGC or ordered by the dispatcher. The unit ramp rate was not considered in this simulation. After the emergency regulation, the system steady-state frequency was restored to 49.95 Hz, which satisfied control requirements.

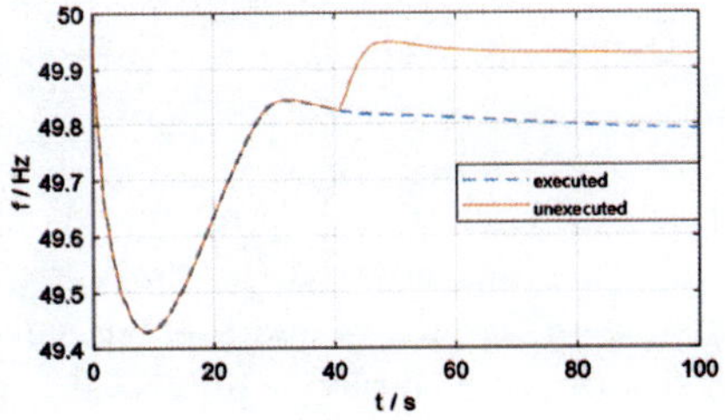

Fig. 4. Frequency curve before and after the implementation of emergency regulation

Voltage decision was made on the basis of frequency decision. After the implementation of frequency strategy, the steady-state voltage of 500kVII bus of RH station was 509.12 kV (the limit value was set as 510 kV). The performance index of measures for voltage was calculated by formula (7), and the results are shown in Table 4.

Table 4. Performance index of measures

Number	Capacitor and reactor	Direction of adjustment	Adjust amount/MW	Index
1	RH #1 reactor	Switch in	+60	98.18
2	RH #2 reactor	Switch in	+50	98.18
3	QS #3 reactor	Switch in	+45	92.56
4	QS #4 capacitor	Switch out	+60	92.56

The reactive power control costs were all set as 0.01 million yuan per Mvar. There were 4 effective measures. There were 16 schemes to be checked after enumerating the combination measures, and the check results of four of them are listed in Table 5. Scheme 3 satisfies the security constraints and has the lowest total cost. Therefore, the outage of two groups of RH reactors and the commission of one group of QS capacitor solved the problem of voltage exceeded the lower limit. A total of 170 Mvar of reactive power was adjusted. After checking, the voltage of 500 kV bus of RH station rose to 512.19 kV, which satisfied control requirements of voltage after the fault. The pre-decision strategy searching of frequency and voltage emergency regulation was completed.

Table 5. Details of different control schemes

Scheme number	Combination	Adjust amount/MW	Total cost/10,000 yuan	Voltage safety margin/%
1	#1/#2/#3	155	1.55	−7.74
2	#1/#3/#4	165	1.65	−1.79
3	#1/#2/#4	170	1.7	1.19
4	#1/#2/#3/#4	215	2.15	21.69

6 Conclusion

The security and stability characteristics of AC/DC hybrid power system are complex. Multiple critical faults that are not covered by the current emergency regulation strategy may occur, resulting in frequency and voltage safety stability problems. In order to make up for the risk of accident plan, and meet the requirements of frequency and voltage online calculation and emergency regulation, an online pre-decision method for emergency regulation of frequency and voltage security and stability is proposed to avoid the risks caused by the insufficient of accident plan. The decision-making model of frequency and voltage aiming at the mini-mum cost of comprehensive control is constructed. The verification schemes are generated by enumerating measures and screened by the control quantity estimation. The pre-decision strategy considering frequency and voltage is carried out by iterative search. After the actual fault occurs, effective regulation action is determined according to the fault and the problem. The effectiveness of this method is verified by an example of a real power grid. This method improves the effectiveness and reliability of the pre-decision strategy and has engineering practicability.

Acknowledgment. The work described in this paper was supported by the Science and Technology Project of China Southern Power Grid Co., Ltd. (066500KK52180014).

References

1. Guoping, C., Mingjie, L., Tao, X.: System protection and its key technologies of UHV AC and DC power grid. Autom. Electr. Power Syst. **42**(22), 1–10 (2018)
2. Guoping, C., Delin, W., Yutao, Q., et al.: Challengers and development prospects of relay protection technology. Autom. Electr. Power Syst. **41**(16), 1–11 (2017)
3. Athari, M.H., Wang, Z.F.: Impacts of wind power uncertainty on grid vulnerability to cascading overload failures. IEEE Trans. Sustain. Energy **9**(1), 128–137 (2018)
4. Zheng, C., Ma, S., Shen, X., et al.: Definition, connotation and form of strong HVDC and weak AC and countermeasures for stable operation of hybrid power grid. Power Syst. Technol. **41**(8), 2491–2498 (2017)
5. Yan Jianfeng, Y., Zhihong, T.F., et al.: Dynamic security assessment and early warning system of power system. Proc. CSEE **28**(34), 87–93 (2008)
6. Cai Bin, W., Sunong, W.S., et al.: Power grid on-line security and stability analysis and forewarning system. Power Syst. Technol. **31**(2), 36–41 (2007)
7. Taishan, X., Jianbing, X., Yanhong, B., et al.: On-line pre-decision based preventive and emergency control system for interconnected power grids. Autom. Electr. Power Syst. **30**(7), 1–4 (2006)
8. Yong, T., Yingtao, W., Fang, T., et al.: Research and development of stability analysis, early-warning and control system for huge power grids. Power Syst. Technol. **36**(7), 1–11 (2012)
9. Voropai, N.I., Kurbatsky, V.G., Tomin, N.V., et al.: Preventive and emergency control of intelligent power systems. In: 2012 3rd IEEE PES Innovative Smart Grid Technologies Europe (ISGT Europe). Berlin, Germany, pp. 1–7. IEEE (2012)
10. Shi, H., Zhang, S.: Optimization modeling of static security online decision support for provincial power grid. Autom. Electr. Power Syst. **39**(20), 98–102 (2015)
11. Yanhong, B., Taishan, X., Hua, Z., et al.: An online pre-decision method for security and stability emergency regulation. Electr. Power **52**(8), 91–97 (2019)
12. Yanhong, B., Changyou, F., Taishan, X., et al.: Online security and stability comprehensive auxiliary decision-making of power system. Autom. Electr. Power Syst. **39**(10), 104–110 (2015)
13. Yongjie, F., Yanhong, B., Wei, X., et al.: A review of on-line calculation methods of preventive control for power system security and stability. Autom. Electr. Power Syst. **39**(11), 199–207 (2015)
14. Yongjie, F., Yanhong, B., Taishan, X., et al.: Coordinated prevention based on manual and automatic emergency control. Autom. Electr. Power Syst. **36**(22), 6–11 (2012)
15. Yao, F., Zhang, B., Zhou, D., et al.: Active power security protection of transmission section and its fast algorithm. Proc. CSEE **26**(13), 31–36 (2006)
16. Qin, X., Bi, T., Yang, Q.: Dynamic state estimator based on WAMS during power system transient process. Proc. CSEE **28**(7), 19–25 (2008)
17. Mingjie, L.: Characteristic analysis and operational control of large-scale hybrid UHV AC/DC power grids. Power Syst. Technol. **40**(4), 985–991 (2016)
18. Zhenya, L.: Innovation of UHVAC transmission technology in China. Power Syst. Technol. **37**(3), 567–574 (2013)
19. Liu, Z., Zhang, Q., Dong, C., et al.: Efficient and security transmission of wind, photovoltaic and thermal power of large-scale energy resource bases through UHVDC projects. Proc. CSEE **34**(16), 2513–2522 (2014)

An Improved Sub-harmonic Analysis Method Based on Negative Frequency Effects

Zhongmin Sun(✉), Xiaoyu Zhang, Pinhao Mo, Chao Zheng, and Qiaogen Gu

NR Electric Co., Ltd., Nanjing 211102, China
sunzm@nrec.com

Abstract. Complex frequency signals are adopted for power harmonic analysis generally, whereas the actual measured signal is always a discrete cosine signal, which contains negative frequency components according to Euler's equation. Taking the negative frequency components into account, an improved multilayer DFT interpolation correction method is proposed based on negative frequency effects. The proposed method is applied to harmonic parameter identification, which can reduce spectral leakage by adding a window function, phase rotation and odd-point interpolation operation. Different frequencies and amplitudes of harmonics are analyzed. The results show that the proposed method has a higher accuracy when the analytical signal is mixed with low-frequency components, and the improved method gets more obvious advantages in sub-harmonic parameter estimation.

Keywords: Harmonic analysis · Negative frequency · Sub-harmonic · Fast fourier transform (FFT)

1 Introduction

The discrete Fourier transform (DFT) and its fast algorithm, the fast Fourier transform (FFT) is the most commonly used analytical tool for the measurement of harmonic parameters considering its clear physical meaning and simple calculation [1].

Due to the difficulty of synchronous sampling of power system signals, the DFT-based harmonic analysis method will cause spectrum leakage in the case of asynchronous sampling [2–4]. In order to reduce the spectrum leakage and improve the measurement accuracy of harmonic parameters, lots of meaningful research work has been done.

The spectrum leakage can be suppressed by adding window functions and the fence effect can be eliminated by interpolation [5–9]. In order to further reduce the influence of spectrum leakage, an odd-frequency interpolation method was proposed in [10], which effectively suppressed the spectrum interference between signals by adjusting the phase of the discrete Fourier transform results. On this basis, the multilayer DFT interpolation correction method is proposed in [11], which is actually an odd-point interpolation correction method based on a cosine combination window. However, the negative frequency component of the signal is not considered in the above methods.

Y. Xue et al. (Eds.): PMF 2020, LNEE 718, pp. 388–397, 2021.
https://doi.org/10.1007/978-981-15-9746-6_31

In harmonic parameter estimation, the signal is usually regarded as a complex frequency signal $Ae^{-j(2\pi ft+\varphi)}$, while in fact the voltage and current signal is a cosine signal as $A\cos(\omega t + \theta)$. According to Euler's identity, the signal function $A\cos(\omega t + \theta)$ can be transformed into a double-band line spectrum with $e^{j\omega t}$ as the core, namely $0.5Ae^{j(\omega t+\theta)} + 0.5Ae^{-j(\omega t+\theta)}$. Thus, the spectral influence is severe between positive and negative components when frequency is close to zero.

The influence of negative frequency on spectrum analysis was studied in [12], which concluded that the negative frequency component not only has practical physical significance, but also must be considered when the signal frequency is low, so as to reduce the analysis error. Some spectrum analysis methods based on DFT, such as FFT phase difference method [13] and full-phase FFT phase difference method [14], effectively improve the accuracy of spectrum analysis by considering the effect of negative frequency [13, 14].

Therefore, in order to further improve the accuracy of harmonic parameter measurement and reduce the interference of negative frequency components, an improved sub-harmonic analysis method based on negative frequency effects was proposed. A unique characteristic, the main lobes are in same direction while the side lobes are in reverse directions after phase rotation, was applied to harmonic and sub-harmonic parameter estimation. Signals with different frequency and amplitude are analyzed subsequently.

2 Influence of Negative Frequency Component

2.1 DFT Analysis

Considering a single discrete cosine signal, and can be expanded by Euler's formula as

$$x_p[k] = \frac{A_p}{2} e^{j(2\pi f_p k\Delta t + \varphi_p)} + \frac{A_p}{2} e^{-j(2\pi f_p k\Delta t + \varphi_p)} \tag{1}$$

where, A_p is the amplitude of the signal, f_p and φ_p are frequency and phase angle respectively, j is a imaginary unit, T_w is the length of the sampling window, $k = 0, 1, \ldots, N - 1$, $N = T_w/\Delta t$, where Δt is sampling period.

Apply discrete time Fourier transform on Eq. (1), the DFT of the discrete signal can be obtained as

$$\begin{aligned} X_p[n] = & \frac{A_p}{2} e^{j(\varphi_p - \pi(n - f_p T_w)\frac{N-1}{N})} \frac{\sin(\pi(n - f_p T_w))}{\sin(\frac{\pi}{N}(n - f_p T_w))} \\ & + \frac{A_p}{2} e^{j(-\varphi_p - \pi(n + f_p T_w)\frac{N-1}{N})} \frac{\sin(\pi(n + f_p T_w))}{\sin(\frac{\pi}{N}(n + f_p T_w))} \end{aligned} \tag{2}$$

where, $n = 0, 1, \ldots, N - 1$.

It can be seen form (2) that, the DFT result of a cosine signal with frequency f_p is relevant to f_pT_w. When f_pT_w is integer, the DFT result is $0.5NA_pe^{j\varphi_p}$ when $n = f_pT_w$, and 0 when $n \neq f_pT_w$, which means the signal is sampled synchronously. However, when

f_pT_w is non-integer, the DFT result has leakage value at each discrete frequency point, and its value is related to the adjacent frequency component and T_w, which is non-synchronous sampling, resulting in spectrum analysis error.

In general, it is difficult to achieve synchronous sampling in practical applications. Spectrum leakage is inevitable, and the negative frequency component will aggravate the effect of leakage and interfere with the original spectrum of other components. Therefore, it is necessary to study how to improve the accuracy of harmonic analysis in the case of asynchronous sampling considering the influence of negative frequency.

2.2 Cosine Window Truncation

To deal with the truncation effect of a signal processed by DFT, specific windows can be used to reduce spectrum leakage. Here the consine window function $w_M[k]$ is adopted, and its time-domain express is as

$$w_M[k] = \sum_{m=0}^{M-1} (-1)^m a_m \cos(\frac{2\pi}{N-1} mk) \tag{3}$$

where, M is the number of terms of the function, a_m is coefficient of each term, and $k = 0, 1, \ldots, N-1$.

According to the frequency domain convolution theorem, time domain multiplication corresponds to frequency domain convolution, and the DFT result after adding window is obtained from Eqs. (2) and (3).

$$\begin{aligned} \bar{X}_p[n] = & \frac{A_p}{4} e^{j(-\pi n\frac{N-1}{N})} e^{j(\varphi_p + \pi f_p T_w \frac{N-1}{N})} \bullet \sum_{m=0}^{M-1} \sum_{i=0}^{1} a_m \frac{\sin[\pi n - \pi(f_p T_w + (-1)^i \frac{N}{N-1} m)]}{\sin[\frac{\pi}{N} n - \frac{\pi}{N}(f_p T_w + (-1)^i \frac{N}{N-1} m)]} \\ & + \frac{A_p}{4} e^{j(-\pi n\frac{N-1}{N})} e^{j(-\varphi_p - \pi f_p T_w \frac{N-1}{N})} \bullet \sum_{m=0}^{M-1} \sum_{i=0}^{1} a_m \frac{\sin[\pi n - \pi(-f_p T_w + (-1)^i \frac{N}{N-1} m)]}{\sin[\frac{\pi}{N} n - \frac{\pi}{N}(- f_p T_w + (-1)^i \frac{N}{N-1} m)]} \end{aligned} \tag{4}$$

The second term of Eq. (4) is the negative component of the signal, which can affect the spectrum analysis.

3 Improved Multilayer DFT Interplation Correction

A multilayer spectral-line-summation method was proposed in [11], which can greatly reduce the spectral interference. After considering the negative frequency component, the signal is weighted in time domain and rotated in frequency domain. The superposition processing as shown in Fig. 1 can effectively reduce the interference of side lobe and negative frequency on the spectral line, so as to achieve the goal of accurate solution.

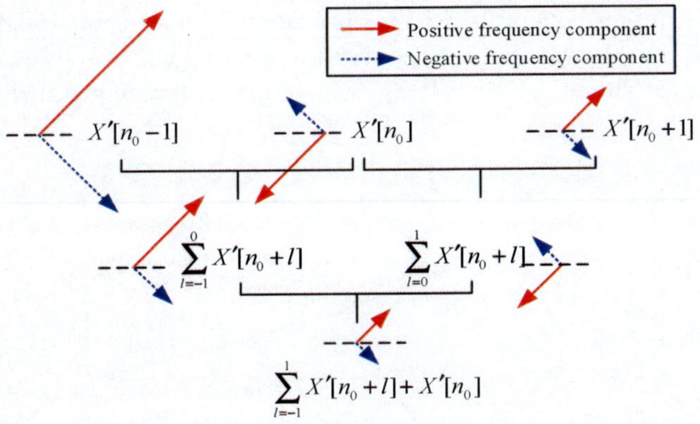

Fig. 1. 3-layer DFT summation calculation after phase rotation adjustment

Figure 1 shows the summation calculation process of 3-layer DFT, where n_0 is the maximum spectral line sequence number in the main lobe corresponding to the concerned frequency component. According to the above characteristics, both the positive and negative frequency sidelobe components of adjacent two DFT values will be greatly reduced after superposition, effectively reducing the influence of sidelobe leakage on spectrum analysis. Of course, the summation of 5-layer and 7-layer can also be adopted, but the number of spectral lines required increases and the algorithm complexity raises. In view of this, the summation calculation of 3-layer DFT is adopted in this paper.

For the signal whose frequency is $f_p = (n_0 - \alpha)/T_w$, the sampling synchronization deviation is $\alpha \in (-0.5, 0.5)$. At the same time, let the concerned component be component p, and the spectral line sequence number corresponding to the maximum spectral line is n_0.

From Fig. 1, according to the spectral lines corresponding to $n_0 - 1$, n_0 and $n_0 + 1$, the summation results of the 3-layer are calculated. According to the second layer, the 3-spectral line optimization ratio r_3 is defined and simplified to

$$r_3 \approx \left| \frac{\sum_{l=0}^{1} \bar{X}'_p[n_0 + l]}{\sum_{l=-1}^{0} \bar{X}'_p[n_0 + l]} \right| = \left| \frac{\sum_{l=0}^{1} w(\alpha, l) + \sum_{l=0}^{1} v(\alpha, l)}{\sum_{l=-1}^{0} w(\alpha, l) + \sum_{l=-1}^{0} v(\alpha, l)} \right| \tag{5}$$

where,

$$w(\alpha, l) = \sum_{m=0}^{M-1} \sum_{i=0}^{1} a_m \frac{\sin[\pi(\alpha + l - (-1)^i m \frac{N}{N-1})]}{\sin[\frac{\pi}{N}(\alpha + l - (-1)^i m \frac{N}{N-1})]} \tag{6}$$

$$v(\alpha, l) = \sum_{m=0}^{M-1} \sum_{i=0}^{1} a_m \frac{\sin[\pi(2n_0 - \alpha + l - (-1)^i m \frac{N}{N-1})]}{\sin[\frac{\pi}{N}(2n_0 - \alpha + l - (-1)^i m \frac{N}{N-1})]} \tag{7}$$

The curve fitting function is used for polynomial fitting approximation, and α can be calculated from the value of r_3. The result of a fitting curve is shown in Fig. 2. The average error of fitting is 2.5084e−9, which meets the requirement of algorithm accuracy.

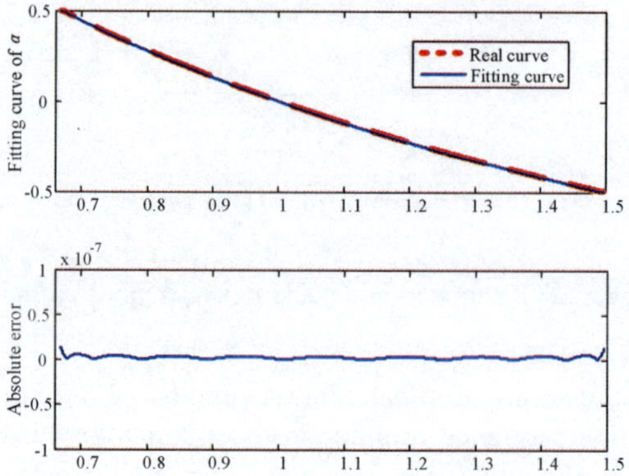

Fig. 2. Fitting curve and absolute error of α

The actual frequency of the concerned component is obtained by α

$$f_p = \frac{n_0 - \alpha}{T_w} \tag{8}$$

The last layer of the 3-layer DFT summation calculation is

$$\begin{aligned}\sum_{l=-1}^{1} \bar{X}'[n_0 + l] + \bar{X}'[n_0] \approx \frac{A_p}{4}\{\mathrm{e}^{\mathrm{j}(\varphi_p + \pi f_p T_w \frac{N-1}{N})}[\sum_{l=-1}^{1} w(\alpha, l) + w(\alpha)] \\ + \mathrm{e}^{-\mathrm{j}(\varphi_p + \pi f_p T_w \frac{N-1}{N})}[\sum_{l=-1}^{1} v(\alpha, l) + v(\alpha)]\}\end{aligned} \tag{9}$$

The calculation of amplitude A_p is affected by the frequency deviation α and phase φ_p, so the accuracy is reduced and the calculation is complicated. Therefore, only the positive frequency component is considered here.

$$A_p = 2\frac{|\sum_{l=-1}^{1} \bar{X}'[n_0 + l] + \bar{X}'[n_0]|}{|\sum_{l=-1}^{1} w(\alpha, l) + w(\alpha)|} \tag{10}$$

From Eq. (9), the expression of phase can be deduced as

$$\varphi_p = \arctan\left(\frac{X}{R}\cdot\frac{[\sum_{l=-1}^{1} w(\alpha,l) + w(\alpha) + \sum_{l=-1}^{1} v(\alpha,l) + v(\alpha)]}{[\sum_{l=-1}^{1} w(\alpha,l) + w(\alpha) - \sum_{l=-1}^{1} v(\alpha,l) - v(\alpha)]}\right) - \frac{N}{N-1}\pi(n_0 - \alpha) \quad (11)$$

where, $R = \mathrm{real}(\sum_{l=-1}^{1} \bar{X}'[n_0 + l] + \bar{X}'[n_0])$, $X = \mathrm{imag}(\sum_{l=-1}^{1} \bar{X}'[n_0 + l] + \bar{X}'[n_0])$.

4 Simulation

4.1 Harmonic Analysis

In order to verify the effectiveness and accuracy of the algorithm, the discrete signal given in Eq. (12) is simulated.

$$x[k] = \sum_{p=1}^{P} A_p \cos\left(\frac{2\pi f_p k}{f_s} + \varphi_p\right) \quad (12)$$

where, A_p, f_p and φ_p are frequency, amplitude and initial phase of each harmonic, f_s is sampling frequency, P is the number of frequency components contained in the signal, and k = 0, 1, …, $N - 1$.

The parameters of each harmonic component are shown in Table 1. The sampling frequency f_s is 1.5 kHz, the fundamental frequency of the signal is 50.1 Hz, and the number of sampling points N is 512.

Table 1. Parameters of each harmonic component

p	Harmonic order	Frequency/Hz	Amplitude/pu	Phase/(°)
1	1	50.1	1.00	340
2	3	150.3	0.02	100
3	5	250.5	0.05	320
4	7	350.7	0.05	50
5	9	450.9	0.02	25

The method (Method 1) in [11] and the method (Method 2) proposed in this paper are used to analyze the above signals. The Blackman window function is taken and the 3-layer DFT interpolation correction is adopted. The specific simulation results are shown in Fig. 3.

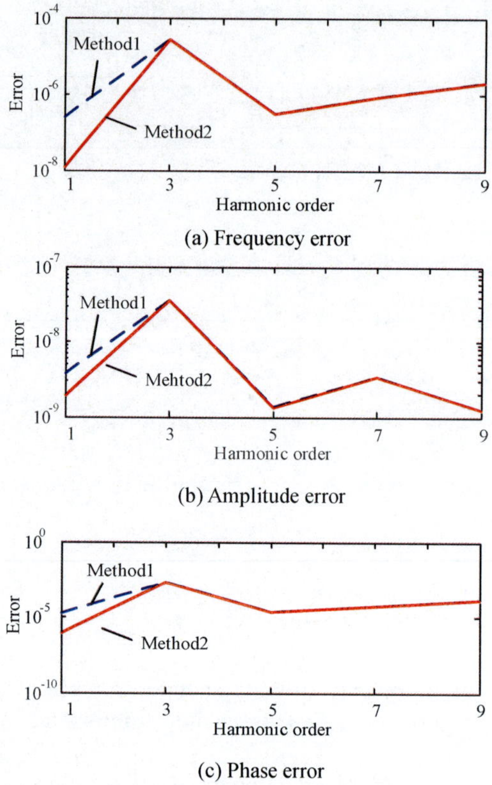

Fig. 3. Comparison of errors in estimating harmonic parameters

It can be seen from Fig. 3 that the calculation error of Method 2 in general better than Method 1, especially the fundamental wave frequency parameter. As for the higher harmonic frequency, Method 2 were similar with Method 1, because the frequency is far from zero, and the negative frequency components of the frequency spectrum leakage has become very small, which has little interference on the precision of parameter calculation.

4.2 Sub-harmonic Analysis

As known from the analysis of the principle, when the harmonic frequency is very low, the negative frequency components can cause great influence on the spectrum analysis. Sub-harmonics with various frequencies are added to Eq. (12), and the amplitude, which gradually increases, is also considered for spectrum analysis. The specific parameters are shown in Table 2.

Table 2. Parameters of each Sub-harmonic component

Sub-harmonic	Frequency/Hz	Amplitude/pu	Phase/(°)
1	8.79	A_p	20
2	8.8	A_p	20
3	17.6	A_p	20
4	26.4	A_p	20
5	35.2	A_p	20

where, A_p = [0.02, 0.05, 0.1, 0.4, 0.7].

The simulation results of absolute errors can be caluculated. Define the Eq. (13) to calculate the difference value of the absolute error of the two methods, as shown in Fig. 4.

$$\Delta E = ||E_1| - |E_2|| \tag{13}$$

where, E_1 and E_2 are absolute errors of Method1 and Method 2, respectively.

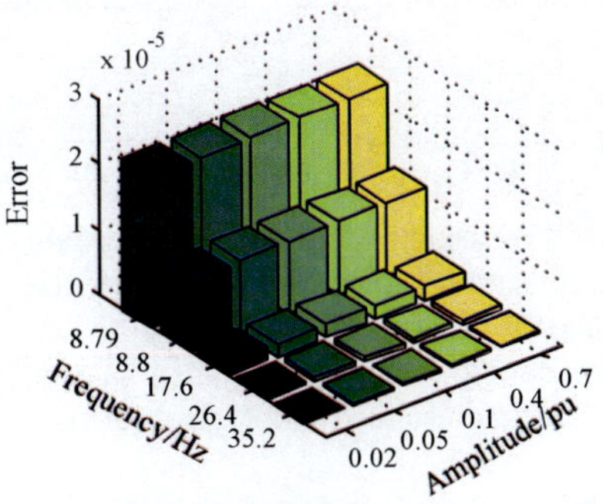

(a) Frequency error

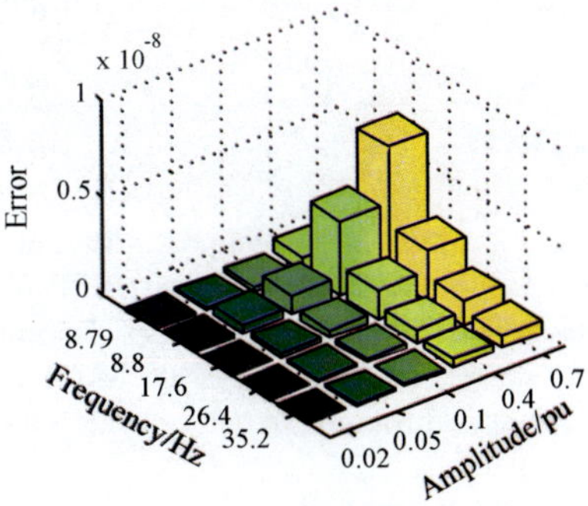

(b) Amplitude error

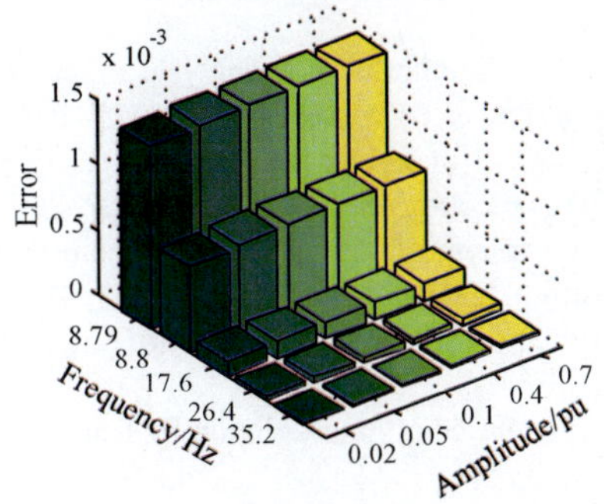

(c) Phase error

Fig. 4. Comparison of the two harmonic measurement methods

As can be seen from Fig. 4, with the reduction of sub-harmonic component, the improved method is better than the original method. This is because when the frequency is small, the influence of negative frequency can no longer be ignored, resulting in relatively serious spectral interference. The correction method considering the negative frequency effect is equivalent to reducing the leakage effect of negative frequency, which can improve the identification accuracy of harmonic parameters efficiently. At the same time, the increase of the amplitude of the sub-harmonic will reduce the error of the harmonic parameters, but it also validates the improvement of the measurement accuracy of the proposed method.

5 Conclusion

The existence of negative frequency components has a certain influence on the measurement of power system harmonic parameters, especially when the signal is with low-frequency components, in which negative part should not be ignored. In order to eliminate the spectral interference, an improved multilayer DFT interpolation method is proposed in this paper. On the basis of adding a window function and interpolation, the negative frequency component interference is reduced. The simulation results show that this method has better measurement accuracy when the concerned frequency is low, and is especially suitable for the analysis of low-frequency sub-harmonics.

References

1. Sun, Z., Chen, Y., Zhang, J.: An improved measurement method for wideband multi-frequency signal. In: 2018 2nd IEEE Conference on Energy Internet and Energy System Integration (EI2), pp. 1–6. IEEE, Beijing (2018)
2. Sun, Z., He, Z., Zang, T., et al.: Multi-interharmonic spectrum separation and measurement under asynchronous sampling condition. IEEE Trans. Instrum. Meas. **65**(8), 1902–1912 (2016)
3. Ming, D., Yuliang, Y., Lin, L., et al.: Synchronized grid-connected parameter measurement based on all-phase discrete Fourier transform with adaptive sampling sequence length. Autom. Electr. Power Syst. **42**(13), 175–181 (2018)
4. Salor, O.: Spectral correction-based method for interharmonics analysis of power signals with fundamental frequency deviation. Electr. Power Syst. Res. **79**(7), 1025–1031 (2009)
5. Zhongmin, S., Zhengyou, H., Tianlei, Z.: A kind of hybrid convolution window and its application in harmonic analysis. Trans. China Electrotech. Soc. **31**(16), 207–214 (2016)
6. Yu, W., Miao, Y., Yonggang, P., et al.: New explanation of frequency domain methods for harmonic analysis and its generalization. Autom. Electr. Power Syst. **41**(20), 70–77 (2017)
7. Jiangfei, L., Longdao, C.: Improved interpolation algorithm applied to power harmonic analysis. Autom. Electr. Power Syst. **43**(8), 138–144 (2019)
8. Hao, Q., Rongxiang, Z.: Interharmonics analysis based on interpolation FFT algorithm. Proc. CSEE **25**(21), 87–91 (2005)
9. Wang, N., Xiao, X., Zeng, S., et al.: An interharmonic estimation method based on interpolation chirp Z transform. Power Syst. Technol. **31**(18), 43–47 (2007)
10. Hui, J., Yang, H.: An approach for harmonic/inter-harmonic analysis based on the odd point interpolation correction. Proc. CSEE **30**(16), 67–72 (2010)

11. Liu, Y., Hui, J., Yang, H.: Multilayer DFT interpolation correction approach for power system harmonic analysis. Proc. CSEE **32**(25), 182–188 (2012)
12. Mao, Y., Tu, Y., Zhang, H., Xiao, W.: Advances and trends in spectrum analysis methodology with negative frequency contribution. Electr. Meas. Instr. **48**(545), 27–32 (2011)
13. Haitao, Z., Yaqing, T.: A new method for phase difference measurement based on FFT with negative frequency contribution. Acta Metrol. Sinica **29**(2), 168–171 (2008)
14. Tan, S., Ren, Z., Sun, C.: Improvement of phase difference correcting spectrum method based on all-phase FFT. Syst. Eng. Electron. **35**(1), 34–39 (2013)

Applied Research on the Non-intrusive Load Identification Based on Improved Manhattan Distance Matching

Qi Li[1,2](✉), Jinyu Xu[1,2], Jian Xiong[1,2], Yong Zuo[1,2], and Hua Gao[1,2]

[1] NARI Group Corporation (State Grid Electric Power Research Institute), Nanjing 211106, Jiangsu, China
liqi2@sgepri.sgcc.com.cn
[2] NARI Technology Co. Ltd., Nanjing 211106, Jiangsu, China

Abstract. NILM is an important technology to realize power consumption behavior perception. In view of the problem that the existing load characteristic model can not match the multi-state load and the huge amount of calculation of identification algorithm can not be applied to smart meter, this paper proposes a method for building a load multi-state model library and a weighted matching algorithm based on Manhattan distance. The effective identification characteristics of each state of electrical equipment are used as characteristic unit to building the multi-state identification model library. Through the multi-time scale window scanning the total power sequence, the load characteristics in the process of load switching and state change are extracted. Using the weighted algorithm based on the improved Manhattan distance, load identification is carried out by calculating the matching distance with the library. The test results show that the proposed method can identify the load category quickly and accurately.

Keywords: Load identification · Manhattan distance · Multi-state identification model

1 Introduction

With the rapid development of economy, the residential users at the end of smart grid have more and more influence on the peak-valley difference of power grid and the comprehensive energy consumption of the whole society [1]. Power consumption behavior perception is an important link to realize the interaction between power grid and users [2]. Load monitoring is the core technology of power consumption behavior perception.

Load monitoring technology can identify the detailed operation status of each kind of electrical appliances in the house by sampling and analyzing the user's electricity data, and obtain the power consumption of each kind of electrical appliances of the user. There are two forms of load monitoring, i.e. intrusive load monitoring (ILM) and non-intrusive load monitoring (NILM) [3]. Compared with ILM, NILM uses advanced

Y. Xue et al. (Eds.): PMF 2020, LNEE 718, pp. 398–416, 2021.
https://doi.org/10.1007/978-981-15-9746-6_32

algorithm to replace the sensors of every electrical appliance in the ILM system. It only monitors the energy information at the user's main entrance, and decomposes the total load data through advanced algorithm, which greatly reduces the economic cost of the equipment and is easy to install and maintain, but also puts forward higher requirements for the identification accuracy of the algorithm.

In recent years, the research focuses on extracting the load characteristics with strong identification ability from the given electrical characteristics database and measured data to realize load decomposition. In reference [4], according to the different characteristics of voltage-current (V-1) curves of different power loads, the hierarchical clustering method is used to analyze the two-dimensional load characteristics in the database to realize load identification. In reference [5], different points of instantaneous current and power waveforms of similar loads are taken as identification features to realize accurate identification of electrical appliances with similar current waveforms. In reference [6], the time-domain transient current is processed by S-transform to obtain the frequency-domain load characteristics, and the 0–1 multidimensional knapsack algorithm is improved to realize the load monitoring decomposition. In reference [7], the characteristics during the transient process of load are analyzed and processed. After obtaining the characteristics of the power changes at the moment of load start and stop, the hidden Markov algorithm is applied to realize load monitoring and identification. In reference [8], by calculating the closeness of unknown samples to the waveform in the transient characteristic database, the fuzzy matching of household load is carried out, and the maximum membership degree is used as the criterion to identify devices. In reference [9], based on the information of steady-state voltage, fundamental current and harmonic current, the constrained incompatible estimation equations are constructed, and the optimal solution of the equations obtained in each period is taken as the result of load identification. In reference [10], the hidden semi-Markov model is applied to the non-invasive load decomposition algorithm to verify the correctness and rationality of the algorithm. In reference [11, 12], the hidden Markov model and its deformation are used to complete the load identification under low frequency sampling. The operation of high-precision appliances can be accurately identified. However, the method has low tolerance for the number of appliances and low efficiency for non-repetitive equipment, which makes applicability very poor. In reference [13], According to the principle that the load capacitance and inductance of power equipment are constant, the steady-state current before transient is extracted. Combined with the operation habits of residential users, the load state is determined by optimization.

The load identification module will be one of the expansion modules of the new generation of smart meter, so it is necessary to realize the extensive promotion of this technology. At present, most of the identification algorithms to deal with the complex domestic power consumption scenarios rely on machine learning, which is computationally large and complex, and cannot be applied to the smart meter. Moreover, the load feature library is mostly a feature corresponding to an electrical appliance. In the face of the actual operation of some electrical appliances, there will be multiple state transitions, which cannot be identified accurately. In view of the above problem, this paper analyzes the waveforms of common household appliances, decomposes the operation process of appliances into multi-states, and takes the effective identification features of each state of electrical appliances as the feature unit, establishes the multi-state identification model of electrical appliances, and constructs the common household load feature library; Through the multi-time scale window scanning the total power sequence, the transient characteristics and the slow characteristic in the process of load switching and state change are separated and extracted. Based on the load characteristic database, the multi-state characteristic matching of load is completed according to the improved Manhattan distance algorithm.

2 Implementation Principles of NILM

2.1 Implementation Structure

The typical architecture of NILM system mainly includes data sampling, data processing, event detection, feature extraction and load identification, as shown in Fig. 1.

Data sampling is the first step of noninvasive load identification. The higher the sampling frequency is, the richer the transient information can be obtained. The sampling rate of this paper is 6400 Hz, which can collect the instantaneous change of energy in the transient process of household appliances.

Data processing includes noise elimination and code value conversion to provide reliable data for subsequent links.

Through event detection, we can know the change of operation state based on the change of load characteristics in a certain period. The load characteristics are determined by the working conditions of the electrical equipment. The load operation characteristics can be divided into three categories: steady-state, transient and operation mode. The steady-state and transient characteristics depend on the characteristics of the components inside the equipment, and the characteristics of the operation mode are determined by the operation control strategy of the equipment.

Feature extraction is to extract a series of different load operation features for load identification from the data before and after the event.

Load identification is the last step of NILM. Based on the establishment of load characteristics database, the features extracted in the previous step are compared with the features in the database to achieve load identification.

The load characteristic database is the data basis of load identification and matching. It is composed of various load identification models. The load identification and characteristic database form a two-way feedback mechanism. Through the identification results, the optimized load identification model can be updated to further improve the identification accuracy.

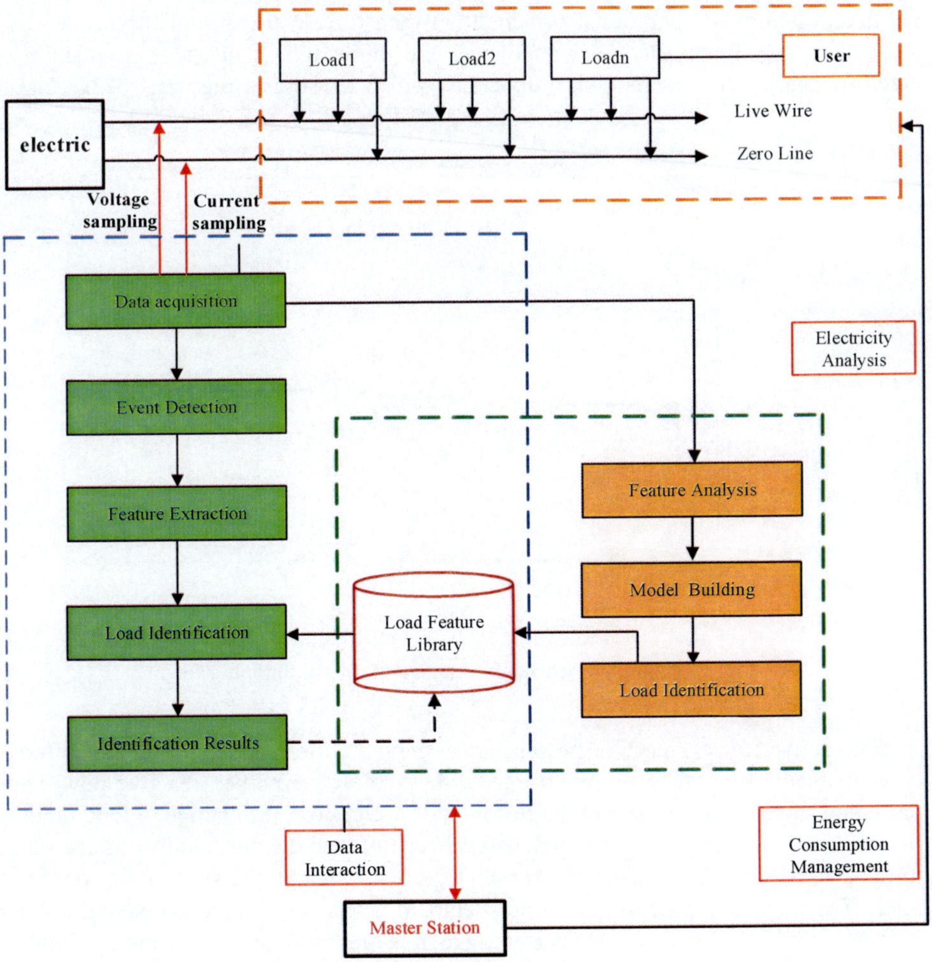

Fig. 1. Implementation structure of non-intrusive load identification

2.2 Load Characteristic

A complete working process of any electrical equipment can be composed of several transition phases (TP) and steady-state phases (SP). TP includes the transition process of startup, shutdown and working state switching, as shown in Fig. 2.

Three kinds of state mutation of TP will cause the change of load operation characteristics, which can be detected in event detection. A certain electric appliance can have multiple steady-state sections, respectively for multiple working states of the corresponding appliance. When the electric appliance operates in the steady-state section, the power fluctuates slightly within a certain range, and the average power basically remains unchanged.

In this paper, the characteristics that can reliably identify the types and working states of electrical appliances are called load characteristic (LC). Due to the different internal structure or operational principle between the electrical appliances, the characteristics of different electrical appliances are more or less different, each kind of electrical appliances has its own unique LC, which constitutes the basis of NILM.

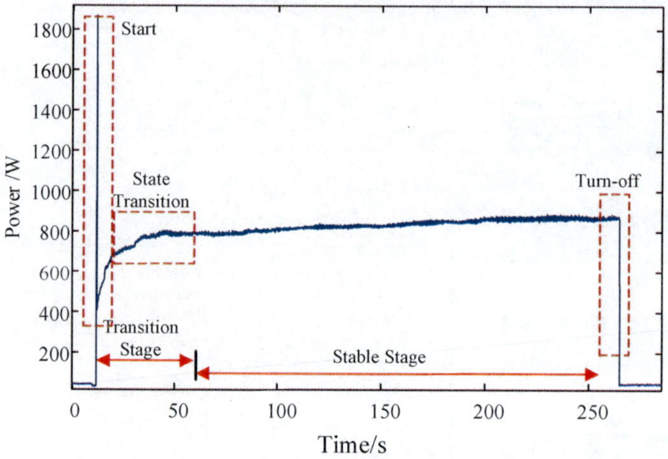

Fig. 2. Complete working sequence of air conditioning load

The active power, reactive power and harmonic current are different for different types of loads. According to the change of the total active power, reactive power and harmonic signal of the system at the power supply entrance, the change of load input or removal in the system can be judged, and the type of load can be determined according to the power. The electromagnetic transient process exists in the start-up of non-linear load. The transient process of some electrical appliances is accompanied by the emergence of 3–9 times impulse current, which is quite obvious. The running duration and mode of household appliances are different. Appliances such as electric heaters also have seasonal running characteristics, which can be used as the load characteristic.

Therefore, load identification is composed of steady power characteristic, steady current harmonic characteristic, transient characteristic and operation characteristic.

2.3 Multi-state Load Identification Model

With the development of science and technology, common household appliances usually have multiple function modes, each function corresponds to a working state, and each working state corresponds to a set of load characteristics. When running in a certain working state, it needs to go through the starting stage, several transition stages and steady-state stages. By statistical analysis of the load characteristics in these three stages, the load identification in the corresponding stage can be obtained. Thus, the multi-state identification model of the electrical appliance can be established. The model structure is shown in Fig. 3.

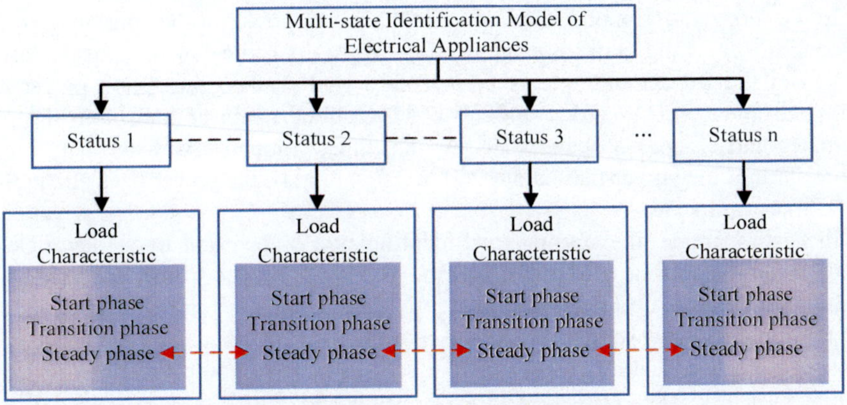

Fig. 3. The structure of multi-state-model

For linear load, because there is no inductive or capacitive component inside the electrical appliance, there is no transient process in the operation process. After starting, the active power rises to the steady-state power step by step, and no harmonic current and reactive power are generated. After starting, the active power rises to the steady-state power step by step, and no harmonic current and reactive power are generated. In Fig. 4, it shows the operation power diagram of electric water heater, so the linear load only has the steady-state load identification.

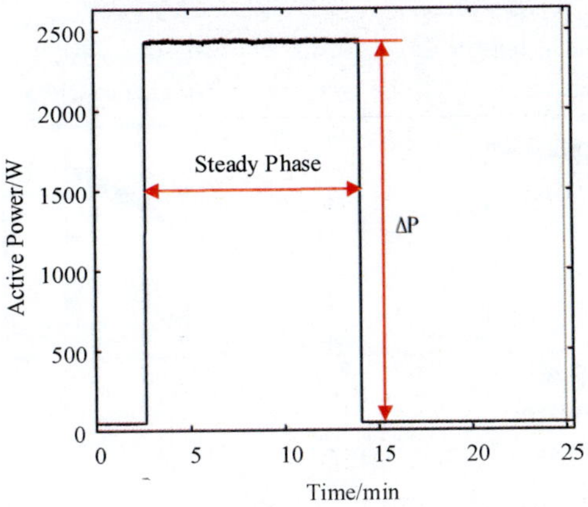

Fig. 4. Operation power diagram of water heater

Nonlinear load usually has many working states, so its identification model is composed of load identification of each operation mode. Take the constant frequency air conditioner as an example, as shown in Fig. 5, the power waveform of the constant frequency air conditioner in the refrigeration mode and heating mode respectively.

In the refrigeration mode, the impact of the active power of the constant frequency air conditioner at the start-up stage is 2000 W(ΔP_{peak}), the impact process lasts for 0.4 s, and the impact multiple is 5($k_P = \Delta P_{peak}/\Delta P_1$), then the active power drops rapidly to about 400 W ($\Delta P_1 = 400$ W). After about 1 min of transition, the power fluctuates and reaches a stable state. At this time, compared with the active power before startup, the power rises about 700 W(ΔP_2), that is, the power rises about 300 W in the transition state.

In heating mode, the starting load identification of constant frequency air conditioner is similar to that of refrigeration mode. However, due to the large indoor and outdoor temperature difference in winter, auxiliary heating is generally started after the air conditioner is first started. When the room temperature reaches the preset temperature, the electric heating stops, and the starting and stopping process of auxiliary heating basically takes on a step type. As shown in Fig. 5(b), the change of active power (ΔP_h) is about 700 W, and then enters the steady state.

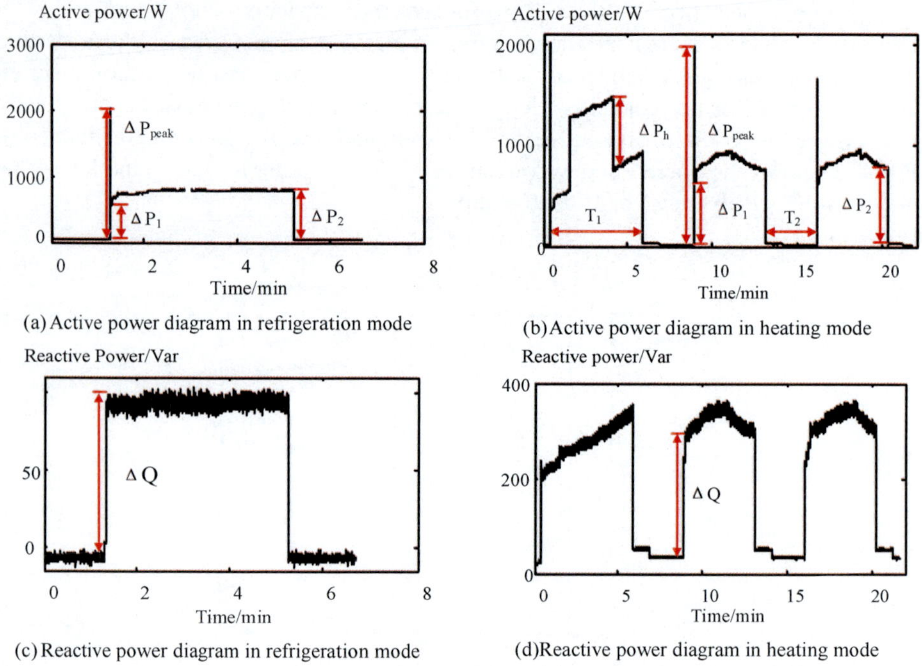

Fig. 5. Operation power diagram of constant frequency air conditioner

The air conditioner will generate 2nd, 3rd and 5th harmonic current during operation. It can be seen from Fig. 6 that the impact of harmonic current will also be generated in the dynamic startup stage. The amplitude of 2nd harmonic current in the steady-state operation stage is significantly larger than other harmonic currents, about 0.6 A, which can be used as one of the load marks.

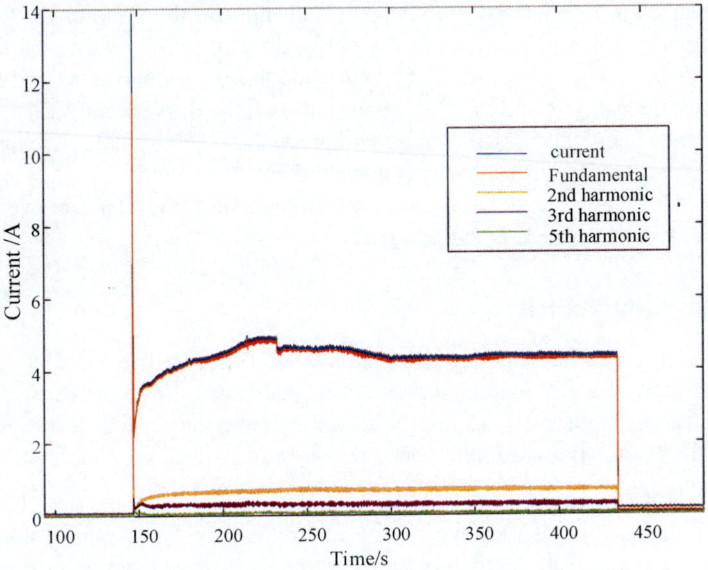

Fig. 6. Harmonic current diagram of constant frequency air conditioner

2.4 Load Feature Extraction

Load feature separation is an important part of load identification. The switching event of a single load usually changes some load characteristics of the sampling sequence at the power supply. The non-parametric CUSUM change point detection algorithm can be used for event detection. The principle of non-parametric CUSUM algorithm is: when the monitored CUSUM is significantly higher than the average level of stable operation, it means that the system has changed. When the power value of the monitoring sequence is higher than the average threshold, it can be judged that the load switching event occurs. The specific algorithm is defined as follows.

$$\begin{cases} h_0^+ = 0 \\ h_t^+ = \max(0, h_{t+1}^+ + p_t - (\alpha_0 + \delta)) \end{cases} \tag{1}$$

$$\begin{cases} h_0^- = 0 \\ h_t^- = \max(0, h_{t-1}^- + p_t - (\alpha_0 - \delta)) \end{cases} \tag{2}$$

It is assumed that the mean value of power sequence $P = \{p(t)\}$ is α, and the noise from outside is δ:

When the power sequence is in steady state, the statistics of power variables (h_t^+, h_t^-)are random variables with 0 as the mean value;

When the sequence changes suddenly, the power increases positively, and the mean value of the power sequence increases to $\alpha_1(\alpha_1 > \alpha_0 + \delta)$, the positive increasing power variable is accumulated to h_t^+, When the absolute value of h_t^+ exceeds the set event threshold θ, it is judged that there is load input in the system. On the contrary, when the power variable increases in a negative direction, the mean value of the power sequence decreases to $\alpha_1(\alpha_1 < \alpha_0 - \delta)$, When the absolute value of h_t^- exceeds the set event threshold value θ, the load in the system is judged to be cut off.

When the change occurs, the power will gradually reach a stable state again. At this time, the average value of the power sequence fluctuates slightly within a certain range. When the statistical value is within the stable threshold σ, the transient event is judged to be over, and the system is stable again.

2.5 Load Identification

Based on the multi-state identification model of common electrical appliances, the characteristic database of load identification is established. Based on the feature library, the optimization model of load matching is established according to the similarity of load identification, so as to realize decompose in real-time.

Manhattan Distance

Manhattan distance is also known as city block distance, as shown in the red line in Fig. 7. In Manhattan with many blocks, the shortest distance from point A to point B is the green route, that is, the Euclidean distance. But in fact, because of the existence of the house, it is impossible to follow this route. It needs to follow the red, blue and yellow routes, that is, the blue and yellow routes are equivalent to the Manhattan distance, so the Manhattan distance is the sum of the absolute values of the wheelbases of the two points on the standard coordinate system.

When Manhattan distance is used to calculate the distance, the smaller component can still contribute the information contained in itself to a certain extent, so as to improve the matching effect.

Manhattan distance is applied to load identification matching, assuming that the load feature library E contains L categories of loads:

$$E = [E_1, E_2, \cdots, E_i, \cdots E_L], i = 1, 2, 3, \cdots, L \tag{3}$$

Each load contains n load characteristics

$$E_i = [e_1, e_2, e_3, \cdots e_n] \tag{4}$$

Suppose D_i is the Manhattan distance between the load to be identified X and the load in the library i.

$$X = [x_1, x_2, x_3, \cdots x_n] \tag{5}$$

$$D_i = f(X, E_i) = \|X - E_i\|_1 = \left(\sum\nolimits_{k=1}^{n} |x_k - e_k|\right) \tag{6}$$

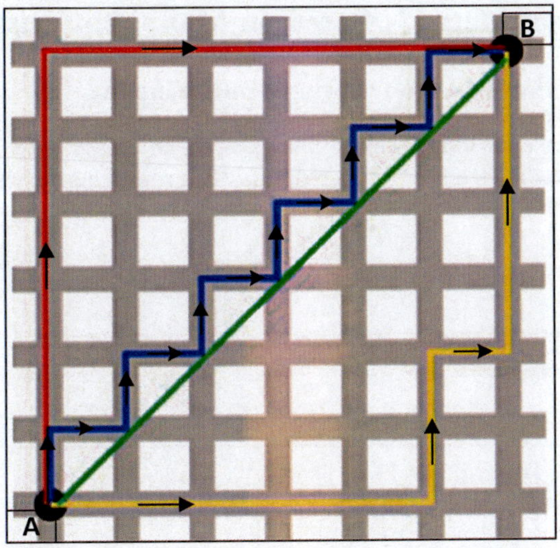

Fig. 7. Schematic diagram of Manhattan distance

Weighted Matching Based on Manhattan Distance
Load characteristic is composed of multi-dimensional load identification features. The dimensions of each feature are different. In order to make the variables of all dimensions play the same role in distance measurement, it must be standardized. Therefore, the Manhattan distance between the load to be identified X and the load in the library i can be improved to:

$$D_i = f(X, E_i) = \|X - E_i\|_1 = (\sum\nolimits_{k=1}^{n} \frac{|x_k - e_k|}{\delta_k}) \quad (7)$$

δ_k is the standard deviation of the k-th load identification of load i in the load feature library, this value can be obtained when the feature library is established.

After standardization, the load characteristics of all dimensions are measured at the same level. However, from a practical point of view, it is necessary to consider that different load characteristics have different impact factors on effective identification. For example, if the active power cannot be matched, other feature quantities are not considered, so weighting coefficient should be assigned to each feature quantity in the load identification. Formula (7) can be modified to:

$$D_i = f(X, E_i) = \|X - E_i\|_1 = (\sum\nolimits_{k=1}^{n} \frac{|x_k - e_k|}{\delta_k}) \times \theta_k \quad (8)$$

θ_k is the weight coefficient of the k-th load identification of load i in the load feature library.

The smaller the value of D_i, the higher the similarity between the load and the load i in the feature database, and the highest degree of matching between the load and the load i in the feature database.

3 Identification Methods Based on Manhattan Distance

3.1 Load Detection Based on Moving Time Window

Load switching and state conversion will cause power changes. Sliding window is used to detect load power at two side change points. The specific process is shown in Fig. 8.

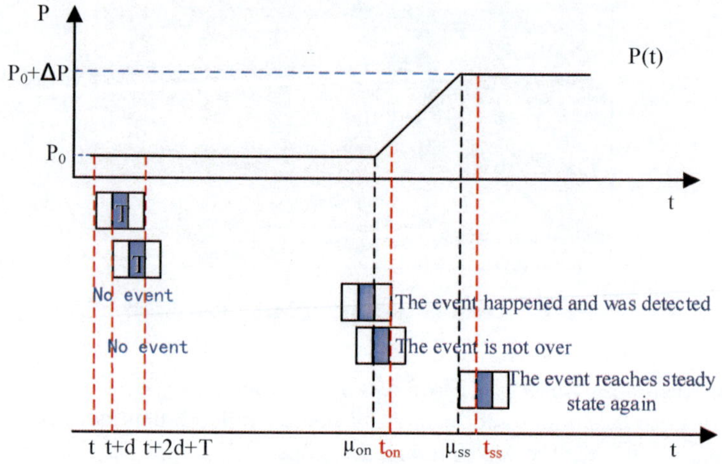

Fig. 8. Process of detection algorithm of bilateral change in sliding window

Assuming that the sliding distance of the sliding window is *d*, the sliding window of a certain time scale is divided into *n* small windows with a width of *d*, and the first and last small windows with a width of *d* are calculated, such as the white lattice of the sliding window in the Fig. 8, and the waiting window is the window with a width of *T*, such as the blue lattice of the sliding window in the Fig. 8, $T = (n - 2)d$.

Compare the difference of power mean value (ΔP) between the head and tail calculation windows:

$$\Delta P = P_{(t,t+d)} - P_{(t+d+T,t+nd)} \quad (9)$$

If $|\Delta P| > \theta$, it is judged that the switching event occurs, and record the time of the rear edge of the tail window as t_{on}. Continue to move the sliding window, ΔP gradually decreases. When $|\Delta P| < \sigma$, it is judged that the event has ended and the system reaches a stable state again, and the time at the back edge of the head window is recorded as t_{ss}.

The transient process of the load is short. The reasonable sliding distance *d* of the sliding window can accurately detect the occurrence and end of each transient event, and the error between the calculated value at the time of the event occurrence and the actual time can also be controlled within a very small error range. For transient load detection, d = 100 ms, n = 2.

The variable-frequency load has a longer time in the transient process, and the power change is slowly fluctuating. Therefore, it is necessary to conduct bilateral change point detection through a long-scale sliding window, and adjust the sliding distance d and the number of split windows n, and increase the length of the window T, so as to know the whole process of load lifting and the amount of active power lifted when it is opened. For the detection of slowly varying load, d = 1 s, n = 10.

If the active power variation of the current two sliding windows exceeds the set threshold, and the trend of load change in the waiting window is consistent with the direction of load change in the window, and then judge the occurrence of slowly changing load. In the process of detecting and extracting slowly changing loads, the influence of transient loads is eliminated by the principle of load superposition. When it is detected that the load change in the sliding window is less than the set stability threshold, it is determined that the change of transient load ends.

3.2 Construction of Load Multi-state Model Library

There are many kinds of household appliances, but the appliances that consume a lot of energy are usually air conditioners and kitchen appliances. In this paper, the multi-state identification model of high-energy appliances is taken as the basic unit of feature library, and the load identification of appliances in different state stages is taken as the basic element of identification model.

In order to ensure that the load characteristic of the feature library is universal for most of the residential users, the main brand electrical appliances in the market are investigated in the early stage, the electric quantity data of the individual operation of electrical appliances are collected for many times, the load characteristics of all kinds of electrical appliances are analyzed statistically, the load characteristic of start state, transition state and steady state are determined, and the multi-state identification model of each electrical appliance is established.

Table 1 is the statistics of load identification of common electrical appliances in various states ('/' means not obvious or not exist). Some electrical appliances have multiple operation modes, the electrical appliances in a certain operation mode are modeled as independent loads. The load characteristics in the start state is the most obvious feature that can be grasped by the appliance at the start time. The load characteristics in the transition state refers to the operating characteristics that will appear within a certain time after the load is started, and has a certain timeliness.

Table 1. Load characteristics of common electrical appliances in various states

Load name	Load state	Load characteristic					
		P/W	Q/Var	Harmonic/A			Feature
				2nd	3rd	5th	
Constant frequency air- conditioning (refrigeration)	Start	2000	200	0.8	0.5	0.2	Intermittent operation
	Transition	350	/	0.3	−0.15	/	
	Steady	750	200	0.6	0.3	/	
Constant frequency air- conditioning (heating)	Start	2000	200	0.8	0.4	0.2	Intermittent operation; Heating only lasts for about 10 min
	Transition	350	150	0.3	−0.15	/	
		700	/	/	/	/	
	Steady	750	350	0.6	0.3	/	
Induction cooker (low-power mode)	Start	1800	220	/	0.65	/	Intermittent operation; Single operation time is only 3–5 s
	Steady	950	150	/	0.28	/	
Induction cooker (high-power mode)	Start	1500–1800	250	/	0.7	/	Continuous operation for a long time
	Transition	± 200	±(25–30)	/	±(0.06–0.1)	/	
	Steady	1500–1800	180–250	/	1	/	
Microwave oven	Start	250	700	/	0.4	0.6	Same interval operation; The duration is 10 s–15 s
	Transition	1000	−400	/	/	/	
	Steady	1300	300–400	/	0.4	0.6	
Electric kettle	Steady	1500–1800	/	/	/	/	Continuous operation for a short time
Electric water heater	Steady	Over 2000	/	/	/	/	Intermittent operation; Single heating time is more than 20 min
Ordinary rice cooker	Steady	500–1300	/	/	/	/	Intermittent operation; The first running time is longer at least 3 min
Electric oven	Steady	1000–1200	/	/	/	/	Same interval operation; Single operation time is about 1 min

3.3 Load Matching Identification

When an event is detected, the extracted load feature change is taken as the input of the load matching algorithm, and the algorithm flow is shown in Fig. 9.

First, determine the direction of the change in load characteristics. If it is a positive increase, it may indicate that there is a new load expenditure. If it is a negative change, it may indicate that the started load may be cut off. However, it is not possible to exclude the possibility that the started load has updated the status, and match by situation.

When the direction of the change in load characteristics is positive increase, the similarity between the positive increase of load characteristics and the dynamic load identification in the load characteristics database is calculated. When the change of load characteristics is positive increase, calculate the similarity between the change and the load characteristic of start-state of each load in the load characteristic library. If it can match successfully, it means that it is indeed the input of a new load, and the operation information of the load is added to the list of started equipment. If it can not match, it may indicate that a certain load condition has been changed, and then the load characteristics of the transition state should be matched.

The load matching of the transition state is carried out in the started device, and the linear load does not have the transition state, so the nonlinear load in the started device is traversed. Index to the transition state load identification of the load in the identification feature library by the information at startup, then match the extracted load characteristics with load characteristics of transition state. If the matching is successful, update the status information of the device in the list of started devices. If it can't be matched, it means that it may be an unknown device, and the operation characteristics can be stored in the unknown device list, which can be transmitted to the master station regularly for big data analysis.

When the direction of the change in load characteristics is negative increase, match the load characteristics with the operation characteristics of each load in the started list, such as steady-state power characteristics, harmonic characteristics, etc. If the match is successful, it means that the load is cut off, change the status of the load in the started list to closed. If the matching is not successful, the load identification of transition state will be matched. If the load identification of transition state fails to match, it will be stored in the unknown equipment list for subsequent analysis and processing.

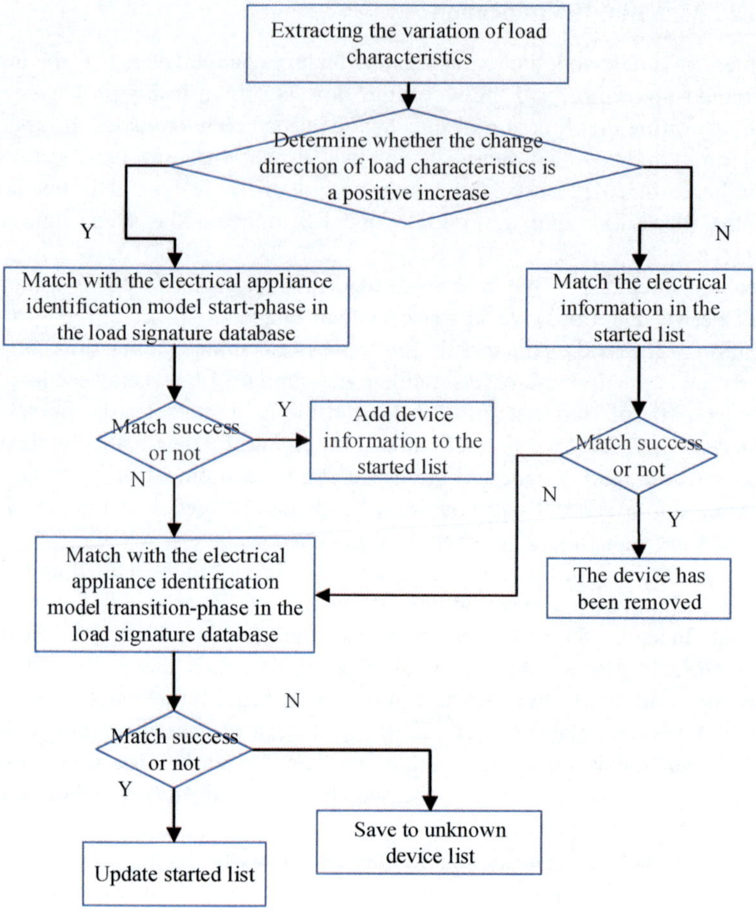

Fig. 9. Load matching algorithm flow

4 Results and Discussion

The load identification algorithm based on Manhattan distance matching needs to verify the accuracy and stability of the algorithm through complex working conditions. There are many types of appliances running in the meal preparation period and the night period in the house, and the appliances are frequently operated. The performance of the algorithm is tested through these two scenarios.

4.1 Example 1

The meal preparation scene is a period of frequent use of various electrical equipment, especially the high energy consumption electric heating equipment and kitchen appliances will be in working state. As shown in Fig. 10, A, B, C, D and E in the figure represent electric kettle, rice cooker, air conditioner, refrigerator and microwave oven.

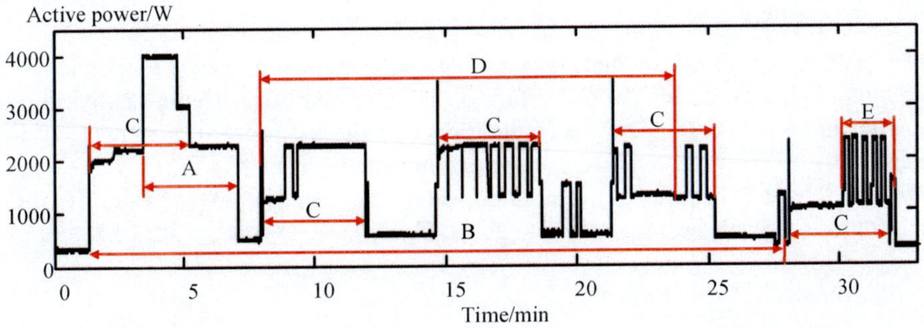

Fig. 10. Active power waveform of family meal preparation scene

The results of load identification are shown in Table 2. The load identification algorithm can identify the switching time of high-power linear load accurately, and the error is basically 0 s. For the judgment of nonlinear load, the delay of switching time is about 1–2 s, and the error rate of partial measurement of electric energy is basically within 5%.

However, the measurement error of microwave oven reaches 9.69%, which is due to the large fluctuation of the operating power of microwave oven, resulting in the large difference between the extracted steady-state power value and the actual power value. In the actual scene, the low-power lights are in the lighting state, but the load characteristics of the low-power electrical appliances are submerged in the changes of the high-power electrical appliances, it is difficult to extract from them, so the algorithm only recognizes the electrical appliances above 300 W at present.

Table 2. Identification results of meal preparation scene

Load name	Load state	Start-stop time (s)			Sub-measure electrical energy (kW·h)		
		Actual	Identify	Error/%	Actual	Identify	Error/%
Electric kettle	Start	255	255	0	0.109272	0.1080444	1.12
	Stop	476	476	0			
Electric cooker	Start	133	134	1	0.420111	0.4130356	1.68
	Stop	1725	1725	0			
Air conditioner	Start	129	130	1	0.048541	0.0478944	1.33
	Stop	362	362	0			
Air conditioner	Start	532	533	1	0.048542	0.047506	2.13
	Stop	765	765	0			
Air conditioner	Start	932	932	0	0.048750	0.047974	1.61
	Stop	1166	1166	0			
Air conditioner	Start	1332	1333	1	0.048750	0.047324	2.93
	Stop	1566	1566	0			
Air conditioner	Start	1732	1732	0	0.048542	0.047636	1.87
	Stop	1965	1965	0			
Microwave oven	Start	1863	1865	2	0.041889	0.037830	9.69
	Stop	1979	1979	0			

4.2 Example 2

The working condition of the night scene is shown in Fig. 11. The electricity consumption behavior in this period is: turn off the air conditioner (A) in the living room, turn on the variable frequency air conditioner in the bedroom (G), start electric rice cooker (b), microwave oven (C) and electromagnetic oven (D) to make breakfast in advance, electric kettle (E) for boiling water, the electric water heater (F) in the bathroom is heated, corresponding to A, B, C, D, E, F, G marked in the figure.

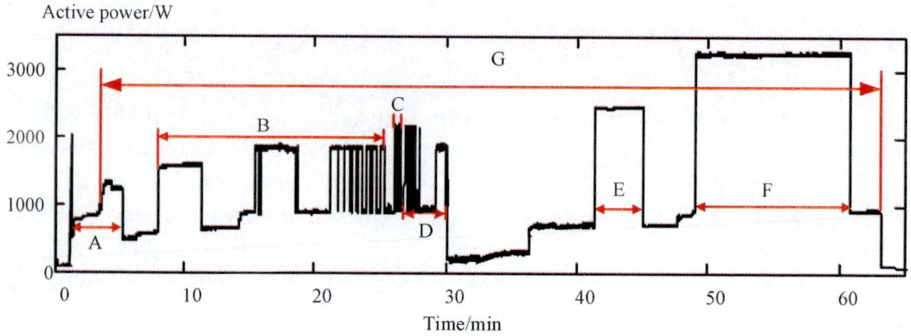

Fig. 11. Active power waveform of night scene

Table 3 is the identification result of night scene. The algorithm can accurately identify the working state of electric heating equipment and kitchen electrical appliances under the condition that the variable frequency air conditioning has been running. After a few minutes of operation, the induction cooker switches to the low gear operation, which causes the power increase before and after the change is too large, resulting in failure to match successfully.

There is a time difference of 20 s between the identified start-up time and the actual start-up time of the variable frequency air conditioner, which is caused by the slow power rise of the variable frequency air conditioner during the start-up. It is more accurate to grab the load at the time of switching, and the error rate of electric energy measurement except the induction cooker is within 5%.

Table 3. Identification results of night scene

Load name	Load state	Start-stop time (s)			Sub-measure electrical energy (kW·h)		
		Actual	Identify	Error/%	Actual	Identify	Error/%
Air conditioner	Start	73	74	1	0.05	0.04921	1.61
	Stop	313	313	0			
Inverter air conditioner	Start	198	218	23	0.850236	0.83023	2.35
	Stop	3799	3799	0			
Electric rice cooker	Start	480	480	0	0.273653	0.26847	1.89
	Stop	1517	1517	0			
Microwave oven	Start	1557	1557	0	0.039325	0.03751	4.63
	Stop	1674	1675	0			
Induction cooker	Start	1747	1751	3	0.257056	/	/
	Stop	2408	/	/			
Electric kettle	Start	2473	2473	0	0.109767	0.10841	1.24
	Stop	2695	2695	0			
Electric water heater	Start	2931	2931	0	0.463472	0.46032	0.68
	Stop	3641	3641	0			

5 Conclusion

The monitoring technology of noninvasive load is a research hotspot at present. The non-invasive load identification algorithm based on Manhattan distance matching proposed in this paper can be seen from the identification results of the experiment, which can effectively extract the characteristics of air conditioning, water heater, hot kettle, rice cooker, microwave oven and other loads, and accurately identify the corresponding start and stop time and power consumption. However, the recognition accuracy of frequency converter and frequently switched appliances is not high enough, which needs further analysis and research to improve the accuracy of the algorithm.

References

1. Li, Y., Wang, B., Li, F.: Outlook and thinking of flexible and interactive utilization of intelligent power. Autom. Electr. Power Syst. **17**, 2–9 (2015)
2. Chen, S., Song, S., Li, L.: Survey on smart grid technology. Power Syst. Technol. **33**(8), 1–7 (2009)
3. Li, P., Yu, Y.: Nonintrusive method for on—line power load decomposition. J. Tianjin Univ. **42**(4), 303–308 (2009)
4. Lam, H., Fung, G., Lee, W.: A novel method to construct taxonomy electrical appliances based on load signaturesof. IEEE Trans. Consum. Electron. **53**(2), 653–660 (2007)
5. Huang, T.D., Wang, W.S., Lian, K.L.: A new power signature for nonintrusive appliance load monitoring. IEEE Trans. Smart Grid **6**(4), 1 (2015)

6. Lin, Y.H., Tsai, M.S.: Development of an improved time-frequency analysis-based nonintrusive load monitor for load demand identification. IEEE Trans. Instrum. Meas. **63** (63), 1470–1483 (2014)
7. Mueller, J.A., Sankara, A., Kimball, J.W., et al.: Hidden Markov models for nonintrusive appliance load monitoring. In: North American Power Symposium, pp. 1–6. IEEE (2014)
8. Gao, Y., Yang, H.: Household load identification based on closeness matching of transient characteristics. Autom. Electr. Power Syst. **37**(9), 54–59 (2013)
9. Yu, Y., Liu, B., Luan, W.: Nonintrusive residential load monitoring and decomposition technology. Southern Power Syst. Technol. **7**(004), 1–5 (2013)
10. Zhang, L.: Non-intrusive house load disaggregation algorithm based on HSMM. Taiyuan University of Science and Technology, Shanxi (2014)
11. Yang, D., Kong, L., Hu, B., et al.: Load identification method based on multi-feature sequence fusion. Autom. Electr. Power Syst. **41**(22), 66–73 (2017)
12. Chen, S., Gao, F., Liu, T., et al.: Load disaggregation method based on factorial hidden Markov model and its sensitivity analysis. Autom. Electr. Power Syst. **40**(21), 128–136 (2016)
13. Han, X., Deng, C., Zhang, Y.: Fast online identification algorithm for delayed feedback load based on V-I characteristics of high frequency data. Autom. Electr. Power Syst. **43**(09), 136–149+223 (2019)

Design of Point-to-Point Power Mutual Transaction Between Residents Based on the Technical Characteristics of Blockchain

Bin Ji[1,2(✉)], Li Chang[1,2], Qia Ding[1], Bin Cao[1,2], Lacheng Pang[1], and Liye Zhu[1,2]

[1] NARI Group Co., Ltd. (State Grid Electric Power Research Institute Co., Ltd.), Nanjing 211100, Jiangsu, China
jibin3@sgepri.sgcc.com.cn
[2] NARI Technology Co., Ltd., Nanjing 211100, Jiangsu, China

Abstract. In order to deepen the reform of electric power market, guide the electric power residents to bring flexible energy resources into the market, and carry out the Point-to-Point(P2P) electric power trade(PEPT)among regional residents. This paper proposes PEPT method based on blockchain and "two-round trade matching" (TRTM). Firstly, based on the blockchain technology framework, a consensus mechanism for Ripple consistency with low computing power requirements is proposed and a P2P operation architecture is designed; Secondly, the TRTM process of PEPT between residents is proposed based on the technical characteristics of blockchain. Purchasers and sellers first declare the energy transaction information for the next period, perform matching and authentication according to the pre-declared power transaction (first round, FRPT) matching mechanism, clear PEPT orders; When there is a deviation in the execution of the transaction, purchasers and sellers are matched according to the deviation power transaction (second round, SRPT) mechanism, and the transaction costs are cleared according to PEPT clearing mechanism; For the problem of power measurement and price between residents, key technology designs, such as the matching mechanism of PEPT, the design of prices with equilibrium prices (EP) and flexible prices (FP). And the clearing method of SRPT are proposed. Finally, this paper verifies the feasibility of the mechanism and model of PEPT between residents through case, and provides a reference for the design of a flexible and open power trading mechanism between distribution network residents.

Keywords: Blockchain technology · PEPT · Market mechanism design · Residents participation

1 Introduction

As the marketization of electricity continues to deepen, residents on the distribution side are included in the power market [1]. In 2017, the competent department of energy issued the notice on carrying out the pilot market-oriented trading of distributed generation, allowing the distributed entities to carry out direct trading of power with

Y. Xue et al. (Eds.): PMF 2020, LNEE 718, pp. 417–432, 2021.
https://doi.org/10.1007/978-981-15-9746-6_33

residents. The direct trading of power for residents has regional characteristics of personalized demand, small scale and large trading volume, and the trading presents the layout of "multi transaction" [2]. However, there is a lack of supporting transaction mechanism and key technology research on PEPT for residents. The research needs to focus on data security, participation threshold, information transparency and others.

Blockchain technology is designed with asymmetric data encryption, consensus mechanism, smart contract, able to implement distributed P2P transactions [3–5]. Blockchain is referring to a traceable data chain which blocks are is serially connected into, according to the time sequence of block formation, and the data cannot be modified through data encryption technology.

At present, some scholars have studied blockchain Technology. Paper [6] introduces the application of blockchain technology in the energy Internet, and explores the power trading methods in the energy Internet; Paper [7] on the basis of the alliance chain, proposed a market credit risk management method for distributed energy trading and distributed power trading system of active distribution network based on blockchain; Paper [8, 9] based on the alliance chain, the method of distributed power transaction in distribution network is proposed, and the model architecture of distributed power transaction system in distribution network is designed; Paper [10] studies the key technologies and architecture model of spot trading based on blockchain technology; Paper [11] discusses the power transaction model based on the blockchain; Paper [12] discusses a flexible price model of distributed active distribution network, which provides the basis of price design for P2P transaction; Paper [13] studies a user oriented power trading mode and method exploration. However, the research on the mechanism and mode of P2P balanced power transaction based on blockchain still needs further study.

This paper studies PEPT mode between residents based on the technical characteristics of blockchain. Under this transaction mode, the decentralized matching mechanism and principle of power transaction, are designed; The ripple consensus verification mechanism, which is suitable for power transaction between residents on the distribution side, is designed; and it simplifies the electric quantity measurement of electric power trade, puts forward the clearing mechanism and value settlement analysis of P2P electric power trade. Finally, the mechanism and model proposed in this paper are analyzed and verified by an example.

2 Blockchain Technology and Ripple Consensus Mechanism

2.1 Technical Characteristics of Blockchain

Blockchain is a chain of data blocks connected by the principle of cryptography. Each data record is stored in the block body in the form of "inverted tree", so the blockchain can be regarded as a distributed ledger database that cannot be tampered with P2P network [14]. In order to solve the problem of distributed data consistency, blockchain technology introduces a consensus mechanism to achieve data consistency authentication. It also introduces the design of smart contract for the personalized needs of various industries, and realizes business logic and business data processing through intelligent algorithm [15]. The P2P transparent transaction application built by blockchain technology has the characteristics of secure data encryption, transparent data information, intelligent mechanism contract, trading environment with consistent rights and responsibilities, personalized trading willingness satisfaction, etc. (Table 1).

Table 1. Blockchain technology features and application scenarios.

Feature	Description
Secure data encryption	Adopt asymmetric hash data encryption method, superior encryption performance, difficult to crack, provide low-cost, high-security data guarantee for PEPT between residents
Transparent data information	Blockchain technology provides decentralized storage of system data copies. Each nodes can obtain system-wide data. Residents can view local data through private keys to provide support for transaction decisions
Smart mechanism contract	Blockchain technology relies on consensus mechanism, and adapts to different application scenarios through smart contracts, providing personalized and intelligent transaction execution mechanisms and constraints for PEPT
Consistent trading environment	Members of the blockchain application perform operations in accordance with the established consensus mechanism and smart contract content. They enjoy consistent power and responsibility for any member of the system, and provide an equal environment for residents
Personalized application design	Blockchain applications change the rules of the application system by modifying the content of smart contracts, and constructing trading rules that meet local needs for residents in different regions

2.2 Ripple Consensus Mechanism

Compared with the PoW, PoS, or DPoS consensus mechanisms which commonly used in blockchain technology, the Ripple consensus mechanism has weaker computing power for each system node. Moreover, the data consistency verification is carried out by the common vote of all members, the validation efficiency of the internal data is almost unchanged [16].

In the Ripple network, the power transaction is initiated by the client, and the power transaction information is broadcast to the network for data through a custom tracking node (TN, light node) or a validating node (VN, heavy node). Among them, TN mainly releases power transaction information request and responds to the account book request of client; In addition to having the function of TN, VN also has the function of adding sample book data as required, that is verifying transaction data.

The Ripple consensus algorithm configures a VN trusted list, the unique node list (UNL), to prevent TN from being unable to participate in power transactions due to their limited computing power. The VN in UNL verifies the data of power transaction according to the established algorithm. The verification process is shown in Fig. 1:

1) Each VN will continuously receive the power transaction data sent from the network. After being verified by the local smart contract algorithm, the illegal transactions are discarded directly, while the legal transaction data will be aggregated into a transaction candidate set (CS). The CS contains transaction information that cannot be determined by consensus verification;
2) Each VN sends its own power transaction CS, as power transaction verification information, to other VNs;
3) After receiving the transaction verification information of other VNs, if the sending node is not in UNL, it will directly ignore the transaction verification information; otherwise, if the transaction verification information is sent by the VN in UNL, it will be compared with the local transaction verification information, the transaction received one vote if the result is the same. In a certain period of time, if a power transaction certification obtains more than 50% of accumulated votes, it will enter the next round. The verification of power transactions that do not exceed a% of the votes will be left to confirm in the next consensus process;
4) The VN sends the power transaction CS with a verification pass rate exceeding a% to other VNs, and sets the corresponding required vote threshold b%, repeating steps 3) and 4) until the verification pass rate exceeds the expected threshold c%;
5) The VN writes the power transaction data confirmed by c% UNL node into the block through an encryption algorithm, and broadcasts to all system nodes for storage.

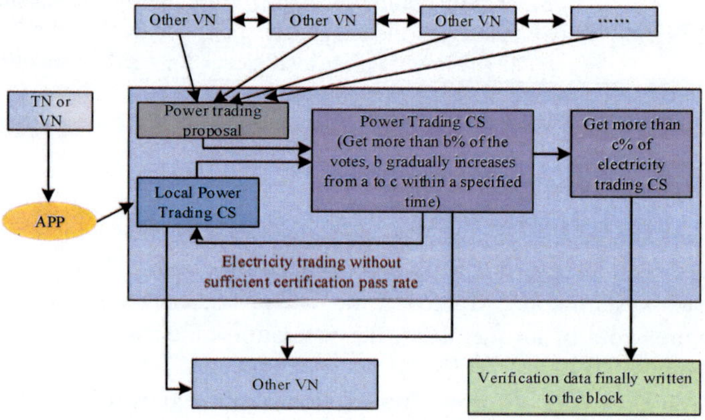

Fig. 1. Ripple consensus mechanism algorithm flow

The Ripple consensus algorithm used by PEPT between residents, classifies nodes combined with actual needs, sets the verification pass rate threshold, and the time limit for each round of verification, so it is more efficient than other consensus mechanisms, and the verification time of general transactions is several seconds, and the requirement of computing power of the node is not high.

3 PEPT Methods

3.1 PEPT Mechanism

Residents proposed in this article include ordinary residents and prosumers who have distributed power generation equipment that integrates generation and consumption. The residents purchase power, and participate in demand-side response, sell their flexible power resources. Prosumers participate in demand side response and distributed generation energy transactions. Considering that residents lack accurate technical support for power generation forecast, they only conduct real-time power transactions. And the transaction cycle is 30 min. Power adopt a continuous bilateral delisting method similar to stock trading. According to the principle of "transaction price first, transaction declaration time first", the buyer will be ranked from high price to low price. If the prices are the same, the earlier the declaration time is, the higher the matching order is; Order of sellers from low to high according to the declared transaction price. In each round of transaction, the buyer's quotation shall be higher than the seller's, and a transaction shall be concluded between the buyer's highest quotation and the seller's lowest quotation, shown in Eq. (1). If the matching of the transaction declaration for a limited time or for a limited number of rounds is unsuccessful, the market subject will purchase or sell power from grid company, the transaction matching is subject to the transaction matching round and time constraint, as shown in Eq. (2).

$$f = (P_{buy,\max}, P_{sale,\min}) \tag{1}$$

$$\begin{aligned} T_i &\le T_{\max} \\ t_i &\le t_{\max} \end{aligned} \tag{2}$$

T represents the number of times; *t* represents the time.

The transaction declaration information of PEPT between residents includes the estimated power consumption and generation, transaction power, price and power response price in the next period. Among them, if the trading volume is positive, it is the selling volume, and if the trading volume is negative, it is the buying volume; The transaction price is the settlement price of the financial contract declared in the transaction, while the deviation electricity response price is the assessment price of the actual power deviation executed in the actual transaction. The quantity and price of the pre-declared transaction is order to be executed, and the quantity and price of the transaction are locked, while the deviation quantity response price is the reward or assessment price of the response quantity obtained in response to its own response or against the market power demand. The deviation quantity response cost is calculated according to the power demand status and actual response quantity of the trading system.

3.2 PEPT Process

The PEPT between residents based on blockchain needs to go through five steps, the flowchart is shown in Fig. 2.

1) According to the efficiency of blockchain technology, the market subjects declare the transaction information 30 min in advance of power transaction, including the information of pre-trading power, trading price and deviation price;
2) The blockchain trading platform performs transaction matching on the declared information according to the consensus mechanism and smart contract content;
3) Determine whether the match is successful, if there is no match, re-declare the transaction information within the specified round and time, until the round and time limit of the matching transaction;
4) Collect measurement data and clear transaction power and transaction prices separately according to the clearance mechanism;
5) Conduct transaction value accounting and transfer.

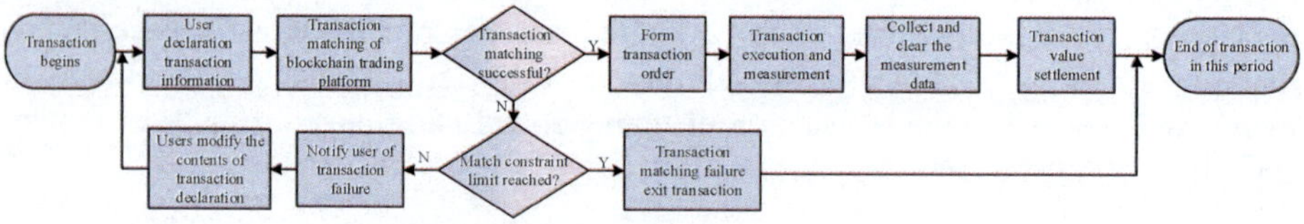

Fig. 2. PEPT flow between users

4 Key Technology Design of PEPT

4.1 Power Transaction Matching Mechanism

PEPT matching principle adopts the basic principle of "price first, time first, limited rounds and time". The purchaser and the seller conduct transaction matching in the way of "many to many" continuous bilateral delisting to the transaction order that must be executed and settled according to the order quantity and price.

The declared price of the purchaser shall be higher than the seller. If not, the declared price of the transaction shall be adjusted within the specified declaration rounds or time limits. Residents who reach the specified declaration rounds or time limits but still do not match successfully are deemed to purchase power directly at the external price. The deviation from actual power consumption adopts the same P2P continuous bilateral delisting matching method. The lowest deviation power response price of the seller matches the highest deviation power response price of the purchaser, but the cost of deviation power is "first power consumption, then transaction matching", so the purchaser's power price can be lower than the seller's, but in order to prevent malicious declaration of deviation power response price transaction mechanism design provisions: When the response price of the purchaser's deviation power is lower than 50% of the seller's, it is deemed that the quotation does not reflect the actual value of the power, so the quotation is invalid. The purchaser's response price of the deviation power is assessed and settled according to the retail price of the external network power, and the seller gets the price subsidy according to the external network power.

PEPT between residents based on blockchain is a power transaction market maintained by both parties. The transaction price needs to reflect the price demands of both parties. In this paper, EP of market purchase and sale is proposed, that is, the average price of the successful quotation of both parties is used as the settlement price, as shown in Formula (3).

$$P_{eq} = \frac{P_{sale} + P_{buy}}{2} \tag{3}$$

4.2 Clearing Mechanism of PEPT

Clearing of Trading Power. The clearing data of P2P multi terminal bilateral power trade between users includes the trading power and trading price. The trading power includes FRPT power, the deviation of the actual power consumption and the network loss allocation power; The trading power corresponding to the trading price is divided into the trading price, the deviation power response price and the network loss price. Thus, the clearing payment fee for residents to participate in the power transaction consists of the product of the above three parts of power and price, as shown in Eq. (4).

$$M = Q_{tra}P_{eq} + Q_{de}P_{de} + Q_{cast}P_{cast} \tag{4}$$

M is the fee for clearing the transaction; Q_{tra} is the declared power quantity of FRPT; Q_{de}, P_{de} are the deviated power quantity and price; Q_{cast}, P_{cast} are the allocated power quantity and price of the network loss; P_{eq} is the actual EP of the power quantity.

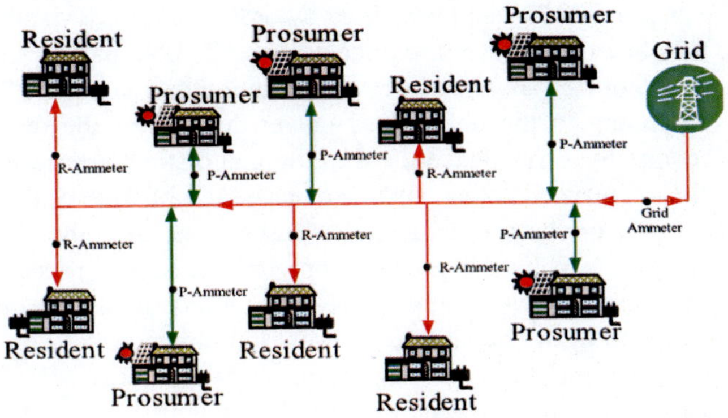

Fig. 3. Schematic diagram of clearing the metering of distribution network power trading

The power transaction between residents shall be executed according to FRPT order, and the deviation power is the difference between FRPT power and the actual power consumption. In Fig. 3, the power consumption measurement diagram of PEPT among residents is shown. The two-way power consumption is measured by the two-way power meter (p-ammeter), the power consumption is measured by the ordinary residents' power meter (r-ammeter), and the power consumption is measured by the grid ammeter of the distribution network merging point. Among them, the difference between the grid ammeter measurement value of the distribution network area and the measurement results of p-ammeter and r-ammeter measurement results of the residents in the area is equally distributed to all residents in the area. And the average value is the network loss that residents need to bear, as shown in formula (5). The deviation of power consumption of ordinary residents is the difference between the declared trading power quantity of FRPT order and the actual measured power quantity, while the deviation power quantity of consumers is the difference between the trading power quantity of FRPT and the actual trading power quantity, as shown in formula (6).

$$Q_{cast,i} = \frac{1}{n}(Q_G + Q_P - Q_R - Q_{R-P}) \tag{5}$$

Q_G is the grid ammeter measurement value; Q_P is the power supply of p-ammeter to the area; Q_R is the actual power consumption of ordinary users measured by r-ammeter; $Q_{R\text{-}P}$is the power consumption of p-ammeter.

$$\begin{cases} Q_{de,R} = Q_{buy} - Q_m \\ Q_{de,P} = Q_{tra} - Q_m \end{cases} \tag{6}$$

$Q_{de,R}$, $Q_{de,P}$ is the deviation power of ordinary users and producers and consumers respectively; Q_{buy}, Q_{tra} is FRPT of ordinary residents and prosumers respectively; Q_m is the metering power of p-ammeter and r-ammeter.

Settlement of Transaction Price. The price of PEPT between residents shall be settled by EP, and the price of FRPT shall be locked by order, and the deviation price shall be settled by the EP of deviation response price declared by both parties. In order to avoid resident's response enthusiasm affected by the deviation response price locked in FRPT order when responding to the unbalanced power demand of the region, the design introduces FP component to balance the deviation, and stimulates residents' response enthusiasm to the unbalanced power, and the matched responding resident shall bear the power deviation cost as the power deviation assessment cost, the FP component is shown in formula (7). When the power supply in the region exceeds the demand, compared with the residents who apply for the transaction order in advance, they are the residents who respond to the unbalanced power, and obtain the corresponding preferential price of FP, the actual deviation power price is the difference between the response deviation EP and FP. When the supply exceeds the demand, it is the opposite, which is shown in Eq. (8). Similarly, the network loss price of PEPT between residents is determined by the relationship between the supply and demand of the regional power. When the power supply is oversupplied, the network loss cost is the average price of the regional deviation price; when the power supply is in short supply, the network loss cost is the retail cost of the external power grid, as shown in formula (9).

$$T_{resp} = \frac{Q_{de} P_{eq,pre}}{\epsilon Q_{eq,pre}} \tag{7}$$

T_{resp} is FP obtained in response to unbalanced power; ϵ is the elastic coefficient, Reflecting the sensitivity of consumers to price, the higher the sensitivity of price, $P_{eq,pre}$ is the actual power consumption and the deviation price respectively.

$$\begin{cases} P_{de} = P_{de,1} + T_{resp}, Q_{need} \leq Q_{tra} \\ P_{de} = P_{de,1} - T_{resp}, Q_{need} > Q_{tra} \end{cases} \tag{8}$$

$P_{de,1}$is the deviation EP; Q_{need} is the actual demand of the region.

$$\begin{cases} P_{cast} = \frac{1}{h} \sum_{k=0}^{h} P_{eq,k}, Q_{need} \leq Q_{tra} \\ P_{cast} = P_{SALE}, Q_{need} > Q_{tra} \end{cases} \tag{9}$$

h is the number of cases of response deviation power in this area; P_{SALE} is the retail price of external network power.

4.3 PEPT Transaction Value Settlement Analysis

Settlement Method. The settlement of PEPT shall be strictly executed and settled in accordance with the order content, and FRPT volume and price shall be strictly executed and settled. The unmatched residents shall trade with the external power grid by default and submit the power purchased by the power seller. The settlement of deviation power transaction matches the deviation trading object according to the established rules. The deviation trading power is the absolute value of the difference between FRPT and the actual measured power consumption. The deviation trading power price is the deviation matching EP, and the power consumers who fail to match have to purchase power from the power sellers. In PEPT, it should be pointed out that the transaction costs are settled strictly according to the transaction orders, and residents who sell and purchase power according to the orders cannot obtain the corresponding power through the deviated power trading market or the power sellers of the external power grid, so there is a risk of transaction right transfer and income transfer in FRPT.

Value Settlement Analysis. The value transfer of PEPT settlement can be divided into three types: 1) creating digital virtual currency of energy within the trading system and establishing virtual currency management fund for regional power transaction. Each resident participating in regional power transaction needs to pay fees, obtain corresponding virtual currency fund shares, and adjust the fund shares owned according to the method of daily and monthly settlement In this way, the price of electric energy can be decoupled from the value of legal currency in a certain period of time, and the value of electric energy in the region can be kept stable, but the exchange ratio of unit fund share and legal currency needs to be adjusted every other period of time to avoid payment disputes when trading with the external power grid; 2) create a virtual value currency of regional electric energy, introduce financial service institutions and connect corresponding settlement applications to obtain the energy/power price of the day, settle the virtual digital currency on a daily basis on the blockchain trading platform and lock the value of the power transaction of the day, and uniformly convert the accumulated power transaction value of the month into legal currency at the end of the month and transfer it, so as to realize the equal value of power and legal currency, avoid frequent modification of the transaction parameters of regional power transaction. Also the system transaction data is limited to external members; 3) do not establish virtual currency among users, directly adopt a more common "daily clearing and monthly settlement" transaction accounting method, transfer the transaction cost through legal currency regularly, which may increase the possibility that other illegal members outside the system will calculate the power trading data between residents according to the transfer amount of power trading, and lead to the result that most of the data of residents are modified and broken, but this method reduces the processing capacity of transaction settlement data.

5 Example

5.1 PEPT Transaction Benefit Model

The PEPT is mainly for the connection between prosumers and ordinary residents, which requires a high proportion of new energy generation from the distribution side. The main benefit index of the power transaction is price. The purchase (sale) price of participating in PEPT needs to be less than (higher than) the retail price (grid connection price) of the grid company, so that residents can participate in the power transaction enthusiastically, otherwise, it will be deemed as direct power transaction with grid company.

In order to verify the advantages of the balanced price, the economic degree of the balanced price is introduced as shown in Eq. (10). And the power consumption satisfaction of residents is evaluated. There are two main factors influence the power consumption satisfaction: 1) the ratio between the actual power consumption and the demand power consumption; 2) the ratio between the actual trading price and the benchmark trading price. The evaluation function of the improvement degree of PEPT satisfaction is constructed in this example. However, in the actual transaction, the comparison object of the customer's PEPT satisfaction is not the price directly traded with the external power grid, but the comparison object of the customer's transaction satisfaction is the transaction price declared by itself, as shown in Eq. (11).

$$\begin{cases} f_{sale} = \frac{P_{eq} - P_{sale}}{P_{sale}} \cdot 100\% \\ f_{buy} = \frac{P_{eq} - P_{buy}}{P_{buy}} \cdot 100\% \end{cases} \quad (10)$$

P_{sale}, P_{buy} are the quotations of the selling purchaser; f_{sale}, f_{buy} are the economic benefits of EP of the selling purchaser.

$$\begin{cases} U = [\alpha U_{com} + \beta U_{eco}] = [\alpha(1 - \frac{\Delta Q}{Q_{need}}) + \beta \frac{P_{p-tra} - P_0}{P_{p-tra}}] \\ P_0' = (Q_{tra} P_{eq}' + Q_{de} P_{de}' + Q_{cast} P_{cast}') \end{cases} \quad (11)$$

U, U_{com}, U_{eco} are respectively represent the PEPT satisfaction, power consumption satisfaction and economic satisfaction; α, β are the user satisfaction bias factor, and $\alpha + \beta = 1$; ΔQ the adjusted power consumption in response to the unbalanced power consumption; ΔQ is the comprehensive power price of the user; P_0 is the cost of direct grid connection. P_{eq}', P_{de}', P_{cast}' are PEPT price, deviation power transaction price and network loss allocation price are respectively included.

5.2 Case Analysis

At present, the domestic Gansu new energy power generation access ratio reaches a maximum of 41%, and its transaction pricing in the ancillary service market is representative. With reference to the Gansu reserve power transaction market subject's declaration of reserve power sale price and retail price, take 0.1 yuan/kWh and 0.6 yuan/kWh respectively, in PEPT, if the prosumers quotes less than or equal to 0.1 yuan/kWh or the price is 0.6 yuan/kWh or more, it means that it does not participate in PEPT. According to residents transaction declaration data in Table 2, the transaction ordering and transaction power graph shown in Fig. 4 can be obtained, where U1 represents resident 1.

Table 2. PEPT declaration and actual execution measurement data (negative sign means power purchase)

Transaction data	Forecast power consumption	Pre declared generation	Pre declared trading capacity	Transaction price	Deviation response price	Actual power consumption	Actual power generation	Actual trading capacity	Deviation trading capacity
U1	10	21	11	0.398	0.4	9	19	10	−1
U2	11	11	0	0.32	0.401	9.8	8	−1.8	−1.8
U3	15	25	10	0.388	0.42	12.7	21	8.3	−1.7
U4	7	0	−7	0.45	0.399	9	0	−9	−2
U5	6	0	−6	0.46	0.45	5.1	0	−5.1	0.9
U6	4	0	−4	0.499	0.42	3.2	0	−3.2	0.8
U7	7	0	−7	0.431	0.4	4.3	0	−4.3	2.7
U8	6	0	−6	0.41	0.41	6.3	0	−6.3	−0.3
U9	9	11	2	0.399	0.399	7	7.6	0.6	−1.4
U10	11	13	2	0.401	0.401	8.21	9.64	1.43	−0.57
U11	12	19	7	0.4	0.419	13.51	15.62	2.11	−4.89
U12	20	27	7	0.4	0.4	15.03	23.87	8.84	1.84
U13	3	0	−3	0.421	0.421	2.97	0	−2.97	0.03
U14	7	0	−7	0.402	0.402	6	0	−6	1
U15	8	0	−8	0.489	0.401	5.3	0	−5.3	2.7
U16	13	14	1	0.42	0.42	11.39	10.69	−0.7	−1.7
U17	15	18	3	0.399	0.401	15.32	16.6	1.28	−1.72
U18	5	0	−5	0.45	0.43	4.59	0	−4.59	0.41
U19	7	0	−7	0.401	0.401	5.31	0	−5.31	1.69
U20	3	0	−3	0.399	0.399	4	0	−4	−1

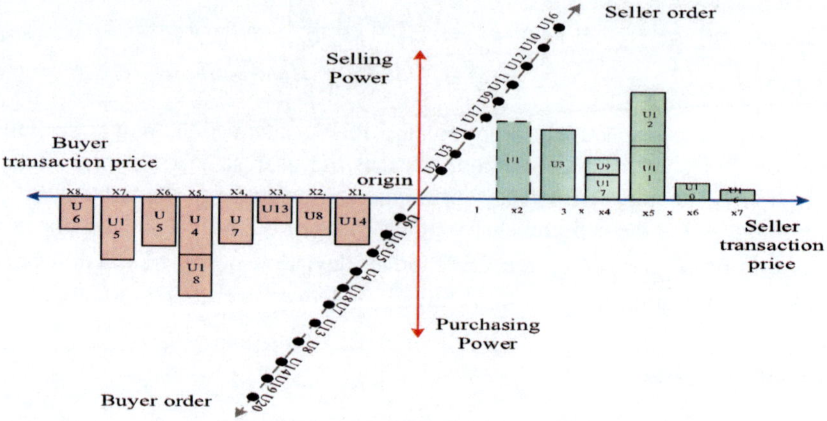

Fig. 4. First round P2P power transaction sequence diagram of the case

According to the transaction declaration data in Table 2 and the transaction mechanisms proposed in this paper, the transaction data analysis chart shown in Figs. 5, 6, 7, 8, 9 and 10.

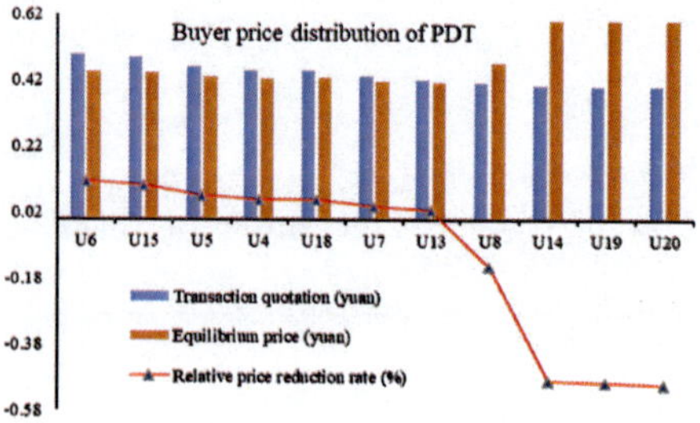

Fig. 5. Price distribution of the first round of purchaser's transaction

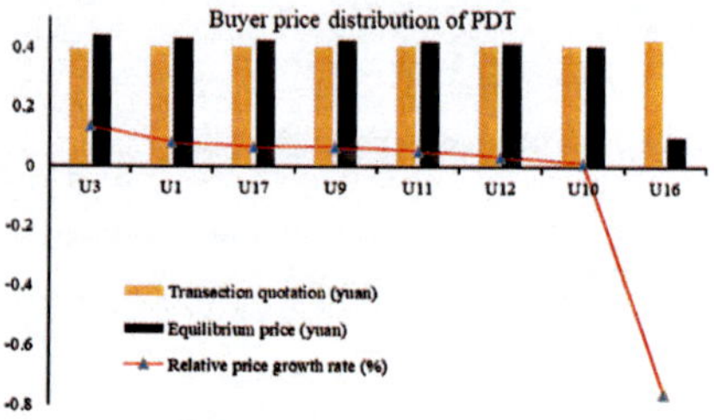

Fig. 6. Price distribution of the first round of seller transactions

From Fig. 5 and Fig. 6, it can be concluded that PDT matching successful EP are greater than the seller's and less than the purchaser's declared transaction prices. Part of the purchaser U8's power demand is obtained through PEPT, and the relative prices of the relative U14, U19 and U20 are slightly lower, but both greater than other purchasers; U16, as a seller, plans to sell the remaining power at the price of 0.1 yuan/kWh, because the offer is too high and does not match the corresponding purchaser in the limited rounds and time, which greatly reduces the revenue. In addition, others have successfully matched the corresponding purchaser and obtained a certain increase in FRPT price.

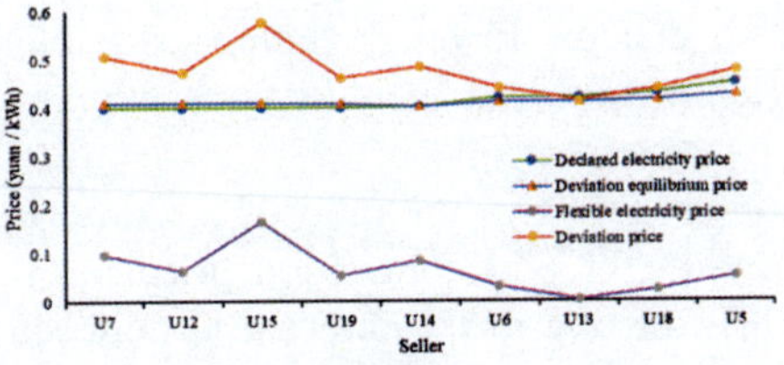

Fig. 7. Price distribution of the second round of purchaser's transaction

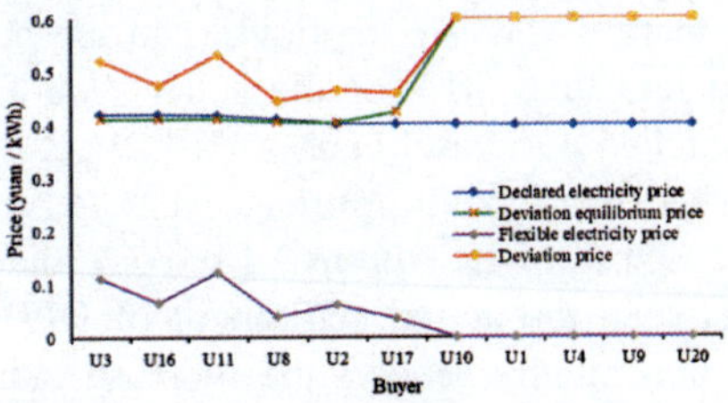

Fig. 8. Price distribution of the second round of seller transactions

Figure 7 and Fig. 8 show the SRT matching data collation between residents. Since SRT between residents only has PDT trading price, and the actual power consumption cannot be fully grasped and pre-set in advance, PDT trading price may be the purchase price or the sale price of SRT of actual deviation power consumption, so the seller's price shown in Fig. 8 will be lower than the purchaser's price, and according to the established rules, the purchaser's price needs to be higher than 50% of the seller's. The deviation price of the actual transaction settlement is the sum of the deviation equilibrium price (DEP) and FP. Through the comparison of Fig. 8 and Fig. 9, it can be concluded that DEP is higher than the declaration and the market subject. And the regional period is the seller's market, and the sellers who respond to the deviation amount of power consumption in the corresponding period in the region obtain higher income, while the purchasers who cause the deviation amount of power are also affected by assessment of high price. In this paper, the elasticity coefficient is taken as 0.5, and all residents participating in the deviation power transaction have FP component.

According to Table 2, the decline and growth rate of the buyer and seller of PEPT relative to the direct grid connection transaction, as shown in Fig. 9, the increase in relative quotation is the ratio of EP of the transaction to the price declared by the seller. From the trend of broken line, it can be seen that residents with low declared price in SRT can obtain a relatively high increase in relative quotation. In particularly, U13 has seen that EP of the transaction is lower than the price of the declared transaction because SRT has been actually implemented. In this case, only the transaction price of

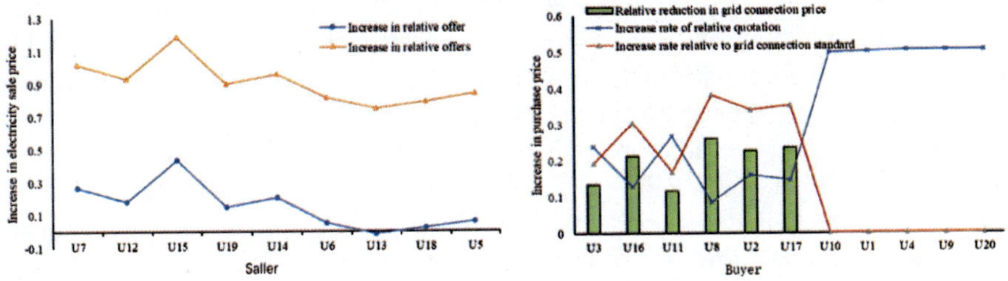

Fig. 9. The decline and growth rate of the trading income of the purchaser and the seller relative to the direct grid connection

deviation power is reported, and the purchaser is allowed to match and rank the price no less than 50% of the seller's quotation, so that the lower deviation price of the matched purchaser causes U13 to get the stretch of FP lower than the transaction price of its own deviation price.

The growth volume of relative bidding object refers to the ratio of the difference between the actual settlement of DEP and the direct grid connection price to DEP, which mainly reflects the increase rate of DEP relative to the direct grid connection transaction price. Figure 9 right, the reduction rate of relative grid connection price refers to the amount of reduction of the price from seller through PEPT compared with the price at which directly traded from the external power grid. Because the regional period is the seller's market, only some of the purchasers successfully match the sellers of PEPT. Therefore, residents whose price is lower than that of U17 fail to match, and they need to purchase power from grid companies at a high price. The increase rate of relative quotation and the increase rate of relative grid connected object are the ratio of the difference between SRPT of DEP and the declared deviation price, respectively, and the ratio of DEP and external retail price. It needs to be pointed out that the emergence of the seller's market is caused by the fact that the power consumption of ordinary residents exceeds the declared plan and the trading power provided by prosumers themselves is less than the declared plan power. Therefore, residents who do not provide the scheduled power according to the plan will spend a higher rate of price purchase deviation power than the expected price.

Fig. 10. Growth rate data of transaction income and relative direct grid connection income

According to the settlement method and formula (10), (11) and (12) in Sect. 3.1, the data of transaction income and growth rate of relative direct grid connection income are shown in Fig. 10. It can be seen from Fig. 10 left that except for U9, U16 and U20, others have obtained certain income. U11 is the prosumer, due to the inaccurate prediction of distributed generation, it is unable to provide sufficient power at SRPT of actual power delivery. Therefore, the prosumer has to purchase retail power with corresponding higher than FRPT of EP from the external power grid to provide it to the purchaser matching the pre transaction, resulting in negative income. U16 failed to match the purchaser in the limited round of quotation in the first round of transaction, resulting in the sale of FRPT price to the external power grid at a very low retail grid connection price. Meanwhile, in SRPT of actual transaction, U16 purchased 1.7 kWh of power at a relatively high price, resulting in a negative return. U20 did not participate in PEPT, the power transaction directly conducted in the external power grid did not generate any increase or decrease in revenue. Correspondingly U19 purchased 7 kWh of power from the external power grid in FRPT, locked the power consumption and price, and the actual power consumption was 5.31 kWh. In the actual transaction process, 1.69 kWh of power that would be locked multiple times was successfully matched with PEPT, and 1.69 kWh*0.141 yuan was obtained, that is because U19 transfered the power locked with 0.6 yuan/kWh in FRPT to the resident with unsuccessful deviation power matching in SRPT, which is equivalent to that U19 purchased power at the cost of 0.459 yuan/kWh and selled it at the price of 0.6 yuan/kWh, and the transaction of financial power sale right. The rest of residents got income from participating in PEPT. The main sources of income are as follows: 1) EP of PEPT among residents is lower than the purchase price of external power grid, higher than the sale price of external power grid; 2) responding to the regional period power imbalance, properly declare the transaction price of deviated power, so as to increase the transaction income or reduce the cost of deviated power consumption.

Since residents use power according to their own needs, when it is 0, it is also expressed as the satisfaction of participating in PEPT. Corresponding to Fig. 10 left, U9, U16 and U20 participate in the power transaction strategy errors, resulting in poor economy (satisfaction).

6 Summary

In this paper, a PEPT scheme based on blockchain technology is proposed for the power transaction. The consensus mechanism of Ripple consistency with low computing power requirements is introduced, and lock the power and electricity price of the first round of transaction through the design of "two round transaction matching" structure. And the second round of deviation electricity price and supporting mechanism will guide the seller (buyer) to report relatively low (high) price and use electricity according to the declared trading plan, so as to improve the income of participating in market transactions. It can be concluded from the example that the mechanism proposed in this paper can improve the satisfaction of transactions. The second round of actual deviation electricity response can transfer the first round of pre-declared trading plan deviation, realize arbitrage through the transfer of power sale right, and will not have an impact on the total electricity trading in local region. The trading income of residents without reasonable planning is negative, so PEPT mechanism designed in this paper has certain market flexibility, guiding the trading game and reasonable planning of electricity consumption.

References

1. Qin, J., Sun, W., Zhu, Y., et al.: Power trading method based on block chain in microgrid. Power Autom. Equip. 1–8
2. Meng, S., Sun, W., Han, D., et al.: Design and implementation of distributed power trading mechanism supporting spot market. Power Syst. Protect. Control **48**(07), 151–158 (2020)
3. Cai, Y., Gu, Y., Luo, G., Zhang, X., Chen, Q.: Green certificate trading platform based on blockchain: concept and practice. Autom. Electr. Power Syst. 1–12
4. Wang, S., Guo, C., Feng, B., Zhang, H., Du, Z.: Application of blockchain technology in power system: prospects and ideas. Autom. Electr. Power syst. **44**(11), 10–24 (2020)
5. Nakamoto S.: Bitcoin: a peer-to-peer electronic cash system. Consulted (2009)
6. Wu, L., Men, K., Xu, S., et al.: Democratic centralism: a hybrid blockchain architecture and its applications in energy internet. In: 2017 IEEE International Conference on Energy Internet (ICE). IEEE (2017)
7. L,i Z., Kang, J., Yu, R., et al.: Consortium blockchain for secure energy trading in industrial internet of things. IEEE Trans. Ind. Inf. 1 (2017)
8. Luo, F., Dong, Z.Y., Liang, G., et al.: A distributed power trading system in active distribution networks based on multi-agent coalition and blockchain. IEEE Trans. Power Syst. 1 (2018)
9. Ping, J., Yan, Z., Chen, S., et al.: Credit risk management method for distributed energy trading market based on blockchain. In: Proceedings of the CSEE, pp. 1–10 (in Chinese)
10. Chen, Q., Wang, K., Chen, S., et al.: A tradable energy system for distributed entities: architecture, mechanism design and key technologies. Power Syst. Autom. **42**(3), 1–7 (2018). (in Chinese)
11. Gong, G., Wang, H., Zhang, T., et al.: Research on power spot trading market based onblockchain. In: Proceedings of the CSEE, vol. 38, no. 23, pp. 6955–6966+7129 (2018). (in Chinese)
12. Ji, B., Mo, J., Tan, J.: Design of power demand response mechanism for high proportion of photovoltaic prosumer. Power Syst. Technol. 1–9 (in Chinese)
13. Ji, B., Tan, J.: Preliminary research of distributed micro-energy trading on power distribution side using blockchain technique. Mod. Electr. Power **1**, 29–36 (2019)
14. Wang, B., Li, Y., Zhao, S., Chen, H., Jin, Y., Ding, Y.: Key technologies of distributed energy trading based on blockchain. Autom. Electr. Power syst. **43**(14), 53–64 (2019)
15. Yang, X., Zhang, Y., Lu, J, et al.: Blockchain-based automated demand response method for energy storage system in an energy local network. In: Proceedings of the CSEE, no.13, pp. 3703–3716 (2017)
16. Zhou, B., Yang, M., Shi, S., Wei, J., Li, Z., Dong, S.: Game model of micro grid market potential based on blockchain. Autom. Electr. Power syst. **44**(07), 15–26 (2020)

An Active Power Control Method for Large-Scale Wind Energy Considering Peak Load and Transmission Constraints

Aimei Liu[1,2](✉), Zemei Dai[1,2], Shiyuan Chen[1,2], Changhai Yu[1,2], and Jiping Wu[1,2]

[1] NARI Group Corporation (State Grid Electric Power Research Institute), Nanjing 211106, Jiangsu, China
{liuaimei, daizemei, chenshiyuan, yuchanghai, wujiping}@sgepri.sgcc.com.cn
[2] NARI Technology Development Limited Company, Nanjing 211106, Jiangsu, China

Abstract. In order to improve the utilization efficiency of wind energy and reduce the wind energy curtailment as much as possible, this paper proposes an active power control strategy based on peaking and tie line constraints, which divides peaking and tie line absorptive capacity into dynamic zones. Through flexible switching of control modes in different zones, the utilization of peaking and tie line capacity can be maximized. Considering the situation that there is the non-energy dissipation of power grid due to peak load regulation or blocked cross-section, this paper proposes an echelon active power control strategy, taken into account the utilization of hours and the completion of trading power, thus, ensuring the fairness of the generating capacity of each station. The practical operation on a power grid system of northern part China demonstrates the feasibility of the proposed method.

Index Terms: Wind energy · Active power control · Peaking constraint · Transmission constraint · Echelon control

1 Introduction

Large-scale of wind power integration brings great challenges to the security and stability of the power grid system. The randomness, fluctuations and peak inversion characteristics of wind power bring great pressure to the dispatch and operation of the power grid [1–6]. The northern part of china is a region which is rich in wind resources. Due to limited power transmission capacity and the lack of peak load regulation, the problem of wind curtailment in some areas is very prominent. Experts and scholars from all over the world have made lots of research on wind power active control

This work is supported by State Grid Corporation of China (Research and Application of Key Technologies for Cross-Level and Multi-Area Frequency Coordinated Control of Southwest China Power Grid under Asynchronous Interconnection, No. 521999190009).

Y. Xue et al. (Eds.): PMF 2020, LNEE 718, pp. 433–445, 2021.
https://doi.org/10.1007/978-981-15-9746-6_34

methods for high-permeability power grids. Reference [7] proposed a coordinated control mode for non-AGC units and AGC units during peak operating hours of the power grid which improves the grid's ability to accept large-scale wind power by coordinating different time scales. Reference [8] proposed a method of active power control that dispatches the grid absorption capacity according to the generation priority of the wind farms, which realizes the optimal dispatching of the wind farm power under peak load regulation constraints. Reference [9] gives the calculation method of maximum wind power absorption capacity from the point of view of peak load regulation capacity and transmission capacity of power grid, but it does not consider allowing partial wind power generation to be limited. Reference [10, 11] have made a lot of research on active power control model when large wind integrated to power grid. Reference [12] uses the method that transfers power generation capacity from wind farm with insufficient wind power to that with sufficient wind power. In condition of the safety of all sections, the maximum utilization of wind energy is achieved as much as possible. From the generation scheduling perspective, References [13–15] propose approaches to overcome the impact of the uncertainty of wind farms output to consume wind power as much as possible through rolling coordinated optimization of power generation plans at different time scales and establishment of regional and provincial two-level optimal dispatch models.

When the wind power is curtailed due to difficulty in peak load regulating or blocked transmission channel, each wind farm generally reduces its output according to the ratio set by the dispatching agency without considering the actual utilization hours of each station and the completion of the transaction power. There has been a situation where the power generation between the stations is seriously unbalanced and the fairness cannot be taken into account. Reference [16] provides a minimum wind abandonment method aiming at consuming wind power as much as possible with fair dispatch. When wind power output increases, wind farms with better evaluation performance and more accurate maximum power output estimation should be given priority for dispatch. To some extent, the above suggestions considers the fairness of dispatch between wind farms, however, the problem of unbalanced power generation between wind farms has not yet been resolved completely.

From the existing research and practical application situation, when wind power participates in dispatching, it generally evaluates the wind power availability margin according to the peak load regulation capacity of power grid, each wind farm is allocated according to the predicted maximum power generation capacity. If the forecast of wind power is inaccuracy, the estimated maximum output of the wind farms will be lower than the actual output. Therefore, this distribution method will lead to considerable loss of abandoned wind. This paper introduces a "Free Generation" mode of wind farms on the base of proposing a method for optimal dispatch of wind active power that adapts to high wind power permeability grids. Based on the on-line analysis of the wind power margin that can be absorbed by the power grid and the elaborate control of the AGC level of the wind farm, under the premise of ensuring the safe and stable operation of the system, the power generation capacity of the station shall be fully utilized to maximize the peak load regulation capacity and transmission capacity. In addition, a method of priority active power control strategy considering station utilization hours and transaction power completion in order to promote the friendly development of the wind power industry and ensure that the fairness of principle is raised in the situation that the power grid has no capacity to absorb wind power when peak shaving or cross section is blocked, and the wind farm needs to be limited.

With the large scale integration and transmission of wind power, peak load regulation capacity and line transmission capacity are the key factors restricting wind power consumption [17]. Wind power dispatching should aim at reducing the loss of abandoned wind, take into account the grid security and meet the requirements of dynamic balance of power generation and consumption, minimum output constraints of conventional units and line transmission capacity, which is a typical multi-objective optimization problem. A new method is proposed in this paper which can be summarized as follows:

A. The common modes of wind farms are "Automatic Control" mode based on current active output superposition adjustment and "Free Generation" mode with installed capacity as the target command. According to the real-time operation data such as rotating reserve, section stability limit and section power flow under thermal power, the power grid peak regulation capacity and section availability margin are dynamically zoned, and the wind farm operation mode is automatically adjusted by the zoning results.
B. According to the utilization hours of wind farm stations and the completion of the guaranteed purchase of electricity and the policies, the priority of power limitation of each wind farm station are divided. When the peak load regulation capacity of the power grid is insufficient or the active power flow of the cross section exceeds the limit, the power limitation strategy is started in time. Based on the principle of reducing the loss of abandoned wind and considering the fairness, the reasonable distribution of curtailment power among wind farms is carried out.

Compared to other methods, the main features of this method are:

a) Wind farms are finely controlled according to the dynamic zoning results of peak load regulation capacity and section transmission capacity constraints. When the wind power accepted margin is sufficient, the target that breaks through the prediction of the maximum power generation capacity is issued to improve the utilization rate of wind resources, which avoids the wind curtailment loss that occurs when the wind power prediction results are used for power allocation because of inaccuracy. When the peak shaving capacity of the power grid is insufficient or the power flow of the transmission cross-section exceeds the limit, the wind abandonment control is started in time. The target instructions from the AGC system fully guarantees the timeliness and safety of the active control of the wind farm.
b) When the wind power is limited, the curtailment control strategy comprehensively considers the utilization hours of each wind farm, guaranteed purchase of electricity and policies. Under the same cross section, priority is given to selecting stations with high average utilization hours throughout the year or completed transaction electricity. Finally, there will be no imbalance of limited power between wind farm stations.

2 Dynamic Consumption Capacity Partition and Mode Setting

Regardless of grid constraints, peak load regulation capacity is the main factor that determines the capacity of wind power consumption in a region. The theoretical power consumption of wind power is the difference between the system load and the minimum technical output of conventional units [13]. In order to balance the random fluctuation of wind power and the deviation of load forecast, and to deal with the possible power failure in the future power grid, the system needs to reserve a certain rotating reserve capacity. Therefore, the calculation formula of the maximum wind power receiving space under the peak shaving constraint is:

$$C_{\max} = P_{load} + P_{tie} + P_{reg} - P_{G\min} - P_{reg}^{dn} \tag{1}$$

where, $C_{\max}$ is the maximum wind power consumption space, P_{load} is the system load, P_{tie} is the plan value for tie line, P_{reg} is the system reserve capacity, $P_{G\min}$ is the minimum technical output, P_{reg}^{dn} is the minimum reserve. When the system load, tie-line plan value and the minimum technical output of the current units have been determined at, the reserve capacity of the system has become an important factor affecting the wind power consumption. For the time when the reserve capacity is abundant, there is sufficient capacity for wind power consumption. The operation mode of "Free Generation" improves the utilization of wind resources. Therefore, according to the surplus of reserve capacity in different periods of the power grid, the peak load regulation capacity can be divided into three operating states: safety, early warning and limited, which are shown in Fig. 1.

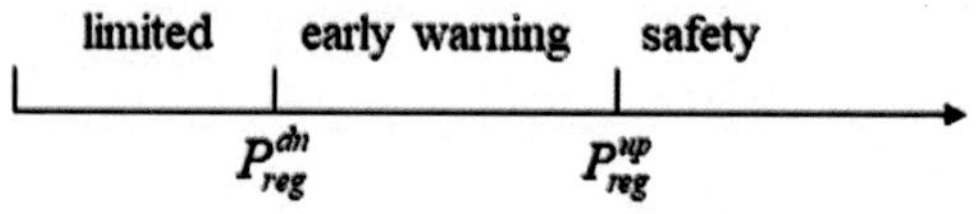

Fig. 1. Peak load capacity zoning diagram

The division method of peak load regulation interval in Fig. 1 is:

$$\begin{cases} safety: & P_{reg} \geq P_{reg}^{up} \\ early\ warning: & P_{reg}^{dn} \leq P_{reg} < P_{reg}^{up} \\ \lim ited: & P_{reg} < P_{reg}^{dn} \end{cases} \tag{2}$$

where P_{reg} is the current actual reserve capacity of the system, P_{reg}^{up} and P_{reg}^{dn} are the upper bound and lower bound of the reserved capacity separately. The determination of P_{reg}^{dn} should take uncertain factors into account such as system load forecast deviation, new energy output forecast deviation, unit failure and other outages, which is calculated generally according to a fixed ratio of the system peak load (5%–10%) or the maximum capacity of a single machine [18]. P_{reg}^{up} can be gained as:

$$P_{reg}^{up} = P_{reg}^{dn} + n\Delta P_{avg} \tag{3}$$

where, ΔP_{avg} is the average value of the increment of active power in a single control cycle which can be obtained according to the new energy ultra-short-term power prediction curve. And n is the number of periods.

Considering the limitation of the cross-section transportation capacity, the new energy acceptance capacity under the cross-section constraint is calculated according to the cross-section stability limit, safety margin and the cross-section actual power flow. If $S_{\max}$ is the stability limit of a section, $S_{m\arg}$ is the safety margin, S_{gen} is the cross-section actual power flow, the safe increment of cross-section can be calculated as

$$S_{reg} = S_{\max}(1 - S_{m\arg}) - S_{gen} \tag{4}$$

It is suggested that there are n wind farms belong to this cross-section. According to the historical operating data, the maximum value of the active power increment in two consecutive control cycles during the period of the largest power generation at the first station is $\Delta P_{\max,i}$, when the actual output of the cross section S_{gen} satisfies the condition of $\sum_{i=1}^{n} \Delta P_{\max,i} + S_{gen} < S_{\max}$, which indicates that the current cross-section capacity is sufficient, and the cross-section completely consumption threshold $S_{sfm\arg}$ is:

$$S_{sfm\arg} = (\sum_{i=1}^{n} \Delta P_{\max,i} + S_{gen})/S_{\max} \tag{5}$$

The consumption capacity of cross-section can be divided into four different cases shown as following (Fig. 2):

$$\begin{cases} \text{sufficient}: \; S_{gen} \le S_{\max}(1 - S_{sfm\arg}) \\ \textit{normal}: \; S_{\max}(1 - S_{sfm\arg}) < S_{gen} < S_{\max}(1 - S_{m\arg}) \\ \text{early warning}: \; S_{\max} > S_{gen} \ge S_{\max}(1 - S_{m\arg}) \\ \text{insufficient}: \; S_{gen} \ge S_{\max} \end{cases} \tag{6}$$

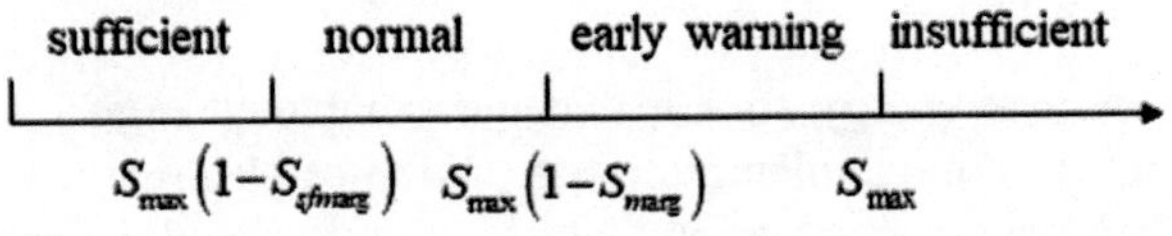

Fig. 2. Transmission consumption capacity zoning diagram

In each control cycle, the operation mode of the wind farm is adjusted according to the changes in the power grid's wind consumption capacity by monitoring the peak load regulation interval and the cross-section operation status. The specific methods are:

a) When the peak load regulation capacity is in a safe state and the cross-sectional consumption capacity is sufficient, we set the wind farm in "Free Generation" Mode.
b) When the peak load regulation capacity is limited, the wind farm should be in output reduction mode, we set the wind farm in "Automatic Control" mode.
c) After the wind farm gets the allocation of peak load regulation and absorption space, it checks the stability limit of the section. The peak load regulation capacity of the power grid is sufficient, but when the output of the wind farm in some areas exceeds the transmission capacity of the power grid in this area, the controlled station under the switching section will be set to "Automatic Control" mode.
d) When the peak load regulation interval is in the early warning state but the cross-section transportation capacity is sufficient, the station will not switch modes and maintain the control mode of the previous cycle.
e) When the peak load regulation interval is in an unrestricted state, and the cross-sectional absorption space is sufficient, the station does not switch modes and maintains the operation mode of the previous cycle.

3 Echelon Control Strategy

The precise management of active energy control of renewable energy requires not only guaranteeing the security of system and economy, but also the fairness between the various stations [19]. In order to ensure the fairness of dispatch, the new energy restriction is based on a priority control strategy based on transaction power. According to the completion of the guaranteed purchase of electricity at each new energy station, the method of determining the priority is as follows:

In order to promote the consumption of new energy, each government has certain requirements on the utilization hours of new energy stations, which is generally called guaranteed electricity purchase; the transaction electricity is to further promote local new energy consumption, and electricity trading contracts signed with external power grids should be completed as much as possible, so the echelon control strategy is based on utilization hours and transaction electricity.

- Group One: A new energy station that has completed a guaranteed purchase of electricity, and there is no transaction electricity or the transaction electricity has been sent out;
- Group Two: A new energy station that has completed a guaranteed purchase of electricity and has not completed the transaction of electricity;
- Group Three: A new energy station that has not completed the guaranteed purchase of electricity and has no trading electricity;
- Group Four: A new energy station that has not completed the guaranteed purchase of electricity and has trading electricity;
- Group Five: The policy protection stations and the poverty alleviation stations.

The Intelligent dispatching management system provides two types of rolling groups which are monthly and daily. Monthly groups are calculated by transaction power and utilization hours provided by scheduling part. And the daily rolling groups are calculated by power supplied and utilization hours provided by the new energy AGC system. The daily rolling group information is sent back to the new energy AGC system. When the grid's absorption capacity is insufficient and it is necessary to abandon the wind, the station output of the previous group shall be controlled to the lower limit first, and then the station of the latter group shall be limited. Before power limitation control, total adjustable downward regulation to ensure the safety of power grid will be calculated. If the down margin of the previous group station meets the current correction requirements, the next group will not participate in the power limit control. The regulation amount of the station broke the step limit which means the target is the lower limit of the station. The following is an example of active power distribution in a limit state. And the specific power distribution description of the algorithm is shown in Fig. 3.

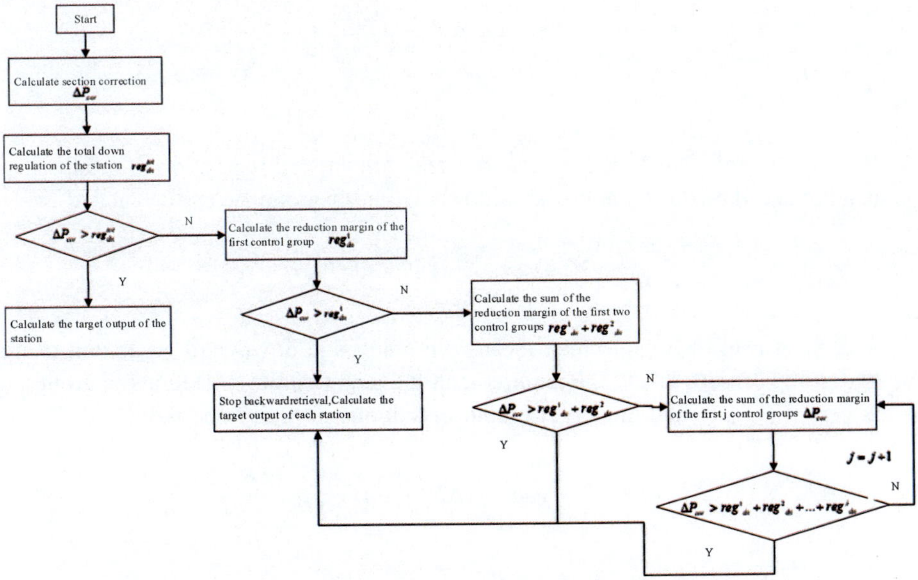

Fig. 3. Power distribution algorithm under the condition of section limitation

a) When the cross-section is overloaded or exceeds the limit, the correction value is calculated as

$$\Delta P_{cor} = P_{gen} - P_{lmt}(1 - m\arg) \quad (7)$$

Where, P_{gen} is actual output of the station, P_{lmt} is the stability threshold of the cross-section, $m\arg$ is the cross-section safety margin;

b) Total regulation of cross-section can be calculated as

$$reg_{dn}^{tot} = \sum_i (gen^i - gen_{dn}^i) \quad (8)$$

Where, gen^i is the output of the station i, gen_{dn}^i is the minimum output of the station i. If $\Delta P_{cor} \geq reg_{dn}^{tot}$, which means the adjustable downward regulation is insufficient, the minimum output will be set as the control objective: $gen_{des}^i = gen_{dn}^i$, If $\Delta P_{cor} < reg_{dn}^{tot}$, then calculate as following:

c) Calculate the total adjustable downward regulation of group one:

$$reg_{dn}^1 = \sum_i (gen^i - gen_{dn}^i) \quad (9)$$

d) If $\Delta P_{cor} \leq reg_{dn}^1$, which means the adjustable downward regulation of group one is sufficient. ΔP_{cor} is allocated as the rated capacity ratio. And each allocation is:

$$reg^i = \Delta P_{cor} \frac{cap^i}{\sum_i cap^i} \quad (10)$$

Then, $gen_{des}^i = gen^i + reg^i$, $gen_{des}^i = gen^i$;

e) If $\Delta P_{cor} > reg_{dn}^1$, which means the adjustable downward regulation of group one is insufficient, the adjustable downward regulation of group two is calculated as

$$reg_{dn}^2 = \sum_i (gen^i - gen_{dn}^i) \quad (11)$$

If $\Delta P_{cor} \leq reg_{dn}^1 + reg_{dn}^2$, which means the adjustable downward regulation of first two group can not satisfy the requirement, the target output of stations in group one is $gen_{des}^i = gen_{dn}^i$ and the target output of stations in group one is

$$gen_{des}^i = gen^i + (\Delta P_{cor} - reg_{dn}^1) \frac{cap^i}{\sum_i cap^i} \quad (12)$$

f) If $\Delta P_{cor} > reg_{dn}^1 + reg_{dn}^2$, then use total demand downward regulation minus the adjustable downward regulation of each group until group j satisfies

$$\Delta P_{cor} \leq reg_{dn}^1 + reg_{dn}^2 + \ldots + reg_{dn}^j \quad (13)$$

The target output of stations from group one to $j - 1$ is $gen_{des}^i = gen_{dn}^i$. And the target output of stations in group j is

$$gen_{des}^i = gen^i + (\Delta P_{cor} - reg_{dn}^1 - reg_{dn}^2 - \ldots - reg_{dn}^{j-1}) \frac{cap^i}{\sum_i cap^i} \quad (14)$$

And the target output of the rest station is

$$gen^i_{des} = gen^i \tag{15}$$

4 Case Study

4.1 Example of Echelon Active Power Control Strategy

The active power dispatching method proposed in this paper has been put into closed-loop operation in power grid of northern part of China. The intermittent energy rated in the power grid accounts for 51.8% of the entire grid power supply, and the maximum output accounts for about 26.9% of the current load which is higher than other cases. The following is an example of a severely restricted section for the analysis of the priority control algorithm in the power limited state.

The actual output, priority and rated capacity of the stations in same cross-section at a certain time are shown in Table 1. The upward regulation step is taken as 10% of the rated capacity, and the downward regulation step is 8% of the rated capacity.

Table 1. Operation and control parameters of a section subordinate substation

Name	Output (MW)	Priority	Rated capacity (MW)
Station 1	16.8	Group 1	35
Station 2	23.2	Group 2	50
Station 3	29.3	Group 3	100
Station 4	36.2	Group 3	97.5
Station 5	17.1	Group 4	49.5
Station 6	18.6	Group 4	49.5
Station 7	38.0	Group 4	148.5

Station 1 has completed the annual utilization hours and the transacted electricity, which is assigned to Group 1; Station 2 has completed the annual utilization hours, but has not completed the transacted electricity, which is assigned to Group 2; Station 3 and Station 4 have not completed the annual utilization hours, but have completed the transacted electricity, which are both assigned to Group3; Station 5, Station 6 and Station 7 have not completed the annual utilization hours, and have not completed the transacted electricity, either, which are assigned to Group 4.

The stability threshold of the cross-section is 700 MW, and the safety and stability margin is 8%. At this moment, the actual output of the section is 697.9 MW, the current actual reserve of the entire network is 1153.5 MW, the upper bound of reserve is 1120 MW, and the lower bound of reserve is 600 MW. The actual reserve is higher than the upper bound of reserve, the peak shaving capacity is sufficient, only the partial cross-section is heavily loaded, and the cross-section correction amount is 53.9 MW. Before the improvement of the algorithm, the limited power distribution of the cross-section is carried out according to the actual output ratio when the power is limited. For the two control strategies, the active commands and adjustments of each field are shown in Table 2.

Table 2. Comparison of adjustment amount between proportional control and priority control

Name	Output (MW)	Target of proportional control (MW)	Target of priority control (MW)	Regulation of proportional control (MW)	Regulation of priority control (MW)
Station 1	16.8	11.7	0	−5.1	−16.8
Station 2	23.2	16.2	0	−7.0	−23.2
Station 3	29.3	20.5	22.3	−8.8	−7.0
Station 4	36.2	25.3	29.3	−10.9	−6.9
Station 5	17.1	12	17.1	−5.1	0
Station 6	18.6	13	18.6	−5.6	0
Station 7	38.0	26.6	38.0	−11.4	0

If we define the limited power increment of the station as the difference between the adjustment amount of the proportional control and the priority control in the limited state, the limited power increment is positive, indicating that the limited power increases after the station uses the priority control strategy, vice versa. And the limited power increment of each station is shown in Table 3.

Table 3. The additional power of each substation under the cascade control strategy

Name	Priority	Increment of power (MW)
Station 1	Group 1	11.7
Station 2	Group 2	16.2
Station 3	Group 3	−1.8
Station 4	Group 3	−4.0
Station 5	Group 4	−5.1
Station 6	Group 4	−5.6
Station 7	Group 4	−11.4

When using the proposed echelon control strategy, the output of Group 1 and Group 2 are controlled to the lower bound. The remaining limited power are allocated to stations in Group 3 according to the rated capacity. And the output of the first three groups of stations have met the power limitation requirements of the cross-section. Hence, the stations in Group 4 can maintain the current actual output. It can be seen from Tables 2 and 3 that the limited power of the stations that have completed the guaranteed purchase of electricity have significantly increased, while the limited power of the stations that have not completed the transaction of electricity or the guaranteed purchase of electricity has decreased, which is beneficial to the balance of power generation of each station.

4.2 Peak Load Regulation Case

Figure 4 shows the operation curve of a certain power grid in North China during the period of winter wind generation. The stability limit of the section is 2000 MW, the safety margin is 2%, the section full absorption threshold parameter is 28%, the reserved reserve lower limit of the power grid is 800 MW, and the upper limit is 1180 MW. Before peak load regulation control, the section power flow was close to the warning limit value, and the section utilization rate was as high as 98%, the controlled stations were set to "Automatic Control" mode. At 09:40, the power grid had difficulty in peak load regulation and adopted wind restriction strategy. At 09:47, the rotating reserve of the system dropped to 558 MW, and the active power of the section decreased by 160 MW. From 09:47 to 09:48, the rotating reserve of the system was gradually restored to the limit value, and the wind power output was gradually and steadily restored. From 09:49 to 09:55, the rotating reserve under the system was restored to 950 MW, and the section power flow was also restored to the level before peak load regulation.

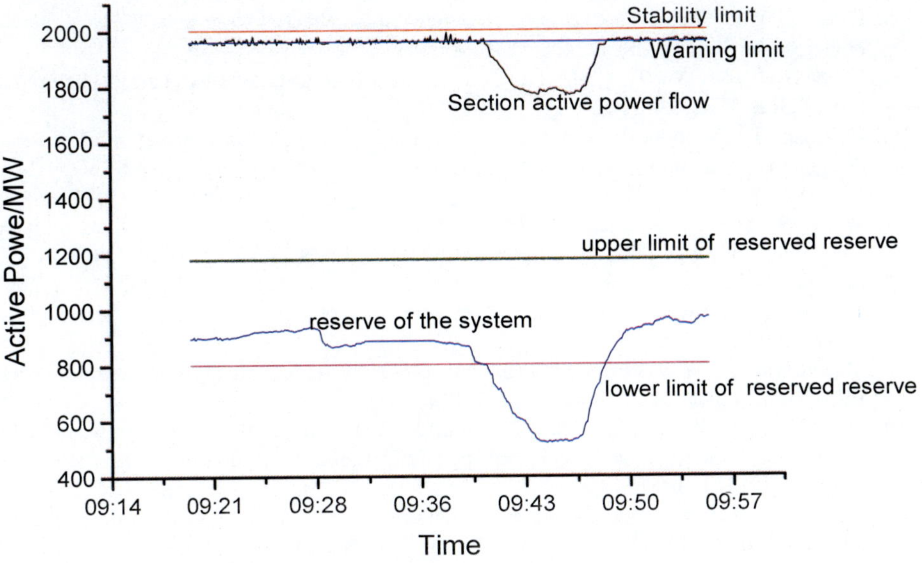

Fig. 4. Section operation curve under peak regulation

5 Conclusions

The active power control strategy mentioned in this paper is safe and efficient. The proposed control strategy of dynamic partition based on peak load regulation capacity and the wind power accepted margin of cross-sections improves the utilization efficiency of wind energy resources, while ensuring the safe and stable operation of the power grid. Through the optimized echelon power limitation strategy, some wind farms that have completed the basic power generate less while the other stations with uncompleted basic power reduced their downward regulation power, thus, achieving fair dispatch of each wind farm. Coordinated control with conventional energy sources such as thermal power and hydropower, as well as large-scale wind power active control with market transactions still need to be further studied.

References

1. Yang, L., Yang, X., Tu, J., et al.: Analysis of wind power output characteristics and its impacts on operation of power system. Smart Grid **2**(5), 18–22 (2014)
2. Liu, J.: Basic issues of the utilization of large-scale renewable power with high security and efficiency. Proc. CSEE **33**(16), 1–8 (2013)
3. Xue, Y., Lei, X., Xue, F., et al.: A review on impacts of wind power uncertainties on power systems. Proc. CSEE **34**(29), 5029–5040 (2014)
4. Ding, M., Wang, W., Wang, X., et al.: A review on the effect of large-scale PV generation on power systems. Proc. CSEE **34**(1), 1–14 (2014)
5. Chen, G., Li, M., Xu, T., et al.: Study on technical bottleneck of new energy development. Proc. CSEE **37**(1), 20–25 (2017)
6. Holttinen, H., Meibom, P., Orths, A., et al.: Impacts of large amounts of wind power on design and operation of power systems, results of IEA collaboration. Wind Energy **14**(2), 179–192 (2011)
7. Wang, S., Yu, J.: Coordinated dispatch regulation strategy between Non-AGC units and AGC units with high wind power penetration. Proc. CSEE **33**(7), 156–163 (2013)
8. Zhai, B., Wang, J., Yang, Z., et al.: Active power control strategy for wind power considering priorities of power generation with peak regulation constraint. Autom. Electr. Power Syst. **41**(23), 83–88 (2017)
9. Matevosyan, J.: Wind power integration in power systems with transmission bottlenecks. IEEE Trans. Power Syst. **17**(4), 1132–1139 (2007)
10. Lin, Z., Chen, Z., Qu, C., Guo, Y., Liu, J., Wu, Q.: A hierarchical clustering-based optimization strategy for active power dispatch of large-scale wind farm. Int. J. Electr. Power Energy Syst. **121**, 106155 (2020)
11. Marannino, G.P., Montagna, G.P., et al.: Differenttime-scale approaches to the real power dispatch of thermal units. IEEE Trans. Power Syst. **5**(1), 169–176 (1990)
12. Zhuo, J., Jin, X., Deng, B., et al.: An active power control method of large-scale wind farm considering security constraints of tie lines. Power Syst. Technol. **39**(4), 1014–1018 (2015)
13. Zhent, T., Feng, L., Wang, S., et al.: An optimized wind power dispatching method considering security constraints in the power grid. Autom. Electr. Power Syst. **34**(15), 71–75 (2010)

14. Wang, X., Li, J., Huang, B., et al.: A two-stage optimal dispatching model for provincial and regional power grids connected with wind farms to promote accommodation of wind power. Power Syst. Technol. **39**(7), 1833–1838 (2015)
15. Wang, K., Zhang, B., Yan, D., et al.: A multi-time scale rolling coordination scheduling method for power grid integrated with large scale wind farm. Power Syst. Technol. **38**(9), 2434–2440 (2014)
16. Wang, B., Sun, Y., Wu, W., et al.: A real-time wind power control method for power systems with high wind power penetration and its implementation. Autom. Electr. Power Syst. **39**(21), 23–29 (2015)
17. Zhu, L., Chen, N., Han, H.: Key problems and solutions of wind power accommodation. Autom. Electr. Power Syst. **35**(22), 29–34 (2011)
18. Su, P., Liu, T., Li, X.: Determination of optimal spinning reserve of power grid containing wind. Power Syst. Technol. **34**(12), 158–160 (2010)
19. Shi, G., Sun, R., Xu, H., et al.: Active power stratification coordination control strategy for large-scale cluster of renewable energy. Power Syst. Technol. **42**(7), 2160–2167 (2018)

An Online Identification Method of Key Transmission Sections for Thermal Stability Considering Cross-Region Faults and Security and Stability Control Measures

Jinyou Meng[1(✉)], Zhihao Tian[1], Yuchen Dai[2], Shibin Bai[1], Shuang Zhang[1], and Feng Luo[2]

[1] State Grid Ningxia Electric Power Co., Ltd., Yinchuan 750001, China
mengjinyou@nx.sgcc.com.cn

[2] NARI Group Corporation (State Grid Electric Power Research Institute), Nanjing 211106, China

Abstract. Accurate and complete key transmission section identification has important significance to reduce power system monitoring range and prevent cascading failures. However, key transmission section identified by analysis of typical power system operating modes may not match power grid real-time operation situation when power grid structure is seriously damaged. Therefore, an online identification method of key transmission section considering cross-region faults and security and stability control measures is proposed in order to identify the weak sections that need to be concerned in power grid monitoring. Firstly, power flow distribution characteristics under different faults are analyzed based on results of online static stability analysis. Secondly, power flow transfer channels with different faults are identified and combined into transmission sections for thermal stability. The key transmission sections for thermal stability are screened by power transfer limit and margin calculation. Finally, the effectiveness of the proposed method is demonstrated by Northwest Power Grid.

Keywords: Transmission section · Overload line · Cross-region fault · Power transfer limit calculation

1 Introduction

With the increasingly prominent characteristics of power electronized and constantly expanding power grid [1, 2], the grid dispatching operators have to deal with the cascading faults caused by electrical components and cross-region faults caused by natural disasters. It is difficult to monitor and analyze all electrical components of the power grid. Therefore, key transmission sections are adopted to reduce power system monitoring range in most dispatching centers at home and abroad [3, 4].

The traditional power system operation monitoring section is a combination of regional connecting lines determined by the experience of dispatchers and geographical location. Affected by the expansion of the scale, the continuous adjustment of operation modes and the integration of power electronic components and natural disasters, the

Y. Xue et al. (Eds.): PMF 2020, LNEE 718, pp. 446–461, 2021.
https://doi.org/10.1007/978-981-15-9746-6_35

traditional operation monitoring sections can no longer cover all the weak sections in the grid. Therefore, the online identification of key transmission sections has become one of the most urgent problems for power grid dispatching operation control.

Based on the optimization and simplification of network partitioning, the combination of regional tie lines is identified to form the transmission section by using graph theory and other calculation methods [5–10]. The key transmission sections of the system are determined according to the calculated power transfer limit. However, most of the sections identified based on geographic divisions have already been monitored by dispatchers. The potential weak sections in non-geographical divisions cannot be identified. The key lines can also be determined by the calculation of load ratios and electric betweenness. The branches closely related to the key lines are searched to form the key transmission sections based on power flow shift factors and clustering index. In reference [11], the parallel transmission lines of overload lines are searched by graph theory method. In reference [12], the line power composition is analyzed based on the graph theory method to identify the power flow shift lines. In reference [13, 14], the key lines is identified by improved electric betweenness of the lines, and power flow shift area are adopted as the search range of the transmission section. This kind of method can accurately identify the transmission section with a single weak component. In order to identify key section with multiple weak components, the post-fault power flow is estimated by multi-branch outage shift factor [15–17]. However, the power flow relationship among multiple weak components is not fully considered.

An online identification method of key transmission sections for thermal stability is proposed in this paper to consider cross-region faults and security and stability control measures. Based on the study of multiple fault components influence on power flow shift, the weak lines are searched and the power flow transfer channels are identified according to the line load ratios and its changes after fault according to the static stability analysis results. Considering the current-carrying capacity of line, generator and load distribution and other factors, the power transfer limit and margin are calculated to identify the key transmission sections for thermal stability.

2 Characteristics of Key Transmission Sections for Thermal Stability

Thermal stability of power system means that the power grid remains stable after a single component fault (such as line N-1, parallel lines N-2, DC bipolar blocking and so on), and other components are not overloaded. Correspondingly, key transmission sections for thermal stability refer to the outage of section components will lead to heavy load or overload of other section components. Such sections have two important characteristics:

1) After one or more components fault, most of the power flow will be transferred to the remaining components.
2) After one or more components fault, there is a risk of overload for other section components.

Power flow transfer of different types of fault have large differences, which can be roughly divided into three categories. The first is transformer/line N-1 and parallel lines N-2; the second is the fault of active components, including generation unit trip, DC fault, and load loss; the third is multiple faults, such as bus fault, N-K fault of non-same pole parallel transmission lines. The power shortage caused by generation unit trip and load loss is usually shared by the whole grid, which has little impact on the power flow of a single transmission section; the power flow in near area receiving channel changes greatly after DC fault, and the relevant sections have already been monitored. Therefore, the power flow transfer characteristics of the first and third categories of faults are mainly analyzed in this paper.

2.1 Characteristics of Power Flow Transfer for a Single Component Fault

A single component fault mentioned in this section refers to transformer/line N-1 and parallel lines N-2. The power flow transfer can be divided into three cases:

1) In Fig. 1(1), the power flow is transferred to the other line after one of the parallel lines fault and such weak transmission channels are usually monitored;
2) In Fig. 1(2), the fault line is in the radial network, there are some problems such as system splitting and unbalanced power after fault, and overload is not the primary concern;
3) In Fig. 1(3), the fault line is in the mesh network, the power flow change of line A-C may be the largest, but not necessarily the weakest. The load rate change of lines C-E-B and C-D-B may be the largest affected by grid structure and distribution of generation and load.

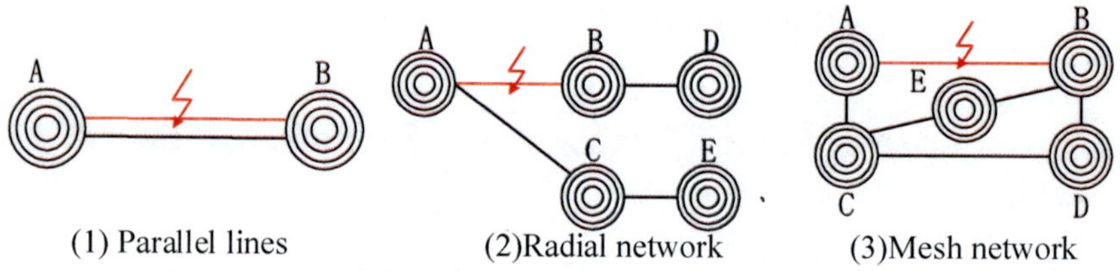

Fig. 1. Flow transferring of N-1 fault.

In addition, the dispatching center needs to formulate a major accident plan for the fault of parallel transmission lines on the same pole that may cause the overload of relevant components and other security and stability problems. Therefore, it is necessary to distinguish power flow transfer caused by fault or fault control measures, and focus on power flow transfer caused by fault.

2.2 Characteristics of Power Flow Transfer for Multiple Faults

Multiple faults can be divided into two categories according to the characteristics of power flow transfer and the electrical connection of fault components.

The electrical distance between components of the first category of fault is relatively long, as shown in Fig. 2. The power flow transfer lines of fault lines A-B and C-D are E-F and G-H respectively, and the power flow transfer between each other is small, which can be regarded as N-1 outage of lines A-B and C-D.

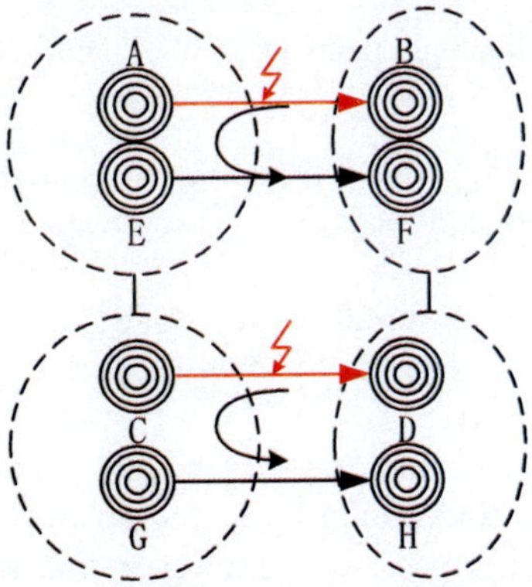

Fig. 2. Multiple faults of fault components with far electrical distance.

The electrical connection between components of the second category of fault is close, mainly including the following power flow transfer forms:

1) There is a series relationship between fault lines A-B and C-D in Fig. 3. The power flow of double circuit lines C-D will greatly decrease after the outage of line A-B. After N-1 fault of double circuit lines C-D, the power flow of the other line will increase, but the increase is far less than the decrease of line A-B outage. The risk of overload after this fault mainly occurs in the fault power flow transfer channel of line A-B;

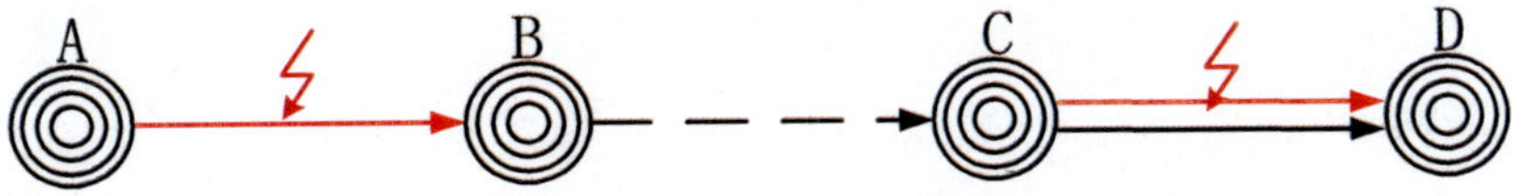

Fig. 3. Multiple faults of series fault components.

2) The fault shown in Fig. 4 is a typical N-K fault of regional transmission channel. The fault lines A-B and C-D are parallel and have a same power flow transfer channel E-F, the simultaneous fault may lead to overload of line E-F. Therefore, dispatchers need to monitor A-C + C-D + E-F regional transmission section;

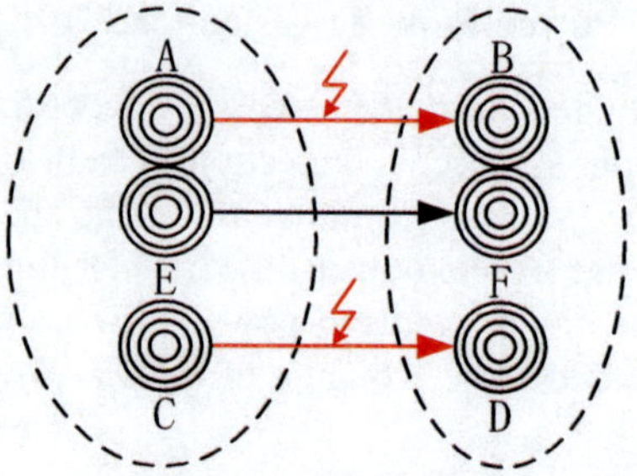

Fig. 4. Multiple faults of parallel fault components.

3) The fault shown in Fig. 5 is a complex cross-region fault. There are series and parallel relationship between fault lines. The power flow transfer has the following cases:
   (1) The fault power flow of parallel lines A1-B1 and A3-C1 will be transferred to line A2-B2. It is necessary to pay attention to the sending section A1-B1 + A2-B2 + A3-C1 of area A;
   (2) The effect of the series lines A3-C1 and C3-D2 outage on B4-C2 is opposite. The power flow of B4-C2 can be increased or decreased, which needs to be determined according to the real-time operation of the power grid;
   (3) The power flow of line B3-D1 is only affected by the outage of line C3-D2, which is less related to the outage of other lines. This characteristic has a great interference on the section identification.

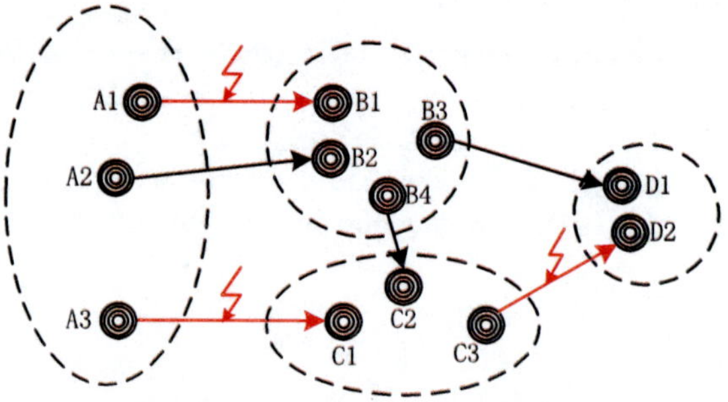

Fig. 5. Multiple faults of parallel and series fault components.

In conclusion, in addition to great changes in power flow, there are the following characteristics of power flow transfer of various faults:

1) Risk of multiple components may occur in the power flow transfer channel;
2) The influence of some fault components on power flow is independent of other fault components, which can form a section independently;
3) Some fault components have little influence on the post-fault power flow, which can be ignored;

4) Part of the power flow transfer is not related to the fault component itself, which is caused by the control measures after the fault.

3 Identification Method of Key Transmission Sections for Thermal Stability

The identification of key transmission sections for thermal stability is divided into the following steps:

1) Combined with the characteristics of power flow transfer under different faults, the power flow transfer lines are screened out;
2) According to the power flow direction and connection relationship of fault line and fault power flow transfer line, the transmission channel is identified;
3) According to the sensitivity between channels, the transmission channel combination that meets the requirements is screened out;
4) The lines of each transmission channel are combined exhaustively to generate a number of transmission sections for thermal stability. Transmission limit of each section is calculated by linear programming method to obtain its transmission margin. Finally, the lower margin is screened out as the key transmission section for thermal stability.

3.1 Power Flow Transfer Lines Filter

Identifying transmission channels is to determine the channel to which the fault line belongs and its related transmission channel. Firstly, it is necessary to screen out the power flow transfer lines of fault lines. Combined with the power flow after and before the expected fault, the power flow transfer line should meet the load rate change constraint, which is as follows:

$$\left(\eta_i^{'} - \eta_i^0\right) > \Delta\eta_{\text{th}} \tag{1}$$

Where, η_i^0 and $\eta_i^{'}$ are the load rate of the line before and after fault, $\Delta\eta_{\text{th}}$ is the threshold value of load rate change, the definition of load rate is as follows:

$$\eta = \frac{I}{I_e} > \eta_{\text{th}} \tag{2}$$

The line load rate η is defined as the ratio of the current I flowing through the line to the rated current I_e of the line. The load rate used in this paper takes the direction of line power flow into account and the direction of power flow after fault is positive. It is helpful to identify the power flow transfer channel with reverse power flow after fault.

3.2 Search of Transmission Channel Consists

According to the power flow consistency of fault line and power flow transfer line, the transmission channel matrix is formed. Each column of the matrix is a transmission channel. Take the radial network in Fig. 1 as an example:

$$S = \begin{bmatrix} \text{A-B} & \text{A-C} & \cdots \\ \text{B-D} & \text{C-E} & \cdots \\ \cdots & \cdots & \ddots \end{bmatrix} \tag{3}$$

The radial network in Fig. 1 contains two transmission channels, A → B → D and A → C → E, respectively. The specific generation method is as follows:

1) The heads of all fault lines and power flow transfer lines form the head node set, and all the ends form the end node set;
2) Determining whether the head node belongs to the end node set. If not, the node is marked as the starting point of the channel;
3) The first line of the transmission channel is the line with the starting point of the channel, which constitutes the initial transmission channel S_1;

$$S_1 = [\text{A-B} \quad \text{A-C}] \tag{4}$$

4) The transmission channel S_2 is obtained by taking the line that is headed by the end node of a line on each channel as the next line of the corresponding channel;

$$S_2 = \begin{bmatrix} \text{A-B} & \text{A-C} \\ \text{B-D} & \text{C-E} \end{bmatrix} \tag{5}$$

5) If the head of multiple lines is the same node, and it is also the end of a power flow transfer line or fault line, it will be split into corresponding number of transmission channels S_3;

$$S_3 = \begin{bmatrix} \text{A-B} & \text{A-C} & \text{A-C} \\ \text{B-D} & \text{C-E} & \text{C-E} \\ \text{D-F} & \text{E-H} & \text{E-I} \end{bmatrix} \tag{6}$$

6) Until the channels of all the power flow transfer lines and fault lines have been identified.

3.3 Identification of Transmission Channel Validity

Due to the interference of cross-region faults and security and stability control measures, some transmission channels are invalid, including two types:

1) Transmission channel of fault line with short electrical distance but no power flow transfer relationship, which has no obvious power flow transfer relationship with other channels.

2) The power flow change of power flow transfer channel caused by security and stability control measures is not the power flow transfer of fault line.

Therefore, it is necessary to screen transmission channels before generating transmission sections for thermal stability. The branch outage shift factor can reflect the power flow relationship between transmission channels, and the calculation formula is as follows:

$$\xi_{12} = \frac{P'_{L2} - P^0_{L2}}{P^0_{L1}} = \frac{X_{21}/x_{L2}}{1 - X_{11}/x_{L1}} \tag{7}$$

Where, ξ_{12} is the branch outage shift factor of line L1 with the largest load rate in transmission channel 1 to line L2 with the largest load rate in transmission channel 2, X_{ij} is the mutual impedance between ports i and j, X_{ii} is the self-impedance of port i, x_{L1} and x_{L2} are the impedance of line L1 and L2 respectively.

There are two criteria for identifying invalid channel, and meeting any one of them is invalid channel:

1) The active sensitivity of the transmission channel to which the fault line belongs and all transmission channels is lower than the threshold value;
2) The active sensitivity of power flow transfer channel and the transmission channels to which all fault lines belong is lower than the threshold value.

3.4 Limit Calculation of Transmission Section

One line of each effective transmission channel is selected to generate one line combination, which can generate several line combinations. Key transmission sections for thermal stability meet the following two principles:

1) Uniqueness principle, any two lines in the transmission section for thermal stability do not belong to the same transmission channel;
2) Integrity principle, all transmission channels have and only one line belongs to transmission section for thermal stability.

The transmission section for thermal stability that the power grid dispatching operation focuses on is the transmission section with low margin, the margin ξ can be calculated according to the following formula:

$$\xi = \frac{P_{\text{lim}} - P^0_{\text{cut}}}{P_{\text{lim}}} \times 100\% \tag{8}$$

Where, P^0_{cut} is the sum of active power under the initial condition of transmission section, P_{lim} is the limit power of transmission section, which can be calculated by linear programming method. The goal of transmission section limit calculation model is to maximize the sum of active power of the section:

$$\max \quad P_{\lim} = P^0_{\text{cut}} + \sum_{Gi} S_{Gi}\Delta P_{Gi} + \sum_{Li} S_{Li}\Delta P_{Li} \tag{9}$$

Where, S_{Gi} is the active power sensitivity of generator i to the transmission section for thermal stability, ΔP_{Gi} is the adjustment amount of generator i, S_{Li} is the active power sensitivity of load i to the transmission section for thermal stability, ΔP_{Li} is the adjustment amount of load i.

The limit of transmission sections for thermal stability refer to the maximum transmission active power of the section when all the lines in the section do not exceed the limit before and after fault. Therefore, the following four constraints need to be met:

1) Under the initial operating condition, all lines do not exceed the limit

$$P^0_k + \sum_{Gi} S^k_{Gi}\Delta P_{Gi} + \sum_{Li} S^k_{Li}\Delta P_{Li} \le P^{\lim}_k, k \in S_{\text{line}} \tag{10}$$

Where, P^0_k is the active power of line k in the composition of section under the initial operating condition, S^k_{Gi} is the active power sensitivity of generator i to line k under the initial condition, S^k_{Li} is the active power sensitivity of load i to line k under the initial condition, $P^{\lim}_k$ is the maximum active power transmitted by line k, S_{line} is the set of all lines in the composition of section.

2) All lines do not exceed the limit after fault

$$P^{'}_k + \sum_{Gi} S^{k'}_{Gi}\Delta P_{Gi} + \sum_{Li} S^{k'}_{Li}\Delta P_{Li} \le P^{\lim}_{k'}, k \in S_{\text{line}} \tag{11}$$

Where, $P^{'}_k$ is the active power of line k in the composition of section under fault, $S^{k'}_{Gi}$ is the active power sensitivity of generator i to line k under fault, $S^{k'}_{Li}$ is the active power sensitivity of load i to line k under fault.

3) Active power balance constraint

$$k_{\text{loss}} \sum_{Gi} \Delta P_{Gi} + \sum_{Li} \Delta P_{Li} = 0 \tag{12}$$

Where, k_{loss} is the active power loss coefficient, which is the ratio of the total active load of whole power to the total active output of generators under the initial condition.

4) Adjust orientation constraint

When calculating the section limit operation mode, it is usually required that the sending end only raises the output of generators and the receiving end only raises the load, as follows:

$$\begin{array}{ll} \Delta P_{Gi} \geq 0, & Gi \in S_{\text{gen}} \\ \Delta P_{Li} \geq 0, & Li \in S_{\text{load}} \\ S_{Gi}\Delta P_{Gi} \geq 0, & Gi \in S_{\text{gen}} \\ S_{Li}\Delta P_{Li} \geq 0, & Li \in S_{\text{load}} \end{array} \tag{13}$$

5) Adjust space constraint

$$\begin{array}{ll} P_{Gi\min} \leq P_{Gi} + \Delta P_{Gi} \leq P_{Gimax}, & Gi \in S_{\text{gen}} \\ P_{Li\min} \leq P_{Li} + \Delta P_{Li} \leq P_{Limax}, & Li \in S_{\text{load}} \end{array} \tag{14}$$

Where, $P_{Gi\,\min}$ and $P_{Gi\,\max}$ are the lower and upper output limit of generator i respectively, $P_{Li\,\min}$ and $P_{Li\,\max}$ are the minimum and maximum value of load i respectively, P_{Gi} is the output of generator i under the initial operating condition, P_{Li} is the magnitude of load i under the initial operating condition.

3.5 Key Transmission Sections Identification Process

Key transmission sections identification process is shown in Fig. 6, which includes three part: search of transmission channel consists, identification of transmission channel validity, limit calculation of transmission section.

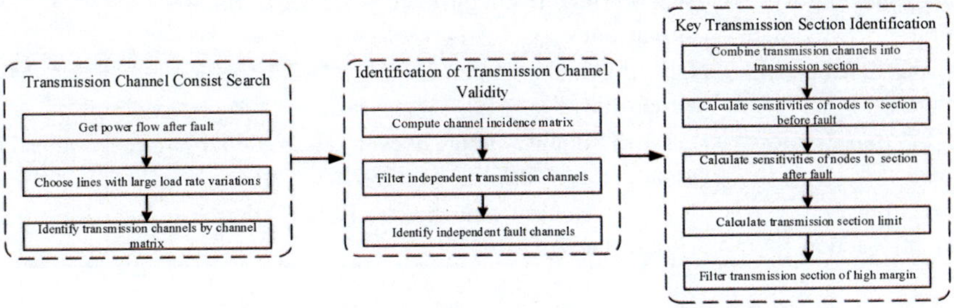

Fig. 6. Flow chart of key transmission sections identification.

4 Example

4.1 Example Description

The online data of Northwest Power Grid in August 2019 is used in the example. The generating capacity is about 110GW, the load is about 85GW, and the UHV DC transmission power is about 25GW. The key transmission sections of online system are shown in Table 1 (Fig. 7).

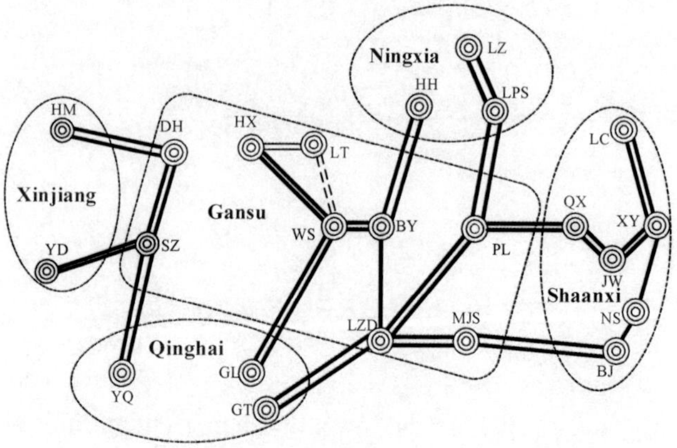

Fig. 7. Simplified geographic wiring diagram of Northwest Power Grid.

Table 1. Key transmission sections of Northwest Power Grid.

Section name	Section composition	Section limit
Heyu section	HX-WS double-circuit lines + HX-LT double-circuit lines + SZ-YQ double-circuit lines	8500 MW
West-to-east power transmission section	PL-QX double-circuit lines + MJS-BJ double-circuit lines	7000 MW
Wubaiguandong section	WS-BY double-circuit lines + GT-LZD double-circuit lines	9500 MW
Xinjiang transmission section	DH-HM double-circuit lines + YD-SZ double-circuit lines	3000 MW
Qinghai transmission section	GT-LZD double-circuit lines + GL-WS double-circuit lines + SZ-YQ double-circuit lines	5000 MW
Ningxia power receiving section	BY-HH double-circuit lines + PL-LPS double-circuit lines	7000 MW

4.2 Transmission Section Identification Analysis of a Single Component Fault

Selecting a peak data of Northwest Power Grid to identify transmission sections for thermal stability. Getting the associated fault results of Heyu section and Ningxia power receiving section, and identifying key sections, the results are shown in Table 2.

Table 2. Results of key transmission section identification.

Fault	Key transmission section	Initial power	Limit power	Margin
HX-WS double-circuit lines outage	HX-WS double-circuit lines HX-LT double-circuit lines	1992	3144	36%
BY-HH double-circuit lines outage	BY-HH double-circuit lines LPS-PL double-circuit lines	6114	73118	16%
BY-HH double-circuit lines outage	BY-HH double-circuit lines BY-LZD single-circuit line	3760	6511	42%
BY-HH double-circuit lines outage	BY-HH double-circuit lines LZ-LPS double-circuit lines	6036	7080	14%

The key transmission section for thermal stability identified by HX-WS double-circuit lines outage is HX-WS double-circuit lines + HX-LT double-circuit lines, which is different from Heyu section and lacks SZ-YQ double-circuit lines. After HX-WS double-circuit lines outage, part of the load rate changes are shown in Table 3. After the fault, the load rate of 330 kV HX-LT double-circuit lines increases from 16% to 70%. According to the calculation, the transmission power limit of the section of HX-WS double-circuit lines + HX-LT double-circuit lines is 3144 MW, and the transmission margin is 36%. The limit of Heyu section is 8500 MW, the current power is 2838 MW, and the margin is 66%. Heyu section cannot accurately reflect the transmission margin of HX-WS double-circuit lines + HX-LT double-circuit lines. It is necessary to pay attention to both the section of HX-WS double-circuit lines + HX-LT double-circuit lines and Heyu section during power grid dispatching operation.

Table 3. Load rate changes after HX-WS double-circuit lines outage.

Line	Load rate before fault/%	Load rate after fault/%	Active power change/MW	Load rate change/%
HX-LT line I	16.79	70.00	464.81	53.21
HX-LT line II	16.42	68.48	454.69	52.05
SZ-YQ line I	11.79	9.01	−409.12	−2.77
SZ-YQ line II	12.02	9.078	−422.94	−2.94

Table 4. Load rate changes after BY-HH double-circuit lines outage.

Line	Load rate before fault/%	Load rate after fault/%	Active power change/MW	Load rate change/%
LPS-PL line I	18.37	60.56	2107	42.20
LPS-PL line II	18.37	60.56	2107	42.20
LZ-LPS line I	28.21	53.62	1783	25.40
LZ-LPS line II	28.18	53.5	1779	25.33
BY-LZD line	7.44	38.3	2382	30.86
LZD-PL line I	14.42	36.13	1247	21.71
LZD-PL line II	14.35	36.05	1245	21.69

There are three transmission sections for thermal stability identified after BY-HH double-circuit lines outage, including Ningxia power receiving section, BY-HH double-circuit lines + BY-LZD single-circuit line section, BY-HH double-circuit lines + LZ-LPS double-circuit lines section. The dispatching operation only monitors Ningxia power receiving section. There is only one power flow transfer channel for BY-HH double-circuit lines outage, which is BY → LZD → PL → LPS → LZ channel. The load rate changes of relevant lines before and after fault are shown in Table 4. Among them, the load rate change of LPS-PL double-circuit lines is the largest. And the load rate after fault is relatively high, reaching 60%. BY - LZD is a single-circuit line, and the active power flow change of single-circuit line is the largest. There may be overload problem when the power flow is further aggravated.

Table 5. Load rate changes after BY-HH double-circuit lines and QX-PL line I outage.

Line	Load rate before fault/%	Load rate after fault/%	Active power change/MW	Load rate change/%
LPS-PL line I	18.37	61.18	2108	42.82
LPS-PL line II	18.37	61.18	2109	42.82
LZ-LPS line I	29.58	54.19	1783	24.61
LZ-LPS line II	29.49	54.07	1779	24.58
BY-LZD line	7.44	38.51	2382	31.07
LZD-PL line I	12.40	37.73	1305	25.33
LZD-PL line II	14.35	37.65	1326	23.29
QX-PL line II	4.57	24.19	1004	19.62

4.3 Transmission Section Identification Analysis of Multiple Faults

Affected by severe weather, it is assumed that BY-HH double-circuit lines and QX-PL line I outage one after another, and the lines with large load rate change are shown in Table 5. Compared with BY-HH double-circuit lines outage, the load rate of BY → LZD → PL → LPS → LZ channel is close. QX-PL line II is added to the lines with large load rate change. And the load rate is increased from 4.57% to 24.19%, an increase of 19.62%, which is the power flow transfer channel of QX-PL line I. However, the load rate of QX-PL double-circuit lines is low before and after fault, and the risk of overload is low. The focus of overload monitoring is still on BY → LZD → PL → LPS → LZ channel after BY-HH double-circuit lines and QX-PL line I outage one after another. Therefore, the identification result is completely consistent with BY-HH double-circuit lines.

4.4 Analysis of the Influence of Security and Stability Control Measures on Identification Results

XY-LC double-circuit lines is an important transmission channel of Shaanxi power grid. Corresponding security and stability control measures have been taken for XY-LC double-circuit lines outage. The measures cut off some generators in Northern Shaanxi, and part of the load of Shaanxi power grid is powered by Gansu generators. Load rate changes after XY-LC double-circuit lines outage are shown in Table 6. The lines with large load rate changes are all related lines of Shaanxi-Gansu section, which are not the power flow transfer lines of fault line. The load rate changes are caused by load transfer. Therefore, transmission section is not identified after XY-LC double-circuit lines outage. It also shows the effectiveness of security and stability control measures, and there is no new line with overload risk in the power grid.

Table 6. Load rate changes after XY-LC double-circuit lines outage.

Line	Load rate before fault/%	Load rate after fault/%	Active power change/MW	Load rate change/%
JW-QX line II	6.43	13.73	563	7.30
MJS-BJ line I	23.19	29.73	348	6.54
PL-QX line I	5.04	11.19	439	6.15
PL-QX line II	5.05	11.18	438	6.14

5 Conclusion

An online identification method of key transmission section considering cross-region faults and security and stability control measures is proposed in this paper. According to the results of online static stability analysis, lines with large load rate change are selected as the power flow transfer branch. Then, the power flow transfer channel of fault line and the fault line channel with less influence of power flow are identified according to the consistency and correlation of power flow. And the lines in each channel cross to form key transmission sections for thermal stability. Finally, considering the distribution and capacity of power grid generator set and load, the limit power and margin of transmission sections for thermal stability are calculated. And sections with low margin are selected as key transmission sections for thermal stability. The example of actual power grid shows that the method can identify not only the sections that have been monitored by the online system, but also the sections that have overload risk. It can provide reference and guidance for power dispatchers to monitor the operation of power grid.

Acknowledgment. This research was supported by the Key Research and Development plan of Ningxia Hui Autonomous Region: Research on Key Technologies and Application of Large-scale New Energy Power Delivery safety Coordinated Control in Ningxia (5229NX18001W).

References

1. Zheng, C., Miao, T.: Study on the dynamic identification method for key branch of AC weak section and stability control. Proc. CSEE **35**(21), 5429–5436 (2015)
2. Hou, K., Shao, G., Wang, H., et al.: Research on practical power system stability analysis algorithm based on modified SVM. Protect. Control Mod. Power Syst. **V3**(1), 1–7 (2018)
3. Agarwal, S., Swetapadma, A., Panigrahi, C., et al.: A method for fault section identification in High voltage direct current transmission lines using one End measurements. Electr. Power Syst. Res. **172**, 140–151 (2019)
4. Fang, Y., Cui, X.: Strong correlated transmission sections and their identification method from perspective of transient stability. Autom. Electr. Power Syst. **43**(05), 75–82 (2019)
5. Zhao, F., Sun, H., Zhang, B.: Electrical zone division based automatic discovery of flowgates. Autom. Electr. Power Syst. **35**(5), 42–46 (2011)
6. He, J., Wang, Z., Zhang, D.: Transmission section identification based on network segmenting and side-path searching. Electr. Power Autom. Equip. **37**(07), 1–7 (2017)
7. Miao, S., Ma, S., Shang, Y., et al.: A fast recognition method of transmission section based on cut-vertex and path search. Autom. Electr. Power Syst. **38**(02), 39–45 (2014)
8. Wang, T., Li, Y., Gu, X., et al.: Identification of the key transmission sections considering optimization of geographical partition boundary for power grids. Trans. China Electrotech. Soc. **29**(04), 220–228 (2014)
9. Yan, Y., Fushuan, W., Jie, W., et al.: A method for automatic discovery of transmission sections with critical branches taken into consideration. Autom. Electr. Power Syst. **36**(14), 6–10 (2012)
10. Jiang, Z.: Studies on key assessment technology of transmission sections and faults in power system. Huazhong University of Science and Technology (2014)
11. Zhou, D., Zhang, B., Yao, F., et al.: Fast search for transmission section based on graph theory. Proc. CSEE **26**(12), 32–38 (2006)
12. Liu, X., Ren, J., Qu, W.: A fast search algorithm of power flow transfer for multi-branches removal event based on breadth first search. Autom. Electr. Power Syst. **39**(13), 69–74 (2015)
13. Xu, Y., Zhi, J.: Identification of transmission section based on key lines. J. North China Electr. Power Univ. Nat. Sci. Edn. **43**(01), 21–26 (2016)
14. Ren, J., Wei, J.: Vulnerable line identification and its transmission section search for power system. Electr. Power Autom. Equip. **35**(07), 37–42 (2015)

15. Wang, B., Guo, W., Xiang, D., et al.: Running section similarity matching based on improved K-means algorithm. Electr. Power Autom. Equip. **37**(09), 166–170 (2017)
16. Ling, X.: Study of recognition of dangerous lines casued by power flow transferring in multi-branches removal event and overload elimination control strategy. North China Electric Power University (2108)
17. Nie, H., Yuan, X., Zhang, H., et al.: Rapid flow transferring identification based on multiple line tripping and key branch set. Power Syst. Protect. Control **42**(17), 38–43 (2014)

Power Quality Prediction of Active Distribution Network Based on Clustering and Neural Network

Junpeng Shu[1], Jie Si[2], Jing Li[3], and Guoqing Weng[1(✉)]

[1] College of Information Engineering, Zhejiang University of Technology, Hangzhou, China
wgq@zjut.edu.cn
[2] Zhejiang Tusheng Power Transmission and Transformation Co., Ltd., Lucheng Branch, Wenzhou, China
[3] State Grid Wenzhou Power Supply Company, Wenzhou, China

Abstract. For the power quality problems that exist in the active distribution network with distributed generation, this paper predicts the power quality data in active distribution combining with various external factors affecting the power quality. Considering that the traditional BP neural network prediction is easy to fall into the local optimum solution, a new prediction approach based on K-means clustering and BP neural network for the steady-state indexes of power quality in an active distribution network is proposed. By adding the clustering step before prediction, the power quality influencing factors such as input time, light intensity, temperature and loads are clustered. After clustering, each type of error-minimum BP neural network is trained separately, and then the influencing factors of the prediction are identified by class markers. At last, the best BP neural network model of the corresponding category is called to predict the power quality of the active distribution network. The analysis of the example shows that compared with the traditional BP neural network prediction method to predict the steady-state indexes of the power quality, the use of clustering and neural network for active distribution network quality prediction can improve the prediction accuracy and has good feasibility.

Keywords: Active distribution network · Power quality · Prediction · Cluster · Neural network

1 Introduction

In recent years, with the increasing penetration rate of distributed generation (DG) and more requirements for better monitoring and control of distribution networks, the form of traditional distribution networks is constantly shifting toward the direction of active distribution networks (AND) [1, 2]. Active distribution networks require a wider range of measurements and more intelligent control analysis among which the monitoring and analysis, situation awareness and optimization decision-support of power quality (PQ) have become the research hot spot in the field of active distribution networks [3]. It is of great significance for both power suppliers and power users to have a research

Y. Xue et al. (Eds.): PMF 2020, LNEE 718, pp. 462–474, 2021.
https://doi.org/10.1007/978-981-15-9746-6_36

on the reasonable prediction of power quality of active distribution network with DG high-permeability access, which can discover the changing trend of power quality and possible hidden problems in the power grid quickly and accurately [4, 5].

In recent years, several scholars at home and abroad have made a research on the field of power quality prediction. Power quality index prediction has certain similarities with power load forecasting. Both of them are based on the characteristics and changing rules of historical data to predict the trend of future time periods. However, the characteristics of power quality indexes change with the influence of external factors. Therefore, different prediction methods and algorithm improvements are usually adopted according to the changing characteristics. At present, the commonly used power quality prediction algorithms mainly include time series algorithm, Monte Carlo method, random forest model, cluster analysis method, neural network method and so on. In [6], based on the mutual differences of typical power quality steady-state indicators, a preferred combination prediction model including linear regression method, random time series method and grey model is established. However, it is preferred that the prediction accuracy of each algorithm of the combined model is not high, resulting in an inaccurate prediction. In [7], PQ prediction was taken as a randomness forecasting problem essentially and a simulation experiment on the probability model can be carried out with random sampling to obtain the statistical characteristics and conclude the approximations of required indexes. However, the Monte Carlo method needs a large number of measured data, which results in more computational complexity and longer forecasting time. Paper [8] quantifies the correlation between typical indicators such as flicker and voltage total harmonic distortion rate in off-grid systems and temperature, energy storage battery status, etc., and establishes a random forest model for training and prediction, but does not consider the impact of power quality in the active distribution network. In [9], a multi-layer neural network with a backward learning model was put forward, which showed good self-learning ability and adaptability. However, it does not consider the complex impact on the power grid in the active distribution network, nor compares the performances of different neural network prediction models. In [10, 11], a main component clustering analysis method was adopted to standardize, reduce dimension and eliminate correlation for the multidimensional data of PQ. It has the advantage of a visual and convenient prediction process, while the dimension reducing process may lead to losing key information and being hard to determine the principal component in the power distribution network.

After taking consideration that BP neural network has the characteristics of being easy to fall into the local optimal solution, and adding the clustering leading step before the prediction is beneficial to reduce the difference of training data to get the global optimal solution, this paper proposes a new prediction method for active distribution network with distributed generation based on clustering and neural network predicting algorithm. The rest of this paper is organized as follows. Section 2 describes active distribution network steady-state PQ indexes as well as the general prediction process. Section 3 describes the characteristics and steps of cluster analysis and the neural network for PQ prediction. Section 4 is dedicated to describe the analysis of the examples and show the property of the neural network and the comparison of the effect adding clustering or not. Finally, conclusions are drawn in Sect. 5.

2 Prediction Process for Power Quality

2.1 Steady-State Power Quality Index

According to the disturbance characteristics, the power quality indexes of the distribution network can be classified into two categories: steady-state indexes and transient indexes. Steady-state power quality indexes mainly include frequency deviation, voltage deviation, voltage fluctuation and flicker, three-phase voltage imbalance, harmonic distortion rate and so on [12].

Since the transient power quality problem has the characteristics of randomness and suddenness, the transient power quality indexes are generally used for post-positioning, detection, and distribution statistics, which is difficult to use for prediction [13, 14]. However, the steady-state power quality indexes have the periodic, aperiodic characteristics of the physical parameters which remain relatively stable for a foreseeable period of time and can be reduced by data analysis [15]. Therefore, this paper only chooses the steady-state power quality index to construct a power quality prediction model for prediction.

2.2 Prediction Process of Steady-State Indexes for ADN Power Quality

Since the application background is an active distribution network, which contains DG, it is necessary to consider the influence of environmental factors on the trend of power quality change. In order to achieve the rapidity and accuracy of the trend prediction of steady-state power quality indicators, it is necessary to establish an active distribution network with DG firstly, and then choose the appropriate prediction algorithm to obtain the highest possible accuracy. The approved prediction process of the steady-state PQ indexes is shown in Fig. 1.

The basic idea of active distribution network power quality prediction based on K-means clustering and BP neural network can be summarized as follows: Firstly, the environmental factors affecting the trend of power quality variation of DG in the active distribution network are divided into several different categories based on K-means clustering, and a cluster-based environment input model is constructed. Then, using the PQ data of the historical period as the output, the BP neural network with the smallest error is trained for each data group in each class, and the influencing factor data of the future time period is input after the prediction checks which category it belongs to. At last, the corresponding BP neural network prediction model is used to predict the power quality of the active distribution network and acquire more accurate and reliable prediction results to achieve accurate prediction of the steady-state PQ data in the target grid.

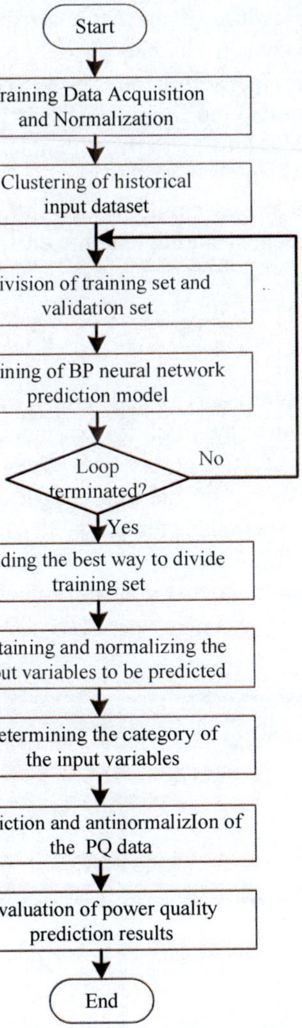

Fig. 1. Power quality index prediction process

3 Power Quality Prediction Model Based on Clustering and Neural Network

3.1 Clustering and Prediction Algorithms

The algorithm of cluster analysis can be divided into Partitioning Methods, Hierarchical Methods, Density-Based Methods, Grid-Based Methods, and Model-Based Methods, etc. This paper uses the Partition-Based K-means clustering algorithm [16]. In terms of prediction, BP neural network is used in this model.

K-means Clustering Algorithm. K-means clustering algorithm is a simple and efficient clustering algorithm, which has been widely used in many research fields. The basic idea of the K-means algorithm is as follows: firstly, we select the initial cluster centers randomly and calculate the Euclidean distance from each sample point to the initial cluster center. Then, according to the principle that the closer the distance is, the larger the similarity is, and they are allocated to the categories represented by the clustering center with the largest similarity. At last, calculate the mean value of all sample points for every category and update the cluster center until the target criterion function converges [17].

A schematic diagram of the clustering process of the K-means algorithm is shown in Fig. 2. In the figure, the upper left corner is the initial state, and each point is the sample point to be predicted. It is assumed that the sample points are divided into three categories. Besides, the three points of red, green, and blue are randomly selected as initial cluster centers. Calculate the Euclidean distance from the remaining points to the clustering center, and classify the points with the smallest distances as a class. The initial clustering result is shown on the upper right, and the points in the same color area are in the same class. Then, the average value of all samples in each class is taken as the new cluster center. The movement of the cluster center is shown in the lower right corner. Repeat the above process until the calculation results of the cluster center converge.

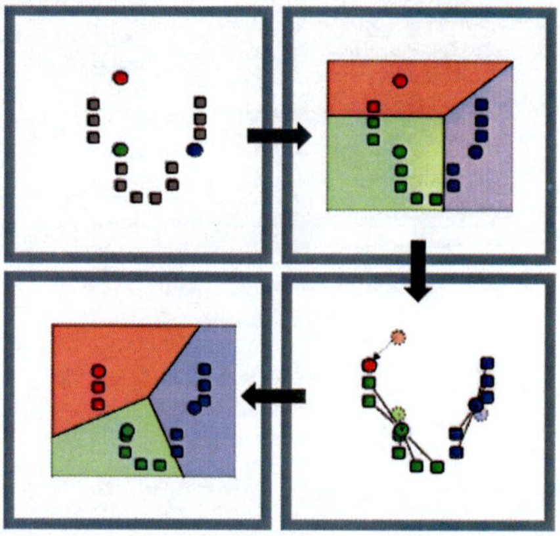

Fig. 2. Clustering process of the K-means algorithm

BP Neural Network Prediction Algorithm. The basic BP algorithm includes two processes, including forward propagation of signals and back propagation of errors. In other words, the calculation of the error output is performed in the direction from the input to the output, and the adjustment weight and the threshold are performed from the output to the input. In the process of forward propagation, the input signal acts on the output node through the hidden layer and undergoes a nonlinear transformation to generate an output signal. If the actual output does not match the expected output, it is transferred to the backpropagation process of the error. Error back propagation is to pass the output error back to the input layer through the hidden layer, and distribute the error to all the units in each layer so as to obtain the error signal from every layer as the basis for adjusting the weight of each unit [18]. By adjusting the threshold value and the connection strength between the input node and the hidden layer node and the connection strength between the hidden layer node and the output node, the error is decreased along the gradient direction. After repeated learning and training, the network parameters corresponding to the minimum error, such as weights and thresholds, are determined, and the training is finished.

According to the analysis of the influencing factors of power quality indexes, we take the time, temperature, illumination data, the load of the public connection point, and the load of the remaining points of the network as the network's input data, and take the BP network constructed with the power quality indexes as the network's output data. Because the training data is relatively small and has volatility, this paper uses the cross-validation method to train the BP neural network as well as find the best number of hidden layer nodes and the best partitioning way of the training set for data prediction.

3.2 Implementation Steps of Power Quality Clustering and Prediction

The prediction model of power quality steady-state indexes of an active distribution network based on K-means clustering and BP neural network can be summarized as follows.

1) Training data collection and normalization processing. In order to obtain predictive performance with sufficiently high accuracy, it is necessary to provide the prediction model with training data covering a sufficiently wide and sufficient number of scenes; therefore, it is necessary to collect data on the power quality indexes items and their influencing factors for the target prediction points firstly, then input the collected data as a historical data set and perform normalization operations.
2) Clustering partition of historical input data sets. On the basis of training data collecting and normalization processing, the k-means clustering algorithm is used to cluster the historical input data sets.
3) Training set and validation set partitioning. In all kinds of data sets that have been clustered, the training set and the verification set are respectively divided. The partition method is to take 10% of all data as the validation set and the remaining 90% is used as the training set.
4) BP neural network prediction model training. Take the training sets after the k-means clustering in step 3 as input signals, which act on the output node through the hidden layer to generate an output signal through nonlinear transformation. If there is an error between the actual output and the expected output, the error is propagated back to the input layer through the hidden layer. By adjusting the threshold value and the connection strength between the input node and the hidden layer node and the connection strength between the hidden layer node and the output node, the error is decreased along the gradient direction. After repeated learning and training, the network parameters corresponding to the minimum error such as weights and thresholds are determined, and the training is finished.

5) Find the optimal training set partition method. In the clustered data set, all types are re-divided into training sets and validation sets. Take 10% of the data that is mutually exclusive with the verification set taken in step 3 as the new verification set and the remaining 90% is taken as the new training set. Repeat steps 3 and steps 4 to record the K error minimum models trained by the optimal number of hidden layer nodes in the optimal division way, and respectively correspond to the K clusters obtained by clustering. The K error minimum models are the minimum error neural network prediction models of every cluster.
6) Predictive input variable value acquisition and normalization. Anti-normalize the obtained prediction data, then compare them with the practical results and analyze the errors. Input the variable value vector of the influencing factor of the power quality prediction node in a certain period of time in the future, and perform normalization processing.
7) Determine the cluster class attribution of the input variables. Calculate the input variable vector to the square of the two norms of each cluster mean vector and find the cluster corresponding to the smallest value and classify it into the cluster.
8) Power quality prediction output and inverse normalization. Take the future input vector into the error minimum neural network prediction model according to its cluster determined in step 5, and obtain the predicted power quality index results, then anti-normalize the output power quality index prediction results.
9) Evaluation of Power Quality prediction results. Collect the real power quality data corresponding to the prediction period of step 6 on the node of the target distribution network. Calculate the prediction error, and evaluate the prediction result.

4 Analysis of Examples

4.1 Construction of Simulation Model and Data Acquisition

In order to obtain a large number of historical data needed for training neural networks, analysis data is obtained by Simulink simulation. Firstly, a model of 10.5 kV IEEE

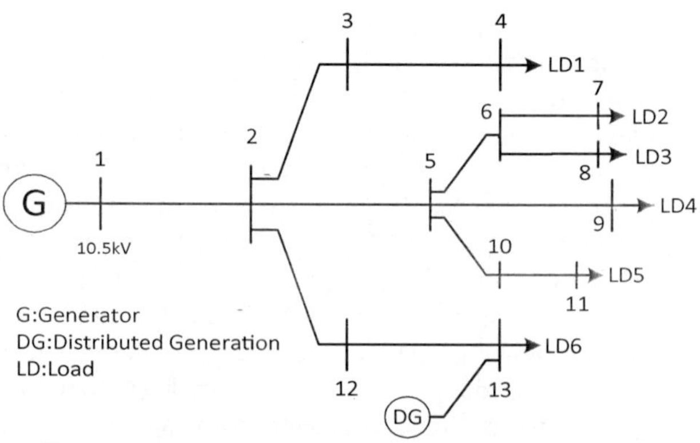

Fig. 3. The topology of 13-node active distribution network

13-node simulation distribution network is constructed as shown in Fig. 3, and the parameters of various circuits, loads and other devices are set up.

This paper constructs a 13-node active distribution network model based on Simulink as shown in Fig. 3. It has six power loads and we can set different load variation curves according to its load characteristics. LD6 is selected as the access point of DG, and the power quality monitoring device is configured here.

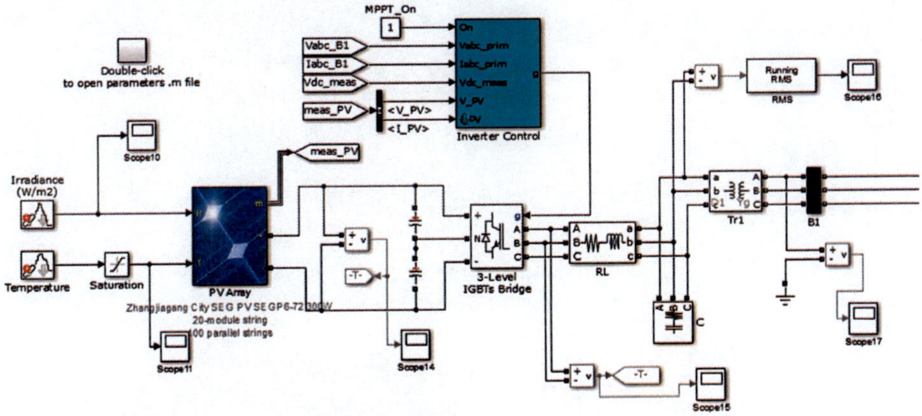

Fig. 4. New energy power generation module

The simulation of the new energy power generation uses the photovoltaic array (PV Array) module as shown in Fig. 4. The module is composed of 100 photovoltaic panels in parallel and 20 in series. The maximum output power of a single photovoltaic panel is 300 W. The daily temperature and illumination curves are input to simulate the changes of environmental factors in a day so that the real-time output power of PV Array can be adjusted. The variation of time-sharing light intensity in 21 days is shown in Fig. 5 and the variation of time-sharing temperature is shown in Fig. 6. In order to broaden the coverage of data, the difference between daily temperatures is enlarged here. The DC output from the photovoltaic array is converted to AC via three-phase IGBT full-bridge inverters. After being filtered by series RL and capacitance, it is lifted to the grid voltage by the transformer for grid connection.

PQ monitoring device is used to collect data of voltage, current and frequency, and then the power quality indexes such as voltage deviation, frequency deviation, three-phase voltage unbalance and harmonic distortion rate are obtained according to the calculation formula. The average values of these indexes in each hour are calculated respectively as the power quality index data in this period. The hourly power quality data of 21 days under different environmental factors can be obtained by many times of simulation with environmental factors changed in each time. Finally, the dimension of historical data is normalized by categories to accelerate the convergence speed of parameter optimization in the prediction process.

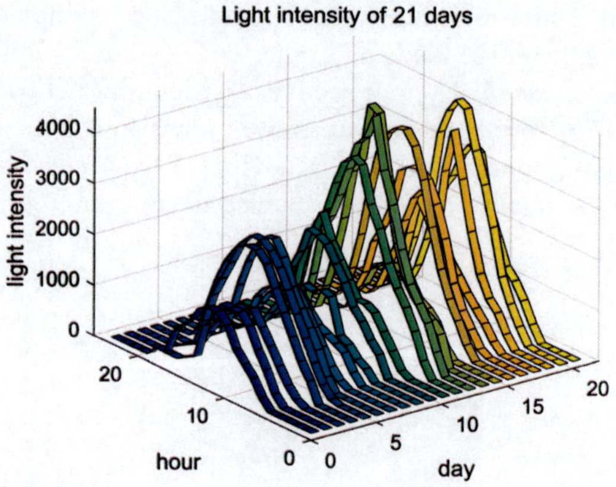

Fig. 5. Variation of time-sharing light intensity in 21 days

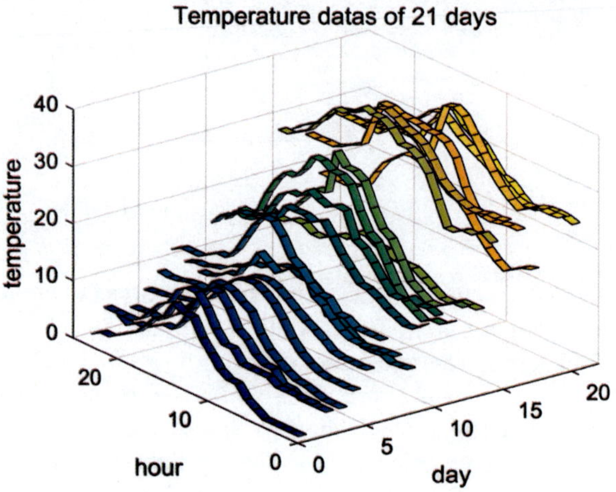

Fig. 6. Variation of time-sharing temperature in 21 days

4.2 Clustering Processing of Input Data

Considering that BP neural network is easy to fall into local optimal solution, this method uses the K-means clustering algorithm to cluster the environmental variables and load variables that is the input of the neural network before prediction, so as to reduce the difference of training set and get the global optimal solution. The steps of clustering are as follows:

1) Select the data of the first 20 days as historical data, and the data of the 21st day as future data to be predicted which is used to be compared with the prediction results. Select four-row vector subset samples randomly from the input variables of historical data, namely time stamp, temperature, light and load, as clustering center.

2) Calculate the Euclidean distance between every row vector subset with every clustering center in the historical data set. Mark the cluster with the minimum Euclidean distance which is the cluster the subset belongs to and put every subset into corresponding clusters.
3) Take the average value in every cluster respectively as the updated clustering center.
4) Repeat steps 2 and 3 until the difference between the updated clustering center and the original clustering center is less than the set error limit, that is, the calculation result converges.

Finally, the input indexes of the BP neural network, such as time mark, temperature, illumination and load data, are divided into four categories.

4.3 Power Quality Indexes Prediction

The key steps of power quality indexes prediction are as follows:

1) Divide the clustered data set into the training set and the validation set: 10% of all data is taken as the validation set, and the remaining 90% is taken as the training set.
2) All categories of training sets clustered by K-means are regarded as input signals of the BP neural network prediction model which act on the output node through the hidden layer, and the output signals are generated by the non-linear transformation. If there are errors between the actual output and the expected output, the errors will be propagated back to the input layer through the hidden layer. By adjusting the joint strength between the nodes of the input layer and the hidden layer, the joint strength as well as the threshold between the nodes of the hidden layer and the output layer, we can make the error decrease along the gradient direction. After repeated learning and training, the weight and threshold corresponding to the minimum error is determined, and the training stops.
3) Re-divide all categories in the clustered data into training sets and verification sets, in which 10% data mutually exclusive from the validation sets in step 2 are taken as validation sets, and the remaining 90% are taken as training sets. Repeat step 2. Record the four models with minimum error trained with the best node number of hidden layers under the optimal partitioning way. The four models are corresponding to four categories clustered. These four models are neural network prediction models with the minimum error of every cluster.
4) Calculate the Euclidean distance between every row vector subset of input data on the 21st day with every clustering center. Mark the cluster where the distance is the smallest and put the subset into the cluster. The input variables to be predicted are input into the neural network prediction models with the minimum error of its cluster determined in step 3, and the prediction results can be obtained. Anti-normalize the output power quality prediction results and compare the results with the future data on the 21st day, then evaluate the predicted results.

Plot the clustered and non-clustered BP neural network prediction results and the relative error waveform using MATLAB. Here, the voltage deviation is taken as an example as shown in Fig. 7 and Fig. 8.

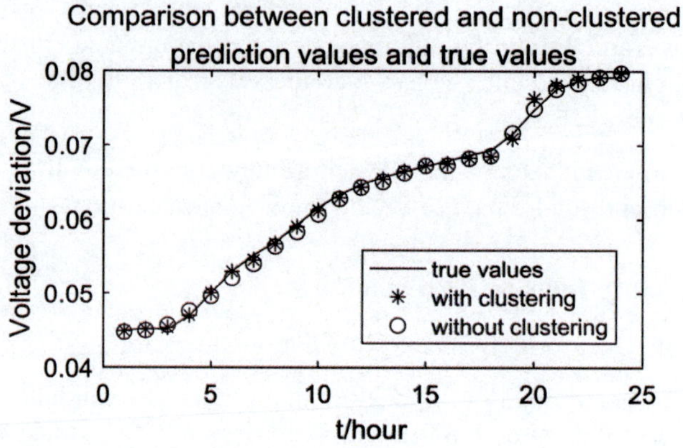

Fig. 7. Comparison between clustered and non-clustered prediction values and true values

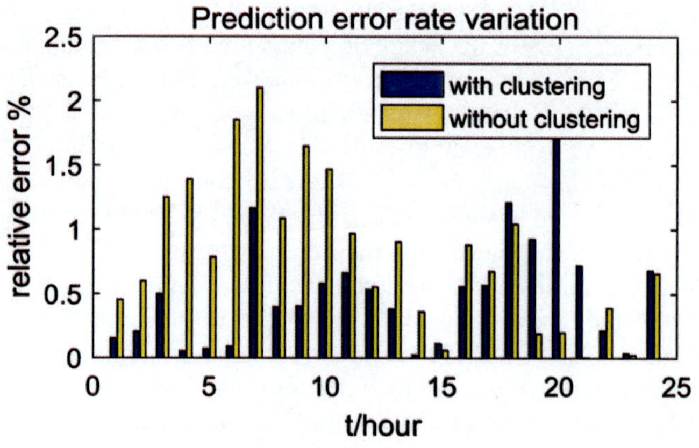

Fig. 8. Prediction error rate variation

According to the simulation process and the error rate variation of the steady-state power quality index shown above, the following conclusions can be drawn:

1) The effect of neural network based on predicting method is good and of high accuracy on power quality indexes. The prediction error can even be less than 1%, which proves the feasibility of using the neural network to predict the power quality of an active distribution network.
2) As can be seen from Fig. 8, the relative error with clustering is smaller than that without clustering in most cases. It indicates that getting the input data clustered

before prediction and training neural networks for every cluster can overcome the shortcomings of easily falling into the local optimum solution when using the BP neural network alone and effectively improve the accuracy of prediction.

5 Conclusion

Active distribution network contains a large number of distributed generations, whose generated energy is influenced by daily temperature, illumination, and load changes with certain fluctuations. Applying clustering technology to the link of power quality data preprocessing before training neural network will provide more pertinence to the training of the network and the selection of prediction model which is helpful to improve the accuracy of prediction and provide with evidence for the warning, diagnosis and treatment as well as ensure the safe and stable operation of the power system. Therefore, it is very important to carry out reasonable data classification before predicting the power quality of the active distribution network. This paper proposes a prediction model based on K-means clustering and BP neural network for steady-state power quality index of active distribution network after considering that the BP neural network is easy to fall into the local optimal solution and adding a clustering step before prediction which is beneficial to reduce the difference of training set so that global optimal solution is more inclined to be obtained. The analysis of examples shows that predicting method combining clustering and the neural network has good feasibility for power quality indexes in the active distribution network and the addition of clustering can make the prediction result more accurate.

References

1. Yi, Y., Dong, L., Weng, W.P.: Technology and its trends of active distribution network. Autom. Electr. Power Syst. **36**(18), 10–16 (2012)
2. Weng, G.Q., Huang, F.T., Tang, Y.F., Yan, J., Nan, Y.R., He, H.B.: Fault-tolerant location of transient voltage disturbance source for DG integrated smart grid. Electr. Power Syst. Res. **144**, 13–22 (2017)
3. Zhao, B., Wang, C.S., Zhou, J.H., Zhao, J.H., Yang, Y.Q., Yu, J.L.: Current situation and future development of active distribution network. Autom. Electr. Power Syst. **38**(18), 125–135 (2014)
4. Liang, Y.W., Hu, Z.J., Chen, Y.P.: A review of distributed generation and its application in power systems. Power Syst. Technol. **2003**(12), 71–75+88 (2003)
5. James, R., Petr, M., Stanislav, M., Lukáš, P., Petr, B., Václav, S.: Towards prediction of photovoltaic power quality. In: 26th IEEE Canadian Conference on Electrical and Computer Engineering, Saskatchewan, pp. 1–4. IEEE (2013)
6. Ding, Z.J., Liu, P., Ou, Y.S.: Mechanism of power quality forecast and early warning and their application. In: Proceedings of the CSU-EPSA, vol. 27, no. 10, pp. 87–92 (2015)
7. Qin, H.T., Li, Q.Z., Liu, Y., Dong, X.: Prediction of the power quality caused by electrified railways based on Monte Carlo. Power Syst. Protect. Control **39**(13), 64–70+77 (2011)

8. Tomas, V., Stanislav, M., Tomas, J., et al.: The power quality forecasting model for off-grid system supported by multi-objective optimization. IEEE Trans. Ind. Electron. **64**(12), 9507–9516 (2017)
9. Jindrich, S., Stanislav, M., Tomas, V., Tomas, B.: A power quality forecasting model as an integrate part of active demand side management using artificial intelligence technique - multilayer neural network with backpropagation learning algorithm. In: IEEE 15th International Conference on Environment and Electrical Engineering, Rome, pp. 611–616. IEEE (2015)
10. Cui, X.F., Ma, Z.Y., Xu, Z., Wen, C.X.: Forecast of power quality index based on the discrete fourier decomposition and AR model. Mod. Electr. Power **30**(06), 38–42 (2013)
11. Lu, M.X.: Power quality prediction based on least squares method. Vibroengineering Procedia **2017**(12), 153–158 (2017)
12. Shao, R.P., Han, Z.W., Lin, J.G.: Power quality index analysis. Electr. Power Syst. Autom. **2007**(03), 118–121 (2007)
13. Sikes, D.L.: Comparison between power quality monitoring results and predicted stochastic assessment voltage sags-"real" reliability for the customer. IEEE Trans. Ind. Appl. **36**(2), 677–682 (2000)
14. Weng, G.Q., Huang, F.T., Yan, J., Yang, X.D., Zhang, Y.B., He, H.B.: A fault-tolerant location approach for transient voltage disturbance source based on information fusion. Energies **9**(12), 1092 (2016)
15. Bai, J.J., Gu, W., Yuan, X.D., Li, Q., Xue, F., Wang, X.C.: Power quality prediction, early warning, and control for points of common coupling with wind farms. Energies **8**(9), 9365–9382 (2015)
16. Sun, J.G., Liu, J., Zhao, L.Y.: Clustering algorithms research. J. Softw. **19**(1), 48–61 (2008)
17. Wang, Q., Wang, C., Feng, Z.Y.: Research review of k-means clustering algorithm. Electron. Des. Eng. **20**(07), 21–24 (2012)
18. Liu, L., Peng, C.J., Luo, Y.: Short-term daily load forecasting of electric power based on BP neural network. Electr. Eng. **2011**(02), 15–17 (2011)

Comprehensive Assessment Method of Natural Disaster Risk for Compact Transmission Corridor

Dawei Wu[1,2,3](✉), Lei Zhang[1,2,3], Hantao Tao[1,2,3], Yue Chen[1,2], and Zhibo Jiang[1,2]

[1] NARI Group Corporation Ltd., Nanjing 211106, China
wudawei@sgepri.sgcc.com.cn
[2] Wuhan NARI Limited Liability Company, State Grid Electric Power Research Institute, Wuhan 430074, China
[3] Hubei Key Laboratory of Power Grid Lightning Risk Prevention, Wuhan 430074, China

Abstract. In order to promote natural disaster protection level of compact transmission corridor, a comprehensive assessment method for evaluating the overall risk of lightning strike, windage, contamination and icing is proposed in this paper. On the basis of existing risk maps, the data is standardized as the risk value for each natural disaster accordingly. Combining with historical tripping data of compact transmission corridor and expert scores for the four natural disasters, the risk weights of lightning strike, windage, contamination and icing are determined by integrating analytic hierarchy process (AHP) with entropy weight. Through multiplying the risk values and risk weights, the overall risk of the four natural disasters is calculated. Using the tripping data of one compact transmission corridor from 2013 to 2016 for risk assessment analysis, the result shows that the accuracy of this comprehensive assessment method is up to 94.4%, which can be used to improve the stability and security of compact transmission corridor.

Keywords: Compact transmission corridor · Natural disaster · Comprehensive risk assessment · Analytic hierarchy process (AHP) · Entropy weight

1 Introduction

Being exposed to atmosphere environment for a long time, transmission lines are vulnerable to different types of natural disasters, which may cause tripping accidents [1]. Each tripping accident will bring huge impact on the operation of power grid. Natural disasters, such as lightning strike, windage, contamination and icing, are the main causes for the unplanned shutdown of transmission lines [2–4]. According to the statistics of line tripping data of State Grid Corporation of China (SGCC) in 2018, the proportion of unplanned shutdown caused by natural disasters is up to 75% for 500 kV transmission lines. In the context of global climate change, the intensity and frequency of extreme

Project Supported by the National Natural Science Foundation of China (51777037), and Science and Technology Project of SGEPRI (524625190050).

Y. Xue et al. (Eds.): PMF 2020, LNEE 718, pp. 475–486, 2021.
https://doi.org/10.1007/978-981-15-9746-6_37

disaster weather are on the increasing trend. Therefore, how to deal with the threat of natural disasters is one of the urgent problems that need to be solved for power grid.

Natural disasters prevention has been a research hotspot in the field of electrical engineering [5, 6]. During the development and construction of power grid, abundant safety ensuring technologies against natural disasters have been accumulated, such as installing protective equipment or improving structural design and so on. However, the existing external insulation designs of transmission line are made by just considering the impact of lightning strike, windage, contamination and icing alone [7]. So, there is a lack of comprehensive understanding about the risk of various natural disasters for transmission lines. As a result, it is difficult to carry out natural disaster prevention in a targeted way, and there will be a series of problems on the actual operation of power grid [8–10]. For example, windage reduces the discharge path, which will make lightning trip-out easier. In order to further improve the ability of natural disasters prevention for power grid, it is necessary to carry out researches on comprehensive assessment.

With the rapid development of extra-high voltages (EHV) and ultra-high voltages (UHV), compact transmission corridors have gradually formed, in which transmission lines are arranged closely and transmission power are huge. These compact transmission corridors have become the hub of power transmission among regional power grids. Because of the long-distance transmission and large-width arrangement, it is inevitable to pass through areas with various natural disasters. The risk of two or more transmission lines tripping caused by natural disasters in one corridor exists objectively. This will cause violent oscillation of the power system, and even lead to blackout events in large area as the condition worsens. For the sake of a better operation level of power grid, it is of great significance to carry out researches on comprehensive assessment of natural disasters risk for compact transmission corridor.

2 Risk Maps

Taking 500 kV transmission lines of one provincial power grid as an example, the causes and outage rates of the line tripping from 2007 to 2016 are analyzed as shown in Table 1.

Table 1. Causes and outage rates of 500 kV transmission lines tripping of one provincial power grid from 2007 to 2016

No.	Causes	Tripping times	Proportion	Outage times	Outage rate
1	Lightning strike	137	43.8%	16	0.028
2	Mechanical touch	45	14.4%	41	0.071
3	Foreign matter	29	9.3%	12	0.021
4	Windage	24	7.7%	18	0.031
5	Bird damage	21	6.7%	2	0.003
6	Mountain fire	12	3.8%	8	0.014
7	Branch discharge	12	3.8%	6	0.010
8	Contamination	7	2.2%	3	0.005

(*continued*)

Table 1. (*continued*)

No.	Causes	Tripping times	Proportion	Outage times	Outage rate
9	Icing	7	2.2%	2	0.003
10	Construction quality	5	1.6%	5	0.009
11	Line galloping	5	1.6%	4	0.007
12	Unknown cause	2	0.6%	2	0.003
13	Other	6	1.9%	6	0.010
14	Improper maintenance	1	0.3%	1	0.002
15	Total	313	100%	126	0.217

It can be seen that the times of line tripping caused by lightning strike, windage, contamination, and icing are 137, 24, 7, and 7 respectively, accounting for 43.8%, 7.7%, 2.2%, and 2.2% of the total. The percentage of line tripping caused by these four natural disasters is up to 55.9%, which means that natural disasters are the main causes of line tripping for this provincial power grid. The conclusion is also applicable to other provincial power grids.

In this paper, lightning strike, windage, contamination, and icing are mainly considered to evaluate the overall risk of natural disaster for compact transmission corridor. Their risk maps in this province drawn by the meteorological data from 2007 to 2016 are shown in Fig. 1. Each risk map is quite irregular itself and the four risk maps vary greatly even in the same place, so it is difficult to form a comprehensive understanding.

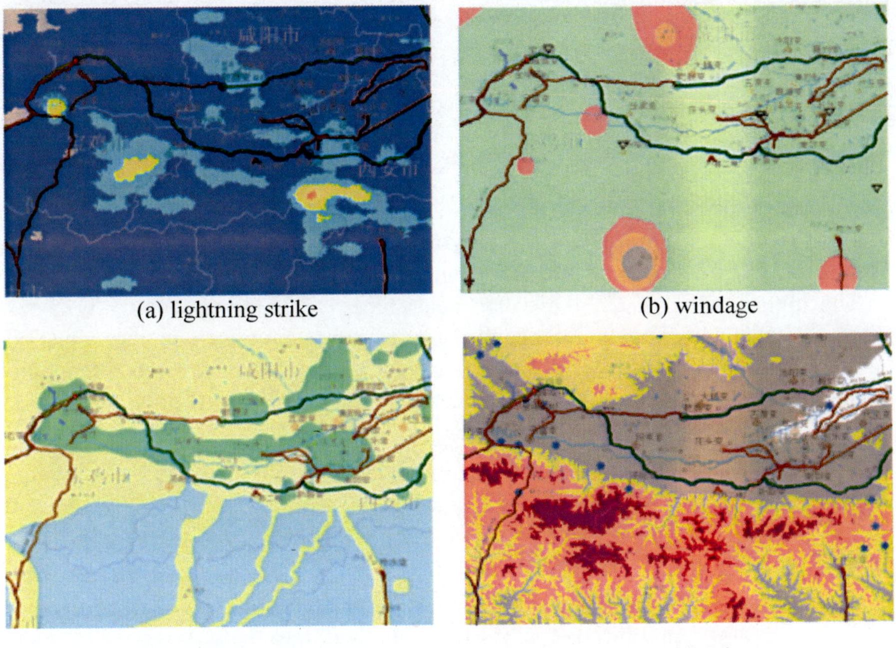

(a) lightning strike (b) windage

(c) contamination (d) icing

Fig. 1. Risk maps of the main natural disasters in one province

3 Risk Weight Determination

3.1 Analytic Hierarchy Process

Analytic hierarchy process (AHP) is a subjective assessment method by combining qualitative analysis and quantitative analysis together, which can be used to solve complex multi-objective decision-making problems. It has the advantages of simplicity, practicality, and less information demand for quantitative data. Its basic idea is to divide the complex problem into several objectives, and decompose each objective to several assessment indexes level by level. At last, qualitative index method and fuzzy quantitative method are adopted to calculate the comprehensive order of assessment index at each level, and the multi-objective decision-making problem is settled on the basis of the order. The basic steps are shown below.

Establishment of Recursive Hierarchy. Relevant indexes of the multi-objective decision-making problem are decomposed into several levels from high to low depending on their attributes, which will result in forming a multi-level index model.

Construction of Judgment Matrix at Each Level. According to the comparison of relative importance of all indexes in this level to the index in upper level, the nine-scale method is usually used to assign values to the elements of judgment matrix.

Consistency Check of the Judgment Matrix. Due to the complexity of multi-level judgment, some values in the judgment matrix may be inconsistent. Therefore, it is necessary to check the consistency of judgment matrix by using the two formulas as shown below.

$$CI = \frac{\lambda_{\max} - n}{n - 1} \tag{1}$$

$$CR = \frac{CI}{RI} \tag{2}$$

Where n represents the order of the judgment matrix, $\lambda_{\max}$ represents the maximum eigenvalue, and RI represents the mean random consistency index.

The higher value of CI, the worse consistency of judgment matrix is checked. When $CR < 0.1$, it means that the consistency of judgment matrix is very well. When the judgment matrix fails to pass the consistency check, it is necessary to reassign the values of relative importance among the assessment indexes until passing the check.

Determination of Hierarchy Weight. If the judgment matrix passes consistency check, eigenvector corresponding to the maximum eigenvalue is the importance ranking of the present indexes relative to the index in upper level.

3.2 Entropy Weight

Entropy weight is an objective assessment method based on information entropy principle [11]. The smaller information entropy of the index indicates the greater information, which means playing a greater importance role in the comprehensive assessment. So, the higher value of weight will be assigned to the index.

Assuming that there is an index system to be assessed, which is consist of m assessment objects and N assessment indexes. The original data can be represented by matrix $R_{\mathrm{m}\times\mathrm{n}}$, then the assessment steps by using entropy weight are shown below.

Dimensionless Processing for the Original Data. In order to eliminate the influence of dimension of the assessment indexes, it is necessary to carry out dimensionless processing to the original data $R_{m\times n}$ by using the extreme value (the maximum or minimum value) of each column. The result after dimensionless processing is recorded as matrix $S = (s_{ij})_{m\times n}$.

Data Normalization. The above dimensionless processing result should be normalized into the range of [0, 1], which will reduce the influence of magnitude of the assessment indexes on the assessment results. After data normalization, the results are recorded as matrix $D = (d_{ij})_{m\times n}$.

Entropy Value Calculation. The entropy value of each assessment index is calculated according to its definition as shown below.

$$H_j = -k\sum_{i=1}^{m} t_{ij}\ln t_{ij} \tag{3}$$

Where $t_{ij} = \frac{d_{ij}}{\sum_{i=1}^{m} d_{ij}} (j = 1, 2, \cdots, n)$, $k = \frac{1}{\ln m}$.

Determination of Entropy Weight. The entropy weight of each assessment index is calculated by the following formula.

$$\varphi_j = \frac{\alpha_j}{\sum_{j=1}^{n} \alpha_j} (j = 1, 2, \cdots, n) \tag{4}$$

Where α_j is the difference coefficient of the jth assessment index, which can be expressed as shown below.

$$a_j = 1 - H_j(j = 1, 2, \ldots, n) \tag{5}$$

3.3 Combinatorial Optimization of AHP and Entropy Weight

AHP is a subjective assessment method, and personal preference has a great influence on the assessment results. It is very easy to make mistakes in decision making. Entropy weight is an objective assessment method, and it is also very easy to be affected by the original data.

Aiming at the shortcomings of single assessment method, least-square is used to optimize the combination of AHP and entropy weight in this paper. The combinatorial optimization can integrate the advantages of the two methods and make up the insufficiency of single method, so as to improve the accuracy of the weights [12].

The weights of each index calculated by AHP and entropy weight method are expressed as ω_j, φ_j respectively. Assuming that the comprehensive weight of each assessment index is μ_j, and the assessment index matrix $R_{m\times n}$ is expressed as $(s_{ij})_{m\times n}$ after dimensionless processing. For the assessment indexes of all objects, the deviation of subjective weight and objective weight calculated by AHP and entropy weight respectively should be as small as possible. So, the combinatorial optimization model of comprehensive weight established by least square can be shown as follows [13].

$$\min F(\mu) = \sum_{i=1}^{m}\sum_{j=1}^{n}\{[(\omega_j - \mu_j)s_{ij}]^2 + [(\varphi_j - \mu_j)s_{ij}]^2\} \tag{6}$$

Where: $\sum_{j=1}^{n}\mu_j = 1, \mu_{ij} \geq 0 (j = 1, 2, \cdots, n)$.

The value of comprehensive weight for each assessment index can be obtained by using Lagrangian method to solve the above model.

4 Establishment of Comprehensive Assessment Model

The risk values of natural disasters such as lightning strike, windage, contamination and icing can be obtained from the existing risk maps. And the comprehensive risk weight can be determined by the combinatorial optimization of AHP and entropy weight. The overall risk can be calculated by multiplying the risk values and risk weights. The processes are shown in Fig. 2.

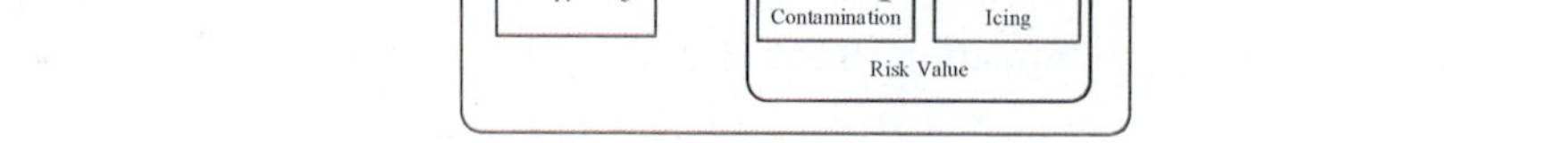

Fig. 2. Processes of the comprehensive assessment method for various natural disasters

The overall risk can be expressed as the following formula [14, 15].

$$P = \sum_{i=1}^{n} u_i \times v_i \tag{7}$$

Where P represents the overall risk of natural disasters for the compact transmission corridor, u_i represents the risk value of the ith natural disaster, and v_i represents the risk weight accordingly.

The final assessment results are output according to the established grading standard of risk level as shown in Table 2.

Table 2. Grading standard of risk level

P	$P < 30\%$	$30\% \leq P < 60\%$	$60\% \leq P < 90\%$	$P \geq 90\%$
Risk level	I	II	III	IV

5 Case Analysis

5.1 Standardized Treatment of Risk Value

The risk maps of lightning strike, windage, contamination and icing can be treated as the risk values for risk assessment. In order to eliminate the influence of dimension and magnitude of the data, standardized treatment should be carried out by the following formula.

$$X = \frac{x - x_{\min}}{x_{\max} - x_{\min}} \tag{8}$$

Where x and X are the risk values before and after standardization respectively, $x_{\max}$ and $x_{\min}$ are the maximum and minimum values of the corresponding parameters. The standardized risk maps of one compact transmission corridor are shown in Fig. 3.

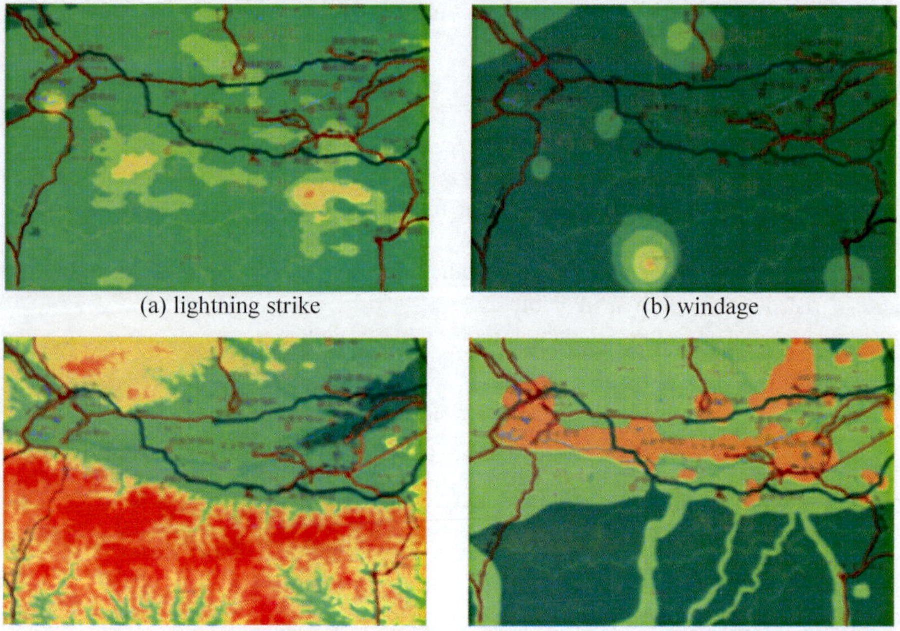

(a) lightning strike (b) windage (c) contamination (d) icing

Fig. 3. Standardized risk maps

5.2 Determination of Comprehensive Risk Weight

Taking the line tripping data of this compact transmission corridor from 2009 to 2013 as samples, the times of line tripping caused by lightning strike, windage, contamination and icing are shown in Table 3.

Table 3. Times of line tripping caused by natural disasters from 2009 to 2013

Year	Lightning strike	Windage	Contamination	Icing
2009	3	2	47	1
2010	2	1	73	1
2011	3	9	31	2
2012	2	6	45	1
2013	2	7	48	4
Total	12	25	244	9

Risk Weight Calculated by AHP. According to the expert scores, the sort of harmfulness brought to the compact transmission corridor by natural disasters is shown as: lightning strike > icing > windage > contamination. The pairwise comparison of the relative harmfulness of natural disasters are conducted by reference to nine-scale method. The results are shown in Table 4.

Table 4. Comparison of natural disasters' relative harmfulness

Natural disaster	Lightning strike	Windage	Contamination	Icing
Lightning strike	1	1/6	1/8	1/2
Windage	6	1	1/3	5
Contamination	8	3	1	7
Icing	2	1/5	1/7	1

Based on this, the judgment matrix is established.

$$P = \begin{bmatrix} 1 & 1/6 & 1/8 & 1/2 \\ 6 & 1 & 1/3 & 5 \\ 8 & 3 & 1 & 7 \\ 2 & 1/5 & 1/7 & 1 \end{bmatrix}$$

The maximum eigenvalue $\lambda_{\max}$ of the matrix is 4.1145, and the corresponding eigenvector ξ is (0.1189, 0.4365, 0.8884, 0.0781).

The consistency of the judgement matrix is verified.

$$CI = \frac{\lambda_{\max} - n}{n - 1} = \frac{4.1145 - 4}{4 - 1} = 0.0382$$

$$CR = \frac{CI}{RI} = \frac{0.0382}{0.89} = 0.0429 < 0.1$$

The results show that the judgment matrix passes the consistency check.

Therefore, the risk weights can be expressed as the normalized eigenvector ω = (0.0781, 0.2868, 0.5837, 0.0513).

Risk Weight Calculated by Entropy Weight. Dimensionless processing for the original data in Table 3 is carried out to obtain the matrix S.

$$S = \begin{bmatrix} 1 & 0.2222 & 0.6438 & 0.25 \\ 0.6667 & 0.1111 & 1 & 0.25 \\ 1 & 1 & 0.4247 & 0.5 \\ 0.6667 & 0.6667 & 0.6164 & 0.25 \\ 0.6667 & 0.7778 & 0.6575 & 1 \end{bmatrix}$$

The matrix S should be normalized into the range of [0, 1].

$$S' = \begin{bmatrix} 0.0808 & 0.0180 & 0.0520 & 0.0202 \\ 0.0539 & 0.0090 & 0.0808 & 0.0202 \\ 0.0808 & 0.0808 & 0.0343 & 0.0404 \\ 0.0539 & 0.0539 & 0.0498 & 0.0202 \\ 0.0539 & 0.0629 & 0.0532 & 0.0808 \end{bmatrix}$$

The entropy values of risk indexes are calculated as H = (0.9873 0.8683, 0.9768, 0.8867).

Finally, the entropy weights of risk indexes can be obtained as φ = (0.0451, 0.4688, 0.0827, 0.4033), which means that the sort of harmfulness caused by natural disasters is shown as: icing > windage > lightning strike > contamination.

Combinatorial Optimization by Least Square. The least square method is used to optimize the combination of the above two methods. And the Lagrangian function for solving the combinatorial optimization model is shown as follows.

$$L = \sum_{i=1}^{m}\sum_{j=1}^{n}\left\{\left[(\omega_j - \mu_j)s_{ij}\right]^2 + \left[(\varphi_j - \mu_j)s_{ij}\right]^2\right\} + 4\lambda\left(\sum_{j=1}^{n}\mu_j - 1\right) \tag{9}$$

The results can be obtained as follows.

$$M = (0.0483, 0.3331, 0.3778, 0.2407)$$

After combinatorial optimization, the sort of harmfulness caused by natural disasters is shown as: lightning strike > icing > windage > contamination.

For comparison, Table 5 shows the sort calculated by the above three methods.

Table 5. Comparison of the sort of harmfulness caused by natural disasters

Assessment method	AHP	Entropy weight	Combinatorial optimization model
Contamination	3	4	4
Icing	2	1	2
Lightning strike	1	3	1
Windage	4	2	3

According to the actual operation experience, the results of the combinatorial optimization model by least square are more reasonable.

5.3 Calculation of Overall Risk

After obtaining the risk values and risk weights, the overall risk can be calculated by formula (7). Tripping data caused by natural disasters from 2013 to 2016 in one compact transmission corridor is used to verify the effectiveness of the proposed method in this paper.

Table 6. Statistics of comprehensive assessment results

No.	Tripping time	Natural disaster	Fault tower	Assessment results	Risk level	Coincidence or not
1	3/4 12:00	Windage	#219	0.45	II	Yes
2	4/11 18:59	Lightning strike	38#	0.57	II	Yes
3	6/5 15:45	Lightning strike	116#	0.53	II	Yes
4	6/12 18:59	Lightning strike	56#	0.86	III	Yes
5	6/30 12:39	Lightning strike	18#	0.92	IV	Yes
6	7/24 10:43	Windage	#19–#20	0.67	III	Yes
7	7/24 10:49	Windage	#20	0.70	III	Yes
8	7/27 16:40	Lightning strike	15#	0.21	I	No
9	7/30 17:18	Windage	#62	0.49	II	Yes
10	7/30 22:43	Lightning strike	73#	0.54	II	Yes
11	7/30 23:06	Lightning strike	73#	0.58	II	Yes
12	7/31 1:03	Lightning strike	195#	0.94	IV	Yes
13	7/31 5:20	Lightning strike	138#	0.59	II	Yes
14	8/4 15:32	Lightning strike	146#	0.91	IV	Yes
15	9/7 17:35	Lightning strike	43#	0.79	III	Yes
16	9/22 5:48	Lightning strike	58#	0.83	III	Yes
17	9/22 5:48	Lightning strike	58#	0.86	III	Yes

It can be seen from Table 6 that in the 17 tripping accidents, there are 16 tripping accidents whose risk level are assessed as II or above. The result shows that the accuracy of this comprehensive assessment method is up to 94.4%, which means a certain degree of accuracy.

6 Conclusions

In this paper, a comprehensive assessment method of natural disaster risk for compact transmission corridor is proposed. On the basis of the existing risk maps of lightning strike, windage, contamination and icing, the data are standardized. Then, the subjective weights are determined by AHP, and the objective weights are determined by entropy weight. Finally, the subjective weights and objective weights are combined and optimized by least square to calculate the comprehensive weights. According to the risk values and risk weights, the overall risk of natural disasters for compact transmission corridor can be obtained by multiplying. This method integrates the assessment of lightning strike, windage, contamination and icing together, and makes up the shortcomings of single assessment method, with an accuracy up to 94.4%. It is of great significance to promote natural disaster protection level of compact transmission corridor.

References

1. Jiang, J.: Study on Risk Assessment and Control for the Operation of Transmission Line. North China Electric Power University, Beijing (2015)
2. Huang, Y., Wei, R., Zhou, E., et al.: Early warning method of transmission line damage under typhoon disaster. Autom. Electr. Power Syst. **42**(23), 142–147 (2018)
3. Wang, J.: Research on Meteorological Disaster risk Analysis and Fault Early Warning Methods for Overhead Transmission Lines. Chongqing University, Chongqing (2016)
4. Hou, H., Geng, H., Huang, Y., et al.: Multi-factor correction based failure prediction method of transmission line with typhoon disaster. Autom. Electr. Power Syst. **43**(23), 193–201 (2019)
5. Chen, B., Li, C., Qin, H., et al.: Evaluation of typhoon resilience of distribution network considering grid reconstruction and disaster recovery. Autom. Electr. Power Syst. **42**(6), 47–52 (2018)
6. Li, Z., Wang, F., Guo, W., et al.: Resilience evaluation of smart distribution network in extreme weather. Autom. Electr. Power Syst. **44**(9), 60–68 (2020)
7. Li, S.: Research on the Evaluation of Overhead Transmission Line Operation Risk Under the Bad Meteorological Disasters. South China University of Technology, Guangzhou (2015)
8. Christodoulow, C.A., Perantzakis, G., Spanakis, G.E., et al.: Evaluation of lightning performance of transmission lines protected by metal oxide surge arresters using artificial intelligence techniques. Energy Syst. **3**(4), 379–399 (2012)
9. Zhao, C., Chen, J., Wang, J., et al.: Research on technology of lightning disaster risk assessment for power system. High Volt. Eng. **37**(12), 3012–3021 (2011)
10. Chen, X.: Jinhua Power Grid Ice and Lightning Disaster Analysis. Zhejiang University, Hangzhou (2008)
11. Feng, J., Wang, Y., Xiao, X.Y.: Synthetic risk assessment of catastrophic failures in power system based on entropy weight method. In: Advanced Materials Research, vol. 446, pp. 3015–3018. Trans Tech Publications (2012)
12. Hu, Z., Liu, J., Zhang, B., et al.: Risk assessment of operation environment for transmission lines based on risk key characteristic. Autom. Electr. Power Syst. **41**(18), 160–166 (2017)
13. Guo, J., Pu, X., Gao, X., et al.: Improved method on weights determination of indexes in multi-objective decision. J. Xidian Univ. **41**(06), 118–125 (2014)
14. Deng, H., Dai, D., Li, S.: Comprehensive operation risk evaluation of overhead transmission line based on hierarchical analysis-entropy weight method. Power Syst. Protect. Control **45** (01), 28–34 (2017)
15. Duan, T., Luo, Y., Shi, L., et al.: A real-time fault probability evaluation model of transmission line considering meteorological factors. Power Syst. Protect. Control **41**(15), 59–67 (2013)

Model and Geometric Characteristic of Multi-quadrants Security Region for Active Distribution Network

Guoqiang Zu[1,2(✉)], Haishen Liang[3], Yanan Zhou[3], Jun Xiao[2], and Yunfei Mu[2]

[1] State Grid Tianjin Electric Power Research Institute, Tianjin, China
zuguoqiang_tju@163.com
[2] School of Electrical and Information Engineering, Tianjin University, Tianjin, China
[3] State Grid Tianjin Electric Power Company, Tianjin, China

Abstract. The region-based method has been applied in transmission systems and traditional passive distribution systems without power source. This paper proposes the model of multi-quadrants security region (MQSR) for active distribution network (ADN) with distributed generation (DG). The MQSR is modeled considering the constraints of state space, normal operation and N − 1 security guideline. Afterwards, the geometric characteristics of MQSR are observed on the case grids with different DG penetration levels. The MQSR can be located in multiple quadrants of the state space. For different DG penetrations, the shape and security feature of MQSR are also different. Finally, the region map is first discovered, which summarizes the features of different kinds of distribution networks.

Keywords: Active distribution network · Multi-quadrants · Security region · Distributed generation · Region map

1 Introduction

The future distribution network, especially in the urban area, will be developed to the active system, which is able to dynamically select the economic and reliable operation mode according to the consecutive changes caused by external environment, such as load fluctuations and intermittent distributed generations (DG). Integrating DGs is the most important features of an active distribution network. Besides, the active controllability of the network is another feature, such as the load transfer among feeders flexibly, which is realized by the distribution automation.

Secure and efficient operations of power systems are always the major objective for power system planners and operators. For ADN with DGs, the security and efficiency of distribution network is complicated. The existing security assessment is based on case-by-case simulating approach [1], which is time-consuming and cannot provide effective quantitative indices. Distribution system security region (DSSR) is a new approach for the security analysis and assessment of distribution systems. The DSSR is

Y. Xue et al. (Eds.): PMF 2020, LNEE 718, pp. 487–497, 2021.
https://doi.org/10.1007/978-981-15-9746-6_38

defined as a closed set of all the N − 1 secure operating points. Compared with the conventional case-by-case simulating approach, the region-based method can give the global information about the security boundary surrounding the operating point of the system and security margin to stability and operating limits [2]. The complicated operational boundary can be approximately expressed by simple hyperplanes, which benefits the optimization problem [3] and probabilistic assessment [4], especially for the ADN with uncertainties due to intermittent DGs.

Most of the existing researches on DSSR are related to traditional distribution systems, including the definition [5], observation [6], model [5], characteristics [6, 7] and application [8]. Since a traditional distribution network passively receives energy from the transmission system and supplies the energy to its load, it is called the load supply network (LSN) in this paper. However, the function of ADN is not only power supply, but also DG accommodation. Because of the bidirectional power flow of branches and nodes, it can be deduced that the security region of ADN will be expanded from only the first quadrant to more quadrants of the state space, turning into the multi-quadrants security region (MQSR). Therefore, the model and geometric characteristics of the region could be also changed greatly and the new region theory for ADN is needed.

This paper proposes the concept and model of MQSR for ADN. 2-dimensional projections of MQSR are observed under 2 different DG penetration scenarios. The characteristics of MQSR are summarized. The region map is first discovered, which summarizes the features of different kinds of distribution networks.

2 Model of MQSR

2.1 Formulation

The MQSR model is compactly formulated as (1)

$$\boldsymbol{\Omega}_{MQSR} = \{\boldsymbol{W} = (S_1, \ldots, S_i, \ldots S_n) \in \boldsymbol{\Theta} | St.A \cap St.B \cap St.C\} \tag{1}$$

where $\boldsymbol{\Omega}_{MQSR}$ represents the multi-quadrants security region. $\boldsymbol{W} = (S_1, \ldots, S_i, \ldots S_n)$ is the n-dimensional operating point, which is defined as the minimum set of the state variables that can determine the security status of the ADN. S_i is the total apparent power of bus i. Considering that one bus is connected to only one object (load or DG), S_i can be expressed as (2)

$$\begin{cases} S_i = S_{L,i}, & i \in \boldsymbol{L} \\ S_i = S_{G,i}, & i \in \boldsymbol{G} \end{cases} \tag{2}$$

where $S_{L,i}$ and $S_{G,i}$ are the apparent power of the load and DG on the bus i, respectively. $\boldsymbol{L}$ is the load set and $\boldsymbol{G}$ is the DG set. The direction is positive when the power flows out of bus i ($S_i > 0$). Thus, $S_{L,i} \geq 0$ and $S_{G,i} \leq 0$. $St.A$, $St.B$ and $St.C$ are three conditions of the set $\boldsymbol{\Omega}_{TQSR}$, which will be illustrated in following sections.

2.2 State Space of Constraints

St.A is the constraints of state space, which represents that the power of each bus should be kept in a reasonable range. For example, the load cannot exceed its capacity of distribution transformer and the DG output cannot exceed its generation capacity. *St.A* is formulated as (3)

$$St.A\begin{cases} S_{L,i}^{\min} \le S_{L,i} \le S_{L,i}^{\max}, & i \in \boldsymbol{L} \\ -S_{G,i}^{\max} \le S_{G,i} \le 0, & i \in \boldsymbol{G} \end{cases} \tag{3}$$

where $S_{L,i}^{\max}$ and $S_{L,i}^{\min}$ are the maximum and minimum apparent power of load i respectively. The default value of $S_{L,i}^{\min}$ is 0. $S_{G,i}^{\max}$ is the capacity of DG i.

2.3 Normal Operational Constraints

St.B is the capacity constraints of branches and substation transformers under normal operating state, which is formulated as (4)

$$St.B\begin{cases} \left|S_{B,i}\right| = \left|\sum\limits_{j\in\boldsymbol{\Omega}_{B,i}} S_j\right| \le c_{B,i}, & \forall i \in \boldsymbol{B} \\ \left|S_{T,i}\right| = \left|\sum\limits_{j\in\boldsymbol{\Omega}_{T,i}} S_j\right| \le c_{T,i}, & \forall i \in \boldsymbol{T} \end{cases} \tag{4}$$

where $S_{B,i}$ is the power of branch i and the set of its downstream buses is $\boldsymbol{\Omega}_{B,i}$. The term of downstream means the direction from a substation transformer towards its feeder terminal. $c_{B,i}$ is the capacity of branch i and $\boldsymbol{B}$ is the branch set. $S_{T,i}$ is the power of substation transformer i, and the set of its downstream buses is $\boldsymbol{\Omega}_{T,i}$. $c_{T,i}$ is the capacity of substation transformer i and $\boldsymbol{T}$ is the transformer set.

2.4 N − 1 Security Constraints

The N − 1 security constraints are formulated as

$$St.C \quad \forall \psi_k \in \boldsymbol{\Psi},$$
$$\begin{cases} \left|S_{B,i}(k)\right| = \left|\sum\limits_{j\in\boldsymbol{\Omega}_{B,i}(k)} S_j\right| \le c_{B,i}, & \forall i \in \boldsymbol{B}, \psi_k \notin \boldsymbol{B} \\ \left|S_{T,i}(k)\right| = \left|\sum\limits_{j\in\boldsymbol{\Omega}_{B,i}(k)} S_j\right| \le c_{T,i}, & \forall i \in \boldsymbol{T}, \psi_k \notin \boldsymbol{T} \end{cases} \tag{5}$$

where $\boldsymbol{\Psi}$ is the contingency set. Ψ_k is the contingency k, which represents a single contingency of a branch or a substation transformer.

After contingency k, the network will be reconfigured to restore the service of non-fault area. The power flow of the branches and transformers could be changed. $S_{B,i}(k)$

and $S_{T,i}(k)$ are the power of branch i and transformer i respectively after contingency k. $\boldsymbol{\Omega}_{B,i}(k)$ and $\boldsymbol{\Omega}_{T,i}(k)$ are the sets of downstream buses of the branch i and transformer i after contingency k, respectively. $\psi_k \notin \boldsymbol{B}$ and $\psi_k \notin \boldsymbol{T}$ means that the calculation of the power flow of the faulted component is not needed.

2.5 Comparison with Existing DSSR Model

The proposed MQSR model and traditional DSSR model [7] is compared here.

(1) The range of state space. Since the traditional DSSR considers loads only and the bus power is always a positive number, the state space is located in the first quadrant only. However, the state space of ADN can cover all the quadrants because the bus power can be either positive or negative with DG considered.
(2) The formulation of normal operating state. In the traditional DSSR, the constraints under normal operating state are usually neglected because the constraints to an operating point are stronger under N − 1 contingency than normal operating state. However, the constraints of MQSR under normal operating state could be more rigorous than N − 1 security and thus it must be formulated.
(3) Contingency set. The contingency set of DSSR contains only substation transformer contingency and feeder outlet contingency. However, the contingency set of MQSR must include the contingencies of all the branches.
(4) Security constraints. For MQSR with bidirectional power flow, the constraints of each branch must be modeled. For a traditional DSSR with radial power flow, the constraints of the feeder outlet branch are modeled only.

3 Case Study

3.1 Scenarios Division

The penetration of DGs will influence the security feature of the distribution systems. The index of CP (Capacity penetration) is used in this paper, which is defined as the ratio of the maximum total DG capacity to the summit loads in the same area within one year.

According to different CPs, two typical scenarios for future ADN is studied in this paper, which are the high penetration supply network (HPSN) and very high penetration balanced network (VHPBN), respectively [9]. The CP of HPSN is around 40%. The power consumption of load is received almost equally from the transmission system and local DGs. The CP of VHPBN is around 100%. All the power consumption of load could be from local DGs within a certain time. Compared to LSN, both HPSN and VHPBN are turned into the system with comprehensive service function of the load supply and DG accommodation.

The geometric characteristics of MQSR are observed on a 4 buses test system under 2 different scenarios including HPSN and VHPBN.

3.2 Scenario 1: HPSN

In the study of HPSN, the security region of LSN and power collection distribution network (PCN) will be used as comparison. The main function of PCN is to collect and deliver the generation of DGs to the upper level transmission system [10]. LSN and PCN can be regarded as the special ADN with the lowest and the highest DG penetration respectively, which is useful to explain how DG penetration affects the characteristics of security region. The topology of 4 buses test system of ADN is shown in Fig. 1. The data of the three case grids are in Table 1. The summit load of the ADN is 2MVA and CP is 40%.

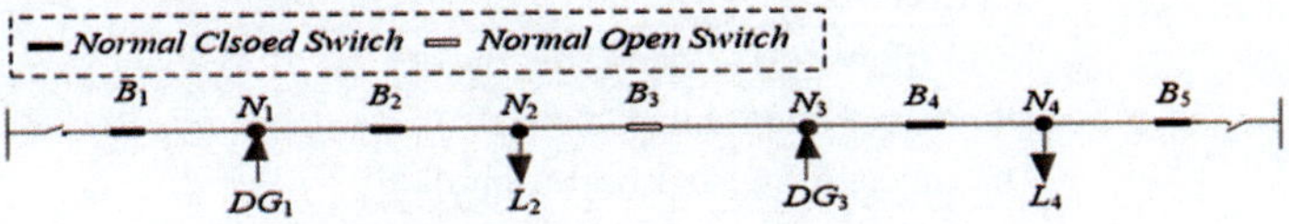

Fig. 1. The 4 buses test system of ADN

Table 1. Basic data of case grids

Case grids	ADN	LSN	PCN
Branch capacities (MVA)	1	1	1
DG output range (MVA)	[− 0.4, 0]	–	[− 0.4, 0]
Load range (MVA)	[0, 1.5]	[0, 0.8]	–

Quadrant. After observing many MQSR projections on 2-D subspace, it is shown that the region can be located in multiple quadrants. The example of quadrant IV is shown in Fig. 2.

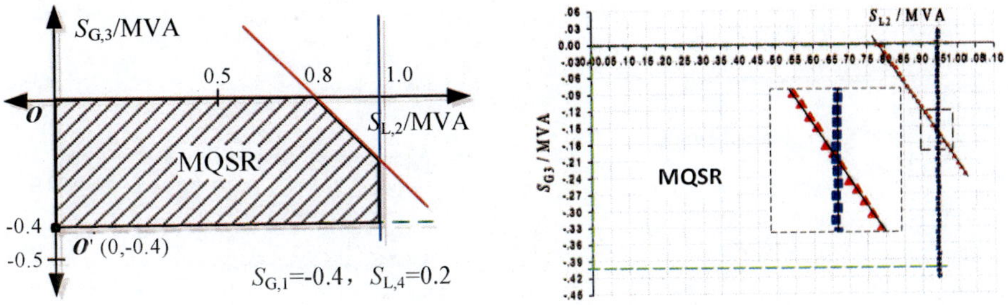

(a)

(b)

Fig. 2. The obtained boundary figures using different power flow models. (a) The boundary figure using DC power flow-based model. (b) The boundary figure using AC power flow-based model.

In Fig. 2(a), the $S_{G,3}$ and $S_{L,2}$ are observed and $S_{G,1}$ and $S_{L,4}$ are frozen at −0.4 MVA and 0.2 MVA respectively. It can be observed that MQSR is closed and bounded by N − 1 security boundary (represented by red solid line), normal operating boundary (blue solid line) and state space boundary (represented by green dot line). If the security boundary is calculated by AC power flow-based model [6], the boundary figure is similar to the Fig. 2 but the boundary fluctuates around the straight line within a very small range. This is shown in Fig. 2(b). Table 2 shows the correspondence between the quadrant and observation variable type.

Table 2. The relationship of the quadrant and observation variable type.

Variable X	Variable Y	Quadrant
Load (flow out)	Load (flow out)	I
DG (inject into)	Load (flow out)	II
DG (inject into)	DG (inject into)	III
Load (flow out)	DG (inject into)	IV

In Table 2, the variable X and Y represent the two observation variables.

Shapes. The shape features of MQSR 2-D projections can be included as follows. In general, 13 shapes can be observed in the 4-buses case of HPSN. 2 shapes are same with that of LSN and 11 shapes are new added. The same region figure in different quadrants will be counted as different shapes.

The 11 new shapes of MQSR are also pentagon, ladder, rectangle or triangle, which is similar to traditional DSSR. The pentagon projection is shown in Fig. 3.

However, the 11 new shapes of MQSR projections differ from the traditional DSSR. They are not in quadrant I. Further, let α be the angle between the inclined boundary and the coordinate axis towards the region, For the traditional DSSR, α is always acute angle (approximately 45°), but it could be obtuse angle (approximately 135°) for MQSR.

Take the pentagon in quadrant IV as an example, as is shown in Fig. 3.

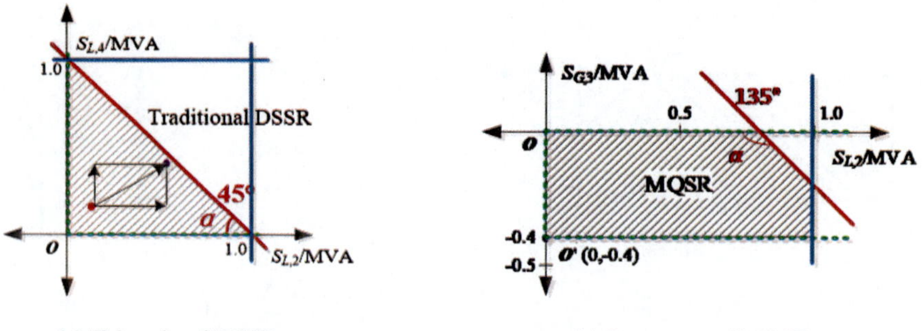

(a) Triangle of DSSR (b) Pentagon of MQSR

Fig. 3. The Comparison of traditional DSSR and MQSR—the angles between the inclined boundary and the coordinate axis

This shape feature will lead to the difference of security feature between DSSR and MQSR, which will be illustrated in following section.

Security. For DSSR, the most secure operating point is the original point (0, 0), which represents the total load of the system is zero. However, the most secure point will be changed in the MQSR. It is corresponding to the maximum output of DGs and the zero loads. This is briefly illustrated in Fig. 4.

In Fig. 4, the security degree of operating point O'' (the purple point) is lower than that of O' (the red point), because $dO_1'' < dO_1', dO_2'' < dO_2'$. Here, d is the security distance of an operating point to the effective security boundary. Thus, the most secure point is the O' which is located in the left bottom of MQSR projection. Also, even if the position of O'' (purple one) varies, the conclusion still remains unchanged.

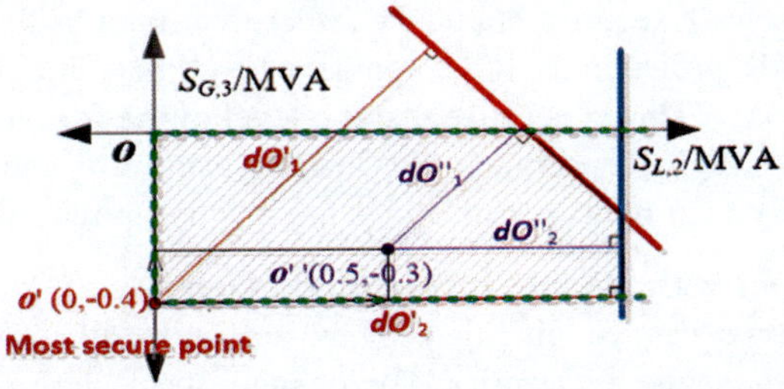

Fig. 4. Most secure operating point of MQSR of HPSN

3.3 Scenario 2: VHPBN

The MQSR of VHPBN is studied in this section and compared with the MQSR of HPSN. The network topology of the 4 buses small case of VHPBN is same as that of HPSN (Fig. 2), only except that the capacity of each DG is increased to 1.5 MVA. The summit load is 3 MVA, and CP is 100%.

Quadrant and Shape
First, the quadrant feature of VHPBN is same as that of HPSN, that is, the MQSR of VHPBN can be located in any quadrant of the state space. Second, as to the shape feature, 24 shapes can be observed in the 4-buses case of VHPBN. 12 shapes are same with that of HPSN and 12 shapes are new added. Here, the same region figure in different quadrants will be counted as different shapes.

Different Shapes from HPSN
12 MQSR shapes of VHPBN are different from the HPSN. The 12 shapes can be divided into two types as follows.

Type 1: The reverse Shapes
In this type, 8 shapes can be obtained. This type is similar to HPSN. The projections can be also pentagon and rectangle. However, the "orientation" of the projections is different. The pentagon projections in quadrant IV of HPSN and VHPBN are compared in Fig. 5 as an example.

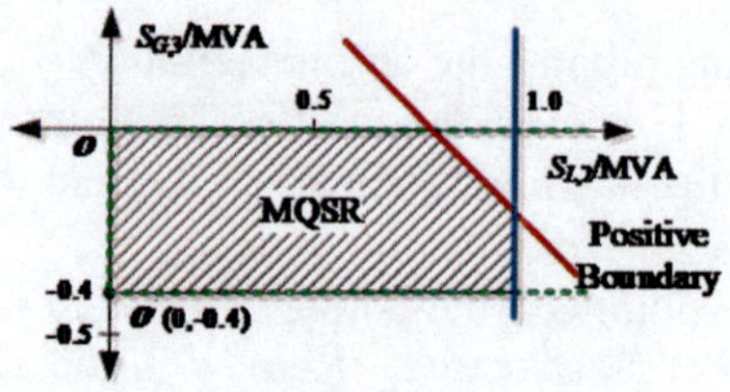

(a) Positive pentagon of HPSN

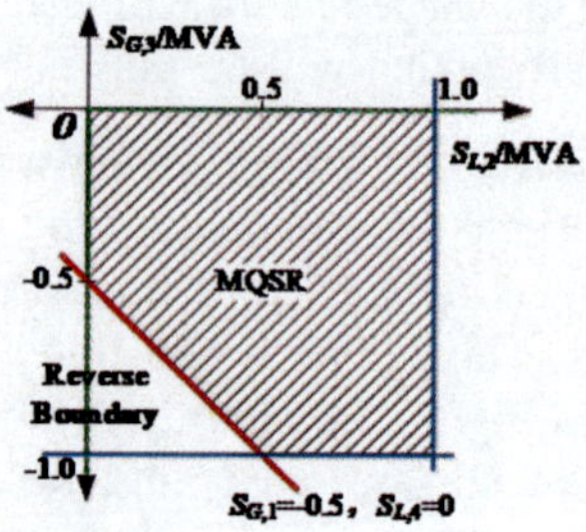

(b) Reverse pentagon of VHPBN

Fig. 5. Comparison of the pentagons of VHPBN and f HPSN

In Fig. 5, the reverse security boundary rather than the positive security boundary exists in the MQSR projections. It is opposite to HPSN and presents the positive security boundary only. This is because that the level of the frozen loads is low and the frozen DG output is high. The branch overloading caused by the reverse power flow happens more easily than the one caused by the positive power flow.

Type 2: The Shapes with Double Incline Boundaries
4 shapes of this type can be obtained. They are quite different from HPSN. The projections can be hexagon and ladder. The hexagon projection is shown in Fig. 6.

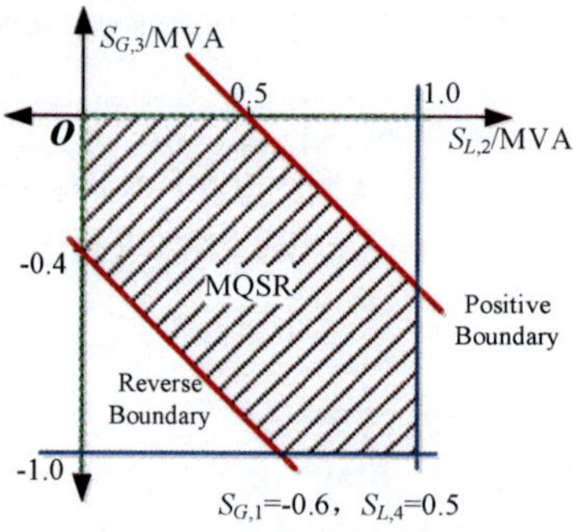

Fig. 6. The 2-D hexagon projection of MQSR in quadrant IV of VHPBN.

In Fig. 6, both the positive and reverse boundaries (red incline boundaries) are presented to form the shape. The reason is that, when the level of the frozen loads and the DG output are moderate, the branch overloading could be caused by both the positive and reverse power flow.

Security
The most secure point for HPSN is corresponding to the maximum output of DGs and the zero loads. However, the most secure point should be the original point for VHPBN. It means that the load is zero or the DG output is also zero.

The reason for the change of security features of VHPBN is as follows. The security feature of MQSR for HPSN is based on the important premise, that is, the total capacity of DGs on a feeder cannot exceed the feeder capacity (40% CP is consistent with this premise). However, in the VHPBN, the capacity of total DGs under 100% CP could be larger than the feeder capacity and the security features will be different. The branch overloading could be caused by both the positive and reverse power flow.

4 Region Map

Through the summary of the region characteristics of LSN, HPSN and VHPBN for many case grids with different DG penetrations, including a modified 114 buses IEEE RBTS test system with DGs [11], the region map is first discovered, as is shown in Table 3 (Fig. 7).

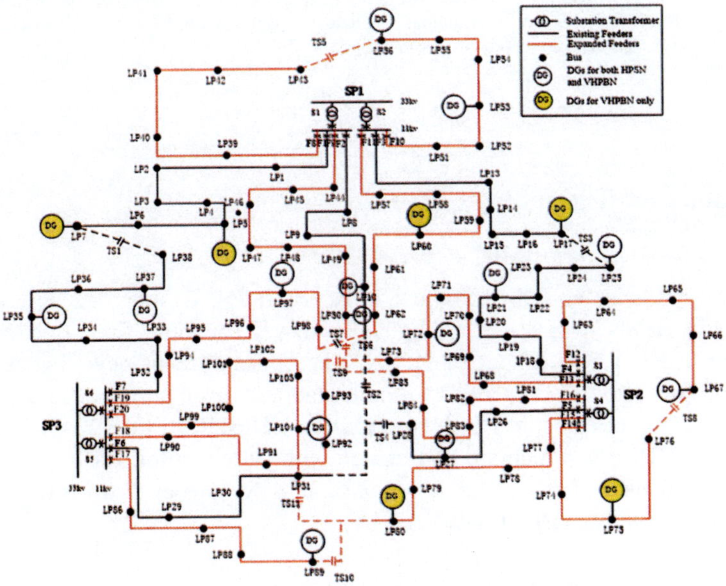

Fig. 7. The modified IEEE RBTS test system with DGs

Table 3. The region map

	Quadrant I				Quadrant II				Quadrant III				Quadrant IV				Total
LSN 0%	1T+	1L+	1P+	1R													4
HPSN 40%	1T+	1L+	1P+	1R	2T+	2L+	2P+	2R	3T+	3L+	3P+	3R	4T+	4L+	4P+	4R	16
VHPBN 100%	1T+	1L+	1P+	1R	2T+	2L+	2P+	2R	3T+	3L+	3P+	3R	4T+	4L+	4P+	4R	40
	1H+-	1L+-	1PL+-		2H+-	2L+-	2PL+-		3H+-	3L+-	3PL+-		4H+-	4L+-	4PL+-		
	1T-	1L-	1P-		2T-	2L-	2P-		3T-	3L-	3P-		4T-	4L-	4P-		

Note: Each symbol is marked with an ID below. In the ID, the first number means the quadrant. The alphabet represents the shape, in which T is triangle; L is ladder; P is pentagon, PL is parallelogram, H is hexagon. "+" means the positive boundary exists in the shape. "−" means the negative boundary exists. The rectangle region does not contain decline boundary so that does not need "+" or "-".

4.1 Features of Region Map

In Table 3, it can be seen that

1) All the 4 shapes exist in LSN also exist in HPSN. Besides, 12 new shapes are added in HPSN, which are highlighted with yellow color. Each quadrant (II, III, IV) provides 4 new shapes of HPSN. The appearance of new shapes is mainly because the region is expanded to total quadrants from single quadrant.
2) All the 16 shapes exist in HPSN also exist in VHPBN. Besides, 24 new shapes are added in VHPBN, which are highlighted with yellow color. Each quadrant (I–IV) provides 6 new shapes. The appearance of new shapes is mainly because the reverse boundary exists in the VHPBN.

4.2 Significance of Region Map

1) The region map summarizes the features of the DSSR and MQSR and shows the maximum possible region shapes that a distribution system could have.
   For instance, the region of LSN has at most 4 shapes according to the map. The pentagon will never be obtained in any given LSN. The region of HPSN has at most 4 shapes according to the map. The hexagon will never be obtained in any given HPSN.
2) The region map can be used to categorize the distribution networks. For example, if a hexagon region projection is observed in quadrant IV, it can be found in the map that the system is a VHPBN.

5 Conclusion

This paper proposes the concept and model of multi-quadrants security region (MQSR) for active distribution networks (ADN), which is useful for the optimization problem and probabilistic assessment of the ADN. The 2-dimensional projections of MQSR are observed under 2 DG penetration scenarios. The geometric characteristics of MQSR are summarized.

1) As to the quadrant, MQSR is transformed from the single-quadrant region to the multi-quadrants region.
2) As to the shape, for HPSN, at most 12 new shapes will be obtained in quadrant II, III and IV compared with LSN, For VHPBN, at most 24 new shapes in all quadrants will be obtained compared with HPSN.
3) As to the security feature, for HPSN, the most secure operating point is the one that the DG output reaches the peak value and the power of load is zero. For VHPBN, the most sécure point is the original point.
4) The region map is first discovered, which provides the maximum possible shapes that a region could have. It can be also used to categorize the distribution networks.

Acknowledgement. The authors gratefully acknowledge the China Postdoctoral Science Foundation 67th General Project and State Grid Science and Technology Program (KJ-20-1-27).

References

1. Morison, K., Wang, L., Kundur, P.: Power system security assessment. In: IEEE Power and Energy Magazine, vol. 2, no. 5, pp. 30–39 (2004)
2. Wu, F.F., Kumagai, S.: Steady-state security regions of power systems. IEEE Trans. Circ. Syst. **29**(11), 703–711 (1982)
3. Yu, Y., Liu, H., Zeng, Y.: A novel optimization method of transient stability emergency control based on practical dynamic security region (PDSR) of power systems. Sci. China Ser. E **47**, 376–384 (2004)
4. Liu, Y.L., Yu, Y.X.: Probabilistic steady-state and dynamic security assessment of power transmission system. Sci. China Technol. Sci. **56**(5), 1198–1207 (2013)
5. Xiao, J., Gu, W., Wang, C., Li, F.: Distribution system security region: definition, model and security assessment. IET Gener. Transm. Distrib. **6**(10), 1029–1035 (2012)
6. Xiao, J., Zu, G., Gong, X., Li, F.: Observation of security region boundary for smart distribution grid. IEEE Trans. Smart Grid **8**(4), 1731–1738 (2017)
7. Xiao, J., Zu, G., Gong, X., et al.: Model and topological characteristics of power distribution system security region. J. Appl. Math. (2014, available online)
8. Zu, G., Xiao, J., Sun, K.: Distribution network reconfiguration comprehensively considering N − 1 security and network loss. IET Gener. Transm. Distrib. **12**(8), 1721–1728 (2018)
9. Xiao, J., Zhang, B., Li, J., et al.: Research idea on high penetration renewable energy generations distribution system planning based on security boundary. Autom. Electr. Power Syst. **41**(9), 1–9 (2017). (in Chinese)
10. Yang, J., Fletcher, J.E., O'Reilly, J.: Multiterminal DC wind farm collection grid internal fault analysis and protection design. IEEE Trans. Power Deliv. **25**(4), 2308–2318 (2014)
11. Billinton, R., Jonnavithula, S.: A test system for teaching overall power system reliability assessment. IEEE Trans. Power Syst. **11**(4), 1670–1676 (1996)

A Method of Generating Group Fault Sets Automatically Taking into Account of Spatial-Electrical Coupling Characteristics

Li Xiang[1], Xiaotong Xu[2(✉)], Zhihao Tian[1], Hailiang Zhong[1], Shuang Zhang[1], and Junjun Yang[2]

[1] State Grid Ningxia Electric Power Co., Ltd., Yinchuan 750001, China
[2] NARI Group Corporation (State Grid Electric Power Research Institute), Nanjing 211106, China
xuxiaotong@sgepri.sgcc.com.cn

Abstract. Taking into account the external natural disaster probability and the space and electrical coupling characteristics between electrical equipment, a combined probability calculation method of group failure and sequential failure under natural conditions is proposed. And a multi-dimensional failure based on grid structure vulnerability, grid state vulnerability and grid accident risk is proposed. Consequence evaluation method provides a basis for effectively screening high-risk expected failure sets; constructs a disaster-problem equipment failure probability state tree model, uses the multiplicity of failures, combined probability threshold, and multi-dimensional failure consequence evaluation indicators as constraints, and uses a backtracking algorithm to achieve natural disasters. The high-risk prediction sets are quickly generated to ensure that small-probability and high-risk faults are included in the calculation, and the real-time and targeted nature of power grid safety analysis and control is improved. The simulation analysis of an actual power grid online example verifies the effectiveness and practicability of the method.

Keywords: Natural disasters · Fault probability · High-risk accident · Multiple faults · Backtracking algorithm

1 Introduction

The expected failure set includes single equipment failure, group failure and sequential failure [1], which is an important data for power grid security analysis and control. With the increasing frequency and degree of damage caused by external natural disasters to the power grid, the occurrence probability of multiple faults such as group failure and sequential failure increases, resulting in the change of fault generation from considering the structure and operation state of the grid to considering the adverse effects of natural disasters on the electrical equipment [2, 3]. In the context of large-capacity long-distance transmission and interconnection of large-scale power grids, the expected failure will bring more serious security risks to the power grid. Both high-probability high-risk failure and low-probability high-risk failure should be included in the active defense failure set [4–6]. At present, the source of the expected failure set is

Y. Xue et al. (Eds.): PMF 2020, LNEE 718, pp. 498–511, 2021.
https://doi.org/10.1007/978-981-15-9746-6_39

the n − 1 fault of the whole network and the fault set by the dispatcher according to experience, which has a large mismatch with the safety characteristics of the grid and cannot fully reflect the real-time operation status of the grid. Therefore, it is necessary to update the expected failure set dynamically in combination with the external impact of natural disasters and the operation structure of the power system.

Based on the fault probability of equipment caused by natural disasters, the set of affected equipment can be acquired. And the equipment combination can be carried out in the set of affected equipment, which can reduce the scale of the expected failure set. Literature [7] adopts the method of exhaustive combination for a single external disaster, and generates multiple fault sets based on probability threshold and multiplicity of failures, which reduces the dimension of fault, but cannot reflect the coupling effects among multiple disasters, and takes a long time. Literature [8] pointed out that analyzing the impact of different natural disasters in isolation cannot reflect the risk of power failure under complex disasters correctly. Literature [9] qualitatively describes the relationship between the probability of "independent n-m fault" and "related n-m fault". However, at present, there are few reports on the quantitative evaluation of the combination probability of multiple failures under natural disasters and the rapid screening method of expected failures. Most of them generate expected group failure sets by exhaustive combination [10].

For large power grid with thousands of units and tens of thousands of nodes, equipment combination is carried out based on the failure probability under natural disasters. Constrained by multiplicity of failures and probability threshold, a large number of low-risk failures are included in the defense range. Due to the large scale of faults, it is difficult to meet the online analysis and control requirements under the condition of limited computing resources, which requires further screening of high-risk failures. The fault set screening algorithm is mainly based on artificial intelligence and transient energy function method, which can filter out the non-serious accidents by defining the performance indexes that reflect the severity of accidents. For example, literature [11] first uses the iterative prediction method of load and power flow, and then uses the improved quadratic curve prediction method to select the static stability faults. Some literatures proposed a method to screen the serious faults set under the current operation mode based on the last round of online evaluation results on the power grid [12], which realized the screening of the expected failure subsets that need to be calculated according to the severity of transient stability [13]. However, the development speed of group failures caused by natural disasters is much faster than the change of power grid operation state. Moreover, it changes dynamically according to the development path of the disaster, the severity of the disaster and the space-electrical coupling characteristics between the affected equipment. The existing fault set screening method is difficult to adapt to the dynamic change and screening of multiple faults under external disasters [14, 15].

Taking the large power grid as the research object, this paper proposes a method for the rapid generation of multiple expected fault sets in the large power grid under natural disasters. First, considering the probability of single equipment failure, the spatial distance between the devices/towers and the electrical coupling characteristics under external natural disasters, a combined probability quantitative evaluation model of group failures and successive failures is established. Then, from the three dimensions of grid structure, grid state and accident risk, an evaluation indicators calculation method of high-risk expected failure set based on the consequences of failures is proposed. Finally, an optimized mathematical model of the combination of failure equipment based on the affected equipment set is constructed. And a backtracking algorithm is used to realize the rapid screening of the expected failure set. The method in this paper can quickly and accurately obtain the expected failure set that not only meets the needs of security and stability prevention and control, but also meets the online real-time calculation speed. It can ensure that low probability and high-risk faults are included in the calculation, which is helpful for the dispatcher to fully grasp the safety and stable operation level of the power grid in real time.

2 Probability Assessment of Failure Combination

From the perspective of the formation mechanism of natural disasters, different types of disasters have different impact ranges, impact times and failure types on power grid equipment. For example, the time scale of ice cover and snow cover impacts ranges from a few hours to a few days, which will cause insulator flashover of the transmission line towers, and eventually leads to disconnection failure and tower failure with the increase of ice cover. Mountain fires cause multiple transmission lines tripping out simultaneously or successively, and even failing to reclose after the trip. Moreover, different types of disasters have induction or coupling effects. A single disaster often induces multiple disasters, and multiple disasters influence each other in the evolution process of time dimension and space dimension, which results in group failures or successive failures of electrical equipment. The uncertainty of external disasters and the complexity of equipment failures that make it difficult to obtain the failure set of affected equipment by mathematical models accurately. However, the coupling coefficient can be introduced according to the disaster probability, the relationship between the disaster and the equipment space and electrical interaction, so as to calculate the combination probability of different types of disasters and different electrical equipment, and carry out the failure equipment combination.

2.1 Failure Probability of a Single Equipment

Under the influence of a single disaster, the failure probability of electrical equipment is the probability of disaster [16–19]. When the same equipment is affected by multiple disasters, each type of disaster is an independent event. According to the disaster-causing, the coupling coefficient of different types of disasters is introduced to modify the failure probability of a single equipment under the multiple disaster as follows:

$$p_l = 1 - \prod_{i=1}^{m} \left(1 - \beta_i P_{l,i}\right) \tag{1}$$

In the above formula, p_l represents the failure probability of the equipment l under multiple disasters, $P_{l,i}$ represents the failure probability of the equipment l under the disaster i, m represents the number of disasters, β_i represents the coupling coefficient between disaster i and other disasters, which is determined according to the disasters reclosing time, takes $0 \leq \beta_i \leq 1$.

Equipment failure is affected by both external disasters and power grid operation [20, 21]. There is a clear cause-and-effect relationship between the current and subsequent faults. According to the superposition effect between the disaster and the equipment state, the modified equipment failure probability is:

$$p_l' = 1 - (1-p_l) \times (1 - \omega p_z) \tag{2}$$

In the above formula, p_l' represents the post-correction failure probability of the equipment l, p_l represents the failure probability of the equipment l under multiple disasters, p_z represents the failure probability of the equipment l caused by system state change, ω represents the influence factor, takes $\omega \geq 1$.

2.2 Combined Probability Calculation

Group failures refer to multiple expected failures of electrical equipment affected by external natural disasters. For example, due to the similar weather and geological conditions in the same area, external extreme disasters often affect many transmission lines and equipment at the same time, resulting in group failures and large-scale power failure. The failure of each equipment in the expected failure set is regarded as an independent event. And the calculation formula of combined probability is as follows:

$$P_K = 1 - \prod_{l=1}^{K} (1 - \alpha_l P_l) \tag{3}$$

In the above formula, P_K represents the combination probability, P_l represents the failure probability of the equipment l, K represents the number of equipment in an expected failure, which is the fault multiplicity for short. α_l represents the coupling coefficient between the equipment l and other equipment in the combination, takes $0 \leq \alpha_l \leq 1$. The larger the value of α_l is, the closer the two equipment is.

According to the spatial distance and electrical distance between the equipment, the coupling coefficient α_l is calculated as follows:

$$\alpha_l = \prod_{j=1, j \neq l}^{K} \left(\frac{L_{\min}}{L_j} \cdot \frac{E_{\min}}{E_j} \right) \tag{4}$$

In the above formula, L_j and E_j represent the spatial distance and the electrical distance between the equipment l and the equipment j respectively, $L_{\min}$ represents the minimum spatial distance between the equipment l and other equipment in the combination, $E_{\min}$ represents the minimum electrical distance between the equipment l and the other equipment in the combination.

During the evolution of natural disasters, the equipment failure probability changes with time. The time series model of the equipment probability during the disaster period is set as $P_a = \{P_{a,t_1}, P_{a,t_2}, \cdots, P_{a,t_i}, \cdots, P_{a,t_n}\}$. P_{a,t_i} represents the failure probability of the equipment a at the moment t_i. When K equipment are selected from the sequence model of the affected equipment set, the subsequent equipment failures are affected by the previous equipment failures. The coupling coefficient of the time dimension between equipment is calculated according to the time series model as follows:

$$\alpha_l = \prod_{a=1,a\neq l}^{K} \frac{\text{cov}(P_l, P_a)}{\psi_{\max}} \tag{5}$$

In the above formula, $\text{cov}(P_l, P_a)$ represents the covariance between P_l of equipment l and P_a of equipment a, $\psi_{\max}$ represents the maximum covariance between P_l of equipment l and other equipment. When the covariance between two equipment is negative, takes $\alpha_l = 0$.

According to the tower failure probability and coupling effect, the combined probability between transmission lines is modified as follows:

$$p_o = 1 - \prod_{i=1}^{G_1} \left(1 - \beta_1 p_{o_1,i}\right) \times \prod_{j=1}^{G_2} \left(1 - \beta_2 p_{o_2,j}\right) \tag{6}$$

In the above formula, p_o represents the combined probability between transmission lines, o_1 represents tightly coupled tower set, o_2 represents general coupled tower set, $p_{o_1,i}$ represents the failure probability of the tower i in o_1, $p_{o_2,j}$ represents the failure probability of the tower j in o_2, G_1 and G_2 represent the number of tower in o_1 and o_2 respectively, β_1 and β_2 represent the coupling coefficient of o_1 and o_2 respectively, $0 \leq \beta_1 < \beta_2 \leq 1$.

3 Evaluation Index of Failure Consequence

In order to reduce the fault scale under natural disasters based on the evaluation indexes of high-risk fault characteristics, this paper selects the vulnerability index of grid structure, the vulnerability index of grid state and the accident risk index to represent the damage degree of grid structure, the safe and stable operation state of grid and accident event levels under natural disasters. The evaluation index of failure consequence constructed from the three dimensions of grid structure, state and risk are complementary and have fast calculation speed. The quantitative evaluation of index is carried out without relying on time domain simulation, which can avoid missing screen serious failure effectively.

3.1 The Vulnerability Index of Grid Structure

The electric betweenness represents the importance of transmission line in carrying power between generation and load. Based on the upper limit of transmission power and its sensitivity to power generation and load nodes, the electric betweenness of transmission line is defined as:

$$B_i = \sum_{g \in \Omega_g} \sum_{l \in \Omega_l} S_{gl} \lambda_{i.g} \quad (7)$$

In the above formula, B_i represents the electric betweenness of transmission line i, Ω_g represents the generator set in the grid, Ω_l represents the load set in the grid, $\lambda_{i.g}$ represents the sensitivity of transmission line i to generator g, S_{gl} represents the minimum transmission capacity between all paths from generator g to load l, that is:

$$S_{gl} = \min\{\frac{S_1}{\lambda_{1.g}}, \ldots, \frac{S_j}{\lambda_{j.g}}, \ldots, \frac{S_n}{\lambda_{n.g}}\} \quad (8)$$

In the above formula, S_j represents the maximum transmission power of the path j from generator g to load l, $\lambda_{j.g}$ represents the sensitivity of path j to generator g, n represents the number of paths from generator g to load l.

In order to facilitate the screening of multiple expected faults with a high degree of damage to the grid structure, the vulnerability index of the grid structure is defined as the minimum value of the transmission line vulnerability index included in the fault. The smaller the vulnerability index is, the more serious the damage of the fault to the grid structure would be. The vulnerability index $C_{net,i}$ of transmission line i is quantified as follows:

$$C_{net,i} = [1 - \frac{(B_i - S_{i,\max})}{S_{i,\max}}] * 100\% \quad (9)$$

In the above formula, $S_{i,\max}$ represents the upper power limit of transmission line i.

3.2 The Vulnerability Index of Grid State

The network overload security, static voltage stability, system frequency security and transient stability after failure can better represent the network state vulnerability. The vulnerability index of grid state $C_{stat,k}$ is defined as:

$$C_{stat,k} = \min(\eta_{s,k}, \eta_{f,k}, \eta_{v,k}, \eta_{t,k}) \quad (10)$$

In the above formula, $\eta_{s,k}$, $\eta_{f,k}$, $\eta_{v,k}$, $\eta_{t,k}$ represent thermal stability margin, static voltage stability margin, frequency safety margin and transient stability margin under fault k respectively.

Through the power flow calculation, sensitivity analysis and modal analysis in the initial test mode, the grid state variables can be quickly estimated to evaluate the grid thermal stability margin, static voltage stability margin, frequency safety margin and transient stability margin.

(1) Equipment overload safety. According to the DC power flow method and the distribution factor method, the power flow of each branch $I_{1,i,k}$ after the failure can be estimated quickly. Then the thermal stability margin of the system is

$\eta_{s,k} = \min_{i \in n}(1 - I_{1,i,k}/I_{\max,i})$, n is the number of monitoring branches, and $I_{\max,i}$ is the thermal stability limit of branch i.

(2) System frequency safety. The quasi-static simulation method can be used to estimate the steady-state frequency of several synchronous networks under the expected fault k quickly, that is:

$$f_{1,i,k} = f_{0,i,k} + f_{N,i} \frac{\Delta P_{0,i} + \Delta P_{1,i,k}}{K_{L,1,i} P_{L,1,i}} \tag{11}$$

In the above formula, $f_{0,i,k}$ and $f_{1,i,k}$ represent the steady-state frequency of synchronous grid i before and after failure respectively, $f_{N,i}$ represents the rated frequency of synchronous grid i, $\Delta P_{0,i}$ represents the active unbalance power of synchronous grid i caused by fault directly, $\Delta P_{1,i,k}$ represents the active unbalance power of synchronous grid i caused by post-fault security measures, $K_{L,1,i}$ and $P_{L,1,i}$ represent the sum of the static frequency characteristic coefficient of active load and active load of synchronous grid i respectively.

(3) Voltage safety and stability. ① Calculate the voltage stability margin $\eta_{v,0}$ based on the initial method, and the participation factor threshold $\delta_{\max}$ is set dynamically according to the margin value. ② Obtain the voltage weak bus, generator, line, transformer and other component participation factors $\delta_{i,k}$, based on the modal analysis results of the limit state mode. Select the equipment with $\delta_{i,k}$ larger than $\delta_{\max}$ to form a dangerous equipment set. ③ If the component equipment of the fault k belongs to the dangerous equipment set, and the electrical distance between the affected equipment and the weak bus is small, then the fault k may cause the static voltage stability problem, takes $\eta_{v,k} = -1.0$. Otherwise, the static voltage is stable, takes $\eta_{v,k} = 1.0$.

(4) Transient stability. Based on the SEEAC algorithm, the transient power angle margin, transient voltage margin, and transient frequency margin can be calculated quickly, and the minimum of the three is taken as the transient stability margin $\eta_{t,k}$.

3.3 The Accident Risk Index

The grid accident risk index under the expected fault i is defined as:

$$\begin{aligned} R_{\mathrm{e}} = k_{g.l} & \sum_{j \in G} \lambda_{g.j}(P_{g.j} - P_{g.i.j}) T_{g.i.j} \\ & + \sum_{j \in L} \lambda_{l.j}(P_{l.j} - P_{l.i.j}) T_{l.i.j} \end{aligned} \tag{12}$$

In the above formula, G and L represent the power plant assembly and load assembly respectively in the initial operating state respectively, $\lambda_{g.j}$ represents the feed-in price of power plant j, $\lambda_{l.j}$ represents the purchasing price of load j, $P_{g.j}$ and $P_{g.i.j}$ represent the active power output of power plant j under the initial operation state and after the failure i respectively, $P_{l.j}$ and $P_{l.i.j}$ represent the load active power of load j under the initial operation state and after the failure i respectively. $T_{g.i.j}$ represents the time from $P_{g.i.j}$ to $P_{g.j}$ of power plant j after the failure i, $T_{l.i.j}$ represents the time from $P_{l.i.j}$ to $P_{l.j}$ of load j after the failure i.

4 Fault Combination Algorithm

The optimization problem of fault equipment combination is expressed as problem W: n tuples $(P_1, P_2, \ldots, P_n)$ composed of the failure probabilities P_i of n disaster-affected equipment, forming a state space E = $\{(P_1, P_2, \ldots, P_n) \mid P_i \in S_i, i = 1, 2, \ldots, n\}$. The constraint set in n-tuples is set as D. Find all n-tuples in E that satisfy the multiplicity constraint of fault combination, the threshold value constraint of combined probability, the vulnerability index constraint of grid structure, the vulnerability index constraint of grid state, and the accident risk index constraint.

The backtracking algorithm can traverse the required path in depth-first order, looking for solutions to problems during the searching process. When it is found that the solution conditions are not satisfied, it will "backtrack" and return to try another path. In order to optimize the search for all solutions satisfying D using the back-tracking method, a state space tree S of $n*n$ is defined. Each element is regarded as a node in the diagram, and the problem W is reduced to the path traversal problem of the diagram.

$$S = \begin{bmatrix} P_1 & P_2 & P_3 & P_4 & \cdots & P_n \\ & P_2 & P_3 & P_4 & \cdots & P_n \\ & & P_3 & P_4 & \cdots & P_n \\ & & & \cdots & \cdots & \cdots \\ & & & & P_{n-1} & P_n \\ & & & & & P_n \end{bmatrix} \quad (13)$$

The elements of each row in the state space tree S are arranged in order of failure probability from large to small. And all combinations of K numbers are selected from n numbers. It is equivalent to finding all paths starting from the [1, 1] position in an $n*n$ upper triangular matrix and searching for any element of the row n as the end in depth-first order. It is stipulated that only the nodes between adjacent rows, and the nodes in the next row must be on the right side of the nodes in the previous row, have paths to be connected, otherwise there are no paths to be connected. Obviously, the number sequence passed by any path corresponds to a combination that meets the requirements.

Cycle through the maximum combination multiplicity K to get all combinations of N − 2, N − 3, …, N − K quickly (Fig. 1).

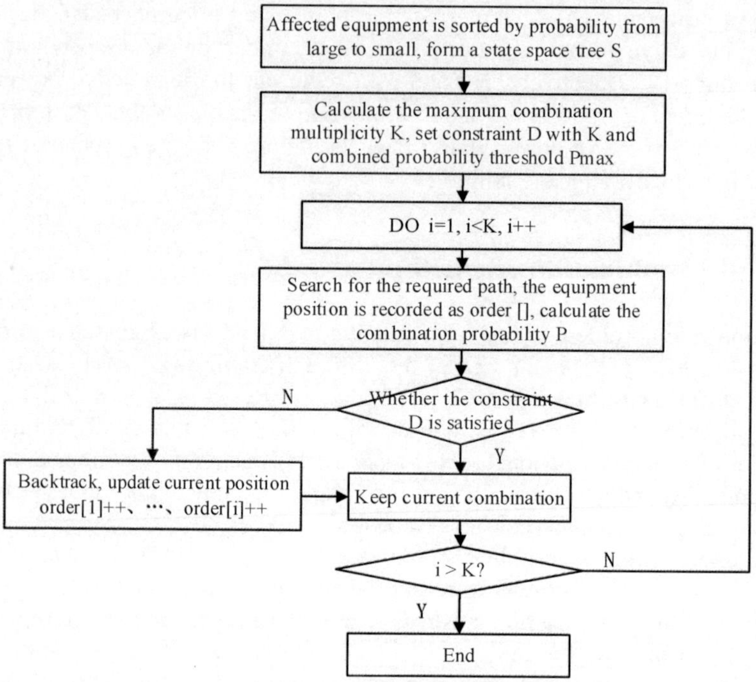

Fig. 1. Flow chart of predictive fault set generation based on backtracking algorithm

The method to determine the maximum combination multiplicity K: the affected equipment is sorted by probability from large to small, the equipment is multiplied one by one until the probability is less than the probability threshold $P_{\max}$. When $P_1 * P_2 * P_3 \cdots P_x \leq P_{\max}$, $K_x = x - 1$, the maximum combination multiplicity is selected as $K = \max(K_x, K_{set})$, K_{set} is the maximum combination multiplicity set before combination.

5 Example Analysis

In order to verify the effectiveness and practicability of the method to generate high-risk accident set in natural disaster, the online calculation data of a certain section of Northwest Power Grid with a scale of 5,872 nodes, 1,032 units, 3,152 AC lines and 4 DC systems were used.

5.1 Combination Probability Calculation

Meteorological disasters of high winds and heavy rains are expected to occur near the Haixi Corridor during the expected period. The equipment affected includes 2 DC systems, 12 transformers and 35 AC lines. The schematic diagram of the power grid is shown in Fig. 2.

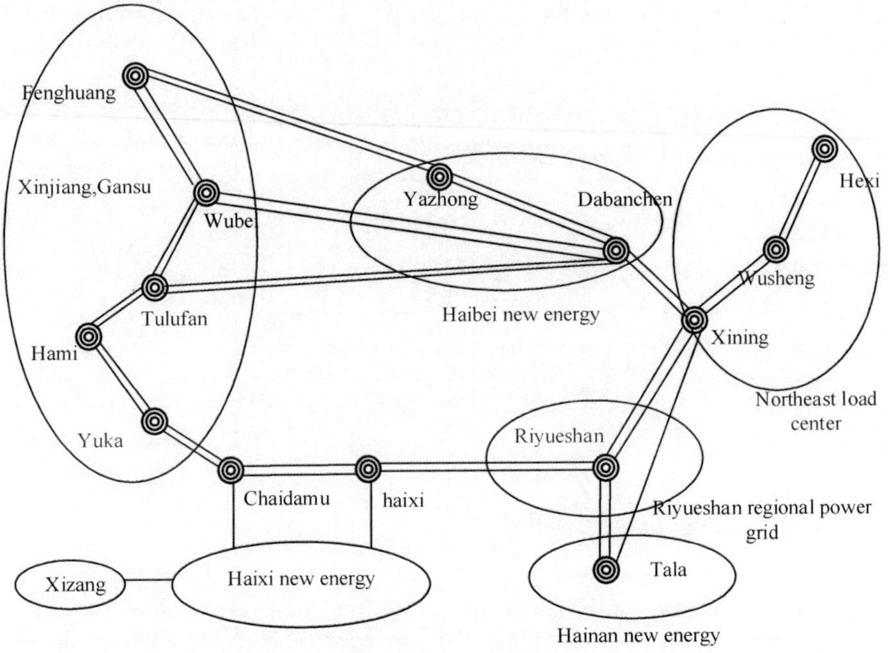

Fig. 2. Schematic diagram of the power grid

According to the probability of equipment failure under a single disaster of high winds and heavy rain, the failure probability of a single equipment under the multiple disaster was calculated. The disaster coupling coefficient was set as 1.0. The calculation results are shown in Table 1. The failure probability of equipment under the multiple disaster increased.

Table 1. Results of failure probability of single equipment

Equipment name	Fault probability		
	High winds	Heavy rains	Multiple disaster
FY Line	0.32	0.56	0.7008
FW Line	0.31	0.52	0.6688
DT Line	0.31	0.42	0.5998
DW II Line	0.28	0.39	0.5608
DY Line	0.28	0.36	0.5392
HW I Line	0.21	0.23	0.3917
QS DC Line	0.18	0.22	0.3604
TZ DC Line	0.14	0.19	0.3034

The coupling coefficient between equipment is calculated according to the spatial distance and the electrical distance, and the results of the combination probability are shown in Table 2. The closer the spatial distance and the electrical distance are, the greater the coupling coefficient is, and the greater the combination probability is.

Table 2. Results of combination probability (units: km/Ω)

Combined equipment	Spatial distance	Electrical distance	Combination probability
FY Line	34.61	36.73	0.577
FW Line	99.49	35.54	0.299
DT Line	42.07	38.43	0.505
DW II Line	27.86	9.99	0.862
DY Line	123.73	42.12	0.043
HW I Line	115.26	52.23	0.037
QS DC Line	135.26	62.28	0.032
TZ DC Line	107.42	43.21	0.042

5.2 Fault Combination Screening

The number of affected equipment is 49. The scale of the fault combination of the exhaustive method and the backtracking method in this paper is shown in Table 3. With the increase of fault multiplicity, the number of fault combination increases rapidly and the time-consuming increases.

Table 3. Results of comparative analysis of fault combination (unit: s)

Fault multiplicity	Exhaustive method		Combination only by probability		Combination only by risk index	
	Fault number	Time-consuming	Fault number	Time-consuming	Fault Number	Time-consuming
2	1176	1.52	113	0.21	52	12.12
3	18424	2.33	384	0.41	127	27.33
4	211876	4.23	876	0.82	336	56.23
5	1906884	12.25	2182	1.18	1237	124.25

Screening fault combination based on the principle of combining by probability first and then combining by risk index, it can not only improve the calculation speed, but also screen out high-risk fault sets. In order to make full use of computing resources, the vulnerability threshold of the grid structure was 0.20, the vulnerability threshold of the grid state was 0.60, and the threshold of accident risk was 100.0. The number of N − 2, N − 3, N − 4 and N − 5 fault combinations was screened out to be 1752.

Some N − 2 faults generated by the combination with FY Line are shown in Table 4. The combination probability of "FY Line + DT Line" fault is 0.577, but it is not satisfied with the risk index of multiple fault, which belongs to the fault with high probability and low risk, so it is not included in the scope of defense.

Table 4. Screening results of fault combination

Fault combination	Vulnerability of grid structure	Vulnerability of grid state	Accident risk	Keep or not
FY Line + FW Line fault	0.13	0.78	0	Yes
FY Line + DT Line fault	0.22	0.74	0	No
FY Line + DW II Line fault	0.13	0.73	0	Yes
FY Line + DY Line fault	0.11	0.72	0	Yes
FY Line + HW I Line fault	0.56	0.55	0	Yes
FY Line + QS DC Line fault	0.87	0.13	210.1	Yes
QS DC Line + TZ DC Line fault	0.91	0.03	412.6	Yes

Take the three-phase permanent fault of FY line and DT line as an example, the power angle stability margin of the system is 100% after fault. And the voltage stability margin is 74%, the key bus is Fenghuang. Therefore, the stability margin of the system is 74%. The simulation results are as follows (Figs. 3 and 4).

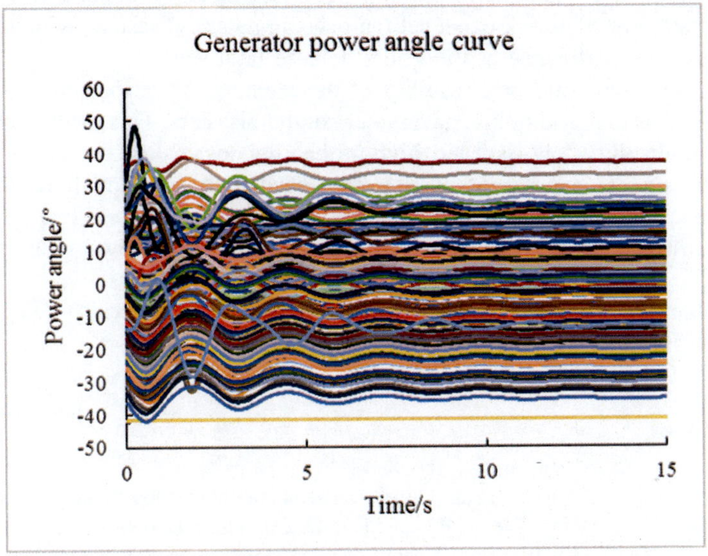

Fig. 3. Generator power angle curve after fault

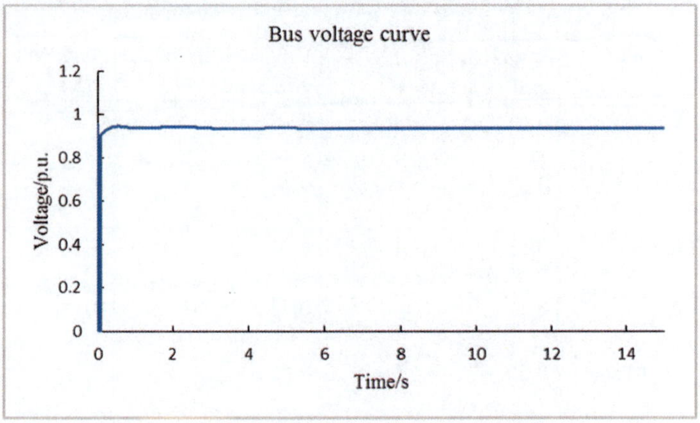

Fig. 4. Bus voltage curve after fault

6 Conclusion

Aiming at the characteristics of uncertainty and high risk of multiple expected fault sets in large power grid under natural disasters, a combined probability assessment model based on disaster probability, equipment space and electrical coupling characteristics is established to provide a basis for affected equipment combination. The vulnerability index of grid structure, the vulnerability index of grid state and the accident risk index are selected, and the evaluation index system of multi-dimensional high-risk expected fault set based on fault consequence is established. Based on the probability of fault combination and risk evaluation index, the backtracking algorithm is used to realize the rapid generation of high-risk expected fault sets in natural disasters, which ensures the effectiveness and pertinence of the grid's defense fault sets.

The effectiveness and practicability of the method are verified by the simulation analysis of the actual grid online data under natural disasters. This method can filter out most low-probability low-risk and high-probability low-risk expected faults, greatly reduces the size of expected fault sets that need to be simulated. It improves the calculation speed of online security analysis and promotes the practical application of online security and stability assessment technology for large power grid.

Acknowledgment. This research was supported by the Key Research and Development plan of Ningxia Hui Autonomous Region (5229NX18001W).

References

1. Bao, Y., Feng, C., Ren, X., et al.: Online transient stability fault screening based on support vector machine. Autom. Electr. Power Syst. **43**(22), 52–58 (2019)
2. Wang, H., Xu, T., Li, B., et al.: Design of coordinated prevention system in self-adapting natural environment for safety and stability of power grid. Autom. Electr. Power Syst. **38**(9), 143–151 (2014)

3. Chen, B., Li, C., Qin, H., et al.: Evaluation of typhoon resilience of distribution network considering grid reconstruction and disaster recovery. Autom. Electr. Power Syst. **42**(6), 47–52 (2018)
4. Li, C., Zhang, Y., Yang, G., et al.: An anticipated multi-contingencies selection method based on fault influence domain. Power Syst. Protect. Control **43**(14), 92–98 (2015)
5. Jiao, H., Wen, Y., Guo, C., et al.: A greedy algorithm for contingency screening based on probability ordered tree and its application in reliability assessment. Proc. CSEE **36**(8), 2068–2076 (2016)
6. Zhang, J., Wei, J., Li, X.: Impact analysis of typhoon on cascading failure for electric power system. Electr. Power Autom. Equip. **39**(10), 157–162 (2019)
7. Yao, K., Yu, J., Xu, T., et al.: A method to generate dynamic accident set for power grid clustered faults caused by tropical cyclone. Power Syst. Technol. **38**(6), 1593–1599 (2014)
8. Xue, Y., Wu, Y., Xie, Y., et al.: Powersystem stability analysis for intercurrent natural disasters. Autom. Electr. Power Syst. **40**(4), 10–18 (2016)
9. Xue, Y., Xie, Y., Wen, F., et al.: A review on cascading failures in power systems. Autom. Electr. Power Syst. **37**(19), 1–9, 40 (2013)
10. Xue, Y., Wu, Y., Xie, Y., et al.: Extension of blackout defense scheme to natural disasters early-warning. Autom. Electr. Power Syst. **37**(16), 18–26 (2013)
11. Zhao, J., Jiang, X., Zhang, B.: A contingency screening and ranking method for power system static stability analysis. Power Syst. Technol. **29**(20), 62–67 (2005)
12. Xu, T., Xue, Y., Li, B., et al.: On-line adaptive contingency screening for early-warning of transient stability. Autom. Electr. Power Syst. **33**(22), 1–4 (2009)
13. Xu, T., Du, Y., Bao, Y., et al.: A classification rolling contingency screening method for on-line transient security and stability assessment. Autom. Electr. Power Syst. **42**(13), 182–188 (2018)
14. Fang, L., Xiong, X., Fang, S., et al.: Power system risk control decision based on cause and effect correlation analysis of fault and meteorology. Power Syst. Protect. Control **42**(17), 113–119 (2014)
15. Xie, M., Yin, Y., Du, Y., et al.: Assessment on available transfer capability considering complex weather risk sources. Electr. Power Autom. Equip. **38**(4), 16–22 (2018)
16. Javanbakht, P., Mohagheghi, S.: A risk-averse security constrained optimal power flow for a power grid subject to hurricanes. Electr. Power Syst. Res. **116**, 408–418 (2014)
17. Wu, F.F., Tsai, Y.K., Yu, Y.X.: Probabilistic steady state and dynamic security assessment. IEEE Trans. Power Syst. **3**(1), 1–9 (1987)
18. Liu, Y., Zhu, Y., Cai, W., et al.: Comprehensive evaluation of power quality based on disaster levels. Electr. Power Autom. Equip. **29**(8), 110–113 (2009)
19. Miki, T.: Consideration of uncertainty factors in search for high risk events of power systems caused by natural disasters. WSEAS Trans. Power Syst. **3**(3), 76–81 (2008)
20. Liu, H., Sun, Y., Cheng, L., et al.: Fast sorting technique considering multi-state component models. Autom. Electr. Power Syst. **32**(1), 16–21 (2008)
21. Xue, Y., Fei, S., Bu, F.: Upgrading the blackout defense scheme against extreme disasters part I new challenges and reflection. Autom. Electr. Power Syst. **32**(9), 1–6 (2008)

Control Method on Photovoltaic Inverter Decoupling Circuit with Electrolytic Capacitor-Less

Xiaodong Liu, Longfei Tang(✉), Sucheng Liu, Wei Fang, and Wenlong Shi

Key Laboratory of Power Electronics and Motion Control, Anhui University of Technology, Ma'anshan, China
llleft@163.com

Abstract. For the problem of the power imbalance between the AC side and DC side of the two-stage single-phase photovoltaic grid-connected inverter, an active power decoupling circuit control method is proposed. While stabilizing the bus voltage, the long-life film capacitors with small capacitance and high stability are used to replace large electrolytic capacitors with poor reliability. Combined with the law of power conservation, the instantaneous energy stored in the decoupling circuit is equal to the energy of the second fluctuation on the output side of the inverter, and the power decoupling can be achieved to calculate the control reference value required by the decoupling circuit. The calculation process of the decoupling circuit parameters and the design process of the controller are given. For a suitable closed-loop control method, the small signal modeling is presented and voltage current double closed-loop control strategy is proposed. Based on the design parameters, an experimental prototype with a power of 1 kW is produced. The experimental results verified the effectiveness of the proposed active power decoupling control strategy.

Keywords: Two-stage inverter · Photovoltaic · Power decoupling · Double loop control

1 Introduction

Single-phase inverters are widely used in distributed power grid-connected systems with power levels less than 10 kW, such as distributed photovoltaic power generation, energy storage units, and fuel cells. The traditional single-phase photovoltaic grid-connected inverter is composed of two stages. The front-stage Boost circuit realizes the boost and MPPT functions to make the photovoltaic panel work at the maximum power point. The latter stage uses a single-phase full-bridge inverter circuit to achieve DC to AC convert. Since the power factor of the inverter is close to 1, the waveform of its output current is the same frequency and phase as the grid voltage. And the output power on the AC side is twice the power frequency ripple power. In this case, the pulsating power can induce a ripple component on the DC side, which reduces the efficiency of the photovoltaic panel [1, 2].

Y. Xue et al. (Eds.): PMF 2020, LNEE 718, pp. 512–525, 2021.
https://doi.org/10.1007/978-981-15-9746-6_40

Therefore, the passive power decoupling circuit uses a large-capacity electrolytic capacitor to achieve power balance. But passive circuits also bring many shortcomings, which not only shorten the life of the inverter but also increase the size of the system [3–5].

There are two ways to reduce the capacitance. One is to increase the DC bus voltage, and the other is to parallel active power decoupling (APD) circuits. For the method of increasing the bus voltage [6], it will increase the voltage stress on the device, which will cause difficulty in device selection. Moreover, the modulation ratio of the inverter will be reduced, affecting the efficiency of the system. Therefore, the mainstream method is to increase the active power decoupling topology. Inductor, as commonly passive devices, can be used in APD circuits [7]. However, the introduction of inductance as an energy storage element will undoubtedly reduce the power density of the system, resulting in an increase in the size of the system. Moreover, the basic topology combination algorithm can also reduce the bus ripple. Based on the traditional bidirectional buck/boost decoupling topology, a new control strategy is proposed in [8]. The weighted control of the output bus voltage fluctuation is introduced based on double closed loop, which can significantly reduce the bus voltage ripple and the decoupling capacitance. And a new type of topology has been proposed in [9]. A small capacitance capacitor is connected in parallel with the symmetric half-bridge decoupling circuit to stabilize the bus voltage. While reducing the decoupling capacitance, the withstand voltage of the decoupling capacitor is reduced and further reduced. The size and cost of the inverter. Besides, many decoupling methods have also been proposed [10, 11], all of which increase the energy storage topology so that when the grid side power is higher than the DC side power, the energy storage circuit releases energy, and when the grid side power is lower than the DC side power, the energy storage The circuit absorbs energy.

The objective of this paper is to propose an active power decoupling method based on the topology mentioned in Reference [9]. While stabilizing the bus voltage and the output current of the inverter, the film capacitor can be used to replace the electrolytic capacitor to achieve power decoupling. Section 2 analyzes the structure and basic working principle of the power decoupling circuit. In Sect. 3, the control strategy of the power decoupling branch is analyzed, including the drawing of the control system block diagram and the design of the double closed loop. A 1 kW prototype has been built and tested, and the simulation and the experimental results are presented in Sect. 4. Section 5 concludes this paper.

2 Circuit Configuration and Operation Principles

2.1 The Topology of the Symmetrical Half-Bridge Decoupling Circuit

The topology of the symmetrical half-bridge decoupling circuit is shown in Fig. 1 below. The topology includes thin film capacitors C_1 and C_2, filter inductance L_f, and switch tubes Q_1 and Q_2. Among them, the capacitors C_1 and C_2 with the same capacitance value are connected in series and then connected in parallel across the DC bus. The midpoint is connected to the switch tubes Q_1 and Q_2 through the filter inductor to form a bidirectional buck/boost circuit so that the fluctuating energy can be exchanged between the film capacitors C_1 and C_2.

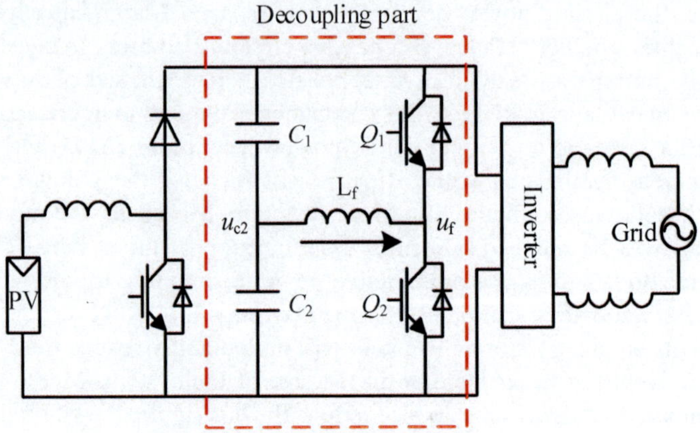

Fig. 1. Novel power decoupling circuit.

There are two working modes of the decoupling circuit, One of the modes is that capacitor C_1 charges the inductor, and the inductor discharges C_2, the other mode is that C_2 charges the inductor, and the inductor discharges C_1. The two working modes of the decoupling circuit are analyzed below:

Mode I. In each switching cycle, the decoupling circuit can be divided into two working steps. During the time of step I, as shown in Fig. 2, the Q_1 is turned on and Q_2 is turned off, The capacitor C_1 charges the inductor L_f, and the voltage of the capacitor C_1 gradually decreases. During the time of step II, the Q_1 is turned off, and Q_2 is turned on, the inductor L_f charges the capacitor C_2, and the voltage of the capacitor C_2 gradually rises. In Mode I, to maintain the stability of the bus voltage, the sum of the average voltage of the capacitors should be equal to the bus voltage.

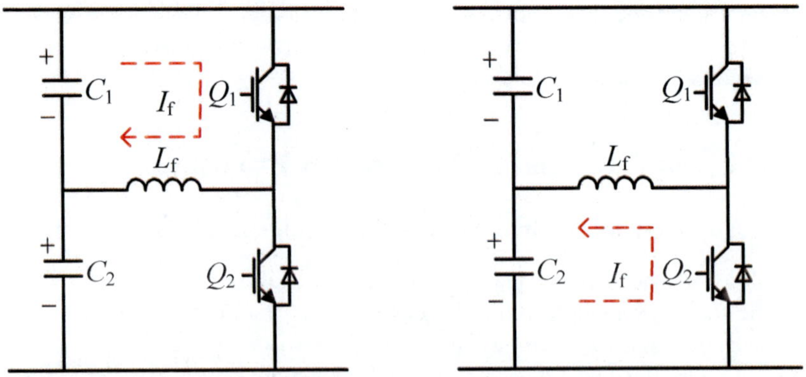

(a) Step I (b) Step II

Fig. 2. Operating mode I of the decoupling circuit.

Mode II. Mode II is similar to Mode I. This mode can also be divided into two steps. as shown in Fig. 3. During the time of step I, the Q_1 is turned off, and the Q_2 is turned on, the capacitor C_2 charges the inductor L_f, and the voltage of the capacitor C_2 gradually decreases. During the period of step II, Q_1 is turned on, and Q_2 is turned off, the inductor L_f charges the capacitor C_1, and the voltage of the capacitor C_1 gradually rises. Similarly, in Mode II, to maintain the stability of the bus voltage, the average capacitance The sum of the voltages should be equal to the bus voltage.

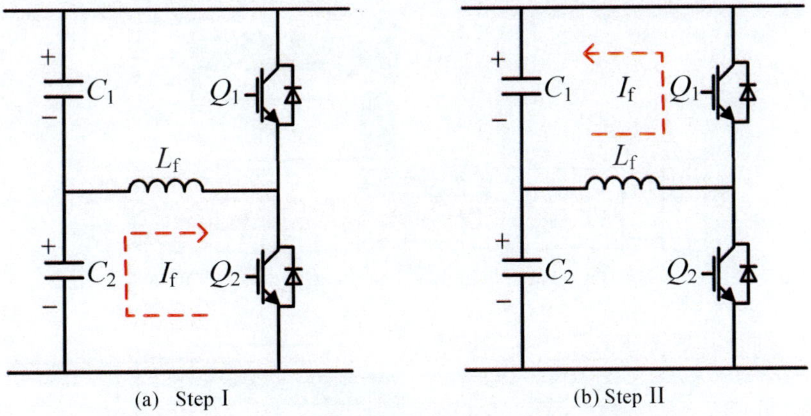

Fig. 3. Operating mode II of the decoupling circuit.

2.2 Operation Principles

When the inverter works at unity power factor, the AC side current of the converter is in the same frequency and phase as the grid voltage, and the grid voltage and grid side current are as shown in Eqs. (1) and (2):

$$u_{ac}(t) = V_g \sin(\omega t) \tag{1}$$

$$i_{ac}(t) = I_g \sin(\omega t) \tag{2}$$

V_g and I_g are the amplitude of the sinusoidal voltage and current on the grid side, respectively, and ω is the power frequency. From the above formula, the power on the AC side is:

$$p_{ac}(t) = \frac{V_g I_g}{2}(1 - \cos 2\omega t) \tag{3}$$

Replace each quantity of the above formula with the following expression:

$$p_{ac}(t) = P_{ac(dc)} + p_{ac(2\omega)} \tag{4}$$

$P_{ac\ (dc)}$ is the average DC power required by the load, and $p_{ac\ (2\omega)}$ is the low-frequency ripple power that needs to be buffered on the DC side.

When the photovoltaic power is higher than the grid-side output power, the decoupling circuit stores energy, otherwise the energy is released by the decoupling circuit, as Fig. 2 shows. If the instantaneous energy of the decoupling circuit is equal to the fluctuating energy on the grid side, the reference values of the decoupling capacitor voltage and inductance current can be obtained (Fig. 4).

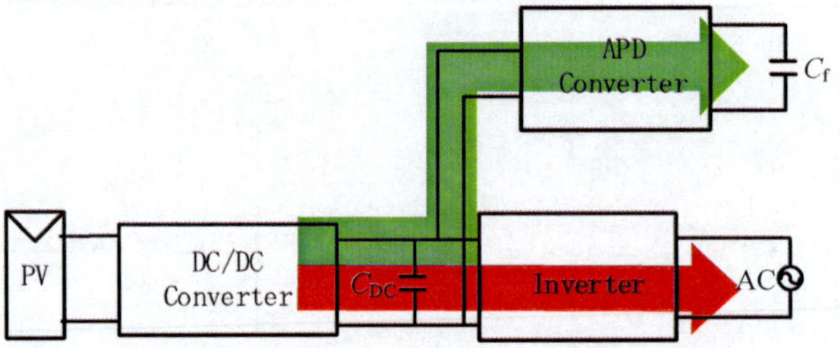

(a). DC power is higher than the grid-side output power

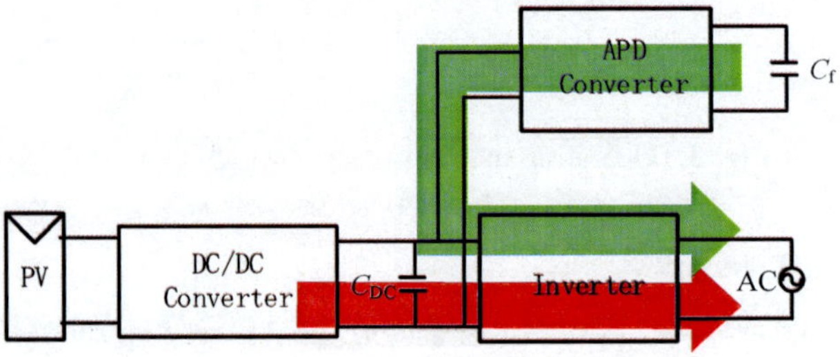

(b). DC power is lower than the grid-side output power

Fig. 4. Schematic diagram of energy flow.

The active power decoupling circuit undertakes the task of eliminating the pulsating power caused by the power imbalance on the AC and DC sides, and at the same time, it is necessary to stabilize the DC side bus voltage. Since the decoupling capacitor $C_1 = C_2 = C_f$, the average voltage of the capacitors is equal and half of the bus voltage. Therefore, the capacitor voltage is obtained by the forward bias of the AC quantity. The capacitor voltage can be set as follows:

$$u_{c1}(t) = \frac{V_{dc}}{2} - V_c \sin(\omega t + \theta) \tag{5}$$

$$u_{c2}(t) = \frac{V_{dc}}{2} + V_c \sin(\omega t + \theta) \tag{6}$$

u_{c1} and u_{c2} are the instantaneous voltage values of the film capacitors C_1 and C_2. Respectively, V_{dc} is the DC-side bus voltage, this paper takes 400 V. Because the capacitor voltage fluctuates with the grid-side power, ω is the power frequency, V_c is the amplitude of the capacitor voltage AC The value, θ is the phase angle of the alternating voltage of the capacitor voltage, which is obtained under the condition that the power of the decoupling circuit is equal to the fluctuating power.

Under the premise of the assumed capacitor voltage, the capacitor current can be obtained as follows:

$$i_{c1}(t) = -\omega C_1 V_c \cos(\omega t + \theta) \tag{7}$$

$$i_{c2}(t) = \omega C_1 V_c \cos(\omega t + \theta) \tag{8}$$

So the power of the decoupling circuit is:

$$p_{APD}(t) = \omega C_f V_c^2 \sin(2\omega t + 2\theta) \tag{9}$$

From Eq. (4), the fluctuating power on the AC side is $p_{ac\ (2\omega)}$. Now let $p_{APD} = p_{ac\ (2\omega)}$:

$$\omega C_f V_c^2 \sin(2\omega t + 2\theta) = \frac{V_g I_g}{2} \cos(2\omega t) \tag{10}$$

To make the real part equal and the imaginary part equal, list the following equations:

$$\begin{cases} \omega C_f V_c^2 = \frac{V_g I_g}{2} \\ 2\omega t + 2\theta - \frac{\pi}{2} = 2\omega t \end{cases} \tag{11}$$

From the above, V_c and θ can be obtained:

$$\begin{cases} V_c = \sqrt{\frac{V_g I_g}{2\omega C_f}} \\ \theta = \frac{\pi}{4} \end{cases} \tag{12}$$

2.3 Design Principles of the Decoupling Capacitor

According to the voltage reference value calculated by the working principle, draw the waveform diagrams of various quantities of the system as shown in Fig. 5. The voltage waveform of V_c can eliminate the secondary fluctuating power on the DC side, to achieve the purpose of power decoupling.

During the period t_1–t_2, the excess power generated by the photovoltaic array, namely the shaded area in Fig. 5, is P_{PV}/ω, where P_{PV} is the constant power output by the photovoltaic array, and ω is the angular frequency of the grid-side voltage. The voltage value of the decoupling capacitor C_1 reaches the maximum value $u_{c2(max)}$ at time t_2, and C_2 reaches the minimum value $u_{c1(min)}$ at the time t_2. Since θ takes the value $\pi/4$, the phase of u_{c1} lags the grid voltage by $\pi/4$. The sum of the instantaneous voltages of u_{c1} and u_{c2} is equal to the bus voltage.

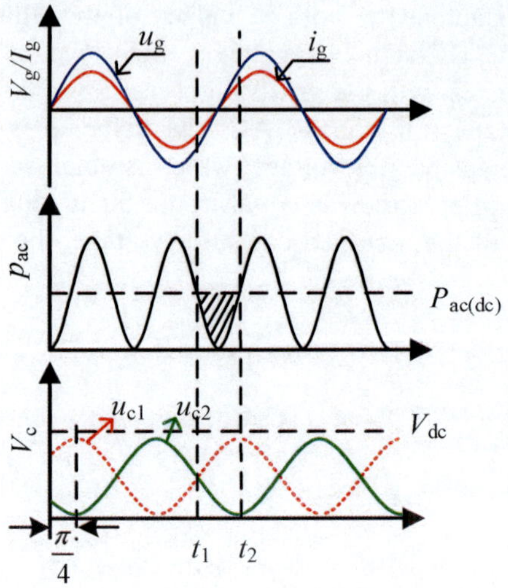

Fig. 5. Control target waveform.

The energy stored in the decoupling capacitor during t_1–t_2 can be obtained:

$$P_{APD} = \frac{1}{2}C_1(u_{c1(\max)}^2 - u_{avg}^2) + \frac{1}{2}C_2(u_{c2(\min)}^2 - u_{avg}^2) \tag{13}$$

Let $P_{PV}/\omega = P_{APD}$, the value of C_f can be got, which is the capacitance of the decoupling capacitor:

$$C_f = \frac{2P_{PV}}{\omega(u_{c1(\max)}^2 - 2u_{avg}^2)} \tag{14}$$

Also, the role of the decoupling capacitor is not only to alleviate the energy imbalance, but its equivalent capacitance $C_1//C_2$ also needs to bear the role of filtering the high-frequency ripple on the DC bus. Therefore, for the capacitance value obtained by Eq. (14), the ability to maintain the stability of the bus voltage needs to be checked. In a switching cycle, it is assumed that the excess energy P_{PV}Ts of the photovoltaic array are all stored by C_1 and C_2, where Ts is the switching frequency of the inverter. The fluctuation of the bus voltage caused by this is Δu, and the Δu obtained by Eq. (15) can be used to check whether the decoupling capacitor meets the requirements.

$$\frac{C_1C_2}{C_1 + C_2}U_{DC}\Delta u = P_{PV}T_s \tag{15}$$

3 System Control Design

The circuit control objectives are the instantaneous voltage of the film capacitors C_1 and C_2 and the current of the filter inductor L_f to eliminate the secondary voltage fluctuation on the DC bus. Therefore, the control circuit adopts the control strategy of the outer loop of the capacitor voltage and the inner loop of the inductor current.

To model the filter inductor current and voltage, from Eqs. (7) and (8):

$$i_{\mathrm{f}} = i_{\mathrm{c1}} - i_{\mathrm{c2}} = -2C_{\mathrm{f}}\frac{\mathrm{d}u_{\mathrm{c2}}}{\mathrm{d}t} \tag{16}$$

$$L_{\mathrm{f}}\frac{\mathrm{d}i_{\mathrm{f}}}{\mathrm{d}t} = u_{\mathrm{c2}} - u_{\mathrm{f}} \tag{17}$$

The Laplace transform can be applied to (16)–(17):

$$G_1(s) = \frac{u_{\mathrm{c2}}}{i_{\mathrm{f}}} = -\frac{1}{2sC_{\mathrm{f}}} \tag{18}$$

$$sL_{\mathrm{f}} = \frac{u_{\mathrm{c2}}}{i_{\mathrm{f}}} - \frac{u_{\mathrm{f}}}{i_{\mathrm{f}}} \tag{19}$$

Bring Eq. (18) into Eq. (19):

$$G_2(s) = \frac{i_{\mathrm{f}}}{u_{\mathrm{f}}} = -\frac{2C_{\mathrm{f}}s}{2C_{\mathrm{f}}L_{\mathrm{f}}s^2 + 1} \tag{20}$$

Draw the control block diagram of the decoupling circuit according to the established model as follows (Fig. 6):

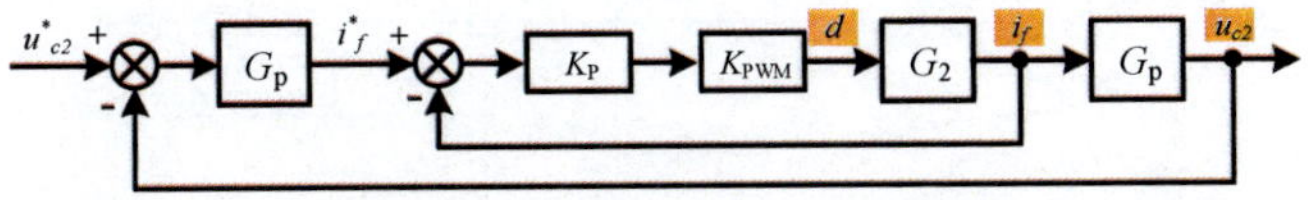

Fig. 6. Control block diagram of the decoupling circuit.

To accelerate the dynamic response of the system, the current inner loop is compensated by proportional adjustment. The Bode diagram before and after compensation is shown in Fig. 7 where K_p is 300 The bandwidth of the system before compensation is small. By increasing the proportional coefficient, the amplitude-frequency characteristic curve is moved up. Increase bandwidth and improve system responsiveness.

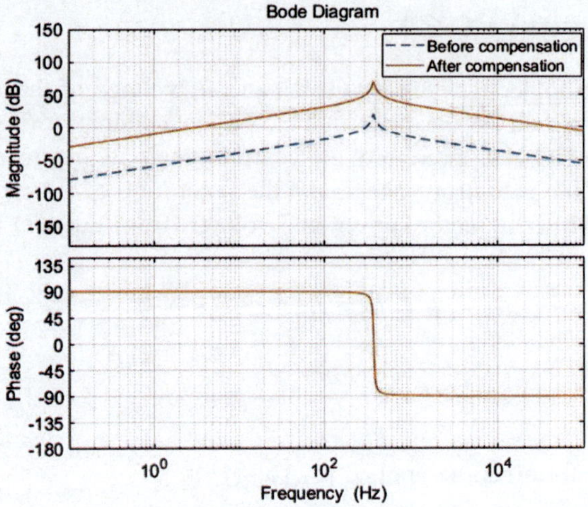

Fig. 7. Bode plot of the current loop compensation

When the outer loop is compensated, the gain of the inner loop can be regarded as approximately 1. The outer loop compensation function is adjusted by PR. The transfer function of the PR regulator is shown in Eq. (15):

$$G_{PR}(s) = K_{\text{fp}} + \frac{2K_{\text{fi}}\omega_{\text{c}}s}{s^2 + 2\omega_{\text{c}}s + \omega_0^2} \tag{22}$$

The open-loop Bode diagram before and after compensation is shown in Fig. 8. The parameters of the PR controller are as follows: $K_{\text{fp}} = 2$, $K_{\text{fi}} = 300$, $\omega_{\text{c}} = 3$, and $\omega_0 = 314$. It can be seen that due to the introduction of quasi-resonant compensation, the gain of the system at 50 Hz is greatly improved.

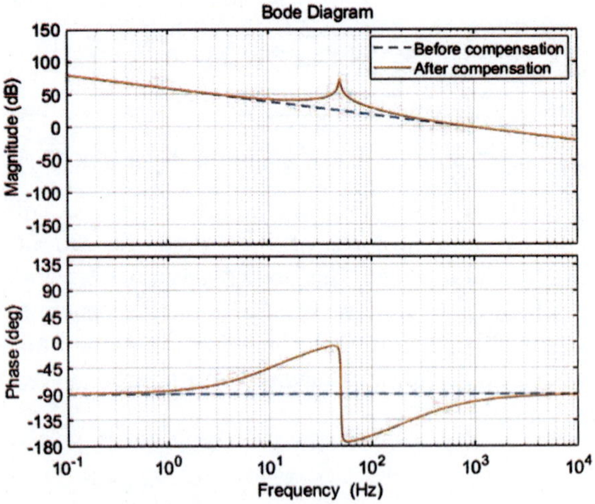

Fig. 8. Bode plot of the voltage loop compensation

4 Simulation and Experimental Results

4.1 Simulation Results

It can be calculated from Eq. (14) that the capacitance of the decoupling capacitor is 79.6 μF, and the simulation circuit is selected to be 90 μF. The larger the capacitance value, the smaller the ripple of the bus voltage. The filter inductance in the decoupling circuit is selected as 1 mH, and other parameters are shown in Table 1:

Table 1. Circuit Parameter.

Parameter	Value
DC-link voltage V_{dc}/V	400
Filter inductor L_f/mH	1
Film capacitor C_1C_2/μF	90
Switching frequency f/kHz	20
AC output RMS voltage V_g/V	220
Grid frequency f_g/Hz	50
Ripple rate of the bus voltage	≤5%

Figure 9 shows the current waveform on the DC side when the power decoupling branch is not added. Since the power on the AC side fluctuates twice the power frequency, the bus voltage is constant and the current should fluctuate twice the power frequency with the power on the AC side, and the simulation waveform conforms to the theoretical derivation. In this case, the fluctuation of the current will undoubtedly affect the working point of the photovoltaic panel, causing system instability and unnecessary waste. Figure 10 shows the waveform after the power decoupling branch is added. The DC bus current is stable at around 2.6A, which significantly suppresses the DC bus current fluctuations.

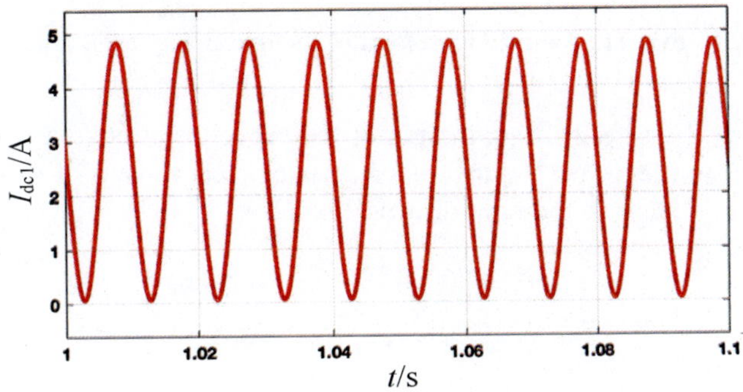

Fig. 9. DC current without decoupling

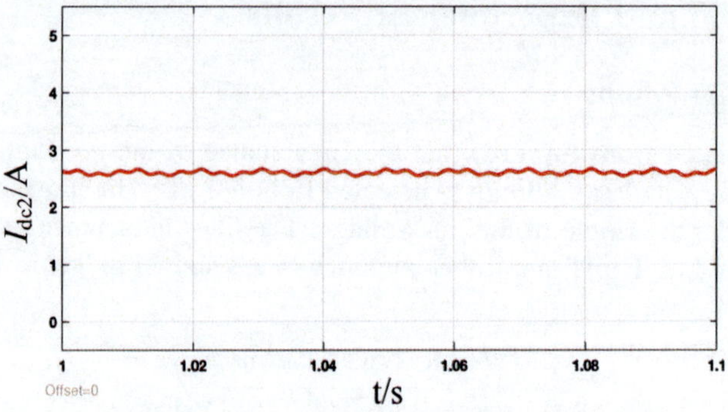

Fig. 10. DC current with decoupling circuit

Figure 11 and 12 shows the decoupling capacitor voltage and DC bus voltage waveform. It can be seen from the waveform that the decoupling circuit waveform follows the given reference value, which embodies the stability of the control loop. The average voltage of the DC bus is 400 V, and the voltage ripple is 4%, which meets the requirements.

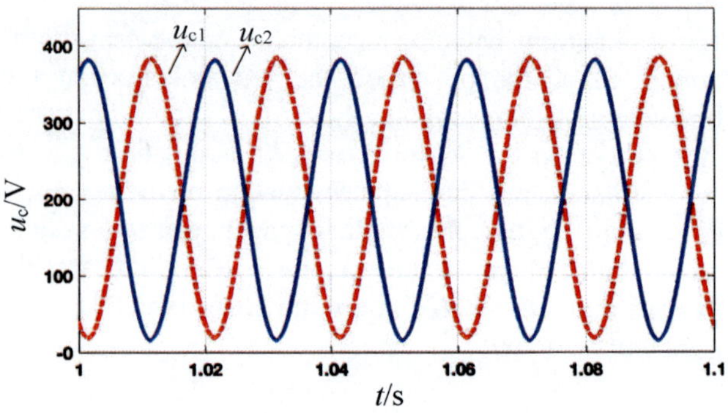

Fig. 11. Voltage waveform of the decoupling capacitor

Considering the safe working margin of the capacitor, it is greater than 0 V at the lowest voltage and less than the theoretically calculated value of 400 V at the highest voltage, so the voltage of the control capacitor C_2 is:

$$u_{c2}(t) = 200 + 190\sin(314t + \pi/4) \tag{23}$$

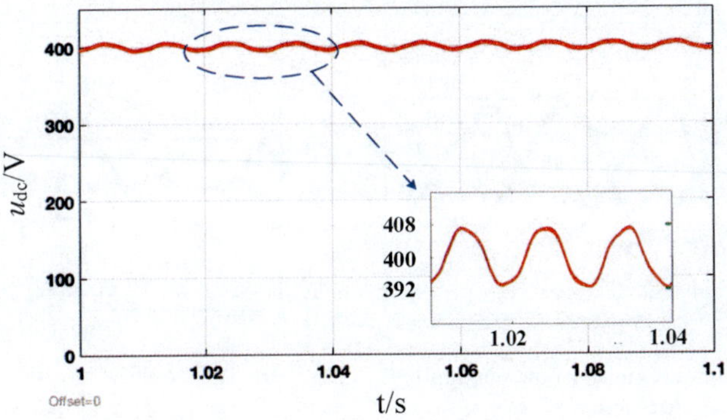

Fig. 12. Voltage waveform of DC bus

Correspondingly, from the results derived above, the voltage of capacitor C_1 should theoretically be:

$$u_{c1}(t) = 200 - 190\sin(314t + \pi/4) \tag{24}$$

4.2 Experimental Results

To verify the integrity of the control strategy, a 1 kW experimental platform was built. The parameters of each device were the same as the simulation parameters in Table 1. Among them, the thin film capacitor is EPCOS B32776E8306K, the capacitance is 30 μF, the withstand voltage is 800 V, and the ESR is 3.0 mΩ.

Figure 13 shows the main experimental waveforms of the converter. The complementary sine wave in Fig. 13(a) is the waveform of the voltages V_{c1} and V_{c2} of the decoupling circuit capacitors C_1 and C_2. The average voltage is 400 V, and the maximum value is less than the bus voltage, and the minimum value is greater than 0, which meets the given requirements of the controller. The fluctuating energy of twice the power frequency is always transferred between the decoupling capacitors. In order to facilitate the observation of the bus voltage ripple ΔV_{dc}, the oscilloscope bus voltage channel is changed to AC coupling measurement. Figure 13(b) is the measurement result. The bus voltage ripple is less than 20 V and the frequency is 100 Hz. Figure 13(c) is the grid voltage and output current waveform. The current and voltage are in the same phase and the frequency is 50 Hz. At the zero crossing point, there will be slight distortion due to the switching of the bridge arm switch. The experimental waveforms are consistent with the simulation results. The control strategy proposed in this paper can replace the electrolytic capacitors with film capacitors to achieve power decoupling while maintaining the stability of the bus voltage and output current.

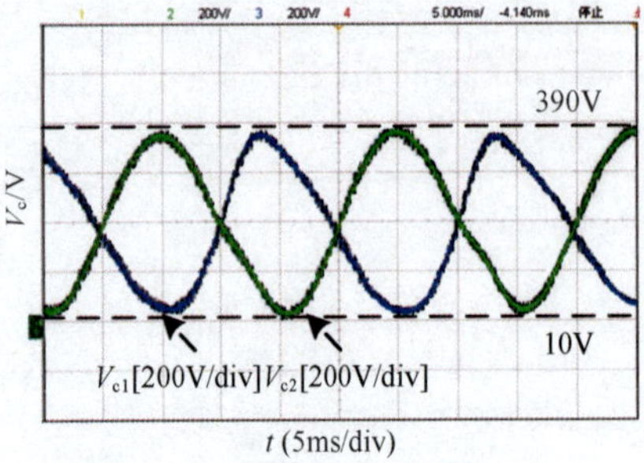

(a) Voltage waveform of the decoupling capacitor

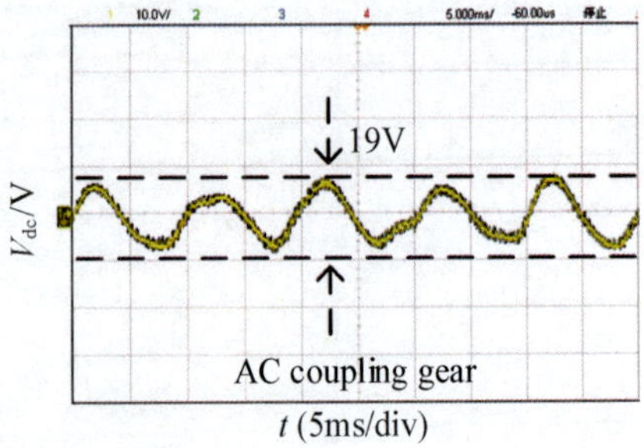

(b) The ripple voltage waveform of the DC bus

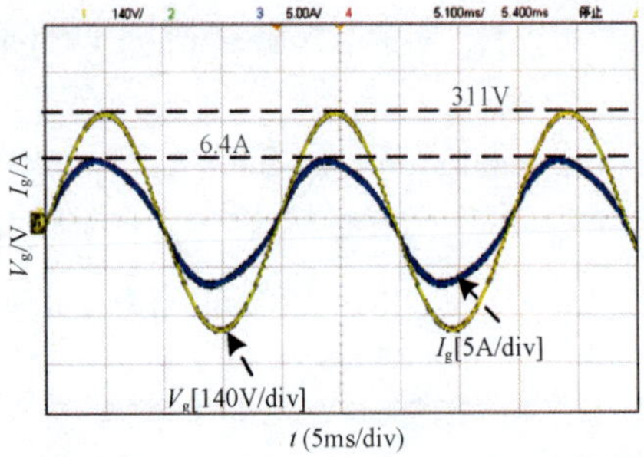

(c) Grid side voltage and inverter output current

Fig. 13. Experimental waveform

5 Conclusion

This paper analyzes the causes of the power imbalance on the AC and DC sides of a two-stage single-phase photovoltaic grid-connected inverter that uses a boost circuit in the front stage and a single-phase full-bridge inverter circuit in the rear stage. Aiming at the disadvantages of poor stability, large size, and low reliability of passive decoupling circuits, based on the symmetrical half-bridge decoupling circuit topology and the idea of power conservation, the control reference value of the decoupling circuit are calculated, and the double loop is given. This method can replace the electrolytic capacitor with a film capacitor while stabilizing the bus voltage, reducing the capacity of the bus capacitor. Finally, a 1 kW inverter simulation and experiment platform were built to verify the effectiveness of the proposed control strategy. Meanwhile, this paper only analyzes the steady-state process of the system. The transient process such as external interference or system faults is not mentioned. Therefore, the impact of the transient process on the photovoltaic system is needed to be studied.

References

1. Latham, J., McIntyre, M.L.: A novel controller for power decoupling in a single-phase grid-tied inverter using a boost converter buffer. In: 44th Annual Conference of the IEEE Industrial Electronics Society, IECON 2018, Washington, DC, pp. 913–918 (2018)
2. Jamatia, A., Gautam, V., Sensarma, P.: Power decoupling for single-phase PV system using Ćuk derived microinverter. IEEE Trans. Ind. Appl. **54**(4), 3586–3595 (2018)
3. Yuan, Y., Zhang, Y., Zhong, Q.: Secondary ripple analysis and suppression of the bus voltage in two-stage inverter. J. Power Supply **16**(1), 37–45 (2018)
4. Xu, Z., Tang, F., Liu, J.: Improved three-phase four-leg topology for active power decoupling control. Autom. Electr. Power Syst. **43**(18), 154–165 (2019)
5. Liu, B., Li, M., Xu, C.: Grid-connected operation control of single-phase photovoltaic inverter under battery input condition in family photovoltaic energy storage system. Autom. Electr. Power Syst. **40**(13), 85–92 (2016)
6. Wang, C.: Research on two-stage PV grid-connected inverter and power decoupling. Nanjing University of Aeronautics and Astronautics, Nanjing (2010). (in Chinese)
7. Li, H., Zhang, K., Zhao, H.: Active power decoupling for high-power single-phase PWM rectifiers. IEEE Trans. Power Electron. **28**(3), 1308–1319 (2012)
8. Ren, Y., Mai, K., Wang, H.: Technology of single-phase PFC output bus voltage fluctuation suppression and ripple power transfer. J. Power Supply **15**(4), 178–182 (2017)
9. Yang, X., Jiang, L., Fen, J.: A new power decoupling circuit for photovoltaic inverter. Trans. China Electrotech. Soc. **30**(16), 42–48 (2015)
10. Xu, S., Chang, L., Shao, R., Mohomad, A.H.: Power decoupling method for single-phase buck-boost inverter with energy-based control. In: 2017 IEEE Applied Power Electronics Conference and Exposition (APEC), Tampa, USA (2017)
11. Sun, Y., Liu, Y., Su, M.: Review of active power decoupling topologies in single-phase systems. IEEE Trans. Power Electron. **31**(7), 4778–4794 (2016)

Optimized Coordinated Control of DAB and Three-Phase Inverter Based Cascaded System with Active Damping

Jicheng Li[1], Xiaofeng Yang[1,2(✉)], Bingkai Zhou[1,2], and Renbiao Nong[1,2]

[1] School of Electrical Engineering, Beijing Jiaotong University, Beijing 100044, China
xfyang@bjtu.edu.cn
[2] State Key Laboratory of Operation and Control of Renewable Energy and Storage Systems, China Electric Power Research Institute, Beijing, China

Abstract. As for the dual active bridge (DAB) converter and three-phase inverter based cascaded system, the stability is affected by the interaction between output and input impedance of the two sub-converters. The small-signal models of both DAB and three-phase inverter are firstly derived for stability analysis. Based on the improved Middlebrook criterion, this paper analyzes the difference in impedance characteristics between forward and reverse power flows, and the parameters that affect the stability of system. Then an optimized coordinated control strategy with active damping is proposed to improve the stability under forward power flow. The theoretical analysis results show that the optimized coordinated control can reduce the voltage fluctuation of the DC-link and achieve smooth transitions during sudden power changes and reversals, thereby improving the dynamic response of the system. Furthermore, robustness of the system is enhanced without changing proportional integral (PI) parameters and the circuit hardware. Finally, the optimized coordinated control is compared with conventional control by simulation results.

Keywords: Cascaded system · Dual active bridge · Three-phase inverter · Active damping · Optimized coordinated control · Impedance characteristic

1 Introduction

With the ever-increasing renewable energy integration in recent decades, the power electronic transformer (PET) has gained extensive attention and study in smart grid application [1–4]. PETs are usually employed for energy conversion, voltage transformation, AC/DC and DC/AC conversion, power flow control, and automatic interconnection to achieve the balance between power supply and demand. Many scholars have studied the topology and control of PETs, and typical AC/DC/AC topology of PET is shown in [5]. In DC/AC conversion, the cascaded structure of isolated dual active bridge (DAB) converter and three-phase inverter with good controllability is widely used [6–8]. DAB realizes DC voltage transformation, and the energy flows bidirectionally between the power grid and the energy storage unit. It has the advantages of flexible voltage level transformation, simple hardware structure, reliable and

Y. Xue et al. (Eds.): PMF 2020, LNEE 718, pp. 526–541, 2021.
https://doi.org/10.1007/978-981-15-9746-6_41

safe performance [9–11]. Three-phase inverter supplies the load and maintains the stable operation of the power grid.

Although multi-module cascaded structure improves the performance and application range of the system, the instability caused by the interaction between the two sub-converters is an important challenge. Reference [12] introduced different control strategies according to different power flows, but they cannot realize the seamless transition of power flow from one mode to another. In reference [13], the cascaded system structure of rectifier, DAB and inverter was studied, impedance models were established and the system stability was improved with the active damping. But this strategy is suitable for the cascaded system with single-phase inverter, and further analysis for three-phase inverter is needed. Moreover, the cascaded system of DAB and three-phase inverter is an important part of renewable energy integration, so it is necessary to analyze and improve the stability of the cascaded system.

The impedance-based stability criterion was firstly proposed by Middlebrook, which requires that the input impedance of load converter is greater than the output impedance of source converter in the full frequency range. However, it is too stringent in practical application. Then the improved Middlebrook criterion with the minor-loop gain for cascaded systems is studied, that is, the ratio of the output impedance of the source converter to the input impedance of the load converter satisfies the Nyquist stability criterion [14–16].

Stability of this cascaded system is affected by the impedance interaction between DAB and three-phase inverter. In this paper, small-signal impedance models of the two sub-converters are derived for stability analysis. Then based on the stability criterion, the minor-loop gains reflecting the impedance interaction under forward and reverse power flows are analyzed. The cascaded system is more instable when power is forward because of the poor impedance characteristics. Moreover, proportional integral (PI) parameters and load affect the system stability. In order to solve the instability issue under forward power flow and to improve dynamic response, an optimized coordinated control with active damping is proposed. The optimized coordinated control is specifically to improve the inverter input impedance characteristics under forward power flow, and to coordinate response difference of the two sub-converters to load changes. Finally, the simulation waveforms of optimized coordinated control and traditional control are compared through MATLAB/Simulink.

2 Impedance Modeling

2.1 Impedance Modeling of DAB with Constant Voltage Control

Figure 1 shows the configuration of cascaded system which composes of DAB and three-phase inverter. The primary and secondary sides of DAB are H-bridges connected by high frequency transformer and inductance L_1, and the transformer transformation ratio is $1/n$. C_1 and C_2 are the DC-link capacitors of primary and secondary sides respectively, L_2 is the filter inductance of inverter. U_{dc1} and U_{dc2} are the DC voltage of primary and secondary sides, $i_{in\text{-}DAB}$ and $i_{out\text{-}DAB}$ are the input and output current of DAB, $i_{in\text{-}INV}$ is the inverter input current, and the positive direction is shown. DAB adopts the constant voltage control in Fig. 2, where U_{dc2}^{*} is the reference output voltage, $K_{p\text{-DAB}}$ and $K_{i\text{-DAB}}$ are respectively the PI parameters in the DAB output voltage controller, h is the voltage gain which affects PI, D is the phase-shift ratio between primary and secondary H-bridges. The DAB output power $P_{\text{out-DAB}}$ can be derived as (1) from Fig. 1 and Fig. 2.

$$P_{\text{out-DAB}} = \frac{U_{\text{dc1}} U_{\text{dc2}} D(1 - D)}{2nf_s L_1} \tag{1}$$

where f_s is the switching frequency.

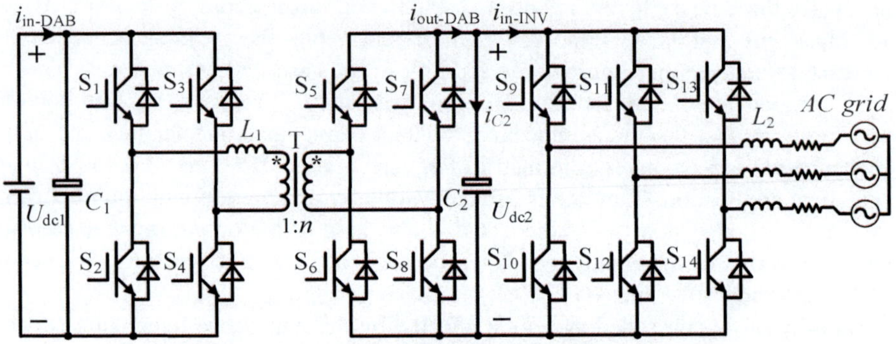

Fig. 1. Configuration of DAB and inverter cascaded system.

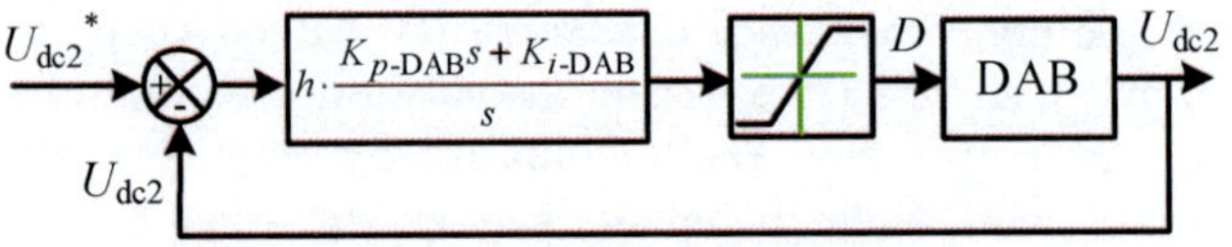

Fig. 2. Constant voltage control of DAB.

In order to obtain the small-signal model of DAB, average output current $I_{\text{out-DAB}}$ within a single switching cycle can be approximatively expressed as (2).

$$I_{\text{out-DAB}} = \frac{P_{\text{out-DAB}}}{U_{\text{dc2}}} = \frac{U_{\text{dc1}} D(1 - D)}{2nf_s L_1} \tag{2}$$

The small-signal model of DAB can be derived as (3) and (4) by small disturbance and linearization of (2).

$$\hat{i}_{\text{out-DAB}} = \frac{\hat{u}_{\text{dc1}} D(1 - D)}{2nf_s L_1} + \frac{U_{\text{dc1}} \hat{d}(1 - 2D)}{2nf_s L_1} \tag{3}$$

$$C_2 \frac{d\hat{u}_{\text{dc2}}}{dt} = \hat{i}_{\text{out-DAB}} - \hat{i}_{\text{in-INV}} \tag{4}$$

where $\hat{i}_{\text{out-DAB}}$ is the disturbance component of $i_{\text{out-DAB}}$, $\hat{u}_{\text{dc1}}$ and $\hat{u}_{\text{dc2}}$ are the disturbance components of U_{dc1} and U_{dc2} respectively, $\hat{d}$ is the disturbance component of D, $\hat{i}_{\text{in-INV}}$ is the disturbance component of $i_{\text{in-INV}}$.

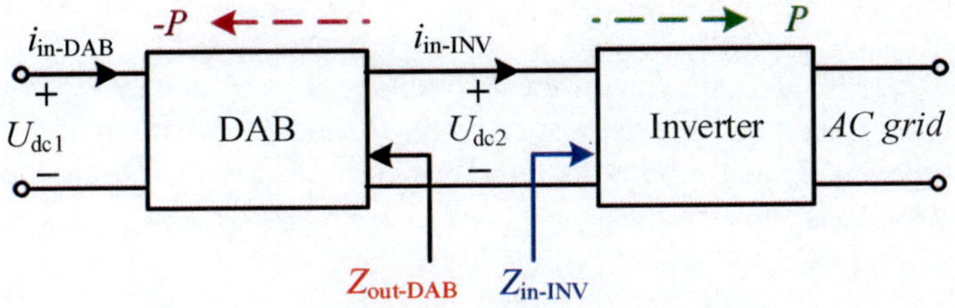

Fig. 3. Simplified model of cascaded system.

The simplified model, as shown in Fig. 3, expresses the impedance model of cascaded system clearly. It is defined that the power flowing from DAB to inverter is forward, while from inverter to DAB is reverse. When the power flows forward, $Z_{\text{out-DAB}}$ is the DAB output impedance and $Z_{\text{in-INV}}$ is the inverter input impedance.

$Z_{\text{out-DAB}}$ is defined as the equivalent impedance observed from the output side. Here, $\hat{u}_{\text{dc2}}$ is consistent with reference voltage direction, and $\hat{i}_{\text{in-INV}}$ is opposite to reference current direction. Therefore, according to (3) and (4) and Fig. 2, $Z_{\text{out-DAB}}$ can be derived as (5).

$$Z_{\text{out-DAB}} = -\frac{\hat{u}_{\text{dc2}}}{\hat{i}_{\text{in-INV}}} = \frac{s}{C_2 s^2 + ahK_{p\text{-DAB}} s + ahK_{i\text{-DAB}}} \tag{5}$$

where $a = \frac{(1-2D)U_{\text{dc2}}}{2nf_s L_1}$.

In fact, the DAB input voltage does not change when the power flows in the reverse direction, while the input current is just opposite in direction [17]. DAB input impedance $\overline{Z}_{\text{in-DAB}}$ can also be derived as (6) when the power flows in the reverse direction with the same voltage control.

$$\overline{Z}_{\text{in-DAB}} = Z_{\text{out-DAB}} = \frac{s}{C_2 s^2 + ahK_{p\text{-DAB}} s + ahK_{i\text{-DAB}}} \tag{6}$$

2.2 Impedance Modeling of Three-Phase Inverter with *PQ* Control

The three-phase inverter adopts *PQ* control shown in Fig. 4 to adjust the power flowing through the inverter in *d-q* rotation frame coordinates.

With the control in Fig. 4, the inverter state equations are given as (7) and (8).

$$L_2\frac{di_d}{dt} = D_d U_{dc2} - U_d + \omega L_2 I_q \tag{7}$$

$$L_2\frac{di_q}{dt} = D_q U_{dc2} - U_q - \omega L_2 I_d \tag{8}$$

where $\omega = 2\pi f$, f is the frequency of the AC voltage, U_d, U_q and I_d, I_q are the voltages and currents on d axis and q axis, respectively. P^* and Q^* are reference active power and reactive power, I_d^* and I_q^* are the reference currents, D_d and D_q are the duty ratios on d axis and q axis.

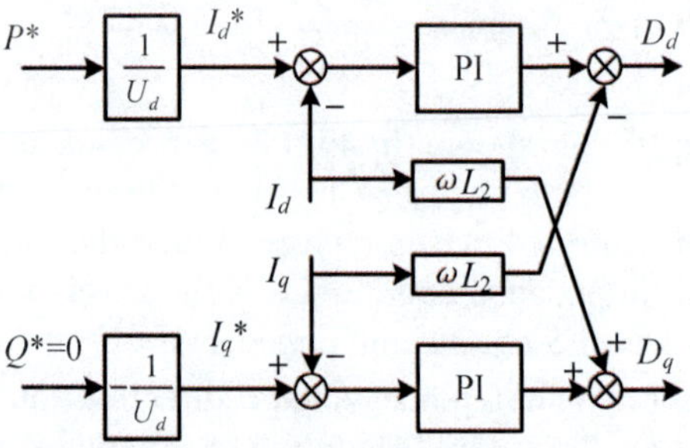

Fig. 4. *PQ* control block diagram.

Without considering the switching loss, the power balancing of AC and DC side can be expressed as (9).

$$P_{\mathrm{INV}} = \frac{3}{2}(U_d I_d + U_q I_q) = U_{dc2} I_{\text{in-INV}} \tag{9}$$

where P_{INV} is the inverter output power, and $I_{\text{in-INV}}$ is the inverter average input current.

Assuming the power grid voltage is constant, in stable state, U_q and I_q are equal to 0. With $\hat{u}_d = 0$, the small-signal model of power balancing is (10).

$$\frac{3}{2}U_d \hat{i}_d = U_{dc2}\hat{i}_{\text{in-INV}} + \hat{u}_{dc2} I_{\text{in-INV}} \tag{10}$$

$$\hat{i}_d = G_{IU}(s)\hat{u}_{dc2} \tag{11}$$

where $\hat{i}_d$ is the disturbance component of d axis current, and its relationship with $\hat{u}_{dc2}$ is (11). $G_{IU}(s)$ is the transfer function with the input DC voltage disturbance and output d axis current disturbance when the power flows forward. It is obtained by (7), and illustrated by the control block diagram in Fig. 5.

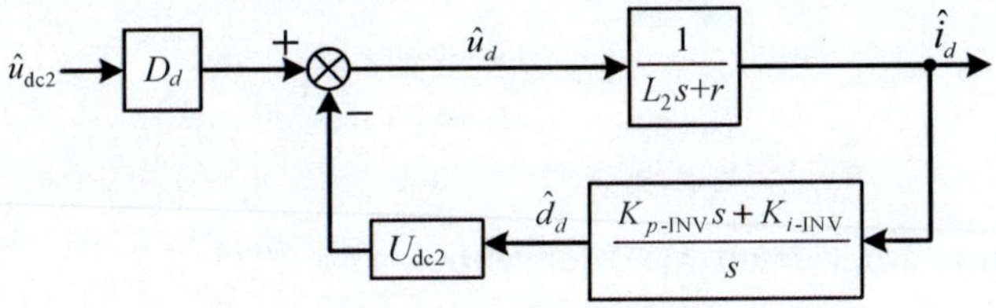

Fig. 5. Block diagram of $G_{IU}(s)$.

where D_d is the duty ratio of d axis voltage, $K_{p\text{-INV}}$ and $K_{i\text{-INV}}$ are the PI parameters in the controller, $G_{IU}(s)$ can be derived as (12). When the power flows forward, according to (10)-(12), the inverter input impedance $Z_{\text{in-INV}}$ is derived as (13).

$$G_{IU}(s) = \frac{D_d s}{L_2 s^2 + (r + U_{dc2}K_{p\text{-INV}})s + U_{dc2}K_{i\text{-INV}}} \tag{12}$$

$$\begin{aligned} Z_{\text{in-INV}} &= \frac{\hat{u}_{dc2}}{\hat{i}_{\text{in-INV}}} = \frac{U_{dc2}}{1.5U_d G_{IU}(s) - I_{\text{in-INV}}} \\ &= \frac{L_2 U_{dc2} s^2 + (rU_{dc2} + U_{dc2}^2 K_{p\text{-INV}})s + U_{dc2}^2 K_{i\text{-INV}}}{-L_2 I_{\text{in-INV}} s^2 + (1.5U_d D_d - rI_{\text{in-INV}} - I_{\text{in-INV}} U_{dc2} K_{p\text{-INV}})s - I_{\text{in-INV}} U_{dc2} K_{i\text{-INV}}} \end{aligned} \tag{13}$$

where $\hat{u}_{dc2}$ and $\hat{i}_{\text{in-INV}}$ are the disturbance components of U_{dc2} and $i_{\text{in-INV}}$ respectively, $I_{\text{in-INV}}$ is the reference input current, which is obtained by the reference power and voltage.

When power flows in the reverse direction, the inverter output voltage does not change, and the output current is just opposite in direction, so its output impedance $\bar{Z}_{\text{out-INV}}$ can also be obtained as (14) with the same control strategy.

$$\bar{Z}_{\text{out-INV}} = -\frac{\hat{u}_{dc2}}{\hat{i}_{\text{out-INV}}} = -\frac{U_{dc2}}{1.5U_d \bar{G}_{IU}(s) - I_{\text{out-INV}}} \tag{14}$$

where $\bar{G}_{IU(s)}$ is the transfer function with the DAB output voltage disturbance and the d axis current disturbance when the power flows in the reverse direction, $\hat{i}_{\text{out-INV}}$ is the output current disturbance component of inverter, $I_{\text{out-INV}}$ is the reference output current of inverter, which can be obtained by the reference power and voltage.

According to [18], DAB is a typical voltage control source converter when the power flows forward, so the forward minor-loop gain T_m of cascaded system for impedance interactions analysis is expressed as (15).

$$T_m = \frac{Z_{\text{out-DAB}}}{Z_{\text{in-INV}}} \tag{15}$$

When power flows in the reverse direction, the inverter is a power control source converter, so the reverse minor-loop gain $\bar{T}_m$ of the cascaded system is expressed as (16).

$$\overline{T}_{\mathrm{m}} = \frac{\overline{Z}_{\text{in-DAB}}}{\overline{Z}_{\text{out-INV}}} \tag{16}$$

2.3 Impedance Interaction with Conventional Control

Figure 6 shows the Nyquist plots of the minor-loop gains described in (15) and (16) when power flows in the forward and reverse directions respectively. It can be seen that under the same parameters and conditions, the minor-loop gain margin of reverse flow is much bigger than that of forward flow, which means that the stability of cascaded system under reverse power flow is more stable, so instability may occur at the moment of power reversal due to the uncertainty of load.

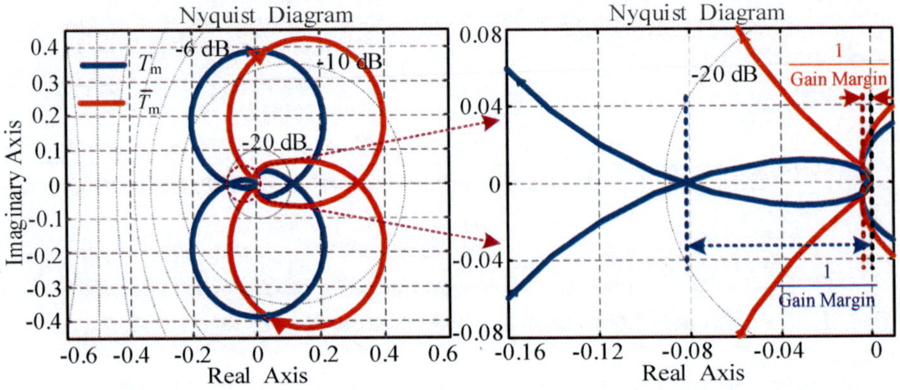

Fig. 6. Nyquist plots of minor-loop gain T_{m} and $\overline{T}_{\mathrm{m}}$ with conventional control.

The above difference is further analyzed in impedance characteristic shown in Fig. 7 when the power is 1 kW with conventional control. The amplitudes of input impedance $Z_{\text{in-INV}}$ and output impedance $\overline{Z}_{\text{out-INV}}$ are very similar, but the phase of $Z_{\text{in-INV}}$ is −180° in the low frequency range, showing as a negative impedance. The phase of $\overline{Z}_{\text{out-INV}}$ is positive in the low frequency when the power flows in the reverse direction. It is shown in Fig. 6 and Fig. 7 that the system is more stable when the power flows in the reverse direction than that in the forward direction, therefore, we should focus on the system stability under forward power flow.

Furthermore, there are other factors that affect the system stability. Fig. 8 is the Bode diagram of DAB output impedance $Z_{\text{out-DAB}}$ and inverter input impedance $Z_{\text{in-INV}}$ under different load levels when the power flows forward. The load is 6 kW, 8 kW, 10 kW and 12 kW respectively.

It can be seen in Fig. 8 that in the low frequency range, the amplitude of $Z_{\text{out-DAB}}$ increases with the increase of load, and the amplitude of $Z_{\text{in-INV}}$ decreases. This phenomenon can also be obtained from (13). When the load is 6kW and 8kW, the two amplitude curves do not intersect. But when the power is 10 kW and 12 kW, the two

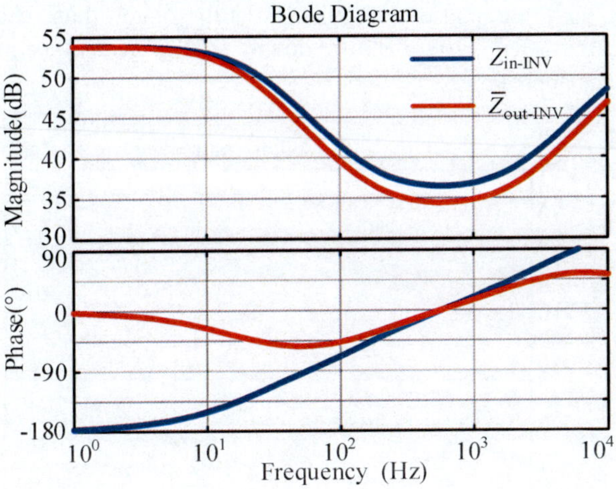

Fig. 7. Three-phase inverter input impedance $Z_{\mathbf{in\text{-}INV}}$ and output impedance $\bar{Z}_{\mathbf{out\text{-}INV}}$ at 1 kW with conventional control.

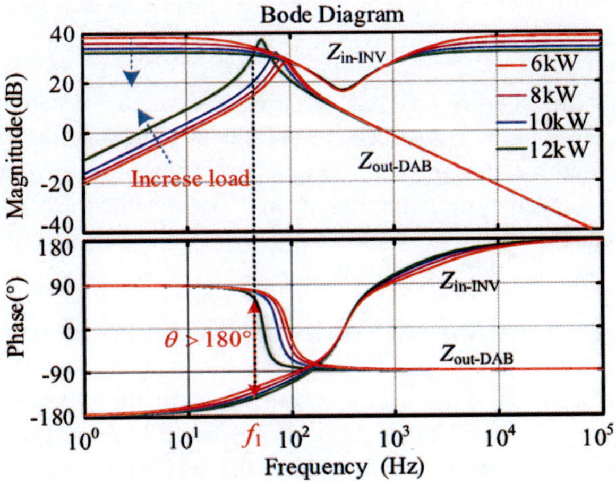

Fig. 8. DAB output impedance $Z_{\mathbf{out\text{-}DAB}}$ and inverter input impedance $Z_{\mathbf{in\text{-}INV}}$ as load increases with conventional control.

amplitude curves intersect at the same load level respectively, and the phase difference is greater than 180° at the intersection frequency f_1, which means that the cascaded system is instable [19]. Therefore, overload may increase the instability of cascaded system.

Moreover, Fig. 9 shows the Bode diagram of $Z_{\mathbf{out\text{-}DAB}}$ and $Z_{\mathbf{in\text{-}INV}}$ at different voltage gains under forward power flow. The amplitude of $Z_{\mathbf{out\text{-}DAB}}$ increases with the decrease of h. When h is reduced to a certain value, the two impedance amplitude curves intersect, and the phase difference is still greater than 180°, the dynamic response of the cascaded system slows down, so the system is instable when PI parameters are not appropriate.

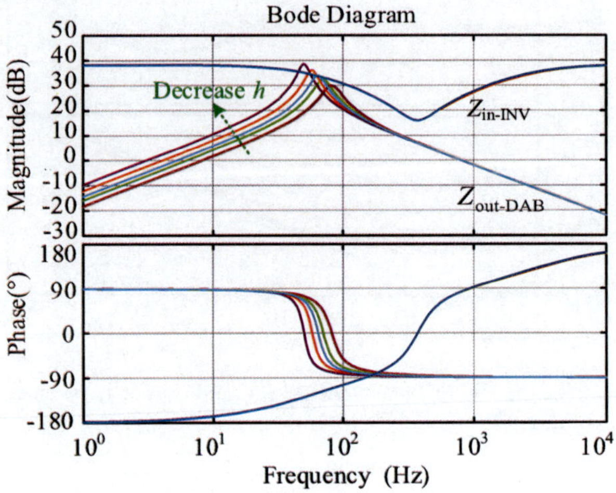

Fig. 9. DAB output impedance $Z_{\text{out-DAB}}$ and inverter input impedance $Z_{\text{in-INV}}$ as h decreases with conventional control.

According to the stability analysis of cascaded system, instability usually occurs when power reverses to the forward flow, and the system stability mainly depends on the size of load and the design of PI parameters. Therefore, in order to improve the stability of this cascaded system, an optimized coordinated control strategy with active damping is proposed.

3 Optimized Coordinated Control with Active Damping

From the above analysis, when power flows forward, the phase difference between inverter input impedance and DAB output impedance is more than 180° in low frequency range, and the amplitude curves intersect when the inverter input impedance amplitude is small. Moreover, the system stability mainly depends on the PI parameters and load. In this paper, active damping is added to improve the phase of the inverter input impedance, while increasing the amplitude to avoid the intersection of two curves in low frequency range. Thus, the influence caused by unreasonable parameters is reduced, the robustness of the system is enhanced, and the anti-interference ability is improved.

According to (13), increasing the amplitude of the inverter input impedance can be considered from two aspects: increasing the numerator or decreasing the denominator. It can be seen that the numerator is constant and the denominator contains variable transfer function $G_{IU}(s)$. Therefore, this paper considers changing $G_{IU}(s)$ to increase the input impedance amplitude. Suppose $G_{IU}(s)$ is updated as (17).

$$G_{IU}^{*}(s) = G_{IU}(s) + X(s) \tag{17}$$

where $X(s)$ is a variable, updated inverter input impedance $Z^{*}_{\text{in-INV}}$ can be obtained as (18) by substituting (17) into (13).

$$Z_{\text{in-INV}}^{*} = \frac{U_{\text{dc2}}}{1.5U_d G_{IU}^{*}(s) - I_{\text{in-INV}}} = \frac{U_{\text{dc2}}}{1.5U_d G_{IU}(s) + 1.5U_d X(s) - I_{\text{in-INV}}} \tag{18}$$

The block diagram of transfer function $G^{*}_{IU}(s)$ is shown in Fig. 10. Since there is already a feed-back path, a feed-forward path is added. And the $X(s)$ is expressed as (19).

$$X(s) = \frac{G_{\text{ex}}(s)}{U_d} \tag{19}$$

It can be seen from Fig. 8 and Fig. 9, at high frequency, the amplitude of DAB output impedance decreases with increasing frequency, and the curves of the two impedances no longer intersect. Therefore, a typical first-order low-pass filter is introduced to improve the low-frequency characteristics, which can be specifically expressed as (20). The new transfer function $G^{*}_{IU}(s)$ can be derived as (21) from Fig. 10 and (20).

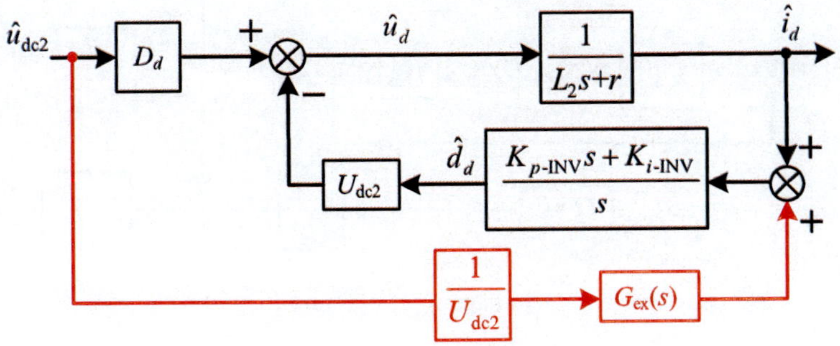

Fig. 10. Block diagram of $G^{*}_{IU}(s)$.

$$G_{\text{ex}}(s) = \frac{K_G \omega_a}{s + \omega_a} \tag{20}$$

$$G_{IU}^{*}(s) = \frac{D_d s}{L_2 s^2 + (r + U_{\text{dc2}} K_{p\text{-INV}})s + U_{\text{dc2}} K_{i\text{-INV}}} + \frac{G_{\text{ex}}(s)}{U_d} \tag{21}$$

where K_G is the proportional coefficient, ω_a is cut-off frequency of the low-pass filter. Therefore, the inverter input impedance $Z^{*}_{\text{in-INV}}$ with active damping is expressed as (22).

$$Z^{*}_{\text{in-INV}} = \frac{\hat{u}_{\text{dc2}}}{\hat{i}_{\text{in-INV}}} = \frac{U_{\text{dc2}}}{1.5U_d G_{IU}(s) + 1.5G_{\text{ex}}(s) - I_{\text{in-INV}}} \tag{22}$$

The phase of the $Z_{\text{in-INV}}$ in the low-frequency range is approximately -180°. According to (22), the appropriate $G_{\text{ex}}(s)$ should be selected, and the $I_{in\text{-INV}}$ should be offset in the low frequency range to reduce the phase difference between the two curves, and at the same time, the amplitude of $Z_{\text{in-INV}}$ can be larger.

Figure 11 is the overall block diagram of optimized coordinated control proposed in this paper. Compared with conventional control, the active damping is added to three-phase inverter. Moreover, the power coordinated response coefficient K_D is added to the DAB voltage control loop. As mentioned above, the system will have better stability when the power flows forward, and at the same time have better dynamic characteristics at the moment of load change.

In order to verify the rationality and feasibility of the proposed control theoretically,

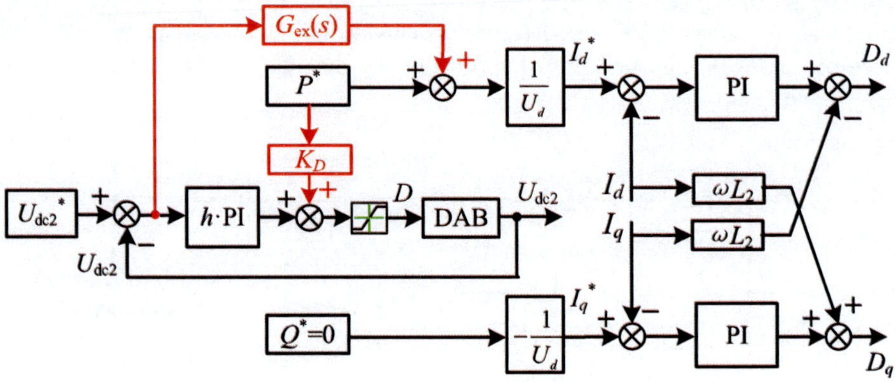

Fig. 11. Overall block diagram of optimized coordinated control.

the Bode diagram with two impedance characteristic curves at 10 kW before and after optimization is shown in Fig. 12. It can be seen that with conventional control, the two impedance amplitude curves intersect, the phase difference is greater than 180°, so the cascaded system is instable. With optimized coordinated control shown in Fig. 11, the amplitude curve of inverter input impedance rises and is separated from the amplitude curve of DAB output impedance in the low frequency range. In addition, the phase of the inverter input impedance is significantly improved in the low frequency range, which is limited to −90° –0°. The phase difference is always less than 180°, which meets the stability conditions. This result is consistent with the theoretical analysis and achieves the expected effect. Next, the variation of the minor-loop gains under different power flows are compared and analyzed from the Nyquist plots.

Similarly, the Nyquist plots of forward minor-loop gain T_{m} and T_{m}^{*} changing with different voltage gain h are shown in Fig. 13. It can be seen in Fig. 13(a) that when h decreases, the DAB output impedance increases, and the Nyquist plots move closer to (−1, j0). The Nyquist plots better illustrate the instability described in Fig. 9. And

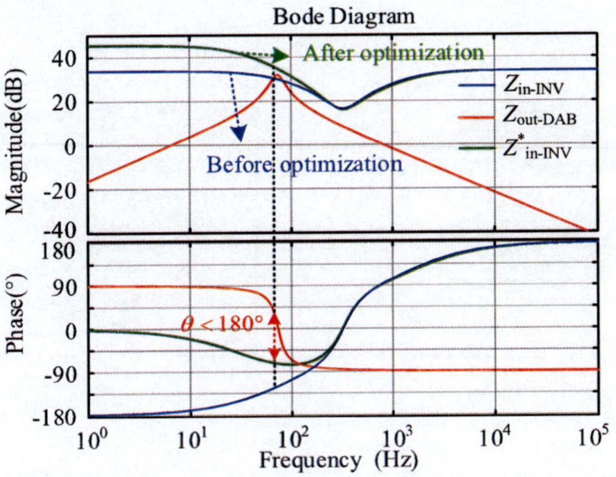

Fig. 12. Impedance characteristic curves at 10 kw before and after optimization.

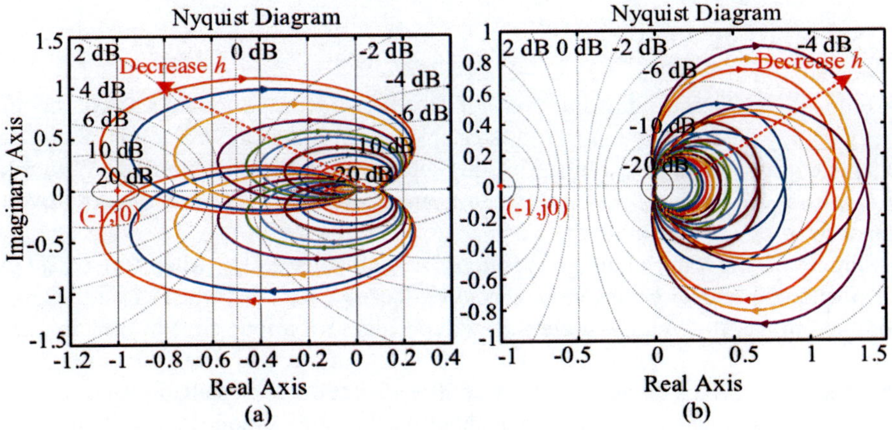

Fig. 13. Nyquist plots of forward minor-loop gain. (a) T_m with conventional control. (b) T_m^* with optimized coordinated control.

Fig. 13(b) are the Nyquist plots of the forward minor-loop gain T_m^* changing with different h with optimized coordinated control. The Nyquist plots move away from (−1, j0) rather than toward it. Therefore, the stability of forward power flow is improved and the system has better robustness.

Then the system under reverse power flow is analyzed. Fig. 14(a) and (b) are the Nyquist plots of the reverse minor-loop gain $\bar{T}_m$ and $\bar{T}_m^*$ with traditional control and optimized coordinated control respectively. The Nyquist plots are gradually away from (−1, j0) with either control strategy, and the optimized coordinated control does not damage the stability of reverse power flow. Therefore, when the stability of forward power flow is improved, a smaller disturbance is expected at the moment of power reversal.

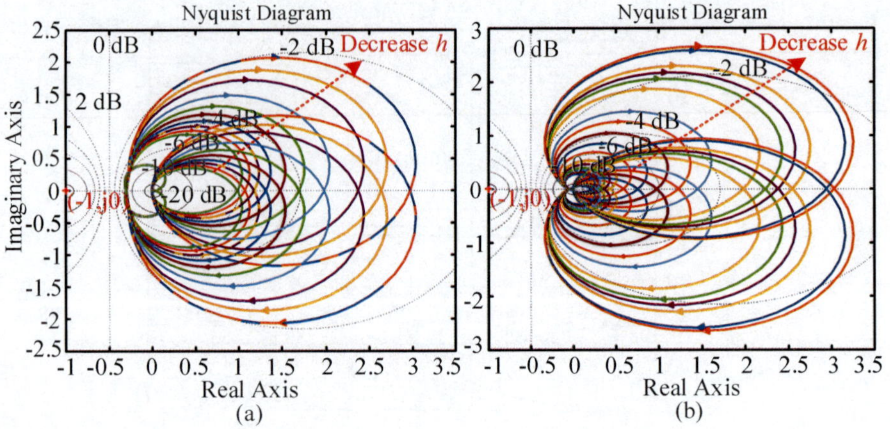

Fig. 14. Nyquist plots of reverse minor-loop gain. (a) $\bar{T}_m$ with conventional control. (b) $\bar{T}_m^*$ with optimized coordinated control.

4 Simulation Analysis

In order to verify the rationality and effectiveness of the proposed control strategy, firstly, a cascaded simulation model is established in MATLAB/Simulink. Simulation parameters are shown in Table 1. The load setting value changes from 100 W to 5 kW at 4 s, from 5 kW to − 5 kW at 6 s and from − 5 kW to 5 kW at 8 s, to observe the response of cascaded converter.

Figure 15 show the simulation result of DC-link voltage U_{dc2} when the voltage gain h is not appropriate with conventional control. It can be seen that when PI parameters are not appropriate, the voltage waveform will oscillate for a long time at the moment of load change. The voltage overshoot reaches 156 V at 4 s, and when the power flow is changed from reverse direction to forward direction at 8 s, the oscillation and overshoot are the most serious. The voltage overshoot reaches 200 V and the regulating time is 500 ms, which reflects the instability of the system. As mentioned above, this is caused by the negative impedance of the inverter with conventional control.

Figure 16 is the simulation result of the response of the cascaded system with conventional control and appropriate voltage gain h, where P_{DAB} is the power of DAB.

Table 1. Simulation parameters.

Parameter	Value	Parameter	Value
DAB U_{dc1}	175 V	Grid voltage	380 V
DAB U_{dc2}	700 V	C_1	2 mF
n	4	L_2	3 mH
f_s	10 kHz	C_1	2 mF
L_1	30 μH	Grid-frequency	50 Hz

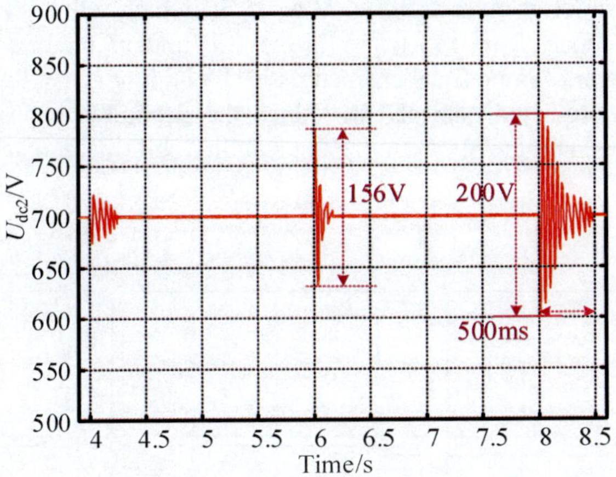

Fig. 15. DC-link voltage U_{dc2} when voltage gain h is not appropriate with conventional control.

Compared with Fig. 15, the oscillation phenomenon is weakened but still obvious, the overshoot is 93 V and 96 V at 4 s and 8 s respectively. The transient oscillation is serious especially when the power flows forward at 8 s, the regulation time is 113 ms, and the dynamic response of the converter is still slow.

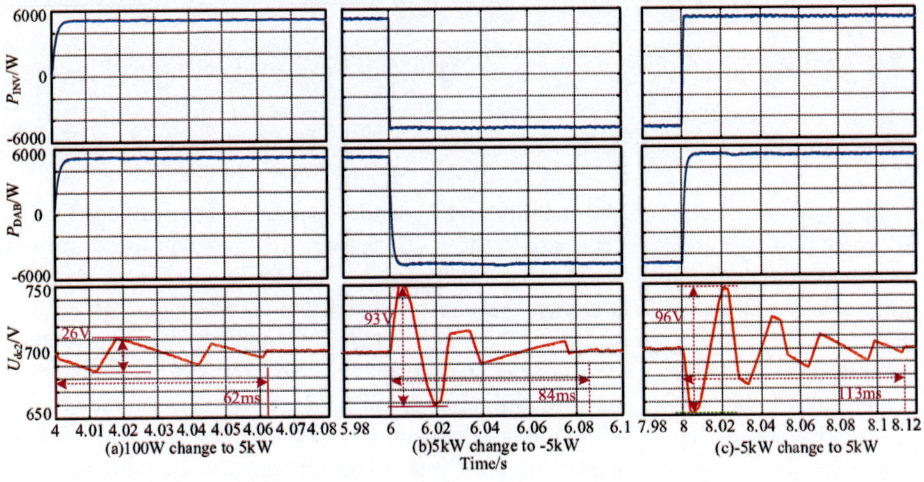

Fig. 16. Response of system with appropriate h with conventional control.

Figure 17 is the simulation result of the cascaded system response when the voltage gain h is not appropriate with optimized coordinated control. Compared with Fig. 16, the oscillation disappears at 6s, the voltage overshoot variable changes from 93 V to 9 V, which reduces by 90.3%, and the regulation time reduces from 84 ms to 22 ms. When the power changes from − 5 kW to 5 kW at 8s, smaller oscillation occurs, and the voltage overshoot changes from 96 V to 21 V, which reduces by 78.1%, and the regulating time reduces from 113 ms to 38 ms. It is undeniable that this is contributed by the improvement of impedance characteristics by adding active damping. Therefore, the optimized coordinated control can reduce the dependence on the sensitivity of parameter changes, and greatly optimize the stability and dynamic response of forward power flow.

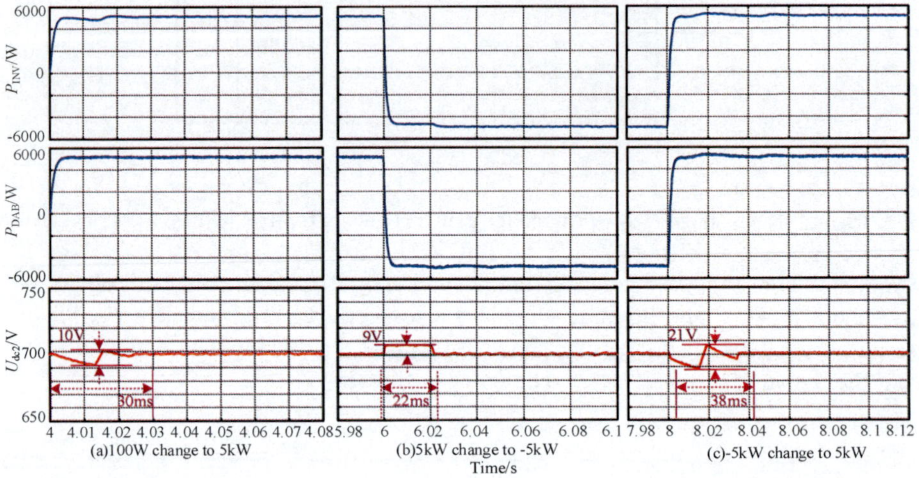

Fig. 17. Response of system when *h* is not appropriate with optimized coordinated control.

5 Conclusion

With conventional control, the stability and dynamic response of the cascaded system are more instable under reverse power flow. Therefore, this paper proposes an optimized coordinated control based on active damping. Through theoretical analysis and simulation comparison, it is shown that the proposed control improves the response difference of the two converters to the load changes. Compared with conventional control, proposed control improves the stability of forward power flow, enhances the robustness of the system, ensures the smoothness of load changes and inversions, effectively reduces the DC-link voltage fluctuation, shortens the regulation time, and optimizes the dynamic response of the system, which verifies the rationality and effectiveness of the proposed strategy.

Acknowledgement. This work was supported by Open Fund of Operation and Control of Renewable Energy & Storage Systems under Grant DGB51201901187.

References

1. Liang, D., Liu, Y., Kou, P., et al.: Analysis of development trend for intelligent distribution transformer. Autom. Electr. Power Syst. **44**(7), 1–4 (2020)
2. Ma, D., Chen, W., Xue, C., et al.: AC-DC power electronic transformer with low-voltage DC bus and strategy of short-circuit fault ride-through. Autom. Electr. Power Syst. **43**(8), 158–166 (2019)
3. Tian, J., Mao, C., Wang, D., et al.: A short-time transition and cost saving redundancy scheme for medium-voltage three-phase cascaded h-bridge electronic power transformer. IEEE Trans. Power Electron. **33**(11), 9242–9252 (2018)
4. Zhang, J., Liu, J., Yang, J., et al.: A modified DC power electronic transformer based on series connection of full-bridge converters. IEEE Trans. Power Electron. **34**(3), 2119–2133 (2019)
5. Zhang, G., Chen, J., Zhang, B., et al.: A critical topology review of power electronic transformers: In view of efficiency. Chinese J. Electr. Eng. **4**(2), 90–95 (2018)
6. Moonem, M.A., Krishnaswami, H.: Analysis of dual active bridge-based power electronic transformer as a three-phase inverter. In: Conference of the IEEE Industrial Electronics Society, pp. 238–243. IEEE,Montreal QC (2012)
7. Zhong, Q., Weiss, G.: Synchronverters: Inverters that mimic synchronous generators. IEEE Trans. Ind. Electron. **58**(4), 1259–1267 (2011)
8. Zhang, H., Jing, L., Wu, X., et al.: Control strategy for autonomous operation of power electronic transformer. Autom. Electric Power Syst. **42**(4), 89–94 (2018)
9. Zhao, B., Song, Q., Liu, W., et al.: Overview of dual-active-bridge isolated bidirectional DC–DC converter for high-frequency-link power-conversion system. IEEE Trans. Power Electron. **29**(8), 4091–4106 (2014)
10. Zhou, B., Yang, X., Zhang, Z., et al.: Backflow power optimization of DAB with gradient descent algorithm based extended-phase-shift control in EER application. In: 3rd International Electrical and Energy Conference (CIEEC), pp. 1737–1742. IEEE, Beijing (2019)
11. Hu, Y., Zhang, Y., Zhang, T., et al.: Optimization method of current stress for dual active bridge converter considering different soft switching modes. Autom. Electric Power Syst. **43** (23), 58–64 (2019)
12. Krishnamurthy, H., Ayyanar, R.: Stability analysis of cascaded converters for bidirectional power flow applications. In: INTELEC 2008 - 2008 IEEE 30th International Telecommunications Energy Conference, pp. 1–8. IEEE, San Diego (2008)
13. Khazraei, M., Prabhala, V., Ahmadi, R., et al.: Solid-state transformer stability and control considerations. In: 2014 IEEE Applied Power Electronics Conference and Exposition - APEC 2014, pp. 2237–2244. IEEE, Fort Worth TX (2014)
14. Sudhoff, S., Glover, S., Lamm, P., et al.: Admittance space stability analysis of power electronic systems. IEEE Trans. Aerospace Electron. Syst. **36**(3), 965–973 (2000)
15. Zhang, X., Ruan, X., Tse, C.: Impedance-based local stability criterion for DC distributed power systems. IEEE Trans. Circ. Syst. **62**(3), 916–925 (2015)
16. Du, Y., Cui, L., Yang, X., et al.: Adaptive control method of impedance for grid-connected inverters. Autom. Electric Power Syst. **42**(6), 120–127 (2018)
17. Wu, T., Ruan, X.: Output impedance analysis of source converters in the DC distributed power system. Chinese J. Electr. Eng. **28**(3), 66–72 (2008)
18. Liu, F., Liu, J., Zhang, B., et al.: Unified stability criterion of bidirectional power flow cascade system. In: 2013 Twenty-Eighth Annual IEEE Applied Power Electronics Conference and Exposition (APEC), pp. 2618–2623. IEEE, Long Beach (2013)
19. Zhang, X., Ruan, X., Zhong, Q.: Improving the stability of cascaded DC/DC converter systems via shaping the input impedance of the load converter with a parallel or series virtual impedance. IEEE Trans. Ind. Electron. **62**(12), 7499–7512 (2015)

A Multi-objective Optimal Orderly Charging Strategy for Electric Vehicles

Tianyun Wang(✉) and Hao Zhang

NARI Group Co., Ltd. (State Grid Electric Power Research Institute), Nanjing 211100, China
649575621@qq.com

Abstract. In order to reduce the adverse effects of grid which is caused by massive unguided charging behavior of electric vehicles and maintain the safe and stable operation of the distribution network, an orderly charging strategy of electric vehicles is proposed in the paper, and the algorithm is based on improve differential evolution algorithm. The main idea of this strategy is to make charging plan through time-of-use (TOU) electricity price, and then electric vehicles can be moved to normal period and valley period for charging. The objective function of the algorithm is to minimize the charging cost of users and the peak-valley difference. Based on this strategy, the paper builds a mathematic model of multi-objective optimization, and improve differential evolution (IDE) algorithm is used to solve the model. In addition, the algorithm optimizes charging continuity,which can effectively prevent load curve mutation and protect charging facilities. The simulation results verify the effectiveness of the algorithm.

Keywords: Electric vehicles · Improve differential evolution algorithm · Orderly charging · Optimization strategy

1 Introduction

Recently, with the continuous development of our economy and science, the happiness index of people's life has improved significantly. However, there are also a series of problems, such as the shortage of fossil energy, global warming and excessive emission of pollutants. What's more, all of these problems have been bringing great challenges to the future of modern society [1]. Compared with traditional fuel vehicles, new energy vehicles, due to the use of clean energy, can effectively relieve the pressure of exhaust emissions to the environment. Therefore, countries around the world begin to support the research of new energy vehicle industry vigorously. In this way, electric vehicles and related industries have ushered in a period of rapid development. At the same time, the charging behavior of large-scale electric vehicles without guidance has had a huge negative impact on distribution network, which embodies in the following aspects: enlarging the peak valley difference of distribution network, generating harmonic pollution and overloading transformers [2–5].

In order to solve the problems caused by disordered charging of electric vehicles, scholars at home and abroad have developed a series of orderly charging strategies by dividing the charging objectives of electric vehicles. Reference [6, 7] used traditional

Y. Xue et al. (Eds.): PMF 2020, LNEE 718, pp. 542–556, 2021.
https://doi.org/10.1007/978-981-15-9746-6_42

algorithms to solve the model. In reference [6], the demand response model was established with the optimization objective that it was necessary to minimize purchase cost of charging station, and in this model, linear programming method was used to deal with the problem. Reference [7] proposed a user guidance based on a matching strategy for EV-charging pile. However, the above traditional optimization algorithms belong to the convex optimization category and have a unique global optimal solution. With the increase of the scale of electric vehicles, the strategy of orderly charging has been transformed into a multi extremum problem, which means that the traditional optimization algorithms are no longer applicable. So, a double-layer model was established in reference [8]. For the upper control layer, the global interests were taken into consideration, and peak load, node voltage and overall electricity cost were adopted as the global optimal objectives. For the bottom layer, considering the high penetrations of EVs, the scalable and flexible alternating direction method of multipliers (ADMM) algorithm was deployed to solve the optimal problem, which blended the benefits of dual decomposition and augmented Lagrangian methods. Reference [9] took charging station as research object and established a two-stage multi-objective optimization model. The model aims at stabilizing the load fluctuation and improving the efficiency of charging station, and in this model, the multi-population genetic algorithm is used to solve the problem. Reference [10] researched the optimal electric vehicle charging and navigating strategy from the following four aspects: electric vehicle users, electric power companies, charging station operators and traffic management departments. And it put forward "vehicle-grid-road-station" multi-objective optimization and orderly charging navigation system. However, the objective functions mentioned in Reference [8–10] are not comprehensive enough, and the interests of the owners of electric vehicles are not considered.

Considering the benefits of the owners of electric vehicles and the safety and stability of the power grid, this paper establishes a multi-objective optimal charging model, which takes the minimum charging cost of electric vehicle users and the minimum peak valley difference of distribution network as the objective function. In this model, the charging behavior of electric vehicles is scheduled on the time level, and the improve differential evolution algorithm is used to solve the problem. By this means, the charging plan can be implemented to a specific vehicle, making the orderly charging strategy more practical.

2 Charging Scenario of Electric Vehicles

Generally speaking, it can be divided into two levels: charging control center and intelligent charging pile. After the electric vehicles drive into the charging station for charging, the Battery Management System (BMS) transmits the battery capacity, State of Charge (SOC), the user's expected end time of charging, the user's expected SOC and other information to the intelligent charging post. Then the intelligent charging piles uploads information from different platform areas to the charging pile control center. After that, the control center, which provides scientific and orderly guidance to the charging behavior of electric vehicles, sends the charging plan to each intelligent charging post. Finally, the concrete plan is carried out by the charging piles.

In this paper, electric vehicles in residential areas are taken as the research object. And most residents of these areas leave in the daytime, and return to charge in the evening. Therefore, this paper only studies the charging behavior from the time they return in the evening to the time they leave in the next morning.

3 Mathematical Model of Orderly Charging Strategy

3.1 Objective Function

In order to take both the benefits of the owners of electric vehicles and the safety and stability of the power grid into account, the optimization model selects the minimum charging cost for the users of electric vehicles and the minimum peak-valley difference of the power grid as the objective function.

(1) Minimize charging costs for EV users

In order to reduce the pressure on the power grid caused by the charging of electric vehicles, users need to be guided to charge in the period of low electricity price according to time-of-use electricity price. On the other hand, it is necessary to prevent users from charging together in the valley period, in case of a new charging peak demand. The mathematical expression of the objective function is,

$$min\ F_1 = \sum_{n=1}^{N} \sum_{t=i}^{i+j-1} c_{ij} \cdot P_{ev} \cdot \alpha_{i,j} \cdot \Delta t \tag{1}$$

Where N is the number of electric vehicles participating in the orderly charging strategy; i is the start time of charging; j the end time of charging; c_{ij} is the electricity price during charging period; P_{ev} is the charging power; $\alpha_{i,j}$ is the charge flag variable; $\alpha_{i,j} = 0$ indicates that the nth electric vehicle is not charged in (i, j) period, $\alpha_{i,j} = 1$ indicates that the nth electric vehicle is charged in (i, j) period; Δt is charging time.

(2) Minimum peak-valley difference of distribution network

Overcentralized charging behavior will lead to over limit of transformer in residential area. Therefore, in order to keep the power supply equipment safe and make the load curve as smooth as possible, it is necessary to clip the peak and fill the valley of the charging power generated by electric vehicles. The mathematical expression of the objective function is,

$$L_{max} = \max_{0 \le t \le 1440} \left(l_t + \sum\nolimits_{n=1}^{N} P_t \right) \tag{2}$$

$$L_{min} = \min_{0 \le t \le 1440} \left(l_t + \sum\nolimits_{n=1}^{N} P_t \right) \tag{3}$$

$$\min F_2 = L_{max} - L_{min} \tag{4}$$

Where l_t is the basic load of residents at time t; P_t is the charging power of vehicle n at time t; L_{max} is the peak load during the day; $L_{\min}$ is the valley load during the day.

When dealing with multi-objective optimization problems, it is necessary to normalize the dimensions. Therefore, the above two objective functions are unified into the following expression.

$$\min F = \lambda_1 \frac{F_1}{F_{1R}} + \lambda_2 \frac{F_2}{F_{2R}} \tag{5}$$

Where F_{1R} is the user's charging cost at the regular electricity price; F_{2R} is the peak-valley difference of power grid during disordered charging; λ_1 and λ_2 are the weighting coefficient of each objective function, and they meet the condition of $\lambda_1 + \lambda_2 = 1$. The values of λ_1 and λ_2 can be selected by AHP according to the actual situation of each region, and can also be weighted objectively according to entropy weight method.

3.2 Constraint Condition

(1) Total charging capacity

The charging load, allocated by the charging post to the electric vehicles, shall meet the SOC value of the batteries as expected by the users at the completion of charging, which is shown as,

$$\eta \sum\nolimits_{t=1}^{1440} P_{n,t} \cdot \Delta t \geq \left(SOC_{n,e} - SOC_{n,0}\right) \cdot Q \tag{6}$$

Where η is the charging efficiency; $P_{n,t}$ is the charging power of the vehicle n at time t; $SOC_{n,e}$ is the expected SOC value of the vehicle n; $SOC_{n,0}$ is the initial SOC value of the vehicle n; Q is the battery capacity of electric vehicle.

(2) Charging time of electric vehicle

$$0 \leq t_n \leq \frac{\left(SOC_{n,e} - SOC_{n,0}\right) \cdot Q}{\eta \cdot P_n} \tag{7}$$

Where t_n is the charging time of the vehicle n, P_n is the charging power of the vehicle n.

(3) Maximum peak load

The peak charging load of the electric vehicles cannot cause transformer overloaded, and the mathematical expression is,

$$l_t + \sum\nolimits_{n=1}^{N} P_{ev} \cdot \alpha_{i,j} < \mu P_M \tag{8}$$

Where P_M is the maximum capacity of transformer, μ is the threshold multiple of maximum capacity of transformer.

(4) Constraint of charging continuity

Frequent start and stop of charging piles will damage the car battery and components of charging pile. Therefore, the minimum start-stop interval of charging pile is set as 5 min in this paper, and the mathematical expression is,

$$\prod\nolimits_{t=t_0}^{t_0+4} \alpha_{i,j} = 1 \tag{9}$$

$$\sum\nolimits_{t=t_0+\Delta t}^{t_0+\Delta t+4} \alpha_{i,j} = 0 \tag{10}$$

Where t_0 is the start time of charging.

4 Optimal Charging Strategy Based on Improve Differential Evolution Algorithm

4.1 Basic Principle of Differential Evolution (dE) Algorithm

Differential evolution algorithm is a widely used evolutionary computation algorithm [11]. Like other evolutionary algorithms, the idea of DE algorithm is also derived from the evolution mechanism of natural organisms, that individuals who adapt to environment (optimal solution) can be preserved after multiple mutation, crossover and selection. This algorithm has the characteristics of self-organization, self-adaptive and self-learning. Therefore, it is a well-developed global method with high robustness and wide applicability, which makes it have the ability to effectively deal with the complex problems, especially when the problems are difficult to be solved by the traditional optimization algorithm [12].

DE algorithm includes initialization, mutation, crossover, selection and other steps. The specific processes are as follows.

(1) Initialization

Like other intelligent optimization methods, DE algorithm also needs to establish initial population first. Set the initial population as $X = \{X_1^t, X_2^t, X_3^t \ldots X_{NP}^t\}$, and $X_i^t = \{x_{i1}^t, x_{i2}^t, x_{i3}^t \ldots x_{iD}^t\}$ is individual vector, where N_P is the population size, suppose N_P = 50. D is the dimension of solution space, t is the number of iterations. Each individual population is generated according to the following formula,

$$x_{ij} = x_{ij}^{min} + rand \cdot \left(x_{ij}^{max} - x_{ij}^{min}\right) \tag{11}$$

Where *rand* is the uniform random number and its value range is [0, 1].

(2) Mutation

Mutations can maintain racial diversity, and the generation of new individuals are as follows.

$$v_{ij}^{t} = v_{r_1j}^{t} + F\left(v_{r_2j}^{t} - v_{r_3j}^{t}\right) \tag{12}$$

Generating three integers (r_1, r_2, r_3) with random function, these integers meet the condition: $r_1 \neq r_2 \neq r_3 \in \{1 \ldots N_P\}$. v_{ij}^{t} is the value of the dimension variable j of the mutant i in the generation population t. F is mutation factor whose range is [0, 2], suppose $F = 0.8$.

(3) Crossover

Randomly generate integer $j_{\text{rand}} \in \{1 \ldots D\}$, and its mathematical expression is as follows.

$$u_{ij}^{t} = \begin{cases} v_{ij}^{t}, if\,(rand_j \leq Cr \text{ 或 } j = j_{rand} \\ \qquad x_{ij}^{t}, \; other \end{cases} \tag{13}$$

Where $rand_j$ is the random number whose range is [0, 1]. Suppose $Cr = 0.8$.

(4) Selection

By comparing the fitness of the experimental individuals, the better ones are selected through the fitness function, and different fitness functions are selected according to the specific situation.

$$x_i^{t+1} = \begin{cases} u_i^t, & f(u_i^t) < f(x_i^t) \\ x_i^t, & other \end{cases} \tag{14}$$

4.2 Basic Principle of Artificial Bee Colony (ABC) Algorithm

Artificial bee colony algorithm is an algorithm proposed by Karaboga in 2005 based on honeybees' behavior of collecting honey [13]. As a swarm intelligence algorithm imitating biological behavior, artificial bee colony algorithm is mainly used to solve mathematical optimization problems. It is inspired by the foraging behavior of bees, so the model contains three parts: employed bees, unemployed bees and food sources. According to the different functions of bees in the process of honey collection in nature, ABC algorithm divides bee colonies into the following three categories: leaders, followers and scouters.

The model can be abstracted as Table 1.

Table 1. Mathematical significance of variables and actions

Behavior and variables	Mathematical significance
Gathering nectar of bees	Process of optimization
Food source	The feasible solution of the optimization problem
The location of the food source	The location of the solution
The amount of nectar from a food source	Fitness in optimization problems
The process of finding and collecting food sources	The process of solving the problem
A food source with the largest amount of nectar	The optimal solution of the problem

The specific search process is as follows.

(1) Set algorithm parameters and initialize the population. Among the $2N$ positions randomly generated, N superior ones are selected as honey source positions.
(2) The leader finds the location of the honey source and continues to search for nearby honey sources. The search algorithm is as follows.

$$V_{ij} = x_{ij} + R_{ij}\left(x_{ij} - x_{kj}\right) \tag{15}$$

Where V_{ij} is a new generated solution, $j \in \{1, 2, 3...D\}$, $k \in \{1, 2, 3...N\}$, R_{ij} is a random number in the range −1 to 1.
(3) According to the fitness value (objective function value of electric vehicles), the newly formed individuals are compared with the contemporary ones, and the best one is selected and marked as initial point.
(4) The leader bee chooses the follow bee through Roulette. The next electric vehicle decides whether to charge at this time or to search for a new time.
(5) The leader selects and marks the one with the largest amount of nectar among the honey sources searched. Followers search for new honey sources in the vicinity and compare them with the marked one. And then, the source with a greater number of nectar is selected for retention.
(6) In order to prevent the algorithm from falling into local optimum, the maximum number of searching limit is set. If the number of search times is greater than the limit, the original honey source will be abandoned and a new search will be selected. Suppose the maximum number of search times is 200.

Repeat these steps until the maximum number of cycles is reached.

4.3 An Improved Differential Evolution Algorithm Based on ABC Algorithm

DE algorithm is a widely used evolutionary algorithm which adopts mutation, crossover and selection operation and has good global search strategy ability. Also, the algorithm has fewer parameters and strong robustness, being easy to implement. However, the local search ability of DE algorithm needs to be improved. The algorithm is easy to fall into the local optimum. Therefore, it is necessary to select the appropriate population size to avoid premature convergence. In order to avoid the problem mentioned above, an improved DE algorithm is proposed, which is called improved differential (IDE) algorithm. The algorithm incorporates the ABC algorithm's local search and global search into the DE algorithm, and by introducing swarm evolution and detection operation, compared with the DE algorithm, the search performance of the IDE algorithm is improved and the population size required by the algorithm is reduced.

The specific implementation processed of IDE algorithm is as follows.

(1) Initialize the parameters required by DE and ABC algorithm and the population size. Randomly generate the charging plan of electric vehicles.
(2) The DE algorithm is used to perform basic evolutionary operations. Through mutation, crossover and selection, the population is renewed and a new generation of individuals is produced. The specific update formula is from (12) to (14). The result in this step is an orderly charging plan for electric vehicles.
(3) The ABC algorithm is used to accelerate evolutionary operation. According to the idea of ABC algorithm, the local search times of individuals can be selectively increased. Besides, the individuals with high fitness are allowed to perform basic DE operations multiple times, so as to generate a new generation of individuals. This step can jump out of the local beat solution and prevents EVs from being charged in a concentrated amount of time.
(4) Determine if there are undesirable individuals. In order to break out of the local minimization cycle, the upper limit of evolution times is defined as $limit = N_P * D$. If an individual has evolved more than limit times but still performs poorly, it will be defined as an undesirable individual. However, if there is no bad individual, go to step (6) for optimization; otherwise, directly go to step (6).
(5) Detect the artificial bee colony. For bad individuals, return to step (1) to regenerate the individuals randomly, replacing the bad individuals with the new individuals, and the number of accelerated evolution of the new individual is zero.
(6) Determine if the termination condition has been met. If the maximum number of iterations is reached, the iteration will be stopped and the objective function value will be calculated. Otherwise return to step (2) and redo the basic differential evolution operation.

4.4 Rolling Optimization Strategy for Electric Vehicles

The orderly charging strategy proposed in this paper updates the charging plan according to the real-time status of EVs, and the update cycle is 10 min. If there is a new EV access at period t, the charging plan will be reformulated by IDE algorithm;

otherwise, execute the charging plan of the previous period. Charging behavior in residential areas is mainly concentrated between 18 pm and 9 am the next day, 900 min in total. Assume that the minimum charging load change cycle is 10 min, the charging plan is a 90-dimension problem, and IDE algorithm can solve the problem effectively.

5 Analysis of Examples

5.1 Parameter Setting

In this paper, load data of a certain area is selected for analysis. Only the charging behavior between users' return in the evening and their departure next day is studied. The specific parameter settings are as follows:

(1) The maximum capacity of the distribution network in this area is set as 15000 kW
(2) The initial SOC of electric vehicle is set to follow a normal distribution with a mean value of 0.7 and a variance of 0.1
(3) The charging time of EVs follows a normal distribution with a mean value of 1140 and a variance of 60
(4) The end charging time of EVs follows a normal distribution with a mean value of 480 and a variance of 30
(5) The uniform electricity price for charging is 1.3 yuan. The TOU price [14] is shown in Table 2.

Table 2. Time-of-use electricity price

Period	TOU (yuan/kW·h)
00:00–08:00	0.4
08:00–12:00	2.0
12:00–15:00	1.2
15:00–21:00	2.0
21:00–24:00	1.2

(6) The original load data of the region is shown in Fig. 1.

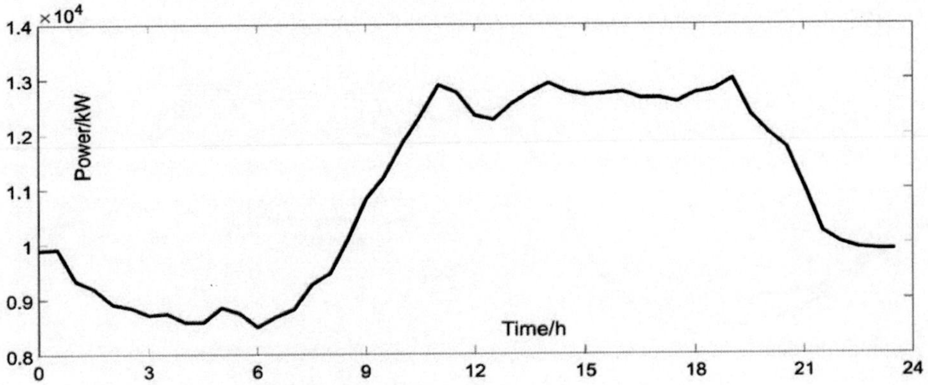

Fig. 1. Original load curve of grid

5.2 Analysis of Simulation Results

The numbers of electric vehicles are selected to be 100, 500 and 1000 respectively for simulation. The optimized scheduling results are shown in Table 3.

Table 3. optimized scheduling results of electric vehicles

The number of EVs	Scheduling mode	Charging expense/yuan	Peak-valley difference/kW
100	Disordered charging	1660.7	4659.7
	Orderly charging	647.9	4236.3
500	Disordered charging	9759.9	5392.1
	Orderly charging	3134.6	3765.3
1000	Disordered charging	22696.2	6879.1
	Orderly charging	6491.4	3849.4

The simulation results of 100 vehicles are shown in Fig. 2.

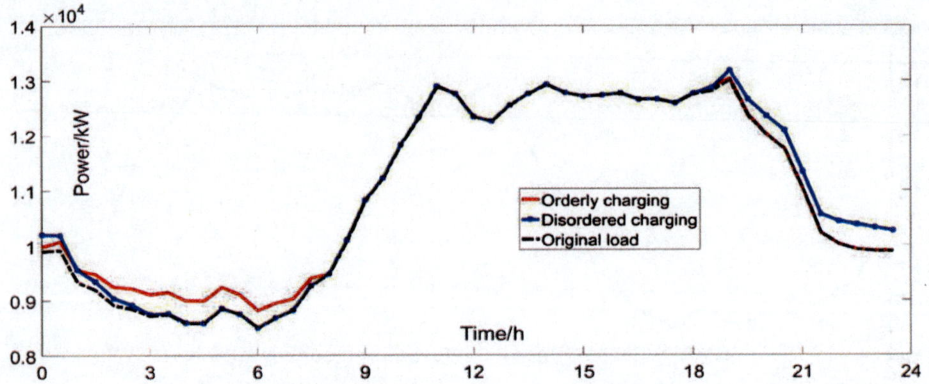

Fig. 2. Charging curve of 100 electric vehicles

As can be seen from Fig. 2, compared with disordered charging, orderly charging can transfer part of the charging load to the low peak period. Therefore, the method plays a certain role in cutting peak and filling valley of power grid. What's more, as can be seen from Table 2, that compared with the disordered charging, the charging cost saved 1012.8 yuan and the peak valley difference of the grid was reduced by 423.4 kW.

The simulation results of 500 vehicles are shown in Fig. 3.

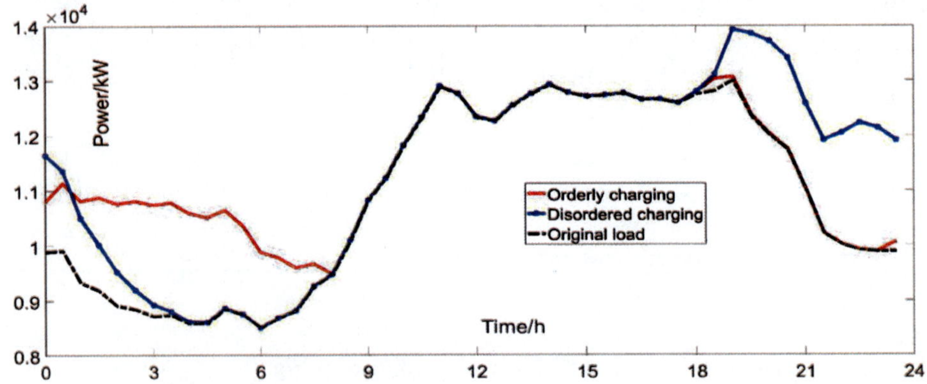

Fig. 3. Charging curve of 500 electric vehicles

As can be seen from Fig. 3, with the increase of the number of electric vehicles, the total load is gradually approaching the limit load value of safe and stable operation of the distribution network. In this moment, the orderly charging algorithm can transfer most of the charging load to the off-peak period. Besides, orderly charging algorithm can reduce load fluctuation rate and make the load curve smoother. What's more, as can be seen from Table 2 that compared with the disordered charging, the charging cost saved 6625.3 yuan and the peak valley difference of the grid was reduced by 1726.8 kW.

The simulation results of 1000 vehicles are shown in Fig. 4.

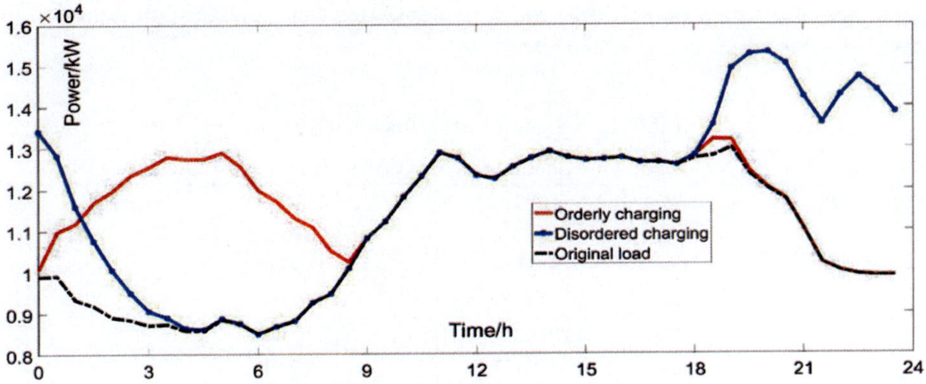

Fig. 4. Charging curve of 1000 electric vehicles

As can be seen from Fig. 4, as the number of electric vehicles increases further, the influence of disordered charging on distribution network becomes more and more intense. And the charging power will exceed the maximum capacity of the distribution network, which will cause a series of problems such as transformer over-limit. However, orderly charging algorithm can maintain the safe and stable operation of distribution network. As can be seen from Table 2 that compared with the disordered charging, the charging cost of orderly charging saved 16 204.8 yuan and the peak valley difference of the grid was reduced by 3029.7 kW. In this situation, some EVs are not transferred to charge at night, because the load at night is close to the limit value. Therefore, some late arriving cars are scheduled to be charged during peak period.

It can be seen from the above three situations that with the increase of the number of electric vehicles, the charging charges reduced by the orderly charging algorithm are 60.98%, 65.87% and 68.63%, respectively, which shows that the orderly charging strategy can significantly reduce the users' charging cost. Thus, the optimization effect of orderly charging strategy on users' charging cost becomes more obvious.

In terms of peak-valley clipping, the orderly charging algorithm can significantly reduce the peak-valley difference of the distribution network. It can be seen from the simulation results that with the increase of the number of electric vehicles, the peak-valley difference of load in the distribution network was reduced by 9.09%, 30.18% and 44.04% respectively by the orderly charging algorithm. Besides, with the increase of the number of electric vehicles, the optimization effect of orderly charging strategy on user charging cost becomes more obvious. As a result,even if the scale of electric vehicles increase rapidly in the future, the orderly charging algorithm is also applicable.

5.3 Performance Analysis of IDE Algorithm

In order to illustrate the convergence effect and convergence precision of the proposed improved IDE algorithm, the results of the IDE algorithm and the DE algorithm are compared in Table 4.

As can be seen from Table 4 that the optimization effect of IDE algorithm is better than DE algorithm in the aspects of the charging cost of EVs users and the peak-valley difference of distribution network, which demonstrates that the algorithm proposed in this paper has better convergence accuracy.

Table 4. Solving model results by algorithm

Number of EVs	Algorithm	Charging expense/yuan	Peak-valley difference/kW
100	DE	1023.5	4356.2
	IDE	647.9	4236.3
500	DE	4658.3	4159.5
	IDE	3134.6	3765.3
1000	DE	10253.8	4203.6
	IDE	6791.7	3849.4

The convergence curve of DE algorithm and IDE algorithm to solve the model is shown in Fig. 5 when 1000 EVs are charging.

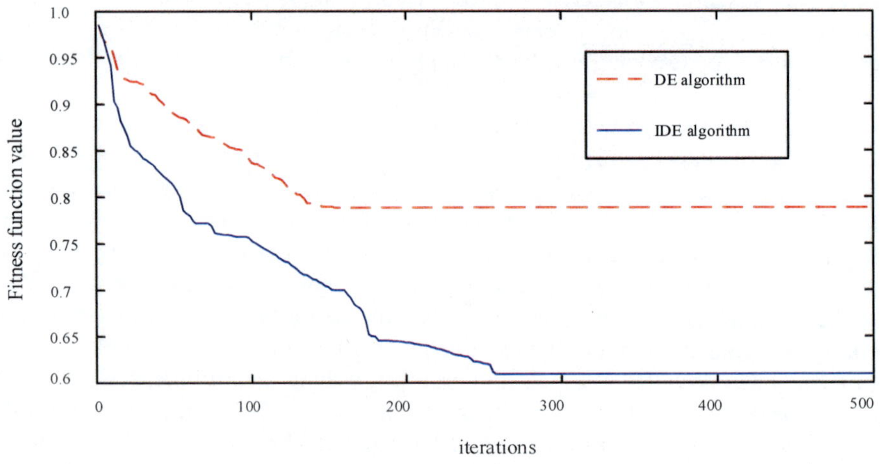

Fig. 5. Convergence curves of the algorithm

As can be seen from Fig. 5 that IDE algorithm can improve local search capability, which can avoid the algorithm to fall into premature, so as to get better search results.

6 Conclusion

In order to alleviate the impact of large-scale electric vehicles charging load on distribution network, this paper proposes an orderly charging strategy of electric vehicles based on improve differential evolution algorithm. To solve the problem that differential evolution (DE) algorithm is lack of local search capability, this paper introduces an artificial bee colony algorithm which improves the local search capability of it, so as to get an improved DE algorithm. The simulation results show that the proposed orderly charging strategy can effectively reduce the charging cost of EVs' users and the peak and valley differences of the load of the distribution network. Besides, with the expansion of the scale of electric vehicles, the optimization effect of the algorithm is more obvious in the aspects of peak load clipping and valley load filling in distribution network and reducing the charging cost of users. Constraints on charging continuity can prevent from load mutation and protect charging facilities. Besides, the rolling optimization strategy can ensure the real-time performance of the charging plan.

References

1. Liu, Z.Y.: Electric Power and Energy of China, 2nd edn. CEPP, Beijing (2012)
2. Li, G., Dong, Y.Z., Wen, F.S., et al.: Charging and discharging behavior prediction of electric vehicles based on mobile social network platform. Autom. Electr. Power Syst. **40**(9), 64–70 (2016)
3. Wang, G.N., Yang, J.F., Wang, S., et al.: Distributed optimization of power grid considering dispatching of electric vehicle battery swapping stations and data storage of blockchain. Autom. Electr. Power Syst. **43**(8), 110–116 (2019)
4. Zhou, T.P., Sun, W.: Electric vehicle-to-grid technology based on microgrid. Autom. Electr. Power Syst. **42**(3), 98–104 (2018)
5. Yao, M., Wang, Y.: Restoration method of electric power supply for isolated island with auxiliary of electric vehicles considering impact of cooling load. Autom. Electr. Power Syst. **43**(7), 144–150 (2019)
6. Chen, Q.F., Liu, N., Zhao, T.Y.: Automatic demand response for PV charging station based on receding linear programming. Power Syst. Technol. **40**(10), 2967–2974 (2016)
7. Li, D.D., Duan, W.Y., Lin, S.F., et al.: User guidance based matching strategy for electric vehicle-charging pile in condition of real-time electricity price. Autom. Electr. Power Syst. **44**(8), 74–82 (2020)
8. He, X., Tu, C.M., Wang, L.L., et al.: Double-layer charging strategy for electric vehicles considering users' driving patterns. Autom. Electr. Power Syst. **42**(3), 64–69 (2018)
9. Chen, K., Ma, Z.L., Zhou, S.Y., et al.: Charging control strategy for electric vehicles based on two-stage multi-target optimization. Power Syst. Protect. Control 1–8 (2020)
10. Hou, H., Fan, H., Su, S., et al.: Multilateral win-win strategy for smart charging service of electric vehicle. Autom. Electr. Power Syst. **41**(19), 66–73 (2017)
11. Peng, C., Sun, H., Guo, J., et al.: Dynamic economic dispatch for wind-thermal power system using a novel bi-population chaotic differential evolution algorithm. Int. J. Electr. Power Energy Syst. **42**(1), 119–126 (2012)
12. Liu, J., Cheng, H.Z., Xiao, J., et al.: A multi-objective reconfiguration strategy for smart distribution network considering N-1 security criterion. Autom. Electr. Power Syst. **40**(7), 9–15 (2016)

13. Bahmani-Firouzi, B., Farjah, E., Azizipanah-Abarghooee, R., et al.: An efficient scenario-based and fuzzy self-adaptive learning particle swarm optimization approach for dynamic economic emission dispatch considering load and wind power uncertainties. Energy **50**(2), 232–244 (2013)
14. Wu, J.W., Qiu, X.Y., Pan, Y.J., et al.: Research on orderly charging strategy of electric vehicle based on improved chicken swarm optimization. Electr. Meas. Instr. **56**(09), 97–103 (2019)

Research on Security of Multi-access Edge Computing Nodes System Integrated by Blockchain Technology

Xuesong Huo[1], Ming Zhang[1], Jiang Zhu[2(✉)], Zi Wang[2], and Hongqin Zhu[1]

[1] State Grid Jiangsu Electrical Power Co., Ltd., Nanjing, China
[2] NARI Group Corporation/State Grid Electric Power Research Institute, Nanjing, China
zhujiang@sgepri.sgcc.com.cn

Abstract. This paper introduces the application mode of the multi-access edge computing nodes system (MECNS), and analyzes the needs of the security research of the MECNS. Aiming at the security problems faced by the MECNS in the application of new energy sources such as power plant-grid-load-store system (PPGLSS), etc., the security edge computing node (ECN) is defined, and the architecture of a MECNS with a security mechanism is designed. It is based on public and private key encryption algorithm. The authorization protocol, registration protocol, cancellation protocol and the address query protocol of ECNs are proposed. In this scheme, the confidentiality and security of communication between ECNs are realized through SSL. And on this basis, the blockchain technology is integrated to propose a trusted protection mechanism for edge computing. This architecture and related protocols can ensure the verifiability and non-repudiation of ECN activities.

Keywords: Multi-access edge computing node · Public key · Private key · Blockchain · Plant-grid-load-store

1 Introduction

With the rapid development of new technologies such as the Internet of Things and 5G communications, the intelligent era of the Internet of Everything is accelerating. In IoT systems, all raw event data are collected from sensors and processed only in the cloud, resulting in unnecessary data redundancy and network overhead. And there is a tighter coupling between events and actions [1]. Edge computing is a new computing model that provides intelligent services at the edge of the network near objects or data sources. It can reduce costs and improve the overall response performance of the system through local data processing techniques.

This work was supported by the Project of State Grid Corporation of China (Blockchain-Based Research and Typical Scenario Application of Trusted Service Support System for Electric Power Business). No: SGJS0000DKJS1900523.

Y. Xue et al. (Eds.): PMF 2020, LNEE 718, pp. 557–568, 2021.
https://doi.org/10.1007/978-981-15-9746-6_43

In recent years, edge computing has entered a period of rapid development, and has quickly implemented application pilots in new energy projects such as power plant-grid-load-store system (PPGLSS). Based on the edge computing model, some researchers have proposed related business solutions. De et al. [2] believe that the development of the Internet of Things system is constrained by the coupling of physical processes and the digital world, and is crucial to the impact of the next industrial revolution. The edge computing paradigm is considered, and a standards-compliant communication architecture is extended to support container-based orchestration mechanisms. Magurawalage et al. [3] introduced a new idea of providing computing services for connected devices. It is based on the key concept that computing resources will be coupled with communication resources in the future, and is used to enhance the user experience and optimize the resources of infrastructure providers. In addition, this coupling is based on the joint / cooperative resource allocation algorithm, which is achieved by integrating computing and communication services and integrating hardware in the network. Wang et al. [4] proposed that the coupling of virtual nodes is one of the reasons hindering the development of sensor cloud [5]. The virtual node is a user visual node composed of physical sensor nodes. It is created, stopped and destroyed with the user's needs [6, 7]. The literature [8, 9] proposes an extended KM algorithm based on the data cache queue to maximize the utilization of resources. Gao et al. [10] proposed a greedy algorithm that uses the KM algorithm to solve the maximum matching problem on the Bipartite Graph, thereby obtaining an effective algorithm for the quasi-optimal solution of the original problem [11]. These algorithms include 2 offline algorithms and one online algorithm, of which 2 offline algorithms return the upper and lower bounds of the original problem [12, 13]. Li et al. [14] proposed to combine the optimal placement of data blocks and the optimal scheduling of tasks in order to reduce the calculation delay and response time of submitting tasks and improve the user experience. The placement of data blocks is considered not only from the popularity of data blocks, but also from the data storage capacity and replacement rate of the edge server that stores these data blocks. The above researches focused on the application analysis of edge computing models.

While conducting the above-mentioned application researches based on the edge computing mode, the security researches of the edge computing have been also gradually carried out. There are three main reasons for studying the security of edge computing models:

1) The environment in which edge computing nodes (ECNs) exist is loose, distributed, and open. ECNs may be distributed on different hosts in the Internet. ECNs may join or exit at any time. Therefore, we need to study how to manage these ECNs to ensure that their behaviors can be carried out safely.
2) ECNs achieve their goals through actions. Any activity of ECNs represents the behavior of users. In order to make the ECNs responsible for their own behaviors, any activities result of the ECNs are required to be unmodifiable and undeniable.
3) In order to avoid the leakage of communication information and prevent malicious attackers, it is necessary to study the communication mechanisms of ECNs.

The development of blockchain technology provides new ideas for solving the above problems. As the underlying technology of Bitcoin, blockchain is essentially a decentralized and trustless distributed database account book. As time goes by, new transactions are continuously generated in the network, and miners verify the transactions according to the consensus algorithm, and store them independently in each node.

The workflow of blockchain specifically includes the following steps [15]:

1) The sending node A broadcasts the newly generated data record information to other nodes in the entire network.
2) All receiving nodes in the entire network verify the received data record information from the sending node A, for example, whether the data record information is legal. If it passes the test, the receiving nodes will store the data record information in a temporary area among the blocks.
3) Each receiving node implements a consensus algorithm on the block, such as Proof of Stake (PoS), Proof of Work (PoW), and Practical Byzantine Fault Tolerance, (PBFT).
4) When a receiving node finds the solution of the consensus algorithm (for example, the proof of work solution), broadcasts the block to all nodes in the whole network.
5) If and only if all transactions in the block are valid and unused, the node can accept that the block is eligible to become a new block.
6) The new block is officially stored at the end of the longest chain of the current blockchain. The nodes of the entire network express the acceptance of the block by using the random hash of the block as the hash value of the latest block.

All nodes in the network regard the longest chain in the blockchain as a valid chain, and add new blocks to the chain, thereby increasing its length [16]. Although two nodes may calculate the proof of work of a block and broadcast a new block at the same time, other nodes in the network have differences in the timing of receiving the new block. Under normal circumstances, the node will receive the first seen block and carry out new mining work on the basis of the block. Because different nodes follow different branch chains, in theory, the computing power of the two branch chains is different, and the growth rate of the blocks is also different. After a period of time, a certain chain will inevitably become relatively longer chain. At this time, the node working on another branch chain will abandon the current chain and continue working on the longer chain. This is how the blockchain prevents forks. Therefore, the final data record information stored on the distributed ledger of the blockchain is the result of joint authentication and confirmation by the nodes of the entire network. Once the nodes of the entire network reach a unified consensus, the data record information has the consistency of the entire network and cannot be tampered with.

In order to study the security of ECNs, we first need to know the various potential unsafe factors in the system. These potential unsafe factors will affect the performance and execution results of the entire system. Typical insecure factors include compromised name services, insecure proxy appointments, insecure communications, etc. In order to solve the above insecure problems, the security architecture of the system is first given. Under this structure, the authorization protocols of the ECNs are defined so that the ECNs can obtain the certificate of joining the system. Then, the communication between ECNs is encrypted by SSL to achieve secure communication between them, and on this basis, the blockchain technology is integrated to propose a trusted protection mechanism for edge computing. Finally, this security mechanism is given Application in a power plant-grid-load-store system (PPGLSS) project.

2 System Structure in Edge Computing Mode

The research on the system structure adopting the edge computing mode has also been the topic of artificial intelligence research. However, regardless of the current cognitive structure or reactive structure, the security factors of ECNs are not considered. Similarly, for ECNs, the more concerned about the coordination and cooperation between them, less consideration of security factors. Therefore, from the perspective of security, the security ECN is defined, and the system structure using the edge computing mode is given.

2.1 Definition of a Secure ECN

If an ECN can meet the following conditions, it can be called a safe ECN.

1) It can indicates its user;
2) It is responsible for the behavior of the node itself;
3) It can ensure that the secret information is not leaked when interacting with the surrounding environment;
4) It can identify the other nodes when interacting with them. It can make correct judgments to other nodes' requests and prevent impersonation and attacks by other nodes.

It is known from the attributes of edge computing nodes that the activities of edge computing nodes are autonomous. Therefore, only the edge computing nodes indicate their users can they truly trace the person responsible for the results of the edge computing nodes' behavior. Even if the edge computing node indicates its user, it still needs to know that a certain behavior is indeed caused by itself or a certain message is indeed sent by itself. That is, the edge computing node is responsible for its own behavior. In addition, in a system that includes multiple edge computing nodes, the addition or departure of edge computing nodes is dynamic. Therefore, when the edge computing node exchanges information with other nodes, it must be able to recognize the requests of other nodes and make correct judgment. Since the system may be an open and distributed system, there must be security measures for the transmission of messages between edge computing nodes. Only in this way can we say that edge computing nodes are safe.

2.2 Business System Structure of Secure ECN

As can be seen from Fig. 1, in addition to the components (such as rule calculation, scene linkage, device access, Collaborative computing, and streaming data analysis) that general ECNs have, the ECNs in this paper also add many security-related components such as encryption, signature, verification.

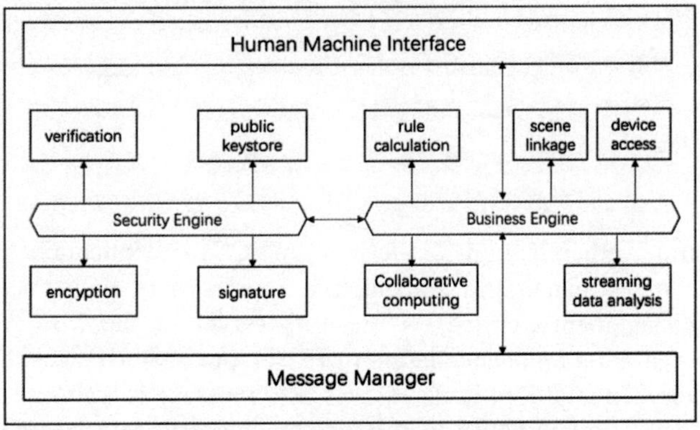

Fig. 1. Structure of secure ECN

From the above definition of the secure ECN, it can be seen that the ECN needs to be bound to its users. Therefore, the ECN needs to confirm the identity of its user through the human-machine interface before joining the system, which is usually realized by the user's signature. The security system in this paper is based on the PKI public key management architecture. Therefore, encryption, signature, and verification components are added to the construction of ECNs. The encryption component is used to encrypt the messages and content. The signature component is used to indicate the identity of the edge computing node, recognize its own behavior results and sent messages. The erification component is mainly to verify the identity of the sender of the message through the consensus mechanism of the blockchain. In addition, in order to reduce the amount of network information transmission and the number of conversations between ECNs during the authentication process, a public keystore component is added to the structure of ECNs to cache the public keys.

When multiple ECNs are combined to complete a goal, a multi-access edge computing nodes system (MECNS) is formed. Currently, a facilitator is usually used to communicate between ECNs. The MECNS architecture framework constructed in this paper is shown in Fig. 2.

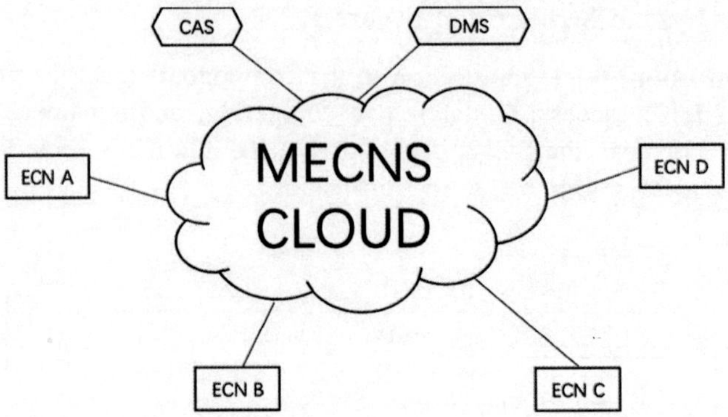

Fig. 2. Structure of MECNS

CAS (Central Authentication Service) is an ECN authentication server, which is responsible for the issuance and authentication services of ECNs certificates. DMS (Distributed Management Service) is the core server of the MECNS. It not only manages the registration and cancellation of ECNs, but also manages the name service and address query. In addition, it also serves as a message roaming service for ECNs. When an ECN needs to join MECNS, it must obtain a certificate to join the system. The certificate can be obtained by applying to CAS. When the ECN holds the certificate, the certificate applies to DMS for registration. After verifying that the ECN's certificate is qualified, DMS assigns an ID to the ECN, so that the ECN is considered a formal member of the system. The following discusses the security authentication protocol and secure communication mechanism for implementing this MECNS.

3 Security Mechanism of MECNS

Convention about notation: Assuming that *A* represents a certain ECN, then *a* represents *A*'s private key, and P_a represents *A*'s public key. The pair of keys for ECN user is $\{d, P_d\}$. The pair of keys of CAS is $\{cas, P_{cas}\}$, the pair of keys of DMS is $\{dms, P_{dms}\}$, $[m]_a$ means that the information *m* is signed by *A* with its private key *a*. *A* Send the message *m* to the ECN B as *AB:m*.

3.1 The Authorization Protocol of ECNs(APE)

In the MECNS structure model(see Fig. 2.), in order to join the system, any ECN first generates a pair of keys *Keygen(A)*, and then applies for its own certificate on the authentication server *CAS*, (*CAS* through the *blockchain* Consensus mechanism) to be recognized by other ECNs. After reviewing the qualification of the applicant, the authentication server *CAS* can issue the certificate *Cergen(A)* to it if it meets the conditions. When the ECN obtains its own certificate, it claims that the applicant has been authorized by *CAS*. It can register with *DMS* and formally join this MECNS. Of course, the ECN can also not join the system after obtaining the certificate.

APE:
A : Keygen(A)= {a, P_a}
A→CAS:m_1, m_1=[[name, P_a]$_d$]$_{Pcas}$
CAS:
①Check if the request in m_1 is legal;
②Check if the signature in m_1 can be verified with P_a;
③Check if the P_a is used by other ECN.
④If ①, ②, ③ are all satisfied, then:
Broadcast m_1 to the other ECNs and receive feedback using blockchain consensus mechanism.
⑤If ④ is satisfied, then C_A= Cergen(A), m_2=(C_A)$_{cas}$
⑥Store C_A in blockchain
CAS→A:m_2, m_2=(C_A)$_{cas}$
A:
①Check if the signature in m_2 can be verified with P_{cas};
②Save (C_A, m_2)
THE END

3.2 The Registration Protocol of ECNs(RPE)

After the ECN obtains the authorization certificate, it does not indicate that the ECN is already a member of the MECNS. Only after registering with DMS, ECNs can be considered as full members and can communicate and cooperate with other nodes.

When registering, the ECN *A* needs to provide his *name, address, port number* and certificate *C*. After receiving the signature registration message from the *A*, DMS must first go to the CAS to verify the validity of the signature. When CAS confirms that the application meets the conditions, it generates an ID number for *A*. This ID is unique in the MECNS. In order to verify whether *A* is at the *address*, DMS sends the newly generated ID to the *port number* of *address*, requests *A* to sign and response this ID. If *A* is not in *address*, *A* will not receive this ID. By this way, you can prevent those denial of service (DOS) attacks.

After the DMS receives the response, it will save the ID, name, address, port number and certificate of *A*, and at the same time, tell *A* registration information and certificates of other ECNs in the MECNS. When *A* interacts with other ECN, it is not necessary to search and verify at the DMS. Thereby it will avoid the bottleneck effect of DMS.

Since DMS has the relevant information of all ECNs registered in the system, when an ECN wants to find the address and other information of other ECNs, it can be obtained by DMS through name and address mapping. In addition to querying the information of other ECNs through DMS, it can also verify whether an ECN is currently in the system.

RPE:
$A \rightarrow DMS: m_1$, $m_1=[register[name, address, port number, C_A]_a]_{Pdms}$
DMS:
①Check if the signature in m_1 can be verified with P_a;
②Check if the name in certificate is right;
③Check if the name is registered;
④Check if the address is right by sending the ID to the address and port number:
$DMS \rightarrow A(address): m_2$, $m_2=(ID, address, port number)_{dms}$
$A \rightarrow DMS: m_3$, $m_3=(ID)_a$
DMS:
if m_3 is valid, then write the C_A, ID, name, address, port number of A to the blockchain
$DMS \rightarrow A: m_4$, $m_4=[registered([ID, address, name, port number, expiration]_{dms}]_{Pa}$
A:
①Check if the signature in m_4 can be verified with P_{dms};
②Save registered information.
THE END

3.3 The Cancellation Protocol of ECNs(CPE)

When an ECN wants to exit the system, it can use the CPE protocol to apply to DMS for exit. When the DMS receives the cancellation request from the ECN, it will first verify the validity of the request. Any registered ECN's cancellation request is valid. Any unregistered ECN's cancellation request is unresponsive. In order to avoid "impostors" revoking an ECN, any ECN's requesting revocation must be signed and confirmed by other ECNs through the blockchain consensus mechanism. DMS only receives signature requests. At the same time, DMS verifies the validity of the signature. The DMS checks whether the request to cancel the registration of the ECN is legal. If it is legal, it deletes the ECN from the registry and uploads its request information to the blockchain for future audit.

CPE:
$A \rightarrow DMS$: m_1, $m_1=[revoke[ID]_a]_{Pdms}$
DMS:
①Check if the signature in m_1 can be verified with P_a;
②Check if the ID is registered;
③Check if the ID is the same as the information in the blockchain;
④If ①, ②, ③ are all satisfied, then:
delete C_A, ID, name, address, et al from blockchain.
$DMS \rightarrow A: m_2$, $m_2=[unregistered([ID]_{dms}]_{Pa}$
A:
Check if the signature in m_2 can be verified with P_{dms};
THE END

3.4 The Address Query Protocol of ECNs(AQPE)

Use AQPE to find the address of an ECN. The AQPE is public, that is, there is no need to sign the address query request. If not found, it indicates that the ECN is not registered in the system, otherwise, the query result is recorded.

AQPE:
A→DMS: m_1, m_1=*Search[ID]*
DMS:
①*If ID is in the blockchain, then get the address from blockchain;*
②*record the query result in the blockchain.*
DMS→A(address):m_2, m_2=*[search result[ID]]*$_{dms}$
A:
①*Check if the signature in* m_2 *can be verified with* P_{dms}*;*
②*Get the address of ID.*
THE END

3.5 Secure Communication Between ECNs

The secure communication protocol SSL is used between the ECNs to establish a connection. Using the SSL layer has two advantages: Firstly, the SSL layer is independent of the specific application, so for different applications, the same method can be used without changing the application. Secondly, because the SSL layer is an independent layer, when more advanced encryption technology appears, we can replace the original encryption technology and use the most advanced encryption technology to encrypt information without changing the system. Therefore, it brings great flexibility to the system.

The business data flow on SSL is as follows:

1) Configure the digital certificate of the software process in ECN *A* and ECN *B* which need to communicate with each other;
2) When communicating between *A* and *B*, the digital certificate is used to perform end-to-end bidirectional identity authentication, and then the data encryption key is generated through negotiation;
3) *A* encrypts and signs the transmitted data through the key generated through negotiation, and then sends it to *B*;
4) During data transmission, decryption and verification are not performed;
5) *B* decrypts and verifies the received encrypted data, and after successful, receives the data.

3.6 Edge Computing Trusted Protection Mechanism

On the basis of MECNS, the edge security trusted network is constructed through the characteristics of the blockchain. It includes regional trust, global trust and trusted sharing. The consensus accounting technology of the blockchain is used to achieve the robust characteristics of system security protection, preventing the destruction of security protection capabilities caused by single-point failures of traditional security protection technologies or DDoS attacks. Private blockchain is used to build a secure and trusted network for edge computing. Multi-factor identity authentication technology or Zero Trust Security is used to ensure heterogeneous edge terminals such as mobile phones, tablets, sensors, smart terminals, computers, buses, cars, unmanned vehicles, drones and unmanned boats which are connected to the edge network is credible. The prior security control of the edge endpoints connected to the access network is achieved. Edge gateways and alliance chains are used to connect to the global trusted cooperative network on the cloud side, so that the trusted security resources of the edge network can be freely combined and orchestrated in the entire edge network domain. The immutable feature of the blockchain is used to realize the supervision of incidents and the traceability of threats afterwards, thereby ensuring the authenticity and integrity of the data. The p2p transmission mechanism and e2e encrypted communication mechanism of the blockchain network are used to to achieve covert communication and important data security in the edge computing network. The public blockchain is used to achieve cross-domain trusted communication and secure data sharing between the edge computing network and the Internet. Through the integration of smart contracts and smart algorithms of the blockchain, it is possible to carry out independent intelligent security protection and risk response according to security threat scenarios. (see Fig. 3).

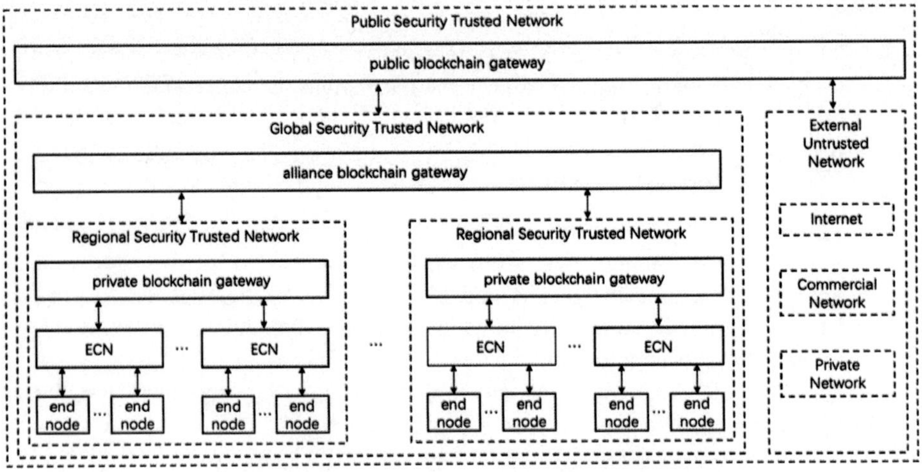

Fig. 3. Edge computing trusted protection mechanism

4 Application in the PPGLSS

In the PPGLSS, we use the above security mechanism to achieve secure collaboration between ECNs. In this project, various functions of power generation side ECN, power grid side ECN, load side ECN, and energy storage side ECN are constructed. (see Fig. 4.)

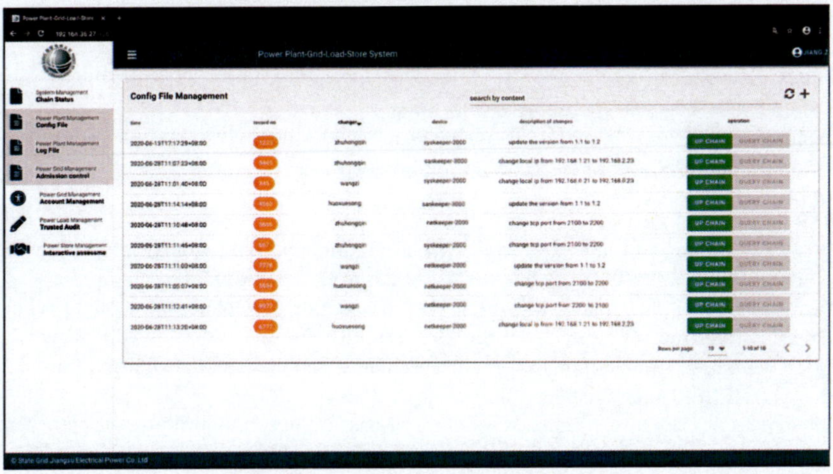

Fig. 4. Application in the PPGLSS

DMS and CAS are used to implement authorization, authentication and public key management of ECNs. In this way, any actions of power plants, power grid companies and load users cannot be denied, and at the same time, the benefits will not be damaged. In the PPGLSS of State Grid Jiangsu Electrical Power Co. Ltd, we verified the security of the proposed architecture, and the feasibility of the architecture is determined.

5 Conclusion

With the gradual application of MECNS in new energy fields such as PPGLSS, its security has become an important research topic. The MECNS architecture with security mechanism, the authorization, registration, cancellation and other protocols and the trusted protection mechanisms designed in this paper can achieve the verifiability and non-repudiation of the behavior of the ECN.

References

1. Wen, Z., Yang, R., Garaghan, P., et al.: Fog orchestration for Internet of things services. IEEE Internet Comput. **21**(2), 16–24 (2017)
2. De Brito, M.S., Hoque, S., Steinke, R., et al.: Application of the fog computing paradigm to smart factories and cyber-physical systems. 10.1002/et.3184. Accessed 20 Aug 2019
3. Magurawalage, C.S., Yang, K., Wang, K.: Aquacomputing: Coupling computing and communications. arXiv preprint, arXiv:1510.07250 (2015)
4. Wang, T., Liang, Y., Jia, W., et al.: Coupling resource management based on fog computing in smart city systems. J. Netw. Comput. Appl. **135**, 11–19 (2019)
5. Li, C., Tang, J., Tang, H., et al.: Collaborative cache allocation and task scheduling for data intensive applications in edge computing environment. Future Gen. Comput. Syst. **95**, 249–264 (2019)

6. Gregory, J.P.: Complexity of a disjoint matching problem on bipartite graphs. Inf. Process. Lett. **116**, 649–652 (2016)
7. Itai, A., Rodeh, M., Tanimoto, S.L.: Some matching problem for bipartite graphs. J. Assoc. Comput. Mach. **25**(4), 517–525 (1978)
8. Wang, T., Li, Y., Jia, W., et al.: Research progress of sensor-cloud security. J. Commun. **39** (3), 3552 (2018)
9. Zeng, J., Wang, T., Jia, W., et al.: A survey on sensor-cloud. J. Comput. Res. Dev. **54**(5), 925–939 (2017)
10. Wang, T., Zhou, J., Liu, A., et al.: Fog-based computing and storage offloading for data synchronization in IoT. IEEE Internet Things J. **6**(3), 4272–4282 (2019)
11. Wang, T., Zhou, J., Huang, M., et al.: Fog-based storage technology to fight with cyber threat. Future Gen. Comput. Syst. **83**, 208–218 (2018)
12. Tian, W., Yang, Li., Guojun, W., et al.: Sustainable and efficient data collectionfrom WSNs to cloud. IEEE Trans. Sustain. Comput. **4**(2), 252–262 (2019)
13. Zhenguo, G., Danjie, C., Peng, S., et al.: KM-based efficient algorithms for optimal packet scheduling problem in celluar/infostation integrated networks. Ad Hoc Netw. **77**, 84–94 (2018)
14. Chunlin, L., Jingpan, B., Jianhang, T.: Joint optimization of data placement and scheduling for improving user experience in edge computing. J. Parallel Distrib. Comput. **125**, 93105 (2019)
15. Ametrano, F.M.: Bitcoin, Blockchain, and Distributed Ledger Technology. Social Science Electronic Publishing (2016)
16. Christidis, K., Devetsikiotis, M.: Blockchains and smart contracts for the internet of things. IEEE Access **4**, 2292–2303 (2016)

Data-Driven Auxiliary Decision-Making Method for Power System Stability

Hongqiang Li[1], Huibiao Yang[1], Lei Zhou[1], Feng Li[2], and Yi Tang[2(✉)]

[1] Ningxia Key Laboratory of Electrical Energy Security, Ningxia Electric Power Research Institute, Yinchuan, China
lhql652@126.com, huibiaoyang@163.com, luxuszl@163.com

[2] Southeast University, Nanjing, China
{lifeng_ee, tangyi}@seu.edu.cn

Abstract. Critical clearing time (CCT) is one of the most important indexes for large-disturbance rotor angle stability margin evaluation. In actual operation, model-driven methods, e.g., integrated extended equal area criterion (IEEAC) methods, are proven to be good methods to calculate CCT. Nevertheless, classical generator models are always used to reduce computation burden, where differences between classical models and detailed models could make CCT predictions useless. To correct error caused by model differences, data-driven method is considered and integrated with classical generator model based IEEAC, making an integrated method. In this paper, extreme learning machine (ELM) is used as the data-driven correction part. Further, cost-sensitive ELM model is constructed by substituting common error summation part with weighted error rate summation part in training objective. Prediction error for cases with small actual value is expected to be lower than that with large actual value by this transformation. Finally, validations on WSCC 9 and New England 39 bus system show effectiveness of proposed integrated method in correcting error caused by model difference and changing error distribution.

Keywords: Large-disturbance rotor angle stability · Critical clearing time · Extended equal area criterion · Extreme learning machine

1 Introduction

Modern electric power systems encounter more challenges than before considering that they transmit more power among wider areas with more economic considerations [1]. To overcome these challenges, more flexible and efficient methodologies and implementations for power system stability evaluation are expected.

Traditionally, power systems operation relies on decision tables formed by off-line analysis results about preset typical faults [2] and then on-line refreshing methods are proposed [3, 4]. With construction of smart grid, advanced measurement instruments and high-performance computing equipment are available for power systems situation awareness. Nowadays, decision table can be made with online system operation

Y. Xue et al. (Eds.): PMF 2020, LNEE 718, pp. 569–581, 2021.
https://doi.org/10.1007/978-981-15-9746-6_44

information and realize quasi real-time preset faults analysis, which could be more adaptive to actual operation.

Large disturbance rotor angle stability is focused in this paper. In this research area, critical clearing time (CCT) is one of the most important indexes recorded in the decision table describing transient stability margin [5]. Methodologies for CCT calculation include four categories: time domain simulation (TDS) method, transient energy function (TEF) method, extended equal area criterion (EEAC) method and machine learning (ML) method. The former 3 methods are model-driven methods and the last is a typical data-driven method. TDS method is accurate but rather time consuming. Both TEF and EEAC methods reduce computational efforts under certain assumptions [6–10]. However, TEF method is essentially conservative and designing a suitable energy function for complex power system models is always difficult.

EEAC method has been widely used in practical engineering for calculating CCTs of preset faults in decision table since 1990s [8]. It includes static extended equal area (SEEAC), dynamic extended equal area criterion (DEEAC) and integrated extended equal area criterion (IEEAC) methods [10]. Former two methods are based on model aggregation, which makes them fast but not accurate enough. IEEAC combines TDS and EEAC theory by trajectory aggregation where accuracy is greatly improved as the sacrifice of computation speed. In IEEAC actual application, classical generator model is always used as the substitution of detailed generator model for reducing the computation burden. Hence, CCT calculation error caused by model simplification could be a problem for classical model based IEEAC method.

ML methods, including artificial neural network (ANN) [11], support vector machine (SVM) [12], extreme learning machine (ELM) [13], have the most promising properties in computation speed and fitting ability. However, performances of ML methods highly rely on quality and quantity of collected training samples. In most ML methods, error tolerance for each training sample is treated equally by setting summation of total training error as part of training objectives in training algorithm. It is good for common regression predictions but not appropriate for CCT predictions. Because the lower the actual CCT is, the severer the fault scenario will be, where accuracy is more important. So error tolerance for each training sample should be treated differently in ML methods for CCT prediction.

Therefore, main objectives of this paper lies in two aspects. One is to reduce CCT calculation error of IEEAC caused by model differences. The other is to improve performance of ML method with consideration of differentiated accuracy importance. To realize former objective, model-driven and data-driven methods integration mode introduced in [14] is referred. While for latter objective, modifications on training model of ML methods are considered.

Remaining of this paper is organized as follows. In Sect. 2, problem description and IEEAC based CCT calculation are introduced in detail. Construction of cost-sensitive ELM model is presented in Sect. 3. The implementing structure is also mentioned in this section. Case studies and conclusion are presented in Sect. 4 and 5 respectively.

2 Problem Description and IEEAC Method

As mentioned in the introduction part, CCT is defined as the smallest fault clearing time by which loss of synchronization can be avoided. To describe CCT physical meaning, electromechanical transient process is analyzed and presented by differential algebraic equation (DAE) based mathematical model.

After a large disturbance, synchronization could be broken in a prompt process, which is academically named with electromechanical transient process. Effects of generator mechanical characteristics, excitation systems, governor and turbine systems, and load characteristics are usually involved in analyzing stability situation with consideration of transient process timescale. Mathematical model for describing this process is abstracted and shown below in form of DAEs.

$$\begin{cases} \frac{dx}{dt} = f(x, y) \\ g(x, y) = 0 \end{cases} \tag{1}$$

Where differential equation sets include differential equations for describing generator sequent and subsequent voltage variation, excitation system control process, governor and turbine control process, induction motor loads transient process, and generator rotor motion process. Algebraic equation sets include algebraic equations for Kirchhoff's law based power network, generator stator winding circuits, and static loads voltage-power relationship.

By solving these DAEs, post-disturbance power systems dynamic process can be described in detail. Most often, first-swing (about 1 s) and several-swing (about several seconds) rotor angle absolute variations after disturbance can reveal stability situation. Rule for stable and unstable cases judgement is that whether any absolute rotor angles gap among generators exceeds 180°.

For model-driven CCT calculation methods, IEEAC is more valuable than TDS method because it can evaluate stability margin with theoretical criterion and lower computation burden for finding critical cases without accuracy sacrifice [10]. It projects trajectories of complete power system, mainly generator rotor angle and power trajectories, onto one machine infinite bus (OMIB) system by complementary-cluster center-of-inertia and relative-motion (CCCOI-RM) method, and then applies equal area criterion (EAC) on OMIB system for stability analysis [15].

Given a power system with n generators, motion equation of generator i can be represented by the following equation.

$$M_i \ddot{\delta}_i = P_{mi} - P_{ei} \tag{2}$$

Where M_i and δ_i are inertia constant and rotor angle of generator i respectively. P_{mi} and P_{ei} are mechanical and electromagnetic power of generator i. To analyze system stability status, problem analysis dimension is reduced from R^n to R^2 and then R^1 by CCCOI and RM transformation respectively. Transformation process is elaborated below.

With rotor angle trajectories figured out by TDS method, generators can be divided into 2 groups, i.e. critical generators (Group S) and non-critical generators (Group A) in Fig. 1.

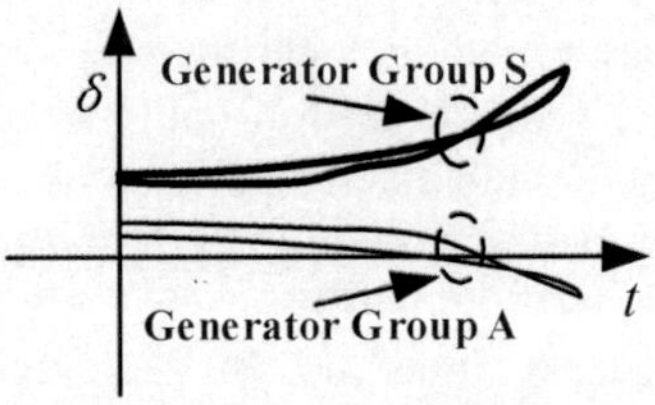

Fig. 1. Illustration of generator clusters

Generators in the same group then can be aggregated together. Equivalent rotor angles (δ_S and δ_A), mechanical power (P_{mS} and P_{mA}) and electromagnetic power (P_{eS} and P_{eA}) can be calculated with following formulas.

$$\begin{aligned} \delta_S &= \sum_{i \in S} M_i \delta_i / \sum_{i \in S} M_i = \sum_{i \in S} M_i \delta_i / M_S \\ \delta_A &= \sum_{j \in A} M_j \delta_j / \sum_{j \in A} M_j = \sum_{j \in A} M_j \delta_j / M_A \\ P_{mA} &= \sum_{j \in A} P_{mj} \\ P_{mS} &= \sum_{i \in S} P_{mi} \\ P_{eS} &= \sum_{i \in S} P_{ei} \\ P_{eA} &= \sum_{j \in A} P_{ej} \end{aligned} \tag{3}$$

Where S and A represents generator sets consisting of Group S and Group A. M_i and M_j are inertia constants of generator i and j. P_{ei} and P_{ej} are electromagnetic power of generator i and j. P_{mi} and P_{mj} are mechanical power of generator i and j. M_S and M_A are defined with following formula.

$$\begin{aligned} M_S &= \sum_{i \in S} M_i \\ M_A &= \sum_{j \in A} M_j \end{aligned} \tag{4}$$

Equivalent double machine power system is constructed with above CCCOI transformation and system dimension is reduced from n to 2. RM transformation is then applied to construct OMIB system reducing system dimension from 2 to 1 with following formula.

$$
\begin{aligned}
\delta &= \delta_S - \delta_A \\
M_T &= \sum_{k=1}^{n} M_k \\
P_m &= M_T^{-1}(M_A P_{mS} - M_S P_{mA}) \\
P_e &= M_T^{-1}(M_A P_{eS} - M_S P_{eA})
\end{aligned} \tag{5}
$$

Where δ, P_m and P_e are equivalent rotor angle, mechanical power and electromagnetic power of OMIB system respectively. Once OMIB system is constructed, observation space can be demonstrated.

Figure 2 shows the trajectories in observation space. With determined acceleration and retardation area, EAC method can be applied in observation space for stability analysis. In common cases, whether dynamic saddle point (DSP) is reached is used as a criterion for stability judgement. If DSP is met, power system will lose synchronization. Otherwise, it will not.

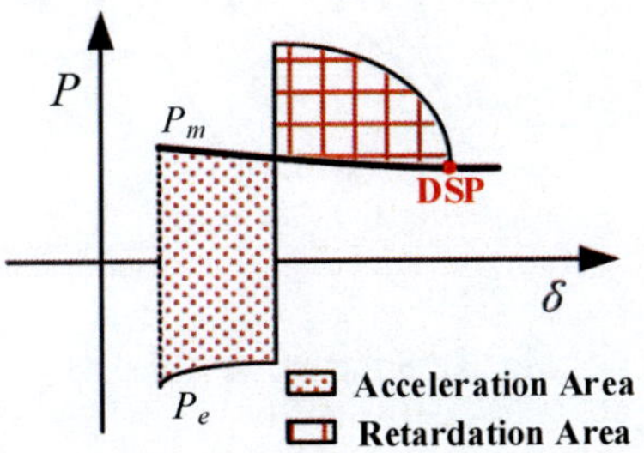

Fig. 2. Illustration of acceleration and retardation area in observation space

To reduce computation burden of CCT calculation under IEEAC method, sensitivity analysis can be made to reduce TDS method calls times. Moreover, detail models in TDS method can be substituted with simplified ones by which a number of differential equations can be eliminated. For example, classical model of a generator assumes both voltage behind transient reactance and power of prime mover to be constants, which means at least $2n$ differential equations could be avoided in comparison with detailed model for n-generator power systems. Therefore, classical model of generator is always useful for reducing computation burden, while use of simplified model could make calculation results inaccurate.

Hence, it is necessary to polish classical model based IEEAC CCT calculation results for better approximation to detail model based IEEAC CCT calculation results in actual applications.

3 Cost-Sensitive ELM and Implementation

In this paper, ML method is utilized as the data-driven part for correcting CCT calculation results of IEEAC method. ELM is chosen as the representative for ML method application with consideration of its fast training process and good generalization ability.

ELM is a kind of single-hidden layer feed forward neural network (SLF-NN) proposed by Huang and it is introduced in detail in [16]. It has good properties in training speed, generalization ability and avoiding complex model parameter selection. Figure below shows the structure of ELM (Fig. 3).

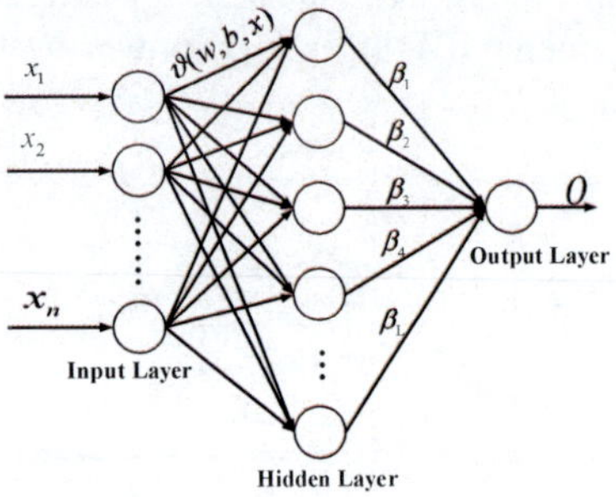

Fig. 3. Structure of SLF-NN

In critical clearing time prediction, $\mathbf{x}_i = [x_{i1}, x_{i2}, \ldots, x_{in}]^{\mathrm{T}} \in \mathbf{R}^n$ is the input vector, including CCT calculated by IEEAC based on classical generator model, pre-disturbance generator active and reactive power, post-disturbance generator bus voltage amplitude and phase short-term variation. $\mathbf{t}_i = [t_{i1}, t_{i2}, \ldots, t_{im}]^{\mathrm{T}} \in \mathbf{R}^m$ is actual CCT results based on detail model.

Conventionally, ELM can also be described as an optimization model showing below [17]:

$$\begin{aligned} \min \quad & f = \frac{1}{2}\|\beta\|^2 + \frac{M}{2}\sum_{i=1}^{N}\varepsilon_i^2 \\ \text{s.t.} \quad & h(x_i)\beta = t_i - \varepsilon_i \quad i = 1, 2, \ldots, N \end{aligned} \tag{6}$$

Where, β is output weight value vector. M is the weight value of training error part. ε_i, $h(x_i)$ and t_i are training error, activation function and actual value of sample i respectively. N is the number of training samples.

So it can be inferred from the model that the optimization objective is to minimize the summation of output weight matrix and training error. While in practical situation, error tolerance for small values is usually low, especially for CCT prediction. Hence, prediction error in cases where actual values are small is expected to be low, while for cases where actual values are large, the error tolerance could be relaxed. For CCT prediction, cases with low CCT values could be severer than that with high CCT values because the larger the CCT is, the more actions can be made to improve results or eliminate risks.

Hence, training error part in optimization objective of common ELM is substituted with weight training error rate part showing below:

$$\begin{aligned} \min \quad & f = \frac{1}{2}\|\beta\|^2 + \frac{M}{2}\sum_{i=1}^{N} \varepsilon_i / t_i^2 \\ \text{s.t.} \quad & h(x_i)\beta = y_i - \varepsilon_i \;\; i = 1, 2, \ldots, N \end{aligned} \tag{7}$$

The modified ELM model, named with cost-sensitive ELM model, can be seen as a quadratic programming problem, where globally optimal output matrix β can be determined by interior point method. Moreover, this transformation can also be expanded to other ML methods because only training objective is modified.

In actual implementation, model-driven method and data-driven method are integrated in serial through decision table. In model-driven calculation part, classical model based IEEAC method is applied as usual to form decision table. After faults, ELM based data-driven correction part is activated and finally output corrected CCT values. The whole process can be easily implemented and tested in actual engineering. Complete working flow is shown in Fig. 4 below.

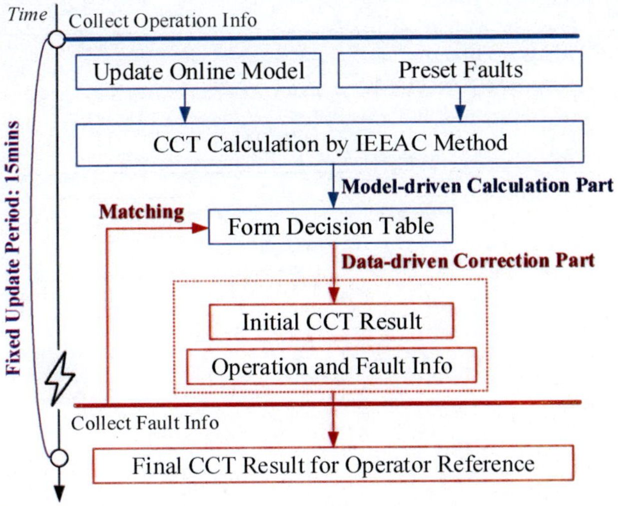

Fig. 4. Online working flow for integrated method

4 Case Study

The proposed method is tested in WSCC 9-bus and New England 39-bus test system. First of all, effects of IEEAC and common ELM integration are demonstrated. Secondly, performance of cost-sensitive ELM model is presented and discussed.

In proposed method, input features for ELM model are chosen as CCTs calculated by IEEAC based on classical generator model, pre-disturbance generator active and reactive power, instantaneous voltage amplitude and phase variation of post-disturbance and pre-disturbance generator buses. Output result is set as the actual CCT value. Simulations are carried out by a computer configured with Inter(R) Core(TM) i5-6200U CPU @2.30 GHz, 16 GB in hardware and Matlab PST v3.0 in software.

4.1 Sample Generation and Evaluation Indexes

Samples are generated by Monte Carlo sampling method based on Matlab PSTv3.0 software. Fault type is set as three-phase short-circuit fault. To simulate different operation scenarios of power system, critical operation parameters are assumed to follow a certain probability distribution. The load level of each bus ranges from 80% to 120% of base load. Fault locations is assumed to be spread over the whole system with uniform distribution. On this basis, 800 samples for WSCC 9-bus test system and 1500 samples for New England 39-bus test system are generated separately.

'10-fold cross-validation' is used for validation and so each test set contains 80 and 150 samples for WSCC 9-bus and New England 39-bus test system. Optimal hidden neurons for ELM models are determined by grid search method. Four evaluation indexes, including mean absolute error (MAE), mean absolute percentage error (MAPE), root-mean squared error (RMSE) and coefficient of determination (R^2), are referred in this paper and their calculation formulas are shown as follows.

$$
\begin{aligned}
E_{\mathrm{MAE}}(f;D) &= \frac{1}{m}\sum_{i=1}^{m}|f(x_i)-y_i| \\
E_{\mathrm{MAPE}}(f;D) &= \frac{1}{m}\sum_{i=1}^{m}\frac{|f(x_i)-y_i|}{|f(x_i)|} \\
E_{\mathrm{RMSE}}(f;D) &= \sqrt{\frac{1}{m}\sum_{i=1}^{m}(f(x_i)-y_i)^2} \\
R^2 &= 1-\frac{\frac{1}{m}\sum_{i=1}^{m}(f(x_i)-y_i)^2}{\frac{1}{m}\sum_{i=1}^{m}(\bar{y}_i-y_i)^2}
\end{aligned}
\tag{8}
$$

Where $D = \{(x_1, y_1), (x_2, y_2, \ldots, (x_m, y_m)\}$ is the test sample set. m is the number of test samples. y_i is the actual value of x_i and $\bar{y}_i$ is the mean value of y_i. f is the prediction function.

4.2 Validation for Proposed Integrated Method

Importance Analysis of IEEAC Results. In this part, common ELM algorithm is applied to verify the performance of proposed integrated method.

For WSCC 9-bus test system, hidden neuron number is set as 100 by grid searching method. The absolute error of IEEAC and proposed integrated method is shown in Fig. 5 below.

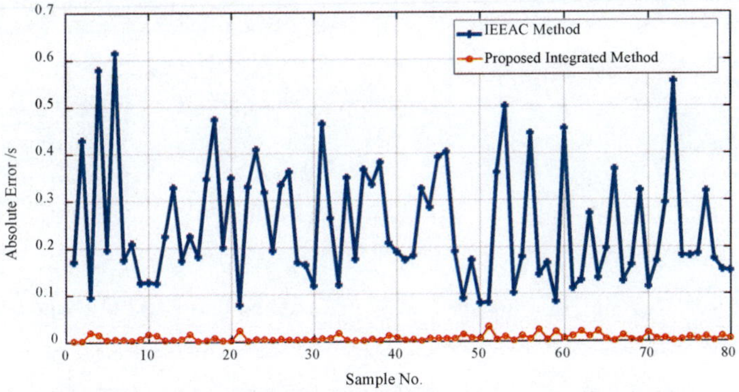

Fig. 5. Error comparison of IEEAC and Integrated methods

It can be inferred from the figure that proposed integrated method is much better than IEEAC method whose error is caused by model difference. Error caused by model difference varies with operation scenarios making it difficult to correct by experience. More specifically, MAE index declines from 0.2456 s for IEEAC method to 0.0076 s for integrated method. It indicates that model difference caused error of IEEAC method can be reduced by ELM based error correction part.

Validation is further tested on New England 39-bus test system where optimal number of hidden neurons is set as 283. Performances of IEEAC and integrated method are compared with four indexes shown in table below.

Table 1. Comparison of IEEAC and integrated method

Method type	MAE/s	MAPE/%	RMSE/s	$R^2/1$
IEEAC method	0.0787	16.55	0.1247	0.5125
Integrated method	0.0209	4.93	0.0342	0.9628

From performance in both WSCC 9-bus and New England 39-bus test system, it can be inferred that error of IEEAC caused by model difference can be corrected by ELM method through integration by constructing the relationship between error of IEEAC and operation scenarios, which shows greatly improved accuracy.

In addition, importance of IEEAC results in the integrated method is also considered. In the integrated method, IEEAC results are used as input features. Both effects of IEEAC results and sample quantity are demonstrated in figure below.

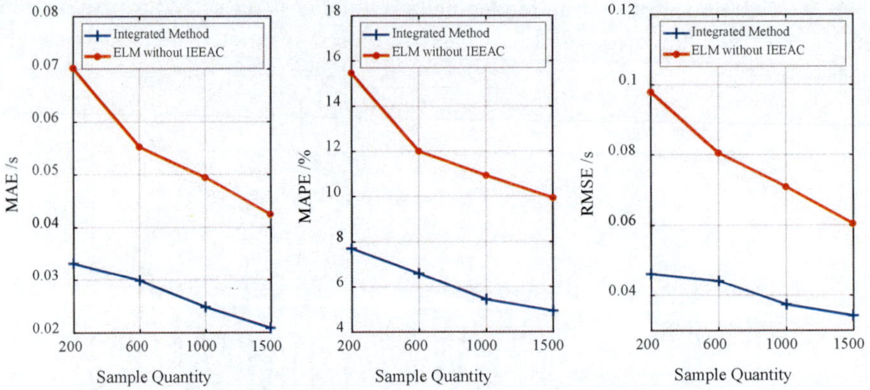

Fig. 6. Comparison between integrated method and ELM only method

It can be inferred from Fig. 6 that overall performance of integrated method where IEEAC result is considered, is better than ELM only method where IEEAC result is not considered. The smaller the sample quantity is, the more obvious the advantage in accuracy will be. Hence, it can be concluded that result of IEEAC method is also an important part for proposed integrated method.

Performance Analysis of Cost-Sensitive ELM Model. From above description, it can be concluded that the overall performance of integrated method based on common ELM model is good. Nevertheless, the accuracy is not always good if results are analyzed in different actual CCT value regions. For example, results can be analyzed based on four regions of actual value distribution. Table 2 shows the specific MAE and MAPE indexes in the four regions.

It can be found in Table 2. that MAPE index for region with low actual CCT values is obviously larger than other regions. The reason lies in that training objective of common ELM model is set as the summation of output weight matrix and training error, i.e., error importance is treated equally.

Table 2. Result analysis for integrated method with common ELM model

CCT region	Sample quantity	MAE/s	MMAPE/%
CCT < 0.3	26	0.0223	10.67
0.3 < CCT < 0.5	49	0.0167	4.03
0.5 < CCT < 0.8	68	0.0194	3.22
CCT > 0.8	7	0.0418	4.46

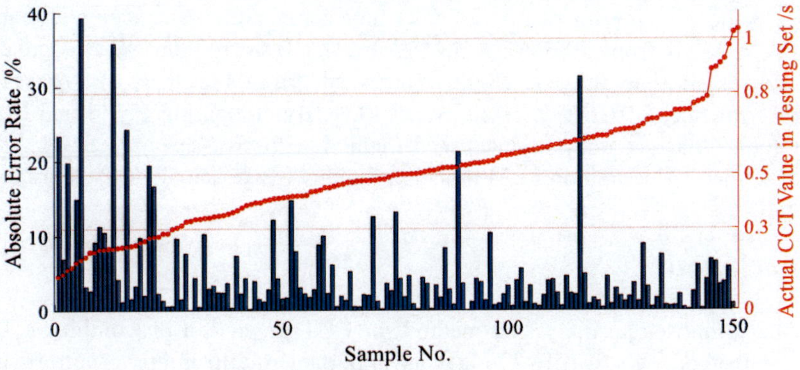

Fig. 7. Error rate distribution of integrated method with common ELM model

Curves in Fig. 7 show that absolute error rate for regions with low actual values is obviously high. Hence, regions with lower actual values should be concentrated more.

Proposed cost-sensitive ELM model, shown in (7), is validated in the same training and testing set. On the whole, MAE, MAPE and RMSE indexes for this model are 0.0172 s, 3.57%, 0.0315 s respectively, which is better than common ELM model by comparing with Table 1. On the local, error for regions with low CCT values is largely reduced with little increase for regions with large CCT values. Detail results are shown in table and figure below.

Table 3. Result analysis for integrated method with cost-sensitive ELM model

CCT region	Sample quantity	MAE/s	MMAPE/%
CCT < 0.3	26	0.0083	4.11
0.3 < CCT < 0.5	49	0.0119	2.97
0.5 < CCT < 0.8	68	0.0158	2.63
CCT > 0.8	7	0.0507	5.24

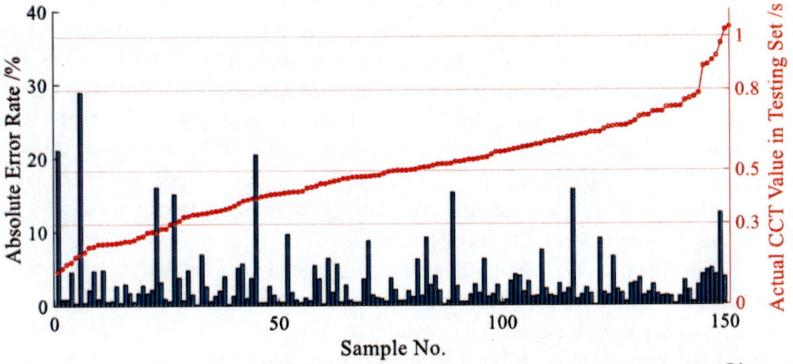

Fig. 8. Error rate distribution of integrated method with cost-sensitive ELM model

Table 3 shows that error rate in regions where actual CCT values are less than 0.3 s is greatly reduced from 10.67% to 4.11%. Error rate in regions where actual CCT values are larger than 0.8 s is slightly increased. From Fig. 8, it can be seen that absolute error rates of most cases are below 10%. By comparing Fig. 7 and Fig. 8, it can be found that accuracy of integrated method with cost-sensitive ELM model is superior to that with common ELM model in regions where actual CCT values are low.

5 Conclusion

This paper proposes an integrated method for CCT prediction by combining IEEAC and ELM method, where IEEAC plays an important role to improve performance of ELM method, and further training objective of ELM training model is modified to improve performance in regions with low actual CCT values. Validations show effectiveness of proposed integrated CCT prediction method and cost-sensitive ELM model. In future, effects of measurement uncertainty, model and topology change should be further considered before actual application.

Acknowledgement. This research is supported by the science and technology project of state grid corporation of China (Project Name: Research and Application of Online Risk Identification and Auxiliary Decision-making Technology in Power Gird Stability Control System, 201935003A).

References

1. Savulescu, S.C.: Real-Time Stability in Power Systems: Techniques for Early Detection of the Risk of Blackout, 2nd edn. Springer, New York (2014)
2. Mandozzi, M., Menditto, V., Moreschini, G., et al.: Recent improvements of emergency automatic control of Enel power system in interconnected and isolated operation. In: International Conference on Large High Voltage Electric Systems (CIGRE Session), Paris, France, pp. 39–302 (1992)
3. Xue, Y., Chen, Y.: Emergency control by using on-line refreshed decision tables and uncertain knowledge on model effects. In: IFAC Symposium on Control of Power Plants and Power Systems, vol. 28, no. 26, pp. 373–378 (1995)
4. Zhao, L., Zhou, M., Fan, Z., et al.: Architecture of online analysis system for power grid oriented to rapid response. Autom. Electr. Power Syst. **43**(18), 95–101 (2019). (in Chinese)
5. Kunder, P.: Power System Stability and Control. McGraw-Hill, New York (1994)
6. Pai, M.A., Padiyar, K.R., Radhakrishna, C.: Transient stability analysis of multi-machine AC/DC power systems via energy-function method. IEEE Trans. Power Apparatus Syst. PAS **100**(12), 5027–5035 (1981)
7. Vu, T.L., Turitsyn, K.: Lyapunov functions family approach to transient stability assessment. IEEE Trans. Power Syst. **31**(2), 1269–1277 (2015)
8. Xue, Y., Van Custem, T., Ribbens Pavella, M.: Extended equal area criterion justifications, generalizations, applications. IEEE Trans. Power Syst. **4**(1), 44–52 (1989)
9. Xue, Y., Lyu, R., Xue, F., et al.: The qualitative criterion of transient angle stability. In: 10th International Conference on Advances in Power System Control, Operation & Management (APSCOM 2015), Hong Kong, China, pp. 1–6 (2015)

10. Xue, Y.: Integrated extended equal area criterion–theory and application. In: 5th Symposium Specialists in Electric Operational and Expansion Planning, Recife, Brazil, pp. 1–6 (1996)
11. Karami, A.: Power system transient stability margin estimation using neural networks. Int. J. Electr. Power Energy Syst. **33**(4), 983–991 (2011)
12. Hu, W., Lu, Z., Wu, S., et al.: Real-time transient stability assessment in power system based on improved SVM. J. Mod. Power Syst. Clean Energy **7**(1), 26–37 (2019)
13. Sulistiawati, I.B., Priyadi, A., Qudsi, O.A., et al.: Critical clearing time prediction within various loads for transient stability assessment by means of the extreme learning machine method. Int. J. Electr. Power Energy Syst. **77**, 345–352 (2016)
14. Wang, Q., Li, F., Tang, Y., et al.: Integrating model-driven and data-driven methods for power system frequency stability assessment and control. IEEE Trans. Power Syst. **34**(6), 4557–4568 (2019)
15. Xue, Y.: Extended equal area criterion: foundations and applications. In: Invited paper in 4th Symposium of Specialists in Electric Operational and Expansion Planning, Brazil (1994)
16. Huang, G.B., Zhu, Q.Y., Siew, C.K.: Extreme learning machine: theory and applications. Neurocomputing **70**(1–3), 489–501 (2016)
17. Huang, G.B., Zhou, H., Ding, X., et al.: Extreme learning machine for regression and multiclass classification. IEEE Trans. Syst. Man Cybern. Part B (Cybern.) **42**(2), 513–529 (2012)

Power System Cascading Outages Risk Assessment Considering Failure of Control Actions

Feng Li and Qi Wang(✉)

Southeast University, Nanjing 210096, China
wangqi@seu.edu.cn

Abstract. In traditional risk assessment method, failure of control actions is always neglected due to its low occurrence probability. However, failure of control actions usually leads to power system cascading outages, where severe consequence has an impact on risk assessment results. Hence, it is necessary to consider control actions failure in power system risk assessment. In this paper, relay protection failure is chosen as an example of control actions failure and considered in power system risk assessment. In the risk assessment procedure, the probability evaluation of relay protection failure and system risk computation method are concerned. Firstly, a method for relay protection failure probability evaluation is presented, where intrinsic property and operating state of relay protection devices are combined. Secondly, an improved state space partitioning (SSP) method, suitable for relay protection failure involved power system cascading outages risk assessment, is put forward. The improved SSP method transforms sampling strategy of sampling process in common SSP method, utilizing enumerated state help low-probability event sampling. It enables to avoid the low-efficiency problem caused by extreme low-probability event in common SSP method based risk assessment. Finally, relay protection failure probability evaluation method and the improved SSP method are validated with simulations and results show effectiveness in risk assessment accuracy and efficiency.

Keywords: Cascading outages · Risk assessment · State space partitioning · Relay protection failure

1 Introduction

Uncertainties and risks in power system operation increase along with power system development, reflected in increase of catastrophic cascading outages occurrence times [1]. The cascading outage usually occurs in a short time, causing a series of devices breaking down and finally resulting in large-scale blackout. Accident analysis shows that most large-scale blackouts are related with failure of control actions [2, 3].

In practical power systems, relay protection unit is the basic equipment for power system stability control. It usually has high reliability but is also affected by some factors, e.g. device aging, extreme weather, intentional cyber-attack. Besides, the cost of consequence is usually high, once failure occurs in relay protection action. Hence, it

Y. Xue et al. (Eds.): PMF 2020, LNEE 718, pp. 582–594, 2021.
https://doi.org/10.1007/978-981-15-9746-6_45

is necessary to involve effect of relay protection action in power system risk assessment [4, 5].

It has been widely recognized that failure of relay protection is sure to aggravate effect of disturbance. In [6] and [7], effects of relay protection hidden failures on cascading outages are analyzed, while the failure probability evaluation for relay protection neglects effect of actual operating state and only considers historical statistical information. Failure probability evaluation of relay protection is further researched in [7], where action logics of relay protection are considered. However, effects caused by external factor are still excluded. In [8], K-means algorithm is used to analyze historical defect data from the perspective of location, cause and severity of defect and quantitative indicators of relay protection defect's characteristics are put forward. In [9], a searching method for high-risk N-k contingencies is proposed based on substation topology and relay protection hidden failure evaluation information. This searching method relies on small probability analysis technique and event tree method, which has the essential of state enumeration method (SEM). It is hardly to be used for power system integral risk assessment. In [10, 11], action logics of relay protection devices are considered to analyze cascading outage propagation mechanism, based on which Monte Carlo sampling method (MCS) is utilized for power system risk assessment. However, the sampling efficiency could be low because of the low occurrence probability of relay protection failure action. In [12], state space partitioning (SSP) method, combing SEM and MCS method, is applied to improve computation efficiency of power system risk assessment. Nevertheless, effects of relay protection and cascading outages are not considered. SSP method divides total state space into two subspaces, i.e. state emulation subspace and state sampling subspace. These two subspaces are totally independent, which challenges application of SSP method in risk assessment that considers low-probability events. Because a lot of useless and repetitive states are sampled in state sampling subspace. Hence, it is meaningful to improve traditional SSP method to adapted to low-probability event involved power system risk assessment. However, few related researches are presented.

In this paper, failure probability evaluation of relay protection action and cascading outages involved risk assessment are concerned. Main contributions of this paper lie in two aspects. On one hand, a failure probability evaluation method for relay protection action is proposed. It utilizes historical statistical information to roughly evaluate failure action probability caused by intrinsic property of relay protection unit, and further polish this probability with actual operating state information. On the other hand, an improved SSP method that considers effect of low-probability event is proposed. In this method, low-probability event sampling, e.g. failure of relay protection action, is also executed in state enumeration subspace to improve sampling efficiency.

Remaining of this paper is organized as follows. In Sect. 2, a probability evaluation method for relay protection failure action is put forward based on failure action factors analysis of relay protection units. The improved SSP method for cascading outages involved power system risk assessment is presented in Sect. 3. Case study and conclusion are presented in Sect. 4 and 5 respectively.

2 Failure Probability Evaluation of Relay Protection Action

Relay Protection is a unit of different function modules that take effects in fault isolation. It usually contains instrument transformers, relays and breakers. Function modules of relay protection and their relationships are presented in following figure.

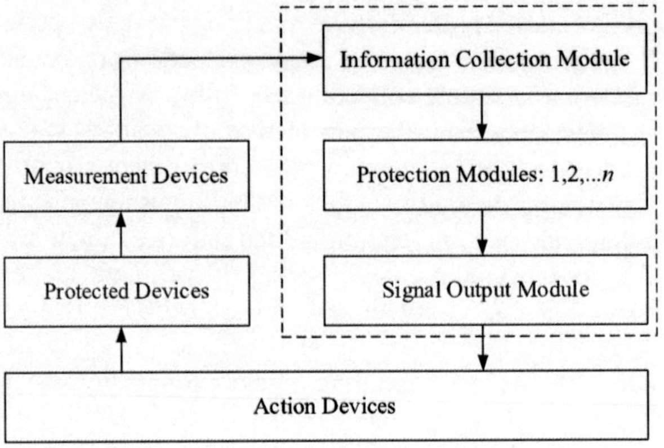

Fig. 1. Components of relay protection unit and their relationships

In the dashed rectangular area, it is the software function part of digital relays. According to constitution of relay protection unit, causes of failure action can be concluded with following two main categories [13, 14].

- The main reason lies in intrinsic property of relay protection unit, e.g. hardware defects, software design flaws, unreasonable setting and etc. Failure action probability caused by this kind of reasons is usually fixed or regular if no further maintenance is made, which can be inferred from historical records.
- The other reason is related with operating state of relay protection unit, e.g. extreme weather condition, abnormal electromagnetic environment, spiteful cyber-attacks and etc. Failure action caused by these reasons is usually rare in historical records. But their quantitive relationship with failure action probability can be estimated by certain theoretical models.

Hence, evaluation of relay protection failure action probability can be realized by two steps. In the first step, intrinsic property is considered and historical information is utilized to describe the fundamental failure action probability P_B. In the second step, actual operating state information provided by sensors are taken into consideration, figuring out correction coefficient C_A, to polish the fundamental failure action probability. So the actual failure action probability can be estimated with following equation.

$$P_{fail} = 1 - C_A \cdot (1 - P_B) \tag{1}$$

2.1 Fundamental Failure Action Probability Calculation

Relay protection unit consists of various components, whose reliability is strongly related with the performance of relay protection unit. In relay protection unit, modules can be divided into 6 categories according to function differences, including power supply module (PSM), measurement module (MM), hardware based data processing module (HBDPM), software based data processing module (SBDPM), communication module (CM) and action execution module (AEM).

From Fig. 1, it can be concluded that relay protection failure action happens if any one module loses effect. Hence, the failure rate of relay protection can be seen as the summation of each module failure rate, shown in equation below.

$$\lambda_B = \lambda_{PSM} + \lambda_{MM} + \lambda_{HBDPM} + \lambda_{SBDPM} + \lambda_{CM} + \lambda_{AEM} \tag{2}$$

Where, λ_B is the equivalent failure rate of relay protection unit, in unit of times/year. λ_{PSM}, λ_{MM}, λ_{HBDPM}, λ_{SBDPM}, λ_{CM} and λ_{AEM} are recorded failure rate of each function module respectively.

Finally, the exponential model is utilized to construct the relationship between failure action probability and failure rate statistical data of relay protection, shown in following equation.

$$P_B = 1 - \mathrm{e}^{-(\lambda_B \cdot t)/\mathrm{T}} \tag{3}$$

Where t is the time scale concerned in risk assessment and T is the total hours of a year.

2.2 Operating State Based Correction Coefficient Calculation

Besides intrinsic property, the external factors also have impact on performance of relay protection. These external factors can be divided into two categories. One is the operating environments, e.g. humidity, temperature and etc., which affects reliability of relay protection components. The other is the operating condition, e.g. power grid operation mode change, protection settings tampering by cyber-attacks and etc., which affects reliability of relay protection action logics.

These two kinds of external factors have the property that the reliability is high if operating states vary in rated ranges. Otherwise, the reliability will sharply decrease. This key property can be described by the super-Gaussian distribution, with following curve shape.

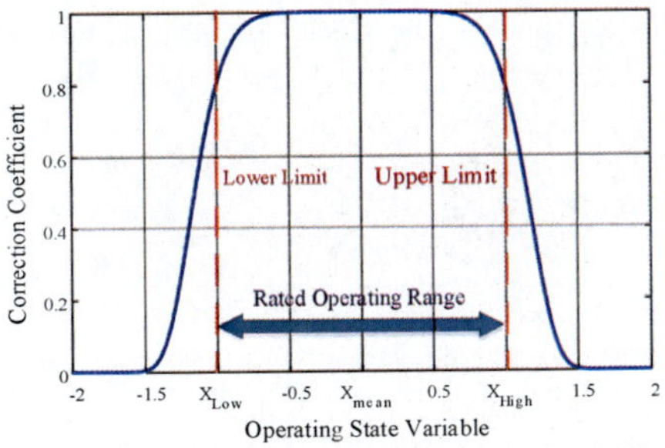

Fig. 2. Response curve of correction coefficient with operating state variable

It can be seen that if operating state is in [X_{LOW}, X_{HIGH}], the reliability is not affected. However, the reliability decreases a lot if operating state goes far beyond requirements. The calculation function has following forms.

$$C_{Ai} = \exp\left(-\left(\frac{\alpha(x_i - m_i)}{w_i}\right)^N\right) \tag{4}$$

Where C_{Ai} is the correction coefficient considering the *i-th* operating state factors. x_i is the measurement from sensors of the *i-th* operating state factors. w_i is the rated operating range of the *i-th* operating state factors and m_i is the middle value. N is the curve shape parameter, usually set as 8 or 10. α can be figured out according to the pre-given correction coefficient value at the operating state bound (Fig. 2).

By determining reasonable rated ranges of operating state, the correction coefficient can reflect the effect of either operating environments and operating conditions.

3 Improved SSP Method for Relay Protection Failure Action Involved Risk Assessment

SSP method divides full state space into two subspaces, where SEM and MCS method are utilized separately to select states. In this way, contingencies with high probability can be screening by SEM method, and multiple contingencies with low probability are sampled by MCS method.

Figure below illustrates the principle of SSP method. The total state space S can be divided into two subspaces, i.e. state space with low risk high probability (left side of the vertical red dash line) and the state space with high risk low probability (right side of the vertical red dash line). Firstly, fast sorting technique (FST) can be used to search possible states, forming state enumeration subspace S_F, underlined by the rectangular area with gird background. In this step, selected states usually have low risk and high probability. The rest states form sampling subspace S_M, and these states usually have high risk and low probability. These states are selected by sampling and the sampled state in common SSP method is presented in form of square black block.

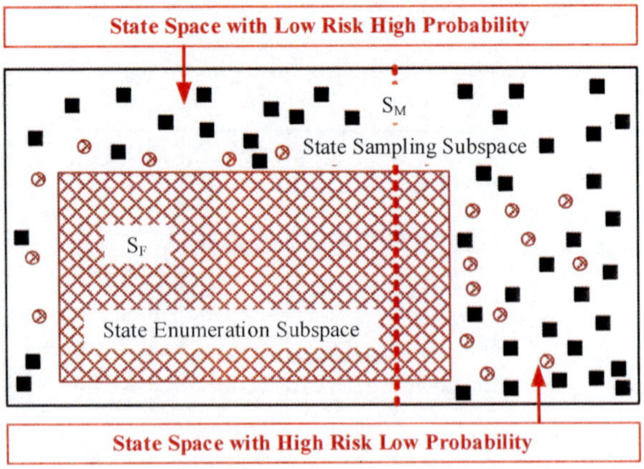

Fig. 3. Illustration of improved SSP method

In common SSP method, common probability events rather than low probability events are considered, whose states can be sampled by MCS method. However, low probability events exist in power system, e.g. relay protection failure action caused cascading outages. For low probability events, the MCS in common SSP method could loss effect. Hence, an improvement is made on common SSP to realize low-probability event involved risk assessment.

The improvement is made in MCS sampling process. When the sampled state belongs to state enumeration subspace, low probability event sampling is further executed based on the sampled state, represented with red circle in Fig. 3. By this improvement, the condition probability increases for low probability event sampling. While the same sampling states will be dropped directly in common SSP method.

3.1 Theoretical Foundation for Improved SSP Method

As mentioned above, the improved SSP method contains two steps to select power system operating states, which is similar to common SSP method.

- **First Step:** FST is utilized to select power system operating states according to operating state probability and evaluate the risks of selected states. Repeat selection and evaluation process until the accumulated probability of selected states reaches preset threshold, by which subspace S_F is formed. The threshold for accumulated probability of enumerated states is set as the upper limit probability α.
- **Second Step:** take MCS method to evaluate rest state space. Repeat sampling process to obtain states excluded by subspace S_F and relay protection failure action state, until computation precision meets requirement. Hence, the accumulated probability of sampled states is set as the rest probability 1-α.

The FST enables to pick out states with high probability without searching the total state space [15]. Considering the low-probability property of cascading outages caused by relay protection failure action, the traditional calculation formula is transformed to following form.

$$\begin{aligned} \alpha &= \sum_{x_i \in S_F} P(x_i)\beta(x_i^j) \\ \beta(x_i^j) &= 1 - \prod_{j \in i} P(x_i^j) \end{aligned} \tag{5}$$

Where $P(x_i)$ is the operating state x_i occurrence probability. $\beta(x_i^j)$ is the correction coefficient, which represents the effect of relay protection failure action. X_i^j is the sampled *j-th* relay protection action state related with operating state x_i.

Expectation of the systematical risk assessment index can be calculated with following condition probability formula.

$$E(F) = \alpha E[F(x)|x \in S_F] + (1 - \alpha)E[F(x)|x \in S_M] \tag{6}$$

Where $F(x)$ is the load loss under operating state x. $E[F(x)|x \in S_F]$ is the risk assessment index calculated by SEM method and $E[F(x)|x \in S_M]$ is the risk assessment index calculated by MCS method. They are determined with following equations respectively.

$$E[F(x)|x \in S_F] = \frac{1}{\alpha} \sum_{x_i \in S_F} F(x_i)P(x_i) \quad (7)$$

$$E[F(x)|x \in S_F] = \frac{1}{N} \sum_{x_i \in S_M} F(x_i) \quad (8)$$

Where N is the total sampling times of MCS method.

3.2 Risk Assessment Scheme for Relay Protection Failure Action Caused Cascading Outages

With improved SSP method, the risk assessment method considering relay protection failure action can be realized by following working flow. It mainly contains two parts.

In the first part, FST is used to enumerate states until preset upper probability is reached. In the second part, Monte Carlo sampling method is used to select states.

If the sampled state has emerged in enumerated subspace, further sampling the state of relay protection action. Sampling relay protection action state until no relay protection failure action happens. After a new state excluded by enumerated subspace is selected, calculate the consequence index, and update consequence index expectation and convergence index of all sampled states. Sampling new operating states until convergence index meet requirement.

Finally, output the risk assessment index (Fig. 4).

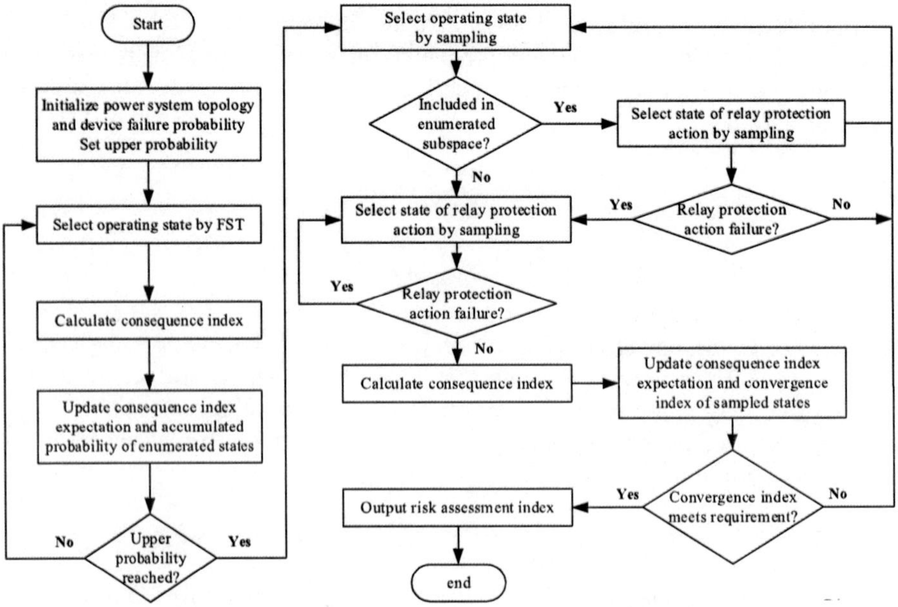

Fig. 4. Working flow of risk assessment considering relay protection failure action

4 Case Study

IEEE RTS-96 system is used as the testbed [16]. It contains 12 generator unit, 33 transmission lines, 5 transformers and 24 buses. Installed capacity of the system is 3405 MW and peak load capacity reaches 2850 MW. System topology is shown in Fig. 5.

In this paper, initial faults at generator, bus and transmission line are concerned. Cascading outages caused by relay protection failure action are considered when initial faults happen at buses or transmission lines.

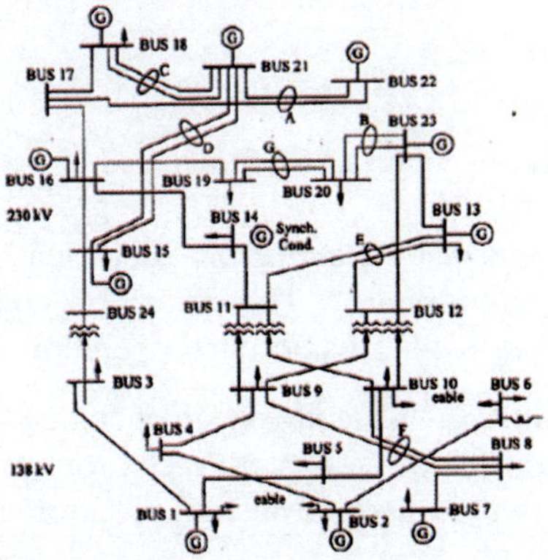

Fig. 5. Topology of IEEE RTS-96 test system

4.1 Probability Evaluation of Relay Protection Failure Action

In [14], annual failure rate for relay protection unit is given as 0.009772 times per year. The fixed failure probability of relay protection then can be figured out according to exponential model, which finally turns 0.0027%. This probability can be seen as the reflection of devices intrinsic property. Further, actual operating state of relay protection unit is considered to polish relay protection failure action probability.

To describe the effect of operating state on relay protection action, the scenarios are divided into categories. One the is effect of operating environment and the other is the operating condition.

Effect of Operating Environment. It is assumed that the rated operating environment ranges in [−1, +1] and the middle value is 0. The curve shape parameter is set as 8 and the correction coefficient at the bound is set as 0.98. Then the correction coefficient can be figured out with following mapping function.

$$C_{AEi} = \exp\left(-\left(\frac{1.2280(x_i)}{2}\right)^8\right) \tag{9}$$

With this function, variation of correction coefficient with operating environment can be described with curves in following figure.

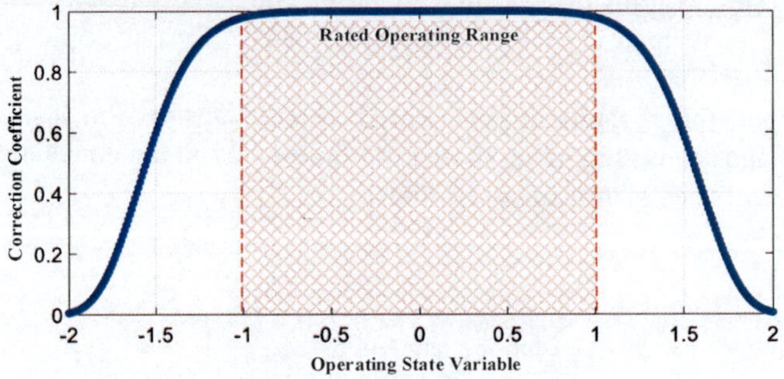

Fig. 6. Relationship between correction coefficient and operating environment variable

It can be seen from Fig. 6 that the correction coefficient is 1 when relay protection works in rated operating environment. The correction coefficient sharply decreases when operating environment is far beyond rated operating environment.

Effect of Operating Condition. Calculation of operating condition correction coefficient is similar to that considered in operating environment. The effect of operating condition mainly considers the effect of operating criterion value in relay protection action.

The rated operating condition range is set as [1] and the middle value is 0.5. The curve shape parameter is set as 8 and the correction coefficient at the bound is set as 0.98. The function for correction coefficient calculation under effect of operating condition is in following form (Fig. 7).

$$C_{AC} = \exp\left(-\left(\frac{2.456\,(x-0.5)}{2}\right)^{8}\right) \tag{10}$$

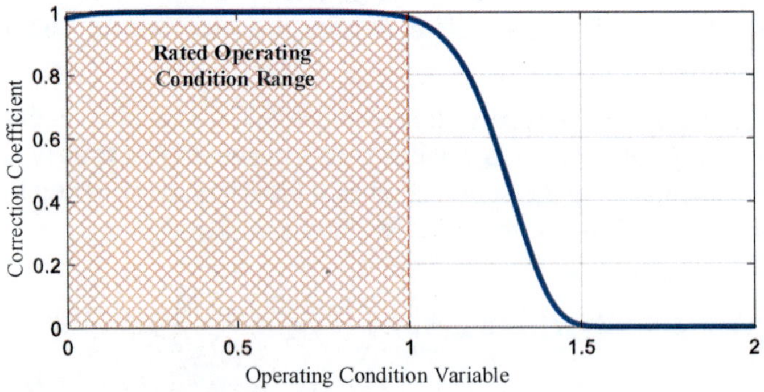

Fig. 7. Relationship between correction coefficient and operating condition variable

From above figure, it shows that the correction coefficient keeps 1 when operation condition is in rated range, while it decreases sharply when operating condition exceeds normal state. It accords with the fact that states at the bound are easier to be affected by operation condition change.

4.2 Performances of Improved SSP Method in Risk Assessment

In this paper, expected power not supplied (EPNS) index is used for risk assessment. In Table 1, performances of SEM, MCS and Improved SSP are summarized. In the improved SSP method, the upper probability α is set as 0.8. The convergence criterion for MCS and improved SSP method is that variance of EPNS is not over 0.1%.

Table 1. Performances comparison between different methods

Method type	EPNS/p.u	Computation times/1
SEM method	0.3013	50001
MCS method	0.7953	102970
Improved SSP method	0.7974	19870

It can be seen from the table that results of SEM Method is different from the other two methods. The reason lies in that the occurrence probability of relay protection failure action becomes lower and lower with the enumerating process, causing enumerated state subspace is not complete. The accumulated probability variation of enumerated states is shown in Fig. 8 below.

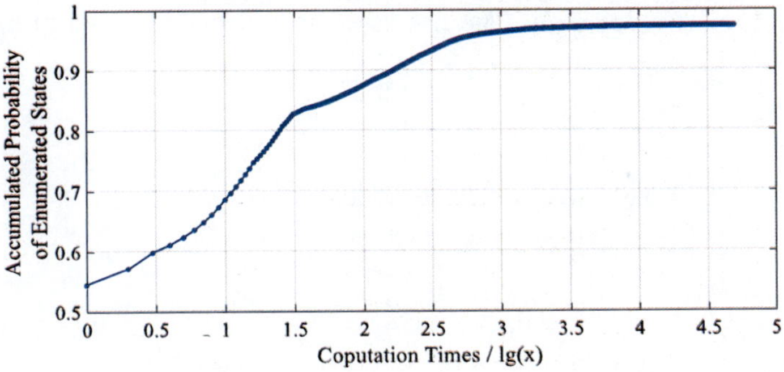

Fig. 8. Relationship between enumerated states accumulated probability and computation times

The above figure shows that the accumulated probability of enumerated states is hardly increase when certain computation times is reached. Hence, SEM method alone is not suitable in this risk assessment scenario.

Meanwhile, MCS and improved SSP method have similar EPNS index, shown in Table 1. However, the computation times of improved SSP is less than one fifth of MCS method. Hence, the improved SSP method is superior to both SEM and MCS methods.

Sensitivity Analysis of Upper Probability α. In this part, effect of upper probability on improved SSP method is tested and results are summarized in Table 2 below.

Table 2. Effect of upper probability on performance of improved SSP method

Upper probability	Enumerated states	Sampled states	EPNS index
0.8	25	19845	0.7974
0.9	168	10348	0.7754
0.95	485	3781	0.7809
0.97	3671	1979	0.7826
0.971	5467	2136	0.7972
0.972	11385	2019	0.7920

Table above shows that the EPNS index are similar even the upper limit probability varies, while the states provided by SEM method increases. The reason lies in that the upper probability determines the combination level of SEM and MCS method.

Sensitivity Analysis of Failure Action Probability Considering the variation of relay protection failure action probability, the EPNS index of risk assessment under different extreme conditions should be different. In this part, four special scenarios are considered, by combining 1.05, 1.10 times operating environment, and normal and bound operating condition. The operating environment and operating condition here represent abstract concepts, and their appropriate correction coefficients can be calculated with Eq. (9) and Eq. (10) separately. The risk assessment result is shown in figure below.

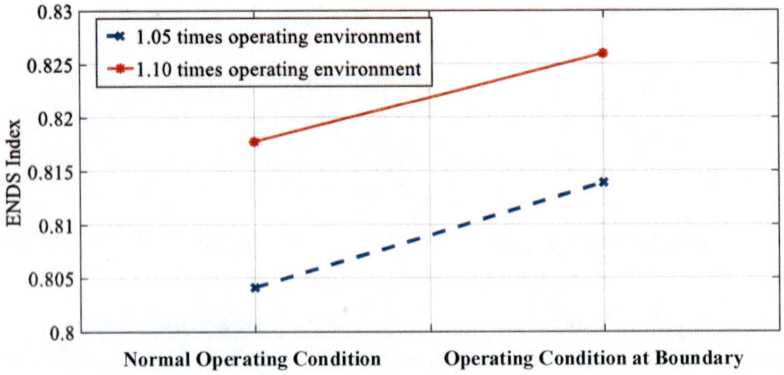

Fig. 9. Effect of failure probability variation on EPNS index calculated by improved SSP method.

In Fig. 9, the EPNS index calculated by improved SSP method in risk assessment shows obvious relation with relay protection failure action probability, even though the failure action probability is not in a large. Hence, it indicates that the improved SSP method is effective to consider the variation of devices failure action probability.

5 Conclusion

This paper proposes a cascading outages risk assessment method considering failure action of relay protection. In this method, failure action probability of relay protection is polished with actual operating state. Moreover, the traditional SSP method is improved to enhance sampling performance when low probability events are considered. This improved SSP method is expected to take good effects in risk assessment scenarios where rate events are involved, e.g. extreme weather and cyber-attack. However, mal-operation of relay protection is not considered in this paper, which deserves further research.

Acknowledgement. This work was supported by the Project of State Grid Corporation of China (Project Name: the Identification Method of Malicious Attack on the Power Grid and Active Defense Method on the Power Grid Side).

References

1. Zeng, B., Ouyang, S., Zhang, J., et al.: An analysis of previous blackouts in the world: lessons for China' s power industry. Renew. Sustain. Energy Rev. **42**, 1151–1163 (2015)
2. Elizondo, D.C., de La Ree, J., Phadke, A.G., et al. Hidden failures in protection systems and their impact on wide-area disturbances. In: 2001 IEEE Power Engineering Society Winter Meeting, pp. 710–714. IEEE, Columbus (2001)
3. Yi, J., Bu, G., Guo, Q., et al.: Analysis on blackout in Brazilian power grid on March 21, 2018 and its enlightenment to power grid in China. Autom. Electr. Power Syst. **43**(2), 1–6 (2019) (in Chinese)
4. Mili, L., Qiu, Q., Phadke, A.G.: Risk assessment of catastrophic failures in electric power systems. Int. J. Crit. Infrastruct. **1**(1), 38–63 (2004)
5. Ning, L.: Relay protection hidden fault monitoring and risk analysis based on the power system. In: International Conference on Advanced Intelligent Systems and Informatics, pp. 991–997. Springer, Cairo (2019)
6. Chen, J., Thorp, J.S., Dobson, I.: Cascading dynamics and mitigation assessment in power system disturbances via a hidden failure model. Int. J. Electr. Power Energy Syst. **27**(4), 318–326 (2005)
7. Yu, X., Singh, C.: A practical approach for integrated power system vulnerability analysis with protection failures. . IEEE Trans. Power Syst. **19**(4), 1181–1820 (2004)
8. Wu, C., Wang, W., Zhang, L., et al.: Analysis of typical modes of relay protection defects based on k-means algorithm. In: 2019 International Conference on Electronic Engineering and Informatics (EEI), pp. 499–505. IEEE, Nanjing (2019)
9. Chen, Q., McCalley, J.D.: Identifying high risk N-k contingencies for online security assessment. IEEE Trans. Power Syst. **20**(2), 823–834 (2005)

10. Yu, X., Singh, C.: Power system reliability analysis considering protection failures. In: IEEE Power Engineering Society Summer Meeting, pp. 963–968. IEEE, Chicago (2002)
11. Liu, X., Shahidehpour, M., Li, Z., et al.: Power system risk assessment in cyber attacks considering the role of protection systems. IEEE Trans. Smart Grid **8**(2), 572–580 (2016)
12. He, J., Sun, Y., Kirschen, D.S., et al.: State-space partitioning method for composite power system reliability assessment. IET Gener. Trans. Distrib. **4**(7), 780–792 (2010)
13. Cheng, Y., Chen, X., Ren, J., et al.: Study on hidden failure of relay protection in power system. In: 2013 IEEE International Conference on Cyber Technology in Automation, Control and Intelligent Systems, pp. 434–439. IEEE, Nanjing (2013)
14. Zhao, L., Li, X., Ni, M., et al.: Review and prospect of hidden failure: protection system and security and stability control system. . J. Mod. Power Syst. Clean Energy **7**(6), 1735–1743 (2019)
15. Liu, H., Sun, Y., Wang, P., et al.: A novel state selection technique for power system reliability evaluation. Electr. Power Syst. Res. **78**(6), 1019–1027 (2008)
16. Grigg, C., Wong, P., Albrecht, P., et al.: The IEEE reliability test system-1996. A report prepared by the reliability test system task force of the application of probability methods subcommittee. IEEE Trans. Power Syst. **14**(3), 1010–1020 (1999)

A Load-Shedding Algorithm for Power System Based on Transfer Learning

Bian Daokuan[1](✉), Wang yong[1], and Yang Ning[2]

[1] School of Electrical Engineering, Shandong University, Jinan 250061, Shandong, China
Biandaokuan@163.com
[2] Huaneng Shandong Power Generation Co. Ltd., Jinan 250014, China

Abstract. In power system fault consequence analysis based on intelligent algorithms, usually set the load removal, as well as load-shedding as the main indicator. In general, it is necessary to use many fault samples generated offline to train the model, which can achieve good results under the support of a certain amount of data and training algebra. However, when the topological structure or operation mode of system changes, the accuracy of the trained model will decrease significantly. To solve this problem, transfer learning is introduced and a fast transfer adaptation method for the trained model is proposed based on the convolutional neural network. Firstly, using the data set obtained by the Monte Carlo simulation method to train the pre-trained model. When the structure of the system changes, the network structure of the pre-trained model is kept unchanged, and some network parameters are transferred to the new model to adapt to the system after changes. Finally, IEEE-RTS79 system was used to verify that the transfer learning method has better adaptability to the system with structural changes, and can effectively update the model and reduce the training time of the new model.

Keywords: Transfer learning · Load cutting · CNN

1 Introduction

The operation of the power system has strong randomness, and the fault types and degrees are different. The analysis of the fault consequences of the power system can better support the operation of the power system.

The modern power system is a high dimensional nonlinear system, the failure generates quickly, the type and the kind are also different. Especially in recent years several large power failure accidents [1], which makes the importance of power system failure consequence analysis is becoming more and more important. The main basis of fault consequence analysis is close to the reliability analysis indicator.

Reference [2] is the first article on the reliability analysis of power generation and transmission systems. After decades of continuous research, a series of index systems have been established, and they have been widely used [2, 3]. When the system fails, mainly consider the selective fault removal and prevention of accident expansion [4], In this process, load removal has higher flexibility and adaptability for failure

Y. Xue et al. (Eds.): PMF 2020, LNEE 718, pp. 595–607, 2021.
https://doi.org/10.1007/978-981-15-9746-6_46

consequence analysis. It is a key problem in fault consequence analysis to get the cutting load quickly. It is necessary to use the optimization algorithm to calculate load removal. Reference [5] transforms the load-shedding problem into the optimal control problem; Reference [6] adopted multi-initial value optimization to improve the convergence of the algorithm. Reference [7] uses an improved particle swarm optimization algorithm to optimize load shedding calculations. All the above references need to set up physical models to simulate and calculate according to the characteristics of the power grid. However, with the increasingly complex structure of power networks, it is more and more difficult to construct the model in this way.

In recent years, artificial intelligence algorithms have been widely used in power system analysis by constructing the mapping relation between input and output, such as load forecasting [8], fault diagnosis [9], and power behavior analysis [10]. In terms of risk control, reference [11] constructs a risk assessment based on deep learning, reference [12] uses deep learning to quickly search for high-risk cascading failures. In terms of the calculation of the load removal, reference [13] proposed a fast cutting load method by using the ANN neural network, which can avoid the information error caused by the simplification and approximation of the simulation process and deal with large-scale systems. In reference [14], deep reinforcement learning is applied to the calculation of load-shedding, and better training results can be obtained in the absence of labels.

The above intelligent algorithm has achieved a lot of results, but the deficiency lies in the lack of feature learning and transfer ability. The neural network relies on training data when learning, each time needs to search all the rules. When the system is becoming complicated, some minor changes will lead to obtaining data for training again. This leads to the high accuracy of the model for a single network, but when a new network needs to use the model, it also requires a lot of new data for retraining. If the data is inadequate, it will be difficult for the network to convergence, and the effect is not ideal with long training time. Therefore, if a model can be trained as a universal model when the network structure changes, it will adapt to the new environment quickly, then this will be very valuable. The transfer learning method [15, 16] can effectively solve the above problems by transferring knowledge from the source domain to the target domain.

In recent years, research on transfer learning in the field of power systems has developed to some extent. Reference [17] effectively improves the positioning accuracy by applying transfer learning to the positioning of synchronous oscillation sources. In reference [18], transfer learning is used in fault detection, fault diagnosis, and short-term load prediction of grid users for power line trip, which improves the accuracy and efficiency of power data mining. While improving the efficiency of network training, transfer learning also improves the performance of network calculation. However, its related research on failure consequence analysis has not appeared. This paper introduced transfer learning into the calculation of the load-shedding when the system failure occurred and proposed a transfer learning method based on a convolution neural network, transferring the useful knowledge of the source domain to the target domain. For systems with structural changes, it can effectively calculate and substantially reducing the training time of the new model, provides a new idea for the analysis of power system failure consequences based on artificial intelligence.

2 Load-Shedding Model Based on CNN

2.1 Introduction of CNN

CNN (convolutional neural network) is a subdivision of ANN (artificial neural network). In general, an artificial neural network includes a feedforward network and a feedback network. From the training method, they can be divided into supervised learning and unsupervised learning. the convolutional neural network is a feedforward artificial neural network model using supervised learning. Classic convolutional neural network models include lenet-5 [19] proposed by LeChun in 1998, AlexNet [20] proposed by Krizhevsky in 2012, VGGNet [21] and GoogleNet [22], etc. The load-shedding model based on the convolutional neural network model has a stronger generalization ability and transfer ability than a simple artificial neural network model.

Based on the input, output, and neuron of the artificial neural network, the convolutional neural network added some layers, including the convolution layer, pooling layer, and full connection layer. The convolutional layer has the characteristics of local connection and weight sharing, which greatly reduces the number of parameters, and it can significantly reduce the complexity of model training. Besides, it will also reduce overfitting. The pooling layer has a lot of advantages. On the one hand, the pooling layer can reduce the features, on the other hand, it can also give the model tolerance for slight deformation so that the generalization ability of the model can be improved. Therefore, compared with simply fully connected artificial neural networks, the convolutional neural network has the advantages of fewer network parameters, short training time, strong generalization ability, and parallel learning ability.

The convolution layer uses the convolution operation to extract the input features. Usually, multiple convolution kernels are selected according to the actual situation. After each convolution kernel is convolved with the feature graph of the upper layer, the output feature graph is obtained. H_i is used to represent the feature map of i th layer of the convolutional neural network layer, which can be represented [23] as Eq. (1):

$$H_i = \mathrm{f}(H_{i-1} \otimes W_i + b_i) \tag{1}$$

Where f is the activation function; $\otimes$ represents the convolution operation; W_i is the weight matrix of the i th layer convolution kernel; b_i is the offset vector of the i th layer. The commonly used activation functions mainly include ReLu, tanh, Sigmoid, and so on. ReLu, as a non-saturated activation function, can greatly accelerate the convergence rate of weight gradient descent, thus it is widely used.

Pooling is a sampling method. After Pooling operation, it can significantly reduce the dimension, reduce the calculation amount, and improve the robustness of the model. The calculation process can be represented as the Eq. (2):

$$H_i = \sigma(H_{i-1}) \tag{2}$$

Where, σ stands for pooling operation, which mainly includes average pooling, maximum pooling, random pooling, etc. The model in this paper uses the method of maximum pooling.

After several convolutional layers and pooling layers, the full connection layer will take the weighted sum of the feature graph of the previous layer, and obtain the output feature graph through the activation function, playing the role of mapping the learned features to the output, which can be expressed as Eq. (3):

$$H_i = \mathrm{f}(H_{i-1}W_i + b_i) \tag{3}$$

Where parameters are the same as those mentioned above. For the convolutional neural network, the training goal is to minimize the loss function. When solving regression problems, MSE (Mean Squared Error) is generally adopted as the loss function. In the training process, the gradient descent method is used for optimization.

2.2 Model Structure

When the convolutional neural network is used for power system load-shedding calculation, it corresponds to the regression problem. The calculation process of the power system load-shedding based on the convolutional neural network is shown in Fig. 1. As can be seen from the figure, there are two parts, part I is offline training and part II is the online application.

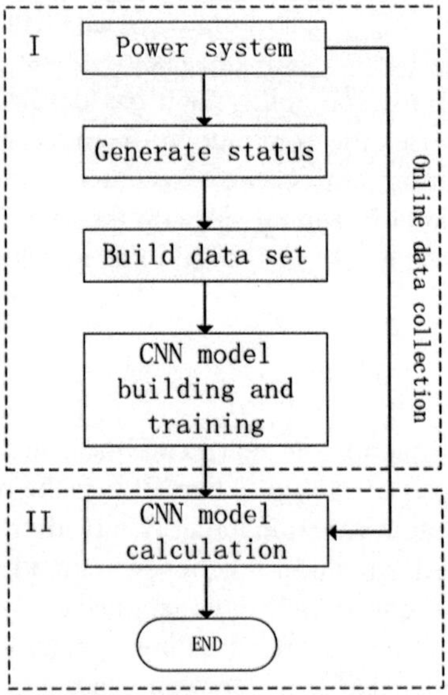

Fig. 1. Power system load-shedding calculation process based on convolutional neural network

Offline training mainly includes the following aspects:

1) Generate status. Generally, the Monte Carlo simulation method is used to randomly set the system status, and a fault sequence is generated for each simulation according to the failure rate of different components.

2) Build a data set. It is necessary to construct the fault consequence data set according to the selected input features. The choice of feature is very important to the effect of the model. If there are many features, the main features may be covered up, resulting in an unsatisfactory learning effect of the model. If the number of features is too small, it may not be able to fully represent the state of the system, and the evaluation effect is not ideal. The model constructed in this paper mainly adopts the input characteristics of switching state of components, the active power output of generators, and the active power of each node. Considering that the input of the convolutional neural network is two-dimensional data, this paper uses the processing method of cutting and merging to directly arrange the one-dimensional features into a two-dimensional matrix in order. As for the number of labels, the output features are set to the total load removal of the system and the number of nodes generating the cutting load.
3) Convolutional neural network model building and training. The convolutional neural network model established in this paper includes 2 convolutional layers, 2 pooling layers, 1 fully connected layer, and 1 dense layer, and they are shown in Fig. 2. Among them, the activation function of neurons uses the ReLu function, and the loss function uses MSE, which is calculated according to the mean squared error of label data and network output data. Each iterates the model was obtained through the Adam algorithm through the supervised training by the gradient descent method.

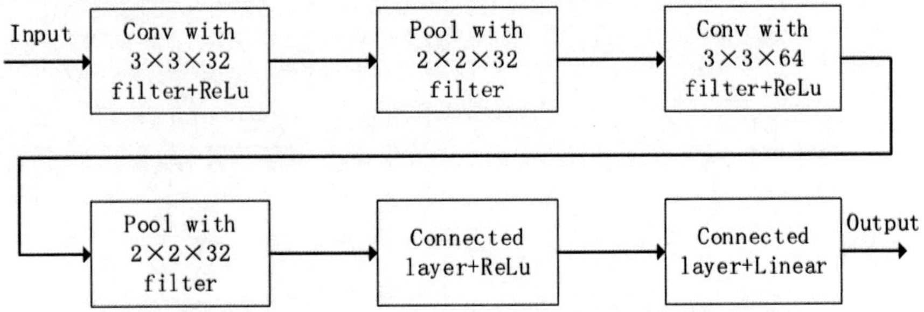

Fig. 2. Model structure of the convolutional neural network

In the online application, the running status is obtained from the system, and the model trained offline is used for data calculation.

3 Load-Shedding Model Based on Transfer Learning

3.1 Transfer Learning Theory

Transfer learning can apply the knowledge learned in one domain to another similar domain and use a machine-learning algorithm to transfer the existing knowledge from the source domain to the target domain.

In the basic concept of transfer learning, the existing knowledge is called the source domain and the new knowledge is called the target domain. We transfer the knowledge of the source domain to accomplish the task of the target domain. In terms of specific methods, there are mainly instance weight method and feature representation method, among which the latter has better transfer ability. In this paper, the source domain is the pre-trained CNN model for a certain system, the target domain is the CNN model after the system has undergone minor changes, and the "knowledge" to be learned is the model structure and parameters, and the aim is to avoid the time loss caused by training a new CNN model. During transfer learning, the probability distribution adaptation method is used to represent the distribution difference between the source domain and the target domain, and the model of the target domain is adjusted accordingly by using the distribution difference.

Transfer learning enables the network to learn feature knowledge from the source domain to fit the network weights and paranoid parameters, and then use the fitted network weights and weight parameters to share with the trained model to achieve sharing and training models. However, there is a gap between the source domain and the target domain, and the predicted output of the pretraining model network is closely related to the data set, so it cannot be copied completely. Each output node corresponds to a specific predicted value, so after network parameters are migrated, the output layer needs to be reconstructed to achieve better adaptation. This paper extracts the features learned by the network on a large number of fault consequence data sets (source domain) of the IEEE-RTS79 system and transfer to the fault consequence data sets (target domain) after the system structure changes. The structure is shown in Fig. 3.

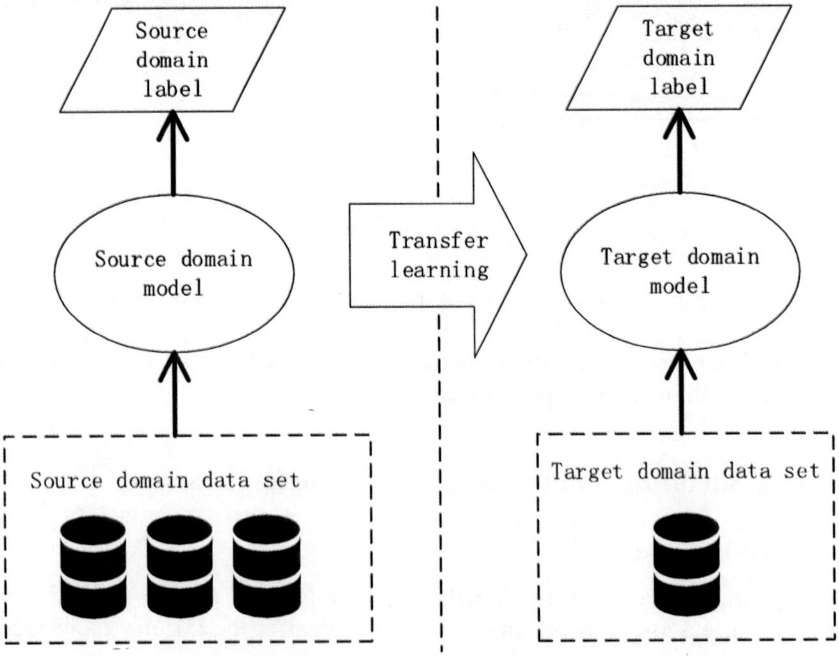

Fig. 3. A simple model of transfer learning

When the topological structure and operation mode of the system is greatly changed, the samples in the current scene are quite different from the original data set, resulting in the unsatisfactory accuracy of the results obtained by the network directly using the features learned from the pre-trained convolutional neural network model for calculation. Therefore, it is necessary to train the pre-trained convolutional neural network model under the data set of the new target domain, so that it can better adapt to the new system structure.

As for the distribution differences between the source domain and the target domain, MMD (the Maximum Mean difference algorithm [24]) is used to measure the adaption degree of different distributions. MMD is an effective method to measure the correlation between the source domain and the target domain. The principle is to find continuous functions in the sample space, find the mean values of the samples with different distributions, and then find the difference between the two means. As the mean difference between the two distributions, the maximum value of the mean difference is the MMD of the two distributions. If the MMD is small enough, then the two distributions are considered the same; If the MMD is large, the two distributions are considered different. MMD is added as a regularization term to the loss function so that the features on both domains are as similar as possible.

In this article, when the system topology changes slightly or the operation mode changes, transferring the pre-trained CNN model is equivalent to applying the problem-solving knowledge in the source domain to another different but close problem in the target domain. To ensure that the transfer can be performed when the above changes occur, the CNN model in the calculation example in this paper sets a certain margin to ensure the adaptability of the algorithm to the changes. From the training of the CNN network to the construction of the transfer model, when the difference between the target domain and the source domain is within a certain range, the theoretically learned features can get ideal results under the current learning conditions.

3.2 Transfer Learning Scheme

To compare the model performance of different transfer schemes, this paper adopts the following schemes to compare the training time of the transfer model and the accuracy of the transfer model when the structure of the power system changes.

Scheme 1: Retrain. Using the same network structure as the pre-trained model, a new convolutional neural network model is trained with the new training data when all parameters are initialized, and then the new model is tested with the test set under the new conditions.

Scheme 2: Transfer the entire network. The network structure and parameters of the pre-trained model are all transfer to the new model. In other words, the parameters of the pre-trained model are used as the initial values of the parameters of the new model, and the new training set is used for training on this basis. After the network parameters of the pre-trained model are adjusted, the new test set is used to test the new model.

Scheme 3: Retain the network structure in the pre-trained model and the network parameters of all the convolutional and pooling layers, and randomly initialize the parameters of the fully connected layer and the output layer. In other words, the main features learned from the convolutional layer and pooling layer are retained, and the new training set is used for training on this basis. After the parameters of the full connection layer and the fates are adjusted, the new test set is used to test the new model.

Scheme 4: Retain the network structure and parameters of the convolutional layer, pooling layer, and the first fully connected layer in the pre-trained model. That is to say, only randomly initialize the parameters of the dense layer. After the parameters of the dense layer were adjusted with the new training set, and then the new test set is used to test the new model.

4 Example Analysis

4.1 Example System Introduction

IEEE-RTS79 is a power generation system and transmission network that can be used to test or compare the reliability analysis of power systems. It includes 32 generators ranging from 12 to 400 MW. The transmission system contains 24 loads or generation buses connected by 38 lines or autotransformers at two voltages, 138 and 230 kV. The structure of the system is illustrated in Fig. 4.

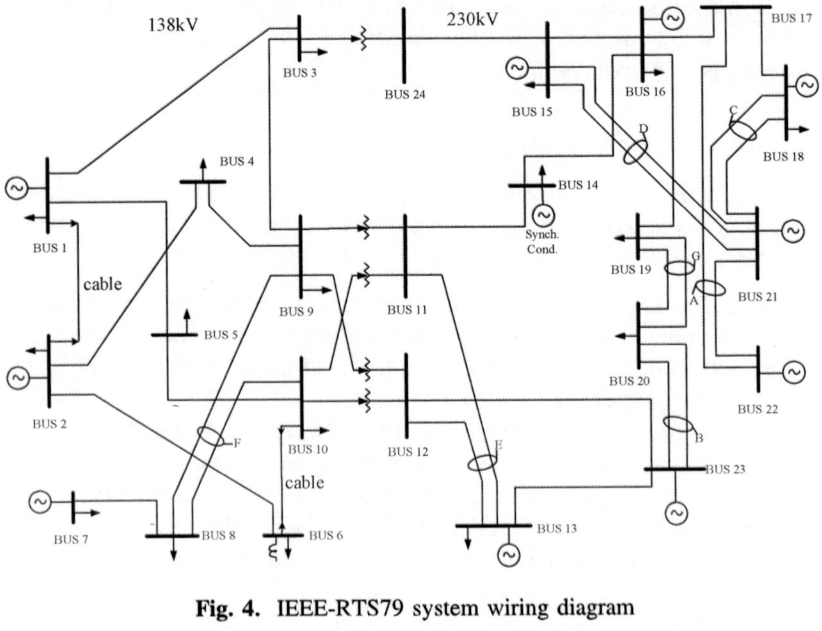

Fig. 4. IEEE-RTS79 system wiring diagram

4.2 Data Set Acquisition

To obtain the data set of load-shedding, we coded based on the Monte Carlo method. For the RTS79 system, the state of the system was randomly assigned according to the failure rate each time. There is no limitation on the number of faults when building the data set. Several conditions may occur, such as a single unit outage, single line failure, multiple units' outage, and multiple lines failure. When the fault is extremely serious, the system will not be able to converge through load cutting. In this case, the data will not be included in the data set. When the fault is light and the load is not cut, the data, in this case, will not be included in the data set. Therefore, the system will judge whether it conforms to the data set standard according to the calculated load-shedding, and then add it to the data set if it does. For the load level of the system, mainly consider the situation of 80% to 120%. Since the load level increases, the consequences of system failures become more and more serious. To ensure that the number in the data set remains balanced under various load conditions, a variable step size scheme is used to adjust the load level. The data conforming to the specifications were written into the input and label of the data set following the above format, and finally, 43800 samples were generated, all of which had a system failure and generated load-shedding. The sample label records the total load-shedding of the system and the number of nodes that generate load-shedding.

4.3 Performance Test of the Pre-trained Model

Randomly scramble 43800 samples, and then randomly select 60% of them, that is, 26280 samples as the training data set. Take 10% of them or 4380 samples as the validation data set. The remaining 30%, that is, 13140 samples will be used as the test data set. The prediction performance of the convolutional neural network mentioned above was compared with that of a simple artificial neural network.

Since the model in this paper deals with a regression problem, the accuracy of the model is not as easy to express as the classification problem. If we set the absolute value difference between the output of the model and the actual value within 0.01, the model result is considered correct. When two tags are included, the aforementioned conditions need to be met at the same time, and the second tag has the same value.

To reduce the error caused by randomness, CNN, and a simple artificial neural network was used for training for 10 times of sampling, and the accuracy is shown in Table 1. As can be seen from Table 1, when only the first training label is considered, both CNN and simple artificial neural networks have better performance, and the accuracy of CNN is slightly higher than that of simple artificial neural networks. However, when considering the two training tags, the performance of an artificial neural network was significantly reduced.

The main reason is that the simple artificial neural network can better fit a single target when modeling this type of problem, but multi labels will generate competition and cause loss of features. By improving the loos function, the CNN network can get better performance. It is worth noting that the training time of CNN is longer than that of the simple artificial neural network, but the characteristics of CNN, especially the transfer ability, are not possessed by the artificial neural network.

Table 1. Performance comparison between CNN and ANN

Test system	Model	Accuracy rate when 1 label (%)	Accuracy rate when 2 labels (%)
IEEE-RTS79	ANN	98.20	83.41
	CNN	98.93	96.87

4.4 Generalization Ability Test of the Pre-trained Model

The generalization ability of the model is a part that cannot be ignored when evaluating the model because the generalization ability represents the adaptability of the model. Only with a strong generalization ability can pre-trained models can cope with different working conditions under the same structure. To test the generalization ability of the CNN pre-trained model, the following new cases are added.

Case 1: As in training, the load level is maintained at 80% to 120%, the topology of the system remains unchanged. Change the output level of the generator, and data calculation is carried out on the RTS79 system under this new working condition.
Case 2: Reduce the load level to 70%, reduce 2 generators, and 3 lines. Among them, the removed generator is unit 4 of bus 1 and unit 1 of bus 16. The removed lines are the line from bus 1 to bus 3, the line from bus 16 to bus 19, and line from bus 21 to bus 22. Data calculation of the new system is carried out under this new working condition.
Case 3: Increase the load level to 130%, add 1 generator, and 2 lines. Among them, add a new unit at bus 7, Add line from bus 2 to bus 7, and line from bus 7 to bus 8. Perform data calculation on the new system under this new working condition.
Case 4: Reduce the load level to 60%, Reduce 6 generators, and 6 lines. The removed generators are unit 2 of bus 1, unit 3 of bus 2, unit 1 of bus 13, unit 4 of bus 15, unit 5 of bus 15, and unit 6 of bus 22. The removed lines are the line from bus 1 to bus 2, the line from bus 4 to bus 9, Line from bus 15 to bus 21, the line from bus 19 to bus 20, the line from bus 20 to bus 23, and line from bus 21 to bus 22. Under this new working condition, the system calculates the data.

The new test sets T1, T2, T3, and T4 are respectively generated in the above four cases using the same method as the previous data set. They are all composed of 5000 samples. For comparison purposes, the following labels will only consider single labels. The generalization ability of the pre-trained CNN model is shown in Table 2. To visually show the difference between each case and the original one, MDD is also listed.

Table 2. The generalization ability of CNN pretraining model

Data set	MDD	Accuracy rate (%)
T1	0.0367	98.82
T2	0.0728	98.36
T3	0.1633	96.38
T4	0.5472	91.44

According to the above tests, it can be seen that the prediction accuracy of the pre-trained model under Cases 1 to 3 is above 96%, indicating that the pre-trained CNN model designed in this paper has a certain generalization ability and a relatively reliable performance the problem that the system structure does not change too much. However, in case 4, the MDD is significantly higher than in the other three cases. There was a significant decline in accuracy, which indicated that when the system structure changed greatly, the adaptability of the pretraining CNN model became worse and the performance decreased, so it was necessary to retrain the model or transfer the pre-training model for this condition.

4.5 Effect Comparison of the Transfer Learning Scheme

Based on the previous section, this section conducts the transfer learning of the pre-trained CNN model for case 4, to improve the accuracy of the model under case 4, and compares the transfer learning scheme proposed in Sect. 2.2.

Due to the change of system structure, no matter what kind of transfer learning method is adopted, new samples need to be generated for the training and testing of the new model. For different schemes, the data sets are generated in the same way, so the comparison in this part will mainly focus on the training time of the new model and the accuracy of the new model. For different schemes, Adam algorithm is also used to inversely adjust the parameters of each layer of the neural network. For the convenience of comparison, the training time of the ANN model and pre-trained CNN model and their accuracy to scene 4 is also listed in the table. The PC configuration used for data acquisition in this paper was: Intel core i5-4200U CPU2.30 GHz/8.00 GB RAM, and no GPU was used for all training.

Table 3. Effect comparison of different transfer learning schemes for case 4

Training scheme	Training time/s	Accuracy rate (%)
Pretraining CNN	3358.66	91.44
Scheme 1	87.34	96.52
Scheme 2	30.72	96.38
Scheme 3	35.43	96.84
Scheme 4	2.12	97.36

As can be seen from Table 3, scheme 1 uses a small new data set for retraining, and the accuracy is significantly improved. However, compared with other schemes, scheme 1 takes a longer time. Since the parameters of the convolution layer and pooling layer are retained in scheme 2 and scheme 3, the speed of scheme 1 is significantly improved. In scheme 4, the calculation performance of the model can be quickly and effectively restored to more than 97% by adjusting the network parameters of the dense layer. And the average training time is significantly reduced compared with other schemes. It is worth noting that this training time is much faster than retraining the simple artificial neural network. According to the comparison of the four schemes, the new model obtained in scheme 4 only through the reconstruction of the dense layer has obvious advantages in terms of training speed and accuracy. It can be seen that the scheme can not only improve the training efficiency of the deep neural network but also make full use of the source domain data to improve the transfer ability of the model so that it can recover to a better performance level in a short time.

5 Conclusion

Based on deep transfer learning, this paper proposes a method for calculating the load-shedding in fault consequence analysis. This method uses training data to model the pre-trained system, and quickly transfer and restore performance when the system structure changes. Through the verification of the IEEE-RTS79 system, it is shown that the model has a good learning effect in the case of multiple tags. During transfer learning, keep the network structure of the pre-trained CNN model and the network parameters of the convolutional layer, pooling layer, and the first full connection layer unchanged, and only the parameters of the dense layer are randomly initialized, to obtain the best transfer effect.

This paper has preliminary verified the feasibility of the application of artificial skills in power system fault consequence analysis and provided a new idea for the reliability evaluation of power systems based on artificial intelligence. However, the performance of this method in dealing with more complex largescale systems needs to be further studied. The application of deep transfer learning to modeling other aspects of power system reliability assessment will be the focus of future research.

References

1. Xue, Y.: Integrated defense evolved from accidental failure to power disaster – warning of "8 · 14" blackout in North America. Autom. Electr. Power Syst. **27**(18), 1–5 (2003). (in Chinese)
2. Billinton, R.: Composite system reliability evaluation. IEEE Trans. Power Apparatus Syst. **88**(4), 276–281 (1969)
3. Billinton, R., Allan, R.N.: Reliability Evaluation of Power Systems. Plenum Press, New York and London (1996)
4. Song, H., Junyong, W., Linfeng, W.: Dynamic partitioning and autonomous disassembly strategy in power system emergency. Trans. China Electrotech. Soc. **27**, 224–230 (2012). (in Chinese)
5. Wang, B., Fang, W., Luo, X.: A fast algorithm for optimal cutting machine load cutting scheme under emergency control. Power Syst. Technol. **35**, 82–87 (2011). (in Chinese)
6. Sun, D., Zhou, H., Xiong, H., Ping Ju, Y., Zhao, J., et al.: Emergency load cutting control optimization method for DC receiver system based on sensitivity analysis. Proc. CSEE (38), 67–75 (2008). (in Chinese)
7. Yi, H., Wang, X., Teng, Y.: Optimal load shedding scheme for AC/DC hybrid receiving-end power grid after UHVDC blocking. Autom. Electr. Power Syst. **42**(22), 98–108 (2018). (in Chinese)

8. Lei, S., Sun, C., Zhou, Q., Zhang, X., Cheng, Q.: Based on RBF neural network and adaptive neural fuzzy system of electric power short-term load forecasting method. Proc. CSEE **22**, 81–85 (2005). (in Chinese)
9. Wang, H., Yang, D., Zhou, B.: Fault diagnosis of multi-terminal HVDC transmission line based on parallel convolutional neural network. Autom. Electr. Power Syst. **44**(12), 84–95 (2020). (in Chinese)
10. Jun, L., Chen, Z., Gong, G., Zhiqiang, X., Qi, B.: Classification and analysis method of residential electricity use behavior based on extreme learning machine. Autom. Electr. Power Syst. **143**, 97–104 (2019). (in Chinese)
11. Li, C., Li, H., Liu, Y.: Intelligent assessment system for dynamic security risk of large-scale power grid. Autom. Electr. Power Syst. **43**(22), 67–75 (2019). (in Chinese)
12. Zhu, Y., Liu, Y.: Fast search for high-risk cascading failures based on deep learning DC blocking judgment. Autom. Electr. Power Syst. **43**(22), 59–67 (2019). (in Chinese)
13. Wang, Y., Wang, Y., Ding, Y., Zhou, Y., Zhang, Z.: A fast load-shedding algorithm for power system based on artificial neural network. In: 2019 International Conference on IC Design and Technology (ICICDT), SUZHOU, China, pp. 1–3 (2019)
14. Liu, W., Zhang, D., Hou, X., Liu, L.: A decision making strategy for generating unit tripping under emergency circumstances based on deep reinforcement learning. Proc. CSEE **38**(1), 109–119+347 (2018). (in Chinese)
15. Zhang, C., Wang, Q., Liu, Z.: Pulmonary nodule auxiliary diagnosis method based on deep transfer learning. Comput. Eng. **46**, 271–278 (2020). (in Chinese)
16. Wang, J., Huang, H., Xu, H.: Low-resource Uyghur speech recognition based on transfer learning. Comput. Eng. (44), 281–285 + 291 (2018). (in Chinese)
17. Chen, J., Du, W., Wang, H.: A Method of locating the power system subsynchronous oscillation source unit with grid-connected PMSG using deep TRANSFER learning. Trans. China Electrotech. Soc. 1–12 (2020). (in Chinese)
18. Wang, Y.: Power Data Mining Technology Based on Deep Learning and Transfer Learning. Zhejiang University (2019). (in Chinese)
19. Lecun, Y., Bottou, L., Bengio, Y., et al.: Gradient-based learning applied to document recognition. Proc. IEEE **86**, 2278–2324 (1998)
20. Krizhevsky, A, Sutskever, I., Hinton, G.E.: Image net classification with deep convolutional neural networks. In: The 25th International Conference on Neural Information Processing systems, pp. 1097–1105. Curran Associates Inc., Lake Tahoe (2012)
21. Parkhi, O.M., Vedaldi, A., Zisserman, A.: Deep face recognition. In: British Machine Vision Conference, pp. 41–42. British Machine Vision Association, Swansea (2015)
22. Szegedy, C., Liu, W., Jia, Y.Q., et al.: Going deeper with convolutions. In: 2015 IEEE Conference on Computer Vision and Pattern Recognition, pp. 1–9. IEEE, Boston (2015)
23. Li, Y., Hao, Z., Lei, H.: Survey of convolutional neural network. J. Comput. Appl. **36**(9), 2508–2515 (2016). (in Chinese)
24. Pan, S.J., Yang, Q.: A survey on transfer learning. IEEE Trans. Knowl. Data Eng. **22**(10), 1345–1359 (2010)

Research on Topology Identification of Distribution Network Based on Multi-source Data Fusion

Rui Mei(✉), Yanan Liu, Chao Yuan, Chenlong Li, and Hongyu Yang

Jiangsu Frontier Electric Technology Co., Ltd., Nanjing 211102, Jiangsu Province, China
merrymr@163.com

Abstract. When the distributed supervisory control and data acquisition (DSACADA) and micro phasor measurement units (μPMU) are installed in the distribution network, a topology identification method based on multi-source data fusion is proposed. The identification model of topology changing time can be constructed with the voltage angle measurements. Based on the change of voltage angle and amplitude, the topology change trend criterion can be constructed for narrowing range of the possible topology. Then based on WLS, the fusion of μPMUs and DSACADA can be realized and all the node voltage angles in every possible topology can be estimated. Finally, the similarity identification model is constructed to identify the real operation topology. he reliable topology identification can be realized even some errors occur in the measurement of μPMU and DSACAD is shown in the example.

Keywords: Distribution network · Topology identification · Multi-source data fusion · DSCADA · μPMU

1 Introduction

As the construction of smart grids accelerates, the structure of distribution network systems becomes increasingly complex, and operating topology changes more frequently. The distribution network collects remote communication data in real time, grasps the network topology changes in time to update the operating topology structure [1]. However, the false and unreported data often occur in the remote communication data, and the data source is relatively single. Thus, the correctness of the generated network topology is unreliability [2].

Most of the power grid topology identification research in China focus on the main network, and there are few studies on the complex structure of the distribution network. In recent years, foreign countries have begun to carry out research on distribution network topology identification. Literature [3] proposed an identification method that a search method is used to reconstruct the operating tree of the distribution network, and combines linearly coupled power flow models to search for possible connected branches to generate a topology. Literature [4] developed an online topology identification algorithm based on the μPMU device, to identify the current switch state changes in the

Y. Xue et al. (Eds.): PMF 2020, LNEE 718, pp. 608–621, 2021.
https://doi.org/10.1007/978-981-15-9746-6_47

distribution network by comparing the actual measurement of the node voltage with time scale and the results of power flow calculation under various possible topologies. Literature [5] combined with the real-time collected node voltage phasors, based on conditional independent testing to identify the real-time topology.

The topology identification of the distribution network just mentioned is based on the new μPMU measurement device which are not suitable for the distribution network with DSCADA installed. Therefore, this paper considers to make full use of the existing distribution automation equipment in China, combined with the new measurement device μPMU, through the fusion of multi-source telemetry data to achieve distribution network operation topology identification, which is of great significance to improve the reliability of identification.

In this paper, when the DSCADA measurement device is installed in the distribution network, a small number of μPMU measurement devices are installed in the nodes at both ends of the tie switch and the multi-branch nodes with a degree greater than two [6]. Firstly, measure the node voltage phase where μPMU installed, construct a topology change time identification model, and determine the time when the topology changes. Then construct the topological change trend criterion, determine the branch range where the running state changes, and enumerate all possible topological situations after reconstruction [7]. Then the telemetry data of μPMU and DSCADA are fused to estimate the state of all possible topologies. A topology similarity identification model can be constructed by the estimated and measured node voltage phase, and identifies the actual running topology. Finally, the influence on the accuracy of topology identification because of the measurement errors of DSCADA and μPMUs, and the applicability of the algorithm in the case of single-contact and multi-contact switch actions should be considered [8].

2 Distribution Network Topology Identification Model Based on Multi-source Data Fusion

The telemetry data fusing of μPMU and DSCADA is considered in two factors. On the one hand, remote measurement data collection frequency of μPMU can reach 8 ms at a time, which can accurately collect the data of DSCADA at any time, and realize the data collection at the same time section, which is more real-time and reliable than other measurement devices. On the other hand, the type of data collected by the μPMU device is relatively rich, which can not only collect the node voltage amplitude and phase, but also the branch current and node injection power. Therefore, this paper fusing the data measured by DSCADA and μPMU to improve the redundancy of the state estimation of the distribution network and provide a data basis for reliable identification of the distribution network topology [9].

2.1 Topology Change Moment Identification Model

When the distribution topology of the distribution network changes, the power flow in the network changes, and the voltage phase change of the tie switch and multi-branch nodes is more obvious [10]. Therefore, consider a phase identification model based on the phase change of the node voltage:

$$delta(t) = \sum_{i=1}^{w} \|\alpha(t) - \alpha(t-i)\|_1, w \in 1, 2, \cdots, \infty \tag{1}$$

Where, vector $\alpha(t)$ represents the vector formed by the voltage phase information of all the nodes measured by μPMU at time t, the dimension of the vector is equal to the number of nodes installed with μPMU. Vector $\alpha(t-i)$ represents the vector formed by the voltage phase values measured by each installed μPMU node at time $t-i$. $delta(t)$ represents a 1-norm function with time t as a variable. Parameter w is equivalent to a time window of the $delta(t)$ function, the greater the w, the greater the difference between the $delta(t)$ and $delta(t-1)$ values when the network operation mode changes at time t. However, if the value of w is set relatively large, the calculation amount of the program will be increased. Therefore, a reasonable value of w should meet the calculation speed requirement, and when the topology changes, the value $delta(t)$ will have a sharp increase spike at time t.

Through the topology change time identification model, the identification problem of the distribution network operation topology at a certain time is converted into the problem of identifying the topology after the reconstruction, which is based on the previous network operation topology after determining the network reconstruction time [11].

2.2 Topological Change Trend Criterion

The switches that are ubiquitous in the distribution network can generally be divided into two categories: one is a normally closed section switch and the other is a normally open contact switch. When selecting the segment switch to be opened in the simplified distribution network, the following two necessary conditions must be met:

(1) In order to avoid island operation mode in the network, only one section switch can be opened in each branch group, which means only one branch is opened.
(2) When N contact switches are closed, N loops appear in the distribution network. In order to build a connected power distribution network with radial operation, only N branches in each branch group under different loops can be disconnected.

According to the necessary conditions for distribution network reconstruction [12], criterion 1 can be obtained.

Criterion 1: When only one tie switch acts, the branch where the segment switch may occur is included in all the branch groups that form a ring network with the tie switch, and only one segment switch acts. When there is an N tie network, during the switching action, the branch where the segment switch may occur is included in all the branch groups of the formed N ring networks, and there is one segmented switching action in each ring network.

Since the μPMU device is installed at both ends of the tie switch, when the branch current is monitored, the tie switch branch can be determined to be in a connected state.

According to criterion 1, the branch group range of all possible action segment switches for radial operation of distribution network can be determined.

The nodes far away from the action switch have smaller changes in power supply distance and smaller voltage fluctuations. Therefore, criterion 2 can be obtained.

Criterion 2: The state of the segment switch of a branch changes, then the voltage amplitude and phase of the nodes at both ends of the branch change more than the previous moment. The criterion formula is derived as follows:

$$\begin{cases} |\theta_i(t) - \theta_i(t-1)| > \varepsilon_\theta \\ \left|\frac{U_i(t) - U_i(t-1)}{U_N}\right| > \varepsilon_U \end{cases} \tag{2}$$

$$\begin{cases} |\theta_j(t) - \theta_j(t-1)| > \varepsilon_\theta \\ \left|\frac{U_j(t) - U_j(t-1)}{U_N}\right| > \varepsilon_U \end{cases} \tag{3}$$

Where, nodes i and j are the nodes where the μPMU measuring device is installed. $\theta_i(t)$ and $\theta_j(t)$ are the voltage phases of nodes i and j at the moment t. $\theta_i(t-1)$ and $\theta_j(t-1)$ are the voltage phases of nodes i and j at time $t-1$. $U_i(t)$ and $U_j(t)$ are the voltage amplitudes of nodes i and j at time t. $U_i(t-1)$ and $U_j(t-1)$ are the voltage amplitudes of nodes i and j at time $t-1$. U_N is the rated voltage. ε_θ is the phase fluctuation constraint value of the node voltage. ε_U is the constraint value of the node voltage amplitude fluctuation.

Due to load fluctuations and other factors, it will cause node voltage fluctuations, thereby expanding the branch range determined by criterion 2. Therefore, criterion 3 is introduced to further narrow the range of branches.

Criterion 3: When the state of a section switches in a branch group changes, the difference in the node voltage amplitude between the two ends of the branch group changes the most.

$$\max M = \left| |U_j(t) - U_i(t)| - |U_j(t-1) - U_i(t-1)| \right| \tag{4}$$

Where, M represents the change value of the node voltage difference at time t and $t-1$. $U_i(t)$ and $U_j(t)$ are the voltage amplitudes of nodes i and j at time t. $U_i(t-1)$ and $U_j(t-1)$ are the voltage amplitudes of nodes i and j at time $t-1$.

2.3 Distribution Network State Estimation Model Based on Multi-source Data Fusion

Weighted Least Squares Model. When the operational topology of the network and the parameter information of the operational branches are known, m measurement information will be collected, and the state estimation of n state variables in the system will be performed by constructing a nonlinear equation between the state quantity and the measurement, the equation is as follows:

$$z = \boldsymbol{h}(\boldsymbol{x}) + \boldsymbol{v} \tag{5}$$

Where: $h(x)$ is the measurement function vector, and $h(x) = [h_1(x), h_2(x), \ldots, h_m(x)]^T$. x is the state vector, and $x = [x_1, x_2, \ldots, x_n]^T$. z is the measurement vector, and $z = [z_1, z_2, \ldots, z_m]^T$. v is the measurement error vector, and $v = [v_1, v_2, \ldots, v_m]^T$.

By setting relevant weights for different measurement values to ensure the accuracy of the estimation result, the matrix expression of the objective function is as follows:

$$J(\boldsymbol{x}) = [z - \boldsymbol{h}(\boldsymbol{x})]^T \boldsymbol{W}[z - \boldsymbol{h}(\boldsymbol{x})] \tag{6}$$

Where, $\boldsymbol{W}$ is the weight matrix, and $\boldsymbol{W} = [w_1, w_2, \ldots, w_m]^T$, where w_i is the weight value of the ith quantity measurement z_i.

Usually the measurement error follows a normal distribution, thus $E(v) = 0$, $E(vv^T) = \boldsymbol{R}$, $\boldsymbol{R}$ is the diagonal matrix of the covariance of the measurement error, and the measurement weight matrix $\boldsymbol{W} = \boldsymbol{R}^{-1}$, as shown in the formula (3).

$$\boldsymbol{W} = diag\left(\frac{1}{\sigma_1^2}, \frac{1}{\sigma_2^2}, \cdots, \frac{1}{\sigma_m^2}\right) \tag{7}$$

Where, σ_i^2 is the variance of the measurement error of the ith measurement z_i.

Thus Eq. (2) can be written as:

$$J(\boldsymbol{x}) = [z - \boldsymbol{h}(\boldsymbol{x})]^T \boldsymbol{R}^{-1}[z - \boldsymbol{h}(\boldsymbol{x})] \tag{8}$$

Since $\boldsymbol{h}(\boldsymbol{x})$ is usually non-linear, $\boldsymbol{h}(\boldsymbol{x})$ should be linearized and substituted into Eq. (5), and the derivative of $\boldsymbol{x}$ is obtained to obtain the state quantity $\hat{\boldsymbol{x}}$. Its iteration can be written as:

$$\begin{cases} \Delta\hat{\boldsymbol{x}}^{(k)} = \left[\boldsymbol{H}^T(\hat{\boldsymbol{x}}^{(k)})\boldsymbol{R}^{-1}\boldsymbol{H}(\hat{\boldsymbol{x}}^{(k)})\right]^{-1}\boldsymbol{H}^T(\hat{\boldsymbol{x}}^{(k)})\boldsymbol{R}^{-1}[z - \boldsymbol{h}(\hat{\boldsymbol{x}}^{(k)})] \\ \hat{\boldsymbol{x}}^{(k+1)} = \hat{\boldsymbol{x}}^{(k)} + \Delta\hat{\boldsymbol{x}}^{(k)} \end{cases} \tag{9}$$

Where, k represents the number of iterations. $\hat{\boldsymbol{x}}^{(k+1)}$ and $\hat{\boldsymbol{x}}^{(k)}$ represent the estimated state quantities calculated at the $k + 1^{th}$ and k^{th} iterations. $\Delta\hat{\boldsymbol{x}}^{(k)}$ represents the amount of state correction in the kth iteration calculation. $\boldsymbol{H}(\hat{\boldsymbol{x}}^{(k)})$ is the Jacobian matrix, which is derived from the measurement function $\boldsymbol{h}(\boldsymbol{x})$ to $\boldsymbol{x}$, and obtained at $\hat{\boldsymbol{x}}^{(k)}$.

The most commonly used convergence criteria in engineering is shown as follows:

$$\max_i \left|\Delta\hat{\boldsymbol{x}}^{(k)}\right| \leq \varepsilon \tag{10}$$

Where, the threshold value ε is usually 10^{-4} to 10^{-6}.

WLS-Based Distribution Network State Estimation Model. Considering the measurement errors of μPMU and DSCADA, the voltage phase of each node of the whole network is estimated by WLS [13]. The corresponding measurement equations are:

(1) Injection power measurement

$$\begin{cases} P_i^m = P_i + e_{P_i} = U_i \sum_{j=1}^{N} U_j (G_{ij}\cos\theta_{ij} + B_{ij}\sin\theta_{ij}) + e_{P_i} \\ Q_i^m = Q_i + e_{Q_i} = U_i \sum_{j=1}^{N} U_j (G_{ij}\sin\theta_{ij} - B_{ij}\cos\theta_{ij}) + e_{Q_i} \end{cases} \tag{11}$$

Where, $\theta_{ij} = \theta_i - \theta_j$, G_{ij} and B_{ij} are the real and imaginary parts of the elements between nodes i and j in the node admittance matrix. P_i^m and Q_i^m are the active and reactive injection measurements for node i. e_{Pi} and e_{Qi} are the injected power measurement errors at node i.

(2) Measurement of branch head end power

$$\begin{cases} P_{ij}^m = P_{ij} + e_{P_{ij}} = U_i^2 g_{ij} - U_i U_j g_{ij}\cos\theta_{ij} - U_i U_j b_{ij}\sin\theta_{ij} + e_{P_{ij}} \\ Q_{ij}^m = Q_{ij} + e_{Q_{ij}} = -U_i^2 b_{ij} - U_i U_j g_{ij}\sin\theta_{ij} + U_i U_j b_{ij}\cos\theta_{ij} + e_{Q_{ij}} \end{cases} \tag{12}$$

Where, P_{ij}^m and Q_{ij}^m are the active and reactive power measurements of the first end of the branch ij. e_{Pij} and e_{Qij} are the measurement errors of the corresponding branch power measurement. g_{ij} and b_{ij} are the real and imaginary parts of the admittance of branch ij.

(3) Power measurement at the end of the branch

$$\begin{cases} P_{ji}^m = P_{ji} + e_{P_{ji}} = U_i^2 g_{ij} - U_i U_j g_{ij}\cos\theta_{ij} + U_i U_j b_{ij}\sin\theta_{ij} + e_{P_{ji}} \\ Q_{ji}^m = Q_{ji} + e_{Q_{ji}} = -U_i^2 b_{ij} + U_i U_j g_{ij}\sin\theta_{ij} + U_i U_j b_{ij}\cos\theta_{ij} + e_{Q_{ji}} \end{cases} \tag{13}$$

Where, P_{ji}^m and Q_{ji}^m are the active and reactive power measurement at the end of branch ij.

(4) Measurement of branch current amplitude

$$(I_{ij}^2)^m = I_{ij}^2 + e_{I_{ij}^2} = (g_{ij}^2 + b_{ij}^2)(U_i^2 + U_j^2 - 2U_i U_j \cos\theta_{ij}) + e_{I_{ij}^2} \tag{14}$$

Where, g_{ij} and b_{ij} are the conductance and susceptance of branch ij respectively. I_{ij}^2 is the square of the current amplitude of branch ij. U_i and U_j are the voltage values of node i and node j. e_{Iij}^2 is the measurement of the current amplitude error.

According to the state variables and measurement equations, the corresponding Jacobian block matrix in the distribution network state estimation model can be obtained:

$$H = \begin{bmatrix} \frac{\partial P}{\partial \theta} & \frac{\partial P}{\partial U} \\ \frac{\partial Q}{\partial \theta} & \frac{\partial Q}{\partial U} \\ \frac{\partial (I^2)}{\partial \theta} & \frac{\partial (I^2)}{\partial U} \end{bmatrix} \tag{15}$$

In the formula: the elements in the matrix $\frac{\partial P}{\partial \theta}$, $\frac{\partial Q}{\partial \theta}$, $\frac{\partial P}{\partial U}$ and $\frac{\partial Q}{\partial U}$ are the node injection power and the partial derivative of the branch power to the phase and amplitude of each node voltage. The elements in the matrices $\frac{\partial(I^2)}{\partial\theta}$ and $\frac{\partial(I^2)}{\partial U}$ are the partial derivatives of the square of the amplitude of the branch current with respect to the phase and amplitude of each node voltage.

2.4 Topological Similarity Identification Model

The vector $\boldsymbol{x}$ = (a_1, a_2,..., a_n), $\boldsymbol{y}$ = (b_1, b_2,..., b_n) are two vectors in the n-dimensional vector space $\mathbf{R}^n$, and the inner product is as follows:

$$(\boldsymbol{x},\boldsymbol{y}) = a_1b_1 + a_2b_2 + \cdots + a_nb_n = \sum_{k=1}^{n} a_kb_k \tag{16}$$

The cosine of the angle between two non-zero vectors $\boldsymbol{x}$ = (a_1, a_2,..., a_n) and $\boldsymbol{y}$ = (b_1, b_2,..., b_n) can be obtained from the relationship between the inner product of the vector and the amplitude, as shown in Eq. (22).

$$\cos\langle\boldsymbol{x},\boldsymbol{y}\rangle = \frac{(\boldsymbol{x},\boldsymbol{y})}{|\boldsymbol{x}||\boldsymbol{y}|} \tag{17}$$

Where $\langle\boldsymbol{x},\boldsymbol{y}\rangle$ represents the angle between vector $\boldsymbol{x}$ and vector $\boldsymbol{y}$. Therefore, the following inequality can be obtained:

$$\left|\frac{(\boldsymbol{x},\boldsymbol{y})}{|\boldsymbol{x}||\boldsymbol{y}|}\right| \leq 1 \tag{18}$$

The equal sign in Eq. (18) holds if and only if the vectors $\boldsymbol{x}$ and $\boldsymbol{y}$ are linearly related.

Considered the state estimation in each topology in turn, according to the inner product principle, estimate the voltage phase based on the nodes in different topology situations, and construct a topology similarity identification model:

$$\max J = \left(\frac{\boldsymbol{\alpha}(t)}{|\boldsymbol{\alpha}(t)|}, \frac{\boldsymbol{\alpha}_i(t)}{|\boldsymbol{\alpha}_i(t)|}\right), i \in \{1, 2, \cdots, M\} \tag{19}$$

Where, J is the similarity value, $\alpha_i(t)$ represents the node voltage phase estimation vector under the ith topology, $\alpha(t)$ represents the node voltage phase measured vector, and the dimensions of the two vectors are equal to the number of μPMU installations, M the number of all possible topologies.

After unitizing $\alpha(t)$ and $\alpha_i(t)$, the measured angle of the node voltage phase vector will be closer to the angle estimation vector of the node voltage phase of a certain topology, its J value tends to 1. And the J value of the node voltage phase estimation vector under other topologies is less than 1. Therefore, the J value closest to 1 is selected as the actual operating topology.

3 Topology Identification Process of Distribution Network

When the operation topology of the distribution network changes at a certain moment, combined with the operation topology structure of the distribution network at the previous moment, the following steps are used to identify the moment when the topology changes and the identification of the new operation topology after reconstruction. The flow chart is shown in Fig. 1.

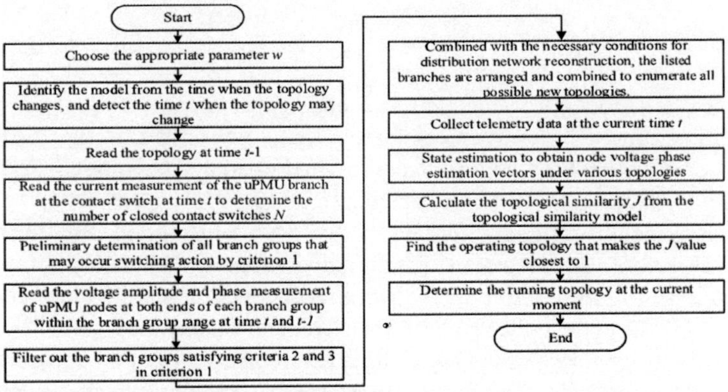

Fig. 1. Flow chart of operation topology identification of distribution network.

In a large-scale distribution network, due to the large number of network nodes, combined with the engineering experience, the number of possible topologies that require state estimation can be effectively reduced, making the identification workload smaller and the identification speed faster.

4 Case Study

This paper simulates the IEEE33 network under the MATLAB environment. The IEEE33 initial network operation topology is shown in Fig. 2, which contains 32 branches and 5 contact switch branches. A total of 13 μPMU measuring devices are installed on the nodes at both ends of the tie switch and nodes 1, 2, and 5 with a degree greater than 2. The calculation example simulates the actual measured value of μPMU based on the results of power flow calculation and considering the influence of random measurement errors. At the same time, it is assumed that DSCADA measurement is installed in the entire network, and the injected power of each node in the entire network can be collected. The reference voltage is 12.66 kV, and the reference power is 10 MVA.

4.1 Single Contact Switch Action

When the network is in the initial state, reconfiguration occurs at a certain moment, the switch TS3 is closed, and the branch 3-4 is opened, forming the operating structure shown in Fig. 3.

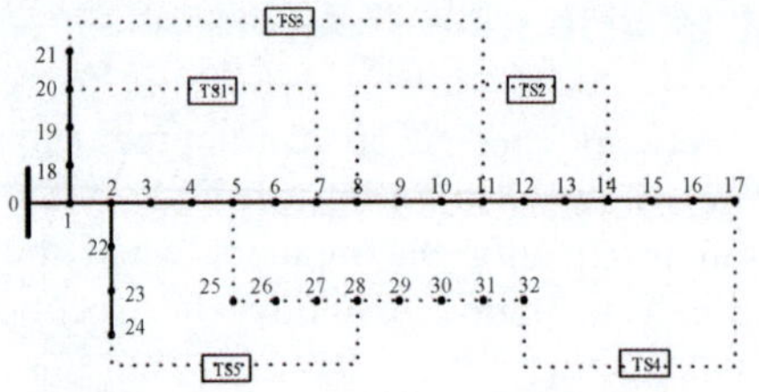

Fig. 2. Initial operation topology of IEEE33 network

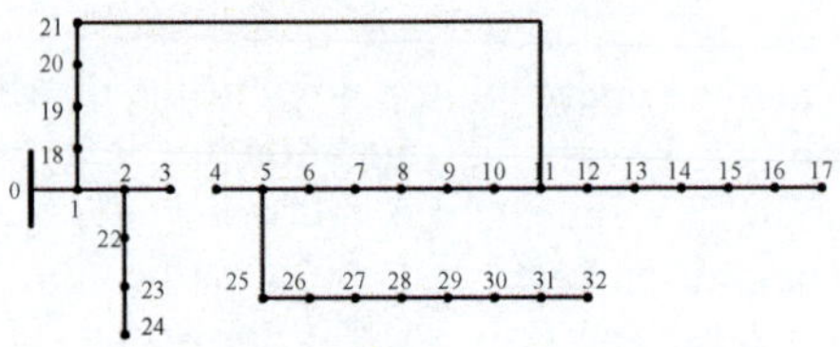

Fig. 3. Operational topology after single contact switch operation

IEEE33 has a topology change at time *t* and keeps this operating mode unchanged for a period of time. The power flow calculation is performed on the IEEE33 initial network operation topology and the reconstructed operation topology respectively, so as to obtain the true value of the voltage phase of 13 μPMU measurement nodes. The actual measurement of μPMU follows a Gaussian distribution and meets the 3σ principle. Therefore, the Gaussian distribution is used to generate the node voltage phase vectors at 40 times before the network reconstruction and 50 times after to form continuous μPMU node voltage phase measurement data. Since the μPMU actually measures the voltage phase of the node, the measurement error of the measured value is set to 5%. By identifying the model at the time of topology change, a *delta*(*t*) time series graph can be formed, as shown in Fig. 4.

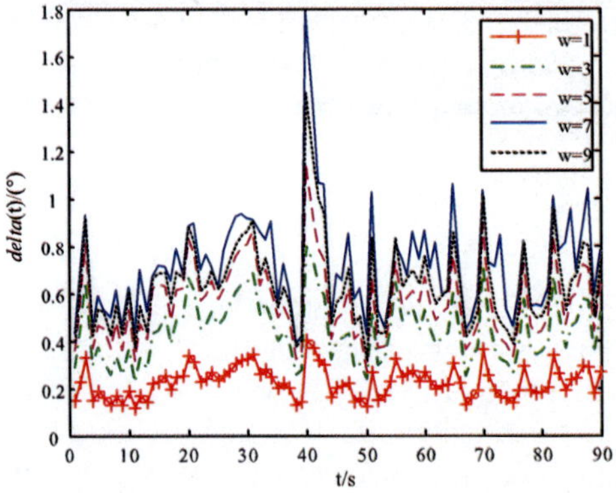

Fig. 4. Time series graphic of *delta*(*t*)

Figure 4 shows that when t = 40 s, the network operation topology changes, the measured vector of the node voltage phase changes compared to the previous time, the value of the *delta*(*t*) function becomes larger, and a sharp peak appears. Thus, the *delta* (*t*) function can identify the moment when the network running topology changes. And when the value of *w* is 7, the change trend is the most obvious, and the effect of identifying the model at the time of topology change is the best.

When the tie switch TS3 switch is closed, since the µPMU is installed at the node of the tie switch TS3, the tie branch current can be monitored to determine the closed tie switch. From the criterion 1 of the topology change trend, it can be seen that there may be a branch where the state of the segment switch changes, within the range of the branches formed by the nodes 1–21, 1–11. Table 1 is the absolute value of the voltage fluctuation of the µPMU measurement nodes in the node interval 1–21, 1–11 before and after reconstruction, and Fig. 5 is the graph of the voltage difference change of the nodes at both ends of the branch group before and after reconstruction.

Table 1. Voltage fluctuations of nodes equipped with µPMUs be-fore and after single contact switch operation.

Node	Amplitude fluctuation/pu	Phase fluctuation/deg	Node	Amplitude fluctuation/pu	Phase fluctuation/deg
1	0.000	0.003	8	0.046	0.473
2	0.160	0.147	11	0.020	0.704
5	0.103	0.645	20	0.055	0.438
7	0.079	0.464	21	0.047	0.853

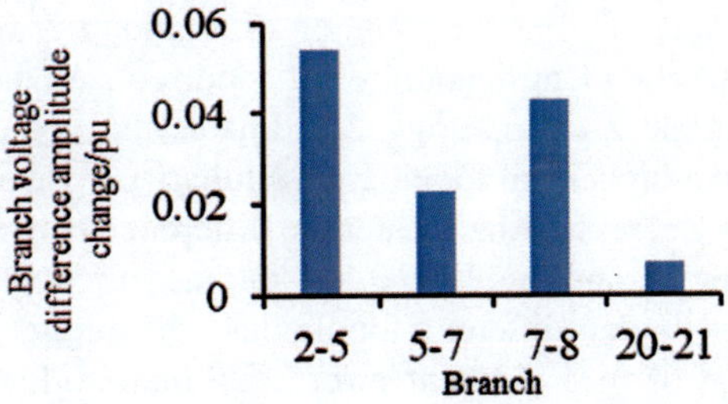

Fig. 5. Changes in branch voltage amplitude before and after single contact switch operation.

Analysis of Table 1 shows that within the range of branches formed by nodes 1–21, 1–11, criterion 2 preliminarily determines that the branch with possible segmented switching action is in branch group 2–5, 5–7, 7–8 and 20–21. Figure 5 shows that the branch 2–5 has the largest voltage difference, and from criterion 3, it can be seen that the branch with possible segment switching action is within the range of branch group 2–5, which can be formed by the enumeration method 3 possible topological situations. They are topology one: branch 2–3 is disconnected; topology two: branch 3–4 is disconnected; topology three: branch 4–5 is disconnected.

The values in the examples are actual measured by µPMU and DSACAD, and the redundancy is greater than 1, regardless of the virtual measurement and pseudo measurement, and the measurement error of different measurement devices is set differently, the accuracy of µPMU measurement has higher measurement accuracy than DSACAD. Discuss 4 kinds of measurement error combinations: (1) $e_1 = 20\%$, $e_2 = 30\%$ (2) $e_1 = 5\%$, $e_2 = 10\%$ (3) $e_1 = 2.5\%$, $e_2 = 5\%$ (4) $e_1 = 0.5\%$, $e_2 = 1\%$. Where, e_1 is µPMU measurement error, e_2 is DSCADA measurement error. In each case of error combination, the measured value is generated by Gaussian distribution. Where, the error cases two, three and four are within the normal measurement error range, and the error combination one is the case where the analog measurement device is unstable and the actual measurement value measurement error is large. The voltage phase of the μPMU node under the three topological situations is estimated, so that the similarity value under the three topologies can be obtained from the topological similarity identification model. Figure 6 is the topological similarity value under different measurement error combinations when the single contact switch operates.

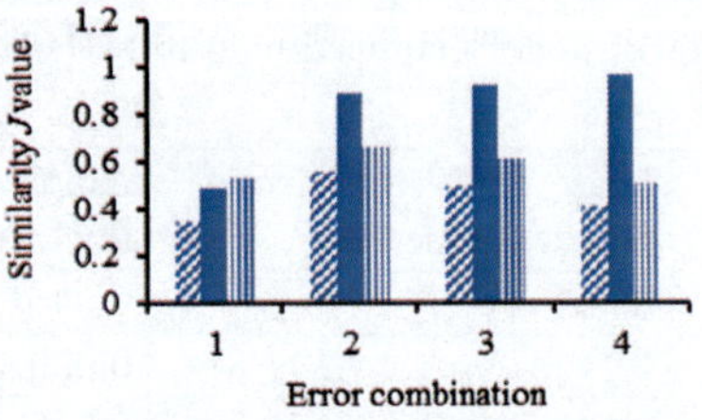

Fig. 6. Topological similarity *J* values under different measurement error combinations.

Figure 6 shows that the similarity value under topology 2 is closest to 1, so the operating topology is identified as the situation of topology 2, which is consistent with the actual topology and the identification result is correct. As the measurement smaller error, the closer the node 2 of topology 2 estimates the voltage phase to the actual measured value and the higher the topological similarity *J* value. At the same time, as the measurement error decreases, the greater the difference between the estimated phase of nodes under topology 1 and topology 3 and the measured value, the greater the gap between the *J* values under different topologies, the higher the sensitivity of the algorithm. In the case of measurement error combination 1, due to the larger measurement error, the measured value deviates from the true value getting more seriously, the topology similarity *J* value in topology 3 is the largest, and topology 3 is recognized as the actual topology and the recognition error.

4.2 Multi-contact Switch Action

When the network is in the initial state as shown in Fig. 2 and a reconstruction occurs at a certain moment. The tie switches TS1 and TS3 are closed, and the branches 3–4 and 8–9 are opened, forming the operating structure shown in Fig. 7. The node voltage of the reconstructed μPMU installation point is obtained by power flow calculation.

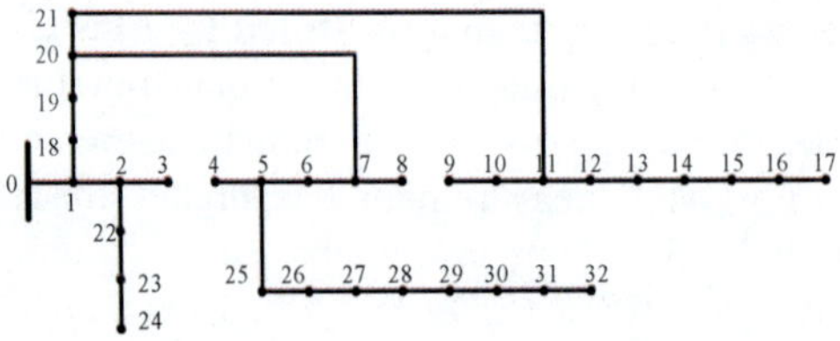

Fig. 7. Operational topology after multiple contact switch operation.

When the tie switches TS1 and TS3 are closed, since μPMU is installed at the nodes of TS1 and TS3, the current of the tie branch can be monitored to determine that TS1 and TS3 are closed. It can be seen from the conditions of the reconstruction that the branches that may have the status change of the segment switch are included in the two ring networks formed by TS1 and TS3, including the branch groups 1–2, 2–5, 5–7, 1–20 and 7–8, 8–11, 20–21 of the branch groups in the TS3 ring. Table 2 shows the absolute value of the voltage fluctuations of the μPMU installation nodes in the above 7 branch groups before and after reconstruction.

Table 2. Voltage fluctuations of nodes equipped with μPMUs before and after multiple contact switch operation.

Node	Amplitude fluctuation/pu	Phase fluctuation/deg	Node	Amplitude fluctuation/pu	Phase fluctuation/deg
1	0.001	0.101	8	0.069	1.917
2	0.046	0.136	11	0.042	2.013
5	0.097	2.047	20	0.003	1.645
7	0.055	1.829	21	0.018	1.759

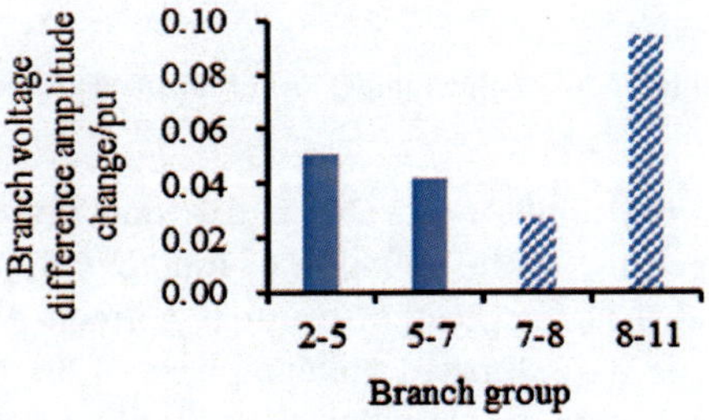

Fig. 8. Changes in branch voltage amplitude before and after multiple contact switch operation.

Table 3 shows that the branch groups satisfying Criterion 2 are 2–5, 5–7, 7–8, and 8–11. Figure 8 shows that in the loop formed by TS1, the voltage difference between the two ends of the branch group 2–5 has the most fluctuating, and in the loop formed by TS3, the voltage difference of the two ends of the branch group 8–11 has the most fluctuating. According to the criterion 3, the branch group 2–5 and the branch group 8–11 are determined as the areas where switching action may occur. According to the necessary conditions for the reconstruction of the distribution network, the branches in the two branch groups are arranged and combined to enumerate 9 possible topologies, as shown in Table 3, where topology 4 is the real topology after reconstruction. As in the case of the single contact switch action, the four measurement error combinations are discussed separately, and the similarity values of different topologies under different measurement error combinations are obtained, as shown in Fig. 9.

Table 3. Possible operational topology after switch operation.

Topology number	Broken branch	Topology number	Broken branch
1	2–3, 8–9	6	3–4, 10–11
2	2–3, 9–10	7	4–5, 8–9
3	2–3, 10–11	8	4–5, 9–10
4	3–4, 8–9	9	4–5, 10–11
5	3–4, 9–10		

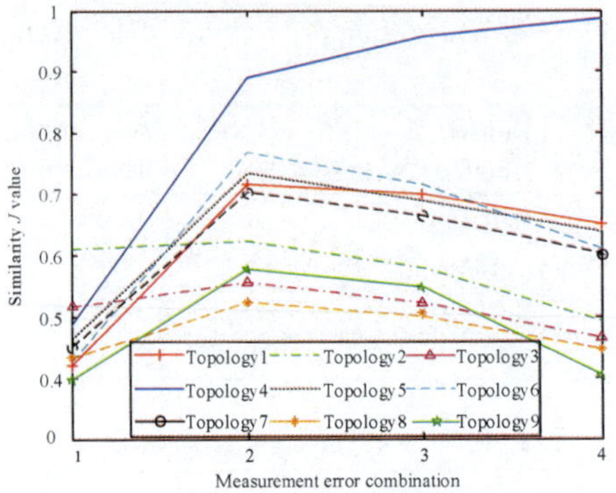

Fig. 9. Topological similarity J values under different measurement error combinations.

Figure 9 shows that the similarity *J* value under topology 4 is closest to 1, and the running topology is identified as the situation of topology 4, which is consistent with the actual topology and the identification result is correct. And as the measurement error decreases, the closer the estimated voltage phase of the node of topology 4 is to the measured value, the higher the topological similarity *J* value. In the case of measurement error combination 1, due to the larger measurement error, the measurement value deviates from the true value getting more seriously, the topology similarity *J* value in topology 2 is the largest, and topology 2 is identified as the actual topology and the identification error.

When combined with Fig. 6 and Fig. 9, it can be seen that the algorithm can achieve the accuracy of the reconstructed topology when the measurement device is within a certain normal error range in the case of single-contact switch action and multi-contact switch action identification, and the smaller the measurement error, the higher the sensitivity. When the measurement device is unstable or fails, and the measurement error is large, the reliability of the algorithm is affected to a certain extent.

5 Conclusion

In this paper, DSCADA measurement device is installed in the distribution network, and a small number of μPMU devices are installed in some nodes. A topology identification method for the distribution network is proposed. The identification model of the topology changing time and the criterion of the topology changing trend are constructed by measuring the voltage of the μPMU node to determine the time when the topology changes and the possible topology situation after the change. The telemetry data of μPMU and DSCADA are fused by WLS, and the voltage phase and the measured voltage phase are estimated by the nodes in various possible topological situations, and a topology similarity identification model is constructed to identify the actual running topology. The simulation of the IEEE33 network in the calculation example shows that the algorithm has a high accuracy when the measurement error of μPMU and DSACAD is within the normal range. When the measurement device fails or unstable, the identification accuracy is affected.

References

1. Liu, K., Zhang, J., Tao, S., et al.: Detection and evaluation method of voltage data quality of distribution network SCADA system based on multi-source and multi-space information. Power Syst. Technol. **39**(11), 3169–3175 (2015)
2. Luo, X., Duan, B., Wu, J., Lai, J.: Quantitative assessment method of SCADA system security vulnerability of wind farm based on evidence reasoning. Autom. Electr. Power Syst. **44**(11), 25–33 (2020)
3. Li, C., Li, X., Tian, S., Yuan, J.: Risk transmission of deep integration of energy internet power information: challenges and prospects. Autom. Electr. Power Syst. **41**(11), 17–25 (2017)
4. Cavraro, G., Arghandeh, R., Poolla, K., et al.: Data-driven approach for distribution network topology detection. In: 2015 IEEE Power & Energy Society General Meeting, pp. 1–5. IEEE (2015)
5. Deka, D., Backhaus, S., Chertkovy, M.: Estimating distribution grid topologies: a graphical learning based approach. In: 2016 Power Systems Computation Conference (PSCC), pp. 1–7. IEEE, Genoa (2016)
6. Si, X., Chen, Q., Gao, Z., et al.: Active distribution network protection based on the direction of current phase angle mutation amount. Power Syst. Autom. **38**(11), 97–103 (2014)
7. Zhai, H.F., Yang, M., Chen, B., et al.: Dynamic reconfiguration of three-phase unbalanced distribution networks. Int. J. Electr. Power Energy Syst. **99**, 1–10 (2018)
8. Amini, M.A., Jalilian, A., Behbahani, M.R.P.: A new method for evaluation of harmonic distortion in reconfiguration of distribution network. Int. Trans. Electr. Energy Syst. **30**(6) (2020)
9. Chen, X., Zhang, Y., Huang, X.: Overview of reactive voltage control methods in the background of active distribution network. Power Syst. Autom. **40**(01), 143–151 (2016)
10. Pappu, S.J., Bhatt, N., Pasumarthy, R., et al.: Identifying topology of low voltage (LV) distribution networks based on smart meter data. IEEE Trans. Smart Grid **PP**(99), 1 (2016)
11. Sudha, R., Murali, D.: Low-cost real-time design and implementation of phasor measurement unit using analog discovery. Microprocess. Microsyst. **76** (2020)
12. Li, Q., Zhang, J.: Solution of voltage over-limit of distribution network with distributed photovoltaic power supply. Power Syst. Autom. **22**, 117–123 (2015)
13. Martínez-Parrales, R., Fuerte-Esquivel, C.R., Alcaide-Moreno, B.A.: Analysis of bad date in power system state estimation under non-gaussian measurement noise. Electr. Power Syst. Res. **186** (2020)

Check for updates

Research on Cross-Regional Optimal Configuration of Emergency Reserve for Fluctuation Risk of Renewable Energy Output

Xiaohu Yang[1(✉)], Chen Yu[1], Shizhi He[2], Haohuai Wang[2], Dongliang Xie[1], and Linjun Cai[1]

[1] NARI Group Corporation (State Grid Electric Power Research Institute), Nanjing 211106, China
somnusyangxh@163.com
[2] Electric Power Dispatching Control Center of CSG, Guangzhou 510623, China

Abstract. In recent years, the centralized interconnection of renewable energy sources has led to a surge in the reserve demand of power grid. However, relying solely on conventional reserve for risk management and control obviously cannot meet the operational requirements of grid. In view of the shortcomings of the existing optimal configuration of reserve, this paper studies the optimal configuration of unconventional reserve, namely emergency reserve, which can participate in short-term regulation of power system and has additional configuration or regulation cost. Aiming at the problem that the existing optimal configuration of reserve is mostly limited to a single region, this paper is on the basis of risk method. With the goal of minimizing the total cost of global risk control, the cross-regional optimal configuration model and algorithm of emergency reserve are established. Through simulation analysis of examples, it is verified that the algorithm can better coordinate the global emergency reserve resources, and play the role of reducing costs, controlling risks and improving efficiency while sharing resources.

Keywords: Emergency reserve · Cross-regional optimal configuration · Risk management and control

1 Introduction

In recent years, the grid-connected capacity of renewable energy has increased year by year, making the cost of relying solely on conventional reserve for risk management and control constantly increasing [1]. At the level of power generation / utilization planning, some power facilities in the form of power source, load, storage or any combination also have a certain degree of adjustability, which can also be called reserve in a broad sense. Among these "unconventional reserve", those that can participate in short-term power system regulation and have additional configuration or regulation cost may be called "emergency reserve". With the increasing level of intelligence in the development of power grid, the technical conditions for emergency reserve participation in risk management and control have gradually matured [2–4].

Y. Xue et al. (Eds.): PMF 2020, LNEE 718, pp. 622–633, 2021.
https://doi.org/10.1007/978-981-15-9746-6_48

At present, there are mainly two types of optimal method applied to the problem of optimal configuration of emergency reserve [5]. Although the probabilistic method takes into account the uncertain impact of output fluctuation of renewable energy, this method only takes into account the reliability constraints of the system, and does not consider the economic benefits of reserve [6]. Compared with the probabilistic method, the risk method does not need to set the index that meets the reliability requirements of system in advance, but converts the reliability requirements into the economic cost or economic value of the stable and safe operation of the system. This type of method combines the economic cost or value of emergency reserve to optimize, seeks the balance between reliability and economy, and adaptively determines the range of probability space.

At the present stage, China's power dispatchers often follow the principle of partition independence when formulating reserve plans. In the independent region mode, the contradiction between the demand for reserve resources and the existing reserve capacity in regions rich in renewable energy resources is increasingly prominent [7]. Therefore, in this paper, based on the risk method, we will address the problem of unreasonable allocation of emergency reserve resources between regions, and jointly optimize conventional reserve and emergency reserve to realize cross-regional optimal configuration of emergency reserve.

2 Optimal Configuration Modeling of Emergency Reserve

2.1 Objective Function

This paper will be discussed on the basis of payment mechanism of the reserve. Under this mechanism, the dispatch department obtains the reserve dispatch right by signing a contract. The configuration of reserve capacity under this mechanism can be called "contract optimization". When performing risk management and control, conventional reserve and emergency reserve are inseparable. When the two types of reserve are involved in contract optimization, there are only difference in compensation methods and means at economic level. Therefore, in the context of contract optimization, the conventional reserve and emergency reserve will be modeled uniformly.

From the view of the instantaneous response, the various type of reserve unit involved in modeling can be roughly divided into two categories, namely, rotating unit and non-rotating unit. For these two types of unit, the control method can be divided into two types: preventive control and emergency control according to the time nodes at which units participate in control [8].

In the cross-regional optimal configuration of emergency reserve, the objective function is established with the goal of minimizing the total cost of global risk control. Taking the optimal configuration of two regions (sending region a_m and receiving region b_n) as an example, the objective function is as follows:

$$\min(Q_{a,m} + Q_{b,n}) \tag{1}$$

Where $Q_{a,m}$ represents the total cost of risk control in the sending region a_m; $Q_{b,n}$ represents the total cost of risk control in the receiving region b_n.

In the model, there is only the difference between the output and input of the cross-regional transmission power in the sending and receiving region, the following will take the sending region a_m as an example to explain the total cost model and constraints of regional risk control. The model and constraints of the receiving region are similar. The total cost model of risk control in the sending region a_m is as follows:

$$Q_{a,m} = \sum_{t=1}^{NT_{a,m}} \sum_{l=1}^{NS_{a,m}} \left\{ \sum_{i=1}^{NR_{a,m}} u_{a,m,i} \cdot (C_{fix}(a,m,i,l,t) + C_{risk}(a,m,i,l,t)) + C_{loss}(a,m,l,t) \right\} \tag{2}$$

Where l, t, i are the risk scenario number, optimization period number and unit number of the region a_m; $NS_{a,m}$, $NT_{a,m}$, $NR_{a,m}$ are the number of risk scenario concentrated in the set of risk scenario, the number of optimization period, and the number of unit participating in the optimization in region a_m; $u_{a,m,i}$ indicates the contract signing status of the $R_{a,m,i}$ (1 indicates signed, 0 indicates unsigned); $C_{fix}(a,m,i,l,t)$ indicates deterministic control cost of the $R_{a,m,i}$ under scenario $s_{a,m,l}$ within the period t; $C_{risk}(a,m,i,l,t)$ represents the risk control cost of the $R_{a,m,i}$ within the period t under scenario $s_{a,m,l}$; $C_{loss}(a,m,l,t)$ represents the control loss in period t under scenario $s_{a,m,l}$.

Among them, $C_{fix}(a,m,i,l,t)$ and $C_{risk}(a,m,i,l,t)$ can be further expressed by the following formula:

$$C_{fix}(a,m,i,l,t) = (q_{a,m,i,t}^{\max} \cdot p_{cap,a,m,i} + v_{a,m,i} \cdot k_{a,m,i} \cdot p_{down,a,m,i} + v_{a,m,i} \cdot (1 - k_{a,m,i}) \cdot p_{up,a,m,i}) \cdot \rho_{a,m,l} \cdot \Delta t \tag{3}$$

$$C_{risk}(a,m,i,l,t) = \{q_{a,m,i,l,t} \cdot p_{ele,a,m,i} + I_{a,m,i}(t-1) \cdot (1 - I_{a,m,i}(t)) \cdot p_{down,a,m,i} + I_{a,m,i}(t) \cdot (1 - I_{a,m,i}(t-1)) \cdot p_{up,a,m,i}\} \cdot \rho_{a,m,l} \cdot \Delta t \tag{4}$$

Where $q_{a,m,i,t}^{\max}$ represents the maximum output of $R_{a,m,i}$ in period t; $p_{cap,a,m,i}$ is the capacity price of $R_{a,m,i}$; $v_{a,m,i}$ is a binary variable in the form of control, 0 represents emergency control after the event, 1 represents preventive control in advance; $k_{a,m,i}$ is a binary variable of the initial operating state of $R_{a,m,i}$, 0 means non-rotating state, 1 means rotating state; $p_{down,a,m,i}$ is the cost of downgrading of $R_{a,m,i}$; $p_{up,a,m,i}$ is the cost of upgrading of $R_{a,m,i}$; $\rho_{a,m,l}$ is the probability of $s_{a,m,l}$ in region a_m; Δt is the duration of each period; $q_{a,m,i,l,t}$ is the output within the period t under $s_{a,m,l}$; $p_{ele,a,m,i}$ is the electricity price of $R_{a,m,i}$; $I_{a,m,i}(t)$ is the running state of $R_{a,m,i}$ in period t, 0 means non-rotating state, 1 means rotating state.

2.2 Constraints

Taking the sending region a_m as an example, the constraints of this model include power balance constraint, regulation rate constraint, output upper and lower limit constraint, minimum continuous output / interruption time constraint, start-stop / interruption interval time constraint, start-stop / interruption time constraint, state of charge constraint. In particular, there are transmission power constraint of transmission line. The specific constraints are as follows:

$$\begin{cases} \sum_{i=1}^{NR_{a,m}} u_{a,m,i} \cdot q_{a,m,i,l,t} = D_{a,m,l,t} \ t \in (1, NT_{a,m}) \\ -q_{a,m,i}^{\max} \cdot \lambda_{down,a,m,i} \cdot \Delta t \le q_{a,m,i,l,t} - q_{a,m,i,l,t-1} \le q_{a,m,i}^{\max} \cdot \lambda_{up,a,m,i} \cdot \Delta t \ t \in (2, NT_{a,m}) \\ q_{a,m,i}^{\min} \le q_{a,m,i,l,t} \le q_{a,m,i}^{\max} \ t \in (1, NT_{a,m}) \\ (d_{a,m,i,t}^{on} - T_{a,m,i,\min}^{on})(I_{a,m,i}(t-1) - I_{a,m,i}(t)) \ge 0 \ t \in (2, NT_{a,m}) \\ (d_{a,m,i,t}^{off} - T_{a,m,i,\min}^{off})(I_{a,m,i}(t) - I_{a,m,i}(t-1)) \ge 0 \ t \in (2, NT_{a,m}) \\ \sum_{t=1}^{NT_{a,m}} |I_{a,m,i}(t) - I_{a,m,i}(t-1)| \le M_{a,m,i} \ t \in (2, NT_{a,m}) \\ SOC_{a,m,i}^{\min} \le SOC_{a,m,i,l,t} \le SOC_{a,m,i}^{\max} \\ L_{m,n,t} \le P_{m,n,t} \end{cases} \tag{5}$$

Where $D_{a,m,l,t}$ is the capacity deficit in period t of $s_{a,m,l}$ in region a_m; $\lambda_{up,a,m,i}$, $\lambda_{down,a,m,i}$ are the upward and downward adjustment of the output rate of $R_{a,m,i}$, respectively; $q_{a,m,i}^{\max}$ is the maximum output constraint of $R_{a,m,i}$; $q_{a,m,i}^{\min}$ is the minimum output constraint of $R_{a,m,i}$; $d_{a,m,i,t}^{on}$ is the actual continuous output time of $R_{a,m,i}$ before period t, if the control measure is interruptible load, it is interruption time; $T_{a,m,i,\min}^{on}$ is the minimum continuous output time of $R_{a,m,i}$; $d_{a,m,i,t}^{off}$ is the time constraint of continuous non-output state of $R_{a,m,i}$ before period t; $T_{a,m,i,\min}^{off}$ is the minimum continuous non-output time of $R_{a,m,i}$; $M_{a,m,i}$ is the limit of switching times between the maximum output state and the non-output state of $R_{a,m,i}$; $SOC_{a,m,i,l,t}$ is the state of charge of the energy storage unit $R_{a,m,i}$ in period t under $s_{a,m,l}$; $SOC_{a,m,i}^{\min}$ is the minimum state of charge of the energy storage unit $R_{a,m,i}$; $SOC_{a,m,i}^{\max}$ is the maximum state of charge of the energy storage unit $R_{a,m,i}$; $L_{m,n,t}$ is the transmission capacity of the transmission line between the region a_m and region b_n; $P_{m,n,t}$ is the planned transmission capacity of the transmission line between region a_m and region b_n in period t.

3 Cross-Regional Optimal Configuration Algorithm of Emergency Reserve

Literature [9] puts forward the concept of cost performance ratio (CPR). The essence of this ratio is reflected and this indicator reflects the risk cost of reserve unit providing unit effective reserve power under the target scenario set. Therefore, taking $R_{a,m,i}$ as an example, the expression of CPR under scenario $s_{a,m,l}$ can be rewritten as the following formula:

$$\phi(R_{a,m,i}) = \frac{\sum_{l=1}^{NS_{a,m}} \sum_{t=1}^{NT_{a,m}} (C_{fix}(a,m,i,l,t) + C_{risk}(a,m,i,l,t) + C_{apportion}(a,m,i,l,t))}{\sum_{l=1}^{NS_{a,m}} \sum_{t=1}^{NT_{a,m}} q_{a,m,i,l,t} \cdot \rho_{a,m,l} \cdot \Delta t} \tag{6}$$

In the above formula, $C_{apportion}(a,m,i,l,t)$ is the cost allocation for the units participating in the optimization in period t under scenario $s_{a,m,l}$.

For each region of the power grid, if it independently carries out risk management and control, some measures with lower CPR may be selected. Based on formula (6), the control measures that are selected later in the round, the worse the cost performance ratio is. Therefore, when the regional power grid suffers from emergency resources without considering cross-regional optimal configuration of resource, in order to deal with possible risk, the economic efficiency of emergency reserve configured in the power grid may be poor. Even when there is a shortage of emergency resources, risk leakage will occur. At this time, the leakage of risk will be controlled by passive means, and the total cost of risk control in the whole process is high. In response to this problem, it may be useful to set the region with abundant emergency reserve resources and connected to the receiving grid region with the connection line as the sending region, and transfer part of the risk in receiving region to the sending region through the transmission line. This process can be called risk transferring. Cover this part of risk with emergency reserve with better cost performance in the sending region, thereby improving the reliability and economy of risk management and control of the entire region in power grid. It should be noted that, the power provided by the emergency reserve in sending region may not be sent to the receiving region in full due to the constraint of transmission capacity. That is, when carrying out risk transferring, due to the capacity constraints of the transmission line, some risk may not be transferred to the sending region and be handed over to the emergency reserve of the sending region for risk management and control. May wish to record this part of the risk that can not be transferred as blocking risk. The blocking risk can only be managed in receiving region. At this time, the total cost of the risk control of the receiving region will not be reduced as expected, which is reflected in the failure of the expensive measures that will be replaced by this round, and the total cost of risk control of sending region will also increase.

Specifically, as shown in Fig. 1, the specific process of the algorithm of cross-regional optimal configuration for emergency reserve is as follows:

(1) Based on the information of the power grid to be optimized, unit parameters and other information, the quick optimal algorithm [8] is applied to complete initial optimal configuration of various types of unit in the sending and receiving regions respectively.

(2) Remove the emergency reserve with the worst CPR in the configuration plan of the receiving region, update the set of risk scenario to be supported after the emergency reserve removed, and combine the constraint of transmission capacity to update the synthetic set of risk scenario of the sending region, then generate an optimal configuration plan for the sending region.

(3) When transferring risk, if the transmission line is not blocked, the control plan after the removal of emergency reserve in the previous step is taken as the unit configuration of the receiving region in this round. If the transmission line is blocked, for the risk that is transferred unsuccessfully, based on the remaining unselected unit measures in the receiving region, optimize the unit under this risk scenario, and generate the optimal configuration plan.
(4) Integrate the configuration plan, conduct global check of risk, and determine whether there is an emergency reserve that can be reduced in the receiving region. If it exists, take the current round of configuration plan in receiving area as the plan to be optimized in the next round, and perform the optimization of next round. If it does not exist, the optimization is ended, and the set of plan with the smallest total cost of risk control is selected as the cross-regional optimal configuration of emergency reserve.

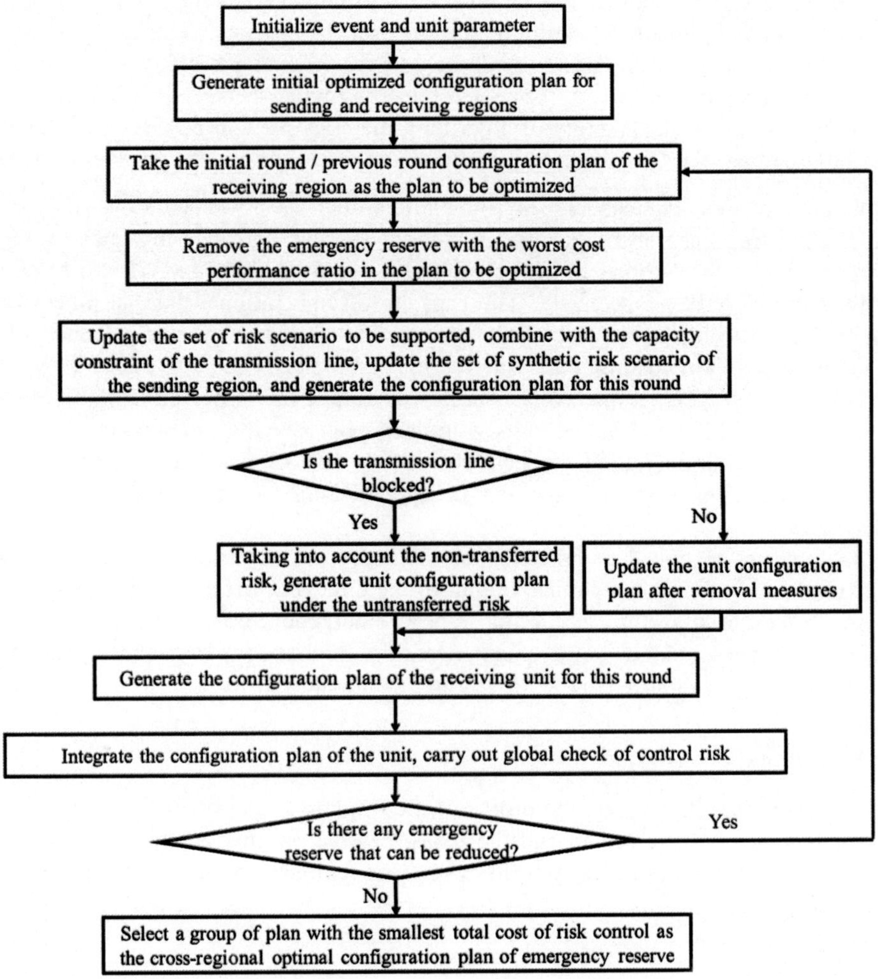

Fig. 1. Flow chart of cross-regional optimal configuration algorithm of emergency reserve

4 Example Analysis

4.1 Parameter Setting

The example simulation is based on the dynamic simulation platform for macro-energy systems (DSMES) [10–12].

Input Scenario Set. Based on the forecast of wind power in a provincial power grid in China in 2019, the output scenario set to be optimized is constructed. The research duration of scenario set is 48h, and the time resolution is 1h. Due to the large number of scenarios, and the reduction of scenarios is not within the scope of the discussion, scenarios in the scenario set of wind power prediction are now aggregated into "strong wind scenario", "moderate wind scenario", and "light wind scenario" according to the degree of output of wind power forecast. Subsequently, a number of scenarios are selected from these three types of scenario and combined to generate an input scenario set for the following study. Any situation can be simulated by adjusting the probability of each type of scenario in the input scenario set. In order to be close to the actual control environment of the power grid, the calculation example will first be discussed based on the input scenario set dominated by strong wind scenario. After that, in order to further verify the adaptability of algorithm, the example will be discussed based on the input scenario set dominated by moderate wind scenario and light wind scenario.

Unit Parameters. The capacity of conventional thermal power units, thermal power units with deep peak shaving function, energy storage, and ILH in receiving region participating in the bidding in the calculation example is 14680 MW, 990 MW, 900 MW, and 400 MW, respectively. The capacity of conventional thermal power units, thermal power units with deep peak shaving function, energy storage, and ILH in sending region participating in the bidding in example is 29760 MW, 2200 MW, 2000 MW, and 1500 MW, respectively. The lower output limits of conventional thermal power units and thermal power units with deep peak shaving function are set to 50% and 40% of maximum output, respectively. 40% to 50% of maximum output of thermal power units with deep peak shaving function belongs to the space of depth peak shaving.

In terms of technical parameters, this example only considers the constraints of maximum and minimum technical output of the emergency reserve. In the subsequent research, constraints such as climbing rate of emergency reserve can be further considered. In terms of economic parameters, the economic parameters of the unit's capacity fee, start-stop fee are set according to literature [5].

4.2 Simulation Analysis

The simulation will compare the results of the single-regional optimal configuration of the emergency reserve region and the cross-regional optimal configuration, and then discuss the advantages of the algorithm of cross-regional optimal configuration.

4.3 Scenario Set Dominated by Strong Wind

Single-Regional Optimal Configuration. Figure 2, Fig. 3 and Fig. 4 show the output of various units under different scenarios when the single-regional optimal configuration is considered. Due to the dominance of strong wind scenarios, the running capacity of conventional thermal power is low, this measure further reduces the amount of wind abandonment in strong wind scenario (see Fig. 2). However, due to insufficient capacity of emergency reserve, when the moderate and light wind scenario unexpectedly appears, as shown in Fig. 3 and Fig. 4, a large number of expensive emergency reserve such as energy storage and ILH are selected.

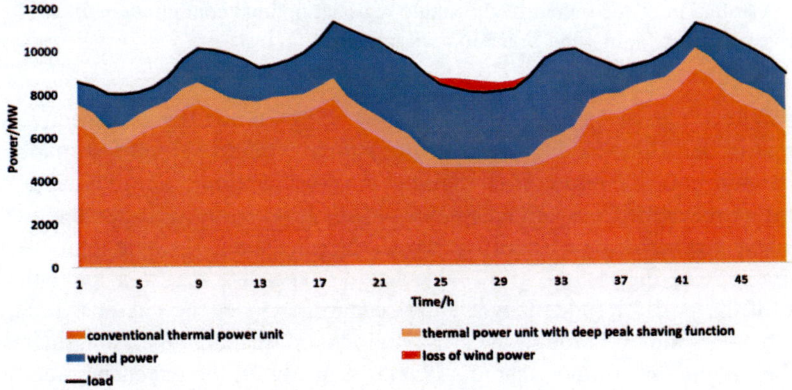

Fig. 2. Distribution of the output of the single-regional optimal configuration under the strong wind-driven forecast (strong wind scenario appears)

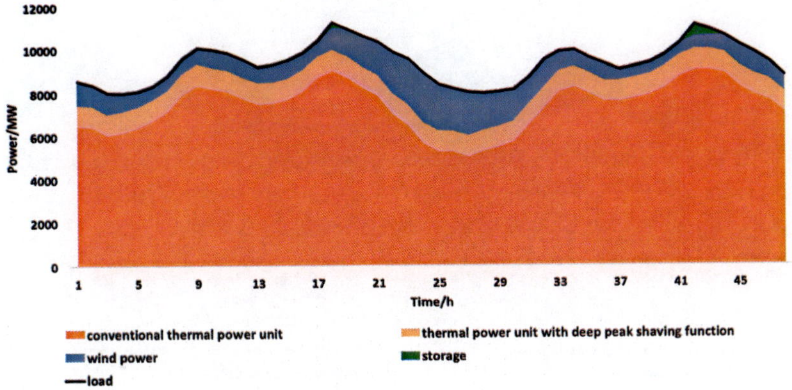

Fig. 3. Distribution of the output of the single-regional optimal configuration under the strong wind-driven forecast (moderate wind scenario appears)

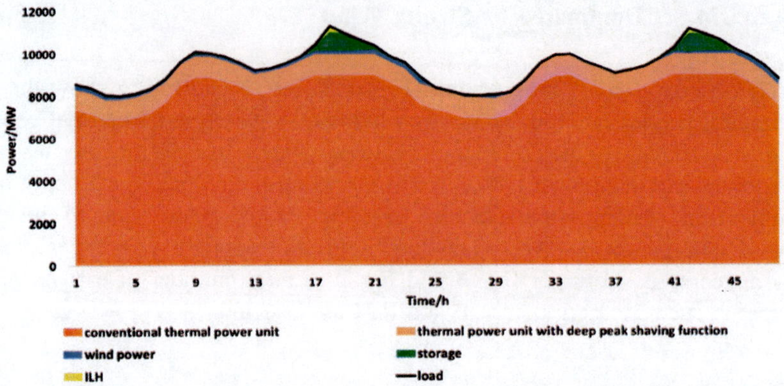

Fig. 4. Distribution of the output of the single-regional optimal configuration under the strong wind-driven forecast (light wind scenario appears)

Cross-Regional Optimal Configuration. Figure 5, Fig. 6 and Fig. 7 respectively show the output of units in receiving region under strong wind-driven input scenario set. As part of resource outside region is configured, the consumption space of wind power has been further released. Comparing Fig. 5 and Fig. 2, it can be seen that the amount of wind abandoned has dropped significantly. Comparing result of single-regional optimal configuration (see Fig. 4), the grid invokes some expensive ILH to achieve the power balance of the grid for sudden light wind scenarios. Under the circumstances, this is really a helpless move. Through the cross-regional optimal configuration, the resources in sending region are further shared. As shown in Fig. 7, the expensive control measures such as the interruptible load originally configured in Fig. 4 are replaced. From the perspective of the total cost of risk control, the advantages of cross-regional optimal configuration are further reflected.

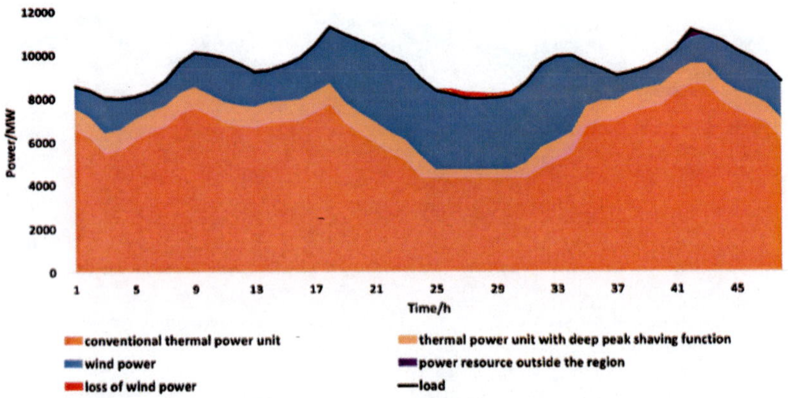

Fig. 5. Distribution of the output of the cross-regional optimal configuration under the strong wind-driven forecast (strong wind scenario appears)

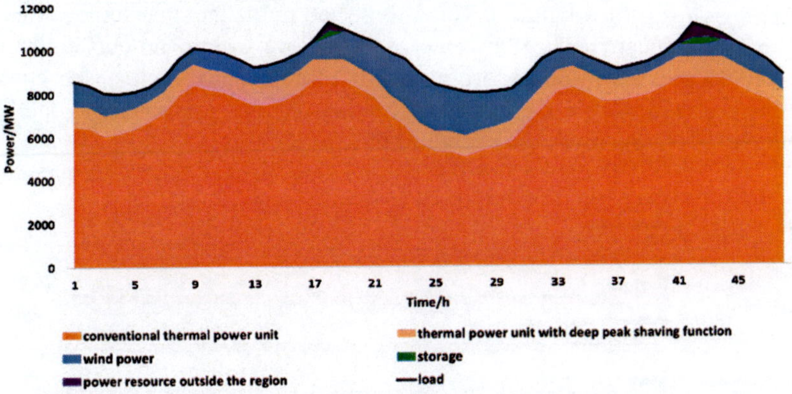

Fig. 6. Distribution of the output of the cross-regional optimal configuration under the strong wind-driven forecast (moderate wind scenario appears)

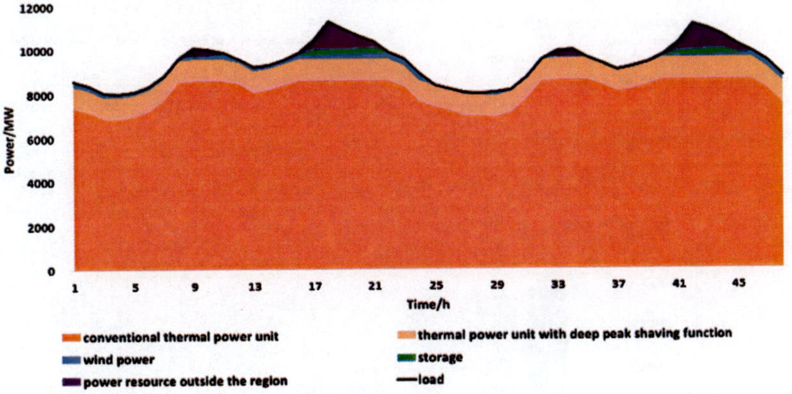

Fig. 7. Distribution of the output of the cross-regional optimal configuration under the strong wind-driven forecast (light wind scenario appears)

Comparison Between Single-Regional Optimal Configuration and Cross-Regional Optimal Configuration. To further illustrate the effect of cross-regional optimal configuration, Fig. 8 and Fig. 9 compare the results of single-regional optimal configuration and cross-regional optimal configuration from the aspects of configuration capacity of various types of unit and total cost of regional risk control.

As shown in Fig. 8 and Fig. 9, no matter what kind of prediction information is input, the algorithm proposed in this paper can adaptively optimize the configuration of various types of unit resource within and outside the region. In the simulation, the receiving region is set to a region with relatively high risk. With the emergency reserve outside access with relatively good CPR, the problem of shortage in resource of emergency reserve has been solved in some scenarios. In addition, by optimizing the configuration across regions, the dependence of conventional thermal power unit in receiving region is further reduced. This is reflected in the fact that some conventional thermal power units are replaced by emergency reserve within and outside the region. At the same time, the resources from outside also partially replace the emergency reserve with relatively poor CPR in the receiving region, such as energy storage and ILH, which further improves the economy of control.

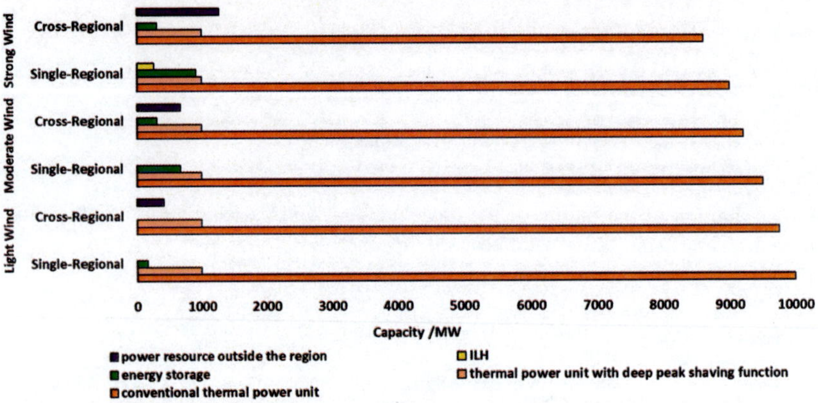

Fig. 8. Comparison of the configuration capacity of various units in the receiving region under different dominant scenario input

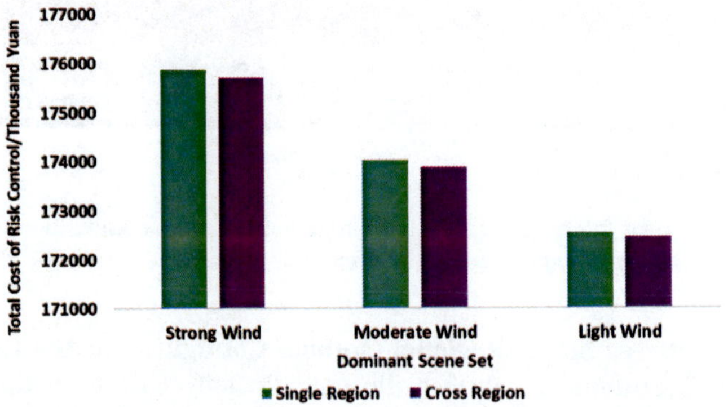

Fig. 9. Comparison chart of the total cost of risk control in the receiving region under different dominant scenario

5 Conclusion

In order to solve the problem of cross-regional optimal configuration of emergency reserve, this paper aims at minimizing the total cost of global risk control, takes the transmission power constraints of transmission line in each period into consideration, puts forward the idea of "risks are transmitted across regions through connection lines, which in turn drives the sharing and mutual benefit of multiple regional reserve resources". Based on this idea, the cross-regional optimal configuration model and algorithm of emergency reserve are established. This algorithm is designed to replace the emergency reserve with poor CPR in the receiving area by the emergency reserve with better CPR in the sending area. The algorithm is implemented based on DSMES. Through the simulation analysis, it is verified that the algorithm can better coordinate the global emergency reserve resources. Also, this algorithm can play a role in reducing costs, controlling risks and improving efficiency when sharing power resources.

Acknowledgements. This work is supported by National Key R&D Program of China (Technology and application of wind power/photovoltaic power prediction for promoting renewable energy consumption, 2018YFB0904200) and Eponymous Complement S&T Program of China Southern Power Grid.

References

1. Tao, R., Li, F., Li, Y., Zhou, S., Gao, G., Fu, L.: Optimal configuration of generalized spinning reverse for power grid based on characteristics of system frequency response. Autom. Electr. Power Syst. **43**(9), 82–91 (2019)
2. Xie, D., Zhang, Y., Wu, J., Liu, C., Wu, J.: Charging optimization strategy and capability analysis as reserve measures for electric vehicles under limited-capacity area. Autom. Electr. Power Syst. **42**(11), 56–63 (2018)
3. Wu, J., Xue, Y., Xie, D.: Optimization of reserve service capability made by electric vehicle aggregator. Autom. Electr. Power Syst. **43**(9), 75–81 (2019)
4. Chen, Z., Zhang, Y., Ma, G., Guo, C., Zhang, J.: Two-stage day-ahead and intra-day robust reserve optimization considering demand response. Autom. Electr. Power Syst. **43**(24), 67–76 (2019)
5. Wu, J., Xue, Y., Shu, Y., Xie, D., Xue, F., Xu, J.: Adequacy optimization for a large-scale renewable energy integrated power system part one spinning-grade reserve optimization. Autom. Electr. Power Syst. **43**(8), 101–110 (2019)
6. Qu, C., Wang, X., Yao, L., Han, J.: Optimal configuration of spinning reserve based on conditional cost/benefit analysis. Autom. Electr. Power Syst. **38**(14), 62–69 (2014)
7. Li, S.: Research on Cross-regional Reserve Sharing in InterconnectedSystem with High Wind Power Penetration. Huazhong University of Science & Technology (2019)
8. Wu, J., Xue, Y., Shu, Y., Xie, D., Song, X., Xu, Y.: Adequacy optimization for a large-scale renewable energy integrated power system part two collaborative optimization of multi-grade reserve. Autom. Electr. Power Syst. **43**(10), 19–28 (2019)
9. Wu, J., Xue, Y., Cai, B., Song, X., Xu, J., Xie, D.: Unified optimization of generation and demand side reserves. In: 2014 International Conference on Power System Technology (POWERCON), pp. 962–968. IEEE, Chengdu (2014)
10. Huang, J., Xue, Y., Xu, J., Xue, F., Zou, Y.: Dynamic simulation platform for power market and power system part one function design. Autom. Electric Power Systems 35(10), 16–22 (2011)
11. Xie, D., Xue, Y., Xue, F., Luo, J.: Dynamic simulation platform for power market and power system part two support layer design. Autom. Electr. Power Syst. **35**(11), 1–7 (2011a)
12. Xie, D., Xue, Y., Xue, F., Luo, J.: Dynamic simulation platform for power market and power system part three application layer design. Autom. Electr. Power Syst. **35**(12), 7–14 (2011b)

Research on Single-Phase Grounding Line Selection in Distribution Network Based on Signal Injection

Jiaxin Yuan and Luqing Sun(✉)

School of Electrical Engineering and Automation, Wuhan University, Wuhan 430072, China
whuslq633@163.com

Abstract. New energy power injection to the distribution network has increased the difficulty of fault line selection. A new approach to solve the problem by using new energy power generation devices is proposed in this paper. And a fault line selection method based on characteristic signal injection in the distribution network is proposed. In the situation of single-phase short-line grounding fault, characteristic signal is injected to the distribution network via inverters in a short period by changing its modulation strategy. With the help of distributed zero sequence current measuring devices, zero sequence currents are analyzed and the content of characteristic signal at every measuring point is compared. Then the fault line is selected and the fault identification is completed. Simulation model of typical active distribution network is set up and the contents of characteristic signal are analyzed when the values of grounding resistances are different. According to the simulation result, compared with traditional signal injection method, the new one proposed in this paper has greater measuring. The accuracy of fault line selection can be improved effectively.

Keywords: Single-phase grounding · Signal injection via inverters · Fault line selection

1 Introduction

Improving the stability and safety of operation in distribution network is an inevitable requirement of society development. At the same time, the distribution network is more complicated because of society development. In addition, potential safety hazards have been increased because of the injection of power electronics devices such as large capacity inverters to distribution network. On the other hand, capacitive currents after fault occurring have been improved significantly because of the increase of both lines' distance and cable lines' proportion. Grounding faults add serious threats to devices and people. The chance of single-phase short-line grounding fault occurring in the distribution network is over 80% [1]. Thus the problem is needed to be solved.

Inverters without isolation transformers can be connected to the distribution network directly. Power electronics devices' actions generate high frequency components of higher randomness. If capacitive currents in certain healthy feeders account for

Y. Xue et al. (Eds.): PMF 2020, LNEE 718, pp. 634–646, 2021.
https://doi.org/10.1007/978-981-15-9746-6_49

higher proportions of those in distribution network, fault line selection methods based on stationary signals are easier to bring about errors. What's more, high frequency components directly exist in transient process after the fault occurring, affecting the selection of energy spectrum of transient currents.

At present, further studies about single-phase short-line grounding fault in power system have made certain achievements. According to different principles, existing fault line selection methods can be divided into passive selection methods and active ones. The former ones are based on stationary signals [2], transient signals [3–5] or transient-stationary signals. Methods based on stationary signals mainly include colony amplitude and phase comparison method [6], zero sequence admittance method [7, 8], and zero-sequence reactive power method [9]. These are easily affected and limited by factors such as relatively weak signals and system disturbances. Methods based on transient signals mainly include first semi-wave method [10] and wavelet transform method [11–13]. As for studies on the devices for fault line selection, combined selection methods are adapt to trend at present in order to increase the accuracy. For example, transient, and other analyzing methods are used in MD-MLA-2B fault line selection device. It was jointly developed by Gansu Mingzhu Electric Power Science Park Co., Ltd. And North China Electric Power University [14]. Active selection methods mainly include medium resistance method [15, 16] and signal injection method. Signal injection method is about to inject certain current signal via devices to select the fault line [17–19]. When high resistance earth fault happening in the system, the effect of this method is relatively poor. Journal article [20] expounds the improving measures for signal injection method.

Certain difficulties have existed in single-phase grounding fault line selection in distribution network for a long time. Especially when single-phase high-impedance grounding fault occurring, the measure of characteristic signal is relatively weak and difficult to detect. Considering that after single-phase short-line grounding fault occurring in the power system, a current signal with certain character is actively injected for a short period via inverters. Then the fault line can be selected after detecting the contents of characteristic signal.

2 Analyses of Single-Phase Short-Line Grounding Fault in Distribution Network with Inverter Injection

Figure 1 shows the schematic diagram of single-phase short-line grounding fault occurring in the typical distribution network. Zero sequence current value in fault line equals to the sum of grounding capacitance current values in non-fault lines. And the direction of zero sequence current in the fault line is opposite to those in non-fault lines. When arc suppression coil is in the state of overcompensation, zero sequence currents flow through the fault line and non-fault lines are of the same direction. Thus the direction of zero sequence currents cannot be the criterion for fault line selection.

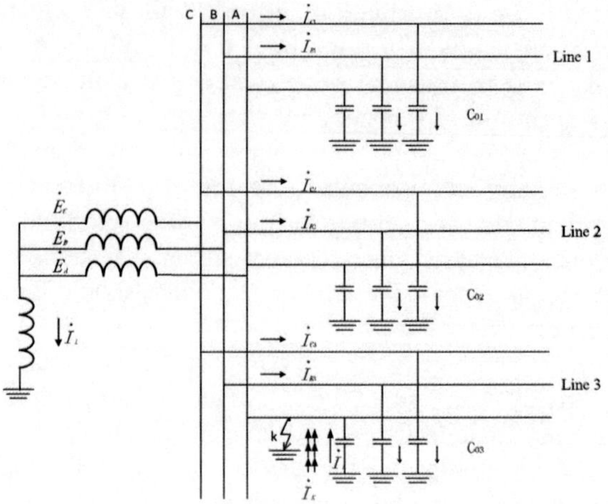

Fig. 1. Schematic diagram of current distribution after single-phase short-line grounding fault occurring in the distribution network with neutral grounding method with an arc suppression coil

2.1 Inverters Topology

Figure 2 shows the topology diagram of a typical three-phase bridge type inverter directly injected to the distribution network. The inverter circuit is consisted of full-controllable electric semiconductor devices IGBT T_1–T_6 and anti-paralleled diodes.

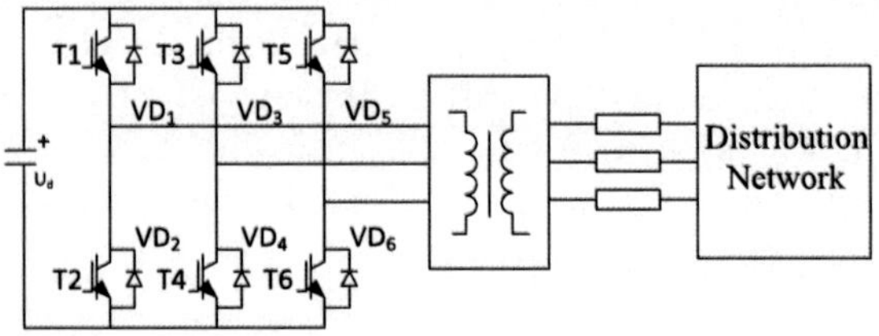

Fig. 2. Topology diagram of the typical three-phase bridge type inverter

2.2 Path Analyses and Equivalent Circuit of Characteristic Signal Injection via Inverters

Once detecting that a permanent fault occurs, a characteristic signal is injected via inverters to the distribution network in a short time. The signal flows through capacitance to earth of lines into ground and flows back via grounding resistance. Then a loop is formed. Characteristic signal current distribution in resonance grounding system is similar to that in isolated neutral system. Then the path of characteristic signal does not have too many differences. Figure 3 shows the schematic diagram of the signal path after it is injected via inverters.

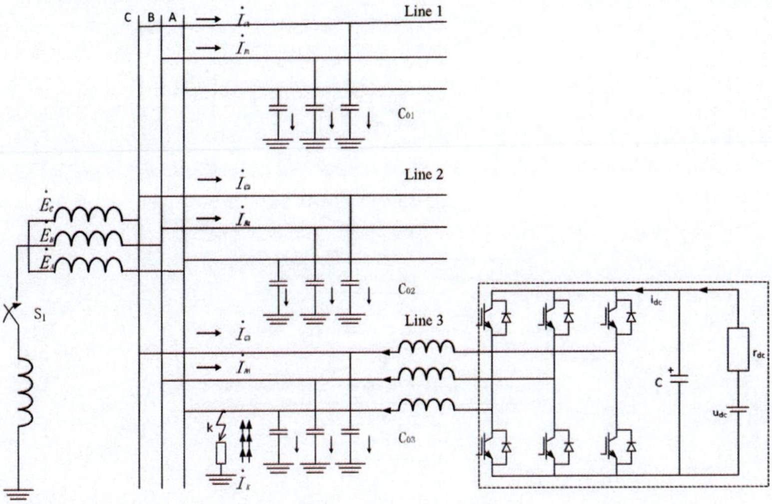

Fig. 3. Schematic diagram of characteristic signal injection via inverters

The schematic diagram of the equivalent circuit is showed in Fig. 4. Switch S_1 is closing means that a characteristic signal is injected via inverters to the distribution network. Switch S_2 is closing means that distribution network is a system with neutral grounding method of an arc suppression coil, on the other hand, it is breaking means that distribution network is a system of isolated neutral one. L and C_0 mean arc suppression coil inductance and distributed capacitance of lines themselves, respectively.

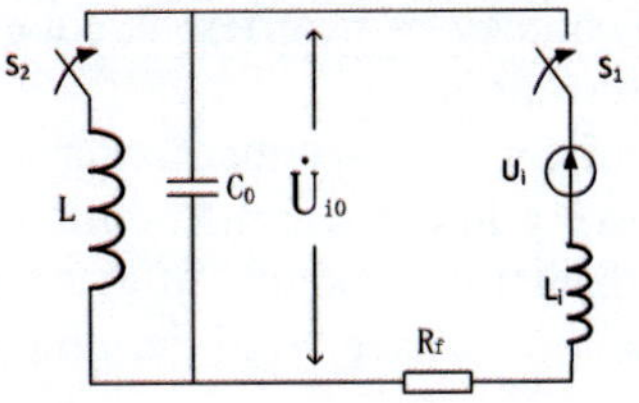

Fig. 4. Equivalent circuit diagram of single-phase grounding fault with characteristic signal injection via inverters

3 Characteristic Signal Injection via Inverters Method for Fault Line Selection

After single-phase short-line grounding fault in distribution network occurring, control strategy of inverters is adjusted to inject characteristic signal directly. And the characteristic current which flows through the fault line is the sum of those flow through non-fault lines. Therefore, the content of characteristic signal in the fault line is the highest. The detailed workflow of fault line selection is showed in Fig. 5.

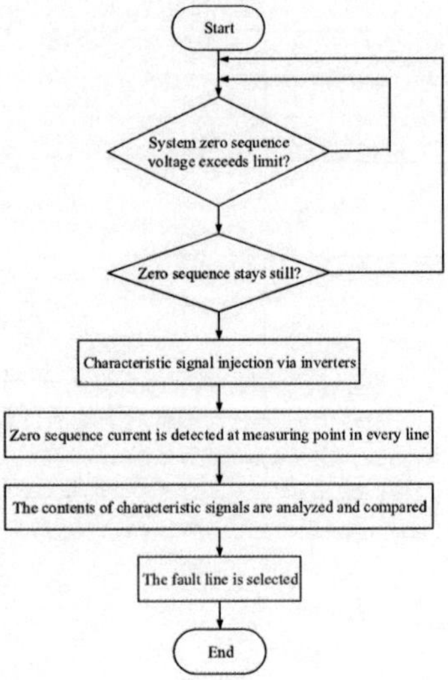

Fig. 5. Flow diagram of single-phase grounding fault line selection with characteristic signal injection via inverters

Fault line selection process is as following:

Step 1: Zero sequence voltage in system is detected by TV that it exceeds the limit in the distribution network. Then a single-phase short-line grounding fault is recognized.
Step 2: The zero sequence voltage stays still. Then a permanent grounding fault is recognized.
Step 3: The modulation strategy of inverters is changed. And it is ready to inject characteristic signal via inverters.
Step 4: Zero sequence current of every measuring point can be gotten via current detecting equipment of every line in the distribution network.
Step 5: The content of characteristic signal of every measuring point is analyzed.
Step 6: The fault line can be selected. And the process is finished.

4 Simulation Analysis of Characteristic Signal Injection via Inverters Method

According to the practical situation, the voltage grade of power module on the system side is 110 kV. And it belongs to the category of the transmission network. The voltage grade of transformer is 110 kV/10 kV. Four lines are set up in the feeder side of 10 kV. These lines are a single overhead line, cable lines and an overhead-cable mixed line, respectively. C_{Li} represents cable lines and O_{Li} represents overhead lines. Single-phase short-line grounding faults occur in the end of Line 1, the middle of Line 3 and the end of Line 4, marked as F1, F2 and F3, respectively. All above are showed in Fig. 6.

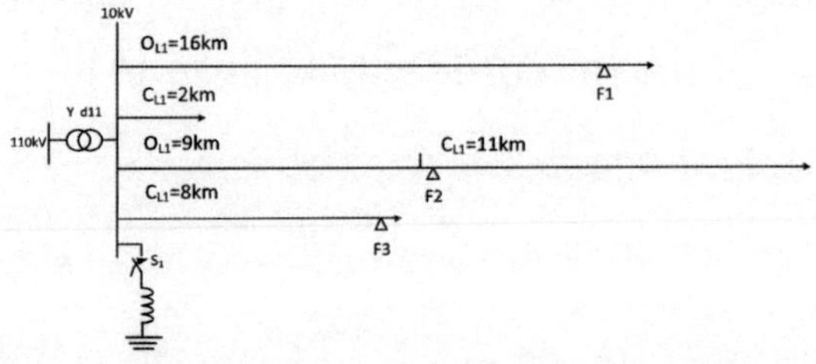

Fig. 6. Schematic diagram of fault position in typical distribution network

Table 1. Parameters of Overhead Lines

Phase sequence	Resistance (Ω/km)	Capacitance (nF/km)	Inductance (mH/km)
Positive sequence	0.1273	9.6900	1.2546
Zero sequence	0.2300	4.7561	4.6553

Table 2. Parameters of Cable Lines

Phase sequence	Resistance (Ω/km)	Capacitance (nF/km)	Inductance (mH/km)
Positive sequence	0.2700	339.0	0.2550
Zero sequence	2.7000	280.0	1.0190

4.1 Simulation Analyses of Faults with Different Grounding Resistances in Different Lines

Three fault types are set up. And simulation analyses are taken in different lines with different grounding resistances. Once the fault occurring, characteristic signal of the fifth harmonic voltage is injected via inverters after certain time delay.

In the fault type of F1, harmonic content table of zero sequence current in every line when metal earth fault occurring is showed in Table 3. And harmonic contents relationship is showed in Fig. 7. The content of characteristic signal in Line 1 is highest. The content of characteristic signal in Line 1 is equal to the sum of those in the other lines. In conclusion, Line 1 is the fault line. It is same as the former theoretical derivation.

Table 3. Content of harmonic after signal injection in F1 with metal earth fault

F1	Fun (rms)	h5(thd)	h5 (rms)
L1	12.090	17.58%	0.6068
L2	1.300	12.16%	0.1148
L3	6.408	20.86%	0.3723
L4	6.409	13.95%	0.4110

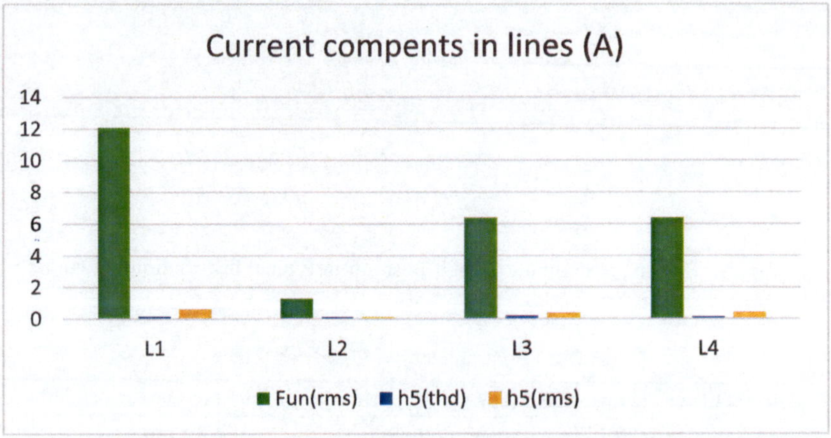

Fig. 7. Schematic diagram of the content of harmonic after signal injection in F1 with metal earth fault

If grounding resistance value is 800 Ω, based on Table 4 and Fig. 8, the relationship of the content of characteristic signal in every line is similar to that with metal earth fault. And the amplitudes of zero sequence currents in the network are decreasing with the increase of grounding resistance. However, the relationship of the content of characteristic signal in every line is still significant enough for fault line selection. In the situation of high resistance grounding, injecting characteristic signal value is adjusted in order to select fault line effectively.

Table 4. Content of harmonic after signal injection in F1 with grounding resistance value is 800 Ω

F1	Fun (rms)	h5 (thd)	h5 (rms)
L1	2.3710	19.35%	1.1220
L2	0.2553	13.41%	0.1245
L3	1.2560	22.89%	0.6245
L4	0.8887	15.45%	0.3753

When metal earth fault occurring in F2, the content of characteristic signal which flows through fault line is highest. Detailed RMSs are displayed is Table 5 and the relationship of harmonic content is displayed in Fig. 9.

If grounding resistance value is relatively big, based on Table 6 and Fig. 10, the relationship of the content of characteristic signal in every line in F3 accords with the conclusion.

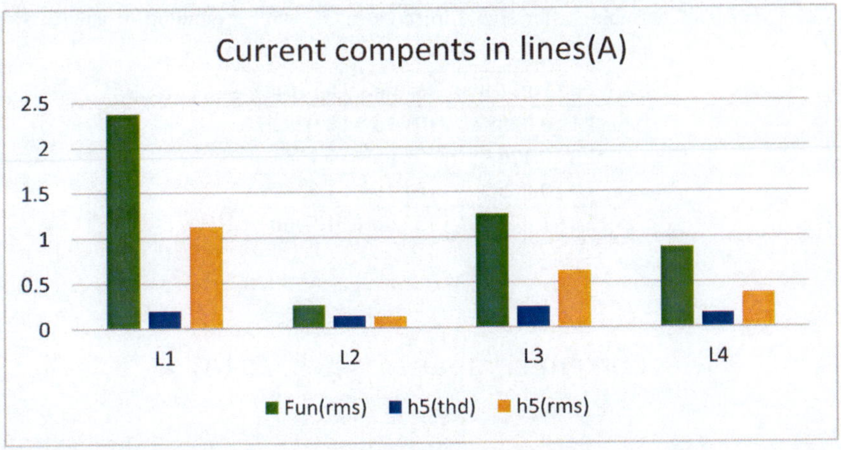

Fig. 8. Schematic diagram of the content of harmonic after signal injection in F1 with grounding resistance value is 800 Ω

Table 5. Content of harmonic after signal injection in F2 with metal earth fault

F2	Fun (rms)	h5 (thd)	h5 (rms)
L1	0.1477	54.73%	0.1059
L2	1.2480	47.92%	0.6574
L3	5.6200	54.73%	3.8620
L4	4.3500	55.15%	3.1260

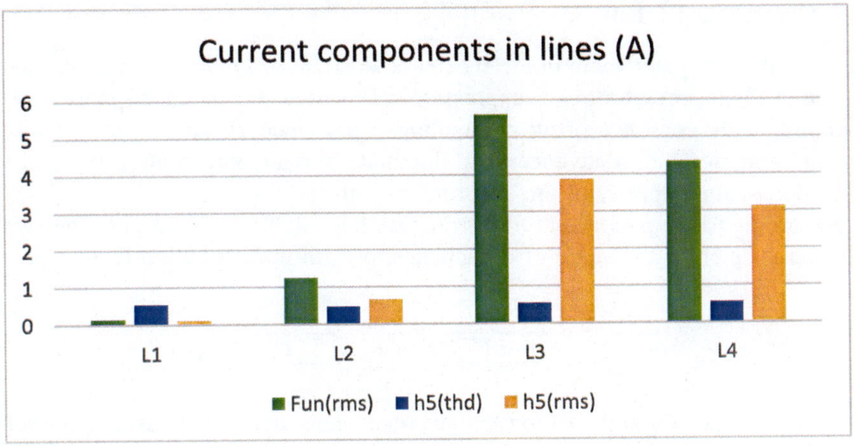

Fig. 9. Schematic diagram of the content of harmonic after signal injection in F2 with metal earth fault

Table 6. Content of harmonic after signal injection in F3 with grounding resistance value is 800 Ω

F3	Fun (rms)	h5 (thd)	h5 (rms)
L1	0.02087	109.76%	0.02226
L2	0.1764	95.88%	0.1683
L3	0.8655	164.16%	1.3580
L4	1.0430	154.40%	1.5460

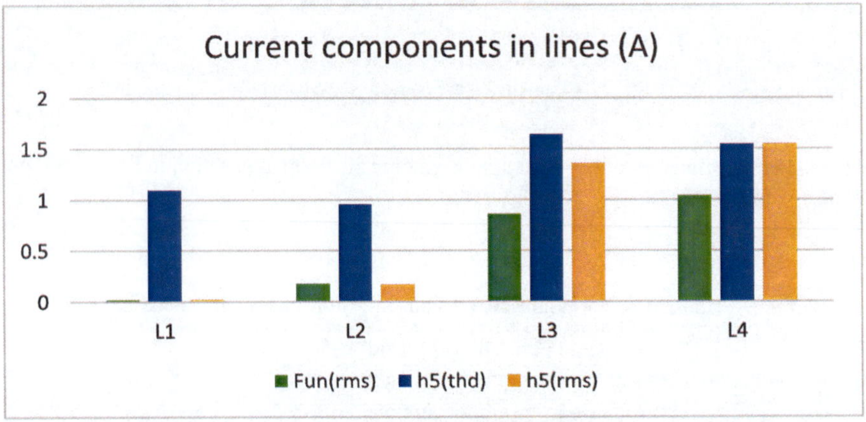

Fig. 10. Schematic diagram of the content of harmonic after signal injection in F3 with grounding resistance value is 800 Ω

4.2 Analyses for Signal Injection via Inverters Method in Resonance Grounding System

Based on the line parameters in the schematic diagram of typical distribution network showed in Fig. 6, presetting arc suppression coil is used. When single-phase short-line grounding fault does not occur, grounding capacitance current in the distribution network is figured out in advance. And the mode of overcompensation is taken to set the compensation capacity of arc suppression coil.

According to the parameters of distribution lines in Table 1 and 2, in the situation of power frequency, the value of capacitance to earth in distribution lines is:

$$\frac{1}{\omega C_{\sum}} = \frac{1}{2\pi f \times (\sum O_L \times c_{ol} + \sum C_L \times c_{OC}))} \tag{1}$$

In equality (1), O_L and C_L represent overhead lines and cable lines, respectively. c_{ol} And c_{oc} represent the value of zero sequence capacitance in per unit length of overhead lines and cable lines, respectively. Based on the model and relating lines parameters displayed in Fig. 8, it can be figured out that,

$$\frac{1}{\omega C_{\sum}} = 176.87\Omega \quad (2)$$

The mode of overcompensation is taken for arc suppression coil in the distribution network and the turn off degree is −0.05. Then,

$$L = \frac{1}{\omega^2 \times C_{\sum} \times (1 + \vartheta)} \quad (3)$$

L = 0.5362 H is gotten.

Arc suppression coil operates near to the point of resonance. And damping ratio increased by arc suppression coil is d ≈ P%. P (%) is the percent of active power loss caused by arc suppression coil in the compensation capacity.

$$\mathrm{P}(\%) = \frac{I_r}{I_L} \times 100\% = \frac{\omega L}{r_L} \times 100\% \quad (4)$$

In the simulation system, P = 2.0% is taken, it can be figured out that r_L = $8.4226 \times 10^3\ \Omega$.

The situation of fault type of F2 with 800 Ω grounding resistance is taken to be analyzed. When the neutral point of the system is grounded through the presetting arc suppression coil, the main part of arc at the grounding point of the fault line is compensated by arc suppression coil. Then the current that flows through every line is just the capacitive current of every line itself. The value and the content of injecting characteristic harmonic are analyzed and displayed in Table 7. The relationship of harmonic currents is showed in Fig. 11. According to the parameters of the distribution network, the rate of RMS of zero sequence current in every line is approximately equivalent to the rate of the capacitance to earth of every line. It can be known that the grounding arc at the point of fault is compensated by arc suppression coil in the operating state of overcompensation.

Table 7. Content of harmonic after signal injection in F2 with grounding resistance value is 800 Ω

F2	Fun (rms)	h5 (thd)	h5 (rms)
L1	0.04878	16.05%	0.01963
L2	0.4126	14.08%	0.1997
L3	1.437	24.27%	0.5848
L4	1.138	16.21%	0.4027

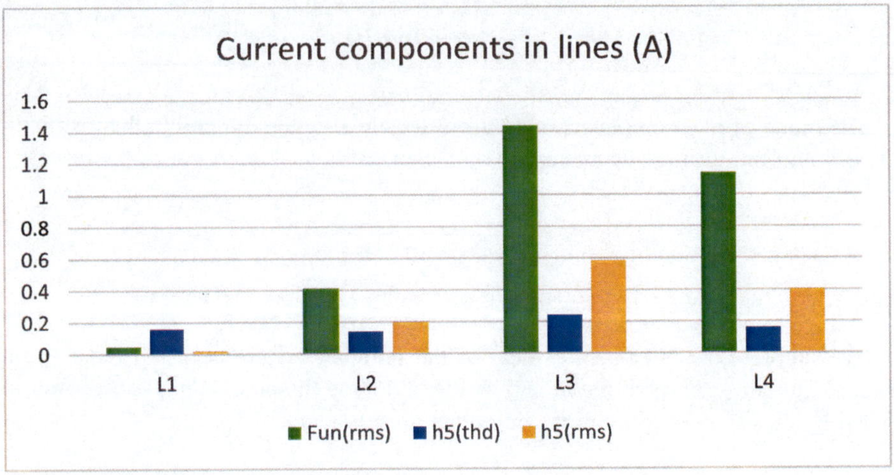

Fig. 11. Schematic diagram of the content of harmonic after signal injection in F2 with grounding resistance value is 800 Ω

Table 8. Content of harmonic after signal injection in F1 with different signal voltages

F1	Signal Source 5 kV			Signal Source 10 kV		
	Fun (rms)	h5 (thd)	h5 (rms)	Fun (rms)	h5 (thd)	h5 (rms)
L1	1.916	31.54%	0.6043	1.915	63.10%	1.209
L2	0.4159	11.17%	0.04645	0.4159	22.33%	0.09289
L3	2.046	19.05%	0.3897	2.046	38.10%	0.7794
L4	1.448	12.86%	0.1862	1.448	25.72%	0.3724

According to Table 7, as for every feeder, RMS of the harmonic current of the zero sequence current in Line 3 is the sum of those in Line 1, Line 2 and Line 4. Thus Line 3 can be selected as the fault line. It accords with the former conclusion. RMS of the characteristic current detected in Line 3 is approximately 2.0218 A. The current of fault point in the cable lines system is no more than 30 A and the current of fault point in the overhead lines system is no more than 10 A. Thus arc reignition at single-phase short-line grounding fault point cannot be caused by the injecting characteristic signal current.

4.3 Analyses for the Effect of Changing Injection Voltages of Inverters on Fault Line Selection

The situation of fault type of F1 with grounding resistance value of 800 Ω is taken to be analyzed. The voltage of the value of 5 kV is injected to the distribution network via inverters. After characteristic signal injection, the content of characteristic signal harmonic in every line is analyzed. Then the value of the voltage that injected to the distribution network is increased to 10 kV. And the content of characteristic signal harmonic in every line is analyzed. The amplitude of characteristic signal in two different levels of signal injection voltage is displayed in Table 8.

According to Table 8, the content of harmonic is double increased thereupon with the voltage level of injecting signal is considered to be double increased. The result of simulation experiment accords with the analysis of theory. And under the situation of the voltage values of signal injection via inverters are 5 kV and 10 kV, respectively, the fault line can be reliably selected without the occurrence of arc at the grounding point.

5 Conclusions

Based on inverter injection to the distribution network, the method of characteristic signal injection via inverters to select fault line is proposed. And the content of characteristic signal of every feeder line is analyzed and compared. Then fault line selection is completed. Conclusions are as following:

(1) If grounding resistance value of the distribution network is relatively small, the measure of characteristic signal current in every line can be significantly improved by the method of characteristic signal injection via inverters.
(2) If grounding resistance value of the distribution network is relatively big, errors are easily caused by traditional zero sequence current amplitude and phase comparison method. The measure of characteristic signal current in every line can be significantly improved by the method of characteristic signal injection via inverters and the accuracy of fault line selection is improved.
(3) Distribution network with neutral grounding method with an arc suppression coil is taken to consideration. The use of presetting arc suppression coil is taken as an example to be analyzed. With presetting arc suppression coil, characteristic signal is injected via inverters to adjust the output voltage so that fault line selection can be completed under the situation without grounding arc reignition at the fault point.

References

1. Ye, R., Liu, R., Liu, J., et al.: Transient stability calculation of power system integrated with direct-drive wind farm with permanent magnet synchronous generators. Trans. China Electrotech. Soc. **29**(6), 211–218 (2014) (in Chinese)
2. Guo, Q., Wu, T.: Survey of the methods to select fault line in neutral point ineffectively grounded power system. Power Syst. Protect. Control **38**(2), 147–152 (2010) (in Chinese)
3. Cheng, S.W., Fu, Z.X.: Single-phase ground fault line selection for distribution network based on frequency domain parameter identification method. IOP Conf. Ser.: Earth Environ. Sci. **267**(4), 147–151 (2019)
4. Liu, M., Wang, Y.: A new fault line selection method based on transient phase current characteristics analysis. Proc. CSU-EPSA **29**(1), 30–36 (2017) (in Chinese)

5. Wei, X., Yang, D.: An adaptive fault line selection method based on wavelet packet comprehensive singular value for small current grounding system. In: 2015 5th International Conference on Electric Utility Deregulation and Restructuring and Power Technologies (DRPT). IEEE (2015)
6. Jiang, J., Bao, G.: Survey of the methods to select fault line in neutral point ineffectively grounded power system. Electr. Eng. (12), 1–5 (2015). (in Chinese)
7. Lin, X., Sun, J., Kursan, I., et al.: Zero-sequence compensated admittance based faulty feeder selection algorithm used for distribution network with neutral grounding through Peterson-coil. Int. J. Electr. Power Ener. Syst. **63**(01), 747–752 (2014)
8. Yuan, J.J.: A study on the algorithm of supply chain protection based on zero-sequence compensation. Int. J. Secur. Appl. **9**(2), 85–96 (2015)
9. Yu, N.H., Li, R., Cao, H.G., et al.: Fault line selection and section location of distribution network based on S-Transform and Zero-Sequence Reactive Power. In: Applied Mechanics and Materials. Trans Tech Publications, Switzerland (2014)
10. Liu, M.Y, Lv, L.P., et al.: Study of the small current neutral grounding fault based on TDFT sampling method of first wave. Electr. Measure. Instrum. 55(23), 22–28+33 (2018). (in Chinese)
11. Jamil, M., Kalam, A., Asari, A.Q., et al.: Generalized neutral network and wavelet transform based approach for fault location estimation of a transmission line. Appl. Soft Comput. **19** (3), 322–332 (2014)
12. Shu, H., Peng, S.: Distribution network fault line detection using the full waveband complex relative entropy of wavelet energy. High Volt. Eng. **35**(7), 1559–1564 (2009) (in Chinese)
13. Silva, K.M., Souza, B.A., Brito, N.S.D.: Fault detection and classification in transmission lines based on wavelet transform and ANN. IEEE Trans. Power Deliv. **21**(4), 2058–2063 (2006)
14. Huang, T.T.: Research on fault line selection in small current grounding system based on injection principle. Jinan University (2019). (in Chinese)
15. Chen, W., Cai, G., Cai, Y.: Neutral grounding mode in 10kV distribution network via Peterson coil with parallel resistance. Power Syst. Technol. **28**(24), 56–60 (2004) (in Chinese)
16. Shi, R.: Development and research on technology of fault line selection based on medium resistance switching in small current neutral grounding system [D].North China Electric Power University, 2015. (in Chinese)
17. Xue, Y., Chen, X., Song, H., et al.: Resonance analysis and faulty feeder identification of high-impedance faults in a resonant grounding system. IEEE Trans. Power Deliv. **32**(3), 1545–1555 (2017)
18. Pan, Z., Zhang, H., Zhang, F.: Analysis and modification of signal injection based fault line selection protection. Autom. Electr. Power Syst. **31**(4), 71–75 (2007) (in Chinese)
19. Guo, X.: Three-phase CH7inverter with a new space vector modulation to reduce leakage current for transformerless photovoltaic systems. IEEE J. Emerg. Select. Topics Power Electron. **5**(2), 708–712 (2017)
20. Guo, X., Wei, B., Zhu, T., et al.: Leakage current suppression of threephase flying capacitor PV inverter with new carrier modulation andlogic function. IEEE Trans. Power Electron. **PP** (99), 1 (2017)

An Online Calculation Data Generation Technology for Power Grid Integrated with Large-Scale Wind Farm

Shibin Bai[1], Minghui Yan[2](✉), Wei Xu[2], Zhihao Tian[1], Xiaoxiang Sun[1], and Haifeng Zhou[2]

[1] State Grid Ningxia Electric Power Co., Ltd., Yinchuan 750000, China
[2] NARI Group Corporation (State Grid Electric Power Research Institute), Nanjing 211106, China
yanminghui@sgepri.sgcc.com.cn

Abstract. The large-scale integration of wind farms leads to more complex stability mode. Online security and stability analysis is necessary to improve the new energy absorption capacity and operation efficiency of power grid. An online calculation data generation technology is proposed to simulate the dynamic characteristics of wind farm accurately. The operating state of the low-voltage network is identified by the fusion of multi-source real-time data from state estimation, SCADA and security and stability control system to form operating mode data integrated with wind farm clusters. The spatio-temporal correlation measurement correction model is adopted to identify and correct bad measurement data unhandled by state estimation. The clustering method based on Analytic Hierarchy Process (AHP) is used to obtain dynamic equivalent model of wind farm based on static characteristic and operating state of wind turbines. Finally, the effectiveness of the method has been verified by the cases of actual power grid.

Keywords: Large-scale wind power · Online calculation · Multi-source data fusion · Low voltage network status identification · Equivalent wind power generators

1 Introduction

The safe operation of power grid and large-scale absorption capacity of renewable energy are facing new challenges [1–3]. Different from conventional generators, wind power equivalent moment of inertia is very small, it lacks the relevant frequency adjustment function and the voltage adjustment ability is limited. The frequency and voltage tolerance ability is insufficient. It is easy to be separated from power grid in the case of large fluctuation of system frequency and voltage, which adversely affects the system's safe and stable operation. Aiming at the actual power grid integrated with large-scale wind power, it can effectively identify the weakness of power grid and improve the operating efficiency of power grid by online calculation [4].

The power and voltage of some wind farms can only be connected to dispatching automation system now. The range of state estimation is difficult to cover the low

Y. Xue et al. (Eds.): PMF 2020, LNEE 718, pp. 647–659, 2021.
https://doi.org/10.1007/978-981-15-9746-6_50

voltage network including wind power cluster. Instead, the corresponding network is equivalent to load or generator on the high-voltage side of 330 kV/220 kV substation [5]. Therefore, the imperfection of online safety analysis calculation [6] data leads to a large difference between the online analysis conclusions and the actual situation of power grid. It is also difficult to consider wind power generators when setting stable sections to adjust mode, the space of preventive control measures and the strategy of generator tripping, which makes it difficult to accurately simulate the dynamic characteristics of wind power when grid-side faults occur.

Although the wind farm data is not included in the state estimation model, with the rapid increase of wind power generator capacity, the control center has access to the measurement information of wind farm to monitor the actual output in real time [7, 8]. The centralized security and stability control management system can directly obtain the measurement information of wind farm internal feeders and some branches of low-voltage network collected by the security and stability control device. However, due to the lack of low-voltage network model parameters, it is impossible to generate online calculation data including wind farms based on the above measured data. In addition, due to the lack of time scale for the measured data, it is impossible to ensure the consistency among the three factors, namely, the wind power generators output, the measured active power of wind farm feeder by security and stability control devices, and the SCADA acquisition output of wind power generators or entire wind farm.

In view of the above problems, an online calculation data generation technology for power grid integrated with large-scale wind farm is proposed in this paper. The operating state of the low-voltage network is identified by the fusion of multi-source real-time data from state estimation, SCADA and security and stability control system to form operating mode data integrated with wind farm clusters. The spatio-temporal correlation measurement correction model is adopted to identify and correct bad measurement data of node power injection. The clustering method based on Analytic Hierarchy Process (AHP) is used to realize online dynamic equivalent model of wind farm based on static characteristic and operating state of wind turbines.

2 Overall Approach to Online Data Generation

In order to ensure the safety, stability and economic operation of large power grid, a large number of security and stability control devices and centralized management system have been constructed at the sending and receiving power grids. The centralized management system of security and stability control devices can directly acquire the measurement information of the internal feeder of wind farms and some branches of low voltage network collected by security and stability control devices [9, 10]. Combined with the typical mode data of power grid security and stability analysis, it can provide the basis for low voltage network state identification with large-scale wind farm (Shown in Fig. 1, where the solid line is the state estimation modeling power grid, and the dotted line is the un-modeled power grid).

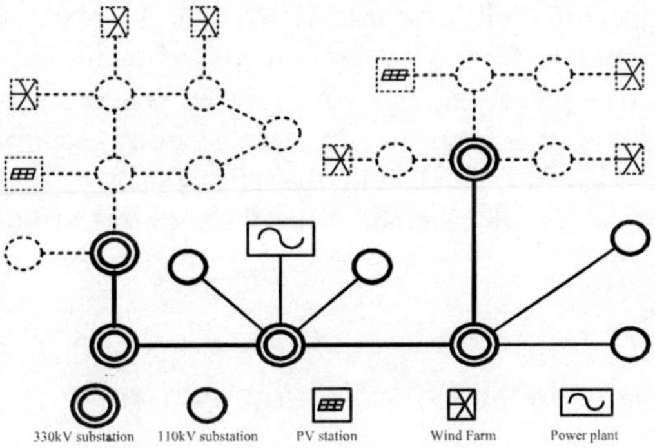

Fig. 1. Diagram for grid structure.

The dynamic model of a wind power generator includes the generator model, voltage ride-through protection model, pitch angle control, active power control, reactive power control and et al. Due to the influence of the amplitude limiting link, the low-voltage ride-through characteristics, active and reactive dynamic responses of wind power generators at different operating points are quite different. There is a big error in using the single-unit multiplication method to equate all wind power generators to multiple generators with the same output. Due to the large uncertainty of wind speed, it is difficult to determine the number of online equivalent groups and equivalent machines of wind farm by offline method. Therefore, it is necessary to carry out online equivalence according to the output of each generator in the wind farm and determine the dynamic model of wind farm matching the current mode. In addition, according to the strategy model and the operating state of power grid, the current strategy of security and stability control system can be identified online. Among them, the wind farm is generally cut off by the feeders, which needs to establish the mapping relationship between the equivalent wind power generators and feeders in the process of dynamic equivalence.

The overall process of online data generation for power grid integrated with large-scale wind farm includes the following steps:

(1) According to the typical mode, the un-modeled low voltage network model in state estimation is established, including topological relationship and branch parameters;
(2) The state of low voltage network is identified by the fusion of typical mode output, SCADA measurement, security and stability control measurement and state estimation equivalent measurement;
(3) The equivalent equipment in the state results is replaced by the corresponding low voltage network, and the power flow of whole power grid is calculated to form the operation data including wind farms;
(4) The output of wind power generator is adjusted according to the feeder active power or the whole output, and the clustering algorithm is used to determine the grouping strategy considering the strategy of security and stability control devices and the response characteristics of wind power generators;

(5) According to the principle of conservatism, the number of generators in each group is determined to form a dynamic model of online safety analysis;
(6) According to the fault form, the wind power generators and control quantity needed to be controlled by security and stability control system under the expected fault are determined, and the grid-related protection action of wind power generators is determined through the electromechanical transient simulation.

3 State Identification of Low Voltage Network

The data that can be used for low voltage network state identification include the power flow of equivalent equipment in state estimation results, the power flow of monitoring equipment in the security and stability control system, and the power flow of wind farms used by SCADA. Because it does not meet the observability requirements, it can't directly obtain the power flow and branch topology state of each wind farm by state estimation. For unmeasured wind farms and branches, the output and topology state can only be determined according to typical mode.

There may be errors or bad data in the measurement data that has not been estimated by the state. It needs to be verified according to the reasonable distribution range of various measurements and the measurement data from different sources.

In order to ensure the convergence of the whole power flow after splicing the low voltage network, the state identification is carried out according to the following principles [11]:

(1) SCADA adopts the transmission mode that combines all data and change data, that is, normal transmission of change data, the whole power grid data is periodically called by master station. The data collected by security and stability control devices adopts the transmission mode of calling by the master station in fixed period (far less than SCADA), and the security and stability control system is generally equipped with multiple sets to ensure reliability. Therefore, the refresh speed and reliability of the data collected by the security and stability control system are higher than that of the data collected by SCADA system, and the data collected by SCADA system should be corrected according to the data collected by security and stability control system.
(2) In order to identify the bad data of measurement data, according to the feasible areas of active power, reactive power and bus voltage of the equipment, the data collected by security and stability control system and SCADA are checked and the wrong data are eliminated. Using the spatial-temporal relationship between measurement data from different measuring points, the suspicious data can be identified and corrected. In the time dimension, the time series method is used to obtain the predicted value of the measurement, and the current measurement is checked and corrected. In the spatial dimension, the data is corrected according to the parameters of the branch and the measurements at both ends.

(3) In order to avoid the contradiction between remote communication data and remote measurement data, according to the principle that the reliability of the remote measurement data is greater than the remote communication data, the remote measurement in security and stability control system acquisition data and SCADA acquisition data shall be taken as the criteria to check and correct the remote communication data of switches in security and stability control system acquisition data and SCADA acquisition data.
(4) In order to ensure that the voltage deviation and unbalanced power of the splicing point before and after the splicing of low voltage network meet the requirements, it is necessary to determine the overall power level of low voltage network according to the state of the equivalent equipment in state estimation results. Among them, the measurement will not participate in the power flow adjustment through the checked and corrected equipment. It is specified that the power transmission from low voltage network to main power grid is in the positive direction. When the power deviation is positive, it shall be adjusted according to the order of reducing the generating output and increasing the active load. When the power deviation is negative, it shall be adjusted according to the order of reducing the active load and increasing the generating output. If there is still unbalanced power after power allocation, the unbalanced power is taken as the equivalent equipment to access the splicing point of state estimation.

4 Online Equivalence of Wind Farm Dynamic Mode

Online safety analysis needs to accurately simulate the dynamic response of wind farm under the expected fault, especially the situation of wind power generators' trip-off from grid under severe faults and its impact on the voltage and frequency in the transient process [12]. The installed capacity of large-scale wind farm may reach thousands. The online safety analysis does not care about the response of each wind power generator in wind farm. The wind power generator in wind farm can be treated with appropriate equivalence. The calculation efficiency of online safety analysis can be improved under the condition that the influence of wind farm on the dynamic characteristics of the system is basically unchanged [13–15].

The predictive fault set of online safety analysis includes different types of AC system faults and DC system faults, and the operating state of wind power generators will change greatly in a short time due to the influence of meteorological environment. According to the dynamic response characteristics of wind farm in the specified fault scenario, wind farm equivalence is carried out, the equivalent model is difficult to cover all fault scenarios [16]. Therefore, different from offline research, the difficulty of online wind farm equivalence is to determine the equivalent model according to the type of wind power generators, topology connection, protection setting value and other static characteristics, as well as the wind speed, pitch angle, control mode and other dynamic characteristics before simulation calculation.

4.1 Analysis of Wind Farm Dynamic Characteristics and Influencing Factors

For the actual wind farm, the wind power generators that are put into production at the same time basically are the same wind power generator. The manufacturer will provide detailed measured parameters in the dynamic model of wind power generators. The method of calculating equivalent wind power generator parameters for a specific fault scenario has large computation. And it is also difficult to meet the accuracy requirements of multiple fault scenarios. Therefore, the output multiple method is used to characterize equivalent wind power generator in this paper. In the online application, the number and wind speed of equivalent wind power generator in the dynamic model are determined according to the measured data of wind power generator output, wind speed and pitch angle. The commercial electromechanical transient simulation software provides a detailed dynamic model of wind power generator. The response of wind power generator control system and electrical control system is closely related to the operation state of wind power generator.

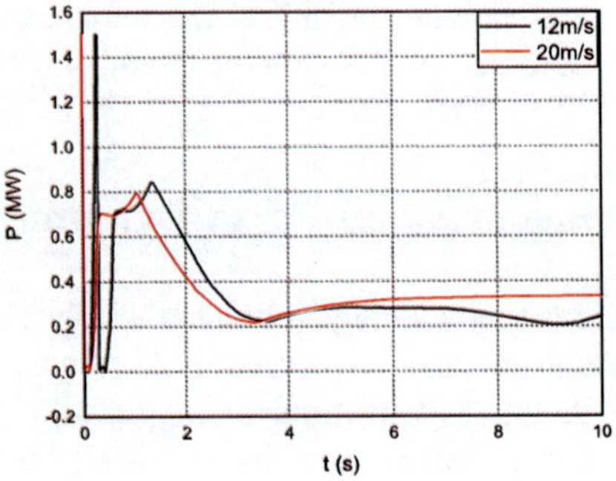

Fig. 2. Wind power active power response curves under different speeds.

For variable-speed wind power generators with variable pitch, the control goal of wind power generators is to get more power from the available wind energy as much as possible without exceeding the rated value of the equipment. When wind speed is higher than rated value, the blade pitch angle is adjusted to control the mechanical power to the rated value. When wind speed is lower than rated value, the pitch angle is set to the minimum value to obtain the maximum power. The active power response of wind power generators under different wind speeds when AC system fails is shown in Fig. 2. Therefore, the control of pitch angle has a great influence on the results of dynamic simulation. And it is necessary to group according to the action of pitch angle.

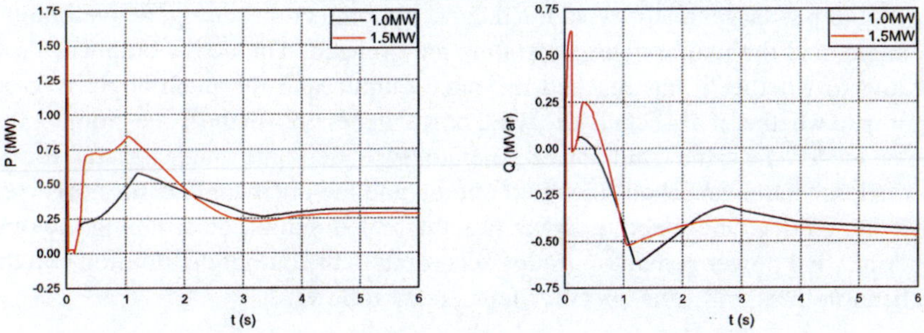

Fig. 3. Wind power response curves under different active power.

Electric control includes electric control model and frequency converter model. Electric control model includes active power control, reactive power control, voltage ride-through control strategy and et al. The control strategies of each control link in transient process are as follows:

(1) The active current control command is calculated by the active power control according to the output power command of wind power generator model and the terminal voltage. And it is limited by the short-time active current output capacity of the converter;
(2) The excitation voltage control command is calculated by the reactive power control according to the difference between the voltage of the node to be controlled and the reference voltage, the reactive power of generator and the terminal voltage. It is limited by the range of reactive power output and excitation current of generator;
(3) Switching control strategy based on voltage state, the voltage ride-through control can control the active and reactive power. The active power control includes calculating current, specifying current and specifying power according to the initial power. The reactive power control includes voltage-based control, specified reactive power, specified current and et al. The limiting link includes the limitation of active and reactive output.

It can be seen that the state variables of each control link are initialized according to the active output of wind power generators and the terminal voltage. The active and reactive power responses are also related to the voltage drop in the near area of the wind farm during fault. Due to the existence of limiting links, there are large differences in the active and reactive power responses of wind power generator with different output levels, as shown in Fig. 3. Therefore, for the generators that do not reach the rated output, it is necessary to group them according to the generators output level.

4.2 Wind Power Generators Clustering Method Based on Hierarchical Analysis

In order to meet the needs of online safety analysis for the dynamic characteristic model of wind farm, in this paper, the hierarchical analysis is carried out according to the order of feeder, wind power generator type, rated capacity, protection setting, active output and pitch angle, and the wind power generators are grouped. The active output is grouped according to whether it has reached the rated output, and the pitch angle is grouped according to whether it has acted. The wind power generators in the same group belong to the same feeder, the same wind power generator type, the same rated capacity, the active power is less than (or reached) the rated output, and the pitch angle is (or not) acted.

For the groups whose active power less than rated output, determining the optimal number of wind power generator groups according to the output distribution and the set classification threshold. The specific steps are as follows:

(1) Each wind power generator is divided into one class according to the output level, the distance between each class is calculated, and the minimum distance between each class is taken as the current combined distance;
(2) The two classes with the smallest distance between classes are combined into a new class, and recalculating the distance between each class (The distance between classes is the largest distance between wind power generators included in two classes).
(3) When the change amount of the merge distance corresponding to the last two merges is less than the specified threshold, the optimal number of groups is obtained.

A similar method is used to determine the optimal number of wind power generator groups based on the distribution of pitch angle for the groups that have reached the rated output.

4.3 Method of Parameter Aggregation of Equivalent Machine

The number m of each group including wind power generators is taken as the number of wind power generators in the equivalent model, and the average wind speed of each wind power generator is taken as the speed of wind power generators in the equivalent model. The equivalent parameters of transformer can be obtained according to (1). The equivalent impedance of trunk and radial collector lines can be obtained according to (2) and (3), respectively.

$$\begin{cases} S_{T\_eq} = mS_T \\ x_{T\_eq} = \dfrac{x_T}{m} \end{cases} \tag{1}$$

$$Z_{eq} = \sum_{i=1}^{m} \left(\sum_{k=1}^{i} \left(Z_k \sum_{j=k}^{n} P_j \right) P_i \right) \Big/ \left(\sum_{i=1}^{m} P_i \right)^2 \tag{2}$$

$$Z_{eq} = \sum_{i=1}^{m} Z_i P_i^2 \Big/ \left(\sum_{i=1}^{m} P_i \right)^2 \tag{3}$$

Where, S_T and x_T are the transformer capacity and reactance respectively; Z_k is the impedance of the k-th cable; P_i is the output power of the i-th wind power generator, and n is the number of main-line connected generators.

5 Example Analysis

The technology proposed in this paper has been applied in Northwest, Ningxia and Gansu power grids. In 2018, the installed capacity of wind power in Northwest power grid was 48914.28 MW, accounting for 19.39%. The proportion of the largest wind power to the load of the day is 33.07%. The technology described in this paper can provide refined wind farm model data when generating online calculation data and improve the accuracy of simulation analysis.

5.1 Data Generation of Whole Power Grid with Low Voltage Network Model

At present, the regulation and control center can monitor the state of wind farm through the real-time monitoring system and the active power prediction system of wind farm and SCADA system. In the state estimation, wind farm is usually equivalent to load on 220 kV or 330 kV side. Taking Yumen station as an example, two main transformers are equivalent to load modeling at 330 kV bus, as shown in Fig. 4. In a certain mode, the low-voltage grid transmits 290.76 MW power to the main grid.

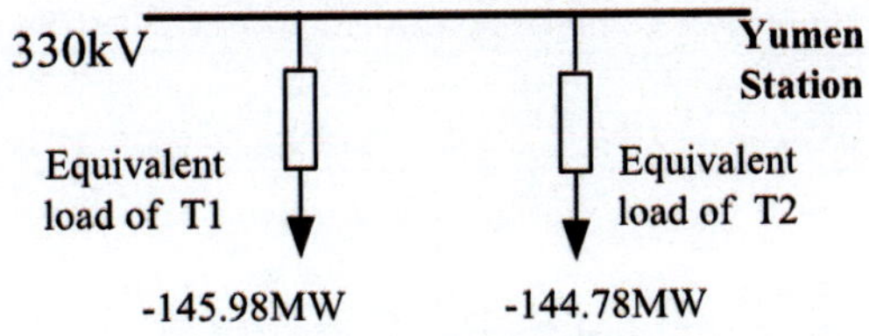

Fig. 4. State estimation model.

After splicing, the grid is as shown in Fig. 5. The complete structure of wind farm and collection station has been established.

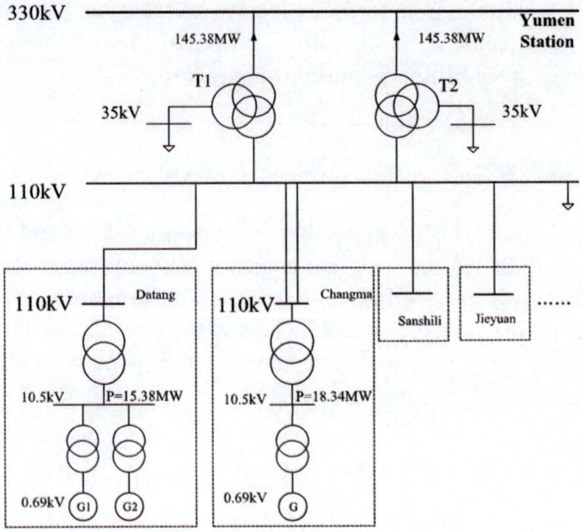

Fig. 5. Network topology after splicing.

The comparison of grid scale before and after the low voltage network is spliced based on the state estimation of the main grid is shown Table 1.

Table 1. Comparison of grid scale.

Data type	Number of compute nodes	Number of lines	Number of windings
State estimation	1533	1532	742
After splicing	4372	2839	2764

5.2 Online Equivalent of Wind Power Generator

Figure 6 is the wind power generator topology diagram of a feeder line of a wind farm. The wind power generators A1-01–A1-07 are out of operation. The actual measurement of other wind power generators in operation are shown in Table 2. The parameters (actual value) of each transformer are R = 0.22 and X = 4.55.

According to the state of wind power generators, the wind power generators in Fig. 6 can be divided into three groups, and the generators in the dotted circle are the same group. The topological diagram after equivalence is shown in Fig. 7, the power of each equivalent generator is shown in Table 3, and the parameters of equivalent transformer (actual value) are R = 0.055, X = 1.14.

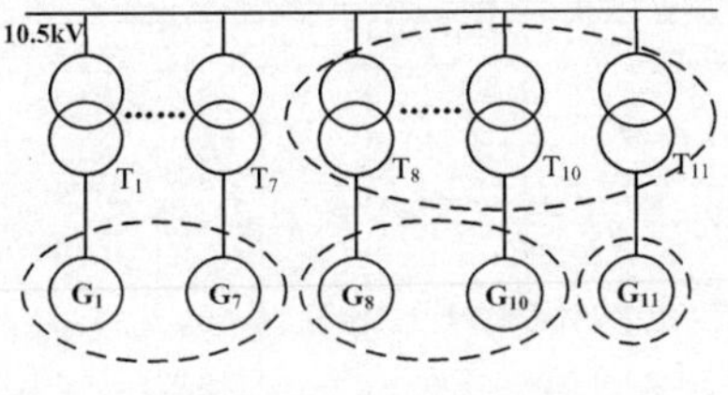

Fig. 6. Wind power generator grouping.

Table 2. Measurement of wind power generator.

Name	Active power (MW)	Wind speed (m/s)	Pitch angle (degree)
A1-08	0.156	6.48	2.8
A1-09	0.154	6.75	2.4
A1-10	0.134	6.23	2.8
A1-11	1.438	2.31	3.6

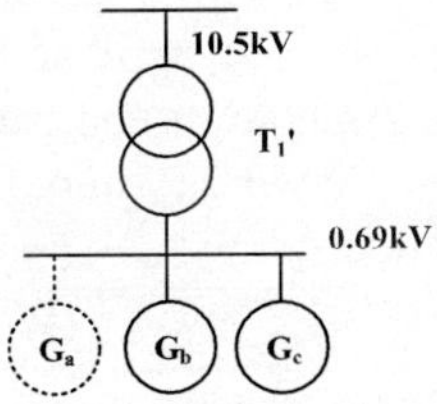

Fig. 7. Equivalent topological diagram.

Table 3. Information of equivalent generator.

Equivalent number	Active power (MW)	Corresponding generator
a	0	A1-1–A1-7
b	0.444	A1-8–A1-10
c	1.438	A1-11

5.3 Simulation Calculation

The analysis is based on the Northwest Power Grid. After DC bipolar block fault, the voltage and frequency response of 330 kV bus in one station are monitored based on the following two calculation models.

Case 1: Regardless of the wind farm model, the wind farm is equivalent on the high-voltage side of the bus.
Case 2: Taking the wind farm model into account, the large-scale wind farm is equivalent online according to the method described in this paper.

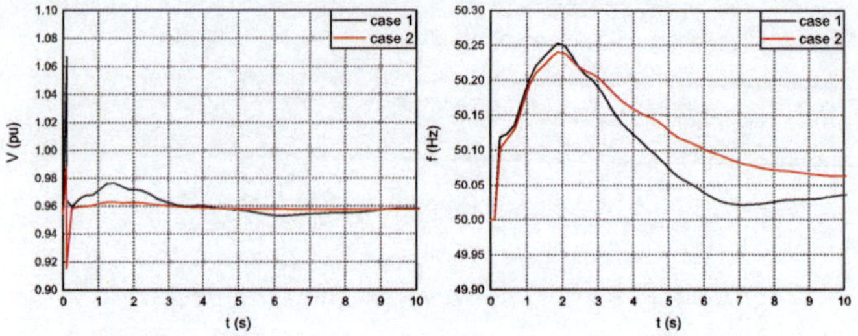

Fig. 8. Bus voltage and frequency curve after the fault.

The simulation results are shown in Fig. 8. Because the grid side of wind power generators has the characteristic of constant reactive power control, the transient overvoltage in case 2 after DC bipolar block fault is smaller than that without wind power generator model, and the bus voltage after the fault is more gentle. The constant active power control characteristic of the rotor side of wind power generators in case 2 makes the maximum transient frequency of the bus after the fault lower than that without wind power generator model, the bus voltage after the fault is gentle, which leads to the gentle change of the active power of wind farm, so the frequency recovery of the bus after the fault is also gentle. The simulation comparison results show that the model considering equivalence of large-scale wind farm online can more accurately reflect the actual operation state of power grid.

6 Conclusion

An online calculation data generation technology for power grid integrated with large-scale wind farm is proposed in this paper. The operating state of the power grid is identified by the fusion of multi-source real-time data such as typical mode data, SCADA and security and stability control system, and the integrity of the underlying model is guaranteed. The spatial-temporal correlation correction model of measurement is established, and bad data recognition and correction of node power injection measurement are realized based on historical measurement data and branch parameters. Through the wind farm equivalence, the data meeting the demand of online safety analysis is established. The results of actual power grid operation verify the correctness and feasibility of this method.

Acknowledgment. This research was supported by Key Research and Development plan of Ningxia Hui Autonomous Region (SGNX0000DKJS1900496).

References

1. Ying, Y., Sun, Y., Yang, J., et al.: Load characteristic analysis of grid fault ride-through process for DFIG based large wind turbine. Autom. Electr. Power Syst. **44**(12), 131–138 (2020)

2. Xu, D., Wang, H.: High risk cascading outage assessment in power systems with large-scale wind power based on stochastic power flow and value at risk. Power Syst. Technol. **43**(2), 400–409 (2019)
3. Wei, W., Wang, Y., Li, X., et al.: Simulation of large scale wind farms integration into Gansu Jiajiu power system. Autom. Electr. Power Syst. **33**(18), 96–101 (2009)
4. Xue, Y., Lei, X., Xue, F., et al.: A review on impacts of wind power uncertainties on power systems. Proc. CSEE **34**(29), 5029–5040 (2014)
5. Ding, P., Xu, Y., Wang, J., et al.: Study of the characteristics of grid model for online dynamic security assessment system in bulk power systems. Power Syst. Protect. Control **41** (19), 132–139 (2019)
6. Sun, S., Xie, C., Lü, Y., et al.: Power system online security and stability analysis application modes. Power Syst. Technol. **39**(10), 2875–2881 (2015)
7. Lin, L., Zhu, C., Zheng, T., et al.: Active power control of wind farm cluster and its strategy. Autom. Electr. Power Syst. **38**(14), 9–16 (2014)
8. Xue, F., Chang, K., Wang, N.: Coordinated control frame of large-scale in-termittent power plant cluster. Autom. Electr. Power Syst. **35**(22), 45–53 (2011)
9. Hou, J., Tao, X., Zhang, J., et al.: A low-voltage ride-through control strategy based equivalence method for wind farms. Power Syst. Technol. **39**(5), 1281–1286 (2015)
10. Roy, B., Hua, C., Ghajar, R.: Time-series models for reliability evaluation of power systems including wind energy. Microelectron. Reliab. **36**(9), 1253–1261 (1996)
11. Lu, J., Wen, B., Jiang, C., et al.: Research of multi-source data and its quality control in energy management system. Autom. Electr. Power Syst. **26**(23), 73–75 (2002)
12. Zhou, H., Xie, H., Wu, T., et al.: Analysis and simulation of wind farm dynamic behavior in high-voltage trip-off accident of Zhangbei. Electr. Power Autom. Equip. **37**(2), 54–60 (2017)
13. Chao, P., Li, W., Jin, X.: Views and discussions on dynamic equivalent of DFIG wind farms. Autom. Electr. Power Syst. **40**(12), 194–199 (2016)
14. Zhou, H., Zhang, M., Xue, Y., et al.: A dynamic equivalent method for doubly-fed induction generator wind farm based on the Thevenin equivalent circuit. Autom. Electr. Power Syst. **36** (23), 42–47, 53 (2012)
15. Tu, J., Zhang, J., Liu, M., et al.: Mechanism analysis of transient overvoltage with large disturbance considering dynamic characteristics of wind generator. Autom. Electr. Power Syst. **44**(11), 197–205 (2020)
16. An, Z., Shen, C., Zheng, Z., et al.: Assessment method for equivalent models of wind farms based on direct-driven wind generators considering randomness. Proc. CSEE **38**(22), 6511–6520 (2018)

Optimal Schedule Strategy of Integrated Energy System Considering Static Voltage Stability

Wei Xu[1(✉)], Hongshui Lu[1,3], Guoxin He[1,2], Yonghua Chen[1,2], and Dongmei Yang[1,2]

[1] State Grid Electric Power Research Institute, Nanjing 211106, China
zhangsan@163.com
[2] NARI Technology Co., Ltd., Nanjing 211106, China
[3] NARI GEIRI Semiconductor Co., Ltd., Nanjing 211106, China

Abstract. The optimization operation of integrated energy system is to achieve coordination between different energy systems such as electricity, heating, cooling, gas, etc. The purpose is to achieve the goals of improving energy efficiency and reduce energy costs and environmental pollution. At present, most relevant studies only consider the operational constraints of various types of equipment, and do not consider network constraints. In this paper, the static voltage stability in the power system is taken into consideration. An optimized scheduling method for integrated energy system operation considering the constraints of the power network is proposed. Simulation examples of the integrated energy system based on the IEEE 13 standard node expansion are verified by simulation. The results show that the proposed method can further promote the global optimization of the integrated energy system.

Keywords: Integrated energy system · Optimal schedule · Voltage stability

1 Introduction

The development of human society has always been faced with major problems in the use of energy. It has became a common choice for people to vigorously develop renewable energy and improve energy efficiency. In the past, people solved problems within a single energy system, such as proposing a smart grid or smart gas. With the rapid development and in-depth combination of new energy technology and information technology, people gradually realize that it is necessary to think about problems from a higher perspective. For example, through the coordinated scheduling of power system and gas system, the flexibility and reliability of energy utilization can be further improved [1].

Under this background, the idea of integrated energy system came into being. The integrated energy system is considered as the physical carrier of the energy Internet. Its characteristics include multiple energy systems such as power, heating, cooling and gas and coordinated and complementary relationship is formed among the energy systems [2]. It is of vital importance to break down the traditional barriers of energy system

Y. Xue et al. (Eds.): PMF 2020, LNEE 718, pp. 660–675, 2021.
https://doi.org/10.1007/978-981-15-9746-6_51

technology, market and management to realize the unified planning, coordination and evaluation of multiple energy systems. It is the key to improve the comprehensive energy utilization efficiency and reduce the energy use costs.

Compared with optimal scheduling of a single energy system, such as smart grid scheduling, scheduling rules of the integrated energy systems are more complicated [3]. The reason lies in the need to balance the different needs of electric energy, heating energy and cooling energy. In addition, due to the expansion of the number of equipment and the scale of the network, the involved constraints become more complicated, which makes optimal scheduling problem of the integrated energy systems more difficult. For example, the Combined Cooling and Heating Power (CCHP) is a typical coupling device, which is a hub device for the coupling of power, heating, cooling and gas, which directly affects the supply and regulation of power, heating, cooling and gas.

In the aspect of optimizing scheduling of the integrated energy systems, preliminary research had been conducted at home and abroad. Literature [4] took the grid-connected micro-grid including wind turbines, photovoltaic cells, CCHP, electric boilers and power storage devices as an example. The optimized software is used to obtain the best output and total operating costs of each micro-power supply in the dispatch cycle, achieving unified and coordinated scheduling of electric heating and reducing micro-grid operating costs. Literature [5] unified modeling of CHP, gas turbine, transformer and other components in the area as a single energy conversion component through an energy hub. Literature [6] proposed the Energy Hub model, which abstracted and classified the supply and demand for Energy from a variety of sources and obtained the best energy supply mode through global optimization. Literature [7] proposed a comprehensive energy model of combined heating and power supply suitable for micro-grid and realized the optimized operation of each equipment under the premise of meeting the requirements of combined heat and power supply. Literature [8] only considers the equipment output constraints in the system and analyzes the optimal output of system when CCHP operating under different electro-thermal ratio operating modes. Literature [9] fully considered the interaction of various energy sources such as electricity, heating, cooling, gas and established a optimization model of regional integrated energy system that fully took into account the key characteristic variables of the electric, heating, cooling and gas.

However, the above studies only considered the equipment components of the integrated energy system and rarely involve the constraints of electricity, heat and gas networks. Literature [10] proposed the analysis method of static voltage stability for integrated energy system and analyzed the static voltage stability index of integrated energy system. Literature [11] proposed the electric-gas coupling based on the static voltage stability index of power grid network static voltage stability index concept. Regarding the optimization problem under the constraints of the integrated energy system network, further research and discussion are needed.

In this paper, the power system static voltage stability is taken into consideration. The γ index are established for research on optimal scheduling of integrated energy systems. An optimized scheduling method for integrated energy system operation considering the constraints of the power network is proposed. The simulation example of the integrated energy system based on the IEEE13 is verified by simulation. The proposed method can further promote the global optimization of the integrated energy system.

2 Basic Model of Integrated Energy System

2.1 Basic Equipment Model

As shown in Fig. 1, the integrated energy systems mainly include power systems, thermal systems, natural gas systems, energy storage systems, distributed energy sources, CCHP, etc.

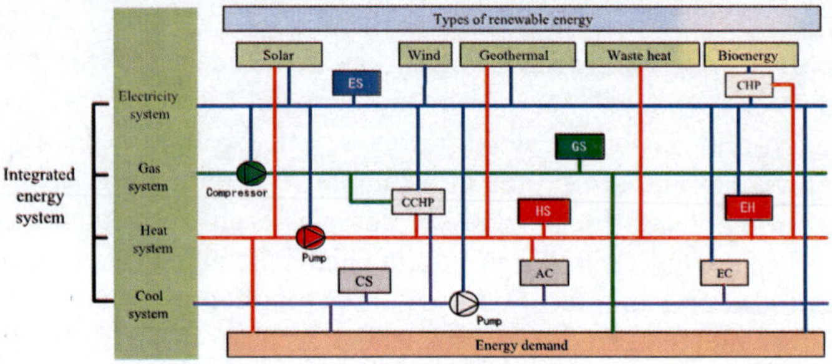

Fig. 1. Integrated energy system

Power Network Model: The parameters of power network mainly includes voltage, current, phase angle, reactance, active power and reactive power. The active power and reactive power of node i satisfy:

$$P_i = V_i \sum_{j=1}^{n} V_j (G_{ij} \cos \theta_{ij} + B_{ij} \sin \theta_{ij}) \tag{1}$$

$$Q_i = V_i \sum_{j=1}^{n} V_j (G_{ij} \sin \theta_{ij} - B_{ij} \cos \theta_{ij}) \tag{2}$$

Active power and reactive power must meet:

$$0 \leq P_i \leq P_{\max} \tag{3}$$

$$0 \leq Q_i \leq Q_{\max} \tag{4}$$

Where P_i and Q_i are the active and reactive power injected into the node. V_i is the voltage amplitude of node i. n is the number of nodes. G_{ij} and B_{ij} are the conductance and susceptance respectively. θ_i is the voltage phase angle of the node. $P_{\max}$ and $Q_{\max}$ are the upper limit of active power and reactive power respectively.

Electric Energy Storage Model: Electric energy storage has two states of charging and discharging. The capacity and power of electric energy storage meet:

$$S_t = S_{t-1}(1-\tau) - (\frac{P_{d\_t}}{\eta_d})\Delta t \tag{5}$$

$$S_t = S_{t-1}(1-\tau) + P_{c\_t}\eta_c\Delta t \tag{6}$$

Charge and discharge power meet:

$$P_{d\_\min} \le P_d \le P_{d\_\max} \tag{7}$$

$$P_{c\_\min} \le P_c \le P_{c\_\max} \tag{8}$$

Where S_t and S_{t-1} are the electric energy storage capacity at time t and time t − 1 respectively. τ is the electric energy storage self-discharge rate. P_{d_t} and P_{c_t} are the discharge and charging power at time t, respectively. η_d and η_c are the discharge and charging efficiency, respectively. Δt is the time interval at time t and time t − 1. P_{d_min} and P_{d_max} are the lower limit and upper limit of the electric energy storage discharge power, P_{c_min} and P_{c_max} are the lower limit and upper limit of the electric energy storage charging power.

Photovoltaic Model: The power of photovoltaic power generation is related to irradiance, temperature, and photovoltaic conversion efficiency:

$$P_{out} = f(I, T, \eta) \tag{9}$$

The output power of photovoltaic power generation must meet the upper and lower power limits:

$$P_{\min} \le P_{out} \le P_{\max} \tag{10}$$

Where P_{out} is the actual output power of photovoltaic power generation. I is the radiance. T is the ambient temperature. η is the conversion efficiency of photovoltaic power generation. $P_{\min}$ and $P_{\max}$ are the lower and upper limits of the power output.

Wind Power Generation Model: The power of wind power generation is related to the environmental wind speed. During the operation of the wind turbine, there are cut in wind speed v_{in}, rated wind speed v_{rate} and cut out wind speed v_{out}. When the wind speed is less than the cut-in wind speed and greater than the cut-out wind speed, the wind power generation is stopped and the rated wind speed corresponds to the rated output power P_{rate}.

$$P_{wind\_out} = \begin{cases} \frac{v - v_{in}}{v_{rate} - v_{in}} P_{rate} & v_{in} < v < v_{rate} \\ P_{rate} & v_{rate} < v < v_{out} \end{cases} \tag{11}$$

The output power of wind power generation need to meet the upper limit and lower limit:

$$P_{\min} < P_{wind\_out} < P_{\max} \tag{12}$$

Where P_{wind_out} is the output power of the wind turbine. $P_{\min}$ and $P_{\max}$ are the lower limit and upper limit of the output power of the wind power generation, respectively.

Combined Cooling Heating and Power: The core devices of CCHP are micro gas turbine and lithium bromide refrigerator. The high-grade thermal energy of natural gas combustion drives the gas turbine to generate electric energy. The emission of high temperature preheated flue gas not only can meet the domestic hot water demand, but also meet the cold energy demand of the life through the lithium bromide refrigerator.

The relationship between the heat generated by the micro gas turbine and the output electric power is:

$$Q_{MT}(t) = \frac{P_e(t)(1 - \eta_e - \eta_1)}{\eta_e} \tag{13}$$

Where $Q_{MT}(t)$ and $P_e(t)$ are the exhaust heat and electric power of the micro gas turbine at time t, respectively. η_e and η_1 are the power generation efficiency and heat loss coefficient of the micro gas turbine, respectively.

The amount of natural gas consumed by the micro gas turbine is:

$$V_{MT}(t) = \frac{P_e(t)\Delta t}{\eta_e L_{NG}} \tag{14}$$

Where $V_{MT}(t)$ is the amount of natural gas consumed by the micro gas turbine at time t. L_{NG} is the low calorific value of natural gas. Δt is the time step.

The lithium bromide refrigerator uses high temperature flue gas to drive work of the refrigerator. The cooling capacity per unit time is:

$$R_a(t) = Q_i(t)C_{op} \tag{15}$$

Where C_{op} is the cooling coefficient of the refrigerator. $R_a(t)$ is the cooling capacity at time t. $Q_i(t)$ is the heat of source heat at time t.

Generally, the CCHP work in a fixed electric-cooling- heating ratio mode. There is a proportional relationship between the electric, cooling and heating energy:

$$P_e : Q : R_a = 1 : \varepsilon_{h\_e} : \varepsilon_{c\_e} \tag{16}$$

$$P_{e\_\min} \le P_e \le P_{e\_\max} \tag{17}$$

Where ε_{c_e} and ε_{c_e} are respectively the cold-to-electric ratio and the thermoelectric ratio. $P_{e\_\min}$ and $P_{e\_\max}$ are the lower limit and upper limit of the combined cold and heat-electricity generation, respectively.

Electric Cooling/Electric Heating Model: Electric cooling and electric heating devices produce cold and heat energy by consuming electrical energy:

$$C_{e\_c} = P\eta_{e\_c} \tag{18}$$

$$H_{e\_h} = P\eta_{e\_h} \tag{19}$$

Where C_{e_c} and H_{e_h} are the cooling power and heating power of electric cooling and electric heating, respectively. η_{e_c} and η_{e_h} are the cooling efficiency and heating efficiency of electric cooling and electric heating respectively.

Cooling/Heating Storage Model: Cooling/heating storage have two operating states: cooling/heating storage and cooling/heating release, cooling/heating storage capacity and power meet:

$$S_h^t - S_h^{t-1} = P_{h\_c}\Delta t \tag{20}$$

$$S_h^{t-1} - S_h^t = P_{h\_f}\Delta t \tag{21}$$

$$P_{h\_c\_\min} \leq P_{h\_c} \leq P_{h\_c\_\max} \tag{22}$$

$$P_{h\_f\_\min} \leq P_{h\_f} \leq P_{h\_f\_\max} \tag{23}$$

Where P_{h_c} and P_{h_f} are the heating storage power and heating release power respectively. S_h^t is heating storage capacity at t time. $P_{h\_c\_\min}$ and $P_{h\_c\_\max}$ are the lower limit and upper limit of the heating storage power. $P_{h\_f\_\min}$ and $P_{h\_f\_\max}$ are the lower limit and upper limit of the heating storage release power.

2.2 Static Voltage Stability

The voltages of two adjacent node i and node j in the power system satisfy:

$$\dot{U}_i = \dot{U}_j + \dot{I}_{ij} * Z_{ij} \tag{24}$$

$$U_i^2 = (U_j + \frac{P_jR_{ij} + Q_jX_{ij}}{U_j})^2 + (\frac{P_jX_{ij} - Q_jR_{ij}}{U_j})^2 \tag{25}$$

The formula is simplified as:

$$\mathrm{U}_j^4 + (2(P_jR_{ij} + Q_jX_{ij}) - U_i^2)U_j^2 + (P_jR_{ij} + Q_jX_{ij})^2 + (P_jX_{ij} - Q_jR_{ij})^2 = 0 \tag{26}$$

If the equation has a solution, satisfy:

$$\Delta = b^2 - 4ac \geq 0 \tag{27}$$

$$4((P_jX_{ij} - Q_jR_{ij})^2 + (P_jR_{ij} + Q_jX_{ij})U_i^2) \leq U_i^4 \tag{28}$$

In the power network, R_{ij} is much smaller than X_{ij}, and meanwhile this paper mainly considers the problem of optimal power system active power scheduling.

The formula is simplified as:

$$P_j \leq \frac{\sqrt{U_i^4 - 4Q_j X_{ij} U_i^2}}{2X_{ij}} \tag{29}$$

For node j with m branches to meet:

$$P_j \leq \sum_{i=1}^{i=m} \frac{\sqrt{U_i^4 - 4Q_j X_{ij} U_i^2}}{2X_{ij}} \tag{30}$$

Define γ indicators to meet:

$$\gamma = \sum_{i=1}^{i=m} \frac{\sqrt{U_i^4 - 4Q_j X_{ij} U_i^2}}{2X_{ij}} - P_j \tag{31}$$

In the process of optimizing scheduling, this paper establishes the γ index by considering the stability of static voltage, and uses it as the inequality constraint in the optimal scheduling model. When $\gamma > 0$, it satisfies Eq. (30). When $\gamma < 0$, the node must not satisfy the static voltage stability, so it must be guaranteed $\gamma > 0$ during the optimization scheduling process.

3 Optimal Scheduling Model of Integrated Energy System

3.1 Objective Function

This paper studies the problem of regional integrated energy system optimization scheduling with the goal of overall economic optimization of the entire system and fully considering the distributed power output as well as the limitations of local environmental factors and equipment itself. Meanwhile considering the γ index of the power network in the system optimization-scheduling model, the optimization objective function is as follows:

$$C_{total} = C_{gas} + C_{ele} \tag{32}$$

Where C_{total} is the total costs of system operation. C_{gas} is the costs of natural gas consumed by system. C_{ele} is the costs of power purchase from the main network.

The costs of natural gas mainly comes from the consumption of CCHP. The gas turbines of the CCHP consume natural gas and drives the generator to generate electrical energy. At the same time, the heat energy generated by gas turbines not only can meet the domestic heat energy demand, but also are used to drive the refrigerator to meet the domestic cold energy demand.

$$C_{gas} = p_{gas} * V_{gas} = p_{gas} * \frac{P_{CCHP}}{C_{g\_e}} \tag{33}$$

Where p_{gas} is the unit price of natural gas. V_{gas} is the consumption of natural gas. P_{CCHP} is the amount of electricity generated by CCHP. C_{g_e} is conversion factor of gas to electricity

The costs of power grid purchases mainly comes from the difference between the amount of system power purchases and system power sales:

$$C_{ele} = C_p - C_s = p_{ele} * (W_p - W_s) \tag{34}$$

Where C_p and C_s are the amount of electricity purchased by the system and the amount of electricity sold by the system, respectively. p_{ele} is the unit price of the electricity. W_p is the amount of electricity purchased. W_s is the amount of electricity sold.

3.2 System Operation Constraints

The comprehensive energy system must not only satisfy the constraints of the equations of electric power balance, cooling power balance and heating power balance, but also satisfy the constraints of equations related to the equipment of electric energy storage, cooling energy storage and heating energy storage. The balance of electric power, cooling power and heating power is to maintain the dynamic balance of the generation and consumption of electric cooling and heating power at each moment. The equality constraints of energy storage is that the total capacity do not change during an optimal scheduling period.

Electric Power Equation Constraints

$$\sum_{i=1}^{n} P_{dg\_i} + P_{grid} + P_{CCHP} - P_{hc} = P_{load} \tag{35}$$

Where P_{dg_i} is distributed power generation. P_{grid} is power of purchases electricity from the main network, P_{CCHP} is the electric energy generated by the CCHP. P_{hc} is the electric energy required of electric cooling and electric heating of the system. P_{load} is electrical load.

Cooling Power Equation Constraints

$$\sum_{i=1}^{n} C_i + C_e + C_{cs} = C_{load} \tag{36}$$

Where C_i is CCHP cooling power. C_e is the electric cooling power. C_{cs} is the cold energy storage. C_{load} is the cooling load.

Heating Power Equation Constraints

$$\sum_{i=1}^{n} H_i + H_e + H_{cs} = H_{load} \tag{37}$$

Where H_i is the CCHP heating power. H_e is the electric heating power. H_{cs} is the thermal energy storage. H_{load} is the thermal load.

Energy Storage Equation Constraints

Constraints on energy storage equations include constraints of electrical energy storage equations, heating energy storage equations and cooling energy storage equations:

$$\sum_{t_c} P_c * t_c = \sum_{t_d} P_d * t_d \tag{38}$$

Where P_c and P_d are the charging and discharging power of energy storage at time t_c and t_d, respectively.

3.3 Optimal Scheduling Model

In this paper, it mainly considers the problem of optimal power system active power scheduling and do not considers the problem of reactive power scheduling, so solving optimal scheduling model is a linear optimization problem. The standard optimal scheduling model includes objective function, equality constraints, and inequality constraints. The standard optimal scheduling model is as follows:

$$\begin{cases} \min C_{total} \\ s.t. A_{eaq}X = b_{eaq} \\ AX \leq b \\ x_{i,\min} \leq x_i \leq x_{i,\max}\ i = 1, 2, 3\ldots \end{cases} \tag{39}$$

Where A_{eaq} is the coefficient matrix of equality constraints. A is the coefficient matrix of inequality constraints. X is the optimization variable, including distributed power, CCHP, and power grid purchase. $x_{i,\min}$ and $x_{i,\max}$ the lower and upper limit of the system equipment output.

In this paper, CPLEX is used to solve the above optimal scheduling model. The optimal scheduling model that does not consider the power network γ index and consider the power network γ index are analyzed.

4 Case Study

4.1 Example Parameters

As shown in Fig. 2, this paper bases on the IEEE13 adding photovoltaic, wind power, CCHP, electric energy storage, thermal energy storage, cold energy storage, electric refrigeration, electric boilers and other key equipment components to form regional integrated energy system. The system can buy electrical energy from the main grid and can sell electricity energy to the main grid. Figure 3 shows the electricity, heating and cooling prediction load curves of a typical day of the system in a summer.

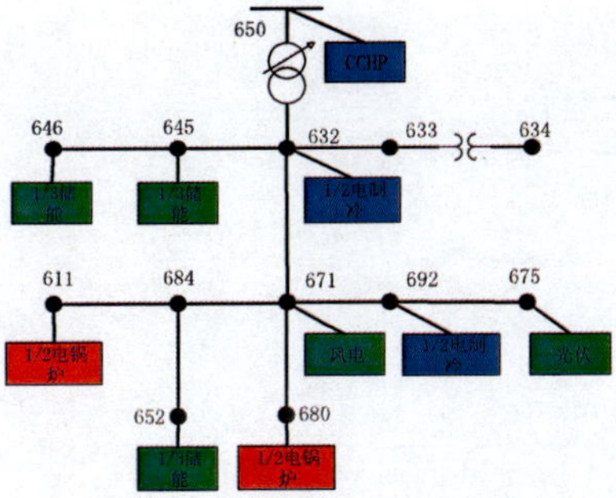

Fig. 2. Power network of IES.

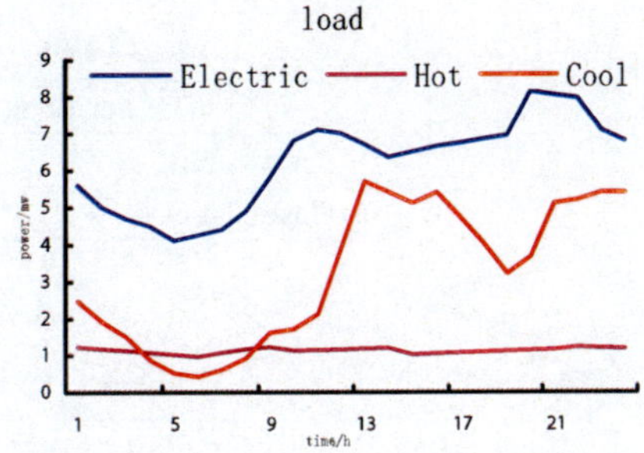

Fig. 3. Electric-cooling-heating load of prediction

Figure 4 shows the real-time electricity price changes of the power grid within a day. Figure 5 shows the typical predicted power values of photovoltaic and wind power at a typical day of the system in a summer.

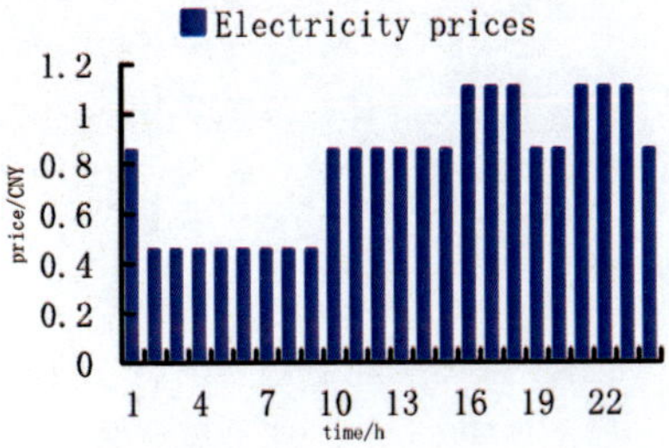

Fig. 4. Real-time electricity price

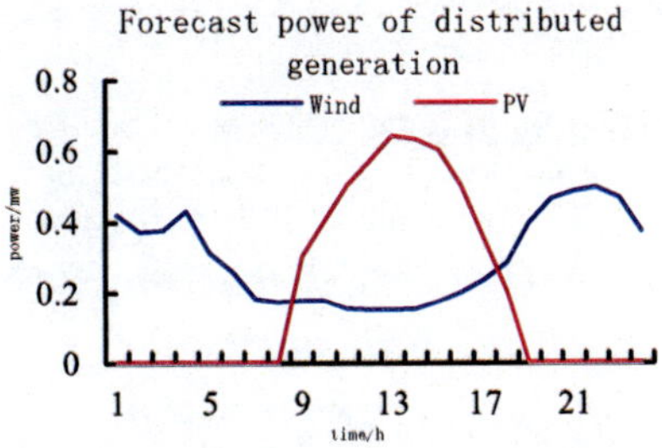

Fig. 5. Predicted power of PV and Wind

As shown in Table 1, two basic scenarios are established in this paper: case1 is an optimal energy system optimal scheduling without considering γ indices; case2 is an integrated energy system optimal scheduling considering γ indices. Discuss and analyze the difference of the optimal dispatching results of the integrated energy system under the two scenarios.

Table 2 shows the upper limit and lower limit of each device power output in the integrated energy system. In addition, the optimized scheduling period of the entire system is 24 h. The scheduling interval is 1 h. The CCHP operate in a fixed electric-heat-cooling ratio of 1: 0.27: 1.08. The price of natural gas is 2.4 yuan per cubic meter. The state of charge of the energy storage is 0.2–0.8.

Table 1. Case

Case	Whether or not consider γ indexs
Case1	No
Case2	Yes

Table 2. Device parameters

Equipment	Lower limit of output/MW	Upper limit of output/MW
Electric heating	0	3
Electric cooling	0	3
Electric storage	−0.5	0.5
Heat storage	−0.3	0.3
Cool storage	−0.3	0.3
CCHP	0.5	4
Photovoltaic	0	0.8
Wind	0	0.8

4.2 Results and Discussion

Figure 6, 7, and 8 show the optimized output of electric power, heating power, and cooling power of case1 and case2, respectively.

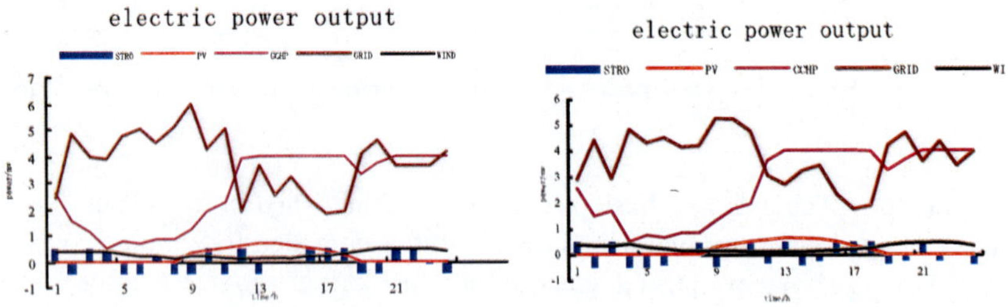

Fig. 6. Electrical power output of case1 and case2

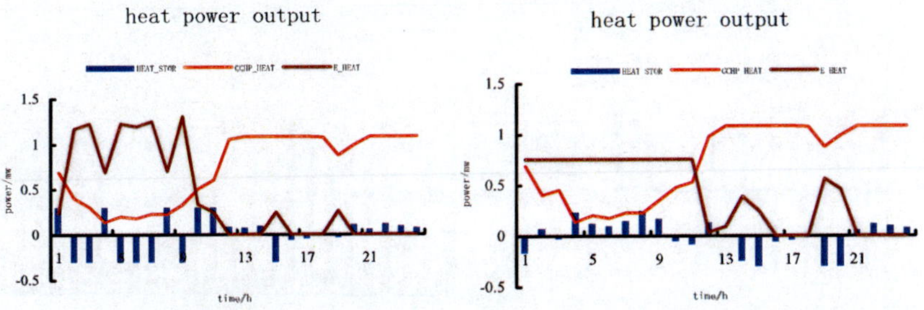

Fig. 7. Heating power output of case1 and case2

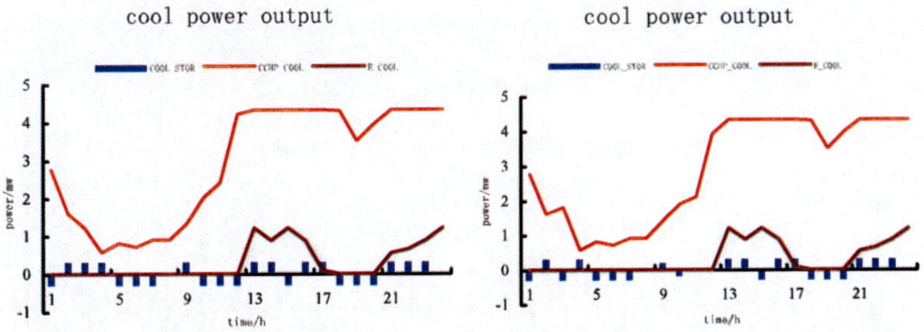

Fig. 8. Cooling power output of case1 and case2

Figure 9 shows the power output of electric cooling, electric heating. Figure 10 shows the power output of electric energy storage and main work in case1 and case2.

Figure 11 and Fig. 12 show the size of the γ indices of each node in case1 and case2, respectively. It can be seen from Fig. 10 that the optimized scheduling results of case1 cannot satisfy all the nodes whose γ indices is greater than 0 in a scheduling cycle. The γ index of the node 652 is less than 0 at times 9, 11, 13, 19, 20 and 24. The γ indices of the node 680 is less than 0 at times 2, 3, 5, 6, 7 and 9.

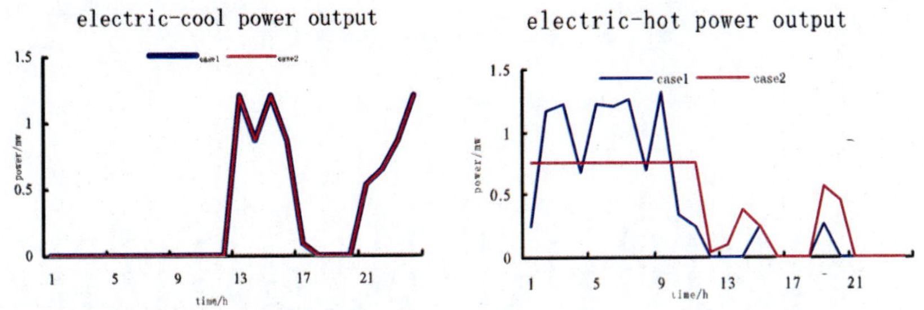

Fig. 9. Electrical-cooling and electrical-heating output of case1 and case2

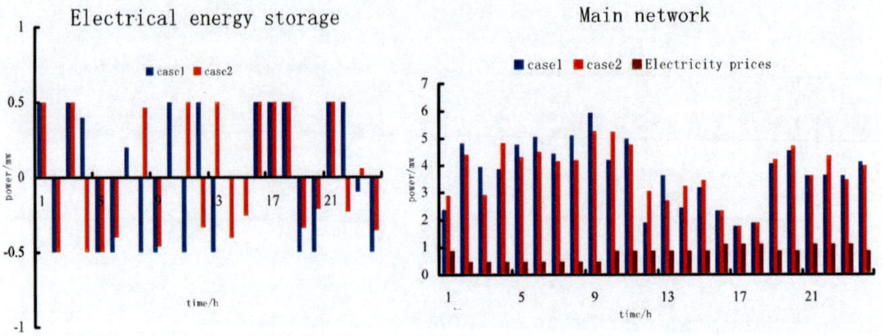

Fig. 10. Electrical energy storage and main network of case1 and case2

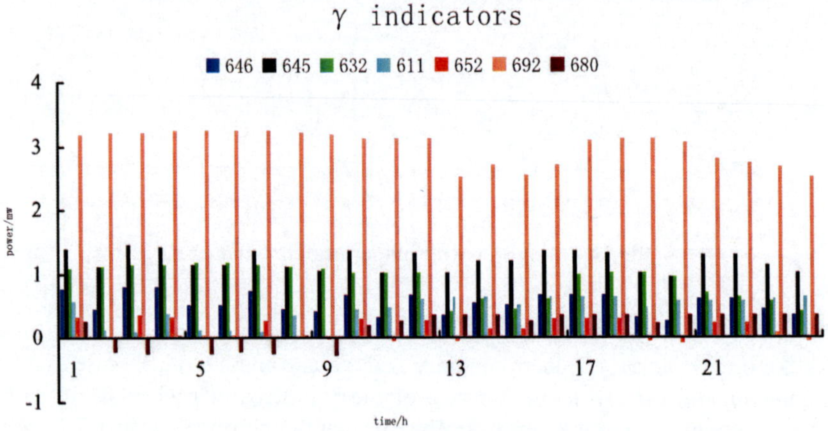

Fig. 11. γ indices of case1

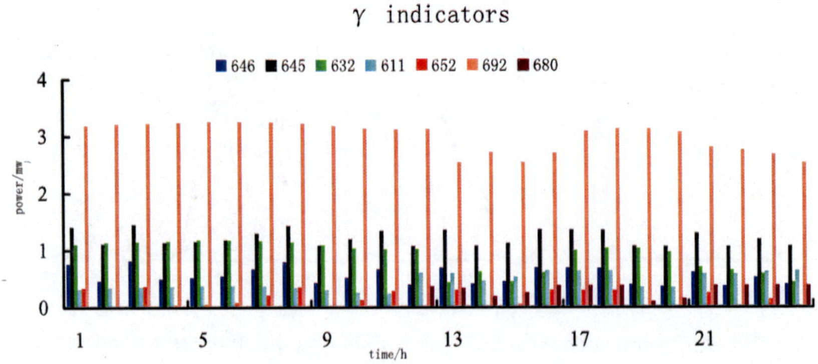

Fig. 12. γ indices of case2

Improve γ indices through coupling devices such as electrical energy storage, electrical heating and electrical cooling:

(1) Comparing Fig. 2 and Fig. 10, 1/3 the energy storage are connected to the node 652. The electrical energy storage in case1 is in full power charging state at six moments of 9, 11, 13, 19, 20 and 24. The γ indices is less than 0, which cannot meet the static voltage stability of the power network. Considering the static voltage stability of the power network in case2, the γ indices of the node 652 are directly used as the inequality constraints of optimization scheduling model of the system. The power output of electrical energy storage at the times 9, 11, 13, 19, 20 and 24 are adjusted during the optimization process. Because node 652 has a low load at the time of 9, 19, 20, 24, it is only necessary to reduce the charging power of the energy storage to make the γ indices greater than 0. However, the load peak period at time 11 and 13 can only change the charging and discharging state of the energy storage at time 11 and 13 to make it in the discharge state, so that the γ indices can be greater than 0.

(2) Comparing Fig. 2 and Fig. 9, 1/2 electric heating are connected to the node 680. The load of electric heating of case 1 is high at times 2, 3, 5, 6, 7, and 9, making the γ indices of node 680 less than 0 at these six times. In case 2, the optimal scheduling strategy is adjusted to reduce the output of electric heating at times 2, 3, 5, 6, 7 and 9 and reduce the electric heating power of the node 680 at times 2, 3, 5, 6, 7 and 9. The reducing of power demand of the electric heating makes the γ indices greater than 0. In addition, all other nodes in case1 satisfy the γ indices greater than 0, so the cooling strategy of case1 and case2 is the same. The power output curves coincide, as shown in Fig. 9.

Table 3 shows the total costs of case1 and case2 in a scheduling optimization period. The total costs in case1 is 112,300 yuan, of which the electricity costs is 65,565 yuan, and the gas costs is 46,735 yuan. The total costs in case2 is 113,790 yuan, of which the electricity costs is 67055, and the gas costs is 46735 yuan.

Table 3. Costs of case1 and case2

Case	Total costs	Electricity purchase costs	Natural gas costs
Case1	112300	65565	46735
Case2	113790	67055	46735

(1) Gas purchasing costs: The gas purchasing costs of case1 and case2 are the same and indicates that the CCHP consumes the same amount of natural gas in the two cases in a cycle. The total electrical energy, total heating energy and total cooling energy produced by CCHP are the same. Because the balance of electrical cooling and heating are satisfied at every time, so purchase of electrical energy of the main network are same in two cases.

(2) Electricity purchasing costs: It can be seen from Fig. 10 that the differences of electricity purchasing costs between the two cases are mainly due to the difference in the electricity purchase amount under different electricity prices. In order to make the γ indices of the node 680 greater than 0 in the low electricity price at

time 2, 3, 5, 6, 7 and 9, the electric heating reduce the power output of electric heating. Meanwhile ensuring the heating power balance at time 2, 3, 5, 6, 7 and 9, the heating energy storage releases a large amount of heating energy. Besides, in order to meet the constant of heating energy storage capacity in a cycle, the electric heating power is increased compared to case1 and Consume more electrical power at time 13, 14, 19, and 20 that the electricity price is higher. It will lead to an increase in the costs of electricity. In addition, for the γ indices of the node 652 greater than 0 at the lower electricity price at the six moments of 9, 11, 13, 19, 20 and 24, the charging and discharging states of electric energy storage are adjusted at the six moments. It also cause an increase in the costs of electricity.

5 Conclusion

This paper first establishes the key equipment model of the integrated energy system, proposes and defines the γ index with considering static voltage stability. Furthermore, a regional integrated energy system optimal scheduling model considering distributed power output characteristics and system operating network constraints was further established. Based on the example of IEEE13, this paper establishes two examples of scenarios that consider the γ index and do not consider the γ index and takes the lowest comprehensive costs of the system as the optimization goal. CPLEX is used to solve the optimal scheduling model. A comparative analysis of the differences in the optimal scheduling results under the two scenarios are carried out. The results show that considering the γ index improves the overall optimization effect of the integrated energy system, and lays the foundation for the follow-up research on comprehensive energy optimization scheduling technology under the comprehensive network characteristics constraints.

References

1. Dong, Z., Zhao, J., Wen, F., et al.: From smart grid to energy internet: basic concept and research framework. Autom. Electric Power Syst. **38**(15), 1–11 (2014)
2. Zheng, Y., Wang, D., Wan, C., et al.: Key technology and application of energy internet oriented to new-type towns. Autom. Electric Power Syst. **043**(014), 1–15 (2019)
3. Hongbin, S., Zhaoguang, P., Qinglai, G.: Energy management for multi-energy flow: challenges and prospects. Autom. Electric Power Syst. (2016)
4. Li, Z., Zhang, F., Liang, J., et al.: Optimization on microgrid with combined heat and power system. Proc. CSEE (2015)
5. Liu, X., Wu, J., Jenkins, N., et al.: Combined analysis of electricity and heat networks. Appl. Ener. **162**, 1238–1250 (2016)
6. Geidl, M., Andersson, G.: Optimal power flow of multiple energy carriers. IEEE Trans. Power Syst. **22**(1), 145–155 (2007)
7. Awad, B., Chaudry, M., Wu, J., et al.: Integrated optimal power flow for the electric power and heat in a microgrid. In: 20th International Universities Power Engineering Conference (UPEC), pp. 1–5 (2009)

8. Zhang, X., Yang, D., Du, W., et al.: Research on operation optimization method of regional integrated energy system. In: 2019 IEEE Innovative Smart Grid Technologies - Asia (ISGT Asia). IEEE (2019)
9. Zhong, Y., Sun, Y., Xie, D., et al.: Multi-scenario optimal dispatch of regional integrated energy system with power-heating-gas-cooling subsystems. Autom. Electr. Power Syst. **43** (12), 76–84 (2019)
10. Huang, Y., Sun, Q., Wang, R., et al.: Static voltage stability analysis for integrated energy system. Proc. CSEE (2019)
11. Wu, F., Lu, Y.: Multi-energy flow calculation and static stability analysis of electric power-LNG regional integrated energy system. Guangdong Electr. Power (2019)

Economic Dispatch Strategy of Solar Thermal-Wind Power System Considering Conditional Value-at-Risk

Zhenyuan Li[1], Yong Sun[1], Fengkai Qiu[2(✉)], Ming Zeng[2], Jialin Lin[2], Dexin Li[1], Baoju Li[1], and Chang Liu[1]

[1] State Grid Jilin Electric Power Co., LTD., Changchun 130022, China
[2] School of Economics and Management, North China Electricity Power University, Beijing 102206, China
602514627@qq.com

Abstract. In view of the influence of the randomness, volatility and anti-peak-regulation characteristics of large-scale grid-connected wind power output on the grid's peak-regulation and dispatching, the technical means of joint operation of concentrated Solar power system with thermal storage and wind power is adopted. The complementarity of solar-thermal and wind power output and the dispatchability of thermal storage are used to suppress the fluctuation of wind power output, reduce the peak-valley difference of power grid load, reduce the cost of thermal power peak shaving, and reduce wind power abandonment. Considering that the uncertainties of wind power and light power bring about the increase of dispatch operation risk, the conditional value-at-risk (CVaR) is used to measure the risk loss caused by uncertainty factors to dispatch operation. Aiming at the lowest total dispatch cost and risk cost, the economic dispatch strategy of solar thermal- wind power system considering CVaR is studied. Finally, the feasibility and rationality of the dispatch strategy and the model are verified by a simulation example.

Keywords: Concentrated solar power · Wind power · Economic dispatch · Conditional value-at-risk

1 Introduction

With the introduction of large-scale wind power into the grid, the randomness, fluctuation and anti peak regulation characteristics of its output indirectly increases the peak valley difference of grid load. At the same time, it not only increases the difficulty of power system dispatching, but also increases the operation and maintenance cost of thermal power units in the system [1]. How to reduce the impact of wind power grid-connection on the grid to maximize the acceptance of wind power has become an urgent problem. Concentrating Solar Power (CSP) system with Thermal Storage (TS) has stable and controllable output, which provides ideas for solving the above problems [2].

At present, the cost and dispatch of the joint operation of CSP and wind power have been studied in related literature at home and abroad. Reference [3] optimized a single

Y. Xue et al. (Eds.): PMF 2020, LNEE 718, pp. 676–689, 2021.
https://doi.org/10.1007/978-981-15-9746-6_52

objective of the combined solar thermal storage and wind power system, such as the lowest generation cost, the maximum benefit of grid-connection and the minimum power fluctuation. Reference [4] established a multi-objective scheduling model with the minimum output power fluctuation and the largest grid-connection benefit of the solar-thermal and wind power generation system. The grid-connection benefit was considered from the generation side, and the maximum benefit of the combined system was obtained by controlling the combined output power. The economic benefits brought by the combined system to the power grid were not analyzed. Reference [2] established a rolling prospective dispatching optimization model of the combined solar thermal power plant and wind power system, and explored the impact of the CSP and wind power capacity ratio on the system operation. However, these literatures did not consider the influence of weather, temperature and other uncertain factors. Due to certain errors in wind power and light power prediction, system dispatchers will face increased risks of dispatching operation when making power generation dispatching plans.

In fact, due to the inaccuracy of wind power and solar thermal power prediction and the non-schedulability of wind power, it may lead to risks such as the increase of system rotating reserve capacity and the transfer of part of load from efficient units with low production cost to units with high production cost. It is difficult to accurately reflect the risk attitude of dispatchers in decision-making and the impact of loss caused by extreme scenario tail risk. Therefore, it is urgent to measure the risk brought by uncertain factors in power generation scheduling process and control the risk at an acceptable level. Compared with risk measurement methods such as the minimum variance method and Value-at-Risk (VaR), conditional value-at-risk has the advantages of satisfying subadditivity, consistency theorem, and more accurate prediction of tail risk. At present, many literatures use Conditional Value-at-Risk (CVaR) to measure the impact of renewable energy uncertainty on generation scheduling. In reference [5], based on the theory of conditional risk value to quantify the uncertain risk of wind power generation system, a stochastic optimal scheduling model of power system based on conditional risk value was constructed. In reference [6], conditional risk value was used to measure the possible risks caused by uncertain factors, and a comprehensive energy system economic scheduling model considering CVaR was established.

In order to solve the economic dispatch problem of risk loss caused by uncertain factors under the background of high proportion of wind power access, according to the operation characteristics of CSP station, CSP system with thermal storage is used to participate in the system peak-regulation operation to solve the peak-regulation problem of the system after the access of wind power. The risk loss of wind abandonment and load loss caused by the uncertainty of wind power and light power is measured by introducing CVaR. Considering the operation economy and risk of the system, an economic dispatching model including CSP system with thermal storage and wind power system is constructed, which takes the condition risk value into account. Finally, the feasibility and effectiveness of the proposed method are proved by simulation.

2 Operation Principle and Characteristics of CSP Station

This paper mainly studies the operating characteristics of tower CSP. The system structure is shown in Fig. 1. As can be seen from Fig. 1, the tower-type CSP system mainly includes four parts, such as a concentrating heat collecting device, a heat storage device, a heat exchange device, and a turbo-generator set. The energy is transferred between each other through the heat transfer working medium, and the tower type heat transfer working medium generally uses molten salt.

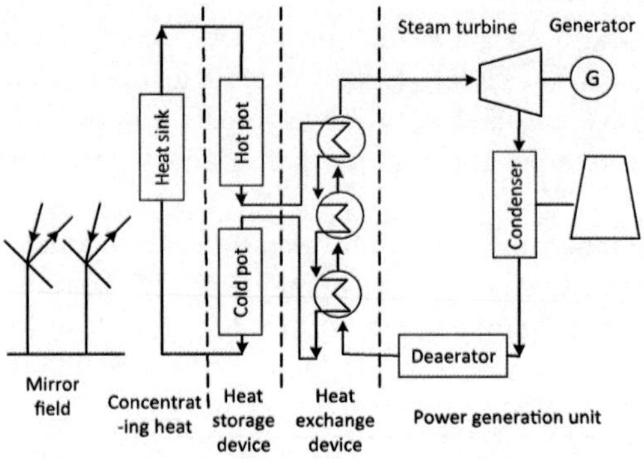

Fig. 1. Schematic diagram of tower CSP plant

The light field uses solar energy to heat the heat-conducting working medium in the heat absorber. The heat-conducting working medium generates steam through the thermal cycle to promote the generation of steam turbine generator set. If the steam generated by the heat-conducting working medium exceeds the steam required for the operation of the turbo-generator set, the heat storage device stores the excess heat energy in the form of high-temperature molten salt in case of insufficient light or peak load. Unlike other renewable energy sources, CSP station can use heat storage to shift their output, thereby achieving peak shifting and valley filling for wind power output. It has dispatching operation characteristics similar to conventional thermal power. The output of the steam turbine generator set can be quickly adjusted by controlling the heat exchange system. The fastest speed of adjustment is 20% of the installed capacity per minute, which is far higher than the conventional thermal power by 2%–5%. It is close to the gas turbine [7], and can quickly respond to the wind power output fluctuation.

3 Economic Dispatch Model of Solar Thermal-Wind Power System Considering CVaR

3.1 Theory of Conditional Value-at-Risk

Value-at-Risk (VaR), as a risk measurement tool widely used in the financial field, provides a comprehensive risk measurement framework for many complex investment portfolios [8]. However, due to the shortcomings of VaR such as lack of subadditivity and inadequate measurement of tail loss, its application is restricted to a certain extent. Conditional value-at-risk (CVaR) can reflect the average potential loss when the loss exceeds the VaR threshold. Compared with VaR, it can reflect the potential value of risk. The relationship between VaR and CVaR is shown in Fig. 2.

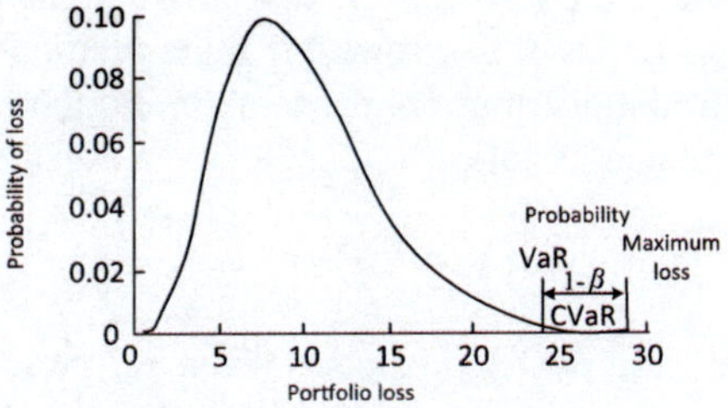

Fig. 2. The relationship between VaR and CVaR

Let $X \subset R^n$, $x \in X$ be the n-dimensional decision-making variables and $y \in R^m$ be the m-dimensional random variables, representing the uncertainty factor related to x. Then, $f(x, y)$ is the loss function related to x and y, and $\varphi(y)$ is the probability density function of y. The probability that the loss $f(x, y)$ caused by y does not exceed the critical value α is as follow.

$$\varphi(x, \alpha) = \int_{f(x,y) \leq \alpha} \rho(y)dy \tag{1}$$

For a given confidence level β, α has a corresponding set, and the minimum value is the value-at-risk.

$$f_{VaR_\beta} = \min\{\alpha \in R : \varphi(x, \alpha) \geq \beta\} \tag{2}$$

The confidence level β indicates the decision-maker's attitude towards risk. The larger level β, the higher the level of safety required by decision makers, which indicates the more risk is excluded.

The expression of conditional value-at-risk $f_{cVaR\beta}$ is as follow.

$$f_{CVaR\beta} = E[f(x, y)|f(x, y) \geq f_{VaR\beta}] = \frac{1}{1-\beta} \int_{f(x,y) \geq f_{VaR\beta}(x)} f(x, y)\rho(y)dy \tag{3}$$

$f_{cVaR\beta}(x)$ is difficult to be solved analytically. An auxiliary function $F_\beta(x, \alpha)$ can be introduced to calculate the value of $f_{cVaR\beta}(x)$.

$$F_\beta(x, \alpha) = \alpha + \frac{1}{1-\beta} \int_{y \in R^m} [f(x, y) - \alpha]^+ \rho(y)dy \tag{4}$$

$$f_{CVaR\beta}(x) = \min F_\beta(x, \alpha) \tag{5}$$

3.2 Risk Measurement of Solar Thermal-Wind Power System Based on CVaR in Probability Scenario

This paper mainly considers the influence of inaccurate prediction of wind power and light power on system operation. It is assumed that the prediction errors of wind power and light power obey normal distribution, the total prediction error of wind power and light power can be calculated as follow.

$$\Delta P_{WS,t} = \Delta P_{W,t} + \Delta P_{S,t} \tag{6}$$

Where, $\Delta P_{W,t}$ is the wind power prediction error in t period, and $\Delta P_{S,t}$ is the light power prediction error in t period.

When the prediction error $\Delta P_{W,t} < 0$, there will be a power shortage in the system. If the positive rotation reserve of the system cannot meet the power shortage, there will be a load loss event. When the prediction error $\Delta P_{W,t} > 0$, there will be a power surplus in the system. If the negative rotation reserve of the system is insufficient, there will be a wind abandonment event [9]. The loss of load and the wind abandonment power are respectively as follows.

$$P_{L,t}^{Loss} = -\Delta P_{WS,t} - UR_t \quad -\Delta P_{WS,t} > UR_t \tag{7}$$

$$P_{W,t}^{Loss} = \Delta P_{WS,t} - DR_t \quad \Delta P_{WS,t} > DR_t \tag{8}$$

Where, UR_t is the positive rotation reserve capacity of the system during t period, DR_t is the negative rotation reserve capacity of the system during t period.

When wind abandonment and load loss events occur, the penalty cost of wind abandonment F_t^{wind} is equal to the product of wind abandonment power $P_{W,t}^{Loss}$ and penalty coefficient of wind abandonment C_{LOWC}, and the penalty cost of load loss suffered by power system F_t^{load} is equal to the product of penalty coefficient of load loss C_{LOLC} and total load loss power $P_{L,t}^{Loss}$ [10]. The penalty cost of wind abandonment and load loss are as follows.

$$F_t^{wind} = C_{LOWC} P_{W,t}^{Loss} \qquad P_{W,t}^{Loss} \geq 0 \tag{9}$$

$$F_t^{load} = C_{LOLC} P_{L,t}^{Loss} \qquad P_{L,t}^{Loss} \geq 0 \tag{10}$$

Where, F_t^{wind} is the risk loss of wind abandonment at time t, F_t^{load} is the risk loss of load loss at time t, C_{LOWC} is the penalty coefficient of wind loss, and C_{LOLC} is the penalty coefficient of load loss.

In this paper, a probabilistic scenario method is used to transform the stochastic optimization problem into a deterministic optimization problem. Considering the total set of wind power and light power scenarios $WS = \{WS_k, k = 1, 2, \ldots, N_Z\}$, where N_Z

is the total number of wind power and light power scenarios [11], the CVaR values of wind abandonment risk and load loss risk are as follows.

$$F_{CVaR,t}^{wind} = \alpha_1 + \frac{1}{N_Z(1-\beta)} \sum_{k=1}^{N_Z} [F_{t,k}^{wind} - \alpha_1]^+ \tag{11}$$

$$F_{CVaR,t}^{load} = \alpha_2 + \frac{1}{N_Z(1-\beta)} \sum_{k=1}^{N_Z} [F_{t,k}^{load} - \alpha_2]^+ \tag{12}$$

Where, α_1 is the critical value of wind abandonment risk loss, α_2 is the critical value of load loss risk loss.

Therefore, the system operation risk cost considering the uncertainty of wind power and light power is as follow.

$$F_{CVaR} = \sum_{t=1}^{T} F_{CVaR,t}^{wind} + \sum_{t=1}^{T} F_{CVaR,t}^{load} \tag{13}$$

Where, T is the scheduling period.

3.3 Objective Function

The objective of optimal dispatching of CSP system with TS is to minimize the total dispatching cost and operation risk under the premise of satisfying the power supply quality. The mathematical description is as follow.

$$\min F = \min(F_G + F_{QT} + \xi F_{CVaR}) \tag{14}$$

Where, F_G is the cost of coal consumption for power generation, F_{QT} is the start-stop cost, and ξ is the risk preference coefficient.

In Eq. (14), the objective function includes two parts [12]. The first part is the total dispatching cost of the system operation, which includes the sum of the power generation cost and the start-stop cost of the system. The second part is the risk cost, and the risk aversion coefficient ξ is used to reflect the decision makers' attitude towards risk, and $\xi \geq 0$. When $\xi = 0$, it means that the decision-maker does not pay attention to the risk loss caused by the uncertain factors of the system. At the same time, the decision-maker does not take measures to avoid the risk. When ξ takes a smaller value, the decision-maker tends to aggressive scheduling, which means that the decision-maker pursues a smaller power generation costs but at the same time faces greater risk losses [13]. When ξ takes a larger value, decision maker tends to be conservative scheduling, indicating that decision makers may undertake higher power generation scheduling costs while pursuing lower risk losses. The decision maker can make reasonable system scheduling decisions based on the degree of risk aversion.

(1) Power generation cost

$$F_G = \sum_{t=1}^{T}\sum_{i=1}^{N_G} a_i(P_{i,t}^G)^2 + b_i P_{i,t}^G + c_i \tag{15}$$

Where, N_G is the number of generating units, a_i, b_i and c_i are the cost coefficients of thermal power unit i, and $P_{i,t}^G$ is the output of thermal power unit i during time t.

(2) Start-stop cost

$$F_{QT} = \sum_{t=1}^{T}\sum_{i=1}^{N_G} u_{i,t}^G(1-u_{i,t-1}^G)S_i + \sum_{t=1}^{T}\sum_{j=1}^{N_{CSP}} x_{j,t}^{CSP}(1-x_{j,t-1}^{CSP})Q_j \tag{16}$$

Where, $u_{i,t}^G$ is the running state of the thermal power unit i at time t, with a value of 1 indicating the running state and 0 indicating the shutdown state. $x_{j,t}^{CSP}$ is the running state of the CSP power unit j at time t, which is a 0–1 variable. 1 represents the start-up state, and 0 represents the outage state. S_i is the start-stop cost of thermal power unit i. Q_j is the start-stop cost of CSP unit j. N_{CSP} is the number of CSP units.

3.4 Constraints

On the basis of meeting the operation constraints of conventional units, the economic dispatch of CSP power system with TS should also meet the following constraints [14].

(1) System power balance constraints

$$\sum_{i=1}^{N_G} P_{i,t}^G + P_t^W + \sum_{j=1}^{N_{CSP}} P_{j,t}^{CSP} = P_{L,t} \tag{17}$$

Where, P_t^W represents the output of the wind turbine, $P_{j,t}^{CSP}$ represents the output of the j th CSP unit at time t, and $P_{L,t}$ is the load value of the system at time t.

(2) Operation constraints of CSP stations

$$P_{j,t}^{CSP} + P_{j,t}^{CSP,Up} \le P_j^{CSP_{\max}} \quad \forall j \in [1, N_{CSP}], t \in [1, T] \tag{18}$$

$$P_{j,t}^{CSP} - P_{j,t}^{CSP,Dw} \ge P_j^{CSP_{\min}} \quad \forall j \in [1, N_{CSP}], t \in [1, T] \tag{19}$$

$$-R_j^D \le P_{j,t}^{CSP} - P_{j,t-1}^{CSP} \le R_j^U \tag{20}$$

Where, $P_{j,t}^{CSP,Up}$ and $P_{j,t}^{CSP,Dw}$ are respectively the up and down reserve of the CSP unit j at time t, $P_j^{CSP_{\min}}$ and $P_j^{CSP_{\max}}$ are respectively the minimum and maximum output of the CSP unit j, R_j^D and R_j^U are the maximum up and down climbing capacity of the unit j respectively.

Equations (18) and (19) are the output limits of the CSP power plant units. Equation (20) is the climbing constraint of the CSP power plant units.

(3) TS operating constraints

$$E_{\min} \leq E_t^{TS} \leq \rho^{FLH} \frac{P_{\max}^{CSP}}{\eta_{TS}} \tag{21}$$

$$0 \leq P_t^C \leq P_{\max}^C x_t^{storage} \tag{22}$$

$$0 \leq P_t^D \leq P_{\max}^D x_t^{storage} \tag{23}$$

$$-\eta_C P_t^C + \frac{P_t^D}{\eta_D} = (1 - \gamma)E_{t-1} - E_t \tag{24}$$

Where, $E_{\min}$ is the minimum energy storage capacity of the heat storage device considering safe operation. E_t^{TS} is the energy stored in the heat storage device at time t. $P_{\max}^{CSP}$ represents the sum of the maximum output of all units of the CSP station. ρ^{FLH} and η^{TS} are the full load hours of the heat storage unit and the thermal conversion efficiency of the generator, respectively. P_{mac}^C and P_{mac}^D are the maximum storage power and release power of the heat storage devices, respectively.. $x_t^{storage}$ represents the storage and release state of the heat storage device. A value of 0 indicates the heat released by the device, and a value of 1 indicates the heat stored. γ, η_C and η_D are the dissipation coefficient, heat storage efficiency and heat release efficiency of the heat storage device respectively.

Equation (21) is the thermal storage capacity constraint of the heat storage device. Equations (22) and (23) are the heat storage and release power constraints [15]. Equation (24) is the relationship between the heat storage capacity and the charge and discharge power.

The economic dispatch model of CSP with TS and wind power system considering CVaR is a mixed integer nonlinear programming problem. The decision variables include the output and start-stop state of conventional units, the heat storage and release power of CSP units, the output and start-stop state of the CSP unit and so on.

4 Example Analysis

4.1 System Parameters

This paper analyzes based on the IEEE39 system and includes 10 thermal power units with a total installed capacity of 1662 MW. For specific parameters of thermal power units are shown in reference [16]. The maximum load of the system is 1549 MW. The installed capacity of the CSP station is 100 MW. Full load generation hours of heat storage device ρ^{FLH} is 12 h. The mirror field area S_{SF} is 2.0×10^6 m^2. For other detailed data of CSP station, see reference [17]. The installed capacity of the wind farm is 400 MW. The data adjusted according to the load data in [18] is shown in Table 1.

The forecast curves of wind power and DNI at 24 h before the date are shown in Fig. 3. It is assumed that the forecast errors of wind power and light power meet the normal distribution [19], and the standard deviations are 10% and 5%. 10 typical scenarios of wind power and light power prediction errors obtained by Monte Carlo simulation and scene reduction technology [20] are shown in attached Fig. 1 and attached Fig. 2 for details. The confidence level β of the example is 0.95. C_{LOWC} is 140 $yuan/(MW \cdot h)$, and C_{LOLC} is 700 $yuan/(MW \cdot h)$ [17]. Risk preference coefficient ξ is 1.0.

Table 1. System load schedule

Time	Load/MW	Time	Load/MW
1:00	1160	13:00	1237
2:00	1110	14:00	1267
3:00	1067	15:00	1328
4:00	1033	16:00	1342
5:00	1143	17:00	1414
6:00	1257	18:00	1468
7:00	1297	19:00	1521
8:00	1328	20:00	1549
9:00	1407	21:00	1385
10:00	1371	22:00	1367
11:00	1341	23:00	1248
12:00	1257	24:00	1220

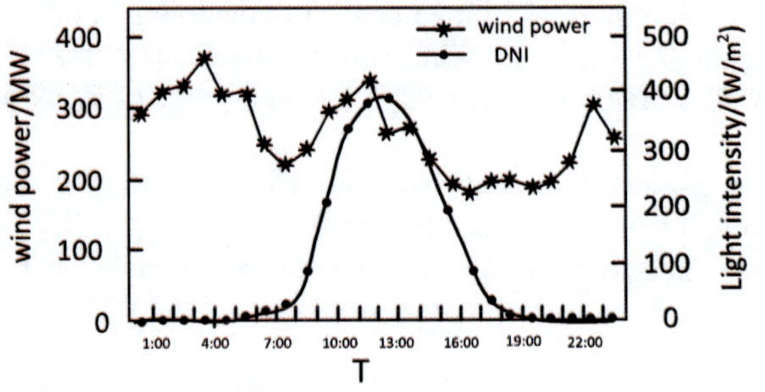

Fig. 3. Wind power and solar power scenarios

According to the established model, this paper calls optimization software CPLEX on the Matlab platform to solve.

4.2 Analysis of Utility of CSP Station Participating in System Dispatching Operation

In order to analyze the utility of the CSP plant with thermal storage in the operation of the wind power system, the economic indicators of the system operation under two solar power generation scenarios are compared under the same light power and power generation installed capacity, as shown in Table 2. Scenario 1 is access to photovoltaic power station; scenario 2 is access to CSP station.

Table 2. Economic indicators under different scenarios. (Unit: Ten thousand yuan)

Scene	Generation cost	Risk cost of wind abandonment	Risk of loss of load	Start stop cost	Total dispatch cost
Scene 1	68.85	14.15	7.11	5.86	74.71
Scene 2	63.32	5.13	1.77	3.02	66.34

As can be seen from Table 2, when the system is connected to CSP station, the total dispatching cost is 11.2% lower than that of light power station, and the risk loss is significantly reduced. The reason is that the heat storage device of CSP station can store solar energy in the form of thermal energy if the solar energy and wind power output exceed the system demand. In the case of large load demand and small solar and wind power output, the stored thermal energy is used to generate electricity, reducing the amount of strong urgent load and thermal power output. The CSP station with heat storage device enables solar power generation to have a certain controllable output, which can compensate for the uncertainty and anti-peak-regulation of wind power output to a certain extent. Therefore, it can effectively reduce the wind abandonment and load loss phenomenon of wind power and light power caused by the uncertainty, and improve the overall economic benefits of the system.

4.3 Output Analysis of CSP Station

This paper analyzes the heat storage, heat release and active power output of the CSP station based on the above example system and data, and obtains the heat storage/radiation power and power generation output curves of CSP station. The curves are shown in Fig. 4.

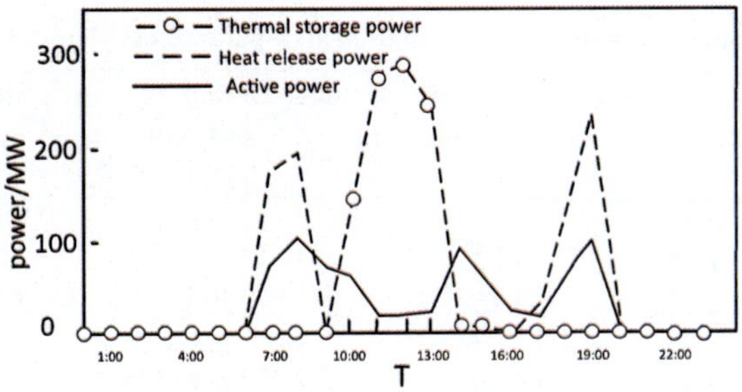

Fig. 4. Curves of CSP power station storage/heat release and power generation

It can be seen from Fig. 4 that in the period of 8:00–9:00 with large load and small wind power and light power, the CSP station uses the stored thermal energy to release the heat so that the unit is in the power generation state. In the period of 11:00–14:00 with large light power and wind power output, the thermal storage system of CSP power station is in heat storage state. During this period, the active power output of the CSP station is small, and the active power output of the wind farm is large, avoiding the wind abandonment phenomenon of the system. In the period of 18:00–20:00, due to the peak load regulation characteristics of wind power, the load is large but the wind power output is small, and the generating units of the photo thermal power station are in the state of large output. The energy stored in the heat storage system is transformed into the high-value power in the peak load demand period of the system, so as to reduce the impact of unit operation with high coal consumption rate and reduce the peak load regulation of wind power. Therefore, the economics of system operation are improved.

4.4 Impact of Risk Aversion Coefficient on Dispatch Operation

In order to analyze the impact of the degree of risk aversion on the optimization of scheduling operations, Table 3 shows the various costs of system operation under different risk aversion coefficients. It can be seen from Table 3 that the total dispatch cost of the system shows a gradual upward trend as the risk aversion coefficient increases. When the risk aversion coefficient is 0, although the power generation cost of the system is low, the cost of wind abandonment risk and loss of load risk are relatively high due to the lack of consideration of risk factors. With the increase of risk aversion coefficient, the cost of power generation has increased, while the cost of wind abandonment risk, loss of load risk, and start-stop costs continue to decrease. When the aversion coefficient increases, decision makers tend to avoid the operational risks caused by wind power and light power fluctuations. Therefore, more controllable thermal power units in the running state are added to the dispatch process, and the power generation load rate of the unit will decrease. So, the increase in fuel consumption per unit of electricity in thermal power units will increase the cost of power generation, and the risk cost will be reduced accordingly. Since the traditional generator set does not need to be started frequently when risk is avoided, the system start-up cost will be reduced.

Table 3. Economic indicators under different risk parameters (Unit: Ten thousand yuan)

Risk aversion coefficient	Generation cost	Risk cost of wind abandonment	Risk of loss of load	Start stop cost	Total dispatch cost
0	60.45	6.53	2.51	3.80	64.25
0.1	61.04	5.92	2.33	3.36	64.40
0.5	62.15	5.45	2.06	3.24	65.39
1.0	63.32	5.13	1.77	3.02	66.34
5.9	64.78	4.79	1.49	2.72	67.50

4.5 Influence of Different Heat Storage Capacity on Dispatch Operation

In order to compare and analyze the influence of different heat storage capacity on the system scheduling operation, this paper sets the heat storage device to generate electricity at full load hours $\rho^{FLH} \in [0, 14]$.The optimized scheduling results under different heat storage capacity are shown in Fig. 5.

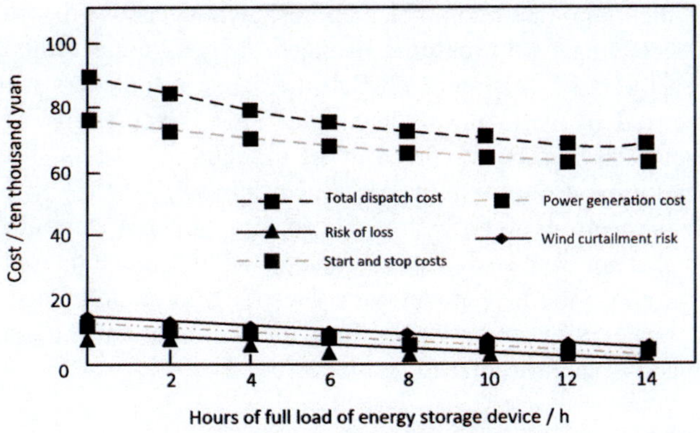

Fig. 5. Economic indicators under different heat storage capacity

In Fig. 5, as the capacity of the heat storage device increases, the total dispatch cost of the system tends to decrease first and then stabilize. When the scale of the heat storage device is small, increasing the heat storage capacity of the CSP station can increase the adjustable capacity of the system and improve the ability to absorb wind power, which reduces the cost of power generation. As the increase in heat storage capacity improves the regulation capability of the system, the impact of wind power and lighting power uncertainty are reduced, and the system's risk of wind abandonment risk, loss of load risk and start-stop costs are also continuously decreased. When the full-load hours of the heat storage device reach 12 h, if it continues to increase, the various economic indicators of the system will basically not change. This is mainly due to the limitation of light resources. The adjustable space of the system has reached saturation, and the increased heat storage capacity has not been utilized, and the economic benefits brought by the heat storage capacity will no longer change accordingly.

5 Conclusions

Based on the research on the output characteristics of CSP stations, this paper proposes an economic dispatch model of CSP plants with thermal storage and wind power systems considering the conditional risk value CVaR. The main conclusions are as follows.

(1) The CSP station with thermal storage has good dispatch controllability. Using the climbing capacity of CSP power station and the adjustable capacity of heat storage system can effectively reduce the phenomenon of power abandonment caused by the anti-peak-regulation characteristics of wind power and the inaccurate prediction of renewable energy output.
(2) Introducing CVaR into the objective function of economic dispatch of CSP stations with thermal storage and wind power systems, dispatchers can determine the impact of uncertain operation risks on dispatch decisions based on their risk aversion. According to the optimal dispatch analysis under different risk coefficients, when the risk aversion coefficient increases, dispatchers will tend to avoid the risks caused by wind power and light power fluctuations, and controllable generator sets will be more involved in the scheduling process. The system's new energy abandonment rate will decrease accordingly.
(3) Reasonable allocation of heat storage capacity of CSP station can reduce the scheduling cost and risk cost of the system, so as to achieve the optimal economic benefits and reduce the investment cost. The study on the influence of heat storage capacity on scheduling operation can provide an important reference for the heat storage capacity planning of CSP station.

Acknowledgments. This work was financially supported by State Grid Technology Project (522300190009):Research and application of the key technologies of source-demand participation in market transactions based on data value discovery.

References

1. Jiang, W., Wang, B., Wang, N., et al.: Research on power output characteristics of large-scale wind power base in multiple temporal and spatial scales. Power Syst. Technol. **41**(2), 493–499 (2017)
2. Santos-Alamillos, F.J., Pozo-Vazque, Z.D., Ruiz-Arias, J.A., et al.: Combining wind farms with concentrating solar plants to provide stable renewable power. Renew. Energy **76**, 539–550 (2015)
3. Sioshansi, R., Denholm, P.: Benefits of colocating concentrating solar power and wind. IEEE Trans. Sustain. Energy **4**(4), 877–885 (2013)
4. Dai, J., Tang, Y., Qu, L., et al.: Research on coordinated optimization control of solar thermal and wind power generation. Comput. Simul. **34**(10), 73–77 (2017)
5. Liu, H., Feng, Z., Wang, J., et al.: Economic dispatch of integrated energy system considering conditional value-at risk. Power Syst. Technol. **42**(5), 1385–1392 (2018)
6. Hemmati, R., Saboori, H., Saboori, S.: Stochastic risk-averse coordinated scheduling of grid integrated energy storage units in transmission constrained wind-thermal systems within a conditional value-at-risk framework. Energy **113**, 762–775 (2016)
7. Chen, R., Sun, H., Li, Z., et al.: Grid dispatch model and interconnection benefit analysis of concentrating solar power plants with thermal storage. Autom. Electr. Power Syst. **38**(19), 1–7 (2014)
8. Fu, B., Wan, X., Xiong, X., et al.: Risk assessment of wind power integrated system based on time-periodic characteristics of wind speed. Power Syst. Prot. Control **46**(19), 43–50 (2018)

9. Xu, W., Jiang, C., Liu, Y., et al.: Decision-making model and method for spinning reserve and risk of power system integrated with large-scale wind farms. Autom. Electr. Power Syst. **28**(13), 64–70 (2014)
10. Liu, X., Wen, B., Jiang, Y.: Study on rotary standby benefit of wind power system with conditional risk value. Trans. China Electrotechn. Soc. **32**(9), 169–178 (2017)
11. Ge, J., Wang, F., Zhang, L.: Spinning reserve model in the wind power integrated power system. Autom. Electr. Power Syst. **34**(6), 32–36 (2010)
12. Santos-Alamillos, F.J., Pozo-Vazque, Z.D., Ruiz-Arias, J.A., et al.: Combining wind farms with concentrating solar plants to provide stable renewable power. Renew. Energy **76**, 539–550 (2015)
13. Zhang, G., Li, F., Zhou, Q., et al.: Day-ahead hierarchical dispatching model considering friendliness of wind power. Power Syst. Prot. Control **47**(6), 118–124 (2019)
14. Wang, S., Lou, S., Wu, Y., Cao, K., Zhou, K.: Robust optimal scheduling of large-scale wind power integration considering the cost of deep peak shaving of thermal power units. Autom. Electr. Power Syst. **44**(01), 118–125 (2020)
15. Liang, H., Pan, Y., Xu, D., Zhou, J.: Autom. Electr. Power Syst. **43**(22), 101–110 (2019)
16. Bie, C., Hu, G., Xie, H., et al.: Optimal dispatch of wind power systems with demand response. Autom. Electr. Power Syst. **38**(13), 115–120, 159 (2014)
17. Cui, Y., Zhang, H., Zhong, W., et al.: Optimized configuration method of wind powerphotothermal combined with external capacity based on time-sharing energy complementation. Power Grid Technology (2018). https://doi.org/10.13335/j.1000-3673.pst.2018.2184
18. Lin, S., Han, M., Zhao, G., et al.: Capacity allocation method of distributed photovoltaic distribution network energy storage system based on random prediction error. Chin. J. Electr. Eng. **33**(4), 25–33, 5 (2013)
19. Vick, B.D., Moss, T.A.: Adding concentrated solar power plants to wind farms to achieve a good utility electrical load match. Sol. Energy **92**, 298–312 (2013)
20. Yu, D., Sun, X., Bingtuan, G., et al.: Coordinated optimization model of wind power integration considering uncertain wind power output. Acta Electrotechn. **31**(9), 34–41 (2016)

Double-Layer Collaborative Optimization Control Architecture and Application for Urban Energy Internet

Guoxin He[1,2,3], Qi Wang[1(✉)], Jinda Zhu[2,3], Dongmei Yang[2,3], and Xiaochen Zhang, Nianchu Zhang[1]

[1] School of Electrical and Automation Engineering, Nanjing Normal University, Nanjing 210023, China
Wangqi@njnu.edu.cn
[2] NARI Group Corporation (State Grid Electric Power Research Institute), Nanjing 211106, China
[3] NARI Technology Co., Ltd., Nanjing 211106, China

Abstract. Given the structural characteristics of the novel urban energy network, the present study complies with "cloud supervision" and "weak centralization" and develops a centralized-distributed collaborative management and control architecture with double-layer optimization. Such architecture is adopted to alleviate the task of centralized optimization by appropriately reducing top-layer management functions. With virtual soft controllers (VSC), the present study also formulates virtual and real (double-layer) controller deployment strategies, allowing the bottom layer with the power of self-decision and self-control. Moreover, the interaction mode and process between the top and bottom layers in the double-layer architecture are analyzed, and the effectiveness of this architecture based on the data harvested from the cases of actual engineering applications is verified.

Keywords: Double-layer · Control architecture · Urban energy internet

1 Introduction

In the conventional urban energy system, the electric power system, heating power system, and gas system are regulated and operated by a range of organizations. Such fragmented energy supply and marketing and have imposed prominent problems (e.g., low overall energy efficiency in urban areas, a low proportion of clean energy, as well as poor supply-demand coordination) [1]. Accordingly, all those who work in relevant fields have come to realize that to achieve interconnection between different energy systems and enhance the overall efficiency of urban energy supply, the strengths of Internet technology should be exploited for research and applications of the energy internet under the premise of risk control [2–4].

On the whole, conventional energy subsystem optimization mostly complies with the model of "centralized management and optimization". With the power system as an example, it is commonly regulated via the Energy Management System (EMS) or the Local Control Center (LCC), so decisions are made after collecting and processing

Y. Xue et al. (Eds.): PMF 2020, LNEE 718, pp. 690–701, 2021.
https://doi.org/10.1007/978-981-15-9746-6_53

data. Such model is employed in considerable studies on the management framework of energy Internet operations. For instance, As revealed from Jin X's study [5], he proposed achieving centralized optimization and scheduling by the optimal mixed power flow algorithm for the regionally integrated energy system, incorporating distribution network reconstruction and complying with the EH (Energy Hub) concept and the principle of whole process (source-network-terminal) optimization. However, with the rise in the number of various entities (e.g., integrated energy hub stations, distributed novel energy stations, energy producers and consumers in the urban energy Internet), the model of centralized management or optimization with a single decision-making body is no longer sustainable. The reasons can be drawn as follows:

(1) As power, heat, and gas supply systems are affiliated with a range of bodies of interest, a centralized control system of the urban energy internet may not be capable of implementing direct control over them. For instance, the air-conditioning system in a commercial building cannot probably to receive any top-down intrusive control:
(2) And that centralized optimization model faces severe challenges in privacy, reliability, cost effect, flexibility, feasibility, etc., for it relies on the complex communication network for data collection and processing. For example, certain industrial, commercial or domestic users are reluctant to report their complete energy usage data for privacy concerns; large-scale communication networks are costly and prone to severe communication delays, and it comes with a high communication failure rate and the risk of cyber-attacks.

To solve the mentioned problems, the idea of distributed control or optimization emerges. Consistent with the internet, distributed optimization allows each resource to communicate with its neighboring resources. Besides, by enabling information exchange between resources, it can perform iterative control with the limited information obtained by each resource and achieve overall stability [6–8]. Over the past few years, the energy internet control based on distributed optimization has attracted growing academic attention for it provides new perspectives to build a novel optimal control system that applies to the urban energy internet system. For instance, Chen G developed a fully distributed operation optimization method for micro-grid groups based on consensus algorithm and Newton algorithm; it was demonstrated that this method is superior to conventional centralized optimization for reliability, flexibility and scalability through simulation verification [9]. Wang Y proposed abstracting the regional energy interconnection system, containing multiple energy entities into a system formed by multiple EH coupled energy supply networks, with each energy entity as a basic decision-making subject; thus, the coordinated control can be achieved over the entire multi-energy interconnection system [10].

Though the distributed optimization model is commonly considered reliable, flexible, easily scalable, low-cost, and highly private, significant limitations remain in its application:

(1) The prerequisite for the overall optimal solution of the distributed optimization model is that each energy subsystem is capable of accurately solving the local optimization problem and successfully implementing the control strategy, whereas

the degree of optimization of each local subsystem varies in actual engineering applications; thus, the consistent control of the network cannot be effectively ensured at all times;

(2) In the distributed optimization mode, a single energy subject may primarily aim to achieve self-optimization, while ignoring overall system requirements, which will adversely affect the integrity of the system and even jeopardize the safe operation of the system.

To address the mentioned problems, the present study proposes a centralized-distributed collaborative management and control architecture with double-layer optimization, to form a three-dimensional management and control structure suitable for the operation and control of urban energy internet, and to achieve consistent overall control that combines the technical advantages of both conventional centralized optimization and the emerging distributed optimization method. Moreover, it was also suggested that applying virtual and real (double-layer) controller deployment strategies with virtual soft controllers (VSC). After comparing the difference between the two deployment strategies (real and virtual) and their applicability is compared in different scenarios, and the mode of top-bottom interactions and specific processes is analyzed under the double-layer optimization architecture, the applicability and validity of which are verified lastly by analyzing the data harvested from cases of actual engineering applications.

2 The Overall Structure of Double-Layer Control

The overall structure of the double-layer coordinated optimization control solution for urban energy internet is illustrated in Fig. 1.

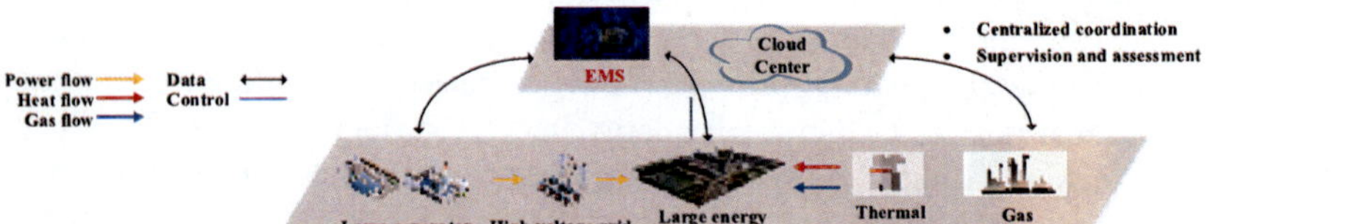

Fig. 1. The overall structure of centralized-distributed double-layer optimization for urban energy Internet.

Figure 1 illustrates that small distributed photovoltaic energy sources, sources of wind power, heat power, gas supply, power-heat-gas supply pipeline network, as well as industrial-commercial-residential multi-energy loads form a small scale integrated energy system corresponding to a virtual soft controller VSC 1; multiple virtual soft controllers (VSC 1, VSC 2 … VSC n) jointly constitute the bottom layer (distributed decision layer) of the dispatch control architecture of the urban energy internet. Moreover, control systems for large centralized generator sets, thermal power plants, natural gas sources, high-voltage transmission grids, heat-gas pipelines and others are linked to the urban energy management system (EMS) and Cloud Center to form the top layer, that is, the supervisory management. All VSCs at the top and bottom layers can communicate with each other, thereby ensuring barrier-free instruction up to and down transmission of instructions.

Note that this architecture inherits the basic functions of conventional centralized management and control system, which is to centrally formulate control strategies and send the end operating directly via remote control. Inconsistent with the conventional model, in this architecture, the top layer EMS can only check the overall volume of on-site decision control operations made by VSCs at the bottom layer, and issue orders for execution after confirmation or making corrections when necessary. The mentioned novel control method eliminates the need for complex, large-scale nonlinear mixed-integer calculations (for optimization) by top layer EMS, and significantly weakens the top layer centralized decision-making function by substituting conventional complex decision-making with checkout decisions or weighted approvals; Besides, since complex optimization decision tools are not required to be used, the top layer only needs to apply conventional decision results to monitor the operation of the entire system, which significantly reduces the centralized optimization task. Such architecture is more consistent with the practical application requirements of the urban energy internet.

3 Deployment Strategy of VSC

The virtual soft controllers (VSCs), the core devices of the two-layer optimization control system, refer to the "decision-makers" or "executors" of the operation optimization of various small distributed integrated energy systems. They can be deployed in two modes, i.e., virtual or real.

Figure 2 suggests that, if all computing resources are located at the top layer EMS, the bottom-level VSCs adopt virtual deployment, in which the top layer EMS allocates computational resources to each VSC to support it in making related calculations for distributed decisions. Under the virtual deployment mode, a VSC is capable of conducting distributed computing and formulating control strategies that will be uploaded to the top layer EMS for verification, but it cannot implement distributed control directly over any small integrated energy system.

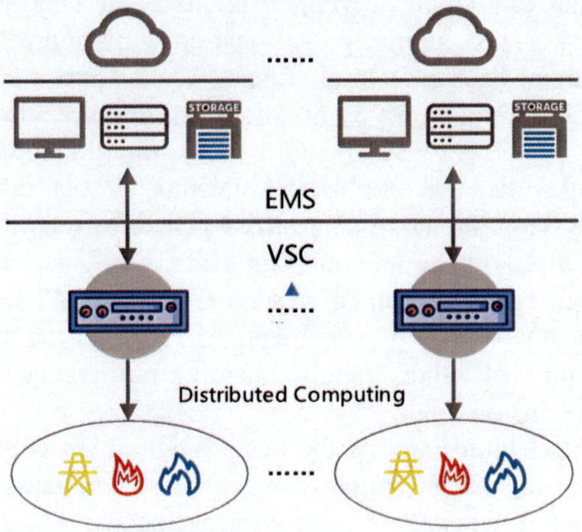

Fig. 2. Virtual deployment method of a controller.

Figure 3 shows the real deployment of controllers, in which the VSCs not only have the capability of distributed computing but also undertakes distributed control functions. In other words, they can perform self-decision and self-control. However, depending on the amount of authority granted, some VSC-generated decisions should still be transmitted to the top layer EMS for verification, and these decisions can only be implemented with the approval of the top layer.

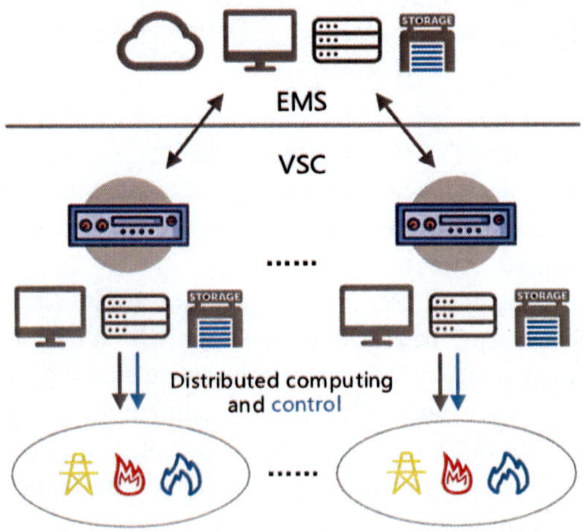

Fig. 3. Real deployment method of a controller.

There currently are three main functions used to make optimal decisions for urban energy Internet. It yields [11]:

(1) The one with minimum operating cost is expressed as:

$$\min F(Cost) = \sum\nolimits_{i=1}^{n} Cost_i \tag{1}$$

Where *F*(*Cost*) is the function representing the total cost of the integrated energy system and $Cost_i$ represents each cost item.

(2) The one with minimum fuel consumption:

$$\min F(Fuel) = \sum\nolimits_{j=1}^{m} Fuel_j \tag{2}$$

Where *F*(*Fuel*) is the function representing the total fuel consumption of the integrated energy system and $Fuel_j$ represents the fuel consumption of each energy subsystem.

(3) The one with the maximum reliability in energy supply:

$$\min F(Lack) = Lack_e + Lack_h + Lack_g \tag{3}$$

Where *F*(*Lack*) is the function representing energy shortage of the integrated energy system; while $Lack_e$, $Lack_h$, and $Lack_g$ represent respectively the power-off, heat-off, and gas-off states of each energy subsystem.

(4) The one with multi-objective optimization decision:

$$\min F(Total) = \alpha F_1 + \beta F_2 + \gamma F_3 \tag{4}$$

Where *F*(*Total*) denotes the function representing the multi-objective optimization of the integrated energy system; F_1, F_2, F_3 represent functions representing different single objectives; α, β, and γ those representing the weight coefficients of the single objectives in the combination function.

Different decision-makers involved in the distributed energy optimization system have different goals. For instance, the planners of integrated energy systems in parks focus on cost effectiveness and environmental protection; the operation and maintenance personal are more concerned with safety and reliability, and the operation managers often stress the optimization of multiple objectives.

Apart from their respective goals, the virtual controller deployment methods are also in close relation to factors e.g., the privacy protection needs of asset owners and users. For instance, as shown in Fig. 4, managers or users who need to control their own integrated energy systems should use the real deployment method, while a third party that does not handle privacy data and provides data value-added services should only adopt the virtual deployment method.

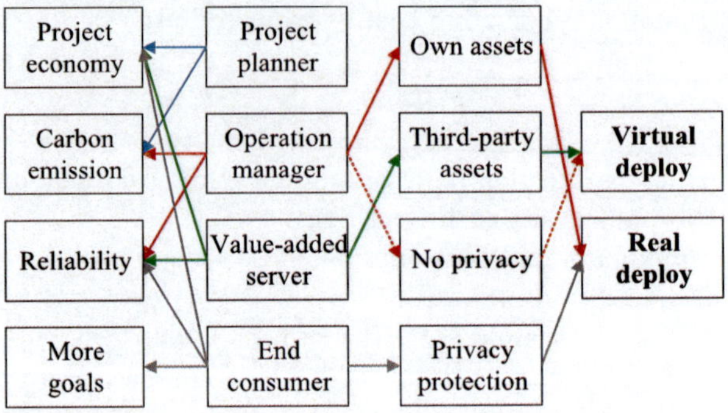

Fig. 4. Deployment options for different subjects.

4 Interactive Process Between EMS and VSC

The present study argues to achieve centralized-distributed collaborative optimization of urban energy Internet is to find the solution to a typical multi-stage decision-making problem [13]:

$$F = maxE\sum\nolimits_{t=0}^{T}\gamma^{t}C_{t}(S_{t},\ x_{t}) \tag{5}$$

$$x_t = X^{\pi}(S_t) \tag{6}$$

Where X^{π} denotes the decision function corresponding to strategy m; S_t represents the state variable of the system; $C_t\ (S_t\ ,x_{t)}$ is the immediate return of the system at a time after adopting decision x_t under state S_t and γ is the attenuation factor.

On the whole, the problem can be solved based on the optimization principle, Bellman principle, that is, to convert the optimization process into a range of single-stage decisions and find the solution to the respective decision by analyzing the transfer and constraint relationships between them. By complying with the Bellman principle, a distributed optimization problem can be considered a dynamic programming problem, meaning that an optimized solution for the entire system can be yielded by gradually accumulating optimal solutions for the local subsystems.

As shown in Fig. 5, in the double-layer collaborative optimization control system for urban energy Internet, the VSCs at the bottom layer and the small system composed of several VSCs are all local systems. Moreover, a large system can be divided into several smaller systems by separately optimizing each VSC subsystem or local system composed of such VSCs. As each original VSC subsystem is considered in the optimization for each local system, the optimal optimization solution for the entire system can be developed by continuously expanding the scope of the local system optimization.

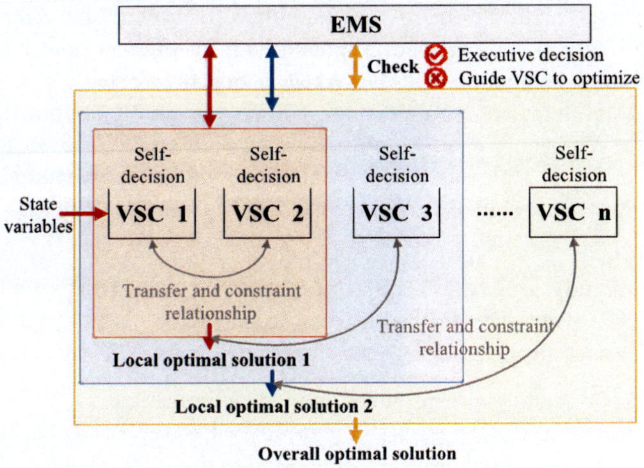

Fig. 5. The interaction process between EMS and VSC.

5 Example Verification

The system structure of the simulation example selected in the present study is shown in Fig. 6. It is a coupled system of two small multi-energy systems:

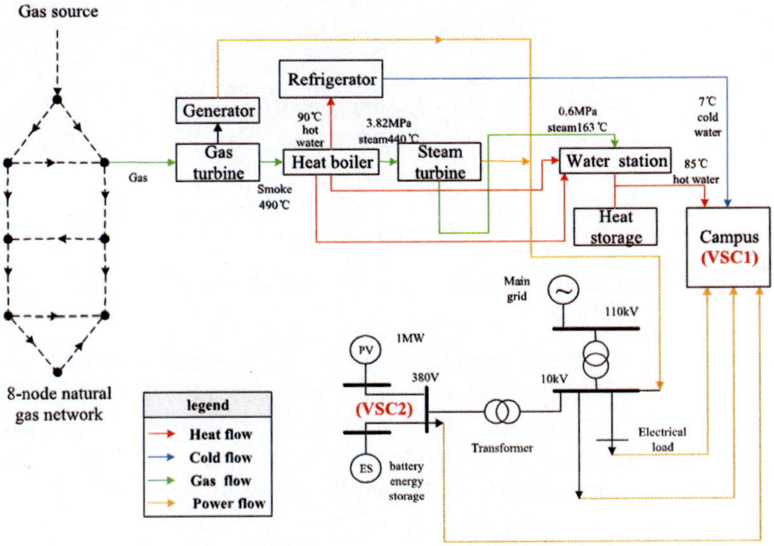

Fig. 6. System structure of the simulation example.

(1) Distributed multi-energy system: It uses the natural gas system with 8-nodes to supply gas and then transmits to the CCHP unit, through which electrical energy,

thermal energy, and cold energy are outputted to provide the campus with electricity, thermal and cooling load. The area with insufficient power is provided by the main grid, as shown in VSC1 marked in Fig. 6.

(2) Micro-grid system: It consists of 1 MW photovoltaic power generation, 500 kW/3MWh battery energy storage system, and 0.6 MW power load. The local power balancing is achieved largely by altering battery energy storage, and the area lacking power is provided by the main grid, as presented in VSC2 marked in Fig. 6.

The present study uses MATLAB for programming and IBM's CPLEX software for optimization solver. The following two scenarios are largely considered in the calculation. For specific parameters, please refer to Zhang X's study [13]:

(1) Scenario 1: Conventional centralized unified optimization
All information in VSC1 and VSC2 is uploaded to the EMS, performing unified centralized optimization. VSC1 and VSC2 fail to make independent decisions; they aim at economy optimization. Their optimization results are illustrated in Fig. 7.

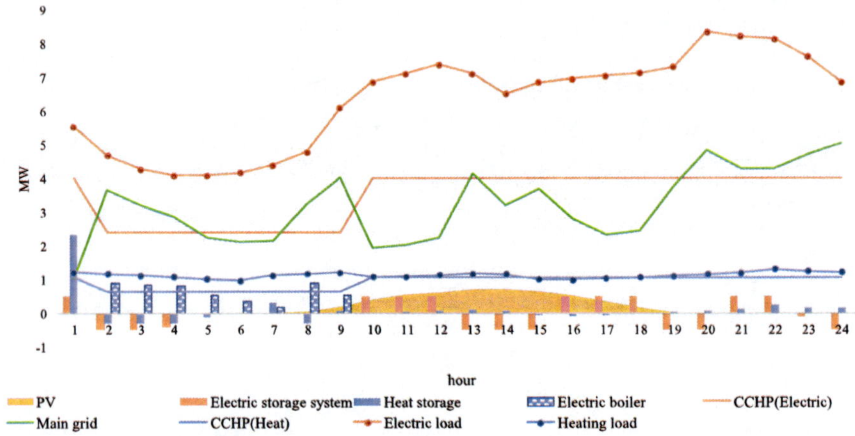

Fig. 7. Optimization results in scenario 1 (only show electricity and heat).

(2) Scenario 2: Centralized-distributed double-layer collaborative optimization
VSC1 and VSC2 are capable of making independent decisions. VSC2's optimization goal is inconsistent with that of Scenario 1. Its energy storage system performs peak-valley arbitrage, while absorbing extra output of photovoltaic during the day. The optimization results are illustrated in Fig. 8 and Fig. 9:

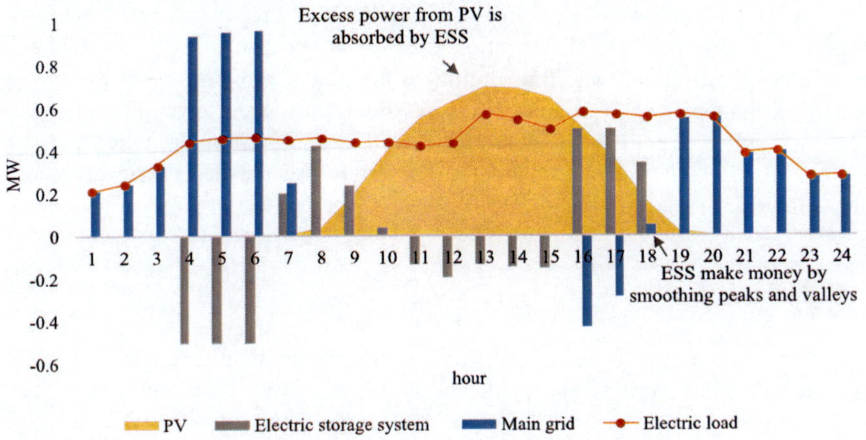

Fig. 8. Optimization results of VSC2 in scenario 2.

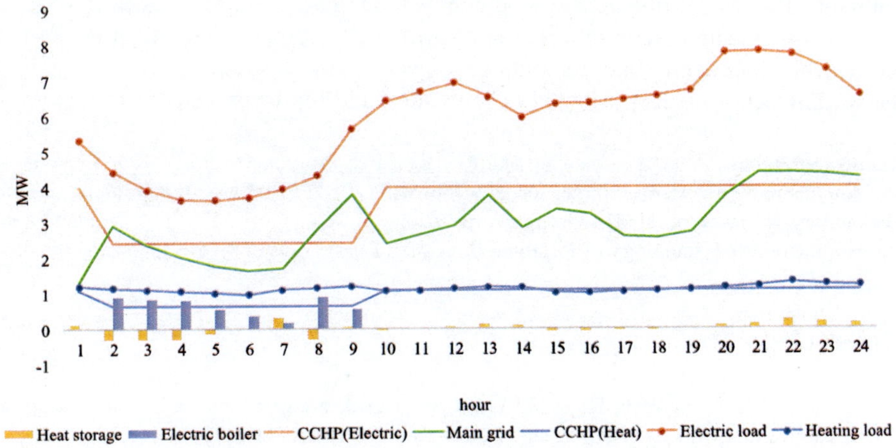

Fig. 9. Optimization results of VSC1 in scenario 2 (only show electricity and heat).

The cost expenditure and the time taken to do the calculations for the two scenarios are listed in Table 1:

Table 1. Comparison results in two scenarios.

Index	Case1	Case2
Electricity purchase cost (CNY)	60751	61293
Natural gas purchase cost (CNY)	60054	60054
Total cost (CNY)	120805	121345
calculating time(s)	3.796	3.055

It is revealed that the cost of scenario 2 is higher than that of scenario 1, whereas the calculation time is shorter. The mentioned result is because in Scenario 1, the EMS obtains global information and thus optimizes the global economy more effectively; in Scenario 2, however, VSC1 and VSC2 consider only their own interests, thereby leading to higher overall cost. Moreover, VSC1 and VSC2 in Scenario 2 can make decisions themselves, and the calculation is faster, and the decision execution is of higher efficiency.

6 Conclusion

The present study proposes the double-layer optimization architecture for urban energy interconnection and verifies, the applicability and validity of the architecture by analyzing the data harvested from cases of real engineering application cases. As the present study is a preliminary study, further in-depth research should be further conducted on the specific deployment patterns of controllers, decision-making scenarios, and methods of the centralized-distributed optimization should be studied in depth. Subsequent technical research will be conducted regarding this architecture (e.g., the study on optimization algorithms based on artificial intelligence, controller deployment mechanisms based on edge computing, as well as the operation scenarios of information interaction with distributed collaboration and top layer supervision).

Acknowledgment. Thanks to the National Key Research Program of China "Key Technologies and Applications of Energy Internet for New Towns" (2018YFB0905000) and the Science and Technology Projects of NARI Group Corporation "Research on Modeling Simulation and Operation Control Technology of Regional Integrated Energy System" for supporting this paper.

References

1. Dong, Z., Zhao, J., Wen, F., et al.: From smart grid to energy internet: basic concept and research framework. Autom. Electr. Power Syst. **38**(15), 1–11 (2014)
2. Shu, Y., Xue, Y., Cai, B., et al.: A review of energy transition analysis part two uncertainties and approaches. Autom. Electr. Power Syst. **42**(10), 1–12 (2018)
3. Huang, J., Xue, Y., Jiang, C., et al.: Carbon market risk analysis and control part : framework design. Autom. Electr. Power Syst. **42**(12), 11–18 (2018)
4. Zheng, Y., Wang, D., Wan, C., et al.: Key technology and application of energy internet oriented to new-type towns. Autom. Electr. Power Syst. **43**(14), 1–15 (2019)
5. Jin, X., Mu, Y., Jia, H., et al.: Calculation of optimal hybrid power flow for integrated community energy system considering electric distribution network reconfiguration. Autom. Elect. Power Syst. **41**(1), 18–24 (2017)
6. Manickavasagam, K.: Intelligent energy control center for distributed generators using multi-agent system. IEEE Trans. Power Syst. **30**(5), 2442–2449 (2014)
7. Qin, X., Sun, H., Shen, X., et al.: A generalized quasi-dynamic model for electric-heat coupling integrated energy system with distributed energy resources. Appl. Energy **251**, 113270 (2019)

8. Wu, W., Zhang, B., Sun, H., et al.: Energy management and distributed energy resources cluster control for active distribution networks. Autom. Electr. Power Syst. **44**(9), 111–118 (2020)
9. Chen, G., Yang, Y., Yang, X., et al.: Distributed optimization scheduling method for microgrid cluster based on distributed newton method. Autom. Electr. Power Syst. **41**(21), 156–162 (2017)
10. Wang, Y., Zhang, N., Kang, C., et al.: Standardized matrix modeling of multiple energy systems. IEEE Trans. Smart Grid **10**(1), 257–270 (2017)
11. Zeng, M., Liu, Y., Zhou, P.: Review and prospects of integrated energy system modeling and benefit evaluation. Power System Technol. **42**(6), 1697–1708 (2018)
12. Heymann, B., Bonnans, J.F., Martinon, P., et al.: Continuous optimal control approaches to microgrid energy management. Energy Syst. **9**(1), 59–77 (2018)
13. Zhang, X., Yang, D., Du, W., et al.: Research on operation optimization method of regional integrated energy system. In: 2019 IEEE Innovative Smart Grid Technologies-Asia (ISGT Asia), pp. 3299–3304. IEEE, Chengdu (2019)

Research on Identification Technology of Weak Links in Power Grid Based on Risk Index Under Extreme Ice Disaster

Yang Yang[1], Xia Zhou[2], Jianqiang Lu[2](✉), Xiangpeng Xie[2], and Chenjie Bu[2]

[1] State Grid Xinjiang Electric Power Research Institute, Wulumuqi 830000, China
[2] Institute of Advanced Technology, Nanjing University of Posts and Telecommunications, Nanjing 210023, China
1098116508@qq.com

Abstract. To prevent the occurrence of large-scale power outages accidents in extreme ice disasters and to evaluate the weak links during power grid operation, a technique for identifying weak links in power grids based on risk indicators under extreme ice disasters is proposed. This technology mainly considers extreme weather conditions such as large temperature difference, multiple strong winds, and thick ice coating when extreme ice disasters occur, calculates the total load on transmission lines and towers, and uses strength-stress interference models to determine transmission line and tower failures Rate, and then determine the overall probability of line failure based on the principle of series reliability. Then, the non-sequential Monte Carlo method and the comprehensive probability of line faults are used to set the expected fault set, and the DC load flow optimization model is used to calculate the load loss of each line. Secondly, comprehensively consider the load level and emergency repair recovery time, determine the economic value of the line load loss, and define it as the fault consequence. Finally, the product of the fault consequence and the comprehensive probability of the line fault is used to obtain the risk index of the line and rank it as the basis for evaluating the weak link. On this basis, the IEEE30 node modified by the actual topology is taken as an example to carry out simulation research to prove the reliability and effectiveness of the technology.

Keywords: Weak links · Extreme disaster · Failure consequences · Risk index

1 Introduction

As one of the frequent extreme weather disasters in China, gale and snow disasters pose severe challenges to the safe and stable operation of the power grid. To prevent the occurrence of large-scale power outage accidents under extreme disasters, how to establish a weak link identification model of the power grid and comprehensively evaluate the weak links of the transmission line are crucial to the safe and stable operation of the power grid.

Y. Xue et al. (Eds.): PMF 2020, LNEE 718, pp. 702–715, 2021.
https://doi.org/10.1007/978-981-15-9746-6_54

In terms of the research on the weak links of the power grid, literature [1–6] track and identify the weak links of the transmission line based on the external environment and the internal operating status, respectively, according to the reliability model; Literature [7–10] proposed a multi-factor comprehensive line outage probability model to evaluate the weak links of the power grid. This type of method can only determine the probability of transmission line outage, and cannot reflect the consequences of line failure; References [11–13] use information resources in the process of production scheduling and use a variety of professional analysis techniques to monitor the trend of power grid operating status changes and propose a method for identifying weak links in the power grid. In the existing related research, the identification model of the weak link of the power grid is mostly from the perspective of the external environment of the transmission line or the internal operating state. The link does not accurately reflect the actual situation. At the same time, the related research is mostly based on the normal power grid, and there are relatively few studies on the identification method of the weak link of the power grid in extreme disaster scenarios.

In most parts of western China, extreme ice disasters are often accompanied by strong winds. To prevent the occurrence of large-scale power outage accidents under extreme disasters, this paper proposes a calculation method of power grid risk indicators that considers the probability of line failure, to identify the weak links of the power grid under extreme disasters. Firstly, by collecting information such as external micro-meteorology, terrain, component aging status, and internal operating status of the power grid, the total load on the ice-covered lines and towers is calculated based on the ice-covering status and weather information. Comprehensively considering the impact of extreme ice disasters on the line failure rate, the probability of occurrence of power disconnection and power tower failure are calculated respectively, and the comprehensive probability of line failure is obtained according to the series reliability model. Further, using the DC power flow optimization model to obtain the line fault load loss amount, a multi-factor grid fault consequence model is established, the impact of the load level and emergency repair time on the fault consequence is considered, and the economic loss of the line load loss is defined as the fault consequence. Finally, the IEEE standard 30 nodes are used as an example to verify the reliability of the weak link identification method.

2 The Identification Process of Weak Links in Power Grid Under Extreme Ice Disaster

When extreme ice disasters occur, they are often accompanied by meteorological environmental characteristics such as large temperature differences, multiple strong winds, and thick ice cover, which have a great impact on the power grid.

To optimize the allocation of emergency repair resources and improve the ability of the power grid to resist disasters, this paper proposes the identification technology of the weak links of the power grid under extreme ice disasters. The overall process is shown in Fig. 1.

1) Collect external micro-meteorology, terrain, component status and other information of the power grid;
2) According to the ice coating state [14–17] and wind speed information, calculate the total load on the ice coating line and the tower;
3) Based on the strength-stress interference model, calculate the probability of power failure and power tower collapse, and use the series reliability formula to obtain the comprehensive probability of line failure;
4) Use the DC power flow optimization model to sequentially obtain the load loss caused by each line fault, and integrate the node load level and fault recovery time to obtain the economic loss of the line load loss as the fault consequence of the corresponding line;
5) The product of the probability of failure of each line and the consequences of the failure is defined as the risk of the transmission line and is used as the basis for the identification of weak links in the power grid.

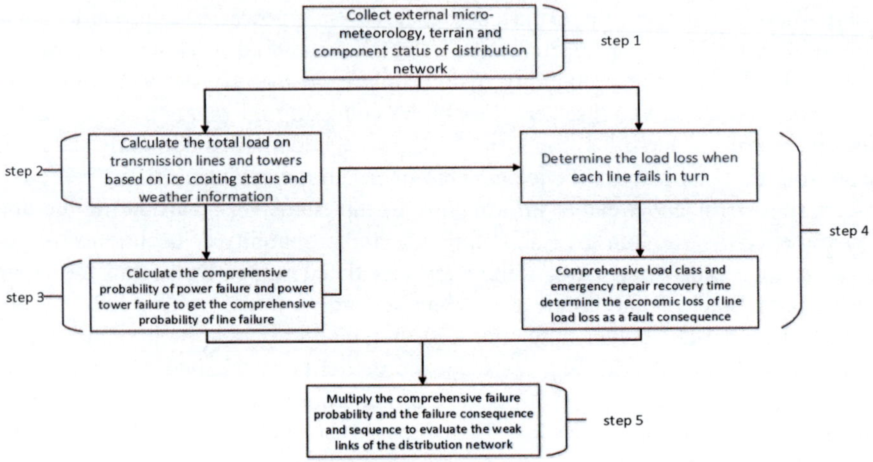

Fig. 1. The identification process of weak links in the power grid under extreme ice disaster.

3 Calculation of Total Load of the Transmission Line (Tower)

Under extreme ice disasters, the transmission line is affected by both the ice load and the wind load, which will not only directly cause power disconnection, but also cause the power to fall into the tower, which will cause power disconnection. The calculation process of the comprehensive failure probability of the transmission line is shown in Fig. 2.

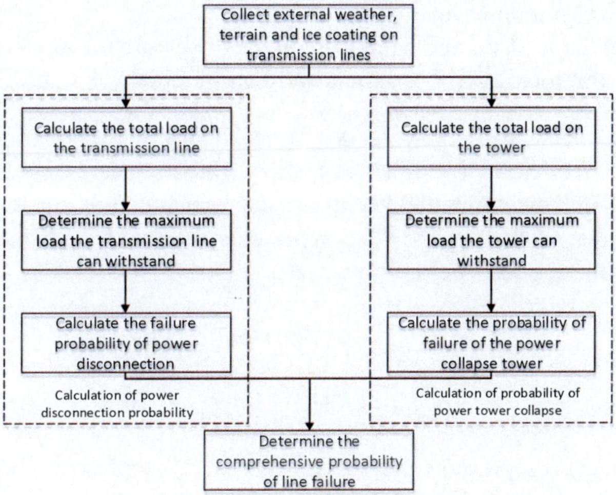

Fig. 2. The calculation process of line comprehensive failure probability.

The increasing load on the transmission line and the tower is the main reason for the power disconnection and power tower collapse.

(1) Calculation of Total Load on Transmission Lines
Under extreme ice disasters, the total load R_l on the transmission line is as shown in Eq. (1):

$$R_l = G_l + F_{lwind} + F_{lice} \tag{1}$$

Among them: G_l is the gravity load of the transmission line itself, F_{lwind} represents the wind load on the line, and F_{lice} represents the ice load on the line [18, 19].

$$F_{lwind} = \alpha \frac{V^2(t)}{1600} \mu_h \mu_{sc} \beta_c d_i B \sin^2 \theta \tag{2}$$

$$F_{lice} = 9.82 \times 10^{-8} \rho_i \pi [(D + 2R^t)^2 - D^2] L_l \tag{3}$$

Among them: α is the wind pressure unevenness coefficient, $v(t)$ represents the wind speed, μ_h is the wind pressure coefficient caused by different heights, μ_{sc} is the shape factor of the wire, β_c is the wind load adjustment coefficient of the transmission line, and d_i is the ice coating the diameter of the line, B is the wind load increase coefficient caused by the increase in the force area during ice coating, θ indicating the angle between the wind direction and the conductor, ρ_i is the ice density, and D is the transmission line diameter; R^t is the ice coating thickness of the transmission line, L_l represents the length of the transmission line between adjacent towers.

(2) Calculation of the total load on the tower
Combining the ice load and wind load on the tower and the unbalanced tension on the tower, the total load R_t on the tower can be obtained.

$$R_t = F_{pull} + F_{twind} + F_{tice} \quad (4)$$

F_{pull} represents the combined lateral resistant generated by the transmission lines on both sides of the tower; F_{twind} represents the wind load on the tower; F_{tice} represents the ice load on the tower [20];

$$F_{twind} = (V(t))^2 \rho C_d(\alpha^t) A_f / 2 \quad (5)$$

$$F^i_{tice} = \pi R^t (h + R^t)\gamma L_p \times 10^{-3} \quad (6)$$

Among them, ρ represents the air density, $Cd(\alpha^t)$ represents the wind direction ∇t drag coefficient of the tower at the angle α_t, A_f represents the effective area of the tower construction to bear the wind pressure. F_{tice} represents the magnitude of the ice load on the ice-covered tower; R^t represents the thickness of the ice-covered tower member with round cross-section; h represents the outer diameter of the tower member; γ represents the weight of the ice coating; L_p represents the total length of the tower member.

Considering the occurrence of uneven ice coating and deicing, it is also necessary to consider the unbalanced tension F_{pull} generated by the transmission lines on both sides of the tower to the tower [21, 22].

$$F_{pull} = \sqrt{(F^1_{pull})^2 + (F^2_{pull})^2 - 2F^1_{pull}F^2_{pull}\cos(\eta^t + \alpha^t)} \quad (7)$$

η^t and α^t are the angles between the sag tension of the transmission lines on both sides of the tower and the vertical direction of the tower during t period; F^1_{pull} and F^2_{pull} are expressed as the tension of the transmission lines on both sides of the tower.

$$F^1_{pull} = \frac{F^1_{lice}}{2\cos\eta^t};\ F^2_{pull} = \frac{F^2_{lice}}{2\cos\phi^t} \quad (8)$$

4 Comprehensive Probability Model of Line Fault

4.1 Probability Model of Power (Tower) Failure

According to the principle of the strength-stress interference model [23], when the total load on the ice-covered line is greater than the limit load that it can withstand, a power failure will inevitably occur. The linear probability is the ratio of the two. The probability of a power disconnection P_l on the transmission line is shown in Eq. (9):

$$P_l = \begin{cases} \frac{R_l}{F_{lmax}}, R_l < F_{lmax} \\ 1, R_l \geq F_{lmax} \end{cases} \quad (9)$$

$$F_{l\max} = 1.0917 \cdot T_d = 1.0917 \cdot \frac{0.6T_m}{k} \quad (10)$$

Among them, R_l represents the total load that the ice-covered line bears, $F_{l\max}$ represents the limit load that the line can withstand, T_d represents the maximum use tension of the line, T_m is the breaking tension, k is the safety factor, and the general value is 2.5.

The probability model of power tower failure is similar to the power failure model, and its calculation formula is as shown in formula (11)

$$P_t = \begin{cases} \frac{R_t}{F_{tmax}}, R_t < F_{tmax} \\ 1, R_t \geq F_{tmax} \end{cases} \quad (11)$$

Among them, P_t represents the probability of power collapse, and F_{tmax} represents the limit load that the tower can bear, which is related to the parameters of the tower itself.

4.2 Comprehensive Probability of Transmission Line Failure

A transmission line can be regarded as a series system formed by a plurality of pole towers and the line support between the pole towers [24]. When a poling tower falls or any power disconnection occurs between the pole towers, the overall line cannot operate normally. Therefore, suppose that a transmission line supports tower numbers 1, 2, 3 … m, and the line number between adjacent towers is set to 1, 2, 3 … m-1. The probability of each tower falling is $P_t^1, P_t^2, P_t^3 \ldots P_t^m$ The probability of power disconnection of the transmission line is $P_l^1, P_l^2, P_l^3 \ldots P_l^{m-1}$, then the comprehensive probability of the transmission line failure P_j, as shown in Eq. (12):

$$P_j = MAX\{[1 - (1 - P_t^1)(1 - P_t^2)\ldots(1 - P_t^m)], [1 - (1 - P_l^1)(1 - P_l^2)\ldots(1 - P_l^{m-1})]\} \quad (12)$$

5 Identification Method of Weak Links in the Power Grid

In the identification method of power grid weak link based on risk index, the risk index is the product of the transmission line failure probability and its corresponding consequence. In this paper, the load loss corresponding to the line fault is determined by the DC power flow model. Based on the economic loss, a failure consequence model considering load factors, the difficulty of an emergency repair, and other factors are established. By ranking the risk of the transmission line of the power grid, Identify the weak links in the power grid during disasters.

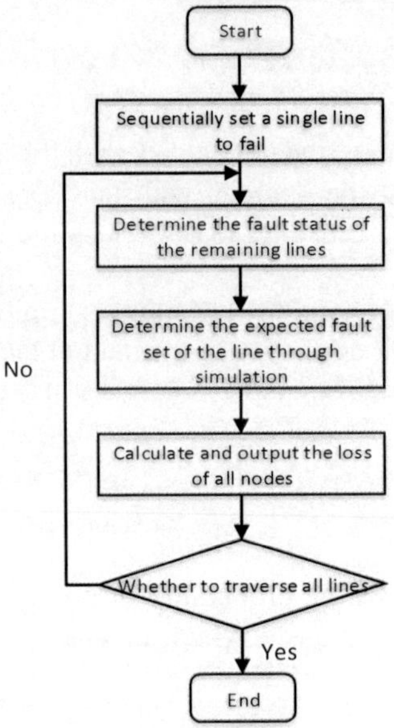

Fig. 3. Flow chart of load loss calculation after line failure.

5.1 Determination of Line Fault Load Loss Based on DC Power Flow Model

The calculation process of load loss after the line fault is shown in Fig. 3. Firstly, a single line is set to fail in sequence, and the fault status of the remaining lines is determined based on the transmission line's comprehensive outage probability P_j. Then, use the DC power flow optimization model to calculate and output the load loss of all nodes.

5.2 Multi-factor Failure Model Considering the Power Grid

The economic loss of load loss is related to many factors. This article mainly considers the load level and the recovery time of emergency repair. Define the consequences of the set line fault as: Economic loss of load loss R_{ens}

$$R_{ens} = \sum_{i=1}^{m} L_i \cdot t_i \cdot C_{ens} \tag{13}$$

Among them, m represents the number of nodes, L_i represents the load value lost by node i; t_i represents the repair time required by load node i; C_{ens} represents the economic loss per unit load per unit time.

Factor one: load rating

$$L_i = L_{lossi} \cdot W_i \tag{14}$$

L_{lossi} represents the actual loss load of node i, W_i represents the load level corresponding to node i, and the W_i corresponding to the first-level load, second-level load and third-level load is 10, 20, 30 respectively. The larger the value of W_i, the more important the node load.

Factor two: node recovery time
The node repair time t_i is related to many factors. This paper introduces the concept of emergency repair resource reserve ratio to reflect the impact on repair time in different scenarios.

$$t_i = \frac{a_i}{g_i} = \frac{a_i}{f_{p_i}/f_{d_i}} \tag{15}$$

a_i represents the time required for unit manual repair of node i per unit time. gi is defined as the ratio of the resources f_{p_i} that can be provided to the resources f_{d_i} required for emergency repair failures, and the value is 0–1 [25, 26].

5.3 Calculation of Line Risk Index

The risk index of the transmission line is defined as the product of the comprehensive probability of the fault and the consequences of the fault [27, 28], and to identify the weak links in the power grid during the disaster. The greater the risk, the weaker the corresponding transmission line.

$$R_f^j = P_j \cdot R_{ens}^j \tag{16}$$

R_f^j represents the risk index of transmission line j; P_j represents the comprehensive probability of failure of line j; R_{ens}^j represents the consequence of the failure of line j.

6 Example Analysis

To analyze the effectiveness of power grid weak link identification technology under strong wind, ice, and snow conditions, this paper uses IEEE30 node as an example to verify. As shown in Fig. 4, the system includes 30 nodes and 41 transmission lines.

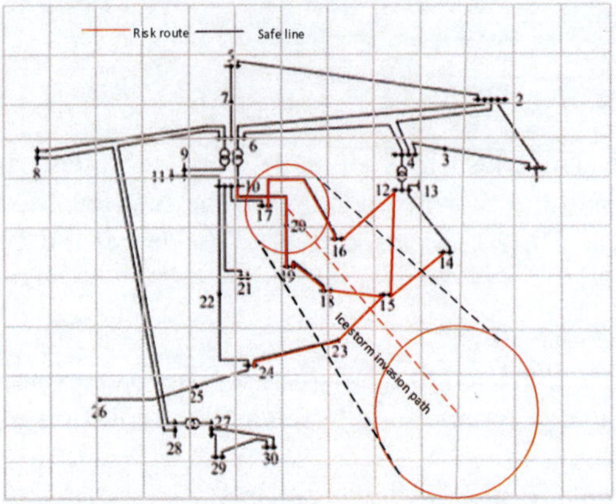

Fig. 4. Ice storm invasion route and IEEE30 node topology structure diagram.

6.1 Calculation of Comprehensive Probability of Line Fault

As shown in Fig. 4, the main lines affected by ice storm weather are 14–15, 12–15, 15–23, 15–18, 18–19, 12–16, 19–20, 10–20, 17–16 And 23–24, this article assumes that the five towers within 5 km have the same probability of falling, and they are equivalent to one and select the line 12–15 to calculate the comprehensive probability of line failure.

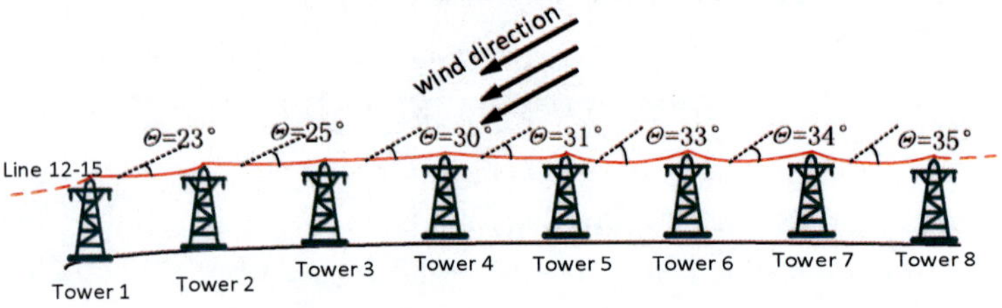

Fig. 5. Line 12–15 schematic

As shown in Fig. 5, the 12–15 line can be equivalent to 8 towers. Calculate the probability of power disconnection and power tower collapse, and the calculation parameters involved are shown in Table 1.

Table 1. Parameter setting of ice load calculation.

Parameter	Value	Parameter	Value
α	0.62	B	1.2
$V(t)$	10 m/s	D	30 mm
μ_{sc}	1.2	d_i	48.7 mm
μ_h	1	ρ_i	0.9 kg/m^3
β_c	1	L_l	1000 m
d	46 mm	ρ	1.293 kg/m^3
R^t	18.7 mm	A_f	10 m^2

Taking the line between Tower 3 and Tower 4 as an example, it can be concluded that the ice load on this section of the line is 0.68kN, the wind load is 0.8kN, and the gravity load is 1kN. 8.787kN, therefore, the probability of power disconnection is 0.307. Taking pole tower 3 as an example, the probability of power collapse is 0.127.

Similarly, the probability of failure of the remaining towers and transmission lines in lines 12–15 are shown in Table 2:

Table 2. Probability of power line disconnection (tower collapse) for line 12-15.

Tower	Power tower collapse probability	Line section	Power disconnection probability
1	0.148	1–2	0.299
2	0.129	2–3	0.304
3	0.127	3–4	0.307
4	0.118	4–5	0.312
5	0.124	5–6	0.314
6	0.162	6–7	0.319
7	0.122	7–8	0.322
8	0.147		

According to the principle of series reliability, the comprehensive failure probability of lines 12–15 is 0.9264. Similarly, the comprehensive probability of failure of the remaining risk lines is shown in Table 3:

Table 3. Comprehensive probability of risk line failure.

Line	Comprehensive failure probability	Line	Comprehensive failure probability
14–15	1.0000	12–16	0.5125
12–15	0.9264	19–20	0.0103
15–23	0.9459	10–20	0.0268
15–18	0.9347	16–17	0.1567
18–19	0.4642	23–24	0.0480

6.2 Analysis of Line Fault Consequences

The failure of a single line is set in turn, and the status of the remaining lines is based on the comprehensive probability of the line fault, determined by the non-sequential Monte Carlo method, and the DC load flow optimization model is used to calculate the load loss caused by the line fault. Then, the comprehensive load class and emergency repair recovery time determine the economic loss R_{ens} of the line load loss, this article sets C_{ens} = 5. The load level and repair resource reserve rate of each node is shown in Table 4.

Table 4. Node load level and emergency repair resource reserve ratio.

Node number	Load level	Emergency repair resource reserve ratio	Node number	Load level	Emergency repair resource reserve ratio
12	30	0.95	18	10	0.80
13	30	0.95	19	20	0.90
14	30	0.95	20	10	0.80
15	10	0.80	21	10	0.80
16	10	0.80	23	20	0.90
17	10	0.80	24	10	0.80

Take the setting of line 12–15 as an example, the remaining faults are lines 15–18, 18–19, and 10–20. The load loss of the node after the line failure is shown in Fig. 5. The total amount of node load loss is 0.3725 MW, the economic loss of load loss R_{ens} is 348,300 yuan, and the risk index is 32.27 (Fig. 6).

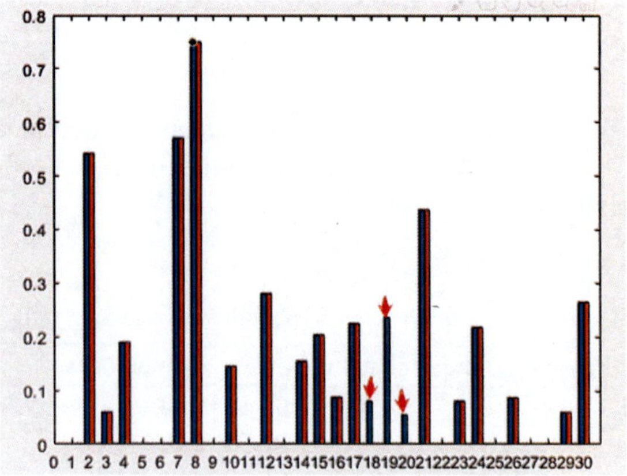

Fig. 6. Loss of load after line 12–15 fault.

Set 14–15, 12–15, 15–23, 15–18, 18–19, 12–16, 19–20, 10–20, 17–16, and 23–24 in sequence, and the economic loss R_{ens}. The line risk indicators are shown in Table 5.

Table 5. R_{ens} and risk indicators of each line fault.

Faulty line	Failure rate	Economic loss of load loss R_{ens}/Ten thousand yuan	Risk index
12–15	0.9264	34.83	32.27
15–18	0.9347	31.99	32.27
15–23	0.9459	13.06	12.35
12–16	0.5125	15.26	7.82
18–19	0.4642	12.83	5.96
14–15	1.0000	3.88	3.88
23–24	0.0480	16.77	0.80
10–20	0.0268	29.83	0.79
16–17	0.1567	2.36	0.36
19–20	0.0103	29.49	0.30

As can be seen from the previous article, due to the impact of ice storm weather, the probability of failure of lines 14–15 is the highest, but when the load level and resource reserve ratio are considered comprehensively, the consequences of failure caused by the failure of lines 14–15 are small, according to Risk definition lines 14–15 have lower risk indicators. The probability of failure of line 12–16 is not high, but due to the higher load level of the line and insufficient node resource reserve rate, the fault consequence caused by the failure of line 12–16 is higher, so the line 12–16 The value at risk is high. A simple assessment of the weak links from the failure rate or the consequences of the failure is easy to cause the waste of repair resources and causes more serious economic losses, and the weak links can be effectively identified by the order of risk value. In the operation of the actual power grid, the operation dispatcher needs to pay enough attention to the weak links with high-risk value, take timely preventive measures, and try to avoid the occurrence of large-scale power outages.

7 Conclusion

Most of the existing weak link identification models are aimed at the power grid in the conventional environment. When an extreme natural disaster occurs, the external environment of the power grid and the running state of the line need to be affected by many factors. The weak link identification technology based on the risk index under extreme ice disasters proposed in this paper comprehensively considers the influence of ice and wind load on the probability of line failure and load level, and the impact of emergency repair recovery time on the fault consequence, and accurately identifies the weak link of the power grid. It is conducive to improving the resilience of the power grid, optimizing the repair strategy and resource allocation of the power grid, and enhancing the initiative of disaster prevention and disaster relief of the power grid.

References

1. Ji, J., Xie, K., Cao, K., et al.: Weak part identification and reliability improvement analysis of Guangdong power grid. Autom. Electr. Power Syst. **35**(13), 98–102 (2011)
2. Chen, L., Liu, M., Yi, J., et al.: The power system caseading outage simulation and vulnerability analysis based on operational reliability model. Power Syst. Technol. **40**(05), 1488–1494 (2016)
3. Zou, L., Chen, L., Sun, Y.: Cascading failure probabilistic assessment of power systems based on operational reliability model for transmission lines. Autom. Electr. Power Syst. **35** (13), 7–11+71 (2011)
4. He, J., Chen, L., Sun, Y., et al.: Power system short-term operational reliability evaluation considering weather forecast. Power Syst. Prot. Control **38**(10), 31–38+51 (2010)
5. Xu, Y., Lou, X., Chen, Z., et al.: Design of weak link identification system for power grid based on topology analysis and association set decomposition. In: 2018 IEEE Power & Energy Society General Meeting (PESGM). IEEE (2018)
6. Wu, Y., Xue, Y., Xie, Y., et al.: Space-time impact of typhoon and rainstorm on power grid fault probability. Autom. Electr. Power Syst. **40**(02), 20–29+83 (2016)
7. Yu, Q., Luan, X., He, Q., et al.: Evaluation of power grid weak links on the basis of comprehensive fault probability model. Sci. Technol. Eng. **16**(30), 92–97+121 (2016)
8. Chen, Q., Mili, L.: Composite power system vulnerability evaluation to cascading failures using importance sampling and antithetic variates. IEEE Trans. Power Syst. **28**(3), 2330–2331 (2013)
9. Sun, F., Chen, L., Sun, Y.: An outage rate model and system adequacy assessment based on adverse weather conditions. Autom. Electr. Power Syst. **33**(13), 7–12+88 (2009)
10. Huang, D.: Classification and identification of power grid weak links. Electr. Power Eng. Technol. **36**(30), 92–97+127 (2017)
11. Chen, L., Liu, M., He, J.: Application and fast simulation of cascading outage based on genetic algorithms. Autom. Electr. Power Syst. **39**(06), 28–33 (2015)
12. Dwivedi, A., Yu, X.: A maximum-flow-based complex network approach for power system vulnerability analysis. IEEE Trans. Ind. Inform. **09**(1), 81–88 (2013)
13. You, L.: Research on short-term prediction models of ice thickness of transmission line. North China Electric University (2017)
14. Xie, Y., Xue, Y., Wen, F., et al.: Space-time evaluation for impact of ice disaster on transmission line fault probability. Autom. Electr. Power Syst. **37**(18), 32–41+98 (2013)
15. Xu, J., Zhen, W., Huang, X., et al.: Transmission line icing prediction model under micro-meteorological conditions. Electr. Power **47**(02), 58–63 (2014)
16. Huang, J., Hu, D., Wang, W., et al.: Research of icing thickness forecast of transmission lines. Hubei Electr. Power **40**(12), 1–4+24 (2016)
17. Yan, M., Zhou, Z., Wen, J., et al.: Assessment method for power grid icing risk based on short-term icing forecasting. Autom. Electr. Power Syst. **40**(21), 168–175 (2016)
18. Yang, H., Huang, L., He, C., et al.: Probabilistic prediction of transmission line fault resulted from disaster of ice storm. Power Syst. Technol. **36**(04), 213–218 (2012)
19. Kudzys, A.: Safety of power transmission line structures under wind and ice storms. Eng. Struct. **28**(5), 682–689 (2005)
20. Sun, Y., Wang, X., Wang, J., et al.: Wind and ice loading risk model and fuzzy forecast for overhead transmission lines. Proc. CSEE **31**(07), 21–28 (2011)
21. Brostrom, E., Ahlberg, J., Soder, L.: Modelling of ice storms and their impact applied to a part of the Swedish transmission network. In: IEEE PowerTech Conference, Lausanne, Switzerland, vol. 7, pp. 1593–1598 (2007)

22. Huang, X., Liu, J., Cai, W., et al.: Present research situation of icing and snowing of overhead transmission lines in China and foreign countries. Power Syst. Technol. (04), 23–28 (2008)
23. Tang, Y., Xu, X., Chen, B., et al.: Research on cumulative damage model of transmission tower based on typhoon parh prediction information. Electr. Power **53**(01), 56–65 (2020)
24. Tang, Y., Xu, X., Chen, B., et al.: Space-time-intensity online early-warning of transmission tower faults by caused rainfall-induced landslides. Electr. Power **52**(07), 69–77 (2019)
25. Pan, B., Zhang, Y., Huang, H.: Modeling for multi-factor rick assessment on ice disaster of power system. Power Syst. Technol. **36**(05), 102–106 (2012)
26. Zhang, Y., Xun, L., Wu, C.: Research on ice disaster risk evaluation model of power system considering multi-factors. Power Syst. Prot. Control **40**(15), 12–17 (2012)
27. Lu, B., Tang, G.: Application of risk-based security assessment in power systems. Autom. Electr. Power Syst. (22), 61–64 (2000)
28. Zhang, Z., Li, Y., He, L., et al.: Application of Monte carlo methods in power system static security risk assessment. Electr. Meas. Instrum. **52**(19), 106–111 (2015)

Research on Index System Construction and Assessment Methods of Distribution Network CPS Security Against Cyber Attack

Xia Zhou, Chengjie Bu(✉), Xiangpeng Xie, and Jianqiang Lu

Institute of Advanced Technology, Nanjing University of Posts and Telecommunications, Nanjing 210023, China
buchengjie1@163.com

Abstract. With development of power distribution automation systems, the distribution network has gradually evolved into a highly coupled distribution network CPS system with information physics. Certain security risks to the distribution network CPS have been brought to the distribution network CPS system when the information system supports the stable operation of the physical system. Such as the presence of a cyber-attack and other methods attack the information system. It may lead the failure of power services or functions, and affect the safe and stable operation of the distribution network CPS in turn. The security definition and security analysis objectives of the distribution network CPS are given in this paper firstly. Then the information security indicators and physical security indicators of the distribution network CPS are constructed. The information physical indicators are integrated and expanded from the perspective of security risk. And information security index and physical security index are constructed, also cyber physical indicators are integrated and expanded from the perspective of security risks. The causal interaction logic of the cyber-attack on the distribution network CPS is analyzed. The cyber-attack is classified according to the impact of the cyber-attack on the distribution network CPS. The impact of the distribution network system under the cyber-attack is deeply analyzed. Finally considering the combination of expected failures, a security assessment process of distribution network under the action of cyber-attack is proposed and verified in the calculation example.

Keywords: Distribution network CPS · Cyber-attack · Security assessment · Cyber-physical combination

1 Introduction

As an important infrastructure closely related to production, living, economic and social development, the distribution network plays an important role in achieving the two strategic goals of smart grid [1]. With the advancement of computing, communication and control technology, a multi-level complex system is gradually becoming

This work was sponsored by the Natural Science Foundation of Nanjing University of Posts and Telecommunications (NUPTSF) under Grant No. NY218146.

Y. Xue et al. (Eds.): PMF 2020, LNEE 718, pp. 716–731, 2021.
https://doi.org/10.1007/978-981-15-9746-6_55

an important feature of information physical systems [2–5]. With the implementation of other strategies such as the smart grid, the smart distribution network already has the key feature of highly coupling information and physics.

Power system cyber-attacks have occurred frequently in recent years. In 2010, the industrial control system in the Iranian nuclear power plant was attacked by the Seismic Virus, resulting in the actual case of centrifuge failure. Malicious attacks against power CPS have gradually attracted widespread attention. At the end of 2015, the large-scale power outage in the Ukrainian power grid was regarded by the outside world as the world's first vicious blackout caused by a cyber-attack [6, 7]. The coupling relationship of power information physics is an important factor for the network security of the distribution network CPS research.

With the coupling of information physics, the distribution network has gradually evolved into a cyber-physical system. Research in this area is mostly based on reliability [8], vulnerability and risk assessment [9]. In 2017, Jiang Zhoushan et al. Started to dig out the risk sources of the information space and physical space of the distribution network [10]. In 2018, based on the self-healing control method of active distribution network failure, the Liu Wenxia team analyzed the impact of the information domain on the information physical system during the failure process and established an information system reliability model based on a hybrid communication network [11].

At present, the existing research on distribution network CPS from the initial architecture to risk assessment [12–14] has become perfect. However, there are few studies on the impact of power grid CPS security. On the physical side of the distribution network CPS, the formation of information physical combination failures will be induced, which will affect the recovery of the distribution network CPS failures and cause secondary loss to the distribution network CPS [3, 15].

Based on the above research background, this paper puts forward the definition and analysis objectives of distribution network security. Constructing the information security indicators and physical security indicators of distribution network CPS. Integrating and expanding the information physical indicators from the perspective of security risks. Then it analyzes the causal interaction logic of the cyber-attack on the distribution network CPS and classifies the cyber-attack according to the impact of the cyber-attack on the distribution network CPS.

2 Distribution Network CPS Security Definition and Analysis Objectives

2.1 Definition of Distribution Network CPS Security

The measurement of the security of the distribution network includes the entire distribution network area from the power supply side to the user. The security research method of the distribution network is to predict the information physical failure of the distribution network and provide guidance for possible failures through the obtained fault treatment plan [16].

The distribution network CPS security is defined as: a time section of the continuous operation of the distribution network CPS. For a set of expected information-physical combination failures, the distribution network physical system maintains normal and stable operation to maintain the ability to supply power to the load.

2.2 The Objective of CPS Security Analysis of Distribution Network

Due to the low voltage level of the distribution network and few dynamic components, it is generally considered that there is no obvious transient process in the distribution network. These determine that the security analysis of the distribution network mainly refers to the static aspect and the criteria of the transmission grid cannot be copied. The static security analysis of the distribution network is based on the k(N+1 − 1) criteria. That is when a line fails and cut off, it is necessary to close a switch to restore power supply [17, 18]. This criterion fully considers the operating characteristics of the distribution network and is the basic criterion for the security analysis of the distribution network. In the process of development from the distribution network to the distribution network CPS, the failure of the information system will threaten the security of the physical system. It is necessary to develop an effective distribution network CPS security analysis method.

3 Construction of CPS Security Index System for Distribution Network Against Cyber Attack

3.1 Distribution Network CPS Information Security Index

Whether the distribution network CPS information system is able to work continuously and can guarantee the integrity of information under the circumstances of cyber-attack is the key of the research on distribution network CPS. This requires an objective assessment of whether the information system has sufficient protection capabilities under various threats [19].

There are three levels of information system security assessment: the first is the physical equipment security level, including physical security, host security and operating system security. Assessing whether existing measures can resist security threats. The second is the level of communication security, including communication topology reliability, communication delay reliability and communication Error reliability, that is the reliability of communication. Which is determined by the topological reliability, delay reliability. The third is the level of data security, including data availability data confidentiality and data integrity. Information system security indicators are shown in Fig. 1.

3.2 Distribution Network CPS Physical Security Index

In the distribution network CPS, the load loss caused by the physical failure can include two parts: the load loss of the faulty section and the load loss of the non-faulty section. The former is fixed because the fault is isolated after the fault occurs. The latter

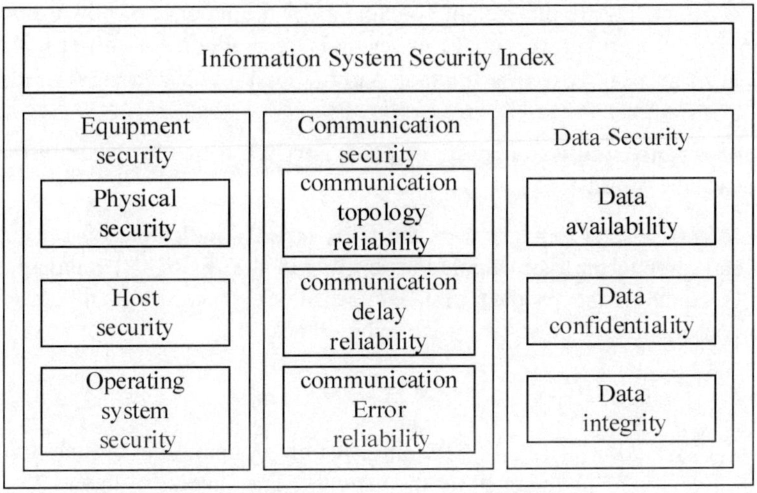

Fig. 1. Information system security index

is not fixed because the radiating mesh structure of the distribution network causes the downstream power supply area to lose power after the fault is isolated. Load loss can be recovered in two ways: on-grid and off-grid. Based on this, this paper puts forward the index of the degree of loss after failure as the index for evaluating the security of distribution network CPS. The loss degree index after failure includes load loss and user loss.

The degree of load loss is the most important index in the security assessment of the distribution network. It refers to the sum of the load reduction of the fault section after the CPS failure of the distribution network and the load reduction of the non-fault section after the fault recovery the ratio between the total load of the network.

$$\rho_{\text{load}} = \frac{\sum\limits_{i \in S_m} w_i P_i + \sum\limits_{i \in S_n} w_i P_i}{\sum\limits_{i \in S} w_i P_i} \times 100\% \qquad (2-1)$$

Where P_i is the active power of node load i; w_i is the level weight of node load i; M, N, S are the fault section, non-fault section and all load node sets.

The degree of user loss ρ_{user} reflects the degree of impact on the user side after the failure of the CPS of the distribution network, including the proportion of the number of power outages and the proportion of user power outage losses.

$$\rho_{user} = (w_1 \frac{X}{Y} + w_2 \frac{\sum\limits_{j \in X} \mu_j}{\sum\limits_{j \in Y} \mu_j}) \times 100\% \qquad (2-2)$$

Where w_1 and w_2 are the weight coefficients of the number of power outages and the proportion of users' power outage losses. μ_j is the grade factor of user j. X and Y are the number of all users lost after the fault and the total number of users in the system.

3.3 Convergence and Expansion of CPS Security Indexes for Distribution Networks Against Cyber Attack

The idea of fusion and expansion of the CPS security indicators of the distribution network can refer to the risk calculation formula in the IEC61850 standard. The risk value R is equal to the product of the cyber-attack success probability P and the physical consequences C caused by the cyber-attack.

$$R = P \cdot C \tag{2-3}$$

In the distribution network CPS information system, the probability of each information device being attacked by the network. The success of the attack is closely related to the level of the attacker and the security of the information node. In this paper, the method of analytic hierarchy process and expert experience scoring is used to calculate the failure risk value of cyber-attack on the information equipment in the distribution network CPS. The specific process is shown in Fig. 2.

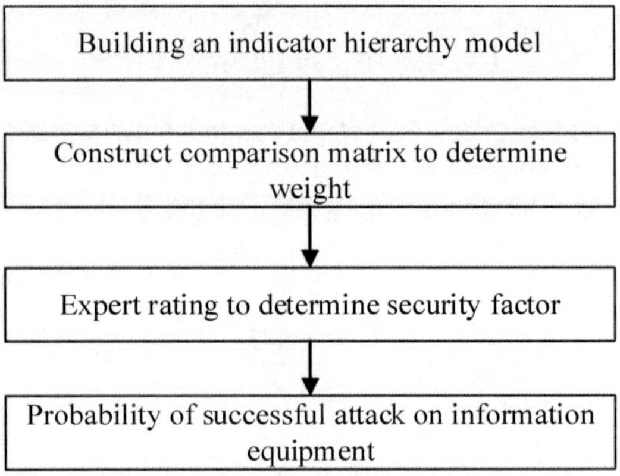

Fig. 2. Information system security risk assessments

Step 1: Establish the assessment model
Considering the characteristics of distribution network CPS and the characteristics of cyber-attack, this paper builds a 3-layer assessment index system. From the three perspectives of attack resources, equipment security and business importance, combine cyber-attack resources with the information system security mentioned in the previous section to establish an AHP hierarchy model for equipment failure risk assessment to comprehensively protect information equipment security assessment, used to quantify the security factor of information equipment. The assessment indicators of equipment

security and communication security are divided into 5 different security levels of extremely low, low, medium, high and extremely high are assigned values of 1, 2, 3, 4 and 5.

Step 2: Determine the weight of the assessment indicators

In order to analyze the security of information systems in conjunction with cyber-attacks, it is necessary to quantitatively evaluate the security of information systems. This paper uses AHP to quantitatively evaluate the security of information equipment. Let us use AHP to establish the judgment matrix of the criterion layer to the target layer and the index layer to the criterion layer from top to bottom, and calculate the normalized feature vector of the matrix:

$$d_i = \sqrt[m]{\prod_{j=1}^{m} a_{ij}} \qquad (2-4)$$

$$w_i = d_i / \sum_{k=1}^{m} d_k \qquad (2-5)$$

Where: a_{ij} is the element of row i and j of the judgment matrix A; d_i is the i component of the feature vector D; w_i is the i component of the normalized feature vector W.

The combined weight of the indicator layer is

$$W^I = (W_1^{IP}, W_2^{IP}, \cdots W_n^{IP}) \cdot W^P \qquad (2-6)$$

Where: W^I is the combined weight vector of the index layer; W^P is the weight vector of the criterion layer to the target layer; W_i^{IP} is the weight matrix of the index layer relative to the element i of the criterion layer; n is the number of elements of the criterion layer. In the calculation of the combined weight of the index layer, it is necessary to verify the consistency index to ensure the credibility of the data.

Step 3: Determine the security factor

Use the method of expert experience scoring to score each indicator in the indicator layer. After normalization, determine the security factor λ_i of each device as:

$$\lambda_i = S_i / \sum_{j=1}^{l} S_j \qquad (2-7)$$

$$S_i = (X_1^i, X_2^i, \cdots X_n^i) \cdot W^I \qquad (2-8)$$

Where: λ_i is the reliability factor of the i device after normalization; S_i is the expert's scoring matrix for the i device; l is the number of devices evaluated; X^i is the expert's scoring vector for the i device; k is the number of elements in the index layer.

Step 4: Information equipment security risk probability

Define the security risk probability of cyber-attack on information equipment as:

$$P_i = 1 - e^{-\frac{C}{\lambda_i}} \quad (2-9)$$

P_i indicates the probability that the information device i is successfully attacked by the network; λ_i indicates the security factor of the information device node i. The greater the security factor, the smaller the probability of successful attack.

C represents the attack resource and reflects the attacker's attack ability. The greater the ability, the greater the probability of successful attack. This article combines the failure loss index and the economic loss to establish a physical security index, including investment losses and maintenance costs for the load loss part. The loss and the economic loss of the user are three parts.

$$C_{j.p} = w_1 C_{\text{invest}} + w_2 C_{service} + w_3 C_{user} \quad (2-10)$$

Where, the investment loss of the load loss part is the purchase cost of the faulty power equipment. The maintenance cost loss refers to the cost of repairing the damaged power equipment. The economic loss is the economic loss caused by the user's power outage.

Based on the dynamic reasoning of the analytic hierarchy process, this paper proposes a comprehensive assessment index of security risk, which includes the security risk probability of information equipment being successfully attacked by the network.

$$R = \prod_{i=1}^{n} P_i \cdot C_{j.p} \quad (2-11)$$

Where P_i is the security risk probability of the information device i suffering a cyber-attack; n is the sum of the information device nodes on the transmission path in the path traversal; $C_{j.p}$ is the overall relative consequence of the failure of the physical device j.

4 Analysis of Causal Interaction Logic of Cyber-Attack on Distribution Network CPS

4.1 Definition and Classification of Distribution Network CPS Cyber Attack

The distribution network CPS cyber-attack is to destroy and reduce the distribution network CPS function. The attack process refers to the attacker's unauthorized positioning and tracking of the communication control system to find the vulnerabilities in the distribution network communication network. Taking the three elements of network security as the dividing standard, CPS cyber-attack in the distribution network can be divided into three attack modes: integrity attacks, availability attacks, and confidentiality attacks.

Confidentiality Attack: In the CPS of the distribution network, certain information needs specific permissions to be obtained. In addition, any illegal and authorized access should be intercepted. Typical attack methods: malicious intrusion, password cracking, internal attacks making confidential information leaked and illegally used.

Usability Attack: All information in the distribution network CPS can be obtained by the authorized party in a reasonable way at any time. The destruction of information availability is reflected in the interruption of communication, which results in the unavailability of information. Typical attack methods include DoS attacks and black hole attacks.

Integrity Attack: In the distribution network CPS, the data must maintain a certain accuracy and any unauthorized means cannot edit or destroy the data. Typical attack methods: false data injection attack, man-in-the-middle attack and replay attack. Injecting false data leads to inaccurate information.

4.2 Failure Analysis of Cyber-Attack on Distribution Network CPS

As an important part of the distribution network CPS, the information system mainly involves advanced distribution automation, network protection and distributed power access services in the Advanced Distribution Operation. At the same time, the introduction of information systems also brings certain security risks, which are easy to be used by malicious attackers.

The spread of cyber-attack can be divided into the following three types as shown in Fig. 3:

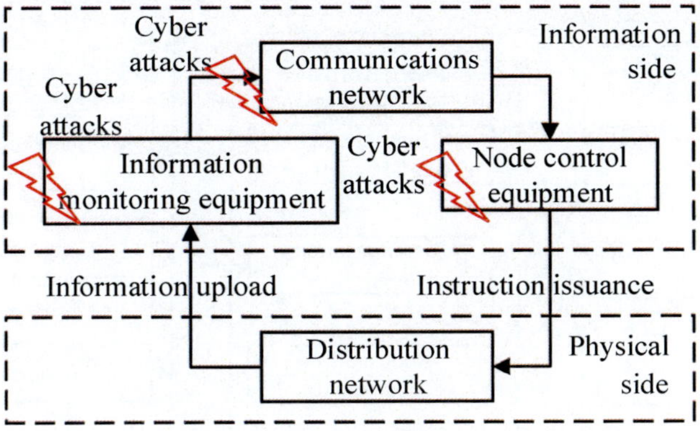

Fig. 3. Cyber-attack on distribution network CPS.

(1) Cyber-attack against secondary equipment (information monitoring equipment);
(2) Cyber-attack against secondary equipment (node control equipment);
(3) Cyber-attack against communication networks.

In the distribution network CPS, the main function of the decision analysis layer is the analysis and processing of information to complete the operations such as self-healing control. The equipment in the decision analysis layer is the core part of the entire information system. The failure of the equipment will directly lead to the failure of the decision and the failure of the self-healing control function. Therefore, the decision analysis layer usually allocates spare hardware resources and puts them into operation at the same time, with high reliability. Studying the impact of cyber-attack on its failures. The research focus of this paper is the impact of cyber-attack on the communication network layer and secondary device layer on the system.

(1) Cyber-attack act on the communication network
The communication network is an important part of the information system, including the backbone network and the access network, network node and transmission channel failure. Message transmission delay and transmission error code will affect its performance.

The impact of cyber-attack on the part of the access network in the communication network is closely related to the type of access network. When the access network is Ethernet based on TCP/IP, the network structure is mostly ring-shaped. After the information element is subjected to a cyber-attack, the link can be smoothed through route conversion. It generally does not cause communication interruption failures and delays in transmission. The influence of line error codes is also closely related to the service load rate and the number of communication nodes. Usually, from the acquisition point to the substation, multiple access is used to ensure data upload. The uplink delay is related to the access method and is relatively fixed. The downlink channel uses broadcast transmission, cyber-attack has little effect on the transmission delay. Bit errors are easily affected by factors such as the quality of the transmission channel, transmission length, signal-to-noise ratio and external interference. When a cyber-attack is encountered, bit errors and even packet loss occur more often, as shown in Fig. 4.

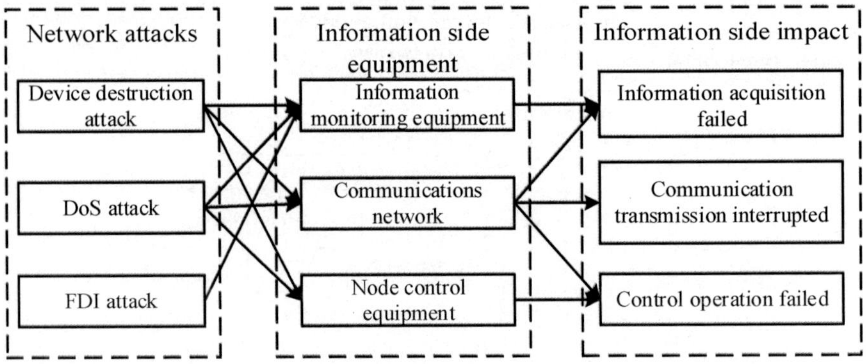

Fig. 4. Schematic diagram of the information failure of the role of cyber-attack.

(2) Cyber-attack act on secondary equipment.
The function of the information monitoring device is to monitor and transmit information such as the voltage, power and switch status of the device. The function of the node control device is to receive and execute the control instructions of the decision analysis center on the physical device. The impact of cyber-attack on secondary device layer failures usually leads to physical system status information monitoring,

transmission failures and the failure of control command execution operations. The failure of the equipment at the secondary device layer will cause the information link to be interrupted and affect the connectivity of the information link. If an error or failure occurs in the information monitoring process, it will affect the accuracy of the system state analysis and control decision. If a failure occurs in the control process, it will cause control failure or control information error, as shown in Fig. 5.

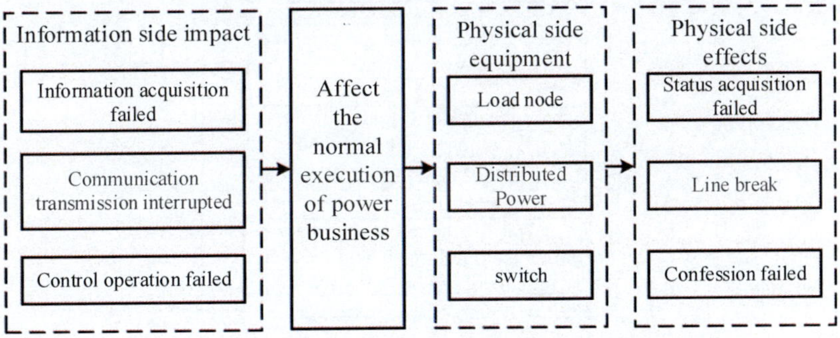

Fig. 5. Schematic diagram of physical faults caused by cyber-attack.

5 Security Analysis and Assessment Method of Distribution Network CPS Considering Combined Expected Failure

5.1 Distribution Network CPS Security Assessment Process

Considering the impact of information system failures in the recovery process of distribution network CPS, a security assessment method for distribution network CPS based on the expected fault set is proposed. The process is shown in Fig. 6.

Step I: Initialization: Enter the distribution network information system and physical system parameters; establish distribution network CPS communication and secondary equipment nodes, network model; establish distribution network physical system node model and topology structure; establish distribution network CPS information - physical correlation matrix model.
Step II: Build expected faults: Establish a set of expected faults in the physical system of the distribution network, which mainly includes the important single components of the physical system in the distribution network. Set the information in the recovery process to anticipate the failure.
Step III: Impact analysis: In view of the physical expected failure, during the failure recovery process, analyzing the impact of the information failure.
Step IV: Index calculation: Calculate the distribution network CPS security assessment indexes such as the loss degree of the distribution network and the fault recovery ability considering the failure of the information system.

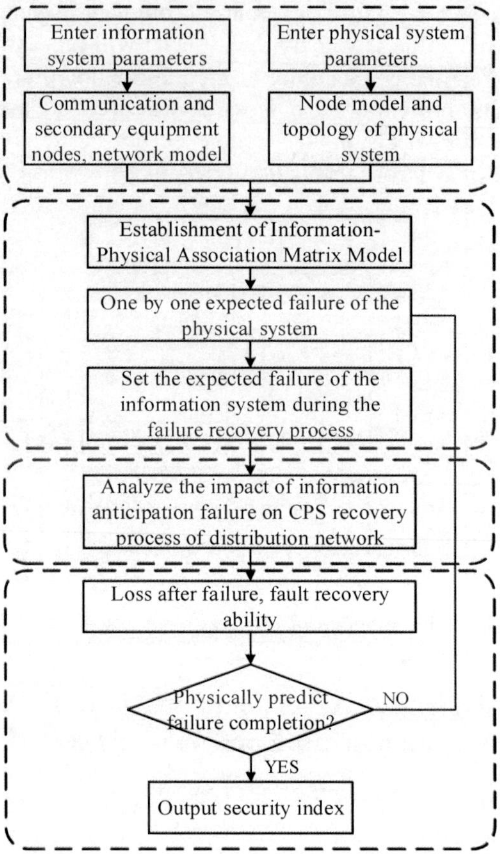

Fig. 6. Distribution network CPS security assessment process.

5.2 Example Analysis

Figures 7 and 8 are structural diagrams of the distribution network 3 feeder system and its corresponding information system. The distribution physical system has a total of 8 distributed power sources to support island operation. There is a tie switch between feeder 2 and feeder 3 for dedicated load supply. At the same time, in order to facilitate calculation, its 33 load nodes are divided into 11 sections, corresponding to 11 kinds of physical expected failures and isolation switches are provided between the sections to ensure the isolation of the faulty section and the non-faulty section. See Table 1, Table 2 and Table 3 for specific parameters and time for self-healing of fault recovery. The communication system adopts the optical fiber communication method based on SDH technology and the system architecture of decision center-information physical coupling-physical entity. The communication terminal nodes are one by one with the secondary equipment nodes of the distribution network load, isolation switch, tie switch and distributed power supply.

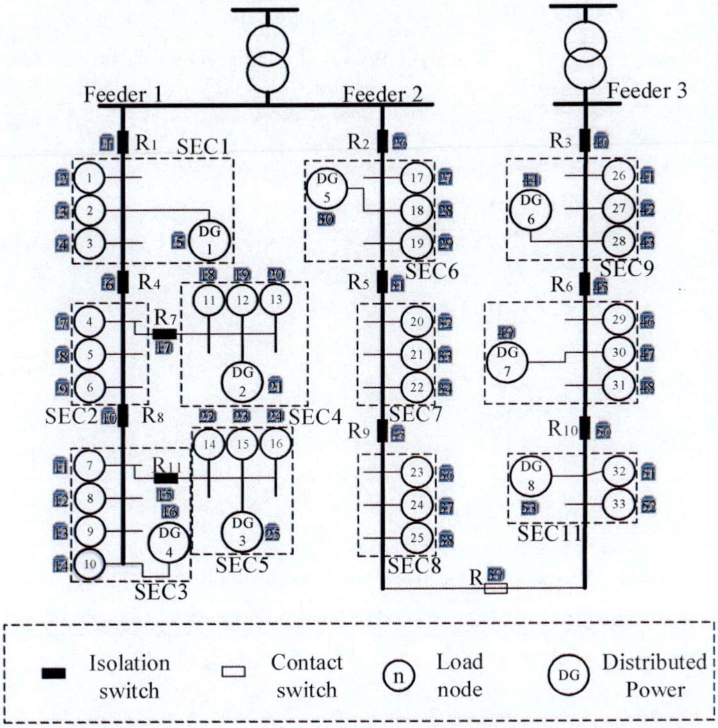

Fig. 7. Physical side structure of distribution network.

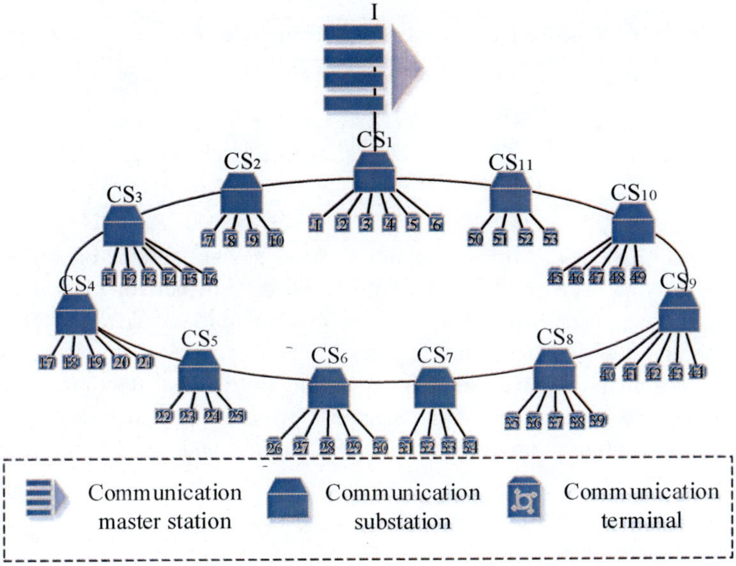

Fig. 8. Information side structure of distribution network.

Table 1. Physical system load parameter configuration.

Load node	Importance	User number	Capacity/MW	Load node	Importance	User number	Capacity/MW
P1–P3	1	1	0.5454	P20–P22	2	4	0.3527
P4–P6	2	2	0.8374	P23–P25	5	5	0.3875
P7–P10	3	8	0.6792	P26–P28	1	1	0.5454
P11–P13	5	5	0.4316	P29–P31	7	9	0.4471
P14–P16	10	6	0.8374	P32–P33	1	1	0.2467
P17–P19	1	1	0.5454				

Table 2. Physical system distributed power parameter configuration.

Distributed power	Capacity/MW	Distributed power	Capacity/MW
1	0.3	5	0.3
2	0.6	6	0.6
3	0.6	7	0.6
4	1.5	8	0.3

Table 3. Time for fault recovery in different operations (normal/fault)

Information system	Fault location	Fault isolation	Load transfer	Island power
Normal	10 s	30 s	60 s	80 s
Fault	30 min	60 min	60 min	60 min

In order to analyze the impact of information system failure on the security of distribution network CPS, the CPS model of distribution network based on the correlation matrix was established for the information upload and command issuance process in the case of system failure recovery of this example. The information of the information system failure was set upload failure (monitoring equipment acquisition failure and communication network transmission failure) and instruction delivery failure (communication network transmission failure and node control equipment control failure).

a. Information upload failure caused by cyber attack

The upload of CPS fault information in the distribution network is mainly two parts: the monitoring equipment collects fault information and the communication network transmits fault information. In this paper, the cyber-attack occurred in SEC1, and setting up the cyber-attack resulted in information monitoring error failure. It is physically expected that the fault occurs between the load nodes P2–P3. But because the information monitoring device of the load node P3 fails, the fault information cannot be transmitted and the monitoring information of the load node P4 is successfully transmitted, the fault section is located at SEC1 and SEC2. Therefore, the security indicators of the expected failure SEC1 in different scenarios are shown in Table 4 below.

Table 4. Calculation result (information upload failure caused by cyber-attack).

Failure information upload status	Security index			
	ρ_{load}/%	ρ_{user}/%	$\rho_{recovery}$/%	T
Normal	2.34	2.40	96.14	120 s
Fault	10.51	4.79	88.16	120 s
Manual positioning	2.34	2.40	90.51	30 min

It can be seen from Table 4: due to the cyber-attack, the information upload error and the fault location range become larger. If the decision control center does not find this information fault, the non-fault section SEC2 is also isolated, resulting in increased load loss and user loss. If this information fault is found in time, manual fault location will be taken, which increases the time for fault recovery.

b. Failure to issue control commands due to cyber attack

The fault information is uploaded to the information center and after a decision is made. A control command is generated to the fault isolation switch and the contact switch for load transfer. If the control command is issued and encounters a cyber-attack that results in a communication network interruption failure or a node control device control failure. The above recovery operation cannot be completed in time. When a fault is expected to occur in feeder 2 and feeder 3, a fault occurs during the issuance of the control command and the contact switch R cannot be closed, resulting in the failure of the load transfer between the feeders, the non-faulty section cannot be restored in time. The security indicators are shown in Table 5 below.

Table 5. Calculation result (command delivery failure caused by cyber-attack)

Anticipated failure set	Security index				Anticipated failure set	Security index			
	ρ_{load}/%	ρ_{user}/%	$\rho_{recovery}$/%	T		ρ_{load}/%	ρ_{user}/%	$\rho_{recovery}$/%	T
SEC1	2.34	2.40	96.14	120 s	SEC7	10.57	18.98	91.27	60 min
SEC2	7.17	4.79	88.16	120 s	SEC8	7.55	11.98	95.43	40 s
SEC3	15.51	16.94	74.38	120 s	SEC9	2.34	2.40	96.59	120 s
SEC4	9.24	11.98	94.92	40 s	SEC10	16.35	18.98	81.43	120 s
SEC5	35.87	19.55	80.26	40 s	SEC11	0.47	1.60	99.74	40 s
SEC6	12.91	21.38	88.06	60 min					

Comparison of Table 4 and Table 5 results can be seen:

Since feeder 1 and feeder 2/3 are not used for load transfer via tie switch R, each expected fault on this feeder has no effect on various indicators.

By comparing the cyber-attack failure SEC6, due to the failure of the node control equipment. When the failure occurs, the failure is isolated. For the downstream SEC7 and SEC8 sections, the load transfer cannot be completed in time. The non-faulty section is powered off, the load The loss of users is greatly affected and the failure recovery rate is also reduced.

Compared with the cyber-attack failure SEC9-11, it is also a contact switch failure, the feeder can also be distributed island power through a distributed power supply, the impact is small. But it is changed from the contact switch to the island power supply to recover from the failure the rate and speed of failure recovery have an impact.

6 Conclusion and Outlook

There are many potential security hazards in the distribution network CPS. The security research of the distribution network is to predict the information physical failure of the distribution network and obtain a solution to the failure to provide guidance for possible failures. This article considers the security definition and analysis objectives of distribution network CPS, constructs a distribution network CPS security index system that targets cyber-attack. The integration of information and physical security indexes is achieved through expert scoring schemes. According to the impact of cyber-attack, the types of cyber-attack are divided and the impact of cyber-attack on the distribution network is analyzed. Finally, the CPS security assessment process of the distribution network is given.

In this paper, only a single information-physical expected fault combination is considered. In actual systems, the impact of multiple faults is greater, so studying the impact of multiple information-physical expected faults is the next step of the work plan. At the same time, the fusion of distribution network CPS security indicators for cyber-attack uses an expert scoring method and subsequent research will expand other methods to integrate security indicators.

References

1. Ma, Z., Zhou, X., et al.: Form and development trend of future distribution system. Proc. CSEE **35**(6), 1289–1298 (2015)
2. Zhao, J., Wen, F., Xue, Y., et al.: Cyber physical power systems: architecture, implementation techniques and challenges. Autom. Electr. Power Syst. **34**(16), 1–7 (2010)
3. Xue, Y.S., Li, M.L., Luo, J.B., et al.: Modeling method for coupling relations in cyber physical power systems based on correlation characteristic matrix. Autom. Electr. Power Syst. **42**(2), 11–19 (2018)
4. Guo, Q., Xin, S., Sun, H., et al.: Power system cyber-physical modelling and security assessment: motivation and ideas. Proc. CSEE **36**(6), 1481–1489 (2016)
5. Falahati, B., Fu, Y., Wu, L., et al.: Reliability assessment of smart grid considering direct cyber-power interdependencies. IEEE Trans. Smart Grid **3**(3), 1515–1524 (2016)

6. US-CERT. Russian government cyber activity targeting energy and other critical infrastructure sectors [EB/OL] (25 July 2018)
7. The Hill. DHS: Russian hackers got into control rooms of US utilities[EB/OL] (25 July 2018)
8. Wang, Z.B., Mu, Y.F., Wang, Y.F., et al.: Cyber Physical system risk assessment of distribution network considering communication failure of multi-agent system for power restoration. Autom. Electr. Power Syst. **40**(17), 51–58 (2016)
9. Yi, T., Qian, C., Mengya, L., et al.: Overview on cyber-attack against cyber physical power system. Autom. Electr. Power Syst. **40**(17), 59–69 (2016)
10. Jiang, Z., Liu, J., et al.: Key technologies for reliability assessment of distribution network cyber physical system. Electr. Power Autom. Equip. **37**(12), 30–42 (2017)
11. Wenxia, L., Qi, G., et al.: Reliability simulation of ADN cyber-physical system based on hybrid communication network. Proc. CSEE **38**(06), 1706–1718+1907 (2018)
12. Song, Y., Liu, X., Li, Z., et al.: Intelligent data attacks against power systems using incomplete network information: a review. J. Mod. Power Syst. Clean Energ. **6**(4), 630–641 (2018)
13. Zhang, Y., Xiang, Y., Wang, L.: Power system reliability assessment incorporating cyber-attack against wind farm energy management systems. IEEE Trans. Smart Grid **8**(5), 2343–2357 (2018)
14. Wang, Q., Pipattanasomporn, M., Kuzlu, M., et al.: Framework for vulnerability assessment of communication systems for electric power grids. IET Gener. Transm. Distrib. **10**(2), 477–486 (2016)
15. Deng, R., Zhuang, P., Liang, H.: CCPA: coordinated cyber-physical attacks and countermeasures in smart grid. IEEE Trans. Smart Grid **8**(5), 2420–2430 (2017)
16. Yi, T., Wang, Q., Ni, M., et al.: Review on the hybrid simulation methods for power and communication system. Autom. Electr. Power Syst. **39**(23), 33–42 (2015)
17. Moussa, B., Akaber, P., Debbabi, M., et al.: Critical links identification for selective outages in interdependent power-communication networks. IEEE Trans. Ind. Inf. **14**(2), 472–483 (2018)
18. Jin, D., Li, Z., Hannon, C., et al.: Toward a cyber resilient and secure microgrid using software-defined networking. IEEE Trans. Smart Grid **8**(5), 2494–2504 (2017)
19. Zhang, H., Cheng, P., Shi, L., et al.: Optimal DoS attack scheduling in wireless networked control system. IEEE Trans. Control Syst. Technol. **24**(3), 843–852 (2016)

Cooperation Scheme of DC 100 Hz Protection and AC Protection Based on Analysis of Converter Fault Characteristics

Lishuai Yin[1(✉)], Guobing Song[1], Junjie Hou[1], Bing Guo[1], Jiangtao Lv[1], Jian Xue[1], and Liyang Yin[2]

[1] Xi'an Jiaotong University, Xi'an 710049, China
yinlishuai@stu.xjtu.edu.cn
[2] Tianjin Electric Power Transmission and Transformation Engineering Corporation, Tianjin 300171, China

Abstract. Aiming at the traditional 100 Hz protection using a long delay strategy without considering the limitation of AC and DC side protection cooperation, an improved 100 Hz protection scheme is proposed. In this paper, according to the transfer characteristics and equivalent impedance of the AC/DC side of the converter, combined with the fault composite sequence network, a calculation method of 100 Hz component fault current is proposed. On this basis, through cooperation with the AC side protection settings and timing, an improved 100 Hz protection scheme with short delay capability is proposed. Simulation verification shows that the protection scheme can effectively identify the asymmetric fault and valve fault on the AC side of the fault, and can be used as remote backup protection for the AC side system.

Keywords: HVDC · Converter · 100 Hz protection · Backup protection

1 Introduction

With the development of power electronic power systems, the nonlinear and time-varying characteristics of AC/DC systems with converter stations as the core are different from traditional power sources [1]. Among them, the establishment of the mathematical model of the converter and the analysis of the fault characteristics have always been the research focus of high-voltage direct current (HVDC) transmission system. For the mathematical modeling of the converter in the traditional HVDC transmission system, the calculation accuracy of the model needs to be considered, and the converter model needs to be simplified to ensure the calculation speed [2–5]. The switching function model based on dynamic phasors can establish a simplified mathematical model of the converter that satisfies the above two conditions at the same time, so that it can communicate the AC harmonics and DC harmonics on both sides of the converter, connecting the AC system and the DC system [6–12].

Project was supported by the National Nature Science Foundation key project (U1766209)

Y. Xue et al. (Eds.): PMF 2020, LNEE 718, pp. 732–748, 2021.
https://doi.org/10.1007/978-981-15-9746-6_56

Relay protection has played an important role in the safe and stable operation of the system. At present, most of AC protection and DC protection have strict protection scope, and lack of cooperation. However, considering that when the fault occurs on one side of the converter station, the characteristic harmonics caused by it will pass through the converter station and affect the other side system and its protection. The nonlinearity and time-varying power electronics make the effect more serious. Therefore, the protection coordination between AC and DC systems needs to be further considered.

Taking DC system protection as an example, if a malfunction occurs, it will have a great impact on the safety and stability of the entire AC/DC system. DC 100 Hz protection is used as a converter valve protection, and at the same time as backup protection for AC asymmetric faults. In particular, attention is needed in the AC and DC systems. Therefore, it is necessary to pay attention to 100 Hz protection. When an asymmetric fault occurs on the AC line, a 100 Hz component will be generated in the DC system. Ref. [13] proposed that the DC 100 Hz component is related to the AC negative sequence component, but no specific quantitative analysis was given. Ref. [14, 15] analyzed the mis-operation of DC 100 Hz protection caused by the AC fault and proposed an improvement plan. Ref. [16] related to the protection principle and operation of 100 Hz protection. The setting value of the 100 Hz protection in the above study was often set based on empirical values, lacking a theoretical basis, and falling into exhaustive simulation verification. Besides, the long-time delay is used in cooperation with AC protection, which often fails to achieve the performance of the 100 Hz protection.

Based on the above research, the simplified model of the converter is analyzed in this paper, and the fault characteristics of the converter are analyzed to obtain its transfer characteristics and equivalent impedance. Assuming that the asymmetrical fault of the AC system occurs on the inverter side, a composite sequence network model is established to calculate the voltage at the fault position under fault steady-state condition, and then use the fault characteristics of the converter to calculate the 100 Hz component on the DC line. Using the 100 Hz component calculation method, the 100 Hz protection setting value in cooperation with the AC line distance protection is calculated, and the time cooperation characteristics in cooperation with the AC protection is given to obtain the 100 Hz protection with short delay characteristics. By comparing with the simulation of the detailed model, it is verified that the proposed 100 Hz protection can operate correctly under the conditions of AC asymmetric fault and valve fault.

2 Analysis of Converter Fault Characteristics

In this paper, the fault transfer characteristics and equivalent impedance characteristics of the converter are analyzed first. It provides a theoretical basis for studying the protection coordination relationship between AC and DC.

2.1 Converter Transfer Characteristics

The transfer characteristic of the converter is the transfer relationship between the AC component and the DC component on both sides of the converter. Based on the switching function theory, the improved dynamic phasor form is used to characterize the transfer relationship of the electrical quantity of the converter under fault conditions.

Traditional Switching Function Model. The basic principle of the switching function is to treat the converter as a non-linear modulation switching circuit. The switching function describes the input and output relationship of the variables in the converter. The basic relationship between AC and DC variables in the form of switching function [6] is shown in (1):

$$\begin{cases} u_d = u_a S_{ua} + u_b S_{ub} + u_c S_{uc} \\ i_a = i_d S_{ia} \\ i_b = i_d S_{ib} \\ i_c = i_d S_{ic} \end{cases} \tag{1}$$

where u_d is DC side voltage, u_a, u_b, u_c and i_a, i_b, i_c are AC side three-phase voltage and current; S_{ua}, S_{ub}, S_{uc} and S_{ia}, S_{ib}, S_{ic} are three-phase voltage switching function and current switching function, and their specific expressions are as shown in (2):

$$\begin{cases} S_a = \sum\limits_{n=1}^{\infty} A_n \cos n(\omega t - \alpha + \varphi_{ua}) \\ S_b = \sum\limits_{n=1}^{\infty} A_n \cos n(\omega t - \alpha + \varphi_{ua} - 2\pi/3) \\ S_c = \sum\limits_{n=1}^{\infty} A_n \cos n(\omega t - \alpha + \varphi_{ua} + 2\pi/3) \end{cases} \tag{2}$$

when representing the voltage switching function and the current switching function, A_n is written as A_{un} and A_{in} as shown in (3):

$$\begin{cases} A_{un} = \frac{4}{n\pi} \sin\frac{n\pi}{2} \cos\frac{n\pi}{6} \cos\left(\frac{n\mu}{2}\right) \\ A_{in} = \frac{4}{n\pi} \sin\frac{n\pi}{2} \cos\frac{n\pi}{6} \frac{\sin\left(\frac{n\mu}{2}\right)}{\frac{n\mu}{2}} \end{cases} \tag{3}$$

Switching Function in The Form of Dynamic Phasors. Compared with the electromagnetic transient simulation, the time-domain switching function has no advantage in calculation speed. In order to improve the calculation speed of the switching function, the switching function in the form of dynamic phasors is adopted.

The dynamic phasor can be defined as the Fourier coefficient obtained by decomposing the time-domain signal in an interval of T on the time axis. When the interval shifts to with time, the Fourier coefficient will change, which is why it is called dynamic phasor. The transformation relationship between the time-domain signal and the dynamic phasor is shown in (4):

$$\langle x\rangle_k(t) = \frac{1}{T}\int_{t-T}^{t} x(\tau)e^{-jk\omega_s\tau}d\tau \tag{4}$$

where $x(t)$is the time domain signal. $\omega_s = 2\pi/T$is the angular frequency. $< x>_k(t)$ is the kth order Fourier coefficient at time t, namely the kth order dynamic phasor at that time. According to the definition of dynamic phasors and the analysis of the improved switching function [8–10], the kth order dynamic phasor of the converter switching function is shown in (5):

$$\begin{cases} \langle S_a\rangle_k = \frac{A_k}{2}e^{jk\left(\varphi_{ua}-\alpha-\frac{\mu_a}{2}-\theta_a\right)} \\ \langle S_b\rangle_k = \frac{A_k}{2}e^{jk\left(\varphi_{ua}-\alpha-\frac{\mu_b}{2}-\theta_b-\frac{2\pi}{3}\right)} \\ \langle S_c\rangle_k = \frac{A_k}{2}e^{jk\left(\varphi_{ua}-\alpha-\frac{\mu_c}{2}-\theta_c+\frac{2\pi}{3}\right)} \end{cases} \tag{5}$$

where μ_m, θ_m (m = a, b, c)are the commutation angle and deviation angle affected by the related two line voltages, respectively. They make the improved switching function applicable to asymmetric faults. Figure 1 shows the waveforms of the components of the improved switching function, where (a) is the basic component, (b) and (c) represent the commutation component and the corrected component, respectively corresponding to μ_m and θ_m.

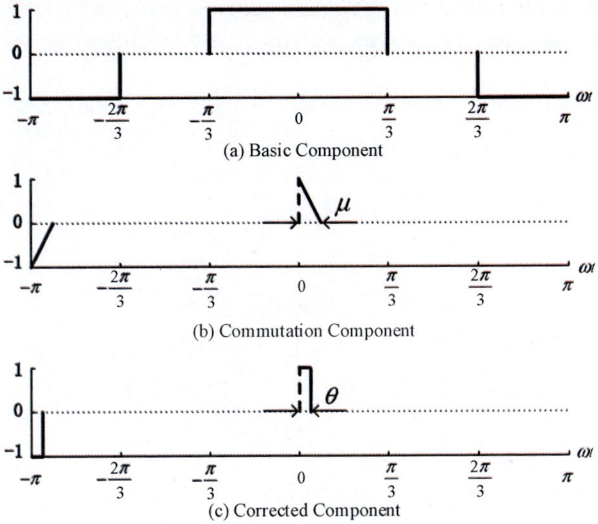

Fig. 1. Components of the switching function

Taking the calculation formula of DC voltage and AC current as an example, the dynamic phasor form is:

$$\begin{cases} \langle u_d \rangle_k = \sum_{i=-\infty}^{+\infty} \sum_{\phi=a,b,c} \langle u_\phi \rangle_{k-i} \langle S_{u\phi} \rangle_i \\ \langle i_\phi \rangle_k = \sum_{i=-\infty}^{+\infty} \langle i_d \rangle_{k-i} \langle S_{i\phi} \rangle_i \end{cases} \tag{6}$$

2.2 Equivalent Impedance of Converter

DC Side Equivalent Impedance of Converter. The equivalent impedance on the DC side of the converter is related to the operating conditions of the converter valve. The converter valve usually works in 2–3 working conditions. The equivalent reactance of the converter is shown in (7) [17].

$$\begin{aligned} Z_{v\_6}(m) &= \frac{\frac{3}{2} Z_{ac}(m) \cdot \mu + 2 Z_{ac}(m) \cdot \left(\frac{\pi}{3} - \mu\right)}{\frac{\pi}{3}} \\ &= \left(2 - \frac{3\mu}{2\pi}\right) Z_{ac}(m) \end{aligned} \tag{7}$$

$Z_{ac}(m)$ is the mth reactance of the AC system. Equation (7) is the equivalent reactance of the 6-pulse converter. When it comes to the 12-pulse converter, its equivalent reactance is two 6-pulse equivalent reactances in series. The impedance corresponding to this reactance is $Z_{v\_12}(m)$. If the DC line impedance $Z_{dcl}(m)$, the line capacitance $Z_c(m)$ and the equivalent impedance $Z_{v\_12}(m)$ of the opposite-end converter valve that have been calculated are taken into account, the equivalent impedance corresponding to the entire DC system Zdc (m) can be obtained as shown in Fig. 2 and (8).

$$Z_{dc}(m) = Z_{dcl1}(m) + \frac{Z_c(m) \cdot (Z_{dcl2}(m) + Z_{v\_12}(m))}{Z_c(m) + (Z_{dcl2}(m) + Z_{v\_12}(m))} \tag{8}$$

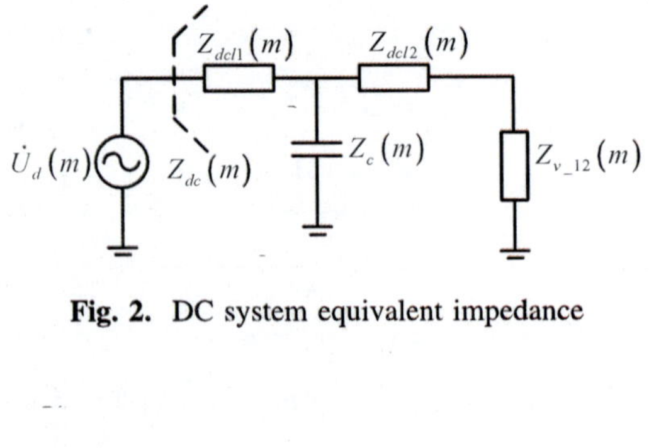

Fig. 2. DC system equivalent impedance

AC Side Equivalent Impedance of Converter. The equivalent impedance on the DC side of the converter is related to the operating conditions of the converter valve. The converter valve usually works in 2–3 working conditions. Using the switching function described above, the modulation calculation is performed on the AC side of the converter. The positive and negative sequence harmonic impedances on the AC side of the 12-pulse converter can be obtained as [18]:

$$Z^{+}_{\text{con12}} = \frac{1}{2}Z_{\text{T}}(\omega) + k^2 / \left(\sum_{n=1}^{\infty} 2(A')^n Z_{\text{ac}}(\omega - 2\omega_1)^{(n-1)} \right) \tag{9}$$

$$Z^{-}_{\text{con12}} = \frac{1}{2}Z_{\text{T}}(\omega) + k^2 / \left(\sum_{n=1}^{\infty} 2(B')^n Z_{\text{ac}}(\omega + 2\omega_1)^{(n-1)} \right) \tag{10}$$

where $A' = 3A_{\text{un}}A_{\text{in}}/(4Z_{\text{dc}}(\omega - \omega_1))$, $B' = 3A_{\text{un}}A_{\text{in}}/(4Z_{\text{dc}}(\omega + \omega_1))$. A_{un} and A_{in} can be obtained by (3). Z_{ac} is the equivalent impedance of the AC system. Z_{dc} is the equivalent impedance of the DC system calculated above.

The research is based on the steady-state condition of the fault. After the fault reaches the steady-state condition, the trigger angle of the converter tends to be stable, and the effect of the converter control strategy on the equivalent impedance no longer changes. Besides, the AC equivalent impedance contains the commutation angle, which can reflect the commutation situation at the steady-state condition of the fault. Therefore, the error caused by the dynamic process of the control system to the calculation of the equivalent impedance is not considered. Considering that when the value of n is too large, the accuracy of calculation cannot be improved any more and the calculation amount will be increased, so n is taken as 4. With this value, the equivalent impedance model can accurately reflect the harmonic characteristics of the AC side impedance of the converter in the low-frequency band below 200 Hz.

3 The Necessity of Improvement for 100 Hz Protection

This section will analyze the causes of the 100 Hz component and the existing problems of the 100 Hz protection to illustrate the necessity of using the converter fault characteristics obtained above to improve the 100 Hz protection.

3.1 Sources of 100 Hz Component

The 100 Hz component of DC line is mainly generated by the fault of the valve not operating, valve misoperating or the AC asymmetric fault [19].

100 Hz Component Generated by AC Asymmetric Fault. AC asymmetric faults will generate a negative sequence component on the AC line, assuming that the negative sequence component is:

$$\begin{cases} u_a = U_h \cos(\omega t + \varphi_{ua}) \\ u_b = U_h \cos(\omega t + \varphi_{ua} + 2\pi/3) \\ u_c = U_h \cos(\omega t + \varphi_{ua} - 2\pi/3) \end{cases} \tag{11}$$

where U_h is the voltage amplitude. Bring (11) into (1) to obtain the harmonic voltage generated on the DC side after the AC negative sequence component passes through the converter:

$$u_{dc}^{-} = \frac{3}{2} A_{u1} U_h \cos[(\omega + \omega_0)t + \varphi_{ua} - \alpha] \tag{12}$$

It can be seen from (12) that the negative-sequence harmonics with a frequency of ω on the AC side will generate harmonic voltage with a frequency of $\omega + \omega_0$ on the DC side. Therefore, the negative-sequence power frequency voltages generated by AC asymmetric faults will generate a 100 Hz voltage on the DC line.

100 Hz Component Generated by Valve Fault. The fault of the valve not operating and valve misoperating can both generate 100 Hz component. Non-opening faults are caused by the loss of converter valve pulses, and misoperating faults are caused by trigger pulses arriving early. The DC voltage is obtained by chopping the AC voltage through the converter. The valve fault causes the converter valve to work abnormally. The modulation effect of the converter valve on the AC voltage is no longer regular, and a variety of frequency harmonics are generated, of which 100 Hz has a large amplitude.

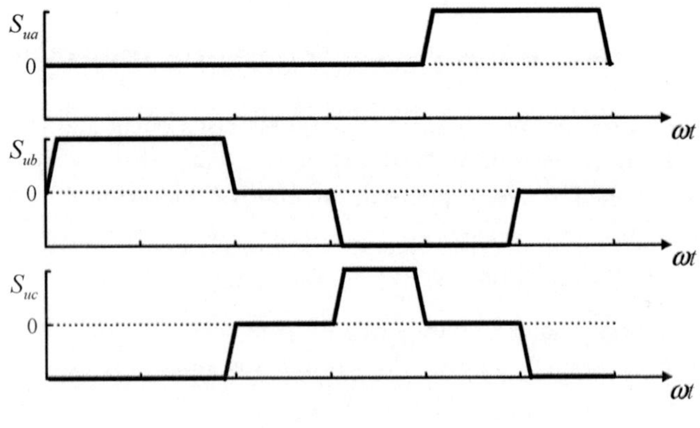

Fig. 3. The Waveform of Voltage Switching Function in the case of the valve not opening fault

The generation of the 100 Hz component can also be figured out from the switching function waveform describing the switching characteristics of the converter. Figure 3 is the waveform of the three-phase switching function when a single valve of the inverter fails to turn on. The a-phase conduction interval becomes smaller, and the c-phase conduction interval becomes larger. The irregular conduction results in harmonics.

3.2 The Necessity for Protection Improvement

The protection criterion of traditional 100 Hz protection is shown in (13) [19]:

$$I_{dc}(100\text{Hz}) > I_{setmin} + K_{set} I_{dc} \tag{13}$$

where I_{setmin} is the minimum starting current, which is mainly used to avoid filter error. The traditional tuning method is an empirical formula. Although it will change with the change of DC current I_{dc}, it still lacks a theoretical basis. 100 Hz protection is considered as backup protection for asymmetric faults on AC lines, but only the DC side electrical components are reflected in (13). Therefore, AC faults should be considered in the protection setting process.

On the other hand, to avoid protection action before the AC protection which may affect the reliability of the DC transmission system, the existing 100 Hz protection delay is set to 3 s according to the tolerance limit of the DC device [13]. Although the long delay can ensure the stability of power supply when the AC protection works normally, it can not play the advantage of 100 Hz protection as the backup protection of the AC system.

Therefore, it is necessary to improve the traditional setting method of 100 Hz protection.

4 Cooperation Scheme of 100 Hz Protection and AC Protection

The pilot protection is usually set as the main protection for the AC line. Distance protection is also provided. The cooperation scheme with AC protection for 100 Hz protection is to be used as the remote backup protection of AC protection. Therefore, the setting value and action delay can be obtained by selecting the close backup zone 3 protection of distance.

First, deduce the calculation formula of the DC 100 Hz component when the AC line faults occur, and then use this calculation method to calculate the 100 Hz protection setting value, and finally determine the time coordination characteristics of the 100 Hz protection according to that of three sections of zone 3 protection of distance.

4.1 DC 100 Hz Current Calculation

Figure 4 shows the AC side simplified circuit diagram of the inverter, where $\dot{U}_i$ is the AC bus voltage, Z_f is the filter bank and compensation capacitor on the AC side of the inverter, and Z_l is the line impedance, $\dot{E}_s$ is the AC power at the receiving end, and Z_s is the internal impedance of the power supply.

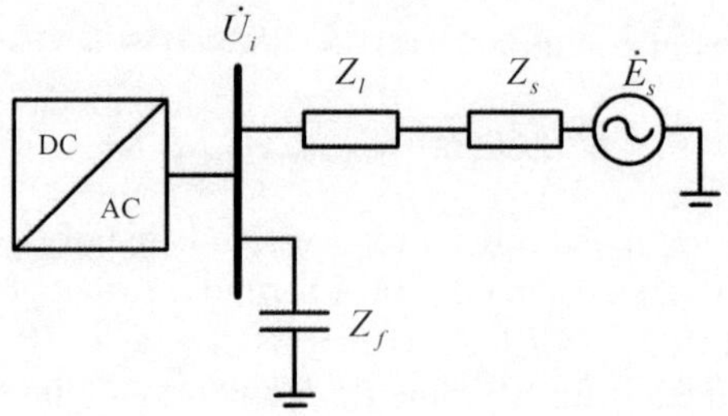

Fig. 4. AC side simplified circuit diagram of the inverter

Replace the inverter with the equivalent impedance of the inverter on the AC side obtained above. Take a single-phase ground fault on the AC line as an example, as shown in Fig. 5.

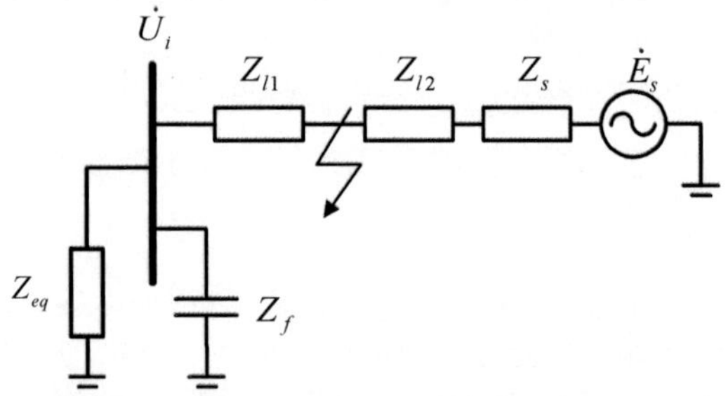

Fig. 5. AC side equivalent fault circuit diagram of the inverter

Z_{l1} and Z_{l2} are the line impedances on both sides of the fault point. Z_{eq} is the AC side equivalent impedance of the inverter. According to the fault circuit diagram, the composite sequence network under the steady-state of the fault is constructed, as shown in Fig. 6:

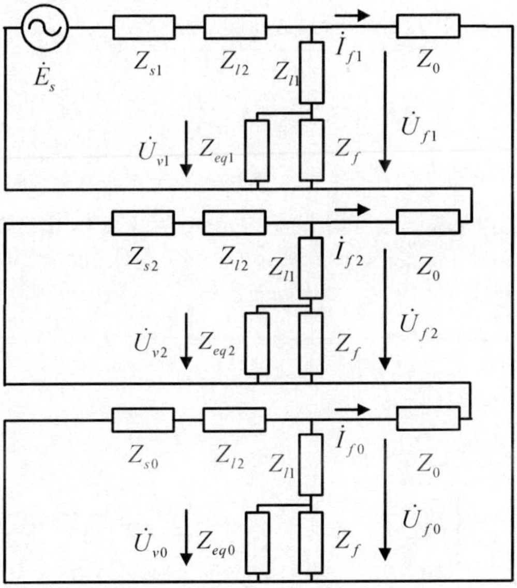

Fig. 6. Composite sequence network for single-phase ground fault of AC line

Among them, the equivalent positive sequence impedance Z_{eq1} and the negative sequence impedance Z_{eq2} on the AC side are the AC side equivalent impedance of the converter calculated above plus the leakage reactance of the converter. For its zero-sequence impedance Z_{eq0}, since there is no zero sequence path in the inverter, there is no zero-sequence equivalent impedance in the inverter part. However, the converter transformer of the 12-pulse converter is grounded at the neutral point of the AC bus side, and there is a zero-sequence path. Therefore, the AC side equivalent zero-sequence impedance of the inverter is a parallel of two zero-sequence 6-pulse transformer impedance. Z_0 is the transition resistance. $\dot{U}_f$ is the voltage of each sequence at fault point. $\dot{I}_f$ is the fault current.

Using Kirchhoff's current law to write the formula for solving the fault point voltage of the composite sequence network:

$$\frac{\dot{U}_{f1} - \dot{E}_s}{Z_{s1} + Z_{l2}} + \frac{\dot{U}_{f1}}{Z_{l1} + \frac{Z_{eq1} \cdot Z_f}{Z_{eq1} + Z_f}} + \dot{I}_{f1} = 0 \tag{14}$$

$$\frac{\dot{U}_{f2}}{Z_{s2} + Z_{l2}} + \frac{\dot{U}_{f2}}{Z_{l1} + \frac{Z_{eq2} \cdot Z_f}{Z_{eq2} + Z_f}} + \dot{I}_{f2} = 0 \tag{15}$$

$$\frac{\dot{U}_{f0}}{Z_{s0} + Z_{l2}} + \frac{\dot{U}_{f0}}{Z_{l1} + \frac{Z_{eq0} \cdot Z_f}{Z_{eq0} + Z_f}} + \dot{I}_{f0} = 0 \tag{16}$$

$$\frac{\dot{U}_{f0} + \dot{U}_{f0} + \dot{U}_{f0}}{3Z_0} - \dot{I}_{f1} = 0 \tag{17}$$

$$\dot{I}_{f1} = \dot{I}_{f2} = \dot{I}_{f0} \tag{18}$$

Solving the above formula can get the sequence voltages at the fault point. Using the voltage at the fault point, it can be calculated back to the DC line to obtain the electrical component of a specific frequency on the DC line under the fault condition.

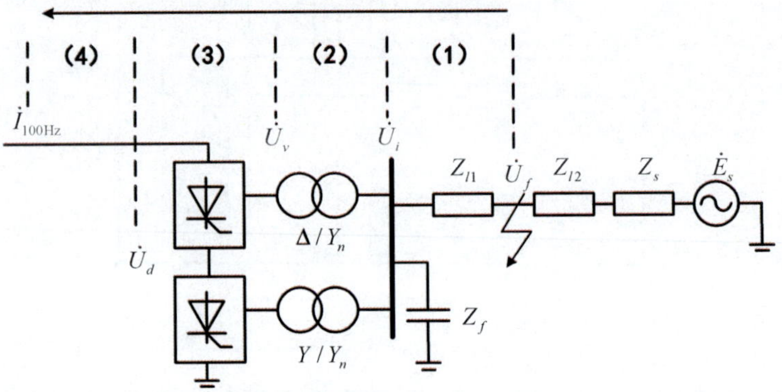

Fig. 7. 100 Hz component calculation circuit diagram

After the fault point voltage is calculated, the DC 100 Hz component at the steady state of the fault can be calculated according to the following steps due to Fig. 6 and Fig. 7:

Step (1): In Fig. 7, according to the composite sequence network, the voltage of the AC line bus $\dot{U}_{i\rho}$ can be obtained from the sequence voltages $\dot{U}_{f\rho}$ at the fault point using the voltage division, which is shown in (19) (where $\rho = 1, 2, 0$ respectively indicate positive, negative, and zero sequences). Convert the sequence voltages of the AC bus $\dot{U}_{i\rho}$ to the voltages of each phase $\dot{U}_{i\varphi}$ ($\phi = a, b, c$) according to the symmetrical component method.

$$\dot{U}_{i\rho} = \frac{Z'_\rho}{Z'_\rho + Z_{l1}} \cdot \dot{U}_{f\rho} \tag{19}$$

Step (2): Considering the different connection methods of the converter transformer, the voltage relationship between the two sides of the transformer with different phase shifts is shown in (20), where $a = e^{2\pi j/3}.\delta$ is the phase shift, which is 0 for Y-Y and π/6 for Y-Δ. k_T is the transformation ratio of the converter transformer. The AC side voltage of the inverter $\dot{U}_{v\varphi}$ can be obtained from $\dot{U}_{i\varphi}$ [20].

$$\begin{bmatrix} \dot{U}_{va} \\ \dot{U}_{vb} \\ \dot{U}_{vc} \end{bmatrix} = \frac{1}{3k_T} \begin{bmatrix} e^{-j\delta} + e^{j\delta} & a \cdot e^{-j\delta} + a^2 \cdot e^{j\delta} & a^2 \cdot e^{-j\delta} + a \cdot e^{j\delta} \\ a^2 \cdot e^{-j\delta} + a \cdot e^{j\delta} & e^{-j\delta} + e^{j\delta} & a \cdot e^{-j\delta} + a^2 \cdot e^{j\delta} \\ a \cdot e^{-j\delta} + a^2 \cdot e^{j\delta} & a^2 \cdot e^{-j\delta} + a \cdot e^{j\delta} & e^{-j\delta} + e^{j\delta} \end{bmatrix} \begin{bmatrix} \dot{U}_{ia} \\ \dot{U}_{ib} \\ \dot{U}_{ic} \end{bmatrix} \tag{20}$$

Step (3): The DC line harmonic voltage $\dot{U}_{dc}(2)$ can be obtained from $\dot{U}_v(2)$ according to the transfer characteristics of the converter, as shown in (21).

$$\dot{U}_{dc}(2) = \dot{U}_{va}(1) \cdot \langle S_{ua} \rangle_1 + \dot{U}_{vb}(1) \cdot \langle S_{ub} \rangle_1 + \dot{U}_{vc}(1) \cdot \langle S_{uc} \rangle_1 \tag{21}$$

Step (4): According to the equivalent impedance of the DC system obtained above, the harmonic current of the DC line $\dot{I}_{dc}(2)$ can be obtained from $\dot{U}_{dc}(2)$, as shown in (22).

$$\dot{I}_{dc}(2) = \dot{U}_{dc}(2)/Z_{v\_12}(2) \tag{22}$$

4.2 Protection Cooperation Scheme

Using the DC 100 Hz component calculation method obtained in the previous section, a 100 Hz protection setting method in cooperation with AC distance protection is proposed. In Fig. 6, the AC line impedance Z_{l1} is replaced by the setting value Z_{set} of the zone 3 protection of distance. Use (14)–(22) to calculate the DC 100 Hz component amplitude I_{dc2} corresponding to Z_{set}. Then calculate the final setting value I_{dc2set} of protection by (23), where K_{rel} is the reliability factor, which is 1.25. By working with the zone 3 protection of distance settings to calculate the setting value of the DC 100 Hz protection, it can ensure that the DC protection can provide backup for the AC protection when an asymmetric fault occurs on the AC line.

$$I_{dc2set} = K_{rel} \times I_{dc2} \tag{23}$$

The time coordination characteristic is also coordinated with zone 3 protection of distance. As shown in (24), a delay difference Δt is added based on the zone 3 protection of distance.

$$t_{dc2set} = t^{\text{III}} + \Delta t \tag{24}$$

The time coordination characteristic of zone 3 protection of distance is considered to avoid the oscillation period [21], t^{III} takes 1.5 s and Δt takes 0.5 s. Therefore, the 100 Hz protection time coordination characteristic is 2 s.

5 Protection Flow Chart

After calculating the 100 Hz protection setting value, the protection process of the 100 Hz protection is shown in Fig. 8.

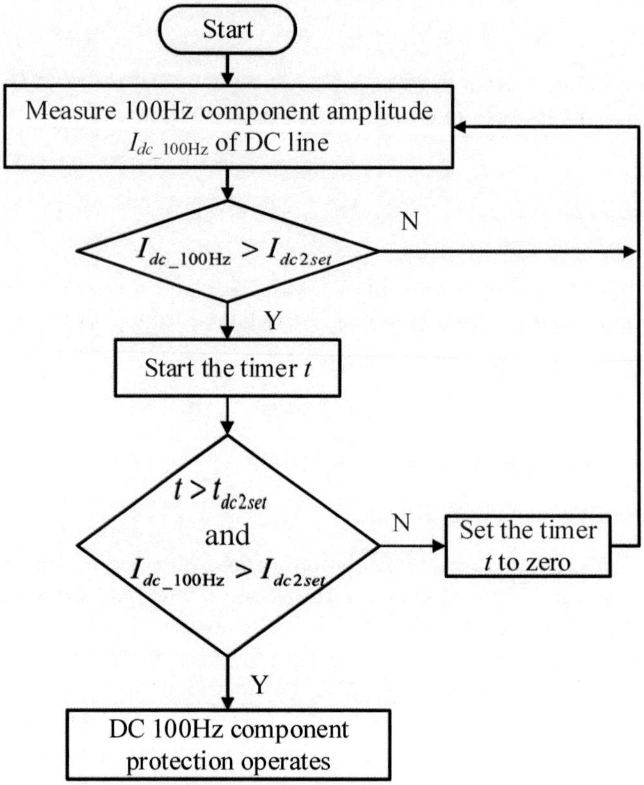

Fig. 8. 100 Hz protection process

6 Simulation

With CIGRE_ Benchmark standardized model, model parameters are shown in Table 1:

Table 1. Model parameters

Parameter name	Value
Rated DC voltage	500 kV
Rated DC current	2 kA
Rated voltage of AC bus	220 kV
AC line impedance per unit length	0.17 + j0.44 Ω/km
AC line length	50 km

6.1 Result of DC 100 Hz Component Calculation

Simulate the occurrence of single-phase grounding faults at different fault distances, record the amplitude of the 100 Hz component on the DC line under the steady-state of the fault, compare it with the calculated value and calculate the error. It can be seen from Table 2 that under different fault distances, the error between the calculated value and the simulated value can be guaranteed to be below 10%. When using the 100 Hz calculated value to calculate the settings, the reliability factor will reduce the effect of the 100 Hz component calculation error.

Table 2. Comparison of simulated values and calculated values of DC 100 Hz component

Fault distance/km	Simulated values of DC 100 Hz component/kA	Calculated values of DC 100 Hz component/kA	Error
1	0.5296	0.5581	5.39%
5	0.5231	0.5427	3.74%
10	0.5163	0.5496	6.45%
20	0.5237	0.5638	7.66%
30	0.4586	0.4821	5.12%
40	0.3527	0.3772	6.96%
50	0.2370	0.2515	6.10%

6.2 Simulation Result of Asymmetric Fault

Take the single-phase ground fault at the end of the AC line as an example. As shown in Fig. 9, the fault occurs at 1.5 s. The solid line is the amplitude of the 100 Hz component measured on the DC line, and the dashed line is the 100 Hz setting value which is calculated by cooperating with zone 3 protection of distance. It can be seen that the measured value of 100 Hz after the fault occurs is much larger than the setting value, and the protection can operate correctly.

For the single-phase ground fault of the AC line, the sensitivity coefficient of the protection setting is:

$$K_{sen} = I_{dc\_100\mathrm{Hz}}/I_{dc2set} = 0.2384/0.0842 = 2.8314 \tag{25}$$

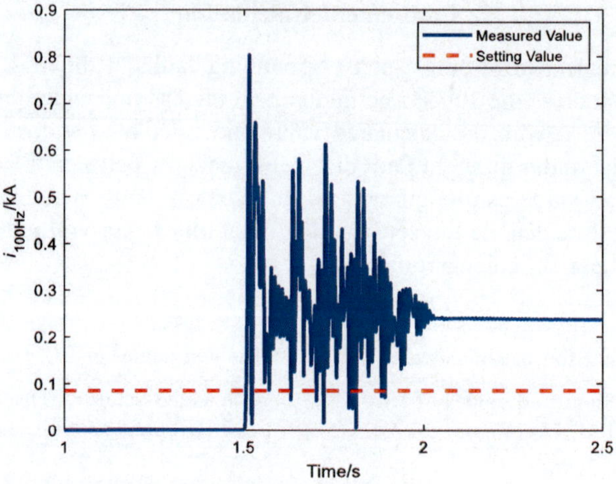

Fig. 9. Comparison of 100 Hz measured value and setting value under asymmetric AC fault.

The calculated sensitivity coefficient is greater than 1.5, which means that the setting value meets the sensitivity requirements.

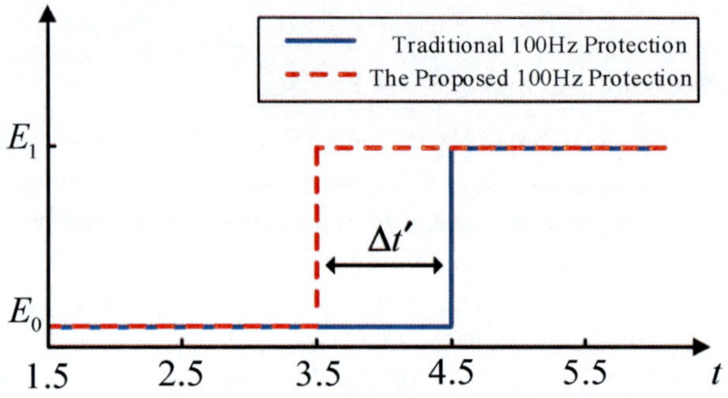

Fig. 10. Comparison of time coordination characteristics

Figure 10 shows the comparison of the time coordination characteristics between the traditional 100 Hz protection and the proposed 100 Hz protection. It can be seen that the proposed 100 Hz protection operate $\Delta t'$ earlier the traditional 100 Hz protection, reflecting its short delay characteristics.

6.3 Simulation Result of Valve Fault

Taking the fault of the valve not opening as an example, a 6-pulse inverter in the simulation model is set to fail to open the valve at 1.5 s. In Fig. 11, the solid line is the amplitude of the 100 Hz component measured on the DC line, and the dashed line is the

100 Hz setting value. It can also be seen that the measured value of 100 Hz after the fault occurs is much larger than the setting value, and the protection can operate correctly.

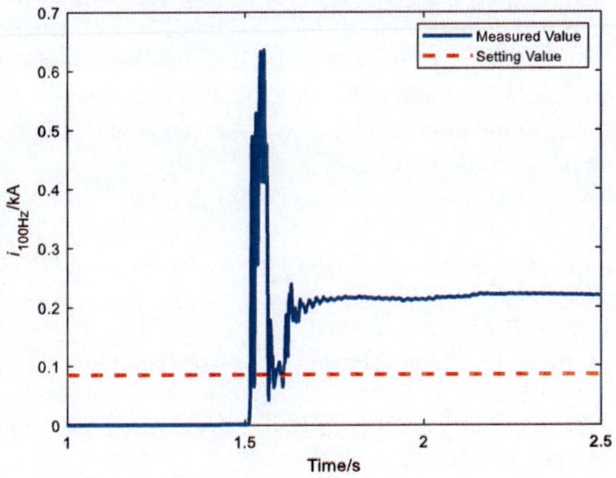

Fig. 11. Comparison of 100 Hz measured value and setting value under valve fault.

For the fault of valve not opening, the sensitivity coefficient of the protection setting is:

$$K_{sen} = I_{dc\_100\mathrm{Hz}}/I_{dc2set} = 0.2203/0.0842 = 2.6164 \tag{26}$$

The calculated sensitivity coefficient is greater than 1.5, which means that the setting value meets the sensitivity requirements.

7 Conclusion

In this paper, the fault characteristics of the converter are analyzed in two parts, including the transfer characteristics of the electrical relationship between the AC side and the DC side of the converter and the impedance characteristics on the single side of the converter. The limitation of traditional 100 Hz protection is analyzed. The calculation method of the DC 100 Hz component under fault steady-state condition and the setting method of the DC 100 Hz protection in cooperation with the zone 3 protection of distance are obtained. The conclusions are as follows:

The calculation method of the 100 Hz component on the DC line under the steady-state of the single-phase ground fault of the AC line based on the transfer characteristics and the equivalent impedance of the converter has high accuracy.

The DC 100 Hz protection setting value in cooperation with the zone 3 protection of distance can operate correctly in the case of an asymmetric fault at the end of the AC line and a fault of the valve not opening. It can meet the sensitivity requirements and improve the limitation of the traditional 100 Hz protection with too long time cooperation characteristics.

References

1. Wanjun, Z.: HVDC Transmission Engineering. China Electric Power Press, Beijing (2009)
2. Li, C., Lin, X., Zhao, Y., et al.: An analytical solution for transient process of HVDC transmission system part 1-mathematical model. Power Syst. Technol. **41**(01), 1–8 (2017)
3. Chen, S.-Y., Li, X.-N., Yu, J., et al.: A method based on the sin-cos components detection mitigates commutation failure in HVDC. Process. CSEE **25**(14), 1–6 (2005)
4. He, C., Yuan, X.: Simulation analysis on commutation failure criteria for HVDC transmission systems. Process. CSEE **28**(7), 51–57 (2008)
5. Xu, Q., Xu, H.: Modeling of static var generator. Autom. Electr. Power Syst. **23**, 37–40 (2001)
6. Hu, L., Yacamini, R.: Harmonic transfer through converters and HVDC links. IEEE Trans. Power Electron. **7**(3), 514–525 (1992)
7. Ma, Y.-L., Xiao, X.-N., Jiang, X.: Analysis of the impact of AC system single-phase earth fault on HVDC. Process. of the CSEE **26**(11), 144–149 (2006)
8. Li, Z., Wang, G., Li, H., Li, X., et al.: An analysis and calculation method of harmonic interaction between AC and DC system under asymmetric operating condition. Autom. Electr. Power Syst. **34**(6), 42–47 (2010)
9. Daryabak, M., Filizadeh, S., Jatskevich, J., et al.: Modeling of LCC-HVDC systems using dynamic phasors. IEEE Trans. Power Deliv. **29**(4), 1989–1998 (2014)
10. Liu, C., Bose, A., Tian, P.: Modeling and analysis of HVDC converter by three-phase dynamic phasor. IEEE Trans. Power Deliv. **29**(01), 3–12 (2014)
11. Wang, T., Jia, K., Bi, T., et al.: Transient characteristics of commutation failure of HVDC system based on dynamic phasor theory. Autom. Electr. Power Syst. **42**(23) (2018)
12. Zhixue, L., Tianqi, L., Xiang, C., et al.: Risk assessment of harmonic instability in HVDC transmission systems considering effect of two-terminal Converters. Autom. Electr. Power Syst. **42**(19) (2018)
13. Xu, M., Cai, Z., Han, K., et al.: Influence analysis of ac system transient invasion on DC protective relaying in AC/DC hybrid power system. High Volt. Eng. **40**(11), 3618–3625 (2014)
14. Yu, J., Zhou, H., Huang, J., et al.: Impact of fault location in AC system to 100 Hz protection of HVDC. Autom. Electr. Power Syst. **32**(3) (2008)
15. Ou, K.-J., Zhu, T.-X., Guo, Q., et al.: Simulation and analysis on the mal-operation of stage 1 of 100 Hz protection in HVDC transmission systems. South. Power Syst. Technol. **2**(5), 36–38 (2008)
16. Xu, H., Ruan, S., Lü, P.: Study of 50 Hz and 100 Hz protection in Gaoling back-to-back HVDC system. Power Syst. Prot. Control **38**(4), 54–57 (2010)
17. Ma, Y., Xiao, X., Jiang, X., et al.: Study on impedance frequency characteristic on HVDC converters. Proc. CSEE **26**(11), 144–149 (1992)
18. Kong, W., Wen, J., Wang, L., et al.: Study on AC-side equivalent harmonic impedances of HVDC converter. J. N. Chin. Electr. Power Univ. **44**(4), 37–43 (2017)
19. Zhi, H., Zexiang, C.: The Principle and Technology of Relay Protection for HVDC Transmission System. China Electric Power Press, Beijing (2013)
20. Wanshun, L.: Power System Fault Analysis. China Electric Power Press, Beijing (2010)
21. Baohui, Z., Xianggen, Y.: Power System Protective Relaying. China Electric Power Press, Beijing (2010)

DQN-Based Voltage Regulation for Active Distribution Network with Distributed Energy Storage System

Wu Yu[1], Gao Huawei[1], Zhang Xiaohai[1], Cui Quansheng[2(✉)], and Zhang Peng[3]

[1] Shuangyashan Power Company, Shuangyashan 155100, Heilongjiang, China
[2] R&D Center, Pinggao Integrated Energy Service Company, Beijing 100052, China
little_cui@163.com
[3] Beijing Zhizhong Energy Technology Development Co., Beijing 100091, China

Abstract. Using the energy storage system to regulate the distribution network voltage can effectively deal with the distribution network voltage level caused by the high intermittence of renewable energies and the fluctuation of load demand. In this paper, the charging and discharging action of energy storage system is modeled as a Markov decision-making process, and a deep-Q-network based (DQN-based) voltage regulation algorithm is proposed for the distribution network with energy storage system. Since the state space of the problem is too large, this DQN-based voltage regulation algorithm uses a neural network to approach the optimal action values of the energy storage, and the neural network is trained by the experience replay strategy. The input of the neural network is the state eigenvector composed of the state of capacity (SOC) of the energy storages, the predicted power outputs of renewable energies and the power load level, and the output of the neural network is the optimal discrete charging and discharging action value used to improve the voltage level. Compared to the traditional voltage regulation methods, the proposed DQN-based voltage regulation method is a learning-based approach, which does not need the explicit uncertainty model and decreases the solution complexity. The proposed DQN-based voltage regulation algorithm is tested in the IEEE 33-bus distribution network and the results are analyzed to prove the effectiveness of the proposed voltage regulation algorithm.

Keywords: Voltage regulation · Deep-Q Network · Energy storage systems

1 Introduction

With the increasing penetration of renewable energies such as photovoltaic and wind power in the distribution networks, the impact of their strong volatility and high uncertainty on the safe and economic operation of the distribution networks is increasing. When a large number of renewable distributed generations (RDGs) are connected to the distribution networks, their power output fluctuation will have a

Y. Xue et al. (Eds.): PMF 2020, LNEE 718, pp. 749–762, 2021.
https://doi.org/10.1007/978-981-15-9746-6_57

negative impact on the voltage operation level of the distribution networks, and even make the voltages exceed the allowable range.

The energy storage technology can effectively achieve the peak load reduction and valley filling, promote the use of the renewable energies, and it also be used as an effective means to assist the voltage regulation of power systems. With the development of the energy storage technology and the rapid decline of its cost, the utilization of energy storage in the voltage regulation of power systems has gradually become another feasible technical means, in addition to the traditional reactive power voltage regulation methods. By allocating the energy storage systems participating in the auxiliary service of the distribution system, the power fluctuation caused by the renewable energies can be restrained and the risk of voltage exceeding the limit can be reduced [1].

Many literatures have studied the energy storage control strategy for suppressing the intermittence of renewable energies and the fluctuation of load output. The paper [2] presents a coordinated control method based on two-stage operation optimization between traditional voltage regulating equipment and distributed energy storage systems, but the modeling of control process is complex and the convergence is poor. In reference [3], based on the power flow solvability, the allocation of energy storages by the static voltage stability index in the distribution networks is studied, but its control strategy is not explained. The traditional model-based methods need to model and analyze the uncertain factors relative to the RDG and load, and the solution process of the energy storage optimal control is complex and poorly practicable.

With the rapid development of the machine learning in recent years, its powerful perceptual-learning ability and data-analysis ability meet the needs of big data application in the smart grid. Reinforcement learning (RL) is a branch of the machine learning, which obtains the environmental knowledge through the continuous interaction between decision-makers and the environment, and takes actions that affect the environment to achieve the preset goals. Deep learning (DL) is another branch of the machine learning, which is completely related to neural network. Deep learning does not rely on any analytical equation, but uses a large number of existing data to describe mathematical problems and their approximate solutions. Applying the deep learning to the reinforcement learning can effectively alleviate the difficulty of value function solving in RL. For example, the paper [4] applies the deep reinforcement learning (DRL) method to real-time energy management and control of the micro-grid, and verifies the calculation efficiency of the strategy; the paper [5] also proposes an intelligent multi-micro grid energy management method, which is based on deep neural network and does not need the model, and proves the effectiveness of the method in solving power system problems with partial observable or uncertain information. In paper [6], a DRL-based uncertainty economic dispatch algorithm is introduced and this kind of algorithm can avoid the difficulty of uncertainty modeling. And in paper [7], the coordinated optimization method of active distribution network based on deep deterministic policy gradient (DDPG) algorithm is proposed, and the effectiveness of the algorithm is verified by an example simulation. DDPG is one of the popular DRL algorithms.

In order to overcome the problems in traditional methods, which comprise the complexity and poor convergence of the uncertainty modeling, and the solution difficulty caused by too large state space, this paper proposes a DRL-based voltage regulation method for the distribution network with energy storage systems to improve the distribution network voltage level.

The algorithm approaches the optimal action value by embedding a deep Q network (DQN), which takes the state eigenvector of the distribution network as the input and takes the optimal charging and discharging action value of the energy storage system as the output, and train the DQN network through the experience replay strategy to obtain optimal charging and discharging strategy. Compared with the traditional model-based method, there is no need to establish the explicit uncertainty model, and the calculation efficiency is higher.

2 Charging and Discharging Characteristics of Energy Storage

The distributed energy storage systems installed at the access nodes of RDG is mainly used to smooth the power fluctuation of the intermittent energy sources. The end of distribution networks can also be equipped with energy storage systems specially participating in the auxiliary services such as voltage regulation, peak shaving and frequency regulation.

In recent years, the electrochemical energy storage technology has been widely used because of its mature technology and cost advantage. In this paper, it is assumed that the type of energy storage system is the battery energy storage system (BESS), and the residual electric quantity of the BESS changes continuously in the process of charging and discharging, which is related to the charging/discharging rate and self-discharging in this period. The recurrence relation of the energy storage system charging is as follows:

$$E^{es}(t) = (1-\sigma)\cdot E^{es}(t-1) + P^{es}(t)\cdot(1-L^{c})\Delta t/E^{es,\,cap} \tag{1}$$

The energy storage discharging process is shown as follows:

$$E^{es}(t) = (1-\sigma)\cdot E^{es}(t-1) + P^{es}(t)\Delta t/[E^{es,cap}\cdot(1-L^{d})] \tag{2}$$

Where, $E^{es}(t)$ is the state of charge (SOC) of the BESS at time t, $P^{es}(t)$ is the charging or discharging power of the BESS at time t;σ is the self-discharge rate of the energy storage medium; L^{c} and L^{d} are the charge and discharge losses of the BESS, respectively; Δt is the duration of each calculation step.

3 Deep Q-Network (DQN)

3.1 Reinforcement Learning

Reinforcement Learning (RL) is a kind of machine learning methods. RL maps from environment states to action, and its goal is to maximize the cumulative rewards for the agent in the interactions of the agent and the environment. Markov decision process (MDP) is used to simplify its modeling, and MDP is usually defined as a four tuple (s, a, r, f) [8].

Where:

1) S is the set of all environment states, and $s_t \in S$ represents the state of the agent at time t;
2) A is the set of executable actions of the agent, $a_t \in A$ represents the action taken by the agent at time t;
3) r is the reward function, $r_t \sim r(s_t, a_t)$ represents the immediate reward value obtained by the agent executing action a_t in the state s_t;
4) f is the probability distribution function of state transition, $s_{t+1} \sim f(s_t, a_t)$ represents the probability of the agent executing action a_t in the state s_t to move to the next state s_{t+1}.

The goal of the Markov model is to find an optimal planning strategy V^{π^*} to maximize the sum of expected rewards after initializing the state s:

$$V^{\pi^*}(s) = \max_{\pi \in \prod} E_\pi[\sum_{t=0}^{step} \gamma_t \cdot r_{t+1} | s = s_0] \tag{3}$$

Where, E_π represents the expectation value under the strategy π; $0 < \gamma < 1$ is a attenuation coefficient in RL that represents the importance of future rewards.

When the scale of the problem is small, the algorithm is relatively easy to solve [8]. However, for practical problems, the state space is usually very large, and there are many kinds of actions in each state. It will take a lot of time to traverse all actions in each state. The cost of the traditional ergodic solutions is too high, and it is difficult to ensure convergence.

3.2 Deep-Q Network

To deal with the above problems, the deep neural network can be combined with the traditional Q-learning algorithm in RL, that is, a neural network with the weight θ can be used to approximate the Q value of all possible actions in each state. Because the deep neural network is used to approximate Q function, it is called deep-Q network (DQN), which is one of the very popular and widely-used DRL algorithms [8].

The update rule of the Q learning is as follows:

$$Q(s,a) = Q(s,a) + \alpha(r + \gamma \max Q(s'a') - Q(s,a)). \tag{4}$$

Where, s denotes the state, s' denotes the next state, $r + \gamma \max Q(s'a)$ is the target value, and $Q(s,a)$ is the predicted value; we tried to minimize the difference between the target value and the predicted value by learning a right policy [8].

Similarly, in DQN, the loss function can be defined as the squared difference between the target value and predicted value, and we try to minimize the loss by updating the weights θ:

$$\begin{aligned} &Loss = (y_i - Q(s,a;\theta))^2 \\ &where\ \ y_i = r + \gamma \max_{a'} Q(s', a'; \theta). \end{aligned} \tag{5}$$

In a nutshell, in DQN, the neural network is used as function approximators for approximating a Q function, and we minimize errors through the gradient descent method [8].

The DQN reinforcement learning algorithm introduces the experience replay mechanism to accelerate convergence and improve the generalization ability. Experience replay is to establish a "memory bank" in the learning process, which comprises the current state s, action a, reward r and the next state s'. Each time the neural network is trained, a batch of memory data is randomly extracted from the memory bank. In this way, the order of the original data is disordered and the correlation of the data is weakened.

4 DQN-Based BESS Charging and Discharging Learning Model

In this DQN-based charging and discharging learning model, The BESS is modelled as an agent, who can interact with the external environments, learn the stochastic and time-varying environments and make use of the learning experience to take optimal charging/discharging actions to better adapt to external environments.

This dynamically changing environment is characterized by the stochastic wind and PV power outputs, the stochastic load fluctuation, and the SOC state of the BESS affected by the BESS charging and discharging action.

The BESS selects the optimal charging and discharging action to get the best reward, and the reward function is defined on the goal of distribution network voltage regulation. The goal of the distribution network voltage regulation includes: minimizing the average network voltage fluctuation, minimizing the voltage fluctuation of weak nodes, and meanwhile make the BESS keeping the best regulation capacity for smoothing the power fluctuation at the next moment.

4.1 Environmental States

For the distribution network with the given to pology parameters, the eigenvector s_t of the environment state sensed by the BESS at time t is defined as:

$$s_t = [E^{es}(t), P_i^{pv}(t), \ldots, P_i^{wt}(t), \ldots, L^{all}(t)]^T \tag{6}$$

Where, $P_i^{\mathrm{pv}}(t)$ is the active power injection prediction value of the PV node i at time t; $P_i^{\mathrm{wt}}(t)$ is the active power injection prediction value of the wind generation node i at time t; $L^{\mathrm{all}}(t)$ is the load of the distribution network at time t. Note that the variables $P_i^{\mathrm{pv}}(t)$, $P_i^{\mathrm{wt}}(t)$ and $L^{\mathrm{all}}(t)$ are random and time-varying, and their values are different for each period t.

4.2 Actions

The power range of the BESS is as follows, $\mathrm{P}_{\max}^{\mathrm{es,d}}$ denotes the maximum discharging power, $\mathrm{P}_{\max}^{\mathrm{es,c}}$ denotes the maximum charging power.

$$-\mathrm{P}_{\max}^{\mathrm{es,c}} \leq P^{es} \leq \mathrm{P}_{\max}^{\mathrm{es,d}} \tag{7}$$

We uniformly discretize the BESS power P^{es}, i.e. the BESS power P^{es} is divided into k discrete charging and discharging options according to its range. The k discrete charging and discharging options comprise the action space A of the BESS.

$$A = \{P^{es,(1)}, P^{es,(2)}, \ldots, P^{es,(k)}\} \tag{8}$$

Where, A is the set of all possible charging and discharging actions; $P^{es,(\mathrm{k})}$ is the k-th charging/discharging action in the uniform discrete action space of the BESS.

When sensing the change of environment state, the BESS will make a response action a_t, and a_t must be of the action space A.

4.3 Rewards

The voltage level of the distribution network is improved by using the flexible charging and discharging characteristics of the BESS. The control objectives of the distribution network voltage regulation include the following ones:

(1) Minimize the average voltage fluctuation of system nodes

$$\min \sum_{i=1}^{N} |V_i(t) - V_i^{\mathrm{base}}|/N \tag{9}$$

(2) Minimize the voltage fluctuation of system weak nodes

$$\min \max_{i \in N} |V_i(t) - V_i^{\mathrm{base}}| \tag{10}$$

(3) The BESS needs to reserve a certain reserve capacity for the subsequent voltage regulation of the system, that is, to maximize the storage energy regulation capacity for smoothing the power fluctuation at the next moment [9]

$$\min |E^{es}(t) - E_0^{es}| \tag{11}$$

Where, N is the total number of nodes in the distribution network; $V_i(t)$ is the static voltage per-unit value of node i at time t, which is calculated under the actual RDG output and the actual load conditions. V_i^{base} is the reference per-unit value of the node i, which is the target of node voltage stability.

V_i^{base} is calculated by power flow, under the condition of the historical average output of each RDG and the average load level of the distribution network. E_0^{es} is the initial SOC value of the BESS involved in the voltage regulation.

The charging and discharging strategy of the BESS is driven by the reward given to it, and the reward should reflect the control goals of the distribution network voltage regulation. So, the reward r_t obtained by the BESS include 3 parts, r_{t1}, r_{t2}, and r_{t3}, and they correspond to formula (9), (10) and (11), respectively. Specifically, r_{t1} and r_{t2} together guarantee to minimize the voltage fluctuation, and r_{t3} ensures the BESS reserve capacity for the subsequent voltage regulation.

$$r_{t1} = \frac{-\sum_{i=1}^{N} |V_i(t) - V_i^{base}|}{N} \tag{12}$$

$$r_{t2} = -\max_{i \in N} |V_i(t) - V_i^{base}| \tag{13}$$

$$r_{t3} = -|E^{es}(t) - E_0^{es}| \tag{14}$$

$$r_t(s_t, a_t) = \begin{cases} \alpha \cdot (r_{t1} + r_{t2}) + (1 - \alpha) \cdot r_{t3}, & 0 \le E^{es}(t) \le 1 \\ r_{\text{done}}, & \text{else} \end{cases} \tag{15}$$

Where, $\alpha \in (0, 1)$ is the reward weight considering the voltage level index; r_{done} is the penalty value, and it is a negative constant with the larger absolute value.

4.4 Algorithm Procedures

The DQN algorithm of the BESS to improve the voltage operation level of the distribution network is as follows:

(1) Build a multi-hidden layer Q network model (see Fig. 1). This neural network is used as function approximator for approximating a Q function. The input of this neural network is the eigenvector s_t sensed by the BESS at time t (see formula (6)), and its output corresponds to the Q value of each action a_t.
(2) Initialize all parameters θ of the Q network, and clear the experience replay set D;
(3) The load level of each node of the distribution system and the RDG power outputs are given according to the historical average values. By the load level of each node and the RDG power outputs, the distribution network power flow and the node voltage per-unit values are calculated;

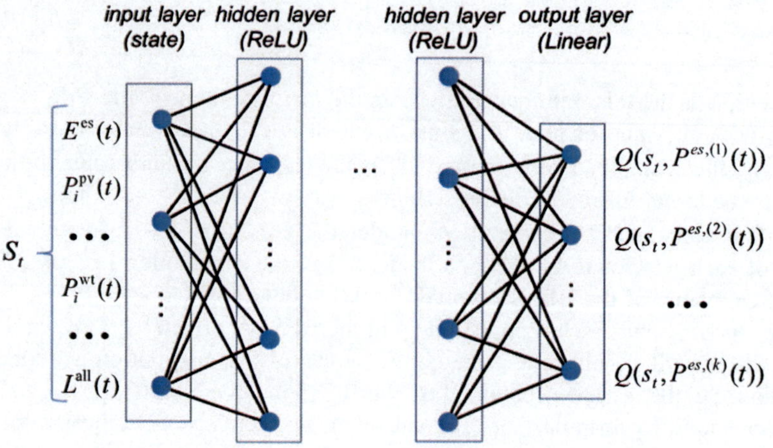

Fig. 1. Structure diagram of deep Q neural network

(4) Initialize the following algorithm parameters, including the total iteration number M, the calculation step T (24 h) at each iteration, and the attenuation factor γ.

(5) Carry out the following iterations:

1. *iter* =1;
2. Initialize the SOC state of the BESS, and the environment state eigenvector is initialized to s_t.
3. In the Q network, s_t is used as the input to get the Q value outputs corresponding to all actions. The greedy method is used to select a corresponding action a_t in the current Q value outputs;
4. Execute the current action a_t in the state s_t, get the new state eigenvector s_{t+1} and the reward r_t(i.e., with s_t and a_t as input parameters, the distribution network power flow is calculated to get s_{t+1} and r_t), and judge whether this iteration is in the termination state;
5. Store s_t , a_t , r_t, s_{t+1} into the experience replay set D;
6. Set the current state $s = s_{t+1}$;
7. Take m samples from D to calculate the current target Q value;
8. All parameters ω of the Q network are updated by the gradient back propagation;
9. If s is in the termination state, the current iteration ends, go to step 10, otherwise, go to step 3;
10. *iter* = *iter* +1; if *iter* > *M,* the algorithm ends, otherwise, go to step 2.

5 Case Study

5.1 Case Conditions and Parameters

The DQN-based algorithm proposed in this paper is implemented with Python 3.5. The IEEE 33-node distribution network is selected as an case study, as shown in Fig. 2. PVs are connected to node 13 and node 24, and the wind power generations are connected to node 8 and node 31, and their corresponding active power outputs are expressed by $P_{13}^{pv}, P_{24}^{pv}, P_{8}^{wt}, P_{31}^{wt}$. The installation location of the BESS should be as close as possible to the end of distribution network [10]. In this case, it is assumed that the BESS is installed at node 16 (Fig 3).

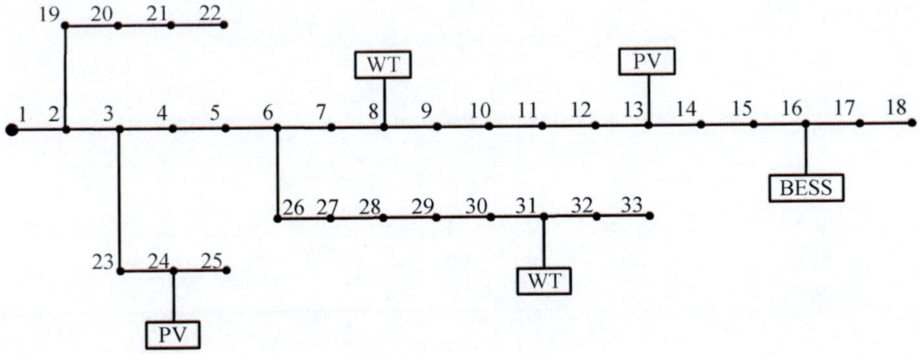

Fig. 2. IEEE 33-bus distribution network

To simplify the analysis process, the parameters of the example are set as follows:

(1) Taking the original load data of IEEE 33 network as the historical average level of the load, the maximum load fluctuation is ±15% (Table 1).
(2) The configuration parameters of the BESS are shown in Table 2, and its charging and discharging power are discretized into 101 actions (k = 101).
(3) In the example, the input layer nodes of the Q network include the SOC of the BESS, the predictive values of the RDG power output and the load level, totally 6-dimension state eigenvector, and the output layer nodes include 101 discrete action values. The double hidden layer architecture is built, and the node number of each hidden layer is 200. The ReLu activation function is selected to accelerate the convergence process. The weight of the normal initialization layer is ω, initialization offset b is set 0.01.
(4) The relevant parameters of the iterative training are set as shown in Table 3. In order to ensure the better convergence of the algorithm, the exploration rate ε gradually reduces with the iteration times, i.e., it changes from $\varepsilon_{\text{ini}} = 0.9$ to $\varepsilon_{\text{final}} = 0.01$.

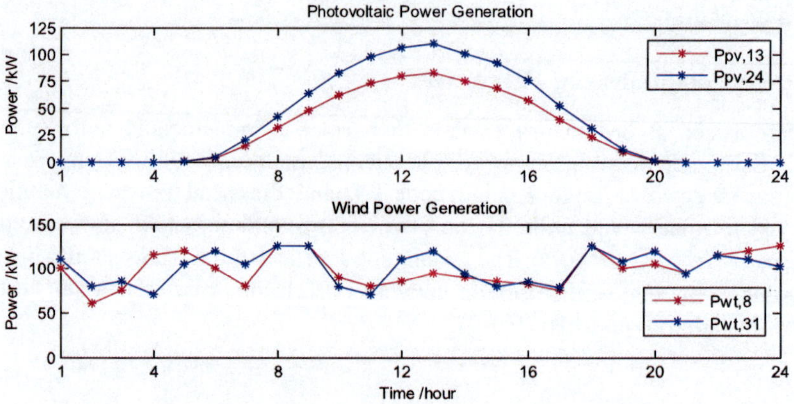

Fig. 3. Actual active power outputs of RDGs

Table 1. Normal distribution parameters of RDG power outputs

Type	Node no.	Actual mean value	Error mean value	Standard deviation
PV	13	50.0	0	7
PV	24	75.0	0	9
WT	8	100.0	0	10
WT	31	100.0	0	15

Table 2. BESS configuration parameters

$P_{max}^{es,c}$/kW	$P_{max}^{es,d}$/kW	L^c	L^d	σ	$E^{es,cap}$/kWh	E_0^{es}
400	400	0.02	0.02	0.01	750	0.5

Table 3. DQN training parameters

T/h	ε_{ini}	ε_{final}	α	m	$\Delta t/h$	r_{done}
24	0.9	0.01	0.7	100	1	−40

5.2 Convergence Analysis of DQN

The voltage target values (per-unit) of 33 nodes in the distribution network are shown in Table 4.

Through the continuous interactive learning with the distribution network environment, the BESS will gradually make an optimal charging and discharging decision to maximize the cumulative reward. In order to verify the convergence of the DQN algorithm, the average cumulative reward R_{mean} is calculated every 100 iterations in the learning process. The total iteration number M is set to 10000. When the initial SOC

Table 4. Voltage target values in the IEEE 33-bus distribution network

Node no.	Base value (per-unit)	Node no.	Base value (per-unit)	Node no.	Base value (per-unit)	Node no.	Base value (per-unit)
1	1.050	10	0.983	19	1.047	28	0.987
2	1.047	11	0.983	20	1.043	29	0.979
3	1.034	12	0.981	21	1.043	30	0.976
4	1.027	13	0.976	22	1.042	31	0.971
5	1.020	14	0.973	23	1.030	32	0.970
6	1.002	15	0.972	24	1.024	33	0.970
7	0.997	16	0.970	25	1.021		
8	0.994	17	0.968	26	1.001		
9	0.989	18	0.968	27	0.996		

value of the BESS is 0.3, 0.5 and 0.7, respectively, the convergence curves of the average cumulative rewards are shown in Fig. 4.

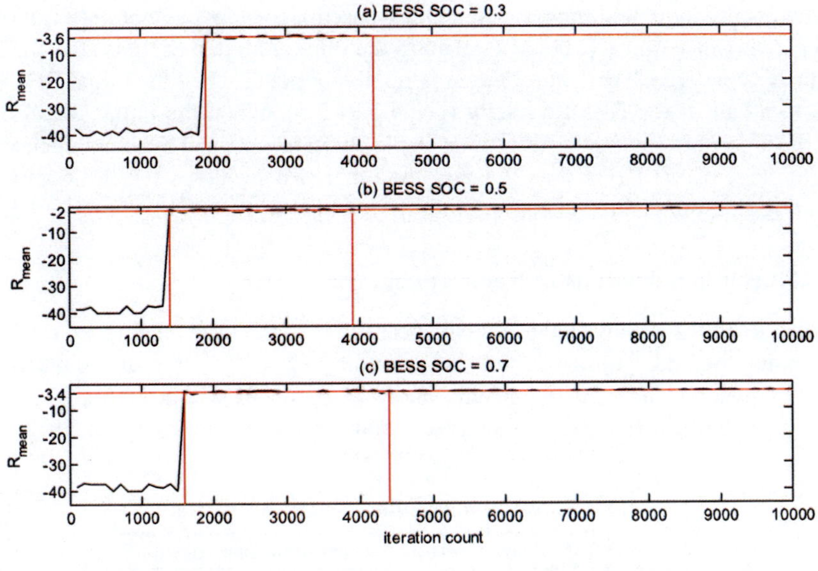

Fig. 4. Average cumulative rewards R_{mean} under different initial values of BESS SOC

It can be seen from Fig. 4 that these curves (when the BESS SOC is 0.3, 0.5 or 0.7, respectively) have the similar pattern. Take Fig. 4(b) as an example, R_{mean} shows a larger volatility with the increase of iteration count in the early stage of training, which indicates that in the early stage ε is larger and the intelligent algorithm is in the exploration stage. When the iteration reaches to the 1300-th iteration, the R_{mean} rapidly increases from −38 to near −2 (the final convergence value) in the subsequent 100 iterations. In the subsequent iteration process, R_{mean} will keep small amplitude oscillation. Finally, after the 3900th iteration or so, R_{mean} reaches the saturation value, and the algorithm is thought to reach the convergence.

Figure 4 describes the average cumulative reward under different initial SOC values.

(a) When the initial BESS SOC = 0.3 (i.e., $E_0^{es} = 0.3$)
The iteration number required to achieve the stable convergence is about 4200, and the R_{mean} finally stabilizes at −3.6, and the average yield is relatively low, that is, the target control effect is relatively low.
(b) When the initial BESS SOC = 0.5
When the number of iterations is close to 3900, the R_{mean} tends to be saturated and fluctuates in a small range, and basically remains stable at about −2. In this case, it has the best control effect.
(c) When the initial BESS SOC = 0.7
In this case, after 1600 iterations, the R_{mean} begins to converge rapidly. When the algorithm finally converges stably, the number of iterations is about 4300. The R_{mean} finally converges to −3.4, and the average yield is relatively low.

The results show that the proposed DQN-based algorithm can not only achieve the good convergence quickly, but also adapt to the different initial SOCs of the BESS and eventually converge. The control effect is relatively better when the initial BESS SOC tends to a half of the BESS capacity (i.e., $E_0^{es} = 0.5$). When the initial BESS SOC = 0.5, the algorithm reaches fastest to the stable convergence, and the convergence value of R_{mean} (i.e. −2) is larger than those in the other two scenarios (i.e. −3.6 and −3.4), indicating that the control effect is the best.

5.3 Fluctuation Analysis of Node Voltages

In order to further analyze the effectiveness of the proposed intelligent method for suppressing the node voltage fluctuation, assuming $E_0^{es} = 0.5$, after the convergence of the DQN method, the voltage per-unit values of all nodes within 24 h are calculated, the voltage fluctuation errors are analyzed, and the results are shown in Table 5.

Table 5. Node voltage fluctuation error analysis

Node no.	Maximum error/p.u.	Maximum relative error/%
14	0.0031	0.32

It can be seen from Table 5 that the maximum node voltage fluctuation error value in all nodes is 0.0031, and the maximum relative error is 0.32%. This two values both corresponds to node 14, which is a PV access node.

The average voltage per-unit values of 33 nodes during one day (24 h) are calculated and compared with the target reference values, as shown in Fig. 5. The two curves in Fig. 5 are very close, which shows that the average voltage values of 33 nodes in a period of time are almost equal to their control target values.

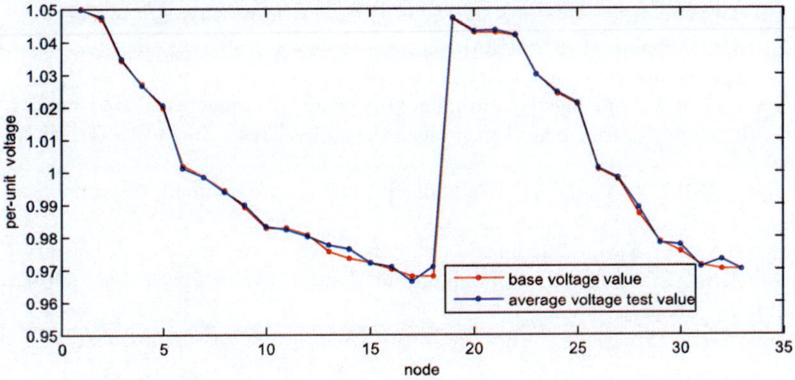

Fig. 5. Comparison of base voltage and average voltage test value

It can be seen from the above results that under the DQN algorithm, the BESS can suppress accurately and reliably the node voltage fluctuation caused by the RDG intermittence and the load fluctuation near the voltage target value, and improve the voltage operation level of the distribution network.

6 Conclusions

In this paper, a DQN-based distribution network voltage regulation algorithm is proposed to improve the voltage operation level of the distribution network. By embedding a deep neural network to approach the optimal action value, the optimal discrete charging and discharging strategy of the BESS is obtained by using the SOC state of the energy storage, the predicted power outputs of renewable energies and the load level as sensing data to drive the convergence. Finally, the effectiveness of the algorithm is verified by a simulation example, and the following conclusions are drawn:

(1) After a certain number of iterations, the algorithm can effectively guide the BESS to learn the optimal charging and discharging strategy which meets the preset goal;
(2) The algorithm can adapt to the different initial SOCs of the BESS, and the control effect is better when the initial SOC tends to a half of the total capacity.

References

1. Mahmud, N., Zahedi, A.: Review of control strategies for voltage regulation of the smart distribution network with high penetration of renewable distributed generation. Renew. Sustain. Energy Rev. **64**, 582–595 (2016)

2. Zhang, J., Zhuang, H., Liu, J., et al.: Two-stage coordinated voltage control scheme of active distribution network with voltage support of distributed energy storage system. Electr. Power Autom. Equip. **39**(5), 15–21 (2019). (in Chinese)
3. Ren, Z.: Research on energy storage configuration method to improve static voltage stability of distribution network. Electr. Switchgear **57**(3), 24–26 (2019). (in Chinese)
4. Bui, V.H., Hussain, A., Kim, H.M.: Q-learning-based operation strategy for community battery energy storage system (CBESS) in micro-grid system. Energies **12**(9), 1789 (2019)
5. Du, Y., Li, F.X.: Intelligent multi-microgrid energy management based on deep neural network and model-free reinforcement learning. IEEE Trans. Smart Grid **11**(2), 1066–1076 (2020)
6. Peng, L., Sun, Y.Z., Xu, J., et al.: Self-adaptive uncertainty economic dispatch based on deep reinforcement learning. Autom. Electr. Power Syst. **44**(9), 33–42 (2020). (in Chinese)
7. Gong, J.X., Liu, Y.M.: Coordinated optimization of active distribution network based on deep deterministic policy gradient algorithm. Autom. Electr. Power Syst. **44**(6), 113–120 (2020). (in Chinese)
8. Ravichandiran, S.: Hands-On Reinforcement Learning with Python. Packt Publishing Ltd. (2018)
9. Ji, Y., Wang, J.H., Xu, J.C., et al.: Real time energy management of a microgrid using deep reinforcement learning. Energies **12**(12), 2291 (2019)
10. Zhang, J., Zhuang, H., Liu, J., et al.: Multi-objective optimal operation scheme based on CS-PSO algorithm of battery energy storage system. Adv. Eng. Sci. **50**(4), 193–200 (2018)

V2G Strategy for Electric Vehicles Considering User Evaluation Based on Fuzzy Control Method

Zhu Chen[1], Renmin Zhang[2], Wenwei Zhou[1](✉), Jie Si[2], and Youbing Zhang[1]

[1] Zhejiang University of Technology, Hangzhou, China
zww@zjut.edu.cn
[2] State Grid Wenzhou Electric Power Supply Company, Wenzhou, China

Abstract. This paper presented an orderly V2G strategy considering user evaluation based on fuzzy control method for massive electric vehicles' real-time V2G control. Satisfying the interests of both users and the power grid is the characteristic of this strategy. Firstly, an orderly V2G management system is proposed, so that V2G strategies could executed on multiple regional control terminals at the same time to reduce the calculation pressure. Then the comprehensive user evaluation model that takes into account battery-wear cost is proposed so as to obtain the user's evaluation of the V2G state in real time for optimal control. Next, the control commands through the fuzzy control method were got to control all the connected electric vehicles through the V2G management system. Finally, the result of simulation experiment with different objectives indicates that this strategy can meet the control demands of massive electric vehicles connected and realizes the peak load transfer of the power grid, while guiding users to participate in V2G responses, it also increases user evaluation.

Keywords: V2G · Electric vehicle · User evaluation · Fuzzy control · Battery wear

1 Introduction

With the increasing awareness of environmental protection and the depletion of fossil fuels, Electric Vehicles (EVs) are becoming more and more popular as an efficient and clean means of transportation [1–3]. As the penetration rate of EVs increases, the benefits that V2G technology can create also increase [4]. V2G technology enables EVs to be used as mobile energy storage units to participate in the energy scheduling of the power grid [5]. To a certain extent, EVs replace peak-shaving power stations and rotating backups to reduce the operating costs of power grids [6, 7]. On the other hand, since EVs are used as a means of daily travel, the time EVs connect to the grid is very random [8, 9]. The load caused by the disorderly charging of EVs will overlap with the peak period of the basic load, which will threaten the safe operation of the equipment in distribution network in the area [10]. If an orderly V2G strategy is implemented for EVs, the above situation can be avoided, thereby reducing peak, maintaining stable

Y. Xue et al. (Eds.): PMF 2020, LNEE 718, pp. 763–780, 2021.
https://doi.org/10.1007/978-981-15-9746-6_58

operation of the power grid, and improving economic efficiency [11, 12]. Therefore, a reasonable orderly V2G strategy is very important.

Zhang [13] adopt the two-stage optimization method with the goal of maximizing the profit of the charging station, and using the second-stage optimization to solve the problem of higher load peaks that the one-stage optimization leading to. Pan [14] formulated a two-layer optimal scheduling scheme from the perspective of transmission and distribution system to reduce network losses, and also to consider the situation of massive EVs connected grid. Zou [15] proposed a multi-objective optimization model based on micro-grid and used the variable charge and discharge rate optimization algorithm to increase the self-consumption of PV power by the orderly V2G strategy. Yang [15] used the decentralized control strategy to avoid from the impact of massive EVs connected grid. Ge [16] divided the parked EVs into two status, control the charge power of the controllable EVs. And use calculated the real-time using margin of EVs to control EVs charge load, so as to achieve the goal of load balance. The fuzzy control methods was used to control EVs to charge and discharge [17], and to avoid the problem of new peak load caused by a large number of EVs during the valley period, but it does not take into account the cost of battery-wear caused by users participating in V2G responses.

The above references do not consider the interests of the user side adequately. If the grid wants users to participate in the V2G strategy, it must meet user demands [18]. Research shows that V2G strategies which can achieve optimal control in theory are difficult for users to accept [19]. The control strategy that fully considers user demands has good practical control effects. Therefore, this paper took massive EVs in a certain area as the research object, designed a reasonable V2G management and control flow, established a targeted user evaluation model by analyzing different demands of users, and encouraged users to participate in V2G strategy by using people's profit-oriented thinking. This paper used fuzzy control methods to optimize the V2G behavior of EVs based on the demands of users and the power grid, and executed the control process through the V2G management system. For different scenarios, various number of EVs were used to simulate the model, and the effectiveness of the proposed model and V2G strategy in handling massive EVs to real-time V2G control is verified.

2 Orderly V2G Management Framework

At present, many areas in China adopt the time-of-use price policy. The purpose is to use low electricity prices to encourage people to make full use of the electricity during the night valley period [20]. However, it is difficult to ensure that users optimize their electricity consumption by simply guiding EV charging through time-of-use price. EV users also cannot actively participate in the V2G response to reduce the peak load of the grid. Massive EVs that are charged at night during periods of low electricity prices may also generate new peak loads. In this context, through the orderly V2G strategy, EVs are allowed to participate V2G response during the peak period during the day, and the charging load is evenly transferred during the valley period during the night.

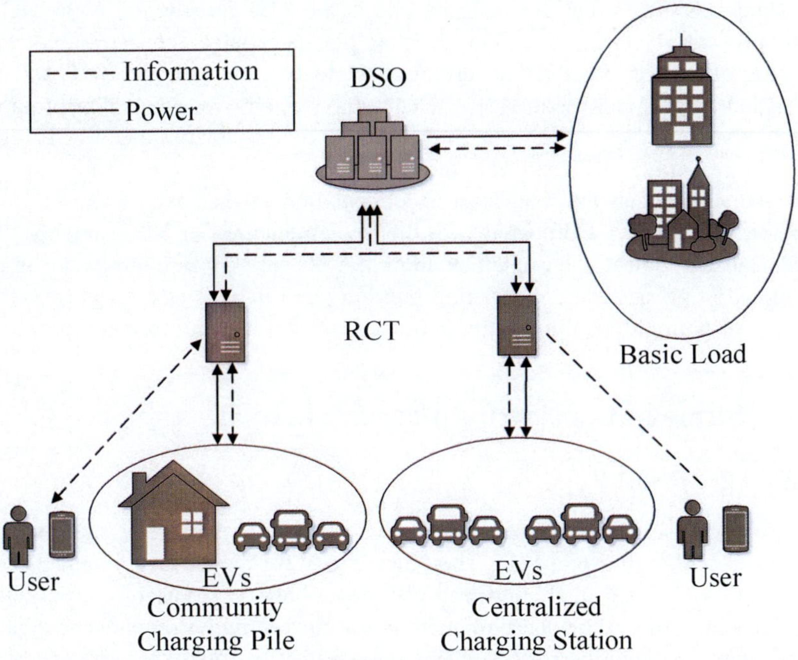

Fig. 1. Orderly V2G management system.

In order to implement the orderly V2G strategy and reduce calculation pressure, this paper proposed an orderly V2G management system, as shown in Fig. 1. It includes EVs, Regional Control Terminal (RCT), and Distribution Network Operator (DSO). Among them, RCT is a management organization of charging equipment in a specific area, and can manage all EVs in an independent charging station, parking lot or community. DSO is responsible for power supply tasks for conventional loads and RCT, and RCT is responsible for controlling the EVs in its jurisdiction to participate V2G response and orderly charging. When EVs participate V2G response, the direction of energy flow is opposite to that during charging. DSO and RCT as nodes of energy flow and information flow, which have the ability to independently process information and calculations. DSO sends conventional load information and electricity price information to RCT, and can also receive and send EV side boundary information. RCT obtains the boundary information on EV-side through the charging piles, exchanges information with the DSO, executes the orderly V2G strategy to obtain a V2G plan for each EV, and issues control commands to the charging pile according to the plan.

This paper assumes that the charging pile in the RCT jurisdiction is the DC smart charging pile with V2G function. The charging pile can control the charging power and V2G power of the connected EV according to the instruction from the RCT, and upload the EV side boundary information. Users can upload their own demand information on the interactive interface of the charging pile or on the smartphone. RCT regularly receives the real-time basic load from the DSO, and can also independently execute the V2G strategy based on the load forecast information in the case of short-term communication interruption. Compared with the implementation of V2G strategy through DSO centralized control, the implementation of V2G strategy through RCT can reduce communication pressure and computing pressure, and has the advantages of reducing the impact of communication failure and protecting information privacy.

3 V2G Strategy Considering User Evaluation

3.1 EV V2G Model

For the power grid, EV can be understood as an energy storage unit when it is connected to a charging pile [21]. Therefore, the V2G model of a single EV can be equivalent to the model of lithium iron phosphate battery. Constant current and constant voltage charging is the charging method for most lithium iron phosphate batteries. The charging voltage of the battery during the constant current phase remains basically unchanged, and this process can be approximated as constant power charging; while the duration of the constant voltage phase is relatively short and can be ignored. In summary, the charging and V2G process of EVs is regarded as a constant power process. Without considering the battery self-discharge and power fluctuations, the charging pile can directly control the charging and discharging power of all connected EVs by controlling the voltage. The V2G model of a single EV is as follows:

$$P_{\mathrm{t}} = \begin{cases} 0 & ,t \notin [\mathrm{t_s}, \mathrm{t_e}] \\ [P_{\min}, P_{\max}] & ,t \in [\mathrm{t_s}, \mathrm{t_e}] \end{cases} \tag{1}$$

$$S_{\mathrm{t}} = S_{\mathrm{in}} + \frac{\sum_{\mathrm{t=t_s}}^{t} \eta_{\mathrm{t}} \Delta t P_{\mathrm{t}}}{E} \tag{2}$$

Where t is the current period, $\mathrm{t_s}$ and $\mathrm{t_e}$ are the period when the EV is connected to the charging pile and the user is scheduled to leave, P_{t} is the power of the EV in t, $P_{\max}$ is the maximum power for V2G, $P_{\min}$ is the maximum power for charge, S_{t} is the SOC of the EV at the current period, and S_{in} is the SOC when the EV is just connected, Δt is the length of a period, E is the capacity of the battery, η_{t} is the V2G power efficiency of EV, the expression is as follows:

$$\eta_{\mathrm{t}} = \begin{cases} \eta_{\mathrm{P}} = 0.9 & P_{\mathrm{t}} > 0 \\ \eta_{\mathrm{V2G}} = 1.1 & P_{\mathrm{t}} < 0 \end{cases} \tag{3}$$

Where η_P is the charging efficiency, and η_{V2G} is the V2G efficiency.

3.2 User Comprehensive Evaluation Model Considering Battery-Wear Cost

The traditional user evaluation model only evaluates V2G results from the perspective of the overall EV users [22], and since the optimization goal is usually the highest overall evaluation mean, it may ignore the demands of a few users.

Otherwise, the traditional user evaluation model cannot evaluate the process of V2G response and charging. Therefore, this paper proposes the evaluation model based on a single user for the V2G process. This model also takes into account the battery-wear cost caused by the V2G process, and make it help to better reflect users' real evaluates.

For users, the demands of V2G process from person to person. The current research mainly divides the user's demands for the V2G process of EVs into two categories, which are the time demand and the cost demand.

On the one hand, users have the demand for the charging time of the EV, and they hope to charge as soon as possible to meet their future travel demands. On the other hand, users hope to participate in orderly V2G strategy to obtain lower electricity prices and V2G subsidies. These two demands together constitute the user's comprehensive evaluation of the V2G process, which in most cases are contradictory.

The model proposed in this paper models these two demands separately. First quantify the two to a unified dimension, and then proposes a user preference coefficient α to represent the user's preference for the two demands.

$$\theta = \alpha\theta_t + (1 - \alpha)\theta_c \tag{4}$$

Where θ is the value of the user evaluation after unified dimension, θ_t is the value of user time evaluation, θ_c is the value of the user cost evaluation. Users determine their preference for these two by determining the user preference coefficient α, and decide how to participate in the orderly V2G strategy.

Time Evaluation Model

For users who want to fully charge the EV as soon as possible or want to use the EV in advance, they expect the battery to reach the highest SOC in the shortest time.

Xu proposed the EV energy boundary model to describe the EV charging process [23]. Based on this, this paper proposed a new EV energy boundary model considering the impact of user participation in V2G response on SOC, as shown in Fig. 2.

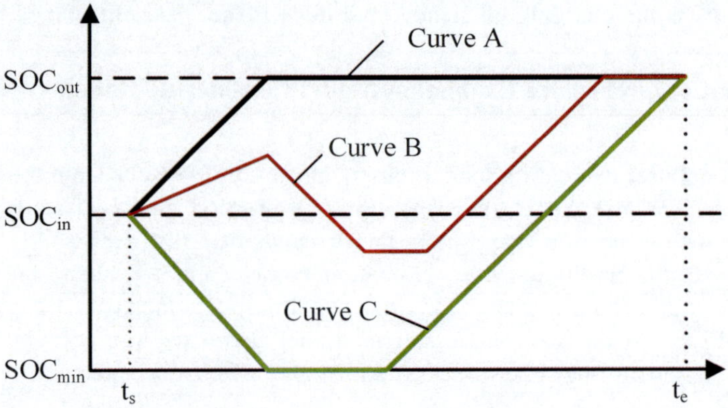

Fig. 2. Electric vehicle energy boundary model considering V2G.

Where SOC_{out} and SOC_{in} represent the SOC when the EV leaves and connected grid. SOC_{min} represents the V2G discharge limit SOC. The slope of the poly line represents the power. Curve A represents that the EV will be directly charged at the highest power after it is connected to the charging pile until it is fully charged and does not participate in the V2G response. Curve C represents that after the EV is connected, it participates in the V2G response immediately, and then delays charging until SOC is just full when leaving. Curve B represents a feasible V2G process. Obviously, charging according to curve A can make the EV fully charged the fastest. When the user needs to use the EV in advance, the EV has enough energy for the next trip. Curve C is completely opposite to the user's time demand, and it takes the longest time to fully charge. It can be known from the model that at a certain moment in the V2G process, the factors that affect user evaluation are the power and SOC.

The higher the SOC of the EV in a certain time, the higher the user time evaluation of the V2G process. Define the SOC parameter as ζ_t.

$$\zeta_t = \frac{S_{max,t} - S_t}{S_{max,t} - S_{min,t}} \tag{5}$$

Where $S_{max,t}$ and $S_{min,t}$ represent the SOC value of curve A and curve C at time t, respectively.

On the other hand, the higher the power, the faster the EV can be fully charged. Define charging power parameter as ζ_p.

$$\zeta_p = \frac{P_{max} - P_t}{P_{max} - P_{min}} \tag{6}$$

Comprehensively consider the impact of these two on θ_t, and get user time evaluation θ_t:

$$\theta_t = 1 - \left(\zeta_t \zeta_p\right)^2 \tag{7}$$

Cost Evaluation Model

This paper assumes that DSO achieves time-of-use prices by selling electricity to users at a fixed electricity price and providing users with charging subsidies during valley period.

For users who are more cared about the cost of fully charging the EV, since DSO sells electricity to users at different subsidies, the main factor affecting its expenses during the valley period is the charging subsidies at this period. During peak period, does DSO provide V2G subsidies and the battery-wear cost of V2G response will affect users' possible income. The power during these periods will affect whether these benefits can be obtained and how much. Therefore, users will want to charge as much as possible during periods of low electricity prices, and participate in V2G responses as much as possible during the periods when the DSO provides V2G subsidies.

This paper establishes a cost evaluation model based on whether or not to provide V2G subsidies. In some cases, this has the advantage of simplifying calculations. The cost evaluation for t can be expressed as:

$$\theta_c = \begin{cases} 1 - \frac{\lambda_{max} P_{max} - \lambda_t P_t}{\lambda_{max} P_{max}} & \omega = 0 \\ 1 - \frac{C_{max} + D_t - C_t}{C_{max} + D_{max}} & \omega \neq 0 \end{cases} \tag{8}$$

$$\begin{cases} \lambda_{max} = \max(\lambda_t) & t_s \leq t \leq t_e \\ C_{max} = P_{max} \varepsilon & \\ C_t = \begin{cases} 0 & P_t \leq 0 \\ P_t \varepsilon & P_t > 0 \end{cases} & \\ D_{max} = -P_{min} \omega & \\ D_t = -P_t(\omega - c) & P_t < 0 \end{cases} \tag{9}$$

Where λ_t is the charging subsidies of the connected EV at time t, C_{max} is the highest charging cost EV may pay at period t, ε is the standard electricity price provided by DSO, which is the highest electricity price, C_t is the charging cost of EV at period t when DSO provides V2G subsidy, D_{max} is the highest V2G subsidy that EV can earn at time t, ω is the V2G subsidy provided by DSO during peak period, D_t is the V2G subsidy earned by EV at period t considering battery-wear cost, c is the unit battery-wear cost of EV participating in V2G response at period t.

It is not difficult to see from the above model that when the EV participates in V2G response with the maximum power during the peak period and charges with the maximum power during the valley period, the user's cost evaluation is the highest. In the opposite case, the cost evaluation is the lowest.

Under normal circumstances, EV participating in V2G response is a win-win situation for users and DSO. Users get subsidies, and DSO reduces peak load. However, due to the high price of replacing the batteries used in EV, the cost of battery-wear for V2G response must be considered. Excessive battery-wear cost will reduce users' enthusiasm for participating in V2G response.

Liu analyzed many factors that affect battery-wear, including depth of discharge, battery temperature, and charge and discharge power [24]. This paper ignored the impact of power and temperature changes on battery-wear, only considered the impact of the depth of discharge, and models of the battery-wear cost.

The relationship between the discharge depth d and the battery SOC value is:

$$d = SOC_1 - SOC_2 \tag{10}$$

Where SOC_1 is the SOC when start participate in V2G response, SOC_2 is the current SOC.

The relationship between the battery's total cycle life and its depth of discharge can be summarized as: the greater the battery's depth of discharge, the less the cycle life. The mathematical expression obtained after fitting according to the experimental results in Reference [25] is as follows:

$$L = ad^b \tag{11}$$

Where L is the number of battery cycle, a and b are fitting coefficients, and the values of a and b are 2151 and −2.301.

The total discharge amount R can be expressed as the product of the number of battery cycles, the discharge depth, and the battery capacity. According to the fitting result, the total discharge capacity of the battery will decrease as the depth of discharge increases. Since the replacement cost of the battery is fixed, the larger the total discharge amount at a certain depth of discharge, the lower the unit battery-wear cost. The formula is as follows:

$$R = LEd \tag{12}$$

$$c = \frac{C_b}{R} \tag{13}$$

Where C_b is the replacement cost of the battery.

After modeling the battery-wear cost, it is possible to clarify the hidden expenditures of users participating in the V2G response, so as to more accurately model the user cost evaluation.

3.3 Fuzzy Control Method

After clarifying the user evaluation, it is feasible to use this parameter to formulate the orderly V2G plan. When massive EVs are connected, focusing the implementation of the strategy on DSO will undoubtedly cause huge communication and computing pressure, and implementing the strategy through distributed RCT is a way to solve this problem. However, RCT may not have sufficient computing power to implement overly complex optimal control strategies in terms of cost, and distributed RCT may encounter problems such as communication interruption.

A well-designed fuzzy controller can achieve control results that satisfy both users and the grid in complex situations even without predicting the behavior of massive EVs. The implementation of fuzzy control on RCT can greatly reduce the amount of calculation and communication [26], so that RCT can implement strategies well in the face of massive EVs.

Since the goal of the strategy proposed in this paper is to achieve a win-win situation for users and the grid, in order to meet the demands of both, the two inputs of the fuzzy controller are the user's comprehensive evaluation U and the current load state of the grid G.

Fuzzification

Set the basic domain of U to $[0, 1]$, and set its fuzzy domain to [2, −1, 0, 1, 2], the corresponding fuzzy subset is (terrible, bad, ok, well, good). The fuzzy subset good and terrible represent the highest evaluation and the lowest evaluation respectively.

Set the basic domain of G to (−1000, 1000) kW, and set its fuzzy domain to (−3, −2, −1, 0, 1, 2, 3), the corresponding fuzzy subset is (NB, NM, NS, ZE, PS, PM, PB). The fuzzy subset NB represents that the current load state of the power grid is close to the maximum value, that is, the load peak state.

In this paper the charging power is divided into 3 levels, the V2G power is divided into 2 levels, plus the level of zero power, there are 6 levels. The levels of power change is the output ΔP of fuzzy control. Set the basic domain of ΔP to $[P_{\min}, \frac{2}{3}P_{\max}]$, and set the fuzzy domain to $[-2, 2]$, the corresponding fuzzy subset is (SR, R, K, I, SI). The fuzzy subset SR represents significantly reduce power, subset K represents keep the current power, and subset SI represents significantly increase power.

According to the principles of consumer psychology, the user's response to the stimulus reaction consists of the insensitive zone (equivalent to the dead zone), the normal response area (equivalent to the linear area) and the response limit period (equivalent to the saturation zone). In order to simplify the problem, this response process is often expressed as a piecewise linear function. Therefore, this paper uses the trapezoid membership function composed of multiple piecewise linear functions to describe the membership of U (Fig. 3).

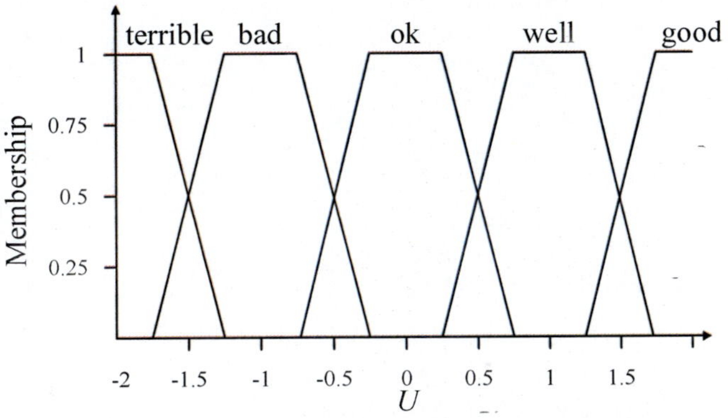

Fig. 3. Membership function for comprehensive user evaluation.

Since the grid does not have the above stimulus response to load pressure, this paper uses the triangular membership function to describe the memberships of G and ΔP. The triangle membership function is relatively simple, so its explanation is omitted, so the description is omitted.

Fuzzy Rules
The designed fuzzy control rules are shown in Table 1.

Table 1. Fuzzy control rules considering user and power grid demands.

U	G						
	NB	NM	NS	ZE	PS	PM	PB
Terrible	SR	SR	SR	K	I	SI	SI
Bad	SR	SR	SR	K	I	I	I
Ok	R	R	K	K	K	I	I
Well	R	R	K	K	K	K	I
Good	K	K	K	K	K	K	K

The reasoning result of fuzzy rules is a fuzzy quantity, which cannot be directly used as a control quantity. It must also be deblurred to output the final control quantity.

In this paper, the area center method is used to defuzzify the output fuzzy set.

3.4 Strategy Flow and Restrictions

Due to the randomness of the connect time of electric vehicles, in this paper, RCT performs unified control on the connected EVs at a fixed time. Divide one day into 96 periods, 15 min each period, update the grid side information at the beginning of each period, and calculate the strategy at that moment. If the connect time of the EV is in the middle of the period, the EV needs to wait and be controlled together with other electric vehicles in the next period. Define the serial number of the above period as T.

The flow of orderly V2G strategy proposed in this paper is shown in the Fig. 4.

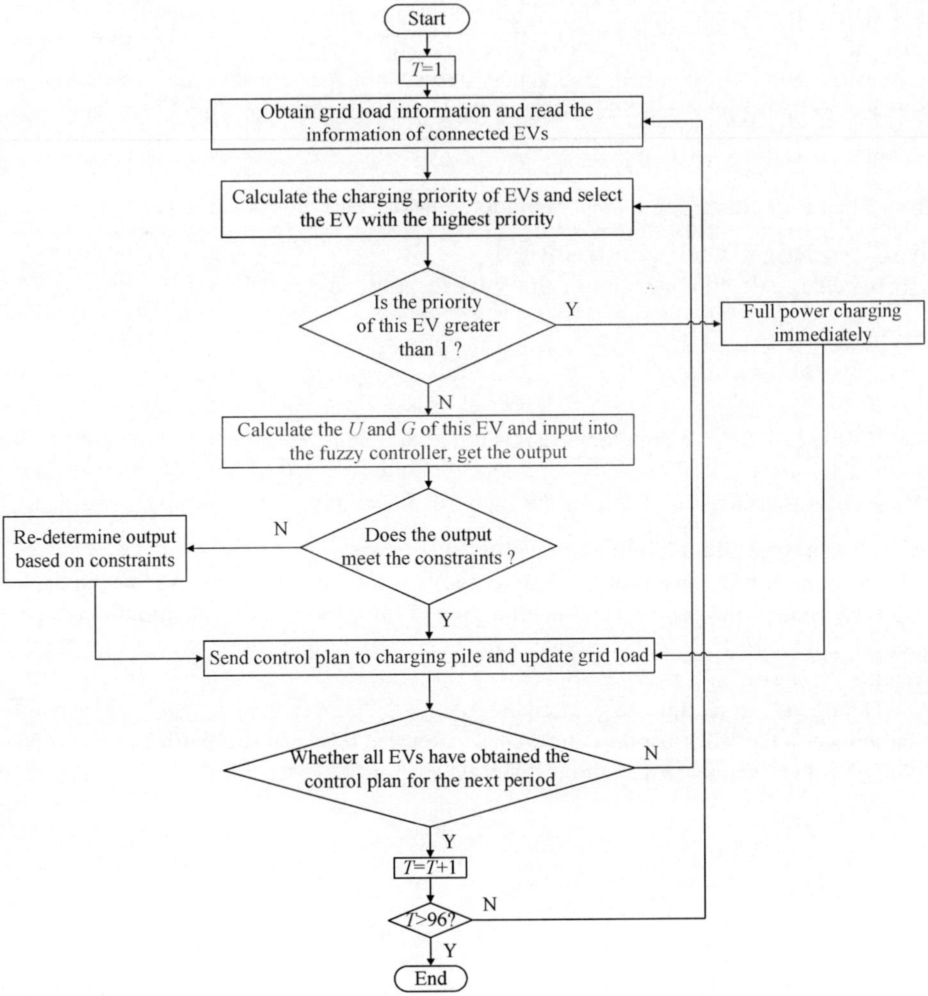

Fig. 4. Orderly V2G strategy flow chart.

EV Charging Priority

Define the charging priority of the EV as F, and the EV with the highest priority as the current electric car that needs to be charged the most. The expression is as follows:

$$T_c = \frac{(S_e - S_t)E}{\eta_p P_{max}} \tag{15}$$

$$F = \frac{T_c}{\Delta t(t_e - t)} \tag{16}$$

Where T_c is the minimum time required to fully charge the EV, and S_e is the SOC that the EV needs to achieve when it leaves, and is generally set to 1. EV with the highest F need to be charged first. When the value of F is greater than 1, it means that the EV must be charged immediately and cannot participate in the orderly V2G strategy.

In order to ensure that the results of the V2G control plan meet the actual needs, this paper has formulated some constraints to achieve better control results.

V2G Discharge Depth Constraint

The discharge depth constraint of the V2G response are as follows:

$$D \leq \delta \tag{17}$$

Where δ is the discharge depth constraint parameter. The higher the parameter, the more V2G subsidies DSO needs to provide, and the longer the EV can participate in the V2G response. Generally, δ is in the range of 0.2 to 0.4.

V2G Discharge Final Time Constraint

The purpose of this constraint is to prevent EVs from participating in V2G responses in the early time of the peak period but charging at full power in the later time of the peak period. This situation will seriously affect the effectiveness of the V2G strategy to reduce peak load.

Define the final time to participate in the V2G response as $\mathrm{t_{lim}}$, EV can only participate in the V2G response before $\mathrm{t_{lim}}$, after this time can still participate in orderly charging, but cannot participate in the V2G response.

$$\mathrm{t_{lim}} = \mathrm{t_e}\Delta t - T_{\mathrm{lim}} \tag{18}$$

$$T_{\mathrm{lim}} = -\frac{\delta E}{\eta_{\mathrm{V2G}} P_{\mathrm{min}}} + \frac{(1 - S_{\mathrm{t}} + \delta)E}{\eta_{\mathrm{c}} P_{\mathrm{max}}} + \beta \tag{19}$$

Where T_{lim} is the V2G response limit time for EVs. This period is composed of three parts: the time required to participate in the V2G response with the maximum power until the discharge depth is limited, the time required to fully charge the maximum power, and the limit period coefficient β.

The larger β is, the lower the possibility of EVs charging at full power during peak period. However, if β is too large, it will lead to too few EVs participating in V2G response. Usually, the value of β is between 0.5 and 2, which is related to the duration of the peak load in this area.

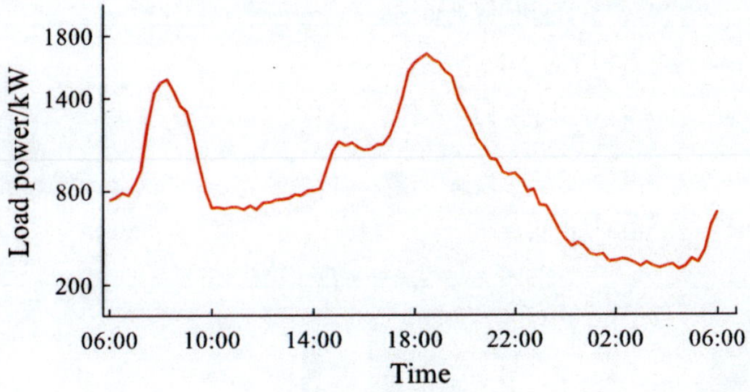

Fig. 5. Grid basic load curve.

4 Simulation and Results Analysis

4.1 Simulation Parameter Settings

Figure 5 shows the basic load curve under a distribution transformer in a certain area. This paper takes this area as an example for simulation analysis. Unless otherwise specified, in the following simulation, the day is divided into 96 periods, the first period starts at 6:00 am, and the time span is 6:00 am to 6:00 am the next day. There is 1 DSO in the simulation, which governs 3 RCTs. Within 1 day of the simulation, a total of 300 EVs are connected to the grid. The user's power demand is calculated from the daily mileage of EV. The probability density functions of the arrival and departure times of the electric vehicle and the daily mileage are based on the analysis results in [27],and made some adjustments according to the actual situation. The daily distribution of the arrival and departure times of EVs is shown in Fig. 6. The evaluation preference coefficient α of the EV users whose parking time is less than and greater than 3 h obey the uniform distribution of $(0, 0.3)$ and $(0.3, 1)$, respectively. The battery capacity of EVs is 32 kW.h, the maximum charging power is 9 kW, the maximum V2G power is −6 kW, the battery replacement price is 78,000 yuan, the discharge depth constraint is 0.4, and the limit time constrain is 0.5. DSO provides V2G subsidies of 1.02 yuan/kW. h for EVs participating in V2G response when the basic load power is higher than 1150 kW. The charging electricity price data for that day is shown in Table 2. The simulation program is implemented in the Matlab environment of a computer with Windows 10, Intel (R) CoreTM i5CPU@3.6 GHz, and 16 GB memory.

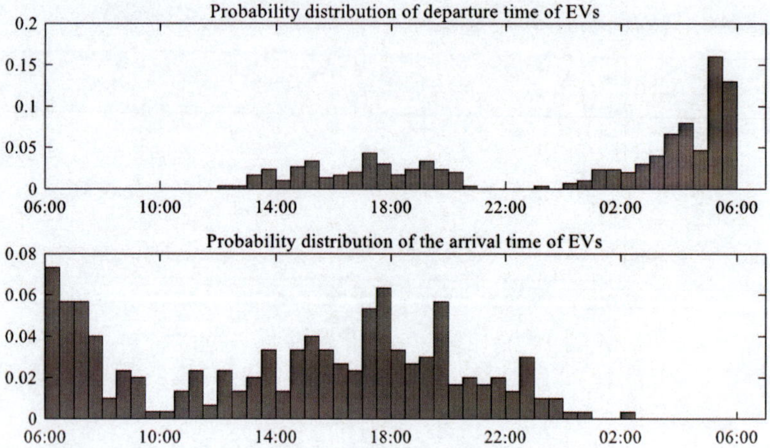

Fig. 6. Driven pattern of electric vehicles.

Table 2. Electric price in each period.

Period	Electric price Yuan/kWh	Period	Electric price Yuan/kWh
07:00–10:00	1.30	18:00–22:00	1.30
10:00–14:00	0.92	22:00–00:00	0.92
14:00–18:00	1.02	00:00–07:00	0.42

Analysis of Simulation Results

Figure 7 is the load curve of disorderly charging and the control strategy proposed in this paper.

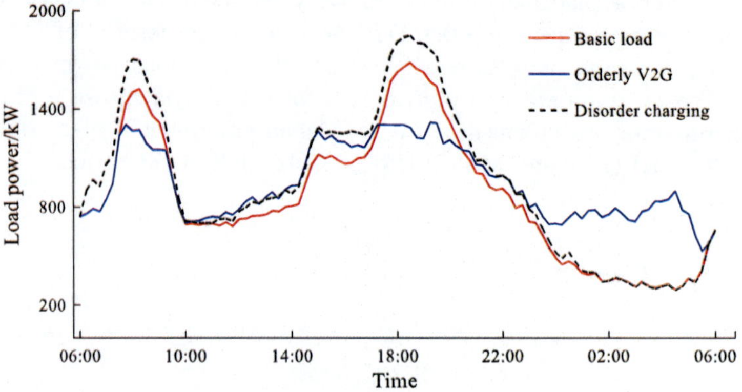

Fig. 7. Grid load curve in two scenarios

It can be seen that due to the coincidence of the time when the EVs are disorderly charged and the peak time of the basic load, disorder charging will lead to serious peak overlap and seriously affect the stable operation of the grid. The V2G strategy proposed in this paper takes into account the comprehensive impact of user evaluation and grid load, and adopts different charging and discharging plans for users with different demands. This strategy reduce the charging situation of EVs during the peak period, and encourage users to participate in V2G response to reduce the peak load. Transfer users to charge during the valley period, make full use of the surplus energy in the valley period, achieve the effect of load transfer, and improve the reliability of the system operation.

Next, the comparative analysis is made in terms of load conditions, user evaluation conditions, the scale of access EVs and the running time of control strategies.

Table 3. Comparison of grid load under different scenarios.

Mode	Load fluctuation/kW^2	Peak load/kW	Load peak-valley difference/kW
Disorderly charging	455.1	1850	1560
Orderly V2G	220.8	1314	784.3
Basic load	394	1679	1389

Table 3 shows the effect of disorder charging and orderly V2G strategy on the load status. It can be seen that after applying orderly V2G strategy, the load situation of the power grid is significantly improved compared to disorderly charging, and the peak load, grid load fluctuation rate and load peak-valley difference are greatly reduced.

Table 4. Comparison of user expenses under different scenarios.

Mode	Charging expenses/yuan	V2G subsidies received/yuan	Battery-wear cost/yuan	Total user expenses/yuan	Average user evaluations
Disorder charging	2419	\	\	2419	0.3918
Consider load	2900.2	1610	222	1512.2	0.4623
Orderly V2G	1954.5	928.7	129.7	1155.5	0.6011

In order to determine the influence of user evaluation on the control effect, this paper simulates the mode that only considers the load of the power grid. In this mode, the user's evaluation is not used as the input of the fuzzy controller. EVs participates in the V2G response when the load is high, and charge when the load is low, and does not have the limit of depth of discharge. This mode is the same as the orderly V2G mode except for the above changes. It can be seen from Table 4 that orderly V2G mode encourages users to participate in V2G response and obtain V2G subsidies during peak periods, and arrange users to charge at valley periods of electricity prices to reduce charging expenses. Compared with the orderly V2G mode, the EVs in the consider load mode takes participate in more V2G response, resulting in increased battery-wear cost. EVs need to charge more due to their participation in additional V2G discharge. Because EVs need to be fully charged before leaving, they may be forced to charge during periods of high electricity prices, which leads to higher charging cost. And too late charging time and too low SOC further reduce user evaluation. As can be seen that the charging expenses of orderly V2G mode is much lower than that of the disorderly charging and consider load mode, the average user evaluations of orderly V2G mode is higher than the disorderly charging and consider load mode.

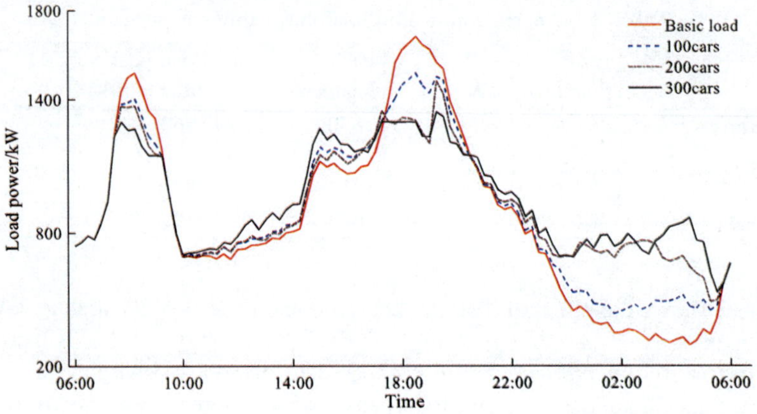

Fig. 8. Grid load curve in different quantity

Table 5. Comparison of grid load and calculation time for different numbers of EVs

Mode	Load fluctuation/kW^2	Peak load/kW	Load peak-valley difference/kW	Strategy calculation time/s	Average calculation time/s
100 cars	328.8	1517	1085	34.82	0.120
200 cars	249.8	1482	994.7	73.06	0.253
300 cars	220.8	1314	784	120.02	0.416

It can be seen from Table 5 and Fig. 8 that orderly V2G strategy can still play a role in suppressing load fluctuations when different numbers of EVs are connected. The larger the number of EVs, the better the effect. It is worth mentioning that the calculation speed of this strategy is relatively fast, and the effect of increasing the number of

EVs on the calculation time of the strategy is linear. Therefore, the calculation time required for each period can be reduced by increasing the number of RCTs in the area, thereby ensuring the rapid execution of the strategy. It proves that this strategy can effectively face the situation of massive and different number of EVs connect.

5 Conclusion

Focusing on V2G strategies that meet the demands of both users and the grid, this paper proposed the user evaluation model consider battery-wear cost, and used fuzzy control to schedule massive EVs to participate in V2G responses and orderly charge.

Simulation results show that this strategy can shift the charging load of massive EVs and guide EVs to participate in V2G response under the premise of user profit, which has certain practical significance. This paper did not do further research on the fuzzy control method itself. The focus of the follow-up work can be on optimizing the fuzzy control method to make the fuzzy control method better meet the actual demands of EVs.

References

1. Zhao, H.J., Wen, F.S., Yang, A., Liu, J.B.: Impacts of electric vehicles on power systems as well as the associated dispatching and control problem. Autom. Electr. Power Syst. **35**(14), 2–10+29 (2011)
2. Chen, L.J., Qing, M., Gu, S.P., Qian, K.J., Xu, X.H.: Battery loss based optimal dispatching strategy of electric bus participating in vehicle-to-grid. Autom. Electr. Power Syst., 1–11
3. Chen, Z., Liu, Y., Chen, X., Liu, T., Xin, Q.: Charging and discharging dispatching strategy for electric vehicles considering characteristics of mobile energy storage. Autom. Electr. Power Syst. **44**(2), 77–88 (2020)
4. Clement-Nyns, K., Haesen, E., Driesen, J.: The impact of charging plug-in hybrid electric vehicles on a residential distribution grid. IEEE Trans. Power Syst. **25**(1), 371–380 (2010)
5. Luo, Z.W., Hu, Z.C., Song, Y.H., Xu, Z.W., Yang, Y.X., Liu, H.: Coordinated charging and discharging of large-scale plug-in electric vehicles with cost and capacity benefit analysis. Autom. Electr. Power Syst. **36**(10), 19–26 (2012)
6. Chen, L.D., Zhang, R., Figueiredo, A.: Overview of charging and discharging load forcasting for electric vehicles. Autom. Electr. Power Syst. **43**(10), 177–197 (2019)
7. Tang, C.H., Zhang, F., Zhang, N., Qu, H.Y., Ma, L.: Day-ahead economic dispatch of power system considering renewable power uncertainty and demand response. Autom. Electr. Power Syst. **43**(15), 18–25+63+26–28 (2019)
8. Zhang, J., Jorgenson, J., Markel, T., Walkowicz, K.: Value to the grid from managed charging based on California's high renewables study. IEEE Trans. Power Syst. **34**(2), 831–840 (2019)
9. Lou, X.S., Ma, G., Guo, C.X.: System and framework design of risk coordination control for whole operation process of power system. Autom. Electr. Power Syst. **44**(5), 161–170 (2020)
10. Zhan, K.Q., Song, Y.H., Hu, Z.C., Xu, Z.W., Jia, L.: Coordination of electric vehicle charging to minimize active power losses. Proc. CSEE **32**(31), 11–18+213 (2012)

11. Xiang, D., Hu, Z.C., Song, Y.H., Ding, H.J.: Business model and day-ahead dispatch strategy to reduce wind power curtailment through vehicle-to-grid. Proc. CSEE **35**(24), 6293–6303 (2015)
12. Wen, G.Q., Huang, T.F., Zhang, Y., Xie, L.Y., Qi, J.: V2G strategy for energy dispatch of island microgrid with EBBG. Electr. Power Autom. Equip. **36**(10), 31–37 (2016)
13. Zhang, L., Yan, Z., Feng, D.H., Xu, S.L., Li, N.H., Jin, L.: Two-stage optimization model based coordinated charging for EV charging station. Power Syst. Technol. **38**(4), 967–973 (2014)
14. Pan, Z.N., Zhang, X.S., Yu, T., Wang, D.: Hierarchical real-time optimized dispatching for large-scale clusters of electric vehicles. Autom. Electr. Power Syst. **41**(16), 96–104 (2017)
15. Zou, T.P., Sun, W.: Electric vehicle-to-grid technology based on microgrid. Autom. Electr. Power Syst. **42**(3), 98–104+117 (2018)
16. Shaoyun, G., Henghui, L., Hong, L.: Timing response capability model and control method of electric vehicle. Autom. Electr. Power Syst. **40**(3), 33–38 (2016)
17. Zhang, Y.B., Liu, Q.H., Hong, C.W., Tang, G.Y.: Charging and discharging dispatch strategy of regional V2G based on fuzzy control. Electr. Power Autom. Equip. **39**(7), 147–153 (2019)
18. Wang, J., Wu, K.H., Liu, Z.Z., Wu, K.Z., Sun, W.: Impact of electric vehicle charging on distribution network load and coordinated control. Electr. Power Autom. Equip. **33**(8), 47–52 (2013)
19. Cui, J.D., Luo, W.D., Zou, N.C.: Research on pricing model and strategy of electric vehicle charging and discharging based on multi view. Proc. CSEE **38**(15), 4438–4450+4644 (2018)
20. Eldjalil, C.D.A., Said, D., Khoukhi, L., Mouftah, H.T.: A novel pricing policy for G2V and V2G services. In: Proceedings of the 2017 14th IEEE Annual Consumer Communications & Networking Conference (CCNC), 8–11 January 2017 (2017)
21. Hu, S.B., Sun, H., Sun, Q.Z.: Risk dispatch of power system considering second-life batteries as flexible loads. Autom. Electr. Power Syst. **43**(10), 78–86 (2019)
22. Pan, Z.N., Yu, T., Wang, K.Y.: Decentralized coordinated dispatch for real-time optimization of massive electric vehicles considering various interests. Proc. CSEE **39** (12), 3528–3541 (2019)
23. Xu, Z., Hu, Z., Song, Y., Wang, J.: Risk-averse optimal bidding strategy for demand-side resource aggregators in day-ahead electricity markets under uncertainty. IEEE Trans. Smart Grid **8**(1), 96–105 (2017)
24. Han, S., Han, S., Aki, H.: A practical battery wear model for electric vehicle charging applications. Appl. Energy **113** (2014)
25. Liu, L., Liu, T., Zhang, T., Liu, J.: Orderly charging and discharging strategy optimization for electric vehicles considering dynamic battery-wear model. Autom. Electr. Power Syst. **40** (5), 83–90 (2016)
26. Singh, M., Kumar, P., Kar, I.: Implementation of vehicle to grid infrastructure using fuzzy logic controller. IEEE Trans. Smart Grid **3**(1), 565–577 (2012)
27. Tian, L.T., Shi, S.L., Jia, Z.: A statistical model for charging power demand of electric vehicles. Power Syst. Technol. **34**(11), 126–130 (2010)

An Event-Detection Algorithm for Non-intrusive Load Monitoring of Residential Appliances

Kangli Zhao[1], Renmin Zhang[2], Youbing Zhang[1(✉)], Qianyi Cai[2], and Junpeng Shu[1]

[1] Zhejiang University of Technology, Hangzhou, China
youbingzhang@zjut.edu.cn
[2] State Grid Wenzhou Electric Power Supply Company, Wenzhou, China

Abstract. Nowadays, the issue of promoting supply-demand balance and energy conservation is urgent by increasingly exhausted traditional fossil energy sources and high penetration of renewable energy resources. Non-intrusive load monitoring (NILM) is one of the key techniques toward energy efficiency and conservation, which provides the guidance for operation scheduling of the power system and demand response strategies of the smart grid. A multi-feature non-intrusive load monitoring method based on event detection algorithm is proposed in this paper to profile residential consumer behavior, which focused on limited applicability and low recognition accuracy of existing NILM techniques. The event detection is conducted to locate the switching of appliances based on the phenomenon of a sliding window that tracks the statistical features of the acquired aggregated load data. To make sure that the model is high-precision, the event detection measurement indexes are applied to obtain the best event detector. On this basis, in order to achieve the comprehensive information of appliance power consumption, the dynamic adaptive particle swarm optimization (DAPSO) is utilized to identify the load type and working state based on the load feature database built by affinity propagation (AP) clustering which combines with the rich electrical characteristics of appliances. A case study using the open database of REDD dataset is applied, and the results validate the effectiveness of the method.

Keywords: Non-intrusive load monitoring · Event detection · Affinity propagation clustering · Dynamic adaptive particle swarm optimization · Load identification

1 Introduction

With the growth of electricity demand and the improvement of living standard, the construction of a "clean, low-carbon, safe and efficient" energy system puts forward urgent demands for the energy production and consumption revolution [1]. In addition, the dual pressure of energy crisis and environmental pollution makes the traditional power system gradually transform to "Grid 2.0" [2]. Smart Grid integrates power plants to end users into a fully automated power transmission network, which achieves a high

Y. Xue et al. (Eds.): PMF 2020, LNEE 718, pp. 781–800, 2021.
https://doi.org/10.1007/978-981-15-9746-6_59

degree of merge of "power flow, information flow, business flow". Intelligent power utilization is an important segment of smart grid and its key technology is Advanced Measurement System (AMI), which primary function is to realize load monitoring [3–5].

NILM is a significant way for big data analysis of end-user energy consumption information. It refers to the installation of intelligent monitoring devices at the user entrance to obtain total power consumption information, according to the operating characteristics of electrical equipment and state data to estimate the power consumptions of the states as well as the usage patterns [6–8]. Personalized energy efficiency services based on the NILM system are considered to be a cost-effective technology for monitoring the details of load power. In the context of energy internet, it can promote two-way interaction of intelligent power utilization as well as dynamic demand response management, optimize the allocation of energy utilization strategy [9–11].

Since Hart pioneered the field of NILM in 1992, many domestic and foreign experts have committed to improving the accuracy and applicability of NILM technology. A spectrum analysis of transient characteristics of voltage waveform is proposed in [12]. S-transform is made for transient characteristics, and then principal component analysis method is used to select feature information to complete identification in [13]. The transient features are easy to handle overlapping events, but the waveform is complex, and the performance requirements of the measurement device are high.

At present, the method based on the characteristics of steady-state signals is gradually extended, which has a bright development future. High-frequency sampling is rich in steady-state information, but requires higher-cost hardware. Therefore, considering the cost problem, an event detection based on probability model method is proposed. In [14], the load identification method based on DTW algorithm and steady-state current waveform is proposed. However, the complexity of the probability model will increase exponentially as the number of appliances increases. In [15], the identification of load state is realized through the extension of Viterbi algorithm in factor hidden Markov problem. In [16], an event detection algorithm in view of sliding window is designed to train multiple BP classifiers, but the selection of event detection parameters is highly subjective and weak in applicability.

Motivated by the above facts, this paper proposes an improved event detection method based on generalized likelihood ratio (GLRT). The power sequence of load is detected by sliding window to realize the time node identification of state change, and the complex process of determining threshold is avoided by using the voting window to establish the event detector. Meanwhile, the event detection metrics is introduced to determine the relevant parameters of the best event detector, thereby increasing accuracy and reducing the computational complexity. Based on cluster analysis of load power, the proposed methodology is extended to apply to the load characteristics with high sample rate, which can realize accurate identification of low-power or similar power appliances. In addition, the multi-objective optimization algorithm is used to identify the load state,and the adaptive weight selection strategy enhances the performance of the algorithm, reduces the computational burden so as to improves the feasibility of the model.

2 Physical Framework of NILM System

The physical architecture of NILM system proposed in this section is shown in Fig. 1.

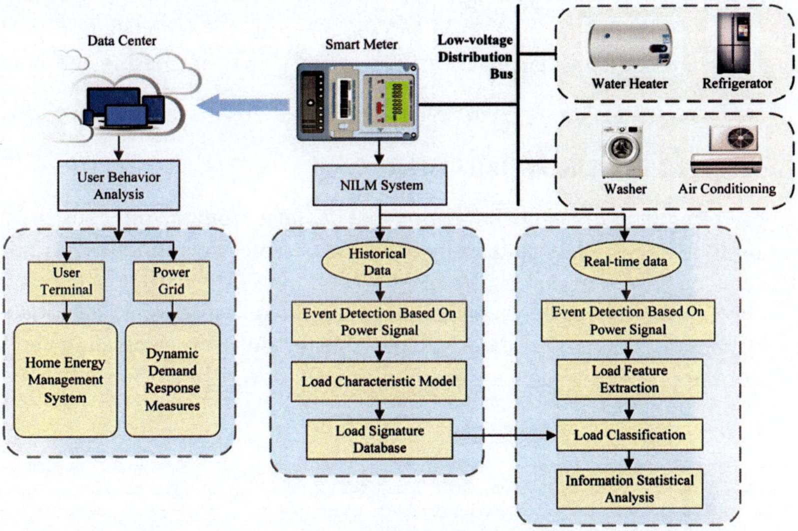

Fig. 1. Physical architecture diagram of NILM system

A typical NILM system consists of event detection, feature extraction as well as device recognition. The precondition of NILM system is to preprocess the load signal (such as denoising, active power calculation, etc.), which solves the uncertainty caused by data measurement. Then, the event detection module detects the state transition of the equipment by analyzing the power change, and determines the electrical switching event. After separating the independent load waveform from the event data, and builds feature database by training state characteristics of event points. Finally, the load characteristics of the test data are compared with the existing feature database to achieve the purpose of appliances identification.

The NILM system can provide detailed feedback of energy consumption on the user side and create itemized energy bills, which is of great significance to users and power companies. For users, they can grasp their own energy demand to adjust energy usage schemes, replace high-energy appliances, etc. [17]. For the power company, it can accurately understand the composition of load, and formulate time-sharing electricity price according to the information of electricity consumption which can provide the basis for distributed flexible energy storage, the electricity price oriented and incentive oriented of demand response. In addition, the power company has the opportunity to tap the potential of user, and rationally allocate elastic loads so as to realize energy saving, consumption reduction and energy cost reduction [18].

3 Event Detection Model

NILM method can be divided into event-based and nonevent-based. The event-based method relies on different edge detection algorithms to capture load features. The calculation efficiency is high since only inferences are performed on the detected events [19]. The nonevent-based method considers various statistical models to directly separate aggregated load data that make up the entire signal, increasing the amount of sample calculation.

3.1 Generalized Likelihood Ratio Detector

The event is a fragment of a signal mutation that deviates from the previous steady state and lasts until the next steady state hasbeen reached. Here, the turning on and off of an appliance are considered as distinct events that have their own starting and ending time, respectively [20]. In this paper, the generalized likelihood ratio event detection method is adopted, which obtains the statistical decision amount by calculating the natural logarithm of the probability density function ratio in two adjacent windows, as shown in Fig. 2.

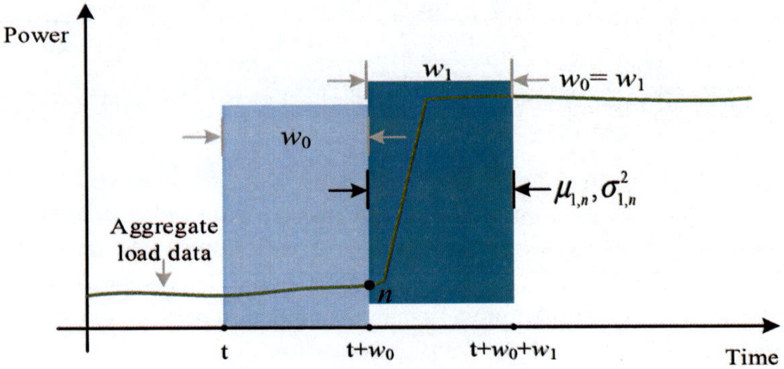

Fig. 2. Event detection diagram

The input power sequence signal is $P[n]$, $n = 1, 2, \cdots, N$, the lengths of the front window and the rear window are ω_0 and ω_1. The starting and ending points of the switching process are judged by determining whether the signal $P[n]$ is located in the front window or the rear window, and define the log-likelihood ratio:

$$S[n] = \ln(\frac{N(P[n])|\mu_{1,n}, \sigma_{1,n}^2)}{N(P[n])|\mu_{0,n}, \sigma_{0,n}^2)}) = \ln(\frac{\sigma_{0,n}}{\sigma_{1,n}}) + \frac{(P[n] - \mu_{0,n})^2}{2 \cdot \sigma_{0,n}^2} - \frac{(P[n] - \mu_{1,n})^2}{2 \cdot \sigma_{1,n}^2} \quad (1)$$

Where, $N(\diamondsuit)|\alpha, \beta$ is a normal distribution with α as mean and β as variance, $\mu_{0,n}, \sigma_{0,n}^2, \mu_{1,n}, \sigma_{1,n}^2$ represents the mean and variance of the front window and back window of the sampling sequence, respectively.

In the case of $n < \omega_0 + 1$ and $n > N - \omega_1$, the window mean and variance cannot be calculated, and the data set is very large, so the part of the window sampling point can be ignored. The full expression is as follows:

$$S[n] = \begin{cases} 0, \ n < \omega_0 + 1 \\ \ln(\frac{\sigma_{0,n}}{\sigma_{1,n}}) + \frac{(P[n] - \mu_{0,n})^2}{2 \cdot \sigma_{0,n}^2} - \frac{(P[n] - \mu_{1,n})^2}{2 \cdot \sigma_{1,n}^2}, \ \omega_0 + 1 \le n \le N - \omega_1 \\ 0, \ n > N - \omega_1 \end{cases} \quad (2)$$

Considering that transient oscillations generally occur before and after switching, a modified log-likelihood ratio $l[n]$ is defined.

$$l[n] = \begin{cases} S[n], \ |\mu_{1,n} - \mu_{0,n}| \ge \theta_p \\ 0, \ \text{others} \end{cases} \quad (3)$$

The spike caused by the surge is easy to incorrectly identify as an event, so the method of introducing the voting window. Slide a length ω_v of voting window on $l[n]$ to traverse all event nodes $e = [e_1, \cdots, e_M]$. Search the node with the largest $|l[n]|$ in the window and obtain the cumulative number of votes superimposed on the sequence $V[n]$, so that the sequence contains the total number of votes obtained by M sampling points. Introduce a vote threshold θ_v, the output event sequence is expressed when $V[n] \ge \theta_v$ is satisfied:

$$\hat{e} = [\hat{e}_1, \cdots, \hat{e}_n, \cdots, \hat{e}_m] \quad (4)$$

3.2 Event Detector Metrics

The main challenge for detection methods based on log-likelihood ratio is to determine the expected threshold θ_p. The setting values for different power sequences vary greatly, and there are still four variables: front window ω_0, rear window ω_1, voting window ω_v, and the number of votes threshold θ_v, which affect the accuracy of event detection. n order to measure the performance of event detector, the measurement indexes of true positive rate and true positive percentage are introduced [21].

True positive rate evaluates the effect of event detector on the working characteristics of subjects, which is usually shown in the ROC curve. The indicator can find the detector with the best trade-off between true positive rate (*TPR*) and false positive rate (*FPR*), which can be defined as the following formula based on Fig. 3:

$$TPR = \frac{N_{\mathrm{TP}}}{N_{\mathrm{TP}} + N_{\mathrm{FN}}} \in [0, 1] \quad (5)$$

$$FPR = \frac{N_{\mathrm{FP}}}{N_{\mathrm{FP}} + N_{\mathrm{TN}}} \in [0, 1] \quad (6)$$

TPR reflects the sensitivity of the detector, *FPR* shows the abnormality of the detector. In any scenario, the change value of *TPR* and *FPR* are directly proportional, so when the *TPR* indicator becomes better, the *FPR* indicator will become worse. Therefore, according to the specific ROC curve, the best detector is defined as $\hat{\psi}_{\text{Rate}}$:

$$\begin{aligned}\hat{\psi}_{\text{Rate}} &= \arg\min\|(0,1)-(TPR,FPR)\|^2 \\ &= \arg\min FPR^2 + TPR^2 - 2TPR + 1\end{aligned} \tag{7}$$

In the energy consumption data of household appliances, the event distribution is sparse. In this case, R_{TP} approaches zero. From Eq. (7):

$$\hat{\psi}_{\text{Rate}} = \arg\min f(TPR) \tag{8}$$

$$f(TPR) = TPR^2 - 2\cdot TPR + 1 \tag{9}$$

The best event detector is the one with the highest true positive rate that $\hat{\psi}_{\text{Rate}}$ tends to 1.

True positive percentage compares the ratio of the number of *TP* and the number of *FP* with the total number of actual events. Define the percentage of true positives (*TPP*) and the percentage of false positives (*FPP*):

$$TPP = \frac{N_{\text{TP}}}{E} \in [0,1] \tag{10}$$

$$FPP = \frac{N_{\text{FP}}}{E} \in [0,1] \tag{11}$$

Where, *E* is the total number of actual events.

Similar to the above metrics, the best detector is the closest point (0,1).

$$\hat{\psi}_{\text{Perc}} = \arg\min_{\psi\in\Psi}\|(0,1)-(FPP,TPP)\|^2 \tag{12}$$

The model under the true positive percentage is equivalent to an optimized model, which only needs to consider true positive and false positive counts.

$$\hat{\psi}_{\text{Perc}} == \arg\min_{\psi\in\Psi}\|(0,E)-(N_{\text{FP}},N_{\text{TP}})\|^2 \tag{13}$$

The detector is a trade-off of actual value, which can more intuitively reflect the actual situation. Hence, it is expected the indicator is better measure the performance than the true positive rate.

Test \ Actual	Event	Non-Event
Event	True Positive(TP)	False Positive(FP)
Non-Event	False Negative(FN)	True Negative(TN)

Fig. 3. The parameter matrix of detector

4 Appliance State Feature Extraction

Early NILM research focused on the disaggregation of high consumption rate and low frequency sampling. Nowadays, more research is mainly concentrating high-frequency sampling data. The high-frequency acquisition mode can better preserve the integrity of the load wave, which gives the opportunity to incorporate more electrical characteristics to accurately classify appliances [22]. However, the transient characteristics of high-frequency sampling exist for a short time, it is difficult to capture its load characteristics once the sampling frequency of the device fails to meet the requirements. Yet the steady-state characteristics obtained by high-frequency sampling are easier to extract and have strong anti-interference ability.

In the paper, based on the active power, FFT is performed on the current signal sampled at high frequency. The amplitude and phase of odd harmonics in the signal can be used for load identification.

In practice, the load waveform extracted from the same appliance apparatus has repeatability, and it is only necessary to select the representative load characteristic waveform of each appliance to establish the load characteristic database. Therefore, the AP clustering is used to cluster the inherent characteristics of load steady-state operation separated from event data. The load types of the independent appliance are distinguished, and each appliance is labeled to complete the adaptive construction of feature database.

For the sample space $X = \{x_1, x_2, x_3, \cdots, x_n\}$, definition $r(i,k)$ represents the attraction of point k to point i, and $a(i,k)$ represents the degree of attribution of point i to point k, which is used to search potential best candidate state features.

$$r(i,k) = s(i,k) - \max_{k' \neq k}\{a(i,k') + s(i,k')\} \tag{14}$$

$$a(i,k) = \begin{cases} \min\left\{0, r(k,k) + \sum_{i' \notin \{i,k\}} \max\left\{0, r(i',k)\right\}\right\}, i \neq k \\ \sum_{i' \notin \{i,k\}} \max\left\{0, r(i',k)\right\}, i \neq k \end{cases} \tag{15}$$

At the end of information dissemination, the clustering center X_k meets the following requirements:

$$\arg\max_{k}\{a(i,k)+r(i,k)\} \tag{16}$$

It is noteworthy that the AP algorithm adaptively obtains the cluster centers based on the similarity of nodes and the minimization of transfer function, which is not bothered by the selection of initial points. Furthermore, the model is a continuous optimization process, and the cluster centers are gradually recognized. In this way, the optimization process is more robust and more practical. In practice, the AP algorithm can obtain more stable results, avoid the influence of abnormal points as state features, and facilitate the automatic establishment of stable load feature data sets [23].

5 Load Identification Model

5.1 Dynamic Adaptive Particle Swarm Optimization

Non-intrusive load identification can be transformed into a knapsack problem in PSO, whose solution is a discrete decision vector, and the discontinuity of the solution space presents a great jump. Therefore, the Dynamic Adaptive Particle Swarm Optimization (DAPSO) is proposed. DAPSO is applied to non-intrusive load decomposition, which shows relatively excellent convergence ability and solution accuracy [24].

DAPSO deviates from the original optimization trajectory by exploring the inertial weight ω to seek a better solution, which represents the ability of the particle to inherit the previous particle velocity at this moment. The dynamic adaptive inertia weight adjustment strategy is as follows:

$$\omega=\begin{cases}\omega_1-(\omega_2-\omega_1)\dfrac{1}{1+H_1\exp(H_2\Delta)}, & f_i^l>f_{\text{average}}^l\\ \omega_3-(\omega_4-\omega_3)\times\left|\dfrac{f_i^l-f_{\text{average}}^{l\,\prime}}{\Delta}\right|, & f_i^l<f_{\text{average}}^{l\,\prime}\\ \omega_2+(\omega_3-\omega_2)\times\left[\dfrac{2l}{T}-\left(\dfrac{l}{T}\right)^2\right], & \text{others}\end{cases} \tag{17}$$

Where, f_i^l, f_{average}^l and $f_{\text{average}}^{l\,\prime}$ are the particle fitness, the average fitness of particles and better particles, respectively.ω_1 and ω_2 are the upper and lower limits of the inertial weight of poor particles. ω_3 is the lower limit of general particle inertia weight. ω_4 is the lower limit of the inertia weight of good particles. H_1 and H_2 are a constant value. T is the maximum number of iterations.

Define the load type with length n and randomly generate N initial population particles $C=\{C_1,C_2,\cdots,C_N\}$. Each particle is recorded as a 0–1 state vector denoted by $\{c_1,c_2,\cdots,c_n\}$ of length n. The power characteristic is expressed as $P=\{p_1, p_2,\cdots,p_n\}$, the current harmonic characteristic is $K=\{K_1,K_2,\cdots,K_m,\cdots,K_M\}$, where M is the total number of harmonics, $K_m = \{k_1, k_2, \cdots, k_n\}$. The objective function expression is:

$$\min_{C_i} d(C, P(t)^*, K(t)^*) \tag{18}$$

$$\min_{C_i} \alpha|P(t)^* - \sum_{i=1}^{n} c_i p_i^*| + (1-\alpha)\sum_{j=1}^{M} |K_j(t)^* - \sum_{i=1}^{n} c_i k_{ij}^*| \tag{19}$$

Where, α represents the weighting factor. $K_j(t)^*$ is the j th current harmonic at time t, and k_{ij}^* is the characteristic of the j th current harmonic of type i appliances in the load characteristic dataset.

5.2 Implementation Steps of NILM

The core of constructing an event-based NILM system lies in the selection of the best detector and the value of important parameters. Ensuring the high precision of event detection can heighten the accuracy of low-power appliance identification, and reduce false detections caused by normal power fluctuations, thus improving the overall performance of the model in identifying appliances. The NILM method proposed in the paper is shown in Fig. 4. The specific implementation steps are as follows:

1) Collect raw aggregated power data to perform generalized likelihood ratio detection, locate the switching point of the appliances and slide the voting window to traverse the sampling point to determine whether an event has occurred.
2) Introduce the true positive rate and true positive percentage indicators to determine the variable parameters of the event detector, weigh the indicator performance to select the best event detector.
3) According to the output event node, the difference between the active power and the current harmonic component of steady operation when the appliances is switched on and off is extracted, and the load feature values calibrated in the training set are AP clustered to construct a load feature database.
4) Considering the electrical characteristics of the load in depth, the current harmonic component is introduced as a feature into the objective function, and the load state sequence of each event node is iteratively optimized through the multi-feature DAPSO algorithm.
5) Identify the performance of each particle, adaptively select the particle weighting factor according to the dynamic adaptive weighting strategy, and fit the load feature sequence of each particle to the feature database to complete the appliances identification.

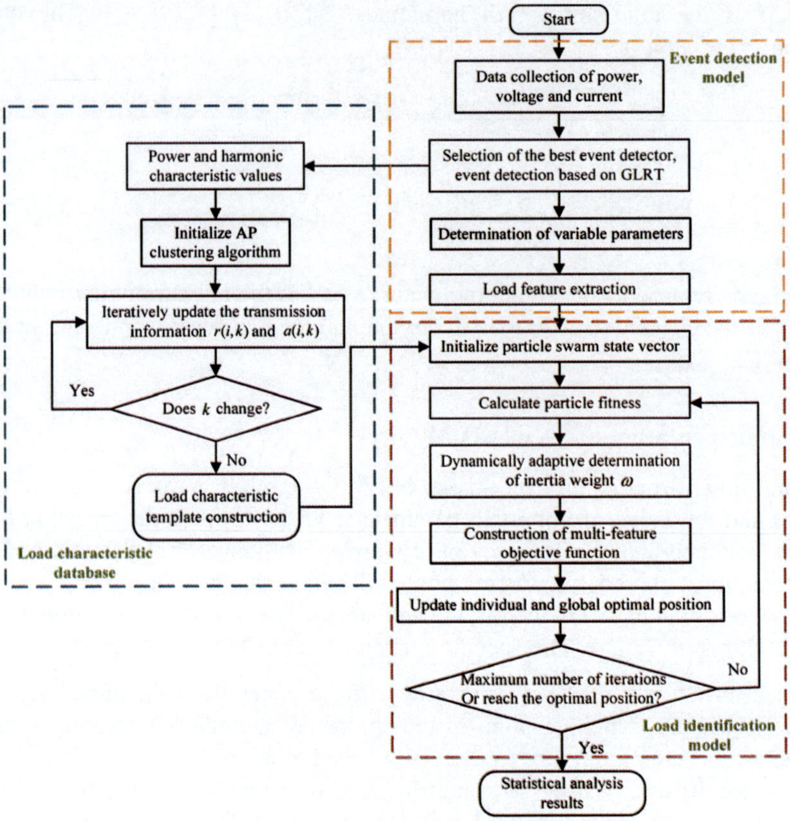

Fig. 4. Flow chart of NILM algorithm

5.3 Performance Evaluation

To evaluate the performance of the model in identifying electrical appliance, two commonly used criteria evaluation indicators are introduced: 1) load recognition accuracy F_R; 2) normalized power disaggregation error (NPDE). F_R reflects the accuracy of the identification for appliance switching activity, NPDE shows the accuracy of energy consumption estimated by the model [25].

$$F_{\mathrm{R}}=\frac{\sum_{i=1}^{L} R_i}{N_s} \tag{20}$$

$$\mathrm{NPDE} = \sqrt{\frac{\sum_{i=1}^{L} (y_i - \hat{y}_i)^2}{\sum_{i=1}^{L} (\hat{y}_i)^2}} \tag{21}$$

Where, R_i is the number of times to correctly identify the state of the electrical switch. N_s is the actual total number of changes in the state of the electrical equipment detected by the event. $\hat{y}_i$ is the power consumption of the i th device estimated by the model. Accordingly, y_i is the actual power consumption of the i th device.

6 Case Study

6.1 Database Setup

In order to verify the accuracy of the proposed model, Matlab R2016b was used to simulate the model, and the measured public data set REDD was selected for verification. REDD contains circuit-level/appliance-level data for several weeks. The appliance-level electricity consumption data could be used as a reference to set up the database in Sect. 4 [26]. This paper uses event detection algorithms to test the non-intrusive load data for 7 types of appliances.

6.2 Non-intrusive Load Identification Based on Event Detection

It is known that the actual number of events in a section of power sequence with event marker is 214. Event detectors composed of different parameter combinations respectively detect it. Further draw the ROC curve using the true positive rate and the percentage of true positives as indicators, as shown in Fig. 5. The best event detector parameters obtained are shown in Table 1.

In the NILM system, too many false positives or the loss of actual events will have a greater impact on other links of the model, and the number of both should be reduced as much as possible. Although the $\hat{\psi}_{\mathrm{Rate}}$ event detector has no the missed inspection, the number of obvious false positives is relatively large, and it is easy to generate more false alarms of the event switch node. At the same time, $\hat{\psi}_{\mathrm{Perc}}$ is a trade-off of actual value, which can better measure the performance of the detector, so the $\hat{\psi}_{\mathrm{Perc}}$ event detector is selected.

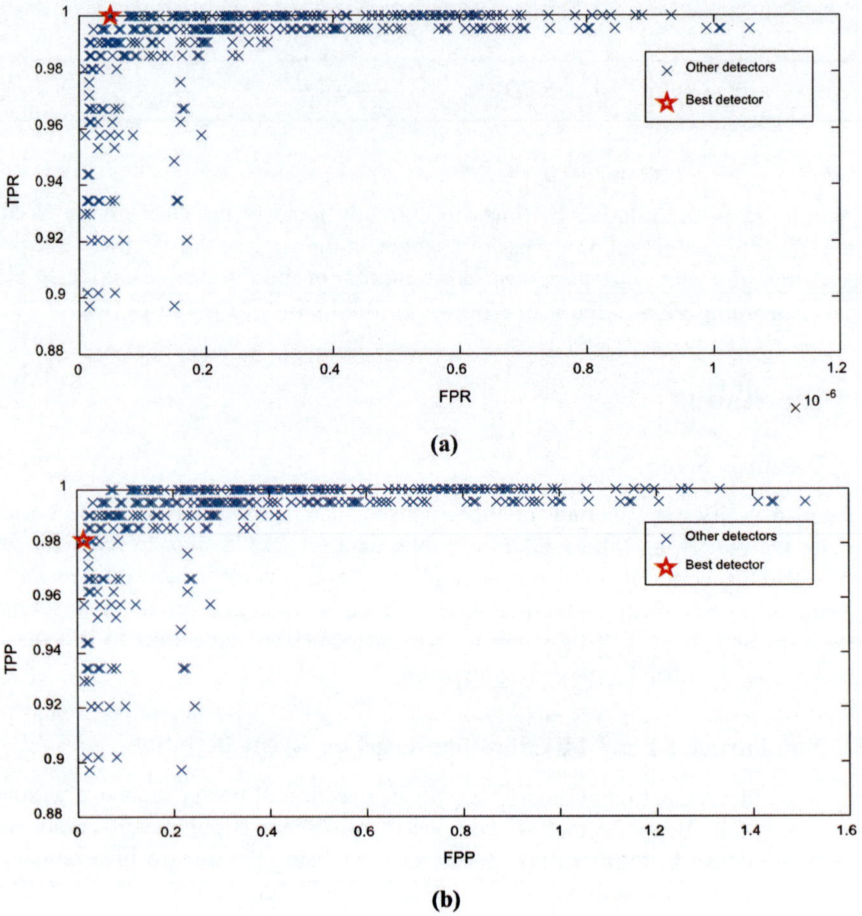

Fig. 5. ROC curve of event detectors

Table 1. Best event detector parameters

Event detector	$\hat{\psi}_{\mathrm{Rate}}$				$\hat{\psi}_{\mathrm{Perc}}$			
Index	TP	FP		M	TP	FP	M	
Number	214	15		0	210	3	4	
Parameters	$\omega_0 = \omega_1$	ω_v	θ_v	θ_p	$\omega_0 = \omega_1$	ω_v	θ_v	θ_p
Value	7	9	5	150	5	7/9	6	190

Use the event detector to perform detection on April 18, 2011 from 20:58:39 to 21:17:03. The red circle in Fig. 6 marks the detected switching position of the appliance, and the voting in Fig. 7 indicates that 14 events occurred during this time period. By comparing the power curve with the sudden change of actual power, it can be seen that the detection accuracy of this event detector is high, and achieves better performance.

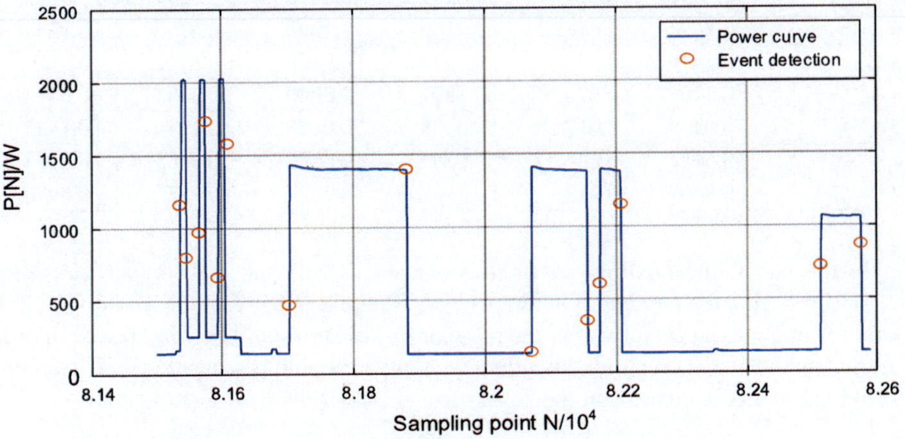

Fig. 6. Schematic of event detection based on power curve

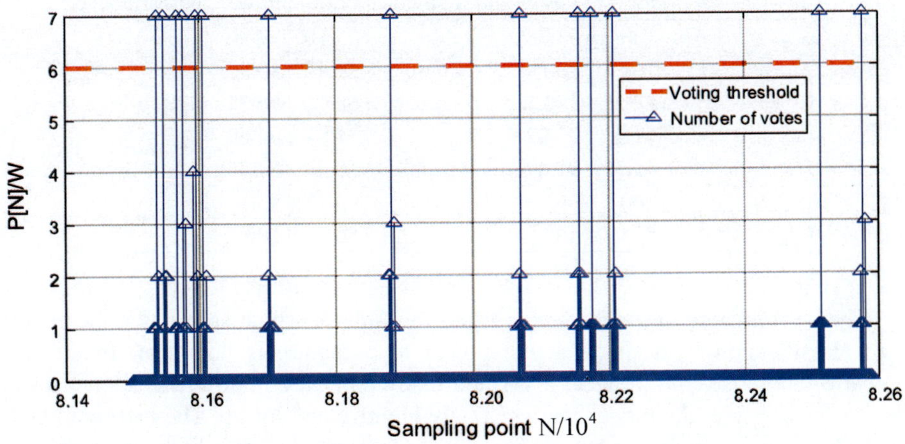

Fig. 7. Sequence diagram of voting window with modified log-likelihood ratio

Extracting the difference between the active power and the current harmonic component during the steady-state operation of the appliances before and after opening, and perform AP clustering as input data to obtain the status characteristics (normalized values) of each appliances, as shown in Table 2. Among, the furnace, dryer, microwave, bathroom GFI, refrigerator, dishwasher, and lights are numbered 1 to 7, respectively.

Table 2. AP clustering result of sample load

Appliance number	Power (W)	1th harmonic	3th harmonic	5th harmonic	7th harmonic	9th harmonic
1	0.2351	0.0347	0.0281	0.0875	0.1064	0.0394
2	0.8693	0.0862	0.1165	0.1262	0.5750	0.0338
3	0.5516	0.9332	0.9342	0.9468	0.8431	0.4850
4	0.5204	0.0732	0.1091	0.0940	0.4193	0.0533
5	0.0409	0.0174	0.0178	0.0465	0.1134	0.1432
6	0.6081	0.0073	0.0289	0.0245	0.0665	0.0554
7	0.0407	0.0731	0.1101	0.2187	0.2881	0.5900

In the paper, the DAPSO parameters are set as follows: the number of particle swarms $N = 100$, the number of iterations $T = 50$, the learning factors $c_1 = c_2 = 1.5$. In order to simulate the actual scene, the influence of noise should be considered. In order to simulate the actual scenario, the influence of noise should be considered. Therefore, a noise load model is created on the basis of load power for load identification.

Table 3. Load identification accuracy of each appliance/%

Appliance type	Non-noise		Noise	
	Power	Power and harmonic	Power	Power and harmonic
1	100	98.78	95.10	97.96
2	100	100	93.47	100
3	88.16	100	68.16	96.33
4	51.02	95.92	31.84	94.69
5	46.53	93.47	30.20	87.76
6	100	96.73	96.73	95.10
7	83.27	100	56.33	100

The performance of the models is listed in Table 3, which shows that the effect of load classification with single active power as a feature is not ideal. In the noisy scenario, obviously the furnace, dryer and dishwasher are high-power appliances and the power is stable, the model can accurately identify and locate. The bathroom GFI is also a high-power appliance but the frequency of power change is high, moreover some of the power is extremely close to the turning-on power of the microwave oven, so the recognition accuracy is low. By adding high-frequency sampling harmonic features, it is more careful to distinguish the types of appliances with similar power, the recognition results of appliances improve obviously.

Table 4. Load identification algorithm indicators

Parameters	Event detection	Non-event detection
F_R	93.88%	70.20%
NPDE	0.1048	0.0574

Table 4 reflects the strong performance and generalization ability of the NILM system model based on event detection, which can accurately identify and locate appliances during working period, and the accuracy rate of identification can reach more than 90%. Due to the overlapping phenomenon of appliances based on the non-event method, the complex combination of appliances in the process of identification leads to the reduction of identification accuracy. The NPDE index of non-event detection is smaller, which is due to the superposition of multiple power and harmonic characteristics in the identification phase. Correspondingly, the pair of high-power appliances can be accurately identified, yet appliances with lower- power are easily submerged under high-power appliances as well as are mistaken for noise, or feature superposition causes appliances to be misjudged as other appliances with similar characteristics. Therefore, the increase in the leakage rate of low-power appliances leads to a reduction in the overall recognition accuracy of the appliances, and a reduction in assembly power errors.

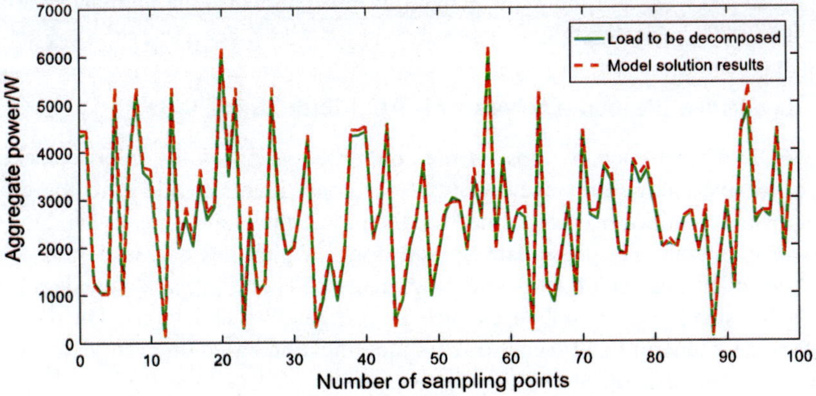

Fig. 8. Comparison of load curve to be decomposed and model solution curve

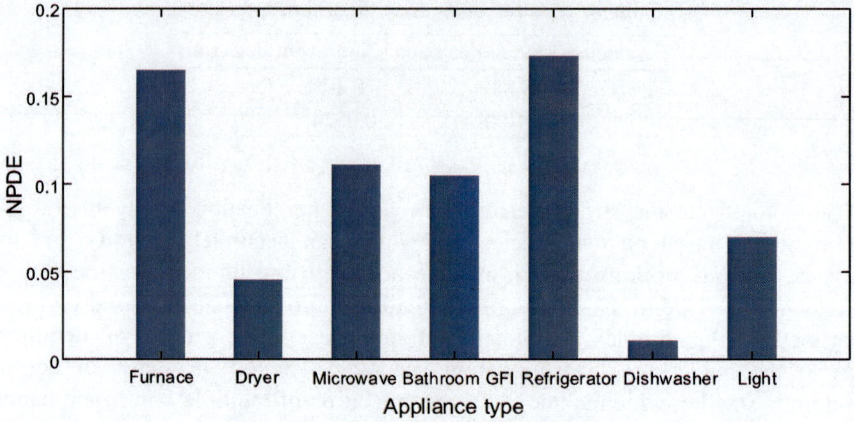

Fig. 9. Power decomposition error diagram

The NPDE evaluation index pursues the highest degree of closeness between the total power consumption to be decomposed and the model results. From the verification results, it can be seen that the actual power and the power estimated by the model have a high degree of fit (see Fig. 8), and the power fitting error of each appliance is small (see Fig. 9). In the paper, the multi-feature NILM method based on event detection has obvious advantages, and the load recognition precision of most appliances attains to more than 90% basically.

6.3 Load Identification Analysis with Air Conditioning Data

The REDD dataset contains most of the common appliances. In order to make more comprehensive energy efficiency analysis, air conditioners is added to the database. When the air conditioner is just turned on, the power fluctuation greatly, and the power after stable operation still oscillate up and down, which has a greater impact on the event detection results of low-power appliances. Therefore, a detailed analysis is needed for small power appliances such as refrigerators and lights. On the basis of power noise, scenario I and scenario II are identification based on event detection when the air conditioner is off and on, respectively.

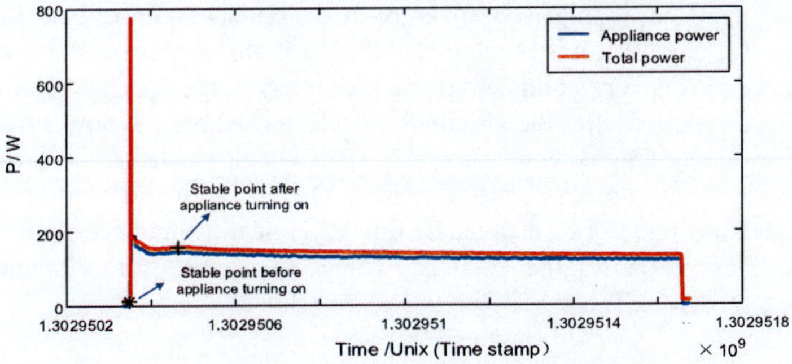

(a). Scenario I: Event detection diagram of the refrigerator

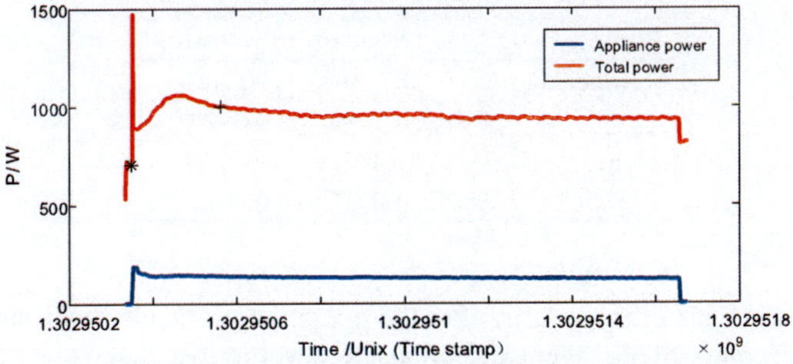

(b). Scenario II: Event detection diagram of the refrigerator

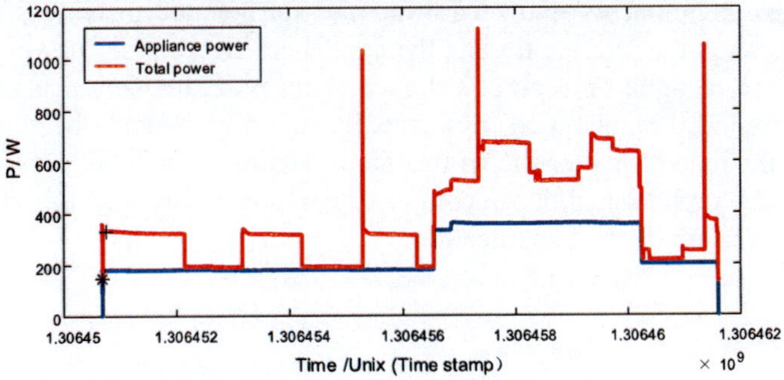

(c). Scenario I: Event detection diagram of the light

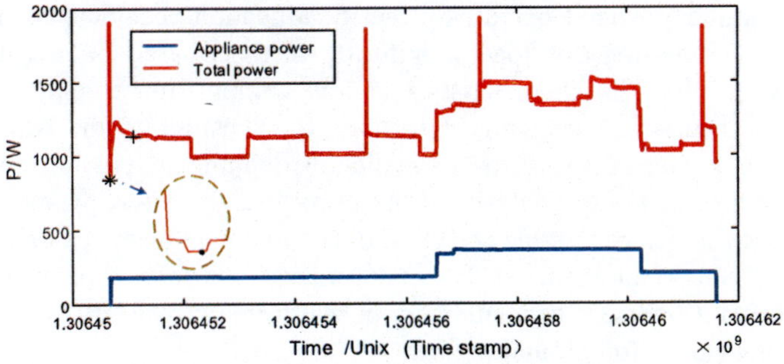

(d). Scenario II: Event detection diagram of the light

Fig. 10. Detecting graphs of appliances in different scenarios

Contrast graph of aggregated power of two appliances under specific working environment (see Fig. 10). It is obvious that the positioning effect of the event detector in the two scenarios is very good, and it can accurately judge the change of the state of the appliance. However, after the air conditioner is turned on, the power fluctuation of the refrigerator and the lights before and after the switching is large and in comparison, it takes a longer time to reach stable operation. There is a great deviation between the load characteristics under the steady-state operation of the appliances before and after opening and those in the original database. The performance test results under the two scenarios are shown in Table 5.

Table 5. Recognition results of the refrigerator and light in different scenarios.

Parameters		Scenario I	Scenario II
Refrigerator	F_R	89.39%	90.20%
	NPDE	0.1088	0.3920
Light	F_R	100%	94.69%
	NPDE	0.0917	0.6915

From the analysis of the results, after the operation of the air conditioner, since the start and end points of the event detection are severely oscillated,the power decomposition error is large. Because of the influence of air conditioner power, the difference of the power components extracted from the refrigerator increases, so the power deviation between the refrigerator and the appliances with similar power enlarge, so as to improve the recognition accuracy slightly. However, the identification accuracy of the lights is reduced by about 5% by comparison, but in general, the current harmonic is added as the load characteristic so that the model has a preferable tracking effect on the state of the appliances. For the complex operating environment, it also has a great performance for appliance identification.

7 Conclusion

This paper aims to improve the accuracy of identifying various appliances with smaller power and similar power. Considering the identification accuracy of multi-objective optimization, a non-invasive load monitoring method based on event detection is proposed, which has the characteristics of low computational complexity and high identification accuracy. Meanwhile it is also very sensitive to the wave motion and peak value of power curve, so as to avoid overlapping of power similar electrical appliances. The sensitivity analysis of the proposed algorithm by introducing event detection index indicates that the selection of optimal parameters can effectively avoid the impact of abnormal events such as false alarms and missed detections caused by system noise and transient oscillation on the performance of NILM system, and has favorable application foreground.

Then, on the basis of power, the load identification part breaks through the limitation of different load feature extraction, separates each load current sampled of the high frequency from the aggregated signal, and combines more electrical features to provide richer information for load identification. Finally, the adaptive search strategy and a dynamic selection strategy are applied to enhance the reliability and robustness of the algorithm and achieve accurate load identification. With the increase of the number of appliances and the influence of environmental factors, the state behavior law of appliances will become complex, and the calculation amount of algorithm will increase. In the future, further research can be carried out in reducing the calculation amount of algorithm and improving the generalization ability of the model

References

1. Yang, T., Zhao, L.Y., Wang, C.S.: Review on application of artificial intelligence in power system and integrated energy system. Autom. Electr. Power Syst. **43**(01), 2–14 (2019)
2. Zhang, D.X., Yao, L.Z., Ma, W.Y.: Development strategies of smart grid in China and Abroad. Proc. CSEE **33**(31), 1–15 (2013)
3. Gao, Y., Yang, H.G., Wang, M.Y.: Household load identification based on closeness matching of transient characteristics. Autom. Electr. Power Syst. **37**(09), 54–59 (2013)
4. Saputro, N., Akkaya, K.: Investigation of smart meter data reporting strategies for optimized performance in smart grid AMI networks. IEEE Internet Things J. **04**(04), 894–904 (2017)
5. Lu, J., Zhu, Y.P., Peng, W.H.: Interactive demand response method of smart community considering clustering of electricity consumption behavior. Autom. Electr. Power Syst. **41** (17), 113–120 (2017)
6. Sun, Y., Cui, C., Lu, J.: A non-intrusive household load monitoring method based on genetic optimization. Power Syst. Technol. **40**(12), 3912–3917 (2016)
7. Wang, K., Zhong, H.W., Yu, N.P.: Nonintrusive load monitoring based on sequence-to-sequence model with attention mechanism. Proc. Chin. Soc. Electr. Eng. **39**(1), 75–83 (2019)
8. Xu, Z., Chen, W., Wang, Q.: A new non-intrusive load monitoring algorithm based on event matching. IEEE Access **7**, 55966–55973 (2019)
9. Song, X.F., Zhou, M., Tu, J.: Non-intrusive load monitoring method based on k-NN and kernel fisher discriminant. Autom. Electr. Power Syst. **42**(6), 73–80 (2018)
10. Li, Y., Wang, B.B., Li, F.X.: Outlook and thinking of flexible and interactive utilization of intelligent power. Autom. Electr. Power Syst. **39**(17), 2–9 (2015)
11. Yao, J.G., Zhang, K.F., Ding, Z.T.: Concept extension and research focus of dynamic demand response. Autom. Electr. Power Syst. **43**(14), 207–215 (2019)
12. Cox, R., Leeb, S.B., Shaw, S.R.: Transient event detection for nonintrusive load monitoring and demand side management using voltage distortion. In: Twenty-First Annual IEEE Applied Power Electronics Conference and Exposition. IEEE, Dallas (2006)
13. Ahmadi, H., Marti, J.R.: Load decomposition at smart meters level using eigenloads approach. IEEE Trans. Power Syst. **30**(6), 3425–3436 (2015)
14. Qi, B., Dong, C., Wu, X.: Non-intrusive load identification method based on DTW algorithm and steady-state current waveform. Autom. Electr. Power Syst. **42**(3), 70–76 (2018)
15. Chen, S.Y., Gao, F., Liu, T.: Load disaggregation method based on factorial hidden markov model and its sensitivity analysis. Autom. Electr. Power Syst. **40**(21), 128–136 (2016)

16. Zhou, M., Song, X.F., Tu, J.: Residential electricity consumption behavior analysis based on non-intrusive load monitoring. Power Syst. Technol. **42**(10), 3268–3274 (2018)
17. Cheng, X., Li, L.Z., Wu, H.: A survey of the research on non-intrusive load monitoring and disaggregation. Power Syst. Technol. **40**(10), 3108–3117 (2016)
18. Cai, J.M., Xie, N., Wang, C.M.: Digitalized techniques and modeling methodologies for distribution network planning-review of CIRED 2017 on power distribution system planning. Power Syst. Technol. **43**(6), 2171–2178 (2019)
19. Xie, Z.Y.: Study of detection methods for non-intrusive load monitoring. North China Electric Power University, Beijing (2018)
20. Attique, U.R., Tek, T.L., Brice, V.: Vent-detection algorithms for low sampling nonintrusive load monitoring systems based on low complexity statistical features. IEEE Trans. Instrum. Meas. 1–9 (2019)
21. Devlin, M.A., Hayes, B.P.: Non-intrusive load monitoring and classification of activities of daily living using residential smart meter data. IEEE Trans. Cons. Electron. **65**(3), 339–348 (2019)
22. Han, X., Deng, C.Y., Zhang, Y.T.: Fast online identification algorithm for delayed feedback load based on V-I characteristics of high frequency data. Autom. Electr. Power Syst. **43**(9), 108–116 + 170 + 117–121 (2019)
23. Xu, Q.S., Lou, O.D., Zheng, A.X.: A non-intrusive load decomposition method based on affinity propagation and genetic algorithm optimization. Trans. China Electrotech. Soc. **33** (16), 3868–3878 (2018)
24. Sun, Y., Zhang, L., Zhao, H.L.: A non-intrusive household load monitoring method based on dynamic adaptive particle swarm optimization algorithm. Power Syst. Technol. **42**(06), 1819–1826 (2018)
25. Ji, T.Y., Liu, L., Wang, T.S.: Non-intrusive load monitoring using additive factorial approximate maximum a posteriori based on iterative Fuzzy c–Means. IEEE Trans. Smart Grid **6**(10), 6667–6677 (2019)
26. Kolter, J.Z., Johnson, M.J.: REDD: a public data set for energy disaggregation research. Sustkdd, 59–62 (2011)

Adaptive Droop Control for Parallel Inverters Power Sharing Based on On-line Identification of Line Impedance

Yang Guoliang[1,2(✉)] and Zhang Yuna[2]

[1] Bayin Guoleng Vocational and Technical College, Korla 841000, China
869874752@qq.com
[2] Yanshan University, Qinhuangdao 066004, Hebei, China

Abstract. A microgrid is mainly composed of renewable energy, energy storage devices and power electronic equipment. The inverter is a power electronic interface that can be connected to the microgrid and plays an important role in the operation and control of the microgrid. In order to reduce or eliminate the circulating current and realize the average output power of each inverter, this paper presents a control method of adaptive adjusting the droop coefficient based on line impedance on-line identification. The realization of this method requires line impedance parameters. Considering that the accuracy of line impedance detection in grid-connected mode is affected by the equivalent impedance of the power grid, therefore, a multi-DG line impedance on-line detection method considering the equivalent impedance of the grid under the grid connected mode is proposed. Simulation and experimental results verify the accuracy and feasibility of the theoretical analysis.

Keywords: Power sharing · Line impedance detection · Microgrid

1 Introduction

Under the dual threats of increasing energy crisis and worsening greenhouse effect, researchers have turned their attention to the field of new energy power generation. The microgrid system composed of distributed power (DG), energy storage, load and converter is the main application of new energy power generation. As the interface between the distributed energy and the AC bus, the inverter's related control technology determines whether the microgrid is stable. At present, the traditional droop control strategy is mainly used to achieve the power of public load according to its corresponding power droop coefficient distribution [1].

However, the application of traditional droop control in low-voltage environment has the following disadvantages: the line impedance of the low-voltage microgrid is mainly resistive, and the line impedance often does not match when multiple micro-sources are connected in parallel. Therefore, it will cause problems such as power coupling and output voltage inconsistent with the theoretical value [2]; It is easy to form a circulating current between the inverters, thereby increasing the loss of the

Y. Xue et al. (Eds.): PMF 2020, LNEE 718, pp. 801–815, 2021.
https://doi.org/10.1007/978-981-15-9746-6_60

inverter, resulting in a reduction in the operating efficiency of the entire parallel system, which in turn affects the stability of the entire microgrid [3].

In response to the above problems, the literature [4–6] introduce a virtual impedance to make the output impedance of the inverter inductive, which reduces the reactive power sharing error to a certain extent, but does not consider the difference in line impedance and ignores the voltage drop due to virtual impedance. The literature [7] to solve the problems of power coupling and line impedance difference, the adaptive virtual impedance is introduced and adjusted by reactive power, and the reactive power is used to adjust the size to achieve the compensation of the output voltage difference without detecting the line impedance parameter. In the case of multiple inverters connected in parallel, it is necessary to ensure the stability of communication.

Using an improved droop control strategy, it is necessary to know the line impedance parameters while achieving power sharing. In addition, according to the current research on line impedance detection, most of them are conducted in the grid-connected mode, and the line impedance and the grid equivalent impedance in this mode play an important role in the operation and control of the grid-connected inverter [8]. And the existence of the equivalent impedance of the power grid directly affects the detection accuracy of line impedance.

In summary, line impedance plays a very important role in system control and power sharing strategy implementation. Taking into account the detection under the conditions of the grid, its accuracy is easily affected by the equivalent impedance of the grid [9]. Therefore, a method for detecting the impedance of the multi-DG grid connected system considering the equivalent impedance of the power grid is proposed. Aiming at the problems existing in the existing circulation suppression and power sharing strategy, in this paper, a control strategy to adaptively adjust the droop coefficient of on-line impedance considering the equivalent impedance of the power grid is proposed to realize the power sharing. Finally, the effectiveness of the control strategy is verified by Matlab/Simulink simulation and experimental results.

2 Adaptive Droop Control Principle of Inverter Based on Line Impedance Identification

The droop control has the advantages of good redundancy, high reliability and good dynamic characteristics, which can realize automatic power distribution by using local communication, so it is widely used in inverter parallel control. The control principle is shown in Fig. 1.

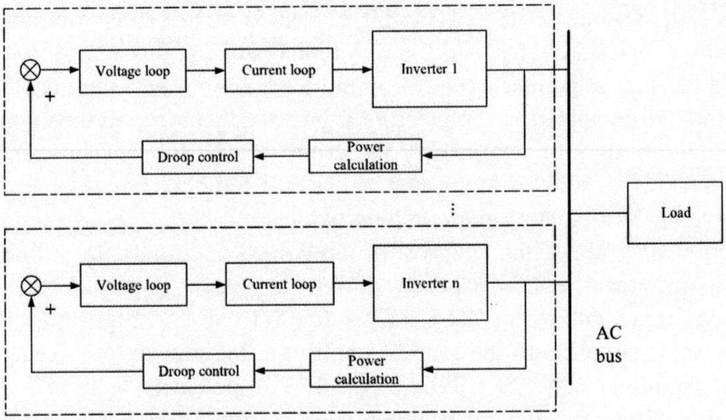

Fig. 1. Block diagram of droop control

The traditional P-V droop control makes the inverter have the external characteristics of the synchronous generator through artificial constraints. The control equation is:

$$\begin{cases} V = V_0 - mP \\ f = f_0 + nQ \end{cases} \tag{1}$$

In the formula, V and f are the inverter output voltage amplitude and frequency; V_0 and f_0 are inverter no-load voltage amplitude and frequency; m and n are active and reactive power droop coefficients; P and Q are the active power and reactive power output by the inverter at the current moment. The corresponding droop curve is shown in Fig. 2.

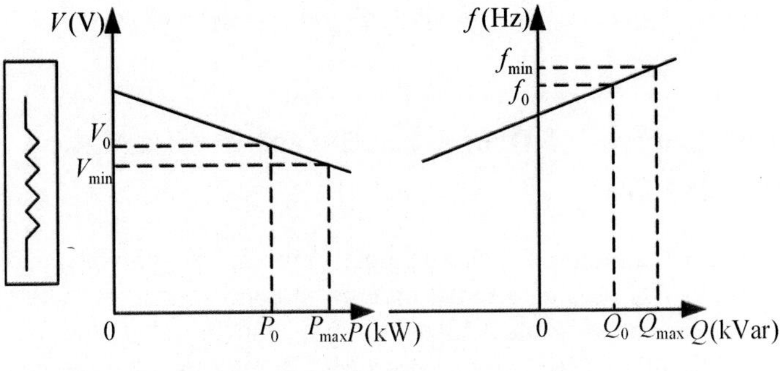

Fig. 2. Droop curve of P-V and Q-f

When two inverters with different capacities are operated in parallel, the difference in line impedance of each inverter will inevitably lead to a circulating current between the systems [10]. The size of the circulating current is related to the output voltage and the impedance of each inverter, and the difference between the output voltage of each inverter and the line impedance should be reduced. At present, the problem of circulating current can be solved by connecting a large inductance, adding a virtual impedance, or using an algorithm to modify the droop coefficient, but these methods have certain disadvantages, such as increasing the size of the device, reducing the accuracy of the voltage, and calculating the complexity.

The control diagram of the adaptive droop control method with online impedance identification proposed in this paper is shown in Fig. 3. The output voltage of the inverter bridge arm is filtered by the LC filter circuit to remove high-order glitches, and then the power is supplied to the load to obtain the inverter output voltage u_a, u_b, u_c, which is transmitted to the grid through the line impedance. Where L, C_f are filter inductance and filter capacitor respectively, i_a, i_b, i_c are the inductor current.

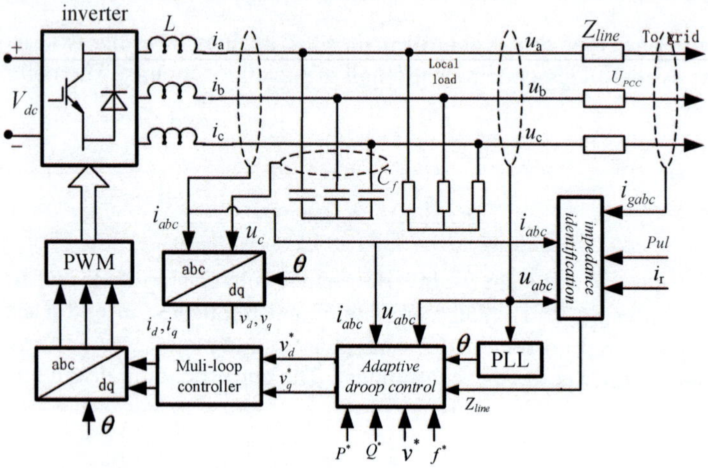

Fig. 3. The control diagram of proposed adaptive droop control method with online impedance Identification

3 Adaptive Control of Droop Coefficient in Circulation Control Strategy

In order to reduce or eliminate the circulating current, this section proposes an adaptive regulation of multi-DG parallel circulating current control strategy with self-adjusting droop coefficient considering local load. As shown in Fig. 4, Z_1 and Z_2 are the line impedance. The traditional P-V droop control is adopted for any one inverter, and the output active power information is transmitted to other inverters which adopt the improved P-V droop control. According to the difference between the output active power of two inverters to adaptive adjust their own P-V droop coefficient, and ultimately to achieve the same output active power, reducing the circulating current of the system [11].

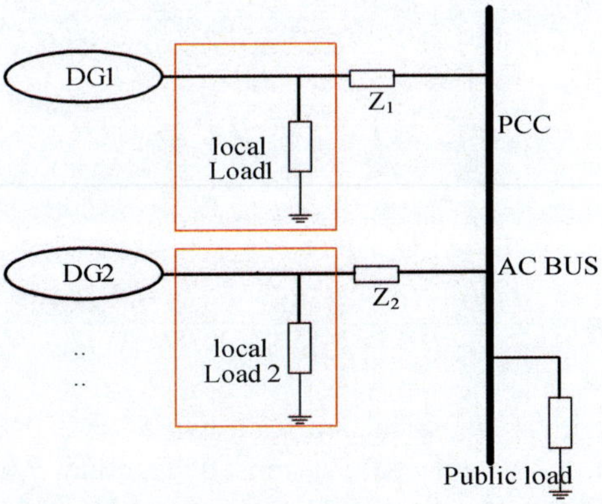

Fig. 4. Equivalent diagram of multi-DGs parallel system with local load

The formula for improving droop control is:

$$V_i - V_{ref} = k_{pi}(P_{ref} - P_i) \tag{2}$$

$$f_i - f_{ref} = -k_{q1}(Q_{ref} - Q_i) \tag{3}$$

$$k_{pi} = k_{p1} - m_i(P_1 - P_i) \tag{4}$$

Suppose the total active power of the system is P, when the power of n inverters is divided equally, it is P/n, which can be obtained according to the P-V droop control:

$$V_1 - V_{ref} = k_{p1}(P_{ref} - P/n) \tag{5}$$

$$V_i - V_{ref} = k_{pi}(P_{ref} - P/n) \tag{6}$$

$$V_1 - E_0 \approx \frac{R_1(P/n - P_{Load1})}{V_{ref}} \tag{7}$$

$$V_i - E_0 \approx \frac{R_i(P/n - P_{Loadi})}{V_{ref}} \tag{8}$$

$$k_{p1}(P_{ref} - P/n) - k_{pi}(P_{ref} - P/n) = \frac{(R_1 - R_i)P/n + R_i P_{Loadi} - R_1 P_{Load1}}{V_{ref}} \tag{9}$$

E_0 is the effective value of AC bus phase voltage, so the Calculation of Adaptive Droop Coefficient k_{pi} is:

$$k_{\mathrm{pi}} = k_{\mathrm{p1}} - m_{\mathrm{i}}(P_1 - P_{\mathrm{i}}) \tag{10}$$

m_i Can be drawn:

$$m_{\mathrm{i}} = \left| \frac{(P/n - P_{\mathrm{Load1}})R_1 - (P/n - P_{\mathrm{Loadi}})R_{\mathrm{i}}}{(P_1 - P_{\mathrm{i}})(P_{\mathrm{ref}} - P/n)V_{ref}} \right| \tag{11}$$

The choice of P_{ref} must be considered from two aspects. From Eq. 9, we can see that if the equation is to be established, P_{ref} must be less than P/n, and K_{pi} is a positive number. When calculating K_{pi} through Eq. 9, it must be meaningful. However, the smaller the P_{ref}, the larger the output voltage error of the inverter or the unstable system. Therefore, P_{ref} should be selected based on the above aspects. The selection of the given value of reactive power generally selects about half of the total reactive power of the system, and the selection of this value is more flexible. The maximum active and reactive power of the system can be estimated according to the system capacity. The minimum value of the voltage is generally 90%–95% of the V_{ref}, and the frequency fluctuation is generally within 2%. According to this, the droop coefficient of the inverter can be calculated.

4 On-Line Identification of Line Impedance

According to the previous introduction to the adaptive adjustment droop control strategy, it can be seen that line impedance plays a very important role in the implementation of system control and power sharing strategy [12]. At the same time, considering that the impedance of the line is detected under grid-connected conditions, its accuracy is easily affected by the equivalent impedance of the power grid. Therefore, a line impedance detection method for multi-DG grid-connected systems considering the equivalent impedance of the power grid is proposed.

When multi-inverter islands are operated in parallel, the line impedance and local load will be inconsistent, which will cause the system to circulate. The realization of the control strategy of adaptively adjusting the droop coefficient needs to know the value of the line impedance. The line impedance detection is mostly done in the grid-connected mode [13]. However, the accuracy of the line impedance detection in the grid-connected mode is affected by the equivalent impedance of the grid, which in turn affects the accuracy of the power sharing strategy implemented in the island mode. It is very necessary to detect the equivalent impedance of the power grid.

At present, the existing impedance detection methods include theoretical calculation method, short circuit test method, load switching method, etc. [14], but the first two methods require accurate and expensive test instruments and reliable protection devices, while the load switching method has relatively high requirements on experimental conditions and measurement environment. These three methods are more difficult to apply in multi-machine systems. The PCC is connected to the power grid through the equivalent impedance Z_g of the power grid, and each DG is connected to the PCC through the corresponding line impedance, and the multi-machine system is equivalent to a virtual single-machine system. Under the virtual stand-alone equivalent model, disturbances are applied to each reference voltage of the grid-connected inverter, and the steady state values of each voltage and current before and after the disturbance are used to calculate each impedance [15]. The impedance network model of the distributed generation system proposed in this paper is shown in Fig. 5.

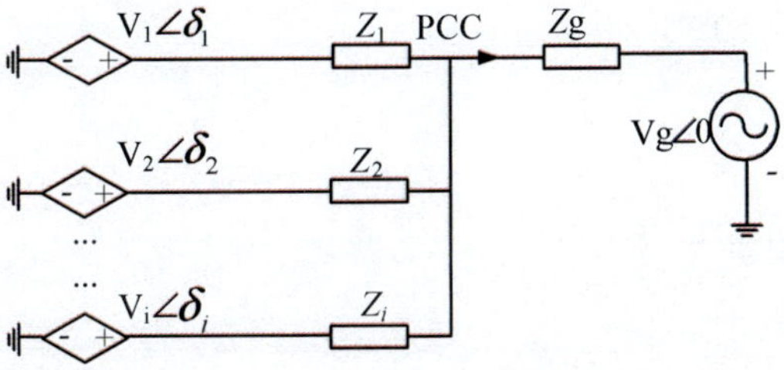

Fig. 5. Impedance model of multi-grid DG system

Apply reference voltage disturbances to multiple inverters with droop control, measure the changes of output voltage current, PCC voltage, and grid current of each inverter before and after the voltage disturbance, and define the first inverter before and after the output voltage change is V_1 and $V_1^{'}$, the amount of change is ΔV_1; the output current is I_1 and $I_1^{'}$ before and after the change, the amount of change is ΔI_1; the output voltage of the i-th inverter is V_i and $V_i^{'}$ before and after the change, the change is ΔV_i; the output current is I_i and $I_i^{'}$ before and after the change, the change is ΔI_i; the PCC voltage is before and after V_{PCC} and $V_{PCC}^{'}$, and the change is ΔV_{PCC}; the change of I_g and $I_g^{'}$ before and after the change of grid current is ΔI_g. The value of k_i can be obtained through the following process:

$$|Z_1| = \left|\frac{\Delta V_{\mathrm{PCC}} - \Delta V_1}{\Delta I_1}\right| \tag{12}$$

$$|Z_i| = \left|\frac{\Delta V_{\mathrm{PCC}} - \Delta V_i}{\Delta I_i}\right| \tag{13}$$

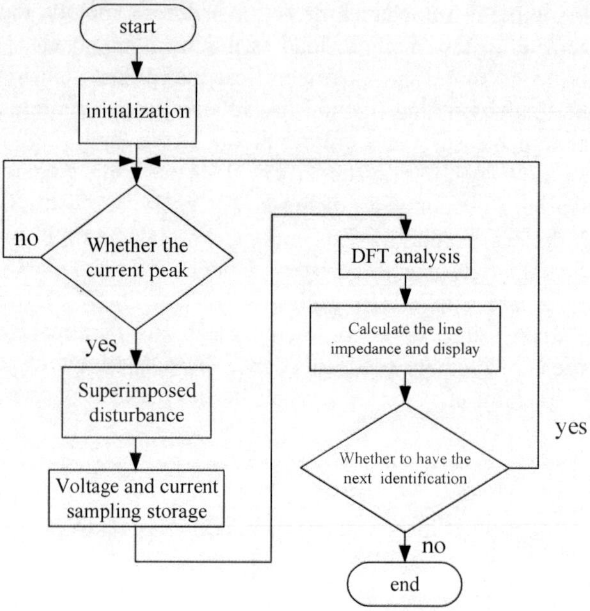

Fig. 6. The flow chart of line impedance online identification

$$|Z_1| = \sqrt{\frac{(\Delta V_{1d} - \Delta V_{PCCd})^2 + (\Delta V_{1q} - \Delta V_{PCCq})^2}{\Delta I_{1d}^2 + \Delta I_{1q}^2}} \tag{14}$$

$$|Z_i| = \sqrt{\frac{(\Delta V_{id} - \Delta V_{PCCd})^2 + (\Delta V_{iq} - \Delta V_{PCCq})^2}{\Delta I_{id}^2 + \Delta I_{iq}^2}} \tag{15}$$

Simplify:

$$L_1 = \frac{\left(\Delta V_{1d} + {}^{\Delta V_{2d}}\!/_{k_2} + \ldots + {}^{\Delta V_{id}}\!/_{k_i}\right)\Delta Igq}{(\Delta Igd^2 + \Delta Igq^2)\omega} - \frac{|Z_1|}{|Z_V|} Lg \tag{16}$$

$$R_1 = \frac{\left(\Delta V_{1d} + {}^{\Delta V_{2d}}\!/_{k_2} + \ldots + {}^{\Delta V_{id}}\!/_{k_i}\right)\Delta Igd}{(\Delta Igd^2 + \Delta Igq^2)} - \frac{|Z_1|}{|Z_V|} Rg \tag{17}$$

Figure 6 is the process of online identification of line impedance. The online identification mentioned in this article can be implemented in order.

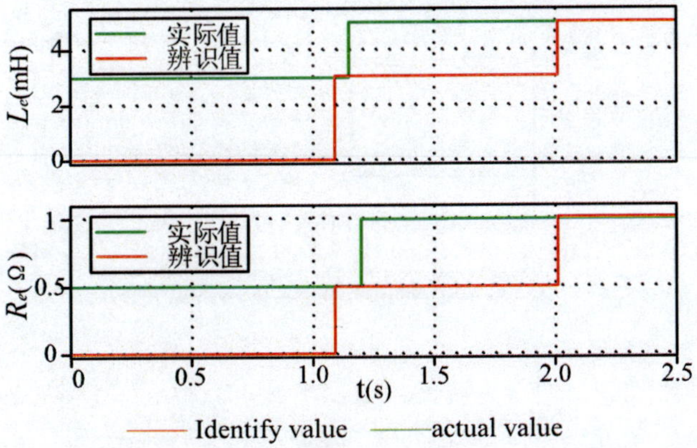

Fig. 7. The identification results of line impedance

5 Simulation and Experimental Results

5.1 Simulation Results of Line Impedance Identification

The identification result of the line impedance is shown in Fig. 7. The initial values of R and L are given as 0.5 Ω and 3 mH respectively. After the simulation, the current disturbance pulse is injected at 0.31 s. After sampling storage and DFT analysis, R_e and L_e can follow the given values of R and L at 1.09 s. At 1.15 s, change the value of R to 1 Ω and the value of L to 5 mH. Both R_e and L_e can follow the value of L again at 2 s. It can be seen that the impedance identification method can accurately track the change value of the line impedance.

5.2 Simulation Verification of Impedance Detection for Three Grid-Connected Inverters

The construction of three grid-connected inverter systems is shown in Fig. 8. Figure 9 shows the measured values of each impedance before and after the sudden change of the equivalent impedance of the power grid. The simulation results show the accuracy and feasibility of the strategy. The accuracy rate is above 97%.

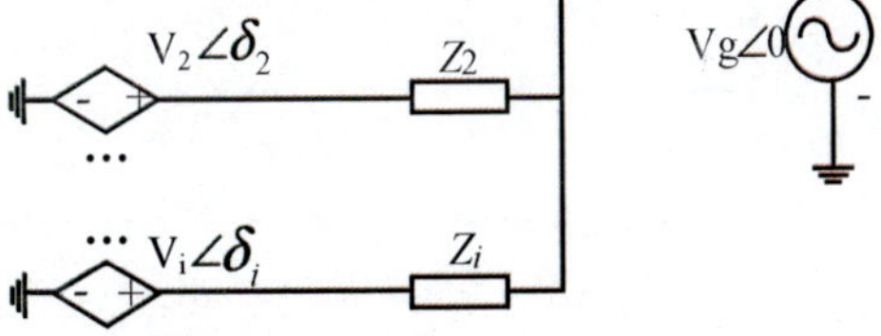

Fig. 8. Impedance model of three grid-connected inverters

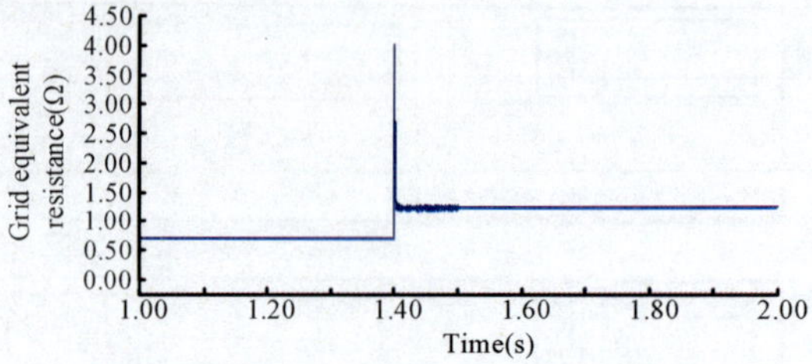

a) Grid equivalent resistance detection value

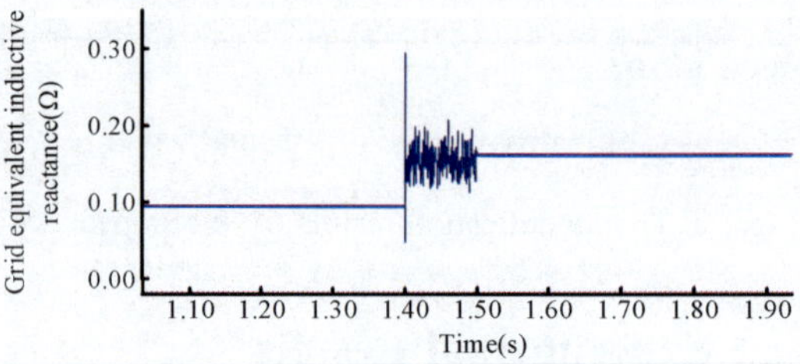

b) Grid equivalent inductance detection value

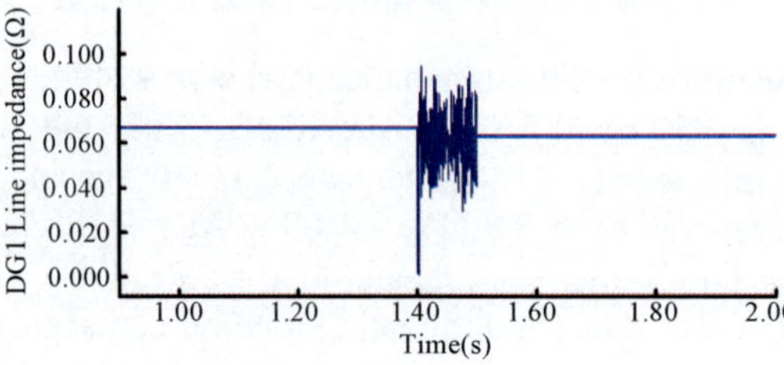

c) DGl line inductance detection value

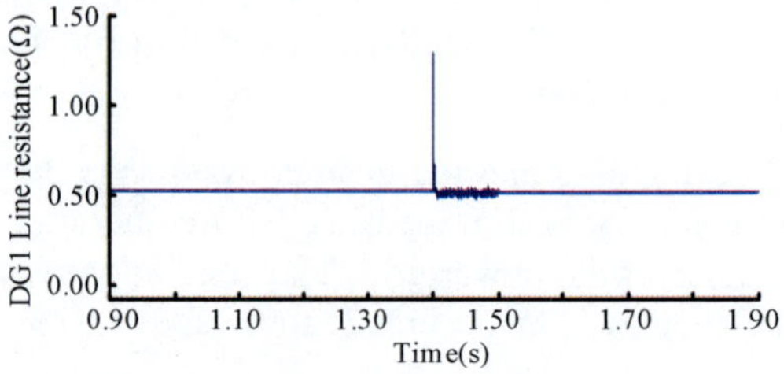

d) DG1 line resistance detection value

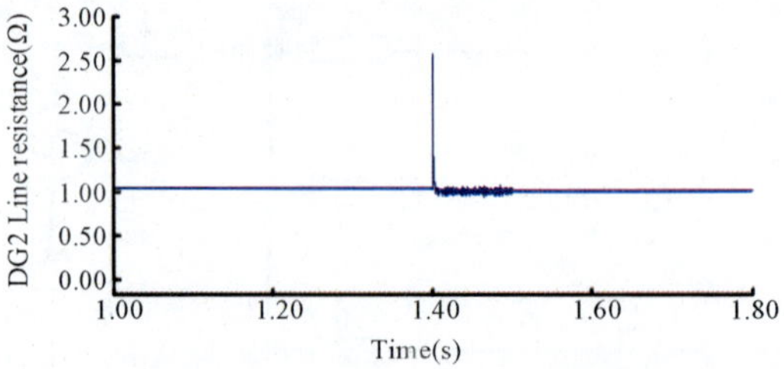

e) DG2 line resistance detection value

Fig. 9. Simulation verification of impedance detection of three grid-connected inverters

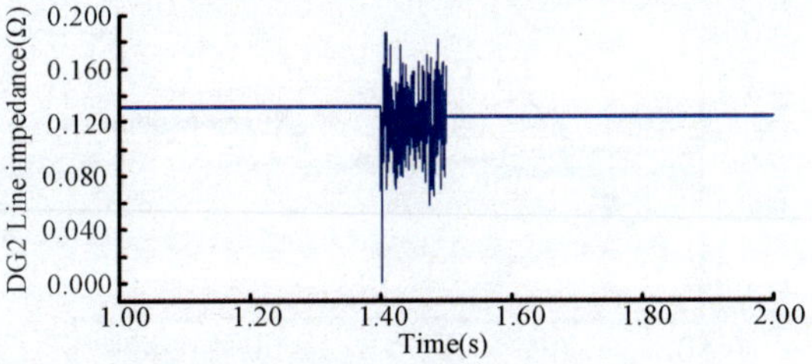

f) DG2 line inductance detection value

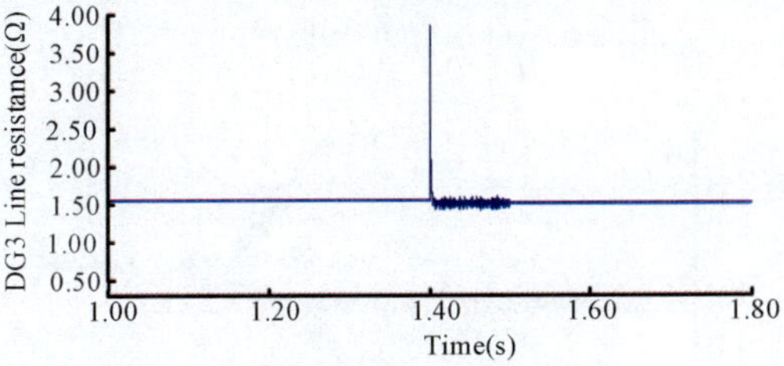

g) DG3 line resistance detection value

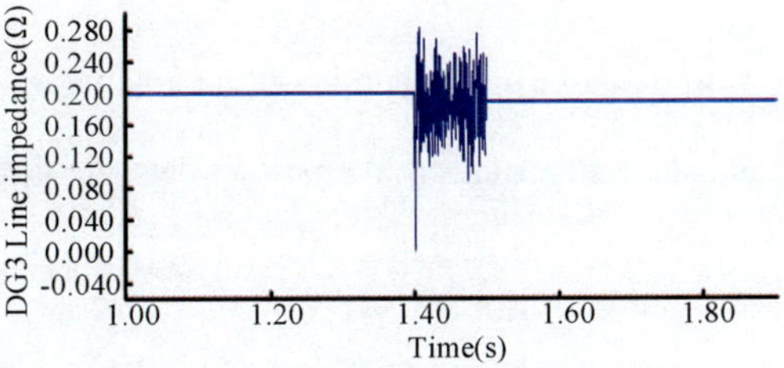

h) DG3 line inductance detection value

Fig. 9. *(continued)*

5.3 Simulation Verification of a Multi-DG Parallel Strategy for Adaptive Adjusting Droop Coefficient

Taking two identical parallel inverters as an example, the system is simulated under the consideration of local load and line impedance $Z_1 = Z_2$, and the traditional P-V droop control and the improved droop control are compared and simulated. Figure 10 is the simulation waveform of the active and reactive power output by the two inverters.

The traditional droop control is used for the inverter before 1.2 s, and the improved control strategy is used after 1.2 s. As can be seen from Fig. 10a), the active power output of the two inverters is not evenly distributed before 1.2 s due to the traditional droop control. But after joining the strategy, the two inverters achieved power sharing.

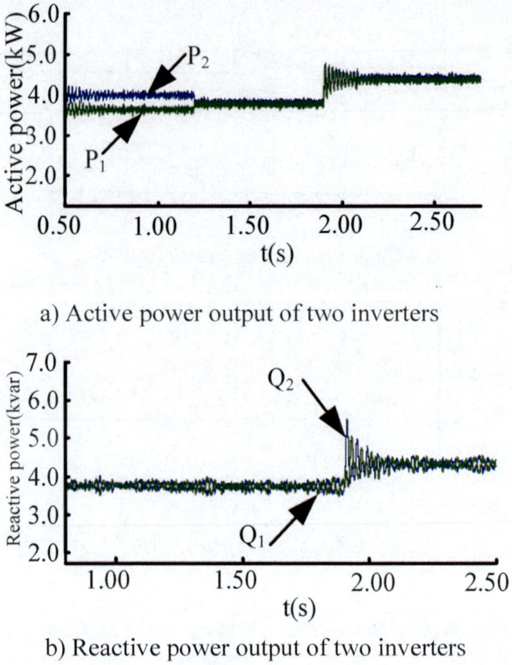

Fig. 10. The inverter outputs active and reactive power when the line impedance is different from the local load

It can be seen from Fig. 10b) that since the frequency of the system is a global signal, the reactive power controlled by Q-f can be divided equally. Then it can be seen that this strategy can also achieve equal sharing of active and reactive power for parallel inverter systems where the line impedance and local load are inconsistent. Under this control strategy, the local load is abrupt at 1.9 s, as can be seen from 1.5 s to 2.5 s in Fig. 10: After the local load is abrupt, the active and reactive power are always divided equally during the dynamic response of the system, and then It is verified that the scheme has good dynamic performance.

5.4 Experimental Verification of a Multi-DG Parallel Strategy for Adaptive Adjusting Droop Coefficient

The two inverters use traditional P-V droop control, that is, the two inverters use the same droop coefficient, and the output voltage, current and power experimental waveforms are shown in Fig. 11.

Based on the first inverter, adjust the droop coefficient of the second inverter. The sag coefficients k_p and m of the second inverter can be obtained. After the inverter output power is divided equally, the public load is abruptly changed. Figure 12a) is the output voltage and current waveform of the inverter when the strategy of adaptively

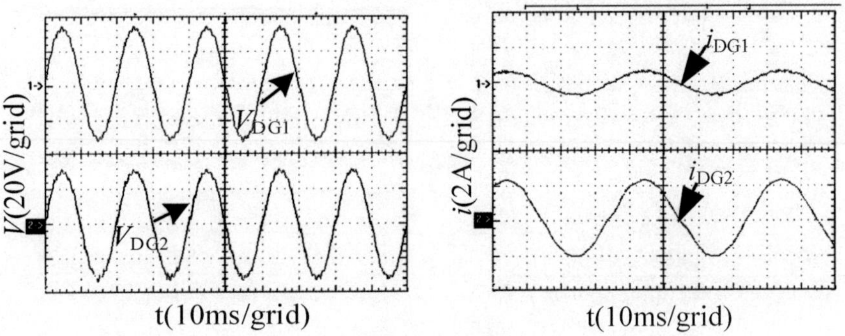

a) Two inverters output voltage and current waveforms

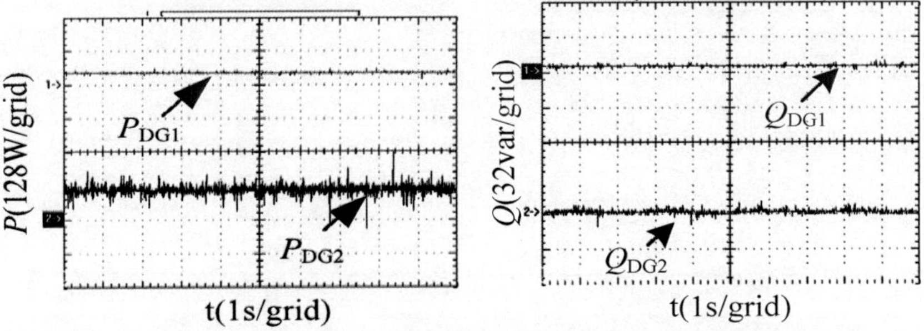

b) Two inverters output active and reactive waveforms

Fig. 11. Inverter output voltage current power waveform after adding strategy

adjusting the droop coefficient is adopted and the public load is not abrupt. Figure 12b) is the active and reactive power waveforms output by the two inverters.

As can be seen from Fig. 11, when the local load is inconsistent with the line impedance, it is verified that the two inverters adopted the traditional P - V droop control, at this time there is circulation in the system, power is not sharing. After adding the adaptive droop coefficient correction strategy, the experimental waveform is shown in Fig. 12. As shown in Fig. 12a), the current output by the two inverters is the same, and the system circulating current is reduced. As shown in Fig. 12b), after adding the strategy, the inverter output power is divided equally and the public load is abruptly changed under this strategy. In the dynamic response process, the inverter output power is still divided, indicating that this strategy can not only achieve power sharing Points, and better dynamic performance.,

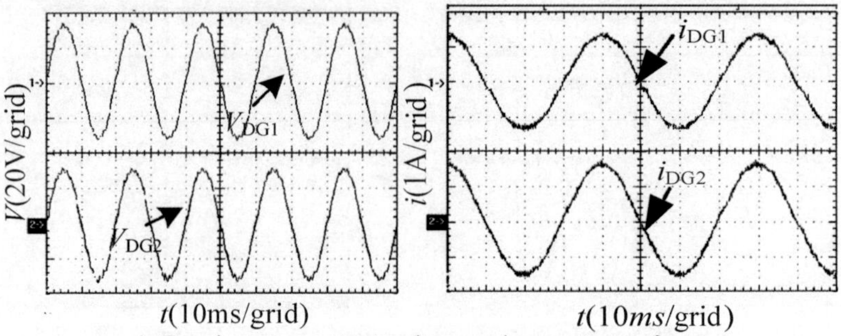

c) Two inverters output voltage and current waveforms (No circulation)

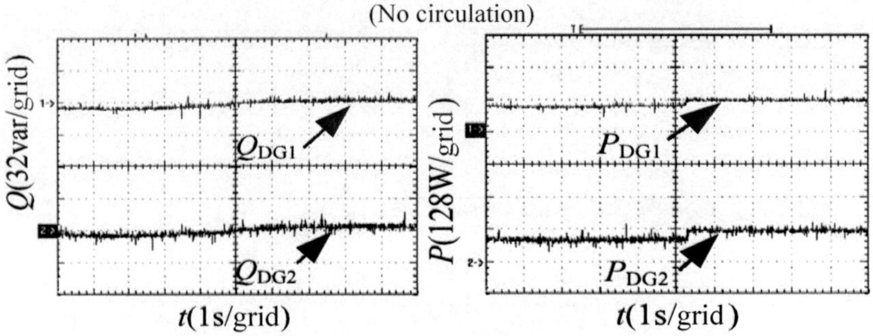

d) Two inverters output active and reactive waveforms (power sharing)

Fig. 12. Output voltage, current and power waveforms of two inverters

6 Conclusion

In this paper, we mainly study the problem of circulating current control and power sharing in parallel inverter in multi-distributed generation microgrid. In view of the existing circulation control strategy and line impedance detection strategy, an adaptive droop coefficient control strategy based on on-line identification of line impedance is proposed. The influence of line impedance and local load on the strategy implementation is analyzed in the paper. Both simulation and experimental results show that this strategy can reduce or eliminate the circulation and can achieve power sharing. The importance of line impedance parameters and the demand calculation of parameters in adaptive adjustment of droop coefficient strategy are discussed. In this paper, an on-line detection method of multi-DG parallel line impedance considering equivalent grid impedance is proposed. The simulation results verify the feasibility and the accuracy of the strategy.

References

1. Fan, Q., Zheng, X., Wang, P., et al.: Decentralized control for autonomous hybrid microgrid with hybrid storage system dynamical regulating. Power Syst. Prot. Control **46**(7), 105–114 (2018)
2. Lu, Z., Sheng, W., Jiang, W., et al.: Frequency dividing droop controllers with the function of voltage stabilization and circulation control. Proc. CSEE **33**(36), 1–9 (2013)
3. Zhu, Y., Zhuo, F., Wang, F., et al.: Virtual impedance optimization method for microgrid reactive power sharing control. Proc. CSEE **36**(17), 4552–4563 (2016)
4. Zhang, P., Shi, J., Li, R., et al.: A control strategy of 'virtual negative' impedance for inverters in low-voltage microgrid. Proc. CSEE **34**(12), 1844–1852 (2014)
5. Zhang, Q., Peng, C., Chen, Y., et al.: A control strategy for parallel operation of multi-inverters in microgrid. Proc. CSEE **32**(25), 126–132 (2012)
6. Guerrero, J.M., Garcia de Vicuna, L., Matas, J., et al.: Output impedance design of parallel-connected UPS inverters with wireless load-sharing control. IEEE Trans. Indus. Electron. **52** (4), 1126–1135 (2005)
7. Bai, X., Miao, H., Zeng, C., Mo site. Improved droop control strategy suitable for inverter reactive power sharing in low voltage microgrid. High Voltage Technol. **46**(04), 1310–1318 (2020)
8. He, J.W., Li, Y.W.: Analysis, design, and implementation of virtual impedance for power electronics interfaced distributed generation. IEEE Trans. Indus. Appl. **47**(6), 2525–2538 (2011)
9. Ciobotaru, M., Agelidis, V.: Line impedance estimation using model based identification technique. In: 14th European Conference on Power Electronics and Applications, Proceedings of the 2011, pp. 1–9 (2011)
10. Tong, Z., Wu, J., Ma, S., Jiang, Y., Luo, X.: A DC microgrid load sharing control strategy based on active voltage disturbance. J. Electrotech. Eng. **34**(24), 5199–5208 (2019)
11. Chen, Q., Zhang, X., Li, M., Guo, Z., Liu, X., Zhao, W.: Isolated island detection method of grid-connected inverter based on impedance identification. Autom. Electr. Power Syst. **44** (07), 123–132 (2020)
12. Tang, Z., Su, M., Liu, Y., Han, H., Wang, H., Li, X.: Active frequency shift island detection technology with load impedance angle feedback. Power Syst. Autom. **42**(07), 199–207 (2018)
13. Pan, P., Chen, W., Hou, K., Wang, Z., Sun, L., Mu, H.: Decentralized unified control strategy for medium and low voltage DC distribution systems. Autom. Electr. Power Syst. **43**(23), 72–85 (2019)
14. Zhu, L., Sun, J., Yang, Y., Lu, Y., Xu, F.: Principle and implementation of dynamic impedance regulator for distribution network lines. J. Electr. Eng. Technol. **34**(S2), 701–708 + 751 (2019)
15. Qiang, Q., Qi, W., Shaojun, X., Jinming, X., Weijian, W.: Multi-frequency impedance model of single-phase grid-connected inverter and its application in resonant circulating current analysis. Autom. Electr. Power Syst. **43**(15), 159–175 (2019)

A New Method of Active Arc Suppression for Single-Phase Grounding Fault in Distribution Network and Its Influencing Factors

Jian Xue[1](✉), Hao Liu[2], Guobing Song[1], Yuanyang Wang[1], and Bing Guo[1]

[1] School of Electrical Engineering, Xi'an Jiaotong University, Xi'an 710049, China
xuejian1996@stu.xjtu.edu.cn
[2] State Grid Shaanxi Electric Power Research Institute, Xi'an 710100, China

Abstract. The distribution network of small current grounding system often has the problem of single-phase grounding fault. The traditional passive arc extinguishment cannot meet the requirement of zero residual current at the fault point. The active arc suppression method uses the zero-sequence current injected by the power electronic equipment to control the voltage of the fault point. This method requires a large-capacity single-phase inverter, which needs high cost in production. This paper introduces the current status of the active grounding arc suppression method of the existing distribution network at first. Then, it proposes a new structure of the active arc suppression device for the distribution network. Also, it analyzes the factors influencing the performance of flexible grounding active arc suppression method. This paper provides a theoretical basis for further improving the control method of the active arc suppression method and enhancing the arc suppression performance.

Keywords: Single-phase grounding fault · Flexible grounding · Active arc suppression · Influencing factors

1 Introduction

The medium voltage distribution network in our country often adopts the neutral point non effective grounding mode. With the development of urban distribution lines, the cable and the flexibility of equipment become more and more common. When a single-phase grounding fault occurs, the grounding fault current increases dramatically. It will produce an arc that is difficult to extinguish [1–4]. If the arc is not handled in time, it may cause overvoltage at the fault. Then it may cause insulation breakdown of non-faulty lines and form interphase faults, which poses a huge challenge to the normal operation of the power system [5].

The single-phase grounding arc suppression method of the distribution network can be divided into two categories: passive arc suppression method and active arc suppression method [6]. The passive arc suppression method mainly generates the reactive

Y. Xue et al. (Eds.): PMF 2020, LNEE 718, pp. 816–827, 2021.
https://doi.org/10.1007/978-981-15-9746-6_61

current of the system when the inductive current compensates the fault in the arc suppression coil. Its method is to reduce the current to make the arc disappear [7–9]. The passive arc suppression method cannot compensate the fault grounding current to 0 [10]. Since the arc suppression coil is usually over-compensated, certain reactive components, active components and harmonic components still exist in the power grid. With the expansion of the distribution network, there may still be excessive residual current after the compensation is completed, so the arc at the fault point still exists. Thus, the active arc suppression method is generated. The active arc suppression method is mainly to inject current through the power electronic device to further reduce the fault current. It is commonly used to cooperate with the inverter and the arc suppression coil. Its aim is to achieve zero residual current at the fault point by injecting a zero-sequence current. This method is also called for the flexible grounding technology of the distribution network. The power electronic devices commonly used in this method are active inverters based on PWM technology, which require a certain capacity and need to determine the fault phase at first.

In reference [11], A new flexible arc suppression method for distribution network is proposed, but it needs to measure the zero sequence network parameters. In reference [12], it proposes an active arc suppression method with double closed-loop control, but the proposed control method is relatively cumbersome and the control accuracy is not high. That is complex to realize. In reference [13], an active arc suppression method based on H-bridge with single continuous current source is introduced. However, it is necessary to measure the parameters of distribution network to the ground, which challenges the real-time performance of arc suppression. In reference [14], a kind of arc suppression technology for ground fault transfer is given. But if another sound phase fails during the fault period, it will cause serious consequences of phase to phase short circuit. In reference [15], an active arc suppression device based on PWM technology inverter is proposed, which aims to control the voltage of fault point and realize arc suppression at the fault point [16].

The current flexible grounding active arc suppression technology of the distribution network is based on the arc suppression coil, which cooperates with an inverter. It can control the fault point voltage to 0. In fact, the injection current is transmitted to the high-voltage side through the transformer. The calculation can find that the inverter output current reaches hundreds of amperes, which makes the inverter often require a large capacity. It also increases the production cost. In addition, the existing methods have not been studied in the case of load access. The effect on arc suppression performance is different under load and no-load conditions. Studying the influencing factors of flexible grounding active arc suppression technology in distribution network is of great significance to the reliable arc suppression of single-phase grounding fault in distribution network.

Based on the introduction of the current active arc suppression technology of flexible grounding in distribution network, this paper proposes a new structure of active arc suppression method. This method can decrease the capacity of inverter and reduce the production cost. At the same time, its arc suppression effect is very good. In addition, this paper analyzes the influencing factors of the active arc suppression method in detail. This paper also makes an in-depth discussion on the residual current and the arc suppression performance at the fault point. Finally, the correctness of the conclusion is verified by PSCAD simulation.

2 Analysis of Current Distribution Network Flexible Grounding Technology

2.1 Current Active Arc Suppression Technology of Flexible Grounding in Distribution Network

The basic device structure of the distribution network through flexible grounding technology is as follows in the Fig. 1 [17]. $\dot{E}_A, \dot{E}_B, \dot{E}_C$ are the three-phase power supply voltage of the distribution network. $\dot{U}_N$ is the neutral point voltage. $\dot{U}_A, \dot{U}_B, \dot{U}_C$ are the three-phase phase voltages of A, B, and C. C_A, C_B, C_C are the capacitance of the distribution network to the ground. r_A, r_B, r_C are the leakage resistance of the distribution network to the ground. In general, $C_A = C_B = C_C = C_0$, $r_A = r_B = r_C = r_0$. Assuming a single-phase ground fault occurs in phase A, the transition resistance at the fault point is R_f. The impedance of the arc suppression coil is Z_L.

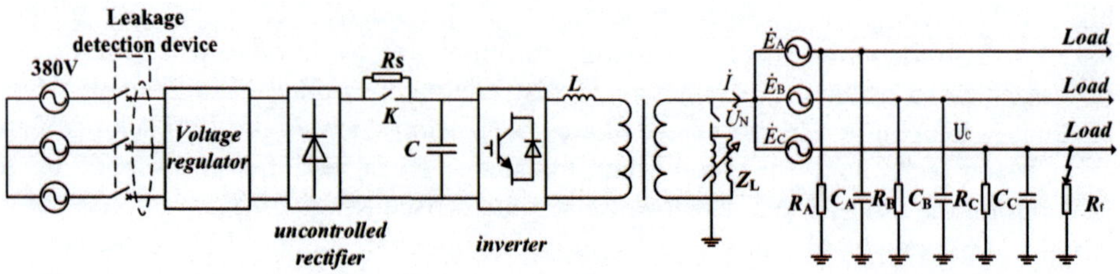

Fig. 1. Structure of flexible grounding active arc suppression device for distribution network.

In the Fig. 1, the station transformer takes the source of 380 V AC. The source passes through the voltage regulator. It outputs the current through the rectifier and inverter. This method cooperates with the arc suppression coil after the transformer to inject the current, which realizes zero residual current arc suppression at the fault point. Leakage detection device is installed in the Fig. 1 to prevent personal electric shock. R_S is a resistance to prevent overcurrent after the equipment is injected with current. The switch K is pressed to cut off R_S after it is stable.

2.2 Analysis of Active Arc Suppression Device with Flexible Grounding in Distribution Network

At present, the arc suppression coil is the main part of the active arc suppression device in the distribution network. It cooperates with the active inverter based on PWM technology, which can control the fault point voltage to 0. Considering that the output voltage level of the general inverter in practice is several hundred volts, the turn ratio of 0.4 kV/10 kV is often adopted for the transformer. In addition, according to the practical experience on site, the injection current $\dot{I}_1$ at the high voltage side of the transformer is usually about 5 to 12A. Therefore, the injection current at the low voltage side can be calculated by the turn ratio:

$$\frac{\dot{I}_2}{\dot{I}_1} = \frac{10}{0.4} = 25 \tag{1}$$

Assuming that the injection current at the high voltage side is $|\dot{I}_1| = 10\text{A}$, there are:

$$|\dot{I}_2| = 25|\dot{I}_1| = 250\text{A} \tag{2}$$

Assuming the output voltage of the inverter is 220 V, the capacity of the inverter can be obtained:

$$P_2 = |\dot{U}_2||\dot{I}_2| = 55000\ \text{kVA} \tag{3}$$

That is to say, the active arc suppression device needs a single-phase inverter with a capacity of about 50 KVA, which is often difficult to achieve in practice. It also requires high cost. At the same time, how to transfer the output current of 250 A to the primary side through the transformer is a problem to be studied. The single-phase inverter with high capacity and high current resistance is the key of this method.

Therefore, the existing flexible grounded active arc suppression methods have the problems of large capacity requirements for power electronic devices (inverters) and high cost of products. How to solve this problem is a research hotspot at home and abroad in recent years.

3 A New Method of Active Arc Suppression with Flexible Grounding in Distribution Network

3.1 New Active Arc Suppression Technology for Distribution Network

The basic idea of the flexible grounding of the existing distribution network is as follows. When a single-phase grounding fault occurs, we judge the fault phase at first. Use an active inverter based on PWM technology to inject a current into the distribution network, controlling the voltage at the fault point to 0 [18]. It can eliminate the reactive power component, active power component and harmonic component to achieve arc extinction at the fault point. Considering that the grid parameters C_0 and r_0 are often difficult to measure in practice, the flexible grounding technology uses closed-loop control. The method controls fault point voltage to be 0 to achieve arc suppression. Since the location of the fault point is often difficult to determine, controlling the voltage at the fault phase bus to0 is usual. The active inverter based on PWM technology is used to control the neutral point voltage equal to the negative value of the fault phase power electromotive force. That is, $\dot{U}_N = -\dot{E}_C$. Therefore, through this idea, a new method of flexible grounding active arc suppression for the distribution network is proposed to reduce the inverter injection capacity.

As shown in Fig. 2, it shows the 220 V voltage used in the station, which has the same phase as the $\dot{E}_C$. The transformer turns ratio is 0.4/10. The new method of flexible grounding active arc suppression is to obtain the opposite value of the station variable power supply $\dot{E}_{C'}$. Then it becomes $-\dot{E}_{C'}$ after passing the transformer. Considering the actual internal resistance of the transformer and other factors, the inverter is connected to the high voltage side of the transformer. Adjust the actual output voltage slightly through the inverter output to make $\dot{U}_N = -\dot{E}_C$, which achieves arc extinction at the fault point. It can be seen from the above figure that the current flowing through the inverter in this method is 5 to 12A. At this time, the capacity of the inverter is greatly reduced. It is also easy to produce. At the same time, the actual production cost is greatly reduced.

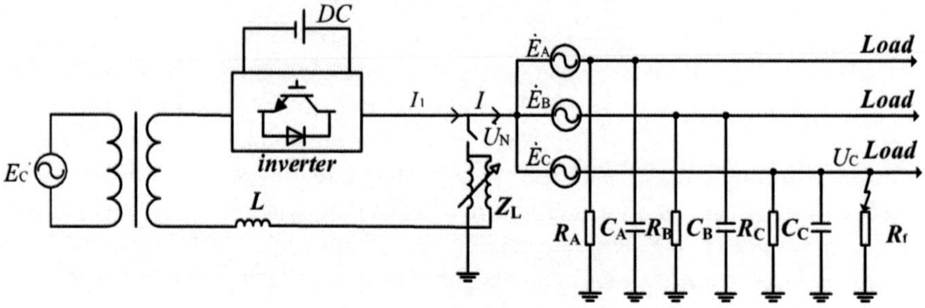

Fig. 2. A new method of active arc suppression for flexible grounding of distribution network.

3.2 Output Voltage and Current Analysis of Inverter

The overall control target of this method is $\dot{U}_N = -\dot{E}_C$. Assuming that the output voltage of the inverter is $\dot{U}_2$. The voltage obtained by passing through the transformer is $\dot{E}_{C''}$, then

$$-\dot{E}_{C'} + \dot{U}_2 = -\dot{E}_C \tag{4}$$

It can be obtained:

$$\dot{U}_2 = -\dot{E}_C + \dot{E}_{C'} \tag{5}$$

The output current value of the inverter is

$$\dot{I}_1 = Y_A \cdot \left(\dot{E}_A + \dot{U}_N\right) + Y_B \cdot \left(\dot{E}_B + \dot{U}_N\right) + \left(Y_C + Y_f\right) \cdot \left(\dot{E}_C + \dot{U}_N\right) \tag{6}$$

Among:
$Y_A = j\omega C_A + \frac{1}{r_A}$, $Y_B = j\omega C_B + \frac{1}{r_B}$, $Y_C = j\omega C_C + \frac{1}{r_C}$, $Y_f = \frac{1}{R_f}$. If the three-phase power supply is symmetrical, there are: $\dot{E}_A + \dot{E}_B + \dot{E}_C = 0$, The above formula can be reduced to:

$$\dot{I} = (Y_A + Y_B + Y_C) \cdot \dot{U}_N + Y_L \cdot \dot{U}_N + Y_F \cdot \left(\dot{E}_C + \dot{E}_N\right) \tag{7}$$

Therefore, there are two methods to realize arc suppression of fault point. (1) Take $\dot{U}_N = -\dot{E}_C$ as the control target of the inverter to realize arc suppression at the fault point. (2) Measure the parameters of the power grid to the ground and take them into Eq. (6) to get the input current of the inverter, which completes arc suppression.

4 Influencing Factors of Active Arc Suppression with Flexible Grounding in Distribution Network

4.1 Analysis of Flexible Grounding Technology in Distribution Network

In fact, it is difficult to determine the location of the fault point, so the fault phase voltage at the control bus is often used. That is to say, the impedance from the bus to the fault point is ignored. Because of the existence of the impedance, there will be voltage drop on the impedance. Thus, the actual fault point voltage U_C is not equal to 0, which will affect the arc suppression performance. It is generally believed that the arc can be extinguished when the fault current is less than 5A, so it is very important to analyze the actual fault voltage and the residual current flowing through the transition resistor.

Assuming that there are m outgoing lines after the bus, a single-phase ground fault occurs on the m-th line. The zero-sequence admittance of the arc suppression coil is Y_L. The sum of all total zero-sequence admittance of the sound line is Y_n. The zero-sequence of the fault line Nano is Y_m. $\dot{U}_C$ is the voltage at the fault point before the fault. $\dot{I}$ is the current injected into the inverter. $\dot{U}_f$ is the zero-sequence voltage at the fault point. The impedance from the bus to the fault point is Z_d [19].

The residual current at the fault point can be deduced from Fig. 2:

$$\dot{I}_f = \frac{\dot{U}_{C'}}{Rf} = \frac{-\dot{E}C + \dot{U}C}{Zd + Rf} \tag{8}$$

4.2 Influence of Load Current on Arc Suppression Effect

The above formula is the arc suppression performance analysis without considering the load current. The influence of load current is considered below. The circuit diagram of fault feeder considering load current is shown in Fig. 3. Due to the injection of the flexible grounded active arc suppression device, the formula $\dot{U}_{OP} = 0$ at the fault bus makes an additional ground point in the line. It forms a new circulation with the actual fault point. This causes the load current to be divided in the process of flowing to the load. Thus, the impedance and transition resistance flowing through the bus to the fault point flow to the load again. Therefore, the fault current is affected by the load current.

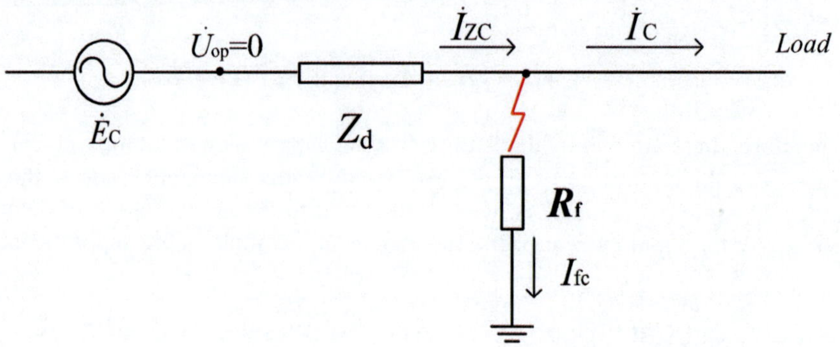

Fig. 3. Equivalent circuit diagram of fault feeder considering load current.

$\dot{I}_C$ in the figure is the load current of the fault line. The current flowing through the line impedance Z_d is $\dot{I}_{ZC}$. The current flowing through the transition resistance R_f is $\dot{I}_{fc}$. This circuit can be simplified as a simple shunt circuit shown in Fig. 4.

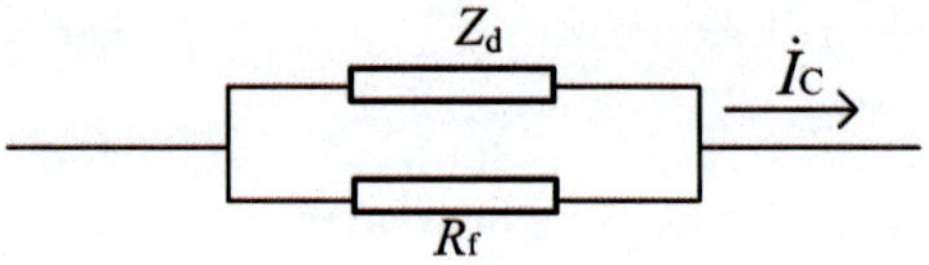

Fig. 4. Equivalent circuit diagram of load current flow direction.

$$\dot{I}_{fc} = \frac{Z_d}{R_f + Z_d}\dot{I}_C \tag{9}$$

The formula (8) adding the formula (9) can be used to obtain the total current flowing through the fault point transition resistance after considering the load current:

$$I_{f'} = \dot{I}_f + \dot{I}_{fC} = \frac{-\dot{E}_C + \dot{U}_C + Z_d\dot{I}_C}{Z_d + R_f} = \frac{-\dot{E}_C + \dot{U}_C}{Z_d + R_f} + \frac{\dot{I}_C}{1 + R_f/Z_d} \tag{10}$$

By analyzing the formula (10), the amount of $-\dot{E}_C + \dot{U}_C$ is small. The influence of load current branch is mainly considered:

(1) When the ground resistance R_f is far greater than the line impedance Z_d, the load current $\dot{I}_{fc}$ is very small, which can be ignored. At this moment, the current flowing through the fault point transition resistance is very small. Also, the fault point voltage is also very small, so the arc suppression performance of flexible grounding technology is good. (2) When the ground resistance R_f is far less than the line impedance Z_d, the load

current $\dot{I}_{fc}$ is very large. At this moment, the current value of the transition resistance through the fault point is very large. The arc suppression performance of the flexible grounding technology is poor [20].

4.3 Analysis of Arc Suppression Effect Under Known Grid-to-Ground Parameters

The analysis in Sect. 4.2 is under control of inverter $\dot{U}_N = -\dot{E}_C$. According to the formula (6), the current of inverter can also be directly controlled to realize zero residual current at fault point by calculating grid parameters. The principle is to obtain the required injected current under grid parameters, so the method can also be classified as current arc suppression method. It does not receive the line from fault phase bus to fault point. The impedance is not affected by the load current, so it has a good arc suppression effect.

5 Simulation Verification

PSCAD/EMTDC software is used to simulate and analyze the distribution network. The simulation model is shown in Fig. 1. The distribution network uses a total of 4 feeders. Lines 1 and 2 are overhead lines, and lines 3 and 4 are cable lines, each with a length of 8 km. Set the fault point to occur on line 3. The fault occurs at 0.4 s and the duration is 1 s. The active arc suppression device is put into use in 0.5 s. Parameters of transmission lines and cable lines are shown in Table 1.

Table 1. Parameters of transmission lines and cable lines.

Line type	R_1(Ω/km)	L_1(mH/km)	C_1(μF/km)	R_0(Ω/km)	L_1(mH/km)	C_1(μF/km)
Overhead line	0.14	1.35	0.008	0.18	4.98	0.005
Cable line	0.09	0.49	0.23	0.30	1.49	0.19

(1) Simulation of the influence of the distance between fault point and bus and transition resistance on arc suppression performance.

After the active arc suppression device is put into operation, the arc suppression performance is affected by the different transition resistance. The simulation verification is based on the residual current and voltage at the fault point. Firstly, the fault at the exit of the bus is verified, which is equivalent to the line impedance Z_d = 0. The steady-state phase after compensation of the flexible grounding active arc suppression device is selected for analysis.

When a fault is set up 5 km away from the bus, the arc suppression performance of the active arc suppression device is simulated.

It can be seen from the simulation diagram that at a certain fault point. With the increase of fault resistance, the current at fault point decreases. It means that the arc suppression performance is better. This is consistent with the conclusion of formula (9).

The current at fault point is inversely proportional to the fault resistance. Furthermore, the farther away from the bus, the smaller the fault current.

(2) Simulation of the influence of load current on arc suppression performance.

It is discussed under the condition that the transition resistance is far greater than the line impedance in Fig. 5 and Fig. 6. The simulation results show that the fault current is very small and the arc suppression performance is good at this moment. In order to consider the influence of load current on arc suppression performance, a metal grounding fault and a single-phase ground fault with a transition resistance of 100 Ω are set up at 5 km away from the bus (Fig. 7 and Fig. 8).

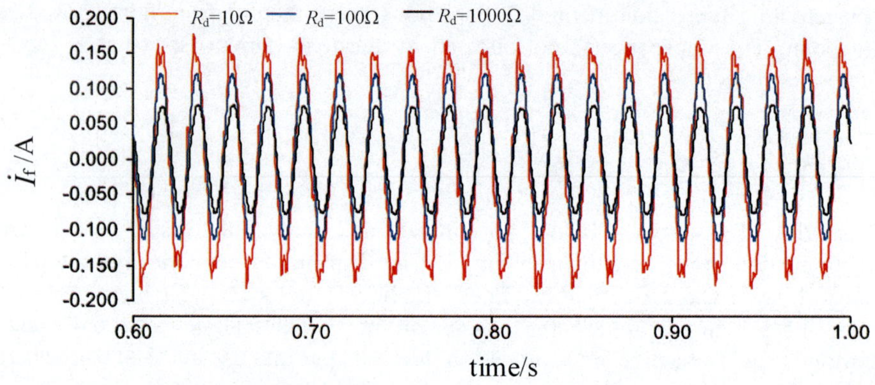

Fig. 5. Current in the fault point after flexible grounding under different transition resistance of single-phase grounding fault at bus outlet.

It can be seen from the simulation results that the metal fault occurs at the distance of 5 km from the bus. The current at the fault point is greatly affected by the load current, which may be detrimental to the effect of the flexible grounding arc extinguishing device. When the grounding resistance is increased to 100 Ω, the fault current decreases rapidly, which indicates that the flexible grounding arc extinguishing device has a good effect.

(3) Simulation of the influence of load current on arc suppression performance.

When the parameters of the power grid are known, the formula (6) is used to simulate the arc suppression performance of the proposed method at this time. At this moment, the method at the bus is consistent with that at the control time. Therefore, the arc suppression performance under the condition that the line impedance is far greater than the transition resistance is mainly considered.

Set R_0 = 8000 Ω, C_0 = 13 μF, set metal fault 5 km away from the bus, and the injection current I_1 = 8.63A.

It can be seen from Fig. 9 that the arc suppression effect is better than that of the control mode when the parameters of the grid to the ground are accurately known. Also, it is not affected by the load current at this time.

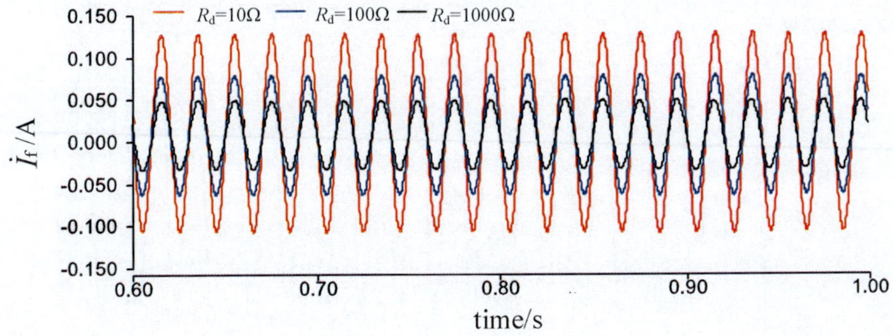

Fig. 6. Current in the fault point after flexible grounding under different transition resistance of single-phase grounding fault 5 km away from bus.

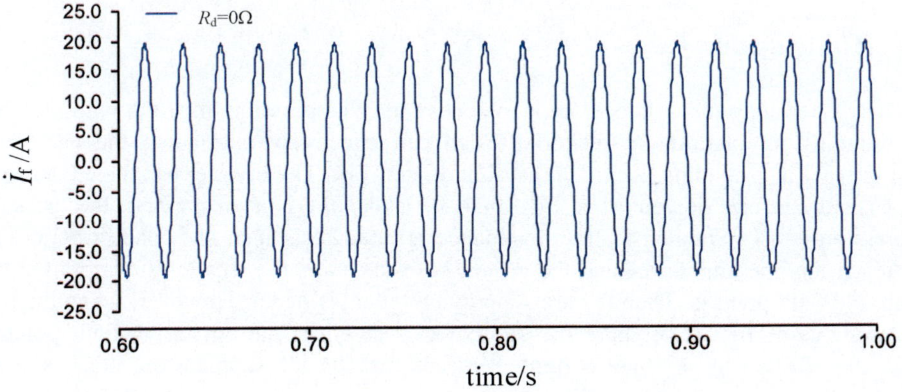

Fig. 7. Current in the fault point when a metallic fault occurs 5 km away from the bus.

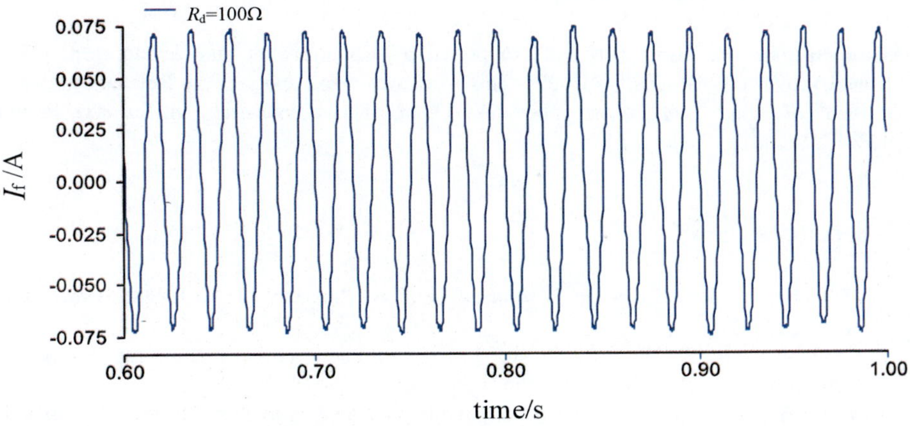

Fig. 8. Current in the fault point when the transition resistance is 100 Ω at 5 km from the bus.

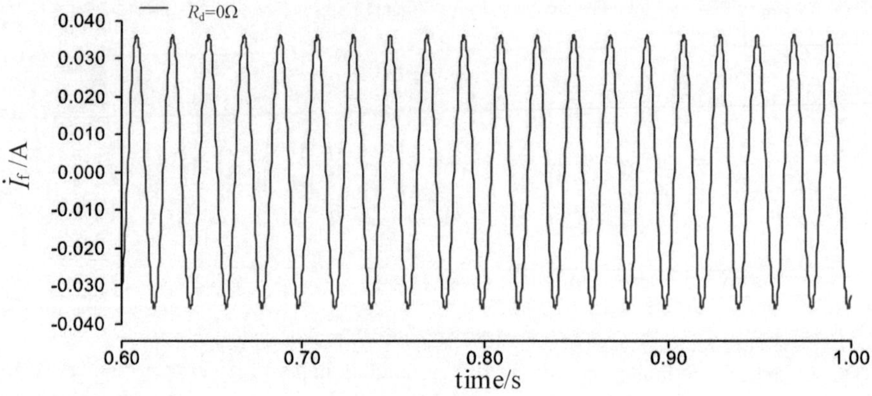

Fig. 9. Current in the fault point when a metallic fault occurs 5 km away from the bus.

6 Conclusion

This paper introduces a new type of active arc suppression method with flexible grounding in distribution network, which can effectively extinguish the arc when single-phase grounding occurs in distribution network. The inverter has the advantages of small capacity, lowercost in this method. Under the control method, the arc suppression effect is related to line impedance, ground resistance and load current. The greater the line impedance and the ground resistance are, the smaller the current at the ground fault point is. Then the arc suppression effect is better. However, when the line impedance is far greater than the ground resistance, the current at the fault point is greatly affected by the load current. It means that the arc suppression effect is poor. When the parameters of the grid to the ground are known, the method is not affected by the transition resistance and load current. The arc suppression effect is good. PSCAD simulation results are also consistent with the theoretical analysis, which proves the correctness of the research content.

Acknowledgements. This work was supported by Key research and development plan of Shaanxi Province (2019ZDLGY18-07). This work was also supported by Relevant projects in the field of smart grid in the innovation chain of key industries in Shaanxi Province (B626KY190001).

References

1. Huannian, Y.: Power System Resonance Earthing, 2nd edn. China Electric Power Press, Beijing (2001)
2. Chen Zhongren, Z., Wu, W., Zhang, Q.: New automatic tuning method for multi-tap arc-suppression coil. Autom. Electr. Power Syst. **41**(5), 75–78 (2005)
3. Wang, Y., Chen, G., Zhang, S.: A continuously magnetic valve arc-suppression coil and its modeling. Trans. China Electrotech. Soc. **29**(5), 239–245 (2014)

4. Tao, L., Xiaopeng, M., Xiaobo, L.: A new tuning method of arc suppression coil based on the constant zero- sequence voltage amplitude. Power Syst. Prot. Control **42**(9), 86–90 (2014)
5. Tang, Y., Wang, T., Luo, J.: Measurement method of the phase-to-earth fault current used in the control of arc-suppression coils. High Voltage Eng. **36**(9), 242–245 (2010)
6. Hänninen, S., Lehtonen, M.: Characteristics of earth faults in electrical distribution networks with high impedance earthing. Electr. Power Syst. Res. **44**(3), 155–161 (1998)
7. Sinclair, J., Gray, I.: Assessing the potential for arc suppression coil technology to reduce customer interruptions and customer minutes lost. In: 20th International Conference and Exhibition on Electricity Distribution, pp. 1–6. CIRED, Prague Czech Republic (2009)
8. Lu, L., Xiang, X., Mei, Z.: A new approach to neutral grounding in distribution network. Power Syst. Technol. **28**(2), 32–35 (2014)
9. Xutao, W., Shaogui, A., Yiping, F.: Field test study on XHG arc extinction, harmonic eliminating and line selection over voltage protective device. Ningxia Electr. Power **7**(2), 14–18 (2007)
10. Janssen, M.: Residual current compensation (RCC) for resonant grounded transmission systems using high performance voltage source inverter. In: Transmission and Distribution Conference and Exposition, pp. 574–578. Transmission and Distribution Conference and Exposition, Toronto (2003)
11. Jianghua, Z., Shanming, W., Yong, Z.: Flexible-grounding arc-suppression method for distribution network based on parameter closed-loop control. High Voltage Eng. **46**(4), 1189–1197 (2020)
12. Rui, C., Duohong, C., Jian, Z.: Research on voltage control method of flexible grounding device in distribution network. J. Electr. Power Sci. Technol. **33**(2), 3–10 (2018)
13. Moufa, G., Jianzhang, Y., Weijun, Z.: Separate-phase flexible arc-suppression method of earth-fault in distribution systems based on three-phase cascaded H-bridge converter. Trans. China Electrotech. Soc. **31**(17), 11–22 (2017)
14. Jian, L., Jun, Q., Zhihua, Z.: Key technologies for intelligent grounding distribution system application. Power Supp. **34**(5), 28–32 (2017)
15. Xiangjun, Z., Yuanyuan, W., Jian, L.: Novel principle of faults arc extinguishing & feeder protection based on flexible grounding control for distribution networks. Proc. CSEE **32**(16), 137–143 (2012)
16. Sepahvand, H., Liao, J., Ferdowsi, M.: Capacitor voltage regulation in single-DC-source cascaded H-bridge multilevel converters using phase-shift modulation. IEEE Trans. Ind. Electron. **60**(9), 3619–3626 (2013)
17. Zeng, X., Hu, J., Wang, Y.: Suppressing method of three-phase unbalanced overvoltage based on distribution networks flexible grounding control. Proc. CSEE **34**(4), 678–684 (2014)
18. Xiangjun, Z., Mingwei, H., Wen, W.: Research on active suppression method of three-phase unbalanced overvoltage for distribution networks. Trans. China Electrotech. Soc. **30**(9), 61–69 (2015)
19. Liu, W., Xue, Y., Xu, B.: An active arc suppression algorithm adaptable to dynamic structure variation of transmission line. Power Syst. Technol. **38**(7), 2008–2013 (2014)
20. Yongrui, X., Gao, X., Yongzhi, S.: Analysis of harmonic in non-solidly earthed network and its influences on harmonic-based earthed network and its influences on harmonic-based earth fault protection. Autom. Electr. Power Syst. **35**(6), 60–64 (2011)

Planning and Analysis of Integrated Energy System Based on Capacity Matching Method

Zhen Li[1], Xichao Zhou[1], Pengxiang Zhao[1], Nan Wang[1], Qiang Yu[2,3,4](✉), Ming Yang[2,3,4], and Huibin Zhu[2,3,4]

[1] State Grid Integrated Energy Service Group CO. LTD., NO. 8, NanHengDong Road, Xicheng District, Beijing 100052, China

[2] Institute of Electrical Engineering, Chinese Academy of Sciences, Beijing 100190, China
yuqiang_21st@sina.com

[3] Key Laboratory of Solar Thermal Energy and Photovoltaic System of Chinese Academy of Sciences, Beijing 100190, China

[4] University of Chinese Academy of Sciences, Beijing 100049, China

Abstract. The coordinated planning of the Integrated Energy System (IES) can not only improve the safety and economy of the whole system, but also achieve the complementary cooperation of various energy flows, and finally improve the overall efficiency of energy utilization. In this paper, the Shanhaitian IES park in Wenchang city, which is an international tourism island, is taken as the research objective Based on the provided data of energy supply and consumption, meteorological conditions, renewable energy resources, building load, and so on, the installed capacities of different renewable energies in this IES park has been planned by using the capacity matching method. The results show that: according to the planned permanent population of about 190,100 people, the installation areas of solar collector can reach 374,000 m^2 which can meet the demands of more than 80% heat load. Besides, the optimal layout area of solar PV system is 878,000 m^2 from the aspect of economic interests, and the remaining power still needs to be supplemented by the large power grid.

Keywords: Coordinated planning · Integrated energy system · Capacity matching method · Building load

1 Introduction

With the rapid growth of China's economy in recent years, the energy demand has been increased year by year. So far, the installed capacity of wind power and photovoltaic (PV) power can reach an average growth rate of more than 30% every year [1]. Therefore, energy strategy has been promoted to be one of important policies in sustainable development agenda. The traditional energy layout is limited to a single energy form such as electricity, gas, heat, cold and so on, and it cannot fully utilize the complementary advantages of different energies. Under this background, Integrated Energy System (IES) [2, 3] has been proposed and studied by many scholars at home and abroad. IES refers to the integrated system including energy production, supply

Y. Xue et al. (Eds.): PMF 2020, LNEE 718, pp. 828–840, 2021.
https://doi.org/10.1007/978-981-15-9746-6_62

and marketing, which are formed through the coordination and optimization of energy generation, transmission, conversion, distribution, storage, consumption in the processes of planning, construction and operation processes [4]. In recent years, IES has attracted more and more attentions from different governments in the world. Compared with previous studies on a single energy system, IES is integrated and converted among different energies, which prolongs the process of energy flows. IES can improve the efficiency of energy utilization, besides, it can also promote the consumption of renewable energy.

So far, most of large renewable energy bases are located in remote areas, which are far away from the large power grid, and they rely on the long-distance transmission technology to solve their power supply problems, however, the construction cost is very high and the operation and maintenance process seem very difficult. Using the local abundant renewable energies, such as solar energy, wind energy and so on, the multi-energy power supply system seems to be a better solution. However, from the operation results of renewable energies, the problems with weak power supply, low power quality and unstable voltage/frequency are still outstanding. So far, several core technical problems have not been well solved yet. The first is the system planning and design: the optimal capacity configuration and operation of the power supply systems have not been organically combined, which results in an unreasonable planning and configuration of IES. Besides, due to the coupling of multiple heterogeneous energy flows in the IES, the coupling relationship is very complex, and the nonlinear characteristics are obvious, which increases the difficulty of coordinated operation of the systems and results in the fluctuation and instability of the IES operation.

Regarding the issues above, many scholars have done a lot of work on the planning of IES park: Xu Weiyan [5] predicted the load during the planning of IES by using the improved adaptive genetic algorithm, and the energy structures of different industries and departments was deeply analyzed. The results show that this algorithm has more accurate prediction capabilities; According to the joint optimization and planning problem of the IES, Yuan Zhiyong et al. [6] reviewed and analyzed the literature of the joint planning such as the electricity-gas system, electricity-gas-heat-cooling system. The results showed that the reasonable planning and coordinated operation of the IES are important means to solve the contradiction in the environmental degradation; Zhao Honggang et al. [7] put forward a management system and planning method based on the IES of cities and towns (named “three layers and two centers”), and the theory was verified by the Xidong science and Technology Park in Wuxi City. Hu Xiao et al. [8] proposed a multi-objective planning model for regional IES with objectives of economy and exergy efficiency. In the model, improving the efficiency of energy utilization is an important task during the planning. The proposed model was finally verified by a certain regional IES in Guangzhou city; Zhang Anan et al. [9] established an electricity-heat-CO2 coupling model of the offshore micro integrated energy system based on the concept of energy hub, and the target of minimizing the total cost and CO2 emission is set as the objective of IES planning. The proposed method is verified by a group of offshore oil and gas platforms and an all-electric vessel; Ammar et al. [10] presented a security constrained optimization problem for co-expansion planning of integrated electric power and natural gas, and the target of minimizing the investment and operation costs as well as voltage stability is set as the objective. The simulation results indicate the importance of voltage stability constraints in long-term energy planning; Wang Qi et al. [11] proposed an expansion planning model of regional integrated energy system considering both short- and long-term uncertainties. The results show that the two types of uncertainty can be effectively managed by adjusting deviation factor and risk parameter. In addition, many scholars have carried out the planning research for different IES areas [12, 13]. From the above, although the pre-scholars have done a lot of work about the planning of IES, however, the matching degree of resource abundance on the source side and demand characteristics of load is not considered comprehensively. Besides, the load calculation of different buildings is not precisely calculated. In this paper, the Shanhaitian IES park in Wenchang city, which is an international tourism island, is taken as the research objective. Based on the provided data of energy supply and consumption, meteorological conditions, renewable energy resources, building load, and so on, the installed capacities of different renewable energies in this IES park has been planned by using the capacity matching method, and the configuration of different energies is analyzed comprehensively.

2 Shanhaitian IES Park

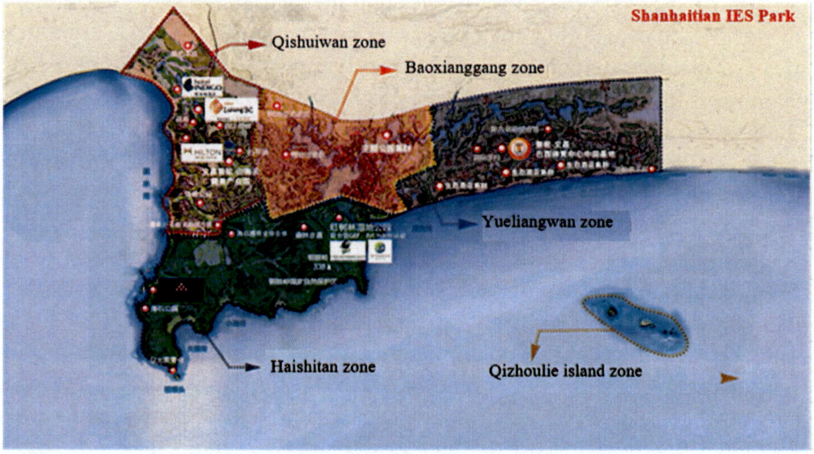

Fig. 1. Schematic diagram of Shanhaitian IES park

Shanhaitian, in Wenchang city of Hainan province, is located in the easternmost part of Hainan island. It is the core project of international tourism island, which is adjacent to the first coastal space city in China. The area map of this project is shown in Fig. 1. According to the plan, the permanent population of long-term planning is about 190,100. In addition, according to the planning scheme of the park, there are mainly two types of buildings in the planning area of this project: residential buildings and public buildings. The total building area is 14.596 million square meters, including 7.6398 million square meters of residential buildings and 6.547 million square meters of public buildings.

2.1 Ambient Temperature

From the temperature distribution of the whole year in Fig. 2, the duration time (the temperature from 20 °C to 30 °C) can reach 5786 h (7.8 months), including 25 °C–30 °C for 3148 h (4.2 months), and 20–25 °C for 2638 h (3.6 months). From the monthly average temperature distribution, the average temperature from October to December and January to April is 15 °C–25 °C, which belongs to a relatively comfortable range.

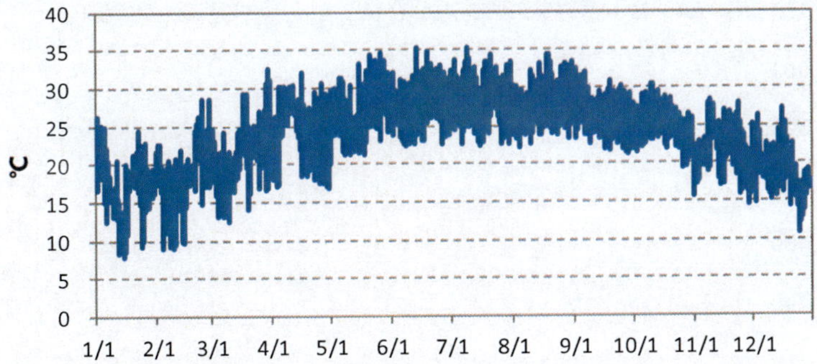

Fig. 2. Temperature distribution in Wenchang city all the year around

2.2 Wind Energy Resource

The annual average wind speed of Shanhaitian IES park is 2.2 m/s. From October to March of the next year, the main wind direction is N–E, and from May to August, the prevailing wind direction is S, besides, the April and September are monsoon transition seasons. In the whole year, the normal wind direction is N–E, and the strong wind direction is N. The maximum wind speed is more than 40 m/s. The wind speed curve of typical year is shown in Fig. 3.

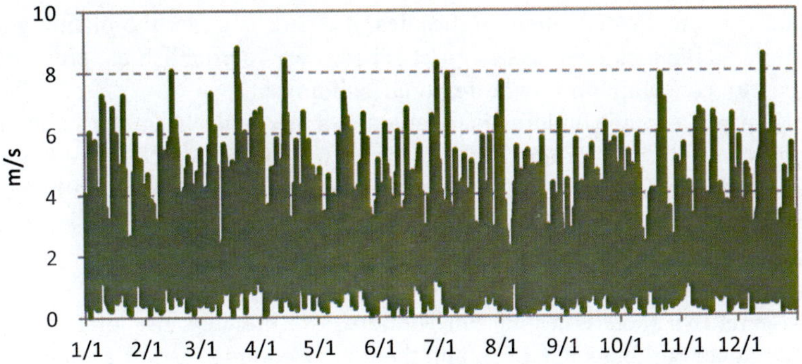

Fig. 3. Wind speed curve in Wenchang city all the year around

2.3 Solar Energy Resource

The variation curve of hourly global solar radiation on horizontal ground for typical years is shown in Fig. 4. It can be seen that the solar radiation for the whole year is relatively well-distributed due to its location (in Hainan province). The yearly global solar radiation on the horizontal ground is about 1386.2 kWh/m^2. In addition, according to the distribution of yearly global solar radiation, the total hours over 300 W/m^2 can reach 697 h, accounting for 21.5% of the total hours (3245 h in total).

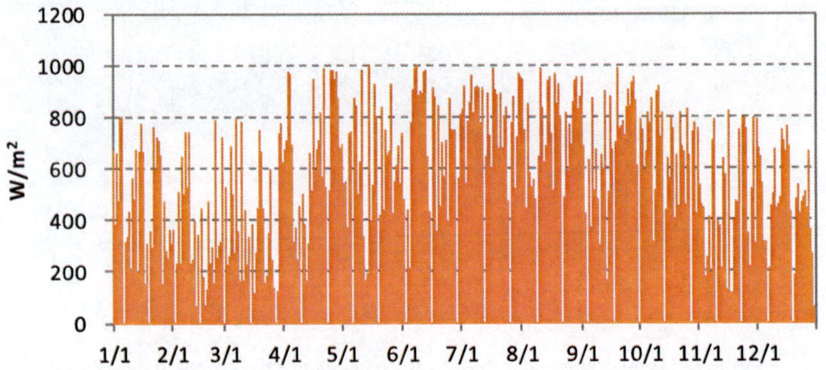

Fig. 4. The yearly global solar radiation on the horizontal ground in Wenchang City

3 Energy Analysis for Energy-Consumption Side

According to the planning of Shanhaitian IES park, there are mainly two types of buildings: residential buildings and public buildings. For these buildings, the energy are mainly used for air conditioning, domestic hot water, cooking, lighting and household appliances. There are many methods to predict the energy consumption of buildings, such as temperature frequency method, equivalent full-load hours method, energy consumption simulation method and so on. Because of the variety of building types and the complicated situation, besides, the park is under the planning stage and there is no specific architectural design, it is necessary to establish the prediction model of the energy consumption for the regional buildings.

In this paper, a single building is regarded as a physical system during the modeling of energy consumption. External disturbances, which are caused by outdoor weather and solar radiation, and internal disturbances, which are caused by the indoor people, indoor equipments and indoor lighting, are considered as inputs to the system, and the load of building is considered as the output.

The modeling process is as follows:

Assuming that there is a total "m" buildings in the area, the total hourly cooling load of regional buildings is $CL(\tau)$, and the hourly cooling load of a single building is $CL_i(\tau)$, so the cooling load of regional buildings is:

$$CL(\tau) = \sum_{i=1}^{m} CL_i(\tau) \quad (1)$$

The maximum hourly cooling load is taken as the planned cooling load of the regional buildings:

$$CL = \max[CL(\tau)] \quad (2)$$

Since the fresh air load of air conditioner $CL_i(\tau)$ is related to the person and using conditions in the building, so the fresh air load can be classified into the internal disturbing load CL_{in}, therefore the Eq. (2) can be written as:

$$CL(\tau) = \sum_{i=1}^{m} CL_{out,i}(\tau) + \sum_{i=1}^{m} CL_{in,i}(\tau) \quad (3)$$

According to the model above, the number of people in the park, the building areas, the living standards of electricity, water and cooking for people in 2018 [14], the calculation results of the annual building energy consumption in the park are as follows (Table 1):

Table 1. Annual cumulative energy consumption statistics of buildings in the park

Building type	Sub-index	Unit	Value
Residential buildings	Air conditioning refrigeration	10^4 kWh/a	44448.4
	Domestic hot water	10^4 kWth/a	6378.7
	Cooking	10^4 kWth/a	7424.1
	Lighting	10^4 kWh/a	4278.3
	Household appliances	10^4 kWh/a	2260.1
Public buildings	Air conditioning refrigeration	10^4 kWh/a	70831.6
	Electricity consumption except air conditioning	10^4 kWh/a	40882.7
	Domestic hot water	10^4 kWth/a	14039.7
	Cooking	10^4 kWth/a	30.4
Summation	Terminal power demand	10^4 kWh/a	162701.1
	Terminal heat demand	10^4 kWth/a	27872.9

According to the statistics above, the annual total power consumption demand in the project area is about 1,627,011,000 kWh/a, and the heat demand is 278,729,000 kWth/a. As seem in this table, the electric power demand is still the main energy consumption in the project area, which is mainly used for air conditioning, and the heat demand is relatively small. Hence, it is important to meet the load requirement of the project area by using the renewable energy.

4 Energy Analysis for Energy-Supplying Side

The energy framework of the energy-supplying side should be rationally distributed according to the local renewable energy resources. At present, the renewable energy in the park mainly includes the solar energy, wind energy, biomass and sewage water.

4.1 Resource Assessment of Solar Thermal Energy

The resource formula of solar thermal power is listed as follows:

$$Q_{sth} = Q_0 \cdot (v/n) \cdot \lambda_{sth} \cdot \gamma_{sth} \cdot \eta_{sth} \cdot A \tag{4}$$

Where, Q_{sth} is the resource of solar thermal power, kWh/a; Q_0 is the annual solar irradiance, kWh/(m^2·a),which can be calculated by the local total horizontal solar irradiance; v is the volume ratio, which is the ratio of the total building areas to the building land areas; n is average number of floors in a building; λ_{sth} is the availability of the roof areas, which is the ratio between the effective areas of the building roof for installing solar collectors and the areas of building roof; γ_{sth} is the ratio of the areas of the solar collector to the areas of the horizontal plane. η_{sth} is average collecting efficiency of solar collector during operation. A is the building areas, m^2;

According to the formula and the provided information above, the total building areas of the four zones is about 14.596,000 m^2. The maximum solar thermal energy resource in the planning area of the project is 578.09 million kWth/a when assuming that the average number of building floors is 7, the utilization rate of roof is 0.8 (ratio of roof areas to roof areas for placing collector), the ratio of solar area to horizontal area is 0.5 and the solar collector efficiency is 0.5.

4.2 Resource Assessment of Solar PV

The resource formula of solar PV is calculated as follows:

$$E_{PV} = Q_0 \cdot (v/n) \cdot \lambda_{PV} \cdot \eta_{PV} \cdot \kappa \cdot A \tag{5}$$

Where, E_{PV} is the resource of solar PV, kWh/a; Q_0 is the annual solar irradiance, kWh/(m^2 · a), which can be calculated by the total horizontal solar irradiance; λ_{PV} is the availability of the roof areas, which is the ratio between the effective areas of the building roof for installing solar PV panels and the areas of the building roof; η_{PV} is the average photoelectric conversion efficiency of solar PV system during the operation; κ is correction coefficient of Solar PV efficiency; A is the building areas, m^2.

According to the provide data, the maximum electric power resource of solar PV in the park is about 115,617,000 kWh/a when assuming that the average building has 7 floors, the roof utilization rate is 0.8, the average photoelectric conversion efficiency of solar PV is 0.1, and the modification coefficient of the photoelectric conversion efficiency is 0.5.

4.3 Resource Assessment of Wind

The annual power output of wind turbines is:

$$E_W = \eta_W \cdot P_V \cdot T_V \tag{6}$$

Where, E_W is the annual power output of wind turbine, kWh/a; η_W is the generation efficiency of wind turbine. At present, the wind turbine for the small wind turbine is about 45%, the generator is about 95%, and the variable speed inverter is about 90% [15]. The total efficiency is the multiple of the three items. P_V is the local wind resource at the effective wind speed V, W/m^2. The local wind energy resources are about 50–100 W/m^2 in China [16]; T_V is the cumulative hours per year at effective wind speed V, h.

Because the project area is very close to the space launch site, there are some certain height restrictions for the buildings. It is difficult to install large wind power in the surrounding areas, and the smaller ones seem much better. When assuming that the efficiency of small wind turbine is 0.38, the wind power is 100 W/m^2, the cumulative number of hours per year is 2159 h (based on the statistics of annual meteorological data), as well as the planned area covers an area of 3537.81 hectares, the resource amount of wind power generation in the planning area of the project is about 816,000 kWh/a, which is only 0.7% of solar PV power, so it can be ignored.

4.4 Evaluation of the Thermal Energy of Sewage Water

The utilization of sewage water heat pump should not affect the treatment process of sewage water. The estimation should take into account the sewage flow per hour. Then the calculation formula is as follows:

$$Q'_{W(h)} = q_w \cdot F_w \cdot t_{c(h)} \tag{7}$$

Where, $Q'_{W(h)}$ is the heat exchange from the sewage, kWh; q_w is the heat exchange of sewage per unit flow, kWh/m^3; F_w is the sewage flow, m^3; $t_{c(h)}$ is the heating or cooling time, h.

The cooling capacity or heating capacity of the sewage source heat pump can be calculated according to the water source heat pump. Considering the daily sewage discharge of 220 L/d/for each person in a typical southern city [17], the sewage capacity is about 41,800 m^3/d based on the population planning of 190,100 in the area. In addition, assuming that the temperature difference between the inlet and outlet of the sewage source heat pump evaporator is 5 °C (COP is 3.0), then, the average daily heat output of sewage source heat pump is 365,000 kWth/d, which can meet the 65.3% requirement of domestic hot water.

Based on the provided data of typical meteorological year, the statistical table of local renewable resources can be obtained as follows (Table 2):

Table 2. Statistics of local renewable resource estimation

Type of energy	Name of the resource	Value (kWh(th)/a)
Heat power	Solar thermal - project area	57808.5
	Solar thermal - residential buildings	30258.0
	Solar thermal - public buildings	25929.9
	Sewage source heat pump	13322.5
Electric power	Solar PV - project area	11561.7
	Solar PV - residential buildings	6051.6
	Solar PV - public buildings	5186.0
	Wind power generation	81.6

According to the renewable energy resources above, the conversion efficiency of the solar thermal system can reach more than 50% due to the high ambient temperature in Hainan province. For some commercial buildings and residential buildings that require domestic hot water all the year round, using solar thermal systems to supply domestic hot water is economically superior to the solar PV systems. According to the amount of renewable energy resources, the maximum available resource of solar PV system can reach 115.617 million kWh/a. If the commercial buildings (the utilization of solar thermal system is considered) is excluded, the possible amount of solar PV can also reach 60.516 million kWh/a. The other renewable energies can reach 81.6 kWh/a. It is seen that the solar PV system is still the main power generation mode in the park, and others can be considered for demonstration.

If the project area can accommodate 190,100 people, the total demand of domestic hot water is 204.184,000 kWth/a, the total demand of daily domestic hot water is 559,000 kWth/d, the average daily output of sewage source heat pump is 365,000 kWth/d, it can meet the 65.3% requirement of domestic hot water.

5 Capacity Matching Analysis and Discussion

5.1 Calculation of Energy Consumption

Due to the low-carbon development trend, the IES park need to make full use of local renewable energy. At the same time, the energy supplying characteristics of different renewable energy systems and the energy demand for different types in the IES park should be seriously studied. The final aim is to achieve areas on able distribution for the supply and demand, and reduce the capacity of energy storage device and finally improve the economy of renewable energy system.

The energy flow process of various renewable energies in the IES park is as follows: solar energy can be converted into electricity and heat (solar PV systems can convert solar energy into electricity, which can be used to provide the power for the buildings, and the surplus electricity can be used to power the grid or stored in batteries. Besides, solar energy can be converted into heat energy and stored in heat storage tanks to heat the buildings). Small wind turbines can convert wind energy into electricity to power the buildings, and the excess energy can be fed into the grid or stored in batteries (Fig. 5).

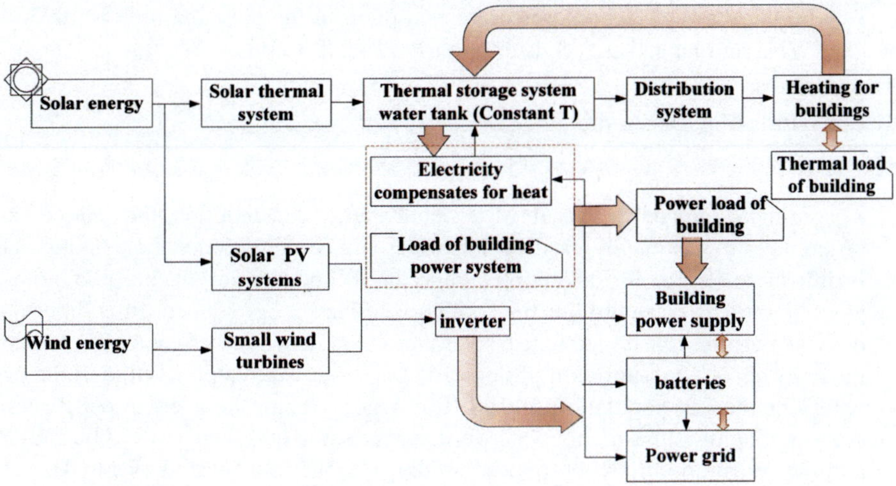

Fig. 5. Energy flow diagram of multi-energy system in the IESpark. (Note: since the total amount of biomass and other energy is very small, it is not considered here. In order to describe the energy flow process, the small wind power is considered here)

According to the annual energy consumption, the annual energy consumption of electricity and gas is obtained, as shown in Table 3. In general, air conditioner uses electricity to drive itself, and the hot water can be heated by the gas or electricity. From the analysis above, we assume that the hot water is supplied by the gas, and the home appliances and electrical equipment are driven by the electricity. Cooking is generally powered by gas.

Table 3. Energy consumption for different building in Shanhaitian IES park

Building type	Sub-index	Power (1×10^4 kWh (th)/a)	Energy types	Energy Consumption (10^4 kWh (th)/a)	Note (Eff)
Residential buildings	Air conditioner	44448.4	Electric power	15874.4	2.8
	Hot water	6378.7	Electric power	6714.4	1.0
	Cooking	7424.1	Fuel gas	12373.5	0.6
	Lighting	4278.3	Electric power	4278.3	1.0
	Household electrical appliances	2260.1	Electric power	2260.1	
Public buildings	Air conditioner	70831.6	Electric power	25297.0	2.8
	Electricity consumption except air conditioner	40882.7	Electric power	40882.7	1.0
	Hot water	14039.7	Electric power	14039.7	1.0
	Cooking	30.4	Fuel gas	50.7	0.6
Summation	Terminal electrical demand	162701.1	Electric power	109346.6	/
	Terminal heat demand	27872.9	Fuel gas	12424.2	/

To sum up, the total annual power consumption in the planning area is 1093.466 million kWh/a and annual gas consumption is 124.242 million kWth/a.

5.2 Analysis and Discussion

(1) Thermal power system
According to the analysis result of terminal energy consumption, the heat demand of the buildings is mainly used for hot water and cooking. Since the cooking habit is difficult to change for the people, while the IES park is very lack of the biomass gas. So, the gas consumption for cooking can hardly be reduced. In addition, the hot water can be considered to use solar thermal system. According to the planning, the daily energy consumption of domestic hot water is 559,000 kWth/d in the park with a permanent population of 190,100. Among them, the average daily energy consumption of domestic hot water for the residential buildings is 175,000 kWth/d, and the consumption for the public buildings is 385,000 kWth/d (Table 4).

Table 4. Area of solar hot water collection system placed on the roof of the buildings

Building type	Area required (10^4 m^2)	Area can be placed (10^4 m^2)
Residential buildings	17.5	43.7
Public buildings	38.5	37.4
Summation	55.9	81.1

Residential buildings have enough areas to place solar collectors to meet the requirements of domestic hot water. However, if the domestic hot water systems are used simultaneously, the efficiency will be very low, so the distributed system should be seriously considered. For the public buildings, if the guarantee rate of solar energy is 120%, the required areas for the solar collectors are about 385,000 m^2. Compared with the areas which can be placed is 374,000 m^2, there is 11,000 m^2 less than the actual solar energy areas. However, the solar collector systems can meet more than 80% demand of domestic hot water, and a small part of the heat demand can be solved by the auxiliary heat sources such as electric boiler. So, the maximum planning areas of solar collector is: 175,000 m^2 for residential buildings and 374,000 m^2 for public buildings.

(2) Power system
According to the statistics in Table 3, the power load is the main load in the IES park. According to the design, two methods can be considered: the solar PV system is adopted for all the roofs of the park or the solar PV system is only arranged on the residential buildings in the park. According to the statistics above, it can be seen that the corresponding areas are 1.668 million m^2 and 0.878 million m^2 respectively, and the annual power generation is 115.617 million kWh/a and 60.516 million kWh/a respectively.

(3) Economic analysis
The initial investment of solar thermal system is about ¥1000/m2 (solar collector areas) [18]. According to the installed areas, the initial investment is ¥175 million for residential buildings, and the annual heat consumption is 63.875 million kwh/a, equivalent to ¥38.32 million/a (assuming average heat price is ¥0.6 yuan/kWth), So, the static investment payback period is 4.6 years. For public buildings, the initial investment is ¥385 million for public buildings, and the annual heat consumption is 163.812 million kwh/a, equivalent to ¥98.29 million/a, So, the static investment payback period is 3.9 years. According to the calculation results of solar PV system, assuming that the installed areas of solar PV panels is 1.668 million m^2, and the initial investment of solar PV panels is ¥700/m^2, so the initial investment is about ¥1.17 billion. The annual power generation is 115.617 million kwh/a. If the local electricity price is ¥0.7/kWh (based on the local data), it is equivalent to be saved with ¥80.932 million every year, and the static investment payback period is about 14.4 years. If the service life of solar PV system is 20 years, the initial investment can be recovered within the life cycle.

As shown in the economic analysis above, it is noted that the recovery period of solar PV is significantly higher than that of solar collector system. Therefore, priority should be given to the application of solar collector system.

6 Conclusion

As an important physical carrier of the energy internet, the IES can not only improve the energy utilization efficiency and promote the consumption of renewable energies, but also ensure the safe and reliable energy supply. Based on the local meteorological data, renewable energy resources, building loads and other data of Shanhaitian IES park, in this paper, the installed capacity of different renewable energies in the IES park is designed by using the capacity matching method. The results show that: based on the planning permanent population of about 190,100 people in the park, in order to meet the energy demand of the park, the maximum wind power generation resources can be planned to be 816,000 kWh/a. For the public buildings, the solar collector areas can be arranged 374,000 m^2 if the assurance rate of solar energy is 120%, which can basically meet the demand of more than 80% heat load, and the left part can be provided by the auxiliary heat sources such as electric boilers. Besides, the optimal layout area of solar PV system is 878,000 m^2 from the aspect of economic interests, and the annual power generation is 60.516 million kWh/a, and the remaining power still needs to be supplemented by the large power grid.

Acknowledgements. The present work is supported by the Science and Technology Project: Simulation Platform Development of Integrated Energy System based on Multi Energy Coupling (Phase I) (SGTYHT/19-JS-215), the Youth Innovation Promotion Association of the Chinese Academy of Sciences, and the Scientific Research Equipment Project of Chinese Academy of Sciences: Development of 100 m^2 Heliostat Shape Measurement Equipment (YJKYYQ 20180052).

References

1. Hongtao, Y: Research of Monitoring System of Wind/Photovoltaic System. Master's thesis, Shandong University (2011)
2. Hongjie, J., Dan, W., Xiandong, X., et al.: Research on some key problems related to integrated energy systems. Autom. Electr. Power Syst. **4**(7), 198–207 (2015)
3. Haozhong, C., Xiao, H., Li, W., Yuquan, L., Qi, Y.: Review on research of regional integrated energy system planning. Autom. Electr. **43**(7), 2–13 (2019)
4. Jianzhong, W.: Drivers and state-of-the-art of integrated energy systems in Europe. Autom. Electr. Power Syst. **40**(5), 1–7 (2016)
5. Weiyan, X.: Research and analysis of comprehensive energy planning based on improved self-adaption genetic algorithms. Power Syst. Big Data **22**(7), 3–16 (2019)
6. Zhiyong, Y.: Research summary of integrated energy systems planning for energy internet. South. Power Syst. Technol. **13**(7), 1–9 (2019)
7. Zhao, H.: Research on urban comprehensive energy planning method based on complex adaptive system theory. Power Des. **21**(4), 35–47 (2019)
8. Xiao, H., Ce, S., Dongwen, C., Li, W., Feng, Z., Haozhong, C.: Multi objective planning method for regional integrated energy systems considering energy quality. Autom. Electr. Power Syst. **43**(19), 22–31 (2019)
9. Anan, Z., Hong, Z., Jianzhong, W., Meysam, Q., Qian, L.: Multi-objective stochastic planning for offshore micro integrated energy systems. Autom. Electr. Power Syst. **43**(7), 129–135 (2019)
10. Ammar, H.M.A., Seyed, H.H., Behrooz, V., Shahab, D.: Security constrained multi-objective bidirectional integrated electricity and natural gas co-expansion planning considering multiple uncertainties of wind energy and system demand. IET Renew. Power Gener. **14**(8), 1395–1404 (2020)
11. Qi, W., Chengmin, W., Ning, X., Haibing, W.: A hybrid CVaR-IGDT expansion planning model for regional integrated energy system. Power Syst. Technol. **44**(2), 505–513 (2020)
12. Lidi, Z., Xun, D., Jun, W., Pan, Z.: Review of distribution network planning in the context of integrated energy. Distrib. Util. **35**(4), 37–45 (2018)
13. Qing, S., Liangliang, Z.: distributed energy planning for electro-thermal coupling system based on integrated power flow. Hubei Power **43**(3), 1–9 (2019)
14. Haizhu, L., Hongtao, D., et al.: Research on the current situation and development trend of energy consumption of civil buildings in China. Constr. Sci. Technol. 10–11 (2018)
15. Haoran, Y., Hua, C.: Power coordination control strategy for small wind and solar complementary power generation system. Bull. Sci. Technol. **35**(12), 105–109 (2019)
16. Heng, X., Ruizhao, Z., Zhenbin, Y., Chunhong, Y.: Assessment of wind energy reserves in China. ACTA Energiae Solaris Sinica **22**(2), 167–170 (2001)
17. Fan, W.: The study on Forecasting of Urban Sewage. Master thesis, East China Normal University(2011)
18. Wu, J.-w.: Analysis and comparison of solar thermal and photovoltaic cooling systems. Chin. J. Refrig. Technol. **36**(3), 42–48 (2016)

Small-Signal Modeling of Modular Multilevel DC Transformer

Kai Hou[1,2,3](✉), Xingyu Wu[4], Fuhong Min[4], Zhigang Wang[1,2,3], and Guoning Wang[1,2,3]

[1] State Key Laboratory of Smart Grid Protection and Control, Nanjing 211106, Jiangsu, China
houkai@sgepri.sgcc.com.cn

[2] NARI Group Corporation (State Grid Electric Power Research Institute), Nanjing 211106, Jiangsu, China

[3] NARI Technology Co. Ltd., Nanjing 211106, Jiangsu, China

[4] School of Electrical and Automation Engineering, Nanjing Normal University, Nanjing 210023, China

Abstract. The dc transformer is considered to be one of the most promising components in future energy systems. The modular multilevel dc transformer inherits many of the characteristics of modular multilevel converter and dual active bridge, which brings many advantages and leads to different working principles from traditional dc transformers. In this paper, the mathematical model is obtained by harmonic decomposition and the small-signal model is obtained by linearization. The parameters of the PI controller are designed according to the small-signal model. The open-loop time-domain simulation verifies the accuracy of the mathematical model. The closed-loop simulation model verifies the reliability of the PI controller. After adding the PI controller, the results of the output voltage show that the model has fast dynamic response with no static errors, which proves that the model obtained by the harmonic decomposition is reliable.

Keywords: Modular multilevel dc transformer · Harmonic model · Small-signal model

1 Introduction

In last several years, with the rapid development of distributed power supply, new energy vehicle technology and flexible dc power grid, dc transformer (DCT) has attracted more and more attention [1, 2]. Dual Active Bridge (DAB) is currently the most representative dc converter [3–5]. The input-series-output-parallel (ISOP) topology DCT due to DAB unit has been a research hotspot in the field of dc microgrid in recent years. This topology owns the advantages of high power density and fast dynamic response, but it also weak in fault handling ability. When the dc short-circuit malfunction happens in the power grid, the malfunction cannot be isolated quickly without over current, and the power grid won't head back into operation immediately after the fault disappears; The topology cannot realize the redundancy technology, so

Y. Xue et al. (Eds.): PMF 2020, LNEE 718, pp. 841–852, 2021.
https://doi.org/10.1007/978-981-15-9746-6_63

the equipment have to stop working when the internal DAB submodule breakdowns. These shortcomings greatly limit its application in practical engineering [6–9].

The DCT based on the modular multilevel converter (MMC) structure overcomes the problem of low voltage stress of the switching device and has good fault handling capabilities [10–12]. When the dc distribution network fails, it can be employed as a dc circuit breaker to cut off the connection to the power grid, which greatly reduces the cost of equipment; MMC can realize the design of redundant modules to enhance the malfunction handling ability of the equipment. The MMC structure is used at the medium-voltage side to improve the voltage level, and the full-bridge modules are connected in parallel at the low-voltage side to increase the current level, which constitutes the modular multilevel dc transformer (MDCT) topology. Literature [13] analyzes the voltage, current, power and soft switching characteristics of the high frequency link under different modulation strategies. Literature [14–17] proposes time domain modeling, state space modeling, and extended description function method modeling. These modeling methods are computationally intensive and relatively complex. In this thesis, the harmonic decomposition is adopted to the modeling of MDCT and the parameters of PI controller are obtained due to the model.

2 Operation Principle of MDCT

The structure of the MDCT is shown in Fig. 1. It is made up of a primary side, an AC high frequency link (ACHFL) and a secondary side. The primary side is composed of two phase MMC, each arm has k SM submodules, SM is the half-bridge submodule, and L_s is the arm inductance of MMC. The ACHFL is composed of h high frequency isolated transformers. The secondary side adopts the topology structure of full-bridges in parallel. L_T is the leakage inductance of the high-frequency transformer; n_T is the transformation ratio of the high-frequency transformer; $V_i$and V_o are the dc voltages at the primary side and the secondary side, respectively; I_i and I_o are the dc currents at the primary side and the secondary side, respectively; $i_{DC1} \sim i_{DCh}$ are the currents at the dc side of the full bridges; $i_{C1} \sim i_{Ch}$ are the capacitor currents of the full-bridges.

In traditional grid-connected MMC, sine wave modulation is often used. Applying sine wave modulation to the DC/DC converter will reduce the utilization rate of the transformer, which is not conducive to improving the power density. But in MDCT, step wave modulation can improve transformer utilization. The primary side uses a step wave modulation strategy, the angle between each sub-module is θ, the secondary side full-bridges are all square wave modulation, and the ideal waveforms of ACHFL are shown in Fig. 2. The working principle of MDCT under single phase shift control is similar to the traditional DAB. The adjustment of the transmission power is achieved by adjusting the voltage phase difference between V_a and V_b.

According to the above definition, the ACHFL voltages V_a and V_b under single phase shift control can be obtained due to Fourier series summation theory, as follow:

$$\begin{cases} V_a(t) = \dfrac{4V_i}{k\pi}\displaystyle\sum_{n=0}^{N}\dfrac{\sin\left[k(2n+1)\frac{-\theta}{2}\right]\sin\left[(2n+1)\left(\omega t-\frac{k-1}{2}\theta\right)\right]}{(2n+1)\sin\left[(2n+1)\frac{-\theta}{2}\right]} \\ V_b(t) = hn_T V_o\dfrac{4}{\pi}\displaystyle\sum_{n=0}^{N}\dfrac{\sin[(2n+1)(\omega t-\delta)]}{2n+1} \end{cases} \tag{1}$$

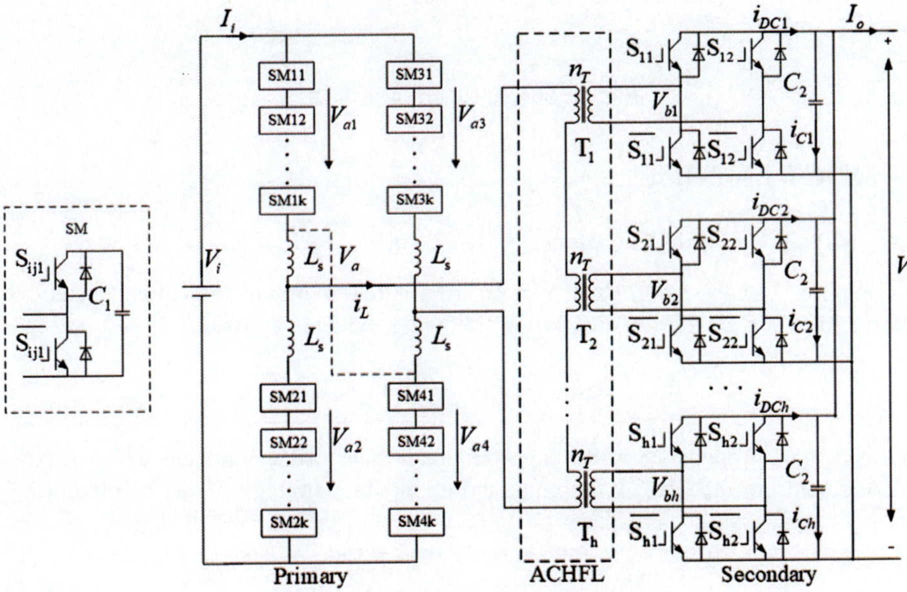

Fig. 1. The circuit diagram of MDCT.

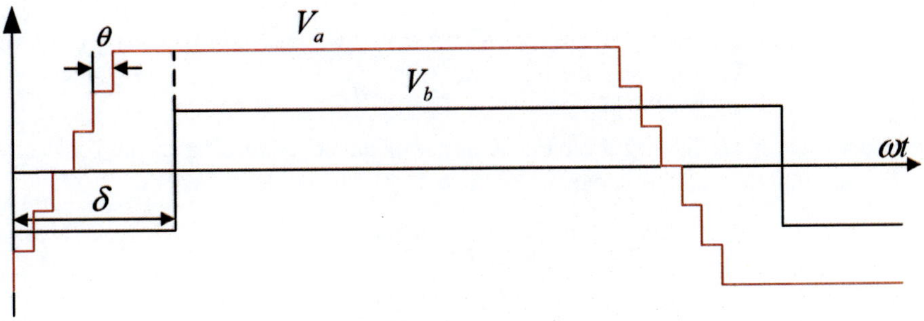

Fig. 2. Ideal waveforms of ACHFL.

According to single phase shift control and the topological structure characteristics of MDCT, ignoring the power semiconductor device losses, the ACHFL can be equivalent to the equivalent structure of V_a and V_b in series with the equivalent leakage inductance $L(L = L_s + hL_T)$, as displayed in Fig. 3.

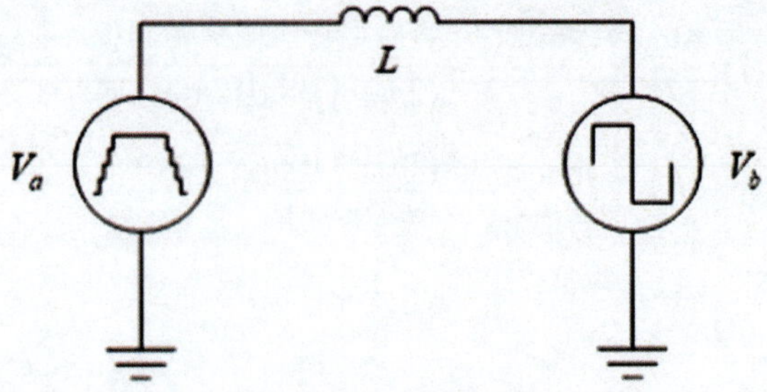

Fig. 3. Simplified circuit of MDCT.

3 MDCT Modeling

3.1 MDCT Dynamic Equations

A Kirchhoff Voltage Loop (KVL) summation is taken around the equivalent circuit, illustrated in Fig. 3. This results in the following dynamic equation.

$$R_L i_L(t) + L\frac{di_L(t)}{dt} = V_a - V_b \tag{2}$$

where $R_L$is the impedance of the high-frequency transformer inductance.

According to the MDCT topology and modulation strategy, it can be obtained:

$$i_{DC1} = \cdots = i_{DCh} = i_{DC} \tag{3}$$

$$i_{C1} = \cdots = i_{Ch} = i_C \tag{4}$$

Kirchhoff Current Law (KCL) is applied to the output node of the MDCT, which gives:

$$i_C(t) = i_{DC}(t) - \frac{I_o(t)}{h} \tag{5}$$

In the case of neglecting the capacitance impedance, according to the basic circuit theory, the equation of the capacitor voltage $V_o(t)$ can be described as:

$$\frac{dV_o(t)}{dt} = \frac{i_C(t)}{C_2} \tag{6}$$

Substituting (5) into (6), the state equation of output voltage, capacitor current and load current can be obtained as:

$$\frac{dV_o(t)}{dt} = \frac{i_{DC}(t)}{C_2} - \frac{I_o(t)}{C_2 h} \tag{7}$$

3.2 Number of Effective Harmonics

In Eq. (1), the value of N determines the number of effective harmonic components contained in the Fourier series expression. Therefore, it is very important to choose the appropriate N, because if the value of N is too small, the model will lack accuracy, but if the value of N is too large, the model will become too complicated. This thesis determines the size of N based on transmitting active power.

According to (1) and (2), the phasor expression of the *(2n + 1)th* component of the voltage and current of ACHFL can be described as:

$$\begin{cases} \dot{I}_{L(2n+1)} = \dfrac{\left\{ \begin{array}{c} \dfrac{2\sqrt{2}V_i}{k} \dfrac{\sin\left[k(2n+1)\frac{-\theta}{2}\right]}{\sin\left[(2n+1)\frac{-\theta}{2}\right]} \angle\left[(2n+1)\left(-\dfrac{k-1}{2}\theta\right)\right] \\ -2\sqrt{2}hn_T V_o \angle[(2n+1)(-\delta)] \end{array} \right\}}{\pi(2n+1)[R_L + j(2n+1)\omega L]} \\ \dot{V}_{a(2n+1)} = \dfrac{2\sqrt{2}V_i}{\pi k(2n+1)} \dfrac{\sin\left[k(2n+1)\frac{-\theta}{2}\right]}{\sin\left[(2n+1)\frac{-\theta}{2}\right]} \angle\left[(2n+1)\left(-\dfrac{k-1}{2}\theta\right)\right] \\ \dot{V}_{b(2n+1)} = \dfrac{2\sqrt{2}hn_T V_o}{\pi(2n+1)} \angle[(2n+1)(-\delta)] \end{cases} \quad (8)$$

According to the theory of steady-state sinusoidal analysis, active power is only generated when the angular frequency of the voltage and current are the same. The unified expression of the *(2n + 1)th* harmonic component of the active power can be derived:

$$\begin{aligned} P_{a(2n+1)} &= \mathrm{Re}\left(\dot{V}_{a(2n+1)} \dot{I}^*_{L(2n+1)}\right) \\ &= \frac{A^2\cos(\varphi_{Zn}) - AB\cos\left[(2n+1)\left(\delta - \frac{k-1}{2}\theta\right) + \varphi_{Zn}\right]}{|Z_n|} \end{aligned} \quad (9)$$

$$\begin{aligned} P_{b(2n+1)} &= \mathrm{Re}\left(\dot{V}_{b(2n+1)} \dot{I}^*_{L(2n+1)}\right) \\ &= \frac{AB\cos\left[(2n+1)\left(\frac{k-1}{2}\theta - \delta\right) + \varphi_{Zn}\right] - B^2\cos(\varphi_{Zn})}{|Z_n|} \end{aligned} \quad (10)$$

where $A = \frac{2\sqrt{2}V_i}{\pi k(2n+1)} \frac{\sin\left[k[2n+1]\frac{-\theta}{2}\right]}{\sin\left[[2n+1]\frac{-\theta}{2}\right]}$, $B = \frac{2\sqrt{2}hn_T V_o}{\pi(2n+1)}$,

$$|Z_n| = \sqrt{R_L^2 + [(2n+1)\omega L]^2}, \quad \varphi_{Zn} = \tan^{-1}\left[\frac{(2n+1)\omega L}{R_L}\right].$$

If ignoring R_L, $P_{a(2n+1)}$ is equal to $P_{b(2n+1)}$, and the harmonic active power summation can be derived,

$$P = n_T h \frac{8V_i V_o}{k\pi^2} \sum_{n=0}^{N} \frac{\sin\left[k(2n+1)\frac{-\theta}{2}\right]}{\sin\left[(2n+1)\frac{-\theta}{2}\right]} \frac{\sin\left[(2n+1)\left(\delta - \frac{k-1}{2}\theta\right)\right]}{(2n+1)^3 \omega L} \tag{11}$$

The harmonic active power summation (11) can be compared with the analytical expression (12) [18]. The number of N is then chosen such that a good match is obtained between these two active power expression.

$$P_o = n_T h \frac{V_i V_o}{\pi\omega L}\left[-\delta^2 + (k-1)\theta\delta + \pi\delta + \frac{(1-2k)(k-1)}{6}\theta^2 - \frac{(k-1)\pi\theta}{2}\right] \tag{12}$$

The simulation parameters are displayed in Table 1. The differences between the harmonic active power summation and the analytical expression are listed in Table 2. When N is equal to 3, the difference between the two power expressions is less than 0.1%, which can be ignored. Therefore, N equal to3 has been used for mathematical modeling in this thesis.

Table 1. Simulation parameters.

Parameter	Value
V_i/kV	10
V_o/V	750
k	8
h	8
θ	0.02π
n_T	1:1
L_T/μH	45
f/Hz	3000

Table 2. Difference between (11) and (12).

Effective harmonics (N)	Difference
0	10.43%
1	2.361%
2	0.514%
3	0.087%
4	0.028%
5	0.034%

3.3 Derivation of Harmonic Model

Ignoring the impedance of the leakage inductance, according to the *(2n + 1)th* harmonic component of the active power expression (10), we have:

$$P_{b(2n+1)} = h\dot{V}^{*}_{o(2n+1)}\dot{I}_{DC(2n+1)} \tag{13}$$

The phasor domain expression of the *(2n + 1)th* harmonic component of i_{DC} can be derived:

$$\dot{I}_{DC(2n+1)} = \frac{AB\cos\left[(2n+1)\left(\frac{k-1}{2}\theta - \delta\right) + \varphi_{Zn}\right] - B^2\cos(\varphi_{Zn})}{hV_o|Z_n|} \tag{14}$$

By replacing (14) in (7), and eliminating high-frequency terms, the mathematical model of MDCT can be described as:

$$\begin{aligned}\frac{dV_o(t)}{dt} &= -\frac{I_o(t)}{C_2 h} + \frac{i_{DC}(t)}{C_2}\\ &= -\frac{I_o(t)}{C_2 h} + \frac{8n_T}{C_2\pi^2}\sum_{n=0}^{N}\frac{1}{(2n+1)^2}\\ &\times\left\{\begin{array}{l}\frac{V_i}{k|Z_n|}\frac{\sin\left[k(2n+1)\frac{-\theta}{2}\right]}{\sin\left[(2n+1)\frac{-\theta}{2}\right]}\cos\left[(2n+1)\left(\frac{k-1}{2}\theta - \delta\right) + \varphi_{Zn}\right]\\ -\frac{hn_T V_o\cos(\varphi_{Zn})}{|Z_n|}\end{array}\right\}\end{aligned} \tag{15}$$

3.4 Small-Signal Disturbance and Linearization

Small-signal disturbances are added around the steady-state operating point of the MDCT, and high-order disturbances are ignored to obtain a small-signal model.

$$\begin{cases}V_o(t) = V_o + \Delta V_o(t)\\ I_o(t) = I_o + \Delta I_o(t)\\ \delta = \delta_0 + \Delta\delta\end{cases} \tag{16}$$

$$\frac{d\Delta V_o(t)}{dt} = a\Delta V_o(t) + b\Delta\delta + c\Delta I_o(t) \tag{17}$$

$$a = -\frac{8hn_T^2}{C_2\pi^2}\sum_{n=0}^{N}\left[\frac{\cos(\varphi_{Zn})}{(2n+1)^2|Z_n|}\right] \tag{18}$$

$$b = \frac{8n_T V_i}{kC_2\pi^2}\sum_{n=0}^{N}\left\{\frac{\sin\left(k[2n+1]\frac{-\theta}{2}\right)}{\sin\left([2n+1]\frac{-\theta}{2}\right)}\frac{\sin\left[\varphi_{Zn} + (2n+1)\left(\frac{k-1}{2}\theta - \delta_0\right)\right]}{(2n+1)|Z_n|}\right\} \tag{19}$$

$$c = -\frac{1}{C_2 h} \tag{20}$$

4 Closed-Loop Control

In order to achieve high performance regulation of MDCT, the system output voltage must keep a good track of the reference command, without steady-state error, and at the same time can achieve fast response. Since the model is first-order in nature, PI controller is sufficient to realize high performance output voltage regulation. Figure 4 shows the basic control strategy of MDCT.

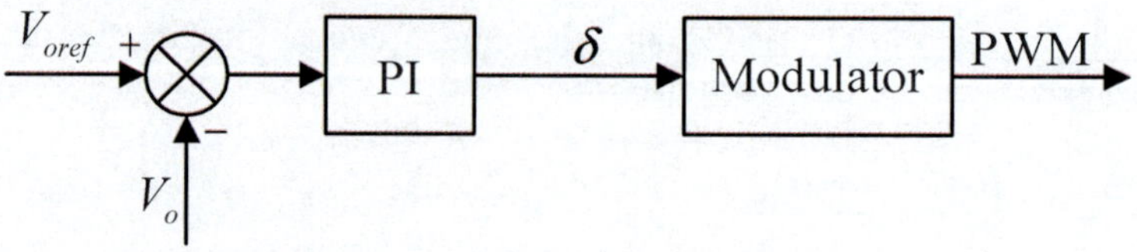

Fig. 4. Control strategy of MDCT.

According to expression (17), when the output voltage becomes state variable and the load current is zero, the transfer function can be described as:

$$G_b(s)\big|_{\Delta I_o(t)=0} = \frac{b}{s-a} \tag{21}$$

Then, when the output voltage becomes the state variable and the phase shift angle is zero, the transfer function that associates the output voltage with load current can be described as:

$$G_c(s)|_{\Delta\delta=0} = \frac{c}{s-a} \tag{22}$$

The bode graph of the forward path before adding PI controller is shown in Fig. 5. The system is a stable system, but the low-frequency gain of the system is small and there is a static error.

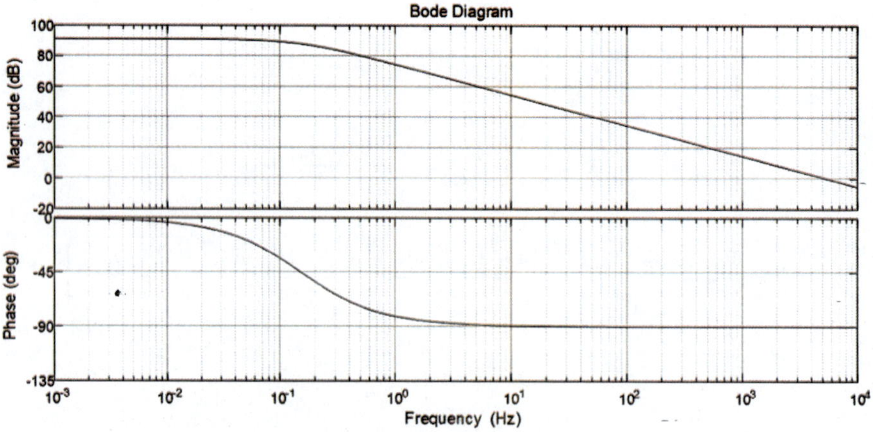

Fig. 5. The Bode diagram of forward path (without PI controller).

This thesis uses SISOTOOL to easily debug accurate PI parameters and complete the control system design. The PID tuning module in SISOTOOL can adjust the bandwidth and phase angle margin of the model. It is easy to get PI (P = 0.06, I = 2.8) parameters. After adding the PI controller with transport delay, the bode diagram of the forward path is displayed in Fig. 6. It can be seen that after adding PI controller, the gain margin is larger than 6 dB and the phase angle margin is larger than 40°, which satisfies the design requirements of the controller.

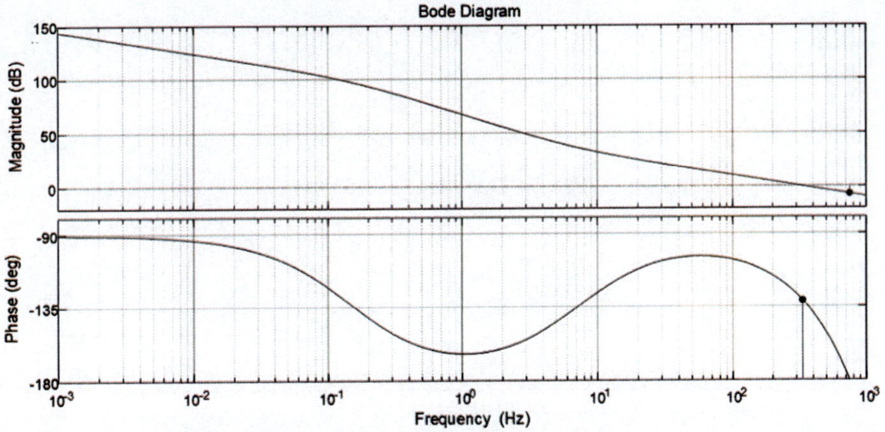

Fig. 6. The Bode diagram of forward path (including PI controller and transport delay).

5 Simulation Results

In this thesis, Matlab/Simulink simulation software is used to build a simulation model according to Fig. 1 to verify the reliability of the model.

5.1 Open-Loop Simulation

The MDCT is open-loop modulated and the phase shift changes from 30° to 35°. The output voltage waveform of Simulink model is compared with the response waveform of the MDCT mathematical model. The results of the mathematical model and Simulink model are displayed in Fig. 7. It can be seen from the voltage waveform that the simulation results are in good agreement with the response of the mathematical model. This proves that the mathematical model is accurate.

5.2 Closed-Loop Simulation

In order to verify the reliability of PI controller, its response to step reference voltage is tested. When the simulation time is 0.3 s, the reference value changes from 750 V till 800 V. Figure 8 shows the simulation results. The good match between the simulation result and the mathematical model proves the reliability of the model. The simulation result shows that the model has short transient time, good dynamic performance and no steady-state error.

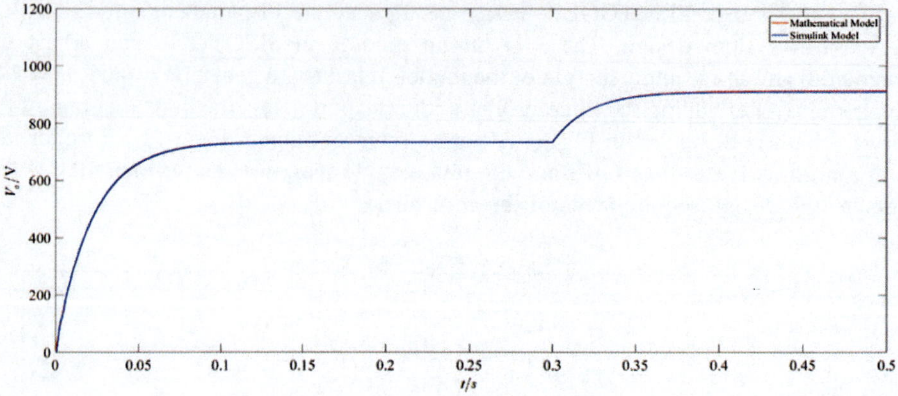

Fig. 7. Open-loop simulation.

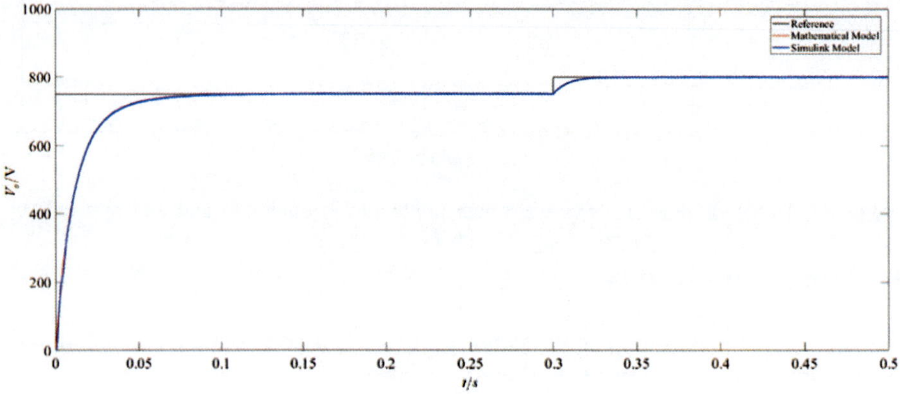

(a) Output voltage.

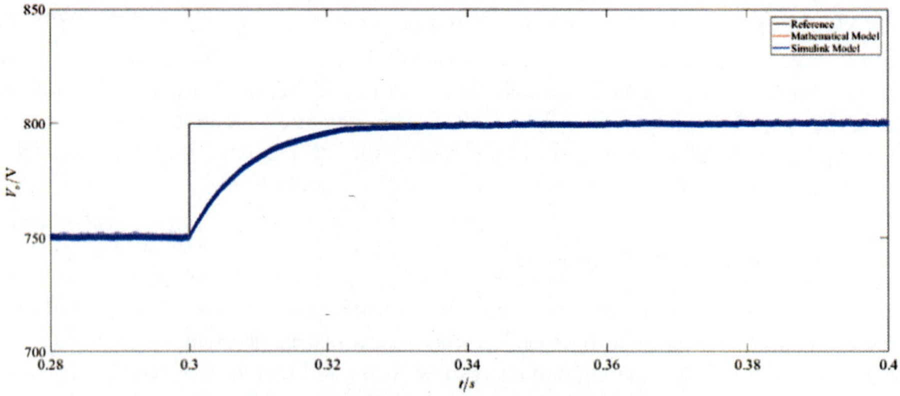

(b) Reference voltage step change from 750V to 800V.

Fig. 8. Closed-loop simulation.

6 Conclusions

In this thesis, the harmonic decomposition is adopted to gain the small-signal model of MDCT. The time domain simulation verifies the accuracy of the mathematical model. This model is used to design the PI controller parameters under the stable operation state. Finally, the simulation test verifies the accuracy of the harmonic modeling strategy and the stability of the controller.

References

1. Huang, A.Q., Crow, M.L., Heydt, G.T.: The future renewable electric energy delivery and management (FREEDM) system: the energy internet. Proc. IEEE **99**(1), 133–148 (2010)
2. Zhao, B., Song, Q., Liu, W.: Overview of dual-active-bridge isolated bidirectional DC–DC converter for high-frequency-link power-conversion system. IEEE Trans. Power Electron. **29**(8), 4091–4106 (2014)
3. Hu, Y., Zhang, Y., Zhang, T.: Optimization method of current stress for dual active bridge converter considering different soft switching modes. Autom. Electr. Power Syst. **43**(23), 58–68 (2019)
4. Hou, N., Song, W., Wu, M.: Minimum-current-stress scheme of dual active bridge DC–DC converter with unified phase-shift control. IEEE Trans. Power Electron. **31**(12), 8552–8561 (2016)
5. Bal, S., Yelaverthi, D.B., Rathore, A.K., Srinivasan, D.: Improved modulation strategy using dual phase shift modulation for active commutated current-fed dual active bridge. IEEE Trans. Power Electron. **33**(9), 7359–7375 (2018)
6. Zhang, H., Li, Y., Gao, F.: High-frequency current fluctuation suppression strategy for isolation stage of cascaded H-bridge based power electronic transformer. Autom. Electr. Power Syst. **44**(07), 130–150 (2020)
7. Zumel, P., Ortega, L., Lazaro, A., Fernandez, C.: Control strategy for modular dual active bridge input series output parallel. In: 14th Workshop on Control and Modeling for Power Electronics. pp. 1–7. IEEE, Salt Lake City, UT (2013)
8. An, F., Song, W., Yang, K.: Virtual power balance control scheme of dual active bridge DC-DC converters with output-parallel structure. Autom. Electr. Power Syst. **42**(12), 106–112 (2018)
9. Zhao, B., Song, Q., Li, J., Liu, W.: Modular multilevel high-frequency-link DC transformer based on dual active phase-shift principle for medium-voltage DC power distribution application. IEEE Trans. Power Electron. **32**(3), 1779–1791 (2017)
10. Kenzelmann, S., Rufer, A., Dujic, D.: Isolated DC/DC structure based on modular multilevel converter. IEEE Trans. Power Electron. **30**(1), 89–98 (2015)
11. Guan, M., Xu, Z.: Modeling and control of a modular multilevel converter-based HVDC system under unbalanced grid conditions. IEEE Trans. Power Electron. **27**(12), 4858–4867 (2012)
12. Debnath, S., Qin, J., Bahrani, B., Saeedifard, M.F.: Operation, control, and applications of the modular multilevel converter: a review. IEEE Trans. Power Electron. **30**(1), 37–53 (2015)
13. Zhao, B., Song, Q., Li, J.: High-frequency-link modulation methodology of DC–DC transformer based on modular multilevel converter for HVDC application: comprehensive analysis and experimental verification. IEEE Trans. Power Electron. **32**(5), 3413–3424 (2017)

14. Wang, Z., Dong, C., Hou, K., Gao, P.: Time domain model of full bridge LLC circuit and its analysis. Autom. Electr. Power Syst. **42**(20), 138–143 (2018)
15. Shi, H., Sun, K., Wu, H., Li, Y.: A unified state-space modeling method for a phase-shift controlled bidirectional dual-active half-bridge converter. IEEE Trans. Power Electron. **35** (3), 3254–3265 (2020)
16. Hebala, O.M., Aboushady, A.A., Ahmed, K.H.: Generalized small-signal modelling of dual active bridge DC/DC converter. In: 7th International Conference on Renewable Energy Research and Applications, pp. 914–919. IEEE, Paris (2018)
17. Hao, K., Lu, J.: Modeling research of dual active bridge DC converter based on double phase shift control. In: 3rd International Conference on Robotics and Automation Sciences, pp. 79–84. IEEE, Wuhan, China (2019)
18. Shao, S., Jiang, M., Zhang, J.: A capacitor voltage balancing method for a modular multilevel DC transformer for DC distribution system. IEEE Trans. Power Electr. **33**(4), 3002–3011 (2018)

A Protection Based on Difference Between DG Impedance and Equivalent Impedance of Grid for Microgrid Transmission Line

Peng Lv, Yiqing Liu(✉), Guangbin Zhu, and Hanxiao Meng

School of Electrical Engineering, University of Jinan, Jinan, China
cse_liuyq@ujn.edu.cn

Abstract. The overcurrent protection applied in the traditional distribution network is applied to the microgrid to protect the microgrid transmission line. However, the application of this method in microgrid transmission line is likely to lead to mal-operation or refuse operation of the protection. In order to solve the problem that mal-operation or refuse operation of the protection in microgrid transmission line, a protection based on difference between DG impedance and equivalent impedance of grid for microgrid transmission line is proposed. In this method, the voltage and current data on both sides of the protected line need to be collected. Then the positive sequence fault component voltage and the positive sequence fault component current are calculated by the two side protections respectively. The measurement impedance is calculated by the positive sequence fault component voltage and the positive sequence fault component current respectively. Finally, according to the relationship between the phase angle and the amplitude of the measured impedance on both sides of line, the internal fault and external fault are determined. The effectiveness of the protection method is proved by simulating internal fault and external fault of the protected line in PSCAD simulation software.

Keywords: Microgrid · Distributed generator · Impedance · Protection · Fault component

1 Introduction

Microgrid is a new power generation and distribution system composed of distributed generation (DG), power electronic devices and loads [1]. Microgrid can give full play to the advantages of photovoltaic power generation, wind power generation and other new energy distributed generation technologies [2, 3]. The structure of microgrid is different from that of traditional distribution network. Microgrid contains a large number of DG. Microgrid is a system with multiple sources. So power flow distribution and short circuit current in microgrid are different from those in traditional distribution network [4, 5]. If traditional overcurrent protection method is directly applied in

This work was supported by Key Research and Development Program of Shandong Province (2019GGX103014) and Science and Technology Program of University of Jinan (XKY2020).

Y. Xue et al. (Eds.): PMF 2020, LNEE 718, pp. 853–867, 2021.
https://doi.org/10.1007/978-981-15-9746-6_64

microgrid, it will easily lead to mal-operation or refuse operation of protection in microgrid transmission line.

To solve the above problems, scholars have conducted a lot of researches on transmission line protection of microgrid, and these achievements can be divided into the several categories:

- Improvement of traditional protection principle based on local information

According to the principle that the voltage amplitude of the node closer to the fault location is lower after the fault, the concept of weighted side length and side voltage are proposed, and then the microgrid protection based on side voltage is proposed in reference [6]. A new microgrid transmission line protection strategy is proposed, in which the setting value of protection changes with the DG power output level and fault current level in reference [7]. However, the setting value solution of this kind of protection strategy is more complex due to the power flow in both directions. And the real-time calculation speed of setting value is affected by hardware conditions.

- Strategy of applying longitudinal current differential protection scheme to microgrid protection

The longitudinal current differential protection distinguishes internal fault and external fault by comparing the amplitude and phase of the current on both sides of the protected line. The scheme of applying longitudinal current differential protection to microgrid transmission line is an improved scheme. The defect of this improved scheme is that the data on both sides must be strictly synchronized.

- Protection based on intelligent algorithm

The protection scheme based on intelligent algorithm can better adapt to the diversity of DG and the flexibility of operation mode. Intelligent algorithms such as data mining, statistical classification and feature extraction are used to obtain the most relevant features to determine the fault location in reference [8]. The protection requires high communication rate and data processing capacity of central processing unit [9, 10].

To sum up, the existing microgrid transmission line protection scheme has many problems, such as the influence of load current, high requirements for hardware equipment and data synchronization. The defects of these protection methods are caused by not considering the characteristics of DG itself. In this paper, the impedance characteristic of DG is used to construct the protection criterion of microgrid transmission line.

The rest of this paper is organized as follows. Section 2 proposes the protection method based on difference between DG impedance and equivalent impedance of grid for microgrid transmission line. In Sect. 3, simulations verify the protection. Section 4 draws a conclusion.

2 Protection Principle Based on Difference Between DG Impedance and Equivalent Impedance of Grid for Microgrid Transmission Line

2.1 Impedance Characteristics of DG During Fault

The protection criterion proposed in this paper makes use of the characteristics of DG impedance. So the impedance characteristics of DG need to be analyzed. The DG in microgrid is usually Inverter-Interfaced Distributed Generator (IIDG). IIDG is widely used and representative in microgrid. So this paper chooses the microgrid with IIDG for research.

The fault characteristics of IIDG are mainly determined by the control strategy of IIDG. So it is necessary to analyze the control strategy of IIDG.

The control strategy of IIDG mainly includes low voltage ride through (LVRT) strategy and output current limiting strategy. When there is a short circuit fault in the transmission line of microgrid, the output reactive current by IIDG can be described as follows:

$$I_q = \begin{cases} 0 & (U_g > 0.9) \\ 1.5 \times (0.9 - U_g) \times I_N & (0.2 \le U_g \le 0.9) \\ 1.05 I_N & (U_g < 0.2) \end{cases} \tag{1}$$

where I_q is the reactive current when fault. U_g is the per unit of the point of common coupling (PCC). I_N is the rated output current of IIDG. In addition, the maximum current shall be limited to 1.2 times of the rated current, so as to avoid the power electronic devices from being burnt out due to excessive current in case of fault. So the output active current by IIDG can be described as follows:

$$I_d = \sqrt{(1.2 I_N)^2 - I_q^2} \tag{2}$$

Due to the influence of LVRT strategy, fault conditions and load current, the phase angle of the equivalent positive sequence impedance of IIDG is not a constant value. In addition, it can be seen from Eq. (1) that I_q changes in a large range from 0 to $1.05 I_N$.

Put Eq. (1) and Eq. (2) into Eq. (3). And phase angle of the equivalent positive sequence impedance of IIDG can be obtained.

$$\begin{cases} \arg(\Delta Z_1) = \arctan \frac{\dot{U}_{|0|} - \dot{U}_1}{\dot{I}_1 - \dot{I}_{|0|}} \\ \dot{U}_1 = k \dot{U}_{|0|} (\cos\delta + j\sin\delta) \\ \dot{I}_1 = I_d (\cos\delta + j\sin\delta) + I_q (\sin\delta - j\cos\delta) \\ \dot{U}_{|0|} = U_N \\ \dot{I}_{|0|} = I_N \end{cases} \tag{3}$$

where δ is positive sequence voltage phase angle difference between fault and normal state, k is degree of voltage drop. $\dot{U}_{|0|}$ is positive sequence voltage before fault, $\dot{I}_{|0|}$ is positive sequence current before fault. $\dot{U}_1$ is positive sequence voltage when occurs fault, $\dot{I}_1$ is positive sequence current when occurs fault. U_N is the positive sequence rated voltage of IIDG, I_N is the positive sequence rated current of IIDG. ΔZ_1 is the positive sequence impedance of IIDG.

Calculate the Eq. (3) by MATLAB, it can be seen that phase angle of the equivalent positive sequence impedance of IIDG is in the range of 0°–360°. Phase angle of the equivalent positive sequence impedance of IIDG as shown in the Fig. 1.

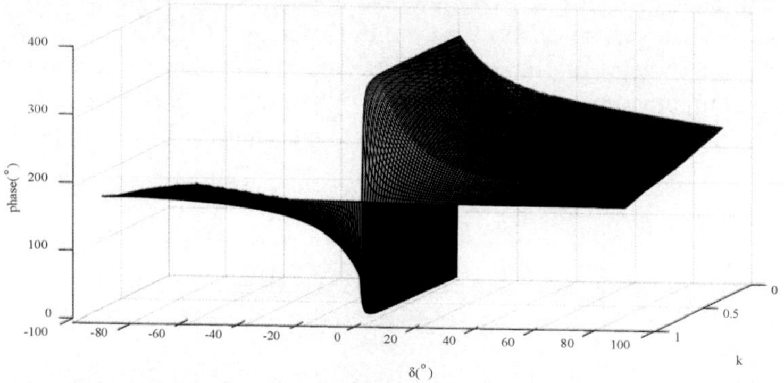

Fig. 1. Phase angle of the equivalent positive sequence impedance of IIDG

IIDG has larger equivalent impedance because of its smaller capacity compared with large power supply and limitation of output current. Therefore, IIDG will provide a small fault current in case of short circuit fault in microgrid.

2.2 Protection Principle Based on Difference Between DG Impedance and Equivalent Impedance of Grid for Microgrid Transmission Line

Based on the standard of IEEE1547 and the definition of microgrid by the consortium for electric reliability technology solutions (CERTS), a simplified model of a microgrid is developed as shown in Fig. 2. As shown in Fig. 2, The voltage level of microgrid is 0.4 kV. The microgrid is connected with the distribution network by PCC. B1 ~ B5 are buses in microgrid. L1 ~ L5 are feeders connected with buses. And Load1 ~ Load7 are three-phase symmetrical loads. E_{DGa}, E_{DGb}, and E_{DGc} are photovoltaic sources. f_1, f_2 and f_3 are different fault positions in microgrid.

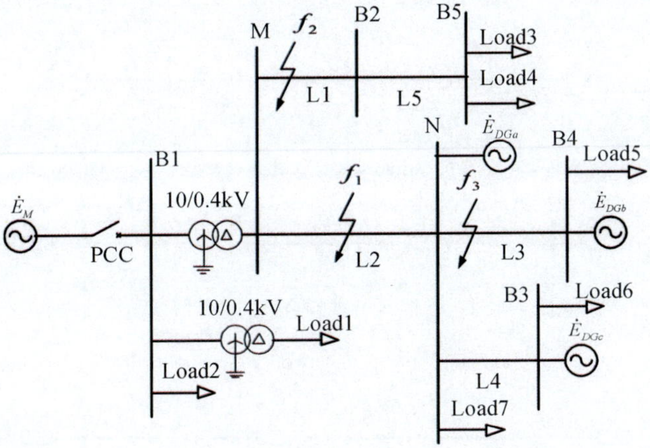

Fig. 2. Typical microgrid structure model

Select line L2 between bus M and bus N in Fig. 2 as the protected line. According to the superposition principle, the microgrid network in Fig. 2 can be divided into normal operation network before fault and component fault positive network. In addition, according to Thevenin equivalence principle, the fault component network of Fig. 2 can be equivalent to Fig. 3.

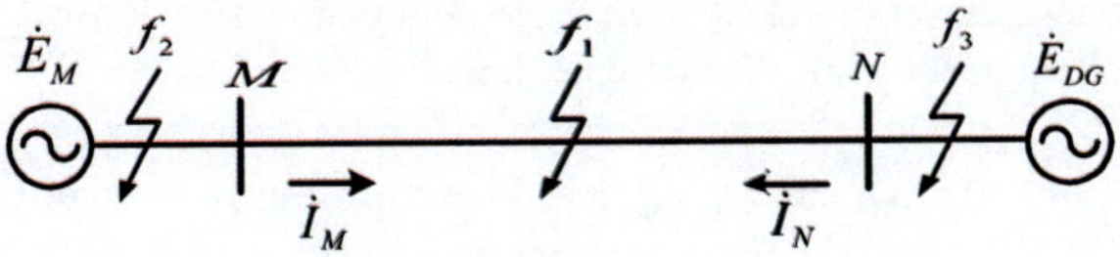

Fig. 3. Equivalent model of fault network

The positive sequence component can reflect all fault types. Using positive sequence component in the protection method can reduce the step of fault phase selection. So it is considered to construct protection criterion by positive sequence component. The equivalent model of positive sequence fault network is shown in Fig. 4 (a). And the component fault networks of different fault locations are shown in Fig. 4 (b) to Fig. 4 (d).

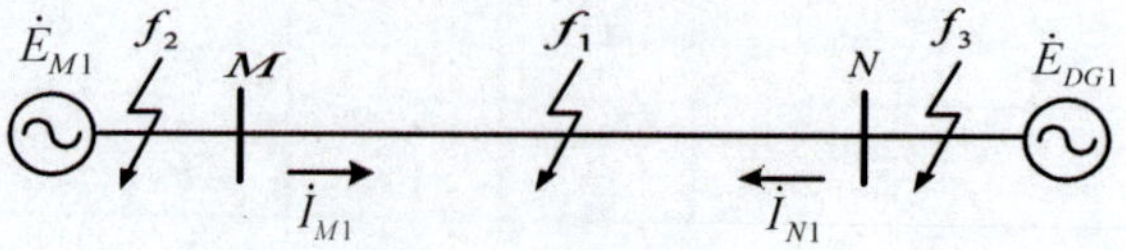

(a) Positive sequence fault network

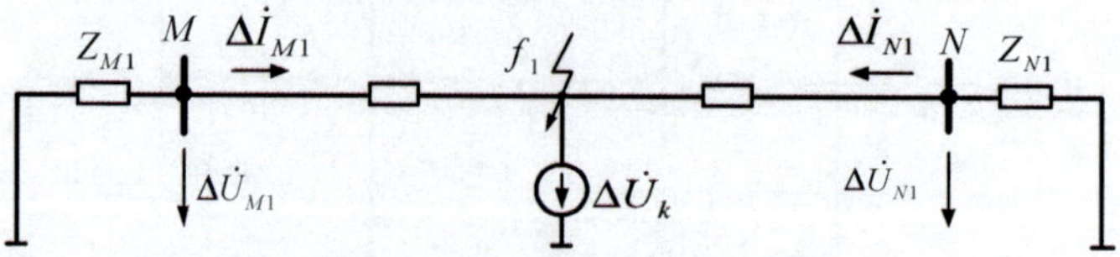

(b) Fault component network of internal fault

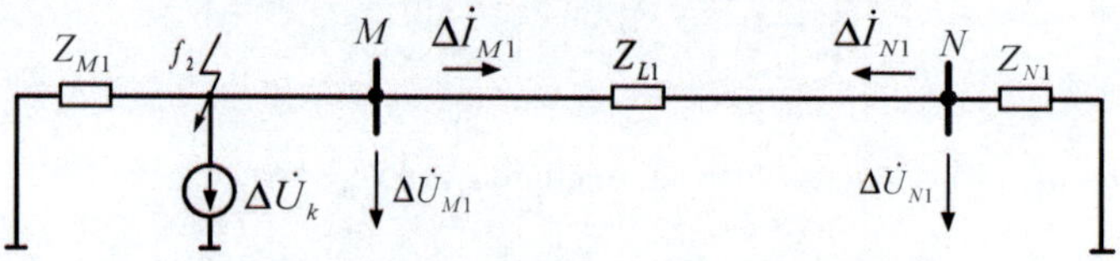

(c) Fault component network of M external fault f_2

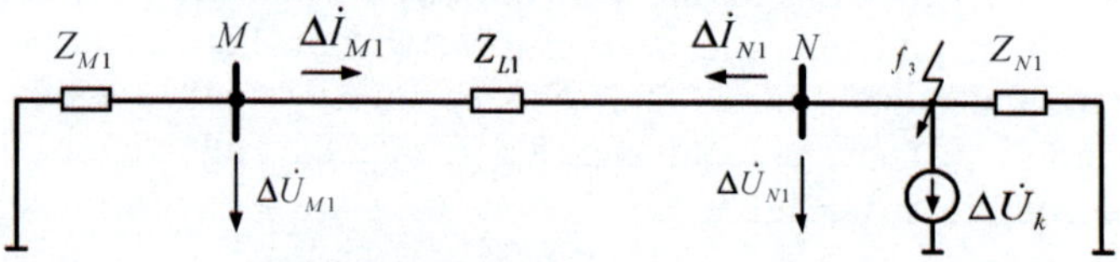

(d) Fault network of N external fault f_3

Fig. 4. The equivalent model of positive sequence fault component network and the component network of positive sequence fault of microgrid line

As shown in Fig. 4, E_{M1} and E_{DG1} are Thevenin equivalent positive potentials of power grid and IIDG. ΔU_k is the component source at the fault point. Z_{M1} is the positive sequence equivalent impedance of the equivalent system in the opposite direction of M side protection. And Z_{N1} is the positive sequence equivalent impedance in the opposite direction of N side protection. It is the positive sequence equivalent impedance of IIDG. Z_{L1} is the positive sequence impedance of the protected line. ΔU_{M1} and ΔU_{N1} are the positive sequence fault component voltage of M side protection and N side protection respectively. $\Delta I_{M1}$and ΔI_{N1} are the positive sequence fault component current of M side protection and N side protection respectively.

Define the positive direction of current as from bus to line. Z_{CM} and Z_{CN} are the measured impedance calculated by the fault component. Z_{CM} and Z_{CN} are expressed as follows:

$$Z_{CM} = -\frac{\Delta\dot{U}_{M1}}{\Delta\dot{I}_{M1}} \tag{4}$$

$$Z_{CN} = -\frac{\Delta\dot{U}_{N1}}{\Delta\dot{I}_{N1}} \tag{5}$$

In microgrid transmission lines, the fault location of microgrid line is divided into: internal, reverse direction of protection on side M, reverse direction of protection on side N. In order to locate these three faults, the impedance characteristics of Z_{CM} and Z_{CN} need to be analyzed.

(1) In case of internal fault in microgrid, the value of Z_{CM} reflects Z_{M1}, and the value of Z_{CN} reflects Z_{N1}.
(2) In case of reverse direction fault of M side protection, the value of Z_{CM} reflects $(-Z_{L1} - Z_{N1})$, and the value of Z_{CN} reflects Z_{N1}.
(3) In case of reverse direction fault of N side protection, the value of Z_{CM} reflects Z_{M1}, and the value of Z_{CN} reflects $(-Z_{L1} - Z_{M1})$.

Combined with the structure and characteristics of microgrid, the protection criterion is constructed by analyzing the results of Eqs. (4) to (5). It can be seen that the phase angle and amplitude of Z_{CM} and Z_{CN} are only related to Z_{M1}, Z_{N1} and Z_{L1}. So the range of Z_{M1}, Z_{N1} and Z_{L1} need to be analyzed. The equivalent system in the opposite direction of M side protection is based on Thevenin equivalent theorem. And the equipment of the equivalent system is mainly resistor inductance. So Z_{M1} is resistor inductance, $\arg(Z_{M1}) \in (0°, 90°)$. According to the previous section, the phase angle of the equivalent positive sequence impedance of IIDG varies in a large range. It is in the range of $(0°, 360°]$. There for, $\arg(Z_{N1}) \in (0°, 360°]$. In addition, according to the characteristics of resistor-inductance of transmission lines in microgrid, It can be concluded that $\arg(Z_{L1}) \in (0°, 90°)$.

In order to determine the internal and external fault, the impedance characteristics of internal and external fault and normal operation need to be analyzed. According to the angle ranges of Z_{M1}, Z_{N1} and Z_{L1}, the angle ranges of Z_{CM} and Z_{CN} in different fault positions can be obtained. The specific analysis process is as follows.

When an internal fault occurs at the protected line, $Z_{CM} = Z_{M1}$, and $\arg(Z_{CM}) \in (0°, 90°)$, $Z_{CN} = Z_{N1}$, and $\arg(Z_{CN}) \in (0°, 360°]$.

When an external fault occurs at reverse direction of protection on side N, $Z_{CM} = Z_{M1}$, and $\arg(Z_{CM}) \in (0°, 90°)$, $Z_{CN} = -Z_{M1} - Z_{L1}$. It can be seen from the phase angle range of Z_{N1} and Z_{L1}, $\arg(Z_{CN}) \in (180°, 270°)$.

When an external fault occurs at reverse direction of protection M, the measured impedance at protection N can be expressed as $Z_{CN} = Z_{N1}$. Owing to $\arg Z_{N1} \in (0°, 360°]$, the range of impedance phase is $\arg(Z_{CN}) \in (0°, 360°]$. Thus, $\arg(Z_{CN})$ changes in a large range. For the convenience of comparison with the case that an external fault occurs at reverse direction of protection N, the range of $\arg(Z_{CN})$ need to be analyzed in two cases of $\arg(Z_{CN}) \in (180°, 270°)$ and $\arg(Z_{CN}) \notin (180°, 270°)$. When $\arg(Z_{CN}) \in (180°, 270°)$, There is $\arg(Z_{N1}) \in (180°, 270°)$. From $Z_{CM} = -Z_{N1} - Z_{L1}$ and the phase angle range of Z_{N1} and Z_{L1}, the range of impedance phase is $\arg(Z_{CM}) \in (0°, 360°]$. When $\arg(Z_{CN}) \notin (180°, 270°)$, that is $\arg(Z_{CN}) \in (0°, 180°]$ and $\arg Z_{CN} \in [270°, 360°]$, $\arg(-Z_{N1}) \in [90°, 360°]$. From $Z_{CM} = -Z_{N1} - Z_{L1}$ and the phase angle range of Z_{N1} and Z_{L1}, The range of impedance phase is $\arg(Z_{CN}) \in [90°, 360°]$.

The angles of Z_{CM} and Z_{CN} in different fault positions are shown in Table 1.

Table 1. The angles of Z_{CM} and Z_{CN} in different fault positions

Fault location	Z_{CM}	Z_{CN}	$\arg(Z_{CM})$	$\arg(Z_{CN})$
Internal fault	Z_{M1}	Z_{N1}	$(0°, 90°)$	$(0°, 360°]$
External fault(f_2)	$-Z_{N1} - Z_L$	Z_{N1}	$(0°, 360°]$	$(180°, 270°)$
			$[90°, 360°]$	$(0°, 180°]$ or $[270°, 360°]$
External fault(f_3)	Z_{M1}	$-Z_{M1} - Z_L$	$(0°, 90°)$	$(180°, 270°)$

According to the results of Table 1, the protection method proposed in this paper uses the phase angle range of Z_{CM} and Z_{CN} to determine internal and external fault. When only the phase angle range of Z_{CM} and Z_{CN} cannot determine the internal and external fault, then use the amplitude relationship of Z_{CM} and Z_{CN} to determine internal and external fault. $\arg(Z_{CN}) \in (180°, 270°)$

In view of the case that external fault at reverse direction of protection on side N, $\arg(Z_{CN})$ has a clear range, that is, $\arg(Z_{CN}) \notin (180°, 270°)$. So this paper divides $\arg(Z_{CN})$ into two parts, which are $\arg(Z_{CN}) \in (180°, 270°)$ and $\arg(Z_{CN}) \notin (180°, 270°)$. Then analyze the situation of these two parts.

According to Table 1, if $\arg(Z_{CN}) \in (0°, 180°]$ or $\arg(Z_{CN}) \in [270°, 360°]$, the location of fault must not be reverse direction of protection on side N. Then use the range of $\arg(Z_{CM})$ to determine the location of fault that internal or reverse direction o protection on side M. When occurs internal fault in line, $\arg(Z_{CM}) \in (0°, 90°)$. When occurs external fault at reverse direction of protection on side M, $\arg(Z_{CM}) \in [90°, 360°]$. It can be seen that according to $\arg(Z_{CM})$, the location of fault can be determined.

However, when $\arg(Z_{CN}) \in (180°, 270°)$, the internal fault and external fault cannot be determined by judging the angle range of Z_{CM}. After analysis, it is considered to determine the internal fault and external fault according to the amplitude relationship between Z_{CM} and Z_{CN}.

When an internal fault occurs in microgrid, there is $|Z_{CM} + Z_{CN}| = |Z_{M1} + Z_{N1}|$. According to the content in SubSect. 2.1, it can be concluded that $|Z_{N1}| \gg |Z_{M1}|$. So the value of $|Z_{CM} + Z_{CN}|$ is very close to the value of $|Z_{N1}|$. In addition, the electricity from DG in microgrid is often used by nearby users and the lines of microgrid are usually short. Therefore, the impedance of the protected line in microgrid is less than the equivalent impedance of IIDG. That is $|Z_{N1}| > |Z_{L1}|$. It can be concluded that $|Z_{CM} + Z_{CN}| > k_{set}|Z_{L1}|$. Where k_{set} is the reliability coefficient. k_{set} is greater than 1. And the optimal value of k_{set} is 1.2.

When an external fault occurs in microgrid, there is $|Z_{CM} + Z_{CN}| = |Z_{L1}|$. Thus, it can concluded that $|Z_{CM} + Z_{CN}| < k_{set}|Z_L|$.

To sum up, according to the phase angle and amplitude relationship between Z_{CM} and Z_{CN}, the process of distinguishing internal fault and external fault is shown in Fig. 5.

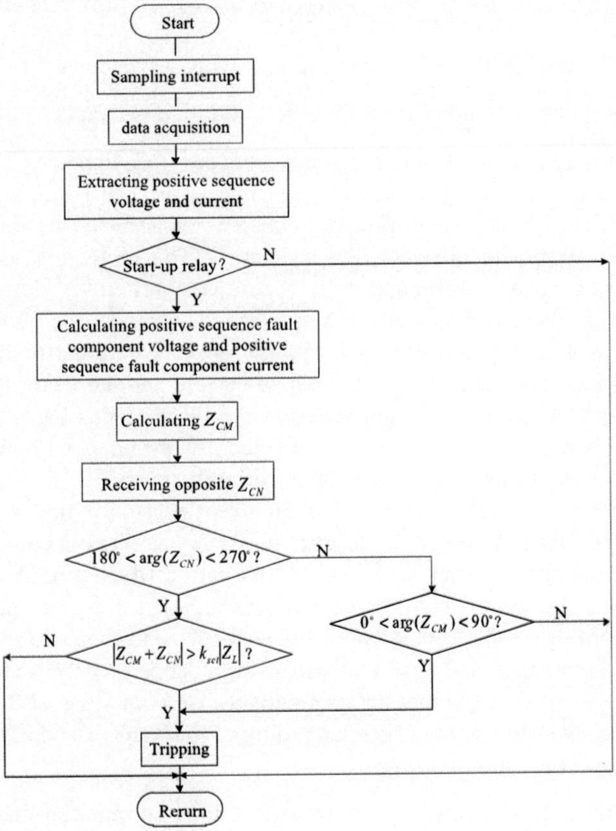

Fig. 5. Flow chart of protection based on the difference between DG impedance and system impedance for microgrid transmission line.

In the flow of Fig. 5, the detailed process is as follows.

1) If $\arg(Z_{CN}) \in (0°, 180°]$ or $\arg(Z_{CN}) \in [270°, 360°]$, then judge the range of arg (Z_{CM}).
a) If $\arg(Z_{CN}) \in (0°, 180°]$, it is judged as internal fault.
b) If $\arg(Z_{CM}) \in [90°, 360°]$, it is judged as external fault.
2) If $\arg(Z_{CN}) \in (180°, 270°)$, then judge the range of $|Z_{CM} + Z_{CN}|$.
a) If $|Z_{CM} + Z_{CN}| > k_{set}|Z_L|$, it is judged as internal fault.
b) If $|Z_{CM} + Z_{CN}| < k_{set}|Z_L|$, it is judged as external fault.

2.3 Advantages of the Proposed Protection Method

The proposed method in this paper has the following advantages.

(1) This method can accurately locate internal and external fault.

In this protection method, the difference between DG impedance and equivalent impedance of grid for microgrid transmission line is used to construct the protection criterion. The criterion of overcurrent protection is constructed according to the difference of fault current in different fault positions. The protection method proposed uses the characteristics of impedance to construct a criterion. Compared with the characteristics of current difference, the impedance can reflect the fault characteristics in essence, so it is better to distinguish the internal and external fault. So protection method of this paper is less affected by the load current and the factor that current flow in two directions. And the protection criterion proposed in this paper is based on the component fault network, which is less affected by the load current and transition resistance. It can determine location of fault correctly.

After analyzing the mal-operation of overcurrent protection under the influence of load current and DG in microgrid, the criterion is constructed according to the characteristics of impedance. After analyzing the structure of microgrid and the characteristics of DG, the equivalent impedance on both sides of the protected line is calculated according to the fault component network, and then a protection criterion based on DG impedance and grid equivalent impedance is proposed. The proposed method needs to calculate the measured impedance on both sides of the protected line according to Eq. (4) and Eq. (5). Then according to the Table 1 and Fig. 5, the internal fault and external fault are determined.

(2) This method is less affected by the factors of load current and that power flow in two directions.

The overcurrent protection mal-operation is affected by load current and DG in microgrid. In the microgrid shown in Fig. 2, the microgrid transmission lines are configured with overcurrent protection. The power grid and IIDG together provide short-circuit current to the fault point, and the output current of IIDG will raise the fault current of protection in L1 when fault occurs at L5. Without considering the injection current of DG, the fault current is greater than the setting value of protection in L1. And the protection which is located at L1 may mal-operate. According to the fault component network, this paper calculate the equivalent impedance of the protection on both sides according to Eq. (4) and Eq. (5), and then put forward the protection criterion based on DG impedance and equivalent impedance of power grid. The fault component network does not contain load current, so the flow direction of power and the flow direction of fault current will not affect the protection method proposed in the manuscript.

(3) This method does not require high data synchronization.

The reason for the low requirement of data synchronization in the protection method proposed is as follow. The Z_{CM} in the protection criterion is obtained by the Eq. (4). The Z_{CN} is calculated in the same way as Z_{CM}. The calculation of Z_{CM} only needs to the synchronized voltage and current data of protection M. The calculation of

Z_{CN} only needs to the synchronized voltage and current data of protection N. Then the protection method proposed needs to accept the opposite impedance to locate the fault. This method does not need the strict synchronization of the voltage and current information on both sides. Compared with the longitudinal current differential protection, the impedance calculation in proposed protection method only needs the synchronization of the voltage and current information collected at the side. However, the real-time current sampling values on both sides are required to be calculated directly in longitudinal current differential protection. So the data synchronization requirement of the proposed method is lower than that of longitudinal current differential protection.

3 Simulation and Analysis

In this paper, a simulation model corresponding to Fig. 2 is built in PSCAD simulation software. Positive sequence impedance of system power is $0.087 + j0.492\ \Omega$. Zero sequence impedance is $0.261 + j1.476\ \Omega$. L2 is the selected protected line. The length of L2 is 0.1 km. The positive sequence impedance of L2 is $0.080 + j0.010\ \Omega$. The Zero sequence impedance of L2 is $2.391 + j0.315\ \Omega$.

3.1 Simulation and Analysis of Internal Fault

In order to verify that the protection method proposed in this paper can accurately distinguish fault, so in the PSACD simulation model, the situation of internal fault and external fault are simulated.

f_1 is fault point of L2, which locate at 50%. The fault type is BC phase fault. And the fault is applied at 1.0 s and removed at 1.2 s. The voltage simulation data and current simulation data of the protection on both sides are imported into MATLAB simulation software. The $\arg(Z_{CM})$ and $\arg(Z_{CN})$ are shown in Fig. 6.

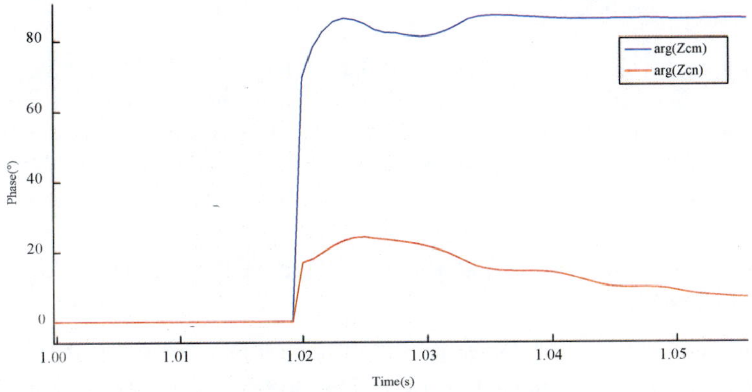

Fig. 6. $\arg(Z_{CM})$ and $\arg(Z_{CN})$ when an internal fault occurs at f_1

Since the Fourier algorithm is used in the calculation of impedance, the calculated impedance after one and a half cycles is accurate. In this simulation, a half cycle is 30 ms. In addition, considering the influence of aperiodic component, the impedance is calculated after 40 ms of fault. As shown in Fig. 6, at 1.04 s, $\arg(Z_{CM}) = 86.0°$, $\arg(Z_{CN}) = 10.2°$. It can be seen that $\arg(Z_{CN}) \notin (180°, 270°)$. According to the analysis of SubSect. 2.2, it is not necessary to calculate the value of $|Z_{CM} + Z_{CN}|$ for the protection of both sides. Because of the existence of $\arg(Z_{CN}) \in (0°, 180°]$, $\arg(Z_{CM}) \in (0°, 90°)$, this method determines that the fault is an internal fault.

3.2 Simulation and Analysis of External Fault in Reverse Direction of Protection M

f_2 is fault point of L1, which locate at 10%. The fault type is BC phase fault. The fault is applied at 1.0 s and removed at 1.2 s. The $\arg(Z_{CM})$ and $\arg(Z_{CN})$ are shown in Fig. 7.

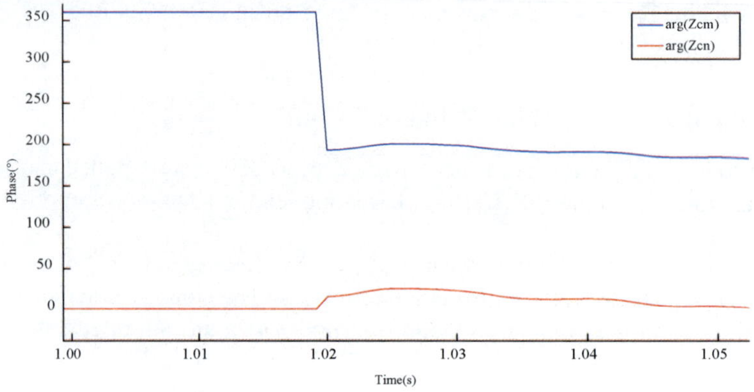

Fig. 7. $\arg(Z_{CM})$ and $\arg(Z_{CN})$ when an external fault occurs at f_2

As shown in Fig. 7, $\arg(Z_{CM}) = 190.2°$, $\arg(Z_{CN}) = 11.0°$. $\arg(Z_{CN}) \notin (180°, 270°)$. According to the analysis of SubSect. 2.2, it is not necessary to calculate the value of $|Z_{CM} + Z_{CN}|$ for the protection of both sides. Because of the existence of $\arg(Z_{CN}) \in (0°, 180°]$, $\arg(Z_{CM}) \in [90°, 360°]$, this method determines that the fault is an external fault.

3.3 Simulation and Analysis of External Fault in Reverse Direction of Protection N

f_3 is fault point of L3, which locate at 10%. The fault type is BC phase fault. The fault is applied at 1.0 s and removed at 1.2 s. The $\arg(Z_{CM})$ and $\arg(Z_{CN})$ are shown in Fig. 8.

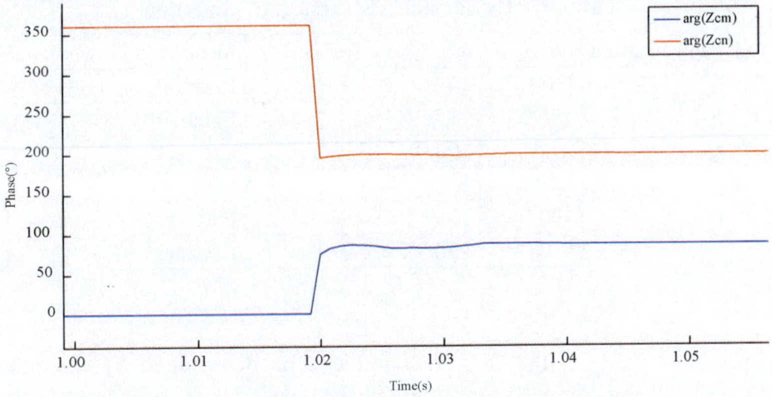

Fig. 8. arg(Z_{CM}) and arg(Z_{CN}) when an external fault occurs at f_3

As shown in Fig. 8, $\arg(Z_{CM}) = 83.4°$, $\arg(Z_{CN}) = 198.0°$. $\arg(Z_{CN}) \in (180°, 270°)$. According to the analysis of SubSect. 2.2, it is necessary to calculate the value of $|Z_{CM} + Z_{CN}|$ for the protection of both sides.

The amplitude of $(Z_{CM} + Z_{CN})$ when an internal fault occurs at f_3 is shown in Fig. 9. As shown in Fig. 9, $|Z_{CM} + Z_{CN}| = 0.080\,\Omega$. The positive sequence impedance of L2 is 0.081 Ω. Because of the existence of $|Z_{CM} + Z_{CN}| < 1.2Z_{L1} = 0.096\,\Omega$, this method determines that the fault is an external fault.

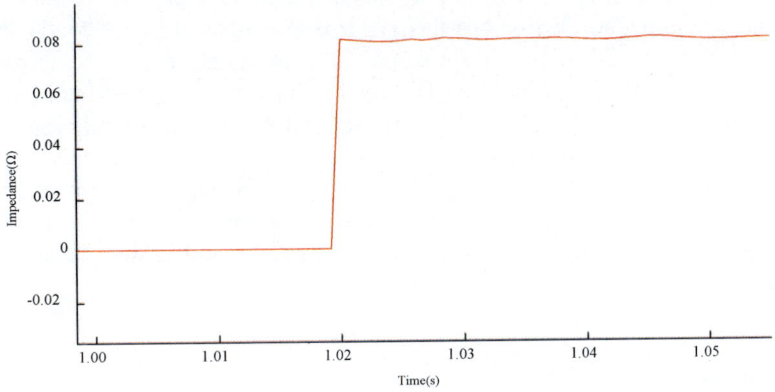

Fig. 9. The amplitude of (Z_{CM} + Z_{CN}) when an external fault occurs at f_3

3.4 Summary of Simulation

In addition, other fault points including f_4, f_5 and f_6 are simulated. And f_4 locates at 60% of L2, f_5 locates at 70% of L1, f_6 locates at 50% of L3.

The identification results of protection in Table 2.

Table 2. The identification results of protection

Fault location	$\arg(Z_{CM})$	$\arg(Z_{CN})$	$\lvert Z_{CM}+Z_{CN}\rvert$	Identification result
f_1	86.0°	10.2°	–	Internal
f_2	190.2°	11.0°	–	External
f_3	83.4°	198.0°	0.080	External
f_4	86.3°	10.5°	–	Internal
f_5	190.0°	10.8°	–	External
f_6	83.7°	198.5°	0.079	External

According to Table 2, the fault location can be determined by the relationship between the phase angle and the amplitude of the measured impedance on both sides of line. In addition, according to the analysis in SubSect. 2.2, the protection method proposed in this paper can also determine other fault points except those listed in Table 2.

4 Conclusion

This paper presents a method based on difference between DG impedance and equivalent impedance of grid for microgrid transmission line to ensure correct action of protection. The fault location is determined by the relationship between the phase angle and the amplitude of the measured impedance on both sides of line. Compared to overcurrent protection and longitudinal current differential protection, there are two advantages: 1) protection method of this paper is less affected by the load current and the factor that current flow in two directions. 2) protection method of this paper does not need the strict synchronization of the voltage and current information on both sides of line. The validity of the method is also verified by PSCAD simulations. And the simulation results show that the method can determine fault location correctly and ensure correct action of protection. Therefore, the application of protection based difference between DG impedance and equivalent impedance of grid for microgrid transmission line can determine fault location correctly, and ensure correct action of protection.

References

1. Ma, Y.W., Yang, P., Wang, Y.W., Zhao, Z.L.: Typical characteristics and key technologies of microgrid. Autom. Electr. Power Syst. **39**(8), 168–175 (2015)
2. Zamani, M.A., Yazdani, A., Sidhu, T.S.: A decision-tree-based method for fault classification in single-circuit transmission lines. Soc. IEEE Trans. Smart Grid **3**(4), 2088–2099 (2012)
3. Shin, D., Lee, K.J., Lee, J.P., et al.: Implementation of fault ride-through techniques of grid-connected inverter for distributed energy resources with adaptive low-pass notch PLL. IEEE Trans. Power Electron. **30**(5), 2859–2871 (2015)

4. Haj-Ahmed, M.A., Illindala, M.S.: The influence of inverter-based DGs and their controllers on distribution network protection. IEEE Trans. Ind. Appl. **50**(4), 2928–2937 (2014)
5. Zamani, M.A., Yazdani, A., Sidhu, T.S.: A communication-assisted protection strategy for inverter-based medium voltage microgrids. IEEE Trans. Smart Grid **3**(4), 2088–2099 (2012)
6. Xu, S.X., Lu, Y.P., Wang, Y.: A section voltage based inversetiming protection method for microgrid. Autom. Electr. Power Syst. **38**(1), 68–73 (2014)
7. Yu, Q., Yu, S., Li, X.H.: Adaptive protection scheme of distribution network with distributed generation. Power Syst. Protect. Control **30**(4), 110–115 (2012)
8. Casagrande, E., Woon, W.L., Zeineldin, H.H., Svetinovic, D.: A differential sequence component protection scheme for microgrid with inverter-based distributed generators. IEEE Trans. Smart Grid **5**(1), 29–37 (2014)
9. Ustun, T.S., Ozansoy, C., Zayegh, A.: Fault current cofficient and time delay assignment for microgrid protection system with central protection unit. IEEE Trans. Power Syst. **28**(2), 598–606 (2013)
10. Saleh, S.A.: Signature-coordinated digital multirelay protection for microgrid systems. IEEE Trans. Power Electron. **29**(9), 4614–4623 (2014)

Research on an Adaptive Weighted Fusion Algorithm Under the Edge Computing Framework of Power Internet of Things

Haiqing Wang[1,2(✉)], Xingshen Wei[1,2], Wei Liu[1,2], and Guo Hu[1]

[1] NARI Group, Nanjing 21003, Jiangsu, China
wanghaiqing@sgepri.sgcc.com.cn
[2] Nanjing NARI Information and Communication Technology Co. Ltd., Nanjing 21003, Jiangsu, China

Abstract. In the power system, each province has nearly millions of sensors, and a large amount of data is generated. In the process of multi-sensor data collection, there are many problems such as inaccurate monitoring accuracy, waste of collection nodes, and zero drift of the sensor itself, which affect the inaccurate data reporting of the central station and the inaccurate execution of feedback information. At the same time, the sensor cannot do it once Acquisition, multiple applications, especially in power distribution rooms, station areas and other areas with high environmental requirements. There are distribution automation, electricity information collection, and dozens of sensor terminals such as temperature and humidity, smoke, video, robots, etc. Highly accurate information collection and environment configuration. To this end, an improved adaptive weighted fusion algorithm is proposed to achieve accurate terminal collection and data fusion upload based on the edge side of the power IoT edge computing framework. The design scheme can improve the accuracy of terminal data collected in power distribution rooms, station areas and other regions. Simulation and experiment proves the effectiveness of the algorithm in terms of acquisition accuracy and data fusion.

Keywords: Adaptive weighted fusion algorithm · Edge calculation · The Internet of Things · Power Internet of Things

1 Introduction

Technologies in emerging fields such as the Internet of Things are becoming more and more popular in remote sensing detection, power monitoring, smart agriculture, and intelligent transportation. [1].

A large number of sensors are deployed in the power monitoring system, especially in the areas with high environmental requirements such as distribution rooms and desk areas. In the process of multi-sensor data collection, there are some problems, such as inaccurate collection accuracy, large number of collection points, zero drift of the sensor itself, interference of external environment noise, etc.An improved adaptive weighted multi-sensor fusion algorithm is proposed based on the multi-sensor data fusion technology, to improve the collection accurate of external environmental

Y. Xue et al. (Eds.): PMF 2020, LNEE 718, pp. 868–881, 2021.
https://doi.org/10.1007/978-981-15-9746-6_65

information in the area where noise interferenceon is high and sensor components are old. Simultaneously, an enterprise-level edge computing framework is designed to convert data format, share data between applications, scene linkage, rule engine and transfer data to IoT management platform [2].

In areas where the power monitoring system has high requirements on the environment, there is high noise interference, the accuracy of the multi-sensor data collection currently used is still very low. At present, multi-sensor acquisition in power monitoring systems often uses arithmetic average algorithm, tangent weighted fusion algorithm, distance-based adaptive weighted fusion algorithm, etc. These algorithms have low accuracy, and are suitable for environment with less interference. [3, 4]

The key of the adaptive weighted fusion algorithm is to determine the weighted value of sensor data fusion in each region First, pre-process the data collected by the sensors in each area, remove the node data with larger errors, perform batch estimation on the basis of the fusion algorithm for each collection node, and group the collection nodes in the batch estimation to obtain the corresponding Fusion data.By comparing the traditional fixed weight fusion algorithm and arithmetic average fusion algorithm, the algorithm we applied was compared and verified. By comparing the relative errors and variances, it was concluded that this algorithm has strong anti-interference ability, better stability and higher accuracy [5].

2 Design of Improved Adaptive Weighted Fusion Algorithm Mode

2.1 Pretreatment of Multi-Sensor Acquisition Information

At present, the acquisition of multi-sensor mainly focuses on the traditional fusion methods such as fixed weighted fusion algorithm, arithmetic average fusion algorithm and least squares weighted fusion algorithm.The key of acquisition accuracy is the weight of each sensor.In the environment of power distribution room and substation where the accuracy of collection is high, the monitoring and coordination of multiple sensors must be required. [6]. A reasonable number of sensors and the accuracy of collection area are the keys of monitoring.

In the power distribution roomwhere we use multiple wireless sensor, affected by the interference of external environmentand the influence of accuracy, the acquisition value of the node may have deviation. The more collection nodes, the more data need to be fused.However, when the external environment has great interference on the acquisition information, it is necessary to preprocess the data and eliminate the acquisition nodes with large deviationIn this paper, the method of fuzzy theory set is used to eliminate sensor nodes with large error.

Take temperature acquisition as an example, suppose that the measurement data collected by a certain temperature sensor in a certain period of time is x1, x2, x3…xn, and it is mainly divided into the following steps in the pretreatment stage:

(1) Based on the corrected distance, taking the i-th measurement data as the base point, calculate the distance between the i-th measurement data and the j-th $(1, 2, 3, 4, 5 \ldots. n)$

measurement data: $d_{[md]ij} = \sqrt{\frac{x_i^2 + x_j^2 + 2x_i^2 x_j^2}{x_i^2 + x_j^2}}$ (i, j = 1, 2, 3,…n), where $d_{[md]ij}$ represents the distance between the node acquisition value x_i and x_j, and calculates the distance between $d_{[md]ij}$ and each acquisition node.

(2) According to the distance between the i point and each node, find the average distance between the collection nodes $\overline{d_{[md]ij}} = \frac{1}{n-1}\sum_{j=1}^{n} d_{[md]ij}$, will $\overline{d_{[md]i}}$ as the final reference variables,

(3) Admissible function is defined as: $R(x_i, x_j) = \frac{2\mathrm{arccot} d_{ij}}{\Pi}$, where d_{ij} represents the distance between any acquisition moments and defines the threshold function as A (xi, xj):

$$A(Xi, Xj) = \begin{cases} 1 & R(x_i, x_j) \leqq \overline{d_{[md]i}} \\ 0 & R(x_i, x_j) \geqq \overline{d_{[md]i}} \end{cases}$$

In this expression, $\overline{d_{[md]i}}$ is the reference variable between any Xi, XJ acquisition nodes which can be calibrated through average value by many experiments,. When A (xi,xj) is 1, it means that the values collected by two nodes are within the allowable range. If A(xi,xj)is 0, it means that there is no allowable value between two collection points, so it should be eliminated.

2.2 Improved Multi-sensor Fusion Algorithm

2.2.1 Multi - data Batch Estimation for Single - Type Sensors

The temperature and humidity in the distribution room is an important index, which is one of the important standards of accurate data acquisition in the moving ring system. The intelligent node can be adapted in the distribution room.,Indoor environment,each Zigbee acquisition node has multiple sensors, including temperature and humidity, light intensity and CO_2 concentration.we take the most typical temperature sensor as an example. First, divide an area into several areas and collect in a certain area.Then, the collected data is divided into five groups: T1, T2, T3, T4, T5. Set the data collected by the I group as $T_{i1}, T_{i2}, T_{i3} \ldots T_{in}$ [7].

The average value of each group is: $\overline{T_i} = \frac{1}{n_i}\sum_{j=1}^{4} T_{ij}$ (j = 1, 2, 3, 4)

The variance of each group is:$\sigma^2_{\overline{T_i}} = \frac{\frac{1}{n_i}\sum_{j=1}^{4} T_{ij} - \overline{T_i}}{n_i}$ (i = 1, 2, 3, 4, 5, j = 1, 2, 3, 4) By using the statistical batching estimation theory, the optimal temperature estimation value T is extracted from the 5 groups of single sensor. Then, the value of T is

$$T = \frac{\sigma^2_{\overline{T}1}\overline{T}_1 + \sigma^2_{\overline{T}2}\overline{T}_2 + \sigma^2_{\overline{T3}}\overline{T}_3 + \sigma^2_{\overline{T4}}\overline{T}_4 + \sigma^2_{\overline{T5}}\overline{T}_5}{\sigma^2_{\overline{T1}} + \sigma^2_{\overline{T2}} + \sigma^2_{\overline{T3}} + \sigma^2_{\overline{T4}} + \sigma^2_{\overline{T5}}}$$

After taking 5 sets of values of the temperature sensor, the optimal variance of the data collected by the sensor is obtained as follows:

$$\sigma = \frac{\sigma^2_{\overline{T1}}\sigma^2_{\overline{T2}}\sigma^2_{\overline{T3}}\sigma^2_{\overline{T4}}}{\sigma^2_{\overline{T_1}}\sigma^2_{\overline{T2}} + \sigma^2_{\overline{T3}} + \sigma^2_{\overline{T4}} + \sigma^2_{\overline{T5}}}$$

2.2.2 Adaptive Weighted Fusion of Multiple Sensors

After the batch estimation of the monitoring data of each sensor, the optimal temperature value T and the optimal variance of a single sensor is obtained. According to the principle of optimal weight allocation, self-adaptive weighted fusion is carried out with in each group, and the optimal weight of each group is obtained. Set n groups of sensors and divide them into m groups. The weight of each group is Wi (i = 1, 2, 3....m). Then the adaptive weighted fusion is performed for each group to obtain the fusion value of each group. Finally, m data of m groups are weighted and fused [8].

Set the weight value of each sensor in each group as Wi:

$$\mathrm{Wi} = \left(\sigma_{\mathrm{i}}^2 \sum_{\mathrm{i}=1}^{k} \frac{1}{\sigma_{\mathrm{i}}^2}\right)^{-1} \quad (\mathrm{k} = 1, 2, 3, 4, \ldots .\mathrm{n/m})$$

Estimated value of sensors in group K after fusion $\mathrm{X_k}$:

$$X_k = \sum_{i=1}^{k} W_i T_i (\mathrm{k} = 1, 2, 3, 4. \ldots \ldots \mathrm{n/m})$$

Similarly, the weighted factor of each group is W_{i}'

$$\mathrm{W}_i' = \left(\sigma_i^2 \sum_{i=1}^{m} \frac{1}{\sigma_i^2}\right)^{-1} \quad (i = 1, 2 \ldots m)$$

Therefore, when the weights of each group are obtained, the final result of n acquisition nodes can be obtained by adaptive weighted fusion as follows:

$$\mathrm{X} = \sum_{\mathrm{i}=1}^{\mathrm{m}} X_i W_i (i = 1, 2, 3, 4 \ldots m)$$

2.3 Test and Analysis of the Improved Adaptive Fusion Algorithm

In order to improve the accuracy of acquisition nodes in low-voltage side branch, meter box, distribution room and other monitoring environment. In this paper, the data of single type acquisition node is first to eliminate the error of measuring distance. Then, the optimal estimation value and the optimal variance at a certain time are extracted by multi-sensor batch estimation, but, the acquisition results in the distribution area are generated by data fusion of multiple sensors. The monitoring area is segmented, andthe value of each area is estimated in batches. The values after batch are fused again to obtain the final fusion value. The batch value and the final fusion value are uploaded to the upper computer for display. In the process of collection, adaptive weighted fusion is carried out in each group to minimize the collection error through the principle of optimal weight allocation.

In order to verify the goodness of the algorithm, a distribution room in lishui city, nanjing is taken as a demonstration site. The system architecture of acquisition and transmission is shown in Fig. 1. Zigbee is used as wireless transmission, and CC2530 and SHT11 are used as acquisition and transmission modules. Each acquisition module is uploaded to the edge iot agent through the coordinator, and the edge iot agent which is an intelligent terminal on the edge, uploads the data to the iot management platform through MQTT protocol. The iot management platform can read the current data in real time [9].

In the aspect of verification simulation, three algorithms are used to compare and verify. Algorithm 1: arithmetic average fusion algorithm. Algorithm 2: fixed weight fusion algorithm. Algorithm 3: improved adaptive weighted fusion algorithm. The fusion data of 50, 100 and 200 are compared. The real environment is set at 25°C.

The system uses matlab to simulate the real environment. Non-correlated zero-mean gaussian white noise is used to simulate the final acquisition result of the sensor. In the real environment, there are some problems in the sensor, such as zero drift, different position of each sensor and indoor environment change, etc. The final fusion result is obtained with the signal-to-noise ratio ranging from 0.1 dB to 1 dB.

2.3.1 Simulated Environment Simulation

In the simulation environment with a SNR of 0.1, a certain temperature monitoring area is selected, and the zero mean value of gaussian white noise is simulated to intercept the values of 200 collection points. The simulation is carried out by Matlab, and the collected results are saved in the excel table. The simulation environment in MATLAB is shown in the following figure:

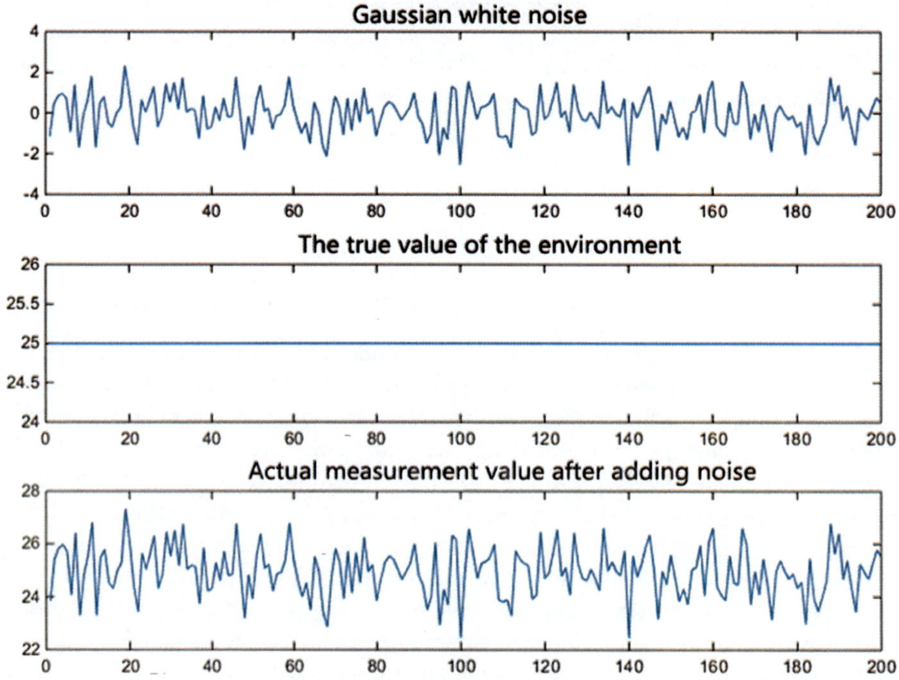

Fig. 1. Matlab simulation of the real environment

2.3.2 Simulation Results Under the Same SNR

Take the data of 200 acquisition nodes,and each data is fused with arithmetic average and fixed weight respectively, and the improved batch estimation adaptive weighted fusion [10]. After collecting ten times, the weight distribution at each time is shown in the following figure (Fig. 2):

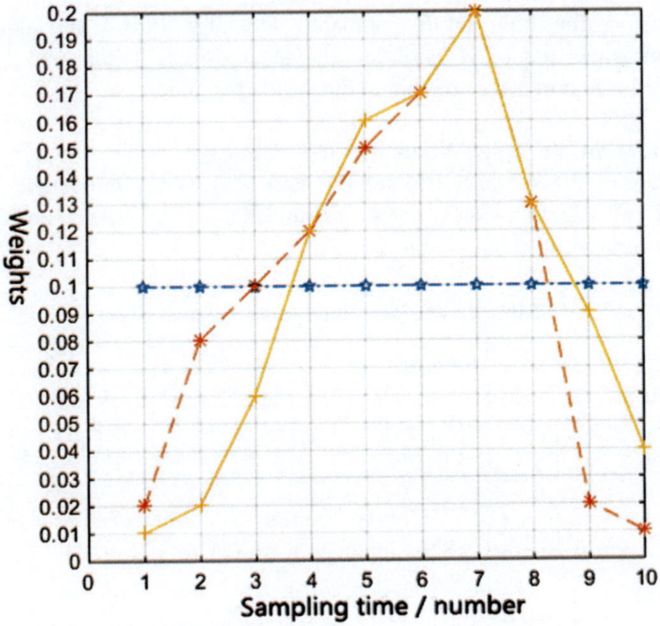

Fig. 2. Weighted values of the three algorithms at each moment

To further verify the effectiveness of the algorithm, we set up the SNR is 0.1 and the real environment is 25°C and the greenhouse environment information is collected. Data fusion is carried out for each acquisition node through three algorithms, and the final results are shown in the following table:

Table 1. Data fusion results of the three algorithms under the same SNR

Sensor(Signal to noise ratio 0.1)		Sampling time							σ^2	Δmax
		t 1	t 2	t 3	t 4	t 5	t 6	t 7		
Fusion result	Arithmetic average	25.02	24.93	24.91	24.83	25.27	25.32	25.05	0.1918	0.32
	Fixed weight fusion	24.84	24.94	24.94	24.87	25.41	24.98	24.88	0.1962	0.41
	This algorithm	24.91	24.95	24.92	24.89	25.28	25.11	24.98	0.1434	0.28

In Table 1,Where, t_m represents the collection times:

$$\sigma 2 = \sqrt{\frac{1}{t_m - 1}\sum_{i=1}^{t_m}(x_i - 25)^2}$$

$$\Delta\max = \max\{|xi - 25|\}\ (i = 1, 2, 3, 4, 5.\ldots..t_m)$$

According to the above table, compared with algorithm 1 and algorithm 2, the population variance and limit deviation of the algorithm in this paper are both the minimum so the fusion effect of this algorithm is the best.

2.3.3 Simulation Analysis When the SNR Changes

The values of 20, 60 and 100 temperature sensors are collected, and the SNR variesfrom 0.1 to 1.Data fusion is performed under three algorithms, and the results are shown in the following table (Table 2):

Table 2. Fusion results under different SNR

Signal to noise ratio	0.1	0.2	0.3	0.4	0.5	0.6	0.7	0.8	0.9	1
Arithmetic average	25.23	25.46	25.37	25.36	25.14	25.33	25.27	25.33	24.85	25.36
Fixed weight fusion	25.19	25.18	25.29	25.21	25.12	25.26	25.18	25.21	24.87	25.34
This algorithm	25.12	25.13	25.15	25.13	25.07	25.13	25.11	25.15	24.94	25.36

In the real environment of 25°C, take 20 groups, and the curve changes as shown in the following figure when the SNR changes from 0.1 to 1 (Fig. 3):

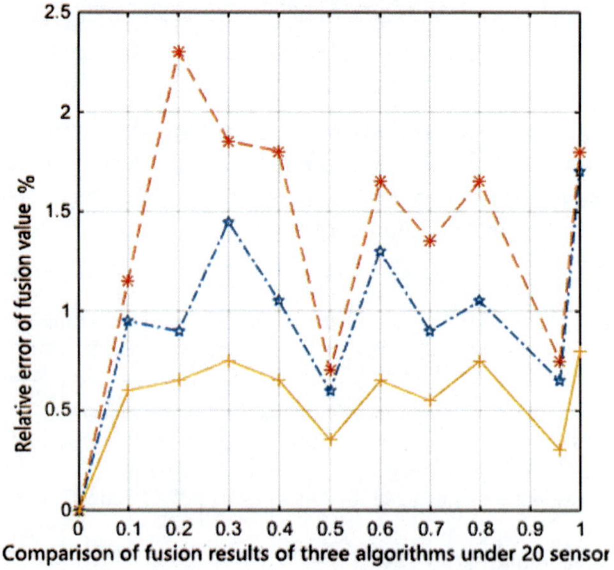

Fig. 3. Comparison of fusion values under different SNR

As we know from the above figure, under the improved weighted fusion algorithm, the fluctuation range, variance and relative error of the fusion value are minimum.

Similarly, the relative error and variance can be obtained when the number of groups is 60 and 100.

The algorithm is placed on the zigee transmission node, and the precise temperature value in the region is obtained through the improved weighted fusion adaptive method. The algorithm can also be applied to the environmental detection of humidity, light intensity, CO_2 concentration, smoke sensation, etc., Finally, it reaches the level of fine and deep layout.

3 Design of Edge Computing Framework Model

3.1 Overall Architecture Description

Firstly, The terminal of the perception layer completes the topology networking, which realizes the multi-dimensional, description of the underlying entity power business, and enhances the depth and breadth of perception. Secondly, the usage of each domain is different, and a cross-domain standardized configuration management method for massive terminals based on the object model is proposed. The configuration management capabilities of business terminals are abstracted from the three dimensions of attributes, events, and services. The JSON format is used to define and describe the object model. The edge side can implement the management configuration interface of the business terminal based on the object model. The object model is issued using the IoT management platform, and the management configuration interface can be called to realize remote configuration, remote upgrade, and remote monitoring. Finally,through the IoT management platform, edge IoT agents and standardized access of various types of terminals to achieve“side logic separation, side management sharing”, promote the unified access of various terminals, edge intelligence and shared sharing [11–15]. Through the edge computing framework, the edge IOT agent is developed to meet the power business system. Functional positioning diagram is shown as follows (Fig. 4):

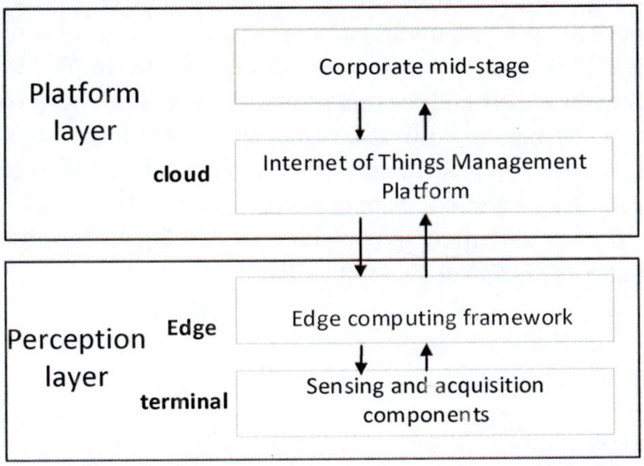

Fig. 4. Functional positioning diagram of edge computing framework

Combined with the mechanism of the improved adaptive weighted fusion algorithm, the optimal fusion data is extracted and uploaded to the edge, and the terminal network topology architecture is deployed to realize intelligent control under a certain environment factor. The corresponding system architecture is shown as follows (Fig. 5):

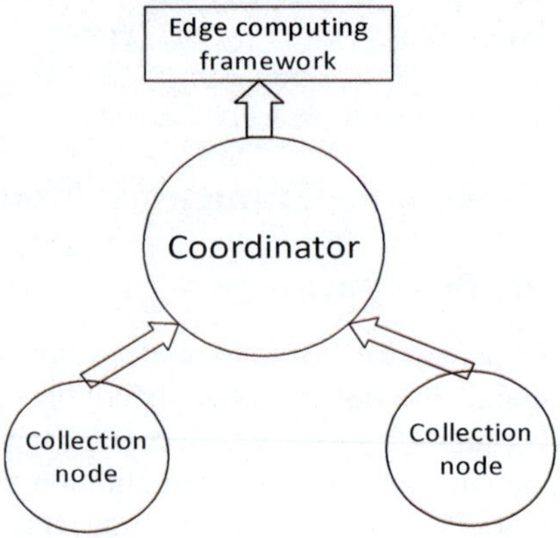

Fig. 5. Basic system architecture diagram

3.2 Basic Framework of Edge Computing

The edge computing framework for Power Internet of Things is divided into operating system layer, basic function layer and edge service layer. The operating system layer includes system monitoring, security access, application isolation, reliability and other functions.The basic function layer includes sub-device access, object model management, message queue and other functions.The edge service layer includes functions such as flow computing and rule engine, and supports cloud edge collaboration of resources, data, intelligence, application management, etc.

The edge computing infrastructure of ubiquitous power Internet of things compatible with multiple programming languages realizes the dynamic migration and reliable operation of software functions on different hardware platforms. It is used for multi type equipment access and message forwarding and provides common industrial fieldbus protocols compatible with standards and realizes interconnection.

Meanwhile, service oriented microservice architecture is adopted to facilitate the instant change and iteration of edge application functions. When fully meeting the edge computing function of ubiquitous power Internet of things, the infrastructure of edge computing framework is shown as follows (Fig. 6):

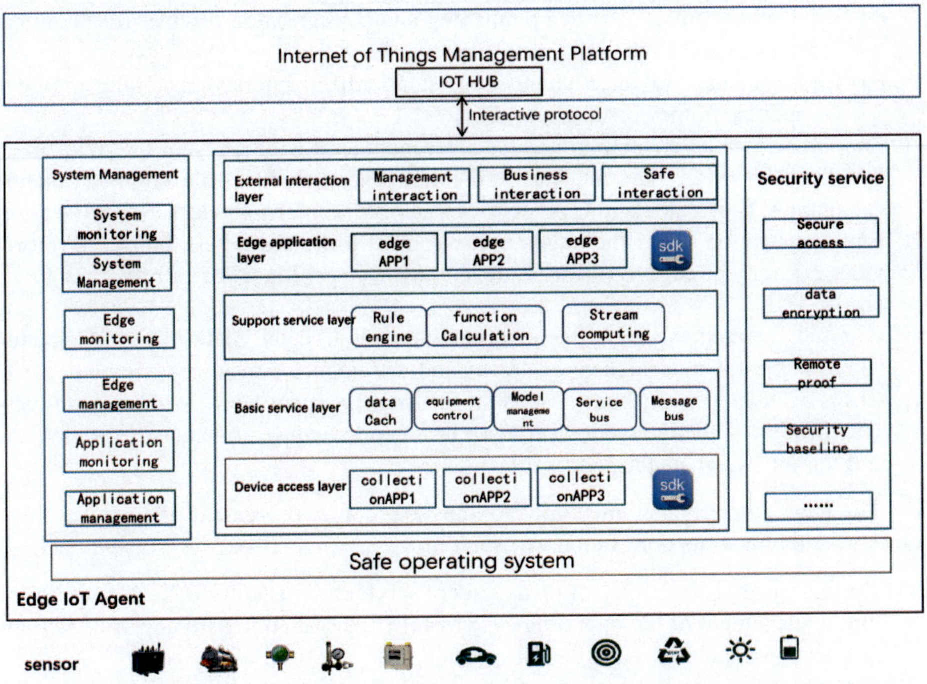

Fig. 6. Basic architecture diagram of edge computing framework

The basic framework of edge computing for Power Internet of Things mainly realizes the following functions:

1) Device access layer: the device access layer is an edge connector that interacts with the power terminal equipment. The edge computing framework supports three application development modes, which are completely based on SDK, partial microservice interface based on edge computing framework, and completely not based on SDK.Among them, edge computing framework provides SDK development kit for C, Go, Java and Python, which facilitates developers to create new acquisition applications quickly and conveniently. Corresponding acquisition applications communicate with devices and sensors through local business access protocol,which can provide services for one or more devices.The collection APP represents specific collection protocol applications such as Modbus and 104 [16–18].
2) Basic service layer: the basic service layer mainly implements the following core components.
   A) data caching: services for persistent storage and related management of data collected by southward objects;
   B) equipment control: realize the service of control request from the north side to the south side;
   C) model management: services for the storage and association management of object metadata connected to the edge computing framework;

D) service bus: provide service publishing and configuration services for the micro services of the edge computing framework;
E) message bus: provide a channel for standardized interaction among the edge computing framework's micro services.

3) Support service layer: support service layer includes a wide range of micro services, which can provide edge analysis and intelligenceincluding rule engine, function calculation, flow calculation, AI services and other micro services.
4) Edge application layer: the edge application layerlike the access service layerprovides convenient edge application development capabilities for third parties based on the SDK.
5) External interaction layer: the external interaction layer realizes the interaction between the iot management platform and the edge iot management platform in terms of management, business and security in accordance with the protocol specification of the interaction protocol between the edge iot management platform and the iot management platform [19–24].

The edge computing framework provides external management and service interfaces in addition to its core functions, such as:

A) System management: system management services are used to provide monitoring and management of the operating systemand edge computing framework itself and application capabilities for the iot management platform.
B) Security services: used to provide security services for the edge computing framework itself and applicationscovering modules such as secure access, data encryption, remote proof and security baseline.

4 Experimental Verification and Analysis

To verify the accuracy of the data uploaded by the sensor terminal and meet the access of large scale equipment, complete the access performance and test the access performance and system performance of the edge framework the maximum number of connections supported by the edge framework, the number of messages that the edge framework can process per second, the number of messages that can be processed by a single node per second, the performance of instructions issued by the edge framework, the performance of instructions issued by a single node, the maximum number of online users, and the average response time of core functionss [25–28],weset up a simulation experiment environment. Through setting up an application scenario simulation environment in the laboratory, carry out end-to-end business verification from terminal, iot agent, iot management platform to business application, and test the connectivity of the edge framework in the south, north and complete links.The results is shown as follows (Fig. 7).

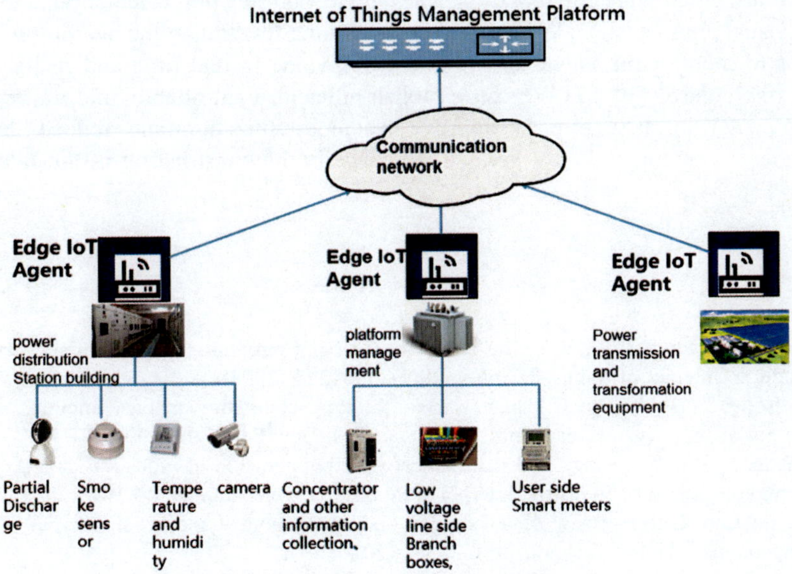

Fig. 7. Experimental scene verification

We deployed several temperature and humidity sensors in the power distribution room at the side of the platform area.Through the improved adaptive weighted fusion algorithm, the high accuracy and deep level calculation at the end side are completed. Compared with other algorithms, the number of sensor nodes could be effectively saved.

Accurate data collection is uploaded to the edge computing framework to complete the edge computing, terminal business acquisition and control capabilities of massive heterogeneous data. The sensor precision has been described in detail in chapter 2.3, and it is finally found that it can meet the test requirements of the above scenario and achieve the expected precision value [29].

5 Conclusion

In this paper, a sensor acquisition accuracy design which is based on edge calculation framework.An improved adaptive weighted fusion algorithm is proposed to realize multiple uses of the data collected at one time.Firstly, a single sensor is collected for a period of time to eliminate large errors caused by zero drift.Secondly,divide the area and calculate the optimal estimate value in this period of time in batches.Thirdly, the weight of the region is calculated and perform weighted fusion of each group according to the principle of optimal weight allocation, and finally integrate the values of each group to obtain The precise measured value at this moment. The measured values are transmitted to the edge IoT generation through wireless or wired methods. The edge IoT agent provides an edge computing framework with an object model, edge computing open capabilities, cloud-edge collaboration, rule engine, and function calculation, and finally gets a visual physics value. We can add deep learning function in the later improvement process to monitor the power environment risk value in real time and realize power safety risk control [30–32].We can establish efficient, controllable, and unified ubiquitous power infrastructure to improve resource utilization and reduce duplicate investment. This method is of great significance in the construction of the Internet of Things [33, 34].

References

1. Jirui, L., Xiaoyong, L., Yali, G., Yunquan, G.: Fang binxing research on data forwarding model in Internet of things. J. Softw. **29**(1), 196–224 (2018)
2. Weisong, S., Hui, S., Jie, C., Quan, Z., Wei, L.: Edge computing: a new computing model in the era of Internet of Everything. J. Comput. Res. Dev. **54**(5), 907–924 (2017)
3. Zhang, J., Wu, X., Yang, Z., Li, J.: Research and application of industrial data acquisition technology based on industrial Internet of things. Sci. Telecommun. **34**(10), 124–129 (2008)
4. Li, S., Luo, G.: Overview of technology and application of industrial Internet of things. Telecommun. Netw. Technol. (03), 26–31 (2014)
5. Zuo, P., Zhou, Q., Dai, X.: A brief analysis of industrial Internet of things technology in intelligent factories. Sci. Technol (08), 88 (2019)
6. Wang, H., Li, Y., Mi, M., Wang, P.: A security data fusion method for industrial Internet of things based on supervision mechanism. J. Instrum. **34**(04), 817–824 (2013)
7. Peng Ying, H., Zenghui, S.H.: A modified distance of evidence. J. Electron. Inf. Technol. **35** (7), 1624–1629 (2013)
8. Wang, H., Wang, D.: An improved adaptive weighted fusion algorithm for multi-wireless sensor data batch estimation. J. Sens. Technol. 2015.08.023
9. Xiaochen, X., Yuanwen, C., Jiangtao, R., et al.: An adaptive weighted data fusion algorithm considering sensor accuracy. Telecommun. Eng. **55**(10), 1079–1086 (2015)
10. Yang, Y., Song, Z.: Research on security and protection technology of industrial Internet of things. Internet Things Technol. **5**(03), 64–66 + 69 (2015)
11. Sha, L., Xiao, F., Chen, W., Sun, J., Wang, R.: Perception method of backdoor privacy leakage in the context of industrial Internet of things. J. Softw. **29**(07), 1863–1879 (2008)
12. Cai, Y., Feng, S., Du, H., Liu, M., Ding, X., Ji, W.: Edge-aware adaptive data processing method for ubiquitous power Internet of Things. High Voltage Technol. (06), 1–8[2019-06-26] (2019). https://doi.org/10.13336/j.1003-6520.hve.20190604005
13. Chen, X., Wan, S., Zhu, Y., Xu, Q.: Analysis of distributed distribution fault handling based on edge computing. Electromechanical Inf. (17), 32–33 (2019)
14. Xu, H.: Realization of edge calculation in motor monitoring system. Electron. Technol. Softw. Eng. (11), 190–192 (2019)
15. Saxena, H., Salem, K.: EdgeX.: edge replication for web applications. In: International conference on cloud computing, pp. 1041–1044 (2015)
16. Zhou, Q.: Five years of GE industrial Internet. China Ind. Inf. (07), 32–38 (2018)
17. Shen, S., Yang, Z.: Architecture of the Internet of things and its standardization. Diss. (2015)
18. PROGRESS INTRODUCES OPENEDGE 12. The Productivity Software **32**(4) (2019)
19. Yang, Z., Mingxia, S., Yuwen, S., et al.: Greenhouse information system based on multi-sensor adaptive weighted fusion. Sens. Microsyst. **33**(6), 100–103 (2014)
20. Shi, W., Cao, J., Zhang, Q., et al.: Edge computing: vision and challenges. IEEE Internet Things J. **3**(5), 637–646 (2016)

21. Shi, W., Dustdar, S.: The promise of edge computing. IEEE Comput. **49**(5), 78–81 (2016)
22. Satyanarayanan, M.: The Emergence of Edge Computing. IEEE Comput. **50**(1), 30–39 (2017)
23. Maheshwari, S., Raychaudhuri, D., Seskar, I, et al.: Scalability and performance evaluation of edge cloud systems for latency constrained applications. Inf. Secur. 286–299 (2018)
24. Ahmed, R., Zaheer, Z., Li, R., et al.: Harpocrates: giving out your secrets and keeping them too. Inf. Secur. 103–114 (2018)
25. Chao, M., Yang, C., Zeng, Y., et al.: F-MStorm: feedback-based online distributed mobile stream processing. Inf. Secur. 273–285 (2018)
26. Patel, P., Patel, M., Panchal, V., et al.: Home automation using internet of things. Imperial J. Interdiscip. Res. **2**(5) (2016)
27. Ai, Y., Peng, M., Zhang, K., et al.: Edge cloud computing technologies for internet of things: a primer. Digital Commun. Netw. **4**(2), 77–86 (2017)
28. Aral, A., Brandic, I.: Dependency mining for service resilience at the edge. In: Information Security, pp. 228–242 (2018)
29. Boutaud, F., Peter, N.E.: Series maxium/minimum function computing devices, systems and methods. U.S. Patent No. 5,072,418 10 December 1991
30. Liu, S., Lin, Z., Li, J., et al.: Review and prospect of situation awareness technologies of power system. Autom. Electr. Power Syst. **44**(3), 229–239 (2020). https://doi.org/10.7500/AEPS20190402008
31. Lou, X., Guang, M.A., Guo, C., et al.: System and framework design of risk coordination control for whole operation process of power system. Autom. Electr. Power Syst. **44**(5), 161–170 (2020). https://doi.org/10.7500/AEPS20190429011
32. Ray, E.T.: Learning XML: Creating Self-Describing Data. O'Reilly Media, Inc., Newton (2003)
33. Mach, P., Becvar, Z.: Mobile edge computing: a survey on architecture and computation offloading. IEEE Commun. Surv. Tutorials **19**(3), 1628–1656 (2017)
34. Hu, YC., et al.: Mobile edge computing—a key technology towards 5G. ETSI white paper **11.11**, 1–16 (2015)

A Self-adaptive Method for Identifying the Last Circuit Breakers in Hybrid AC/DC Power Grids

Guangxiao Zhang(✉), Xiaoyang Tong(✉), Jiale Fan, Wenwen Li, and Xiaolin Zhou

School of Electrical Engineering, Southwest Jiaotong University, Chengdu, China
gxzhang@my.swjtu.edu.cn, xytong@swjtu.cn

Abstract. The last circuit breaker (LCB) is the key circuit breaker in the hybrid AC/DC power grids for avoiding severe overvoltage of AC/DC equipment caused by LCBs tripping. A self-adaptive LCBs identification scheme based on the Dijkstra algorithm is proposed. The LCBs identification is formulated as the single-source shortest path problem, which can be solved by the Dijkstra algorithm. If a circuit breaker in the shortest path from each converter valve to the combined AC lines is removed and there are no paths between them, then the circuit breaker can be considered an LCB. The simulation results reveal the effectiveness of the proposed scheme in Hybrid AC/DC power grids for identifying LCBs under different operating modes of AC field.

Keywords: Last circuit breaker · Hybrid AC/DC power grids · Single-source shortest path · Dijkstra algorithm

1 Introduction

The hybrid AC/DC power grid is a novel power network architecture of the future power grids [1, 2]. The new issues and challenges for the protection of hybrid AC/DC power grids are emerging with the interwoven of AC and DC transmission grids [3-5]. The last circuit breaker (LCB) is the key circuit breaker to connect AC and DC power systems. Once the LCB is tripped, the current coming from the converter valve would flow rapidly into the compensator in AC field, causing the overvoltage of the AC and DC equipment in the inverter station and endanger their safety. Therefore, the LCB identification will be of great significance for ensuring the safe and stable operation of the power grid [6-8].

The schemes developed for LCBs protection can be traditionally classified into two groups: arrester-based [9, 10] and switch states-based [11, 12] schemes. The arrester-based schemes are utilizing the energy consumption characteristic of arresters to protect AC/DC equipment during the overvoltage. However, the impact of overvoltage is still a significant problem during this period, which may lead to the destruction of AC/DC equipment. The switch states-based schemes that utilizing topology data from the AC

Y. Xue et al. (Eds.): PMF 2020, LNEE 718, pp. 882–891, 2021.
https://doi.org/10.1007/978-981-15-9746-6_66

field are more secure than arrester-based schemes because of the effect of overvoltage is eliminated [13].

In this paper, a self-adaptive LCBs identification method in hybrid AC/DC power grids is proposed. The identification of LCBs is converted to the shortest path problem, which can be solved by the Dijkstra algorithm. As is well-known, the connection path between the starting and ending points means that there is at least one path between them. If a vertex in the path between the starting point and the ending point is removed and the connection path no longer exists, then the vertex removed is the key vertex which is a bridge connecting the starting point and ending point. The definition for the key vertex is consistent with the LCB and located on the shortest path. Therefore, we can obtain the LCBs of each converter valve conveniently.

The organization of this paper is as follows. Section 2 summarizes the challenges for the LCBs identification scheme. Section 3 discusses the modeling for the AC field based on graph theory, Dijkstra's algorithm and its implementation for LCBs identification. In Sect. 4, simulation cases and results are given to prove the effectiveness of the scheme. Section 5 is the conclusion of this paper.

2 Problem Formulation

Figure 1 shows the schematic of switch state-based LCBs protection. As described in the figure, the two main stages of this scheme are presented, namely the LCBs identification stage and the blocking converter valve stage. In the first stage, the Pole Control Host (PCH) collects the real-time states of circuit breakers in the AC field and identifies the LCBs from all circuit breakers. In the second stage, the Converter Valve Control System (VVCS) would firstly issue an ESOF signal for blocking converter valve corresponding before the Relay tripped LCBs.

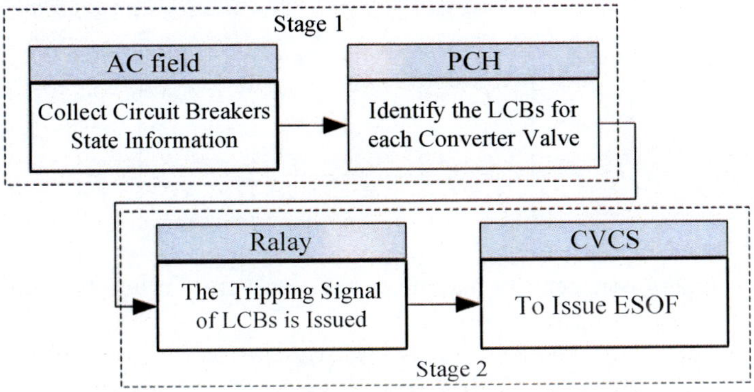

Fig. 1. Schematic of switch state-based LCBs protection.

The LCBs identification in a small-scale network can be easily completed through implement a look-up table approach [14]. Unfortunately, this is a significant setback because it requires the acquaintance of all possible structures of AC field beforehand.

Besides, any new deployment, which is very common for the HVDC project at the early stage, requires that the whole table should be re-written [15]. This scheme above cannot adapt to the changing topology structures, especially for a large-scale AC field. Some of the challenges to LCBs identification are summarized:

- The designs of the AC field are various in different converter stations. A general modeling method with portability should be developed to meet various network topologies of AC field.
- The power delivery mode, fault isolation, and circuit breakers maintenance would form different combinations for the states of circuit breakers [16], So the LCBs identification should be dynamic and updated frequently.
- The new deployments, such as AC transmission lines and intervals were being put into/out of service, are very common for the HVDC project at the early stage. It requires that the model for the identification of LCBs should be easily modified.

In this paper, we aim at developing a self-adaptive LCBs identification scheme to overcome all these problems.

3 Proposed LCBs Identification Scheme

3.1 Modeling for the AC Field

The electric wiring configuration of AC field can be abstracted into an undirected connected graph $G = (V, E)$ with nonnegative weights [17, 18]. The $V = \{v_1, v_2, \ldots, v_N\}$ is the set of vertices. The elements in set V consist of vertices abstracted by converter valves, AC lines, buses, circuit breakers, and teed nodes. The $E = \{(v_i, v_j), i, j = 1, 2, \ldots, N\}$ is a set of edges. The element (v_i, v_j) in the set E represents the edge between vertex v_i and v_j. The connection relationship between any two vertices can be expressed by the adjacency matrix [19].

$$C = \begin{bmatrix} c_{11} & c_{12} & \cdots & c_{1j} & \cdots & c_{1N} \\ c_{21} & c_{22} & \cdots & c_{2j} & \cdots & c_{2N} \\ \cdots & \cdots & \cdots & \cdots & \cdots & \cdots \\ c_{i1} & c_{i2} & \cdots & c_{ij} & \cdots & c_{iN} \\ \cdots & \cdots & \cdots & \cdots & \cdots & \cdots \\ c_{N1} & c_{N2} & \cdots & c_{Nj} & \cdots & c_{NN} \end{bmatrix} \quad (1)$$

Where c_{ij} is binary connectivity parameter between vertex v_i and v_j defined as:

$$C_{ij} = \begin{cases} 1 & \text{If } (v_i, v_j) \in E \\ 0 & \text{If } i = j \\ \infty & \text{Otherwise} \end{cases} \quad (2)$$

Where $c_{ij} = 0$ represents that the vertex v_i is the same as v_j while $c_{ij} = 1$ represents that there is an edge between the vertex v_i and v_j and the $c_{ij} = \infty$ represents that there is no edge between them.

It is worth noting that the states of vertices for circuit breakers are divided into the closed and open while the remaining vertices are all regarded as closed. As previously mentioned, the adjacency matrix should be dynamically updated due to the changing operating modes of AC field. For example, the state of circuit breaker changes from open to closed or from closed to open, the corresponding elements of adjacency matrix should be changed from "∞" to "1" or from "1" to "∞".

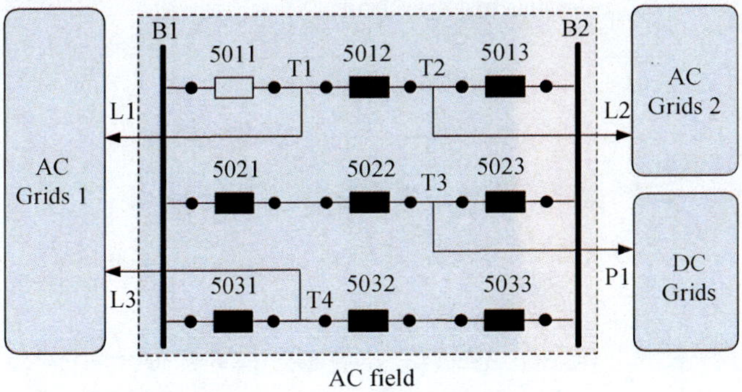

Fig. 2. A sketch map of the AC field with three intervals.

Figure 3 shows a connected undirected graph represented by the AC field in Fig. 2. As it is very time-consuming exercise to identify LCBs by searching for each path between the converter valves and the AC lines, the AC lines in this study are combined into one vertex, e.g., the vertices L1, L2, and L3 in the dashed box in Fig. 3 are combined into one vertex L [15]. In this way, we can quickly identify LCBs by only searching for the path from each converter valve to vertex L.

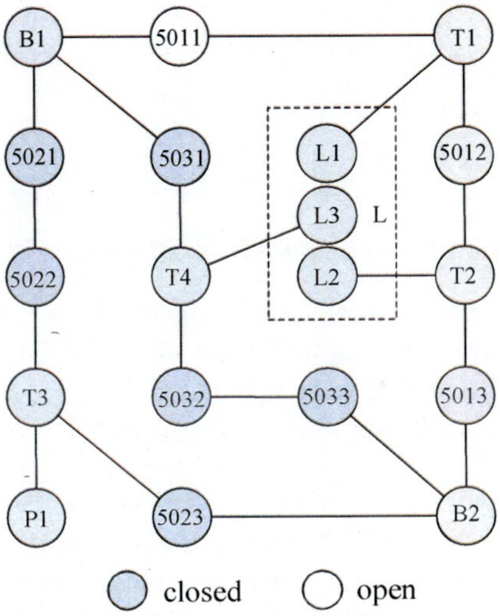

Fig. 3. Graphic representation of the AC field.

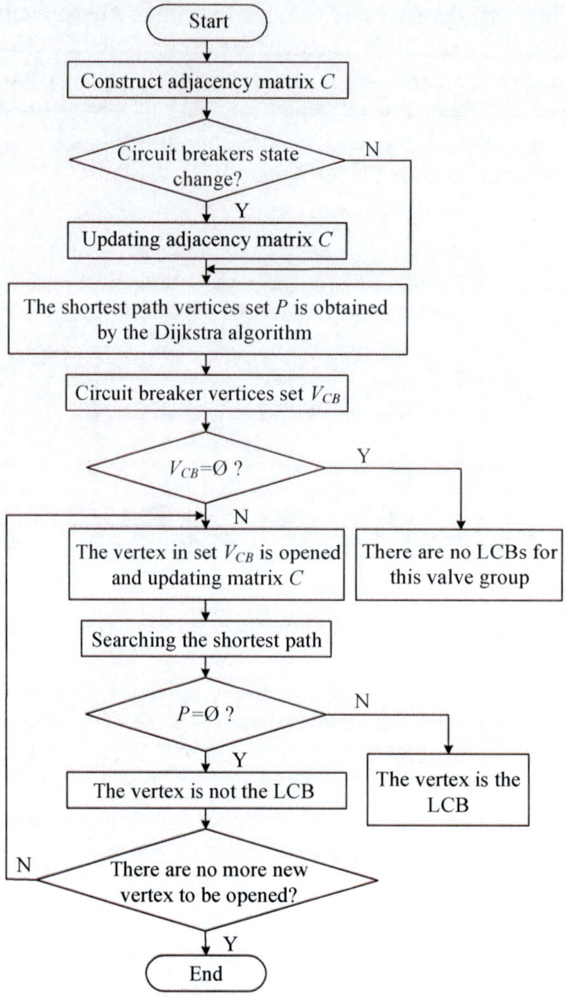

Fig. 4. Flowchart of the LCBs identification.

3.2 LCBs Identification Scheme Based on Dijkstra Algorithm

If a circuit breaker is opened in the shortest path from the converter valve to the combined AC lines vertex L leading to there is no path between them, then the opened circuit breaker is an LCB. Otherwise, it is not the LCB. Using this method, we can conveniently identify all LCBs for each converter valve. As we all know, the shortest path can be solved by the Dijkstra algorithm, which is a well-known algorithm for a given nonnegative graph to solve the single-source shortest path [20]. The Dijkstra algorithm employing the principle of greedy starts from the source vertex to find the shortest path to a neighbor vertex at every step and then moves on to its neighbors to keep track of the shortest path until the target vertex is visited. Thus, the shortest path from the source vertex to the target vertex is obtained. The identification of LCBs in the AC field can be formulated as the solution to the single-source shortest path problem. Now to utilize the algorithm aforementioned to our problem, the flowchart of the proposed LCBs identification method is given in Fig. 4. The specific process is explained as follows:

1) The AC field is firstly abstracted into the connected undirected graph and the real-time states of circuit breakers are collected to update adjacency matrix C for responding to current topology of AC field.
2) Let the combined AC lines vertex L as source vertex and the converter valve as target vertex. Next, the Dijkstra Algorithm is used to obtain the shortest path vertices set P and the non-circuit breaker vertices in the set P should be removed, yielding:

$$V_{CB} = P \cap CB \tag{3}$$

   Where V_{CB} is the set consists of vertices for circuit breakers. CB represents the vertices set of all circuit breakers in the AC field.
3) To identify LCB for each converter valve. If the set V_{CB} is empty, then there are no LCBs for the corresponding converter valve. Otherwise, a vertex in the set V_{CB} is open and the adjacency matrix C is updated accordingly. Meanwhile, we rerun the Dijkstra algorithm to search for the path between the source and target vertex, and to judge whether there is a path between them. If there is a path, the vertex opened in set V_{CB} is the not LCB. Otherwise, it is the LCB. Repeat the above process for identifying all LCBs until there is no more un-opened vertex in the set V_{CB}.

4 Simulation Results

Figure 5 represents a typical main electrical wiring of AC field with nine intervals. The network consists of four converter valves (P1 ~ P4), eight AC transmission lines (L1 ~ L8), four bus-tie circuit breakers (5101 ~ 5104), and 27 circuit breakers (5011 ~ 5093) among the six sectionalized bus (B1 ~ B6). In this figure, the white and black rectangle represents the open and closed circuit breaker, respectively.

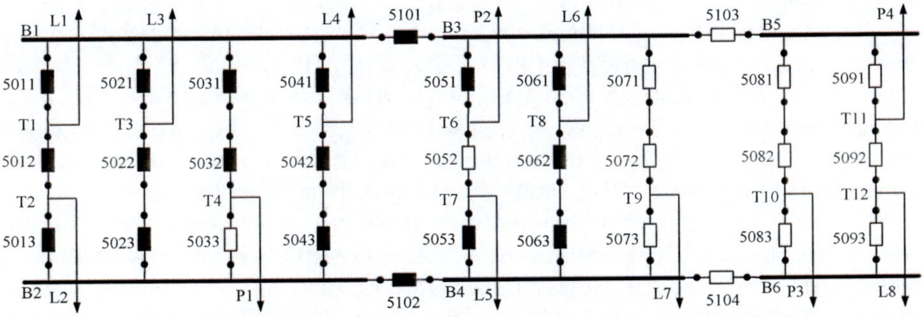

Fig. 5. Typical wiring diagram of the AC field.

Assume that the six intervals in the AC field in Fig. 5 were put into operation at the early stage. The six-interval AC field has 33 vertices after all AC lines are combined as one vertex. For the first case, let all circuit breakers be closed in the six-interval AC field. The graphic representation of AC field should be completed as illustrated in Sect. 3.1. And then, the connection relationship between any two vertices is stored in the adjacency matrix C as shown in Fig. 6.

$$
\begin{array}{c}
 \\ \text{L} \\ \text{P1} \\ \text{P2} \\ \text{B1} \\ \text{B2} \\ \text{B3} \\ \text{B4} \\ \text{T1} \\ \vdots \\ \text{T8} \\ 5011 \\ 5102 \\ 5011 \\ \vdots \\ 5063
\end{array}
\begin{array}{c}
\begin{array}{cccccccccccccc}
\text{L} & \text{P1} & \text{P2} & \text{B1} & \text{B2} & \text{B3} & \text{B4} & \text{T1} & \cdots & \text{T8} & 5011 & 5102 & 5011 & \cdots & 5063
\end{array} \\
\left[\begin{array}{ccccccccccccccc}
0 & \infty & \infty & \infty & \infty & \infty & \infty & 1 & \cdots & 1 & \infty & \infty & \infty & \cdots & \infty \\
\infty & 0 & \infty & \infty & \infty & \infty & \infty & \infty & \cdots & \infty & \infty & \infty & \infty & \cdots & \infty \\
\infty & \infty & 0 & \infty & \infty & \infty & \infty & \infty & \cdots & \infty & \infty & \infty & \infty & \cdots & \infty \\
\infty & \infty & \infty & 0 & \infty & \infty & \infty & \infty & \cdots & \infty & 1 & \infty & 1 & \cdots & \infty \\
\infty & \infty & \infty & \infty & 0 & \infty & \infty & \infty & \cdots & \infty & \infty & 1 & \infty & \cdots & \infty \\
\infty & \infty & \infty & \infty & \infty & 0 & \infty & \infty & \cdots & \infty & 1 & \infty & \infty & \cdots & \infty \\
\infty & \infty & \infty & \infty & \infty & \infty & 0 & \infty & \cdots & \infty & \infty & 1 & \infty & \cdots & 1 \\
1 & \infty & \infty & \infty & \infty & \infty & \infty & 0 & \cdots & \infty & \infty & \infty & 1 & \cdots & \infty \\
\vdots & \vdots & \vdots & \vdots & \vdots & \vdots & \vdots & \vdots & & \vdots & \vdots & \vdots & \vdots & & \vdots \\
1 & \infty & \infty & \infty & \infty & \infty & \infty & \infty & \cdots & 0 & \infty & \infty & \infty & \cdots & \infty \\
\infty & \infty & \infty & 1 & \infty & 1 & \infty & \infty & \cdots & \infty & 0 & \infty & \infty & \cdots & \infty \\
\infty & \infty & \infty & \infty & 1 & \infty & 1 & \infty & \cdots & \infty & \infty & 0 & \infty & \cdots & \infty \\
\infty & \infty & \infty & 1 & \infty & \infty & \infty & 1 & \cdots & \infty & \infty & \infty & 0 & \cdots & \infty \\
\vdots & \vdots & \vdots & \vdots & \vdots & \vdots & \vdots & \vdots & & \vdots & \vdots & \vdots & \vdots & & \vdots \\
\infty & \infty & \infty & \infty & \infty & \infty & 1 & \infty & \cdots & \infty & \infty & \infty & \infty & \cdots & 0
\end{array}\right]
\end{array}
$$

Fig. 6. Adjacent matrix C for the six-interval AC field.

When the Dijkstra algorithm is executed, it will start from source vertex, (i.e., the combined AC lines vertex L), to get the shortest path from source vertex to target vertex, i.e., the converter valve P1. Thus, we can obtain the shortest path vertices set P = {P1, T4, 5032, 5031, B1, 5011, T1, L}.

Once the set above is obtained, the next step is to remove the non-circuit breaker vertices, i.e., the set P and CB = {5011, 5012, 5013, 5021, 5022, 5023, 5031, 5032, 5033, 5041, 5042, 5043, 5051, 5052, 5053, 5061, 5062, 5063, 5101, 5102} are intersected by (3) to obtain the set V_{CB} = {5011, 5031, 5032}. Then, we assume that the vertex 5011 in the set V_{CB} is open, and the adjacency matrix C is updated, i.e., the value of element corresponding 5011 should be changed from "1" to "∞".

Next, we rerun this algorithm and find the shortest path source vertex to target vertex (i.e., L→T4→5032→5031→B1→5021→T3→P1), we can conclude that the vertex 5011 is not the LCB. However, the vertex 5031 and 5032 are both LCB because there is no shortest path when they are open, respectively. The results for each converter valve are shown in Table 1.

Table 1. Results of LCBs for each converter valve in the six-interval AC field.

Converter valve	Set *P*	Vertex opened	Path	LCB
P1	P1, T4, 5032, 5031, B1, 5011, T1, L	5011	P1 → T4 → 5032 → 5031 → B1→ 5021 → T3 → L	No
		5031	–	Yes
		5032	–	Yes
P2	P2, T6, 5051, B3, 5061, T8, L	5051	–	Yes
		5061	P2 → T6 → 5051 → B3 → 5101→ B1 → 5011 → T1 → L	No

For the second case, it is assumed that all intervals in the AC filed in Fig. 6 are put into operation and circuit breakers 5011, 5031, 5032, 5051, 5071, 5083, and 5092 are open whereas the others remain closed. The new rows and columns would be added to the adjacency matrix *C* which elements are assigned values by Eq. (2). Thus, a real-time adjacency matrix *C* is constructed according to the updated principle described in Sect. 3.1. The results of LCBs identification for each converter valve at the nine-interval AC field are presented in Table 2.

Table 2. Results of LCBs for each converter valve in the nine-interval AC field.

Converter valve	Set *P*	Vertex opened	Path	LCB
P1	P1, T4, 5033, B2, 5013, T2, L	5013	P1 → T4 → 5033 → B2→ 5023 → 5022 → T3 → L	No
		5033	–	Yes
P2	P2, T6, 5052, T7, L	5052	–	Yes
P3	P2, T10, 5082, 5081, B5, 5103, B3, 5061, T8, L	5061	P3 → T10 → 5082 → 5081 → B5 → 5103 →B3 → 5101 → B1 → 5021 → T3 → L	No
		5081	–	Yes
		5082	–	Yes
		5103	–	Yes
P4	P2, T11, 5091, B5, 5103, B3, 5061, T8, L	5061	P4 → T11 → 5091 → B5 → 5103 → B3 →5101 → B1 → 5021 → T3 → L	No
		5091	–	Yes
		5103	–	Yes

As Table 2 shows, the 5033, 5052, 5081, 5082, 5091, and 5103 are all LCBs in view of the combined AC lines vertex L would be disconnected from the corresponding converter valve when one of those is open. While the circuit breakers 5013 and 5061 are both not the LCBs due to the combined AC lines vertex L is still connected to the corresponding converter valve when they are open. It should be noted that a circuit breaker may be the LCB of multiple converter valves, such as 5103.

5 Conclusion

A shortest path-based LCBs identification scheme for the converter valve in the hybrid AC/DC power grid is proposed. The topology of AC field is firstly converted into an undirected connected graph and the adjacency matrix is updated by the real-time states of circuit breakers. The identification of LCBs is formulated as the single-source shortest path problem and thus form a practical principle, i.e., if there is no path between the combined AC lines to the converter valve when a circuit breaker is open in the shortest path, the opened circuit breaker is an LCB. Simulation studies show that the proposed scheme is a simple, effective, and portable approach compared to the look-up table scheme with a network structure predetermined and it can provide reliable LCBs identification results.

References

1. Chen, G., Li, M., Xu, T.: System protection and its key technologies of UHV AC and DC power grid. Autom. Electr. Power Syst. **42**(22), 2–10 (2018)
2. Meyer-Huebner, N., Suriyah, M., Leibfried T.: Distributed optimal power flow in hybrid AC–DC grids. IEEE Trans. Power Syst. **34**(4), 2937–946 (2019)
3. Chai, R., Zhang, B., Dou, J., Hao, Z., Zheng, T.: Unified power flow algorithm based on the NR method for hybrid AC/DC grids incorporating VSCs. IEEE Trans. Power Syst. **31**(6), 4310–4318 (2016)
4. Qi, C., Wang, K., Fu, Y., Li, G., Han, B., Huang, R., Pu, T.: A decentralized optimal operation of AC/DC hybrid distribution grids. IEEE Trans. Smart Grid **9**(6), 6095–6105 (2018)
5. Fan, J., Tong, X., Zhao, J.: Unified optimal power flow model for AC/DC grids integrated with natural gas systems considering gas-supply uncertainties. J. Mod. Power Syst. Clean Energy **6**(6), 1193–1203 (2018)
6. Li, M., Li, Y., Tong, C., Lin, X., Ding, S., Sun, J., Tian, F.: Last breaker protection strategy based on arrester energy consumption characteristics in light loaded system. 2014 IEEE PES Asia-Pacific Power and Energy Engineering Conference (APPEEC), pp. 1–4. IEEE, Hong Kong (2014)
7. Tian, F., Li, M., Lin, X., Yang, P., Zheng, P.: The protection strategy analysis of the last circuit breaker in the 1 + 1/2 operation mode. 2014 IEEE PES Asia-Pacific Power and Energy Engineering Conference (APPEEC), pp. 1–4. IEEE, Hong Kong (2014)
8. Jiang, Y., Xia, Y.: Suggestions to improve the last circuit breaker and line protection at Shenzhen converter station. South. Power Syst. Technol. **3**(1), 69–71 (2009)
9. Lin, X., He, Y., Xie, Z., Li, Z., Tong, C., Tan, A., Khalid, M.S.: Study of improved arrester based last circuit breaker protection strategy based on voltage-energy inverse proportional characteristics. Proc. CSEE **37**(12), 3400–3408 (2017)
10. Lin, X., Liu, C., Wang, Z., Xie, Z., He, Y., Li, Z., Tong, C., Deng, K., Huang, J.: Tripping strategy of multi criteria of last breaker protections based on dynamic weight correction D-S evidence theory. Proc. CSEE **38**(9), 2609–2621 (2018)
11. Yu, H., Zeng, L., Yang, Y., Dai, G., Li, Y., Zhang, A.: Self-adaptive strategy for the last breaker in UHVDC project based on graph theory. Autom. Electr. Power Syst. **40**(5), 121–126 (2016)

12. Yu, H., Fu, Y., Yu, F., Zeng, L., Hu, H., Li, Y.: Intelligent judgment method for the last breaker in HVDC project including bus-tie switch. Power Syst. Prot. Control **46**(5), 152–158 (2018)
13. Wu, J., He, J., Peng, G., Li, Q., Guo, J., Yang, G.: Last breaker protection of converter based on virtual current network. Autom. Electr. Power Syst. **44**(1), 168–174 (2020)
14. Cui, X.,Li, X., Zhang, M., Cui, X.: Design of the RRAM-based polymorphic look-up table scheme. IEEE J. Electron Devices Soc. **7**, 949–953 (2019)
15. Zhang, G., Tong, X.: The last circuit breakers identification in hybrid AC/DC power grids based on improved Tarjan algorithm. IEEE Trans. Power Delivery https://doi.org/10.1109/tpwrd.2020.2971664
16. Huang, S., Dinavahi, V.: Fast distribution network reconfiguration with graph theory. IET Gener. Transm. Distrib. **12**(13), 3286–3295 (2018)
17. Ishizaki, T., Chakrabortty, A., Imura, J.: Graph-theoretic analysis of power systems. Proc. IEEE **106**(5), 931–952 (2018)
18. Srivastava, A.K., Ernster, T.A., Liu, R., Krishnan, V.G.: Graph-theoretic algorithms for cyber-physical vulnerability analysis of power grid with incomplete information. J. Mod. Power Syst. Clean Energy **6**(5), 887–899 (2018)
19. Xu, Y., Yu, T., Yang, B.: Reliability assessment of distribution networks through graph theory, topology similarity and statistical analysis. IET Gener. Transm. Distrib. **13**(1), 37–45 (2019)
20. Barbehenn, M.: A note on the complexity of Dijkstra's algorithm for graphs with weighted vertices. IEEE Trans. Comput. **47**(2), 263 (1998)

Check for updates

Research Overview of the Protection and Location Technologies for a Flexible DC Distribution Network

Zilong Hu[2(✉)], Tonghua Wu[1,2,3], Xindong Li[1,2], Feng Hong[1,2], Xiaojiang Zheng[1,2], and Kuncheng Zheng[1,2]

[1] State Key Laboratory of Smart Grid Protection and Control, Nanjing 211106, China
[2] NARI Group Corporation (State Grid Electric Power Research Institute), Nanjing 211106, China
577016781@qq.com
[3] College of Energy and Electrical Engineering, Hohai University, Nanjing 210098, China

Abstract. Due to the advantages of flexible control, good power quality, and convenient connection to distributed energy, flexible DC distribution networks have become a hot research topic around the world. However, the protection and location technologies for DC distribution networks are one of the main technical difficulties that confine their large-scale application, so it is necessary to overcome these difficulties as a matter of urgency. This paper presents an overall introduction to the research statuses of the fault features, protection principles, and fault location technologies of a DC distribution network. First, the fault features of a flexible DC distribution network are analyzed based on two-level VSC converter and modular multilevel converter. Second, the protection principles and fault location technologies of a DC distribution network as well as the main principles, advantages, and disadvantages of different technologies are analyzed. Third, the outlook for the development of protection and location technologies for a DC distribution network is determined.

Keywords: DC distribution network · Fault feature · Protection principle · Fault location · Pilot protection

1 Introduction

With the development of the economy and the rapid expansion of the urban scale, the demand for electricity has increased, and traditional AC distribution networks are now facing a series of challenges related to urban power supply. With the maturity of power electronic technology, DC distribution networks have been able to provide a new power supply scheme for urban power supply. The advantages of DC distribution networks over AC distribution networks are mainly embodied in the following three points:

(1) A DC distribution network can reduce the conversion steps in a system and facilitate the connection to a distributed power supply and a DC load [1]. (2) A DC

Y. Xue et al. (Eds.): PMF 2020, LNEE 718, pp. 892–907, 2021.
https://doi.org/10.1007/978-981-15-9746-6_67

distribution network can effectively alleviate the shortage of urban transmission line corridors due to its higher transmission capacity at the same voltage level [2]. (3) In terms of power quality, there are no problems such as rotor angle stability, frequency deviation, and three-phase asymmetry with DC distribution. In addition, once a breakthrough has been made in energy storage technology, the technology can be widely configured in DC distribution networks as needed, which will effectively solve the power quality problems such as voltage flicker [3, 4].

With the ability to build a flexible, strong, and efficient distribution network, DC distribution systems will become an important form of urban distribution networks in the future. However, the existing AC system relay protection technology is difficult to apply to DC distribution networks, which are characterized by complex fault features quite different from those of AC systems, multistage and strong transient processes, non-fundamental components, rich harmonic content, and multiple attenuation components [5]. Furthermore, the protection principle of a conventional HVDC transmission system is mainly applied to the both-terminal system with the boundary of the smoothing reactor, so it is difficult to directly apply this principle to a DC distribution network [6–8]. DC distribution networks have complex topologies, short transmission lines, and weak overcurrent resistance, which results in strict requirements for the rapidity and selectivity of fault detection and location [9]. Thus, fault detection and location technology is one of the technical difficulties in confining the large-scale application of a DC distribution network and one of the key components of DC distribution network research [10].

This study presents a review and analysis of the current research results around the world based on the aspects of the fault analysis, fault detection, and fault location of DC distribution networks. The prospects of the key problems and the development direction in future research are also described.

2 Analysis of the Fault Features of a Flexible DC Distribution Network

The analysis of the transient features of a DC distribution network fault is the basis of fault detection and location. Because various converters differ greatly from each other with respect to the fault features of a flexible DC distribution network, the corresponding analysis methods and protection requirements are also different. At present, there are two relatively typical converter topologies. The research statuses of their fault transient features are described here.

2.1 Fault Features of a Two-Level VSC Type DC Grid

It was indicated in Reference [11] that the single-pole ground fault features of a system are related to the grounding mode. This was based on research about the transient features of a single-pole ground fault of two-level VSC type converter DC grids with theoretical analysis and simulation verification for the condition of unblocking the IGBT and transformer connections via different grounding modes. Reference [12] pointed out that for the case of a single-pole ground fault on the DC side, the capacitance on the grounding pole side formed an RLC second-order oscillating fault circuit through a fault line and ground resistance, resulting in the rapid discharge of the capacitance and the rapid increase of the current amplitude. After the research about the unsymmetrical fault at the AC side of a DC system in Reference [13], it was found that the zero sequence component formed a path through the ground branch line of the DC side energy-storage capacitance.

The bipolar short-circuit fault at the DC side of a DC distribution network imposes the most serious harm to a system. Reference [14] analyzed the response features of a fault circuit from the perspective of power electronic technology and presented computing methods for fault voltage and current. According to the assumption in Reference [15], the immediate shut-off of IGBT at the instant of a fault caused the short-circuit of two poles, resulting in the three-stage fault transient processes shown in Fig. 1: (a) A DC capacitance discharge stage, for which the fault current was mainly represented by the DC capacitance discharge to the fault point; (b) the natural diode commutation and conduction stage, for which the AC power supply and the capacitance discharged to the fault point simultaneously; (c) the concurrent diode conduction stage, for which the DC side could be equivalent to a first-order discharge circuit, while the AC side could be equivalent to a three-phase short-circuit. It was shown in Reference [16] that the capacitance discharge stage was an under-damped oscillation process due to the small damping of a DC line, and the DC capacitance voltage oscillated, resulting in zero-passage triggering. At that time, the converter was subject to a sudden overcurrent flow that was jointly contributed by the AC side three-phase short-circuit current and the DC side short-circuit reactance follow current, so the protection needed to act quickly.

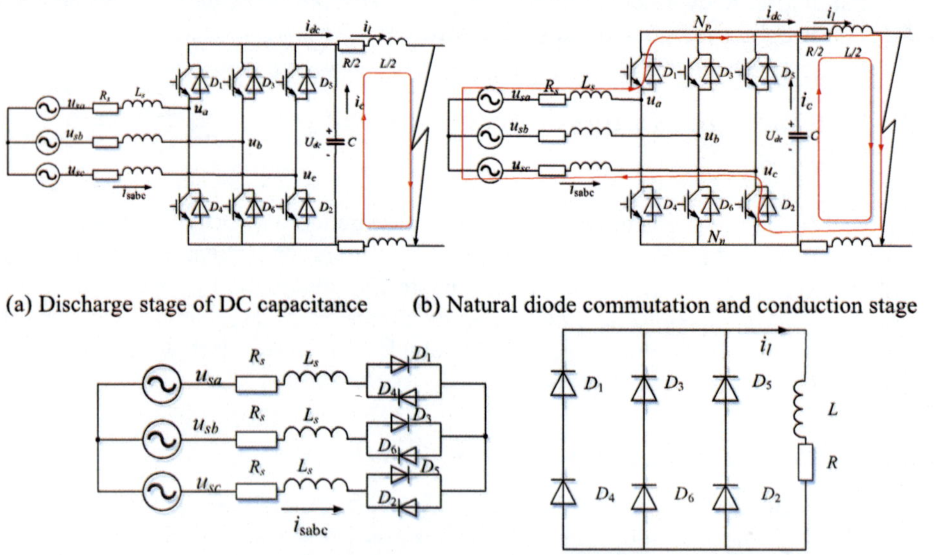

(a) Discharge stage of DC capacitance (b) Natural diode commutation and conduction stage

(c) Concurrent diode conduction stage

Fig. 1. Bipolar short-circuit fault process of a two-level VSC type DC grid.

2.2 Fault Features of a Modular Multilevel Converter (MMC)-Type DC Grid

An MMC adopts the cascade connection of submodules generally characterized by half-bridge types, full-bridge types, and clamping double types, which are shown in Fig. 2. Compared with a two-level VSC type converter, an MMC has numerous advantages such as the easy expansion of voltage level and low-level harmonic content as well as low switching frequency and low operating loss, so MMCs have increasingly been used in the flexible DC transmission and distribution fields. Due to different topologies and modulation strategies, an MMC-based DC distribution network differs from a two-level VSC type DC distribution network in terms of the fault features when a fault is found at the DC side.

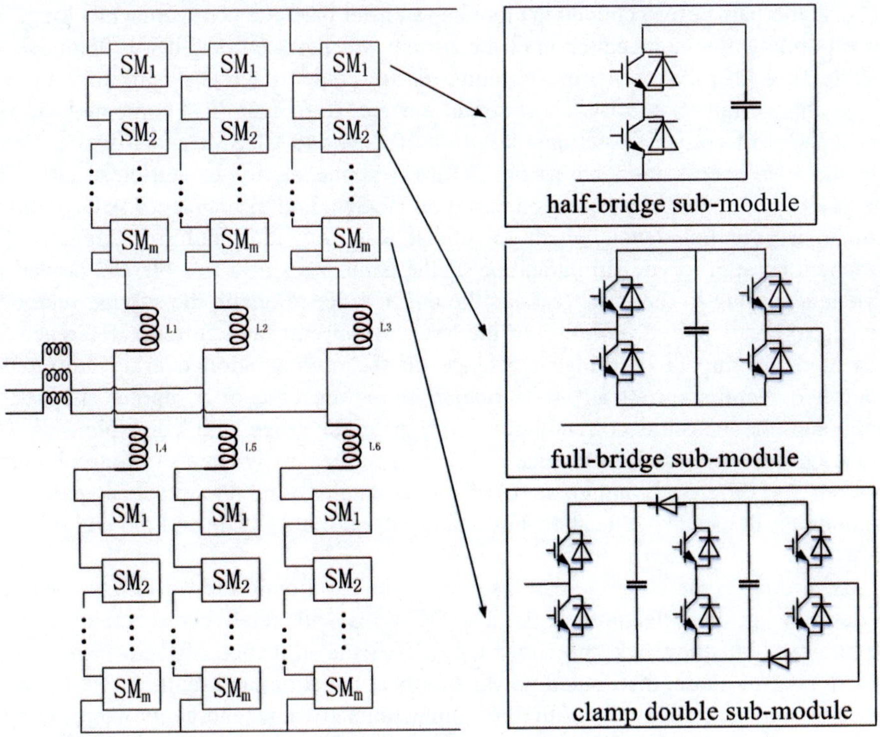

Fig. 2. An MMC topology.

Reference [17] presented research about the single-pole ground fault features of an MMC-based DC distribution network and the analysis of the impact of the AC/DC-side grounding modes on the single-pole ground fault features of a DC system. Reference [18] presented the computing method for single-pole ground short-circuit current and the analysis of the law of the impact of parameters such as the current-limiting reactor inductance on short-circuit current. Reference [19] analyzed the characteristics of a half-bridge MMC-MTDC single-pole-to-ground short-circuit fault, and it was pointed out that when the DC-side grounded MMC was single-pole short-circuited, the sound line voltage was twice the original and the system power could be transported normally. In addition, the converter station did not need to be urgently locked, but the fault had to be removed before the line withstood the voltage limit. Reference [20] showed that when a single-phase grounding fault occurs on an AC side, because the control strategy of the MMC converter station was constant voltage control, the voltage between the DC lines remained constant. However, there were short-circuit fundamental frequency fluctuations in the DC line voltage, which were not conducive to the stable operation of the system. Reference [21] presented a simulation analysis for an inter-pole short-circuit fault between single-terminal and multiterminal systems. It was found that the rates of increase of the fault currents of multiterminal systems were higher. In References [22, 23], the fault transient process was divided into two parts, namely, the part before converter blocking and the part after converter blocking. The corresponding analytical equation of the current was deduced, and it was demonstrated in Reference [22] that the current-limiting reactor could effectively inhibit the increase of the change rate of the DC short-circuit current. Reference [24] compared the fault mechanisms of converter stations based on HDSM and CDSM, and it was indicated that the converter station based on CDSM had the ability to self-clear DC faults compared with the converter station based on HDSM, and the capacitor voltage did not drop to a very low level, which facilitated restarting after failure. Reference [25] presented an appropriate simplification of the parameters of a flexible DC power grid system according to the basic features thereof in order to obtain the solving method of the approximate analytic solution of the line short-circuit fault current. Reference [26] presented the impact of bridge reactance on the commutation overlap during non-controlled rectification steady-state operation for the case of a bipolar short-circuit fault, and the possible commutation overlap angle range and corresponding fault scenarios were analyzed. Reference [27] analyzed the steady-state short-circuit current features for different conduction overlapping angles, and the practical calculation methods for the steady-state short-circuit current on the AC side and the DC side were deduced.

According to the above research studies, scholars around the world have been able to describe the fault features of flexible DC grids with relatively accurate analytical expressions, but they have not fully considered the influence of factors such as the system control mode, distributed power supply connection, and fault current limitation on fault transient features. With the connections to new energy power generation systems in the future, significant consideration should be given to the influence of intermittent power sources such as photovoltaics and wind turbines on fault output. In addition, equal attention should also be given to the fault mechanisms and transient features of multiterminal flexible DC distribution networks with respect to fault ride-through control strategies such as fault current limiting and fault clearing.

3 Protection Principles of a Flexible DC Distribution Network

For the case of a DC distribution network fault, the amplitude of a fault current will rise rapidly to cause serious damage to a system, so fast and reliable protection shall act. According to the different protection principles, protections can be divided into voltage and current protection, fault location protection, boundary protection, and pilot protection.

3.1 Voltage and Current Protection

Voltage and current protection is the most basic form of protection in electric power systems. It uses the voltage, current amplitude, or rate of change to detect faults. In spite of this simple principle, it is difficult to meet the requirements of a DC distribution network for acting speed and selectivity [28]. Therefore, scholars around the world have made improvements in traditional voltage and current protection principles.

Reference [29] presented an inverse time-limit protection for DC distribution network voltage to achieve coordination between the upper and lower levels, but it did not provide the calculation method of the voltage setting value. Reference [30] presented a main protection that integrated the differential value of the fault current with the set detection time limit according to the difference in the internal and external fault current differential values during the capacitance discharge in case of a fault. Reference [31] presented an approach to enable fault detection and to identify a specific fault line by detecting the voltage change rate on the current-limiting reactor at both terminals of a line after a fault. In addition, the comprehensive design of the protection scheme could also be carried out. For example, the multiterminal flexible DC system fault identification method using the voltage change rate and current change rate was presented in Reference [32].

The main problems in the application of voltage and current protection in a DC distribution network are as follows: (1) Fixed value setting mostly depends on simulation computing, and there is a lack of a universal setting method; (2) In a distributed DC distribution network, it is difficult to coordinate in the protection setting between adjacent lines; (3) The protection criterion that is composed only of the amplitude and differential of the electric quantity is susceptible to noise interference.

3.2 Fault Location Protection

Fault location protection enables protection and fault location based on the transient fault information of a DC distribution network and the computation of the distance between the protection installation place and the short-circuit point.

Fault location protection can be divided into traveling wave fault location protection and parameter identification protection. Traveling wave fault location protection was used to calculate a fault distance by measuring the travel time of a traveling wave between a fault point and a protection installation place [33]. Reference [34] presented a traveling wave location method based on mathematical morphology for the simultaneous detection of the mutation moment and the direction of the fault voltage signal. Parameter identification protection was used to calculate the fault distance by computing the line impedance from the protection installation place to the fault point. Reference [35] presented an improved scheme for the computing error and numerical stability of the algorithm of an RL model algorithm in a DC distribution network. An RL model was used to analyze the fault circuit, set up time-domain differential equations, solve the fault distance, and identify the fault location. Reference [36] presented an approach to setting up the time-domain equations of an equivalent fault circuit and solving the fault distance with simultaneous equations according to the values of the voltage and current measured at the protection installation place and the given line parameters.

Fault location protection can be used to accurately locate a fault location, but the following problems exist: (1) There are high requirements for the sampling frequency of the hardware due to the susceptibility of traveling wave fault location protection to noise interference and the limited ability to withstand transition resistance; (2) There are problems concerning the unreliability of the locating accuracy because the parameter identification protection is susceptible to a load.

3.3 Boundary Protection

Boundary protection is used to enable system protection via different fault features on both sides of the boundary elements such as the current limiting reactor at the line output for the case of a fault.

Reference [37] presented an approach for enabling fault identification according to the difference of transient energy for the case of external and internal faults by using the blocking of transient high-frequency components via boundary elements to distinguish the external and internal faults. The advantage of the protection principle lies in the fast acting speed and the ability to withstand a certain transition resistance, but the disadvantage lies in the fact that the protection setting value needs to be obtained through simulation. Reference [38] presented a fault identification method using current limiting reactor voltage. It was found in the research that the initial values of the reactance voltage were different for the case of internal and external faults, so the initial value of the reactance voltage was used to identify the fault. Reference [39] detected faults using voltage gradients, determined the fault interval using the smoothing effect of current-limiting reactors on the fault voltage signal, and determined the fault pole based on the difference between the zero-mode and one-mode voltage propagation characteristics. References [40, 41] presented the design for a protection principle applicable to a DC grid protection based on the voltage change rate of a current-limiting reactor. In addition, Reference [40] presented discussions about the selection principle of the setting value of the voltage change rate at the line side. Reference [41] presented an approach to enabling DC grid protection using low voltage as the criterion of protection activation, judged the fault interval using the voltage change rate and the polarity of the current-limiting reactor, and confirmed the fault pole with the zero-mode fault component. Reference [42] proposed a new full-line quick-action protection method based on the boundary characteristics of HVDC transmission lines. This method identified internal and external faults based on the difference in the amplitude-frequency characteristics of the input impedance, and it did not require high sampling frequency.

Boundary protection features the advantages of fast acting speed and no need for communication, but determining whether to install a current limiting reactor in an actual DC distribution network requires further research, and such methods have a protection dead zone problem for DC distribution networks with multilevel lines. Therefore, the application of this method in DC distribution networks is limited.

3.4 Pilot Protection

Pilot protection is based on the protection principle of both-terminal electrical quantities, mainly including current differential pilot protection and directional pilot protection.

Reference [43] presented a fast directional pilot protection criterion based on the amplitude ratio of the voltage transient components on both sides of a line DC reactor. Additionally, the judgment of the fault poles based on the amplitude ratio of the positive and negative transient voltages was not affected by the distributed capacitance of the line. Reference [44] presented the design for a set of protection principles based on a current state matrix. In fact, in addition to the ability to judge fault type, the protection criterion could be also used to judge fault lines by the current direction information at both terminals of the line. Reference [45] proposed a distributed area protection method for a DC distribution network based on the current differential state quantity. This method used the current differential at both ends of the line as the fault transient characteristic quantity and determined the fault by exchanging the information of the current differential status word between the protection devices. Therefore, this method did not require strict data synchronization. Reference [46] presented an approach for detecting a fault using the short-time energy of a transient differential current at both terminals of a line for the case of a single-pole grounding fault for a symmetric single-pole DC system with a low-current grounding mode. Reference [47] presented a current differential protection based on a frequency band distribution based on the research on bipolar short-circuit faults. It was found that the energy of the line differential current during external and internal faults was concentrated in different frequency bands. Reference [48] presented a pilot protection method based on the full-current correlation test for faults, with the use of a t-test algorithm to conduct a correlation test of the full-current signal on both sides, namely, the united description of the amplitude, direction, polarity, and mutation features of the current. Based on the distribution characteristics of the transient high-frequency components in a DC distribution network after a fault, a DC line pilot protection method based on high-frequency power was proposed in reference [49]. This method used the difference of the linear relationship and correlation coefficient between the high-frequency power on both sides of a line for different fault conditions, achieving fault identification and pole selection.

Pilot protection has absolute selectivity and high reliability, but it has the problems of channel delay and data synchronization as well as poor quickness. In addition, a DC distribution network imposes strict requirements for the fast action of protection. Thus, if pilot protection is applied to a DC distribution network, it is necessary to optimize the data transmission channel, improve the sampling rate, and improve the operation capacity of the processor in order to achieve the purpose of the rapid removal of protection.

3.5 Other Protections

In addition to the common methods mentioned above, scholars around the world have proposed protection principles such as traveling wave protection [50] and protection based on an intelligent algorithm [51] for DC distribution networks. Traveling wave protection depends on the accurate identification of a traveling wave front, but a DC distribution network has short lines, so it is difficult to accurately identify a traveling wave front. Therefore, the application of traveling wave protection in DC distribution networks remains to be studied. Tremendous data training is required to build a rule base for intelligent algorithm-based protection, which is difficult to implement.

4 Fault Location Method for a Flexible DC Distribution Network

The fault location methods for a DC distribution network can be divided into three types: the fault location method based on local information, the fault location method based on multipoint information, and the fault location method based on signal injection.

4.1 The Fault Location Method Based on Local Information

The fault location method based on local information is used to implement fault location by collecting post-fault information for local electrical quantities combined with line parameters and solving for the equivalent impedance.

By targeting a bipolar short-circuit fault, Reference [52] presented a fault location method based on the sixth harmonics to achieve fault location by solving the equivalent impedance of a fault circuit. In particular, a fault location could be identified by improving the discrete Fourier transform algorithm to extract the sixth harmonics of the voltage and current, as well as computing the equivalent impedance to confirm the fault location. Reference [53] presented a fault location method for single-pole grounding that featured a strong ability to withstand transition resistance. Specifically, the least square method was adopted to estimate the equivalent inductance between the measured point and the fault point based on the instantaneous voltage, current, and current change rate of the measured point in order to confirm the fault location. Reference [54] extracted transient voltage oscillation characteristics with the Prony algorithm and then calculated the distance to the fault. This method could locate a DC fault within a few milliseconds, and it had good resistance to transition resistance. However, the method had a certain dead zone range.

For the fault location method based on local information, it is enough to only make use of local information without additional equipment. The method features better economy and a faster fault location speed. However, there are still two problems with this method: 1) Less fault data can be used, and the location accuracy is affected by the transition resistance, line parameters, and other factors. 2) Due to the short length of each section of a DC distribution network, the difference in fault features is very small, and it is difficult to accurately identify fault sections.

4.2 Fault Location Method Based on Multipoint Information

The fault location method based on multipoint information is used to achieve fault location by integrating the information for a double-terminal DC circuit breaker, a section switch, and a multipoint.

Reference [55] presented a method for identifying fault areas based on the fact that the polarities of the current mutation at both terminals of a DC line were positive only for the case of an internal fault. With this method, only the communication state information was needed and there was a low requirement for communication synchronization. It was suggested in reference [56] that differential protection should be adopted as the main location criterion of a DC distribution network, and multipoint current direction information should be used as the auxiliary location method. This criterion and method could be reliably used to locate fault sections. Reference [57] proposed an accurate fault location method based on the transient current in the initial stage after a sub-module was blocked. By establishing an equivalent model for a flexible DC distribution network, the time domain expression of the DC current was derived, and the fault distance and transition were included. The time domain equations of two unknown parameters of the resistance were solved using the least square method. This method is suitable for short-circuit faults between poles under different grounding methods, and for single-pole grounding faults, the system grounding method needs to be considered. Reference [58] presented a method for achieving fault location by extracting the high-frequency components of the fault transient current on both sides of a DC line and deducing the relationship among the fault distance, system parameters, and harmonic current based on the fact that the voltage was the same between each output terminal of the converter and the transition resistance at the fault point. Reference [59] presented a method for enabling sparse measurement-based fault location by building a high-frequency impedance equivalent model based on a transient high-frequency current circuit, extracting the transient high-frequency voltage at the sparse measurement points of a DC distribution network with wavelet transform, and setting up high-frequency voltage and current equations at sparse measurement points for a solution.

The fault location method based on multipoint information can reduce the influence of the error of transition resistance and fault feature extraction to some extent and increase the accuracy of fault location.

4.3 Fault Location Method Based on Signal Injection

The aim of the fault location method based on signal injection is to compute and confirm a fault location using additional signals injected by an additionally mounted injection device after tripping a DC circuit breaker.

Reference [60] presented a method for achieving fault location by connecting an LC injection device to a fault line to form an oscillation circuit along with a fault line, and then computing the fault distance according to the measured oscillation frequency and the parameters of capacitance and inductance. Reference [61] presented a method for achieving fault location using a signal injector made up of a capacitor in parallel with a chopper circuit to inject a triangular wave current signal into a fault point, and then computing the high-frequency impedance of the fault circuit. However, the above schemes had the problem of high sampling frequency. Reference [62] presented a method for computing fault distance by extracting oscillation frequency and waveform envelope-based attenuation coefficients after the formation of an RLC circuit based on a signal injector and a DC network. However, this method featured a lower ability to withstand transition resistance. Reference [63] presented the design of a kind of fault location module composed of a DC voltage source, capacitance, adjustable inductance, and switches. According to this design, when the DC line fault occurred, it was possible to compute the equivalent series resistance of an RLC second-order circuit according to the zero input response features of the circuit, set up the energy conservation equation, and compute the fault distance and the transition resistance. The module featured a strong ability to withstand transition resistance.

With better accuracy, the signal injection method can be properly used to solve the two problems of less available fault data and complex fault transient features. This method usually considers the impact of transition resistance, which is included in the equation set for the solution to eliminate the influence of the transition resistance and improve the location accuracy. However, additional fault location devices have to be added, resulting in the increase of the location cost.

4.4 Development Prospects for Flexible DC Distribution Network Protection

Flexible DC distribution technology has attracted extensive research from scholars around the world due to its excellent distribution performance. According to the summary of the advantages and disadvantages of the protection and location technologies of existing DC distribution networks, the following three suggestions have been proposed:

1) The analysis of the fault features of a DC distribution network should be conducted by incorporating the influence of a control strategy. A DC distribution network contains a large number of controlled power electronic devices whose fault features have been changed to a nonlinear process dominated by control features. While analyzing the fault features of a DC distribution network, the control strategy of the converter should be fully considered. At the same time, more specific and detailed analytical expressions for the fault transient and the method for extracting a transient with engineering application value should be studied.
2) The development of DC distribution network protection has the aim of achieving higher rapidity and reliability. The fault transient quantity contains abundant fault information, such as the time, location, and specific type of a fault. As one future research direction, the transient information-based transient quantity protection can

be used to rapidly detect faults and achieve the ultra-high speed action of DC distribution network protection. Pilot protection features absolute selectivity, strong reliability, and broad development prospect. However, such protection needs to focus on solving the problems of data synchronization and channel delay to enhance the protection speed. At present, the research of scholars around the world about the time coordination and setting value coordination of the main and backup protection of DC lines in flexible DC distribution networks needs to be improved, so the protection principles covering the main and backup protection should be further studied.

3) A DC distribution network has a short length of line and it imposes high requirements for location accuracy. Fault location based on local information and signal injection is confirmed by computing impedance, but the small errors in measurement and computing therein are likely to lead to inaccurate location. Differential protection is absolutely selective, and it can be used to accurately locate fault sections, but it will exit directly in the case of a high channel bit error rate. The fault location method based on multipoint information direction only provides directional information interaction and it imposes low requirements for communication. The location method combining multipoint information direction protection with differential protection can make full use of the advantages of the two protection methods and it is worthy of further research.

In the research of DC distribution network protection and location technologies, it is necessary to consider the features of the structure and control of DC distribution networks, get rid of the shackles of the traditional relay protection concept of an AC grid, and explore new protection principles and positioning technologies in line with the features of a DC distribution network.

5 Conclusion

Currently, China has established DC distribution network demonstration projects in Shenzhen, Suzhou, Hangzhou, Zhuhai, and other places successively. Preliminary exploration has been carried out for related technologies such as voltage grades, technical architecture, and control strategies, but the protection principles and fault location technologies are not mature enough, and it is difficult for a DC distribution network to be applied on a large scale. This paper presents an overall classification and a description of the worldwide research status of the key technologies such as fault feature analysis, protection principles, and fault location in flexible DC distribution networks, and it provides suggestions for future research direction. At present, the protection principles and location technologies for a DC distribution network are still mainly limited to the theoretical research level. With the construction of DC distribution network demonstration projects, the viability of the existing principles needs to be verified by practical projects and improved according to onsite operating situations.

Acknowledgments. This work was supported by the National Key R&D Program of China (No. 2018YFB0904700), the Science and Technology Program of the State Grid Corporation of

China: Research on the transient protection of an AC/DC hybrid grid with high-proportion renewable energy.

References

1. Jiang, D.Z., Zheng, H.: Research status and developing prospect of DC distribution network. Autom. Electric Power Syst. **36**(08), 98–104 (2012)
2. Starke, M., Li, F.X., Tolbert, L.M., et al.: AC VS. DC distribution: maximum transfer capability. In: Proceedings of IEEE Power and Energy Society General Meeting Conversion and Delivery of Electrical Energy in the 21st Century, pp. 1–6. IEEE, Pittsburgh (2008)
3. Lu, W.X., Ooi, B.T.: Premium quality power park based on multi-terminal HVDC. IEEE Trans. Power Deliv. **20**(2), 978–984 (2005)
4. Wang, D., Mao, C.X., Lu, J.M., et al.: Technical analysis and design concept of DC distribution system. Autom. Electric Power Syst. **37**(8), 82–88 (2013)
5. Yang, J., Fletcher, J.E., O'reilly, J.: Short-circuit and ground fault analyses and location in VSC-based DC network cables. IEEE Trans. Ind. Electron. **59**(10), 3827–3837 (2012)
6. Wei, D.H., Miao, S.H., Liu, Z.W., et al.: A fault identification method for HVDC transmission with long line based on boundary characteristics. Power Syst. Protect. Control **46**(17), 75–82 (2018)
7. Zhou, J.P., Zhao, C.Y., Li, C.Y., et al.: Boundary protection scheme for multi-terminal flexible DC grid based on voltage of DC reactor. Autom. Electric Power Syst. **41**(19), 89–94 (2017)
8. Willem, L., Dirk, V.H.: Classification of fault clearing strategies for HVDC grids. In: Cigre Lund Symposium, Paris, France (2015)
9. Yan, S., Luo, X., He, Z.Y.: Prospect of core equipment and key technology for DC power grid. Autom. Electric Power Syst. **43**(03), 205–216 (2019)
10. Wang, S.X., Liu, Q., Xue, S.M., et al.: Key technologies and prospect for coordinated control and protection in DC distribution system. Autom. Electric Power Syst. **43**(23), 23–30 (2019)
11. Dai, Z.H., Ge, H.B., Yan, S.Q., et al.: Effects of grounding mode on fault characteristics in flexible DC distribution system. Power Syst. Technol. **41**(07), 2353–2364 (2017)
12. Hu, J.J.: Fault analysis and protection technology of DC distribution system. Zhejiang University (2014)
13. Carmintim, M., Ragaini, E., Tironi, E.: DC and AC ground fault analysis in LVDC micro-grids with energy storage systems. In: Environment and Electrical Engineering, pp. 1047–1054. IEEE, Rome (2015)
14. Hu, J.J., Gao, Y.B., Yan, Y.T., et al.: DC side short-circuit fault analysis of VSC. J. Mech. Electr. Eng. **31**(04), 512–516+544 (2014)
15. Li, B., He, J.W.: DC fault analysis and current limiting technique for VSC-based DC distribution system. Proc. CSEE **35**(12), 3026–3036 (2015)
16. Li, B., He, J.W.: Research on the DC fault isolating technique in multi-terminal DC system. Proc. CSEE **36**(01), 87–95 (2016)
17. Wu, J., Wu, D., Zhu, J.D., et al.: Grounding method design of multi-terminal flexible DC distribution. Proc. CSEE **37**(09), 2551–2561 (2017)
18. Chen, N., Qi, L., Cui, X., et al.: Calculation method for line-to-ground short-circuit currents of VSC-HVDC grid. Electric Power Constr. **40**(04), 119–127 (2019)

19. Zhang, X.N., Zhao, C.Y., Pang, H., et al.: A control and protection scheme of multi-terminal DC transmission system based on MMC for DC line fault. Autom. Electric Power Syst. **37** (15), 140–145 (2013)
20. Jia, Y.Y., Gao, S., Si, X.Y., et al.: Protection method of flexible DC power distribution network based on grounding mode with high resistor and arrester in parallel. Autom. Electric Power Syst. **44**(10), 166–173 (2020)
21. Zhu, T., Zhao, Y.M., Yu, Z.Q., et al.: Fault transient simulation of middle-voltage MMC-HVDC distribution system. High Voltage Apparatus **51**(11), 26–32+40 (2015)
22. Su, J.S., Guo, J.D., Jin, T.: DC fault characteristics and line fault recovery strategy in flexible DC power network. Trans. China Electrotech. Soc. **34**(S1), 352–359 (2019)
23. Zhang, Z.R., Xu, Z.: Short-circuit current calculation and performance requirement of HVDC breakers for MMC-MTDC system. IEEE Trans. Electr. Electron. Eng. **11**(2), 168–177 (2016)
24. Ding, Y.Z., Su, J.H., Zhou, J.: Analysis on fault current limitation and self-recovery of MMC based on clamp double sub-module. Autom. Electric Power Syst. **38**(1), 97–103 (2014)
25. Tan, L.X., Dong, X.Z.: An approximate method for the calculation of transmission line fault current in MMC-HVDC grid. Proc. CSEE **39**(02), 490–498+646 (2019)
26. Li, B., Li, Y., He, J.W., et al.: Stable fault characteristic analysis of the DC system based on modular multilevel converter. Power Syst. Protect. Control **44**(21), 1–8 (2016)
27. Zhang, G.J., Qi, X.C., Chen, Y., et al.: Characteristic analysis of modular multilevel converter under DC pole-to-pole short-circuit fault. Autom. Electric Power Syst. **40**(12), 151–157+199 (2016)
28. Gao, R.D., Wu, Z.J., Fan, W.C., et al.: Inverse time current variance protection scheme for two-terminal DC distribution system. Power Syst. Technol. **42**(09), 2849–2859 (2018)
29. Hajmaharsi, M.Y.: Novel DC ring topology and protection system-a comprehensive solution for mega city power grids. North Carolina State University, Raleigh (2009)
30. Zhou, J.Y., Li, F.T., Chen, W.W., et al.: Line protection schemes for flexible DC distribution network based on capacitor discharge. Power Syst. Protect. Control **47**(08), 42–48 (2019)
31. Zheng, X., Tai, N., Wu, Z., et a1.: Harmonic current protection scheme for voltage source converter-based high-voltage direct current transmission system. IET Gener. Transm. Distrib. **8**(9), 1509–1515 (2014)
32. Wang, J., Berggren, B., Linden, K., et al.: Multi-terminal DC system line protection requirement and high speed protection solutions. In: CIGRE Symposium, pp. 26–30. Cape Town, South Africa (2015)
33. Su, W., Gong, Y.F., Li, Y.: Traveling-wave-based line fault location in star-connected multi-terminal HVDC systems. IEEE Trans. Power Deliv. **27**(4), 2286–2294 (2012)
34. Yu, X.Y., Xiao, L.Y.: Fault location scheme for multi-terminal MMC-HVDC system based on mathematical morphology. Electric Power Autom. Equip. **40**(01), 93–100 (2020)
35. Li, M., Jia, K., Bi, T.S., et al.: Fault distance estimation-based protection for DC distribution networks. Power Syst. Technol. **40**(03), 719–724 (2016)
36. He, J.H., Zhou, L., Luo, G.M., et al.: Transient protection based on single-end electrical signals for multi-terminal flexible DC distribution system. Electric Power Autom. Equip. **37** (08), 158–165 (2017)
37. Li, B., He, J.W., Li, Y., et al.: Single-ended protection scheme based on boundary characteristic for the multi-terminal VSC-based DC distribution system. Proc. CSEE **36**(21), 5741–5749+6016 (2016)
38. Dai, Z.H., Huang, M., Su, H.B., et al.: Single-terminal quantity based line protection for ring flexible DC distribution system. Proc. CSEE **38**(23), 6825–6836+7117 (2018)
39. Wang, Y.T., Zhang, B.H., Fan, X.K.: Fast protection scheme for overhead transmission lines of VSC-based HVDC grid. Autom. Electric Power Syst. **40**(21), 13–19 (2017)

40. Sneath, J., Rajapakse, A.D.: Fault detection and interruption in an earthed HVDC grid using ROCOV and hybrid DC breakers. IEEE Trans. Power Deliv. **31**(3), 973–981 (2016)
41. Yu, X.Y., Xiao, L.Y., Lin, L.Z., et al.: Single-ended fast fault detection scheme for MMC-based HVDC. High Voltage Eng. **44**(02), 440–447 (2018)
42. Song, G., Chu, X., Gao, S., et al.: A new whole-line quick action protection principle for HVDC transmission lines using one-end current. IEEE Trans. Power Deliv. **30**(2), 599–607 (2015)
43. He, J.W., Li, B., Li, Y., et al.: A fast directional pilot protection scheme for the MMC-based MTDC grid. Proc. CSEE **37**(23), 6878–6887+7078 (2017)
44. Dai, Z.H., Ge, H.B., Chen, B.Y., et al.: Line protection schemes for flexible medium voltage DC distribution networks. Autom. Electric Power Syst. **41**(17), 78–86 (2017)
45. Lyu, J.L., Yang, J.G., Chen, Q., et al.: Current differential state quantities based distributed area protection for DC distribution network. Autom. Electric Power Syst. (2020)
46. Bi, T.S., Wang, S., Jia, K., et al.: Short-term energy based approach for monopolar grounding line identification in MMC-MTDC system. Power Syst. Technol. **40**(03), 689–695 (2016)
47. Jia, K., Li, M., Bi, T.S., et al.: Energy distribution-based differential protection for VSC-DC distribution lines. Power Syst. Technol. **41**(09), 3058–3065 (2017)
48. Wang, C.B., Jia, K., Zhao, Q.J., et al.: Pilot protection for flexible-DC distribution line based on correlation test of DC current. Trans. China Electrotech. Soc. **35**(08), 1764–1775 (2020)
49. Chen, S.Y., Huang, W.T., Tai, N.L., et al.: High-frequency power pilot protection method for multi-terminal flexible DC distribution network. Autom. Electric Power Syst. (2020)
50. Chen, F.F., Yang, Y., Song, G.B., et al.: Adaptability analysis of single terminal traveling wave protection in mid-voltage DC distribution based on VSC. Power Syst. Protect. Control **44**(22), 50–55 (2016)
51. Ramesh, M., Laxmi, A.J.: Fault identification in HVDC using artificial intelligence-recent trends and perspective. In: Proceedings of International Conference on Oower, Signals, Controls and Computation, pp. 1–6. IEEE, Thrissur (2012)
52. Li, M., Jia, K., Bi, T.S., et al.: Sixth harmonic-based fault location for VSC-DC distribution systems. IET Gener. Transm. Distrib. **11**(14), 3485–3490 (2017)
53. Feng, X.Y., Qi, L., Pan, J.P.: A novel fault location method and algorithm for DC distribution protection. IEEE Trans. Ind. Appl. **53**(3), 1834–1840 (2017)
54. Jia, K., Li, M., Bi, T.S., et al.: A voltage resonance-base single-ended online fault location algorithm for DC distribution networks. Sci. China: Technol. Sci. (05), 721–729 (2016)
55. Cai, X.L., Song, G.B., Gao, S.P., et al.: Research on applicability of current fault component polarity pilot directional protection for VSC-HVDC cable lines. Power Syst. Protect. Control **41**(05), 10–16 (2013)
56. Wu, T.H., Dai, W., Li, X.D., et al.: Protection scheme and device development of flexible DC distribution system. Autom. Electric Power Syst. **43**(23), 123–135 (2019)
57. Gao, X.P., Huang, W.T., Tai, N.L., et al.: Transient current based fault location method of pole-to-pole short-circuit for modular multilevel converter low voltage direct current. Autom. Electric Power Syst. (2020)
58. Xu, F., Fan, C.J., Sun, Q., et al.: Dual-terminal fault location method for DC distribution lines with power electronic transformer. Proc. CSU-EPSA **31**(10), 73–80 (2019)
59. Jia, K., Feng, T., Zhao, Q.J., et al.: Bipolar short-circuit fault location for DC distribution network based on sparse measurement of fault high-frequency voltage. Autom. Electric Power Syst. **44**(01), 142–151 (2020)
60. Jia, K., Li, L., Xuan, Z.W., et al.: Study on fault location method and simulation for VSC-DC distribution network based on disturbance injection. Power Syst. Protect. Control **47**(04), 99–106 (2019)

61. Jai, K., Christopher, E., Thomas, D., et al.: Advanced DC zonal marine power system protection. IET Gener. Transm. Distrib. **8**(2), 301–309 (2014)
62. Park, J.D., Candelaria, J., Ma, L., et al.: DC ring-bus micro-grid fault protection and identification of fault location. IEEE Trans. Power Deliv. **28**(4), 2574–2584 (2013)
63. Yang, Y.C., Huang, C., Wang, X.Y., et al.: A novel line fault location method for DC distribution network. Power Syst. Technol. **43**(05), 1787–1794 (2019)

Scheme for Turn-to-Turn Fault Protection of Stepped Controllable Shunt Reactor Control Winding

Liwei Zhu[1,2,3(✉)], Tonghua Wu[1,2], Wei Dai[1,2], Kunxian Zhang[4], Xiang Zhu[1,2], and Saisai Ni[5]

[1] State Key Laboratory of Smart Grid Protection and Control, Nanjing 211106, Jiangsu, China
bbzhuliwei@outlook.com

[2] Nari Group Corporation, Nanjing 211106, Jiangsu, China

[3] NARI School of Electrical and Automation Engineering, Nanjing Normal University, Nanjing 210023, Jiangsu, China

[4] State Grid Gansu Electric Power Company, Lanzhou 730000, Gansu, China
hnzhangkunxian@163.com

[5] Electric Power Research Institute of State Grid Gansu Electric Power Company, Lanzhou 730000, Gansu, China

Abstract. This paper analyzes the negative-sequence network diagram of stepped controllable shunt reactor (SCSR) in case of turn-to-turn short-circuit fault and turn-to-ground fault of the SCSR control winding, as well as the line fault out of the SCSR area and open-phase running of the system based on the "T"-type equivalent circuit, thus deducing the discriminant of the negative-sequence power direction. The paper also configures the negative-sequence power directional protection for the SCSR control winding and derives the relationship of negative-sequence current between line side winding and control winding. The occurrence of fault is identified by the negative-sequence current of the control winding and the position of fault is determined by the negative-sequence power direction angle. For small turn-to-turn fault, the paper proposes a negative-sequence current amplitude adaptive starting threshold detection method based on the capacity to determine the starting threshold of the negative-sequence current. It also puts forward a negative-sequence phase angle smoothing method based on the dynamic mean to deal with the large phase angle fluctuations of the negative-sequence voltage during the small turn-to-turn short-circuit. The simulation results verify that the method can judge the turn-to-turn fault of the SCSR control winding reliably and effectively.

Keywords: Stepped Controllable Shunt Reactor (SCSR) · Turn-to-turn fault · Negative-sequence power directional protection · Small Turn-to-turn protection of adaptative threshold

Y. Xue et al. (Eds.): PMF 2020, LNEE 718, pp. 908–924, 2021.
https://doi.org/10.1007/978-981-15-9746-6_68

1 Introduction

The development of ultra/extra-high voltage (UHV/EHV) AC power transmission system brings problems such as violent tidal change and large capacitive reactive power; the power generation of photovoltaic, wind energy and other new energy sources has the intermittent, fluctuating and random characteristics; the access of power generation of new energy intensifies the change of tide on power transmission channel and fluctuation of reactive voltage, and increases the control difficulty of reactive voltage [1, 2]. Adding reactors to UHV/EHV transmission lines can reduce overvoltage to a certain extent [3–5], however, conventional shunting reactors cannot provide smooth power regulation, hence cannot satisfy the requirements of excessively-high voltage long-distance transmission networks [6]. The controllable shunt reactor (CSR) can adjust its own capacity according to different reactive compensation demands, adjust reactive power on power transmission line dynamically, reduce transmission of reactive power on power transmission line, and limit the power frequency overvoltage [7, 8].

The stepped controllable shunt reactor (SCSR) is adopted as one of important type of controllable shunt reactor, and it has the advantages of simple principle, reliable running, rapid response speed and low cost. The SCSR mainly consists of high-resistance transformer and reactors, and the difference between this transformer and common transformer lies in that the leakage reactance of the transformer is very high and is close to 100% [9], so the SCSR is also called high-resistance type controllable shunt reactor, and its topological structure is as shown in Fig. 1.

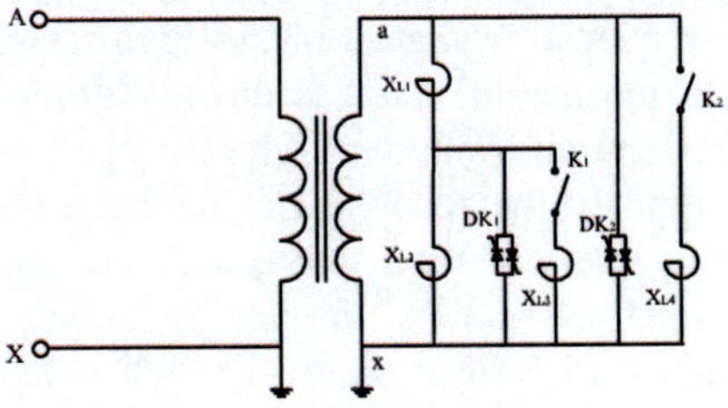

Fig. 1. Wiring of SCSR single-phase principle

In Fig. 1, the SCSR input capacity includes three levels, i.e. 33%, 67% and 100%; in case of accessing X_{L1} and X_{L2} only, the input capacity is 33%; in case of accessing X_{L3}, the input capacity is 67%; in case of accessing X_{L3} and X_{L4} at the same time, the input capacity is 100%. The thyristors DK_1 and DK_2 can realize automatic turn-on or turn-off of corresponding branches rapidly [10, 11].

Protections for conventional FSRs (Fixed Shunt Reactors) have been well documented [12–15]. However, the research on the protection for CSR, has rarely been studied. The SCSR turn-to-turn short-circuit fault is a kind of common internal fault; the transformer has relatively high leakage reactance, and its short-circuit resistance is much greater than that of common transformer. This makes the current change on the line side and control-side relatively small when there is any turn-to-turn short-circuit fault, and the fault characteristics are relatively weak. The problem of SCSR turn-to-turn short-circuit protection also becomes an important part of research, and some experts and scholars have proposed different protection schemes to solve this problem.

The reference [7] proposed a kind of new method specially for turn-to-turn short-circuit protection of SCSR control winding; the differential protection element and zero-sequence overcurrent element constitute fault detection, and the abnormal locking element out of the region as well as abnormal elements of voltage transducer and current transducer constitute locking element to prevent maloperation. Adding the abnormal locking out of the region can enhance protection reliability, but cannot improve the protection sensitivity.

The reference [16] proposed a kind of new turn-to-turn protection principle, took the standard deviation which represented variation of the equivalent total leakage inductance as action component, and took the stability of variation of each side leakage inductance as restraint component, thus constituting a kind of new protection criterion of percentage restraint characteristic. The reference [9] proposed that the zero-sequence power directional protection was adopted as main criterion, with the fault component and steady criterion as auxiliary criteria, which improved the reliability of turn-to-turn protection. The reference [17] proposed the turn-to-turn protection principle based on equivalent inductance, which conducted energy accumulation to adjacent period before and after fault, took the difference between both of them as criterion of occurrence of turn-to-turn fault. This method constituted two criteria only depending on its own electrical quantity, so as to enhance protection reliability. The reference [18] proposed that the negative-sequence power directional protection was adopted to replace the zero-sequence power directional protection, and it could improve the sensitivity of protection. For the zero-sequence reactance of bus high-resistance equals to negative-sequence reactance when the neutral point is directly grounded, this method cannot show obvious effect when the bus is under high-resistance.

The reference [10] proposed the absolute value comparation for negative-sequence and zero-sequence composite direction protection criterion, when the negative-sequence current was relatively small, in case of multiplying a relatively large negative-sequence resistance to form a relatively high voltage, and then conducting judgment, the possibility of error would be reduced. Meanwhile, when the negative-sequence power direction protection and zero-sequence power direction protection were used at the same time, the protection sensitivity would be improved.

The reference [19] referred to configuring the negative-sequence power directional for the line side winding, control winding and compensation winding of Magnetically Controlled Shunt Reactor (MCSR), so as to realize less-turn fault detection of turn-to-turn short-circuit.

This paper takes the “T”-type equivalent circuit of transformer as foundation, analyzes the action criteria of faults in different positions in the negative-sequence network diagram, configures the negative-sequence power directional protection for control winding. The adaptive threshold of negative-sequence current detection is set according to the change of capacity. With regard to the problem that the fluctuation of voltage phase angle of negative-sequence is relatively large in case of small turn-to-turn short-circuit fault, different detection thresholds are set according to different access capacities. The mean value of fluctuating waveform takes the place of real waveform value, so as to speed up the fault detection. The protection acts when the difference of several times of continuous mean values is within the allowable scope, which can guarantee the protection reliability.

2 Protection Principle in SCSR Negative-Sequence Power Direction

2.1 Analysis on SCSR Sequence Network Diagram

The line side winding of SCSR transformer adopts "Y"-type connection, and the neutral point is grounded via small reactor, and its control winding adopts "Y"-type connection, with the neutral point grounded directly, as shown in Fig. 2.

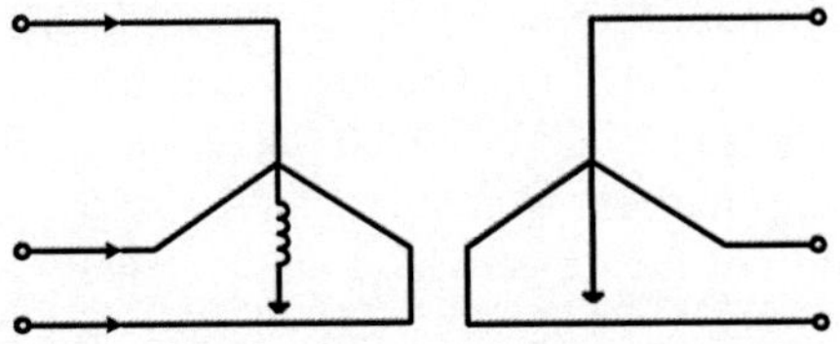

Fig. 2. YNyn joint transformer with neutral point grounded via reactor

The zero-sequence and negative-sequence equivalent circuits of transformer are the same as the positive-sequence equivalent circuit in form, whether there is zero-sequence impedance in the transformer and the magnitude of the zero-sequence impedance depend on the connection of the transformer winding and the iron core structure of the transformer. In the zero-sequence equivalent circuit, only the elements which the zero-sequence current flows through have zero-sequence resistance, while the parts where no zero-sequence current flows through is not calculated into the zero-sequence resistance.

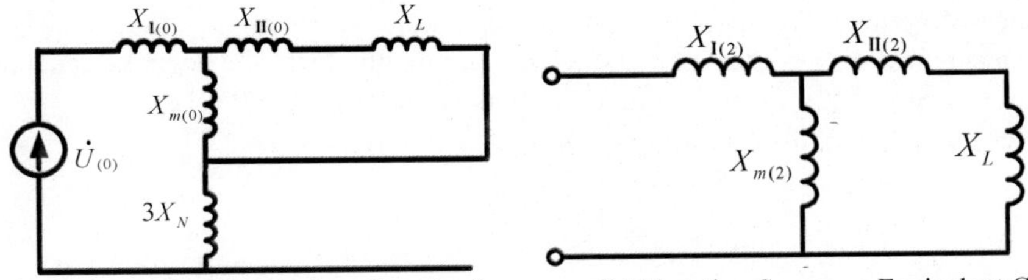

(a) Zero-Sequence Equivalent Circuit of External Circuit with Grounding Point

(b) Negative-Sequence Equivalent Circuit

Fig. 3. Zero-sequence and negative-sequence equivalent circuit

When the external circuit of transformer has grounding point, there will be path of zero-sequence current; at this time, the zero-sequence equivalent circuit is as shown in Fig. 3(a) and the negative-sequence equivalent circuit is as shown in Fig. 3(b).

Each phase of magnetic circuit of three-phase transformer group consisting of three single-phase transformers is independent mutually. The main flux is closed in the iron core, with relatively large magnetic conductance, and very large sequence excitation reactance. Compared with leakage reactance of transformer, the sequence excitation reactance can be treated as infinite [20].

The equivalent reactance of zero- and negative-sequence is:

$$X_{(0)} = X_{\mathrm{I}(0)} + X_{m(0)} // (X_{\mathrm{II}(0)} + X_L) + 3X_N \approx X_{\mathrm{I}(0)} + X_{\mathrm{II}(0)} + X_L + 3X_N \tag{1}$$

$$X_{(2)} = X_{\mathrm{I}(2)} + X_{m(2)} // (X_{\mathrm{II}(2)} + X_L) \approx X_{\mathrm{I}(2)} + X_{\mathrm{II}(2)} + X_L \tag{2}$$

The zero-sequence and negative-sequence equivalent leakage reactance of primary and secondary windings of transformer equal to equivalent leakage reactance of primary and secondary windings respectively. The zero-sequence equivalent reactance also includes neutral point reactance with relatively large value. Therefore, in case of any fault, the negative-sequence current is greater than the zero-sequence current, which is more helpful for detection of fault electrical quantity.

2.2 Protection Principle in SCSR Negative-Sequence Power Direction

The negative-sequence voltage of the negative-sequence power directional protection is from the lead-out line circuit of SCSR, and the negative-sequence current is from the negative-sequence current at the head terminal of SCSR; the negative-sequence current takes the direction of current flowing to SCSR as positive direction.

The criterion [21] of negative-sequence power direction is:

$$-90^{\circ} \leq \arg \frac{\dot{U}_{(2)} e^{-j\varphi_m}}{\dot{I}_{(2)}} \leq 90^{\circ} \tag{3}$$

where, φ_{m} represents the maximum sensitive angle of negative-sequence directional element, and the maximum sensitive angle is decided by the negative-sequence resistance of system. To guarantee that the negative-sequence power directional protection can reflect turn-to-turn fault and internal grounding fault, the negative-sequence resistance is considered as pure electric reactance, then $\varphi_{\mathrm{m}} = -90°$ [18].

After put $\varphi_{\mathrm{m}} = -90°$into formula (3), then

$$-180^{\circ} \leq \arg \frac{\dot{U}_{(2)}}{\dot{I}_{(2)}} \leq 0^{\circ} \tag{4}$$

Whether it is a short-circuit fault inside or outside the SCSR area, there will be unbalanced three-phase voltage and current, resulting in negative-sequence voltage and current. The occurrence of the fault will generate a significant negative-sequence current. Therefore, this paper uses the negative-sequence current amplitude of the SCSR control winding as the criterion to judge the occurrence of faults. The principle of negative-sequence power direction was used to judge whether the fault was located inside or outside the SCSR area.

This section takes the 33% access capacity for example, and based on the "T"-type equivalent circuit of transformer [22], it analyzes the negative-sequence equivalent circuit of turn-to-turn short-circuit fault and turn-to-ground fault of SCSR transformer control winding, as well as the external circuit fault of SCSR region and open-phase running. Under different running situations, there are different relationships between negative-sequence voltage and current.

Turn-to-Turn Short-Circuit Fault and Turn-to-Ground Fault of SCSR. When the SCSR winding occurs to turn-to-turn short circuit, it is equivalent to adding a negative-sequence voltage source with reversed direction inside the winding, as shown in Fig. 4. When it occurs to turn-to-ground fault, it is equivalent to the situation that a negative-sequence voltage power source is added between fault point of winding and ground, which is as shown in Fig. 5.

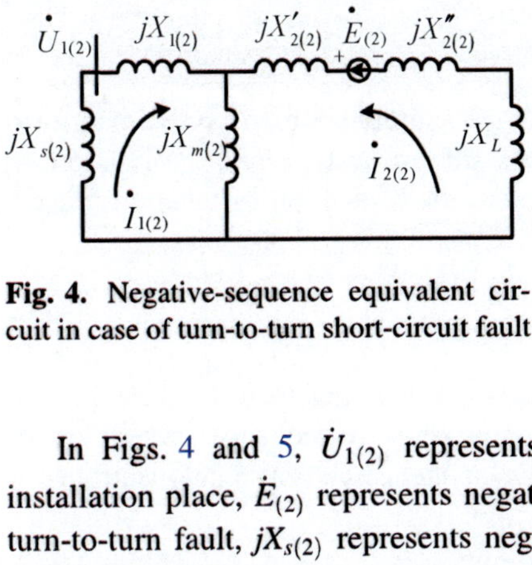

Fig. 4. Negative-sequence equivalent circuit in case of turn-to-turn short-circuit fault

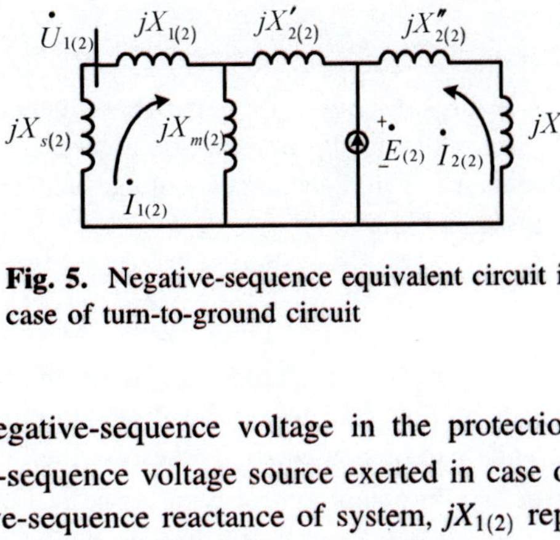

Fig. 5. Negative-sequence equivalent circuit in case of turn-to-ground circuit

In Figs. 4 and 5, $\dot{U}_{1(2)}$ represents negative-sequence voltage in the protection installation place, $\dot{E}_{(2)}$ represents negative-sequence voltage source exerted in case of turn-to-turn fault, $jX_{s(2)}$ represents negative-sequence reactance of system, $jX_{1(2)}$ represents negative-sequence reactance of line side winding, $jX'_{2(2)}$ and $jX''_{2(2)}$ represent negative-sequence reactance on both sides of fault point respectively, $jX_{m(2)}$ represents negative-sequence excitation reactance of transformer, jX_L represents equivalent load reactance, $\dot{I}_{1(2)}$ and $\dot{I}_{2(2)}$ represent negative-sequence current of line side and control windings.

From Figs. 4 and 5, it can be known that the relationship between negative-sequence voltage and current in the protection installation place is shown as below:

$$\dot{U}_{1(2)} = -jX_{s(2)} \cdot \dot{I}_{1(2)} \tag{5}$$

From formula (5), it can be known that when the SCSR control winding occurs to turn-to-turn short-circuit or turn-to-ground fault, the negative-sequence voltage in the protection installation place lags the negative-sequence current.

External Fault of SCSR and Open-Phase Running Analysis of System. In case of external fault of SCSR, the negative-sequence equivalent circuit is as shown in Fig. 6. When the system refers to open-phase running, its equivalent circuit of the SCSR is as shown in Fig. 7.

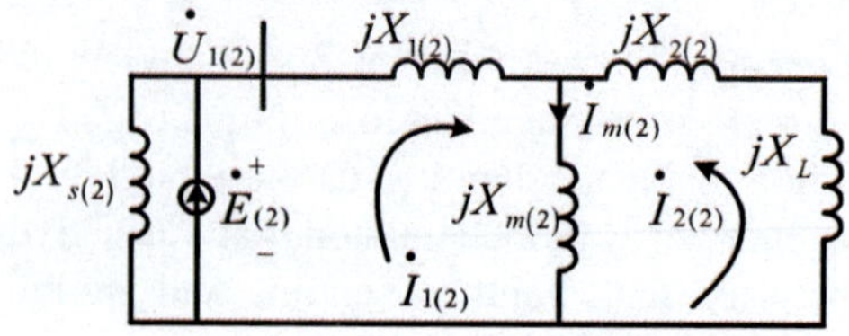

Fig. 6. Negative-sequence equivalent circuit in case of SCSR external fault

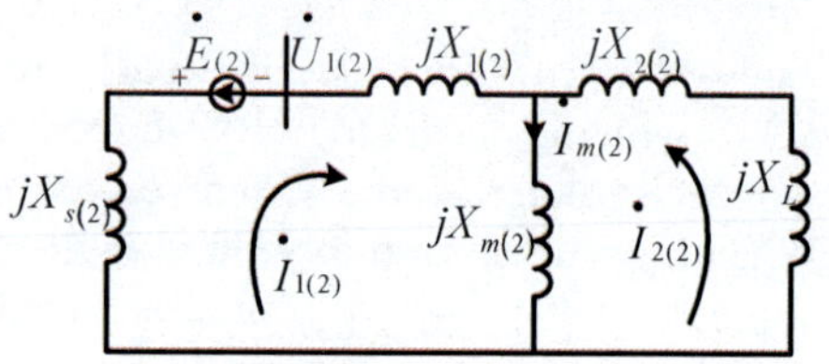

Fig. 7. Negative-sequence equivalent circuit in case of open-phase running

In Figs. 6 and 7, the negative-sequence voltage source is located at the exterior of SCSR, close to the negative-sequence reactance side of system, and the relationship between negative-sequence voltage and current in the protection installation place is shown as below:

$$\dot{U}_{1(2)} = jX_{1(2)} \cdot \dot{I}_{1(2)} - j(X_{2(2)} + X_{L(2)}) \cdot \dot{I}_{2(2)} \tag{6}$$

For the negative-sequence excitation reactance is much larger than the leakage reactance of line side and control windings, and the negative-sequence current on the branch can be neglected. Therefore, the negative-sequence current relationship of line side winding and control winding is $\dot{I}_{1(2)} \approx -\dot{I}_{2(2)}$.

Formula (7) can be further simplified as below:

$$\dot{U}_{1(2)} = jX_{1(2)} \cdot \dot{I}_{1(2)} - j(X_{2(2)} + X_{L(2)}) \cdot \dot{I}_{2(2)} \approx j\dot{I}_{1(2)} \cdot \left(X_{1(2)} + X_{2(2)} + X_{L(2)}\right) \tag{7}$$

From formula (8), it can be known that when there is any fault at the exterior of the SCSR or when the system refers to open-phase running, the negative-sequence voltage in the protection installation place is ahead of the current.

In above four kinds of running situations, the phase relationship between the negative-sequence voltage and the negative-sequence current in the protection installation place is shown as below: when there is turn-to-turn short-circuit fault and turn-to-ground fault, the negative-sequence voltage lags the current; in case of external fault and open-phase running, the negative-sequence voltage is ahead of the current. Therefore, it can differentiate the position of fault occurrence, thus deciding whether the protection acts.

For the turn-to-turn short-circuit faults in the SCSR control winding, the winding negative-sequence current can be obtained from Fig. 4, as shown below:

$$I_{1(2)} = \frac{E_{(2)}}{\left[X_{s(2)} + X_{1(2)}\right]//X_{m(2)} + X'_{2(2)} + X''_{2(2)} + X_L} \cdot \frac{X_{m(2)}}{X_{m(2)} + X_{s(2)} + X_{1(2)}} \tag{8}$$

The negative-sequence current of the control winding is:

$$I_{2(2)} = \frac{E_{(2)}}{\left[X_{s(2)} + X_{1(2)}\right]//X_{m(2)} + X'_{2(2)} + X''_{2(2)} + X_L} \tag{9}$$

It can be known from Formula (8) and Formula (9) that $I_{1(2)} < I_{2(2)}$. When the same fault occurs, the negative-sequence current of the control winding has more significant fault characteristics.

For the turn-to-ground faults in the SCSR control winding, the negative sequence current of line side winding can be obtained from Fig. 5, as shown below:

$$I_{1(2)} = \frac{E_{(2)}}{\left[X_{s(2)} + X_{1(2)}\right]//X_{m(2)} + X'_{2(2)}} \tag{10}$$

The negative-sequence current of the control winding is:

$$I_{2(2)} = \frac{E_{(2)}}{X''_{2(2)} + X_L} \tag{11}$$

In Formula (10), the excitation impedance $X_{m(2)}$ is much larger than other impedance of the system, so it can be ignored during the analysis. The transformer has a higher leakage impedance evenly distributed on both sides of line side winding and control winding, $X_{1(2)} = X'_{2(2)} + X''_{2(2)}$. The leakage impedance on both sides is greater than the system impedance and the load impedance. The following can be obtained:

$$\left[X_{s(2)} + X_{1(2)}\right]//X_{m(2)} + X'_{2(2)} > X''_{2(2)} + X_L$$

Furthermore, $I_{1(2)} < I_{2(2)}$.

It can be seen from the above analysis that when the same turn-to-turn fault occurs in the SCSR control winding, the negative-sequence current of the control winding is greater than that of the line side winding. The control winding has a higher sensitivity in the detection of system faults.

When the access capacity is 67% and 100% respectively, the equivalent load impedance will be changed, but this will not change its nature. The analysis above is also applicable when the access capacity was 67% and 100% respectively.

3 Small Turn-to-Turn Short-Circuit Protection of Adaptive Threshold

3.1 Capacity-Based Negative-Sequence Current Amplitude Adaptive Starting Threshold Detection Method

When the SCSR access capacity is different, the load impedance is also different. The larger the access capacity, the smaller the load impedance. The characteristic value of the same fault is different for different access capacities. Therefore, it is necessary to set different fault electrical quantity detection thresholds in accordance with different capacities. When the capacity is switched, the detection threshold should be adaptively adjusted based on the switched capacity. The analysis of the control winding negative-sequence current amplitude threshold is as follows.

In Formula (9), the excitation impedance is much larger than the leakage impedance, which can be ignored, Formula (9) can be simplified as follows:

$$I_{2(2)} = \frac{E_{(2)}}{X_{s(2)} + X_{1(2)} + X'_{2(2)} + X''_{2(2)} + X_L} \tag{12}$$

The load impedance X_L in Formula (12) changes along with the input capacity. When the same fault occurs, the fault source, the system impedance and the leakage impedance remain unchanged, only the load impedance X_L changes. The negative-sequence current changes along with the load impedance.

When the starting threshold of the negative-sequence current is set in accordance with a certain capacity, the threshold will not be applicable when the same fault occurred in another capacity. It is essential to set the starting threshold for the negative-sequence current based on different capacities.

When the negative-sequence current amplitude of the control winding $|I_{(2)}|_2$ is greater than the set threshold $K_i|I_{(2)}|_{2.set}$, namely, $|I_{(2)}|_2 > K_i|I_{(2)}|_{2.set}$, it can be determined that a fault occurs. In the formula, K_i is an adaptive adjustment coefficient, which can be set according to different access capacities.

3.2 Negative-Sequence Phase Angle Smoothing Method Based on Dynamic Mean

When the number of turn-to-turn short-circuit turns is small, the amplitude of the negative-sequence voltage is small, which will result in fluctuations in the phase angle of the negative-sequence voltage to a certain extent, making it impossible to identify the position of the fault effectively. Further treatment is needed for this case.

When the number of short-circuit turns is very small, the negative-sequence voltage phase angle will have certain fluctuations. The amplitude of the fluctuations can be identified by setting the threshold of the negative-sequence voltage phase angle variation $\angle V_{(2)TH}$. When the fluctuation of the negative-sequence voltage phase angle exceeds the threshold, the average value within the $2\Delta t$ time period $[t_i - \Delta t, t_i + \Delta t]$ before and after the time point t_i is calculated for the calculation of the power direction angle. With the passage of time t_i, the average value of the waveform within the $2\Delta t$ time period is continuously calculated. When the calculated average value meets the criterion for multiple times, it can be determined that the detection is reliable.

The generated negative-sequence voltage and current are also different when the same fault occurs in different capacities. The phase angle of the negative-sequence voltage will fluctuate accordingly. Therefore, it is vital to set different detection thresholds for the negative-sequence voltage phase angle, which can be adaptively adjusted in accordance with different input capacities.

When the system is in three-phase symmetrical operation, the negative-sequence voltage is zero, the phase angle of the negative-sequence voltage will also fluctuate greatly at this time. When a turn-to-turn fault occurs, the negative-sequence current will increase significantly. The amplitude of the negative-sequence current can be set as the criterion used to judge the fault occurrence. The phase angle can be set as the criterion used to distinguish the fault position. The criterion for the negative-sequence voltage phase angle and that for the negative-sequence current amplitude are set as "AND" logically so as to avoid the maloperation generated by the fluctuations of the negative-sequence voltage phase angle when the system operates normally.

The judgement flow chart of small turn-to-turn short-circuit faults is shown in Fig. 8. It shows the procedures to judge small turn-to-turn short-circuit faults of the SCSR control winding. The specific steps are as follows.

Step 1: Take the head three-phase voltage and current of the SCSR line side winding as well as the three-phase current of the control winding.

Step 2: Calculate the negative-sequence current amplitude $|I_{(2)}|_2$ of the control winding. Determine whether it reaches the detection threshold $K_i|I_{(2)}|_{2.set}$ for the fault occurrence. In the criterion $|I_{(2)}|_2 > K_i|I_{(2)}|_{2.set}$, K_i is the adaptive adjustment coefficient of the negative-sequence current, which is adaptively valued in accordance with different access capacities. When the criterion is not satisfied, it is determined to be three-phase symmetrical operation; when the criterion is satisfied, it performs Step 3.

Step 3: Determine the fluctuation degree of the negative-sequence voltage phase angle $\Delta\angle V_{(2)}$. $\angle V_{(2)\max}$ and $\angle V_{(2)\min}$ are the maximum and minimum values of the negative-sequence voltage phase angle in $[t_i - \Delta t, t_i + \Delta t]$ respectively.

Step 4: Determine whether the fluctuation of the negative-sequence voltage phase angle $\Delta\angle V_{(2)}$ reaches the threshold $K_0\angle V_{(2)TH}$ (K_0 is the adaptive adjustment coefficient of the negative-sequence voltage phase angle, which is adaptively valued in accordance with different capacities). When the threshold is reached, calculate the mean value of the negative-sequence voltage phase angle $\overline{\angle V_{(2)i}}$ within the time period. Otherwise, directly use the measured value of the voltage phase angle $\angle V_{(2)}$ to calculate the negative-sequence power angle φ. Then perform Step 8.

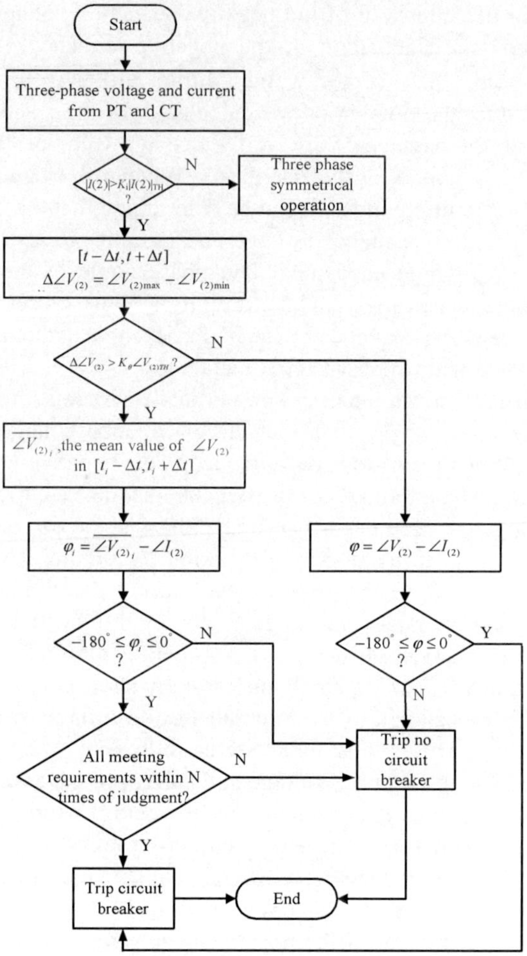

Fig. 8. Judgment flow chart of small turn-to-turn short-circuit faults

Step 5: Calculate the negative-sequence power direction angle φ_i by subtracting the negative-sequence current phase angle $\angle I_{(2)}$ from the average value of the negative-sequence voltage phase angle $\overline{\angle V_{(2)i}}$.

Step 6: Determine whether the negative-sequence power direction angle φ_i meets the criterion $-180° \leq \varphi_i \leq 0°$. If the criterion is satisfied, record the number; otherwise, it is determined that the protection does not act.

Step 7: Determine whether the criterion is satisfied for multiple (n) times. If YES, perform Step 9. Otherwise, it is determined that the protection does not act.

Step 8: Determine whether the negative-sequence power angle φ meets the criterion $-180° \leq \varphi \leq 0°$. If YES, perform Step 9. Otherwise, it is determined that the protection does not act.

Step 9: Output the protection action signal.

Compared with the time required to judge the stable waveform, this method can still shorten the time after multiple judgments. For the SCSR with different capacities, the negative-sequence current amplitude threshold $K_i|I_{(2)}|_{TH}$ and the negative-sequence voltage phase angle threshold $K_\theta\angle V_{(2)TH}$ are set to be different values adaptively in accordance with different capacities.

4 Simulation Verification

The SCSR model is built on the simulation software. The internal resistance of the power supply is set very small, and the line side zero-sequence and negative-sequence current amplitudes with different compensation capacities and turn-to-turn short-circuits are shown in Table 1.

Table 1. Current amplitudes (/A) of turn-to-turn fault under different capacities

Capacity	33%		67%		100%	
Short-circuit proportion	Zero-sequence	Negative-sequence	Zero-sequence	Negative-sequence	Zero-sequence	Negative-sequence
30%	29.24	39.42	9.25	15.69	0.49	1.00
20%	20.64	27.82	6.08	10.30	0.29	0.60
10%	10.96	14.78	2.99	5.07	0.13	0.27
5%	5.66	7.63	1.49	2.52	0.06	0.13

When 10% turn-to-turn short-circuit occurs, the amplitudes of zero-sequence and negative-sequence currents of line side winding are shown in Fig. 9.

It can be known from Table 1 and Fig. 9 that when the same turn-to-turn short-circuit occurs, the negative-sequence current amplitude of the line side winding is greater than that of the zero-sequence current amplitude. So, the sensitivity of the negative-sequence power direction protection is relatively high.

When a 10% turn-to-turn short-circuit fault occurs, the negative-sequence current amplitudes of the line side winding and the control winding are shown in Fig. 10.

It can be found from Fig. 10 that when the same turn-to-turn short-circuit fault occurs under different capacities, the negative-sequence current amplitude of the control winding is greater than that of the line side winding. The detection of faults using the negative-sequence current of the control winding has a higher sensitivity.

In the case of 10% turn-to-turn short-circuit with 33% compensation capacity, the changes in the negative-sequence power direction angle are shown in Fig. 11.

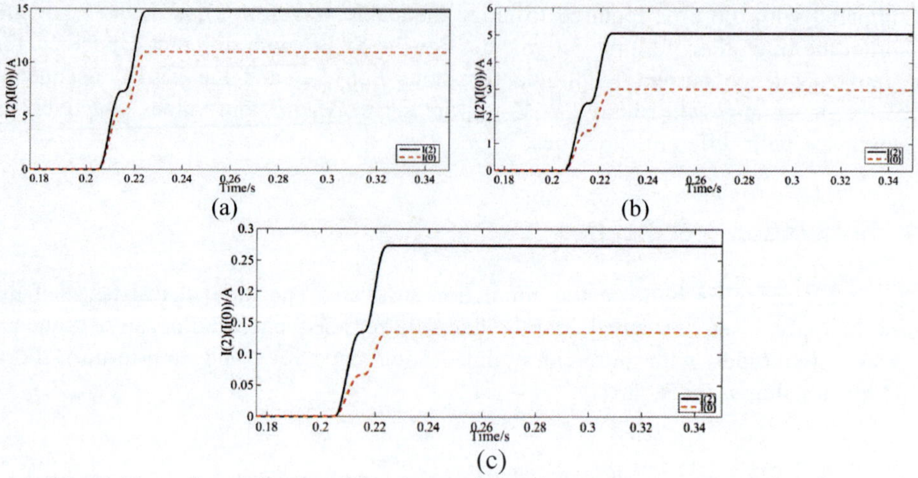

(a), (b) and (c) are the amplitudes of the zero- and negative-sequence currents when the capacity is 33%, 67% and 100%, respectively.

Fig. 9. Zero- and negative-sequence current amplitudes at 10% turn-to-turn short-circuit

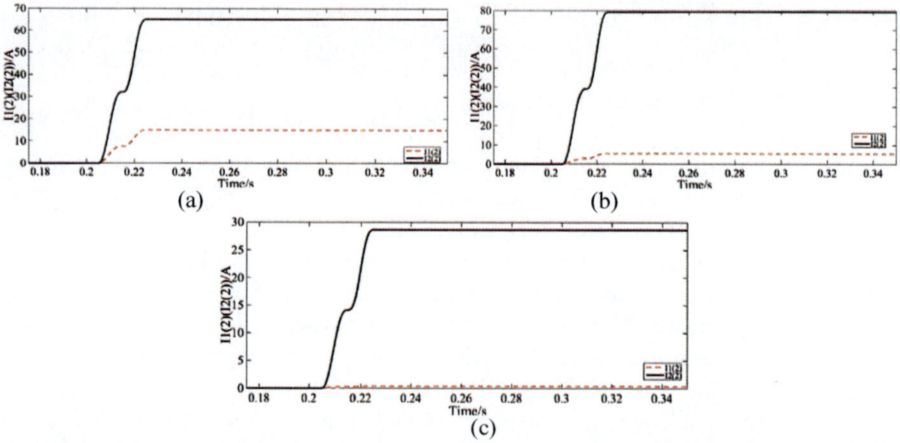

(a), (b) and (c) are the amplitudes of negative-sequence current at line side and control winding of turn-to-turn short-circuit when the capacity is 33%, 67% and 100%, respectively.

Fig. 10. Negative-sequence current amplitudes of line side winding and control winding at 10% capacity with turn-to-turn short-circuit

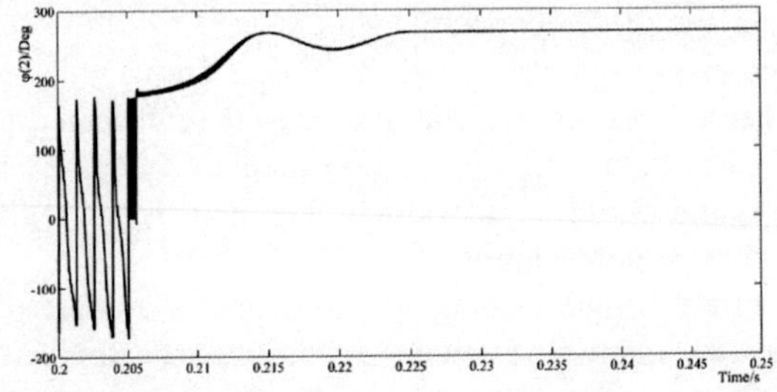

Fig. 11. 10% Turn-to-turn short-circuit power direction angle at 33% capacity

In Fig. 11, the turn-to-turn short-circuit fault occurred at 0.205 s. The fluctuation before 0.205 s resulted from the negative-sequence voltage phase angle fluctuation generated during the normal operation when the voltage amplitude was small.

Under the capacities of 33%, 67% and 100%, the test sets 30%, 20%, 10%, 5%, 3%, 2% and 1% turn-to-turn short-circuit faults and turn-to-ground faults respectively. The values of the negative-sequence power direction angle under different conditions were recorded. In addition, the external faults under different conditions (phase A of circuit is grounded, phase A and phase B are short-circuited and phase A and phase B are short-circuited to ground) were set. The negative-sequence power direction angle was recorded, and the simulation verification results are shown in Table 2.

Table 2. Negative-sequence power direction angle of turn-to-turn fault under different capacities

Fault types	Short-circuit proportion	Capacity			Internal fault or external?
		33%	67%	100%	
Turn-to-turn short circuit	30%	−90.2011°	−90.2004°	−90.1976°	Internal fault
	20%	−90.201°	−90.2011°	−90.1968°	Internal fault
	10%	−90.2005°	−90.2009°	−90.2145°	Internal fault
	5%	−90.1999°	−90.1979°	***−90.2049°***	Internal fault
	3%	−90.2015°	−90.203°	***−90.201°***	Internal fault
	2%	−90.2028°	−90.203°	***−90.2093°***	Internal fault
	1%	−90.1975°	−90.191°	***−90.0968°***	Internal fault
Turn-to-ground fault	30%	−90.2007°	−90.2005°	−90.2003°	Internal fault
	20%	−90.2006°	−90.2008°	−90.1997°	Internal fault
	10%	−90.2007°	−90.2007°	−90.2017°	Internal fault
	5%	−90.2007°	−90.2005°	−90.2009°	Internal fault
	3%	−90.2007°	−90.2007°	−90.2005°	Internal fault
	2%	−90.2008°	***−90.2007°***	−90.2006°	Internal fault
	1%	−90.2008°	***−90.2007°***	−90.2007°	Internal fault
External A-G		87.63°	89.12°	89.56°	External fault
External A-B		87.63°	89.13°	89.57°	External fault
External AB-G		87.65°	89.13°	89.55°	External fault

In Table 2, under the 100% capacity, with regard to turn-to-turn short circuit of 10% below, the negative-sequence power direction angle will have relatively large value fluctuation when a fault occurs. The phase angle represented by bold italics in the table indicates the case where the negative-sequence voltage phase angle fluctuates greatly. This value is the phase angle value calculated by averaging the value of the negative-sequence voltage phase angle.

The fluctuation of the negative-sequence voltage phase angle of the system can be better reflected when the internal resistance of the power supply is considered. When the 10% turn-to-turn short-circuit occurs at 100% capacity, the fluctuation of the negative-sequence power direction angle is shown in Fig. 12.

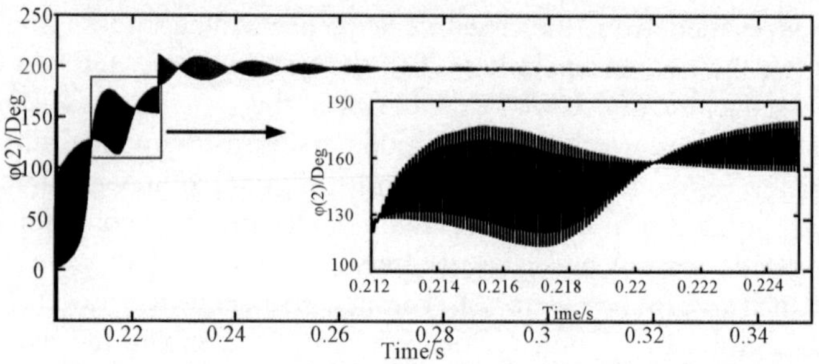

(a)Negative-sequence power direction angle when the negative-sequence voltage phase angle is the actual value

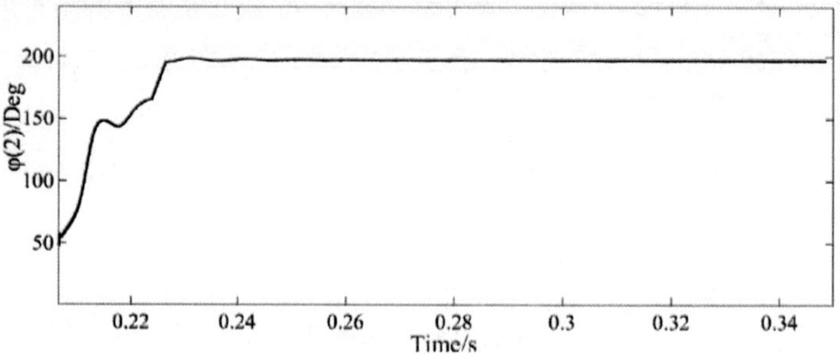

(b)Negative-sequence power direction angle when negative-sequence voltage phase angle is averaged

Fig. 12. Direction angle of negative-sequence power of 100% capacity and 10% turn-to-turn short circuit

It is known from Fig. 12(a) that after the fault occurs, the negative-sequence power direction angle fluctuates greatly, which is caused by the small negative-sequence voltage, and the internal fault or external fault cannot be accurately determined in the fluctuation period. Figure 12(b) shows the waveform obtained by negative-sequence phase angle smoothing method based on dynamic mean in this paper. The waveform has been significantly improved, which can reflect the negative-sequence power direction angle better in case of system fault.

5 Conclusion

This paper analyzes the applicability of the negative-sequence power directional protection in the SCSR control winding turn-to-turn faults based on the "T"-type equivalent circuit of the transformer, and configures the SCSR control winding with the negative-sequence power directional protection. The occurrence of fault is judged through the control winding negative-sequence current while the position of fault is identified through the negative-sequence power direction angle. The simulation verification shows that when the same fault occurs, the negative-sequence current of the control winding changes greatly with a higher sensitivity.

When small turn-to-turn fault occurs in the control winding, different negative-sequence current detection thresholds are set based on different access capacities. This paper proposes a capacity-based negative-sequence current amplitude adaptive starting threshold detection method whose negative-sequence current starting threshold can adaptively change in accordance with different input capacities.

When the number of turns with short-circuit faults in the control winding is small, there will be a small negative-sequence voltage resulting in large fluctuations in the negative-sequence voltage phase angle. Therefore, this paper proposes a negative-sequence phase angle smoothing method based on dynamic mean, which can set different negative-sequence voltage phase angle comparison thresholds corresponding to different capacities. The threshold angle can adaptively change in accordance with the input capacity. When the fluctuation of the negative-sequence voltage phase angle is greater than the threshold, a phase angle with small fluctuations can be obtained by processing the phase angle through dynamic averaging. After the calculation of the negative-sequence power direction angle and multiple judgments, it can ensure reliable actions of protection. The simulation verifies the correctness of this method. This method can better prevent the occurrence of turn-to-turn faults in the SCSR control winding.

Acknowledgments. This work is supported by Science and Technology Program of State Grid Corporation of China: Research of the fast protection technology of power grid with distributed generation.

References

1. Zheng, T., Zhao, Y.J.: Overview of key techniques of EHV/UHV controllable shunt reactor. Autom. Electric Power Syst. **38**(7), 127–135 (2014)
2. Zhang, Y.X., Cui, D.F., Xu, W.: Protection scheme of control winding for 500 kV magnetically controlled shunt reactor. Autom. Electric Power Syst. **42**(19), 148–154 (2018)
3. Nashawati, E., Fischer, N., Le, B., Taylor, D.: Impacts of shunt reactors on transmission line protection. In: 38th Annual Western Protective Relay Conference, pp. 1–16. IEEE, Spokane (2011)
4. Basha, F.K., Thompson, M.: Practical EHV reactor protection. In: 66th Annual Conference for Protective Relay Engineers, pp. 408–419. IEEE, College Station (2013)

5. Das, S., Sidhu, T.S., Zadeh, M.R.D., Zhang, Z.Y.: A novel hybrid differential algorithm for turn to turn fault detection in shunt reactors. IEEE Trans. Power Deliv. **32**(6), 2537–2545 (2017)
6. Tian, M.X., Li, Q.F., Li, Q.F.: A controllable reactor of transformer type. IEEE Trans. Power Deliv. **19**(4), 1718–1726 (2004)
7. Mo, P.H., Wen, J.F., Bao, B., Zhang, X.Y., Lv, H., Gu, Q.G.: Inter-turn protection for control winding of high voltage step controlled shunt reactor. Autom. Electric Power Syst. **40**(14), 105–109 (2016)
8. Zhang, J.K., Su, X.H., Hu, Y., Du, D.X.: Impact of large-capacity controllable shunt reactor on line differential protection and solutions. Autom. Electric Power Syst. **38**(1), 115–120 (2014)
9. Tu, L.M., Su, Y., Yu, K.S., Zou, D.X., Zou, W.H., Yue, L., Liu, J.F.: Development of microcomputer protection for a controllable high-voltage shunt reactor. Autom. Electric Power Syst. **31**(24), 94–98 (2007)
10. Yao, Q.L., Li, R.S., Su, X.H., Zhang, J.K., Zhang, X.G., Cao, L.L.: Configuration and principle analysis of microcomputer protection for classified high-voltage controllable shunt reactor. Autom. Electric Power Syst. **33**(21), 58–65 (2009)
11. Zhang, L., Liu, B., Shen, H.: Voltage control strategy with stepped controllable shunt reactor in large-scale wind power system. In: 2011 IEEE Power Engineering and Automation Conference (PEAM), pp. 78–81. IEEE, Wuhan (2011)
12. IEEE Guide for the Protection of Shunt Reactors, IEEE Std. C37.109–2006, April 2007
13. The shunt reactor protection guide working group of the IEEE power system relaying committee: summary of the 'guide for the protection of shunt reactors' ANSI C37.109. IEEE Trans. Power Deliv. **6**(1), 116–118 (1991)
14. Nishida, S., Sakaguchi, T.: A new algorithm of digital protection for shunt reactors. IEEE Trans. Power Apparatus Syst. **103**(10), 2934–2942 (1984)
15. Zheng, T., Zhao, Y.J.: Microprocessor-based protection scheme for high-voltage magnetically controlled shunt reactors. In: 12th IET International Conference on Developments in Power System Protection (DPSP), pp. 1–5. IET, Copenhagen (2014)
16. Zheng, T., Shu, Y.Z., Dong, S.H., Cao, Y.H.: Interturn fault protection for transformer type controllable shunt reactor based on leakage inductance change. Autom. Electric Power Syst. **35**(12), 65–69 (2011)
17. Xin, Z.T., Yin, X.G., Yang, J.C., Cai, S.L., Huang, X.: A novel principle of the EHV parallel connection reactor interturn protection based on an equivalent inductance. Autom. Electric Power Syst. **28**(10),73–75+87 (2004)
18. Li, B., Li, Y.L., Chen, J., Zhang, Y.: Turn-to-turn faults protection of EHV shunt reactors. J. Tianjin Univ. **38**(8), 717–721 (2005)
19. Khan, M.A., Zheng, T., Liu, X.X., Wei, J.Q.: A turn-to-turn fault detection and inter-windings identification scheme for EHV MCSR. In: 22nd International Conference on Electrical Machines and Systems (ICEMS), pp. 1–6. IEEE, Harbin (2019)
20. Sun, S.Q., Li, A., Li, Z.H.: Power System Analysis, 2nd edn. China Machine Press, Beijing (2018)
21. Zhang, C.H., Gao, X., Shen, Q.R.: Analysis of inter-turn protection in shunt reactor in ultra-high-voltage substation-used power system. Power Syst. Protect. Control **37**(20), 53–56+60 (2009)
22. Zheng, T., Liu, X.X., Wei, J.Q.: A novel protection scheme against turn-to-turn fault of MCSR based on equivalent magnetizing inductance. Electr. Power Energy Syst. **19**(4), 1–11 (2020)

Capacitance Current Compensation Method for Cable-Overhead Hybrid Transmission Lines

Gang Yao[1,2(✉)], Tonghua Wu[1,2], Yuping Zheng[1,2], Xianliang Teng[1], Feng Hong[1,2], and Guozhou Chen[1]

[1] NARI Group Corporation, Nanjing 211106, China
yaogang@sgepri.sgcc.com.cn
[2] State Key Laboratory of Smart Grid Protection and Control, Nanjing 211106, China

Abstract. At present, the capacitance current compensation technology for transmission line differential protection is based on uniform parameters of entire lines, so it cannot be adapted to cable-overhead hybrid transmission lines. In this paper, a steady-state compensation method based on segment parameters, a time-domain compensation method based on whole line equivalent parameters and a time-domain compensation method based on segment parameters are proposed, which can adapt to different application scenarios, such as hybrid lines with shunt reactance at the ends and junctions. A PSCAD/EMTDC simulation shows that when a hybrid transmission line is short, the time-domain compensation method based on whole line equivalent parameters is simple to implement, and it can effectively compensate the capacitance current of hybrid transmission lines. The time-domain compensation method based on segment parameters can accurately compensate the capacitance current of hybrid transmission lines, taking into account the reliability and sensitivity of differential protection.

Keywords: Hybrid transmission line · Longitudinal differential protection · Capacitance current compensation · Time-domain compensation

1 Introduction

Power cable lines can cross rivers and oceans, and they can be adapted to other scenarios without poles and towers. With less influence from natural elements, relatively low maintenance workload, safe power supply, and reliable operation, cable-overhead hybrid transmission lines have been gradually applied in the urban trunk network, cross-river rack gas insulated line (GIL) transmission channels, offshore wind power transmission, and extra high voltage (EHV) electric transmission projects involving voltages of 220 kV and above. Compared with overhead lines, cable lines have a smaller distance between cable cores and line cores and between line cores and sleeves [1], and they have a much larger capacitance per unit line length, resulting in a striking distributed capacitance current. Accordingly, EHV line protection requires

Y. Xue et al. (Eds.): PMF 2020, LNEE 718, pp. 925–939, 2021.
https://doi.org/10.1007/978-981-15-9746-6_69

more research concerning the influences of hybrid transmission lines on differential protection.

The traditional method of increasing the differential current setting to avoid capacitance current influence will reduce the sensitivity of the protection and may fail to operate in the case of high resistance grounding fault of transmission lines, which will affect the reliability of differential protection. Existing capacitance current compensation technologies, such as the steady-state compensation method, the time-domain compensation method, the transient compensation method, and the Bergeron-based compensation method, are based on the premise that the parameters of the whole line are evenly distributed, However, the aforesaid research methods are not suitable for cable-overhead hybrid lines with uneven parameters throughout the line. In this paper, the steady-state compensation method based on segment parameters, the time-domain compensation method based on the whole line equivalent parameters, and the time-domain compensation method based on segment parameters are proposed for cable-overhead hybrid lines. Among them, the time-domain compensation method based on segment parameters can accurately compensate the capacitance current of the hybrid lines, taking into account the reliability and sensitivity of the differential protection. Since both ends of cable-overhead hybrid lines and the middle of the cable-overhead lines may have shunt reactors (referred to as shunt reactance), the corresponding correction schemes of various compensation methods for shunt reactance on the sides and in the middle of the lines are given. Finally, EMTDC simulation software is used to simulate and compare the differential criterion based on various compensation methods on 500 kV cable-overhead lines, and a simplified scheme based on equivalent parameters of the whole line is proposed for short cable-overhead lines.

2 Steady-State Compensation Method Based on Segment Parameters

2.1 Hybrid Transmission Lines Without Shunt Reactors

The equivalent lumped parameter Π model of cable-overhead hybrid lines without shunt reactors is shown in Fig. 1, where the M side is the cable side, the N side is the overhead line side and the K point is the junction of the cable and overhead line.

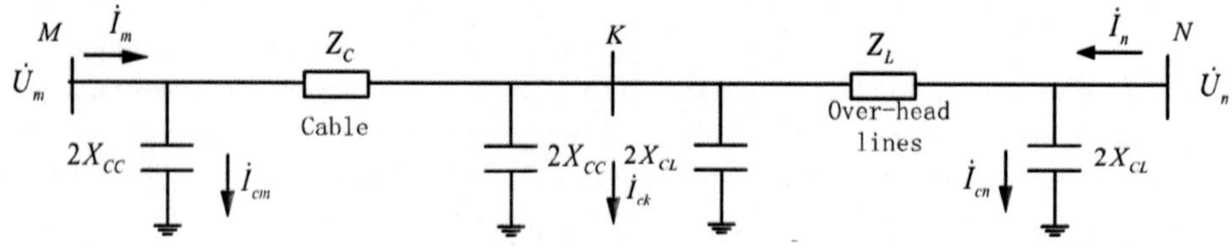

Fig. 1. Lumped parameter segment Π model of hybrid lines without shunt reactance.

As demonstrated by the semi-compensation method of hybrid lines based on a type Π network, the equivalent network of hybrid lines can be taken as the connection of the type Π network based on cable line parameters and connected with the type Π network based on overhead line parameters. The capacitance current to be compensated by the line differential protection, including that on the M side $\dot{I}_{cm}$, that at the K point $\dot{I}_{ck}$ and that at the N point $\dot{I}_{cn}$, can be calculated with the following formulas:

$$\begin{cases} \dot{I}_{cm} = \frac{\dot{U}_m - \dot{U}_{m0}}{-j2X_{cc1}} + \frac{\dot{U}_{m0}}{-j2X_{cc0}} \\ \dot{I}_{ck} = \frac{\dot{U}_k - \dot{U}_{k0}}{-j2X_{cc1}} + \frac{\dot{U}_{k0}}{-j2X_{cc0}} + \frac{\dot{U}_k - \dot{U}_{k0}}{-j2X_{cl1}} + \frac{\dot{U}_{k0}}{-j2X_{cl0}} \\ \dot{I}_{cn} = \frac{\dot{U}_n - \dot{U}_{n0}}{-j2X_{cl1}} + \frac{\dot{U}_{n0}}{-j2X_{cl0}} \end{cases} \tag{1}$$

where $\dot{U}_m$ and $\dot{U}_n$ are the voltages at both ends of the lines; $\dot{U}_k$ is the voltage at a junction between the cable and the overhead line; $\dot{U}_{m0}$, $\dot{U}_{n0}$ and $\dot{U}_{k0}$ are the zero sequence voltages; X_{cc1} and X_{cc0} are the positive sequence reactance and zero sequence reactance of the cable line; and X_{cl1} and X_{cl0} are the positive sequence reactance and zero sequence reactance of the overhead line.

In a practical engineering application, the K-point voltages at the junctions of hybrid lines can be calculated by the voltage, current and line parameters on the sides. Exampled by the M side of the cable line, the calculation formula of K-point voltage is

$$\dot{U}_{k\varphi} = \dot{U}_{m\varphi} - Z_{c1}(\dot{I}_{m\varphi} + k_c 3\dot{I}_{m0}) \tag{2}$$

where $\varphi = a, b,$ c phase, Z_{c1} is the positive sequence reactance of the cable line, and k_c is the zero sequence compensation coefficient of the cable line $k_c = (Z_{c0} - Z_{c1})/3Z_{c1}$. Similarly, for the N side of the overhead line, the K-point voltage can be obtained according to the N side voltage, current and overhead line parameters.

2.2 Hybrid Transmission Lines with Shunt Reactors

Compared with overhead lines, cable lines have lower reactance and higher charging power per unit length, so it is necessary to mount shunt reactors on both sides of hybrid lines or at the junctions of hybrid lines to suppress over-voltage and potential supply current.

(1) Hybrid lines with shunt reactance on both sides
The capacitance current of a hybrid line with shunt reactance compensation on both sides is the capacitance current of the line minus the shunt reactance current on both sides. The calculation formula of the current to be compensated is as follows:

$$\begin{cases} \dot{I}_{cm} = \dfrac{\dot{U}_m - \dot{U}_{m0}}{-j2X_{cc1}} + \dfrac{\dot{U}_{m0}}{-j2X_{cc0}} + \dfrac{\dot{U}_m - \dot{U}_{m0}}{jX_{Lm1}} + \dfrac{\dot{U}_{m0}}{jX_{Lm0}} \\ \dot{I}_{ck} = \dfrac{\dot{U}_k - \dot{U}_{k0}}{-j2X_{cc1}} + \dfrac{\dot{U}_{k0}}{-j2X_{cc0}} + \dfrac{\dot{U}_k - \dot{U}_{k0}}{-j2X_{cl1}} + \dfrac{\dot{U}_{k0}}{-j2X_{cl0}} \\ \dot{I}_{cn} = \dfrac{\dot{U}_n - \dot{U}_{n0}}{-j2X_{cl1}} + \dfrac{\dot{U}_{n0}}{-j2X_{cl0}} + \dfrac{\dot{U}_n - \dot{U}_{n0}}{jX_{Ln1}} + \dfrac{\dot{U}_{n0}}{jX_{Ln0}} \end{cases} \tag{3}$$

where $X_{Lm1} = \omega L_{pm}$, $X_{Lm0} = \omega(L_{pm} + 3L_{fm})$, $X_{Ln1} = \omega L_{pn}$, $X_{Ln0} = \omega(L_{pn} + 3L_{fn})$, L_{pm} and L_{pn} are shunt reactance phase inductances and L_{fm} and L_{fn} are low-reactance neutral point inductances.

(2) Hybrid lines with shunt reactance at junctions
The capacitance current of a hybrid line with shunt reactance compensation at junctions is the line capacitance current minus the shunt reactance currents at the junctions. The calculation formula of the current to be compensated is as follows:

$$\begin{cases} \dot{I}_{cm} = \frac{\dot{U}_m - \dot{U}_{m0}}{-j2X_{cc1}} + \frac{\dot{U}_{m0}}{-j2X_{cc0}} \\ \dot{I}_{ck} = \frac{\dot{U}_k - \dot{U}_{k0}}{-j2X_{cc1}} + \frac{\dot{U}_{k0}}{-j2X_{cc0}} + \frac{\dot{U}_k - \dot{U}_{k0}}{-j2X_{cl1}} + \frac{\dot{U}_{k0}}{-j2X_{cl0}} + \frac{\dot{U}_k - \dot{U}_{k0}}{jX_{Lk1}} + \frac{\dot{U}_{k0}}{jX_{Lk0}} \\ \dot{I}_{cn} = \frac{\dot{U}_n - \dot{U}_{n0}}{-j2X_{cl1}} + \frac{\dot{U}_{n0}}{-j2X_{cl0}} \end{cases} \tag{4}$$

where $X_{Lk1} = \omega L_{pk}$, $X_{Lk0} = \omega(L_{pk} + 3L_{fk})$, L_{pk} is the shunt reactance phase inductance at the junction and L_{fm} and L_{fn} are low-reactance neutral point inductances.

2.3 Differential Criterion of Steady-State Compensation Based on Segment Parameters

The formula of the steady-state compensation differential protection of hybrid lines based on segment parameters is as follows:

$$\begin{cases} I_{cd} = \left|\dot{I}_m - \dot{I}_{cm} + \dot{I}_n - \dot{I}_{cn} - \dot{I}_{ck}\right| > I_{set} \\ \left|\dot{I}_m - \dot{I}_{cm} + \dot{I}_n - \dot{I}_{cn} - \dot{I}_{ck}\right| > k * \left|\dot{I}_m - \dot{I}_{cm} - \dot{I}_n + \dot{I}_{cn}\right| \end{cases} \tag{5}$$

where I_{set} is the differential current setting value, generally 480 A to 600 A for 220 kV and above, and k is the differential ratio restraint coefficient, ranging from 0.6 to 0.75.

The steady-state compensation can effectively compensate the steady-state capacitance current. As the transient-state capacitance for high-voltage hybrid lines is obvious, the steady-state compensation method cannot compensate the transient capacitance current. Although the influence of capacitance current caused by external fault and line energization can be avoided by raising the setting value of differential current, the sensitivity of differential protection may be decreased, and the differential protection may fail to operate in the high reactance grounding fault.

3 Time-Domain Compensation Method Based on Segment Parameters

The compensation idea of time-domain capacitance current [2] is to calculate the capacitance current according to the relationship among the capacitance current, the capacitance voltage and the capacitance value (i.e., $i_c = C\frac{du_c}{dt}$). It is applicable to high-frequency components in the transient process and can accurately compensate for the steady-state capacitance current and the transient capacitance current of transmission lines.

The parameters of hybrid lines are not evenly distributed in the whole line (although they are evenly distributed in each segment), making the traditional time-domain capacitance current compensation method not directly applicable. The time-domain compensation method based on segment parameters can divide hybrid transmission lines into segments with even parameters. For example, cable-overhead hybrid lines can compensate for capacitance current to cable segment and overhead line segment equivalently.

The time-domain compensation calculation method based on segment parameters is used to deduce cable-overhead hybrid lines without shunt reactance, hybrid lines with shunt reactance on the sides, and hybrid lines with shunt reactance at junctions.

3.1 Hybrid Lines Without Shunt Reactance

The segment parameter equivalent circuit of hybrid lines without shunt reactance is shown in Fig. 1. According to the calculation method of the time-domain capacitance current of the whole line even parameter type Π equivalent circuit presented by Wu [2], the capacitance current to be compensated on the M side, the junction, and the N side of the line based on the segment parameters are, respectively, I^a_{cm}, I^a_{ckm}, I^a_{ckn}, and I^a_{cn} (demonstrated by Phase A), and the corresponding formulas are as follows:

$$\begin{cases} I^a_{cm} = \dfrac{C_{cpg}}{2}\dfrac{du^a_m}{dt} + \dfrac{C_{cpp}}{2}\dfrac{du^{ab}_m}{dt} + \dfrac{C_{cpp}}{2}\dfrac{du^{ac}_m}{dt} \\ I^a_{ckm} = \dfrac{C_{cpg}}{2}\dfrac{du^a_k}{dt} + \dfrac{C_{cpp}}{2}\dfrac{du^{ab}_k}{dt} + \dfrac{C_{cpp}}{2}\dfrac{du^{ac}_k}{dt} \\ I^a_{ckn} = \dfrac{C_{lpg}}{2}\dfrac{du^a_k}{dt} + \dfrac{C_{lpp}}{2}\dfrac{du^{ab}_k}{dt} + \dfrac{C_{lpp}}{2}\dfrac{du^{ac}_k}{dt} \\ I^a_{cn} = \dfrac{C_{lpg}}{2}\dfrac{du^a_n}{dt} + \dfrac{C_{lpp}}{2}\dfrac{du^{ab}_n}{dt} + \dfrac{C_{lpp}}{2}\dfrac{du^{ac}_n}{dt} \end{cases} \quad (6)$$

where C_{cpg} and C_{cpp} are the ground capacitance and interphase capacitance of the cable line; and C_{lpg} and C_{lpp} are the ground capacitance and interphase capacitance of the overhead line. The ground capacitance and interphase capacitance of a cable line or overhead line can be obtained according to the relationship between the positive sequence capacitance and the zero sequence capacitance of the lines [4].

For hybrid lines without shunt reactance, the capacitance current to be compensated $I_{cm}^{\varphi''}$ and $I_{cn}^{\varphi''}$ can be:

$$\begin{cases} I_{cm}^{\varphi''} = I_{cm}^{\varphi} + I_{ckm}^{\varphi} \\ I_{cn}^{\varphi''} = I_{cn}^{\varphi} + I_{ckn}^{\varphi} \end{cases} \quad (7)$$

3.2 Hybrid Lines with Shunt Reactance

(1) Hybrid lines with shunt reactance on both sides

The segment parameter equivalent circuit of the hybrid lines with shunt reactance on the sides is shown in Fig. 2.

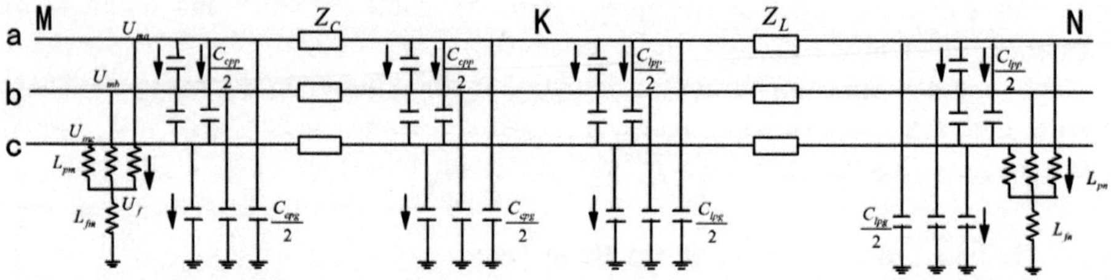

Fig. 2. Segment parameter equivalent circuit of hybrid line with shunt reactance on the sides.

In the figure, the calculation method of capacitance current compensation is consistent with formula (6), and the calculation method of inductance current on both sides of the hybrid line is as follows. According to $\frac{di_L^{\varphi}}{dt} = \frac{U_{L\varphi}}{L_p}$, the calculation formula of real-time inductance current $I_L^{\varphi}(t)$ is as follows:

$$I_L^{\varphi}(t) = I_L^{\varphi}(t_0) + \int_{t0}^{t} \frac{U_{L\varphi}}{L_p} dt \quad (8)$$

where $U_{L\varphi} = U_{\varphi} - U_f$, $\varphi = A, B, C$, U_{φ} is phase voltage, $U_{L\varphi}$ is the inductive voltage, and U_f is the inductive voltage at the neutral point: $U_f = (U_a + U_b + U_c)/(L_p/L_f + 3)$.

For lines with shunt reactance on both sides, the capacitance current to be compensated $I_{cm}^{\varphi''}$ and $I_{cn}^{\varphi''}$ can be expressed as:

$$\begin{cases} I_{cm}^{\varphi''} = I_{cm}^{\varphi} + I_{ckm}^{\varphi} - I_{Lm}^{\varphi} \\ I_{cn}^{\varphi''} = I_{cn}^{\varphi} + I_{ckn}^{\varphi} - I_{Ln}^{\varphi} \end{cases} \quad (9)$$

(2) Hybrid lines with shunt reactance at junctions

The segment parameter equivalent circuit of the hybrid lines with shunt reactance at junctions is shown in Fig. 3.

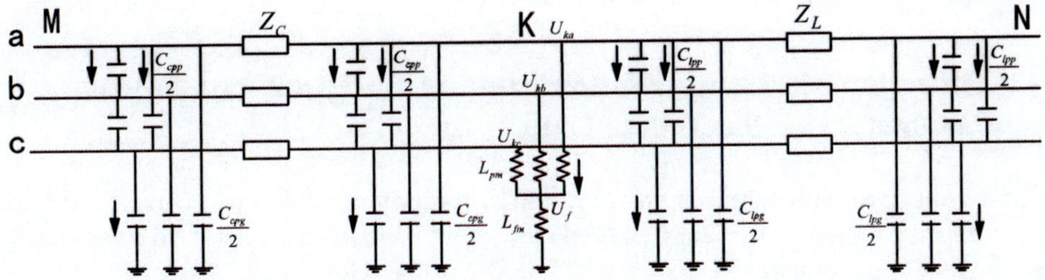

Fig. 3. Segment parameter equivalent circuit of the hybrid lines with shunt reactance at junctions.

The voltage of each K point at the junctions of the hybrid lines can be obtained from the M-side voltage & current and the N-side voltage & current of the hybrid lines, and the first segment parameters and last segment parameters of the cable line and the overhead line. The K-point voltage obtained from the M side is expressed as:

$$\dot{U}_k^{\varphi} = \dot{U}_m^{\varphi} - Z_{C1}(\dot{I}_m^{\varphi} + k_{C0}3\dot{I}_{m0}) \tag{10}$$

The K-point voltage obtained from the N side is:

$$\begin{cases} \dot{U}_k^{\varphi} = \dot{U}_m^{\varphi} - Z_{C1}(\dot{I}_m^{\varphi} + k_{C0}3\dot{I}_{m0}) \\ \dot{U}_k^{\varphi} = \dot{U}_n^{\varphi} - Z_{L1}(\dot{I}_n^{\varphi} + k_{L0}3\dot{I}_{n0}) \end{cases} \tag{11}$$

According to the obtained voltage on both sides and formula (7), the shunt reactor current can be obtained on both sides in real time. The calculation formula of capacitance current on both sides of the hybrid lines is consistent with formula (6).

For hybrid lines with shunt reactance at the junctions, the capacitance current to be compensated, $I_{cm}^{\varphi''}$ and $I_{cn}^{\varphi''}$, can be expressed as:

$$\begin{cases} I_{cm}^{\varphi''} = I_{cm}^{\varphi} + I_{ckm}^{\varphi} - I_{Lk}^{\varphi}/2 \\ I_{cn}^{\varphi''} = I_{cn}^{\varphi} + I_{ckn}^{\varphi} - I_{Lk}^{\varphi}/2 \end{cases} \tag{12}$$

3.3 Differential Criterion of Time-Domain Compensation Based on Segment Parameters

The current differential protection formula of hybrid lines using the time-domain compensation based on segmented parameters is as follows:

$$\begin{cases} I_{cd}^{\varphi} = \left| \dot{I}_m^{\varphi} - \dot{I}_{cm}^{\varphi''} + \dot{I}_n^{\varphi} - \dot{I}_{cn}^{\varphi''} \right| > I_{set} \\ \left| \dot{I}_m^{\varphi} - \dot{I}_{cm}^{\varphi''} + \dot{I}_n^{\varphi} - \dot{I}_{cn}^{\varphi''} \right| > k * \left| \dot{I}_m^{\varphi} - \dot{I}_{cm}^{\varphi''} - \dot{I}_n^{\varphi} + \dot{I}_{cn}^{\varphi''} \right| \end{cases} \tag{13}$$

where $\varphi = A, B, C$, and the differential current setting value I_{set} and the ratio restraint coefficient k are the same as the values of steady-state compensation.

4 Time-Domain Compensation Method Based on Equivalent Parameters of the Whole Line

For hybrid lines with short length and small proportion of cable line or overhead line, the lumped parameter segment Π model (Fig. 1) of cable-overhead hybrid lines can be equivalent to the equivalent parameter Π model of the whole line, as shown in Fig. 4.

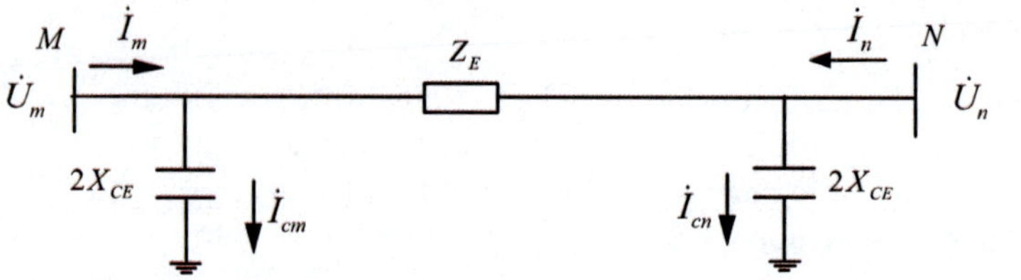

Fig. 4. Equivalent parameter Π model of the whole line of a hybrid line.

In the figure, the equivalent capacitive reactance on each side of the line is $X_{CE} = \frac{1}{j\omega} \frac{C_c + C_L}{C_c * C_L}$, where C_c is the capacitance value of the cable line, C_L is the capacitance value of the overhead line, and Z_E is the equivalent reactance of the line: $Z_E = (R_c + R_l) + j(X_c + X_l)$.

If the hybrid lines have shunt reactance on the sides, the time-domain capacitance current compensation method presented by Wu [2] can be used to accurately compensate the capacitance current.

If the hybrid lines have shunt reactance at the junction between the cable line and overhead line, the shunt reactance at the junction shall be equivalent to that at both ends of the lines for further treatment according to the reactance distribution ratio of the cable line and overhead line, and the post-equivalence calculation method shall be the same as that with shunt reactance at the sides of the lines.

If the total length of the lines is l and the ratio of the reactance of the cable line to that of the overhead line is a, $Z_C = a * Z_L$, and it can be deduced that the equivalent shunt reactor reactance value of the cable line side is $X_{LCequ} = (1 + a) * X_L$ and that of the overhead line side is $X_{LLequ} = (1 + a) * X_L / a$. The line equivalent parameters are shown in Fig. 5. According to the equivalent capacitance reactance of the lines and the equivalent reactance on both sides, the problem of capacitance current compensation of hybrid lines with shunt reactance at the junction can be solved.

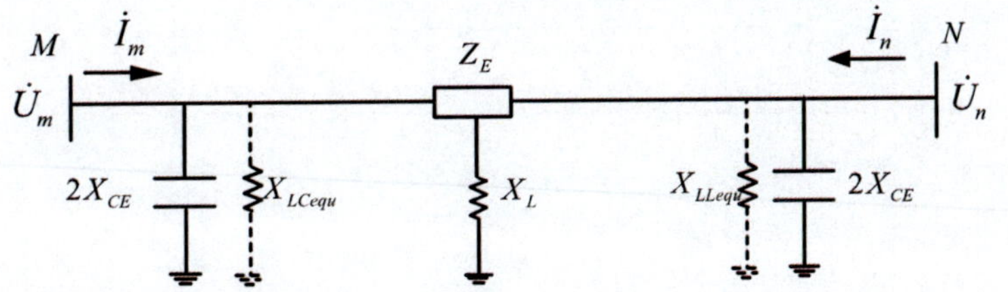

Fig. 5. Equivalent parameter model of the hybrid lines with shunt reactance at the junctions.

5 Simulation Analysis

In this paper, a PSCAD/EMTDC simulation model is built based on the Bergeron line form with the typical parameters of 500 kV overhead line and cable line, and simulation analysis is conducted for hybrid long lines composed of 150 km overhead line and 50 km cable line as well as hybrid short lines composed of 40 km overhead line and 10 km cable line. The shunt reactors are set with reference to typical engineering applications, and the reactive power compensation degree of the lines is 80%. The hybrid long line model simulation verifies the compensation effect of the shunt reactance at junctions of the overhead line and cable line (3*254 Mvar at K1 point), and that on both sides of the lines (3*127 Mvar at K2 and K3 point). The hybrid short line model simulation verifies the compensation effect of the shunt reactance at the junctions (3*55 Mvar at K1 point). The simulation system model is shown in Fig. 6, and the line parameters are shown in Table 1.

Table 1. System reactance and line parameters.

	Item	Parameter	Unit
Overhead line	Positive sequence resistance	0.0064	Ω/km
	Positive sequence inductive reactance	0.2529	Ω/km
	Positive sequence shunt reactor	0.2264	MΩ*km
	Zero sequence resistance	0.1647	Ω/km
	Zero sequence inductive reactance	0.8764	Ω/km
	Zero sequence shunt reactor	0.3935	MΩ*km
Cable line	Positive sequence resistance	0.0263	Ω/km
	Positive sequence inductive reactance	0.0265	Ω/km
	Positive sequence shunt reactor	0.0161	MΩ*km
	Zero sequence resistance	0.0938	Ω/km
	Zero sequence inductive reactance	0.069	Ω/km
	Zero sequence shunt reactor	0.0161	MΩ*km
Long line with shunt reactance on sides	M-side shunt reactance capacity	3*127	Mvar
	M-side shunt reactance capacity	3*127	Mvar
Long line with shunt reactance at junctions	shunt reactance capacity	3*254	Mvar
Short line with shunt reactance at junctions	shunt reactance capacity	3*55	Mvar

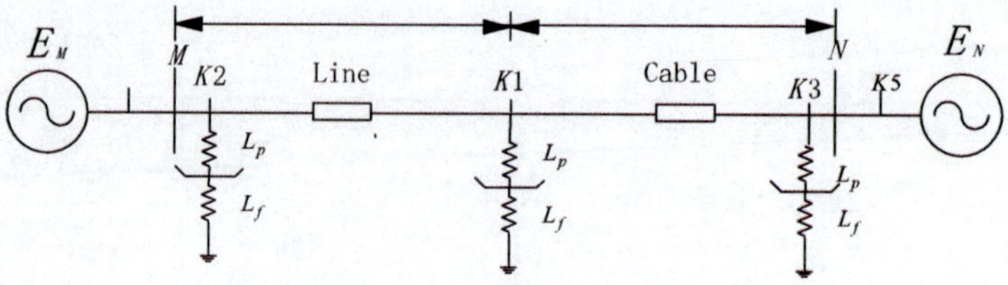

Fig. 6. 500 kV system model.

Simulation analysis calculation comparison is conducted on the criterion of the uncompensated differential, steady-state compensation differential based on segment parameters, time-domain compensation based on segment parameters, and time-domain compensation differential based on whole line equivalent parameters. The voltage and current data are sampled at 24 points per cycle, and the difference filter and the full cycle Fourier are used for numerical calculation.

5.1 Simulation of the Hybrid Long Lines with Shunt Reactance on Sides

5.1.1 No-Load Closing and External Fault Removal

Figure 7 and Fig. 8 respectively show the effective value of differential current of each criterion with no-load closing and external fault removal. It can be found that all three compensation methods can compensate the capacitance current when the system is in stable operation, among which the time-domain compensation method based on the segment parameters has the best performance and the smallest post-compensation differential current.

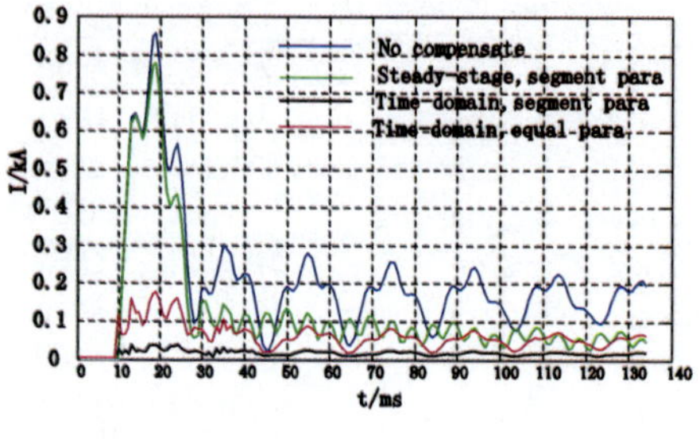

Fig. 7. No-load closing.

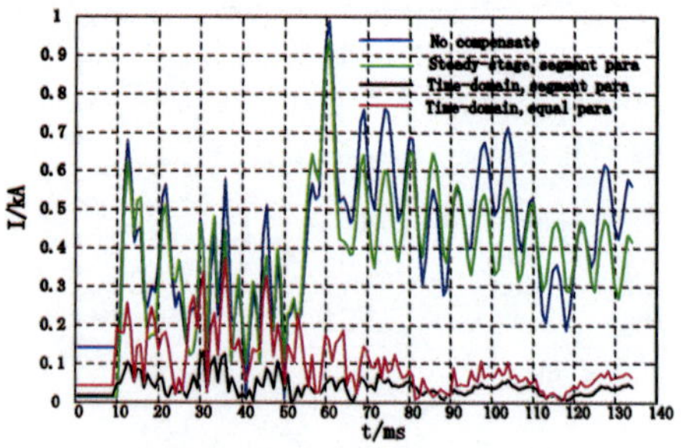

Fig. 8. External fault removal.

In the transient process of no-load closing or fault removal, the steady-state compensation method based on segment parameters has a poor compensation effect and cannot compensate the transient capacitance current. The time-domain compensation method based on segment parameters has the best effect; the time-domain compensation method based on the whole line equivalent parameters only using one type Π network is less accurate in transient components than the time-domain compensation method based on segment parameters using two type Π networks, making the compensation effect worse than the time-domain compensation method based on segment parameters. If multiple type II networks are used in series, the result will be more accurate, but the calculation will be more complex and possibly unsolvable. See Table 2 for the maximum differential current effective values of different compensation methods in the transient process.

Table 2. Criterion maximum differential current effective values when no-load closing and external fault removal.

Differential criterion	No-load closing/kA	External fault and removal/kA
Non-compensation	0.85	0.99
Steady compensation based on segment	0.84	0.96
Time-domain compensation based on segment parameters	0.11	0.14
Time-domain compensation based on equivalent parameters	0.27	0.37

5.1.2 Internal Fault

In the event of internal fault, the influence of the action threshold of the differential criterion of various compensations on the sensitivity of protection is mainly discussed. In this simulation, the value of differential current of each criterion is set to above maximum unbalanced differential current as no-load closing and external fault removal, in which the high setting value of the steady-state compensation criterion and the time-domain compensation criterion takes the reliability coefficient of 1.5, and the low setting value of the steady-state compensation criterion takes the 75% maximum unbalanced differential current. See Table 3 for the action threshold of each criterion set according to the above principle. Simulation analysis is conducted for the internal metal fault and transit resistance fault.

Table 3. Action threshold of each criterion.

Segment steady-state compensation criterion		Segment time-domain compensation criterion/kA	Equivalent time-domain compensation criterion/kA
High setting value	Low setting value		
1.44	0.72	0.21	0.55

The differential current when there is an internal metallic fault is shown in Fig. 9. The differences among steady-state compensation based on segment parameters, time-domain compensation based on segment parameters, time-domain compensation based on equivalent parameters and uncompensated differential current are small. The differential protections formed by various compensation methods are almost the same in action speed.

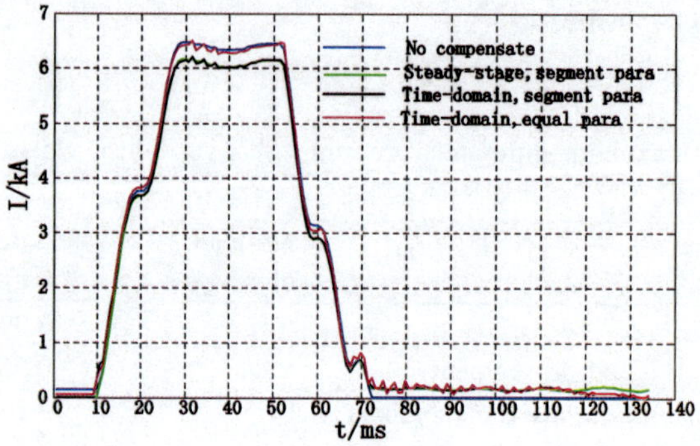

Fig. 9. Internal metallic fault.

When the potential angle difference of the power supply on the sides of the line is 5°, the differential current resulting from the simulation of the internal fault through 300 Ω and 500 Ω transition resistance is shown in Fig. 10. When the transition resistance is 300 Ω, the high setting value of the steady-state compensation based on segment parameters cannot act, whereas the low setting value criterion can reflect the grounding fault of 300 Ω, but the current setting should be above the capacitance current in the transient process. Both the time-domain compensation criterion based on segment parameters and the time-domain compensation criterion based on equivalent parameters allow the transition resistance of more than 500 Ω, and the former criterion is higher in sensitivity and quick in action without delay, thus greatly improving the action speed when there is an internal grounding fault with shunt reactance.

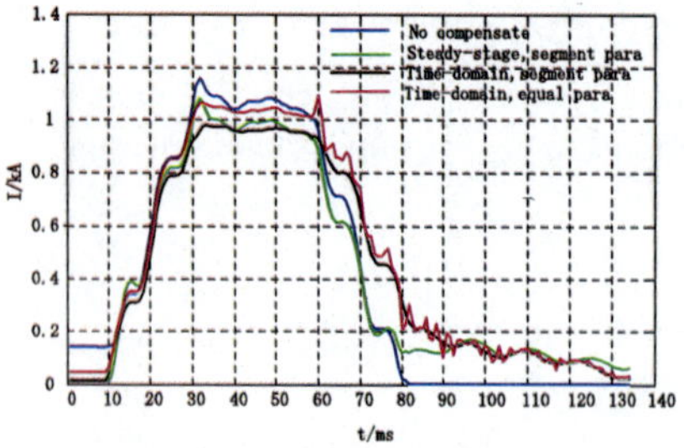

(a) Internal phase-a fault through 300 ω transition resistance.

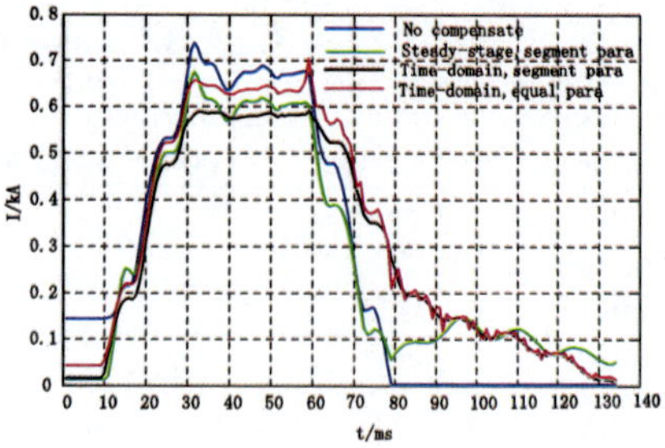

(b) Internal phase-a fault through 500 ω transition resistance.

Fig. 10. Internal fault through transition resistance.

5.2 Simulation of Hybrid Long Lines with Shunt Reactance at Junctions

In order to verify whether the segmented compensation method is applicable to hybrid lines with shunt reactance at junctions, simulation analysis is also carried out for lines with shunt reactance at the K2 point, showing that the compensation degree of the lines is consistent with that of the system with reactance on the sides. During no-load closing and external fault removal, the differential current from the time-domain compensation based on segment parameters can completely compensate the transient capacitance current and steady-state capacitance current with the best effect. The simulation waveforms are shown in Fig. 11.

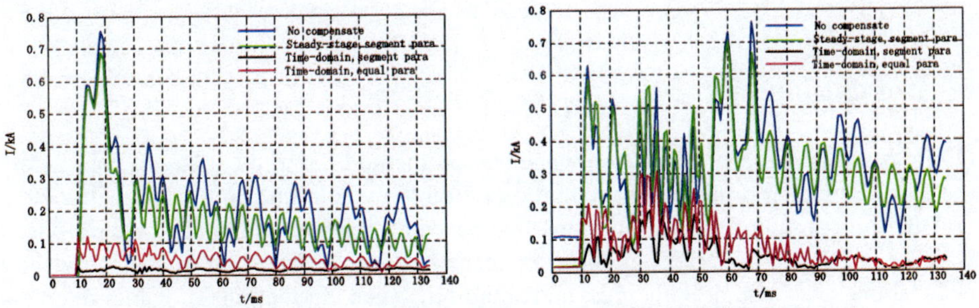

(a) No-load closing. (b) External fault removal.

Fig. 11. Hybrid lines with shunt reactance at junctions.

5.3 Simulation of Hybrid Short Lines with Shunt Reactance at Junctions

Figure 12 shows the effective value of differential current of each criterion during no-load closing and external fault removal. Through simulation, it can be found that the differential current calculated by the time-domain compensation method based on equivalent parameters is almost the same as that calculated by the time-domain compensation method based on segment parameters in the steady-state and transient process when the lines are short. The difference between the equivalent parameters using one type Π network and the segment parameters using two type Π networks is small. The differential protection of the working line device can easily calculate the setting values of the reactance and shunt reactor of the hybrid lines by the methods based on whole line equivalent parameters without any change.

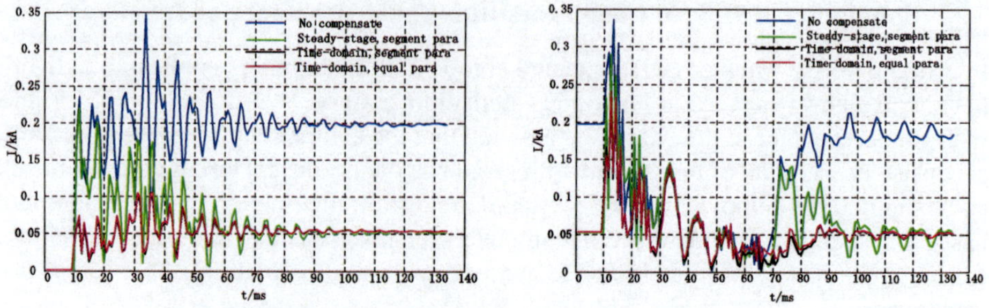

(a) No-load closing. (b) External fault removal.

Fig. 12. Short lines with shunt reactance at junctions.

6 Conclusion

(1) In this paper, the steady-state compensation method based on segment parameters, the time-domain compensation method based on the whole line equivalent parameters, and the time-domain compensation method based on segment parameters are proposed, which can adapt to different application scenarios such as no shunt reactance at both ends, shunt reactance at both ends, and shunt reactance at junctions of cable–overhead hybrid lines.

(2) The time-domain compensation method based on segment parameters can accurately compensate the capacitance current of hybrid transmission lines and has a better effect than the steady-state compensation method based on segment parameters and the time-domain compensation method based on segment parameters, improving the sensitivity and reliability of line differential protection.

(3) For short hybrid transmission lines, the time-domain compensation method based on segment parameters is equivalent to the time-domain compensation method based on the whole line equivalent parameters in terms of its compensation effect. It is not necessary to change the compensation scheme based on the whole line parameters of the existing operation device protection, but only to do the equivalent calculation of the reactance setting value and the shunt reactor setting value of the hybrid lines using methods based on whole line equivalent parameters, which is simple in engineering realization.

Acknowledgments. This work was supported by Science and Technology Program of State Grid Corporation of China: "Research on fault coupling characteristics and new protection principle of AC/DC hybrid power grid."

References

1. Xu, B., Xu, N.: High voltage power cable's influence on the relay protection equipment. J. Elect. Power Sci. Technol. **6** (2010). 25-2

2. Wu, T., Zheng, Y., Zhu, X.: Current differential protection based on transient capacitance current compensation. Autom. Elect. Power Syst. **12**, 61–67 (2005)
3. Zheng, Y., Wu, T., Ding, Y., Chen, Y.: Applied criterion of current differential protection based on Bergeron model. Autom. Elect. Power Syst. **23**, 50–55 (2004)
4. Suonan, J., Zhang, Y., Qi, J., Jiao, Z.: Study of current differential protection using time-domain capacitive current compensating algorithm on π-model. Proceed. CSEE **5**, 12–18 (2006)
5. Suonan, J., Zhang, Y., Qi, J., Jiao, Z.: Current differential protection based on time-domain algorithm of capacitive current compensating. J. Xi'An Jiao Tong University. **12**, 1370–1374 (2005)
6. Zheng, Y.: Technology and Application of Power Network Relay Protection. China Electric Power Press (2020)
7. Ge, Y.: Principle and Technology of New Type Relay Protection and Fault Location. Xi'an Jiaotong University Press (1996)
8. Su, B., Dong, X., Sun, Y.: Distributed capacitive current compensation algorithm for current differential relay of UHV transmission line. Autom. Elect. Power Syst. **8**, 36–40 (2005)
9. Li, Y., Chen, D., Zhang, Z., Yin, X.: The emulation analysis for the influence of capacitance current of UHV transmission line on differential current protection and compensating countermeasure. Relay **6**, 6–9 (2001)
10. Sang, B., Wang, X.: Adaptive compensation scheme of capacitance current to differential protection in UHV lines. Relay **8**, 1–5 (2010)
11. Suonan, J., Yang, C., Yang, Z., Song, G.: Study of time-domain compensation algorithm of capacitive current for parallel transmission lines protection. Proceed. CSEE **1**, 77–81 (2010)
12. Bi, T., Yu, Y., Huang, S., Yang, Q.: Novel accurate compensation method of capacitance current of UHV transmission line. Autom. Elect. Power Syst. **15**, 30–34 (2005)
13. Miao, S., Liu, P., Cheng, S., Meng, L.: Research and simulation of capacitance current compensating countermeasure for phase differential protection of 750 kV transmission lines. Power Syst. Technol. **6**, 54–60 (2008)
14. Zang, H., Lin, F., Rong, Y., Jiao, J., Zhang, J.: Time-domain capacitance current compensation algorithm with shunt reactors. High Volt. Eng. **12**, 3694–3671 (2016)
15. Tai, N., Guo, P., Yu, Z., Fan, C.: Current differential protection using capacitive current compensating algorithm on pi-model for four-parallel lines on same towers. Power Syst. Protect. Control **12**, 34–12 (2015)

Capacity Configuration of Energy Storage for Photovoltaic Power Generation Based on Dual-Objective Optimization

Linfeng Li[1], Shenjun Hou[1], Hui Gu[2], and Changcheng Xu[2](✉)

[1] Jiangsu Nantong Power Generation Co. Ltd., Nantong, China
[2] Nanjing University of Posts and Telecommunications, Nanjing, China
sutong_lf@163.com

Abstract. Capacity configuration is the key to the economy in a photovoltaic energy storage system. However, traditional energy storage configuration method sets the cycle number of the battery at a rated figure, which leads to inaccurate capacity allocation results. Aiming at this problem, this paper proposes a mixed integer programming model to optimize capacity and power of energy storage which the number of cycles as one of optimization parameters. First, the best daily battery charging and discharging strategy, best capacity, power configuration, and best battery cycle number are obtained under the object of economy then we analyze the relationship between battery cycle life and system economy by simulation. According to the results, we can get the range of the number of battery cycles for the sake of better economy. Furthermore, a dual-objective optimization model for economy and photovoltaic self-consumption rate is established as the dual-objective function. The results show that when the PV self-consumption rate exceeds a certain level, the system economy will decrease accordingly. In summary, the proposed method can better guarantee the economy of system.

Keywords: PV power generation · Energy storage · Capacity configuration · Dual-objective optimization · Number of cycles

1 Introduction

In recent years, photovoltaic (PV) power generation has been increasingly affected by its huge resource reserves and small geographical restrictions. Energy storage for PV power generation can increase the economic benefit of the active distribution network [7], mitigate the randomness and volatility of energy generation to improve power quality [8], and enhance the schedulability of power systems [9]. Investors in industrial photovoltaic microgrids can purchase electricity from the grid to charge energy storage (ES) batteries during periods of low electricity prices, and supply stored energy to loads while electricity prices are high. At noon, excess PV can also be stored in ES batteries or connected to the grid. In existing PV power generation, reasonable battery capacity and power allocation is crucial to arrangement photovoltaic energy storage systems [1–6]. If the capacity is too small, the problem of high peak load can't be solved

Y. Xue et al. (Eds.): PMF 2020, LNEE 718, pp. 940–949, 2021.
https://doi.org/10.1007/978-981-15-9746-6_70

effectively. In contrast when the capacity is too large, the investment cost of the battery will increase.

At present, there are many literatures on energy storage allocation. Paper [10] and [11] respectively use genetic algorithm and linear programming to solve capacity optimization, but they do not consider the impact of PV access on the economic benefits of energy storage. In paper [12], a linear programming model for capacity and power allocation of energy storage is established to optimize economic benefits throughout the year, but the effect of battery cycle numbers on system economy is not considered. Paper [13] builds a multi-objective optimization model for the optimization of the energy storage capacity, including economic goals and PV self-consumption rate, which also does not consider the impact of excess PV grid connection and battery cycle numbers on the system. Paper [14] uses mixed integer linear programming with the goal of reducing total annual cost, which decouples the capacity optimization and operation optimization. But this method ignored the impact of PV self-consumption rate on the economy.

In summary, most existing research work focus on the economy of PV energy storage system. But none of them take the number of battery cycles into consideration. In reality, as the actual number of battery cycles increases, the capacity and power of energy storage to be configured will increase accordingly, which result in increased revenue caused by the peak valley arbitrage and high investment cost. In this paper, we establish a mixed integer programming model of battery capacity and power configuration which sets both system economy and PV consumption rate as the objective function and takes battery number of cycles as one of the decision variables. We can reveal the relationship among PV consumption rate, battery cycle numbers and system economy, and obtain the more accurate capacity and power configuration by optimization.

The rest of the paper is organized as follows: Sect. 2 analyzes the characteristics of an actual industrial microgrid. Section 3 introduces the system model, which consists of the dual-objective function, the constraints of system and the model solving method. In Sect. 4, the simulation is carried out to verify the effectiveness of the proposed method. Section 5 concludes the paper.

2 Analysis of PV Characteristics

Here we average hourly value of PV and load data in a typical month to get the data of typical day. The PV output and load in an actual industrial microgrid are shown in Fig. 1, as well as electricity price at different times of the typical day. It can be seen that PV output has evident unimodal distribution. From 11a.m. to 4p.m, PV power is greater than the load, so the redundant PV output can be stored in the battery or sold to the grid. Battery can be discharged to meet the load instead of buying electricity from the grid while the electricity price is high in peak hours—the time from 9a.m.to 12p.m. and 6p.m. to 9p.m. At other times, the system purchases electricity for the sake of economy.

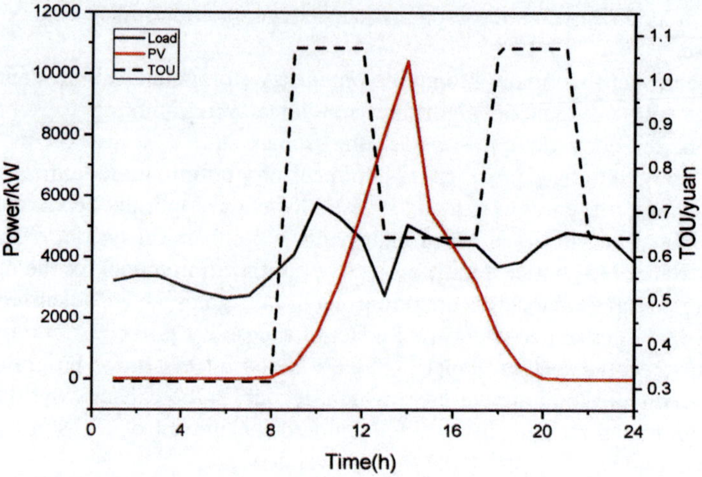

Fig. 1. The PV output, load and electricity price of an industrial PV microgrids

3 Optimal Model of System

3.1 Power Flow in System

The photovoltaic output P_{PV} of the system at each moment is divided into three parts, the power $P_{PV.dmd}(t)$ from PV to load, the power $P_{\mathrm{PV.charge}}(t)$ from PV to battery and the power $P_{PV.export}$ from PV to grid. We can get Eq. (1)

$$P_{PV}(t) = P_{PV.dmd}(t) + P_{\mathrm{PV.charge}}(t) + P_{PV.export}(t) \tag{1}$$

Purchasing power P_{buy} from the grid is divided into $P_{buy.dmd}$ to meet the load and $P_{buy.charge}$ to charge the battery as follows:

$$P_{buy}(t) = P_{buy.dmd}(t) + P_{buy.charge}(t) \tag{2}$$

At each moment, the load of the system is partially satisfied by three parts, which can be obtained by Eq. (3)

$$P_{dmd}(t) = P_{PV.dmd}(t) + P_{buy.dmd}(t) + P_{discharge}(t) \tag{3}$$

Here, $P_{discharge}(t)$ denotes battery discharges power.

3.2 Cost and Benefit Analysis of PV Energy Storage System

The system cost in this paper mainly includes the investment cost of battery and the annual electricity purchase cost due to charging for energy storage. The system benefits are primarily from the peak-valley arbitrage of energy storage and PV grid-connected profit.

The cost of configuring capacity $C_{battery.cap}$ is the product of the battery capacity and the investment cost per unit capacity $C_{unit.battery.cap}$. Considering the battery discharge depth DOD, then we get:

$$C_{battery.cap} = \frac{S_{\max}}{DOD} \times C_{unit.battery.cap} \tag{4}$$

The cost generated by configuring the power is $C_{battery.power}$, which is mainly the investment of the PCS bidirectional converter that meets the power requirements. The power of the PCS bidirectional converter is the same as the maximum battery charge-discharge power $P_{\max}$ selected, and the cost of the configuration power $C_{battery.power}$ is the product of the converter and the unit price, that is:

$$C_{battery.power} = P_{\max} \times C_{unit.battery.power} \tag{5}$$

The annual electricity purchase cost:

$$C_{electricity.buy} = D_r \times \sum_{t=1}^{24} C_e P_{buy.charge} \tag{6}$$

Here, D_r is the operation days of the system. C_e denotes peak and valley time price.

The annual revenue of the system mainly comes from the peak-valley arbitrage of energy storage and PV grid-connected profit, which is defined as:

$$R = D_r \times \sum_{t=1}^{24} (C_e P_{discharge} + C_{sell} P_{PV.export}) \tag{7}$$

Here, C_{sell} is the feed-in tariff for PV.

3.3 Dual-Objective Optimal Function

We select the power allocation from PV and battery charge-discharge power as optimal parameters, in addition to energy storage capacity and power. In this paper, the cycle number is specially introduced to be optimized due to its effect on the system economy. Since the system economy and PV Photovoltaic absorption rate are two important targets in PV energy storage system, we design the dual-objective function as follows:

(1) *Economic benefits*

$$\max F = R - C_{electricity.buy} - \frac{r(1+r)^{\frac{n}{D_r m}}}{(1+r)^{\frac{n}{D_r m}} - 1} (C_{battery.cap} + C_{battery.power}) \tag{8}$$

The energy battery investment is generally a one-time expenditure, and that is amortized annually over the battery cycle life in order to calculate annual benefit. In the above Eq. (8), $\frac{r(1+r)^{\lceil \frac{n}{D_{rm}} \rceil}}{(1+r)^{\lceil \frac{n}{D_{rm}} \rceil}-1}$ is the capital recovery factor converted into the annual investment cost, where n is the actual number of cycles of the energy storage battery to be optimized, r is the discount rate, and m is the battery charge-discharge times every day, which can be configured according to the real scenario, $\lceil\rceil$ denotes round up.

(2) *PV self-consumption rate*

The self-consumption ability of PV power generation in the system is measured by the parameter SCR which can be defined as Eq. (9)

$$\max SCR = \frac{\sum P_{PV.dmd} + \sum P_{PV.charge}}{\sum P_{PV}} \tag{9}$$

3.4 Restrictions

There are some restrictions of parameters above to be optimized, including battery charge-discharge power constraints, battery capacity constraints and the number of cycles constraint. They are expressed respectively as follows.

3.4.1 Battery Charge-Discharge Power Constraints

The linear inequality constraints of each power decision variables can be obtained by Eq. (10) to (12)

$$0 \le P_{PV.dmd}(t) \le P_{dmd}(t) \tag{10}$$

$$P_{PV.dmd}(t) + P_{PV.charge}(t) \le P_{PV}(t) \tag{11}$$

$$P_{PV.dmd}(t) + P_{discharge}(t) \le P_{dmd}(t) \tag{12}$$

Because the charge-discharge power of the battery is within the rated range at each moment, it does not exceed the rated maximum charge-discharge power $P_{\max}$, which can be denoted as Eq. (13) to (16):

$$P_{buy.charge}(t) + P_{PV.charge}(t) \le P_{\max} \tag{13}$$

$$0 \le P_{PV.charge}(t) \le P_{\max} \tag{14}$$

$$0 \le P_{buy.charge}(t) \le P_{\max} \tag{15}$$

$$0 \le P_{discharge}(t) \le P_{\max} \tag{16}$$

3.4.2 Battery Capacity Constraints

For convenience, the battery capacity is equivalent to zero when it discharge to the lowest limit, and the upper limit of charging is equivalent to the battery capacity being fully charged. When the battery is charged and discharged, its capacity of each interval meets Eq. (17):

$$S(t) = S(t-1) + \eta_c \times P_{PV.charge}(t) + \eta_c \times P_{buy.charge}(t) - P_{discharge}(t)/\eta_d \quad (17)$$

Here, η_c and η_d are charge-discharge efficiency respectively.
Since the battery capacity is always within the rated range, we get:

$$0 \leq S(t) \leq S_{\max} \quad (18)$$

In order to eliminate the impact that remain capacity have on the battery operation on next day, the battery capacity should be reset to zero at the end of one day. The daily battery charge-discharge capacity should be equal in one day, which can be expressed as follows:

$$\sum_{t=1}^{24} \left\{ \eta_c \times \left[P_{PV.charge}(t) + P_{buy.charge}(t) \right] - \frac{1}{\eta_d} \times P_{discharge}(t) \right\} = 0 \quad (19)$$

3.4.3 Number of Cycles Constraint

In the economic objective function mentioned above, n is the actual number of cycles of the energy storage battery, which meets the following constraint

$$n_{\min} \leq n \leq n_{\max} \quad (20)$$

The upper and lower bounds of the cycle number n can be set in the optimization. Here, the upper bound $n_{\max}$ is the rated number of cycles.

The optimization model can be applied to calculate the system annual net income, the best operation strategy, the best battery capacity and power configuration, and the optimal number of battery cycles. This problem is the typical mixed integer programming with linear inequality constraints, linear equality constraints, and upper and lower bounds of decision variables. GA solver is applied for single-objective optimization to obtain the best number of battery cycles, and GAMULTIOBJ solver for dual-objective optimization can solve the models.

4 Case Analysis

The case used in this paper is a PV energy storage system of an enterprise in Nantong, Jiangsu. The PV and load data are acquired from historical data, where the load is far higher during daytime than at night as shown in Fig. 1. In this paper, lead-carbon batteries are selected as energy storage batteries (Table 2). Parameters of the energy storage system are set as Table 1.

Table 1. Parameters in the energy storage system

Unit capacity cost	600 yuan/kWh
Unit Power Cost(PCS)	150 yuan/kW
Charge-discharge Efficiency	90%
Discharge depth	75%
Battery cycle times	2000 ~ 2500 times
On-grid price	0.4515 yuan/kWh
Discount rate	0.06
Charging/discharging times	0:00 ~ 8:00, 13:00 ~ 17:00/9:00 ~ 12:00, 17:00 ~ 24:00

Table 2. Time-of-use (TOU) price

Time period	0:00 ~ 8:00	8:00 ~ 12:00 17:00 ~ 21:00	12:00 ~ 17:00 21:00 ~ 24:00
Electricity price(yuan/kWh)	0.3139	1.0697	0.6418

The relationship between system annual net profit and battery cycle number is shown as Fig. 2. It can be seen that the system annual net profit achieves maximization while the numbers of cycle are 2348 rather than the rated number. Furthermore, in order to evaluate the economic and energy performance of the system with energy storage of the optimal capacity, peak-load shifting rate (PSR) is introduced as follows.

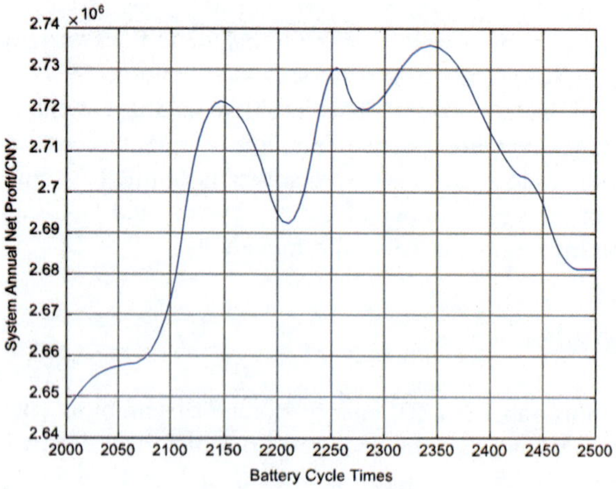

Fig. 2. Relationship between battery cycle times and system economics

$$PSR = (P'_{grid,buy} - P_{grid,buy})/C'_{grid,buy} \times 100\% \quad (21)$$

Here $P'_{grid,buy}$ is the power bought from the grid in the system without energy storage.

To analyze the effect of PV energy storage on the system, the capacity configuration, power configuration and two metrics mentioned above are calculated separately under three scenarios including the system without ES, the system with ES under the rated number of battery cycles (2500), and the system with ES under the optimal number of battery cycles (2348). The results are shown in Table 3. Compared to the system without ES, the total annual profit of system with ES has been greatly improved due to peak valley arbitrage by ES. It can be seen that the system with ES under the optimal number of battery cycles (2348) can achieve best profit. Moreover, the SCR with ES also has greatly increased which means more PV consumption. The capacity configuration, power configuration and PV SCR for the optimal number of cycles is slightly reduced compared to the rated number of cycles.

Table 3. Results of energy storage configuration

	No ES	With ES(2500)	With ES(2348)
Configured capacity	0	13.72MWh	13.19MWh
Power allocation	0	4.48 MW	4.27MWh
Annual profit	310,608 yuan	2,681,372 yuan	2,746,615 yuan
Capital recovery factor	/	0.1803	0.1898
SCR	62.61%	80.58%	80.48%
PSR	/	31.27%	30.86%

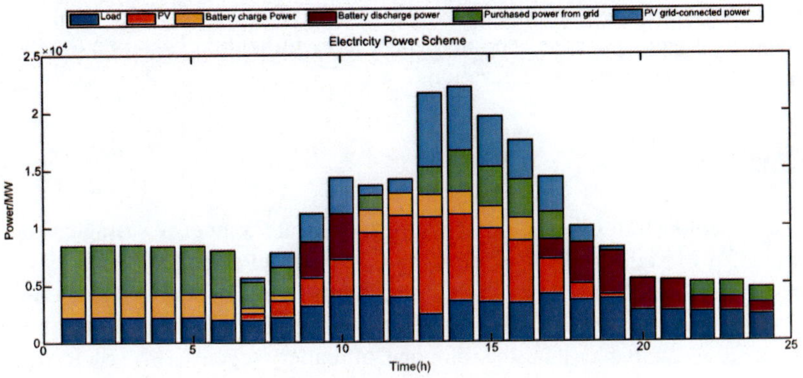

Fig. 3. Electrical power schedule

The electrical power schedule obtained by the proposed optimal method for a typical day is shown in Fig. 3. The simulation results show that the ES can reduce the cost of the system through valley charging and peak discharge, and ultimately achieve profitability. Therefore, the proposed method can obtain the optimal capacity, optimal power and optimal scheduling scheme to enhance the energy efficiency and economy of the system.

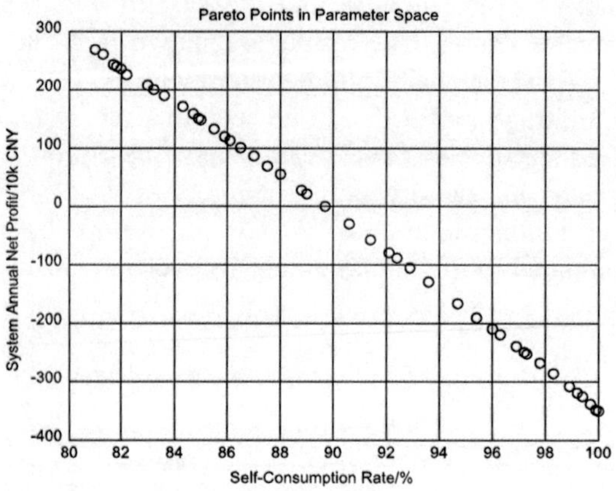

Fig. 4. Pareto optimal frontier

Pareto optimal frontier obtained by GA is shown in Fig. 4. It can be seen that the system economy and the PV self-consumption rate are a set of opposing goals. When the PV self-consumption rate is between 80.9% and 89.4%, the annual net profit is positive. More capacity and power needs to be configured in order to improve SCR, which leads to declining revenue in turn. Investors of energy storage in microgrid can comprehensively consider the economy and energy utilization based on the results, and finally make a reasonable configuration.

5 Conclusion

This paper demonstrates the optimization of industrial PV energy storage systems with heavy load. A Mixed Integer Programming (MIP) model of battery capacity and power configuration is established which set both system economy and PV consumption rate as the objective function and take battery number of cycles as one of the decision variables. It is analyzed that the number of battery cycles, PV SCR and system economy have a balance problem, so we can get the best solution according to the target requirement by optimization. The case simulation results verify that the proposed method can effectively improve the economy and SCR of PV energy storage system.

References

1. Jiantao, L., Jiancheng, Z., Jie, M., et al.: Analysis of energy storage technology applied in grid-connected PV system. Power Syst. Clean Energy **7** (2011)
2. Shuanghui, L., Shengxia, C., Shouxiang, W.: Economy evaluation and development suggestions for distributed PV-Energy storage system in China. Electr. Power **48**(2), 139–144 (2015)
3. Jian, S., Limei, Z., Rui, L.: Cost-benefit analysis of distributed grid-connected photovoltaic power generation. Proceed. CSEE **33**(34), 50–56 (2013)
4. Xiaogang, W., Zongqi, L., Liting, T., et al.: Optimized capacity configuration of photovoltaic generation and energy storage device for stand-alone photovoltaic generation system. Power Syst. Technol. **38**(5), 1272–1276 (2014)
5. Guanqun, L., Yue, Y., Min, W., et al.: Energy storage capacity determining of PV plant considering economic cost. Renew. Energy Res. **32**(01), 1–5 (2014)
6. Jie, W., Chen-Yang, W., Shan, L., et al.: Optimal allocation of PV and accumulator system based on time-of-use price. Adv. Technol. Electr. Eng. Energy **37**(01), 23–30 (2018)
7. Nick, M., Cherkaoui, R., Paolone, M.: Optimal allocation of dispersed energy storage systems in active distribution networks for energy balance and grid support. IEEE Trans. Power Syst. **29**(5), 2300–2310 (2014)
8. Santos, S.F., Fitiwi, D.Z., Cruz, M.R.M., et al.: Impacts of optimal energy storage deployment and network reconfiguration on renewable integration level in distribution systems. Appl. Energy **185**, 44–55 (2017)
9. Kerdphol, T., Qudaih, Y., Mitani, Y.: Optimum battery energy storage system using PSO considering dynamic demand response for microgrids. Int. J. Electr. Power Energy Syst. **83**, 58–66 (2016)
10. Xiong, X., Lin, Y., Rengang, Y.: Optimal allocation and economic benefits analysis of energy storage system on power demand side. Autom. Electr. Power Syst. **39**(17), 42–48 (2015)
11. Yitong, Z., Huifang, W., Benteng, H., et al.: Optimization strategy of configuration and operation for user-side battery energy storage. Autom. Electr. Power Syst. **44**(6), 121–128 (2020)
12. Chihao, N., Xuezhi, L.: Battery capacity configuration and economic analysis of photovoltaics energy storage system. Zhejiang Electr. Power **38**(01), 1–10 (2019)
13. Jian, X., Nian, L., Lei, Y., et al.: Optimal allocation of energy storage system of PV microgrid for industries considering important load. Power Syst. Protect. Control **44**(9), 29–37 (2016)
14. Qianwen, Z., Qiqiang, L., Bingying, Z., et al.: Capacity optimization for electrical and thermal energy storage in multi-energy building energy system. Energy Proc. **158**, 6425–6430 (2019)

Structural Modeling and Identification Method for System Protection Control Strategy of Power Grid

Lin Liu[1], Xiongfeng Xu[2(✉)], Jianyu Luo[1], Shengming Wang[2], Ting Zhou[1], and Xiongfei Wang[2]

[1] State Grid Jiangsu Electric Power Co., Ltd., Nanjing 210024, Jiangsu, China
[2] NARI Group Corporation/State Grid Electric Power Research Institute, Nanjing 211106, Jiangsu, China
xuxiongfeng@sgepri.sgcc.com.cn

Abstract. According to the requirement of system protection control strategy simulation of UHV AC and DC power grid, this paper analyzes the control structure and execution logic of system protection devices, establishes structural strategy models based on formula configuration, and puts forward the duty strategy identification method after considering the operation condition of power grid and the running status of devices. It realizes the automatic identification, as well as online monitoring and early warning, of system protection duty strategy, which timely discover possible safety risks of power grid after system protection actions finished under the circumstances of serious multiple DC fault. It also provides technical support for refined evaluation of system protection strategies. Finally, the effectiveness of the proposed method is verified by combining an example of system protection control strategy of DC receiving-end grid.

Keywords: System protection control device · Control strategy modeling · Duty strategy identification · Online monitoring and early warning · Practicability assessment

1 Introduction

With the rapid development of UHV AC and DC power grid, the features of "strong DC & weak AC" for power grid is prominent. The safety and stability control system and devices of traditional power grid (hereinafter referred to as "stability control system and device") have gradually shown the disadvantages of single control measures, lacking overall consideration, lacking coordinated control and being difficult in effectively dealing with cascading faults, etc. in response to Ultra-HVDC blocking and other serious faults. There are great risks in the safe operation of power grid. In order to ensure the safety of the transitional power grid, it is necessary to construct "UHV AC-DC power grid system protection" (referred to as "system protection") to realize the coordinated control of wide area multi-class control resources in multi-spatial dimension and multi-time scale [1–6].

Serious AC/DC faults, especially serious multiple DC faults, will lead to large unbalanced power at the sending-end and receiving-end of power grid. The offline

Y. Xue et al. (Eds.): PMF 2020, LNEE 718, pp. 950–961, 2021.
https://doi.org/10.1007/978-981-15-9746-6_71

configuration strategies may be mismatched by the change of operation mode, which may lead to equipment overload, voltage over-limit, frequency over-limit and other problems after the action of various control and protection systems or devices such as system protection [7–9]. Therefore, it is very important to accurately perceive the running status of the system protection control devices and the control strategy simulation of the system protection, so as to assist the dispatcher to control the safe and stable operation of the power grid online [10, 11].

Research on related technologies of online monitoring and management of stability control devices has been carried out in China [12–17]. The research shows that is closely related to the safe and stable operation of the power grid whether the duty strategy of the stability control system is implementable (whether it can be executed sufficiently) in practical operation. The stability control system may have the risk of "under control" or "over control" under special operating conditions such as device exceptions and channel exceptions, and in serious cases, these control results may even expand the power grid accident [18–20]. At the same time, in order to ensure the requirement of real-time response [21], the stability control devices themselves cannot conduct real-time evaluation at the implementation of duty strategy for all monitoring contingency faults. In general, the online warning function of the stability control duty strategy is established in the automatic system of the control center, so as to automatically identify whether the duty strategy can be fully implemented.

At present, the duty strategy early warning function of the stability control system that has been constructed does not fully consider the influence of the running status of multi-level control devices in the system protection or the control mode of each DC pole in the DC system on the controllable quantity and control-required quantity of the contingency fault. It is mainly reflected in the following two aspects: 1) The influence of the channel on-and-off state between control devices, the switching on and off state of the control object's associated functional pressing plate, the running status of the pumping storage unit and other equipment on the controllable quantity of system protection is not fully considered; 2) The influence of the contingency fault, including the control mode of each DC pole in each DC system, on the control-required quantity of the DC contingency fault is not fully considered.

This paper mainly aims at simulating the action of the system protection devices. It does not involve the specific calculation model of power grid security and stability analysis, nor does it include the online identification of specific primary equipment fault types of power grid. The unified modeling method of system protection control strategy and duty strategy identification technology are studied on the dispatching end. Based on the power grid operating condition and the system protection devices running status of pre-contingency state, the duty control strategy of system protection and the control quantity and specific action object of various control measures are identified, which provides technical support for the accurate simulation of the duty strategy of system protection.

2 General Framework

The system protection duty strategy identification function is specifically composed of control strategy modeling, formula parsing, duty strategy matching, control-required quantity calculation, controllable quantity identification, control measures selection, and practicability assessment module. The control strategy modeling module realizes the unified design and maintenance of system protection device and its strategy model; The formula parsing module realizes the unified parsing of various configuration expressions in the control strategy model; The duty strategy matching module realizes the analysis and matching of all kinds of constraints of the control strategy that meet the specified operating conditions of the power grid, and calculates the control-required quantity of control measures for the contingency faults under the specified operating conditions of the power grid; The controllable quantity identification module realizes the controllable state recognition and controllable quantity statistics of each control measure; The control measures selection module realizes the distribution of control quantity of all kinds of control measures and the combination of control measures of two sets of devices under different operation modes and outlet connection modes; The practicability assessment module realizes the implementable status calculation of the duty strategy of the contingency fault containing various control measures. See Fig. 1 for details.

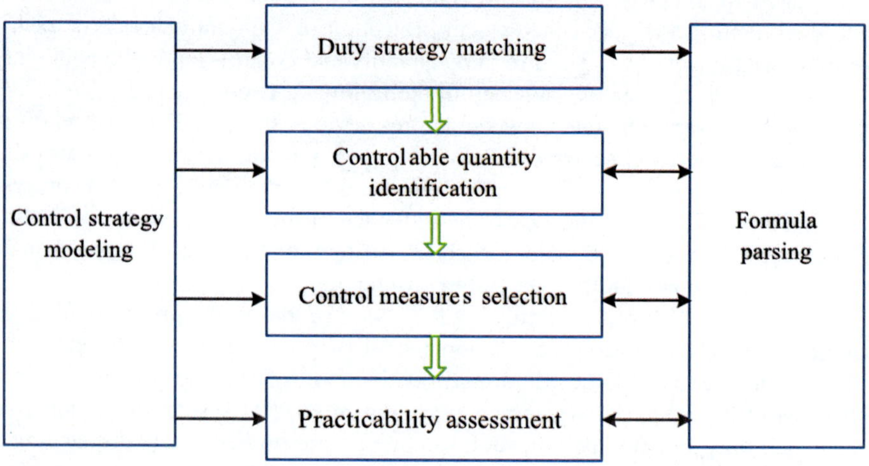

Fig. 1. Framework of system protection duty strategy identification

The system protection duty strategy identification function provides unified publish-subscribe access interface and request-reply access interface for external application functions through service encapsulation, so as to meet the requirements of strategy simulation under different application scenarios of online operation and offline study.

3 Key Technology

3.1 Control Strategy Modeling

According to the different function of co-control master station, control primary station, control substation and executive station in typical system protection systems, the control strategy is logically divided. Combined with the characteristics of easy understanding of object-oriented modeling and flexible configuration of expression definition [17, 22–24], the strategy model of system protection is divided into three types: control strategy, execution strategy and set of control measures, which logically corresponds to the strategy functions of the co-control master station, control primary station and control substation, control substation and execution station respectively, as shown in Fig. 2.

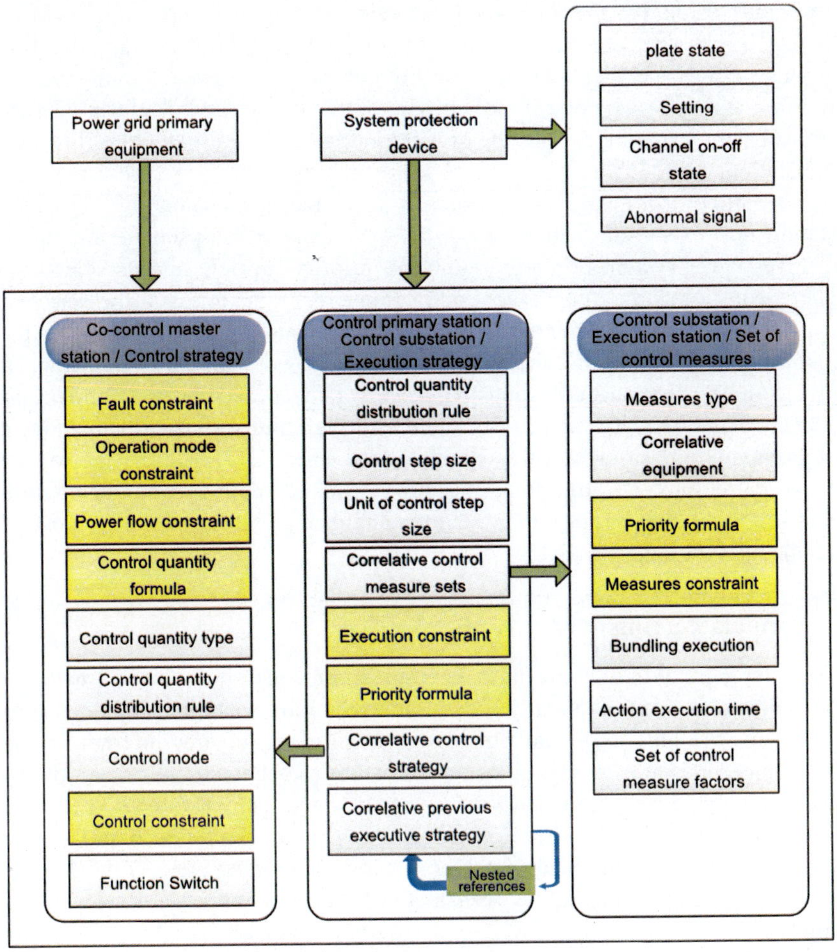

Fig. 2. Model of system protection control strategy

The control strategy corresponds to the strategy logic of the co-control master station, which is composed of fault constraint, operation mode constraint, power flow constraint, control quantity formula, control quantity type, control quantity distribution rule, control mode, control constraint and function switch; In which the control mode specifically includes three modes: minimum over-cut, minimum under-cut, and the closest cut, and the control quantity distribution rule includes three modes: priority, controllable quantity proportion, and optimal combination; The control quantity type mainly includes active power, amount (number of units or load cycles) and logical quantity (whether the line is separated or the reactor is removed); Function switch mainly defines whether the control strategy is put into control execution.

The execution strategy corresponds to the control quantity distribution logic of the control primary station and control substation, which is composed of control quantity distribution rule, control step size, unit of control step size, correlative control measures, execution constraint, priority formula and its selection rule, correlative control strategy and correlative previous-level execution strategy. Control commands issued between various control primary stations and control substations are described by executing strategy nesting method. The control primary station mainly defines the distribution logic of such control measures among multiple control substations, and the control substation mainly defines the distribution logic of multiple control measures in the substation.

The set of control measures corresponds to the measures selection logic of control substation and execution station, which is specifically composed of measures type, correlative power system primary equipment, priority formula and its selection rules, measures constraint, bundling execution, action execution time and the set of control measure factors. In which the control measures type mainly includes removing the unit, removing the load, removing the transformer, DC power modulation, line separation, removing the reactor and adjusting the power of the new energy unit; The primary equipment of the power system associated with the control measures includes generators, pumping and storage power stations, load lines, DC systems, reactor devices, new energy stations, etc. Bundling execution primarily defines whether control measures with the same priority in a set of candidate control measures are executed simultaneously or sequentially.

3.2 Formula Parsing

The specific logic of fault constraint, operation mode constraint, power flow constraint, control constraint, control quantity constraint, execution constraint, measures constraint and priority formula of system protection strategy model are different in different devices. In order to meet the requirement of customized characteristic description of the control strategy, a series of keywords, logical operators, arithmetic operators and relational operators, can be used for unified description combined with the operation condition of primary equipment of power system, the running state of the devices such as pressing plate and channel, as well as the formula composed of relevant strategy settings.

In which the fault constraint mainly defines the fault equipment and fault type; Operation mode constraint mainly defines the operation mode corresponding to the control strategy, which is mainly composed of the running state of the power grid primary equipment and the switch state of the device pressing plates; Power flow constraint mainly defines the power tap position where the power flow of control strategy is located in, which is mainly composed of the switching state of the power grid primary equipment, active power, current and strategy settings; The priority formula mainly defines the priority of control measures, which is mainly composed of the switching state of the power grid primary equipment, active power, the priority settings of device.

3.3 Duty Strategy Matching

For each DC contingency fault in the system protection contingency fault set, formulas of operation mode constraint, power flow constraint and control constraint associated with its control strategy are parsed in turn, and a set of control strategies that match the specified grid operation condition and device operation state are searched out.

The control-required quantity of DC contingency fault is determined by the fault loss power and the strategy setting, in which the fault loss power is not only related to the operating power of the fault DC system, but also related to the control mode of the DC pole of each DC system in fault [25]. Specifically, it is calculated according to Formula (1):

$$P_{lost} = \sum_{i=1}^{f} \left(\max\left\{ \sum \mathrm{P}_{ri} - \sum P_{mi} - \sum P_{ni}, 0 \right\} \right) \tag{1}$$

Where, P_{lost} is fault loss power of fault DC system, f is the number of DC systems included in the contingency fault, P_{ri} is the operating power sum of all converters operating of the i th DC system under the specified power grid operating conditions; P_{mi} is the operating power sum of all the converters without power transfer capability after the fault of the i th DC system under the specified power grid operating conditions; P_{ni} is the ultimate operating power sum of all the converters with power transfer capability after the fault of the i th DC system under the specified power grid operating conditions.

For a single DC pole of a DC system, the power transfer logic of converter that local pole is working normally under different control modes is shown as follows:

(1) If the DC pole is in the bipolar power control mode, whether the local pole converter or remote pole converter breaks down or not, the local pole sound converter has the ability of power transfer.

(2) If the DC pole is in the constant power control mode, only when the local pole converter fails, the local pole sound converter has the ability of power transfer for the same-pole fault converter.

(3) If the DC pole is in the constant current control mode, the local pole converter is not capable of power transfer.

3.4 Controllable Quantity Identification

When a candidate control measure is under control, it not only requires the execution station device of the candidate control measure is in operation, but also meets the following four conditions at the same time under general circumstances [15, 17]:

(1) The command channel between the control device of the duty strategy and the execution device of the candidate control measure is in normal communication state;

(2) The allowable cutting pressing plate corresponding to the correlative equipment of the candidate control measure is in the active state;

(3) The output pressing plate corresponding to the correlative equipment of the candidate control measure is in the active state;

(4) The active power of the correlative equipment of the candidate control measure is greater than the threshold power setting which is allowed to be controlled.

According to the controllable state and operating condition of a single candidate control measure, the controllable quantity of the control measure can be counted. On this basis, the total controllable quantity of various candidate control measures included in the duty strategy can be further counted.

3.5 Control Measures Selection

Aiming at the control-required quantity of each contingency fault of duty strategy, according to the priority and total controllable quantity of various control measures, the control-required quantity of various control measures is determined, and then the action set of specific control measures of each control substation is determined according to the controllable quantity distribution rules and priorities of various control primary stations [17].

The actual control measures of the dual-configuration devices are not only related to the action measures identified by each device, but also closely related to the operation mode and the connection mode of the action outlet of the dual-configuration devices. The operation mode of the dual-configuration devices includes the main-auxiliary operation and parallel operation [26, 27], and the merging of the duty strategy results is shown below.

Main-Auxiliary Operation Mode

For the dual devices in the main-auxiliary operation mode, the action outlet is connected in parallel. If the set of action measure of the main operation device is not null, the action measure of the main-operation device shall be taken as the actual action measure whether the set of the action measures of the auxiliary operation device is null or not. If the main operation device does not operate and the set of action measures of the auxiliary operation device is not null, the set of action measures of the auxiliary operation device is taken as the actual action measures [26]. See Table 1 for details.

Where, $N_{c1,Fj}$ and $N_{c2,Fj}$ are the control-required quantities of the c th class control measures of the duty strategy F_j of the first and second set of control devices; $X_{c1,Fj}$ and $X_{c2,Fj}$ are the actual control quantities to be executed of the c th class control measures of the duty strategy F_j of the first and second set of control devices.

Table 1. The combined logistic of duty strategy with dual devices on main-auxiliary and parallel operation mode.

Operation mode and outlet connection mode		First set of devices		Second set of devices		Merged result	
		Control-required quantity	Action measures	Control-required quantity	Action measures	Control-required quantity	Action measures
Main-auxiliary	First set of main operation	$N_{c1,Fj}$	$X_{c1,Fj}$	$N_{c2,Fj}$	Random	$N_{c1,Fj}$	$X_{c1,Fj}$
			Ø		$X_{c2,Fj}$	$N_{c2,Fj}$	$X_{c2,Fj}$
					Ø	$N_{c1,Fj}$	Ø
	Second set of main operation		Random		$X_{c2,Fj}$	$N_{c2,Fj}$	$X_{c2,Fj}$
			$X_{c1,Fj}$		Ø	$N_{c1,Fj}$	$X_{c1,Fj}$
			Ø			$N_{c2,Fj}$	Ø
Parallel	Parallel	$N_{c1,Fj}$	$X_{c1,Fj}$	$N_{c2,Fj}$	$X_{c2,Fj}$	$\max\{N_{c1,Fj}, N_{c2,Fj}\}$	$X_{c1,Fj} \cup X_{c2,Fj}$
	Series						$X_{c1,Fj} \cap X_{c2,Fj}$

Parallel Operation Mode

For the dual devices in parallel operation mode, if their action outlet is connected in parallel, the actual action result is the union of the action measure set of the first and the second set of devices; If its action outlet is connected in series, the actual action result is the intersection of the action measures set of the first and the second set of devices [26]. See Table 1 for details.

3.6 Practicability Assessment

The practicability of control measures of contingency fault duty strategy is calculated according to the practicability of various class control measures it contains, and the practicability of each class control measures are calculated based on the actual controlled quantity and control-required quantity of the control measures.

When the control-required quantity is 0, it means that it does not require control. When the control-required quantity is greater than 0 and the actual controlled quantity is greater than or equal to the control-required quantity, it means that the control quantity is sufficient. When the control-required quantity is greater than 0 and the actual controlled quantity is less than the control-required quantity, it means that the control quantity is insufficient [17].

4 Application Examples

Taking the system protection of a typical DC receiving-end grid as an example, it is composed of a co-controlled master station, three kinds of control primary stations of DC, pumping and storage and interruptible load, three kind of control substations of DC, pumping and storage and interruptible load, and interruptible load execution stations. The specific configuration is shown in Fig. 3.

Taking DC system A and DC system B bipolar lock control strategy as an example, under the operation condition of a specified power grid, it is assumed that: 1) All the

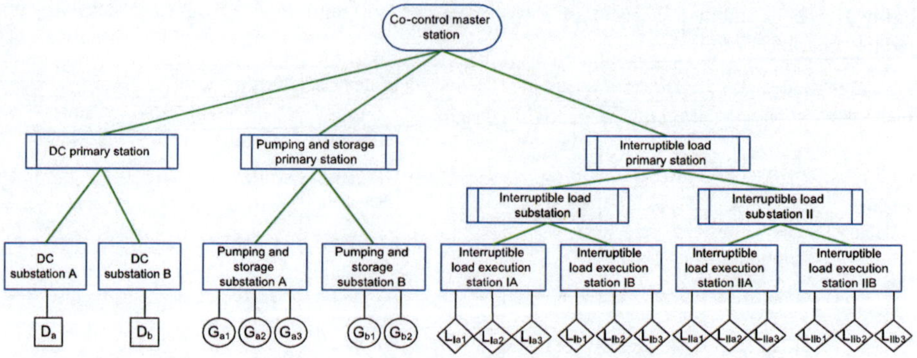

Fig. 3. Configuration diagram of system protection sample

candidate control measures are in the allowable controlling state, and the priority rule is the lesser value first, among which the priority of all control measures are DC modulation < pumped storage unit cutting < interruptible load shedding. The DC modulation priority is A DC < B DC, and the increasable power of each DC is 360 MW and 270 MW respectively. The control strategy of pumped storage unit cutting is first arranged according to the priority of pumped storage plant and then according to the priority of each unit in a single plant; The cutting priority of pumped storage plant is A plant < B plant, the priority of units in a single pumped storage plant is $G_{a1} < G_{a2} < G_{a3}$ and $G_{b1} < G_{b2}$, assume that the units G_{a1}, G_{a2} and G_{b1} are in the pumping state, its operating power is 300 MW. The interruptible load control strategy is firstly distributed according to the controllable quantity of each load control substation in proportion, and then shedding in turn according to the round priority of the execution station, the shedding priority of the load execution station is IA station < IB station and IIA station < IIB station; The shedding priority order of specific load line is $L_{Ia1} < L_{Ib1} < L_{Ia2} < L_{Ib2} < L_{Ia3} < L_{Ib3}$ and $L_{IIa1} < L_{IIb1} < L_{IIa2} < L_{IIb2} < L_{IIa3} < L_{IIb3}$, assume that each power is 20 MW, the load control mode is the minimum undercut, and the other control modes are the minimum overcut. 2) According to the loss power of DC system A and DC system B bipolar lockout fault, the total DC power required to be increased by the duty strategy is calculated as 630 MW, the cutting power of the pumped storage unit is 370 MW, and the interruptible load requires no control; Under the condition that the first and second control devices are in normal operation, the action measures corresponding to the duty strategy in the main-auxiliary operation mode and the parallel operation mode are to increase DC D_1 and D_2, and cut off the pumping storage units G_{a1} and G_{b1}.

It is assumed that there is no information exchange between the two sets of configuration devices in the same station. If the channel between DC master station and DC substation B in the first set of system is abnormal, the D2 DC increase measures of DC substation B in the first set of system will be invalid. If the first system and the second system do not exchange data, the duty strategy in different operation modes is shown in Table 2:

Table 2. Example of practicability assessment of duty strategy for dual devices.

<table>
<tr><th colspan="2" rowspan="2">Operation mode and outlet connection mode</th><th colspan="2">The first set</th><th colspan="2">The second set</th><th colspan="2">Merged result</th></tr>
<tr><th>Candidate measures</th><th>Action measures</th><th>Candidate measures</th><th>Action measures</th><th>Practicability assessment</th><th>Action measures</th></tr>
<tr><td rowspan="2">Main-auxiliary</td><td>First set of main operation</td><td rowspan="4">D_1; G_{a1}, G_{a2}, G_{b1}; L_{Ia1}, L_{Ib1}, …, L_{Ia3}, L_{Ib3}, L_{IIa1}, L_{IIb1}, …, L_{IIa3}, L_{IIb3}</td><td rowspan="4">D_1; G_{a1}, G_{a2}, G_{b1}</td><td rowspan="4">D_1, D_2; G_{a1}, G_{a2}, G_{b1}; L_{Ia1}, L_{Ib1}, …, L_{Ia3}, L_{Ib3}, L_{IIa1}, L_{IIb1}, …, L_{IIa3}, L_{IIb3}</td><td rowspan="4">D_1, D_2,G_{a1}, G_{b1}</td><td>Sufficient</td><td>D_1; G_{a1}, G_{a2}, G_{b1}</td></tr>
<tr><td>Second set of main operation</td><td>Sufficient</td><td>$D1$, D_2; G_{a1}, G_{b1}</td></tr>
<tr><td rowspan="2">Parallel</td><td>Parallel</td><td>Sufficient</td><td>D_1, D_2; G_{a1}, G_{a2}, G_{b1}</td></tr>
<tr><td>Series</td><td>Insufficient</td><td>D_1, G_{a1}, G_{b1}</td></tr>
</table>

Due to the first set of system and the second set of system do not exchange data, there are differences in candidate control measures set and duty strategies between the two sets of systems. The duty strategy of the first set of system includes 360 MW quantity to be increased of DC and 640 MW quantity of pumping storage units to be controlled; the second set of system includes 630 MW quantity to be increased and 370 MW quantity of pumping storage units to be controlled. In the main-auxiliary mode, it will not lead to insufficient controllable quantity of duty strategy. However, in parallel mode, if the outlets are connected in series, the final action measures are to increase the DC D_1 and cut off the pumping storage units G_{a1} and G_{a2}, namely, the total DC power is increased by 360 MW and the power of pumping storage units is removed by 600 MW, which leads to the phenomenon of "under cutting" of the controllable quantity of the duty strategy; In the parallel mode, if the outlets are in parallel, the final action measures are to increase the DC D_1 and D_2, cut off the pumping storage units G_{a1}, G_{a2} and G_{b1}, namely, the total DC power is increased by 630 MW and the power of the pumping storage units is removed by 900 MW, which leads to the phenomenon of "over cutting" of the controllable quantity of the duty strategy.

To avoid the problem of "under-control" or "over-control" problems caused by inconsistencies in input data of the two configuration devices due to single device or channel exception, which further leads to inconsistent duty strategies, in practical engineering applications, the data exchange between the two sets of devices in the same station is adopted to ensure the consistency of decision input data in the two sets of control systems [26, 27]. The abnormal channel of a single set of devices in a station will not lead to the difference of the candidate control measures set between the two systems, thus ensuring the final actual action measures of the two systems are consistent.

5 Conclusion

The system protection and control strategy modeling and duty strategy identification method proposed in this paper comprehensively considers the influence of the operation mode and outlet connection mode of the two sets of devices on the control logic. It is a system protection control strategy simulation method based on the power grid equipment model and its specified operating conditions, which can provide the control-required quantity, actual control object and practicability of all kinds of control measures included in the system protection duty strategy under the specified power grid operation condition. The research results of this paper can provide technical support for online monitoring of operation status and action simulation of UHV AC-DC system protection. It should be pointed out that the method proposed in this paper is only applicable to the identification and early warning of the duty strategy for system protection, and the decision support technology in the scenario where the duty strategy cannot be fully executed needs further study.

Acknowledgements. This work was funded by State Grid Jiangsu Electric Power Co., Ltd. (The Science and Technology Project of "Research on intelligent analysis and control decision-making technology in the whole process of system protection action", Contract No.: SGJS0000DKJS1900572).

References

1. Chen, G., Li, M., Xu, T.: System protection and its key technologies of UHV AC and DC power grid. Autom. Electr. Power Syst. **42**(22), 2–10 (2018)
2. Xu, T., Li, G., Yu, Z., et al.: Design and application of emergency coordination control system for multi-infeed HVDC receiving-end system coping with frequency stability problem. Autom. Electr. Power Syst. **41**(8), 98–104 (2017)
3. Luo, Y., Chen, D., Li, Y., et al.: Design of system protection scheme for North China multi-UHV AC and DC strong coupling large receiving-end power grid. Autom. Electr. Power Syst. **42**(22), 11–18 (2018)
4. Dong, Y., Zhang, X., Yu, R., et al.: Stability control and system protection scheme for power grid with hydropower pool and multi-DC weak sending-end. Autom. Electr. Power Syst. **42** (22), 19–25 (2018)
5. Shao, G., Hou, K., Wang, K., et al.: System protection scheme for over frequency of sending-end power grid caused by ultra-HVDC blocking. Autom. Electr. Power Syst. **42** (22), 26–31 (2018)
6. Wang, Y., Wang, J., Feng, C., et al.: Online monitoring for large power system protection and its optimization strategy. Electr. Power Eng. Technol. **38**(5), 63–70 (2019)
7. Cao, Y., Zhang, H., Xie, Y., et al.: Configuration scheme of emergency load shedding for HVDC receiving-end power grid based on transient low-frequency and high-voltage features. Autom. Electr. Power Syst. **43**(6), 156–162 (2019)
8. Zhou, Y., Liu, F., Chang, H., et al.: Accurate control strategies of hydropower group for frequency emergency control. Autom. Electr. Power Syst. **44**(11), 121–126 (2020)
9. Li, C., Yan, Y., Xiong, J., et al.: Design and application of AC/DC coordinated control system in southwest power grid of China. Autom. Electr. Power Syst. **44**(8), 200–206 (2020)

10. Xu, H., Yao, J., Yu, Y., et al.: Architecture and key technologies of dispatch and control system supporting integrated bulk power grids. Autom. Electr. Power Syst. **42**(6), 1–8 (2018)
11. Shan, X., Wang, Y., Jin, Y., et al.: Scheme and application of integrated monitoring and fault co-disposal technology of UHV interconnected power grid. Autom. Electr. Power Syst. **42** (2), 84–91 (2018)
12. Wu, G., Wang, Y., Yang, Y., et al.: Management station design and development of security-stability control system. Autom. Electr. Power Syst. **36**(10), 118–122 (2012)
13. Xu, T., Li, F., Zhang, J., et al.: Online risk assessment and coordinated decision scheme for emergency load shedding control. Autom. Electr. Power Syst. **39**(20), 91–97 (2015)
14. Qi, Z., Shi, Z., Li, F., et al.: Development and application of the security stability control management system. Power Syst. Prot. Control **44**(1), 122–127 (2016)
15. Wang, F., Shen, F., Li, J., et al.: Design of load shedding risk monitoring function for security and stability control devices in power grid of large city. Autom. Electr. Power Syst. **40**(20), 161–167 (2016)
16. Shen, F., Zhang, W., Zhang, H., et al.: Design and practice of management system for safety and stability control devices adapted to integrated mode of power grid dispatching and control. Autom. Electr. Power Syst. **41**(23), 143–149 (2017)
17. Wang, S., Xu, T., Chen, G., et al.: Practicability assessment method of duty strategy for security and stability control system in power grid. Autom. Electr. Power Syst. **43**(24), 126–133 (2019)
18. Xue, Y.: Space-time cooperative framework for defending blackouts: Part I from isolate defense lines to coordinated defending. Autom. Electr. Power Syst. **30**(1), 8–16 (2006)
19. Fang, Y.: Lessons from September 8, 2011 Southwest America blackout for prevention and control of cascading outages. Autom. Electr. Power Syst. **36**(15), 1–7 (2012)
20. Fang, Y.: Application of emergency control to reduce risk of system collapse triggered by power transmission interface tripping: thinking on the india power blackouts. Autom. Electr. Power Syst. **37**(4), 1–6 (2013)
21. Dong, X., Luo, J., Cui, X., et al.: Whole control time and its constitution of security and stability control system. Autom. Electr. Power Syst. **42**(5), 163–168 (2018)
22. Wu, G., Song, X., Tang, Y., et al.: Modeling of security and stability control system for power system dynamic simulation. Autom. Electr. Power Syst. **36**(3), 71–75 (2012)
23. Shao, W., Zhou, X., Wang, S., et al.: An object-oriented stability control strategy modeling method. Power Syst. Clean Energy **30**(9), 19–25 (2014)
24. Xie, M., Yan, J., Xin, Y., et al.: Application of constraint and the control strategy description language in system analysis. In: International Conference on Power System Technology, Chengdu, China, pp. 1053–1057 (2014)
25. Li, D., Luo, J.: Typical design of security and stability control system for UHVDC transmission. Autom. Electr. Power Syst. **40**(14), 151–157 (2016)
26. Xiao, S.: Security Stability Control Application Technology of Power Grid. China Electric Power Press, Beijing (2011)
27. Li, H., Tang, Y., Li, X., et al.: Analysis of the typical schemes for security and stability control systems of electric power transmission and the devices' main and auxiliary operation configuration principles. Power Syst. Prot. Control **39**(4), 141–145 (2011)

LSTM Network-Based Method for Flexibility Prediction of Aggregated Electric Vehicles in Smart Grid

Huayanran Zhou[1], Yihong Zhou[1], Haijing Zhang[1], Junjie Hu[1](✉), Lars Nordströmd[2], and Guangya Yang[3]

[1] State Key Laboratory of Alternate Electrical Power System with Renewable Energy Sources, North China Electric Power University, Beijing 102206, China
junjiehu@ncepu.edu.cn

[2] Division of Electric Power and Energy Systems, School of Electrical Engineering and Computer Science, KTH Royal Institute of Technology, Stockholm, Sweden

[3] Center for Electric Power and Energy Department of Electrical Engineering, Technical University of Denmark, Kongens Lyngby, Copenhagen, Denmark

Abstract. The flexibility of Demand Response (DR) resources in grid operations has become a valuable solution to respond to the several problems brought about by the growth of intermittent renewable generation. However, the flexibility prediction of the DR resources has not yet been fully addressed in the available literature. This paper trained a long short-term memory (LSTM) recurrent neural network to predict the aggregated flexibility of electric vehicles (EVs). The prediction is based on the historical charging behavior of EVs and the DR signal (DS) which is proposed to facilitate prediction and DR management. Both the size and the maintaining time of the aggregated flexibility can be obtained from the prediction results. The accuracy of the flexibility prediction is verified through the simulation of case studies. The simulation results reveal that the size of flexibility changes under different maintaining time. The proposed flexibility prediction method may be of great assistance for DR management as well as the reserve of the gird.

Keywords: Load flexibility · Demand response (DR) · Electric vehicles (EVs) · Long short-term memory (LSTM) · Deep learning (DL)

1 Introduction

The integration of renewable generation and distributed resources in the power grid tends to introduce problems such as power unbalance between demand and supply. With the evolution of the smart grid and electricity markets, the demand side re-sources are expected to play active roles in addressing the problem [1]. Demand response (DR) has received increased attention in recent years since it can support the power grid operation in a cost-effective way [2, 3].

The DR program usually employs the flexibility of demand response resources (DR resources) to modify their own electricity consumption by offering price compensation

Y. Xue et al. (Eds.): PMF 2020, LNEE 718, pp. 962–974, 2021.
https://doi.org/10.1007/978-981-15-9746-6_72

or economic incentives, which is known to have significant potential for energy and cost savings, as well as load shifting and peak power reserves [4]. Com-pared to conventional measures, the DR resources have the advantages of environment-friendliness and fast-response time, albeit they are typically small and need to be aggregated by an aggregator to participate in system operation.

Up to now, there have been many researches on how to make full use of those DR resources in the power grid. In [5], the peak load shifting effect of aggregated EVs is achieved with the aim of minimizing charge cost. Literature [6] designed the strategy for aggregated residential heating, ventilation, and air conditioning (HVAC) loads to provide regulating reserve service with certain ramping rate. The frequency control strategy of DR for a multi-area power system is particularly designed in [7]. The approach presented in [8] rewards the DR participant to provide the flexibility power in realizing load shifting as well as the voltage improvement in the supply feeder.

The effectiveness of DR largely depends on the flexibility of an aggregated group of DR resources [9]. However, the aggregated flexibility of DR resources is uncertain due to different characteristics of various distributed DR resources. Therefore, many risk control strategies are widely used in the DR operation. Reference [10] takes into account the conditional value at risk when aggregated electric vehicles (EVs) providing reserve service to the gird. A robust optimization based EV scheduling model is established in [11] to develop a conservative frequency regulation strategy. A model predictive control based rolling horizon optimization strategy is proposed in [12] to deal with the deviation between planned and actual dispatching capacity of grid connected EVs.

Compared to the aforementioned strategies [10–12], an accurate flexibility prediction can help the aggregator to deal with the uncertainty better when participating in the grid operation. Yet, few researches regarding flexibility prediction have been carried out. Reference [13] proposed a two artificial neural network (ANN) based model to predict the scale of the DR power from the total load curve, but it cannot give the specific DR flexibility. There is a recurrent neural network (RNN) based flexibility prediction method in [14] which can generate the flexibility of aggregated domestic hot water system (DHWS), and the following day-ahead and real-time load shift scheduling can be carried out based on the flexibility prediction results. However, it relies on knowing the specific load shift signal of the next day in advance. What's more, it could only give the size of the aggregated flexibility, but could not provide the corresponding flexibility maintaining time.

Building forth on the previously developed RNN based flexibility prediction method in [14], in this paper, a long short-term memory (LSTM) recurrent neural network based deep learning method is proposed to predict the aggregated flexibility of aggregated EVs which is a typical DR resources, considering the physical characteristics as well as the user habits. The main contribution of this study is twofold. First, a LSTM network based flexibility prediction method for aggregated EVs is proposed. Second, both the size and maintaining time of the aggregated flexibility can be obtained from the prediction results. The accuracy of the flexibility prediction is verified through the case study. The flexibility prediction results can support DR planning in power grid daily operation.

The remainder of this paper is organized as follows. Section 2 is a preliminaries of our work including the problem description, system framework and some important definitions. Subsequently, the LSTM based prediction method is detailed in Sect. 3. Section 4 presents simulation results and discussions. Finally, the conclusion is given in Sect. 5.

2 Preliminaries

2.1 Problem Description and System Framework

The main aim of this paper is to predict the flexibility of aggregated EVs for the next day by processing realistic electricity consumption data based on the recorded end-users' behavior in the past.

We predict the EV flexibility in the form of aggregation as it reduces the uncertainty of the actual flexibility of the demand side and the flexibility profile of aggregated DR resources is more predictive than the profile of individual customers. An EV aggregators (EVA) are used to aggregate distributed EVs.

$$F_{EVA}(t) = \hat{P}_{EVA}(t) - P_{EVA}(t) \tag{1}$$

The aggregated flexibility of EVA at time step t can be denoted as $F_{EVA}(t)$ which could be derived as (1). Where $P_{EVA}(t)$ represents the aggregated power of EVs at time-step t when not engaging in DR program. $\hat{P}_{EVA}(t)$ represents the aggregated power of EVs at time-step t when engaging in DR program.

The system framework is shown in Fig. 1. It is assumed that the positive direction of power is from the grid to the demand side. The EVA have the ability to transmit instructions i.e., DR signal (see Sect. 2.3) for coordinating and scheduling the power of all the distributed EVs, respectively, and offer attractive tariffs or discounts in the

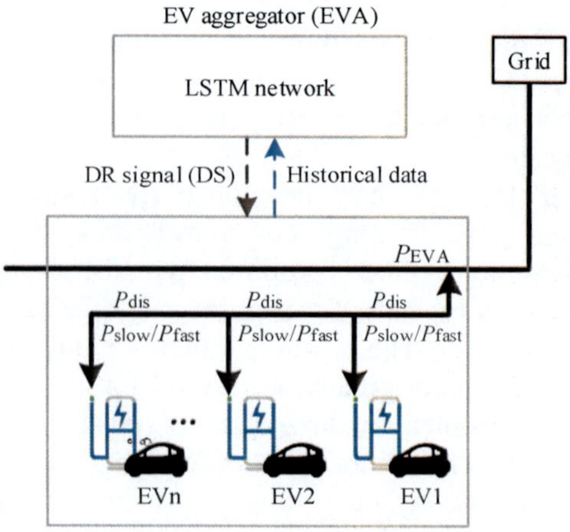

Fig. 1. System framework.

electricity bill to the DR users who shift their loads according to the instructions. There are an LSTM networks in EVA to predict the flexibility of EVs. The prediction details of LSTM network can be seen in Sect. 3.

2.2 EV Operation Model

1. The DR management undertaken by the aggregator must be constrained to the characteristics of each DR resource and the satisfaction of each end-user.

For EV, the end-user's requirement is to get a certain amount of battery state of charge (SOC) when departure as formulated in (6). We consider three charging/discharging modes of each EV, i.e., slow-charging with rated slow-charging power P_{slow}, fast-charging with rated fast-charging power P_{fast} and discharging with rated discharging power P_{dis}. The operation model of EV is as follows:

$$P_{dis} \leq P_{EV}(t) \leq P_{fast} \tag{2}$$

$$E(t+1) = E(t) + \eta P_{EV}(t)\Delta t, \quad if\ t \in \left[t_{start}, t_{dep}\right) \tag{3}$$

$$E(t) = 0, \quad if\ t \notin \left[t_{start}, t_{dep}\right] \tag{4}$$

$$E(t) = E_{start}, \quad if\ t = t_{start} \tag{5}$$

$$E(t) \geq E_{exp}, \quad if\ t = t_{dep} \tag{6}$$

$$E(t) \leq E_{max} \tag{7}$$

$$P_{EV}(t) \geq P_{slow}, \quad if\ E(t) < E_{ms} \tag{8}$$

Where the power at time-step t of EV could be denoted as $P_{EV}(t)$. P_{fast} refers to fast-charging power, $P_{ds} < 0$ refers to discharging power and P_{slow} refers to slow-charging power. $E(t)$ represents the state of charge (SOC) of EV at time step t, η indicates the efficiency, Δt represents the sampling time scale. cc refers to the initial SOC of EV when EV plugs-in at the staring time of charging t_{start}, E_{exp} refers to EV owner's expected SOC at the time of departure t_{dep} and E_{max} refers to the maximum SOC limitation in charging. In response to the sudden departure of EV owners, we set a minimum SOC level E_{ms} to ensure the EV charging demand, so (8) is to guarantee that the EV could only be charged when its energy is lower than that level.

2.3 Instructions in DR Management

To facilitate DR management and flexibility prediction, we introduce demand response signal (DS) to represent the instructions in DR management. The DS is the demand signal sent from the EVA to all end-users to switches their states. Here, we consider three options for the DS as can be seen in Table 1. All the managed EVs would engage in different types of DR depends on the value of DS. To activate the power of EVs to the greatest extent when DS $\neq 0$, and respecting their characteristics and the requirement of end-users when DS = 0, the setting of the responses of EV can also be seen in Table 1. To better show the power consumption and state variations of a single EV under different DS, a detailed illustration can be seen in Fig. 2.

Table 1. Three options of DS and the corresponding reaction of EV

DS value	Demand	EV response
1	Absorb more power from the grid	$P_{EV}(t) = P_{fast}$
−1	Release more power to the grid	$P_{EV}(t) = P_{dis}(<0)$
0	Not engage in DR program	$P_{EV}(t) = P_{slow}$ till the charging requirements are met

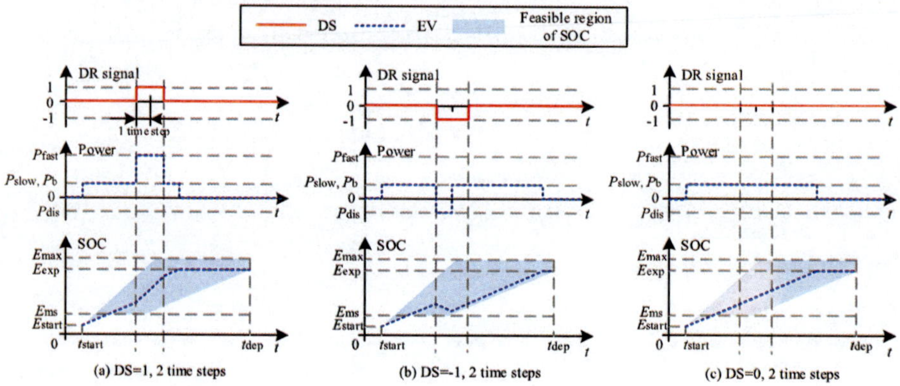

Fig. 2. The power consumption and state variations of a single EV under DS lasting for two time steps; (a) DS = 1; (b) DS = −1; (c) DS = 0.

Please note that, to explore the maximum flexibility, the EVA would send the same DS to all the EVs instead of sending different DS to each one. The reason is explained in Sect. 2.4.

2.4 DS-Based Flexibility Formulation

After the setting of DS, we can notice that, the $\hat{P}_{EVA}(t)$ under DS = 0 in (1) is equal to $P_{EVA}(t)$. So, the (1) can be rewritten as (9).

$$F_{EVA}(t) = \hat{P}_{EVA}(t)\big|_{DS\neq 0} - \hat{P}_{EVA}(t)\big|_{DS=0} \tag{9}$$

Where $\hat{P}_{EVA}(t)\big|_{DS\neq 0}$ represents the aggregated power of EVs under DS ≠ 0, $\hat{P}_{EVA}(t)\big|_{DS=0}$ represents the aggregated power of EVs under DS = 0.

In DR programs, the flexibility provided by DR resources usually need to be maintained for a period of time. In order to derive the flexibility under different maintaining time, we proposed a DS-based calculation method as: if the aggregator wants to derive the flexibility of increasing electricity consumption that can remain at least k time steps, the EVA can send $DS(t) = 1, DS(t+1) = 1, \ldots, DS(t+k) = 1$ to EVs, which means all the EVs would be fully activated from then on. Then $F_{EVA}(t), F_{EVA}(t+1), \ldots, F_{EVA}(t+k)$ can be derived based on the DS signal. After that, the flexibility that can remain at least k time steps F_{EVA}^{k} can be derived by finding the minimum value of $F_{EVA}(t), F_{EVA}(t+1), \ldots, F_{EVA}(t+k)$ as illustrated in (10).

$$F_{EVA}^{k} = \min[F_{EVA}(t), F_{EVA}(t+1), \ldots, F_{EVA}(t+k)] \tag{10}$$

Based on the above analysis and modeling, we can calculate the feasible DR power of EV, and obtain the flexibility of aggregated EVs from the calculated DR power. A detailed illustration of the power consumption and state variations of a single EV under different DS can be seen in Fig. 2, where the action time of DS lasts for two time steps.

According to Eq. (9), the example EV in Fig. 2 can provide scheduling flexibility of $P_{fast} - P_{slow}$ with duration of 2 time steps when DS = 1. Due to the limitation of charging demand of EV users, it can provide scheduling flexibility of $P_{dis} - P_{slow}$ with duration of 1 time step when DS = −1. So, the flexibility of EV maintaining for 2 time steps is 0 According to Eq. (10). It is noted that although the power consumption of EVs will also change after the action of DS, since we only utilize the flexibility data during the time step of DS action for prediction, this would not affect the flexibility prediction results.

3 Methodology and Algorithm Design

3.1 Long Short-Term Memory (LSTM) Neural Network-Based Flexibility Prediction Method

As shown in Eq. (9–10), to predict the flexibility of EVs, we need to predict $\hat{P}_{EVA}(t)$ under DS = 0 and DS ≠ 0. In this study, considering that $\hat{P}_{EVA}(t)$ is related to the previous DSs and power, the prediction of power can be defined as a multi-step-ahead time-series prediction problem [15]. Generally, there are two ways to achieve multi-step-ahead time-series prediction. The first way is to repeat one-time-step-ahead prediction based on the previous prediction at each time-step, which is a recursive way. The second is to create a direct mapping from the previous states of system to multi-step-ahead values. The first way may suffer low accuracy since the error could accumulate in each recursion. Therein, we select the second way to model the problem in this study.

As an updated variant of Recurrent Neural Network (RNN), Long Short-Term Memory (LSTM) is one of the most advanced deep learning architectures for time-series prediction problem [16, 17]. The LSTM network can model temporal correlation of time-series using feedback connections to the internal nodes (i.e., the LSTM units in LSTM layer). The unique LSTM unit structures, including forget gate, input gate and output gate, have significantly improved the performance of LSTM network in multi-step-ahead time-series prediction problem.

3.2 Training Details

In this article, the LSTM based network is composed of an input layer, LSTM layers and fully-connected layers. Before the training process, the hyperparameters including the number of layers, the number of hidden units and the learning criterion must be specified. During the training process, the weights of the network are continuously adjusted according to the error between the network outputs and the target output in training dataset, until the network converges. In this section, we illustrate some training details for EVA.

Inputs and Outputs of the LSTM Network. As mentioned above, the outputs of the network should be $\hat{P}_{EVA}(t)$ under different DS (including DS = 0 and DS $\neq$ 0). Considering that the power of EVA shows similarity between each day and each week, we choose the EVA power one day and seven days before the predicted day, plus DS at the next day to predict the EVA power under different DS at the next day. In addition, there are significant difference between workday and weekend in terms of the power of EVs, so we add a day type indicator *DayType* as one of the inputs. Please note that the aggregated power in the inputs is that did not engage in DR since there would be huge difference among the power under different types of DR. Overall, the prediction can be formulated as (11), where k is the number of time-steps in one day, F represents the mapping ability of LSTM network, and the number 7 means that the data is from a week ago.

$$
\begin{aligned}
&\left[\hat{P}_{EVA}(t+k), \hat{P}_{EVA}(t+k-1), \ldots, \hat{P}_{EVA}(t+1)\right] \\
&= F[P_{EVA}(t-1 \times day+k), P_{EVA}(t-1 \times day+k-1), \ldots, P_{EVA}(t-1 \times day+1) \\
&P_{EVA}(t-7 \times day+k), P_{EVA}(t-7 \times day+k-1), \ldots, P_{EVA}(t-7 \times day+1), \\
&DS(t+k), DS(t+k-1), \ldots DS(t+1), DayType]
\end{aligned} \tag{11}
$$

Network Structure and Hyper-parameters Settings. After deciding the inputs and outputs of the network, we need to specify the hyper-parameters of the network so that it can be trained to make feasible prediction. The overall structure of the LSTM for EVA is given in Fig. 3. There are two LSTM layers with 50 hidden units in both networks. The sequence input layer is used to scale the input values to the range of [0, 1] by min-max normalization to improve accuracy of prediction. The Fully-connected layer is used as the output layer, while the regression layer is used to calculate the Root Mean Squared Error (RMSE), which could reflect the convergence of network when training. We use Adam optimizer to train this network since Adam is considered as a good training algorithm that can be applied in many tasks with its default settings [18]. We set the initial learning rate to 0.01 to reduce the convergence time, and after every

60 epochs, the learning rate would multiple 0.1 to avoid fluctuation so that make the network converge better.

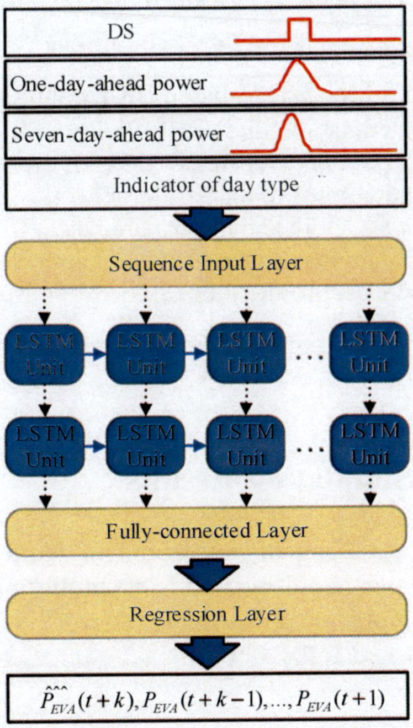

Fig. 3. The overall structure of the LSTM for EVA.

Training Dataset. After setting up the overall structure and hyper-parameters of the network, it can be trained to generate feasible prediction of aggregated power day-ahead under different DS. The training dataset for EVA is introduced as follows. To train the EVA LSTM network, we assume that the EVA could collect the historical information of each EVs including the battery capacity, SOC, the arriving time and the departing time of all the EVs, after which EVA could derive the EV charging/discharging power on each day as given in Eq. (2–8), (9) as well as Table 1 to train the network.

3.3 Prediction

- After training, the LSTM can map inputs and outputs well. Once we input a new set of EVA data to the input layer in the trained EVA LSTM network, the network can generate the corresponding power of EVA under different DSs (including DS = 0 and DS $\neq$ 0) for the next day. Then the flexibility of the next day can be easily obtained by Eq. (9–10).

3.4 Algorithm Summary

Based on the afore-described process, the overall procedure of our proposed LSTM network-based flexibility prediction algorithm is summarized in Algorithm 1. It should be noted that after the LSTM network is trained, the EVA only needs to repeat steps 5 in application.

Algorithm 1: Algorithm for the LSTM-based DR flexibility prediction

1. Prepare an LSTM networks to predict the flexibility of EVA
2. Get historical power dataset under different DS of EVA
2. Specified the inputs and outputs of the LSTM network
3. Set the LSTM network structure and hyper-parameters
4. Train the LSTM and obtain the trained network parameters
5. Input a new set of EVA data to the trained LSTM network, and the flexibility for the next day can be predicted.

4 Case Study and Simulation Results

In this section, we provide a series of simulations to demonstrate the effectiveness of our proposed LSTM network-based method in flexibility prediction of EVs. All simulations were run on a computer with an Intel® Core^{TM} i7-7500U CPU @2.70 GHz–2.90 GHz, 8 GB of RAM and RTX2060.

4.1 Data and Parameters

The case of a residential area with 1000 eV users is studied. The EV parking activities data and parameters come from [19]. The value of some parameters we used are listed in Table 2. We selected the data of 256 weekdays in a year for our research. The first 196 days were selected as the training dataset of the LSTM network, and the 197–226 days were selected as validation dataset to avoid over-fitting of the network. Finally, we randomly select one day from the 227–256 days to test the performance of our proposed algorithm.

Table 2. Some parameters of EVs.

Parameters	Value	Parameters	Value
P_{fast}	9 kW	E_{max}	0.9 × 30 kWh
P_{slow}	3.3 kW	E_{exp}	0.85 × 30 kWh
P_{dis}	−3.3 kW	E_{ms}	0.5 × 30 kWh

The studied horizon is from 12:00 p.m. to 12:00 p.m. the next day. We took 1 time-step as 15 min; thus, we can divide one day into 96 time-steps.

4.2 Convergence Performance of the LSTM Network

Firstly, the learning performance of the LSTM network is evaluated. The loss function of the LSTM network during iterative training is shown in Fig. 4. In this study, the loss function is the RMSE between the network outputs and the target values in the training dataset. Figure 4 shows that over the first 30 epochs, the validation loss is less than training loss, while as the network converges, the training loss becomes less than validation loss. After 200 epochs, the training loss reaches to less than 10^{-4}, and the validation loss does not get lager, so we can conclude that the network converges well and does not suffer overfitting. The overall training process costs only 4 min 16 s to converge, since our proposed network is not so complex. The low convergence time and high convergence performance also prove that our proposed network structure achieve a balance between complexity and performance.

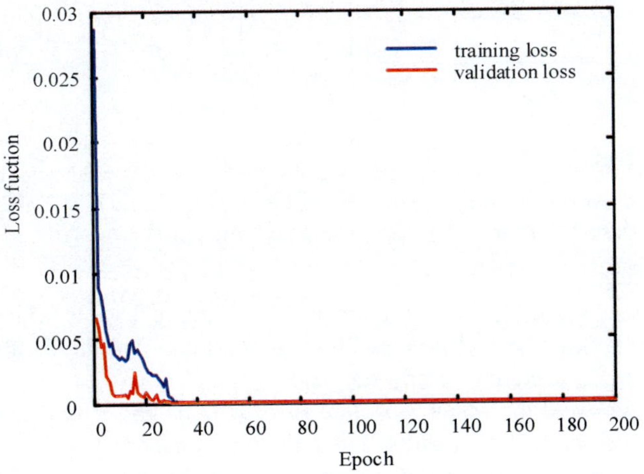

Fig. 4. Loss function of the LSTM network during the training process.

4.3 Flexibility Prediction Results

The performance of the proposed flexibility prediction method is evaluated here. In DR programs, the flexibility provided by DR resources usually needs to be maintained for a period of time. In order to test the prediction performance under different maintaining time, we predict the flexibility of EVA in 6 scenarios: (a) DS = −1 and lasts for 15 min; (b) DS = −1 and lasts for 30 min; (c) DS = −1 and lasts for 1 h; (d) DS = 1 and lasts for 15 min; (d) DS = 1 and lasts for 30 min; (d) DS = 1 and lasts for 1 h. The flexibility prediction results of EVA under 6 studied scenarios are shown in Fig. 5.

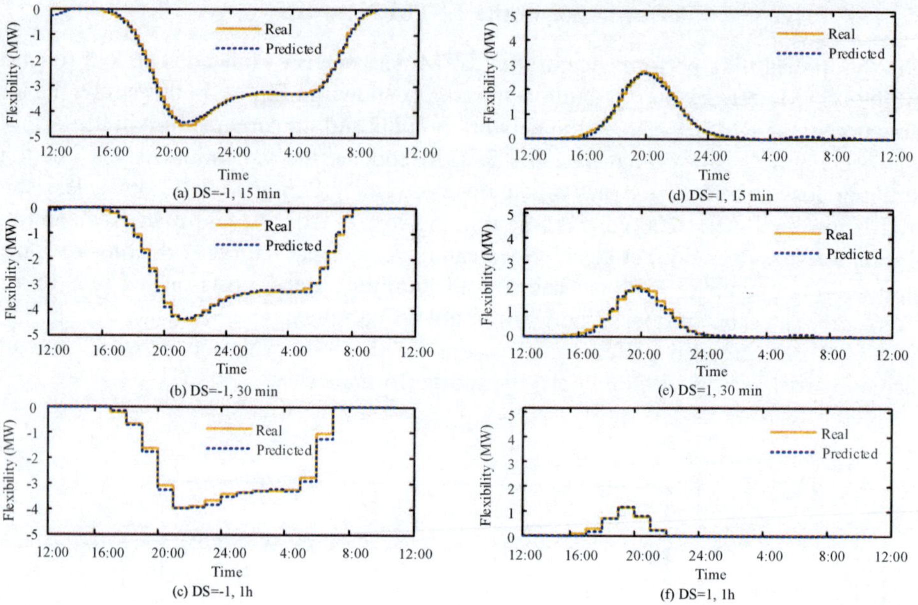

Fig. 5. The flexibility prediction results of EVA under (a) DS = −1 and lasts for 15 min; (b) DS = −1 and lasts for 30 min; (c) DS = −1 and lasts for 1 h; (d) DS = 1 and lasts for 15 min; (e) DS = 1 and lasts for 30 min; (f) DS = 1 and lasts for 1 h.

It can be seen from Fig. 5 that, compared with the real value which is derived by sending DS to EVs at the real time as illustrated in Sect. 2.4, the flexibility prediction results have certain accuracy in any scenario.

We also compared the Mean Absolute Error (MAE) and Root Mean Square Error (RMSE) of EVA prediction results under different scenarios which can be seen in Table 3 and Table 4, respectively. Compared to the magnitude of flexible power, it can be concluded that the predicted flexibility is close to the real value with high accuracy.

Table 3. MAE of the flexibility prediction results.

		MAE under different maintaining time (MW)		
Resources	DS	15 min	30 min	1 h
EVA	1	0.046519	0.019663	0.002512
EVA	−1	0.054673	0.021356	0.01199

In addition, from Fig. 5, it can be concluded that the smaller the maintaining time is, the greater the scheduling flexibility is. The reason is that the EVs are decentralized and with uncertainty, so it is difficult to require them to provide stable scheduling flexibility during a long time. A DR program with smaller maintaining time can activate more potential load shifting ability of EVs.

Table 4. RMSE of the flexibility prediction results.

		RMSE under different maintaining time (MW)		
Resources	DS	15 min	30 min	1 h
EVA	1	0.063376	0.047121	0.011221
EVA	−1	0.080346	0.045909	0.038910

When DS = −1, there is a negative flexibility (the ability to release power to the grid) between 16:00 and 8:00 the next day. When DS = 1, the positive flexibility is distributed between 16:00 and 24:00. When DS = −1, EVA can provide flexibility of about −4.5 mw/15 min at most, while when DS = 1, EVA can provide flexibility of nearly 3 MW/15 min at most. Based on the driving behaviors of electric vehicles in weekdays, when an EV plugs in and is charged over the minimum SOC, it would have a great potential of discharging, but if it is close to the departure time, the negative flexibility would no longer exist to meet the power demand. When DS = 1, due to the fast charging mode, the EV SOC increases rapidly, and due to the limitation of the maximum battery capacity, the EV cannot continue to provide flexibility after reaching the maximum battery capacity.

Based on the prediction, the flexibility power of aggregated EVs can be obtained day-ahead. Through the analysis of flexibility prediction results, the EVA can trade reserve capacity in reserve market, and can also adjust the level of incentives or tariffs offered to DR resources in time to increase involvement of the users in DR actions.

5 Conclusion

To predict both the size and maintaining time of EV flexibility in DR grogram, an LSTM network-based algorithm is proposed in this paper. The prediction is based on the historical power consumption behaviors of EVs and DR signals (DSs) which is proposed to facilitate prediction and DR management. Through the case studies of 1000 aggregated EVs, the accuracy of the flexibility prediction is verified. The results show the characteristics of EV flexibility under end-users' behaviors. The size of flexibility would decrease with the increase of the maintaining time. The flexibility prediction results can support DR planning in power grid daily operation.

Funding Information.. This work was supported by the State Key Laboratory of smart grid protection and operation control open project (SGNR0000KJJS1907535).

References

1. Obaid, Z.A., Cipcigan, L.M., Abrahim, L., et al.: Frequency control of future power systems: reviewing and evaluating challenges and new control methods. J. Mod. Power Syst. Clean Energy **7**, 9–25 (2019)

2. Xue, Y., Yu, X.: Beyond smart grid—cyber–physical–social system in energy future [point of view]. Proc. IEEE **105**(12), 2290–2292 (2017)
3. Chen, Z., Sun, Y., Ai, X., Malik, S.M., Yang, L.: Integrated demand response characteristics of industrial park: a review. J. Mod. Power Syst. Clean Energy **8**(1), 15–26 (2020)
4. Chai, Y., Xiang, Y., Liu, J., et al.: Incentive-based demand response model for maximizing benefits of electricity retailers. J. Mod. Power Syst. Clean Energy **7**, 1644–1650 (2019)
5. Li, Z., Guo, Q., Sun, H., Wang, J.: Storage-like devices in load leveling: complementarity constraints and a new and exact relaxation method. Appl. Energy **151**, 13–22 (2015)
6. Sanandaji, B.M., Vincent, T.L., Poolla, K.: Ramping rate flexibility of residential HVAC loads. IEEE Trans. Sustain. Energy **7**(2), 865–874 (2016)
7. Bao, Y., Li, Y., Wang, B., Hu, M., Chen, P.: Demand response for frequency control of multi-area power system. J. Mod. Power Syst. Clean Energy **5**(1), 20–29 (2017)
8. Vivekananthan, C., Mishra, Y., Ledwich, G., Fangxing, L.: Demand response for residential appliances via customer reward scheme. IEEE Trans. Smart Grid **5**(2), 809–820 (2014)
9. Wu, J., Xue, Y., Xie, D., et al.: Evaluation and simulation analysis of reserve capability for electric vehicles. Autom. Electr. Power Syst. **42**(13), 101–107 (2018)
10. Wang, J., Jia, Y., Mi, Z., et al.: Reserve service strategy of electric vehicles based on double-incentive mechanism. Autom. Electr. Power Syst. **44**(10), 68–76 (2020)
11. Yao, W., Zhao, J., Wen, F., et al.: Frequency regulation strategy for electric vehicles with centralized charging. Autom. Electr. Power Syst. **38**(9), 69–76 (2014)
12. Zhang, B., Xu, G.: Rolling horizon optimization for grid-connected electric vehicles considering demand difference. Autom. Electr. Power Syst. **44**(13), 106–114 (2020)
13. Ponoćko, J., Milanović, J.V.: Forecasting demand flexibility of aggregated residential load using smart meter data. IEEE Trans. Power Syst. **33**(5), 5446–5455 (2018)
14. Paridari, K., Nordström, L.: Flexibility prediction, scheduling and control of aggregated TCLs. Electr. Power Syst. Res. **178** (2020)
15. Bao, Y., Xiong, T., Hu, Z.: Multi-step-ahead time series prediction using multiple-output support vector regression. Neurocomputing **129**, 482–493 (2014)
16. Hochreiter, S., Schmidhuber, J.: Long short-term memory. Neural Comput. **9**(8), 1735–1780 (1997)
17. Gers, F.A., Schmidhuber, J., Cummins, F.: Learning to forget: continual prediction with LSTM. Neural Comput. **12**(10), 2451–2471 (2000)
18. Kingma, D.P., Ba, J.: Adam: a method for stochastic optimization. arXiv:1412.6980 [cs.LG]
19. Luo, Z., Hu, Z., Song, Y., Xu, Z., Lu, H.: Optimal coordination of plug-in electric vehicles in power grids with cost-benefit analysis—part II: a case study in China. IEEE Trans. Power Syst. **28**(4), 3556–3565 (2013)

BESS Configuration for Power Shortage of Regional Power System by Slice Cutting Approach

Qian Xu[1], Jianmin Zhang[2(✉)], Yikai Sun[1], Lijun Zhang[1], and Qianzhi Zhang[3]

[1] Research Institute of Economy & Technology, State Grid Zhejiang Electrical Power Ltd., Hangzhou 310020, China

[2] School of Automation, Hangzhou Dianzi University, Hangzhou 310018, China
zhangjmhzcn@hdu.edu.cn

[3] Department of Electrical and Computer Engineering, Iowa State University, Ames, IA 50011, USA

Abstract. To tackle with the unusual or unexpected regional short-term and long-term power shortages without adding new generating units and power importation from outside, a proposal to coordinate the potential power system resources tapping in that region and equipping the battery energy storage system (BESS) in substations is provided. To ease the computational complexity, a novel fast slicing approach based on power and energy balancing approach has been designed instead of the NP-hard configuration optimization. At first, the peak-regulating power plants are converted into base load operation and their daily operation curve is calculated to obtain a whole-region total daily energy storage operation curve (WESOC). According to the space distance from power shortage load, the substation site selection sequence with available capacity for BESS is worked out; then, the residual load curve (RLC) which is the remaining part of WESOC, is sliced and cut according to this sequence to obtain the corresponding BESS daily operation curve. The RLC is cut in sequence until all available locations are traversed or the RLC becomes impossible for further cut. The proposed method has been verified with a real BESS planning project in Zhejiang province.

Keywords: Battery energy storage system · Power shortage · Power and energy balance approach · Whole-region total energy storage operation daily curve (WESOC) · Residual load curve (RLC) · Slice cutting

1 Introduction

Electrical energy storage has been recognized as core element in the future power and energy system [1–6]. Unlike other energy storage techniques, such as the pumped-hydro plants have strict installation requirements, battery energy storage system (BESS) can be distributed more widely where it is in need in different scenario [5, 6]. BESS has been largely equipped at "variable renewable energy (VRE)" generation side [7, 8], customer side [9–14], transmission network [15–21], distribution network [22–25] for different

Y. Xue et al. (Eds.): PMF 2020, LNEE 718, pp. 975–992, 2021.
https://doi.org/10.1007/978-981-15-9746-6_73

purposes. Because grid connected BESS will change the power flow of power system, the impact of such BESSs should be explored [26–32]. BESS is also considered as an efficient method to tackle with the increasing of valley-to-peak difference [33–35]. As a matter of fact, valley-to-peak difference indicates a spatio-temporal power and energy unbalance between supply and consumption, which can be sorted as one type of shortage of power and energy.

As described in IEEE Std 2030.2.1-2019 [6], basic and advanced functions of BESS in power system include frequency regulation, voltage regulation, emergency back-up/black start, optimization of electric capacity, electric energy time-shift, power quality/electric service reliability improvement. However, in regional transmission network planning or regional power system planning, there are two unusual or unexpected situations of regional power shortage may bring serious problems to power system planners:

(1) The short-term power shortages are caused by the delay of generation plants project construction, and/or delay of outside importation, due to policy change or financial crisis or some nature disaster, etc.
(2) The long-term power shortages are caused by ceasing the expansion planning including conventional & renewable power generations (C&RGs) and transmission/distribution network development, due to policy change or financial crisis or some nature disaster, etc.

In these circumstances, generation and substation resources and transmission/distribution networks are all fixed and there are no any available generation resources or power import resources from outside. Above two power shortage situations are practical problems to power system planners, but no literature has studied so far.

BESS seems a promising solution, e.g., using mobile BESS devices for short-term power shortages and installing stationary BESS for long-term power shortage situation. However, such power shortage belongs to large unbalance of load and generation and can't be met simply by demand response as well as conventional BESS functions, therefore, we provide a proposal to configure new BESSs along with full utilization of existed generation and network resources.

The computational complexity of optimal configuration of BESS in transmission network is mainly caused by the spatial temporal interaction between installation location and multi-period operation of BESSs, so that conventional NP-hard optimization method would face difficulties, thus many measures had been taken to decrease the complexity [15–21]. Most of studies used DC power flow model [15, 17–21]; hierarchical computing or multi-stage model was used [17–19]. In optimization method, linear programming [15, 17], mixed-integer linear programming [17, 20, 22], convex semi-definite program SDP [16], column-and-constraint generation algorithm [17], were used. As for the system scale of case studies, 14 bus [16], 24 bus [21], 30 bus benchmark systems [15, 16], 96 bus [18], 118 bus [15, 21] were studied. ISO New England system was studied but only considered a prospective wind generation portfolio [19].

Compared to a complex conventional NP-hard BESS configuration optimization method which is still impractical in real application, an efficient BESS reconfiguration method based on power and energy balance approach is proposed in this paper to discover the possibility of applying BESS to solve power shortage situations.

The contributions of the paper can be summarized as follows:

(1) A feasible adjustment and potential digging for full utilization of existed power system resources and BESS functions is suggested.
(2) A practical and fast calculation method for BESS configuration is provided and testified by real application.

2 Methodology

2.1 Strategy to Match the Power Shortage

Since all generation and substation resources including networks are already determined, and there are no any available generation resources or power import resources from outside, so that only way out is to full utilization of existed generation power plants as well as substation resources. In this paper, we only consider the BESS site in substation where extra capacity is available to equip a BESS.

Assuming the daily load curve of supply area i with power shortage can be predicted as $P_{gr,i}(t)$ where i is the supply area number, t is hour of a day. Such daily load curve can be divided into two parts, i.e., base load curve, and variable load curve, and let peak regulation power plants (PRPPs) to take the base load, and let BESSs to take the variable load curve.

Above idea is also a scheme to keep the roles of the original resources unchanged as possible, only the conventional peak-regulation power plants are taken to meet the base load, and suitable substations are selected to install BESSs. The maximum powers of a BESS to be equipped in a substation will be limited by the operational margins of that substation, and those BESSs will be dispatched to work with conventional PRPPs to meet short-term or long-term power energy shortages of the considered supply area.

Whole region total daily energy storage operation curve (WESOC) is a demand curve to meet the power shortage. If WESOC is calculated, the next step is to assign that demand load curve to each available BESS sites to have a power and energy configuration. Therefore, the work can be summarized as follows:

Task 1: Obtain the WESOC of the supply area by calculating the power shortage curve and the initial daily peak curve of PRPPs of the supply area.

Task 2: Slice the WESOC of the supply area according to the comprehensive order of the substation power margin and the shortage load distance to configure the corresponding BESS

2.2 Calculation of WESOC

As shown in Fig. 1, the daily regional shortage load curve can be predicted as $P_{r,i}(t)$, where i is the supply area number, t is hour of a day. It is assumed that the original peak load curve operated by conventional PRPPs of the supply area i is $P_{pk,i}(\mathrm{t})$. By adding $P_{r,i}(\mathrm{t})$ and $P_{pk,i}(\mathrm{t})$ together, an equivalent total load curve $P_{gr,i}(\mathrm{t})$ is formed as:

$$P_{gr,i}(\mathrm{t}) = P_{r,i}(\mathrm{t}) + P_{pk,i}(\mathrm{t}) \quad (1)$$

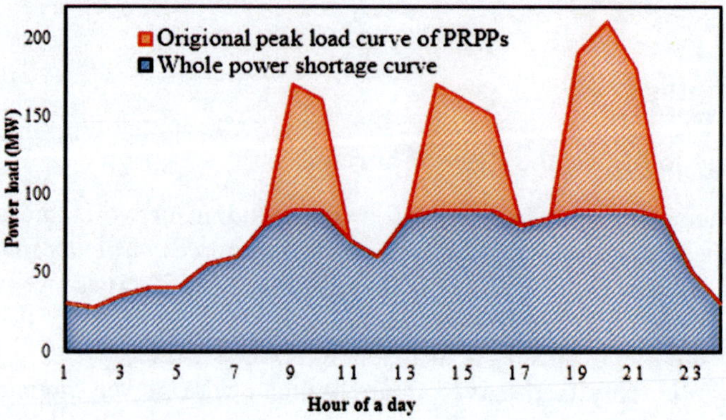

Fig. 1. Equivalent power shortage curve formed by whole-region total power shortage curve plus original peak load curve of PRPPs

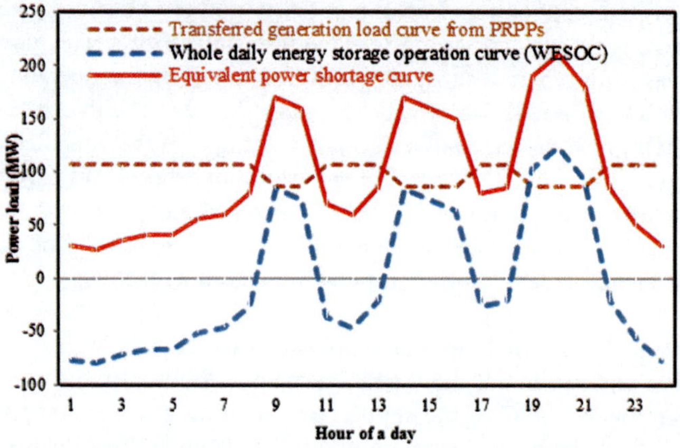

Fig. 2. Decomposition principle for shortage power daily curve

As shown in Fig. 2, if the conventional PRPPs can supply the base load power generation curve $P_{conv,i}(t)$ and the whole-region BESSs can operate as $P_{ess,i}(t)$, it will have:

$$P_{ess,i}(t) = P_{gr,i}(t) - P_{conv,i}(t) \tag{2}$$

To maintain $P_{esse,i}(t)$ the daily regulation, it needs:

$$\sum_{t \in D} \left\{ \frac{\Delta t}{\eta_d} P^{+}_{esse,i} \right\} = \sum_{t \in D} \left\{ \eta_c P^{-}_{esse,i} \Delta t \right\} \tag{3}$$

where η_d and η_c are BESS equivalent efficiency of discharge and charging; $P^{+}_{esse,i}$, $P^{-}_{esse,i}$ are BESS discharge power and charge power, respectively.

As observed from Fig. 2, the base load power generation curve and the original PRPPs can be used to calculate the straight line curve $P^{*}_{conv,i}(t)$:

$$P^{*}_{conv,i}(t) = \frac{1}{T_{day}} \int_{t=0}^{T_{day}} P_{gr,i}(t)dt = P_{const} \tag{4}$$

where T_{day} is the number of hours in a day. If the Eq. (4) cannot meet the requirements of Eq. (3), then the corrected power generation curve $P_{conv,i}(t)$ of PRPPs can be obtained as following equation:

$$P_{conv,i}(t) = \begin{cases} P_{const}, \ when \ P_{gr,i}(t) > P_{const} \\ P_{const} \cdot \eta_d \cdot \eta_c, \ otherwise \end{cases} \tag{5}$$

Therefore, the total BESS operation curve and WESOC can be obtained from Eq. (2). According to above calculations, the original problem will be transformed into the configuration problem of the BESS substation group that satisfies the total BESS operation curve.

2.3 BESS Location with Maximum Power and Sorting

In general, the BESSs are either installed near the substations with extra power capacity or near the load with power shortage. In this work, a priority-based method is proposed to determine the installation locations of BESSs:

(i) Mandatory places: the substations that locate inside the supply area with power shortage will have the highest priority to install BESS.
(ii) Preferred places: the substations close to the supply area with power shortage will have the secondary priority to install BESS.

Assume that in the supply area i, the total number M substations are selected in following maximum power vector:

$$P^{i}_{subavl}(t) = [j, P_{av,j}], j \in [1, M] \tag{6}$$

$$\max\{P_{esse,i}(t)\} < \sum_{j=1}^{M} P_{av,j} \tag{7}$$

where $P_{av,j}$ is the maximum power available in substation for the installation of BESS and j is the priority order to plan for BESS installation.

2.4 Daily Operation Calculation Based on Horizontal Slice Cutting

Figure 3 shows the process of "horizontal slice cutting" for WESOC.

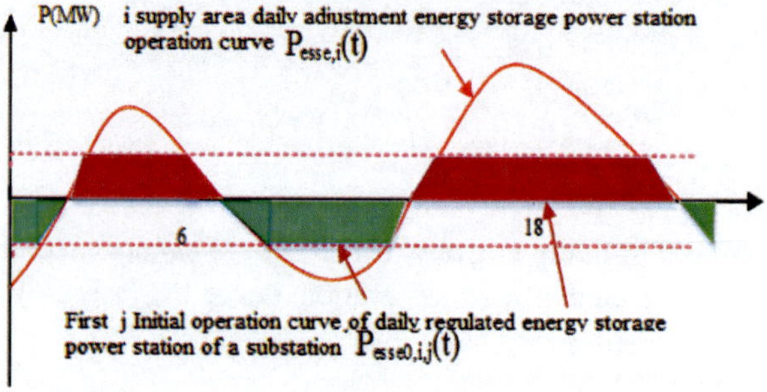

(a) Initial slice cutting with the available maximum power $P_{av,j}$ for BESS operation

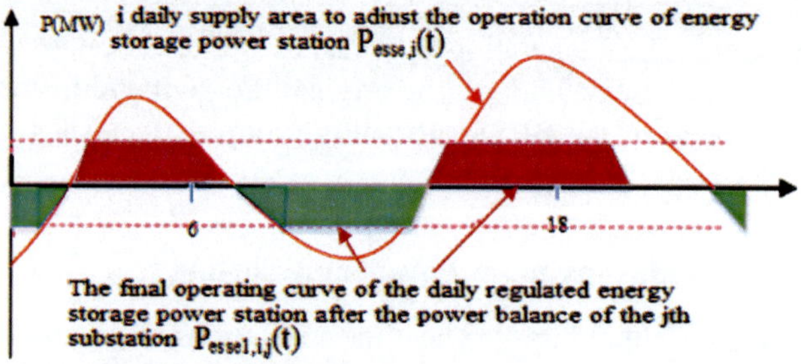

(b) Energy balancing to maintain daily operating characteristics

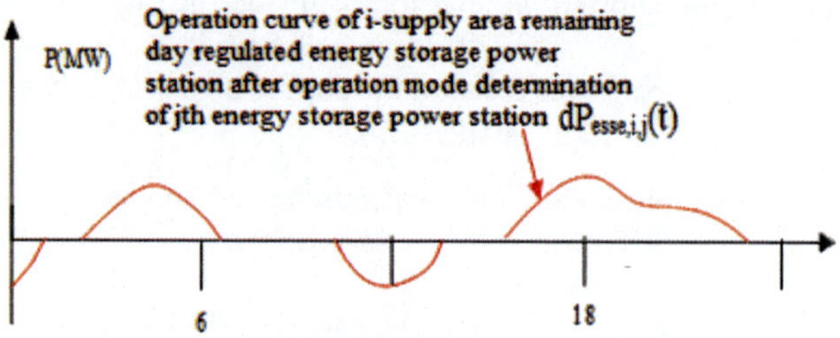

(c) Obtain the new RLC for next turn slice cutting

Fig. 3. Horizon slice cutting

At first, the initial slice cutting is made with the available maximum power $P_{av,j}$ for BESS operation (Fig. 3(a)). Such BESS operation curve can't meet the daily balance of BESS. Thus the operation curve should be revised as shown in Fig. 3(b), where the sequential charge or discharge should be kept to maintain the charging with sufficient space for storing the energy and discharge with sufficient stored energy. The SOC at the designed end of a day will return to the initial SOC state of the designed beginning of a day.

Residual load curve (RLC) is another important curve which is the remaining part of the *(j−1)*-th BESS participating in the apportionment of the WESOC. After the correction as shown in Fig. 3(c), a new RLC will be obtained for next turn slice cutting by the followed BESS site.

The calculation of the RLC $P_{res,i,j-1}(t)$ can be summarized as follows:

(1) When j = 1: the RLC for the first BESS selection is exact the same as WESOC of the supply area, i.e.,

$$P_{res,i,j-1}(t) = P_{esse,i}(t) \tag{8}$$

(2) When j > 1: the RLC for the none-first BESS, i.e., $P_{res,i,j-1}(t)$ is determined by the following formula:

$$P_{res,i,j-1}(t) = P_{esse,i}(t) - \sum_{k=1}^{j-1} P_{essel,i,k}(t) \tag{9}$$

where $P_{essel,i,k}(t)$ is the final operating curves of the *k*-th BESS. The final operating curve of the *j*-th BESS $P_{res,i,j}(t)$ is determined by the following formula:

$$P_{res,i,j}(t) = P_{esse,i}(t) - \sum_{k=1}^{j} P_{essel,i,k}(t) \tag{10}$$

2.5 j-th BESS Operation Mode Based on Horizontal Slice Cutting

The flowchart of horizontal slice cutting algorithm is shown in Fig. 4, the calculation steps are summarized as follows:

(1) Initial operation mode of the *j*-th BESS.
The specific operation mode is determined as:

$$P_{esse0,i,j}(t) = \begin{cases} \eta_d P_{av,i,j}, & \text{when } P_{res,i,j}(t) > 0 \\ 0, & \text{when } P_{res,i,j}(t) = 0 \\ -P_{av,i,j}/\eta_c, & \text{when } P_{res,i,j}(t) < 0 \end{cases} \tag{11}$$

The positive energy $A^+_{i,j}$ d negative energy $A^-_{i,j}$ are calculated as follows, respectively:

$$A^{+}_{i,j} = \int_{t=0}^{Tday} P^{+}_{esse0,i,j}(t)dt \tag{12}$$

$$A^{-}_{i,j} = -\int_{t=0}^{Tday} P^{-}_{esse0,i,j}(t)dt \tag{13}$$

where $P^{+}_{esse0,i,j}(t)$ the positive (charging) power process and $P^{-}_{esse0,i,j}(t)$ the negative (discharge) power process. If $A^{+}_{i,j}$ is not equal to $A^{-}_{i,j}$, it means that the initial run curve does not ensure the day-running characteristics of BESS. In this case, some adjustments are needed here.

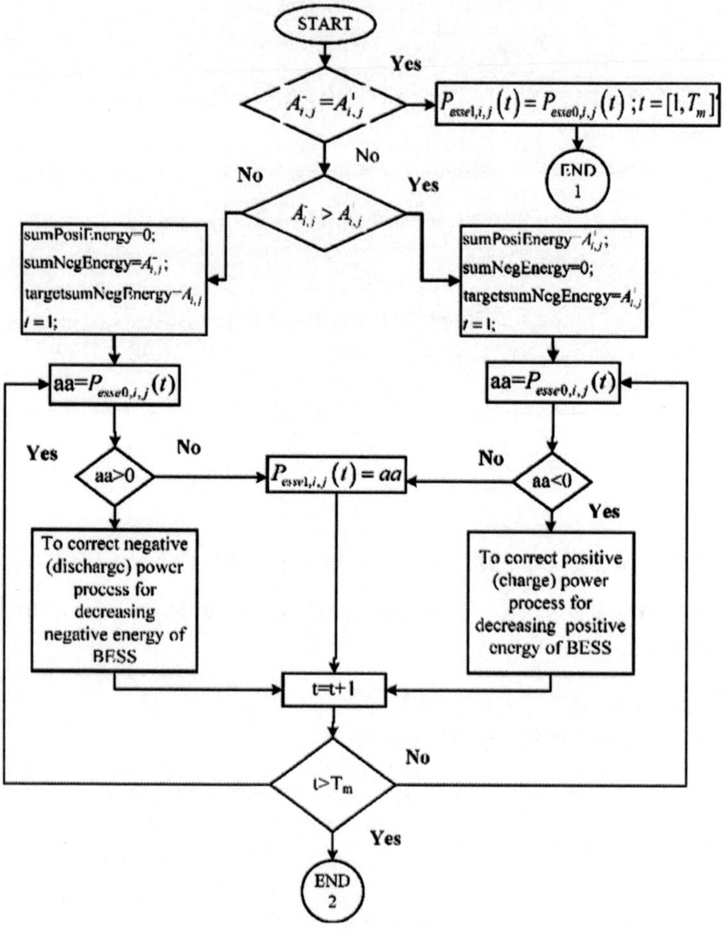

Fig. 4. Flowchart of horizontal slice cutting algorithm

(2) Final operation mode of the *j*-th BESS:

a) If $A_{i,j}^{+} = A_{i,j}^{-}$, then the final operating curve of the *j*-th BESS is $P_{esse0,i,j}^{-}(t)$
b) If $A_{i,j}^{-} > A_{i,j}^{+}$, then the negative (discharge) power process should be adjusted so that the sum of the energy of the negative power process is equal to $A_{i,j}^{+}$.
c) If $A_{i,j}^{-} < A_{i,j}^{+}$, then the positive (charge) power process should be adjusted so that the sum of the energy of the positive power process is equal to $A_{i,j}^{-}$.

(3) Energy rating of *j*-th BESS
Assuming energy rating of j-th BESS is Eng_j and its battery remaining capacity $S_j(t)$ operates between 10% and 90% of rated energy:

$$\begin{cases} S_{\min,j} = 0.1 \cdot Eng_j \\ S_{\max,j} = 0.9 \cdot Eng_j \end{cases} \tag{14}$$

And assume that the energy storage at the beginning of the day and the end of the day $S_{\min}$ is:

$$S_j(t = 0) = S_j(t = T_m) = S_{j,\min} = 0.1Eng_j \tag{15}$$

The equation of battery remaining capacity $S_j(t)$ is:

$$S_j(t) = \begin{cases} S_j(t-1) - \eta_c P_{esse1,i,j}(t)\Delta t, & \text{when } P_s(t) \le 0 \\ S_j(t-1) - \frac{P_{esse1,i,j}(t)\Delta t}{\eta_d}, & \text{when } P_s(t) > 0 \end{cases} \tag{16}$$

where

$$\eta_c = \eta_c = 0.90, \Delta t = 1h \tag{17}$$

Since $p_{esse1,i,j}(t)$ is known, initial value S_j can also be calculated, then rated energy of *j*-th BESS can be got based on (14).

(d) Condition for further slice cutting
The indexes $RES_{i,j}^{+}$ and $RES_{i,j}^{-}$ can be calculated as:

$$RES_{i,j}^{+} = \int_{t=0}^{T_{day}} P_{res,i,j}^{+}(t)dt \tag{18}$$

$$RES_{i,j}^{-} = -\int_{t=0}^{T_{day}} P_{res,i,j}^{-}(t)dt \tag{19}$$

If $RES_{i,j}^{+}$ or $RES_{i,j}^{-}$ is equal to 0 or close to 0, it means that the total residual load curve can no longer be met by BESS, and it is not possible to do further slice cutting for rest of BESSs.

2.6 The Result of the Calculation of the Sorted Substation

The total residual load curve can be calculated as:

$$P_{res,i,j+1}(t) = P_{esse,i}(t) - \sum_{k=1}^{j} P_{esse1,i,k}(t) \tag{20}$$

$$RES^{+}_{i,j+1} = \int_{t=0}^{T_{day}} P^{+}_{res,i,j}(t)dt \tag{21}$$

$$RES^{-}_{i,j+1} = -\int_{t=0}^{T_{day}} P^{-}_{res,i,j}(t)dt \tag{22}$$

If $RES^{+}_{i,j}$ or $RES^{-}_{i,j}$ is greater than 0, it indicates the total selected M substations is insufficient and more candidate substations need to be involved.

2.7 Total Investment Calculation of BESS

The investment of BESS is related to the rated power, rated capacity, technical and economic characteristics of the selected energy storage system, which includes energy storage battery pack, battery management system, energy storage converter and monitoring system. The investment of BESS can be calculated:

$$C_0 = \gamma_p \sum_{j=1}^{M} P_{av,i,j} + \gamma_E \sum_{j=1}^{M} Eng_j \tag{23}$$

where γ_p the power unit cost and γ_E is the energy unit cost.

3 Case Study

In the case study, the test system is a district grid in Zhejiang province, where the initial peak regulation power plants with total capacity 108 MW and the peak load of shortage power is around 107 MW. An equivalent power shortage curve with 215 MW peak demand is formed, as shown in Fig. 5. Those peak regulation power plants will change their roles to take a base load in 108 MW, and a WESOC curve with its maximum power of 107 MW can be also worked out accordingly.

8 substations with total capacity 116 MW are taken as the candidates for BESS installation configuration. In Table 1, the BESSs' available capacities are listed and selected based on the priority-base method and the distances of substations to the loads. The same priority can be exchanged for their preferred order. Therefore, 2 * 3 * 2 * 3 * 2 = 72 optional order sequences can be studied to choose the best configuration of the BESSs.

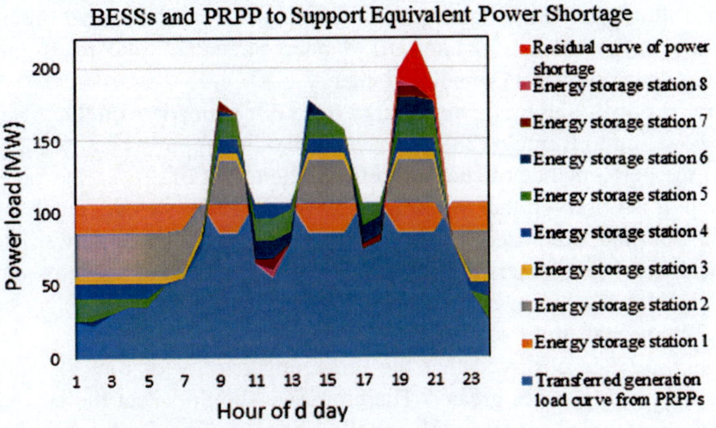

Fig. 5. Area chart of conventional power generation and BESSs of combination 64

To ensure the smooth completion of the slice cutting, 10:00 PM is considered as the starting hour of daily operation. Lead-carbon battery energy storage battery is selected as BESS type with charging and discharging efficiency of 0.9 and the unit price of power is assumed as 8.4 million RMB/MW, the energy unit price is assumed as 900,000 RMB/MWh.

Table 1. Available capacities of 8 sites with priorities in 2 schemes

Priorities	2	2	2	3	3	3	1	1
Available capacities (MW)	10	16	5	15	12	8	30	20

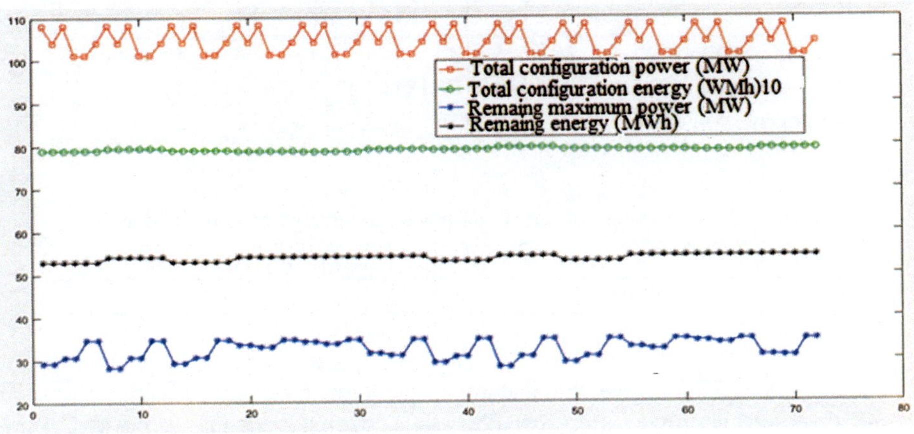

Fig. 6. Four important indexes comparison for 72 optional order sequences in case study

Figure 6 shows four important indexes comparison for 72 optional order sequences in case study, in names as, ① total BESS rated power,② total rated energy, ③remaining maximum power, ④remaining energy.

Total rated power and total rated energy have direct impacts on the investment cost of BESS according to formula (23). Remaining maximum power and remaining energy will reflect the performance of success to fulfill the target of power shortage supply by BESS group. It is observed that the total rated energy is not varying much with the predefined 72 optional order sequences. However, the total rated power is obviously different with total rated energy. Furthermore, it is found that 3 groups of total rated power can be classified in average distribution that means there are many best choices from that 72 optional order sequences.

From Fig. 6, it is observed that the remaining energy of that 72 optional order sequences are varying much greater. Therefore, we should select the best choice from the smaller remaining energy with small remaining maximum power, from the investment smaller group. According to this rule, 64th order is selected as the relative best choice. Figure 5 is an area chart of conventional power generation and BESSs of the selected order, where 215 MW is the peak demand.

Table 2 shows the configuration results of the best capacity selection of the selected order. It is observed that the BESS in front row will have large rated energy with a long time for operation, the daily work hour distribution for the selected order are [2, 6, 8, 16, 18, 22]. It means that no direct relationship with the rated power. However, the BESSs in the front row of selected order will take higher portion in the curve of WESOC, and leave less space for the BESSs in the rear row of selected order. Therefore the work hours of BESS seem a proximate linearly decay as shown in Table 2. It is also observed that only the last one or 8th BESS does not fully utilize the available capacity of that substation, it only uses 4.5 MW out of the available 15 MW.

Table 2. The configuration results of the best capacity selection of the selected order

SN	1	2	3	4	5	6	7	8
P Conf. (MW)	20	30	5	10	16	12	8	4.5(out of 15)
E Conf. (MWh)	224.97	337.4	42.7	76.18	67.47	33.96	4.07	4.5
Hour/Day	22	22	18	16	8	6	2	2

Table 3. The feature of remaining power and energy of combination 64

Posi. power (MW)	Pos. energy (MWh)	Nega. power (MW)	Nega. energy (MWh)
30.405	47.1091	0	0

Table 3 and Fig. 8 give the feature of the remaining power and energy. The residual power shortage after 8 BESS slice cutting remains a positive power 30.405 MW and 47.1091 MWh positive energy, negative power and negative energy are all zero. That means WESOC can't be satisfied in combination 64 after 8 BESS being installed. Because the residual negative power and negative energy being zero,

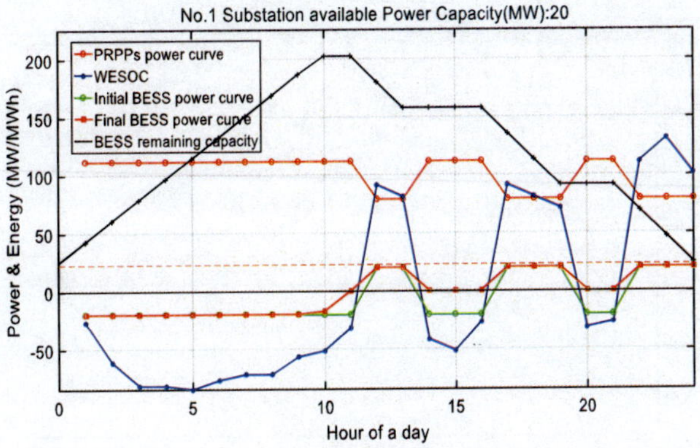

(a) 1st BESS

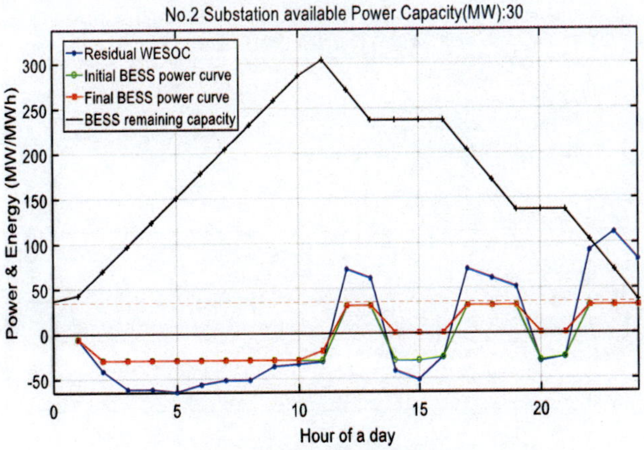

(b) 2rd BESS

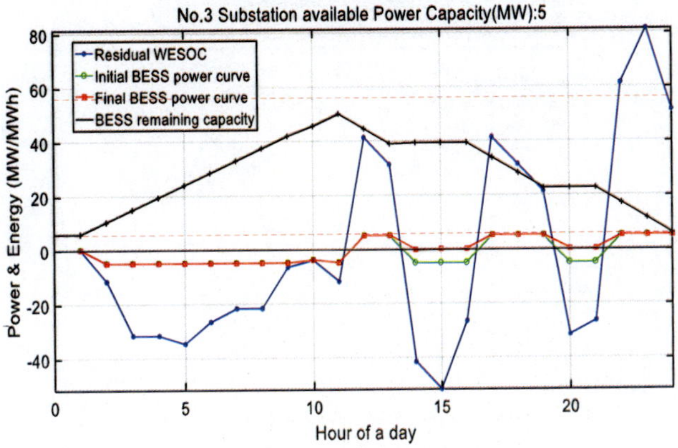

(c) 3rd BESS

Fig. 7. Final operation curves of first eight BESSs

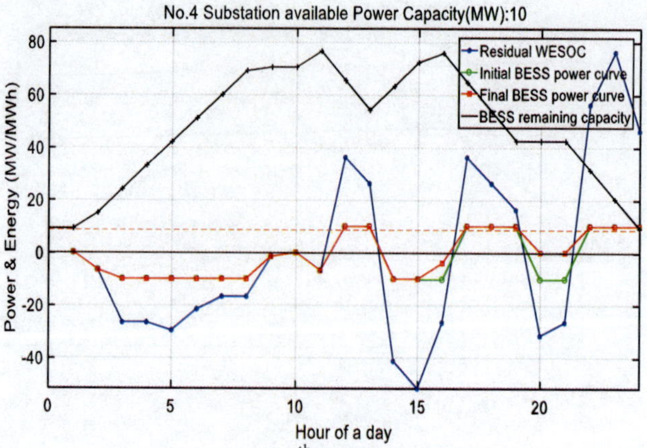

(d) 4th BESS

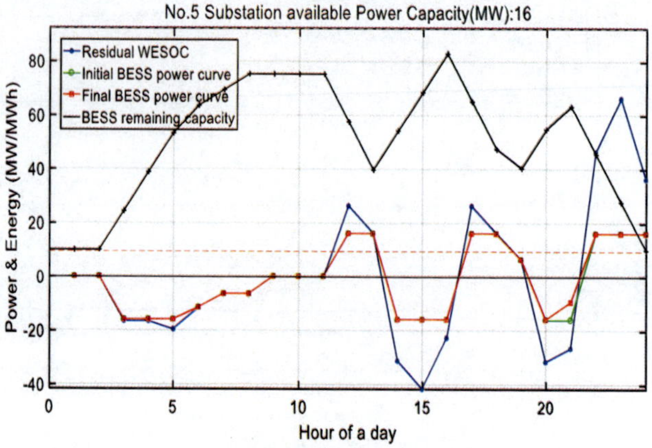

(e) 5th BESS

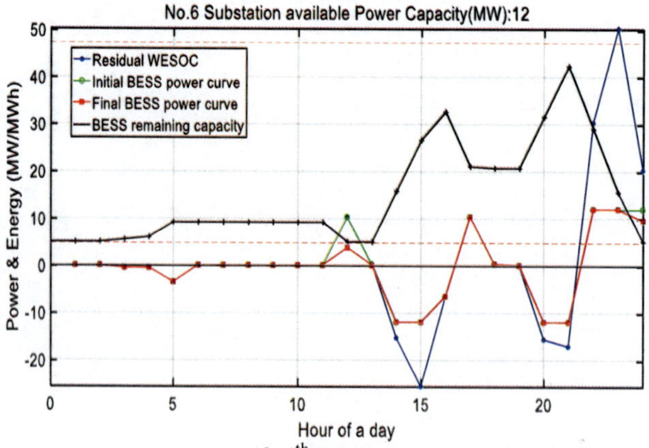

(f) 6th BESS

Fig. 7. (*continued*)

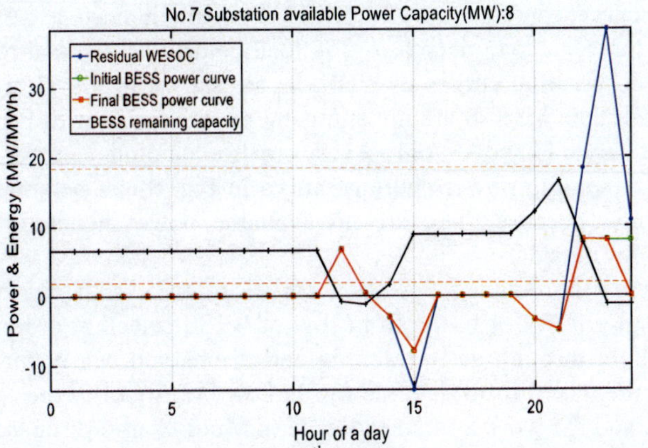

(g) 7th BESS

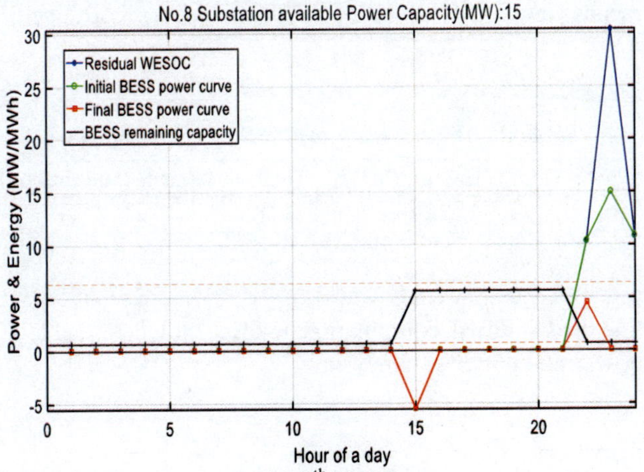

(h) 8th BESS

Fig. 7. ***(continued)***

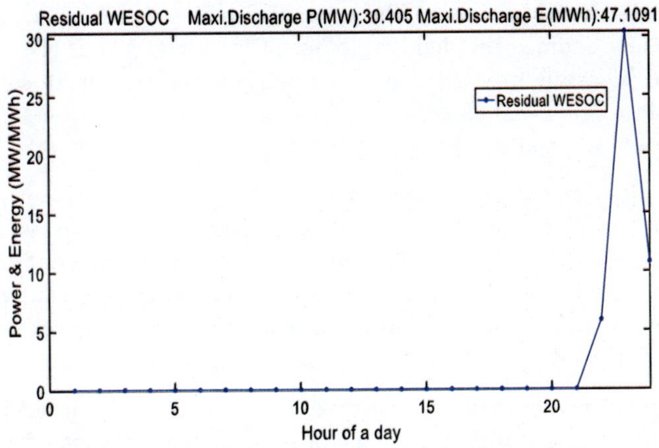

Fig. 8. Remaining residual load curve after 8 BESSs configuration

those residual curve can't be met by BESS element, it need a generation plant to generate such curve, or load decreasing by load control or demand response.

The detailed operation curves of 8 BESSs are shown in Fig. 7 (a)–(h).

It can be seen that 8 BESS are not sufficient to coordinate with PRPPs to cover the power shortage curve. However, only a very small remaining shortage is left, as shown in Fig. 8. The remaining power shortage curve in Fig. 8 can be supplied by demand response measure, because there are no available power generations or no power importations from outside.

Figure 7 gives the detailed operation curves of BESSs. It can be observed that BESSs with high priority or in the front row of BESS selection order will be charged from the late night through early morning and discharged nearly for whole daytime. The charging time will shift to daytime for the low priority or in the rear row of BESS selection order, and the work hours become less. Multi-charge-discharge cycle happens for the middle priority or in the middle row of BESS selection order, like 3rd to 6th BESSs. That means, the BESS operation mode changes according to its place in the selection order.

Table 4. The summary of BESS of combination 64

Power (MW)	Energy (MWh)	Total investment (100 million)
101	786.79	15.565

The investment of optimal combination is also shown in Table 4, which can be used for further techno-economic analysis.

4 Conclusion

A simple and fast power/ energy balance approach of BESS configuration is proposed to tackle the short-term and long-term power shortages of a regional power system, under the conditions of infrequent changing of dispatching of existed power generations and without adding and changing other primary equipment resources. The proposed method is testified by Zhejiang provincial power system to show its satisfied performance in solving real cases.

The proposed method has following merits:

(1) It has a clear and foundational principle, and it can work out the necessary parameters for planning or configuration of BESSs, in names as, the BESS site selection, the BESS power configuration, and BESS energy configuration.
(2) It is useful and practical in real power system planning problem, while the BESS configuration optimization is still impractical to solve real power system planning due to its NP-hard difficulties in numerical computation.

The proposed method has its own disadvantages. Because no grid network constraint model has been introduced, it cannot promise the violation of bus power flow limitation for all nodes of power grid. However, the configuration result from proposed method can provide a good initial solution for the numerical optimization model which can search a much more accurate solution further. In the future, we will develop a hierarchical computing or multi-stage model using DC or AC power flow network modeling to observe and check how to avoid the violation of bus power flow limitation.

Acknowledgement. Thanks for support from the State Grid Zhejiang Electric Power Company Sci. & Tech. Project 5211JY17000N and National Natural Science Foundation of China 51767047.

References

1. Luo, X., Wang, J., Dooner, J.M., Clarke, J.: Overview of current development in electrical energy storage technologies and the application potential in power system operation. Appl. Energy **137**, 511–536 (2015)
2. Li, J., Tian, L., Lai, X.: Outlook of electrical energy storage technologies under energy internet background. Autom. Electric Power Syst. **39**(23), 15–25 (2015)
3. Akhil, A., Huff, G., Currier, A., et al.: DOE/EPRI Electricity Storage Handbook in Collaboration with NRECA. Report Sand (2015)
4. Lawder, M.T., et al.: Battery energy storage system (BESS) and battery management system (BMS) for grid-scale applications. Proc. IEEE **102**(6), 1014–1030 (2014)
5. Qian, X., Sun, Y., Liu, L., Zhang, J., et al.: Review of function and typical application scenario analysis for energy storage station. Zhejiang Electric Power **38**(5), 3–11 (2019)
6. IEEE guide for design, operation, and maintenance of battery energy storage systems, both stationary and mobile, and applications integrated with electric power systems. In: IEEE Std 2030.2.1-2019, pp. 1–45, 13 December (2019). https://doi.org/10.1109/ieeestd.2019.8930450
7. Jiang, Q., Hong, H.: Wavelet-based capacity configuration and coordinated control of hybrid energy storage system for smoothing out wind power fluctuations. IEEE Trans. Power Syst. **28**(2), 1363–1372 (2013)
8. Hessami, M.A., Bowly, D.R.: Economic feasibility and optimization of an energy storage system for Portland wind farm. Appl. Energy **88**, 2755–2763 (2011)
9. Giraud, F., Salameh, Z.M.: Steady-state performance of a grid-connected rooftop hybrid wind–photovoltaic power system with battery storage. IEEE Trans. Energy Conversion **16**(1), 1–7 (2001)
10. Jesse, M., Gantz, S., Massoud, A., Anthony, M.G.: Optimal capacity partitioning of multi-use customer-premise energy storage systems. IEEE Trans. Smart Grid **5**(3), 1292–1299 (2014)
11. Liu, J., Zhang, N., Kang, C., et al.: Cloud energy storage for residential and small commercial consumers: a business case study. Appl. Energy **188**, 226–236 (2017)
12. Choi, S., Min, S.-W.: Optimal scheduling and operation of the ESS for prosumer market environment in grid-connected industrial complex. IEEE Trans. Ind. Appl. **54**(3), 1949–1957 (2018)
13. Hu, X., Xu, G., Shang, C., et al.: Joint planning of battery energy storage and demand response for industrial park participating in peak shaving. Autom. Electric Power Syst. **43**(15), 116–123 (2019)
14. Sun, W., Xiang, W., Pei, L., et al.: Generalized energy storage control strategies on user side in power ancillary service market. Autom. Electric Power Syst. **44**(2), 68–75 (2020)
15. Oh, H.: Optimal planning to include storage devices in power systems. IEEE Trans. Power Syst. **26**(3), 1118–1128 (2011)

16. Gayme, D., Topcu, U.: Optimal power flow with large-scale storage integration. IEEE Trans. Power Syst. **28**(2), 709–717 (2013)
17. Jabr, R.A., Dzafic, I., Pal, B.C.: Robust optimization of storage investment on transmission networks. IEEE Trans. Power Syst. **30**(1), 531–539 (2015)
18. Pandzic, H., Wang, Y., Qiu, T., Dvorkin, Y., Kirschen, D.S.: Near-optimal method for siting and sizing of distributed storage in a transmission network. IEEE Trans. Power Syst. **30**(5), 2288–2300 (2015)
19. Dvorkin, Y., Fernandez-Blanco, R., Kirschen, S., et al.: Ensuring profitability of energy storage. IEEE Trans. Power Syst. **32**(1), 611–623 (2016)
20. Blanco, R.F., Dvorkin, Y., Xu, B., et al.: Optimal energy storage siting and sizing: a WECC case study. IEEE Trans. Sustain. Energy **8**(2) (2017)
21. Zhang, X., Conejo, A.J.: Coordinated investment in transmission and storage systems representing long- and short-term uncertainty. IEEE Trans. Power Syst. **33**(6), 7143–7151 (2018)
22. Awad, A.S.A., El-fouly, T.H.M., Salama, M.M.A.: Optimal ESS allocation for benefit maximization in distribution networks. IEEE Trans. Smart Grid **8**(4), 1668–1678 (2017)
23. Xu, B., Wang, Y., Dvorkin, Y., et al.: Scalable planning for energy storage in energy and reserve markets. IEEE Trans. Power Syst. **32**(6), 4515–4527 (2017)
24. Jia, Z., Zhang, F., Ding, L.: Optimal allocation strategy of energy storage in distribution network considering power four-quadrant output. Autom. Electric Power Syst. **44**(2), 104–112 (2020)
25. Wu, X., Liu, Z., Tian, L., et al.: Energy storage device locating and sizing for distribution network based on improved multi-objective particle swarm optimizer. Power Syst. Technol. **38**(12), 3405–3411 (2014)
26. Oudalov, A., Chartouni, D., Ohler, C.: Value analysis of battery energy storage applications in power systems. In: IEEE Power System Conference and Exposition, Atlanta, Georgia, USA, pp. 2206–2211 (2006)
27. Pudjianto, D., Aunedi, M.: Whole-systems assessment of the value of energy storage in low-carbon electricity systems. IEEE Trans. Smart Grid **5**(2), 1098–1109 (2014)
28. Sperstad, I.B., Korpås, A.H.M.: Valuation of stored energy in dynamic optimal power flow of distribution systems with energy storage. In: 2016 International Conference on Probabilistic Methods Applied to Power Systems (PMAPS) (2016)
29. Sidhu, A.S., Pollitt, M.G., Anaya, K.L.: A social cost benefit analysis of grid-scale electrical energy storage projects: evaluating the smarter network storage project. https://www.eprg.group.cam.ac.uk/wp-content/uploads/2017/06/1710-Text.pdf. Accessed 11 June 2017
30. Ma, Y., Verbič, G., Chapman, A.C.: Estimating the option value of grid-scale battery systems to distribution network service providers. In: 2019 IEEE Milan PowerTech, Milan, Italy (2019)
31. Sun, W., Pei, L., Xiang, W., et al.: Method of system value for energy storage in power system. Autom. Electric Power Syst. **43**(8), 47–55 (2019)
32. Li, J., Niu, M., Wang, S., et al.: Operation and control analysis of 100 MW class battery energy storage station on grid side in Jiangsu power grid of China. Autom. Electric Power Syst. **44**(2), 28–35 (2020)
33. Yang, R.: Nine reviews: technology route of mass energy storage for the peak load regulation of power grid in China. Power Syst. Clean Energy **28**(4), 1–12 (2012)
34. Xiu, X., Li, J., Hui, D.: Capacity configuration and economic evaluation of energy storage system for grid peak load shifting. Electric Power Constr. **34**(2), 1–5 (2013)
35. Li, J., Wang, S.: Optimal combined peak-shaving scheme using energy storage for auxiliary considering both technology and economy. Autom. Electric Power Syst. **41**(9), 44–50 (2017)

Design and Auto-Generation of Full Phase Wiring Diagram for Situational Awareness of Household PV Integrated Low-Voltage Distributed Transformer Service Area

Li Yao[1], Jiangming Zhang[1], Jianmin Zhang[2](✉), Lina Ni[1], and Yan Xia[2]

[1] Electric Power Research Institute of State Grid Zhejiang Electric Power Co., Ltd., Hangzhou 310014, China
[2] School of Automation, Hangzhou Dianzi University, Hangzhou 310018, China
zhangjmhzcn@hdu.edu.cn

Abstract. Due to single phase household photovoltaic (HPV) being gradually and widely applied, corresponding distributed transformer service area becomes an active distribution network, thus it is necessary to have a good on-line visualized monitoring. In addition to the basic function of electricity metering and billing, smart meters can also be used as sensors to realize the comprehensive perception and situation visualization of power metering quality, power supply quality, line loss, etc., thus forming a modern supervision means for the distributed transformer service area. Based on the wiring characteristics of the low-voltage distribution network, the design and automatic generation of the full phase wiring diagram of the low-voltage distribution network for situational awareness is proposed, so as to serve as the base map of the situational awareness of the distributed transformer service area. The diagram design and auto-generation software for voltage situation awareness visualization has been verified by practical applications and demonstrated the good situation awareness performance.

Keywords: Single phase household photovoltaic · On-line visualized monitoring · Distribution transformer service area · Full phase wiring diagram · Situational awareness visualization · Diagram auto-generation

1 Introduction

Smart meters equipped in the distribution transformer service area (DTSA) not only do the basic functions of electricity metering and billing, but also do as sensors which can be used to monitoring the quality of electric energy metering, power supply quality, distribution network comprehensive economic operation evaluation, etc. Moreover, a dynamic visualization display based the wiring diagram of low voltage distribution network (LVDN) by rendering with data collected from smart meters, can form a modern situation awareness visualization platform for the DTSA of LVDN [1, 2].

The China State Grid Corporation Sci. & Tech. Project 5600-201919168A-0-0-00 and National Natural Science Foundation of China 51767047.

Y. Xue et al. (Eds.): PMF 2020, LNEE 718, pp. 993–1006, 2021.
https://doi.org/10.1007/978-981-15-9746-6_74

The LVDN at DTSA has unique operation characteristics, and its power users can be divided into three-phase users and single-phase users. Even for three-phase users, there are single-phase electrical appliances in the house, therefore, three-phase unbalance is the basic mode of LVDN; unbalance operation of LVDB will increase line loss several times as much as the line loss in symmetrical operation, resulting a total power loss 50%–60% of whole distribution network [3, 4]. The main line of LVDN adopts three-phase four-wire system, and there are a large number of repeated grounding devices in neutral line [5, 6]; the measurement of LVDN mainly comes from the smart meters equipped in the low-voltage side of distribution power transformer and in the entry of the end user, so that the state estimation and power flow calculation has its special considerations or features [7–10]. The situation perception and situation visualization based on the phase-separated single-line diagram of LVDN is necessary for situation awareness visualization of LVDN.

Household photovoltaic (HPV) has been gradually and widely applied in LVDN. Due to its small capacity, HPV is connected in a single-phase mode, so it will aggravate the three-phase imbalance and has an impact on the power quality such as over voltage [11]. In addition, unbalanced regulation devices has been widely used in LVDN with different types and control mode [11–15], which is also necessary to perceive their effects and have a good visualization monitoring means.

The situation perception of medium and high voltage distribution network based on intelligent side has been relatively fully studied [16–20]; literature [17] proposed a substation-centered distribution network (SCDN) concept; the proposed SCDN single line diagram presents a SCDN feeder radiation tree layout with 10 kV busbar of high voltage substation as the graph center, which has better situation awareness ability than horizontal vertical single line diagram [17]; literature [18–20] have studied SCDN in a detailed way, especially for the automatic generation of single line diagram of SCDN. However, the situation awareness of LVDN, especial for the design and auto-generation of single-line diagram of LVDN to support a good situation awareness has rarely been studied, so that this paper fills in this blank.

2 Single Line Diagram Modeling for Situation Visualization

2.1 Features of LVDN Wiring in DTSA

The wiring of the LVDN in DTSA has the following characteristics:

(1) The main line is laid with three-phase four-wire, and branch boxes are set in the middle to lead out the three-phase four-wire branches, or lead out the three-phase four-wire user's entry line, or lead out the some single phase line together with neutral line to form the single-phase branches, or lead out the single-phase user's entry line.

(2) Local centralized meter boxes with user meter group and a meter data collector are generally installed at the end branch boxes.

Normally there is no interconnection switch, therefore the topology of LVDN is completely radial, which is far simpler than that of medium-voltage distribution network [18–20]. However, the unbalance operation caused by topology asymmetry as well as load asymmetry are the core feature of LVDN, therefore, the single-line diagram of LVDN need to have a full phase wiring form.

2.2 Operation Characteristics of LVDN

The LVDN in the DTSA has two types of symmetry and asymmetry:

(1) The symmetry and asymmetry of the phase wiring
According to the power supply business rules, the user adopts single-phase 220 V power supply or 380 V power supply according to the total capacity of the electric appliance. Therefore, the user's entry line may be single-phase two-line or three-phase four-line.
(2) Symmetry and asymmetry of phase load

Even if the electrical supply entry system to the user's place is using three-phase four-wire, the unbalance of phase load is still absolute because most of the user's electrical appliance is a single-phase load.

2.3 Single Line Layout Design

(1) Display method of phase single-line diagram in a parallel sector shape
Figure 1 shows one display mode of different phase being arranged in a parallel sector shape, so we can observe the common connection part, e.g., the main trunk and entry line for the three phase users; the different connection part will be observed for single phase line branch and entry line for the single phase users.

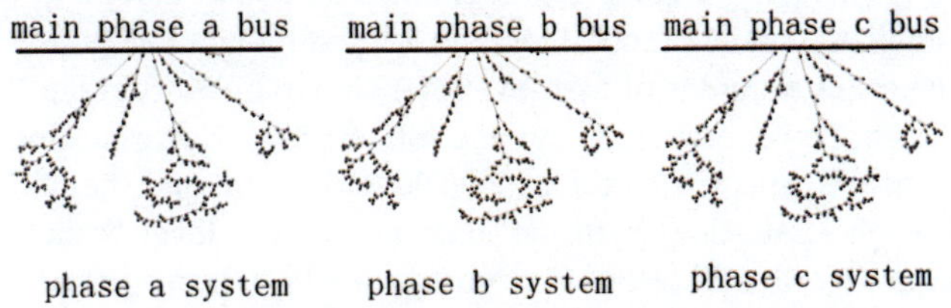

Fig. 1. Display mode in parallel sector shape

(2) Display method of phase single-line diagram in a circle sector shape

Figure 2 shows other display mode of different phase being arranged in a circle sector shape. It has the same advantages as the parallel sector shape (Fig. 1). However, the rendering performance of Fig. 2 will better than that of Fig. 1.

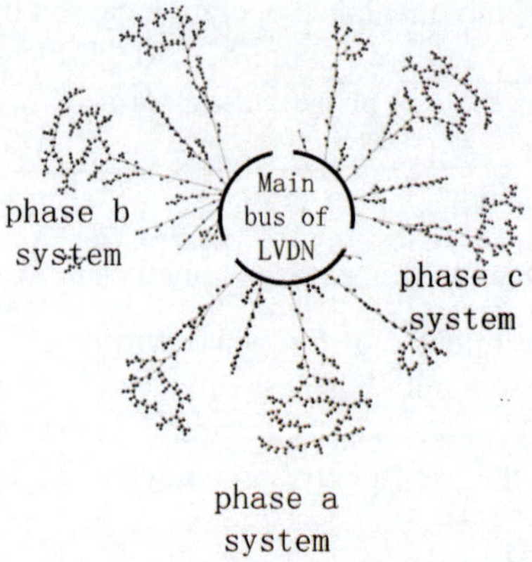

Fig. 2. Display mode in circle sector shape

3 Principle and Algorithm of Auto-Generation of Single-Line Diagram in Straight-Line

The three-phase four-wire bundle separated from the low-voltage bus of distribution transformer is called the low-voltage outlet wire bundle. Suppose there is M bundles, and its sequence is denoted by m, namely m ∈ [1, M]; its three-phase sub-bundle is assumed as a_m, b_m and c_m corresponding to the outgoing line m bundle.

3.1 Automatic Composition of Single-Phase Single Beam

As shown in Fig. 3, taking bound a_m as example. First to carry out static topology analysis and establish a branch model; i.e., the longest branch line is selected as the first level branch, the order of nodes of first level branch is ranged from the low-voltage bus to the end of that first level branch; again, the longest branch line starting from the node of first level branch is selected as the second level branch, and the node order of that second level branch is starting from the node at the first level branch; likewise, line grade table and node order table can be formed. See Fig. 3, the wide line in horizontal is the first level branch, the second level branches are the straight lines whose starting point are the nodes in that horizontal line.

The low-voltage bus of the distribution transformer is set as a start point, whose coordinate is (0,0). And set the initial distance between nodes, such as *dis* = 200; the first level branch line is fixed in the positive direction of X-axis. In order to make the sector formed by the feeder smaller, starting from the origin (0,0), the first node is selected to be 5 times dis from the origin. Subsequently, the abscissa of each node accumulates *dis* in sequence, while the ordinate is always 0.

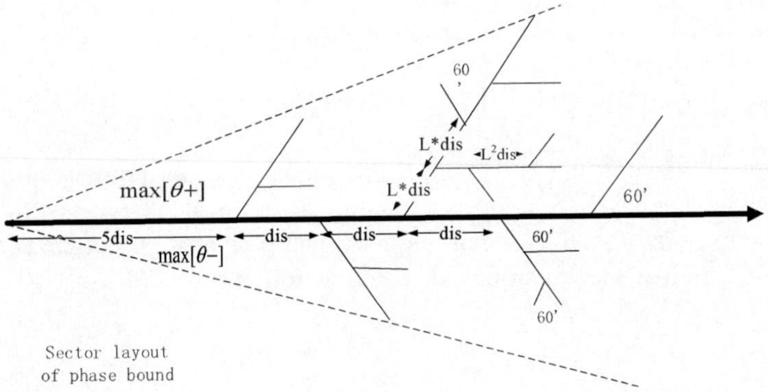

Fig. 3. Composition principle of horizontal single feeder and the formed sector

Coordinates of nodes of second-level branch lines is designed as: from the coordinates of nodes of first-level branch lines, to get the coordinates (x_i, y_i) as the starting nodes, then the coordinates of j-th node (x_{ij}, y_{ij}) of the i-th second-level branch can be calculated as:

$$\begin{aligned} x_{ij} &= x_{i(j-1)} + l \times \text{dis} \times \cos(k \times \pi/3) \\ y_{ij} &= y_{i(j-1)} + l \times \text{dis} \times \sin(k \times \pi/3) \end{aligned} \quad (1)$$

where k is calculated as:

$$k = \begin{cases} 1; & \textit{when i is odd number} \\ -1; & \textit{otherwise} \end{cases} \quad (2)$$

and l is assumed as:

$$l = 0.8 \quad (3)$$

Coordinates of nodes of three-level branch lines and branches above three levels: from the node coordinates of the higher-level branch lines to get the starting node coordinates (x_i, y_i) of the up level branch line, the coordinates of j-th node (x_{ij}, y_{ij}) of the i-th second-level branch can be calculated as:

$$\begin{aligned} x_{ij} &= x_{i(j-1)} + l^2 \times \text{dis} \times \cos(k \times \pi/3) \\ y_{ij} &= y_{i(j-1)} + l^2 \times \text{dis} \times \sin(k \times \pi/3) \end{aligned} \quad (4)$$

3.2 Splicing Calculation of the Whole Map Based on Single-Phase Single Line Beam Sector

The single-phase wiring of all phase bundles will be calculated according to the above node coordinates, and the polar coordinates (r_{ik}, θ_{ik}) of all nodes can be calculated as follows:

$$\begin{cases} (x_{ik}, y_{ik}) = r_{ik}e^{j\theta_{ik}} \\ r_{ik} = \sqrt{x_{ik}^2 + y_{ik}^2} \\ \theta_{ik} = \arctan(\frac{y_{ik}}{x_{ik}}) \end{cases} \tag{5}$$

As shown in Fig. 3, the horizontal single feeder diagram presents a regular hierarchical branch line layout and forms a measurable and scalable sector. The maximum and minimum angle $[\theta_{mi}^{+}, \theta_{mi}^{-}]$, sector angle ϑ_{i0} and pole angle θ_{ci} of the center line of single-phase original sector can be calculated as follows:

$$\begin{cases} \theta_{mi}^{+} = \max(\theta_{ik,+}) \\ \theta_{mi}^{-} = \max(-\theta_{ik,-}) \end{cases} \tag{6}$$

$$\vartheta_{i0} = \theta_{mi}^{+} + \theta_{mi}^{-} \tag{7}$$

$$\vartheta_{ci} = \theta_{mi}^{+} - \theta_{mi}^{-} \tag{8}$$

where, $[\theta_{ik,+}, \theta_{ik,-}]$ are the pole angle of the I and V quadrant nodes.

As shown in Fig. 4, to scale and rotate the sector, and finally form the overall layout; the sector of each phase is 120°, then the scaling factor $\delta_{\theta i}$ of the original single-phase sector, the polar angle $\theta_{ci}^{'}$ of the center line of the sector after scaling can be calculated as:

$$\delta_{\theta i} = \frac{120 - \vartheta_{i0}}{\vartheta_{i0}} \tag{9}$$

$$\theta_{ci}^{'} = \theta_{ci} \tag{10}$$

Since scaling is only performed on the polar angle, the polar diameter is kept unchanged and the polar angle of the center line is kept unchanged; Assuming the polar coordinates of each node after scaling are $(r_{ik}^{'}, \theta_{ik}^{'})$, then:

$$\theta_{ik}^{'} = \begin{cases} \theta_{ik} - \delta_{\theta i} & when\{\theta_{ik} - (\theta_{ik} + \theta_{ik})/2\} \geq 0 \\ \theta_{ik} + \delta_{\theta i} & \text{otherwise} \end{cases} \tag{11}$$

$$r_{ik}^{'} = r_{ik} \tag{12}$$

Calculation of sector splicing can be done after above process. Assuming that the scaled n sectors are arranged one by one counterclockwise from the X-axis in order, $(r_{ik}^{''}, \theta_{ik}^{''})$ are the pole diameter and angle of the *k*-th node of the *i*-th phase feeder, and $(x_{ik}^{''}, y_{ik}^{''})$ are its Cartesian coordinates, then the angle $\vartheta_{ci}^{''}$ of the center line of the final *i*-th phase feeder can be calculated as:

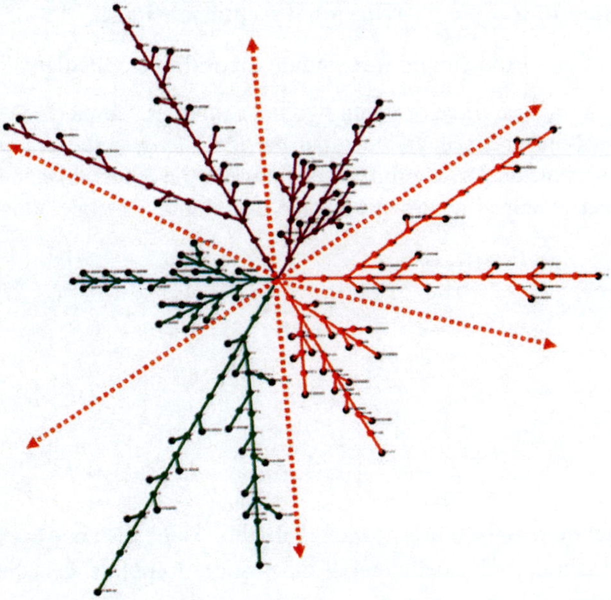

Fig. 4. A complete image composed of the sector-shape of each phase line bundle

$$\begin{cases} \vartheta''_{c1} = 0 \\ \vartheta'_{ci} = \sum_{j=1}^{i} \left(\frac{1}{2}\vartheta_{js} + \theta_{r0}\right) \quad i \geq 2 \end{cases} \tag{13}$$

Furthermore, the Cartesian coordinates of the 3-phase feeder on the whole plane can be calculated according to the following formula:

$$\begin{cases} r''_{ik} = r'_{ik} \\ \theta''_{ik} = \theta'_{ik} - \vartheta_{c1} + \vartheta''_{ci} \\ x''_{ik} = r''_{ik} \times \cos(\theta''_{ik}) \\ y''_{ik} = r''_{ik} \times \sin(\theta''_{ik}) \end{cases} \tag{14}$$

where $(r''_{ik}, \theta''_{ik})$ and (x''_{ik}, y''_{ik}) are the polar coordinates and Cartesian coordinates of k-th node of i-th feeder.

4 Single Line Diagram Beautification Calculation

For the single-line diagram in Fig. 4, the line segments are directly connected in straight-line, so the aesthetic degree is not strong. Therefore, a force guide method is adopted to beautify the generated diagram in Fig. 3.

4.1 Calculation of Simple Gravitational Repulsion Force

The repulsion force of the simple force guide algorithm is calculated as:

(1) There is an attraction force between two nodes that are connected. When the actual distance between nodes is greater than the ideal length, the gravitational force is positively proportional to the difference between the actual distance between nodes and the ideal length. The attraction force calculation of node V_i is calculated as:

$$f_r(V_i) = \sum_{j=1}^{m-1} f_a(V_i, V_j) \tag{15}$$

where

$$f_a(V_i, V_j) = \begin{cases} 0, & d_{V_iV_j} \leq L \\ k_a \times (d_{V_iV_j} - L), & d_{V_iV_j} > L \end{cases} \tag{16}$$

m is the total number of nodes of the graph, $d_{V_iV_j}$ is the distance between nodes and, k_a is the attraction force coefficient of other nodes to points, L is the ideal length of the spur line.

(2) There is a repulsive force between any nodes, and the magnitude of the repulsive force is inversely proportional to the distance between nodes; the repulsion force calculation of node V_i is as followed:

$$f_r(V_i) = -\sum_{j=1}^{m-1} \frac{k_r}{d_{V_iV_j}} \tag{17}$$

where k_r is the repulsion weight of other nodes against that node.

4.2 Calculation of Complex Gravitational Repulsion Forces

(1) Introducing point-edge, edge-edge repulsion force and total repulsion force
For distribution network structure, literature [19] introduces a new repulsion forces of node-edge avoidance and edge-edge avoidance, thus greatly reducing the probability of point edge crossing and side edge crossing. Therefore, in this paper, when dealing with node repulsion forces of a branch of a feeder, the method in literature [19] is simplified, that is, the total repulsion forces of nodes are calculated as follows:

$$f_r(V_i) = -\left(\sum_{j=0}^{m1} \frac{k_{r,1}}{d_{V_iV_j}} + \sum_{j=0}^{n1} \frac{k_{r,2}}{d_{V_iV_j}} + \sum_{j=0}^{q} \frac{k_{r,3}}{d_{V_iV_j}}\right) \tag{18}$$

where $m1$ is the number of nodes of non-native feeder; n1 and q are the number of nodes and the remaining nodes on the branch of the current node of this feeder. $k_{r,1}$ is the repulsion weight of nodes at the point V_i; $k_{r,2}$ is the repulsion weight of nodes on a level branch at the point V_i; $k_{r,3}$ is the repulsion weight of remaining nodes on the point V_i.

(2) Gravity force action and calculation
Only taking the above repulsion force, it is easy to make the topology diverges and stay away from the midpoint, and increase the effort to pull the node to make the topology converge to the center of gravity of the graph (0,0).
The gravity force added by any node V_i is:

$$\begin{cases} f_g(V_i) = k_g \times d_{V_i} \\ f_{gx}(V_i) = k_g \times x_i \\ f_{gy}(V_i) = k_g \times y_i \end{cases} \tag{19}$$

where f_g, f_{gx}, f_{gy} are gravity force and its component forces on horizontal and vertical axis. d_{V_i} is the distance from the node (x_i, y_i) to the center, and k_g is the gravity coefficient.

4.3 Node Motion Equation

Each node in the single-line diagram is affected by four forces, and then the improved motion equation of the node is as follows:

$$x^{k+1}(P_i) = x^k(P_i) + f_x^k(P_i) - g_f \times \frac{x^k(P_i)}{d^k} \tag{20}$$

$$y^{k+1}(P_i) = y^k(P_i) + f_y^k(P_i) - g_f \times \frac{y^k(P_i)}{d^k} \tag{21}$$

$$f_x^k(P_i) = f_{x,a}^k(P_i) + f_{x,r}^k(P_i) \tag{22}$$

$$f_y^k(P_i) = f_{y,a}^k(P_i) + f_{y,r}^k(P_i) \tag{23}$$

$$d^k = \sqrt{x^k(P_i)^2 + y^k(P_i)^2} \tag{24}$$

Where, k is the number of iterations of the node movement; and $f_x^k(P_i)$, $f_{x,a}^k(P_i)$, $f_{x,r}^k(P_i)$, $f_y^k(P_i)$, $f_{y,a}^k(P_i)$, $f_{y,r}^k(P_i)$ are the corresponding attraction and repulsive force and their projections on the X-axis or Y-axis; d^k is the distance of the node from the center point in iteration k, and g_f is the gravity coefficient.

5 Case Study

5.1 Original Information and Parameters of Case LVDN

Figure 5 give the original single line diagram of case LVDN, where three trunk lines in three-phase four-wire are taken from low voltage bus, marked as AB1-A2-…-A15,

AB1-B1-1-.....B1-1-7, AB1-B2-....-B4-E-5, respectively. Line type as 4x70, 4x50 are also marked in Fig. 5. Length of line segment is also marked between two nodes. Table 1(a) and (b) give the topology definition of that case LVDN. The end customer

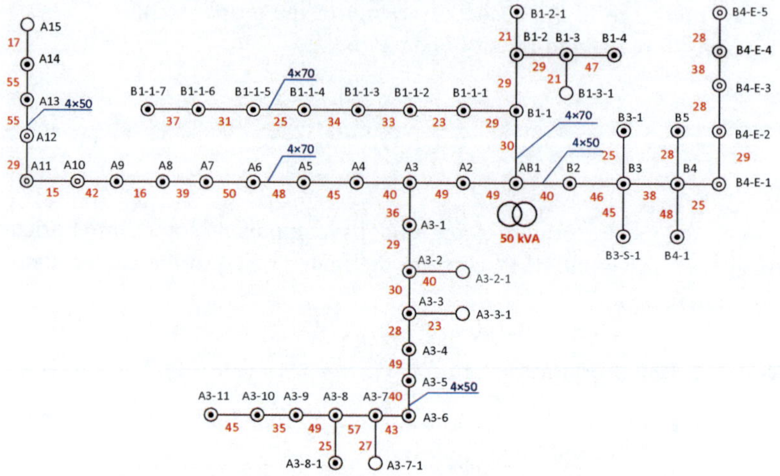

Fig. 5. Original single line diagram of case LVDN

Table 1. Topology definition of LVDN

(a) Node definition (part)		(b) LineSegment definition (part)		
NodeName	NodeID	LineSegment ID	ForontNodeName	BackNodeName
AB1	1	1	1	2
A1	2	2	2	9
A2	9	3	9	10
A3	10	4	10	26
A4	26	5	26	27
A5	27	6	27	28
A6	28	7	28	29

nodes will equip a smart meter, which will supply the data sheet as shown in Table 2.

5.2 Auto-Generated Single-Line Diagram in Straight-Line and Beautified

The auto-generated single-line diagram in straight-line of case LVDN in Fig. 5 is shown in Fig. 4 and beautified one is shown in Fig. 6 (a).

Table 2. Customer meter data at 2018/8/27 14:15 (part)

CustormerNodeName	ActivePower (kW)	PowerFactor	Voltage (V)	Current (A)
3011167101	0.2657	0.933	236.7	1.205
3011167102	0.1384	0.836	233.9	0.707
3011167103	−4.0311	0.963	242.8	−16.818
3011167104	0.022	0.713	230	0.132
3011167105	0.1077	0.646	238.5	0.697
3011167106	0.1478	0.863	229	0.75
3011167107	0.0987	0.742	239.2	0.551
3011167108	0.2933	0.785	228.1	2.036
3011167109	0.2166	0.644	245.1	1.37
3011167110	−4.7996	0.993	242	−20.256
3011167111	−2.253	0.998	245.1	−10.043
3011167112	1.8655	1	223.3	8.628
3011167117	0.1534	0.961	232.6	0.682
3012001607	0.0562	0.545	236	0.437
3012003840	2.6531	0.985	235.2	9.737

5.3 Power Flow Calculation

Based on the characteristics of the low-voltage distribution network, especially the accuracy of power flow calculation and redundancy characteristics, a power flow calculation method based on the current measurement and power measurement is proposed in our research team [20] taking the same case study. The result of the case using the actual data of a low-voltage distribution network shows that the accuracy of the current measurement is better than that of power measurement. At the same time, many erroneous data can be found by comparing the voltage and current data calculated from power flow with the voltage and current data from the consumer's smart meter, which can be analyzed to find measurement abnormality consumption abnormality of customers. Table 2 show the part of customer metering data at 2018/8/27 14:15, where customer 3011167103、3011167110、3011167111 with household PV have PV power generation supply to the LVDN, which form three areas with high voltage than nearby surrounds which is shown in Fig. 6 (c).

5.4 Performance Analysis for Real Application

Figure 6(a) shows the topology of that case LVDN. It can be seen that the trunks of each phase is basically the same, and the circled area in blue color is the asymmetric area of wiring. Figure 6(b), (c) and (d) show the voltage distribution of the LVDN at 1:45 am, 14:15 pm and 21:00 pm. It can be seen that the voltage distribution at Fig. 6 (b) has a certain symmetry, while Fig. 6(c) shows a high voltage around the node installed with household photovoltaic power generation.

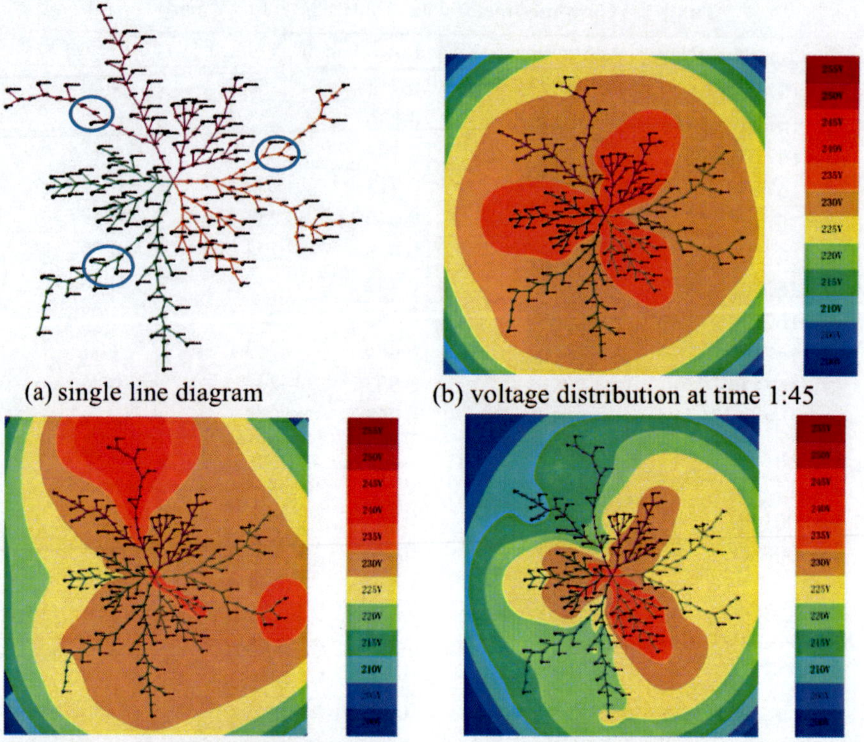

(a) single line diagram (b) voltage distribution at time 1:45

(c) voltage distribution at time 14:15 (d) voltage distribution at time 21:00

Fig. 6. Single line diagram and voltage situation diagram of LVDN

6 Conclusion

In this paper, two kinds of wiring diagram layouts suitable for situation visualization of LVDN are proposed, and the algorithm of automatic generation of full phase single line diagram is also provided. The calculation is fast and the final separate phase single line diagram is beautiful, providing a good base diagram for situation rendering based on the measurement values of smart meters. The time-series voltage situation pictures of the real application can perceive a dynamic changing of the load, especially the PV generation impacts on LVDN can be easily observed and detected. The impact of asymmetry of topology as well as unbalance of phase load on LVDN can also be easily observed and detected. Therefore, such single line diagram design as well as rendering method is a good for situation awareness for LVDN or DTSA.

Acknowledgement. This project has been funded by The State Grid Corporation of China Science and Technology Project 600-201919168A-0-0-00 and The National Natural Science Foundation of China 51677047.

References

1. Luan, W., Wang, B., Zhou, N., et al.: Modeling of LV distribution network based on metering data. Power Syst. Technol. **39**(11), 3141–3146 (2015)
2. Luan, W., Yu, Y., Wang, B.: AMI data analytics. Proc. CSEE **35**(1), 29–36 (2015)
3. Zhang, M., Xie, S., Luo, Y.: Research on three-phase load imbalance optimization model of low-voltage distribution network. J. Wuhan Univ. Sci. Technol. **38**(1), 59–62 (2015)
4. Fang, H., Sheng, W., Wang, J., et al.: Research on real-time online treatment method of three-phase load imbalance in power distribution station area. Proc. CSEE **35**(9), 2185–2193 (2015)
5. Gao, Y., Miao, H., Wu, W.: Impedance compensation based forward/backward sweep flow algorithm for three-phase four-wire distribution networks. Autom. Electric Power Syst. **39** (7), 47–51 (2015). https://doi.org/10.7500/AEPS20140811002
6. Zhao, X., Luo, L., Wang, F., Yan, W., Yu, J.: Three-phase power flow model of medium and low voltage distribution network with ungrounded inverter. Proc. CSEE **36**(20), 5421–5430 (2016)
7. Wang, S., Huang, L., Wang, C., et al.: Unbalanced three-phase power flow calculation for distributed generation system. Electric Power Autom. Equip. **27**(8), 11–15 (2007)
8. Pappu, S.J., Bhatt, N., Pasumarthy, R., Rajeswaran, A.: Distribution networks based on smart meter data. IEEE Trans. Smart Grid **9**(5), 5113–5122 (2018)
9. Dehghanpour, K., Wang, Z., Wang, J., Yuan, Y., Fankun, B.: A survey on state estimation techniques and challenges in smart distribution systems. IEEE Trans. Smart Grid **10**(2), 2312–2322 (2019)
10. Zhang, X., Xia, Y., Zhang, J., Zhang, J., et al.: Power flow calculation and application of low-voltage distribution network for redundancy measurement. Zhejiang Electric Power **39** (4), 22–28 (2020). https://doi.org/10.19585/j.zjdl.202004004
11. Cai, Y.J., Tang, W., Xu, Y., et al.: Research review on voltage control of low-voltage distribution networks with high proportion of household photovoltaic. Power Syst. Technol. **42** (1), 220–229 (2012)
12. Wan, Y., Xiang, S., Liu, H., et al.: Study on selection of three-phase unbalance control device for the three-phase four-wire low voltage distribution network. Power Capacitor React. Power Compensation **38**(6), 113–118 (2017)
13. Alam, M.J.E., Muttaqi, K.M., Sutanto, D.: Community energy storage for neutral voltage rise mitigation in four-wire multi-grounded LV feeders with cooling solar PV allocation. IEEE Trans. Smart Grid **6**(6), 2845–2855 (2015)
14. Alam, M.J.E., Muttaqi, K.M., Sutanto, D.: Potential rise of rebound-to-ground under current allocation of rooftop PV using distributed energy storage. IEEE Trans. Sustain. Energy **6**(3), 889–898 (2015)
15. Worthmann, K., Kellett, C.M., Braun, P., et al.: Distributed and decentralized control of residential energy systems incorporating storage. IEEE Trans. Smart Grid **6**(4), 1914–1923 (2015)
16. Zhang, J., Chen, H., Chen, J., et al.: Smart grid situation awareness diagram modeling and conceptual design of situation awareness visualization. Autom. Electric Power Syst. **8**(9), 168–176 (2014)
17. Zhang, J., Chen, H., Zhou, M., et al.: Generation of uniformly-distributed distribution network connection diagram and perception measurement for situation awareness diagram. Autom. Electric Power Syst. **38**(13), 166–173 (2014)

18. Zhan, J., Zhang, J., Chen, S., et al.: Single-line diagram of substation centralized medium voltage distribution network: part I diagram-data model based on generic switching station. Autom. Electric Power Syst. **43**(18), 102–110 (2019). https://doi.org/10.7500/aeps20180826004
19. Zhang, J., Qiu, C., Wang, F., et al.: Single-line diagram of substation centralized medium voltage distribution network: part II initial layout auto-generation. Autom. Electric Power Syst. **43**(19), 124–131 (2019). https://doi.org/10.7500/aeps20180827010
20. Qiu, C., Zhang, J., Liu, L., et al.: Single-line diagram of substation centralized medium voltage distribution network: part III beautified layout calculation. Autom. Electric Power Syst. **43**(20), 170–175 (2019). https://doi.org/10.7500/aeps20180827011

Author Index

Y. Xue et al. (Eds.): PMF 2020, LNEE 718, pp. 1007–1010, 2021.
https://doi.org/10.1007/978-981-15-9746-6